Chalcogen Chemistry
Fundamentals and Applications

Chalcogen Chemistry
Fundamentals and Applications

Edited by

Vito Lippolis
University of Cagliari, Italy
Email: lippolis@unica.it

Claudio Santi
University of Perugia, Italy
Email: claudio.santi@unipg.it

Eder J. Lenardão
Universidade Federal de Pelotas, Brazil
Email: lenardao@ufpel.edu.br

and

Antonio L. Braga
Universidade Federal de Santa Catarina, Brazil
Email: braga.antonio@ufsc.br

ROYAL SOCIETY OF **CHEMISTRY**

Print ISBN: 978-1-83916-422-4
PDF ISBN: 978-1-83916-738-6
EPUB ISBN: 978-1-83916-739-3

A catalogue record for this book is available from the British Library

The Royal Society of Chemistry is a charity, registered in England and Wales, Number 207890, and a company incorporated in England by Royal Charter (Registered No. RC000524), registered office: Burlington House, Piccadilly, London W1J 0BA, UK, Telephone: +44 (0) 20 7437 8656.

Visit our website at www.rsc.org/books

Printed in the United Kingdom by CPI Group (UK) Ltd, Croydon, CR0 4YY, UK

Foreword

Group XVI elements (oxygen, sulfur, selenium, tellurium, and polonium—the chalcogens) possess fascinating properties. Of these elements, oxygen usually takes center-stage as it is the most widely-studied and its importance in biological processes is common knowledge. In contrast, the chemistries of sulfur, selenium, and tellurium are lesser known. However, new discoveries have meant that oxygen is increasingly sharing the spotlight with other Group XVI members. Inorganic chemists are continuously discovering interesting new chalcogen complexes, organic chemists are expanding their synthetic toolboxes with chalcogen-containing reagents, and biochemists are finding new roles for the chalcogens in biological processes. This book focuses on the chemistries of three important chalcogen elements—sulfur, selenium, and tellurium—areas which are developing rapidly.

Sulfur occurs in its elemental form in many parts of the world, and has been known since prehistoric times. Under the ancient name of brimstone, it is frequently mentioned in the Bible. In fact, many religions and traditions use this element to depict their version of hell—imagery inspired by the vile odors and toxicity of the readily formed hydrogen sulfide, sulfur dioxide, and other foul-smelling sulfur compounds. Organic compounds containing sulfur are also famously malodorous. It was therefore perhaps somewhat surprising that sulfur was detected in plants in 1781, and then later in the bile and blood of animals in 1813. Of course, many of us will have discovered firsthand the presence of hydrogen sulfide in rotten eggs! Nowadays, the importance of sulfur in metabolic processes is firmly established. Since the beginning of the nineteenth century sulfur compounds have been widely used in the chemical industry. In addition, many pharmaceutical compounds contain sulfur atoms within their structures and several reagents

Chalcogen Chemistry: Fundamentals and Applications
Edited by Vito Lippolis, Claudio Santi, Eder J. Lenardão and Antonio L. Braga

Published by the Royal Society of Chemistry, www.rsc.org

used in organic synthesis are sulfur-containing compounds. As a result, the field of sulfur chemistry is well established, being thoroughly described in undergraduate inorganic and organic chemistry textbooks. Overall, the importance of this element is clearly apparent.

In contrast to sulfur, selenium was only identified in 1817, when Berzelius discovered selenium deposits formed in the lead chambers of his sulfuric acid plant in Gripsholm, Sweden. Berzelius named the element selenium to honor Selene, the goddess of the moon in Greek mythology. Berzelius realized the similarity between sulfur and selenium—organosulfur compounds have a bad smell, but selenium analogs smell even worse! Then, for more than a century, selenium chemistry was a neglected field. It was only in 1955 that the first extensive review work on selenium chemistry was published, as a chapter of Houben-Weyl, by Professor H. Rheinboldt, a German chemist who came to Brazil in 1934 to escape Nazi persecution. He was contracted by the recently founded São Paulo University. During his lifetime in Brazil, Rheinboldt trained a number of PhD students in the fields of selenium and tellurium chemistry. In 1973, I started my graduate studies in the laboratory of Petragnani in São Paulo, and witnessed firsthand the difficulty in manipulating organic selenium compounds. This was due to not only their bad smell, but also the poor reproducibility of reported results, owing to the limited knowledge on the chemistry of this element at that time.

Another problem associated with selenium compounds is its reputation as being highly detrimental to health. Around 1930, it was observed that when selenium is present in the soil, it is taken up by the vegetation, causing damage to animals grazing on the toxic plants. Around the same time, it was shown that feeding selenium-containing cereals to hens caused malformation or death in their chicks. These findings deterred, and continues to deter, chemists from studying selenium chemistry.

Finally, in 1957, something positive about selenium was discovered — selenium was identified to be an essential trace element in the animal diet, and the absence of selenium causes disease in both animals and humans. In 1973, it was discovered that selenium is present in the reactive site of mammalian glutathione peroxidase (GPX-1) enzyme, as a component of the 21st amino acid, selenocysteine. GPX-1 is involved in removing peroxides and serves as part of the antioxidant defense system of organisms. In this book, Professor Flohé describes this important discovery in more detail.

The above-mentioned discoveries garnered the attention of the scientific community to the chemistry of selenium. There is currently intense activity in the area of selenium compounds and how they relate to biological systems.

In 1970, the serendipitous discovery of the selenoxide *syn* elimination drew the attention of the organic chemistry community toward the synthetic potential of organoselenium compounds, and this branch of chemistry enjoyed an explosive growth in popularity during the years that followed.

Various synthetic transformations using selenium were found to be superior to the corresponding sulfur versions, clarifying the chemistry of selenium as a field in its own right, and not a mere extension of the chemistry of sulfur. As a direct result, selenium-containing intermediates have been intensely used in the synthesis of complex molecules. Nowadays, chemical transformations using selenium compounds can be found as standard in undergraduate organic chemistry textbooks.

The history of the element tellurium is older than that of selenium. Tellurium was discovered by von Reichenstein in 1782, in mineral ores in the gold district of Transylvania. Its name is derived from the Latin "Tellus", meaning earth. The preparation of the first organotellurium compound, the volatile and malodorous diethyl telluride, was reported by Whöler in 1840. Whöler described the smell of this compound as "obnoxious". Indeed, the low molecular weight tellurium compounds possess a terrible and nauseating smell, which I experienced firsthand when dimethyl telluride was prepared in my laboratory. Whöler's comments contributed to a long hibernation period in the study of the organic chemistry of tellurium. For example, only about 50 papers were published on tellurium chemistry between 1910 and 1950. Around a century after Whöler's original paper, the first comprehensive account of the inorganic and organic chemistry of this element appeared in the literature as part of the aforementioned chapter in Houben-Weyl by H. Rheinboldt. Rheinboldt's group showed that several classes of organic tellurium are high melting point solids. I can myself testify, from as early as the beginning of my graduate studies, that several organic derivatives of tellurium, especially those containing Te(IV), are almost odorless.

After the growth of selenium chemistry during the 1970s and 1980s, the attention of the chemistry community turned to the chemistry of tellurium. It was soon realized that the reactivities of the organic compounds of selenium and tellurium are quite different. For example, the telluroxide *syn* elimination is much more difficult than the corresponding transformation of selenoxides. In this sense, it is more interesting to explore the differences in reactivity between the compounds of these elements, than to explore their similarities. A more synthetically useful transformation of tellurides proved to be tellurium/metal exchange, which occurs more readily than selenium/metal exchange. Many organometallic intermediates used in organic synthesis can be prepared using this reaction. The first total synthesis of a complex natural product using a telluride as an intermediate was reported in 2002. Enantiomerically-enriched hydroxyl tellurides were obtained by enzymatic kinetic resolution, and were subsequently transformed into the corresponding dianions and then used in the synthesis of enantiomerically-enriched natural products. Contrary to early statements, organotellurium compounds can be handled in the presence of the air and under ceiling light, although the exposure of organotellurium compounds to air for prolonged periods should be avoided, especially when in solution.

There are few reports in recent literature on the interaction of tellurium compounds with biological systems, although it has been described that organic tellurides possess antioxidant properties, and that tellurium(IV) derivatives are potent inhibitors of some important enzymes, such as cathepsins. Recent studies on the toxicity of tellurium compounds are scarce.

A more detailed account on the history of the development of selenium and tellurium chemistry has been published (J. V. Comasseto, J. Braz. Chem Soc., 21 (2010) 2027–2031).

It is with great pleasure that I introduce this book—whose chapters are written by distinguished scientists—describing the recent advances in the area of sulfur, selenium, and tellurium chemistry. I hope you enjoy reading about these latest discoveries.

João V. Comasseto
São Carlos, Brazil

Preface

The idea for a new book on chalcogen chemistry, to be published by the Royal Society of Chemistry (RSC), came about in the middle of the COVID pandemic's acute phase and in the middle of the three years occurring between the ICCST-14 (International Conference on the Chemistry of Selenium and Tellurium) organized in Italy in 2019 and the ICCST-15 due to be held in Brazil in 2022. As organizers of the two events we wished to build a cultural bridge between the two conferences and scientifically project chalcogen chemistry into the future pulling together, in a period of isolation, acclaimed researchers around the world. We strongly desired to edit a book that would provide an update and one-stop resource for newcomers in the chemistry of chalcogens (S, Se, Te), but also a guide to the latest developments for established experts; a book organized for both beginners and experts, gathering in the right balance, fundamentals with results in frontier research fields stemming from the similarities and differences between the chemistry of these elements. We aimed at a manageable book for readers, highlighting the impact of chalcogens in many aspects of research; a book that could attract young chemists toward chalcogen chemistry highlighting the increasing role of chalcogen elements in multidisciplinary fields of modern applied chemistry and hot topics related to supramolecular chemistry, materials science, biology, pharmaceuticals medicine, and environmental science, but also giving the necessary fundamentals on the inorganic, organic, analytical, synthetic, structural, and theoretical chemistry of chalcogens.

To reach these goals, we asked both emerging young scientists and acclaimed experts from different countries and continents to contribute. We asked authors not only from different countries, but also from different continents to work together united by their expertise in specific fields and

Chalcogen Chemistry: Fundamentals and Applications
Edited by Vito Lippolis, Claudio Santi, Eder J. Lenardão and Antonio L. Braga

Published by the Royal Society of Chemistry, www.rsc.org

topics of chalcogen chemistry. While many books on the chemistry of each chalcogen have been published, books that consider in a comparative way the chemistry of these elements are very few and they have not recently been published.

This book comprises 27 chapters dealing either with the synthetic, structural, and multidisciplinary applicative chemistry of particular classes of compounds (*e.g.*, chalcogen-based metal clusters, metal-chalcogenolates, pincer and macrocyclic ligands, chalcogen heterocycles, chiral chalcogen-containing derivatives, therapeutic molecules, molecular probes, chalcogen-containing proteins, and nucleic acids), or emerging synthetic approaches involving new nucleophilic and electrophilic chalcogen-based reagents and redox processes. Particular attention has been paid to both the structural features, supramolecular chemistry, and synthetic applications of chalcogen bonding, and the biological role of chalcogen elements including physiological, metabolic, nutritional, and food analysis processes.

We believe this book is a candidate to become a point of reference for beginners and experts for many years thanks to its multidisciplinary approach toward hot topics in biological, material, and supramolecular chemistry, starting from basic aspects of chalcogen chemistry.

We are greatly indebted to all authors for their commitment and to the RSC for believing in our project in support of the chemistry of chalcogens.

Vito Lippolis
Università degli Studi di Cagliari, Cagliari, Italy
Claudio Santi
Università degli Studi di Perugia, Perugia, Italy
Eder J. Lenardão
Universidade Federal de Pelotas, Pelotas, Brazil
Antonio L. Braga
Universidade Federal de Santa Catarina, Florianópolis, Brazil

Contents

Chalcogen Chemistry: Fundamentals and Applications
Edited by Vito Lippolis, Claudio Santi, Eder J. Lenardão and Antonio L. Braga

Published by the Royal Society of Chemistry, www.rsc.org

CHAPTER 1

Carbonyl Chalcogenide Cluster-based Polymers

MINGHUEY SHIEH* AND YU-HUEI LI

Department of Chemistry, National Taiwan Normal University, Taipei 116325, Taiwan, Republic of China
*Email: mshieh@ntnu.edu.tw

1.1 Introduction

In the modern era, coordination polymers have gained a lot of attention from chemists and materials scientists owing to their versatile structures and topologies, unusual physical and chemical properties, and a wide range of applications as sensing, catalytic, and semiconducting materials.[1–3] These coordination frameworks usually consisted of main-group element-containing ligands that coordinated to more electropositive metal atoms, complexes, or clusters, leading to the formation of various coordination polymers.[1] Despite the most used pnictogenides as bridging ligands, the chalcogenides (O, S, Se, and Te) could also bind to metal fragments to produce chalcogen–metal coordination polymers.[2] The strong coupling effect of the heavier chalcogens (S, Se, and Te) and transition metals usually resulted in versatile structures with unique magnetic or electrical properties.[2a]

The introduction of metal clusters into polymeric frameworks was one of the challenging issues in the chemistry of coordination polymers.[3] Metal clusters usually possess intrinsic physical and chemical properties, such as magnetic and semiconducting properties, which effectively make functionalized polymers suitable for use in materials science. Among them, metal carbonyl clusters have long been known as electron reservoirs, which could

Chalcogen Chemistry: Fundamentals and Applications
Edited by Vito Lippolis, Claudio Santi, Eder J. Lenardão and Antonio L. Braga

Published by the Royal Society of Chemistry, www.rsc.org

further modulate the electronic properties of the building units for the construction of multifunctional polymers.[3d] Recent studies showed that introducing carbonyls into polymers could delicately give rise to numerous secondary interactions in the solid state that enhanced the stability and facilitated electron transport of their polymeric frameworks, making them potential semiconductors.[3a] Given these intriguing facts, this chapter will mainly focus on the recent progress on carbonyl heavier chalcogenide cluster-based coordination polymers, ranging from group 6 to group 8 metals, covering the syntheses, structural features, and potential applications. Some representative metal carbonyl complexes will also be introduced as building units for polymeric frameworks.

1.2 Group 6 Carbonyl Chalcogenide Coordination Polymers

Following a survey in the CCDC database, only a few group 6 metal carbonyl chalcogenide polymers were covered in this category. This section will discuss syntheses, structures, and some unique potential applications of several mono- or dinuclear group 6 metal carbonyl-containing polymers.

In 1981, Cooper and McPartlin *et al.* reported the syntheses and structural characterizations of SH-containing metal carbonyl complexes [Na(18-crown-6)][$M(CO)_5(SH)$] and [Na(18-crown-6)][$M_2(CO)_{10}(SH)$] (M = Cr, Mo, W). These complexes could be obtained from UV irradiations of $M(CO)_6$, Na_2S, and 18-crown-6 in ethanol at ambient temperature.[4] Compared to cations $[Et_4N]^+$ used as counterions, the bulky cations $[Na(18\text{-crown-}6)]^+$ enhanced the stability of these complexes in the solid state. X-ray structural analysis of the tungsten analogues, [Na(18-crown-6)][$W(CO)_5(SH)$] and [Na(18-crown-6)][$W_2(CO)_{10}(SH)$], showed that anions $[W(CO)_5(SH)]^-$ and $[W_2(CO)_{10}(SH)]^-$ were each further linked by the cation $[Na(18\text{-crown-}6)]^+$ *via* Na–O(carbonyl) bonds to form 1D polymeric chains (Figure 1.1a and b). This type of tight Na–O(carbonyl) ion-pairing led to stabilization of the unstable metal carbonyl clusters and to the construction of infinite supramolecular frameworks.

In addition, the SCN-containing Bi–Mo carbonyl polymer $[Bi(SCN)\{Mo(CO)_3(Cp)\}_2]_n$ was synthesized from the reaction of $Bi(SCN)_3$ with two equiv. of K[$Mo(CO)_3(Cp)$] in THF solution.[5] The X-ray analysis showed that $[Bi(SCN)\{Mo(CO)_3(Cp)\}_2]_n$ contained a linear polymeric array built through intermolecular Bi···NCS–Bi interactions (Figure 1.1c). The Lewis-acidic bismuth centre and the Lewis-basic nitrogen of the thiocyanate ligand provided a good Lewis acid–base interaction to form the polymeric framework. The zinc-containing mono-thiolate-bridged tungsten carbonyl polymer [Zn(tbtp-dach)$W(CO)_5]_n$ (tbtp-dach = *N*-(3-thiabutyl)-*N*′-(3-thiapentanoate)-1,4-diazacycloheptane) was obtained from the reaction of the *S*-donor metallo-ligand Zn(tbtp-dach) with (THF)$W(CO)_5$.[6] The [Zn(tbtp-dach)$W(CO)_5$] molecules were connected together by intermolecular Zn···O(carboxylate) interactions to form the $W(CO)_5$-decorating $[Zn(tbtp\text{-}dach)]_n$ zig–zag chain (Figure 1.1d). The strong attraction of the Zn···O(carboxylate)

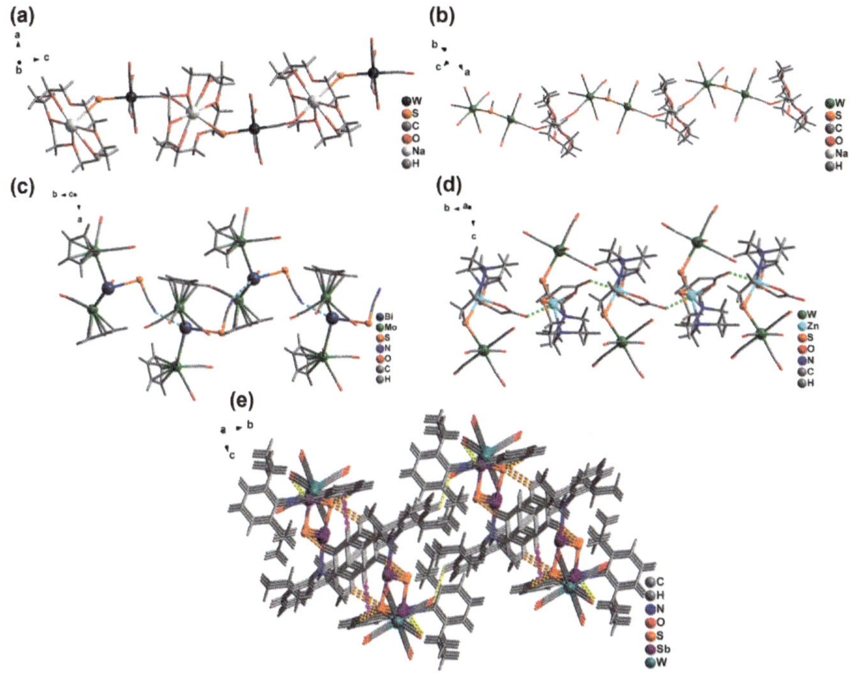

Figure 1.1 A portion of (a) [Na(18-crown-6)][W(CO)$_5$(SH)] (ref. code NACRWA), (b) [Na(18-crown-6)][W$_2$(CO)$_{10}$(SH)] (NACRWB), (c) [Bi(SCN){Mo(CO)$_3$(Cp)}$_2$]$_n$ (WETRAO), (d) [Zn(tbtp-dach)W(CO)$_5$]$_n$ (RULXIG), and (e) *cyclo*-[{2-(2′,6′-^{i}Pr$_2$C$_6$H$_3$N=CH)C$_6$H$_4$}SbS]$_2$[W(CO)$_5$] (EPOCIX). Bright-blue, bright-green, yellow, pink, and orange dashed lines represent Bi···N, Zn···O, O···H, C–H···π, and S···H interactions, respectively.

interaction was attributed to the formation of the (CO)$_5$W–thiolate bond that reduced the electron-donating character of the thiolate and enhanced the electrophilicity of the Zn atom. In addition, the organoantimony-bridged S–W carbonyl complex, *cyclo*-[{2-(2′,6′-^{i}Pr$_2$C$_6$H$_3$N=CH)C$_6$H$_4$}SbS]$_2$[W(CO)$_5$], was synthesized from the reaction of the heterocyclic sulfide, *cyclo*-[{2-(2′,6′-^{i}Pr$_2$C$_6$H$_3$N=CH)C$_6$H$_4$}SbS]$_2$, with [W(CO)$_5$(THF)] in a 1 : 1 molar ratio.[7] The dimeric units of the S–Sb–W compound were formed *via* weak S–H and C–H···π(Ph$_{centroid}$) contacts. These dimeric moieties were further linked through intermolecular O···H and C–H···π(Ph$_{centroid}$) interactions to give a polymeric structure (Figure 1.1e).

1.3 Group 7 Carbonyl Chalcogenide Coordination Polymers

Several group 7 metal carbonyl chalcogenide polymers and supramolecules were documented in the CCDC database. Several paramagnetic and low-energy-gap semiconducting Mn telluride carbonyl supramolecules will be discussed. Some rhenium carbonyl compounds with rectangular geometry showed unique host–guest chemistry in the solid state.

1.3.1 Te–Mn Cluster-based Carbonyl Polymers

Recently, a new series of small- to medium- to large-sized manganese telluride carbonyl clusters have been synthesized. They were obtained from the facile one-pot reaction of $Mn_2(CO)_{10}$ and Te powder with various metathesis trapping reagents (Et_4NBr, TMBACl, PPNCl) in different molar ratios, temperatures, and concentrations of the base. These one-pot reactions led to the formation of a mono spirocyclic $[TMBA]_2[Mn_4Te(CO)_{16}]$, four-membered ring $[TMBA]_2[Mn_2Te_2(CO)_8]$, hydride-containing square pyramidal $[PPN]_2[HMn_3Te_2(CO)_9]$, and dumbbell-shaped $[Et_4N]_4[Mn_6Te_6(CO)_{18}]$ and $[Et_4N]_4[Mn_6Te_{10}(CO)_{18}]$ compounds (Scheme 1.1).[8] It was worth noting that two large-sized, electron-precise clusters $[Mn_6Te_6(CO)_{18}]^{4-}$ and $[Mn_6Te_{10}(CO)_{18}]^{4-}$ possessed unusual paramagnetic properties with $S=1$, derived from the presence of two types of Mn atoms in different oxidation states in the compounds. Their detailed magnetic properties will be further discussed in Section 1.5.3. In addition, interesting structural and spin-state transformations were revealed in these systems. It was found that the small-sized anions $[Mn_4Te(CO)_{16}]^{2-}$ and $[Mn_2Te_2(CO)_8]^{2-}$ could convert into the larger anionic cluster $[Mn_6Te_{10}(CO)_{18}]^{4-}$. The reversible off/on magnetic-switched transformation between $[HMn_3Te_2(CO)_9]^{2-}$ and $[Mn_6Te_{10}(CO)_{18}]^{4-}$ was successfully achieved, and the magnetic transformation between the two anionic magnetic complexes $[Mn_6Te_6(CO)_{18}]^{4-}$ and $[Mn_6Te_{10}(CO)_{18}]^{4-}$ was also observed. These hexamanganese complexes represented the first examples of electron-precise paramagnetic manganese telluride carbonyl complexes. Furthermore, the hydride complex anion $[HMn_3Te_2(CO)_9]^{2-}$ could undergo the reversible dehydridation/hydridation processes upon the treatment with $HCl/NaBH_4$, and also could proceed with a reductive coupling to form the dimeric hydride complex $[\{HMn_3Te_2(CO)_{10}\}_2]^{2-}$ upon the addition of CO followed by O_2.

Importantly, X-ray structural analysis revealed that these Te–Mn anionic clusters were closely packed to each other, and their cations $TMBA^+$, Et_4N^+, and PPN^+ were further connected to the neighbouring anions *via* non-classical C–H···O(carbonyl) hydrogen bonds. Therefore, $[TMBA]_2[Mn_4Te(CO)_{16}]$ and $[PPN]_2[HMn_3Te_2(CO)_9]$ could be extended to supramolecular 1D chains, while $[TMBA]_2[Mn_2Te_2(CO)_8]$, $[Et_4N]_4[Mn_6Te_6(CO)_{18}]$, $[Et_4N]_4[Mn_6Te_{10}(CO)_{18}]$, and $[PPN]_2[\{HMn_3Te_2(CO)_{10}\}_2]$ formed supramolecular 2D sheets in the solid state (Figure 1.2). These 1D and 2D Te–Mn supramolecules possessed low-energy-gap semiconducting characteristics, which will be discussed in Section 1.5.1.

1.3.2 S–Re Rectangular Carbonyl Polymers

Several representative cases of the Re carbonyl supramolecules and polymers are mentioned below. In 2006, the attractive thiolato-bridged Re(I) molecular rectangles, $[\{(SR)_2Re_2(CO)_6\}_2(dpy)_2]$ ($R=C_4H_9$, C_8H_{17}; dpy = 4,4′-dipyridyl), were obtained from the solvothermal reaction of $Re_2(CO)_{10}$, dpy, and

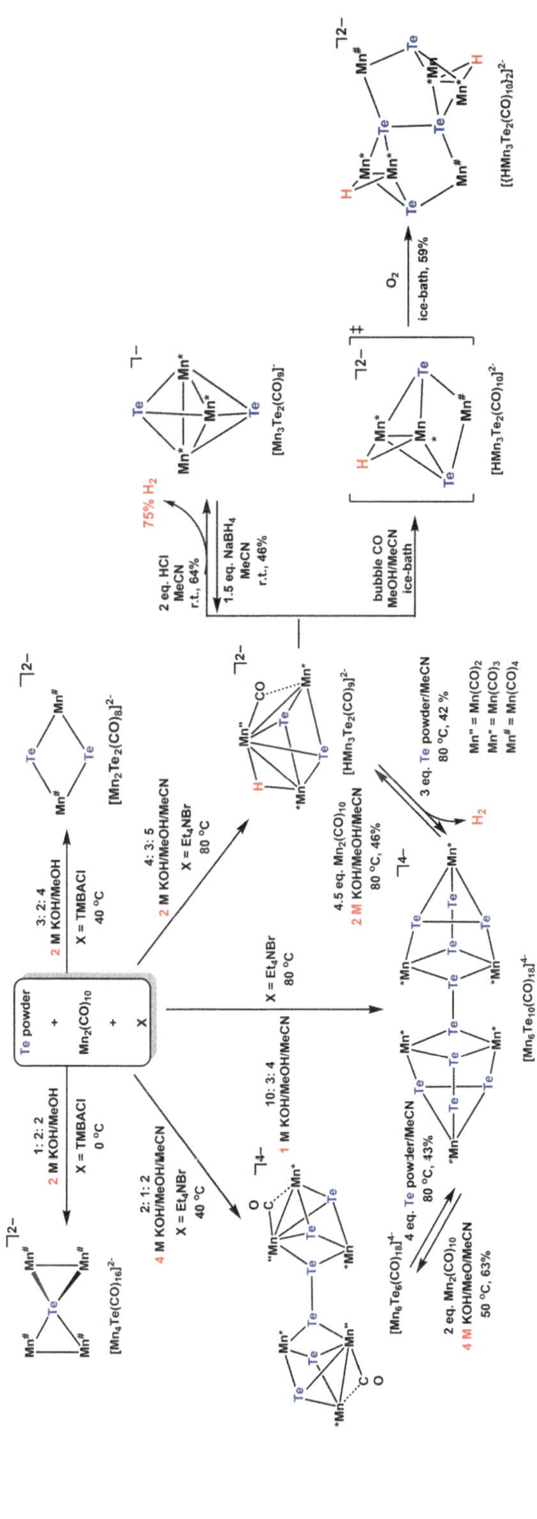

Scheme 1.1 Synthesis and structural transformations of a series of Mn telluride complexes.

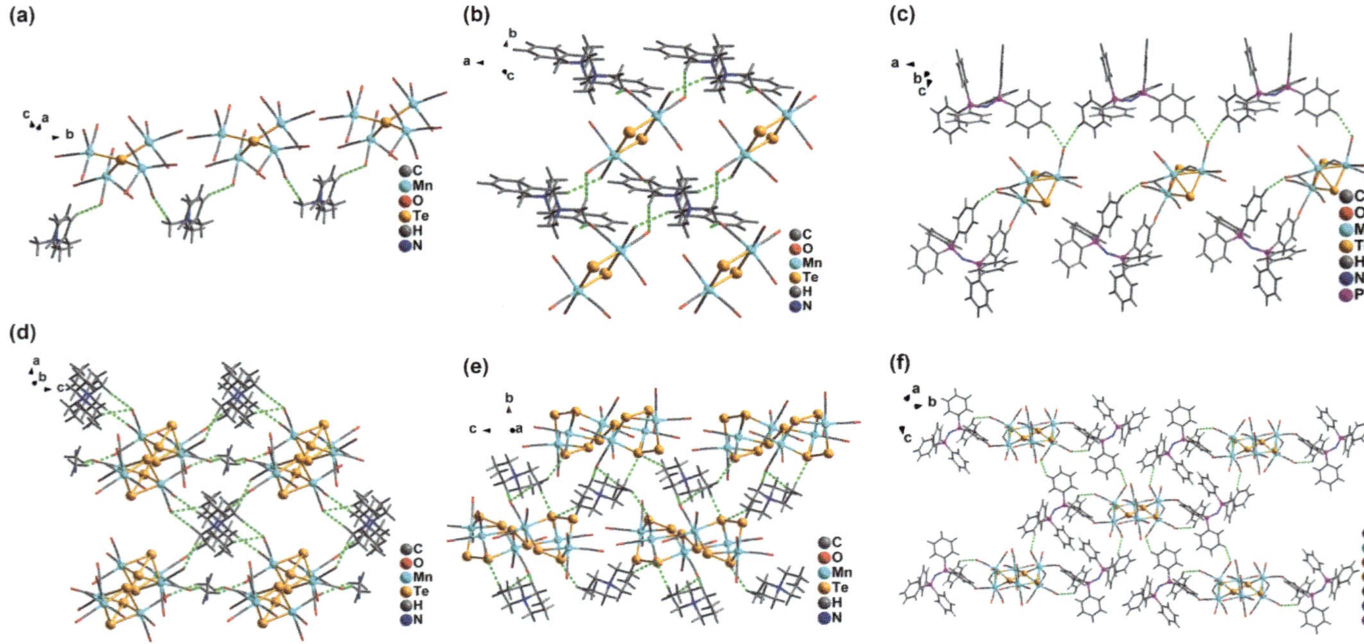

Figure 1.2 A portion of (a) an 1D chain in crystal $[TMBA]_2[Mn_4Te(CO)_{16}]$ (ref. code VUGKOA), (b) a 2D plane in crystal $[TMBA]_2[Mn_2Te_2(CO)_8]$ (VUGKUG), (c) an 1D chain in crystal $[Et_4N]_2[HMn_3Te_2(CO)_9]$ (VUGLAN), (d) a 2D plane in crystal $[Et_4N]_4[Mn_6Te_6(CO)_{18}]\cdot 2MeCN$ (VUGLER), (e) a 2D plane in crystal $[Et_4N]_4[Mn_6Te_{10}(CO)_{18}]$ (VUGLIV), and (f) a 2D plane in crystal $[PPN]_2[\{HMn_3Te_2(CO)_{10}\}_2]$ (VUGLOB). Light-green dashed line represents the O···H interaction.

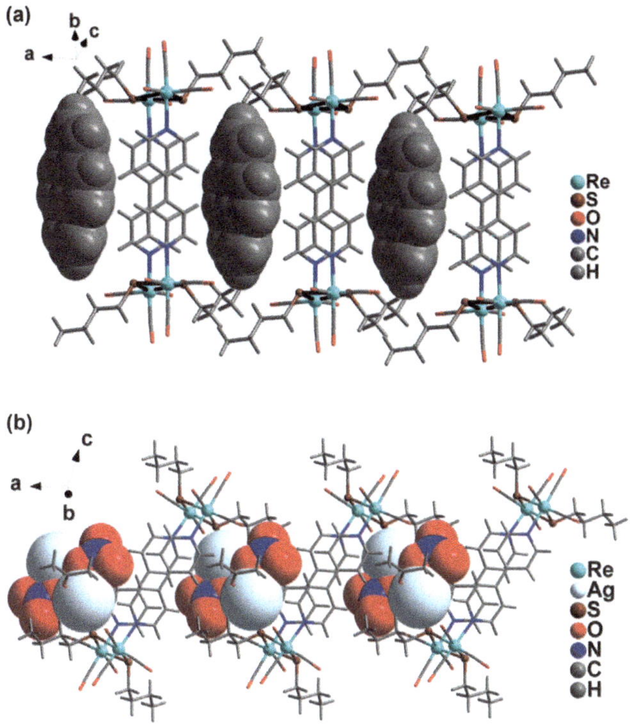

Figure 1.3 A portion of (a) $[\{(CO)_3Re(SC_4H_9)_2Re(CO)_3\}_2(dpy)_2(pyrene)]_n$ (ref. code HEPLOE) and (b) $[\{(CO)_3Re(SC_4H_9)_2Re(CO)_3\}_2(dpy)_2(AgNO_3)_2(C_3H_6O)_2\}\cdot C_3H_6O]_n$ (HEPLUK). Light-green dashed line represents the O···H interaction.

respective alkyl thiol.[9] These rectangles and their OR-substituted derivatives ($R = C_8H_{17}$, $C_{12}H_{25}$) showed good host–guest interactions toward the pyrene molecule and Ag^+ ion upon the addition of pyrene or $AgNO_3$, to give polymers $[\{(SC_4H_9)_2Re_2(CO)_6\}_2(dpy)_2(pyrene)]_n$ and $[\{(SC_4H_9)_2Re_2(CO)_6\}_2(dpy)_2(AgNO_3)_2$–$(C_3H_6O)_2\}\cdot C_3H_6O]_n$, respectively. The X-ray analysis of the pyrene-encapsulated polymer revealed that the two face-to-face dpy ligands interacted with pyrene *via* CH···π interactions to form a supramolecular stair-like array (Figure 1.3a). The S···Ag···ONO_2 interaction was found in the framework for its Ag-intercalated polymer, giving a 1D supramolecular chain (Figure 1.3b). These rectangular complexes further revealed important Ag^+- and pyrene-recognitions and absorptions properties, which will be described in detail in Section 1.5.5.

1.4 Group 8 Metal Carbonyl Chalcogenide Coordination Polymers

In the group 8 metal system, many carbonyl chalcogenide polymers and supramolecules have been reported. The representative triiron carbonyl chalcogenide cluster anions $[EFe_3(CO)_9]^{2-}$ [E = Te ($\mathbf{A}^{Te}$); Se ($\mathbf{A}^{Se}$); S ($\mathbf{A}^{S}$)] were reported to readily react with metal salts either in the presence or in the absence of organic ligands. These reactions would form a series of coordination polymers or supramolecular frameworks. In addition, the di-copper-based compounds $[EFe_3(CO)_9Cu_2(MeCN)_2]$ [E = Te ($\mathbf{B}^{Te}$); Se ($\mathbf{B}^{Se}$)] could also serve as predesigned building units to form a variety of coordination polymers *via* solution- or mechanochemical-based reaction. This section will describe the related syntheses and their fascinating structures.

1.4.1 S–Fe Cluster-based Carbonyl Polymers

The aryl sulfonate-bridged, di-thiolate-$Fe_2(CO)_6$-based complex, $Na\{(SCH_2N(C_6H_4SO_3)CH_2S)\}[Fe(CO)_3]_2$, was synthesized from the reaction of $(SH)_2Fe_2(CO)_6$ with sulfanilic acid and $NaHCO_3$.[10] This compound adopted an S_2Fe_2 butterfly geometry, with the $CH_2N(C_6H_4SO_3)CH_2$ moiety coordinated to the two S atoms. The SO_3 fragment of the $CH_2N(C_6H_4SO_3)CH_2$ moiety was further connected to the Na^+ ion *via* Na–O linkages, giving a 2D network. When an aqueous slurry of this di-thiolate iron complex was added to β-cyclodextrin (β-CyD), a di-β-CyD-hosted polymer, $\{Na\{(SCH_2N(C_6H_4SO_3)CH_2S)\}[Fe(CO)_3]_2{\cdot}2\beta\text{-CyD}\}_n$, was obtained.[10b] X-ray analysis showed that the polymer unit contained an S_2Fe_2 anionic guest cluster encapsulated by two host β-CyD moieties. The Na^+ cations further connected these anionic host–guest units *via* Na–O interactions to form a 1D array. This novel β-CyD-hosted complex represented a model of a first-generation artificial protein with the [FeFe]-hydrogenase enzyme active site. Besides, the dinuclear iron(II)–cyanocarbonyl polymer $\{[Na{\cdot}2.5H_2O][(\mu\text{-SEt})_3Fe_2(CO)_4(CN)_2]\}_n$ was obtained from the reaction of $[PPN][Fe(CO)_4(CN)]$ with I_2 and Na(SEt).[11] The anionic unit $[(\mu\text{-SEt})_3Fe_2(CO)_4(CN)_2]^-$ was found to interact with the hydrated Na^+ cations to create a polymeric framework *via* CN–Na linkages.

1.4.2 EFe_3 (E = Te, Se, S) Cluster-based Carbonyl Polymers

In 2008, the important Cu-linked Te–Fe polymers $[\{\mathbf{A}^{Te}\}Cu^-]_n$ and $[\{\mathbf{A}^{Te}\}Cu_2(dpy)_{1.5}]_n$ were obtained from the stoichiometric one-pot reaction of $[TeFe_3(CO)_9]^{2-}$ ($\mathbf{A}^{Te}$) and $[Cu(MeCN)_4]^+$ in the absence or in the presence of dpy in THF at 0 °C, respectively (Scheme 1.2).[12] These two Te–Fe–Cu polymers could also be synthesized from the reactions of the neutral cluster compound $TeFe_3(CO)_9Cu_2(MeCN)_2$ ($\mathbf{B}^{Te}$) with 1 equiv. of $[Et_4N]_2[\mathbf{A}^{Te}]$ or 1.5 equiv. of dpy in THF, respectively. Polymer $[\{\mathbf{A}^{Te}\}Cu^-]_n$ contained tetrahedral $TeFe_3$ metal cores that were bridged by Cu^+ ions across the Te–Fe edges, forming a 1D zig–zag chain (Figure 1.4a). The dpy-linked 1D polymer consisted of Cu-capped trigonal bipyramidal $TeFe_3(CO)_9Cu_2$ units alternately linked by a single and a double dpy ligand bridge to give a 1D zig–zag-like framework (Figure 1.4b). The intramolecular π–π interactions were found between dpy linkers in the double bridge, implying efficient electron communications.

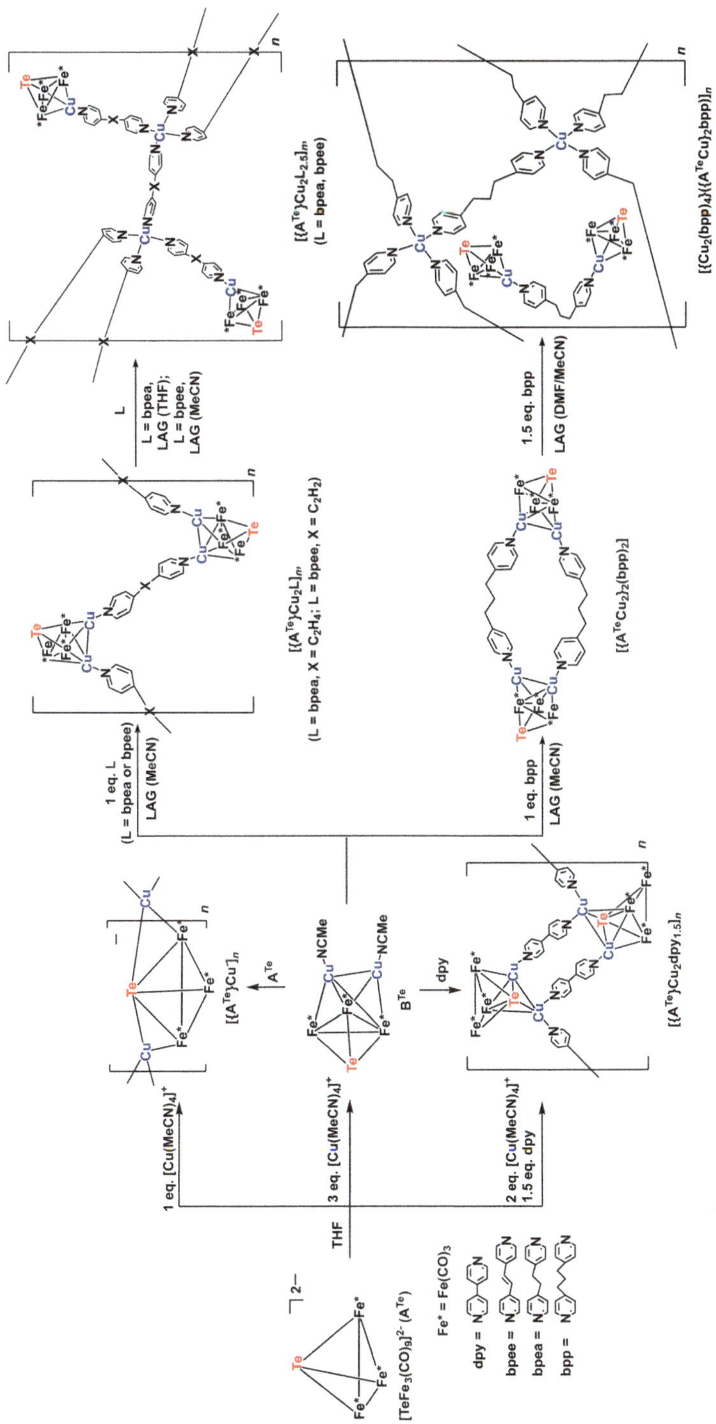

Scheme 1.2 Synthesis of the Te–Fe–Cu cluster-based polymers.

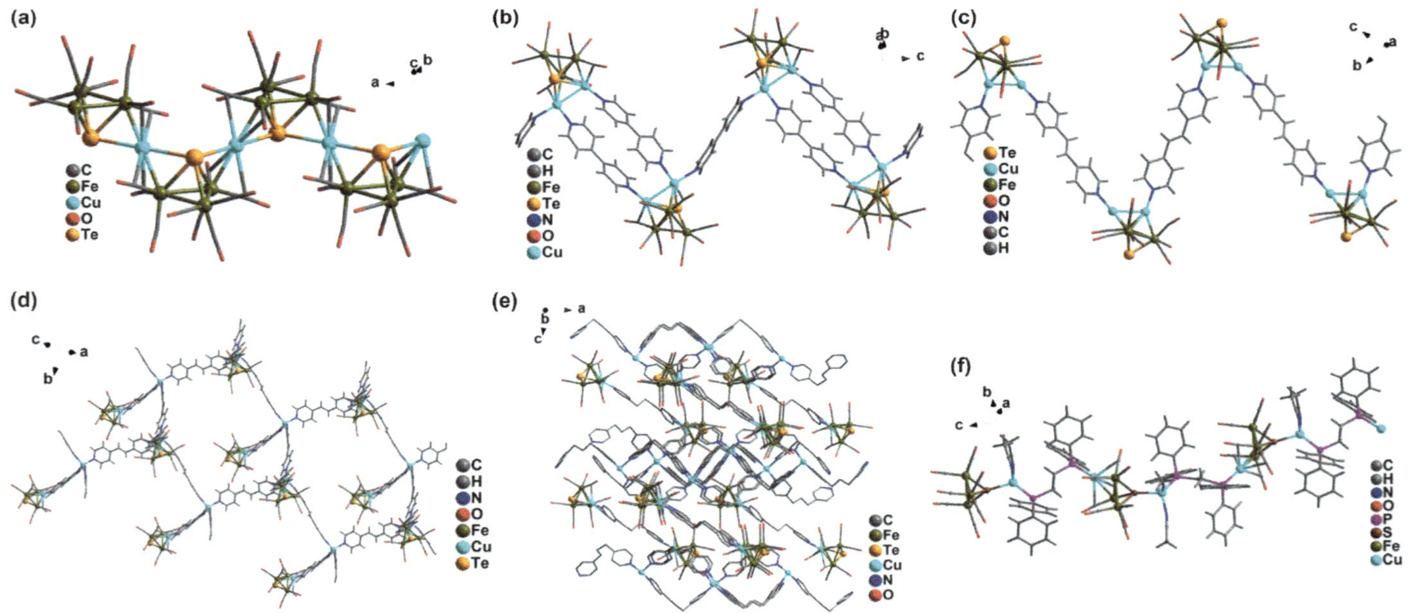

Figure 1.4 A portion of (a) $[\{\mathbf{A}^{\mathbf{Te}}\}Cu^{-}]_n$ (ref. code XOMSEY), (b) $[\{\mathbf{A}^{\mathbf{Te}}\}Cu_2(dpy)_{1.5}]_n$ (XOMSIC), (c) $[\{\mathbf{A}^{\mathbf{Te}}\}Cu_2bpee]_n$ (WAYQOF), (d) $[\{\mathbf{A}^{\mathbf{Te}}\}Cu_2L_{2.5}]_n$ (L = bpee, WAYRAS), (e) $[\{Cu_2(bpp)_4\}(\{\mathbf{A}^{\mathbf{Te}}Cu\}_2bpp)]_n$ (WAYRIA), and (f) $[(\mathbf{A}^{\mathbf{S}})Cu_2(dppe)(MeCN)_2]_n$ (PINTAJ). The H atoms in (e) were omitted for clarity.

In 2017, a new family of 1D and 2D dipyridyl-bridged Te–Fe–Cu-based polymers were mechanochemically synthesized from the predesigned cluster $\mathbf{B}^{Te}$ with a series of conjugated and conjugation-interrupted dipyridyl ligands (Scheme 1.2).[13] Among them, the 1D zig-zag chains $[\{\mathbf{A}^{Te}\}Cu_2L]_n$ (L = 1,2-bis(4-pyridyl)ethane (bpea), 1,2-bis(4-pyridyl)ethylene (bpee)), were found to be quantitatively synthesized *via* liquid-assisted grinding (LAG)[14] of $\mathbf{B}^{Te}$ with stoichiometric amounts of the corresponding dipyridyl ligands (Scheme 1.2).[13] These 1D isostructural chain polymers, $[\{\mathbf{A}^{Te}\}Cu_2L]_n$ (L = bpea, bpee), each contained the $TeFe_3(CO)_9Cu_2$ units that were connected by two dipyridyl ligands *via* the Cu–N bonds to form the 1D zig–zag chain polymer (Figure 1.4c). Interestingly, these 1D $[\{\mathbf{A}^{Te}\}Cu_2L]_n$ (L = bpea, bpee) polymers and the molecular cycle $[\{\mathbf{A}^{Te}Cu_2\}_2(bpp)_2]$ (1,3-bis(4-pyridyl)propane (bpp)) could further react by grinding with the corresponding dipyridyl ligands to afford the 2D honeycomb-like polymers $[\{\mathbf{A}^{Te}\}Cu_2L_{2.5}]_n$ (L = bpea, bpee) and the 2D wave-like cation–anion polymer $[\{Cu_2(bpp)_4\}(\{\mathbf{A}^{Te}Cu\}_2bpp)]_n$ in quantitative yields (Scheme 1.2).[13] The 2D polymers, $[\{\mathbf{A}^{Te}\}Cu_2L_{2.5}]_n$ (L = bpea, bpee), are isostructural. They consisted of the tetrahedrally coordinated Cu nodes, each of which was bound to three bridging L (L = bpea or bpee) ligands and coordinated with one pendant $TeFe_3(CO)_9Cu(L)$ ligand to give a honeycomb layer with (6,3) nets (Figure 1.4d). These 2D layers were closely stacked with pendant $TeFe_3(CO)_9Cu(L)$ (L = bpea or bpee) fragments penetrating upward and downward to construct a 3D supramolecular framework *via* numerous C–H···O(carbonyl) contacts between COs and CH moieties of the dipyridyl linkers. Besides, the cation–anion polymer $[\{Cu_2(bpp)_4\}(\{\mathbf{A}^{Te}Cu\}_2bpp)]_n$ contained a cationic (4,4) 2D net $\{[Cu_2(bpp)_4]^{2+}\}_n$ and anionic bpp-bridged clusters $[\{\mathbf{A}^{Te}Cu\}_2bpp]^{2-}$. These anionic clusters were intercalated between two cationic bilayers *via* several intermolecular interactions, *i.e.* aromatic C–H···π interactions and C–H···O(carbonyl) hydrogen bonds, giving a supramolecular 3D network (Figure 1.4e). These Te–Fe–Cu polymers exhibited semiconducting properties with narrow energy gaps,[12,13] which will be mentioned in Section 1.5.1.

As far as S–Fe–Cu-based polymers are concerned, a bisphosphine-bridged $SFe_3(CO)_9Cu_2$-based neutral cluster $[(\mathbf{A}^S)Cu_2dppe]$ was synthesized from the reaction of $[SFe_3(CO)_9]^{2-}$ ($\mathbf{A}^S$), $[Cu(MeCN)_4]^+$, and bis(diphenylphosphino)-ethane (dppe) in CH_2Cl_2 (Scheme 1.3).[15] Surprisingly, the polymeric complex $[(\mathbf{A}^S)Cu_2(dppe)(MeCN)_2]_n$ was readily produced when the neutral complex $[(\mathbf{A}^S)Cu_2dppe]$ was exposed to MeCN vapour (Scheme 1.3). X-ray analysis revealed that this dppe-linked polymer contained the $\mathbf{A}^S$ cluster that was connected to the $Cu(dppe)Cu(MeCN)_2$ moiety *via* Cu–Fe_2 and S–Cu linkages, giving a 1D zig–zag chain (Figure 1.4f). The depolymerization of this polymer chain to the neutral cluster was successfully achieved *via* grinding with CH_2Cl_2 or THF.[15] These results showed a novel solid-state transformation between the $\mathbf{A}^S$-based cluster and its 1D polymer *via* the rare vapor-diffusion-mediated and mechanochemical approaches.

For the Se–Fe–Cu based polymers, a series of inorganic–organic hybrid Cu polymers were obtained *via* LAG of the three-components $[Cu(MeCN)_4]^+$, the

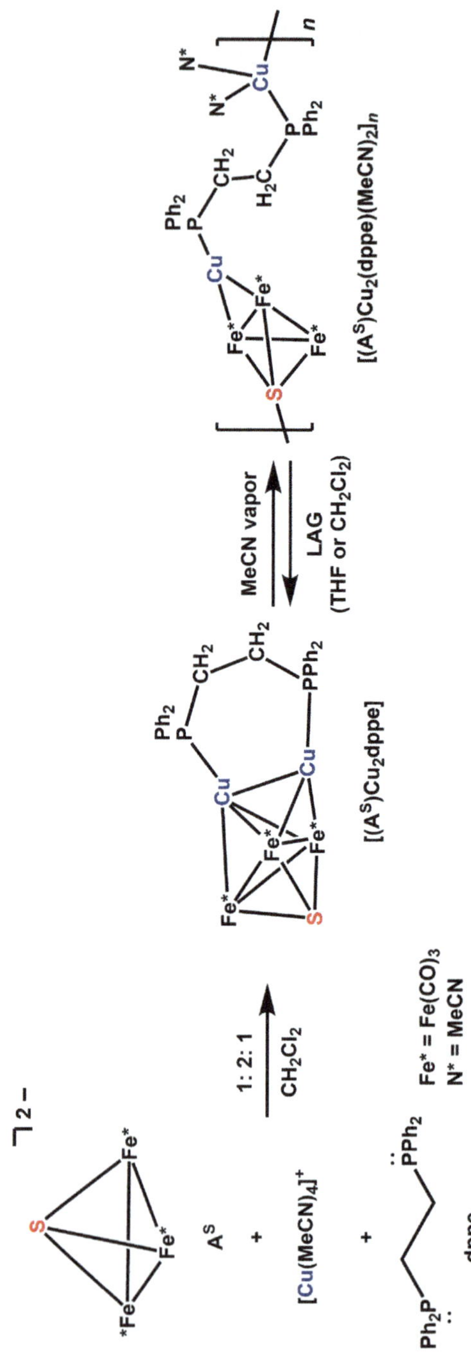

Scheme 1.3 Synthesis and the reversible structural transformation of the dppe-linked S–Fe–Cu-based cluster and polymer.

inorganic anionic ligand $[SeFe_3(CO)_9]^{2-}$ ($\mathbf{A^{Se}}$), and the conjugated or conjugation-interrupted dipyridyl ligands.[16] X-ray analysis showed that the $\mathbf{A^{Se}}$ cluster units of these synthesized polymers were connected to the Cu atom by three distinct bonding patterns, cluster-linked, -blocked, and -pendant coordination modes. Hence, this series of new 1D and 2D polymers were classified based on their structural features as: cluster-linked 2D polymers $[(\mu_4\text{-}\mathbf{A^{Se}})Cu_2(MeCN)(dpy)_{1.5}]_n$ and $[(\mu_4\text{-}\mathbf{A^{Se}})Cu_2(bpp)_2]_n$, cluster-blocked 1D polymers $[(\mu_3\text{-}\mathbf{A^{Se}})Cu_2(L)]_n$ (L = bpee, bpea), as well as the cluster-pendant 1D chain $[(\mu_3\text{-}\mathbf{A^{Se}})Cu_2(dpy)_3]_n$ (Figure 1.5). Interestingly, reversible dimensionality transformations between 2D and 1D polymers accompanied by the change in the cluster bonding pattern were achieved by the LAG addition of the corresponding dipyridyls or $\mathbf{A^{Se}}/[Cu(MeCN)_4]^+$. Notably, the 2D dpy-connected cluster-linked polymer, $[(\mu_4\text{-}\mathbf{A^{Se}})Cu_2(MeCN)(dpy)_{1.5}]_n$, revealed a 5-fold interpenetrating 3D framework with numerous C–H···O(carbonyl) and C–H···π interactions in the solid state (Figure 1.5b).[16] The relationship between these polymeric structures and their resulting properties, *i.e.* semiconducting and photodegradation properties, will be described in Sections 1.5.1 and 1.5.2.

Furthermore, a series of *N*-heterocyclic carbene (NHC)-incorporated compounds $[\mathbf{A^{Se}}\{Cu(NHC)\}_2]$ (NHC = 1,3-dimethylimidazol-2-ylidene (Me_2-imy), 1,3-dimethylbenzimidazol-2-ylidene (Me_2-bimy), 1,3-diisopropylbenzimidazol-2-ylidene ($^iPr_2$-bimy), 1,3-dimethyl-4,5-dichloroimidazol-2-ylidene (Me_2-Cl_2-imy)), could be synthesized *via* one-pot stoichiometric reactions of $[SeFe_3(CO)_9\{Cu(MeCN)\}_2]$ ($\mathbf{B^{Se}}$), imidazolium salts, and KO^{t}Bu in THF.[17] These NHC compounds contained the Cu_2-$\mathbf{A^{Se}}$ cores with the Cu atoms further coordinated by NHC ligands. In the solid state, their structures could be further extended into 1D chains (NHC = Me_2-imy and Me_2-bimy) or 2D frameworks (NHC = $^iPr_2$-bimy and Me_2-Cl_2-imy) through C–H···O(carbonyl) hydrogen bonds between CO groups and CH moieties of the NHC ligands. Surprisingly, these Se–Fe–Cu NHC compounds had catalytic activities for the homocoupling of arylboronic acids with high catalytic yields, which will be described in Section 1.5.4.

1.4.3 Mixed Group 15 and 16 Fe Carbonyl Supramolecules

Recently, a rare class of mixed pnictogen (Bi and Sb) and chalcogen (Te, Se, or S) iron carbonyl clusters were systematically synthesized.[18,19] In the case of the Bi system, when $\mathbf{A^{Te}}$ or $\mathbf{A^{Se}}$ was treated with $BiCl_3$, square pyramidal anionic complexes $[EBiFe_3(CO)_9]^-$ (E = Te, Se) were obtained (Scheme 1.4).[18] Complexes $[EBiFe_3(CO)_9]^-$ (E = Te, Se) displayed a $BiEFe_2$ basal plane with an apical $Fe(CO)_3$ fragment, in which both the Bi and chalcogen atom possessed an outward stereoactive lone pair. These square pyramidal complexes can undergo methylation and metalation *via* competitive 6s/5s and 6s/4s lone pairs selectivity, to give a 4s-methylated compound $\{(Me)SeFe_3(CO)_9\}Bi$, 6s-$Cr(CO)_5$-decorated complexes $[BiEFe_3(CO)_9\{Cr(CO)_5\}]^-$ (E = Te, Se) (Scheme 1.4). The novel nucleophilicity of the 6s lone pair of the naked Bi

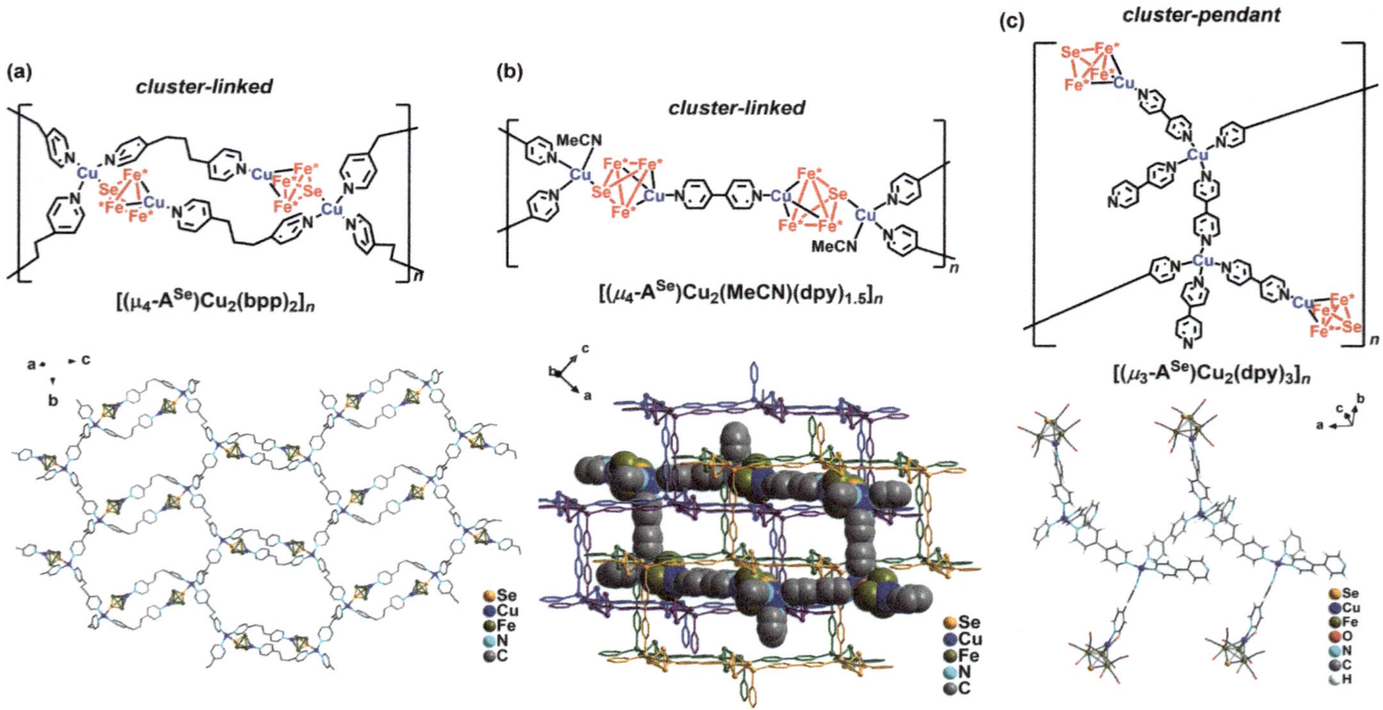

Figure 1.5 A portion of (a) $[(\mu_4\text{-}\mathbf{A}^{Se})Cu_2(bpp)_2]_n$ (ref. code SALJAU), (b) $[(\mu_4\text{-}\mathbf{A}^{Se})Cu_2(MeCN)(dpy)_{1.5}]_n$ (SALHUM), and (c) $[(\mu_3\text{-}\mathbf{A}^{Se})Cu_2(dpy)_3]_n$ (SALJEY). The H atoms and COs in (a) and (b) were omitted for clarity.

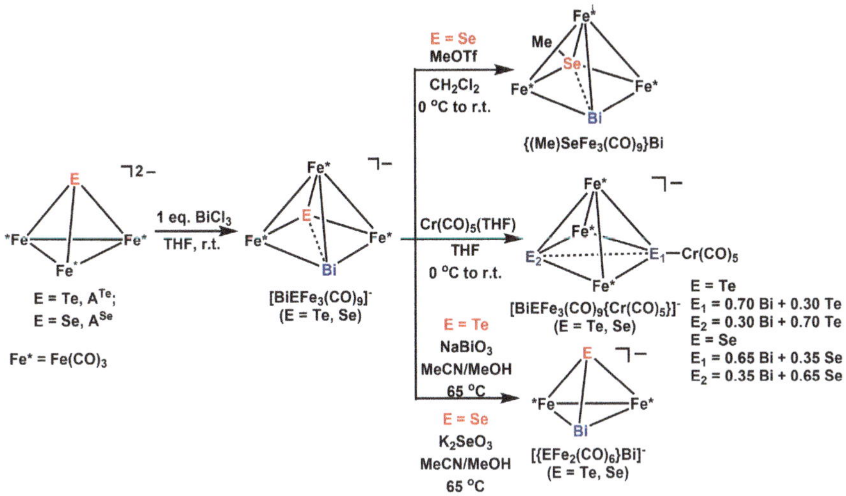

Scheme 1.4 Syntheses and reactivities of mixed Bi–chalcogen iron carbonyl complexes.

atom can be well explained by XPS, XANES, and DFT calculations. These square pyramidal complexes also can be demetalated to give tetrahedral anionic complexes $[\{EFe_2(CO)_6\}Bi]^-$ (E = Te, Se), in which the reactive sites can be understood by their corresponding HOMO orbitals.[18] Importantly, these square pyramidal, tetrahedral, and $Cr(CO)_5$-decorated complexes were found to have intra- and intermolecular Bi···E interactions to give ···Bi···E··· chains or Bi···E···E···Bi (E = Te or Se) dimers. These chains or dimeric units were further connected to their counterions through C–H···O(carbonyl) hydrogen bonds to give 2D supramolecular networks (Figure 1.6).

Similarly but differently, when reactions of **A** (**A** = $\mathbf{A^{Te}}$, $\mathbf{A^{Se}}$, $\mathbf{A^{S}}$) with $SbCl_3$ were carried out under similar conditions, three novel pendant cluster-coordinated square pyramidal-based anionic complexes $[\{SbTeFe_3(CO)_9\}\{Te_2Fe_3(CO)_9\}]^-$, $[\{SbSeFe_3(CO)_9\}\{Se_2Fe_2(CO)_6\}]^-$, and $[\{SbSFe_3(CO)_9\}\{SFe_3(CO)_9\}]^-$, were successfully obtained.[19] These complexes each contained the $ESbFe_3(CO)_9$ (E = Te, Se, S) square pyramidal core that was further coordinated to the respective pendant fragment, $Te_2Fe_3(CO)_9$, $Se_2Fe_2(CO)_6$, or $SFe_3(CO)_9$, *via* the Sb atom. They showed high electrophilicity toward metal carbonylates, producing a series of transmetalated products. These metalated products included "spiked" square pyramidal anionic complexes $[\{SbEFe_3(CO)_9\}\{M(CO)_x\}]^-$ ($M(CO)_x = Fe(CO)_4$, E = Te, Se, S; $Cr(CO)_5$, E = Se, S), and $Mn(CO)_4$-bridged di-square pyramidal clusters $[\{SbEFe_3(CO)_9\}_2Mn(CO)_4]^-$ (E = Se, S). Significantly, these ternary and quaternary complexes revealed O···O(carbonyl) or C–H···O(carbonyl) interactions to construct 2D and 3D supramolecular frameworks in the solid state (Figure 1.7). These rarely seen mixed pnictogens (Bi, Sb) and chalcogen (Te, Se, S) iron carbonyl clusters possessed important intra- and inter-molecular interactions in the solid state, facilitating efficient

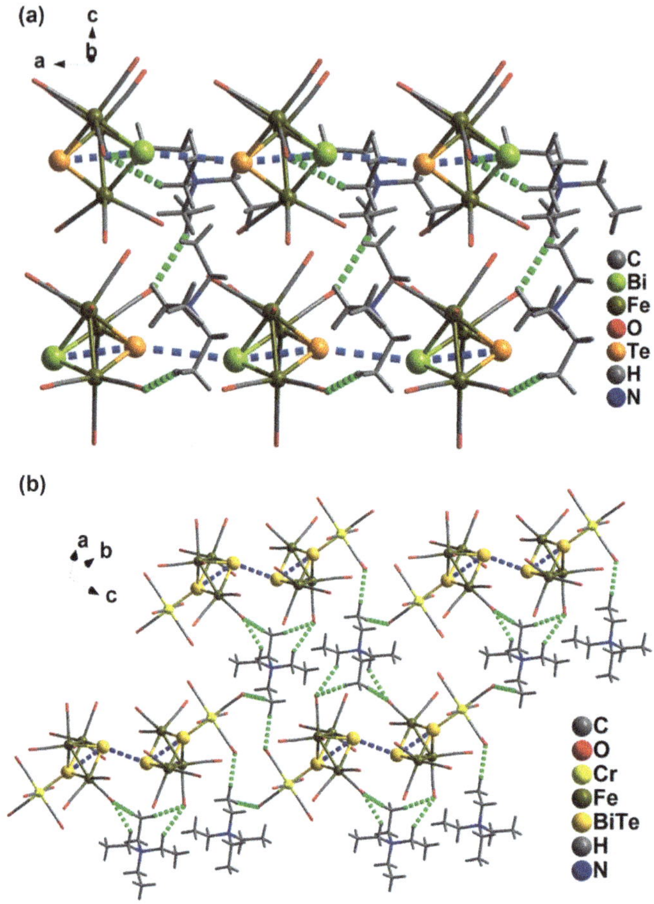

Figure 1.6 A portion of (a) $[Et_4N][TeBiFe_3(CO)_9]$ (ref. code COFFEL) and (b) $[Et_4N]$-$[BiTeFe_3(CO)_9\{Cr(CO)_5\}]$ (COFFOV).

electron transports to result in tunable low energy gaps. The details will be discussed in Section 1.5.1.

1.4.4 Ru Telluride Carbonyl Polymer

The first ternary Te–Ru–Cu cluster-based polymeric chain, $\{[PPh_4]_2[Te_2Ru_4(CO)_{10}Cu_4Br_2Cl_2]\cdot THF\}_n$, was synthesized from the reaction of K_2TeO_3 and $Ru_3(CO)_{12}$ with PPh_4Br in MeOH followed by the addition of CuCl in MeCN. Alternatively, this polymer could also be synthesized upon the treatment of the CuX-bridged octahedral complexes $[PPh_4]_2[Te_2Ru_4(CO)_{10}(CuX)_2]$ (X = Br, Cl) with the linking agent CuX (X = Cl, Br), respectively.[20] X-ray analysis showed that polymer $\{[PPh_4]_2[Te_2Ru_4(CO)_{10}Cu_4Br_2Cl_2]\cdot THF\}_n$ contained the octahedral $Te_2Ru_4(CO)_{10}$ units which were connected by two Cu_2BrCl fragments, in which the opposite Ru–Ru edges each were bridged by

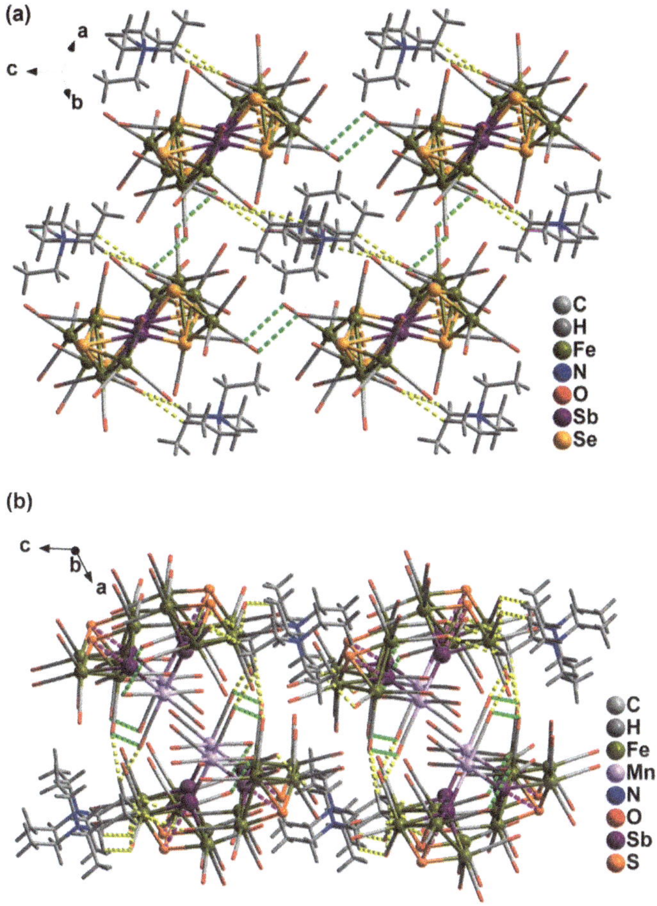

Figure 1.7 A portion of (a) $[Et_4N][\{SbSeFe_3(CO)_9\}\{Se_2Fe_2(CO)_6\}]$ (ELEZAZ) and (b) $[Et_4N][\{SbSFe_3(CO)_9\}_2Mn(CO)_4]$ (ELIBUZ).

the Cu_2Br moiety of the bridging Cu_2BrCl, to give a 1D infinite chain. The bridging chloride ions could further connect with the phenyl H atoms from the PPh_4^+ cations *via* weak hydrogen bonds to form a supramolecular network (Figure 1.8), which enhanced the stability in the solid state. This chain polymer showed surprising semiconducting behaviours with a small optical energy gap of about 0.37 eV. The related discussion will be shown in Section 1.5.1.

1.5 Potential Applications

On the basis of the versatile structural features and unique properties, metal carbonyl chalcogenide polymers are believed to have some potential applications. This section will describe some representative polymers regarding

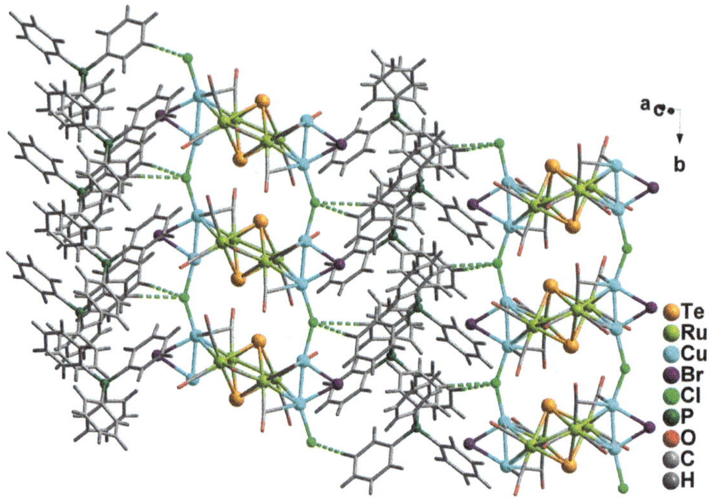

Figure 1.8 A portion of Te–Ru–Cu polymer $\{[PPh_4]_2[Te_2Ru_4(CO)_{10}Cu_4Br_2Cl_2]\cdot THF\}_n$ (GIHSEW).

semiconducting behaviours, photodegradation, magnetism, catalysis, and adsorption/desorption properties.

1.5.1 Semiconducting Behaviors

The discovery of new semiconductors is one of the advanced targets of material science for chemists to pursue. Although quite a few conductive polymers have been reported,[1d,1e] the conductive/semi-conductive metal carbonyl chalcogen-containing polymers or supramolecules were much less investigated. In 2007, the first Te–Ru–Cu polymer $\{[PPh_4]_2[Te_2Ru_4(CO)_{10}Cu_4Br_2Cl_2]\cdot THF\}_n$ was synthesized and found to exhibit semiconducting behaviours with the dc conductivity of $(1–5)\times10^{-2}\ \Omega^{-1}\,cm^{-1}$ and a small optical band gap of *ca.* 0.37 eV.[20] The electron transport was determined by the electron density propagation along the chain units as supported by DFT calculations. Notably, its energy gap was found to be smaller than that of Te_2Ru (0.491 eV),[21] indicating that the incorporation of Cu would efficiently enhance the conductivity.

In 2008, the first $\mathbf{A}^{Te}$-based Te–Fe–Cu polymers, the mono-Cu-bridged polymer $[Et_4N]_n[\{\mathbf{A}^{Te}\}Cu]_n$ and the single- and double-dpy-linked polymer $[\{\mathbf{A}^{Te}\}Cu_2(dpy)_{1.5}]_n$, were rationally synthesized.[12] They were shown to exhibit semiconducting behaviours with low energy gaps (0.59 and 0.41 eV) and high dc conductivities (2×10^{-2} and $5\times10^{-2}\ \Omega^{-1}\,cm^{-1}$).[12] The more efficient electron transport of the single- and double-dpy-linked polymer $[\{\mathbf{A}^{Te}\}Cu_2(dpy)_{1.5}]_n$ was proven *via* DFT calculations to be aided by the $\pi\cdots\pi$ interactions between the double dpy linkers. Later in 2017, a series of dipyridyl-linked $\mathbf{A}^{Te}$-based Te–Fe–Cu polymers, $[\{\mathbf{A}^{Te}\}Cu_2L]_n$, $[\{\mathbf{A}^{Te}\}Cu_2L_{2.5}]_n$ (L = bpea, bpee), and $[\{Cu_2(bpp)_4\}(\{\mathbf{A}^{Te}Cu\}_2bpp)]_n$ were also reported to

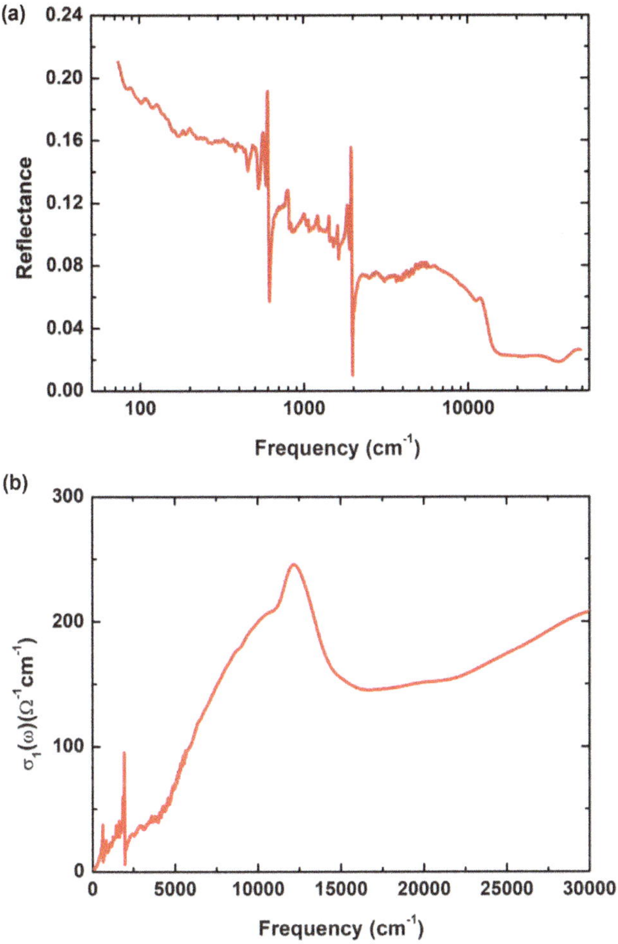

Figure 1.9 (a) Room-temperature optical reflectance spectrum and (b) frequency-dependent optical conductivity of polymer $[\{Cu_2(bpp)_4\}(\{\mathbf{A}^{Te}Cu\}_2bpp)]_n$.

possess semiconducting behaviours with tunable optical energy gaps (1.43–1.52 eV) and dc conductivities (2.5×10^{-3}–$2.2 \times 10^{-2}\ \Omega^{-1}\ cm^{-1}$).[13] The efficient electron transport of these dipyridyl-linked Te–Fe–Cu polymers was supported by the change of the oxidation state of the Cu atoms using XPS and XANES. These results confirmed the electron propagation from electron-rich metal clusters $\mathbf{A}^{Te}$ to the Cu centres, then to the dipyridyl linkers and polymeric frameworks.[13] It was surprising to find that the most conjugated bpp-linked polymer possessed the lowest optical energy gap (1.43 eV) and pronounced dc conductivity ($1.5 \times 10^{-2}\ \Omega^{-1}\ cm^{-1}$) (Figure 1.9). These results can be attributed to the significant C–H···O(carbonyl) and C–H···π(bpp) interactions within the framework. These results strongly demonstrated that COs played an essential role in facilitating the electron communication of metal carbonyl-based polymeric frameworks.

The dppe-bridged $\mathbf{A^S}$-based S–Fe–Cu polymer, $[(\mu_4\text{-}\mathbf{A^S})Cu_2(MeCN)_2(dppe)]_n$, was found to be synthesized from its discrete precursor, $[(\mu_3\text{-}\mathbf{A^S})Cu_2(dppe)]$, *via* vapochemical polymerization. This polymer could undergo unique mechanochemical depolymerization back to its discrete complex.[15] Surprisingly, the S–Fe–Cu–dppe polymer possessed a lower energy gap (1.69 eV) than the discrete complex (1.71 eV). The subtle difference was attributed to the bridging dppe unit, although dppe is usually considered as a poor electron-transfer ligand.

Recently, a series of inorganic–organic hybrid $\mathbf{A^{Se}}$-based Cu polymers with variable cluster bonding modes were mechanochemically synthesized as described in Section 1.4.2.[16] These Se–Fe–Cu polymers were found to possess low energy gaps of 1.49–1.72 eV that increased based on their structure modes following the order: cluster-linked, cluster-blocked, and cluster-pendant coordination fashion. Notably, the most conjugation-interrupted bpp-linked polymer $[(\mu_4\text{-}\mathbf{A^{Se}})Cu_2(bpp)_2]_n$ had a relatively lower energy gap (1.58 eV), mainly ascribed to numerous C–H(bpp)···O(carbonyl) hydrogen bonds and C–H···π(bpp) interactions within the polymeric framework with the aid of X-ray analysis, the density of state (DOS) and band structure calculations. The electrical conductivity of three representative polymers, $[(\mu_4\text{-}\mathbf{A^{Se}})Cu_2(bpp)_2]_n$, $[(\mu_3\text{-}\mathbf{A^{Se}})Cu_2(bpea)]_n$, and $[(\mu_3\text{-}\mathbf{A^{Se}})Cu_2(dpy)_3]_n$, were measured to be 3.13×10^{-7} (Figure 1.10), 2.92×10^{-7}, and 2.30×10^{-7} S cm^{-1}, respectively. These electrical conductivities were parallel to the trend of their energy gaps, showing semiconducting behaviours.

Apart from the semiconducting chalcogen–metal carbonyl polymers mentioned above, some discrete clusters with supramolecular networks were

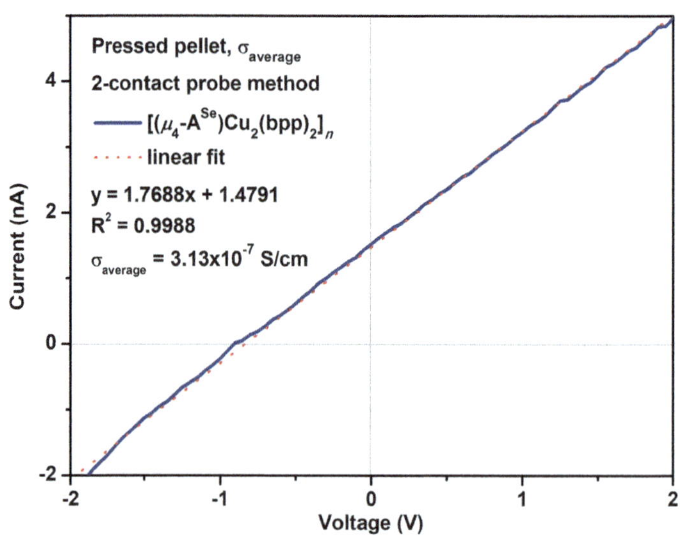

Figure 1.10 The electrical conductivity of the pressed pellet sample of polymer $[(\mu_4\text{-}\mathbf{A^{Se}})Cu_2(bpp)_2]_n$.

also found to have low energy gaps. In the Te–Mn system, a series of small- to large-sized polynuclear Te–Mn complexes have been synthesized and structurally characterized, as mentioned in Section 1.3.1. These manganese complexes possessed a wide range of energy gaps (1.06–1.62 eV).[8] Besides, the rare class of pnictogen (Bi, Sb) and chalcogen (Te, Se, S)-containing iron carbonyl clusters also possessed low-energy-gap characteristics of 1.01–1.21 eV in the Bi series and 0.84–1.48 eV for the Sb series.[18,19] Although structurally characterized as discrete molecular systems, these complexes can be extended into supramolecular frameworks *via* weak but significant C–H(counterion)···O(carbonyl) and O···O(carbonyl) interactions, as evidenced by X-ray analysis and DFT calculations. The formation of supramolecular frameworks is responsible for their unexpected low energy gaps.

1.5.2 Photodegradation

Recently, water/light-stable Se–Fe–Cu semiconducting polymers, $[(\mu_4\text{-}\mathbf{A}^{Se})Cu_2(bpp)_2]_n$, $[(\mu_3\text{-}\mathbf{A}^{Se})Cu_2(bpea)_{2.5}]_n$, and $[(\mu_3\text{-}\mathbf{A}^{Se})Cu_2(dpy)_3]_n$, were shown to have photodegradation activities toward *para*-nitrophenol (PNP) under the Xe irradiation in water (Figure 1.11a).[16] Their degradation rate obeyed the pseudo-first-order kinetics with k_{obs} being 6.46×10^{-3} ($[[(\mu_4\text{-}\mathbf{A}^{Se})Cu_2(bpp)_2]_n$), 6.08×10^{-3} ($[[(\mu_3\text{-}\mathbf{A}^{Se})Cu_2(bpea)_{2.5}]_n$), and 1.21×10^{-3} min^{-1} ($[[(\mu_3\text{-}\mathbf{A}^{Se})Cu_2(dpy)_3]_n$) (Figure 1.11b). The results indicated that cluster-linked $[(\mu_4\text{-}\mathbf{A}^{Se})Cu_2(bpp)_2]_n$ possessed a higher degradation efficiency in comparison with cluster-pendant polymers $[(\mu_3\text{-}\mathbf{A}^{Se})Cu_2(bpea)_{2.5}]_n$ and $[(\mu_3\text{-}\mathbf{A}^{Se})Cu_2(dpy)_3]_n$, in agreement with their increased energy gaps (discussed above in the section of semiconducting properties). Furthermore, the photodegradation of 2,4-dinitrophenol (2,4-DNP) and organic dyes (methylene blue (MB) and sunset yellow (SY)) by the most efficient polymer $[(\mu_4\text{-}\mathbf{A}^{Se})Cu_2(bpp)_2]_n$ was also successfully achieved with good pseudo-first-order kinetics.[16] These polymers were shown to serve as efficient agents for the photodegradation of nitroaromatics and organic dyes, implying their potential uses as catalysts for recycling dye wastewater.

1.5.3 Magnetic Properties

Significantly few metal carbonyl chalcogenide polymers showed magnetic properties due to the strong-field CO ligands. Two electron-precise Te–Mn anionic clusters $[Mn_6Te_6(CO)_{18}]^{4-}$ and $[Mn_6Te_{10}(CO)_{18}]^{4-}$ were shown to exhibit unexpected paramagnetic properties with $\mu_{eff} = 2.60$ and 2.45 μ_B at 300 K (Figure 1.12a).[8] These hexamanganese complexes represent the first examples of electron-precise paramagnetic manganese telluride carbonyl complexes. XPS and EPR analyses of these Te–Mn complexes further established the oxidation states of the Mn atoms, exchange couplings among the magnetic centres, hyperfine splittings of ^{55}Mn ($I = 5/2$), and zero-field splitting parameters for the triplet state (Figure 1.12b).[8]

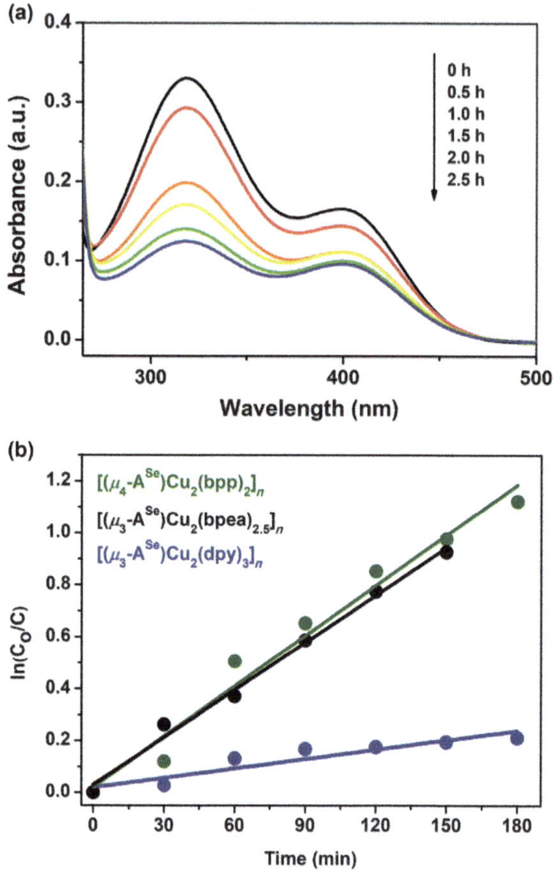

Figure 1.11 (a) The absorption spectra of the photodegradation of PNP by polymer $[(\mu_4\text{-}\mathbf{A}^{Se})Cu_2(bpp)_2]_n$ and (b) kinetic data for the photodegradation of PNP by polymers $[(\mu_4\text{-}\mathbf{A}^{Se})Cu_2(bpp)_2]_n$ (green), $[(\mu_3\text{-}\mathbf{A}^{Se})Cu_2(bpea)_{2.5}]_n$ (black), and $[(\mu_3\text{-}\mathbf{A}^{Se})Cu_2(dpy)_3]_n$ (blue).

1.5.4 Catalysis

Due to the electron-rich nature and high stability, chalcogen-containing metal carbonyl polymers and supramolecules were believed to have potential applications in catalysis. In 2019, a novel family of NHC-functionalized ternary Se–Fe–Cu compounds, $[\mathbf{A}^{Se}\{Cu(NHC)\}_2]$ (NHC = Me_2-imy, Me_2-bimy, $^iPr_2$-bimy, Me_2-Cl_2-imy), were obtained and structurally characterized.[17] These NHC compounds could serve as catalysts in the Suzuki homo-coupling coupling reaction of arylboronic acids with high turnovers (80–87) under mild conditions. Importantly, these Se-containing NHC compounds further represented excellent models for studying chalcogen effects compared to their Te-analogues,[22] which was demonstrated by their catalytic performances and electrochemical behaviours, supported by DFT calculations.

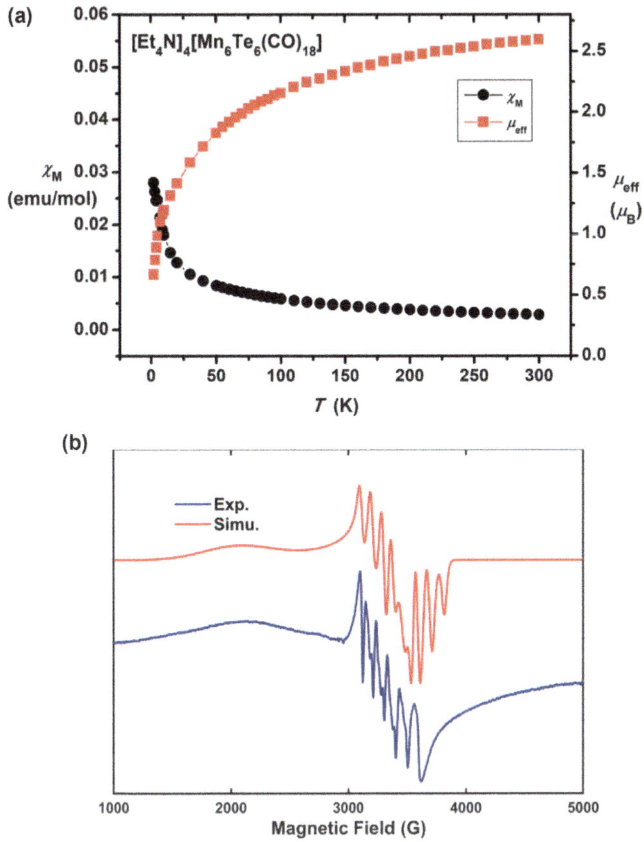

Figure 1.12 (a) Temperature-dependent magnetic susceptibility data for $[Et_4N]_4[Mn_6Te_6(CO)_{18}]$ and (b) experimental (blue line) and simulated (red line) 100 K EPR spectra of a frozen MeCN solution of $[Et_4N]_4[Mn_6Te_6(CO)_{18}]$.

1.5.5 Adsorption/Desorption Properties

Host–guest chemistry was an important topic for the coordination polymer due to various secondary interactions between organic/inorganic guest molecules and host polymers that could modulate their physical and chemical properties. Some reports of metal carbonyl chalcogenide polymers regarding adsorption/desorption properties have been reported. In 2006, the rectangular S–Re carbonyl compound $[\{(SC_4H_9)_2Re_2(CO)_6\}_2(dpy)_2]$ was found to be capable of encapsulating the pyrene molecule or $AgNO_3$.[9] These reactions would lead to formations of the pyrene-introduced polymer $[\{(SC_4H_9)_2Re_2(CO)_6\}_2(dpy)_2(pyrene)]_n$ and the Ag-intercalated polymer $[\{(SC_4H_9)_2Re_2(CO)_6\}_2(dpy)_2(AgNO_3)_2(C_3H_6O)_2\}\cdot C_3H_6O]_n$. The interaction of the pyrene- and $AgNO_3$-intercalated products in solutions was studied by ^1H NMR, UV–vis, and fluorescence spectroscopies. The π–π, CH···π, and S···Ag···O interactions of these intercalated products were further found in the solid state, which could help stabilize their resulting frameworks. In 2010, the S_2Fe_2 cluster-incorporated, β-cyclodextrin-hosted polymer $Na\{(SCH_2N(C_6H_4SO_3)CH_2S)\}[Fe(CO)_3]_2\cdot 2\beta$-CyD has been synthesized and reported.[10b] The included S_2Fe_2 cluster $[(SCH_2N(C_6H_4SO_3)CH_2S)Fe(CO)_3]^-$ had good intermolecular interactions with β-CyD, as evidenced by NMR spectra. The inclusion polymer further exhibited the negative-shifted electrochemical peaks for the catalytic H_2 production compared to the free cluster, reflecting the hydrophobic characteristics of the β-CyD. This result provided a new host–guest approach to a biomimetic cavity with a hydrogenase active site.

1.6 Concluding Remarks

This chapter is intended to give an overview of the recent progress of carbonyl chalcogenide cluster-based coordination polymers. They can be obtained *via* various synthetic methods, such as solution-based, hydro(solvo)thermal, mechanochemical, UV irradiated, and vapochemical reactions. These polymers were constructed mainly by metal–chalcogen linkages, while some supramolecular frameworks could be formed by intra- or intermolecular C–H···O(carbonyl) hydrogen bonds, which can also enhance their stabilities in the solid state. Due to the electron-rich nature and abundant redox properties of the incorporated metal carbonyl cluster units, their resulting polymers were demonstrated to have a wide range of applications, *i.e.* semiconductors, photodegradation activities, magnetism, catalysis, and guest molecule absorption/desorption. Although this research area has received growing interest, the structural characterizations of these polymers still relied on the conventional X-ray analysis of single crystals that were difficult to obtain mainly due to their poor solubility in most organic solvents. In the future, the study of metal carbonyl cluster-based chalcogenide polymers should focus on discovering systematic synthetic routes and their special properties/applications as well as rationalizing their intriguing structure–property relationships.

Acknowledgements

Minghuey Shieh wishes to thank the Ministry of Science and Technology of Taiwan (MOST 110-2113-M-003-014) for the continuous support of her research. She also thanks all graduate students in her research group for their efforts on this topic. She is also grateful to the National Center for High-Performance Computing and National Synchrotron Radiation Research Center of Taiwan for the technical support.

References

1. (a) H. Furukawa, K. E. Cordova, M. O'Keeffe and O. M. Yaghi, *Science*, 2013, **341**, 1230444; (b) P. Thanasekaran, C.-H. Su, Y.-H. Liu and K.-L. Lu,

Coord. Chem. Rev., 2021, **442**, 213987; (c) J.-K. Sun, X.-D. Yang, G.-Y. Yang and J. Zhang, *Coord. Chem. Rev.*, 2019, **378**, 533; (d) L. S. Xie, G. Skorupskii and M. Dincă, *Chem. Rev.*, 2020, **120**, 8536; (e) A. E. Thorarinsdottir and T. D. Harris, *Chem. Rev.*, 2020, **120**, 8716.

2. (a) J. Xie, L. Wang and J. S. Anderson, *Chem. Sci.*, 2020, **11**, 8350–8372; (b) Q. Yue and E.-Q. Gao, *Coord. Chem. Rev.*, 2019, **382**, 1.
3. (a) M. Shieh, Y.-H. Liu, Y.-H. Li and R. Y. Lin, *CrystEngComm*, 2019, **21**, 7341; (b) K. H. Whitmire, *Coord. Chem. Rev.*, 2018, **376**, 114; (c) E. Peresypkina, A. Virovets and M. Scheer, *Coord. Chem. Rev.*, 2021, **446**, 213995; (d) M. Shieh, C.-Y. Miu, Y.-Y. Chu and C.-N. Lin, *Coord. Chem. Rev.*, 2012, **256**, 637; (e) M. Shieh and C.-C. Yu, *J. Organomet. Chem.*, 2017, **849–850**, 219.
4. (a) M. K. Cooper, P. A. Duckworth, K. Henrick and M. McPartlin, *J. Organomet. Chem.*, 1981, **212**, C10–C12; (b) M. K. Cooper, P. A. Duckworth, K. Henrick and M. McPartlin, *J. Chem. Soc., Dalton Trans.*, 1981, 2357.
5. A. Crispini, R. J. Errington, G. A. Fisher, F. J. Funke, N. C. Norman, A. G. Orpen, S. E. Stratford and O. Struve, *J. Chem. Soc., Dalton Trans.*, 1994, 1327.
6. E. Almaraz, W. S. Foley, J. A. Denny, J. H. Reibenspies, M. L. Golden and M. Y. Darensbourg, *Inorg. Chem.*, 2009, **48**, 5288.
7. A. M. Preda, C. I. Raţ, C. Silvestru, H. J. Breunig, H. Lang, T. Rüffer and M. Mehring, *Dalton Trans.*, 2013, **42**, 1144.
8. M. Shieh, Y.-H. Liu, T.-S. Lin, Y.-C. Lin, W.-K. Cheng and R. Y. Lin, *Inorg. Chem.*, 2020, **59**, 6923.
9. B. Manimaran, L.-J. Lai, P. Thanasekaran, J.-Y. Wu, R.-T. Liao, T.-W. Tseng, Y.-H. Liu, G.-H. Lee, S.-M. Peng and K.-L. Lu, *Inorg. Chem.*, 2006, **45**, 8070.
10. (a) M. L. Singleton, D. J. Crouthers, R. P. Duttweiler, III, J. H. Reibenspies and M. Y. Darensbourg, *Inorg. Chem.*, 2011, **50**, 5015; (b) M. L. Singleton, J. H. Reibenspies and M. Y. Darensbourg, *J. Am. Chem. Soc.*, 2010, **132**, 8870.
11. W.-F. Liaw, W.-T. Tsai, H.-B. Gau, C.-M. Lee, S.-Y. Chou, W.-Y. Chen and G.-H. Lee, *Inorg. Chem.*, 2003, **42**, 2783.
12. M. Shieh, C.-H. Ho, W.-S. Sheu, B.-G. Chen, Y.-Y. Chu, C.-Y. Miu, H.-L. Liu and C.-C. Shen, *J. Am. Chem. Soc.*, 2008, **130**, 14114.
13. M. Shieh, C.-C. Yu, C.-Y. Miu, C.-H. Kung, C.-Y. Huang, Y.-H. Liu, H. L. Liu and C. C. Shen, *Chem. - Eur. J.*, 2017, **23**, 11261.
14. (a) T. Friščić, C. Mottillo and H. M. Titi, *Angew. Chem., Int. Ed.*, 2020, **59**, 1018; (b) G. K. Kole and J. J. Vittal, *Chem. Soc. Rev.*, 2013, **42**, 1755.
15. C.-N. Lin, W.-T. Jhu and M. Shieh, *Chem. Commun.*, 2014, **50**, 1134.
16. Y.-H. Liu, K.-T. Huang, W.-C. Chen, Y.-W. Li, W.-M. Ke, B.-R. Ho, M.-C. Hsu, Y.-H. Li and M. Shieh, *Inorg. Chem.*, 2021, **60**, 18270.
17. M. Shieh, Y.-H. Liu, C.-C. Wang, S.-H. Jian, C.-N. Lin, Y.-M. Chen and C.-Y. Huang, *New J. Chem.*, 2019, **43**, 11832.
18. M. Shieh, Y.-H. Liu, C.-Y. Huang, S.-W. Chen, W.-K. Cheng and L.-T. Chien, *Inorg. Chem.*, 2019, **58**, 6706.

19. H.-H. Yeh, M.-C. Hsu, Y.-H. Li, Y.-N. Hsu, F.-Y. Shr and M. Shieh, *J. Organomet. Chem.*, 2021, **937**, 121717.
20. M. Shieh, M.-H. Hsu, W.-S. Sheu, L.-F. Jang, S.-F. Lin, Y.-Y. Chu, C.-Y. Miu, Y.-W. Lai, H.-L. Liu and J.-L. Her, *Chem. – Eur. J.*, 2007, **13**, 6605.
21. P. C. Liao, J. K. Huang, Y. S. Huang and T. R. Yang, *Solid State Commun.*, 1996, **98**, 279.
22. C.-N. Lin, C.-Y. Huang, C.-C. Yu, Y.-M. Chen, W.-M. Ke, G.-J. Wang, G.-A. Lee and M. Shieh, *Dalton Trans.*, 2015, **44**, 16675.

CHAPTER 2

Chalcogenocarbonyl and Chalcogenonitrosyl Metal Complexes

YUICHIRO MUTOH

RIKEN Center for Sustainable Resource Science, 2-1 Hirosawa, Wako, Saitama 351-0198, Japan
Emails: yuichiro.mutoh@riken.jp; yuichiro.mutoh@gmail.com

2.1 Introduction

Transition-metal compounds that contain carbonyl (CO) and thiocarbonyl (CS) ligands have been extensively studied, and their chemical properties have been disclosed. In contrast, the heavier chalcogen isologues, namely selenocarbonyl (CSe) and tellurocarbonyl (CTe) complexes have received less attention. Similarly, although nitrosyl (NO) complexes of transition-metals are well studied, their sulfur and selenium congeners, *i.e.*, thionitrosyl (NS) and selenonitrosyl (NSe) complexes are less explored. These limitations should be related to the lack of suitable reagents and efficient synthetic methods available for synthesizing them.

Recent notable achievements in this field include among others the discovery of CSe and CTe ligands that can adopt various coordination modes to bridge multiple metals, the synthesis of hitherto unknown CTe complexes with a half-sandwich structure that enables the systematic investigation regarding a series of half-sandwiched CE complexes, and the increasing number of selenonitrosyl (NSe) complexes, only one of which existed ten years ago. This chapter mainly describes the recent-ten-year

Chalcogen Chemistry: Fundamentals and Applications
Edited by Vito Lippolis, Claudio Santi, Eder J. Lenardão and Antonio L. Braga

Published by the Royal Society of Chemistry, www.rsc.org

developments in the chemistry of the transition-metal complexes with heavier chalcogenocarbonyl, CE (E = Se, Te) and heavier chalcogenonitrosyl, NE (E = S, Se) ligands. Some review articles on the chemistry of CE and NE complexes have been reported and will be mentioned in each section.

2.2 Chalcogenocarbonyl (CE) Complexes

While the chemistries of CO and CS complexes are well documented, those of selenocarbonyl (M–CSe) and tellurocarbonyl complexes (M–CTe) have been less studied as mentioned above. However, various CSe and CTe complexes have been reported in the last decade or so, and the field of such heavier group 16 CE complexes has gradually been developing. The systematic study of the properties of a series of chalcogenocarbonyl (CE = O, S, Se, Te) complexes is desirable as it would deepen the understanding of the chemistry of free CE (E = Se, Te) molecules, which cannot be detected even under cryogenic conditions.

2.3 New Thiocarbonyl (CS) Complexes Reported after 2008

Since the first sulfur analog of carbonyl complex, *i.e.*, the thiocarbonyl (CS) complex $[Rh(CS)Cl(PPh_3)_2]$ which was prepared in 1966,[1] various transition metal complexes with the CS ligand have been extensively studied and well established as described in review articles up to 2008.[2] Easy-to-handle sulfur compounds such as CS_2 and $Cl_2C{=}S$ have been employed as the CS source for synthesizing CS complexes. Various new CS complexes have been reported since 2008, and only CS complexes characterized by a single-crystal X-ray diffraction study are listed below:

Rhenium: $[CpRe(CO)_2(CS)]$;[3]

Iron: $[Fe\{B\{o\text{-}(i\text{-}Pr_2P)C_6H_4\}_3)(CS)]$[4] and $[Na(OEt_2)_2][Fe\{B\{o\text{-}(i\text{-}Pr_2P)C_6H_4\}_3)(CS)]$;[4]

Ruthenium: $[Ru(CPh{=}CHPh)Cl(CS)(PPh_3)_2]$,[5] $[Ru\{C(C{\equiv}CPh){=}CHPh\}Cl(CS)(PPh_3)_2]$,[5] $[Fe\{\eta^5\text{-}C_5H_4CO_2Ru\{C(C{\equiv}CPh){=}CHPh\}(CS)(PPh_3)_2\}_2]$,[5] $[Ru\{C(C{\equiv}CPh){=}CHPh\}(\kappa^2\text{-}O_2CFc)(CS)(PPh_3)_2]$ ($Fc = CpFeC_5H_4$),[5] $[Ru(CS)Cl_2(H_2IMes)(PCy_3)]$ (H_2IMes = 1,3-dimesitylimidazolin-2-ylidene),[6] $[Ru(CS)Cl_2(H_2IMes)(4\text{-}Me_2Npy)_2]$,[6] $[Ru(CS)H\{(\mu\text{-}H)_2BH_2\}(PCy_3)_2]$,[7] and $[Ru(CS)Cl_2(H_2IMes)_2]$,[8] and $[Ru(CNCH_2Ts)(CS)Cp(H_2IMes)][BAr^F_4]$;[9]

Rhodium: $[Rh(CS)(PEt_3)_2(SGePh_3)]$,[10] $[Rh_2(\mu\text{-}CS)Cl_2(\mu\text{-}dppm)_2]$,[11] and $[Rh_2(\mu\text{-}CS)Cl_2(OH_2)(\mu\text{-}dmad)(\mu\text{-}dppm)_2]$;[11] **Iridium:** $[Ir(CS)(N\{2\text{-}(P\text{-}i\text{-}Pr_2)\text{-}4\text{-}MeC_6H_3\}_2)]$.[12]

2.4 Selenocarbonyl (CSe) Complexes

In 1975, two kinds of transition metal complexes with the seleonocarbonyl (CSe) ligand were reported independently. Initially, the half-sandwich CSe complexes $[M(CO)_2(ring)(CSe)]$ (M = Mn, Cr; ring = Cp, $\eta^6\text{-}C_6H_5CO_2Me$) were prepared using carbon diselenide (CSe_2) as the CSe source.[13] The octahedral

CSe complexes $[Ru(CO)(CSe)X^1X^2(PPh_3)_2]$ ($X^1 = X^2 = Cl$, Br, I; $X^1 = I$, $X^2 = OH$), one of which was also characterized by X-ray crystallographic analysis for the first time, were also obtained using CSe_2.[14] Since then, various CSe complexes of transition metals have been reported; in them the CSe ligand is coordinated to the metals through the carbon atom. Only a terminal coordination mode of CSe was known until 2008.[15] Since around 2018, bridging CSe complexes of transition metals have been reported.

The C–Se bond distances in terminal coordinated CSe ligand (1.61–1.79 Å) are similar to the sum of the double- and triple-bond covalent radii (C=Se, 1.74 Å; C≡Se, 1.67 Å),[16] but much shorter than that calculated for the C–Se single bond length (C–Se, 1.91 Å). Regarding a bridging coordinated CSe ligand, the C–Se lengths fall in a range of 1.75–1.88 Å. Absorption bands assigned to νCSe in terminal coordinated CSe complexes appear in a range of 1005–1201 cm^{-1}, whereas the ^{13}C NMR signals of a terminal and a bridging coordinated CSe ligand are observed at 279–368 ppm and 255–369 ppm, respectively. Since 2012, the ^{77}Se NMR spectra of CSe complexes have been reported (see below).

Only recent progress on CSe complexes is summarized in this section, as some reviews are available.[2c,h,17,18]

2.4.1 Complexes Featuring Terminal Coordinated CSe

2.4.1.1 Synthesis

Before 2001, nearly all CSe complexes were prepared using CSe_2 as the starting material (Scheme 2.1). The C–Se bond activation of the initially formed CSe_2-η^2C,Se complexes was achieved by reaction with a phosphine[13] or methylation followed by treatment with hydrogen halide.[14] This method allowed the synthesis of CSe complexes of various transition metals, such as chromium,[13,19,20] manganese,[13,19,21] rhenium,[13,19] ruthenium,[14,22] cobalt,[23] iridium,[24] and platinum,[25] though toxic and foul-smelling CSe_2 was necessary as the CSe source.

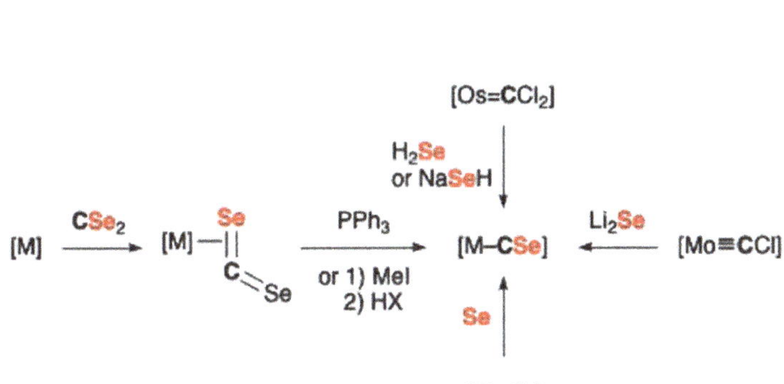

Scheme 2.1 Synthetic pathways to transition-metal complexes featuring terminal coordinated CSe.

The reaction of the dichlorocarbene complexes $[M(CO)(=CCl_2)Cl_2(PPh_3)_2]$ (M = Ru, Os) with H_2Se or NaSeH affords the CSe complexes $[M(CO)(CSe)Cl_2(PPh_3)_2]$ (M = Ru, Os).[26,27] In this case, however, the toxic organomercury reagent $[Hg(CCl_3)_2]$ is required for the preparation of the starting complex. Later, a similar strategy was developed to synthesize the anionic CSe complex $[Et_4N][Mo(CO)_2(CSe)(Tp^{Me_2})]$ (Tp^{Me_2} = hydrotris(3,5-dimethylpyrazolyl)borate),[28] where a reaction of the chlorocarbyne complex $[Mo(CO)_2(CCl)(Tp^{Me_2})]$ with Li_2Se is carried out. The recently utilized strategy aims at avoiding the use of such disagreeable reagents and it consists in the reaction of a terminal coordinated carbido ligand with elemental selenium (Scheme 2.1). In this manner, the anionic dimer of the CSe complex $[\{Mo(CSe)(\mu\text{-}K)(N\{C(CD_3)_2CH_3\}3,5\text{-}Me_2C_6H_3)_3\}_2]$ has been obtained from the reaction of the corresponding carbido complex with selenium.[29] Monomeric anionic CSe complexes of molybdenum and tungsten $[Et_4N][M(CO)_2(CSe)(Tp^{Me_2})]$ (M = Mo, W) can also be prepared[30] and employed as very nice precursors of bridging coordinated CSe complexes.[18] From the reaction of the carbido complex $[Ru(C)Cl_2(H_2IMes)(PCy_3)]$,[31] which is derived in one step using the commercial Grubbs second-generation catalyst, the neutral CSe complexes $[Ru(CSe)Cl_2(H_2IMes)(PCy_3)]$ and $[Ru(CSe)Cl_2(H_2IMes)(4\text{-}Me_2Npy)_2]$ have also been synthesized.[6] The octahedral $[Ru(CSe)Cl_2(H_2IMes)(4\text{-}Me_2Npy)_2]$ can be transformed into the half-sandwiched complex $[Ru(CSe)ClCp(H_2IMes)]$.[32] The adopted multi-step synthetic procedure does not require the use of CSe_2 that is required for the synthesis of related Cp/CSe complexes $[CpM(CO)_2(CSe)]$ (M = Mn, Re).[13,19]

2.4.1.2 Reactions of Complexes Featuring Terminal Coordinated CSe

The reactivity of transition metal CSe complexes has also been explored mainly for the substitution of coordinated ligands other than CSe, and halide abstraction reactions leading to cationic derivatives. These reactivity profiles of CSe complexes have been reviewed.[17,18] Therefore, the most representative and recent examples are summarized in this section.

As far as ligand displacement reactions are considered, when $[Ru(CSe)Cl_2(H_2IMes)(4\text{-}Me_2Npy)_2]$ is reacted with CpLi in the presence of Et_3B, the half-sandwiched $[Ru(CSe)ClCp(H_2IMes)]$ is obtained.[32] Reactions of the anionic $[Et_4N][W(CO)_2(CSe)(Tp^{Me_2})]$ complex with Diazald® (*N*-methyl-*N*-nitroso-*p*-toluenesulfonamide) affords $[W(CO)(CSe)(NO)(Tp^{Me_2})]$,[33] a very unique monomeric tungsten compound that contains coordinated CO, NO, and CSe ligands.

The chlorido ligand abstraction is achieved by the treatment of the complex $[Ru(CSe)ClCp(H_2IMes)]$ with $Na[BAr^F_4]$ ($Ar^F = 3,5\text{-}(CF_3)_2C_6H_3$) in the presence of tosylmethyl isocyanide ($CNCH_2Ts$), and in this way the cationic CSe complex $[Ru(CNCH_2Ts)(CSe)Cp(H_2IMes)][BAr^F_4]$ is formed.[9] This 'chiral-at-metal' complex has been studied by VT NMR spectroscopy to confirm its high configurational stability.

The C–Se bond cleavage reactions of a coordinated CSe ligand has also been observed. As exemplified in the reaction of [Ru(CSe)Cl_2(H_2IMes)(4-Me_2Npy)$_2$] with phosphines,[6] extrusion of the selenium atom has been observed with concomitant regeneration of the carbido complex [Ru(C)Cl_2(H_2IMes)(4-Me_2Npy)$_2$]. Alternatively, [Re(CO)$_2$Cp(thf)] can abstract a selenium atom from [W(CO)-(CSe)(NO)(Tp^{Me_2})] to afford the μ-carbido complex [W(CO)(NO)(Tp^{Me_2})(μ-C)Re(CO)$_2$Cp] along with a μ-selenido complex [{Re(CO)$_2$Cp}$_2$(μ-Se)],[33] albeit both are isolated in low yield.

2.4.1.3 *List of Complexes Featuring Terminal Coordinated CSe and Characterized by X-ray Diffraction Studies*

Spectroscopic and X-ray data for transition metal complexes featuring a terminal coordinated CSe ligand, which have been characterized by single-crystal X-ray diffraction analysis, are listed in Table 2.1.

2.4.2 Complexes Featuring Bridging Coordinated CSe

As expected from the chemistry of CO and CS ligands, a CSe ligand could also adopt various coordination modes (Scheme 2.2). Other than the terminal mode, there are several other possible coordination modes regarding bimetallic systems, such as (1) symmetric bridging mode (μ-κC:κC), (2) semi-bridging mode, (3) end-on/side-on (σ–π, μ-κC:η^2-C,Se bridging mode), and (4) end-on/end-on (isoselenocarbonyl-type, μ-κC:κSe bridging mode). However, although the first complexes featuring a terminal coordinated CSe were isolated in 1975, only [Mo(CO)$_2${μ-CSe}{Pt(C≡C*t*-Bu)(PPh$_3$)$_2$})(Tp^{Me_2})] featuring a bridging CSe ligand in an end-on/end-on fashion was reported until 2011.[15] Other complexes with bridging CSe were only proposed as putative intermediates in several reactions of terminal coordinated CSe complexes.[34]

The latest review article published in 2020 summarized many reactions of CSe complexes with transition-metal complexes that lead to bridging CSe complexes.[18] Therefore, in this section, such complexes featuring bridging CSe are only briefly mentioned. The highest number of bridging CSe complexes synthesized so far belongs to the type end-on/end-on (M–CSe–M′), whereas a semi-bridging coordination mode is still not known. All the complexes featuring bridging CSe have been obtained from the anionic complexes [Et_4N][M(CO)$_2$(CSe)(Tp^{Me_2})] (M = Mo, W) except for complexes containing a symmetric CSe bridge. Consequently, [Et_4N][M(CO)$_2$(CSe)(Tp^{Me_2})] would be an excellent precursor for complexes characterized by new coordination modes of the CSe ligand.

2.4.2.1 *Complexes Featuring Symmetric Bridging CSe*

Analogously to the reaction of a terminal coordinated carbido ligand with elemental selenium that forms a terminal coordinated CSe ligand

Table 2.1 Spectroscopic and X-ray data for selected CSe complexes with a terminal coordinated CSe ligand.

CSe complexes	νCSe (cm^{-1})	C–Se (Å)	M–CSe (Å)	M–C–Se (°)	δ_{CSe} (ppm)	δ_{Se} (ppm)	Ref.
Chromium							
[$Cr(CO)_2(\eta^6$-$C_6H_5CO_2Me)(CSe)$]	1063[a]	1.728(10)	1.786(11)	179.04(65)	368.0[b]	—	13, 19, 20*a* and 20*b*
mer-[$Cr(CO)_2(CSe)\{P(OMe)_3\}_3$]	1018[a]	1.750(3)	1.785(3)	176.9(2)	356.4[c,d]	—	20*c* and *d*
[K(benzo-15-crown-5)$_2$][$Mo(CSe)(N\{C(CD_3)_2CH_3\}3,5$-$Me_2C_6H_3)_3$]	—	1.750(12)	1.835(12)	—	279.1[c,e]	—	29
Molybdenum							
[Et_4N][$Mo(CO)_2(CSe)(Tp^{Me_2})$]	1005[f]	1.786(2)	1.857(2)	171.23(15)	328.9[g]	—	28 and 30*b*
	1009[f]						
Tungsten							
[Et_4N][$W(CO)_2(CSe)(Tp^{Me_2})$]	1030[h]	1.781(11)	1.872(11)	170.9(7)	311.3[g]	542[g]	30*c*
[$W(CO)(CSe)(NO)(Tp^{Me_2})$]	1088[b]	—	—	—	367.2[i]	1188[i]	33
	1078[h]						
Ruthenium							
[$Ru(CO)(CSe)Cl_2(PPh_3)_2$]	1125[j]	1.668(25)	1.828(25)	176.1(2)	—	—	14 and 22
[$Ru(CNXy)_2(CSe)I_2(PCy_3)$]	1102[f]	1.682(12)	1.826(5)	172.7(7)	334.2[i]	—	17
		1.648(13)	1.841(5)	171.9(8)			
[$Ru(CNCH_2Ts)(CSe)Cp(H_2IMes)$][$BAr^F_4$]	—	1.693(4)	1.827(4)	174.2(2)	322.4[i]	886[i]	9
[$Ru(CNXy)_2(CSe)Cl_2(H_2IMes)$]	1104[f]	1.694(2)	1.841(5)	165.53(16)	326.9[i]	—	17
[$Ru(CSe)ClCp(PCy_3)$]	1104[f]	1.703(8)	1.806(8)	168.5(4)	322.5[i]	—	17
		1.714(10)	1.787(10)	170.3(6)			
$Ru(H_2IMes)(CSe)Cl_2(py)$]	1132[f]	1.718(6)	1.735(6)	171.4(4)	293.7[i]	—	17
	1155[f]						
[$Ru(CSe)ClCp(H_2IMes)$]	1090[h]	—	—	—	319.6[i]	804[i]	9 and 32
[$Ru(CSe)Cl_2(H_2IMes)(4$-$Me_2Npy)_2$]	1119[f]	1.720(3)	1.766(3)	165.4(2)	317.3[i]	—	6
$Ru(CSe)Cl_2(H_2IMes)(PCy_3)$]	1150[f]	1.723(4)	1.729(4)	176.9(2)	287.9[i]	—	17
[$Ru(CSe)Cl_2(H_2IMes)(4$-$NMe_2py)$]	1159[f]	1.724(4)	1.718(4)	169.8(3)	292.7[i]	—	17
	1119[f]	1.720(5)	1.728(5)	169.4(3)			

$[Ru(CSe)Cl_2(H_2IMes)(py)_2]$	1123[f]	1.720(20)	1.766(2)	164.20(12)	314.3[i]	—	17
$[Ru(CSe)Cl_2(PCy_3)_2]$	1151[f]	1.726(6)	1.726(6)	179.6(3)	291.2[i]	—	17
$[Ru(CSe)Cl_2(H_2IMes)(PPh_3)]$	1146[f]	1.731(7)	1.710(7)	172.0(4)	289.1[i]	—	17
Osimium							
$[Os(CO)(CSe)Cl_2(PPh_3)_2]$	1156[j]	1.609(15)	1.913(5)	175.3(9)	278.8[k]	—	26 and 27*b*
$[Os(CS)(CSe)Cl_2(PPh_3)_2]$	1130[j]	—	—	—	—	—	27*b*

[a] CS_2.
[b] CH_2Cl_2.
[c] C_6D_6.
[d] In the presence of a small amount of $[Cr(acac)_3]$.
[e] Data for $[\{Mo(CSe)(\mu\text{-}K)(N\{C(CD_3)_2CH_3\}3,5\text{-}Me_2C_6H_3)_3\}_2]$.
[f] KBr.
[g] CD_3CN.
[h] Neat, ATR method.
[i] $CDCl_3$.
[j] Nujor.
[k] $(CDCl_2)_2$.

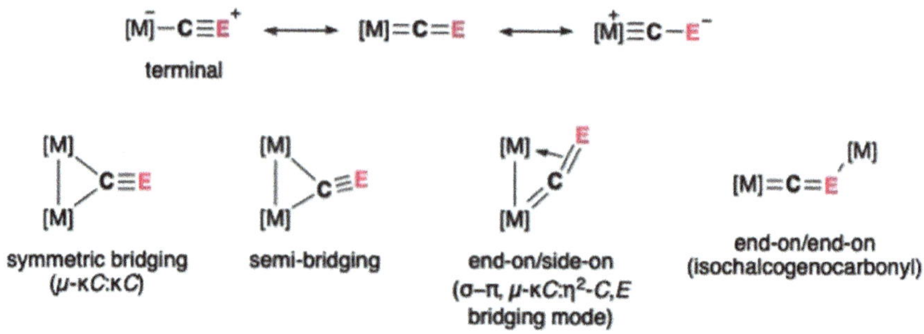

Scheme 2.2 Possible coordination modes of a CE ligand.

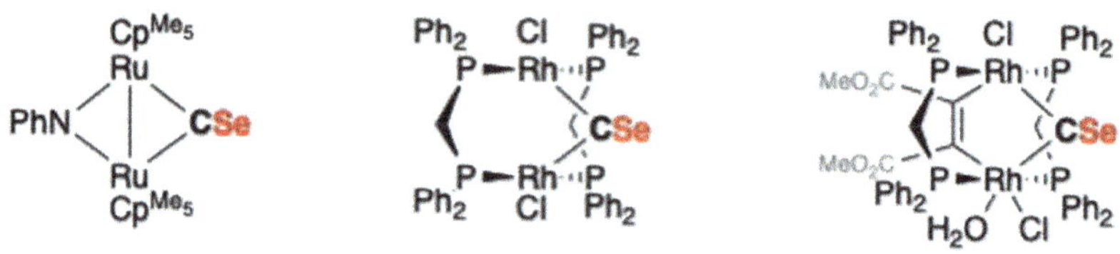

Scheme 2.3 Symmetric bridging CSe complexes (μ-κC:κC).

(Scheme 2.1), the bridging carbido complex $[(RuCp^{Me_5})_2(\mu\text{-}C)(\mu\text{-}NPh)]$, which is generated *in situ* at −78 °C, also undergoes selenation with elemental selenium to afford the complex $[(RuCp^{Me_5})_2(\mu\text{-}CSe)(\mu\text{-}NPh)]$ in which the CSe ligand adopts a symmetric bridging mode (Scheme 2.3).[35] On the other hand, the reaction of the isolated μ-carbido complexes $[Rh_2(\mu\text{-}C)Cl_2(dppm)_2]$ (dppm = $Ph_2PCH_2PPh_2$)[36a] and $[Rh_2(\mu\text{-}C)(\mu\text{-}dmad)Cl_2(dppm)_2]$ (dmad = dimethylacetylenedicarboxylate)[36b] with selenium affords the complexes $[(RhCl)_2(\mu\text{-}CSe)(\mu\text{-}dppm)_2]$ and $[(RhCl)\text{-}(\mu\text{-}CSe)(\mu\text{-}dmad)(\mu\text{-}dppm)_2\{RhCl(OH_2)\}]$, respectively,[36c] which represent rare examples of dinuclear rhodium complexes with symmetric bridging CSe. Both molecular structures have been unambiguously determined by a single-X-ray diffraction study to confirm that the CSe ligand coordinates to each rhodium center through the carbon atom. The C–Se bond lengths at 1.75 Å (mean) are shorter than those observed in the aforementioned μ-CSe (μ-κC:κC) dinuclear complex of ruthenium [1.82 Å (mean)]. The first observation of the infrared absorption bands for a symmetric bridging CSe (957 and 1000 cm^{-1} for $[(RhCl)_2(\mu\text{-}CSe)(\mu\text{-}dppm)_2]$ and $[(RhCl)(\mu\text{-}CSe)(\mu\text{-}dmad)(\mu\text{-}dppm)_2\{RhCl(OH_2)\}]$, respectively) represents a significant achievement in this field. However, these bands weakly appear as compared with those assigned to a terminal coordinated CSe ligand. These reactions of complexes with a coordinated carbido ligand resembles those between N-heterocyclic carbenes and elemental selenium, in which a C=Se bond is smoothly formed.[37]

2.4.2.2 Complexes Featuring End-on/Side-on Bridging CSe

There are two complexes in which the CSe ligand binds two metal centers in an end-on/side-on bridging fashion (Scheme 2.4). The reaction of

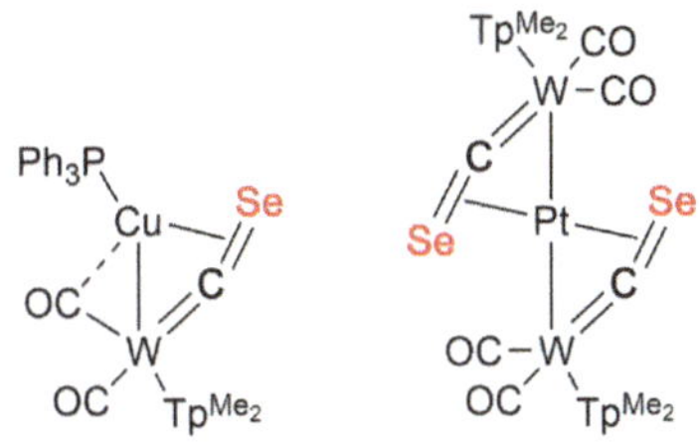

Scheme 2.4 Complexes featuring end-on/side-on (σ–π, μ-κC:η^2-C,Se bridging mode) bridging CSe ligand.

[Et_4N][W$(CO)_2$(CSe)(Tp^{Me_2})] with [Cu$(NCMe)_4$][PF_6] and PPh_3 afforded the first complex featuring a σ–π bridging CSe ligand [WCu$(CO)_2$(μ-CSe)(PPh_3)(Tp^{Me_2})],[30c] and this complex can be further converted into the complex [W$(CO)_2${μ-CSeCu$(PPh_3)_3$}(Tp^{Me_2})] featuring an endo-on/endo-on bridging CSe. Another complex containing a σ–π bridging CSe ligand [W_2Pt(μ-CSe)$_2$(CO)$_4$(Tp^{Me_2})$_2$] has readily been obtained by heating [W_2Pt$(CO)_4$(μ-CSe)$_2$$L_2(Tp^{Me_2})_2$] {$L_2$ = cod (η^4-cycloocta-1,5-diene), and nbd (η^4-norborna-2,5-diene)} in which an end-on/end-on (μ-κC:κSe) bridging coordination mode of CSe is observed to afford the trimetallic moiety W–(CSe)–Pt–(SeC)–W.[38]

2.4.2.3 Complexes Featuring Bridging End-on/End-on CSe

As mentioned above, complexes featuring end-on/end-on bridging CSe, are currently the most common. [Mo$(CO)_2${μ-CSePt(C≡C*t*-Bu)$(PPh_3)_2$}(Tp^{Me_2})] is the first complex of this type reported in 2008.[15] The anionic complexes of molybdenum and tungsten, [Et_4N][M$(CO)_2$(CSe)(Tp^{Me_2})] (M = Mo, W) featuring a terminal coordinated CSe can afford various dinuclear complexes of the type [Mo–CSe–M] (M = Fe, Ru)[30b] and [W–CSe–M] (Pt, Cu, Ag, Au)[30c,38] with end-on/end-on bridging CSe if treated with appropriate transition-metal complexes. The structures of almost all these complexes have been determined by X-ray analysis and are listed in Table 2.2.

2.4.2.4 Complexes Featuring Bridging μ_3- and μ_4-CSe

The CSe ligand can bridge more than two metal centers, and trinuclear and tetranuclear CSe complexes are also known (Scheme 2.5). Two types of μ_3-CSe ligands can be found in [$W_2PtAu_2(CO)_4$(μ_3-CSe)$_2$$Cl_2$L(Tp^{Me_2})$_2$] (L = cod, nbd)[38] and [{Mo$(CO)_2$(Tp^{Me_2})(μ_3-CSe)Rh(cod)}$_2$].[39] In [$W_2PtAu_2(CO)_4$(μ_3-CSe)$_2$$Cl_2$L(Tp^{Me_2})$_2$] the coordination mode of CSe can be regarded as a combination of semi-bridging and end-on/end-on modes (an overall μ_3-κC:κC:κSe coordination). In [{Mo$(CO)_2$(Tp^{Me_2})(μ_3-CSe)Rh(cod)}$_2$], CSe adopts a μ_3-κC:κSe:κSe coordination mode, which has not been found yet in CO and CS coordination chemistry.

Only one μ_4-CSe complex is known so far, [WCu_3(μ_4-CSe)$(CO)_2$([9]-aneS_3)$_3$(Tp^{Me_2})][PF_6]$_2$ {[9]aneS_3 = 1,4,7-trithiacyclononane}.[30c] The CSe ligand

Table 2.2 Spectroscopic and X-ray data for selected CSe complexes with a bridging CSe ligand(s).

CSe complexes	νCSe (cm^{-1})	C–Se (Å)	M–CSe (Å)	M–C–Se (°)	δ_{CSe} (ppm)	δ_{Se} (ppm)	Ref.
Symmetric bridging							
[(RuCp^{Me_5})$_2$(μ-CSe)(μ-NPh)]	—	1.844(12) 1.802(13)	1.934(12) 1.931(12) 1.954(12) 1.949(12)	138.3(6) 138.5(7) 139.0(7) 139.0(7)	368.5^b	—	35
[(RhCl)$_2$(μ-CSe)(μ-dppm)$_2$]	957^c	1.746(9) 1.727(11)	1.929(8) 1.911(10) 1.957(10) 1.931(10)	134.1(5) 134.1(5) 135.8(5) 132.3(5)	—	—	36*c*
[(RhCl)(μ-CSe)(μ-dmad)(μ-dppm)$_2${RhCl(OH$_2$)}]	1000^c	1.785(6)	1.927(6) 1.935(6)	133.6(3) 102.1(3)	—	—	36*c*
End on/side on							
[WCu(CO)$_2$(μ-CSe)(PPh$_3$)(Tp^{Me_2})]	—	1.798(14)	1.895(15)^d	177.0(7)^e	278.4^b	—	30*c*
[W$_2$Pt(μ-CSe)$_2$(CO)$_4$(Tp^{Me_2})$_2$]		1.820(9)	1.892(9)^d	177.8(5)^e	—	502	38
End on/end on							
[Mo(CO)$_2${μ-CSeRuCp(PPh$_3$)$_2$}(Tp^{Me_2})]	—	1.822(6)	1.829(6)^f	173.0(3)^g	—	—	30*b*
[Mo(CO)$_2${μ-CSeRuCp^{Me_5}(PPh$_3$)$_2$}(Tp^{Me_2})]	—	1.825(4)	1.841(4)^f	172.9(2)^g	—	—	30*b*
[Mo(CO)$_2${μ-CSeFeCp(PPh$_3$)$_2$}(Tp^{Me_2})]	—	—	—	—	296.8^h	—	30*b*
[Mo(CO)$_2${μ-CSeRuH(CO)$_2$(PPh$_3$)$_2$}(Tp^{Me_2})]	—	1.838(7)	1.840(7)^f	159.2(4)^g	308.6^h	—	30*b*
[Mo(CO)$_2$({μ-CSe}-*cis*-{Pt(C≡C*t*-Bu)(PPh$_3$)$_2$})(Tp^{Me_2})]	—	1.836(3)	1.832(3)	166.5(2)	297.2^b	—	15
[Mo(CO)$_2$({μ-CSe}-*trans*-{Pt(C≡C*t*-Bu)(PPh$_3$)$_2$})(Tp^{Me_2})]	—	1.828(4)	1.834(4)	163.1(2)	302.5^b	—	15
[W(CO)$_2${μ-CSeCu(PPh$_3$)$_3$}(Tp^{Me_2})]	—	1.798(5)	1.850(5)^d	167.0(3)^e	278.6^b	—	30*c*
[W(CO)$_2$(μ-CSeAgPPh$_3$)(Tp^{Me_2})]	—	—	—	—	—	—	30*c*
[W(CO){μ-CSeAg(PPh$_3$)$_3$}$_2$(Tp^{Me_2})]	—	1.786(6)	1.864(6)^d	167.4(3)^e	—	—	30*c*
[W(CO)$_2$(μ-CSeAuPPh$_3$)(Tp^{Me_2})]	—	1.866(6)	1.826(6)^d	170.8(4)^e	268.2^b	—	30*c*
[W(CO)$_2${μ-CSeAu(PPh$_3$)$_3$}(Tp^{Me_2})]	—	1.770(9)	1.879(9)^d	167.2(5)^e	266.7^i	394^i	30*c*
[W(CO)$_2$(μ-CSeAuPCy$_3$)(Tp^{Me_2})]	—	—	—	—	270.8^b	488^a	30*c*
[W$_2$Pt(cod)$_2$(CO)$_4$(μ-CSe)$_2$(Tp^{Me_2})$_2$]	—	—	—	—	266.6^i	597^i	38
[W$_2$Pt(nbd)$_2$(CO)$_4$(μ-CSe)$_2$(Tp^{Me_2})$_2$]	—	1.867(8)	1.813(8)^d	168.5(5)^e	268.8^i	660^i	38

$[W_2Pt(CN\textit{t}\text{-}Bu)_2(CO)_4(\mu\text{-}CSe)_2(Tp^{Me_2})_2]$	—	1.849(6)	1.828(6)^d	166.4(3)^e	289.0^i	—	38
$[W_2Pt(CNMes)_2(CO)_4(\mu\text{-}CSe)_2(Tp^{Me_2})_2]$	—	—	—	—	—	—	38
μ_3-CSe							
$[W_2PtAu_2(CO)_4(\mu_3\text{-}CSe)_2Cl_2(cod)(Tp^{Me_2})_2]$	—	1.838(8) 1.840(8)	1.934(7)^d 1.914(7)^d	148.8(4)^e 152.2(4)^e	257.1^i	545^i	38
$[W_2PtAu_2(CO)_4(\mu_3\text{-}CSe)_2Cl_2(nbd)(Tp^{Me_2})_2]$	—	—	—	—	255.2^i	587^i	38
$[\{Mo(CO)_2(Tp^{Me_2})(\mu_3\text{-}CSe)Rh(cod)\}_2]$	—	1.854(3)	1.815(3)^j	173.07(18)^k	—	521^h	39
μ_4-CSe	—	—	—	—	—	—	
$[WCu_3(\mu_4\text{-}CSe)(CO)_2([9]aneS_3)_3(Tp^{Me_2})][PF_6]_2$	—	1.881(3)	1.879(3)^d	155.6(2)^e	—	—	30c

[a] Distances and bond angles of both molecules per unit cell are shown.
[b] CD_2Cl_2.
[c] Neat, ATR method.
[d] W–CSe bond length.
[e] W–C–Se bond angle.
[f] Mo–CSe bond length.
[g] Mo–C–Se bond angle.
[h] C_6D_6.
[i] $CDCl_3$.
[j] Rh–CSe bond length.
[k] Rh–C–Se bond angle.

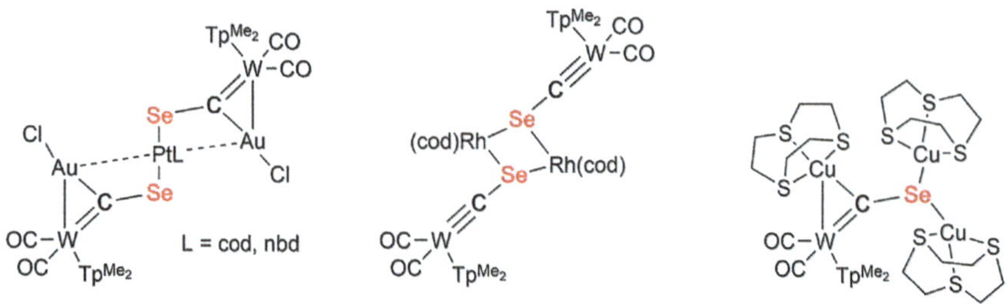

Scheme 2.5 Complexes featuring μ_3- and μ_4-CSe coordination modes.

can be viewed as a combination of semi-bridging and end-on/end-on coordination modes (μ_4-κC:κC:κSe:κSe).

2.4.2.5 List of Complexes Featuring Bridging CSe

Spectroscopic and selected X-ray data for complexes featuring bridging CSe are listed in Table 2.2.

2.5 Tellurocarbonyl (CTe) Complexes

The first tellurocarbonyl (CTe) complex [Os(CO)(CTe)$Cl_2(PPh_3)_2$] was reported in 1980.[26] Since then, only a limited number of CTe complexes have been reported. This situation is probably due to the limited variety of methods available for the synthesis of CTe complexes. As described above, various synthetic methods can be applied to obtain CS complexes.[2] For example, commercially available thiophosgene ($Cl_2C{=}S$) and carbon disulfide (CS_2) are usually employed to synthesize CS complexes. In contrast, their tellurium homologs are difficult to be produced and utilized, particularly tellurphosgene ($Cl_2C{=}Te$) and carbon disulfide (CTe_2), which are still unknown.

The first CTe complex with a half-sandwich structure [Ru(CTe)-ClCp(H_2IMes)] was synthesized in 2017.[32] Any bridging coordinated CTe ligands were not known until 2019,[40] the most outstanding achievements in this field. Since around 2017, the ^{125}Te NMR signals of CTe complexes have also been reported.

The C–Te bond lengths in CTe complexes fall in the range of 1.91–2.02 Å for complexes featuring a terminal coordinated CTe ligand, which are longer than the sum of triple-bond covalent radii and similar to the sum of double-bond covalent radii (C≡Te, 1.81 Å; C=Te, 1.95 Å).[16] On the other hand, complexes featuring bridging CTe ligands show C–Te bond lengths of 2.03–2.10 Å, and these distances are close to the single-bond distance (C–Te, 2.11 Å).[16]

Spectroscopically, terminal coordinated CTe ligands exhibit a νCTe band in the range of 951–1065 cm^{-1} and the ^{13}C NMR signals attributable to the

CTe ligand in the range 284–359 ppm. ^{13}C NMR signals due to bridging coordinated CTe ligands are observed in the range 261–276 ppm. The ^{125}Te NMR spectra show signals at around 973–1248 ppm assigned to a terminal CTe ligand and at around 299–491 ppm due to a bridging coordinated CTe ligand. A couple of reviews summarize these data.[17,18]

2.5.1 Complexes Featuring Terminal Coordinated CTe

2.5.1.1 Synthesis

Synthesis of CTe complexes has been achieved in only a few limited ways due to the scarcity of suitable CTe sources (Scheme 2.6). The first CTe complex $[Os(CO)(CTe)Cl_2(PPh_3)_2]$ was synthesized by the reaction of the dichlorocarbene complex $[Os(CO)(=CCl_2)Cl_2(PPh_3)_2]$ with NaTeH.[26,27] Later, a similar strategy was applied to the synthesis of an anionic CTe complex $[Et_4N][Mo(CO)_2(CTe)Tp^{Me_2}]$,[28] where the chlorocarbyne complex $[Mo(CO)_2(CCl)Tp^{Me_2}]$ was used as the starting material.

The most commonly used reaction, in recent years, is that of terminal carbide complexes, which have been reviewed recently,[31d] with elemental tellurium. This method does not require the highly sensitive NaTeH and Na_2Te reagents. From the reaction of the anionic carbido complex $[Mo(C)(\mu\text{-K})(N\{C(CD_3)_2CH_3\}3,5\text{-}Me_2C_6H_3)_3\}_2]$ and $Li[M(C)(CO)_2(Tp^{Me_2})]$ (M = Mo, W), the corresponding CTe complexes $[Mo(\mu\text{-K})(CTe)(N\{C(CD_3)_2CH_3\}3,5\text{-}Me_2C_6H_3)_3\}_2]$[29] and $[Et_4N][M(CO)_2(CTe)(Tp^{Me_2})]$ (M = Mo, W)[41] are obtained, respectively. Tellurium atom transfers to the neutral carbido complex $[Ru(C)Cl_2(H_2IMes)(PCy_3)]$[31] in the presence of 4-Me_2Npy to afford the CTe complex $[Ru(CTe)Cl_2(H_2IMes)(4\text{-}Me_2Npy)_2]$.[6]

By ligand substitution reactions starting from $[Ru(CTe)Cl_2(H_2IMes)(4\text{-}Me_2Npy)_2]$, hitherto unknown half-sandwich-type ruthenium CTe complexes have been obtained. When CpLi is reacted with $[Ru(CTe)Cl_2(H_2IMes)(4\text{-}Me_2Npy)_2]$ in the presence of Et_3B, the Cp/CTe complex $[Ru(CTe)ClCp(H_2IMes)]$

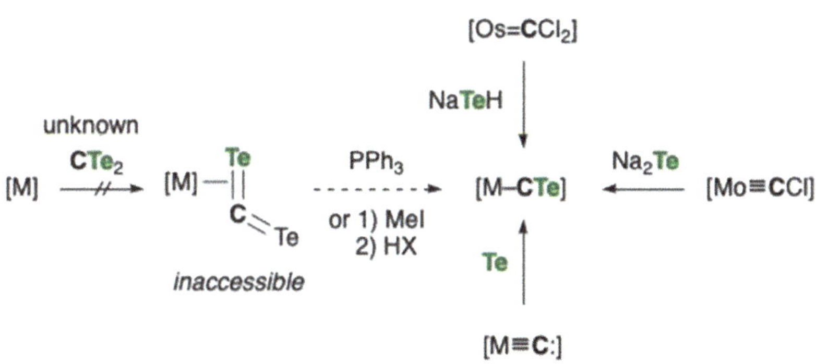

Scheme 2.6 Synthetic pathways to transition-metal complexes featuring terminal coordinated CTe.

is prepared.[32] Moreover, the obtained CTe complex can be transformed into the cationic CTe complex $[Ru(CNCH_2Ts)(CTe)Cp(H_2IMes)][BAr^F_4]$ ($Ar^F = 3,5\text{-}(CF_3)_2C_6H_3$) with a tosylmethylisocyanido ($CNCH_2Ts$) ligand.[9]

2.5.1.2 Reactions of Complexes Featuring Terminal Coordinated CTe

The reactivity profiles of the anionic CTe complexes $[Et_4N][M(CO)_2(CTe)(Tp^{Me_2})]$ (M = Mo, W) have been summarized very recently,[18] and various derivatives can be obtained. In addition, a previous review already described some reactions of previous CTe complexes.[17] Therefore, only the most recent activity of this kind of complexes is described in this section.

When $[Ru(CTe)ClCp(H_2IMes)]$ is treated with an excess amount of PPh_3 at 100 °C, no reaction is observed.[32] This result contrasts with that observed in the case of $[Ru(CTe)Cl_2(H_2IMes)(4\text{-}Me_2Npy)_2]$, which reacts smoothly with tertiary phosphines at room temperature and regenerates the carbide complexes $[Ru(C)Cl_2(H_2IMes)(PR_3)]$ (R = Cy, Ph).[6,17] Treatment of such a stable $[Ru(CTe)ClCp(H_2IMes)]$ complex with S_8 at room temperature furnishes the CS complex $[Ru(CS)ClCp(H_2IMes)]$ *via* a tellurium–sulfur exchange.[32] Furthermore, in the presence of $Et_3B{\cdot}DMAP$, a tellurium/oxygen exchange reaction of $[Ru(CTe)ClCp(H_2IMes)]$ with O_2 has been observed, yielding the corresponding CO complex $[Ru(CO)ClCp(H_2IMes)]$.[32]

2.5.2 Complexes Featuring Bridging Coordinated CTe

Analogously to CSe, CTe can also adopt various coordination modes, such as (1) symmetric bridging (μ-κC:κC), (2) semi-bridging, (3) end on/side-on (σ–π, μ-κC:η^2-C,Te bridging mode), and (4) end-on/end-on (μ-κC:κTe bridging mode, isotellurocarbonyl) (Scheme 2.2).

As mentioned in the introduction section regarding CTe complexes, some new coordination modes of a CTe ligand were reported in 2019 (Scheme 2.7).[40] Reactions of $[Et_4N][W(CO)_2(CTe)(Tp^{Me_2})]$ with appropriate transition-metal complexes afford not only the dinuclear μ-CTe complexes $[WCu(\mu\text{-}CTe)(PCy_3)(CO)_2Tp^{Me_2}]$ and $[W(CO)_2(\mu\text{-}CTe)(Tp^{Me_2})AuPCy_3]$ but also the formal trinuclear μ_3-CTe complex cation $[WAu_2(\mu_3\text{-}CTe)(dppm)(CO)_2Tp^{Me_2}]^+$ in solution, which forms the trimer $[\{WAu_2(\mu\text{-}CTe)(dppm)(CO)_2(Tp^{Me_2})\}_3][PF_6]_3$ in the solid state.

2.5.3 List of all CTe Complexes

Spectroscopic and selected X-ray data for all complexes featuring a terminal coordinated CTe are listed in Table 2.3.

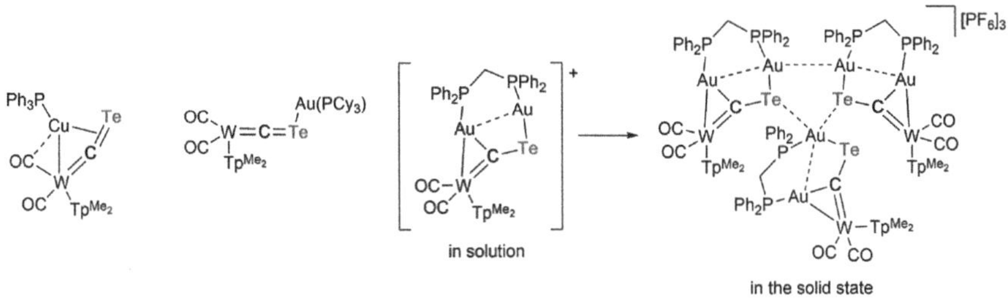

Scheme 2.7 Complexes featuring μ- and μ_3-CTe coordination modes.

2.6 Comparison of a Series of CE (E = O, S, Se, Te) Complexes

Some complete sets of chalcogenocarbonyl (CE; E = O, S, Se, Te) complexes of the type $[Os(CO)(CE)Cl_2(PPh_3)_2]$ (E = O,[26,27,42] S,[26,27,43] Se,[26,27] Te[26,27]) $[Et_4N]$-$[Mo(CO)_2(CE)(Tp^{Me_2})]$, (E = O,[44] S,[28] Se,[28,30b] Te[28,41]) and $[Ru(CE)Cl_2(H_2IMes)$-$(4\text{-}Me_2Npy)_2]$,[6] $[Ru(CE)ClCp(H_2IMes)]$,[32] and $[Ru(CNCH_2Ts)(CTe)Cp(H_2IMes)]$-$[BAr^F_4]$[9] as well as the calculated set $[Ru(CO)_4(CE)]$[45] are known (Table 2.4). A systematic comparison of some of them has already been reviewed.[17,18] The *trans* influence, which is related to the σ-donor properties, and π-acceptor abilities of the CE ligands is evaluated by comparing the bond lengths in the fragment [L–M–CE]. Both abilities of the CE ligands appear to increase in the order CO < CS < CSe < CTe. For example, the latest complete series $[Ru(CNCH_2Ts)(CTe)Cp(H_2IMes)][BAr^F_4]$[9] has been investigated by X-ray dif fraction and NMR spectroscopy as well as detailed computational analyses. The study indicates the importance of the contribution of the [Ru]=C=E and $[Ru^+]{\equiv}C{-}E^-$ canonical structures, which reflects the π-acidic nature of the ligand that increases in the order CO < CS < CSe < CTe. A similar trend is observed in the anionic series $[Et_4N][Mo(CO)_2(CE)(Tp^{Me_2})]$ and $[Et_4N][W(CO)_2(CE)(Tp^{Me_2})]$ (E = O,[44a,46] Se,[30c] Te[41]), albeit not for a complete set. Bridging CE series, $[(RuCp^{Me_5})_2(\mu\text{-}CE)(\mu\text{-}NPh)]$ (E = O,[47] S,[35] Se[35]) and $[(RhCl)_2(\mu\text{-}CE)(\mu\text{-}dppm)_2]$ (E = O,[48] S,[36] Se[36]) are also listed in Table 2.4 for comparison purposes.

2.7 Chalcogenonitrosyl (NE) Complexes

Many nitric oxide (NO) compounds of transition-metals, nitrosyl complexes, have been studied in-depth and periodically reviewed.[49] In contrast, the chemistry of the heavier isologues of NO, *i.e.*, nitric sulfide (NS), nitric selenide (NSe), and nitric telluride (NTe) has been less explored, as these species can only be detected in harsh environments such as low-temperature matrixes.[50] The corresponding transition-metal complexes that contain NE (E = S, Se, Te), *i.e.*, thionitrosyl (M–NS), selenonitrosyl (M–NSe), and telluronitrosyl complexes (M–NTe) should be prepared as stable compounds. Studying the

Table 2.3 Spectroscopic and X-ray data for all CTe complexes.

CTe complexes	νCTe (cm^{-1})	C–Te (Å)	M–CTe (Å)	M–C–Te (°)	δ_{CTe} (ppm)	δ_{Te} (ppm)	Ref.
Molybdenum							
$[Et_4N][Mo(CO)_2(CTe)(Tp^{Me_2})]$	951[a] 952[b]	—	—	—	—	1248[c]	28 and 41
$[\{Mo(CTe)(\mu\text{-}K)(N\{C(CD_3)_2CH_3\}3,5\text{-}Me_2C_6H_3)_3\}_2]$	—	—	—	—	252.6[d]	—	29
Tungsten							
$[Et_4N][W(CO)_2(CTe)(Tp^{Me_2})]$	982[a]	2.018(7)	1.855(7)	171.7(4)	302.3[e]	1031[f]	41
$[WCu(\mu\text{-}CTe)(PCy_3)(CO)_2(Tp^{Me_2})]$	—	2.033(5)	1.876(5)[g]	175.4(2)[h]	276.2[i]	491[j]	40
$[W(CO)_2\{\mu\text{-}CTe\ AuPCy_3\}(Tp^{Me_2})]$	—	2.059(4)	1.841(4)[g]	171.1(3)[h]	260.6[i]	299[j]	40
$[\{WAu_2(\mu\text{-}CTe)(dppm)(CO)_2(Tp^{Me_2})\}_3][PF_6]_3$	—	2.066(11) 2.096(8) 2.073(8)	1.895(10)[g] 1.882(8)[g] 1.908(9)[g]	159.5(6)[h] 150.7(5)[h] 155.8(5)[h]	—	—	40
Ruthenium							
$[Ru(CNCH_2Ts)(CTe)Cp(H_2IMes)][BAr^F_4]$	1007[k]	1.908(4)	1.813(4)	171.4(2)	343.7[j]	1055[j]	9
$[Ru(CTe)Cl_2(CNXy)_2(H_2IMes)]$[l]	1137[a]	1.914(6) 1.925(6)	1.825(6) 1.817(5)	171.2(2)	350.5[i]	—	17
$[Ru(CTe)Cl_2(H_2IMes)(4\text{-}Me_2Npy)]$	1065[a]	1.948(3)	1.701(3)	170.53(17)	284.0[i]	—	17
$[Ru(CTe)Cl_2(H_2IMes)(4\text{-}Me_2Npy)_2]$	1024[a]	1.952(5)	1.748(3)	177.4(4)	327.9[m]	—	6
$[Ru(CTe)ClCp(H_2IMes)]$	990[k]	1.9572(16)	1.7719(16)	161.92(9)	331.0[j]	973[j]	32
$[Ru(CNXy)_2(CTe)I_2(PCy_3)]$	1006[a]	—	—	—	358.9[i]	—	17
Osimium							
$[Os(CO)(CTe)Cl_2(PPh_3)_2]$	1046[n]	1.923(12)	1.813(12)	175.3(7)	297.7[j]	—	26 and 27
$[Os(CO)_2(CTe)(PPh_3)_2]$	—	1.947(20)	1.857(19)	—	—	—	27*a*
$[OsCl(NCMe)(CO)(CTe)(PPh_3)_2][ClO]$	1048[n]	—	—	—	—	—	27*b*

[a] KBr.
[b] CH_2Cl_2.
[c] $(CD_3)_2SO$.
[d] C_6D_6.
[e] $CD_3CN/CD_3C_6D_5$ or $CDCl_3$.
[f] CD_3CN.
[g] W–CTe bond length.
[h] W–C–Te bond angle.
[i] CD_2Cl_2.
[j] $CDCl_3$.
[k] Neat, ATR method (corrected).
[l] Distances and bond angles of both molecules per unit cell are shown.
[m] CD_2Cl_2 in the presence of excess $4\text{-}Me_2Npy$.
[n] Nujor.

Table 2.4 Spectroscopic and X-ray data for a series of CE complexes after the 2012 review.

CE complexes	νCE (cm^{-1})	C–E (Å)	M–CE (Å)	L–MCE (Å)	δ_{CE} (ppm)	Ref.
$[Et_4N][Mo(CO)_3(Tp^{Me_2})]$	1981[a] 1751[a]	1.165(8) 1.173(9) 1.163(9)	1.938(7) 1.940(8) 19.46(8)	2.269(5) 2.260(6) 2.260(6)	—	44
$[Et_4N][Mo(CO)_2(CS)(Tp^{Me_2})]$	1139[b]	—	—	—	—	28
$[Et_4N][Mo(CO)_2(CSe)(Tp^{Me_2})]$	1005[b] 1009[b]	1.786(2)	1.857(2)	2.307(18)	328.9[c]	28 and 30*b*
$[Et_4N][Mo(CO)_2(CTe)(Tp^{Me_2})]$	951[b] 952[d]	—	—	—	—	28 and 41
$[Et_4N][W(CO)_3(Tp^{Me_2})]$	1881[a] 1744[a] 1874[b] 1736[b]	—	—	—	—	44*a* and 46
$[Et_4N][W(CO)_2(CSe)(Tp^{Me_2})]$	1030[e]	1.781(11)	1.872(11)	2.293(7)	311.3[c]	30*c*
$[Et_4N][W(CO)_2(CTe)(Tp^{Me_2})]$	982[b]	2.018(7)	1.855(7)	2.308(5)	302.3[f]	41
$[Ru(CO)ClCp(H_2IMes)]$	1993[g]	—	—	—	202.3[h]	32
$[Ru(CS)ClCp(H_2IMes)]$	1232[g]	—	—	—	300.7[h]	32
$[Ru(CSe)ClCp(H_2IMes)]$	1090[g]	—	—	—	319.6[h]	32
$[Ru(CTe)ClCp(H_2IMes)]$	990[g]	1.9572(16)	1.7719(16)	161.92(9)	331.0[h]	32
$[Ru(CNCH_2Ts)(CO)Cp(H_2IMes)][BAr^F_4]$	2000[g]	1.141(4)	1.879(3)	1.896(1)[i]	197.2[h]	9
$[Ru(CNCH_2Ts)(CS)Cp(H_2IMes)][BAr^F_4]$	1309[g]	1.5478(19)	1.8347(19)	1.922(1)[i]	299.3[h]	9
$[Ru(CNCH_2Ts)(CSe)Cp(H_2IMes)][BAr^F_4]$	—	1.693(4)	1.827(4)	1.928(2)[i]	322.4[h]	9
$[Ru(CNCH_2Ts)(CTe)Cp(H_2IMes)][BAr^F_4]$	1007[g]	1.908(4)	1.813(4)	1.941(2)[i]	343.7[h]	9
$[(RuCp^{Me_5})_2(\mu\text{-CO})(\mu\text{-NPh})]$	1744[b]	1.201(9)	2.002(8)	140.2(6) 138.6(6)	258.7[j]	47
$[(RuCp^{Me_5})_2(\mu\text{-CS})(\mu\text{-NPh})]$[k]	—	1.629(11) 1.627(10)	1.968(12) 1.976(12) 1.981(10) 1.989(11)	139.2(8) 139.6(8) 139.9(7) 139.4(7)	352.3[l]	35

Table 2.4 (*Continued*)

CE complexes	νCE (cm^{-1})	C–E (Å)	M–CE (Å)	L–MCE (Å)	δ_{CE} (ppm)	Ref.
$[(RuCp^{Me_5})_2(\mu\text{-}CSe)(\mu\text{-}NPh)]$[k]	—	1.844(12)	1.934(12)	138.3(6)	368.5[l]	35
		1.802(13)	1.931(12)	138.5(7)		
			1.954(12)	139.0(7)		
			1.949(12)	139.0(7)		
$[(RhCl)_2(\mu\text{-}CO)(\mu\text{-}dppm)_2]$	1745[l]	1.22(3)	1.90(3)	136(2)	—	48
			1.90(3)	133(2)		
$[(RhCl)_2(\mu\text{-}CS)(\mu\text{-}dppm)_2]$[k]	1093[e]	1.614(6)	1.912(6)	135.1(4)	—	11
		1.613(6)	1.920(6)	135.1(4)		
			1.906(6)	135.5(3)		
			1.922(6)	134.0(3)		
$[(RhCl)_2(\mu\text{-}CSe)(\mu\text{-}dppm)_2]$[k]	957[e]	1.746(9)	1.929(8)	134.1(5)	—	11
		1.727(11)	1.911(10)	134.1(5)		
			1.957(10)	135.8(5)		
			1.931(10)	132.3(5)		

[a] CH_3CN.
[b] KBr.
[c] CD_3CN.
[d] CH_2Cl_2.
[e] ATR method.
[f] $CD_3CN/CD_3C_6D_5$ or $CDCl_3$.
[g] Neat, ATR method (corrected).
[h] $CDCl_3$.
[i] Ru–Cp (centroid) bond.
[j] $CD_3C_6D_5$.
[k] Distances and bond angles of both molecules per unit cell are shown.
[l] CD_2Cl_2.

properties of chalcogenonitrosyl (NE; E = S, Se, Te) complexes would allow us to understand further the chemistry of transient NE (E = S, Se, Te) molecules.

In contrast to the aforementioned chalcogenocarbonyl complexes (M–CE; E = O, S, Se, Te),[2,17,18] the chemistry of a series of NE complexes has been much less developed so far, probably due to their scarcity, particularly in the case of NSe and NTe complexes. The latter class of complexes is currently still unknown.

2.8 Thionitrosyl (NS) Complexes

The first thionitrosyl (NS) complex $[Mo(NS)(S_2CNR_2)_3]$ was reported in 1974.[51] Groups 5–9 metal NS complexes have been reported and many review articles are available for the NS complexes.[2e,52] The majority of the crystallographically determined NS complexes are rhenium and osmium complexes.

The N–S bond lengths in NS complexes fall in the range of 1.38–1.59 Å, similar to the sum of triple- and double-bond covalent radii (N≡S: 1.49 Å; N=S: 1.54 Å).[16] The M–N–S linkages are sometimes a bit bent but essentially linear in all NS complexes, unlike NO complexes, where the linear or bent coordinated NO ligand exists. The NS complexes exhibit an νNS absorption band in a range of 1070–1318 cm^{-1}, which is usually assigned by comparison with the band observed for their ^{15}N congener, ^{15}NS complexes, or starting metal complexes.

2.8.1 Synthesis

There are several routes for the synthesis of NS complexes (Scheme 2.8). The first synthesis of an NS complex was achieved by the reaction of the nitrido complex $[Mo(N)(S_2CNR_2)_3]$ and elemental sulfur to afford $[Mo(NS)(S_2CNR_2)_3]$ (R = Me, Et, pentyl).[51] Recently, NS complexes such as $[V(NS)(nacnac)\{O(2,6\text{-}^iPr_2C_6H_3)\}]$,[53] $[(Ir\{N(CHCHP\textit{t}\text{-}Bu_2)_2\}(NS)][PF_6]$,[54] *mer*-$[RuCl_3(NS)(AsPh_3)_2]$,[55] *cis*-$[RuClL_2(NS)]$ (L = 2-[(2,6-diisopropylphenyl)imino]methyl-4,6-dibromophenolate),[56] $[Ph_4P][Ru(hybeb^{R_1,R_2})(NS)(4\text{-}Me_2Npy)]$ (hybeb = *N*,*N*′-(1,2-phenylene)bis(2-hydroxybenzamide), R^1 = H, Cl; R^2 = H, Cl, CF_3)[57] have also been prepared

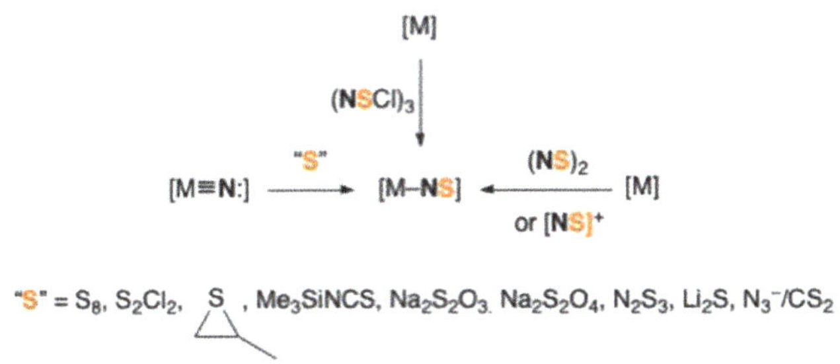

Scheme 2.8 Synthetic pathways to transition-metal complexes featuring terminal coordinated NS.

using this method, that is, sulfurization of the nitrido ligand in a transition-metal complex with elemental sulfur. In place of elemental sulfur, S_2Cl_2,[58] propylene sulfide,[58b,59] Me_3SiNCS,[60] thiosulfate,[61] hydrosulfite,[62] Na_2S_4[63] and Li_2S[64] as sulfur sources can be used for this purpose. While CS_2 is frequently used for the synthesis of CS complexes, NS complexes have rarely been prepared using CS_2.[65] For example, the nitrido complex $[OsCl_2(N)(Tp)]$ (Tp = hydrotris(pyrazoryl)borato) is treated with CS_2 and $[PPN]N_3$ {PPN = $(PPh_2)_2N$} to furnish $[OsCl_2(NS)(Tp)]$. In rare cases, NS complexes can be formed from nitrido complexes and organosulfur ligands, such as dithiocarbamato (S_2CNEt_2)[66] and dithioimidodiphosphinato {N(*t*-$Bu_2P = S)_2$} ligands.[55]

The first "organometallic" NS complexes $[Cr(CO_2)Cp(NS)]$ were synthesized using trithiazyl chloride ($N_3S_3Cl_3$) as the NS source,[67] and various NS complexes have also been prepared by using this reaction system.[68] Although already considered in a previous review,[52f] N_2S_2[69] and $[NS]^+$[70] have also been employed to obtain NS complexes. Reaction of the NSCl complex $[Ph_4P][ReCl_4(NSCl)_2]$ with excess $N(SiMe_3)_3$ affords the dinuclear complex $[Ph_4P]_2[(SN)ReCl_3(\mu\text{-N})(\mu\text{-NSN-}\kappa N,\kappa N')ReCl_3(thf)]$.[71]

Recently, $[Cr(NS)(L)_5][PF_6]_2$ (L = OH_2, dmso, *N*-methylformamide) have been prepared by ligand substitution starting from $[Cr(NS)(CH_3CN)_5][PF_6]_2$ and adding the equivalent amount of the appropriate solvent molecule.[72] In the reaction with NaCN, $[Cr(NS)(CN)_5]^{3-}$ has been obtained. Derivatization of *mer*-$[RuCl_3(NS)(AsPh_3)_2]$ with $K[N(R_2P{=}S)_2]$ (R = Ph, i-Pr, *t*-Bu) has also proceeded to afford *trans*-$[RuCl(NS)\{N(R_2P{=}S)_2\}_2]$.[55]

2.8.2 Reactions

The reactivity of the NS complex $[OsCl_2(NS)(tpm)]X$ (tpm = tris(1-pyrazolyl)methane; X = PF_6, BF_4) has been well studied in recent publications.[65c] As already described,[52f,59] the N–S bond cleavage reaction of the NS complexes $[OsCl_2(NS)(tpm)][PF_6]$ with PPh_3 proceeds to furnish the phosphiniminato complex $[OsCl_2(NPPh_3)(tpm)][PF_6]$.[65c] Similar reactions of $[RuCl_2(CpCo\{P(O)(OEt)_2\}_3)(NS)]$[55,61b] and *trans*-$[RuCl(NS)\{N(Ph_2P{=}S)_2\}_2]$[55] with PPh_3 have also been reported, and the reaction of the latter have afforded the nitrido-briged dinuclear complex $[Ru\{N(Ph_2P{=}S)_2\}_2\text{-}(\mu\text{-N})RuCl\{N(Ph_2P{=}S)_2\}_2]$. Desulfurization of *trans*-$[RuCl(NS)\{N(t\text{-}Bu_2P{=}S)_2\}_2]$ using $[Ni(cod)_2]$ has also been achieved, and $[(Ru\{N(Ph_2P{=}S)_2\}_2)_2(\mu\text{-N})]$ is formed. On the other hand, reaction of $[RuCl(NS)(ttp)]$ with PMe_3 smoothly results in the formation of the substitution product $[Ru(PMe_3)_2(ttp)]$.[68b]

In the presence of air, chemical reduction of $[OsCl_2(NS)(tpm)][PF_6]$ with Zn(Hg) in HCl_{aq} also induces the N–C bond cleavage to afford $[OsCl_2(NH_3)(tpm)][PF_6]$ with concomitant generation of H_2S.[65c] Moreover, in the reaction of $[OsCl_2(NS)(tpm)][BF_4]$ with excess $NO[BF_4]$, abstraction of a sulfur atom from the NS ligand and NS/NO exchange reaction take place, the nitrosyl complex $[OsCl_2(NO)(tpm)][BF_4]$ and the nitrido complex

$[OsCl_2(N)(tpm)][BF_4]$ are formed in a 1 : 4 molar ratio. The NS/NO exchange has also been observed in the reaction of [RuCl(NS)(ttp)] with NO gas.[68b]

The NS ligand in $[OsCl_2(NS)(tpm)][PF_6]$ can be oxidized with trimethylamine *N*-oxide to form the thiazato complex $[OsCl_2(NSO)(tpm)]$. When [RuCl(NS)(ttp)] is reacted with $AgNO_2$, the thiazato/nitrosyl complex [Ru(NO)(NSO)(ttp)] is obtained.[68b] Halogenation of the NS ligand in *trans*-$[Re(CO_4)(NCMe)(NS)][AsF_6]_2$ with CsF has also been known to lead to the formation of the thiazyl fluoride complex *trans*-$[Re(CO_4)(NCMe)(NSF)][AsF_6]$.[73]

Photochemical generation of an NS molecule from the NS complex $[Cr(NS)(CH_3CN)_5][PF_6]_2$ has been developed and summarized in a review article.[52i]

2.8.3 List of NS Complexes Characterized by an X-ray Diffraction Study

NS complexes previously reviewed have also been included in Table 2.5 for comparison purposes.[74] Only NS complexes determined by a single-crystal X-ray diffraction study are listed in Table 2.5.

2.9 Selenonitrosyl (NSe) Complexes

Since the first NS complex $[Mo(NS)(S_2CNMe_2)_3]$ was prepared in 1974,[51] many NS complexes have been studied in detail.[52] However, we had to wait for 14 years before the first selenium isologues of the NO complex, the selenonitrosyl (NSe) complex of the type $[OsCl_2(NSe)Tp]$, was reported in 1988.[59] Since then, only seven examples of NSe have been determined. Five out of seven NSe complexes are characterized by a single-crystal X-ray diffraction study. The N–Se bond lengths in NSe complexes fall in the range of 1.63–1.68 Å, which are slightly longer than the sum of triple-bond covalent radii and similar to the sum of double-bond covalent radii (N≡Se: 1.61 Å; N=Se: 1.67 Å).[16] Similar to that observed in NS complexes, the [M]–NSe linkage in NSe complexes is slightly bent (M–N–Se bond angles, 165–178°). The NSe complexes exhibit an νNSe absorption band at around 1134–1137 cm^{-1}.

2.9.1 Synthesis

Six out of seven NSe complexes have been prepared using the same method, *i.e.*, reaction of a nitrido complex with elemental selenium to afford $[OsCl_2(NSe)Tp]$,[59] *mer*-$[Ru(AsPh_3)_2Cl_2(NSe)]$,[55] $[Ph_4P][Ru(hybeb^{R_1,R_2})(NSe)\text{-}(4\text{-}Me_2Npy)]$ ($R^1/R^2 = Cl/Cl$, H/CF_3, Cl/CF_3),[57] and $[Ir\{N(CHCHP\textit{t}\text{-}Bu_2)_2\}\text{-}(NSe)][PF_6]$[54] (Scheme 2.9). This approach may be less generic for NSe complexes than for NS complexes. For example, a couple of nitrido complexes, such as *cis*-$[RuClL_2(N)]$ (L = 2-[(2,6-diisopropylphenyl)imino]methyl-4,6-dibromophenolato)[56] and $[RuCl_2(CpCo\{P(O)(OEt)_2\}_3)(N)]$,[55] did not undergo the selenation. The selenium atom transfer also depends on the substituents in the nitrido complex $[Ph_4P][Ru(hybeb^{R_1,R_2})(N)]$.[57] While a

Table 2.5 Spectroscopic and selected X-ray data for NS complexes.

NS complexes	νNS (cm^{-1})	MN–S (Å)	M–NS (Å)	M–N–S (°)	Ref.
Vanadium					
[V(NS)(nacnac){O(2,6-i-$Pr_2C_6H_3$)}]	1070	1.5757(15)	1.6817(14)	170.25(10)	53
Chromium					
[CrCp$(CO)_2$(NS)]	1180	1.551(2)	1.694(2)	176.8(1)	67*b*
Molybdenum					
[Mo(NS)$(S_2CNMe_2)_3$]	∼1100	1.592(11)	1.738(11)	172.0(7)	74*a*
[Mo(NS)$(S_2CNEt_2)_3$]	∼1100	1.568(4)	1.765(4)	174.88(2)	74*b*
Technetium					
[$TcCl_2$(NS)$(S_2CNEt_2)_2$][a]	1248	1.49(2)	1.80(2)	177(2)	74*c*
		1.54(2)	1.70(2)	174(2)	
[$TcBr_2$(NS)$(S_2CNEt_2)_2$]		1.504(9)	1.754(9)	177.2(7)	74*d*
[$TcCl_3$(NS)(PMe_2Ph)(O=PMe_2Ph)]	1240	1.521(5)	1.746(5)	179.9(1)	74*e*
trans-[$TcCl_3$(NS)$(PMe_2Ph)_2$]	1220	1.523(4)	1.761(4)	176.2(3)	58*f*
[TcCl(NS)$(phen)_2$][PF_6][b]	1173	1.538(6)	1.782(6)	170.0(4)	58*c*
mer-[$TcCl_3$(NS)(4-Mepy)]	1173	1.54(1)	1.73(1)	176.0(1)	62
mer-[$TcCl_2$(NS)$(PMe_2Ph)_3$]	—	1.548(4)	1.747(3)	178.8(2)	74*f*
Rhenium					
$[Ph_4P]_2$[$ReBr_4$(NS)(NSBr)]	1225	1.38(2)	1.86(2)	178.2(8)	74*g*
[Re$(CO)_4$(NS)][AsF_6]		1.45(2)	1.87(2)	178(2)	73
$[Ph_4As]_2$[$ReCl_4$(NS)$(NSCl)_2$]	1222	1.456(4)	1.837(4)	177.09(25)	69
[Re(NS)$(NCS)_2(PMe_2Ph)_3$]	1176	1.49(2)	1.85(3)	177(1)	60
$[Ph_4P]_2$[(SN)$ReCl_3$(μ-N)(μ-NSN-κN,κN′)$ReCl_3$(thf)]	1208	1.498(8)	1.792(8)	175.6(5)	71
[$ReCl_2$(NS)(NSCl)(py)]	1230	1.520(6)	1.772(5)	176.3(4)	74*h*
trans-[Ph_4P][$ReCl_4$(NS)(py)]	1219	1.525(3)	1.771(3)	172.8(1)	58*h*
trans-[$ReBr_3$(NS)$(PMe_2Ph)_2$]	1229	1.530(7)	1.762(7)	174.8(6)	58*g*
cis-[Ph_4P][Re(NS)Cl_4(py)]	1235	1.535(5)	1.742(5)	174.2(4)	58*h*
trans-[$ReCl_3$(NS)$(PMe_2Ph)_2$]	1225	1.544(4)	1.759(4)	175.7(3)	58*d* and 58*g*
		1.545(4)	1.759(4)	180.0	
mer-[$ReCl_2$(NS)$(PMe_2Ph)_3$]	1178	1.570(5)	1.751(5)	179.4(4)	58*d*
[ReCl(NS)$(PMe_2Ph)_2(S_2CNEt_2)$][a]	1160	1.59(1)	1.72(1)	179.0(7)	58*e*
		1.55(1)	1.795(9)	177.4(6)	

Ruthenium					
$[Ph_4P]_2[\{RuCl_3(\mu\text{-}Cl)(NS)\}_2]$	1280	1.466(7)	1.752(6)	177.3(5)	74*i*
trans-$[RuCl(NS)\{N(Ph_2P{=}S)_2\}_2]$	1281	1.478(9)	1.745(9)	177.0(6)	55
[RuCl(NS)(ttp)]	1271	1.489(5)	1.768(4)	169.1(3)	68*b*
mer-$[Ru(AsPh_3)_2Cl_3(NS)]$	1310	1.502(4)	1.753(4)	180.0	55
$[Ph_4P]_2[\{RuBr_4(NS)\}_2(\mu\text{-}N_2S_2\text{-}\kappa N,\kappa N')]$	1263	1.51(3)	1.69(3)	175(2)	74*j*
$[Ph_4P][Ru(hybeb^{H,CF_3})(NS)(4\text{-}Me_2Npy)]$[a]	—	1.511(4)	1.746(4)	171.0(3)	57
		1.516(4)	1.746(4)	172.2(3)	
$[Ph_4P][RuCl_4(NS)(OH_2)]$	1318	1.540(4)	1.729(4)	170.9(3)	74*k*
Osmium					
[Os(NS)Me(oep)][b]	1196	1.433(13)	1.999(8)	163.0(8)	68*c*
trans-$[OsCl_2(NS)(terpy)][SCN]$	1295	1.459(8)	1.834(7)	178.5(5)	65*b*
$[Ph_4As][OsCl_4(NS)(NSCl)]$[b]	1298	1.46(1)	1.842(8)	169.9(5)	74*l*
	1215				
trans-$[Hpy][OsCl_4(NS)(py)]$	1289	1.47(1)	1.79(1)	168.0(7)	58*h*
$[OsCl_2(NS)(tpm)][BF_4]$	1316	1.489(15)	1.780(14)	179.3(9)	65*c*
$[OsCl_2(NS)Tp]$[c]	1284	1.493(8)	1.800(8)	166.9(5)	65*c*
trans-$[Ph_4P][OsCl_4(NS)(py)]$	1282	1.498(1)	1.770(3)	175.0(2)	58*h*
[OsCl(NS)(oep)][b]	1270	1.501(12)	1.832(13)	174.5(14)	68*c*
$[OsCl_3(NS)(PPh_3)_2]$	1310	1.503(10)	1.779(9)	180.0(1)	74*m*
$[Ph_4P][Os(NS)(OH_2)(S_4)_2]$	1249	1.513(12)	1.752(12)	171.2(7)	63
$[Ph_4P][OsCl_4(NS)(OH_2)]$	1285	1.514(5)	1.731(4)	174.9(3)	74*n*
Iridium					
$[Ir\{N(CHCHPt\text{-}Bu_2)_2\}(NS)][PF_6]$	1285	1.522(2)	1.768(2)	176.97(14)	54

[a]Distances and bond angles of both molecules per unit cell are shown.
[b]Data for major part of the disordered models are shown.
[c]As a cocrystal with 11% of $[OsCl_3Tp]$.

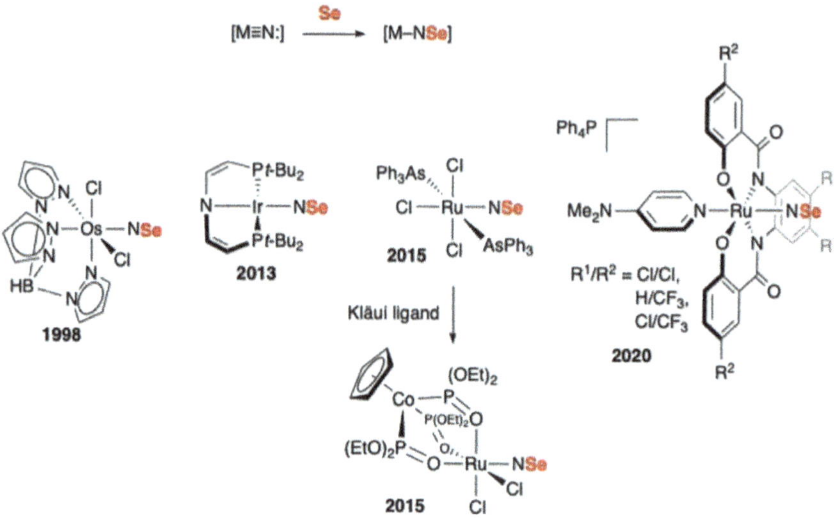

Scheme 2.9 Complexes featuring terminal coordinated NSe.

combination of $R^1/R^2 = H/H$ or H/Cl in $[Ph_4P][Ru(hybeb^{R_1,R_2})(N)]$ has not furnished the corresponding NSe complexes, the reaction of $[Ph_4P]$-$[Ru(hybeb^{Cl,Cl})(N)]$ with selenium and 4-Me_2Npy leads to the formation of an equilibrium mixture between the starting nitrido complex and the product $(Ph_4P)[Ru(NSe)(hybeb^{Cl,Cl})(4\text{-}Me_2Npy)]$.

Following an exchange ligand procedure, the reaction of *mer*-$[Ru(AsPh_3)_2Cl_2(NSe)]$ with the Kläui's tripodal ligand, $[Na][CpCo\{P(O)(OEt)_2\}_3]$, allows us to obtain the NSe complex $[RuCl_2(CpCo\{P(O)(OEt)_2\}_3)(NSe)]$.[55]

2.9.2 Reactivity

Reactivity of NSe complexes is scarcely reported. Similar to the reactions of NS complexes, the NSe complex $[RuCl_2(CpCo\{P(O)(OEt)_2\}_3)(NSe)]$ is reacted with PPh_3, and the corresponding phosphiniminato complex $[RuCl_2(CpCo\{P(O)(OEt)_2\}_3)(NPPh_3)]$ is formed.[55] Anionic NSe complexes of the type $[Ph_4P][Ru(hybeb^{R_1,R_2})(NSe)(4\text{-}Me_2Npy)]$ ($R^1/R^2 = Cl/Cl$, H/CF_3, Cl/CF_3) decompose in solution at room temperature, even under an atmosphere of argon.[57] When a $CDCl_3$ solution of the NSe complexes was kept standing at room temperature, the gradual formation of the parent nitrido complex $[Ph_4P][Ru(hybeb^{R_1,R_2})(N)]$ and 4-Me_2Npy is observed.

2.9.3 List of NSe Complexes Characterized by an X-ray Diffraction Study

Spectroscopic and selected X-ray data for NSe complexes are listed in Table 2.6.

Table 2.6 Spectroscopic and selected X-ray data for NSe complexes.

NSe complexes	νNSe (cm^{-1})	N–Se (Å)	M–N (Å)	M–N–Se (°)	Ref.
Ruthenium					
mer-$[Ru(AsPh_3)_2Cl_2(NSe)]$	1137	1.650(3)	1.756(3)	171.2(2)	55
$[RuCl_2(CpCo\{P(O)(OEt)_2\}_3)(NSe)]$	—	1.651(4)	1.731(4)	178.4(3)	55
$[Ph_4P][Ru(hybeb^{H,CF_3})(NSe)(4\text{-}Me_2Npy)]^a$	—	1.670(4)	1.734(4)	171.2(3)	57
		1.671(4)	1.732(4)	169.9(3)	
Osmium					
$[OsCl_2(NSe)Tp]$	1156	1.629(10)	1.800(10)	164.7(6)	59
Iridium					
$[Ir\{N(CHCHP\textit{t}\text{-}Bu_2)_2\}(NSe)][PF_6]$	1134	1.678(4)	1.756(4)	177.4(4)	54

^{a}Distances and bond angles of both molecules per unit cell are shown.

2.10 Telluronitrosyl (NTe) Complexes

The heaviest chalcogen congener of nitrosyl complex, *i.e.*, telluronitrosyl complexes of transition-metals, are not known so far. All attempts similar to those considered in the synthesis of NSe complexes failed.[55,57,59] The reaction of nitrido complexes with elemental tellurium did not proceed, probably due to the low solubility of tellurium in organic solvents. A suitable tellurium transfer reagent is highly desirable for accessing hitherto unknown NTe complexes.

Some hypothetical NTe complexes, $[Cr(CO)_2Cp(NTe)]$[75] and *mer*-$[ReCl_2(NTe)(PH_3)_3]$,[76] have been compared with their lighter NE (E = O, S, Se) complexes by theoretical calculations.

2.11 Comparison of Chalcogenonitrosyl (NE; E = O, S, Se) Complexes

Regarding a series of chalcogenonitrosyl (NE; E = O, S, Se) complexes of the type $[Cr(NE)(OH_2)_5]^{2+}$,[52i,72b] $[Cr(CO)_2Cp(NE)]$,[67,76] and $[Ir\{N(CHCHP\textit{t}\text{-}Bu_2)_2\}(NE)][PF_6]$,[54,77] a systematic comparison of the structures with the theoretical calculations has been reviewed.[52i] The σ-donor and π-acceptor abilities of the NSe and NS ligand are evaluated by comparison with the lighter NO ligand. For example, in the $[Cr(NE)(OH_2)_5]^{2+}$ series, the π-accepting ability of the NE ligand follows the order NS ≈ NSe < NO. The NS and NSe ligands are stronger σ-donors than the NO ligand. The studies on $[Ir(NE)\{N(CHCHP^tBu_2)_2\}][PF_6]$ indicate that the π-acceptor ability of the NS ligand is lower than that of the NO and NSe ligands, *i.e.*, NS < NSe < NO. The NE complexes *mer*-$[Ru(AsPh_3)_2Cl_3(NE)]$ (E = O,[78] S,[55] Se[55]), $[Ph_4P][Ru(hybeb^{H,CF_3})(NE)(4\text{-}Me_2Npy)]$ (E = O, S, Se),[57] $[OsCl_2(NE)Tp]$ (E = O,[59] S,[59,65c] Se[59]) are summarized in this section (Table 2.7).

For neutral series of the type *mer*-$[Ru(AsPh_3)_2Cl_3(NE)]$ and $[OsCl_2(NE)Tp]$, the M–NO bond lengths are shorter than the M–NS and M–NSe distances, which suggests that the order of the π-accepting property should be NS ≈ NSe < NO. This order is in good agreement with that in the

Table 2.7 Spectroscopic and selected X-ray data for a series of NE complexes.

NE complexes	νNE (cm^{-1})	N–E (Å)	M–N (Å)	L–M(NE)[a] (Å)	Ref.
mer-$[Ru(AsPh_3)_2Cl_3(NO)]$	1869	1.151(9)	1.729(7)	2.346(2)	78
mer-$[Ru(AsPh_3)_2Cl_3(NS)]$	1310	1.502(4)	1.753(4)	2.3937(10)	55
mer-$[Ru(AsPh_3)_2Cl_3(NSe)]$	1137	1.650(3)	1.756(3)	2.3953(10)	55
$[Ph_4P][Ru(hybeb^{H,CF_3})(NO)(4\text{-}Me_2Npy)]$[b]	1810	1.144(3)	1.752(2)	2.1325(19)	57
		1.149(3)	1.744(2)	2.1327(18)	
$[Ph_4P][Ru(hybeb^{H,CF_3})(NS)(4\text{-}Me_2Npy)]$[b]	—	1.511(4)	1.746(4)	2.155(4)	57
		1.516(4)	1.746(4)	2.157(4)	
$[Ph_4P][Ru(hybeb^{H,CF_3})(NSe)(4\text{-}Me_2Npy)]$[b]	—	1.670(4)	1.734(4)	2.173(4)	57
		1.671(4)	1.732(4)	2.173(4)	
$[OsCl_2(NO)Tp]$	1832	1.19(4)	1.65(3)	2.085(5)	59
$[OsCl_2(NS)Tp]$[c]	1284	1.493(8)	1.800(8)	2.086(5)	59 and 65*c*
$[OsCl_2(NSe)Tp]$	1156	1.629(10)	1.800(10)	2.101(7)	59

[a]M–L bond lengths located *trans* to the NE ligands are shown.
[b]Distances and bond angles of both molecules per unit cell are shown.
[c]As a cocrystal with 11% of $[OsCl_3Tp]$.

aforementioned complexes $[Cr(NE)(OH_2)_5]^{2+}$ and $[Ir(NE)\{N(CHCHP^tBu_2)_2\}][PF_6]$. In contrast, in the cationic $[Ph_4P][Ru(hybeb^{H,CF_3})(NE)(4\text{-}Me_2Npy)]$ series, the π-acidity would follow the trend, NO ≈ NS < NSe, as the Ru–NSe bond length is the shortest among the cationic NE complexes. On the other hand, in all NE complexes listed in Table 2.7, the NSe ligand would exhibit a better σ-donating ability than the lighter NO and NS ligands. The length of the L–M(NE) bonds increases on descending the chalcogen group, which appears to reflect the order of *trans* influence of the NE ligands, NO < NS < NSe.

References

1. M. C. Baird and G. Wilkinson, *Chem. Commun.*, 1966, 267.
2. (a) I. S. Butler and A. E. Fenster, *J. Organomet. Chem.*, 1974, **66**, 161; (b) P. V. Yaneff, *Coord. Chem. Rev.*, 1977, **23**, 183; (c) I. S. Butler, *Acc. Chem. Res.*, 1977, **10**, 359; (d) P. V. Broadhurst, *Polyhedron*, 1985, **4**, 1801; (e) M. Herberhold, *Nachr. Chem., Tech. Lab.*, 1981, **29**, 365; (f) E. K. Moltzen, K. J. Klabunde and A. Senning, *Chem. Rev.*, 1988, **88**, 391; (g) I. S. Butler, *Pure Appl. Chem.*, 1988, **60**, 1241; (h) W. Petz, *Coord. Chem. Rev.*, 2008, **252**, 1689; (i) L. Busetto and V. Zanotti, *Inorg. Chim. Acta*, 2008, **361**, 3004.
3. R. D. Young, A. F. Hill, G. E. Cavigliasso and R. Stranger, *Angew. Chem., Int. Ed.*, 2013, **52**, 3699.
4. M. M. Deegan and J. C. Peters, *Polyhedron*, 2021, **209**, 115461.
5. A. R. Cowley, A. L. Hector, A. F. Hiil, A. J. P. White, D. J. Williams and J. D. E. T. Wilton-Ely, *Organometallics*, 2007, **26**, 6114.

6. Y. Mutoh, N. Kozono, M. Araki, N. Tsuchida, K. Takano and Y. Ishii, *Organometallics*, 2010, **29**, 519.
7. O. Blacque, T. Lubica and H. Berke, *CSD Communication*, 2017, CCDC 1572572, DOI: 10.5517/ccdc.csd.cc1psd40.
8. O. Blacque, T. Lubica and H. Berke, *CSD Communication*, 2017, CCDC 1572580, DOI: 10.5517/ccdc.csd.cc1psdd8.
9. A. Suzuki, Y. Mutoh, N. Tsuchida, C.-W. Fung, S. Kikkawa, I. Azumaya and S. Saito, *Chem. – Eur. J.*, 2020, **26**, 3795.
10. T. Ahrens, B. Schmiedecke, T. Braun, R. Herrmann and R. Laubenstein, *Eur. J. Inorg. Chem.*, 2017, **2017**, 713.
11. H. J. Barnett and A. F. Hill, *Chem. Commun.*, 2020, **56**, 12593.
12. M. T. Whited and R. H. Grubbs, *Organometallics*, 2009, **28**, 161.
13. I. S. Butler, D. Cozak and S. R. Stobart, *J. Chem. Soc., Chem. Commun.*, 1975, 103.
14. G. R. Clark, K. R. Grundy, R. O. Harris, S. M. James and W. R. Roper, *J. Organomet. Chem.*, 1975, **90**, C37.
15. L. M. Caldwell, A. F. Hill, J. Wagler and A. C. Willis, *Dalton Trans.*, 2008, 3538.
16. P. Pyykkö and M. Atsumi, *Chem. – Eur. J.*, 2009, **15**, 12770.
17. Y. Mutoh, N. Kozono, K. Ikenaga and Y. Ishii, *Coord. Chem. Rev.*, 2012, **256**, 589.
18. B. J. Frogley, A. F. Hill and L. J. Watson, *Chem. – Eur. J.*, 2020, **26**, 12706.
19. I. S. Butler, D. Cozak and S. R. Stobart, *Inorg. Chem.*, 1977, **16**, 1779.
20. (a) J.-Y. Saillard and D. Grandjean, *Acta Crystallogr., Sect. B: Struct. Crystallogr. Cryst. Chem.*, 1978, **34**, 3772; (b) D. Cozak, I. S. Butler and I. M. Baibich, *J. Organomet. Chem.*, 1979, **169**, 381; (c) A. A. Ismail, I. S. Butler, J.-J. Bonnet and S. Askenazy, *Acta Crystallogr., Sect. C: Cryst. Struct. Commun.*, 1985, **41**, 1582; (d) A. A. Ismail and I. S. Butler, *J. Organomet. Chem.*, 1988, **346**, 185.
21. I. S. Butler, D. Cozak, S. R. Stobart, K. R. Plowman, P. V. Yaneff and R. O. Harris, *Inorg. Synth.*, 1979, **19**, 193.
22. G. R. Clark and S. M. James, *J. Organomet. Chem.*, 1977, **134**, 229.
23. O. Kolb and H. Werner, *J. Organomet. Chem.*, 1984, **268**, 49.
24. W. R. Roper and K. G. Town, *J. Organomet. Chem.*, 1983, **252**, C97.
25. M. Ebner and H. Werner, *Chem. Ber.*, 1988, **121**, 1449.
26. G. R. Clark, K. Marsden, W. R. Roper and L. J. Wright, *J. Am. Chem. Soc.*, 1980, **102**, 1206.
27. (a) W. R. Roper, *J. Organomet. Chem.*, 1986, **300**, 167; (b) G. R. Clark, K. Marsden, C. E. F. Rickard, W. R. Roper and L. J. Wright, *J. Organomet. Chem.*, 1988, **338**, 393.
28. T. Desmond, F. J. Lalor, G. Ferguson and M. Parvez, *J. Chem. Soc., Chem. Commun.*, 1984, 75.
29. (a) J. B. Greco, J. C. Peters, T. A. Baker, W. M. Davis, C. C. Cummins and G. Wu, *J. Am. Chem. Soc.*, 2001, **123**, 5003; (b) J. C. Peters, PhD thesis, Massachusetts Institute of Technology, 1998. https://dspace.mit.edu/handle/1721.1/9855 (accessed Oct 2022).

30. (a) R. L Cordiner, A. F. Hill and J. Wagler, *Organometallics*, 2008, **27**, 5177; (b) I. A. Cade, A. F. Hill and C. M. A. McQueen, *Dalton Trans.*, 2019, **48**, 2000; (c) B. J. Frogley, A. F. Hill and L. J. Watson, *Dalton Trans.*, 2019, **48**, 12598.
31. (a) R. G. Carlson, M. A. Gile, J. A. Heppert, M. H. Mason, D. R. Powell, D. V. Velde and J. M. Vilain, *J. Am. Chem. Soc.*, 2002, **124**, 1580; (b) P. E. Romero, W. E. Piers and R. McDonald, *Angew. Chem., Int. Ed.*, 2004, **43**, 6161; (c) S. R. Dubberley, P. E. Romero, W. E. Piers, R. McDonald and M. Parvez, *Inorg. Chim. Acta*, 2006, **359**, 2658; (d) A. Reinholdt and J. Bendix, *Chem. Rev.*, 2022, **122**, 830.
32. A. Suzuki, T. Arai, K. Ikenaga, Y. Mutoh, N. Tsuchida, S. Saito and Y. Ishii, *Dalton Trans.*, 2017, **46**, 44.
33. A. F. Hill and L. J. Watson, *Chem. Commun.*, 2020, **56**, 2356.
34. (a) I. A. Cade, A. F. Hill and C. M. A. McQueen, *Organometallics*, 2009, **28**, 6639; (b) L. M. Caldwell, R. L. Cordiner, A. F. Hill and J. Wagler, *Organometallics*, 2010, **29**, 1526.
35. S. Takemoto, J. Ohata, K. Umetani, M. Yamaguchi and H. Matsuzaka, *J. Am. Chem. Soc.*, 2014, **136**, 15889.
36. (a) H. J. Barnett and A. F. Hill, *Chem. Commun.*, 2019, **55**, 1734; (b) H. J. Barnett and A. F. Hill, *Angew. Chem., Int. Ed.*, 2020, **59**, 4274; (c) H. J. Barnett and A. F. Hill, *Chem. Commun.*, 2020, **56**, 12593.
37. (a) B. L. Benac, E. M. Burgess and A. J. Arduengo, III, *Org. Synth.*, 1986, **64**, 92; (b) D. J. Williams, M. R. Fawcett-Brown, R. R. Raye, D. VanDerveer, Y. T. Pang, R. L. Jones and K. L. Bergbauer, *Heteroat. Chem.*, 1993, **4**, 409; (c) N. Kuhn, G. Henkel and T. Kratz, *Z. Naturforsch., B: J. Chem. Sci.*, 1993, **48**, 973; (d) A. Doddi, M. Peters and M. Tamm, *Chem. Rev.*, 2019, **119**, 6994.
38. B. J. Frogley, A. F. Hill and L. J. Watson, *Chem. Commun.*, 2019, **55**, 14450.
39. A. F. Hill and C. M. A. McQueen, *Organometallics*, 2012, **31**, 2482.
40. B. J. Frogley, A. F. Hill, C. S. Onn and L. J. Watson, *Angew. Chem., Int. Ed.*, 2019, **58**, 15349.
41. B. J. Frogley, A. F. Hill, R. A. Manzano and M. Sharma, *Chem. Commun.*, 2018, **54**, 1702.
42. J. P. Collman and W. R. Roper, *J. Am. Chem. Soc.*, 1966, **88**, 3504.
43. K. R. Grundy, R. O. Harris and W. R. Roper, *J. Organomet. Chem.*, 1975, **90**, C34.
44. (a) S. Trofimenko, *J. Am. Chem. Soc.*, 1969, **91**, 588; (b) C. P. Marabella and J. H. Enemark, *J. Organomet. Chem.*, 1982, **226**, 57.
45. T. Ziegler, *Inorg. Chem.*, 1986, **25**, 2721.
46. S. Thomas and C. G. Young, *Inorg. Synth.*, 2002, **33**, 218.
47. S. Takemoto, T. Kobayashi and H. Matsuzaka, *J. Am. Chem. Soc.*, 2004, **126**, 10802.
48. (a) M. Cowie and S. K. Dwight, *Inorg. Chem.*, 1980, **19**, 2500; (b) L. Gelmini, D. W. Stephan and S. J. Loeb, *Inorg. Chim. Acta*, 1985, **98**, L3.
49. (a) T. C. Berto, A. L. Speelman, S. Zheng and N. Lehnert, *Coord. Chem. Rev.*, 2013, **257**, 244; (b) *Nitrosyl Complexes in Inorganic Chemistry,*

Biochemistry and Medicine I, ed. D. M. P. Mingos, Springer Berlin Heidelberg, Berlin, Heidelberg, 2014, vol. 153; (c) *Nitrosyl Complexes in Inorganic Chemistry, Biochemistry and Medicine II*, ed. D. M. P. Mingos, Springer Berlin Heidelberg, Berlin, Heidelberg, 2014, vol. 154; (d) L. Li and L. Li, *Coord. Chem. Rev.*, 2016, **306**, 678; (e) N. Lehnert, E. Kim, H. T. Dong, J. B. Harland, A. P. Hunt, E. C. Manickas, K. M. Oakley, J. Pham, G. C. Reed and V. S. Alfaro, *Chem. Rev.*, 2021, **121**, 14682; (f) J. B. Harland, E. C. Manickas, A. P. Hunt and N. Lehnert, Reactivity and Structure of Complexes of Small Molecules: Nitric Oxide, in *Comprehensive Coordination Chemistry III*, Elsevier, 2021, pp. 806–874.

50. R. T. Boeré and T. L. Roemmele, in *Comprehensive Inorganic Chemistry II*, ed. J. Reedikj and K. Poeppelmeier, Elsevier, 2013, vol. 1, pp. 375–411.
51. J. Chatt and J. R. Dilworth, *J. Chem. Soc., Chem. Commun.*, 1974, 508.
52. (a) H. W. Roesky and K. K. Pandey, *Adv. Inorg. Chem. Radiochem*, 1983, **26**, 337; (b) P. F. Kelly and J. D. Woollins, *Polyhedron*, 1986, **5**, 607; (c) T. Chivers and F. Edelmann, *Polyhedron*, 1986, **5**, 1661; (d) M. Herberhold, *Comments Inorg. Chem.*, 1988, **7**, 53; (e) K. K. Pandey, *Prog. Inorg. Chem.*, 1992, **40**, 445; (f) T. Chivers and R. S. Laitinen, in *Handbook of Chalcogen Chemistry*, ed. F. A. Devillanova, Royal Society of Chemistry, Cambridge, 2007, pp. 223–285; (g) T. Chivers and R. Laitinen, in *Handbook of Chalcogen Chemistry: New Perspectives in Sulfur, Selenium and Tellurium*, ed. F. A. Devillanova and W.-W. du Mont, Royal Society of Chemistry, 2nd edn, 2013, vol. 1, pp. 191–237; (h) A. Døssing, *Trends Inorg. Chem.*, 2010, **12**, 1; (i) A. Døssing, *Coord. Chem. Rev.*, 2016, **306**, 544.
53. B. L. Tran, R. Thompson, S. Ghosh, X. Gao, C.-H. Chen, M.-H. Baik and D. J. Mindiola, *Chem. Commun.*, 2013, **49**, 2768.
54. M. G. Scheibel, I. Klopsch, H. Wolf, P. Stollberg, D. Stalke and S. Schneider, *Eur. J. Inorg. Chem.*, 2013, 3836.
55. H.-Y. Ng, W.-M. Cheung, E. K. Huang, K.-L. Wong, H. H.-Y. Sung, I. D. Williams and W. H. Leung, *Dalton Trans.*, 2015, **44**, 18459.
56. H.-Y. Ng, N.-M. Lam, M. Yang, X.-Y. Yi, I. D. Williams and W.-H. Leung, *Inorg. Chim. Acta*, 2013, **394**, 171.
57. C.-W. Fung, G. Fukada, Y. Mutoh, N. Tsuchida and S. Saito, *Dalton Trans.*, 2020, **49**, 613.
58. (a) M. W. Bishop, J. Chatt and J. R. Dilworth, *J. Chem. Soc., Chem. Commun.*, 1975, 780; (b) M. W. Bishop, J. Chatt and J. R. Dilworth, *J. Chem. Soc. Dalton Trans.*, 1979, 1; (c) J. Lu and M. J. Clarke, *J. Chem. Soc., Dalton Trans.*, 1992, 1243; (d) U. Abram and S. Ritter, *Z. Anorg. Allg. Chem.*, 1994, **620**, 1223; (e) S. Ritter and U. Abram, *Z. Anorg. Allg. Chem.*, 1996, **622**, 965; (f) U. Abram, E. S. Lang, S. Abram, J. Wegmann, J. R. Dilworth, R. Kirmse and J. D. Woollins, *J. Chem. Soc., Dalton Trans.*, 1997, 623; (g) A. Voigt, U. Abram and R. Kirmse, *Z. Anorg. Allg. Chem.*, 1999, **625**, 1658; (h) M. Reinel, T. Höcher, U. Abram and R. Kirmse, *Z. Anorg. Allg. Chem.*, 2003, **629**, 853.
59. T. J. Crevier, S. Lovell, J. M. Mayer, A. L. Rheingold and I. A. Guzei, *J. Am. Chem. Soc.*, 1998, **120**, 6607.

60. R. Hübener, U. Abram and J. Strähle, *Inorg. Chim. Acta*, 1994, **216**, 223.
61. (a) A. Wu, A. Dehestani, E. Saganic, T. J. Crevier, W. Kaminsky, D. E. Cohen and J. M. Mayer, *Inorg. Chim. Acta*, 2006, **359**, 2842; (b) X.-Y. Yi, T. C. H. Lam, Y.-K. Sau, Q.-F. Zhang, I. D. Williams and W.-H. Leung, *Inorg. Chem.*, 2007, **46**, 7193.
62. J. Lu and M. J. Clarke, *Inorg. Chem.*, 1990, **29**, 4123.
63. B. Vollmer, S. Wocadlo, W. Massa and K. Dehnicke, *Z. Anorg. Allg. Chem.*, 1996, **622**, 1306.
64. C. F. Leung, D. T. Y. Yiu, W. T. Wong, S. M. Peng and T. C. Lau, *Inorg. Chim. Acta*, 2009, **362**, 3576.
65. (a) K. D. Demadis, E.-S. El-Samanody, T. J. Meyer and P. S. White, *Inorg. Chem.*, 1998, **37**, 838; (b) K. D. Demadis, T. J. Meyer and P. S. White, *Inorg. Chem.*, 1998, **37**, 3610; (c) E. S. El-Samanody, K. D. Demadis, L. A. Gallagher, T. J. Meyer and P. S. White, *Inorg. Chem.*, 1999, **38**, 3329.
66. S. B. Seymore and S. N. Brown, *Inorg. Chem.*, 2002, **41**, 462.
67. (a) B. W. S. Kolthammer and P. Legzdins, *J. Am. Chem. Soc.*, 1978, **100**, 2247; (b) T. J. Greenhough, B. W. S. Kolthammer, P. Legzdins and J. Trotter, *Inorg. Chem.*, 1979, **18**, 3548.
68. (a) K. K. Pandey and U. C. Agarwala, *Inorg. Chem.*, 1981, **20**, 1308; (b) K. K. Pandey and K. Hemlata Garg, *Polyhedron*, 1995, **14**, 1987; (c) D. S. Bohle, C. Hung, A. K. Powell, B. D. Smith and S. Wocadlo, *Inorg. Chem.*, 1997, **36**, 1992; (d) L. Cheng, L. Chen, H.-S. Chung, M. A. Khan, G. B. Richter-Addo and V. G. Young, *Organometallics*, 1998, **17**, 3853.
69. J. Anhaus, Z. A. Siddiqi, H. W. Roesky, J. W. Bats and Y. Elerman, *Z. Naturforsch. B*, 1985, **40**, 740.
70. R. Mews, *Angew. Chem., Int. Ed. Engl.*, 1976, **15**, 691; R. Mews and C. Liu, *Angew. Chem., Int. Ed.*, 1983, **22**, 162.
71. A. Dietrich, B. Neumüller and K. Dehnicke, *Z. Anorg. Allg. Chem.*, 2000, **626**, 1268.
72. (a) J. W. Dethlefsen and A. Døssing, *Inorg. Chim. Acta*, 2009, **362**, 259; (b) J. R. Dethlefsen, A. Døssing and E. D. Hedegård, *Inorg. Chem.*, 2010, **49**, 8769.
73. C. Ruf, U. Behrens, E. Lork and R. Mews, *Chem. Commun.*, 1996, **2**, 939.
74. (a) M. B. Hursthouse and M. Motevalli, *J. Chem. Soc., Dalton Trans.*, 1979, 1362; (b) T. A. Coffey, G. D. Forster and G. Hogarth, *Acta Crystallogr., Sect. C: Cryst. Struct. Commun.*, 1996, **52**, 2157; (c) J. Baldas, J. Bonnyman, M. F. Mackay and A. G. Williams, *Aust. J. Chem.*, 1984, **37**, 751; (d) J. Baldas, S. Colmanet and G. Williams, *Aust. J. Chem.*, 1991, **44**, 1125; (e) L. Kaden, B. Lorenz, R. Kirmse, J. Stach, H. Behm, P. T. Beurskens and U. Abram, *Inorg. Chim. Acta*, 1990, **169**, 43; (f) W. Hiller, R. Hübener, B. Lorenz, L. Kaden, M. Findeisen, J. Stach and U. Abram, *Inorg. Chim. Acta*, 1991, **181**, 161; (g) H.-G. Hauck, W. Willing, U. Müller and K. Dehnicke, *Z. Naturforsch., B: Anorg. Chem., Org. Chem.*, 1986, **41**, 825; (h) H.-G Hauck, W. Willing, U. Müller and K. Dehnicke, *Z. Anorg. Allg. Chem.*, 1986, **534**, 77; (i) W. Willing, U. Müller, U. Demant and K. Dehnicke, *Z. Naturforsch., B: Anorg. Chem., Org. Chem.*, 1986, **41**, 560;

(j) U. Demant, W. Willing, U. Müller and K. Dehnicke, *Z. Anorg. Allg. Chem.*, 1986, **532**, 175; (k) J. W. Bats, K. K. Pandey and H. W. Roesky, *J. Chem. Soc., Dalton Trans.*, 1984, **6**, 2081; (l) R. Weber, U. Müller and K. Dehnicke, *Z. Anorg. Allg. Chem.*, 1983, **504**, 13; (m) H. W. Roesky, K. K. Pandey, W. Clegg, M. Noltemeyer and G. M. Sheldrick, *J. Chem. Soc., Dalton Trans.*, 1984, **2**, 719; (n) K. K. Pandey, H. W. Roesky, M. Noltemeyer and G. M. Sheldrick, *Z. Naturforsch., B: Anorg. Chem., Org. Chem.*, 1984, **39**, 590.

75. C. M. Teague, T. A. O'Brien and J. F. O'Brien, *J. Coord. Chem.*, 2002, **55**, 627.
76. S. F. Vyboishchikov and G. Frenking, *Theor. Chem. Acc.*, 1999, **102**, 300.
77. K. K. Pandey and P. Patidar, *Polyhedron*, 2014, **68**, 87.
78. D. H. F. Souza, G. Oliva, A. Teixeira and A. A. Batista, *Polyhedron*, 1995, **14**, 1031.

CHAPTER 3

Metal–Chalcogenolates: Synthesis and Applications in Material Chemistry

VIMAL K. JAIN*[a] AND ROHIT SINGH CHAUHAN*[b]

[a] UM-DAE Centre for Excellence in Basic Sciences, University of Mumbai, Kalina Campus, Santacruz (E), Mumbai-400098, India; [b] Department of Chemistry, K. J. Somaiya College of Science and Commerce Vidyavihar, Mumbai-400077, India
*Emails: jainvk@cbs.ac.in; rohit.chauhan@somaiya.edu

3.1 Introduction

Metal chalcogenolate complexes have been known for a long time. Siemens[1] and Roemer[2] reported mercury complexes, $Hg(SeEt)_2$ and $Hg(SR)_2$ as early as 1847 and 1873, respectively. During the last quarter of the 19th century and early 20th century thiolate complexes of several elements (*e.g.*, Zn, Cd, Hg, Co, Rh, Pd, Bi) were synthesized.[3] Over the last several decades the chemistry of metal thiolates has been extensively explored owing to the ease of their synthesis. These complexes find numerous applications in widely different areas like pharmaceuticals, structural models for metalloproteins, hydrodesulfurization reactions, *etc.* The chemistry of analogous selenolates and tellurolates, however, remained sluggish until the early 1990s.[4,5]

The relevance of metal chalcogenolates was quickly realized with the emergence of nano-science and nanotechnology in the 1990s making the proverb "*necessity is the mother of invention*" quite pertinent for these complexes. Although the thermal and photolytic decomposition of thiolates has

Chalcogen Chemistry: Fundamentals and Applications
Edited by Vito Lippolis, Claudio Santi, Eder J. Lenardão and Antonio L. Braga

Published by the Royal Society of Chemistry, www.rsc.org

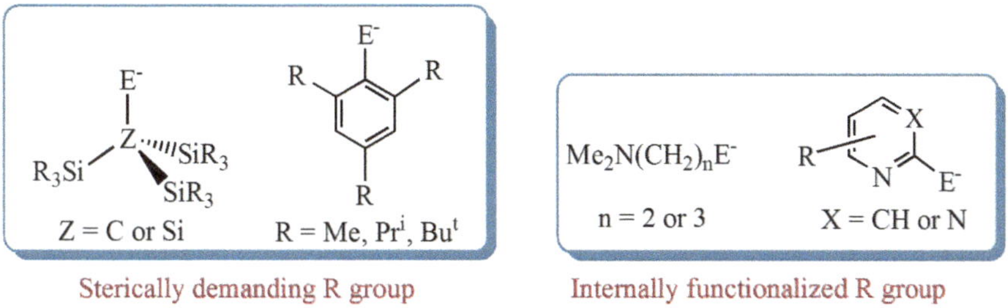

Scheme 3.1 Representative example of chalcogenolate ligands.

been investigated since the early 20th century[6] and even thin films of layered MoS_2 were prepared during the 1960s by exfoliation using adhesive tape,[7] their promising potential was only exploited in the last two–three decades for the design and development of numerous inorganic chalcogenide materials.

Metal chalcogenides constitute an important family of inorganic materials of great technological relevance. Their chemical, physical, electrical, optical, and magnetic properties are magnanimously influenced by their composition, shape, size, and phase. Among various preparative methods for metal chalcogenides for nano-science and nanotechnology applications, a single-source molecular precursor route has emerged as one of the versatile synthetic strategies. This strategy has an edge over other methods in terms of controlling the composition, phase, shape of the nanostructures, and their morphology. Ever increasing interest in nano-crystals and thin films of metal chalcogenides has led to rapid development of single-source molecular precursor chemistry.[5,8–10] Metal chalcogenolates, $M(ER)_n$ were amongst the earliest and simplest examples of single source molecular precursors, but over time several other anionic organochalcogen ligands[10–12] and chalcogenethers[13] have been used to design single-source molecular precursors for the preparation of metal chalcogenide nano-crystals and for the deposition of thin films. In view of limited space, we have focused on metal chalcogenolates derived from ligand systems as given in Scheme 3.1, with reference to their utility as single-source molecular precursors for metal chalcogenides.

3.2 Synthesis of Chalcogenolate Complexes

Metal chalcogenolate complexes are synthesized in several ways through reactions employing different chalcogen sources, *viz.* direct reaction of chalcogen; diorganodichalcogenides, chalcogenolate ion, chalcogenol, (Scheme 3.2) and chalcogeno ethers.[14] Some representative reactions in each category are described here.

3.2.1 Reactions Involving Chalcogen Insertion into the Metal–Carbon Bond

Insertion of chalcogens into the metal–carbon bond of s-block organometallic compounds (*e.g.*, organolithium, Grignard reagent) is frequently

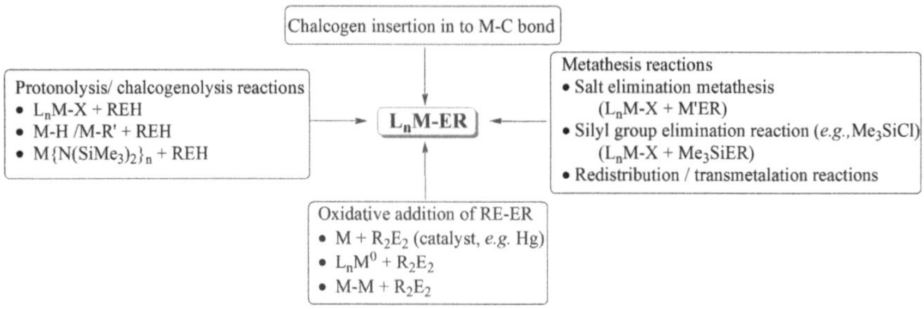

Scheme 3.2 General synthetic routes for the preparation of metal chalcogenolate complexes.

employed for the preparation of organochalcogenolates (eqn (3.1)).[5,15,16] Direct insertion of chalcogens into the main group[17] as well as transition metal–carbon bonds is also well documented. Reactions of gray selenium powder with Mes_3In in refluxing toluene results in insertion of Se in the In–C bond to give $[Mes_2In(\mu\text{-}SeMes)]_2$.[18] Insertion of chalcogens into the Zr–C bond of $(\eta^5\text{-}R'C_5H_4)_2ZrR_2$ has been reported (eqn (3.2)). The rate of insertion of chalcogen decreases in the order $S > Se > Te$.[19] Trialkylphosphine tellurides, due to their solubility in organic solvents and facile dissociation to tellurium and phosphine, are often preferred over elemental tellurium for insertion reactions. For instance, treatment of a toluene solution of $[MnBz(CO)_5]$ with R_3PTe readily yields insertion products, $[Mn(TeBz)(CO)_3(PR_3)_2]$ ($R = Me$ or Et), although the parent manganese compound is inert toward metallic tellurium at RT/100 °C.[20]

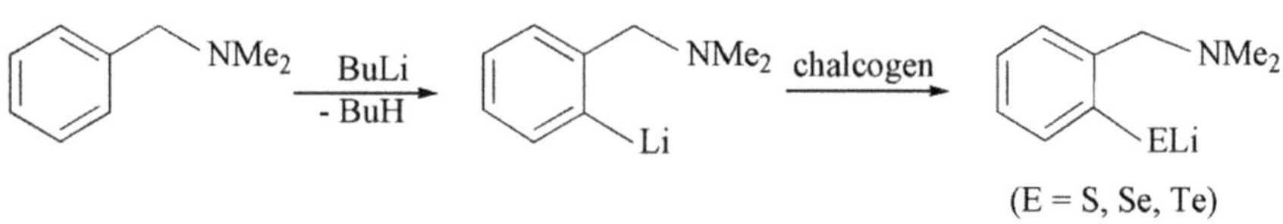

(3.1)

$$[\eta^5\text{-}R'C_5H_4)_2ZrR_2] + 2E \rightarrow [(\eta^5\text{-}R'C_5H_4)_2Zr(ER)_2] \quad (3.2)$$

$$(R' = H \text{ or Me}; E = S, Se, Te; R = Me, Bz, Ph)$$

3.2.2 Oxidative Addition Reactions of Diorganodichalcogenides

Oxidative addition of diorganodichalcogenides to free metal, low valent metal complexes (L_nM), or even metal–metal bonded compounds has been a successful strategy to prepare metal chalcogenolates. Several diorganodichalcogenides oxidatively add to elemental mercury in toluene/methanol to afford $Hg(ER)_2$ complexes.[21] Freshly activated metal powders are quite reactive to insert across the E–E bond. For example activated germanium powder reacts readily with Ph_2E_2 in THF at low temperatures to give homoleptic chalcogenolates, $[Ge(EPh)_4]$ (E = Se or Te).[22] Dipyridyl diselenides react with powdered copper, indium, and other elements, often in the presence of a catalytic amount of mercury, in refluxing toluene to afford homoleptic 2-pyridyl selenolates, $[Cu\{SeC_5H_3(R\text{-}3)N\}]_4$ (R = H or $SiMe_3$, R is at the third position of the pyridyl ring) and $[In\{SeC_5H_4N\}_3]$.[23] Similarly, reduction of Ph_2Se_2 with indium metal in the presence of a catalytic amount of mercury yields $[In(SePh)_3]$.[24] Reduction of diorganodiselenides (R'_2Se_2; R′ = Me or Ph) with triorgano indium, R_3In, yields $[R_2In(\mu\text{-}SeR')]_2$ (eqn (3.3)).[18,25]

$$R_3In + R'_2Se_2 \rightarrow [R_2In(\mu\text{-}SeR')]_2 + RSeR' \quad (R = Me \text{ or } Mes) \tag{3.3}$$

The reactions of Na_2PdCl_4 or other Pd(II) precursors with R_2E_2 (E = S, Se, Te) yield palladium(II) chalcogenolates of composition $[PdCl(ER)]_n$ formation which is believed to involve a Pd(IV) complex, $[PdCl_2(ER)_2]$ generated by oxidative addition of diorgano-dichalcogenides on a Pd(II) precursor. The Pd(IV) intermediate after reductive elimination of RECl yields $[PdCl(ER)]_n$.[26]

Oxidative addition of diorganodichalcogenides to low valent metal phosphine (eqn (3.4))[11,27,28] and olefin[29] complexes is another important synthetic strategy to prepare metal chalcogenolates. DFT calculations on reaction pathways and the trends in controlling the factors of R_2E_2 oxidative addition on $[M(PR'_3)_2]$ (M = Pd or Pt) have been reported.[30] In these reactions, the E–E bond activation energy decreases in the order S > Se > Te which is also the sequence of E–E bond energy. Activation of the RE–ER bond is more facile with Pd(0) complexes than the corresponding Pt(0) analog.[30]

$$[Pt(CH_2{=}CH_2)(PPh_3)_2] + Ph_2Se_2 \xrightarrow[-\,CH_2=CH_2]{} cis\text{-}[Pt(SePh)_2(PPh_3)_2] \xrightarrow{\text{isomerization}} trans\text{-}[Pt(SePh)_2(PPh_3)_2] \tag{3.4}$$

Unlike diorgano-disulfides/-diselenides, oxidative addition reactions of diorganoditellurides are more complex giving several products formed by the cleavage of the Te–Te bond as well as the Te–C bond.[28,31–34] The reaction of $[Pt(PPh_3)_4]$ with dipyridiyl ditelluride in toluene not only yields the expected tellurolate complex, $[Pt\{TeC_5H_3(R\text{-}3)N\}_2(PPh_3)_2]$ but also produces the serendipitous product $[Pt(Te)\{C_5H_3(R\text{-}3)N\}_2(PPh_3)]$ (~15% yield) containing formally Te^0 as a ligand (eqn (3.5)).[32] The reaction of $[Pt(PPh_3)_4]$ with Th_2Te_2, on the other hand, affords a tellurido-bridged trinuclear complex, $[Pt_3(\mu\text{-}Te)_2Th(PPh_3)_5]Cl$ together with *trans*-$[PtTh(Cl)(PPh_3)_2]$.[31] In contrast, the reaction involving platinum(0) complexes of chelating phosphines gives a variety of products.[33] For example, the reaction of $[Pt_2(dppm)_3]$ with dipyridyl ditelluride results in the double deprotonated methylene group of the dppm ligand together with the expected tellurolate complex (eqn (3.6)).[33] The addition of diorganodichalcogenides across the metal–metal bond has also been employed for the synthesis of metal chalcogenolate complexes.[35,36]

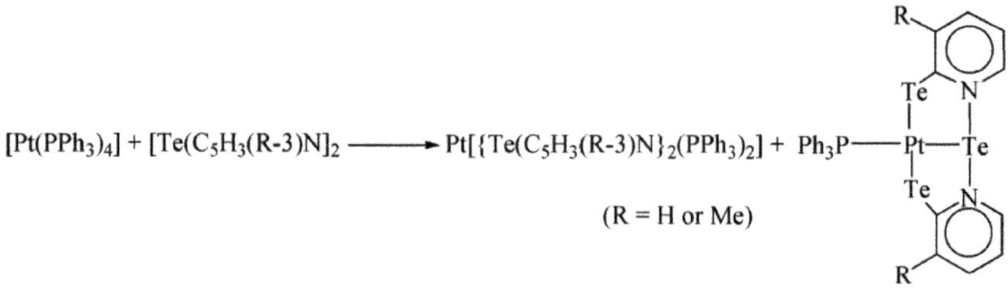

(R = H or Me)

(3.5)

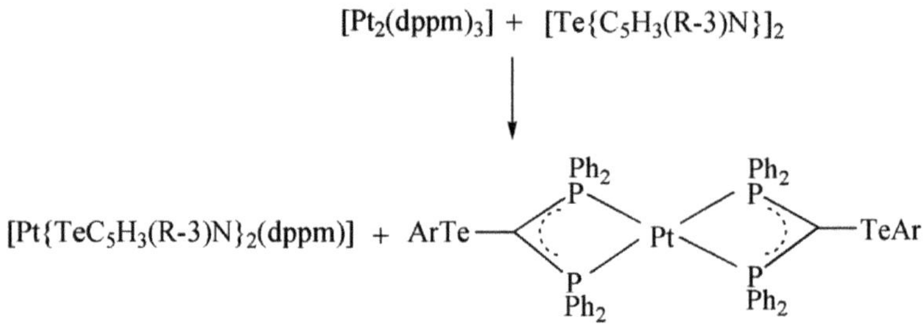

(Ar = C_5H_3(R-3)N; R = H or Me)

(3.6)

3.2.3 Reactions with Nucleophilic Organochalcogenolate Reagents

Organochalcogenolate anions (RE^-), usually used as Li or Na salts generated *in situ*, are versatile nucleophilic reagents commonly employed for the synthesis of a variety of metal chalcogenolate complexes (eqn (3.7) and (3.8)).[28,37–41] These reagents are obtained either by the reaction of chalcogenol with alkali metal compounds (*e.g.*, BuLi, NaOMe, *etc.*) or by reductive cleavage of dichalcogenides using a variety of reducing agents such as $NaBH_4$ and $LiBEt_3H$, *etc.*

$$[PtCl_2(PR'_3)_2] + 2NaER \rightarrow [Pt(ER)_2(PR'_3)_2] + 2NaCl \quad (3.7)$$

$$[Pt_2(\mu\text{-}Cl)_2Cl_2(PR'_3)_2] + 2NaER \rightarrow [Pt_2(\mu\text{-}ER)_2Cl_2(PR'_3)_2] + 2NaCl \quad (3.8)$$

Metathesis reactions, often referred to as transmetalation or redistribution reactions, involving $Pb(ER)_2$,[39] $Hg(ER)_2$,[42] *etc.* with metal halides have been successfully employed for the preparation of metal chalcogenolates. Redistribution of main group (eqn (3.9))[43] and transition metal chalcogenolates (eqn (3.10))[44] are quite facile producing redistributed products in nearly quantitative yields. Metathesis reactions of metal halides with silyl reagents like Me_3SiER proceed cleanly with the elimination of volatile Me_3SiCl with concomitant formation of metal chalcogenolate in high yields.[4]

$$R_2SnCl_2 + [R_2Sn\{SeC_4H(Me_2\text{-}4,6)N_2\}_2] \rightarrow 2[R_2SnCl\{SeC_4H(Me_2\text{-}4,6)N_2\}] \quad (3.9)$$

$$[M_2(\mu\text{-}SR)_2Cl_2(PR'_3)_2] + [M_2(\mu\text{-}Cl)_2Cl_2(PR'_3)_2] \rightarrow 2[M_2(\mu\text{-}SR)(\mu\text{-}Cl)Cl_2(PR'_3)_2] \quad (3.10)$$

3.2.4 Chalcogenolysis/Protonolysis Reactions

These reactions are also commonly employed for the synthesis of metal chalcogenolate complexes. Direct reaction of metal salts L_nMX (X = halide, carboxylate, hydroxide, acetylacetonate, *etc.*) with REH either in the absence (eqn (3.11))[45] or in the presence of a base (*e.g.*, Et_3N, py, *etc.*)[27,46] readily yield metal chalcogenolates.

$$[Pt_2Cl_2(\mu\text{-}Cl)_2(PR'_3)_2] + 2RSH \rightarrow [Pt_2(\mu\text{-}SR)_2Cl_2(PR'_3)_2] + 2HCl \quad (3.11)$$

Chalcogenolysis of metal hydrides, (*e.g.*, alkali metal hydrides), metal alkyls/aryls (eqn (3.12))[47,48] and metal amides (eqn (3.13))[49] proceeds with the elimination of hydrogen, alkane/arene, and amine, respectively, with concomitant formation of metal chalcogenolates. Both di- and tri-valent lanthanide complexes of $(Me_3Si)_3SiE^-$ ligands are conveniently isolated by the protonolysis reaction between an amide derivative and $(Me_3Si)_3SiEH$ (eqn (3.14)).[50,51]

$$R_3M + R'EH \rightarrow \tfrac{1}{2}[R_2M(\mu\text{-}ER')]_2 + R'H \quad (3.12)$$

(M = Ga or In; R = Me, Pr^i, Bu^n, Bu^s, Bu^t, R'E = SMe, SeMe, SPh, SePh, SEt, SeEt, SPr^n, SPr^i, SBu^t, SHex)

$$[M\{N(SiMe_3)_2\}_2] + 2HESi(SiMe_3)_3 \rightarrow [M\{ESi(SiMe_3)_3\}_2] + 2(Me_3Si)_2NH$$
$$(M = Sn, Pb;\ E = S, Se, Te) \quad (3.13)$$

$$[Ln\{N(SiMe_3)_2\}_2(THF)_2] + 2\,(Me_3Si)_3SiEH \xrightarrow[\text{TMEDA}]{\text{Excess}} [Ln\{ESi(SiMe_3)_3\}_2(TMEDA)_2] + 2\,(Me_3Si)_2NH \quad (3.14)$$
$$(Ln = Sm, Eu, Yb;\ E = Se \text{ or } Te)$$

3.3 Structural Aspects of Metal Chalcogenolates

Monofunctional chalcogenolate ligands (RE^-) can act in terminal, μ_2-bridging, and μ_3-bridging modes (Scheme 3.3). The doubly bridging (μ_2-) chalcogenolate generally adopts pyramidal stereochemistry at the chalcogen atom while the triply-bridging (μ_3-) ligand usually acquires a distorted tetrahedral configuration.[52] Monofunctional chalcogenolate, due to its high

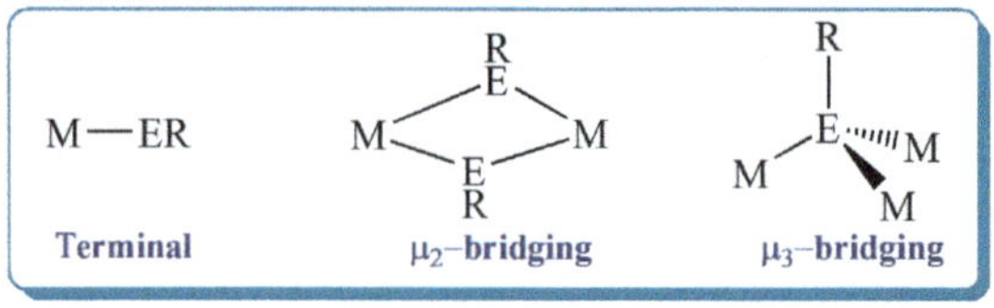

Scheme 3.3 Different bonding modes of monofunctional chalcogenolate ligands.

propensity to bridge, frequently results in aggregation. The overall structure of homoleptic neutral metal chalcogenolate complexes of general type $M(ER)_n$ is governed by all three constituents, *viz.* both metal and chalcogen atoms and the steric and electronic nature of the organic group (R). Structures ranging from discrete monomeric to bi- and tri-nuclear, oligomeric to polymeric and cyclic cages are observed. To suppress oligomerization/polymerization *via* bridging chalcogenolate ligands, several strategies have been developed. These include the use of either a sterically demanding R group (*e.g.*, silyl substituted groups[53] and bulky aryl groups such as $R_3C_6H_2$ (R = Me, Pr^i, Bu^t),[54–56] or internal functionalization of organic substituents (*e.g.*, *N*,*N*-dimethylaminoalkyl,[57] 2-pyridyl/2-pyrimidyl[11,12,57] attached to the chalcogen atom (Scheme 3.1).

Oligomeric/polymeric structures are frequently found for the homoleptic transition metal chalcogenolates of the general composition $[M(ER)_2]$ (R = non-coordinating organic group). In general, smaller R groups favor the formation of one- and two-dimensional coordination polymers while bulkier R groups facilitate the formation of cyclic structures. Oligomers can adopt a three-dimensional (3D) adamantoid network, two-dimensional (2D; *e.g.*, $^{2}_{\infty}[Ag(SePh)]$),[58] and one-dimensional (1D) polymeric chain to oligomeric ring type structures. The interplay of electronic and steric requirements of the R group has a great influence on the overall structure. Thus, depending on R, structures may vary from discrete monomeric two-coordinate geometry with an almost linear S–M–S angle (*e.g.*, $[M\{SC_6H_3\text{-}2,6\text{-}(C_6H_2Pr^i_3\text{-}2,4,6)_2\}_2]$ (M = Mn, Co, Ni),[59] a dimeric with a planar M_2S_2 core (*e.g.*, $[M\{SC_6H_2Bu^t_3\text{-}2,4,6\}_2]_2$ (M = Mn, Co),[60] a three-dimensional adamantoid cage (*e.g.*, $^{3}_{\infty}[Mn(SPh)_2]_4$)[61] to a one-dimensional chain (*e.g.*, $^{1}_{\infty}[Mn(SMes)_2]$)[61] (Figure 3.1). Similar structural diversity can also be noted for cadmium chalcogenolate complexes, *viz.* dimeric (*e.g.*, $[Cd\{EC_6H_2Bu^t_3\text{-}2,4,6\}_2]_2$ (E = S,[62] Se[55]), 3D adamantoid network (*e.g.*, $^{3}_{\infty}[Cd(EPh)_2]_4$ (E = S,[63] Se[64]) to a 1D polymeric chain (*e.g.*, $^{1}_{\infty}[Cd(TeMes)_2]$).[65] Rich structural diversity is also exhibited by d^{10} metal chalcogenolate complexes.[66]

The nature of the chalcogen atom, to some extent, also influences the structure of the resulting metal chalcogenolate. For instance, manganese phenylthiolate adopts a 3D adamantanoid structure ($^{3}_{\infty}[Mn(SPh)_2]_4$)[61] whereas analogous phenylselenolate ($^{1}_{\infty}[Mn(SePh)_2]$)[67] exists as a 1D polymeric chain, although the mesityl derivative irrespective of chalcogen atom, $^{1}_{\infty}[Mn(EMes)_2]$ (E = S,[61] Se or Te[68]) possesses a 1D polymeric chain. "Tiara"-like cyclic structures are frequently encountered for late transition metal

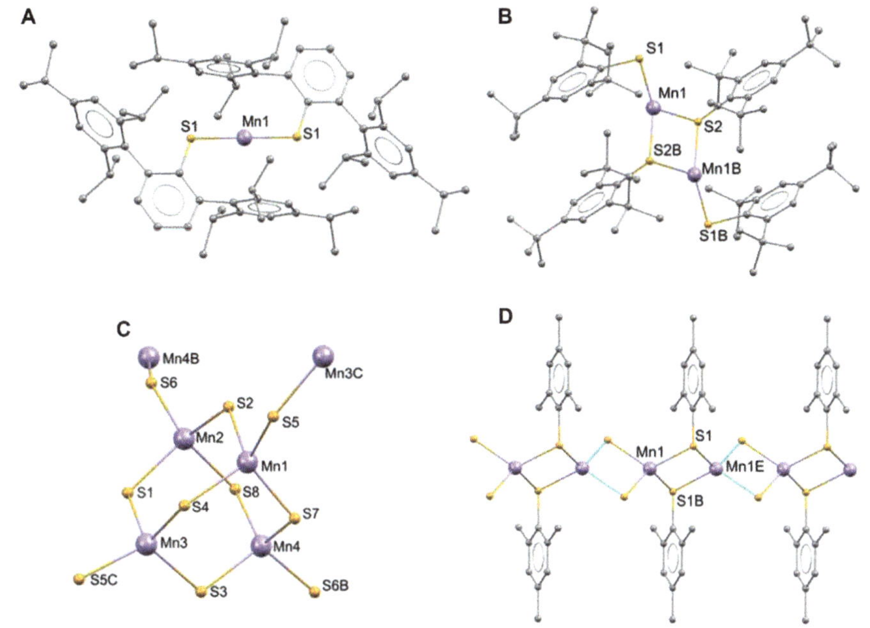

Figure 3.1 Molecular structures of (A) $[Mn\{SC_6H_3\text{-}2,6\text{-}(C_6H_2Pr^i_3\text{-}2,4,6)_2\}_2]$, (B) $[Mn\{SC_6H_2Bu^t_3\text{-}2,4,6\}_2]_2$, (C) ${}^3_\infty[Mn(SPh)_2]_4$ (phenyl group are omitted for clarity), and (D) ${}^1_\infty[Mn(SMes)_2]$. Redrawn from the data from ref. 95–97.

chalcogenolates, *e.g.*, *cyclo*-$[Ni(SR)_2]_n$ ($n = 4$, 5, 6, 8, 9, 11)[69] and *cyclo*-$[Pd(ER)_2]_n$ (n/ER = 6/SR (R = Et, Pr^n, Bu^n, $C_{12}H_{25}$, CH_2COOMe);[70,71] 6/SeR (R = $CH_2CH_2CH_2NMe_2$, CH_2CH_2COOMe);[72,73] 8/SPr^n[74] and 12/TePh[75]). These rings are comprised of square planar metal atoms bridged by four chalcogenolate groups.

Chalcogenolates of main group elements also have a variety of structures. Germanium(IV),[76] tin(IV) and group 15 chalcogenolates[9] exist as monomers. However, tin(II) chalcogenolates adopt several coordination geometries, *viz.* monomeric (*e.g.*, $[Sn(SC_6H_2Bu^t_3\text{-}2,4,6)_2]$),[54] dimeric (*e.g.*, $[Sn\{ESi(SiMe_3)_3\}_2]_2$),[49] trimeric (*e.g.*, $[Sn\{SC_6H_3Pr^i_2\text{-}2,6\}_2]_3$)[54] and 1D linear polymers (*e.g.*, ${}^1_\infty[Sn(EPh)_2]$ (E = S, Se, Te).[77] Likewise for lead(II) chalcogenolates several coordination geometries are possible.[78]

3.4 Applications of Metal Chalcogenolates to Materials Chemistry

3.4.1 Precursors for Main Group Metal Chalcogenides

3.4.1.1 Group 13 Chalcogenides

Group 13 chalcogenides exhibit rich structural and compositional diversity. Gallium and indium chalcogenolate complexes have been primarily used as precursors for deposition of thin films on various substrates, like GaAs(100), Si(100), Si(111), silica, and glass, and only a few of them have been employed for the preparation of metal chalcogenide nanocrystals.[8] Both organometallic and tris-chalcogenolate complexes have been used for the synthesis of gallium and indium chalcogenides. The diorganometal complexes, $[R_2M(\mu\text{-}ER')]_2$ result in the formation of monochalcogenides (ME) on decomposition; whereas mono-organometal chalcogenolates and tris complexes usually afford M_2E_3 materials.

Thin films of the hexagonal phase of GaS have been deposited using $[Bu^t_2Ga(\mu\text{-}SBu^t)]_2$ on KBr single crystals, n- or p-type GaAs(100), and glass substrates at 400 °C.[79] The films comprise nearly spherical particles of 30 nm diameter which upon annealing at 475 °C result in crystallization of the films. Dibutylindium thiolates, $[Bu^x_2In(\mu\text{-}SR')]_2$ ($Bu^x = Bu^n$, Bu^i, Bu^t; $R' = Pr^n$, Pr^i or Bu^t) upon decomposition above 300 °C yield InS,[80,81] mono-organoindium bis(thiolates), and $[RIn(\mu\text{-}SR')(SR')]_2$ ($R/R' = Me/Bu^t$, Bu^n/Pr^n, Bu^n/Pr^i) have been used for deposition of In_2S_3 films.[80–82]

Di-organoindium selenolates as a single source precursor on thermolysis lead to different stoichiometric or selenium deficient indium selenide.[24,83–85] Phase pure hexagonal InSe films have been deposited on GaAs(100) using $[Me_2In(\mu\text{-}SeMe)]_2$ at 320 °C and showed properties comparable to bulk InSe.[83,85] Gysling and co-workers[24] used a toluene solution of $[Me_2In(\mu\text{-}SePh)]_2$ for the deposition of InSe films on a GaAs(100) substrate by spray MOCVD.

Indium tris selenolates such as $[In(SePh)_3]$, $[In\{SeC(SiMe_3)_3\}_3]$ and $[In\{SeC_5H_3\text{-}(Me\text{-}3)N\}_3]$ have been used for deposition of In_2Se_3 films.[24,86,87] The 2-pyridylselenolate complex, $[In\{SeC_5H_3(Me\text{-}3)N\}_3]$ upon thermolysis at 330 °C in HDA, TOPO, or OA yields nanocrystals of different morphologies and phases of In_2Se_3. The hexagonal phase of In_2Se_3 nanocrystals as agglomerated spheres and spherical particles is formed in HDA and TOPO, respectively, whereas in OA, a rhombohedral phase of In_2Se_3 with sheet-like structures is formed.[87]

3.4.1.2 Group 14 Chalcogenides

The large family of binary compound semiconducting Group 14 chalcogenide materials finds numerous applications in optoelectronic,[88] thermoelectric,[89] and photovoltaic (PV)[90] devices and as a catalyst in hydrogen evolution reactions[91] as well as electrode materials in lithium-ion batteries.[92] Among them PbE has been explored extensively since the mid-1980s. However, the lead toxicity restricts their environmentally-friendly utilization.

Molecular precursors for germanium chalcogenide are rare. Germanium tellurolate complexes, $[Ge\{N(Me_2SiCH_2CH_2SiMe_2)\}_2(Pr^i)(TePr^i)]$ and $[Ge(TeBu^n)_4]$, have been used to deposit thin films of GeTe by LPCVD (0.2 mm Hg).[14,93]

A variety of tin and lead chalcogenolate complexes have been employed as precursors for the synthesis of metal chalcogenides. Tin complexes in both +2 and +4 oxidation states have been explored. Divalent tin and lead chalcogenolates (*e.g.*, $^{1}_{\infty}$[Sn(EPh)$_2$] (E = S, Se, Te),[77] [M(Sepy)$_2$],[94] [M{ESi-(SiMe$_3$)$_3$}$_2$] (M = Sn or Pb; E = S, Se, Te),[49] [Pb(ECH$_2$CH$_2$R)$_2$] (E/R = S/NMe$_2$, Se/NMe$_2$, Se/COOMe)[95]) undergo a single step decomposition to yield ME with concomitant elimination of chalcogeno ethers. Thermolysis of [Pb(SeCH$_2$CH$_2$NMe$_2$)$_2$] in HDA at 120 °C yields quantum confined PbSe nanocrystals with an average diameter of 15 nm.[95]

Both classical and organotin(IV) chalcogenolates have been used as molecular precursors for the preparation of tin chalcogenides. For instance, bis(chalcogenolates), [Sn(EPh)$_2${N(SiMe$_3$)$_2$}$_2$], obtained by oxidative addition of Ph$_2$E$_2$ on stannylene, [Sn{N(SiMe$_3$)$_2$}$_2$], undergo single step decomposition (TGA) and upon thermolysis in oleylamine at 210 °C yields nanocrystals of SnE (E = S, Se, Te). The size of the nanocrystals can be controlled by the duration of thermolysis.[96]

Jain and co-workers[43,97–101] developed diorganotin(IV) 2-pyridyl/2-pyrimidyl chalcogenolates as molecular precursors for the preparation of tin chalcogenides (SnE and SnE$_2$). Diorganotin thiolates, [R$_2$Sn(Spy)$_2$] (R = Me, Et, Bu^t),[43,98–101] [Bu$^t_2$Sn(Spyz)$_2$] (Spyz = 2-pyrazinylthiolate)[100] and [R$_2$SnCl(SAr)] (R = Me, Et, Bu^t; Ar = 2-py or 4,6-Me$_2$pym)[98,101] on thermolysis in oleylamine afford orthorhombic SnS nano-sheets of varying thickness (30–80 nm) with an optical direct band gap of 1.61–0.90 eV.

Diorganotin 2-pyridyl, 3-methyl-2-pyridyl, and 5-methyl-2-pyridyl selenolates undergo clean thermolysis in oleylamine at 225 °C to give different morphologies of orthorhombic SnSe or hexagonal SnSe$_2$ nanoparticles depending on the nature of the precursor.[97,99] The optical band gaps of these nanostructures are blue-shifted relative to the bulk material indicative of quantum confinement. These nanocrystals showed a photo-response.[97,99] The complex, [Bu$^t_2$Sn(Sepy)$_2$] has been employed to deposit films of orthorhombic SnSe on glass and p-type silicon(100) substrates at 490 °C.[97] Diorganotin 4,6-dimethyl-2-pyramidyl selenolates on thermolysis in oleylamine afford a nano-crystalline hexagonal phase of SnSe$_2$. The complex [Bu$^t_2$Sn(SepymMe$_2$)$_2$] has been used to deposit films of SnSe$_2$ on silicon wafers at 375 °C by AACVD.

3.4.1.3 Group 15 Chalcogenides

Binary chalcogenides of antimony and bismuth of composition M$_2$E$_3$ are highly anisotropic semiconductor materials and crystallize as a layered structure parallel to their growth directions. These materials find several applications in photovoltaic and room temperature thermoelectric devices.[102] Schulz[9] reviewed the chemistry of antimony and bismuth compounds derived from selenium and tellurium ligands.

Antimony and bismuth thiolates, [M(SR)$_3$] (M/R = Sb/Bu^t,[103] Sb/CH$_2$CF$_3$,[103] Bi/CH$_2$Ph[104]) have been used as precursors for the synthesis of M$_2$S$_3$. Antimony thiolates have been employed for the deposition of phase pure thin films of orthorhombic Sb$_2$S$_3$ (stibnite) by an LPCVD process on glass and silicon substrates at 300–450 °C. Films deposited using [Sb(SBu^t)$_3$] exhibit photoactivity with an optical band gap of 1.6 eV[103] whereas films deposited from $[Sb(SCH_2CF_3)_3]$ show contamination by traces of metallic antimony.

Thermolysis of $[M\{SeC_6H_2Bu^t_3\text{-}2,4,6\}_3]$ at 200 °C yields Sb_2Se_3 when M = Sb but a mixture of bismuth and Bi_2Se_3 when M = Bi with the elimination of selenoether.[105] In contrast, thermolysis of less bulky chalcogenolates $[M\{EC_6H_2R_3\text{-}2,4,6\}_3]$ (M = Sb or Bi; E = S or Se, R = Me or Pr^i) results in elemental antimony/bismuth with the elimination of mainly diaryldichalcogenides (Ar_2E_2).[105]

Jain and co-workers developed 2-pyridyl and 2-pyrimidyl selenolates of antimony and bismuth, $[M(SeAr)_3]$ (M = Sb or Bi, Ar = 2-C_5H_4N, 2-C_5H_3(Me-3)N; 2-$C_4H(Me_2\text{-}4,6)N_2$) for the preparation of phase pure M_2Se_3.[106,107] Pyrolysis of $[M\{SeC_5H_3(Me\text{-}3)N\}_3]$ in a furnace at 400 °C under an argon atmosphere yields orthorhombic Sb_2S_3 and the hexagonal phase of BiSe.[106] Thermolysis of $[M\{SeC_5H_3(Me\text{-}3)N\}_3]$ in HDA at 200 and 230 °C, and $[M\{SeC_4H(Me_2\text{-}4,6)N\}_3]$ in OA at 200 °C gives orthorhombic Sb_2Se_3 and the rhombohedral phase of Bi_2Se_3. The TEM image of Sb_2Se_3 revealed the formation of single crystalline nano-rods with a diameter of 40–60 nm and length of 0.6–1.6 μm.[106,107] The complexes $[M\{SeC_5H_3(Me\text{-}3)N\}_3]$ have also been used for the deposition of metal selenide thin films on a glass substrate by AACVD in the temperature range of 375–500 °C.[106] The antimony complex-deposited films of orthorhombic Sb_2Se_3 which grew preferentially in the (221) direction and was comprised of nanowires like structures (240–500 nm diameter and several micron in length).[106] In contrast, the bismuth complex produced films of different compositions of bismuth selenide depending on the deposition temperature. Thus, rhombohedral Bi_3Se_4 (at 350 °C), hexagonal BiSe (at 375 °C), rhombohedral Bi_4Se_3 (at 400–425 °C) and hexagonal Bi_3Se_2 (at 475 °C) films could be deposited. Clearly with increasing deposition temperature films with selenium deficiency are formed.[106]

Methylantimony bis-chalcogenolates, $[MeSb(EBu^n)_2]$ (E = Se or Te) have been shown to be effective precursors for the deposition of high-quality p-type Sb_2E_3 films on silica substrate through a low pressure (0.05 mm Hg) CVD process at a substrate temperature of 500 °C for Sb_2Se_3[108] and 450 °C for Sb_2Te_3.[108,109] Diphenylantimony tellurolate $[Ph_2Sb(TeR)]$ (R = Et or Ph) on decomposition in oleylamine at 300 °C yields nanoparticles of phase pure rhombohedral Sb_2Te_3. These particles are almost hexagonal in shape with an edge length of ~0.4–2.0 μm and thickness of 20–50 nm.[110] However, decomposition of $[Ph_2SbTePh)]$ in the absence of any solvent in a sealed glass ampule at 300 °C yields only antimony particles. Diorganobismuth tellurolate, $Et_2BiTeEt$ has been used for the deposition of Bi_2T_3 films on Si(100) substrate by LPCVD (10 mbar) at 230 °C. At higher temperatures (245 °C) bismuth rich films (Bi_4Te_3) are formed.[111]

3.4.2 Precursors for Transition Metal Chalcogenides

Since the first mechanical exfoliation of graphene in 2004 by Geim and Novoselov,[112] 2-dimensional transition metal dichalcogenides (TMDs) have attracted tremendous attention due to their technological applications.[113] Besides these layered dichalcogenides, these metals, in particular platinum group[114] and coinage metals, also form several stoichiometric and non-stoichiometric chalcogenides and show polymorphism. Depending on the composition and structural configurations TMDs exhibit a wide range of electrical properties ranging from metallic (*e.g.*, TiS_2, NbS_2, VSe_2, *etc.*), semi-metallic (*e.g.*, $MoTe_2$), semiconducting (*e.g.*, MoS_2, $MoSe_2$, *etc.*), to super-conducting (*e.g.*, $NbSe_2$, TaS_2).

Homoleptic titanium thiolate, $[Ti(SBu^t)_4]$, an air-sensitive deep-red crystalline solid, has been employed for the preparation of titanium sulfide. An initial report revealed the formation of TiS films by an LPCVD process,[115] but subsequent detailed study showed that at low pressure it produces blue-gray/purple films of TiS_2.[116] However, the use of *in situ* prepared $[Ti(SBu^t)_4]$ gives gold colored reflective TiS_2 films on glass substrate at 200–300 °C using an AACVD process.[117] Homoleptic zirconium tellurolate, $[Zr\{TeSi(SiMe_3)_3\}_4$ on heating (200–375 °C) under vacuum yields zirconium telluride.[48]

Metallocene complexes, $[Cp_2M(EBu^t)_2]$ (M = Ti, Zr, Hf; E = S, Se, Te), have been used as single source molecular precursors for deposition of metal chalcogenide films by LPCVD method.[118,119] The thiolate complexes produce sulfur deficient films of composition $TiS_{1.3}$ and ZrS in the temperature range 400–650 °C.[118] Selenolate complexes yield good quality films of MSe_2 at a substrate temperature of 450 °C for $TiSe_2$ and 600 °C for $ZrSe_2$ and $HfSe_2$.[119] The tellurolate complexes under similar conditions in the temperature range 450–600 °C however, deposit gray elemental tellurium rather than metal telluride films.[119]

$[Cp_2V(SeBu^t)_2]$, an extremely moisture sensitive compound, has been employed for deposition of vanadium selenide films on silica substrate at 500 or 600 °C by LPCVD (0.05 mm Hg). EDX measurements indicated Se-deficient compositions, $VSe_{1.55}/VSe_{1.4}$.[120] Homoleptic niobium thiolate, $[Nb\{SC_6H_3Me_2\text{-}2,6\}_5]$, when heated at 450 °C in a dynamic vacuum in a glass ampule gives black films of NbS rather than NbS_2.[121] The latter, however, could be obtained by a sol–gel process.[121] Carmalt and co-workers subsequently used *in situ* generated metal thiolates, which are believed to be $[Nb(SBu^t)_5]$ and $[Ta(NMe_2)_3(SBu^t)_2]$ formed in the reaction between $[M(NMe_2)_5]$ and excess Bu^tSH.[122] These thiolates produced thin films of NbS_2 (at 250–350 °C) and TaS_2 (225 and 450 °C) on glass substrate by AACVD.[122]

Most of the work on TMDs has been centered around Mo and W systems for which multi-source precursors are frequently used.[123] The first single source precursor for MoS_2 was $[Mo(SBu^t)_4]$,[116a] although several 1,1-dithiolate complexes were developed later. Through the LPCVD process (10^{-4} torr), MoS_2 films were deposited on quartz, silicon, and stainless-steel substrates in the temperature range 110–135 °C. The films were polycrystalline and composed of closely packed granules of 30–90 nm size.[116a] Homoleptic chromium and molybdenum arylthiolates, $[M(SC_6H_4X\text{-}4)_3]$ (M/X = Cr/H, Cr/Bu^t, Mo/Bu^t) decompose in the temperature range 300–350 °C to give MS_2 (M = Cr or Mo).[124]

Decomposition of manganese octanethiolate, $[Mn(SC_8H_{17})_2]$ in 1-octadecene at 200–240 °C produces nano-crystals of MnS of ~3.5 nm.[125] Manganese tellurolate complexes, $[Mn(TeBz)(CO)_3(PEt_3)_2]$ and $[Mn(\mu\text{-}TeR)(CO)_4]_2$ (R = Me or Et) on pyrolysis in a stream of hydrogen in a tube furnace at 300 °C yield polycrystalline MnTe.[20] A trinuclear rhenium thiolate, $[Re_3(\mu\text{-}SPr^i)_3(SPr^i)_6]$ has been employed to deposit thin films of ReS_2 by AACVD on a glass substrate at 475 °C.[126] The films grow preferentially in the (001) direction.

Polymeric homoleptic iron selenolate, ${}^1_\infty[Fe(SePh)_2]$ undergoes a two-step decomposition under vacuum to give the tetragonal phase of FeSe at 250 °C, but at higher temperatures (at 550 °C) a mixture of tetragonal $Fe_{1+x}Se$ and hexagonal Fe_7Se_8 phases is formed.[127] Similarly, homoleptic ruthenium thiolates/selenolates, $[Ru(EC_6H_4X\text{-}4)_3]_n$ (E/X = S/OMe, S/Pr^i, S/Bu^t, Se/H, Se/Bu^t) on thermolysis in the temperature range 300–350 °C yields RuE_2 (E = S or Se).[124] Pyrolysis of $[Ru(SePh)_3]_n$ at 450 °C in vacuum resulted in the formation of nano-crystals of $RuSe_2$ with a diameter in the range 3–15 nm.[124]

Polynuclear homoleptic palladium chalcogenolates have been used for the synthesis of palladium chalcogenides. A hexanuclear palladium thiolate, $[Pd(SC_{12}H_{25})_2]_6$ in refluxing diphenylether affords monodispersed PdS nanoparticles of 2.87 ± 0.51 nm in diameter, which are protected by a layer of thiolate species.[128] Palladium complexes with internally functionalized selenolate ligands[5] undergo a two-step decomposition in the temperature range 200–350 °C with the formation of $Pd_{17}Se_{15}$. The binuclear platinum complex, $[PtCl(\mu\text{-}SeCH_2CH_2COOMe)(PPr^n_3)]_2$ on pyrolysis affords Pt_5Se_4.[72] Mononuclear palladium tellurolates, $[PdCl\{TeC_5H_3(Me\text{-}3)N\}(PR_3)]$ ($PR_3 = PEt_3$ or PPh_3) on pyrolysis under an argon atmosphere at 340 °C yield aggregate of nanocrystalline PdTe (Figure 3.2).[129]

Chalcogenato-bridged allylpalladium complexes undergo clean thermolysis to yield palladium chalcogenides, the composition of which depends on the nature of the chalcogenolate and allyl groups.[42,130–132] Thermolysis of

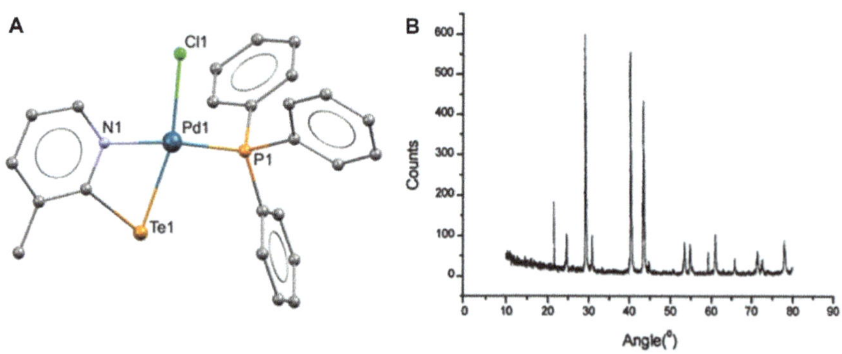

Figure 3.2 (A) Molecular structure of $[PdCl\{Te(3\text{-}MeC_5H_3N)\}(PPh_3)]$ and (B) XRD pattern of PdTe obtained from this complex on pyrolysis at 340 °C under an argon atmosphere. Reproduced from ref. 129 with permission from John Wiley & Sons, Copyright © 2002 WILEY-VCH Verlag GmbH & Co. KGaA, Weinheim.

allylpalladium complexes, $[Pd(\mu\text{-}SeCH_2CH_2COOEt)(\eta^3\text{-}C_3H_5)]_n$ ($n=2$ or 3) either in diphenyl ether or hexadecyl amine at 180 °C produces nano-crystals of Pd_7Se_4, but at higher temperature (220 °C) $Pd_{17}Se_{15}$ is formed.[131] The methylallylpalladium complexes, $[Pd(\mu\text{-}ER)(\eta^3\text{-}C_4H_7)]_2$ (ER = SBu^t, SC_6F_5, SPh, SMes, SePh, SeMes) undergo a three step decomposition and in refluxing xylene/HDA yield nanocrystals of Pd_4E (E = S or Se).[42,130] The tellurolate complex $[Pd(\mu\text{-}TePh)(\eta^3\text{-}C_4H_7)]_2$ tends to decompose during its preparation giving a black powder identified as Pd_3Te_2.[42]

Although thermal decomposition of coinage metal thiolates, $[M(SC_6F_5)]_n$ (M = Cu, Ag, Au) was reported by Peach in 1973,[133] their potential as molecular precursors could be realized only in the past two decades or so. Copper dodecane thiolate, $[Cu(SC_{12}H_{25})]_n$ on thermolysis in dodecane thiol at 200 °C gives the hexagonal phase of Cu_2S nanocrystals which could be isolated in different morphologies like nanodisks, spheres, and hexagon nanoplates by adjusting the experimental conditions.[134] Interestingly, the analogous silver complex, $[Ag(SC_{12}H_{25})]_n$ on thermolysis at 180 °C yields nanodisks of fcc-silver.[135] Alkylthiolates $[Cu(SR)]_n$ ($R = C_nH_{2n+1}$; $n = 4, 8, 12, 16$) shows liquid–crystalline mesophasic behavior which plays an important role in templating the morphology of Cu_2S nanocrystals. The short alkyl chain precursor (*e.g.*, $[Cu(SBu^n)]_n$) on thermolysis produces two dimensional nanosheets of Cu_2S whereas the longer chain (*e.g.*, $[Cu(SC_{12}H_{25})]_n$) yields nanodisks of Cu_2S exclusively.[136] Non-mesogenic thiolates (*e.g.*, $[Cu(SCMe_2C_9H_{19})]_n$, $[Cu(SBz]_n$) in the temperature range 140–200 °C produce nanoparticles of irregular size and composition.[136]

Copper selenolates, $[Cu(SeR)]_\infty$ (R = Me, Ph, $C_6H_4Bu^t\text{-}3$),[137] $[Cu(SeC_6F_5)(py)_2]$,[138] $[Cu(Sepy)]_4$,[23] $[Cu\{SeC_5H_3(SiMe_3\text{-}3)N\}]_4$,[23] and $[Cu\{SeC_5H_3(Me\text{-}3)N\}]_4$[139] on thermolysis affords orthorhombic or cubic phases of α-CuSe, $Cu_{2-x}Se$, and Cu_2Se depending on the decomposition temperatures. In contrast, thermolysis of $[Ag(SePh)]_\infty$ at 180 °C under vacuum results in the formation of metallic silver.[137]

Jain and co-workers[139–141] developed pyridyl and pyrimidyl chacogenolates of copper and silver for the synthesis of metal chalcogenides. The complexes $[Ag(\mu\text{-}Spy\text{-}4)(PPh_3)_2]_2$ and $[Ag(Spy\text{-}4)]_n$ either in the solid state at 270 °C or in hot hexadecyl amine yield monoclinic Ag_2S.[141] Similarly selenolate complexes, $[Ag(Sepy\text{-}4)(PPh_3)_3]$, $[Ag_3(\mu\text{-}Sepy\text{-}4)_3(PPh_3)_4]$, and $[Ag(Sepy\text{-}4)]_n$ in HDA or TOPO at 270 °C give the orthorhombic phase of Ag_2Se.[141] Homoleptic 2-pyrimidyl selonolate complexes, $[M\{SeC_4H(Me_2\text{-}4,6)N_2\}]_6$ (M – Cu or Ag) in dodecylthiol at 150 °C gives nano-crystals of cubic Cu_7Se_4 and orthorhombic Ag_2Se.[140] The complexes $[Cu(SepyMe\text{-}3)]_4$ and $[Cu\{SeC_4H(Me_2\text{-}4,6)N_2\}]_6$ have also been used as precursors to deposit thin films of copper selenides. The 2-pyridyl selenolate on glass substrate in the temperature range 360–400 °C deposited uniform films of orthorhombic phase[139] of Cu_5Se_4 whereas the 2-pyrimidyl selonolate complex at 400 °C substrate (glass or silicon) temperature produced the cubic phase of Cu_7Se_4 films.[140] Copper 2-pyridyl tellurolates, $[Cu\{TeC_5H_3(R\text{-}3)N\}]_4$ (R = H or Me) on thermolysis in TOPO at 150 °C produces the hexagonal phase of $Cu_{2-x}Te$ (for R = H) and tetragonal phase of $Cu_{2-\delta}Te$ (for R = Me).[139] $[Cu\{TeC_5H_3(Me\text{-}3)N\}]_4$ has been employed to deposit thin films of $Cu_{1.85}Te$ on a glass substrate at 350 °C by AACVD.[139]

3.4.2.1 *Precursors for Group 12 Chalcogenide Materials*

Remarkable shape and size dependent optical and electronic properties of binary chalcogenides of group-12 relative to the bulk form make them useful candidates for a variety of applications in optoelectronics and photovoltaic devices and as a fluorescent probe in biological imaging. The relevance of group 12 metal chalcogenolates as molecular precursors was realized in the 1980s and they have played an important role in the progress of single source molecular precursor chemistry.

Pyrolysis of oligomeric zinc and cadmium chalcogenolates, $[M(ER)_2]_n$ and mercury tellurolate $[Hg(TePh)_2]$ in the solid-state yields nearly quantitatively ME with concomitant elimination of chalcogeno ethers.[38,142] Dimeric cadmium complexes, $[Cd\{EC_6H_2Bu^t_3\text{-}2,4,6\}_2]_2$, are volatile solids and can be sublimed under vacuum (10^{-2} torr). These complexes have been used to deposit polycrystalline thin films of CdS and CdSe at 450 and 350 °C, respectively at 10^{-2} torr on a glass substrate.[62] In contrast, mercury complexes $[Hg\{EC_6H_2R_3\text{-}2,4,6\}_2]_n$ (E/R/n = S,Se,Te/Me/∞; S,Se,Te/Pr^i or Bu^t/1) on decomposition afford metallic mercury and the corresponding diorgano dichalcogenides.[21]

Thermolysis of oligomeric group-12 metal chalcogenolates often results in contaminated and/or phase impure products. Mononuclear Lewis base adducts with nitrogen donor ligands proved to be better candidates. Thus thermolysis of $[Zn(TeMes)_2(py)_2]$ either in the solid state or in paraffin oil at 270–320 °C results in the formation of cubic ZnTe whereas the parent polymeric complex $[Zn(TeMes)_2]_n$ in refluxing 4-ethylpyridine decomposes slowly to yield nano crystalline ZnTe contaminated with crystalline tellurium.[143] Similarly, injection of a trioctylphospine solution of $[Zn(EPh)_2(TMEDA)]$ (E = Se,[144] Te[145]) in hot trioctylphosphine oxide (for Se)/dodecylamine (for Te) results in the formation of spherical nano crystals of ZnE. The size of these particles depends upon their growth temperature, smaller particles being formed at higher temperatures.[144,145] Both ZnSe and ZnTe nano crystals exhibit photoluminescence with blue shifted band maxima. Thermolysis of $[Zn(ESiMe_3)_2(TMEDA)]$ (E = S, Se) in refluxing toluene generated TMEDA protected nanoparticles of ZnE of 2–3 nm size, but in the absence of a solvent at 250 °C resulted in unligated bulk ZnE (E = S or Se).[146]

An oligomeric organo-zinc thiolate, $[EtZn(SEt)]_{10}$, obtained by insertion of sulfur in the Zn–C bond of Et_2Zn, comprises of three 6-membered Zn–S rings in chair confirmation resembling the wurtzite-like core of bulk ZnS, has been used as a molecular precursor. Interestingly, thermolysis at different temperature leads to the formation of cubic ZnS (sphalerite) rather than wurtzite form existing in the core of the molecule.[147]

Group 12 metal complexes derived from an internally functionalized chalcogenolate ligand, *viz.* $Me_2NCH_2CH_2E^-$ (E = Se or Te),[148–150] $Me_2NCH_2CH_2CH_2Se^-$ [151] and 2-pyridyl/2-pyrimidyl chalcogenolate,[152–155] have been used as precursors for ME materials. Mononuclear complexes, $[Cd\{SeC_5H_3(R\text{-}3)N\}_2(TMEDA)]$ (R = H or Me)/$[Hg(Sepy)_2(TMEDA)]$ and $[Hg\{SeC_5H_3(Me\text{-}3)N\}_2]$ undergo a two-step decomposition.[154] Cadmium complexes in hot HDA/TOPO gives red colored nano-crystals of hexagonal CdSe. Thermolysis of $[Cd(Sepy)_2(TMEDA)]$ at 200 °C produced rectangular shaped nano-crystals of CdSe with a width of 5–7 nm and length 5–11 nm.[154] The mercury complexes on thermolysis in HDA afford nano-crystals of cubic HgSe.[154] Similarly 2-pyrimidyl selenolate complexes, $[Cd\{SeC_4H(Me_2\text{-}4,6)N_2\}_2(TMEDA)]$ and $[Hg\{SeC_4H(Me_2\text{-}4,6)N_2\}_2]$ in OA gives a mixture of spherical and elliptical shaped nano-crystals of hexagonal CdSe and rod shaped particles of cubic HgSe.[155] Cadmium and mercury complexes of 2-pyridyl tellurolate, $[M(Tepy)_2]$ and $[Cd(Tepy)_2(TMEDA)]$ on thermolysis either in the solid state or in solution yield MTe. Thermolysis of $[Cd(Tepy)_2]$ in HDA affords red colored HDA capped hexagonal CdTe nanoparticles of size ~5 nm, whereas the mercury complex in TOPO at 145 °C yields cubic HgTe nanoparticles of 13 nm (by TEM).[153]

3.4.3 Precursors for Lanthanide Chalcogenides

Lanthanides form mono (LnE) and sesqui-chalcogenides (Ln_2E_3) with electronic properties ranging from insulator to semiconducting to conducting depending on the nature of the lanthanide ion. Polymorphism in Ln_2E_3 is quite common. Conventional high temperature synthesis of lanthanide chalcogenides is a tedious process and is often encountered by incorporation of impurities owing to the high oxophilicity of the lanthanide metal center.

Controlled pyrolysis of inorganic and organometallic lanthanide chalcogenolates of the type $[Ln(ER)_x]$ (E = S, Se, Te; R = alkyl or aryl; $x = 2$ or 3), usually isolated as Lewis base adducts of pyridine, THF or DME (DME = 1,2-dimethoxyethane), has emerged as an important strategy for the synthesis of Ln_xE_y binary materials. In general, divalent lanthanide chalcogenolates, $[Ln(ER)_2L_x]$ decompose to yield monochalcogenides (LnE) whereas trivalent lanthanide chalcogenolates, $[Ln(ER)_3L_x]$ on pyrolysis give Ln_2E_3. In all these cases, the thermolysis is accompanied by elimination of coordinated neutral ligand (L) and chalcogenoether (R_2E).

Both di- and tri-valent lanthanide complexes of $(Me_3Si)_3SiE^-$ ligands have been used for the synthesis of lanthanide chalcogenides.[50,51] The homoleptic trivalent lanthanide tellurolate complexes are thermally unstable and tend to decompose to telluride–tellurolate clusters, $[Ln_5Te_3\{TeSi(SiMe_3)_3\}_9]$ (Ln = La, Ce). The latter are believed to be intermediate species in the conversion of molecular precursors to metal chalcogenides.[50]

Pyrolysis of $[Ln(SPh)_3(py)_3]_2$ at >550 °C yields crystalline Ln_2S_3 (Ln = Ho, Tm).[156] The analogous selenolate complexes, $[Ln(SePh)_3(py)_3]_2$ also yield single phasic Ln_2Se_3 (Ln = Tm, Yb), but the corresponding holmium complex under similar pyrolytic conditions affords a metastable mixture of HoSe and $HoSe_2$.[157] Ytterbium tellurolate complex, $[Yb(TeMes)_3(THF)_x]$ either in refluxing toluene or heating *in vacuo* at 250–300 °C gives a black powder which on annealing at 900 °C under argon affords crystalline YbTe.[158]

3.4.4 Precursors for Ternary Chalcogenide Materials

Both ternary and quaternary chalcogenide semiconductor materials find application in photovoltaic and thermoelectric devices.[159] For instance, $CuInSe_2$ has been investigated since 1975 as a solar cell material with solar conversion efficiency in laboratory devices[160] now exceeding 21%.

Copper-based ternary materials, *viz.* chalcopyrite compounds, ($CuInE_2$), and diamond-like chalcogenide (Cu_2SnE_3) have been prepared by molecular precursor routes. Kanatzidis and co-workers[161] reported the synthesis of binuclear chalcogenolate-bridged complexes of the general formula, $[(PR'_3)_2Cu(\mu\text{-}ER)_2In(ER)_2]$. Subsequently, other preparative routes including a one-pot synthesis of these complexes has been described (Scheme 3.4).[162–166]

Direct heating in a Pyrex tube and % weight loss in TG analysis of $[(PR'_3)_2M(\mu\text{-}ER)_2M'(ER)_2]$ ($PR'_3 = PBu_3$, PPh_3; ER = SEt, SPr^n, SBu^i, SPh, SeEt; M = Cu, Ag; M′ = Al, Ga, In) is consistent with the formation of $MM'E_2$ materials.[161,165,167] Thermolysis of $[(PPh_3)_2Cu(\mu\text{-}SeBz)_2In(SeBz)_2]$ in oleylamine or hexadecylamine at 300 °C results in the formation of the body centered tetragonal phase of $CuInSe_2$ nanoparticles of irregular shape with sizes varying in the range of 20–30 nm (Figure 3.3).[168]

Thermolysis of a mixture of dual chalcogenolate precursors has also been successfully used to prepare copper based ternary materials. Thus thermolysis of a 1 : 1 mixture of either $[Me_2In\{SeC_5H_4N\text{-}4\}]_n$ and $[Cu(SeC_5H_4N\text{-}4)\text{-}(PEt_3)_2]$ in oleylamine[84] at 300 °C or $[In\{SeC_5H_3(Me\text{-}3)N\}_3]$ and $[Cu(SeC_5H_3\text{-}(Me\text{-}3)N\}]_4$ in oleylamine/HDA/TOPO[87] at 330 °C yields the tetragonal phase of $CuInSe_2$.

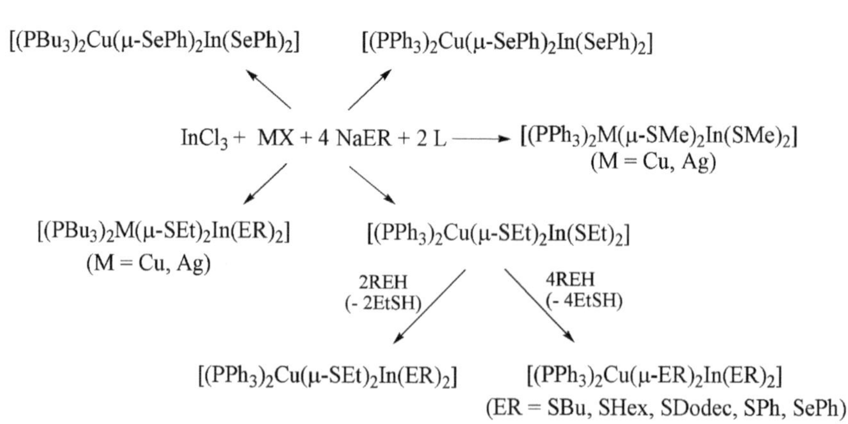

Scheme 3.4 General preparative routes for $[(PR'_3)_2M(\mu\text{-}ER)_2In(ER)_2]$ complexes.

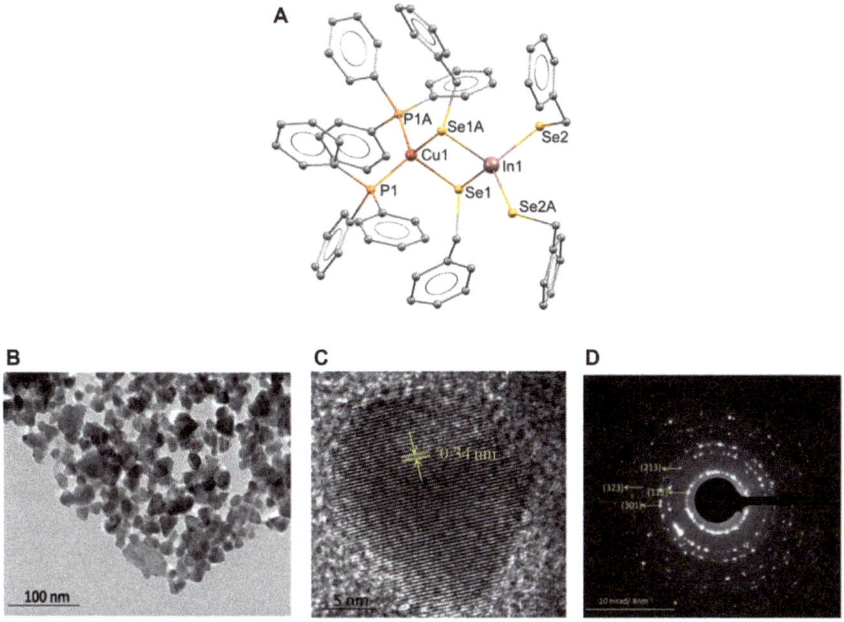

Figure 3.3 (A) Molecular structure of $[(PPh_3)_2CuIn(SeCH_2C_6H_5)_4]$, (B) TEM image, (C) HRTEM image, and (D) SAED pattern of the $CuInSe_2$ nanoparticles obtained by thermolysis of $[(PPh_3)_2CuIn(SeCH_2C_6H_5)_4]$ in oleylamine at 300 °C. Reproduced from ref. 168 with permission from the Royal Society of Chemistry.

Similarly, nanostructures of Cu_2SnE_3 (E = S or Se) have been prepared by using a mixture of dual chalcogenolate precursors.[100,169]

Acknowledgements

The authors are grateful to Mr. Amey P. Wadawale for redrawing molecular structures from the X-ray data deposited in CCDC. We thank the Department of Atomic Energy for funding our research activity. We sincerely acknowledge the contributions of our collaborators, colleagues, and students whose names appear as co-authors in the publications. Permission from publishers to reproduce figures is gratefully acknowledged.

References

1. C. Siemens, *Ann. Chem.*, 1847, **61**, 360.
2. H. Roemer, *Chem. Ber.*, 1873, **6**, 784.
3. (a) P. Claesson, *J. Prakt. Chem.*, 1877, **15**, 205; (b) K. A. Hofmann and W. O. Rabe, *Z. Anorg. Allg. Chem.*, 1897, **14**, 293; (c) F. G. Mann and D. Purdie, *J. Chem. Soc.*, 1935, 1549.
4. J. Arnold, *Prog. Inorg. Chem.*, 1995, **43**, 353.

5. V. K. Jain and G. Kedarnath, in *Selenium and tellurium reagent: In chemistry and material science*, ed. R. S. Laitinen and R. Oilunkamieni, DeGrutyer, 2019.
6. (a) H. Lecher, *Chem. Ber.*, 1915, **48**, 1425–1432; (b) R. J. Kern, *J. Am. Chem. Soc.*, 1953, **75**, 1865.
7. R. F. Frindt, *J. Appl. Phys.*, 1966, **37**, 1928–1929.
8. M. A. Malik, M. Afzal and P. O'Brien, *Chem. Rev.*, 2010, **110**, 4417.
9. S. Schulz, *Coord. Chem. Rev.*, 2015, **297**, 49.
10. M. D. Khan, M. A. Malik and N. Revaprasadu, *Coord. Chem. Rev.*, 2019, **388**, 24.
11. G. Kedarnath and V. K. Jain, *Coord. Chem. Rev.*, 2013, **257**, 1409.
12. R. Cargnelluti, R. F. Schumacher, A. L. Belladona and J. C. Kazmierzak, *Coord. Chem. Rev.*, 2021, **426**, 213537.
13. W. Levason, S. D. Orchard and G. Reid, *Coord. Chem. Rev.*, 2002, **225**, 159.
14. T. Chen, W. Hunks, P. S. Chen, G. T. Stauf, T. M. Cameron, C. Xu, A. G. Dilasquale and A. L. Rheingold, *Eur. J. Inorg. Chem.*, 2009, 2047.
15. M. D. Janssen, D. M. Grove and G. van Koten, *Prog. Inorg. Chem.*, 1997, **46**, 97.
16. G. Mugesh and H. B. Singh, *Acc. Chem. Res.*, 2002, **35**, 226.
17. A. H. Cowley, R. A. Jones, P. R. Harris, D. A. Atwood, L. Contreras and C. J. Burek, *Angew. Chem., Int. Ed. Engl.*, 1991, **30**, 1143.
18. H. Rahbarnoohi, R. Kumar, M. J. Heeg and J. P. Oliver, *Organometallics*, 1995, **14**, 3869.
19. G. Tainturier, M. Fahim, G. Trouve-Bellan and B. Gautheron, *J. Organomet. Chem.*, 1989, **376**, 321.
20. A. P. Coleman, R. S. Dickson, G. B. Deacon, G. D. Fallon, M. Ke, K. McGregor and B. O. West, *Polyhedron*, 1994, **13**, 1277.
21. M. Bochmann and K. J. Webb, *J. Chem. Soc., Dalton Trans.*, 1991, 2325.
22. S. Schlecht and K. Friese, *Eur. J. Inorg. Chem.*, 2003, 1411.
23. Y. Cheng, T. J. Emge and J. G. Brennan, *Inorg. Chem.*, 1996, **35**, 7339.
24. H. J. Gysling, A. A. Wernberg and T. N. Balton, *Chem. Mater.*, 1992, **4**, 900.
25. J. Y. Chao, H. C. Jeong, K. S. Kin, D. H. Kang, H. K. Kim and H. W. Shim, *Bull. Korean Chem. Soc.*, 2003, **24**, 645.
26. N. Ghavale, A. Wadawale, S. Dey and V. K. Jain, *Indian J. Chem., Sect. A: Inorg., Bio-inorg., Phys., Theor. Anal. Chem.*, 2009, **48**, 1510.
27. V. K. Jain, S. Kannan and E. R. T. Tiekink, *J. Chem. Res. (S)*, 1994, 85.
28. V. K. Jain and R. S. Chauhan, *Coord. Chem. Rev.*, 2016, **306**, 270.
29. V. G. Albano, V. DeFelice, M. Monari, G. Roviello and F. Ruffo, *Eur. J. Inorg. Chem.*, 2005, 416.
30. J. M. Gonzales, D. G. Musaev and K. Morokuma, *Organometallics*, 2005, **24**, 4908.
31. R. Oilunkanieni, R. S. Laitinen and M. Ahlgren, *J. Organomet. Chem.*, 2000, **595**, 232.
32. R. S. Chauhan, G. Kedarnath, A. Wadawale, A. Muñoz-Castro, R. Arratia-Perez, V. K. Jain and W. Kaim, *Inorg. Chem.*, 2010, **49**, 4179.

33. R. S. Chauhan, G. Kedarnath, A. Wadawale, A. L. Rheingold, A. Muñoz-Castro, R. Arratia, Perez and V. K. Jain, *Organometallics*, 2012, **31**, 1743.
34. R. S. Chauhan, G. Kedarnath, A. Wadawale, A. M. Z. Slawin and V. K. Jain, *Dalton Trans.*, 2013, **42**, 259.
35. S. Heimann, A. Kuczkowski, D. Bläser, C. Wölper, R. Haack, G. Jansen and S. Schulz, *Eur. J. Inorg. Chem.*, 2014, 4858.
36. (a) I. D. Sadekov, A. I. Uraev and A. D. Garnovskii, *Russ. Chem. Rev.*, 1999, **68**, 415; (b) N. Begum, M. I. Hyder, M. R. Hassan, S. E. Kabir, D. W. Bennett, D. T. Haworth, T. A. Siddiquee, D. Rokhsana, A. Sharmin and E. Rosenberg, *Organometallics*, 2008, **27**, 1550.
37. M. Iwaoka, in *Organoselenium Chemistry: Synthesis and reactions*, ed. T. Wirth, Wiley-VCH Verlag & Co. KGaA, Weinheim, 2012.
38. V. K. Jain and G. Kedarnath, *Phys. Sci. Rev.*, 2018, 20170127.
39. V. K. Jain and S. Kannan, *J. Organomet. Chem.*, 1992, **439**, 231.
40. V. K. Jain, S. Kannan, R. J. Butcher and J. P. Jasinski, *Polyhedron*, 1995, **14**, 3641.
41. (a) R. S. Chauhan, D. B. Cordes, A. M. Z. Slawin, S. Yadav and C. Dash, *Inorg. Chim. Acta*, 2018, **478**, 125; (b) R. S. Chuahan, A. Kumar and P. Prabhu, *Inorg. Chim. Acta*, 2019, **487**, 395; (c) R. S. Chauhan, D. Oza, S. Yadav, C. Dash, A. M. Z. Slawin and N. Shivram, *New J. Chem.*, 2019, **43**, 2381.
42. A. Singhal, V. K. Jain, R. Mishra and B. Varghese, *J. Mater. Chem.*, 2000, **10**, 1121.
43. A. Tyagi, G. Kedarnath, A. Wadawale, A. Y. Shah, V. K. Jain and B. Vishwanadh, *RSC Adv.*, 2016, **6**, 8367.
44. (a) V. K. Jain and G. S. Rao, *Inorg. Chim. Acta*, 1987, **127**, 161; (b) V. K. Jain, *Inorg. Chim. Acta*, 1987, **133**, 261.
45. H. C. Clark, V. K. Jain and G. S. Rao, *J. Organomet. Chem.*, 1985, **279**, 181.
46. (a) V. K. Jain, R. P. Patel and K. Venkatsubramanian, *Polyhedron*, 1991, **10**, 851; (b) V. K. Jain, S. Kannan, R. J. Butcher and J. P. Jasinski, *J. Organomet. Chem.*, 1994, **468**, 285.
47. (a) G. E. Coates and R. G. Hayter, *J. Chem. Soc.*, 1953, 2519; (b) R. Nomura, S. Inazawa, K. Kamaya and H. Matsuda, *Polyhedron*, 1989, **8**, 763; (c) G. G. Hoffman and R. Faist, *J. Organomet. Chem.*, 1990, **391**, 1; (d) M. B. Power, J. W. Ziller, A. N. Tyler and A. R. Barron, *Organometallics*, 1992, **11**, 1055; (e) A. N. MacInnes, M. B. Power, A. F. Hepp and A. R. Barron, *J. Orgnaomet. Chem.*, 1995, **449**, 95.
48. V. Christou and J. Arnold, *J. Am. Chem. Soc.*, 1992, **114**, 6240.
49. A. L. Seligson and J. Arnold, *J. Am. Chem. Soc.*, 1993, **115**, 8214.
50. (a) D. R. Cary and J. Arnold, *J. Am. Chem. Soc.*, 1993, **115**, 2520; (b) D. R. Cary, G. E. Ball and J. Arnold, *J. Am. Chem. Soc.*, 1995, **117**, 3492.
51. D. R. Cary and J. Arnold, *Inorg. Chem.*, 1994, **33**, 1791.
52. (a) I. G. Dance, *Polyhedron*, 1986, **5**, 1037; (b) P. J. Blower and R. J. Dilworth, *Coord. Chem. Rev.*, 1987, **76**, 121.

53. A. Pladzyk, D. K. Zedler, A. Ciborska, A. Schnepf and A. Dolega, *Coord. Chem. Rev.*, 2021, **437**, 213761.
54. P. B. Hitchcock, M. F. Lappert, B. J. Samways and E. L. Weinberg, *J. Chem. Soc., Chem. Commun.*, 1983, 1492.
55. M. Bochmann, K. J. Webb, M. B. Hursthouse and M. Mazid, *J. Chem. Soc., Dalton Trans.*, 1991, 2317.
56. (a) J. R. Dilworth and J. Hu, *Adv. Inorg. Chem.*, 1993, **43**, 411; (b) I. D. Sadekov and A. Z. Zakharov, *Russ. Chem. Rev.*, 1999, **68**, 909.
57. V. K. Jain, *Dalton Trans.*, 2020, **49**, 8817.
58. H. L. Cuthbert, A. I. Wallbank, N. J. Taylor and J. F. Corrigan, *Z. Anorg. Allg. Chem.*, 2002, **628**, 2483.
59. T. Nguyen, A. Panda, M. M. Olmstead, A. F. Richards, M. Stender, M. Brynda and P. P. Power, *J. Am. Chem. Soc.*, 2005, **127**, 8545.
60. P. P. Power and S. C. Shoner, *Angew. Chem. Int. Ed. Engl.*, 1991, **30**, 330.
61. A. Eichhöfer and S. Lebedkin, *Inorg. Chem.*, 2017, **57**, 602.
62. M. Bochman, K. J. Webb, M. Harman and M. B. Hurthouse, *Angew. Chem., Int. Ed. Engl.*, 1990, **29**, 638.
63. I. G. Dance, R. G. Garbutt, D. C. Craig and M. L. Scudder, *Inorg. Chem.*, 1987, **26**, 4064.
64. K. S. Anjali and J. J. Vittal, *Inorg. Chem. Commun.*, 2000, **3**, 708.
65. M. Bochmann, A. P Colemann, K. J. Webb, M. B. Hurthouse and M. Mazid, *Angew. Chem., Int. Ed. Engl.*, 1991, **30**, 973.
66. O. Veselska and A. Demessence, *Coord. Chem. Rev.*, 2018, **355**, 240.
67. A. Eichhöfer, P. T. Wood, R. Vishwanath and R. A. Mole, *Eur. J. Inorg. Chem.*, 2007, 4794.
68. M. Bochmann, A. K. Powell and X. Song, *J. Chem. Soc., Dalton Trans.*, 1995, 1645.
69. S. I. Ivanov, M. A. Kozee, W. A. Merrill, S. Agarwal and L. F. Dahl, *J. Chem. Soc., Dalton Trans.*, 2002, 4105.
70. (a) N. R. Kunchur, *Acta Cryst.*, 1968, **B24**, 1623; (b) I. Schneider, M. Hörner, R. N. Olendzki and J. Strähle, *Acta Cryst.*, 1993, **C49**, 2091; (c) A. I. Stash, T. I. Perepelkova, Y. G. Noskov, T. M. Buslaeva and I. P. Romm, *Russ. J. Coord. Chem.*, 2001, **27**, 585; (d) A. I. Stash, V. V. Levashova, S. A. Lebedev, Y. G. Hoskov, A. A. Mal'kov and I. P. Romm, *Russ. J. Coord. Chem.*, 2009, **35**, 136.
71. Z. Yang, A. B. Smetana, C. M. Sorensen and K. J. Klabunde, *Inorg. Chem.*, 2007, **46**, 2427.
72. L. B. Kumbhare, V. K. Jain, P. P. Phadnis and M. Nethaji, *J. Organomet. Chem.*, 2007, **692**, 1546.
73. S. Dey, V. K. Jain, A. Klein and W. Kaim, *Inorg. Chem. Commun.*, 2004, **7**, 601.
74. J. D. Higgins and J. W. Suggs, *Inorg. Chim. Acta*, 1988, **145**, 247.
75. B. Tirloni, E. S. Lang, G. M. deOliveira, P. Piquini and M. Hörner, *New. J. Chem.*, 2014, **38**, 2394.
76. B. Kersting and B. Krebs, *Inorg. Chem.*, 1994, **33**, 3886.

77. A. Eichhöfer, J. J. Jiang, H. Sommer, F. Weigand, O. Fuhr, D. Fenske, C. Y. Su and G. Buth, *Eur. J. Inorg. Chem.*, 2010, 410.
78. R. L. Davidovich, V. Stavila and K. H. Whitnire, *Coord. Chem. Rev.*, 2010, **254**, 2193.
79. A. N. MacInnes, M. B. Power and A. R. Barron, *Chem. Mater.*, 1993, **5**, 1344.
80. A. N. MacInnes, M. B. Power, A. F. Hepp and A. R. Barron, *J. Orgnaomet. Chem.*, 1995, **449**, 95.
81. R. Nomura, S. Inazawa, K. Kanaya and H. Matsuda, *Appl. Organomet. Chem.*, 1989, **3**, 195.
82. R. Nomura, K. Konishi, S. Futenma and H. Matsuda, *Appl. Organomet. Chem.*, 1990, **4**, 607–610; R. Nomura, K. Konishi and H. Matsuda, *Thin Solid Films*, 1991, **198**, 339.
83. J. Y. Chao, H. C. Jeong, K. S. Kin, D. H. Kang, H. K. Kim and H. W. Shim, *Bull. Korean Chem. Soc.*, 2003, **24**, 645.
84. M. K. Pal, S. Dey, A. P. Wadawale, N. Kushwah, M. Kumar and V. K. Jain, *ChemistrySelect*, 2018, **3**, 8575.
85. I. H. Choi and P. Y. Yu, *J. Appl. Phys.*, 2003, **93**, 4673.
86. J. Cheon, J. Arnold, K. M. Yu and E. D. Bourret, *Chem. Mater.*, 1995, **7**, 2273.
87. R. K. Sharma, G. Kedarnath, N. Kushwah, M. K. Pal, A. Wadawale, B. Vishwanadh, B. Paul and V. K. Jain, *J. Organomet. Chem.*, 2013, **747**, 113.
88. M. A. Franzman, C. W. Schlenkar, M. E. Thompson and R. L. Brutchley, *J. Am. Chem. Soc.*, 2010, **132**, 4060.
89. L. D. Zhao, S. H. Lo, Y. Zhang, H. Sun, G. Tan, C. Uher, C. Wolverton, V. P. Dravid and M. G. Kanatzidis, *Nature*, 2014, **508**, 373.
90. P. D. Antunez, J. J. Buckley and R. L. Bructchey, *Nanoscale*, 2011, **3**, 2399.
91. M. D. Khan, M. Aamir, M. Sohail, M. Sher, N. Baig, J. Akhtar, M. Z. Malik and N. Revaprasadu, *Dalton Trans.*, 2018, **47**, 5465.
92. J. Choi, J. Jin, I. G. Jung, J. M. Kim, H. J. Kim and S. U. Son, *Chem. Commun.*, 2011, **47**, 5241.
93. S. L. Hawken, R. Huang, C. H. DeGroot, A. L. Hecker, M. Jura, W. Levason, G. Reid and G. B. G. Stenning, *Dalton Trans.*, 2019, **48**, 117.
94. Y. Cheng, T. J. Emge and J. G. Brennan, *Inorg. Chem.*, 1996, **35**, 342.
95. G. Kedarnath, L. B. Kumbhare, S. Dey, A. P. Wadawale, V. K. Jain and G. K. Dey, *Polyhedron*, 2009, **28**, 2749.
96. J. R. Thompson, I. Y. Ahmet, A. L. Johnson and G. Kociok-Köhn, *Eur. J. Inorg. Chem.*, 2016, 4711.
97. R. K. Sharma, G. Kedarnath, A. Wadawale, C. A. Betty, B. Vishwanadh and V. K. Jain, *Dalton Trans.*, 2012, **41**, 12129.
98. A. Tyagi, G. Kedarnath, A. Wadawale, V. K. Jain, M. Kumar and B. Vishwanadh, *RSC Adv.*, 2015, **5**, 62882.
99. A. Tyagi, G. Karmakar, A. Wadawale, A. Y. Shah, G. Kedarnath, A. P. Srivastava, V. Singh and V. K. Jain, *J. Organomet. Chem.*, 2018, **873**, 15.

100. A. Tyagi, G. K. Kole, A. Y. Shah, A. Wadawale, A. P. Srivastava, M. Kumar, G. Kedarnath and V. K. Jain, *J. Organomet. Chem.*, 2019, **887**, 24.
101. A. Tyagi, G. Karmaker, B. P. Mandal, D. D. Pathak, A. Wadawale, G. Kedarnath, A. P. Srivastava and V. K. Jain, *Dalton Trans.*, 2021, **50**, 13073.
102. J. R. Sootsman, D. Y. Chung and M. G. Kanatzidis, *Angew. Chem., Int. Ed.*, 2009, **48**, 8616.
103. J. Rodriguez-Castro, P. Dale, M. F. Mahon, K. C. Molloy and L. M. Peter, *Chem. Mater.*, 2007, **19**, 3219.
104. B. Boudjouk, M. P. Remignton, D. G. Grier, B. J. Jarabek and G. J. McCarthy, *Inorg. Chem.*, 1998, **37**, 3538.
105. M. Bochmann, X. Song, M. B. Hursthouse and A. Karaulov, *J. Chem. Soc., Dalton Trans.*, 1995, 1649.
106. R. K. Sharma, G. Kedarnath, V. K. Jain, A. Wadawale, M. Nalliath, C. G. S. Pillai and B. Vishwanadh, *Dalton Trans.*, 2010, **39**, 8779.
107. R. K. Sharma, G. Kedarnath, B. Vishwanadh and V. K. Jain, *Int. J. Chem*, 2013, **2**, 423.
108. S. L. Benjamin, C. H. DeGroot, A. L. Hector, R. Huang, E. Houkharenko, W. Levason and G. Reid, *J. Mater. Chem. C*, 2015, **3**, 423.
109. R. Huang, S. L. Benjamin, C. Gurnani, Y. Wang, A. L. Hector, W. Levason, G. Reid and C. H. DeGroot, *Sci. Rep.*, 2016, **6**, 27593.
110. G. Gupta and J. Kim, *Dalton Trans.*, 2013, **42**, 8209.
111. G. Bendt, S. Gassa, F. Rieger, C. Jooss and S. Schulz, *J. Cryst. Growth*, 2018, **490**, 77.
112. K. S. Novoselov, A. K. Gein, S. V. Morozov, D. Jiang, Y. Zhang, S. V. Dubonos, I. V. Grigorieva and A. A Firsov, *Science*, 2004, **306**, 666.
113. M. Chhowalla, Z. Liu and H. Zhang, *Chem. Soc. Rev.*, 2015, **44**, 2584.
114. S. Dey and V. K. Jain, *Platinum Met. Rev.*, 2004, **48**, 16.
115. M. Bochmann, I. Hawkins and L. M. Wilson, *J. Chem. Soc., Chem. Commun.*, 1988, 344.
116. (a) J. Cheon, J. E Gozum and G. S. Girolami, *Chem. Mater.*, 1997, **9**, 1847; (b) C. J. Carmalt, C. W. Dinnage, I. P. Parkin, A. J. P. White and D. J. Williams, *Dalton Trans.*, 2001, 2554.
117. C. J. Carmalt, S. A. O'Neill, I. P Parkin and E. S. Peters, *J. Mater. Chem.*, 2004, **14**, 830.
118. F. Senocq, H. Viguier and A. Gleizes, *Eur. J. Solid State Inorg. Chem.*, 1996, 1185–1197.
119. A. L. Hector, W. Levason, G. Reid, S. D. Reid and M. Webster, *Chem. Mater.*, 2008, **20**, 5100.
120. A. L. Hector, M. Jura, W. Levason, S. D. Reid and G. Reid, *New J. Chem.*, 2009, **33**, 641.
121. C. J. Carmalt, C. W. Dinnage, I. P. Parkin, A. P. J. White and D. J Williams, *Inorg. Chem.*, 2002, **41**, 3668.
122. E. S. Peters, C. J. Carmalt, I. P. Parkin and D. A. Tocher, *Eur. J. Inorg. Chem.*, 2005, 4197.
123. M. Chhowalla, H. S. Shin, G. Eda, L. J. Li, K. P. Loh and H. Zhang, *Nat. Chem.*, 2013, **5**, 263.

124. S. L. Chen, K. H. Low, G. K. M. So, S. S. Y. Chui and C. M. Che, *Chem. Commun.*, 2011, **47**, 8808.
125. A. Bera and B. L. V. Prasad, *Bull. Mater. Sci.*, 2018, **41**, 125.
126. N. Al-Dulaimi, D. L. Lewis, X. L. Zhang, M. A. Malik and P. O'Brien, *J. Mater. Chem. C*, 2016, **4**, 2312.
127. A. Eichhöfer, G. Buth, F. Dolci, K. Fink, R. A. Mole and P. T. Wood, *Dalton Trans.*, 2011, **40**, 7022.
128. Z. Yang, A. B. Smetana, C. M. Sorensen and K. J. Klabunde, *Inorg. Chem.*, 2007, **46**, 2427.
129. S. Dey, V. K. Jain, J. Singh, V. Trehan, K. K. Bhasin and B. Verghese, *Eur. J. Inorg. Chem.*, 2003, 744.
130. N. Ghavale, S. Dey, A. Wadawale and V. K. Jain, *Organometallics*, 2008, **27**, 3297.
131. L. B. Kumbhare, A. Wadawale, S. S. Zade and V. K. Jain, *Dalton Trans.*, 2011, **40**, 7957.
132. K. R. Chaudhary, D. K. Paluru, A. P. Wadawale and S. Dey, *Inorg. Chim. Acta*, 2017, **467**, 221.
133. M. E. Peach, *J. Inorg. Nucl. Chem.*, 1973, **35**, 1046.
134. X. S. Du, M. Mo, R. Zheng, S. H. Lim, Y. Meng and Y. W. Mai, *Cryst. Growth Des.*, 2008, **8**, 2032.
135. Y. B. Chen, L. Chen and L. M. Wu, *Inorg. Chem.*, 2005, **44**, 9817.
136. W. Bryks, M. Wette, N. Velez, S. W. Hsu and A. R. Tao, *J. Am. Chem. Soc.*, 2014, **136**, 6175.
137. K. H. Low, C. H. Li, A. V. Roy, S. S. Y. Chui, S. F. L. Chan and C. M. Che, *Chem. Sci.*, 2010, **1**, 515.
138. K. Holligan, P. Rogler, D. Rehe, M. Pamula, A. Y. Korienko, T. J. Emge, K. K. Jespersen and J. G. Brennan, *Inorg. Chem.*, 2015, **54**, 8896.
139. R. K. Sharma, G. Kedarnath, V. K. Jain, A. Wadawale, C. S. S. Pillai, M. Naliath and B. Vishwanadh, *Dalton Trans.*, 2011, **40**, 9194.
140. R. K. Sharma, A. Wadawale, G. Kedarnath, D. Manna, T. K. Ghanty, B. Vishwanadh and V. K. Jain, *Dalton Trans.*, 2014, **43**, 6525.
141. G. K. Kole, K. V. Vivekananda, M. Kumar, R. Ganguly, S. Dey and V. K. Jain, *CrystEngComm*, 2015, **17**, 4367.
142. M. L. Steigerwald and C. R. Sprinkle, *J. Am. Chem. Soc.*, 1987, **109**, 7200.
143. M. Bochmann, G. C. Bwembya, A. K. Powell and X. Song, *Polyhedron*, 1995, **14**, 3495.
144. Y. Jun, J. E. Koo and J. Cheon, *Chem. Commun.*, 2000, 1243.
145. Y. Jun, C. S. Choi and J. Cheon, *Chem. Commun.*, 2001, 101.
146. M. W. DeGroot, C. Khadka, H. Rösner and J. F. Corrigan, *J. Cluster Sci.*, 2006, **17**, 97.
147. D. Zeng, M. J Hampden-Smith and E. N. Duesler, *Inorg. Chem.*, 1994, **33**, 5376.
148. S. Dey, V. K. Jain, S. Chaudhary, A. Knoedler and W. Kaim, *Polyhedron*, 2003, **22**, 489.
149. G. Kedarnath, S. Dey, V. K. Jain, G. K. Dey and B. Varghese, *Polyhedron*, 2006, **25**, 2383.

150. G. Kedarnath, S. Dey, V. K. Jain, G. K. Dey and R. M. Kadam, *J. Nanosci. Nanotechnol.*, 2008, **8**, 4500.
151. G. Kedarnath, S. Dey, V. K. Jain and G. K. Dey, *J. Nanosci. Nanotechnol.*, 2006, **6**, 1031.
152. Y. Cheng, T. J. Emge and J. G. Brennan, *Inorg. Chem.*, 1994, **33**, 3711.
153. G. Kedarnath, V. K. Jain, A. Wadawale and G. K. Dey, *Dalton Trans.*, 2009, 8378.
154. R. K. Sharma, G. Kedarnath, A. Wadawale, V. K. Jain and B. Vishwanadh, *Inorg. Chim. Acta*, 2011, **365**, 333.
155. R. K. Sharma, A. Wadawale, G. Kedarnath, B. Vishwanadh and V. K. Jain, *Inorg. Chim. Acta*, 2014, **411**, 90.
156. J. Lee, D. Freedman, J. H. Melman, M. Brewer, L. Sun, T. J. Emge, F. H. Long and J. G. Brennan, *Inorg. Chem.*, 1998, **37**, 2512.
157. J. Lee, M. Brewer, M. Berardini and J. G. Brennan, *Inorg. Chem.*, 1995, **34**, 3215.
158. A. R. Strzelecki, P. A. Timinski, B. A. Helsel and P. A. Bianconi, *J. Am. Chem. Soc.*, 1992, **114**, 3159.
159. (a) F. J. Fan, L. Wu and S. H. Yu, *Energy Environ. Sci.*, 2014, **7**, 190; (b) C. Coughlan, M. Ibanez, O. Dobrozhan, A. Singh, A. Cobat and K. Ryan, *Chem. Rev.*, 2017, **17**, 5865.
160. B. J. Stanbery, *Crit. Rev. Solid State Mater. Sci.*, 2002, **27**, 73.
161. W. Hirpo, S. Dhingra, A. C. Sutorik and M. G. Kanatzidis, *J. Am. Chem. Soc.*, 1993, **115**, 1597.
162. K. K. Banger, M. H. C. Jin, J. D. Harris, P. E. Fanwick and A. F. Hepp, *Inorg. Chem.*, 2003, **42**, 7713.
163. J. S. Gardner, E. Shrudha, C. Wang, L. D. Lau, R. G. Rodriguez and J. J. Pak, *J. Nanopart. Res.*, 2008, **10**, 633.
164. K. R. Margulieux, C. Sun, L. N. Zakharov, A. W. Holland and J. J. Pak, *Inorg. Chem.*, 2010, **49**, 3959.
165. K. R. Margulieux, C. Sun, M. T. Kihara, A. C. Colson, L. N. Zakharov, K. H. Whitmire, A. W. Holland and J. J. Pak, *Eur. J. Inorg. Chem.*, 2017, 2068.
166. C. Sun, R. D. Westover, K. R. Margulieux, L. N. Zakharov, A. W. Holland and J. J. Pak, *Inorg. Chem.*, 2010, **49**, 4756.
167. K. K. Banger, J. Cowen and A. F. Hepp, *Chem. Mater.*, 2001, **13**, 3827.
168. M. K. Pal, S. Dey, S. Neogy and M. Kumar, *RSC Adv.*, 2019, **9**, 18302.
169. A. Tyagi, A. Y. Shah, G. Kedarnath, A. Wadawale, V. Singh, D. Tyagi, C. A. Betty, C. Lal and V. K. Jain, *J. Mater. Sci.: Mater. Electron.*, 2018, **29**, 8937.

CHAPTER 4

Chalcogenones and Pincer Ligands: Coordination Chemistry and Interaction with Transition and Heavy Main Group Elements

ALEXANDRA POP, ANCA SILVESTRU AND CRISTIAN SILVESTRU*

Department of Chemistry, Supramolecular Organic and Organometallic Chemistry Centre, Faculty of Chemistry and Chemical Engineering, Babeş-Bolyai University, 400028 Cluj-Napoca, Romania
*Email: cristian.silvestru@ubbcluj.ro

4.1 Introduction

The chemistry of chalcogenones – compounds containing a heavier chalcogen–carbon double bond, C=E (E = S, Se or Te), *i.e.* a thione, selone, or tellone functional group – has been developed extensively over a number of years, *e.g.*, chalcogenoketones,[1] chalcogenoamides,[2a,b] chalcogenoureas, and related *N*-substituted derivatives,[2a,b,3] as well as a number of other related classes of compounds,[4] including the so-called "*heavy ketones*" – species with a double bond between tetrels (Si, Ge, or Sn) and chalcogens (S, Se, and Te).[5] The chemistry of compounds with a heavier chalcogen–phosphorus double bond, P=E (E = S, Se, or Te) was also developed

Chalcogen Chemistry: Fundamentals and Applications
Edited by Vito Lippolis, Claudio Santi, Eder J. Lenardão and Antonio L. Braga

Published by the Royal Society of Chemistry, www.rsc.org

considerably during recent years as revealed by several book chapters dedicated to this topic.[6] Moreover, for most of these classes of compounds the coordination chemistry toward transition or main group metals was quite thoroughly investigated. In many cases excellent reviews are available, *e.g.*, p- and d-block metal complexes with thio-, seleno-, or telluroureas and related heavy chalcogen–carbon multiple-bond compounds,[3b,c,7] metal complexes with species containing P=E (E = S, Se, or Te) bonds,[6a,d] also emphasizing their potential as catalysts, precursors in materials chemistry, or compounds with biological activity.

Starting with the pioneering works of Shaw[8a] and van Koten and Noltes[8b] on monoanionic tridentate ligands of the types shown in Scheme 4.1, the chemistry of these so-called aryl "pincer" ligands has been developed continuously and gained more and more impact, especially in catalysis, as revealed by several books and reviews.[9] From the initial anionic 2,6-substituted phenyl ligands, abbreviated as PCP- and NCN-pincer ligands based on the atoms coordinated to a metal center, several other classes of pincer ligands were created through changes such as those summarized in Scheme 4.1,[10] affecting either the central core containing the X atom [including development of non-symmetric and non-cyclic ligands, species with C(sp^3) instead of C(sp^2) central atom, or neutral pincer ligands], or the side arms containing the donor E atoms and their Z spacer (including chiral ligands).

This chapter is dedicated to organochalcogen(II) compounds which can behave as so-called "pincer" ligands, being connected through three atoms to a metal center, in particular to symmetric species containing either C=E

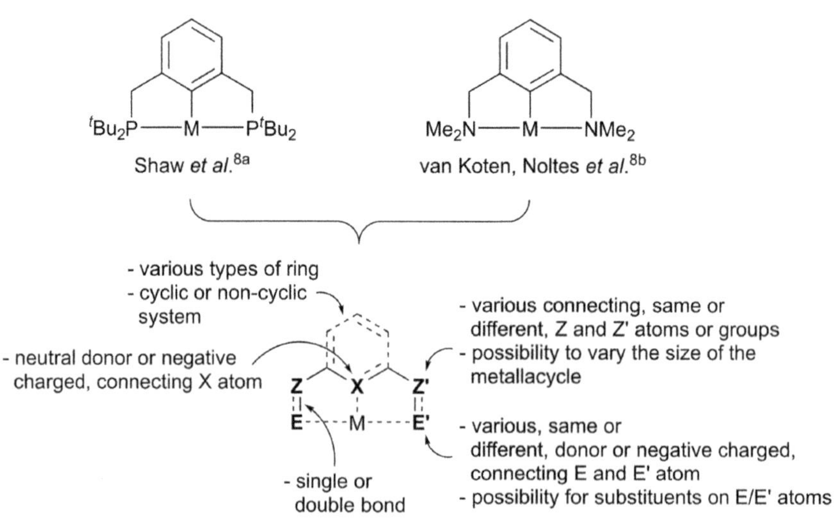

Scheme 4.1 Diversity of changes on original pincer ligands to design new potentially six electron donor EXE′ species.

or P=E (E = S, Se, Te) double bonds. A search of the original literature on such compounds was performed using the SciFinder database, while the Cambridge Structure Database was used to identify pincer–metal complexes for which structure determinations by single-crystal X-ray diffractometry are available. Due to the fast development achieved in recent years, a large diversity of ligands which fit in the above definition are now used in pincer–metal chemistry and a comprehensive treatment of this topic is not the aim of this contribution. Only the most recent, significant illustrative examples will be discussed with respect to the synthetic routes available both for the proligands and the pincer–metal complexes, their structure, stability, and reactivity as well as some relevant applications. Literature older than 2010 will be cited mainly to emphasize some important particular compounds. However, for readers interested in this topic, relatively recent, more comprehensive, reviews and book chapters[6a–c,11] are available on the synthesis and properties of proligands and pincer–metal complexes containing at least one Z=E (Z = C, P; E = S, Se) double bond, also including their applications in catalysis and other areas.

To distinguish easily between the different types of compounds which fit in this template we have introduced a new convention to designate a particular class of pincer ligands, thus indicating not just only the atoms of the ligand moiety bound to the metal center, but also the atoms to which the donors are attached and the size of the fused metallacycles, *e.g.*, instead of just SCS-pincer for the six electron donor ligands of both metal complex fragments (**I**) and (**IV**) in Figure 4.1, we have used [(S=C)/C/(C=S)]-5,5 and [(S=P)/C/(P=S)]-5,5, respectively. The advantage of this designation applies not only when discussing subclasses of pincer ligands with the same donor atoms in the side arms, but also for non-symmetrical ligands with different sizes of the fused metallacycles.

4.2 Symmetric Pincer Ligands of Type [(E=Z)/X/(Z=E)]-5,5

The central moiety of the first representatives of pincer ligands[8] was a 2,6-substituted phenyl ring, so that a large majority of later developed species maintained this structural rigid feature and the formally monoanionic pincer ligands suffered changes just on the side arms, *e.g.*, symmetric pincer ligands as 2,6-bis(thioamide)phenyl [types (**I**) and (**II**)] and 2,6-bis(thiophosphinyl)phenyl [type (**IV**)]. Later changes were applied to this cyclic central fragment by additional phenyl ring substitution, by changing this aromatic ring with azulene [type (**III**)], indene [types (**V**) (although dissymmetrical in nature, this structural type is mentioned here for description reasons), and (**VI**)] or pyrrole rings [types (**VII**)–(**IX**)], or by replacement of the C(sp^2)–metal bond with a P(v)–metal bond in species of type (**X**) (Figure 4.1). On the other hand, neutral pincer proligands with a central pyridine core, including the unique selenium derivative, were also

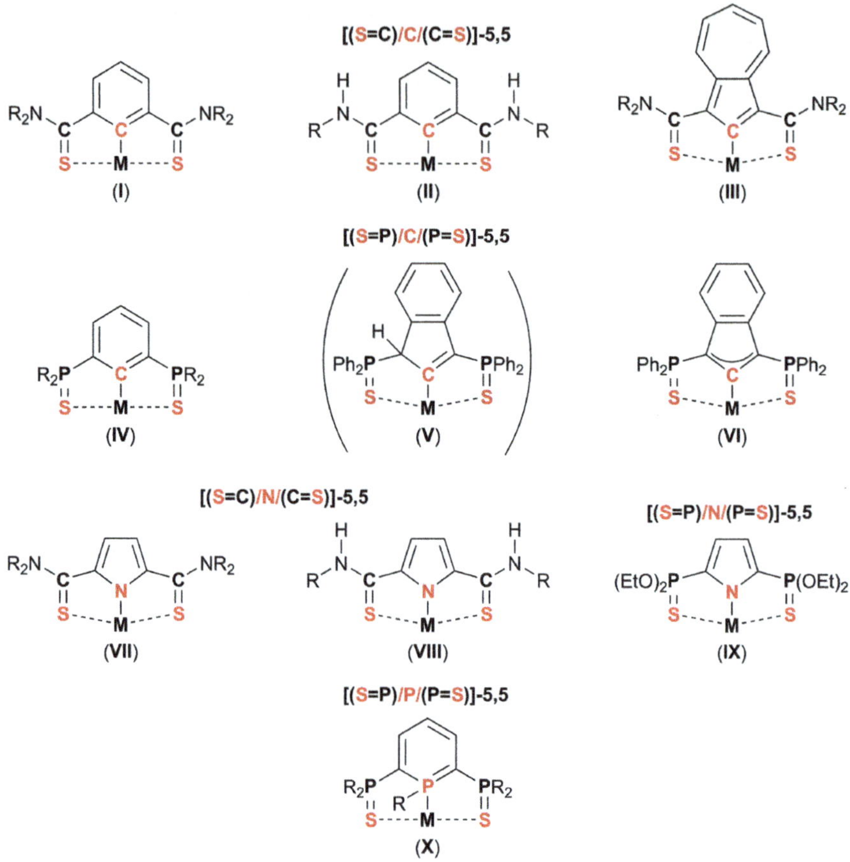

Figure 4.1 General representation of symmetric pincer complexes with two fused five-membered metallacycles, based on anionic SC(sp^2)S, SNS and SPS tridentate ligands with a central cyclic core.

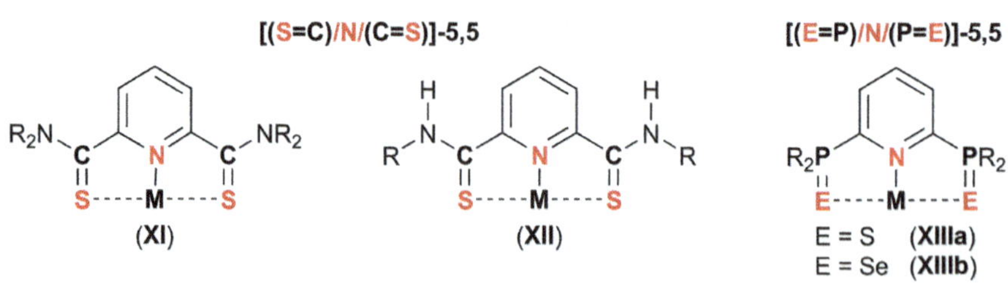

Figure 4.2 General representation of symmetric pincer complexes with two fused five-membered metallacycles based on neutral ENE (E = S, Se) tridentate ligands with a central cyclic core.

designed and metal complexes with fragments of types (**XI**)–(**XIII**) were reported (Figure 4.2). Excellent reviews on these topics summarize earlier results reported before 2016.[11]

4.2.1 Anionic Ligands

The synthesis of symmetrical proligands and their metal complexes containing monoanionic ligands of type {[(S=Z)/X/(Z=S)]-5,5}$^-$ can be easily achieved and, based on this reason, they were extensively used in catalysis.

The main strategies to prepare the 1,3-bis(thioamide) proligands with tertiary (**HL^{I}**) and secondary (**HL^{II}**) thioamide units are based either on (i) thionation of a 1,3-bis(carboxamide) precursor with the Lawesson's reagent,[12] or (ii) a modified Willgerodt–Kindler reaction of isophthalaldehyde with a secondary amine (including heterocyclic species) or a primary amine, in the presence of elemental sulfur and acetic acid (iii)[12a,13] (Scheme 4.2). The thionation with the Lawesson's reagent was also successful in preparing bis(thioamide) pincer proligands functionalized on the central aromatic ring[12d] (*e.g.*, **L^1**Cl, in Scheme 4.3) or polythioamide macrocycles[12b] (*e.g.*, H$_2$**L^2**, in Scheme 4.3). A particular method was used for the preparation of proligand **HL^3** (Scheme 4.3), based on the reaction of Mg with 3,5-dibromotoluene, followed by treatment with phenylisothiocyanate.[14]

Pincer proligands containing two thiophosphinyl side-arms (**HL^4**, **HL^5**, Me**L^6**) were obtained in two steps: (i) Pd-catalyzed phosphination of an aromatic dihalide reagent, then (ii) sulfurization with elemental sulfur (Scheme 4.4).[15] Step-by-step treatment of indene with ^{n}BuLi and quenching with R$_2$PCl (R = Ph, ^{i}Pr), followed by sulfurization with elemental sulfur gave the related proligands 1,3-bis(diorganothiophosphinyl)indene (H$_2$**L$^{7\text{-Ph}}$** and H$_2$**L$^{7\text{-iPr}}$**).[16]

Pincer proligands based on a central pyrrole fragment were designed with both thioamide and thiophosphoryl side-arms. Thus, 2,5-bis(piperidinothiocarbonyl)pyrrole (**HL^8**)[17a] and 2,5-bis(benzylthioamide)pyrrole (**HL^9**)[17b] were prepared by reacting pyrrole-2,5-dicarbaldehyde with piperidine or benzylamine and elemental sulfur, while other pyrrole-based pincer ligands were obtained by treating the corresponding amides with Lawesson's reagent.[17c] Functionalization of pyrrole with two *O*,*O*-diethylthiophosphoryl groups was achieved by two sequential [1,2] base-induced

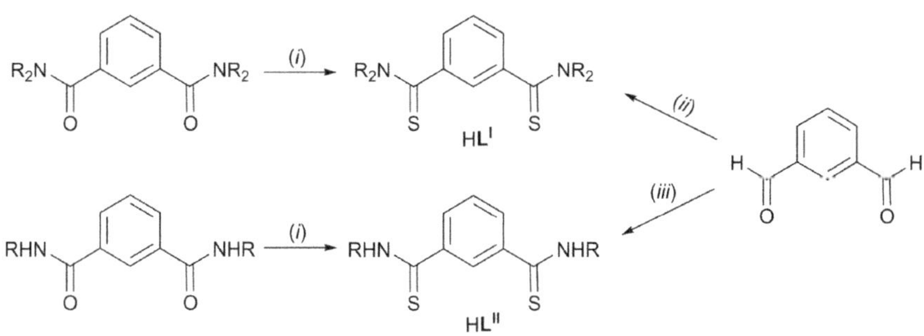

Scheme 4.2 Synthesis of symmetric 1,3-bis(thioamide) pincer proligands of type HL^{I} and HL^{II}: (i) Lawesson's reagent, [*p*-MeOC$_6$H$_4$P(S)S]$_2$; (ii) S$_8$, Me$_2$NH·HCl, NaOAc, or S$_8$, R$_2$NH, AcOH [R = Me, Et, Ph, CH$_2$Ph; R$_2$N = (CH$_2$CH$_2$)$_2$N, CH$_2$(CH$_2$CH$_2$)$_2$N, O(CH$_2$CH$_2$)$_2$N]; (iii) S$_8$, RNH$_2$, AcOH (R = Ph, CH$_2$Ph).

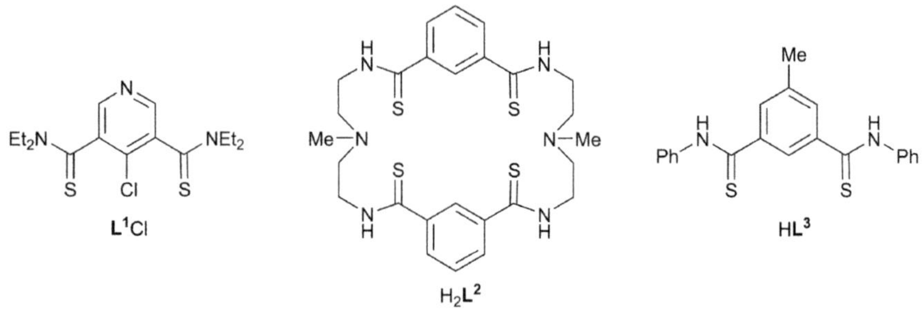

Scheme 4.3 Unusual symmetric bis(thioamide) pincer proligands $\mathbf{L}^1$Cl, $H_2\mathbf{L}^2$ and $H\mathbf{L}^3$.

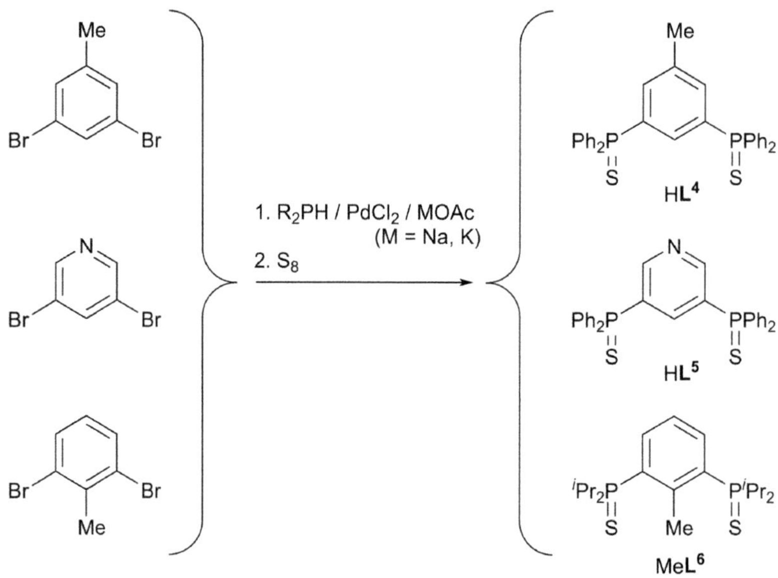

Scheme 4.4 Synthesis of bis(thiophosphinyl) pincer proligands $H\mathbf{L}^4$, $H\mathbf{L}^5$ and $Me\mathbf{L}^6$.

rearrangement with a shift of the thiophosphoryl group from nitrogen to the adjacent carbon atoms, thus leading to 2,5-bis(*O*,*O*-diethylthiophosphoryl)pyrrole ($H\mathbf{L}^{10}$) (Scheme 4.5).[18]

Of particular interest as a unique pincer proligand with a phosphinine central unit is 2,6-bis(diphenylthiophosphinyl)-3,5-diphenylphosphinine ($\mathbf{L}^{11}$), obtained in good yields by treatment of 2,6-bis(diphenylphosphino)-3,5-diphenylphosphinine with elemental sulfur, in toluene.[19] The selective sulfurization of the phosphorus atoms from the side arms over the phosphorus atom of the central heterocycle is the result of a lower basicity of the sp^2 phosphorus atom compared to the sp^3 ones. The phosphinine $\mathbf{L}^{11}$ is a valuable precursor for the synthesis of anionic pincer ligands when reacted with RLi (R = Me, ^{n}Bu, ^{t}Bu, CH_2Py-2), MCl_2(cod) (M = Pd, Pt; cod = 1,5-cyclooctadiene) or $NiBr_2$(DME)

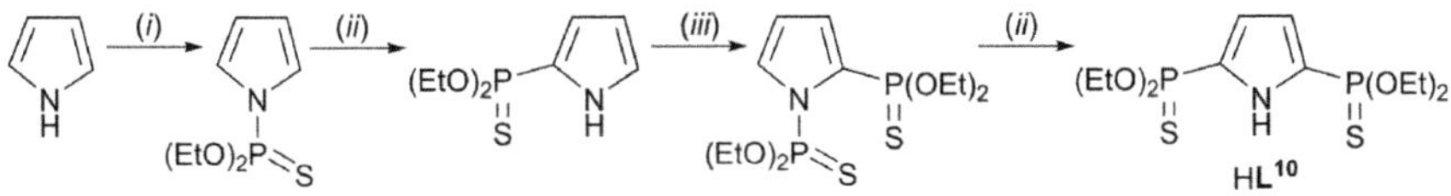

Scheme 4.5 Synthesis of bis(thiophosphoryl)pyrrole pincer proligand HL^{10}: (i) ^{n}BuLi, then $(EtO)_2P(S)Cl$; (ii) ^{n}BuLi, then hydrolysis; (iii) NaH, then $(EtO)_2P(S)Cl$.

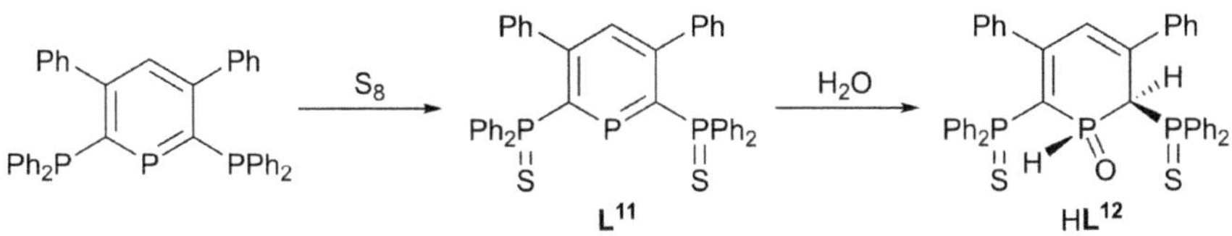

Scheme 4.6 Synthesis of 2,6-bis(diphenylthiophosphinyl)-3,5-diphenylphosphinine ($\mathbf{L^{11}}$) precursor and its oxidation product HL^{12}.

(DME = dimethoxyethane).[19,20] The compound $\mathbf{L^{11}}$ is easily oxidized with water in THF to the 1,2-dihydrophosphinine oxide derivative ($\mathbf{HL^{12}}$)[20] (Scheme 4.6), which can also be used to afford a monoanionic ligand of type $\{[(S{=}P/P/(P{=}S)]\text{-}5,5\}^-$.

The results reported until 2016 regarding the synthesis and applications of pincer metal complexes based on symmetric ligands of type [(S=Z)/X/(Z=S)]-5,5 were quite comprehensively reviewed,[11b] and here we will emphasize some more recent examples. One of the simplest and most efficient methods to prepare pincer metal complexes containing a C(sp^2)–M bond in the structural motif (**I**)–(**VI**), as shown in Figure 4.1, is the C–H activation/cyclometallation of the corresponding proligand using different metal sources. This method is dependent on the electronic character and bulkiness of the side arm ligands. Thus, for example, a ligand of type $\mathbf{HL^{I}}$ (R_2N = 1-pyrrolidinyl, Scheme 4.2) reacted with anhydrous NiI_2, in refluxing toluene for three days, gave the pincer nickel complex **1** (Scheme 4.7).[13d] Proligands $\mathbf{HL^4}$ and $\mathbf{HL^5}$, containing two thiophosphinyl side-arms, were also metalated by treatment with K_2MCl_4 (M = Pd, Pt) and $PdCl_2(PhCN)_2$, respectively, to afford isolation of the corresponding pincer metal complexes ($\mathbf{L^4}$)MCl [M = Pd (**2**), Pt (**3**)][15a] or ($\mathbf{L^5}$)PdCl (**4**).[15b] Based on the complexation ability of the pyridine nitrogen atom, heterodinuclear complexes **5** and **6** were prepared (Scheme 4.8).[15b]

Using a similar C–H activation/cyclometallation procedure, pincer proligands with secondary thioamide groups of type $\mathbf{HL^{II}}$ (Scheme 4.2) were metalated with different transition metals as Ru(II), Pd(II), Pt(II), or Ni(II).[11b] For example, the reaction of the proligand $\mathbf{HL^3}$ with K_2PtCl_4 afforded the Pt(II) complex **7** (Scheme 4.9).[14] Its crystallization from a mixture of DMSO and EtOH resulted in HCl elimination and formation of an organometallic complex with the remaining iminothiolate moiety. In the solid state trimetallic species **8** was formed through Pt–S–Pt bridges (Scheme 4.9).[21]

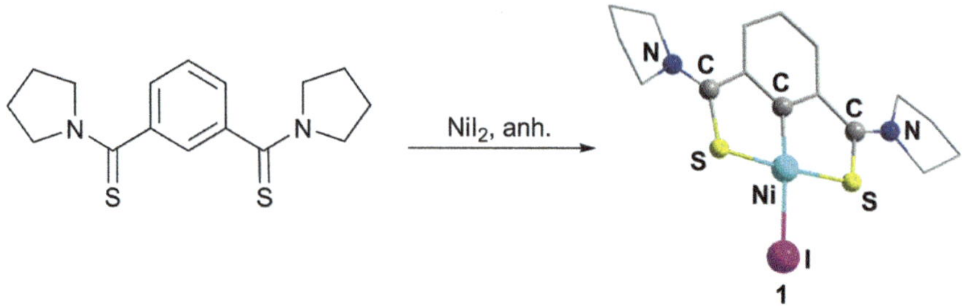

Scheme 4.7 Synthesis of organonickel pincer complex **1** with a monoanionic ligand of type $\{[(S{=}C)/C(sp^2)/(C{=}S)]\text{-}5,5\}^-$.

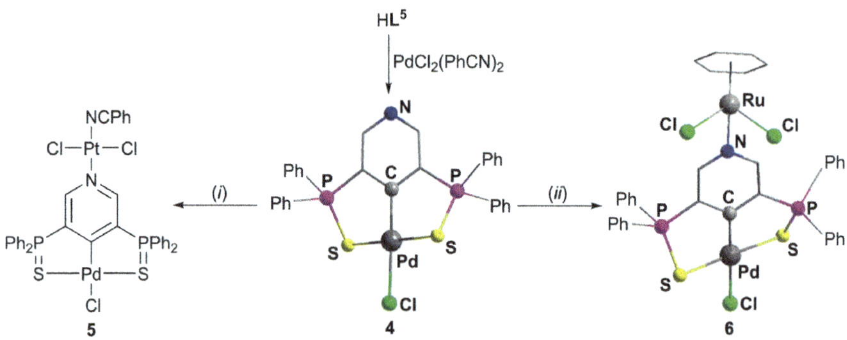

Scheme 4.8 Synthesis of organopalladium pincer complex **4** and hetero-binuclear complexes **5** [(i) $PdCl_2(PhCN)_2$] and **6** [(ii) $[Ru(\eta^6\text{-}C_6H_6)(\mu\text{-}Cl)Cl]_2$].

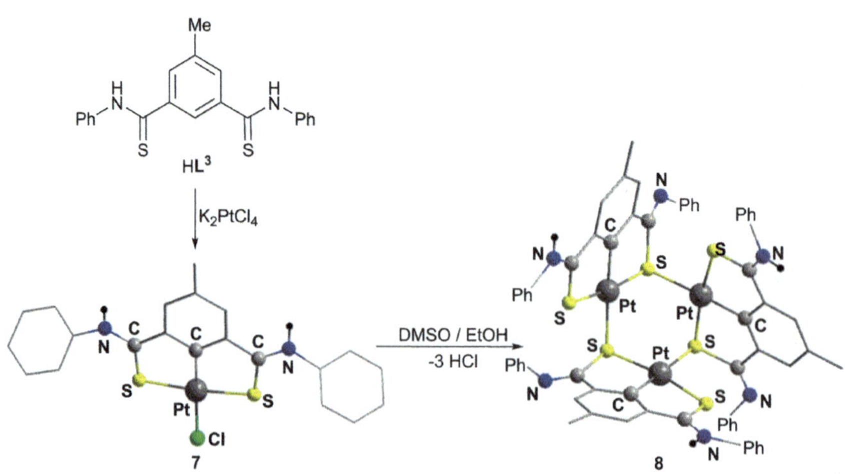

Scheme 4.9 Synthesis of organoplatinum pincer complex **7** and of the trinuclear species **8**.

A two-step, reversible deprotonation–protonation process of the one or both secondary thioamide units in pincer Pt(II) and Ru(II) complexes containing 2,6-bis(benzylaminothiocarbonyl)phenyl was achieved under treatment with different bases (DBU = 1,8-diazabicyclo[4.4.0]undec-7-ene, DBN = 1,5-diazabicyclo[4.3.0]non-5-ene, *etc.*) or methanesulfonic acid, respectively.[22]

Reaction of the polythioamide macrocycle $H_2\mathbf{L}^2$ with 2 equiv. of K_2PdCl_4, in DMSO, resulted in double cyclopalladation and isolation of a ditopic double pincer palladacycle, $(\mathbf{L}^2)(PdCl)_2$ (**9**).[23]

Treatment of a proligand with an azulene central core bearing two thioamide moieties, $H\mathbf{L}^{13}$ (NR_2 = 1-piperidinyl), obtained from the corresponding amide precursor and the Lawesson's reagent ($R^{Lawesson}$), with $PdCl_2$/LiCl in methanol or $PtCl_2(PhCN)_2$ in acetonitrile afforded the isolation of the square planar complexes of type **III**, $(\mathbf{L}^{13})$MCl [M = Pd (**10**), Pt (**11**)] (Scheme 4.10).[17a]

A considerable interest raised the use of 1,3-bis(diorganothiophosphinyl)indene ($H_2\mathbf{L}^{7\text{-}Ph}$ and $H_2\mathbf{L}^{7\text{-}iPr}$) as pincer proligands. The presence of these two $R_2P(S)$ side arms with coordination potential facilitates the rare in-plane σ coordination to various transition metals (Zr, Pd, Pt). Metalation of proligand $H_2\mathbf{L}^{7\text{-}Ph}$ at the carbon atom in position 2 using either $PdCl_2$(cod) or $Zr(NMe_2)_4$ was first reported by Bourissou *et al.* to afford the organometallic pincer complexes **12**, with a non-symmetric monoanionic SCS-tricoordinate 1,3-bis(thiophosphinyl)-2-indenyl ligand (Scheme 4.11), and **13**, featuring a dianionic SCS-tricoordinate 1,3-bis(thiophosphinyl)-2-indenylidene moiety (Scheme 4.12).[24] Analysis by DFT calculations is consistent with a strong σ-bonding and a negligible (if any) π interaction between C2 and the palladium atom. Treatment of complex **12** with amines as $^iPr_2EtN$ or with polystyrene-supported diisopropylethylamine (PS-DIEA) in the presence of tetrabutylammonium chloride resulted in ionic species as **14**, while treatment with 2 equiv. of Cy_2NH (dicyclohexylamine) or 5 equiv. of PS-DIEA afforded isolation of neutral mononuclear **15** and trinuclear **16** indenediide-based species (Scheme 4.11).[24,25] The chloropalladate **14** was used as a ligand for various metal fragments, resulting in homo- and heteropolymetallic species.[25a] Its reactivity towards organic electrophiles was also investigated and its non-innocent behavior was emphasized.[26] Similar

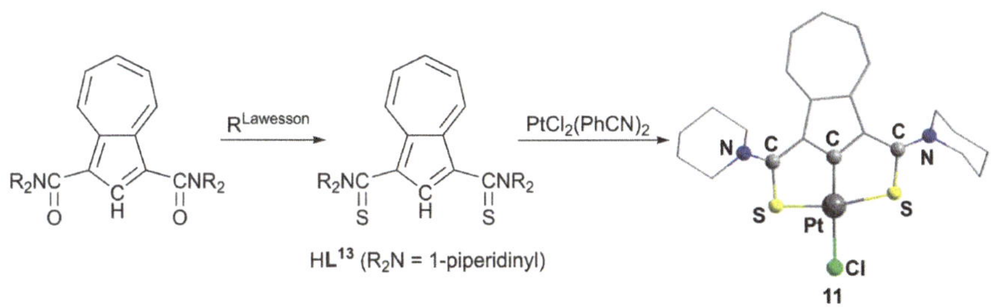

Scheme 4.10 Synthesis of organoplatinum pincer complex **11** with the azulene central unit.

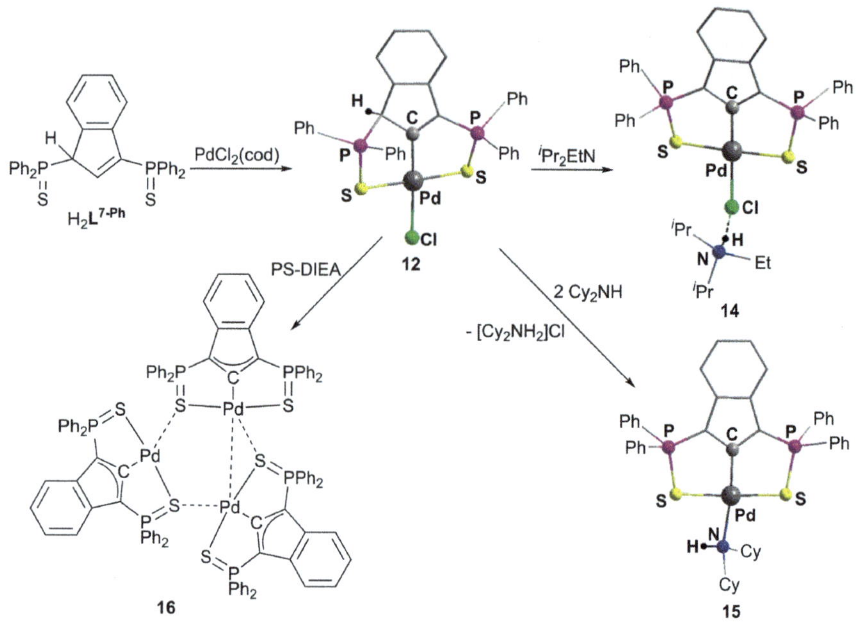

Scheme 4.11 Synthesis of organopalladium pincer complexes with 2-indenyl, 2-indenylidene, and indenediide central units.

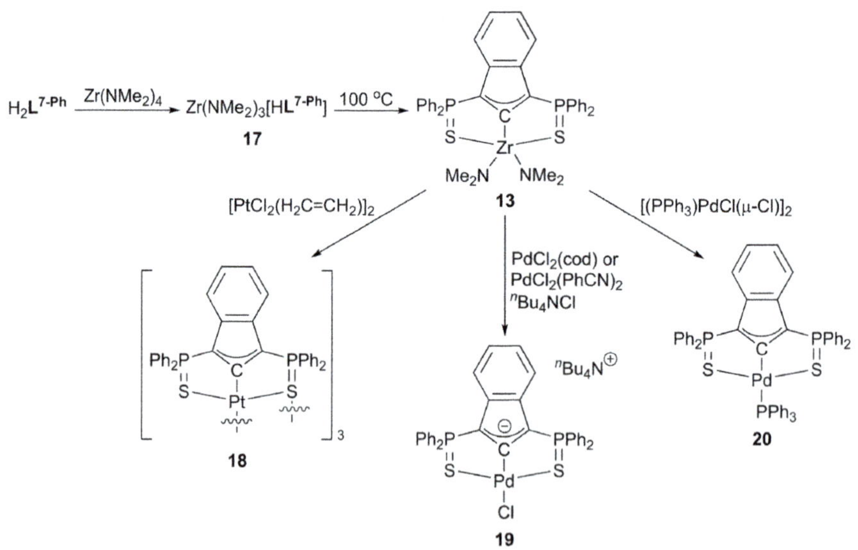

Scheme 4.12 Synthesis of organozirconium pincer complex **13** and transmetallation reactions.

organopalladium pincer complexes were prepared with the related $H_2L^{7\text{-iPr}}$ and their reactivity was investigated.[16b] In this case an indenediide-based palladium(II) compound was crystallized as a dimer species. In the search for increased efficiency in the synthesis of unsaturated δ- and ε-lactones and lactams by catalytic cycloisomerization several organoplatinum(II) pincer complexes were prepared using the indene-based proligands $H_2\mathbf{L}^{7\text{-Ph}}$ and $H_2\mathbf{L}^{7\text{-iPr}}$.[27] The organozirconium pincer complex **13**, generated by thermolysis from the more stable intermediate $Zr(NMe_2)_3[C_9H_5(Ph_2P{=}S)_2]$ (**17**), was successfully used as a precursor in transmetallation reactions for the preparation of SCS-indenediide pincer complexes of Pd(II) and Pt(II) (Scheme 4.12).[28]

The simplest and efficient method to prepare pincer metal compounds containing a C(sp^2)–M bond would be the oxidative addition of M(0) complexes to the carbon–halogen bond of the pincer proligand. However, this method was sparingly used to obtain metal complexes with the fragments (**I**)–(**VI**) exhibited in Figure 4.1. An example is shown in Scheme 4.13: treatment of the chloride $\mathbf{L}^1$Cl with $Ni(cod)_2$ resulted in isolation of the pyridine-based nickel pincer complex **21** as a dark-red solid (*ca.* 70% yield). While direct methylation of pyridine nitrogen of **21** failed, the pyridinium-based nickel pincer complex **22** was obtained following a similar oxidative addition reaction between the preformed $\mathbf{L}^{1\text{-Me}}$Cl proligand and $Ni(cod)_2$ as a black solid (*ca.* 90% yield).[12d]

An uncommon oxidative addition of a C(sp^2)–C(sp^3) was exclusively achieved when the Me$\mathbf{L}^6$ proligand was treated with the Rh(I) precursor $[Rh(coe)_2(acetone)_2]BF_4$ (coe = cyclooctene), at room-temperature, in acetone, with the formation of complex **23**. No cleavage of a more accessible

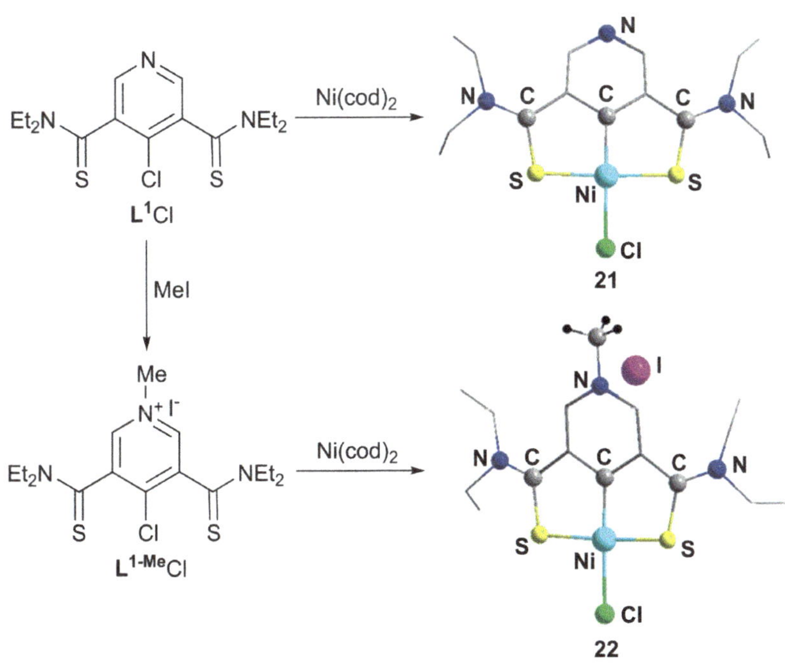

Scheme 4.13 Synthesis of pyridine-based nickel pincer complexes **21** and **22**.

C(sp^3)–H bond was observed (Scheme 4.14).[15c] The reactivity of compound **23** toward carbon monoxide was investigated. Reaction in solution (solvent X = acetone, methanol, or acetonitrile) with 1 equiv. of CO afforded the insertion of CO with the formation of the acetyl derivative **24** which, under subsequent treatment with an excess of CO, gave the carbonyl complex **25**. Compound **24** could be easily recovered by purging a solution of **25** with argon or by treating **25** with an excess of acetonitrile (Scheme 4.14).[29]

Metal complexes containing pincer ligands based on a central 1*H*-pyrrole core 2,5-substituted by tertiary or secondary thioamide units or by thiophosphoryl side-arms [types (**VII**)–(**IX**) in Figure 4.1] are quite rare and can be easily obtained by direct deprotonation of the pyrrole nitrogen under treatment with metal halides, usually in the absence of an additional base. Thus, only one proligand with tertiary thioamide side arms on the 5-membered heterocycle, *i.e.* 2,5-bis(piperidinothiocarbonyl)pyrrole (**HL^8^**), was reacted

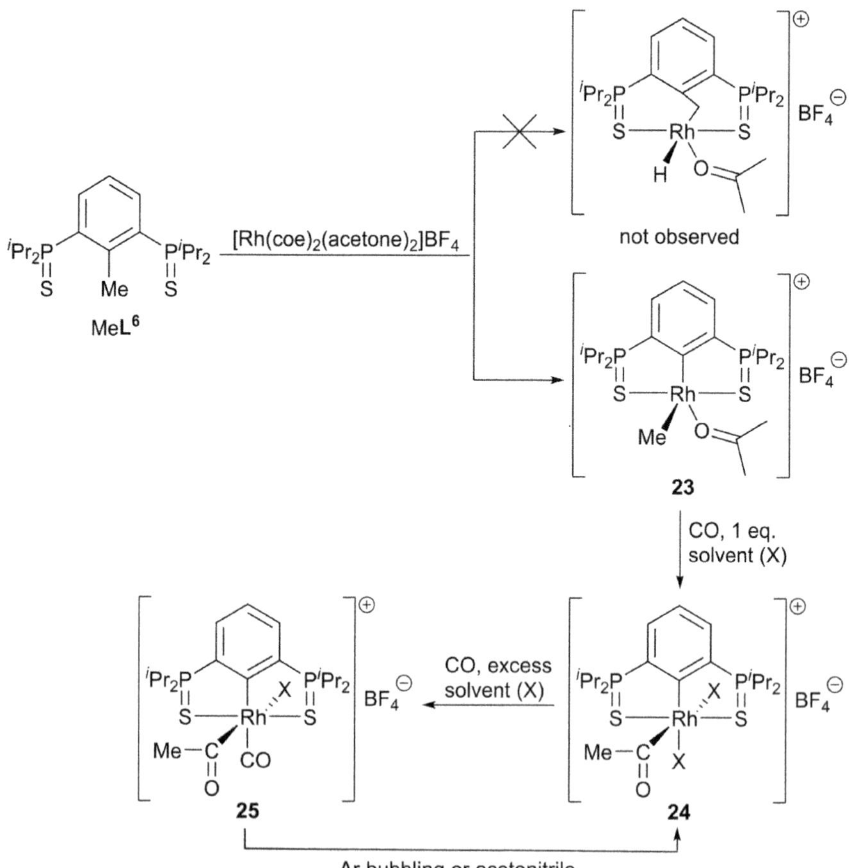

Scheme 4.14 Synthesis of an organorhodium(III) pincer complex *via* C(sp^2)–C(sp^3) bond activation.

with a 1:2 molar mixture of $PdCl_2$ and LiCl, in MeOH, to afford a square planar complex of type (**VII**), (**L**^{8})PdCl (**26**).[17a]

Pincer–metal complexes were reported for only three 1*H*-pyrrole-based proligands with secondary thioamide units as side arms, *i.e.* 2,5-bis(benzylthioamide)pyrrole (**HL**^{9}),[17b,30] 3,4-diphenyl-2,5-bis(butylcarbothioamide)-pyrrole (**HL**$^{14\text{-Bu}}$), and 3,4-diphenyl-2,5-bis(phenylcarbothioamide)pyrrole (**HL**$^{14\text{-Ph}}$).[21,31] Thus, **HL**^{9} was reacted with NiX_2 in ethanol or K_2MCl_4 (M = Pd, Pt) in acetonitrile with the formation of square planar complexes of type (**VIII**), *i.e.* (**L**^{9})NiX [X = Cl (**27**), Br (**28**), I (**29**)],[17b,30] and (**L**^{9})MCl [M = Pd (**30**), Pt (**31**)].[17b] These complexes exhibit an electron-delocalized resonance structure, as shown in Scheme 4.15, and their electrochemical properties were investigated in comparison with those of related complexes with an M–C bond, obtained by using pincer proligands with secondary thioamide groups of type **HL**^{II} (R = CH_2Ph).

While complexes of type (**L**^{9})MX (**27**–**31**) with Ni, Pd, or Pt exhibit a monometallic square planar structure, the Cu(II) analog with the same pincer ligand, (**L**^{9})CuCl (**32**), obtained by cyclometallation of **HL**^{9} with $CuCl_2$, at room temperature, was found to form a dimer association through intermolecular Cu–S interactions [2.809(2) Å], leading to a square pyramidal coordination geometry around the copper atom (Scheme 4.16).[21] Similar dimers, with albeit longer intermolecular Cu–S interactions [3.159(1) Å and 3.142(1) Å, respectively], were reported to be formed in the solid state for the related Cu(II) complexes, (**L**$^{14\text{-Bu}}$)CuCl (**33**) and (**L**$^{14\text{-Ph}}$)CuCl (**34**).[31] It was also found that both Cu(I) and Cu(II) cations exhibit a tendency to promote the formation of benzothiazole side arms following a ring closing reaction when proligand **HL**$^{14\text{-Ph}}$ was used. This process was not observed when Cu(II) was substituted with Co(II), under the same conditions.[31]

The 2,5-bis(*O*,*O*-diethylthiophosphoryl)pyrrole (**HL**^{10}) proligand reacts with $PdCl_2(MeCN)_2$ in THF, at room temperature and in the presence of Et_3N, to give the expected pincer metal complex of type (**IX**), (**L**^{10})PdCl (**34**). The reaction between **HL**^{10} and Ag_2O in CH_2Cl_2 also resulted in deprotonation of the proligand; crystallization afforded the isolation of (**L**^{10})Ag (**35**), which proved to exist as a tetramer based on bimetallic triconnective and trimetallic tetraconnective monoanionic ligands with bridging sulfur atoms [Ag–S distance range: 2.465(1)–3.448(1) Å] and argentophilic interaction [Ag···Ag 3.0382(3), 3.0903(4) Å] (Scheme 4.16).[18] Compound **35** is one example of a metal complex in which the ligand does not act as a classic pincer moiety of type {[(S=P/N/(P=S)]-5,5}$^-$ but as an asymmetric polymetallic bridging ligand.

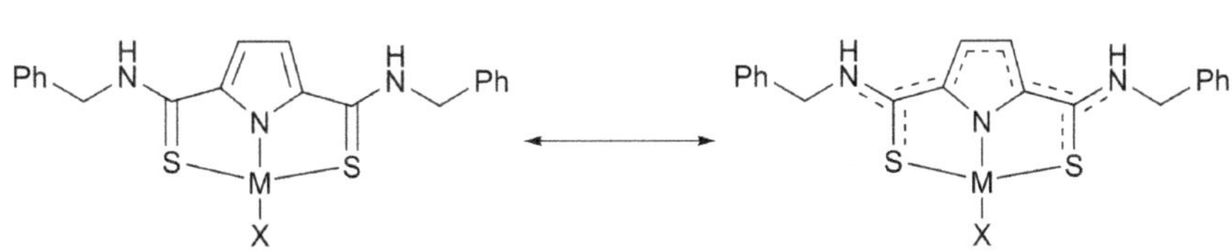

Scheme 4.15 Delocalization of electrons in the (**L**^{9})MX (M = Ni, Pd) complexes.

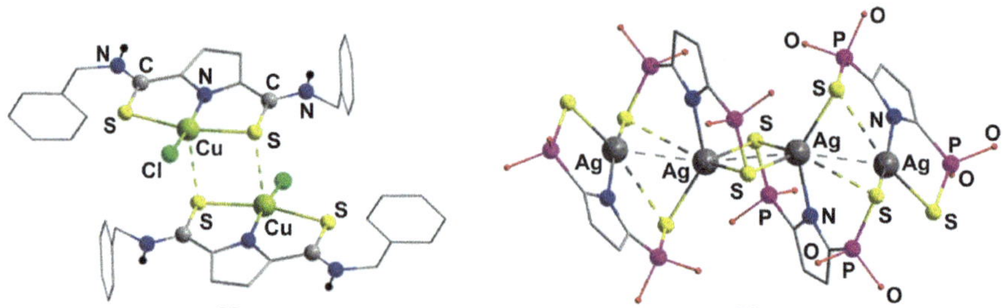

32 35

Scheme 4.16 Polymetallic associations in the crystal of ($\mathbf{L^9}$)CuCl (**32**) and ($\mathbf{L^{10}}$)Ag (**35**).

The anionic pincer ligands of type {[(S=P/P/(P=S)]-5,5}$^-$ [type (**X**) in Figure 4.1] generated by treatment of the unique $\mathbf{L^{11}}$ precursor with organolithium or organomagnesium reagents,[19] were shown to be versatile ligands for several d-transition metals (Mn, Rh, Ir, Ni, Pd, Pt, Cu, Au), as well as for f-elements (Ce, Nd, U). Their rich coordination chemistry was reviewed in excellent book chapters, strongly recommended to those interested in this topic.[11a,32] Some results on Pd(II) and U(IV) complexes reported in publications which are not cited in these reviews are summarized in Scheme 4.17.[33]

4.2.2 Neutral Ligands

Neutral *SNS*-pincer proligands were designed by replacing the central benzene core for a pyridine ring. Such thione proligands such as those used to obtain pincer metal complexes with fragments of types (**XI**) and (**XII**) (Figure 4.2) are obtained by reacting the corresponding 2,6-di(carboxamide)pyridine precursors containing tertiary or secondary amide units with P_2S_5 in benzene,[34] or with Lawesson's reagent in toluene,[12c,35] or following a one-pot synthesis using a mixture of 2,6-di(carboxyaldehyde)pyridine, primary amine, and elemental sulfur in anhydrous dimethoxyethane.[36] Similarly, oxidation of 2,6-bis(diphenylphosphanyl)pyridine by elemental sulfur or selenium, in toluene or methylene dichloride, resulted in the corresponding 2,6-bis(diphenylchalcogenophosphinyl)pyridine derivatives (Scheme 4.18).[37] Note that compound $\mathbf{L^{16}}$ is one of the very few symmetric ligands containing P=Se bonds which was used to obtain metal complexes.

Metal complexes containing a fragment of type (**XI**) were reported for Co(II), Ni(II), and Cu(II).[34] Most of the reported compounds are ionic in nature, with square planar mononuclear [(*SNS*-L)CuX]$^+$ cations, dinuclear [(*SNS*-L)CuCl]$_2^{2+}$ cationic units with either asymmetric Cu–S···Cu or Cu–Cl···Cu double bridges, or polynuclear associations of monocations [(*SNS*-L)CoCl]$_n^{n+}$ with asymmetric Co–S···Co double bridges between staggered cations. Discrete neutral mononuclear (*SNS*-L)CoCl$_2$ (**41**) and dinuclear (*SNS*-L)$_2$Ni$_2$(μ-NCS)$_2$(NCS)$_2$ (**42**) were also described.

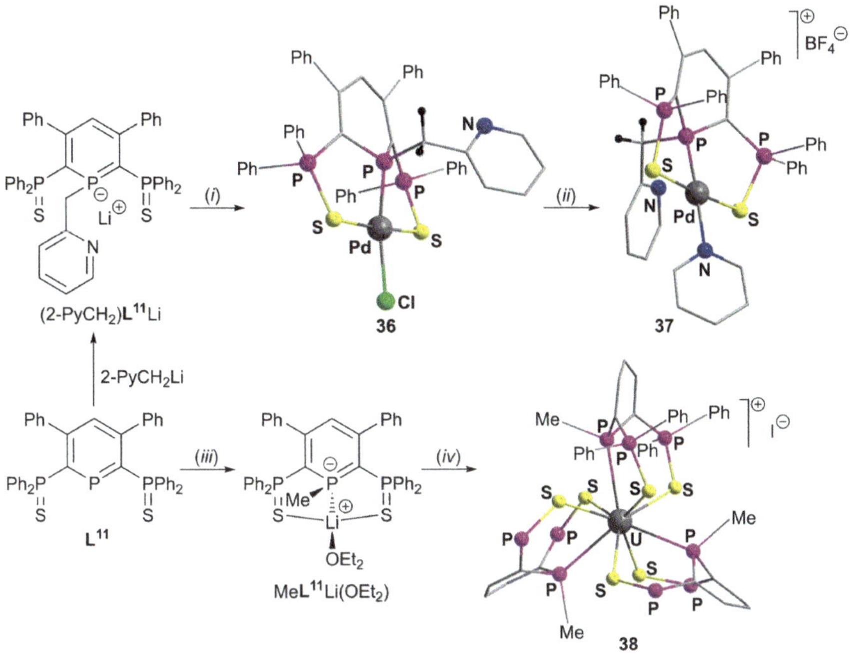

Scheme 4.17 Synthesis of metal pincer complexes based on anionic $[RL^{11}]^-$ ligands: (i) $PdCl_2(cod)$, THF; (ii) 1. $Ag[BF_4]$, CH_2Cl_2; 2. Py; (iii) MeLi, Et_2O; (iv) $UI_3(THF)_4$, THF (for compound **38** the phenyl substituents on phosphorus atoms and on the central phosphinine core are shown only for one of the three ligands).

The acidity of secondary thioamide groups in neutral proligands incorporated in fragments of type (**XII**) (Figure 4.2) might easily result in mono- or di-anionic ligands due to the loss of NH-protons even in the absence of a base. Changes in organic groups attached to nitrogen atoms in a proligand as well as changes of the metal precursor can afford spontaneous deprotonation with isolation of different complexes containing pincer ligands based on a central pyridyl group and either symmetrical side arms (neutral and dianionic ligands) or non-symmetric side arms (thioamido/iminothiolate monoanionic ligands) (see, for example, complexes **43–46** in Scheme 4.19).[12c,35a]

A similar step-by-step deprotonation of the secondary thioamide groups, with successive elimination of two HCl molecules, was reported to take place for the cation of the ruthenium(II) complex **47**, with formation of neutral mononuclear **48** and dinuclear **49** species (Scheme 4.20) and increased electron density at the metal center. A change of the organic group on the nitrogen of the side arms, for example from phenyl to benzyl, was found to influence considerably the deprotonation ability and the oxidation potential of the Ru center.[36]

Using bulkier substituents, such as 2,6-$(Ph_2CH)_2$-4-$^iPrC_6H_2$ or 2,6-$^iPr_2C_6H_3$, on the nitrogen of both secondary thioamide side arms grafted on the

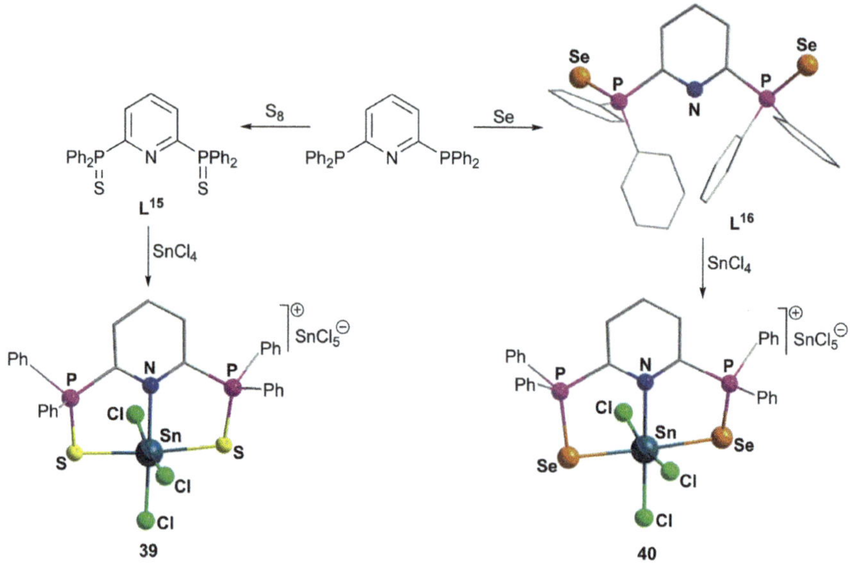

Scheme 4.18 Synthesis of neutral pincer proligands L^{15} and L^{16} and their trichlorotin(IV) pincer complexes **39** and **40**.

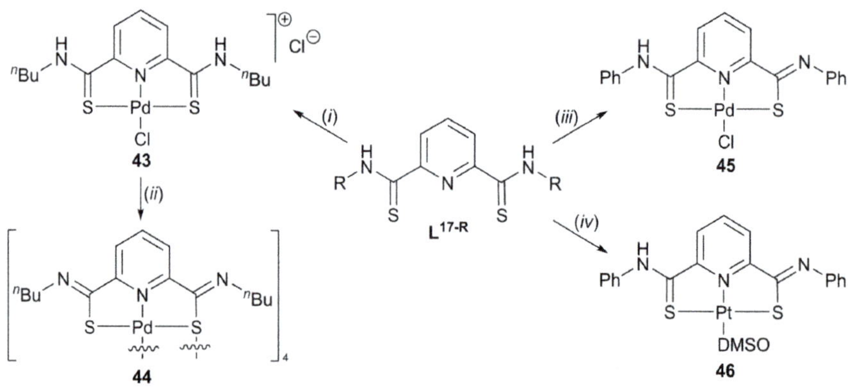

Scheme 4.19 Synthesis of palladium and platinum pincer complexes from neutral proligands with a central pyridyl group and secondary thioamide side arms: (i) $PdCl_2(MeCN)_2$, MeCN; (ii) CyZnCl, THF; (iii) $PdCl_2(PhCN)_2$, DMF/CH_2Cl_2; (iv) $PtCl_2(PhCN)_2$, DMSO/CH_2Cl_2 (R = Ph, ^{n}Bu).

pyridine central core of the pincer proligand, different iron(II) pincer complexes of type (**XII**) were obtained (Scheme 4.21).[35c,d] The reactivity of complexes **50** and **51** related either to deprotonation of the thioamide pincer ligand or to substitutions at the metal center was investigated and several compounds were reported.

Rhenium(I) complexes of type $[Re(CO)_3(\mathbf{L^{15}})]X$ [X = $[Re_2(CO)_6(\mu\text{-}Br)_3]^-$ (**52**), $[SbF_6]^-$ (**53**)] were reported in an earlier work.[37b] The reaction between neutral proligands $\mathbf{L^{15}}$ or $\mathbf{L^{16}}$ and $SnCl_4$, in a 1 : 2 molar ratio, in CH_2Cl_2,

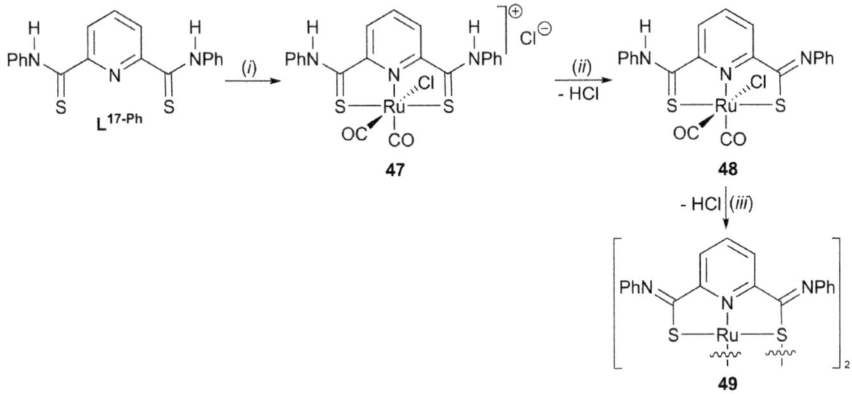

Scheme 4.20 Synthesis of ruthenium pincer complexes based on neutral proligand $L^{17\text{-}Ph}$: (i) $[Ru(CO)_2Cl_2]_n$, anhydrous MeOH; (ii) $CHCl_3$/hexane in air; (iii) Al_2O_3, $CHCl_3$/EtOH.

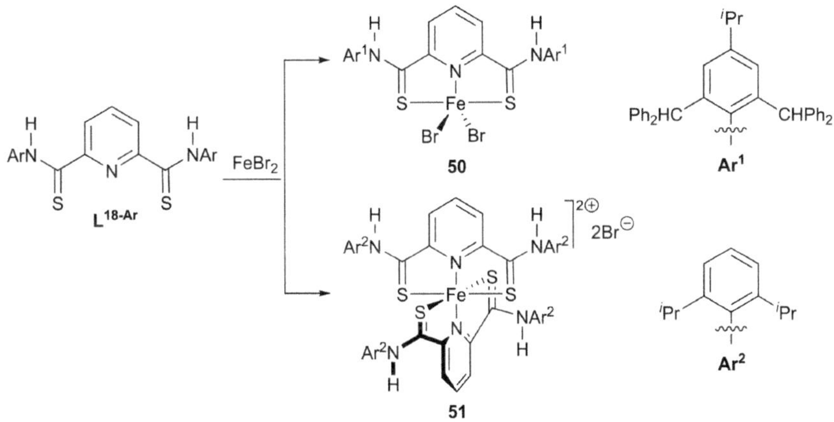

Scheme 4.21 Synthesis of iron pincer complexes based on neutral proligands $L^{18\text{-}Ar}$.

afforded the isolation of two of the very few main group metal complexes with symmetric pincer ligands with a Z=E (Z = C, P; E = S, Se) double bond in the side arms. Both compounds **39** and **40** (Scheme 4.18) contain an octahedral cation with *mer* coordination of the pincer ligand and a trigonal pyramidal $[SnCl_5]^-$ anion.[37c]

4.3 Symmetric Pincer Ligands of Type [(E=Z)/X/(E=S)]-6,6

While a relatively large variety of symmetric pincer proligands of type [(S=Z)/X/(Z=S)]-5,5 was developed in recent decades and used in coordination chemistry of transition metals, related symmetric proligands of type [(E=Z)/X/(E=S)]-6,6 able to act either as monoanionic or neutral pincer ligands and

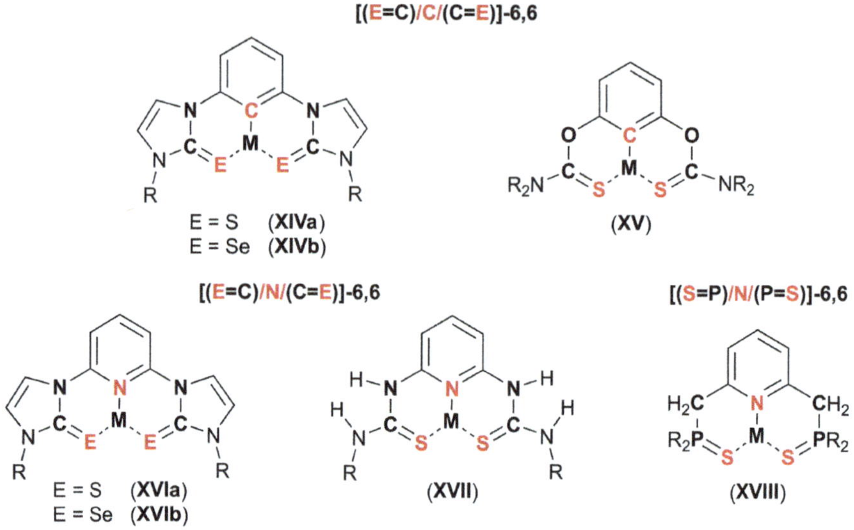

Figure 4.3 General representation of symmetric pincer complexes with two fused six-membered metallacycles based on anionic EC(sp^2)E and neutral ENE (E = S, Se) tridentate ligands with a central cyclic core.

thus to provide fused six-membered metallacycles are very rare. Only proligands based on a 1,3-substituted benzene or a 2,6-substituted pyridine as central cores were used to generate anionic pincer ligands as those in metal complexes of type (**XIV**) and (**XV**) or to act as neutral ligands as those in metal complexes of type (**XVI**), (**XVII**), and (**XVIII**), respectively (Figure 4.3). Note also that the system of fused metallacycles afforded by pincer proligands of type [(E=Z)/X/(E=S)]-6,6 exhibit a twisted conformation, while the system of fused metallacycle afforded by pincer proligands of type [(S=Z)/X/(Z=S)]-5,5 discussed above is planar.

4.3.1 Anionic Ligands

With the aim to provide an accessible pathway to organometallic pincer complexes with fused six- or seven-membered metallacycles, several bis(chalcogenone) proligands were designed. Thus, the synthesis of 1,3-bis(3′-methylimidazolyl-2′-thione)benzene ($\mathbf{HL^{19\text{-}S\text{-}Me}}$) and its seleno analog ($\mathbf{HL^{19\text{-}Se\text{-}Me}}$) was achieved in a one-pot method by reacting phenylene-bridged imidazolium dihalide (X = Br, I) derivatives with sulfur or selenium powder in the presence of K_2CO_3, in hot ethanol (Scheme 4.22).[38] A similar procedure was used to prepare several other related bis(chalcogenone) species, *e.g.* $\mathbf{HL^{19\text{-}S\text{-}R}}$ (R = Et, ^{n}Bu, allyl),[39] or $\mathbf{HL^{20\text{-}S\text{-}R}}$ (R = Me, Et, ^{n}Bu, allyl).[40] Using an alternative method, *i.e.* treatment with *in situ* prepared disodium dichalcogenides and KO^{t}Bu in dry THF, bis(chalcogenone) derivatives $\mathbf{HL^{21\text{-}E\text{-}Bu}}$ (E = Se, Te) and $\mathbf{L^{21\text{-}Se\text{-}R}}$Br (R = ^{i}Pr, ^{n}Bu, ^{t}Bu, pyridyl) (Scheme 4.22) were also obtained with the aim to be used as potential

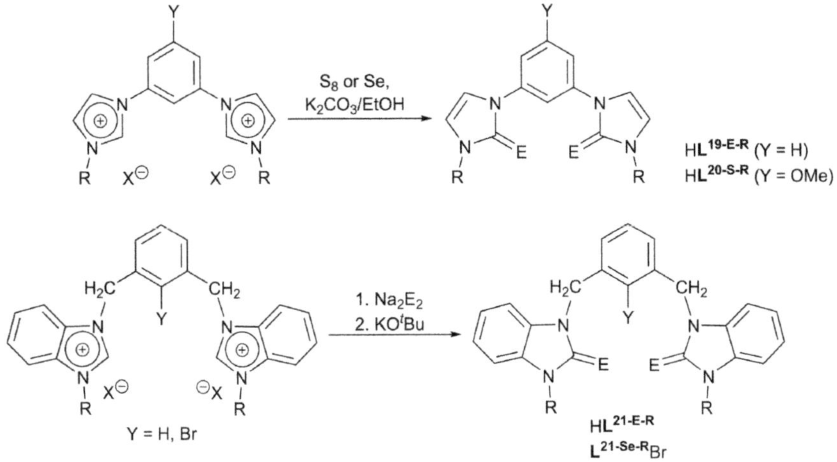

Scheme 4.22 Synthesis of pincer proligands HL^{19-E-R}, HL^{20-E-R}, HL^{21-E-R} and $L^{21-Se-R}Br$.

symmetric proligands which allowed the formation of fused 7-membered metallacycles in pincer complexes.[38b]

Other proligands such as 1,3-$[Me_2NC(S)O]_2$-4-RC_6H_3 [R = H, Me, C_5H_{11}, O(O)CMe] (**HL**$^{22-R}$), which, when acting as pincer moieties, might provide complexes of type (**XV**), were obtained by treating the sodium salts of the resorcinol or its 4-substituted derivatives with 2 equiv. of dimethylthiocarbamoyl chloride in dimethylformamide.[41a]

The C–H activation/cyclometallation of various **HL**$^{19-S-R}$ (R = Me, Et, ^{n}Bu, allyl) proligands was successfully achieved under treatment with different palladium reagents, in a 1 : 1 molar ratio, affording quantitatively first neutral pincer complexes **54** that utilize N-heterocyclic thione donor groups to provide two fused 6-membered metallacycles (Scheme 4.23).[39] Using an appropriate molar ratio of reactants, the complexes **55** (R = Et, ^{n}Bu, allyl) containing two neutral **HL**$^{19-S-R}$ ligands coordinated to the metal center, could be isolated; further reaction with additional Pd reagents afforded conversion of complexes **55** to pincer species **54**.[39] The treatment of compounds **54** with N-heterocyclic carbene precursors, in the presence of Cs_2CO_3 in MeCN, resulted in ionic, mononuclear (*e.g.*, **56**, R = Me, Et), and dinuclear (**57**, R = Me, ^{n}Bu) palladium complexes.[42]

Few organoruthenium(II) and rhodium(III) compounds were also prepared by C–H activation/cyclometallation of proligands with N-heterocyclic thione donor groups. Thus, reaction of **HL**$^{19-S-Me}$ with [Ru(η^6-*p*-cymene)(μ-Cl)Cl]$_2$, followed by treatment with KPF_6, afforded the ionic complex [(**L**$^{19-S-Me}$)Ru(η^6-*p*-cymene)]PF_6 (**58**),[38a] while reaction of **HL**$^{20-S-R}$ (R = Me, Et, ^{n}Bu, allyl) with [Rh(η^5-C_5Me_5)(μ-Cl)Cl]$_2$ gave [(**L**$^{20-S-R}$)Rh(η^5-C_5Me_5)]Cl (**59**).[40]

The proligand **HL**$^{22-H}$ was reported to be efficiently palladated at the benzene ring during liquid–liquid extraction with the formation of a stable compound, [2,6-$[Me_2NC(S)O]_2C_6H_3$]PdCl (**60**), exhibiting a fused

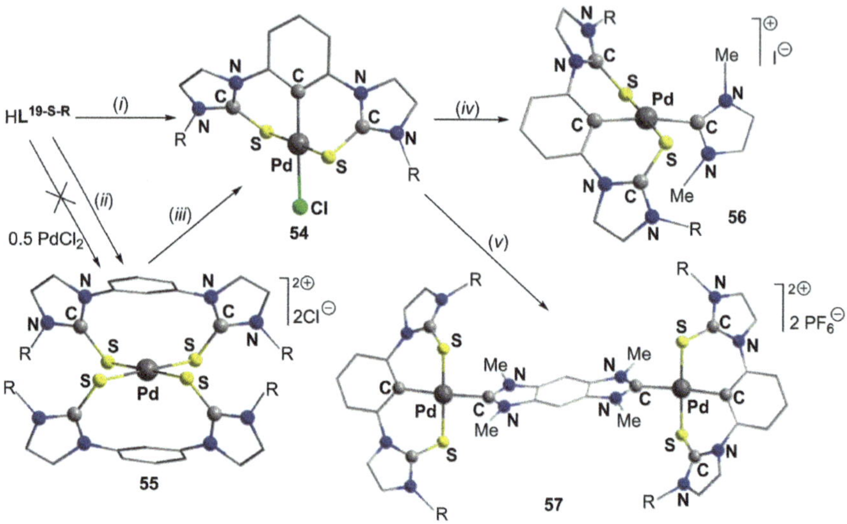

Scheme 4.23 Synthesis of organopalladium pincer complexes based on reaction with $HL^{19\text{-}S\text{-}R}$: (i) $PdCl_2(MeCN)_2$ or $PdCl_2$; (ii) 0.5 $PdCl_2(MeCN)_2$; (iii) 0.5 $PdCl_2(MeCN)_2$; (iv) $[NHC]^+I^-$, and (v) $[NHC]^{2+}2I^-$, then KPF_6.

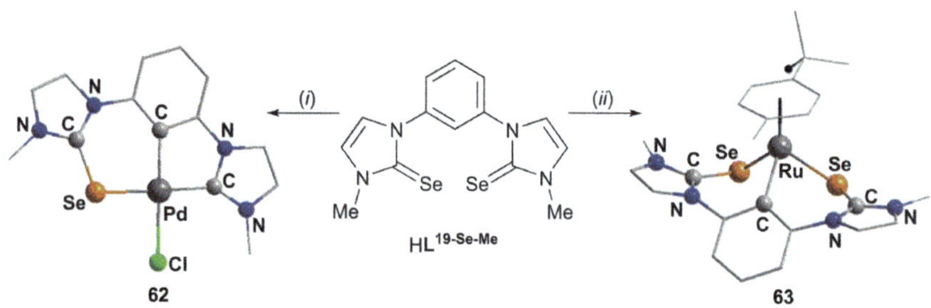

Scheme 4.24 Synthesis of organometallic pincer complexes based on reaction with $HL^{19\text{-}Se\text{-}Me}$: (i) 0.5 $[\{2\text{-}(Me_2NCH_2)C_6H_4\}Pd(\mu\text{-}Cl)]_2$, and (ii) 0.5 $[Ru(\eta^6\text{-}p\text{-}cymene)(\mu\text{-}Cl)Cl]_2$.

six-membered chelate rings system.[41b] The preparation of the same organopalladium pincer complex was also achieved for first time under mechanochemical conditions from $HL^{22\text{-}H}$ and $PdCl_2(PhCN)_2$.[41c]

So far, in contrast to its thio analog, the Se-based pincer proligand $HL^{19\text{-}Se\text{-}Me}$ was seldom used in cyclometallation reactions. Reaction of $HL^{19\text{-}Se\text{-}Me}$ with $PdCl_2(PhCN)_2$ and NH_4PF_6 (2 : 1 : 2 molar ratio) in refluxing acetonitrile gave the complex **61**, a Se-analog of thio-complex **55**. When a reaction mixture of $HL^{19\text{-}Se\text{-}Me}$ and $[\{2\text{-}(Me_2NCH_2)C_6H_4\}Pd(\mu\text{-}Cl)]_2$ (2 : 1 molar ratio) was refluxed in benzene/glacial acetic acid, the C–H activation was achieved with simultaneous elimination of one selenium atom. A non-symmetric, almost planar, organopalladium pincer complex **62** is thus formed, with fused 5- and 6-membered metallacycles as a result of a strong coordinated N-heterocyclic carbene side arm (Scheme 4.24).[38b] On the other hand, the reaction of $HL^{19\text{-}Se\text{-}Me}$ with $[Ru(\eta^6\text{-}p\text{-cymene})(\mu\text{-Cl})Cl]_2$ (2:1 molar ratio), followed by treatment with KPF_6, afforded the expected ionic, organoruthenium pincer complex **63** (Scheme 4.24).[38a] Note the flexibility of this pincer ligand which can coordinate to the metal atom either in *rac*- (as in **54**) or *meso*-conformation (as in **63**).

Attempts to use heavier chalcogene-containing proligands $HL^{21\text{-}E\text{-}Bu}$ (E = Se, Te) and $L^{21\text{-}Se\text{-}R}Br$ (R = ^{i}Pr, ^{n}Bu, ^{t}Bu, pyridyl) in the synthesis of organopalladium pincer complexes with two fused 7-membered metallacycles failed. Reactions with $[\{2\text{-}(Me_2NCH_2)C_6H_4\}Pd(\mu\text{-Cl})]_2$, $Pd(OAc)_2$, or $[Pd(MeCN)_4](BF_4)_2$ to achieve C–H activation/cyclometallation, or $Pd_2(dba)_3$ (dba = dibenzylideneacetone) to promote oxidative addition of the C–Br bond were unsuccessful, leaving generally only complex mixtures of compounds which could not be separated.[38b]

4.3.2 Neutral Ligands

Three types of symmetric proligands have been reported to act as neutral pincer ligands, thus providing fused six-membered metallacycles; all of them being based on a central 2,6-disubstituted pyridine core (Figure 4.3). Various proligands $L^{23\text{-}E\text{-}R}$ with 3′-organoimidazolyl-2′-thione (E = S; R = Me, Et, ^{i}Pr, ^{n}Bu, ^{t}Bu, neopentyl)[43] or -selone (E = Se; R = Me, Et, ^{i}Pr)[44] side arms attached to a pyridine core, 2,6-bis(3′-organoimidazolyl-2′-chalcogenone)-pyridine, analogous to $HL^{19\text{-}E\text{-}R}$ (Scheme 4.22) have been prepared using similar methods, *i.e.* a one-pot procedure *via* reactions of pyridine-bridged imidazolium dibromide with elemental sulfur or selenium, in the presence of a base (K_2CO_3, NaOAc, or KO^{t}Bu, in refluxing methanol or acetonitrile). An adapted method was employed for the synthesis of 1,1′-(pyridine-2,6-diyl)-bis(3-methylthiourea) (L^{24}) when 2,6-diaminopyridine was treated with excess of methyl isothiocyanate.[45] A third type of proligand which can provide fused six-membered metallacycles are those containing diorganothiophosphinomethyl side arms grafted on a pyridine core. Such proligands, $2,6\text{-}[R_2P(S)CH_2]_2C_5H_3N$ ($L^{25\text{-}R}$), were obtained by treatment of 2,6-bis(diphenylphosphinomethyl)pyridine with elemental sulfur (R = ^{i}Pr, Ph)[46] or by adding 2,6-bis(chloromethyl)pyridine to a reaction mixture of RLi/$(EtO)_2P(S)H$ (R = ^{n}Bu, tolyl).[46a]

Treatment of $L^{23\text{-}S\text{-}R}$ with organometallic precursors, *i.e.* $[M(\eta^5\text{-}C_5Me_5)(\mu\text{-Cl})Cl]_2$ (M = Ir, Rh) and $[Ru(\eta^6\text{-}p\text{-cymene})(\mu\text{-Cl})Cl]_2$, afforded the ionic, pincer metal complexes $[M(L^{23\text{-}S\text{-}R})(\eta^5\text{-}C_5Me_5)]Cl_2$ (M = Rh, Ir; R = Me, ^{i}Pr, ^{t}Bu) (**64**, for M = Ir, R = Me)[43a] and $[Ru(L^{23\text{-}S\text{-}Me})(\eta^6\text{-}p\text{-cymene})][PF_6]_2$ (**65**).[38a] For both types of compounds the half-sandwich organometallic cation exhibits a "three-legged piano-stool" structure with two fused six-membered chelate rings (Scheme 4.25). Reaction of the same Ru(II) precursor with proligand L^{24} gave complex $[Ru(L^{24})(\eta^6\text{-}p\text{-cymene})]Cl_2$ (**66**) which could be converted into $[Ru(L^{24})(\eta^6\text{-}p\text{-cymene})][BF_4]_2$ (**67**) after treatment with $AgBF_4$. The structure determination of **67** revealed a similar "piano-stool" cation as found for **65**, due to a *meso*-conformation of the neutral pincer ligand.[45]

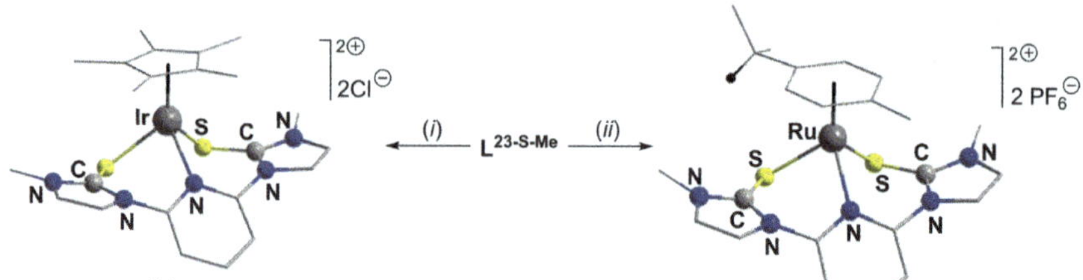

64

65

Scheme 4.25 Synthesis of organometallic pincer complexes based on reaction with $\mathbf{L}^{\mathbf{23\text{-}S\text{-}Me}}$: (i) 0.5 $[Ir(\eta^5\text{-}C_5Me_5)(\mu\text{-}Cl)Cl]_2$, and (ii) 0.5 $[Ru(\eta^6\text{-}p\text{-cymene})\text{-}(\mu\text{-}Cl)Cl]_2$.

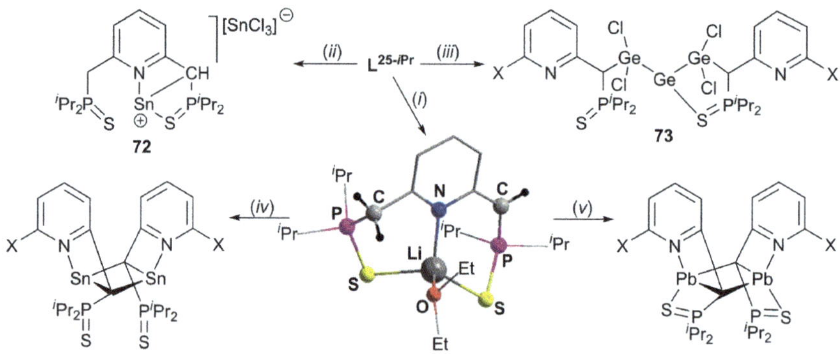

72

73

74

71

75

Scheme 4.26 Synthesis of organometallic complexes based on reaction with $\mathbf{L}^{\mathbf{25\text{-}iPr}}$: (i) ^nBuLi, THF/Et_2O; (ii) excess $SnCl_2$; (iii) 1.5 $GeCl_2$·dioxane; (iv) $Sn[N(SiMe_3)_2]_2$, and (v) $Pb[N(SiMe_3)_2]_2$ ($X = CH_2\text{–}^iPr_2P(S)$).

Pincer, ionic Zn(II) complexes, $[(\mathbf{L}^{\mathbf{23\text{-}S\text{-}R}})ZnCl][Zn(H_2O)Cl_3]$ (**68**),[43b] and neutral Cu(II) complexes, $(\mathbf{L}^{\mathbf{23\text{-}S\text{-}R}})CuCl_2$ (**69**),[48] were obtained by reacting metal dichlorides with the $\mathbf{L}^{\mathbf{23\text{-}S\text{-}R}}$ proligand (R = ^{i}Pr, ^{n}Bu, neopentyl).

A neutral complex containing a (**XVIII**) fragment, *i.e.* $(\mathbf{L}^{\mathbf{25\text{-}Ph}})Ni(NO_3)_2$ (**70**), was obtained by reacting $(2,6\text{-}[Ph_2P(S)CH_2]_2C_5H_3N$ ($\mathbf{L}^{\mathbf{25\text{-}Ph}}$) with $Ni(NO_3)_2{\cdot}6H_2O$. The compound revealed a *mer* coordination of the pincer ligand, an octahedral polyhedron around the metal center being completed by oxygen atoms from bidentate and monodentate nitrate anions, respectively.[46a] The coordination chemistry of the analogous $\mathbf{L}^{\mathbf{25\text{-}iPr}}$ proligand was investigated towards heavier tetrels (Ge, Sn, Pb).[46b] The reaction of $\mathbf{L}^{\mathbf{25\text{-}iPr}}$ with ^{n}BuLi in THF/Et_2O gave the monoanionic lithium complex **71**, with a non-symmetric pincer ligand due to C–H activation at the CH_2 group of one of the side arms of the neutral proligand. Reactions of $\mathbf{L}^{\mathbf{25\text{-}iPr}}$ or treatment of **71** with tetrels(II) precursors, *i.e.* MCl_2 (M = Ge, Sn) or $M[N(SiMe_3)_2]_2$ (M = Sn, Pb), afforded only isolation of ionic and neutral, monometallic, or polymetallic species which exhibit an M–C bond and no pincer structure (Scheme 4.26).

Note here that some of the proligands described in this section were also found to be coordinated only through the chalcogen atoms of both thione or selone side arms, thus acting as monometallic biconnective ligands in discrete species, *e.g.*, $[(S,S\text{-}\mathbf{L}^{\mathbf{23\text{-}S\text{-}R}})AuCl_2]Cl$ (R = Me, Et, ^{i}Pr) (**76**),[43c] $[(Se,Se\text{-}\mathbf{L}^{\mathbf{23\text{-}Se\text{-}R}})AuCl_2][AuCl_2]$ (R = Me, Et, ^{i}Pr) (**77**),[44] $[Pd(dppe)(S,S\text{-}\mathbf{L}^{\mathbf{24}})]Cl[PF_6]$ (**78**) [dppe = 1,2-bis(diphenylphosphino)ethane],[45] or as bimetallic biconnective (bridging) ligands in polynuclear species, *e.g.*, $[\{Cu(S,S\text{-}\mathbf{L}^{\mathbf{23\text{-}E\text{-}R}})\}_8(PF_6)][PF_6]_7$ (E = S, Se; R = ^{i}Pr) (**79**).[47]

4.4 Applications

Metal complexes with pincer ligands found applications as catalysts for various transformations in organic synthesis, luminescent moieties in optoelectronic devices, or as biologically active species. In addition, their complexation ability made them attractive for analytical applications, *i.e.* proligand $\mathbf{HL}^{\mathbf{22\text{-}H}}$ was reported as a performant agent for selective extraction of Pd from a mixture of metal cations in acid solution.[41b]

4.4.1 Catalysis

Various strategies were used to finely tune the properties of pincer complexes in catalytic processes. By far, most of the reported compounds are palladium-containing derivatives. Complex **19** and related species bearing SCS-pincer ligands with a central 2-indenyl, 2-indenylidene, and indenediide unit [pincer complexes of type (**V**) and (**VI**) in Figure 4.1] showed that by only slightly changing the structure of the ligand backbone, a significant variation of the electron density at the metal atom can be achieved, while the structural and steric features remain very similar. The catalytic performance of the Pd complexes in imine allylation reactions showed that the activity increased with increasing electron density at palladium.[25c]

Spectacular progress was achieved in metal–ligand cooperative catalysis using pincer complexes which successfully combine an electrophilic metal center and an electron-rich ligand backbone.[25b] Such ligands, with a non-innocent character, besides the influence on the electronic environment around the metal, participate actively in the catalytic process. Indenediide Pd and Pt pincer complexes **14**,[16b] or **19**,[25b] and **18**,[27,28] respectively, were employed as efficient and selective catalysts in the cycloisomerization of alkynoic acids, giving access to 5–7 membered alkylidene lactones.[25b,27]

Other selective applications involve the use as catalysts of (i) half-sandwich Rh(III) complexes with monoanions from phenylene-bridged bis(thione) ligands $\mathbf{HL}^{\mathbf{20\text{-}S\text{-}R}}$ (R = Me, Et, ^{n}Bu, allyl) [pincer complexes of type (**XIVa**) in Figure 4.3, *e.g.*, **59**] for transfer hydrogenation of ketones to the corresponding alcohols;[40] (ii) half-sandwich Ru(II) complexes with monoanions from phenylene-bridged bis(chalcogenone) ligands $\mathbf{HL}^{\mathbf{19\text{-}E\text{-}Me}}$ (E = S, Se) or neutral pyridine-bridged bis(thione) ligand $\mathbf{L}^{\mathbf{23\text{-}S\text{-}Me}}$ [pincer complexes of type (**XIVa**), (**XIVb**), and (**XVIa**), in Figure 4.3, *e.g.*, **58**, **63**, or **65**] for tandem dehydrogenation of $NH_3 \cdot BH_3$ and hydrogenation of nitro derivatives to the corresponding amines;[38a] (iii) mononuclear and dinuclear Pd(II) complexes **56** (R = Me, Et) and **57** (R = Me, Bu) with N-heterocyclic carbene and monoanion from **HL$^{19\text{-S-Me}}$** [pincer complexes of type (**XIVa**)] for regioselective reduction of quinolines to the corresponding 1,2,3,4-tetrahydroquinolines,[42] or (iv) various Pd(II) complexes of symmetric SCS-, SNS-, or SPS-pincer ligands in C–C bond formation in Heck,[12c,19a,23,38b,39a] Suzuki–Miyaura,[19a,39a] Sonogashira,[39a] or Negishi[35a,49] coupling reactions.

4.4.2 Luminescence

Various Pt(II) and Pd(II) pincer complexes were found to exhibit interesting optical properties.[11b] Although not light-emissive in solution at room temperature, compounds (**L^4**)MCl [M = Pd (**2**), Pt (**3**)] [pincer complexes of type (**IV**) in Figure 4.1] showed photoluminescence in the glassy frozen state (CH_2Cl_2–EtOH–MeOH, at 77 K). The platinum complex **3** was highly light-emissive even in the solid state, at room temperature.[15a] Photochemical properties were also reported for several Pt(II)[13b,14,22a,50a] and Pd(II)[13c,50b] complexes of type (**I**) and (**II**) (see Figure 4.1). Interestingly, the ionic complex [(**L$^{\text{II-Bn}}$**)Pt(PPh_3)]Cl (**80**), based on a monoanion from a 1,3-bis(secondary thioamide) proligand (**HL^{II}**) with benzyl (Bn) substituents on nitrogen atoms, was reported to exhibit aggregation-induced emission (AIE) activity in different solvent mixtures induced by hydrogen bonding and interionic interactions, which differs from the conventional AIE activity occurring by metal–metal interactions.[50a]

Organopalladium(II) pincer complexes bearing secondary thioamide units [pincer complexes of type (**II**) in Figure 4.1], *e.g.*, (**L$^{\text{II-Bn}}$**)PdCl (**81**), were found to exhibit strong phosphorescence at room temperature when incorporated into a poly(vinylpyrrolidone) matrix. By exposing a hybrid film to volatile organic compounds, a reversible switching of luminescence was observed.[50b] Complex **81** showed variation in the intensity of the photoluminescence in a crystalline state, depending on the molecular arrangement and crystal packing. Enhanced intensity was observed for a densely-packed arrangement, which determines AIE, whereas no emission was observed in solution or for non-crystalline powder. The enhanced luminescence at room temperature was explained by a higher density of the crystal due to hydrogen bonding.[50c]

Organopalladium and -platinum compounds (**L^{13}**)MCl [M = Pd (**10**), Pt (**11**)], with an azulene central unit [pincer complexes of type (**III**) in Figure 4.1] were found to feature interesting optical properties. Based on DFT calculations their optical activity was assigned to the electron transition from the metal center to the extended π system of the azulene unit.[17a]

4.4.3 Biological Activity

Nickel(II) and palladium(II) complexes with SCS ligands containing pyrrolidine, piperidine, and cyclohexylamine functionalities [pincer complexes of type (**I**) in Figure 4.1] showed interesting antiproliferative activity against estrogen-responsive human breast cancer cells (MCF7 and MC4L2) and against triple-negative breast cancer (4 T1).[13d,51] The most active Ni(II) compound was **1** (Scheme 4.7); it showed *in vitro* antiproliferative activity against estrogen-responsive human breast cancer cells (MCF7) and significantly inhibited *in vivo* the tumor growth of MC4L2 cancer cells in mice.[13d] Comparative studies indicated that the palladium derivatives were generally much safer anticancer agents than nickel analogous, and they proved to be effective for inhibiting TNBC 4 T1 cell adhesion and migration.[51]

The complex ($\mathbf{L}^1$)NiCl with a pyridinium-based pincer SCS-ligand framework [type (**I**) in Figure 4.1] mediated the dehydrogenation of alcohols with ligand participation in the hydride transfer and it was designed as a model for lactate racemase,[12d] while ionic Zn(II) pincer complexes [($\mathbf{L}^{23\text{-S-R}}$)ZnCl][Zn($H_2O$)$Cl_3$] (**68**) [type (**XVIa**) in Figure 4.3] were investigated as models for the zinc-active site in liver alcohol dehydrogenase (LADH).[43b]

4.5 Concluding Remarks

The chemistry of pincer proligands and their pincer–metal complexes with heavier chalcogen–carbon or –phosphorus double bond, Z=E (Z = C, P; E = S, Se)

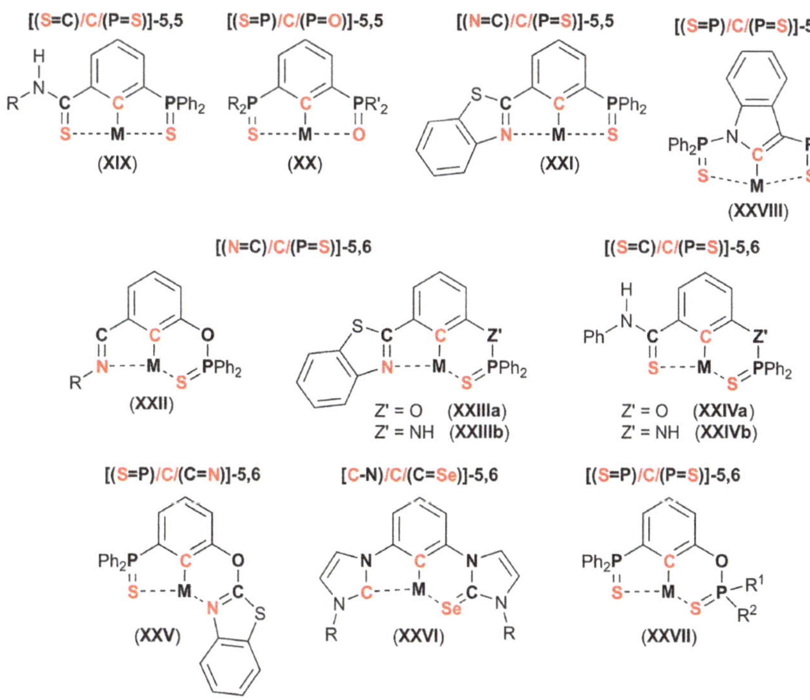

Figure 4.4 General representation of non-symmetric pincer complexes with two fused five-membered or fused mixed five- and six-membered metallacycles based on anionic SC(sp^2)E (E = N, O, S) and CC(sp^2)Se tridentate ligands with a central cyclic core.

in the side arms was developed considerably in the context of the increased interest regarding the use of anionic or neutral pincer ligands in coordination chemistry and the significant potential of the metal complexes for catalysis, new materials, or biological applications. Symmetric species with a central cyclic core and Z=E double bond-containing side arms were discussed within this chapter. Those which are interested in the field should be aware that, in addition, rich literature dedicated to hybrid pincer–metal complexes based on a diversity of non-symmetric proligands with a central cyclic core (Figure 4.4) as well as non-cyclic proligands (*i.e.* without a central cyclic core) (Figure 4.5) (both types containing at least one Z=E double bond involved in the pincer system), and some book chapters including aspects of this chemistry are recommended.[6a,11]

An overview of the scientific literature revealed that there are also topics scarcely developed. Thus, (i) pincer proligands containing at least one element–selenium double bond and their metal complexes are rare; (ii) pincer proligands with a tellone functional group or related element–tellurium

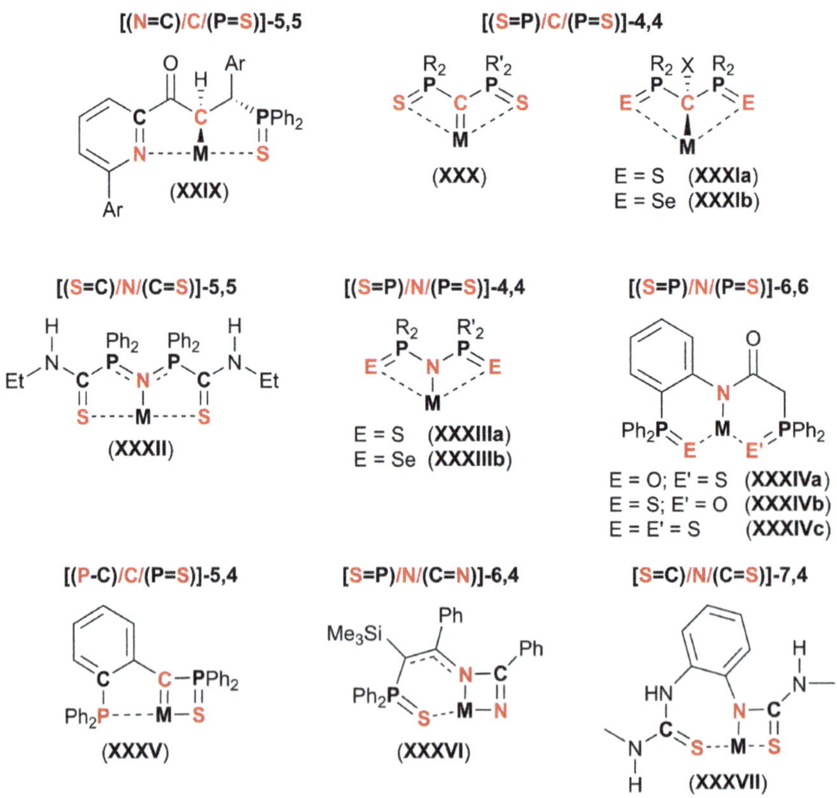

Figure 4.5 General representation of symmetric and non-symmetric, non-cyclic pincer complexes without a central cyclic core, generating fused metallacycles of the same or different size.

double bond, analogous to the largely investigated sulfur species (Figures 4.1–4.5), are not yet available; (iii) the chemistry of main group metal complexes based on pincer ligands with potential Z=E (Z = C, P; E = S, Se, Te) is poorly developed; (iv) even for transition metals most studies are concentrated on the coordination pincer chemistry of palladium, platinum, and a few other metals, leaving a large area to be investigated. All these topics and others which are not mentioned here are expected to receive much more attention in the future.

References

1. (a) N. Kuhn and G. Verani, in *Handbook of Chalcogen Chemistry – New Perspectives in Sulfur, Selenium and Tellurium*, ed. F. A. Devillanova, Royal Society of Chemistry, Cambridge, 2007, ch. 2.3, pp. 107–144; (b) G. Verani and A. Garau, in *Handbook of Chalcogen Chemistry – New Perspectives in Sulfur, Selenium and Tellurium*, ed. F. A. Devillanova and W.-W. du Mont, Royal Society of Chemistry, Cambridge, 2nd edn, 2013, ch. 2.3, pp. 118–146.
2. (a) M. Koketsu and H. Ishihara, in *Handbook of Chalcogen Chemistry – New Perspectives in Sulfur, Selenium and Tellurium*, ed. F. A. Devillanova, Royal Society of Chemistry, Cambridge, 2007, ch. 2.4, pp. 145–194; (b) M. Koketsu, in *Handbook of Chalcogen Chemistry - New Perspectives in Sulfur, Selenium and Tellurium*, ed. F. A. Devillanova and W.-W. du Mont, Royal Society of Chemistry, Cambridge, 2nd edn, 2013, ch. 2.2, pp. 94–117.
3. (a) V. Lippolis, in *Organoselenium Compounds in Biology and Medicine – Synthesis, Biological and Therapeutic Treatments*, ed. V. K. Jain and K. I. Priyadarsini, Royal Society of Chemistry, London, 2018, ch. 4, pp. 122–149; (b) J. S. Ritch, in *Selenium and Tellurium Reagents in Chemistry and Materials Science*, ed. R. S. Laitinen and R. Oilunkaniemi, Walter de Gruyter GmbH, Berlin/Boston, 2019, ch. 6, pp. 207–241; (c) B. S. Stadelman and J. L. Brumaghim, in *Biochalcogen chemistry – The biological chemistry of sulfur, selenium, and tellurium*, ed. C. A. Bayse and J. L. Brumaghim, American Chemical Society, Washington, DC, 2013, ch. 3, pp. 33–70.
4. (a) T. Murai and S. Kato, *Topics in Current Chemistry*, in *Organoselenium Chemistry – Modern Developments in Organic Synthesis*, ed. T. Wirth, Springer, Berlin Heidelberg, 2000, vol. 208, pp. 177–199; (b) T. Murai, in *Organoselenium Chemistry – Synthesis and Reactions*, ed. T. Wirth, Wiley-VCH, Weinheim, 2012, ch. 6, pp. 257–286; (c) *Patai's Chemistry of Functional Groups – The Chemistry of Organic Selenium and Tellurium Compounds*, ed. S. Patai and Z. Rappoport, John Wiley & Sons, Inc., New York, NY, 1986, vol. 1; (d) *Patai's Chemistry of Functional Groups – The Chemistry of Organic Selenium and Tellurium Compounds*, ed. S. Patai, John Wiley & Sons, Inc., New York, NY, 1987, vol. 2; (e) *Patai's Chemistry of Functional Groups – The Chemistry of Organic Selenium and Tellurium Compounds*, ed. Z. Rappoport, J. F. Liebman and I. Marek, John Wiley &

Sons, Ltd., Chichester, 2012, vol. 3; (f) *Patai's Chemistry of Functional Groups – The Chemistry of Organic Selenium and Tellurium Compounds*, ed. Z. Rappoport, J. F. Liebman and I. Marek, John Wiley & Sons, Ltd., Chichester, 2014, vol. 4.

5. (a) N. Takeda, N. Tokitoh and R. Okazaki, in *Handbook of Chalcogen Chemistry – New Perspectives in Sulfur, Selenium and Tellurium*, ed. F. A. Devillanova, Royal Society of Chemistry, Cambridge, 2007, ch. 3, pp. 195–222; (b) N. Takeda and N. Tokitoh, in *Handbook of Chalcogen Chemistry – New Perspectives in Sulfur, Selenium and Tellurium*, ed. F. A. Devillanova and W.-W. du Mont, Royal Society of Chemistry, Cambridge, 2nd edn, 2013, ch. 3, pp. 160–190.
6. (a) R. Davies, in *Handbook of Chalcogen Chemistry – New Perspectives in Sulfur, Selenium and Tellurium*, ed. F. A. Devillanova, Royal Society of Chemistry, Cambridge, 2007, ch. 5, pp. 286–343; (b) R. Davies and L. Patel, in *Handbook of Chalcogen Chemistry – New Perspectives in Sulfur, Selenium and Tellurium*, ed. F. A. Devillanova and W.-W. du Mont, Royal Society of Chemistry, Cambridge, 2nd edn, 2013, ch. 5, pp. 238–306; (c) G. Hua and J. D. Woollins, in *Selenium and Tellurium Reagents in Chemistry and Materials Science*, ed. R. S. Laitinen and R. Oilunkaniemi, Walter de Gruyter GmbH, Berlin/Boston, 2019, ch. 5, pp. 151–206; (d) A. Nordheider, J. D. Woollins and T. Chivers, *Chem. Rev.*, 2015, **115**, 10378; (e) C. M. Timperley and N. Cooper, in *Best Synthetic Methods: Organophosphorus(V) Chemistry*, ed. C. M. Timperley, Elsevier Ltd., Oxford, 2015, ch. 5, pp. 563–632; (f) G. Hua and J. D. Woollins, in *Best Synthetic Methods: Organophosphorus(V) Chemistry*, ed. C. M. Timperley, Elsevier Ltd., Oxford, 2015, ch. 6, pp. 633–720.
7. (a) Y. Mutoh, in *Patai's Chemistry of Functional Groups – The Chemistry of Organic Selenium and Tellurium Compounds*, ed. Z. Rappoport, J. F. Liebman and I. Marek, John Wiley & Sons, Ltd., Chichester, 2014, ch. 21, vol. 4, pp. 1433–1500; (b) K. Srinivas, P. Suresh, C. Naga Babu, A. Sathyanarayana and G. Prabusankar, *RSC Adv.*, 2015, **5**, 15579, and references cited therein; (c) M. Vaddamanu, K. Velappan and G. Prabusankar, *New J. Chem.*, 2020, **44**, 129, and references cited therein; (d) A. Arora, S. Singh, P. Oswal, D. Nautiyal, G. K. Rao, S. Kumar and A. Kumar, *Coord. Chem. Rev.*, 2021, **438**, 213885; (e) A. Schlachter, K. Tanner and P. D. Harvey, *Coord. Chem. Rev.*, 2021, **448**, 214176.
8. (a) C. J. Moulton and B. L. Shaw, *J. Chem. Soc., Dalton Trans.*, 1976, 1020; (b) G. van Koten, K. Timmer, J. G. Noltes and A. L. Spek, *J. Chem. Soc., Chem. Commun.*, 1978, 250.
9. For selected books and reviews, see: (a) *The Chemistry of Pincer Compounds*, ed. D. Morales-Morales and C. M. Jensen, Elsevier, Amsterdam, 2007; (b) *Pincer Compounds – Chemistry and Applications*, ed. D. Morales-Morales, Elsevier, Amsterdam, 2018; (c) *Organometallic Pincer Chemistry*, ed. G. van Koten and D. Milstein, *Topic in Organometallic Chemistry*, Springer, Berlin–Heidelberg, 2013, vol. 40; (d) The Privileged Pincer-Metal Platform: Coordination Chemistry and Applications, in

Topic in Organometallic Chemistry, ed. G. van Koten and R. A. Gossage, Springer, Cham–Heidelberg, 2016, vol. 54; (e) *Pincer and Pincer-Type Complexes – Applications in Organic Synthesis and Catalysis*, ed. K. J. Szabó and O. F. Wendt, Wiley-VCH, Weinheim, 2014; (f) M. Albrecht and G. van Koten, *Angew. Chem., Int. Ed.*, 2001, **40**, 3750; (g) N. Selander and K. J. Szabó, *Chem. Rev.*, 2011, **111**, 2048; (h) A. Kumar, G. K. Rao, F. Saleem and A. K. Singh, *Dalton Trans.*, 2012, **41**, 11949; (i) A. Kumar, G. K. Rao and A. K. Singh, *RSC Adv.*, 2012, **2**, 12552.

10. (a) G. van Koten, Organometallic Pincer Chemistry, in *Topic in Organometallic Chemistry*, ed. G. van Koten and D. Milstein, Springer, Berlin–Heidelberg, 2013, vol. 40, pp. 1–20; (b) P. A. Chase, R. A. Gossage, G. van Koten, The Privileged Pincer-Metal Platform: Coordination Chemistry and Applications, in *Topic in Organometallic Chemistry*, ed. G. van Koten and R. A. Gossage, Springer, Cham–Heidelberg, 2016, vol. 54, pp. 1–15.
11. (a) N. Mézailles and P. Le Floch, in *The Chemistry of Pincer Compounds*, ed. D. Morales-Morales and C. M. Jensen, Elsevier, Amsterdam, 2007, ch. 11, pp. 235–271; (b) D. V. Aleksanyan and V. A. Kozlov, The Privileged Pincer-Metal Platform: Coordination Chemistry and Applications, in *Topic in Organometallic Chemistry*, ed. G. van Koten and R. A. Gossage, Springer, Cham–Heidelberg, 2016, vol. 54, pp. 209–238.
12. For selected articles, see: (a) S. Takahashi, M. Nonoyama and M. Kita, *Transition Met. Chem.*, 1995, **20**, 528; (b) M. A. Hossain, S. O. Kang, J. M. Llinares, D. Powell and K. Bowman-James, *Inorg. Chem.*, 2003, **42**, 5043; (c) R. A. Begum, D. Powell and K. Bowman-James, *Inorg. Chem.*, 2006, **45**, 964; (d) T. Xua, M. D. Wodrich, R. Scopelliti, C. Corminboeuf and X. Hu, *Proc. Natl. Acad. Sci. U. S. A.*, 2017, **114**, 1242.
13. For selected articles, see: (a) Y. Nojima, M. Nonoyama and K. Nakajima, *Polyhedron*, 1996, **15**, 3795; (b) T. Kanbara, K. Okada, T. Yamamoto, H. Ogawa and T. Inoue, *J. Organomet. Chem.*, 2004, **689**, 1860; (c) M. Akaiwa, T. Kanbara, H. Fukumoto and T. Yamamoto, *J. Organomet. Chem.*, 2005, **690**, 4192; (d) M. Hosseini-Kharat, D. Zargarian, A. M. Alizadeh, K. Karami, M. Saeidifar, S. Khalighfard, L. Dubrulle, M. Zakariazadeh, J.-P. Cloutier and Z. Sohrabijam, *Dalton Trans.*, 2018, **47**, 16944.
14. K. Okamoto, T. Kanbara, T. Yamamoto and A. Wada, *Organometallics*, 2006, **25**, 4026.
15. (a) T. Kanbara and T. Yamamoto, *J. Organomet. Chem.*, 2003, **688**, 15; (b) H. Meguro, T. Koizumi, T. Yamamoto and T. Kanbara, *J. Organomet. Chem.*, 2008, **693**, 1109; (c) M. Montag, I. Efremenko, Y. Diskin-Posner, Y. Ben-David, J. M. L. Martin and D. Milstein, *Organometallics*, 2012, **31**, 505.
16. (a) M. Stradiotto, C. M. Kozak and M. J. McGlinchey, *J. Organomet. Chem.*, 1998, **564**, 101; (b) N. Á. Espinosa-Jalapa, D. Ke, N. Nebra, L. Le Goanvic, S. Mallet-Ladeira, J. Monot, B. Martin-Vaca and D. Bourissou, *ACS Catal.*, 2014, **4**, 3605.
17. (a) J. Kuwabara, G. Munezawa, K. Okamoto and T. Kanbara, *Dalton Trans.*, 2010, **39**, 6255; (b) K. Okamoto, T. Kanbara and T. Yamamoto,

Chem. Lett., 2006, **35**, 558; (c) G. W. Bates, P. A. Gale, M. E. Light, M. I. Ogden and C. N. Warriner, *Dalton Trans.*, 2008, 4106.
18. A. Fraix, M. Lutz, A. L. Spek, R. J. M. Klein Gebbink, G. van Koten, J.-Y. Salaün and P.-A. Jaffrès, *Dalton Trans.*, 2010, **39**, 2942.
19. (a) M. Doux, N. Mézailles, M. Melaimi, L. Ricard and P. L. Floch, *Chem. Commun.*, 2002, 1566; (b) M. Doux, C. Bouet, N. Mézailles, L. Ricard and P. L. Floch, *Organometallics*, 2002, **21**, 2785; (c) M. Doux, P. Thuéry, M. Blug, L. Ricard, P. L. Floch, T. Arliguie and N. Mézailles, *Organometallics*, 2007, **26**, 5643.
20. M. Doux, N. Mézailles, L. Ricard and P. L. Floch, *Eur. J. Inorg. Chem.*, 2003, 3878.
21. K. Okamoto, J. Kuwabara and T. Kanbara, *J. Organomet. Chem.*, 2011, **696**, 1305.
22. (a) K. Okamoto, T. Yamamoto, M. Akita, A. Wada and T. Kanbara, *Organometallics*, 2009, **28**, 3307; (b) T. Teratani, T. Koizumi, T. Yamamoto, K. Tanaka and T. Kanbara, *Dalton Trans.*, 2011, **40**, 8879.
23. A. Hossain, S. Lucarini, D. Powell and K. Bowman-James, *Inorg. Chem.*, 2004, **43**, 7275.
24. (a) P. Oulié, N. Nebra, N. Saffon, L. Maron, B. Martin-Vaca and D. Bourissou, *J. Am. Chem. Soc.*, 2009, **131**, 3493; (b) N. Nebra, J. Lisena, N. Saffon, L. Maron, B. Martin-Vaca and D. Bourissou, *Dalton Trans.*, 2011, **40**, 8912.
25. (a) N. Nebra, N. Saffon, L. Maron, B. Martin-Vaca and D. Bourissou, *Inorg. Chem.*, 2011, **50**, 6378; (b) N. Nebra, J. Monot, R. Shaw, B. Martin-Vaca and D. Bourissou, *ACS Catal.*, 2013, **3**, 2930; (c) J. Lisena, J. Monot, S. Mallet-Ladeira, B. Martin-Vaca and D. Bourissou, *Organometallics*, 2013, **32**, 4301.
26. P. Oulié, N. Nebra, S. Ladeira, B. Martin-Vaca and D. Bourissou, *Organometallics*, 2011, **30**, 6416.
27. D. Ke, N. Á. Espinosa, S. Mallet-Ladeira, J. Monot, B. Martin-Vaca and D. Bourissou, *Adv. Synth. Catal.*, 2016, **358**, 2324.
28. J. Monot, N. Merceron-Saffon, B. Martin-Vaca and D. Bourissou, *J. Organomet. Chem.*, 2017, **829**, 37.
29. M. Montag, I. Efremenko, G. Leitus, Y. Ben-David, J. M. L. Martin and D. Milstein, *Organometallics*, 2013, **32**, 7163.
30. T. Koizumi, T. Teratani, K. Okamoto, T. Yamamoto, Y. Shimoi and T. Kanbara, *Inorg. Chim. Acta*, 2010, **363**, 2474.
31. L. E. Karagiannidis, P. A. Gale, M. E. Light, M. Massi and M. I. Ogden, *Dalton Trans.*, 2011, **40**, 12097.
32. C. S. MacNeil, T. K. K. Dickie and P. G. Hayes, in *Pincer Compounds – Chemistry and Applications*, ed. D. Morales-Morales, Elsevier, Amsterdam, 2018, ch. 7, pp. 133–172.
33. (a) M. Blug, M. Doux, X. Le Goff, P. Maître, F. Ribot, P. Le Floch and N. Mézailles, *Organometallics*, 2009, **28**, 2020; (b) T. Arliguie, P. Thuéry, P. Le Floch, N. Mézailles and M. L. Ephritikhine, *Polyhedron*, 2009, **28**, 1578.

34. (a) R. Kapoor, A. Kataria, P. Venugopalan, P. Kapoor, G. Hundal and M. Corbella, *Eur. J. Inorg. Chem.*, 2005, 3884, and references cited therein; (b) R. Kapoor, A. Kataria, P. Kapoor, A. P. Singh Pannu, M. Singh Hundal and M. Corbella, *Polyhedron*, 2007, **26**, 5131.
35. (a) H. Wang, J. Liu, Y. Deng, T. Min, G. Yu, X. Wu, Z. Yang and A. Lei, *Chem. – Eur. J.*, 2009, **15**, 1499; (b) Q.-Q. Wang, R. A. Begum, V. W. Day and K. Bowman-James, *Inorg. Chem.*, 2012, **51**, 760; (c) T. Suzuki, Y. Kajita and H. Masuda, *Dalton Trans.*, 2014, **43**, 9732; (d) T. Suzuki, J. Matsumoto, Y. Kajita, T. Inomata, T. Ozawa and H. Masuda, *Dalton Trans.*, 2015, **44**, 1017.
36. Y. Komiyama, J. Kuwabara and T. Kanbara, *Organometallics*, 2014, **33**, 885.
37. (a) H. C. E. McFarlane, W. McFarlane and A. S. Muir, *Polyhedron*, 1990, **9**, 1757; (b) P. J. Heard and A. E. Aliev, *Polyhedron*, 1998, **17**, 3981; (c) R. Sevcik, M. Necas and J. Novosad, *Polyhedron*, 2003, **22**, 1585.
38. (a) W.-G. Jia, T.-T. Du, L.-L. Gao and J. Du, *Appl. Organomet. Chem.*, 2020, **34**, e5651; (b) N. Ghavale, S. T. Manjare, H. B. Singh and R. J. Butcher, *Dalton Trans.*, 2015, **44**, 11893.
39. (a) G. E. Tyson, K. Tokmic, C. S. Oian, D. Rabinovich, H. U. Valle, T. K. Hollis, J. T. Kelly, K. A. Cuellar, L. E. McNamara, N. I. Hammer, C. E. Webster, A. G. Oliver and M. Zhang, *Dalton Trans.*, 2015, **44**, 14475; (b) W.-G. Jia, L.-L. Gao, Z.-B. Wang, L.-Y. Sun and Y.-F. Han, *Organometallics*, 2019, **38**, 1946.
40. W.-G. Jia, L.-L. Gao, X.-T. Zhi, X.-D. Li, Z.-B. Wang and Y. Sun, *Polyhedron*, 2021, **195**, 114978.
41. (a) H. J. Kurth, U. Kaatz and F. Korte, *Chem. Ber.*, 1973, **106**, 2419; (b) M. R. Gandhi, M. Yamada, Y. Kondo, A. Shibayama and F. Hamada, *RSC Adv.*, 2016, **6**, 1243; (c) D. V. Aleksanyan, S. G. Churusova, R. R. Aysin, Z. S. Klemenkova, Y. V. Nelyubina and V. A. Kozlov, *Inorg. Chem. Commun.*, 2017, **76**, 33.
42. W.-G. Jia, L.-L. Gao, Z.-B. Wang, J.-J. Wang, E.-H. Sheng and Y.-F. Han, *Organometallics*, 2020, **39**, 1790.
43. (a) W.-G. Jia, Y.-B. Huang and G.-X. Jin, *J. Organomet. Chem.*, 2009, **694**, 4008; (b) J. R. Miecznikowski, W. Lo, M. A. Lynn, B. E. O'Loughlin, A. P. DiMarzio, A. M. Martinez, L. Lampe, K. M. Foley, L. C. Keilich, G. P. Lisi, D. J. Kwiecien, C. M. Pires, W. J. Kelly, N. F. Kloczko and K. N. Morio, *Inorg. Chim. Acta*, 2011, **376**, 515; (c) W.-G. Jia, Y.-C. Dai, H.-N. Zhang, X. Lu and E.-H. Sheng, *RSC Adv.*, 2015, **5**, 9491.
44. H.-N. Zhang, W.-G. Jia, Q.-T. Xu and C.-C. Ji, *Inorg. Chim. Acta*, 2016, **450**, 315.
45. M. L. Soriano, J. T. Lenthall, K. M. Anderson, S. J. Smith and J. W. Steed, *Chem. – Eur. J.*, 2010, **16**, 10818.
46. (a) X. Gan, E. N. Duesler, S. Parveen and R. T. Paine, *Dalton Trans.*, 2003, 4704; (b) W.-P. Leung, K.-W. Kan and T. C. W. Mak, *Organometallics*, 2010, **29**, 1890.
47. S. Katam and P. Ganesan, *Dalton Trans.*, 2017, **46**, 16615.

48. J. R. Miecznikowski, M. A. Lynn, J. P. Jasinski, W. Lo, D. W. Bak, M. Pati, E. E. Butrick, A. E. R. Drozdoski, K. A. Archer, C. E. Villa, E. G. Lemons, E. Powers, M. Siu, C. D. Gomes, N. A. Bernier and K. N. Morio, *Polyhedron*, 2014, **80**, 157.
49. J. Liu, H. Wang, H. Zhang, X. Wu, H. Zhang, Y. Deng, Z. Yang and A. Lei, *Chem. – Eur. J.*, 2009, **15**, 4437.
50. (a) H. Honda, Y. Ogawa, J. Kuwabara and T. Kanbara, *Eur. J. Inorg. Chem.*, 2014, 1865; (b) Y. Ogawa, A. Taketoshi, J. Kuwabara, K. Okamoto, T. Fukuda and T. Kanbara, *Chem. Lett.*, 2010, **39**, 385; (c) J. Kuwabara, Y. Ogawa, A. Taketoshi and T. Kanbara, *J. Organomet. Chem.*, 2011, **696**, 1289.
51. M. Hosseini-Kharat, R. Rahimi, A. M. Alizadeh, D. Zargarian, S. Khalighfard, L. P. Mangin, N. Mahigir, S. H. Ayati and A. A. Momtazi-Borojeni, *Bioorg. Med. Chem. Lett.*, 2021, **43**, 128107.

CHAPTER 5

Chalcogen Heterocycles Containing Main Group Elements: Synthesis and Reactivity

PIM PUYLAERT, JENS BECKMANN* AND EMANUEL HUPF*

Institut für Anorganische Chemie und Kristallographie, Universität Bremen, Leobener Str. 7, 28359 Bremen, Germany
*Emails: j.beckmann@uni-bremen.de; hupf@uni-bremen.de

5.1 Introduction

This chapter covers the compound class of chalcogen heterocycles containing other main group elements. The scope of this chapter is to present recent developments in the synthetic protocols, reactivity, and structure of selected chalcogen heterocycles with the p-block elements of groups 13, 14, and 15. The main focus is on the heavy chalcogens selenium and tellurium. Heterocycles containing chalcogens as sole heteroatoms and transition metals containing chalcogen heterocycles are covered in different chapters of this book. For a more comprehensive overview of chalcogen heterocycles we recommend recently published reviews and books.[1–9]

5.2 Group 13 Chalcogen Heterocycles

Amongst the first reported chalcogen heterocycles containing group 13 elements were pentamethylcyclopentadienylaluminium and -gallium

Chalcogen Chemistry: Fundamentals and Applications
Edited by Vito Lippolis, Claudio Santi, Eder J. Lenardão and Antonio L. Braga

Published by the Royal Society of Chemistry, www.rsc.org

chalcogenides with a cubane structure of the type $[Cp^*ECh]_4$ (E = Al, Ga; Ch = Se, Te; $Cp^* = Me_5C_5$)[10,11] as well as the dimeric $[Me_3N(H)Al(\mu\text{-}Ch)]_2$ possessing a 4-membered Al_2Ch_2 core (Ch = Se, Te).[12]

Chalcogen heterocycles containing boron are a rather scarce class of compounds and only recently such species were reported, featuring interesting bonding motifs. Braunschweig and co-workers reacted a cyclic alkyl(amino)carbene (CAAC) stabilized diborene with elemental sulfur and selenium in different stoichiometric ratios to give rise to 3-, 4-, and 5-membered heterocycles **1–3** (see Scheme 5.1).[13] Other examples of 3-membered diboraseleniranes and -telluriranes include the N-heterocyclic carbene (NHC) stabilized $[R(IMe)B]_2(\mu\text{-}Ch)$ (IMe = 1,3-dimethylimidazol-2-ylidene, R = 5-$(SiMe_3)C_4H_2S$, Ch = Se, Te),[14] and the phosphine stabilized $[An(Me_3P)B]_2(\mu\text{-}Se)$ (An = 9-anthryl).[15]

The reaction of the NHC stabilized manganese borylene $Cp(CO)_2Mn{=}B(IMe)^tBu$ with tellurium gave rise to boradichalcogenanes **4**, whereas the reaction with the lighter chalcogenes sulfur and selenium led to the formation of heterocycles of type **5**.[16] Compounds **4** and **5** are valuable synthetic building blocks for unsymmetrically substituted 4-membered B_2Ch_2 heterocycles, such as **6** and **7** as well as the 5-membered trichalcogenadiborolanes **8** (see Scheme 5.2).[17] The boradiselenane **5b** can also be converted into a boraselenone $^tBu(IMe)B{=}Se$ by treatment with PMe_3, which itself decomposes at elevated temperatures into the 4-membered heterocyclic $[^tBu(IMe)B]_2(\mu\text{-}Se)_2$.[16] Chalcogen heterocycles containing boron can also be

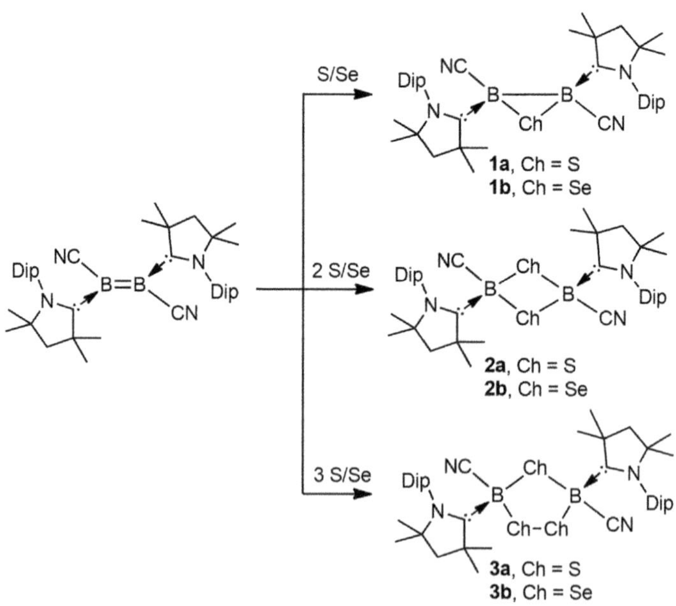

Scheme 5.1 Different B/Se heterocycles reported by Braunschweig *et al.* Dip = 2,6-$(^iPr)_2C_6H_3$.

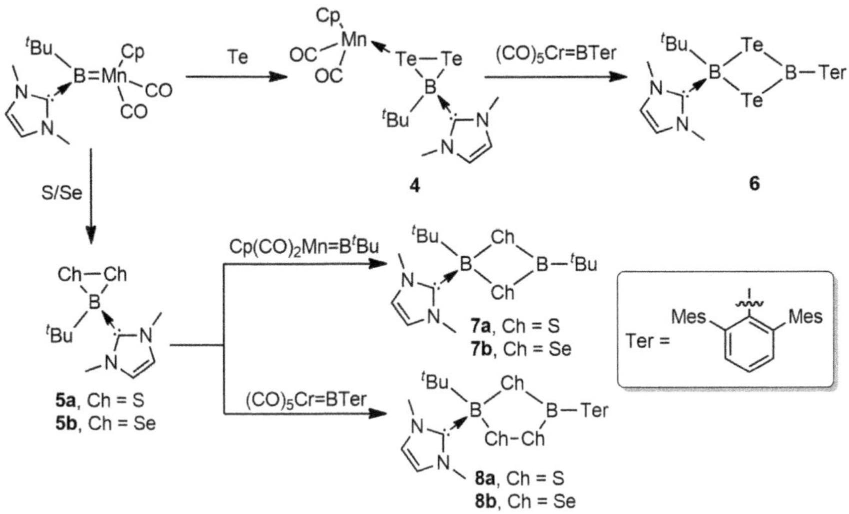

Scheme 5.2 Unsymmetrically substituted B/Ch heterocycles.

generated by using 1,4-diborabenzene, 9,10-diboraanthracene, as well as 1,4,2,5-diazadiboronine building blocks, which form different B_2Ch_n ring sizes.[18,19]

Whereas the presented B/Ch (Ch = S, Se or Te) containing heterocycles were mainly prepared by the reaction with elemental chalcogens, Stephan *et al.* reported the reaction of telluride $(PhC{\equiv}C)_2Te$ with borane $B(C_6F_5)_3$ resulting in the formation of the 6-membered B/Te heterocycle **9**,[20] which can further be modified at the B-side by the reaction with various alcohols, such as benzyl or cyclohexyl alcohol at elevated temperatures.[21] Notably, those alkoxy substituted heterocycles of type **10** can be converted into unsymmetrically substituted B/Te heterocycles **11** *via* alkyne exchange even when treated with an excess of alkyne of more than five equivalents (see Scheme 5.3). This is in contrast to the parent heterocycle **9**, which undergoes double-alkyne exchange leading to symmetric heterocycles of type **12**, which was assigned to the higher Lewis-acidity at the B-side in **9**.[22] The perhydrogenated parent of **9** (TeC_4H_4BH) was investigated computationally by Shuford *et al.* to study the degree of aromaticity and resonance contributions in 6-membered Te/B heterocycles. Those heterocycles indeed show Lewis-pairing character between Te and B in conjunction with increased charge-transfer being shifted from the Te lone-pair to the boron, supporting an aromatic character.[23]

A rather commonly observed structural feature of the heavier group 13 members Al, Ga, and In is the formation of 4-membered chalcogen heterocycles with a E_2Ch_2 core. However, to the best of our knowledge no compound with a Tl_2Ch_2 core has been reported yet. Roesky and co-workers reported the synthesis of 4-membered heterocycles of type Al_2Ch_2 **13** and Ga_2Ch_2 **14** by using an azaallyl ligand $((Me_3Si)_2C(Ph)C(SiMe_3)N)$ with

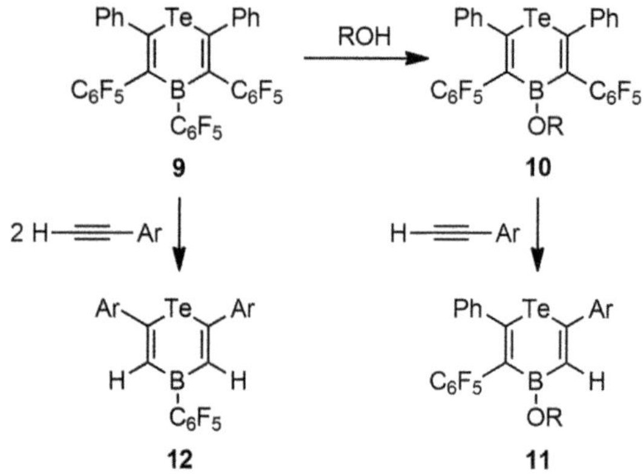

Scheme 5.3 6-Membered B/Te heterocycles reported by Stephan *et al.* R = *e.g.*, Bn, Cy; Ar = 4-BrC_6H_4, 3-C_4H_3S.

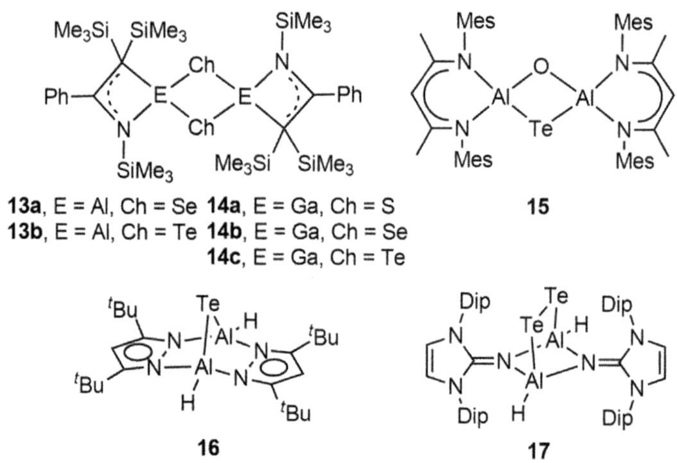

Scheme 5.4 4-Membered E_2Ch_2 heterocycles as well as telluride and ditelluride bridged heterocycles; Dip = 2,6-(^iPr)$_2C_6H_3$.

sterically demanding trimethylsilyl groups (see Scheme 5.4).[24,25] Furthermore, similar structures could be obtained by choosing the bidentate NacNac ligand HC{C(Me)N(Dip)}$_2$ (Dip = 2,6-(^iPr)$_2C_6H_3$).[26] Moya-Cabrera *et al.* reported the Al/Te heterocycle **15** bearing a mixed 4-membered Al_2(μ-O)(μ-Te) core by applying the sterically less demanding NacNac ligand HC{C(Me)N(Mes)}$_2$ (see Scheme 5.4).[27] Considerably less explored are compounds with bridging Al–Ch–Al or Al–Ch–Ch–Al structural motifs and include Roesky's Al/Te heterocycle **16** with a mono-bridging telluride and a N_4Al_2Te core[28] as well as Inoue's ditelluride **17** featuring a $N_2Al_2Te_2$ motif (see Scheme 5.4). Notably, compound **17** can be converted to a non-bridged $L_2L'AlTe$ species containing a Al=Te double-bond, which, at elevated temperatures forms a 4-membered Al_2Te_2 heterocycle.[29] The first 3-membered

$AlSe_2$ heterocycle, comparable to the BSe_2 motif in **5** was reported by Coles and co-workers in the aluminyl anion $[Al(NON^{Dip})(\kappa^2\text{-}Se_2)]^-$ **18** ($NON^{Ar} = [O(SiMe_2NAr)_2]^{2-}$) (see Figure 5.1) obtained from the reaction of the corresponding aluminyl selenide $[Al(NON^{Dip})(Se)]$ with elemental selenium.[30]

Jambor and Bouška showed that 4-membered gallium and indium chalcogenides can also be generated by the use of $L^{N,C,N}$ pincer ligands, such as 2,6-$(Me_2NCH_2)_2C_6H_3$, examples of which are the Ga_2Ch_2 species **19** as well as the indium telluride **20**. The gallium telluride **19b** is light-sensitive and gives rise to the 5-membered Ga_2Te_3 heterocycle **21**. The $L^{N,C,N}$-ligand is also suitable to stabilize 5-membered ECh_4 heterocyclic chalcogenides of type $L^{N,C,N}Ga\text{-}(\kappa^2\text{-}Se_4)$ (**22**)[31] and $L^{N,C,N}In\text{-}(\kappa^2\text{-}Ch_4)$ **23** (see Scheme 5.5).[32]

Chivers, O'Brien, and co-workers demonstrated that the PNP- and PCP-bridged anions of type $[(ChPR_2)_2N]^-$ and $[(ChPR_2)_2CH]^-$ (Ch = Se, Te; R = ^iPr, Ph) are extremely valuable synthetic building blocks in terms of the generation of group 13-containing chalcogen heterocycles, often leading to multiple rings within one compound.[33,34] Furthermore, these building blocks are also suitable to form Tl/Se and Tl/Te heterocycles.[35,36] Scheme 5.6 presents selected examples of 4-, 5-, and 6-membered group 13 heterocycles **24** to **29** containing PNP- and PCP-ligands, respectively. Lastly, other group 13 containing chalcogen heterocycles with interesting structural features include a BFTe moiety as reported by Gabbaï *et al.*[37] or a BP_2Ch (Ch = S, Se) fragment from Ragogna and co-workers.[38]

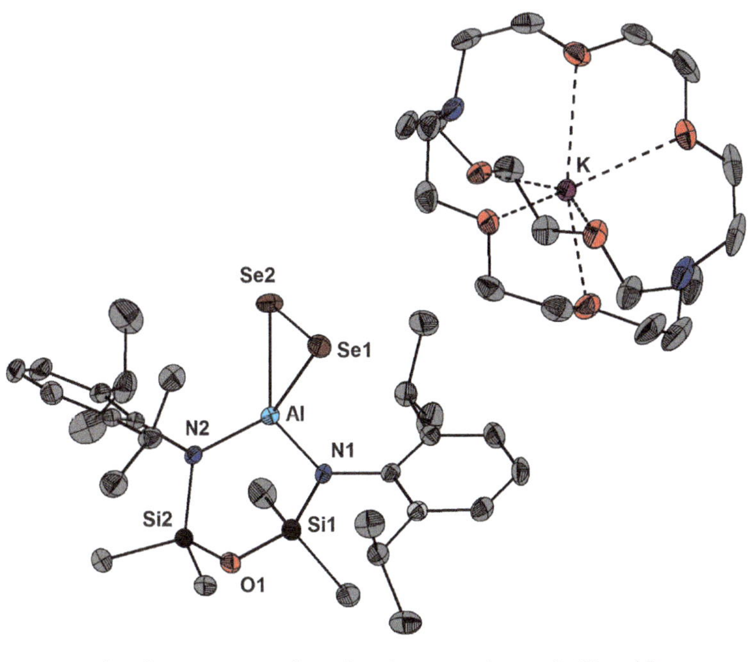

Figure 5.1 Molecular structure of **18** showing 50% thermal ellipsoids. H-atoms and THF solvate molecule omitted for clarity.

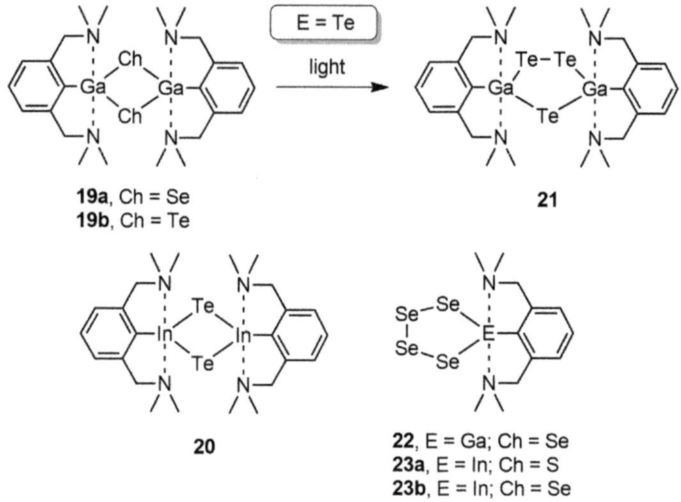

Scheme 5.5 N,C,N-Pincer ligand stabilized four- and 5-membered gallium and indium chalcogenides.

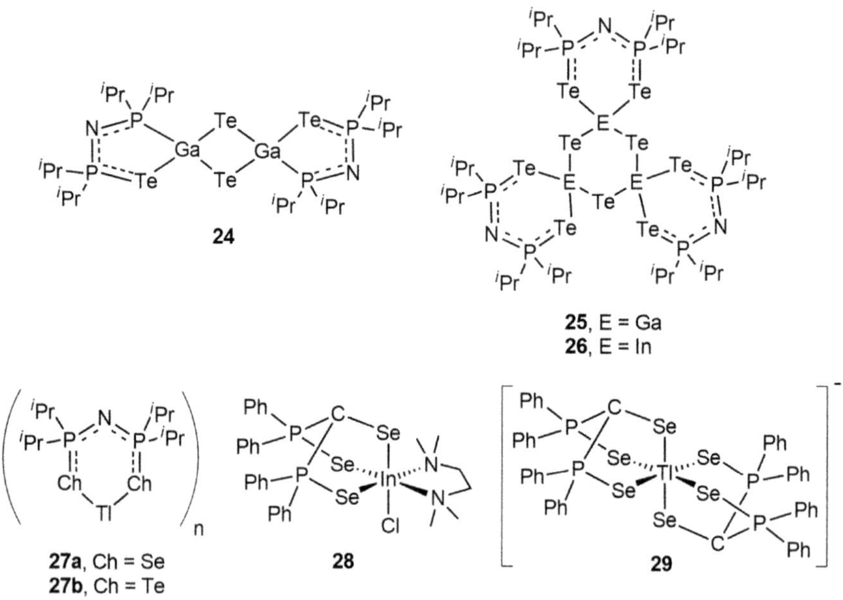

Scheme 5.6 Selected examples of Ga/In/Tl chalcogen heterocycles containing PNP- and PCP-ligands.

5.3 Group 14 Chalcogen Heterocycles

A large portion of mixed group 14/16 ring systems originate from the interest in silylenes and stannylenes and their synthetic potential to react with chalcogens to form tetrel–chalcogen double bonds as "heavy ketone" analogues. The heavier tetrelenes germylenes and plumbylenes are comparatively less studied in that regard. This type of chemistry is nicely reviewed by Chivers in Chapter 10, Section 7 of his 2009 book.[39] With an emphasis on heavier chalcogens the most striking examples are briefly revisited, and expanded to recent developments and examples in the field. Reaction of the tetrel precursors R_2ECl_2 with H_2Ch (or the related Li and Na salts, which are preferable over the toxic H_2Ch gases) lead to the *in situ* formation of $R_2E{=}Ch$, which oligomerize to the heterocycles of type $[R_2E(\mu\text{-}Ch)]_3$ **30** and $[R_2E(\mu\text{-}Ch)]_2$ **31** (See Scheme 5.7). The selectivity of the ring sizes is governed by the steric bulk of the substituents R on the tetrel element: methyl or phenyl groups tend to favor 6-membered heterocycles **30**, whereas the sterically more demanding *t*-butyl groups favor the formation of 4-membered rings such as **31**. As the stability of the "E=Ch" unit is largely influenced by steric effects, bulky substituents have been installed more frequently with the goal of isolating stable ketone analogues and as a consequence the formation of 4-membered E_2Ch_2 has been more commonly encountered.

The reaction of diorganosilanes and -stannanes R_2EH_2 with a base, such as *n*- or *t*-butyllithium followed by the reaction with elemental chalcogens gives rise to the formation of $R_2E\text{-}(\kappa^2\text{-}Ch_4)$ **32**, with a similar 5-membered ECh_4 core as the group 13 heterocycles **22** and **23** (see Scheme 5.7). Instead of elemental chalcogens, various studies have employed trialkyl- or triarylphosphine chalcogenides as a chalcogen source.[39]

Various haptotropic rearrangement reactions of decamethylsilicocene, $Cp^*{}_2Si$, with chalcogen reagents were reported as early as 1989 by Jutzi *et al.* in an attempt to obtain compounds featuring a Si=Ch double-bond (see Scheme 5.8). The reaction of $Cp^*{}_2Si$ with elemental sulfur gave rise to the 4-membered Si_2S_2 heterocycle $[Cp^*{}_2Si(\mu\text{-}S)]_2$ **33**, which was also formed when using cyclohexane sulfide as the sulfur source, but additionally the partially

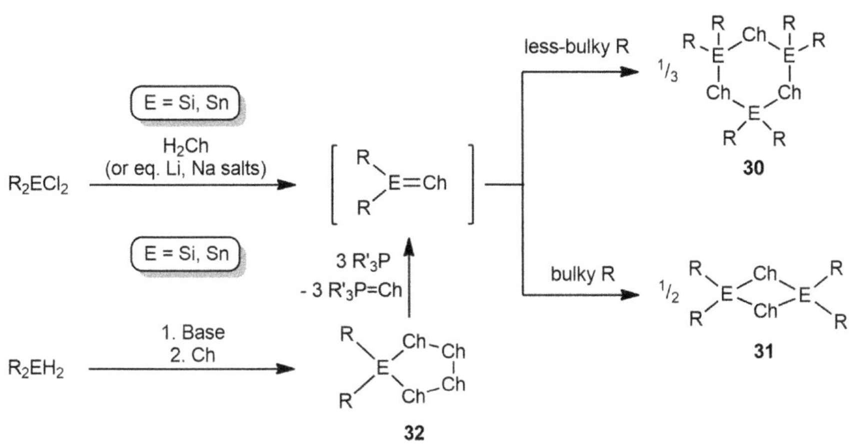

Scheme 5.7 General preparation and reactivity scheme of E=Ch doubly bonded species to E/Ch mixed heterocycles.

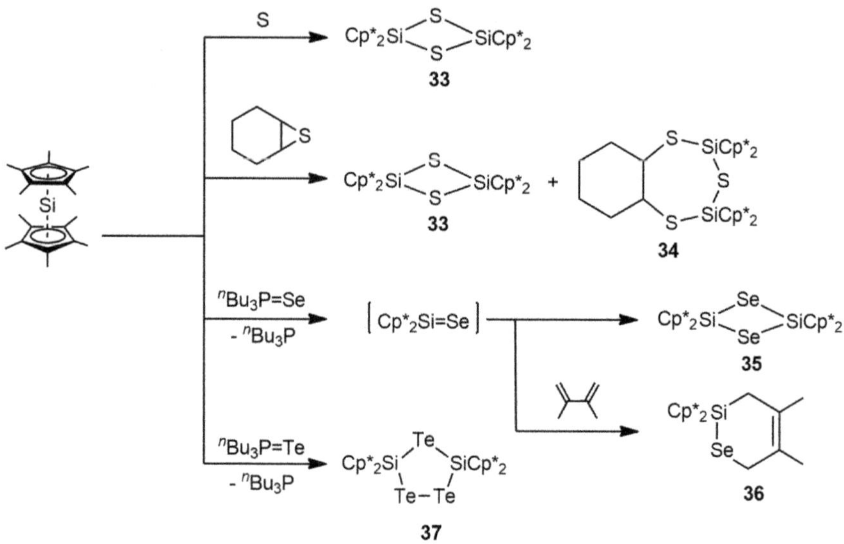

Scheme 5.8 Reactivity of $Cp^*{}_2Si$ with various chalcogen sources.

organic heterocycle **34** with a $C_2Si_2S_3$ core was obtained. Analogously, the reaction with tributylphosphine selenide gives the diselenadisiletane **35**. Notably, using 2,3-dimethylbutadiene as a trapping reagent gives rise to the C_4SiSe heterocycle **36**, indicating the formation of $Cp^*{}_2Si{=}Se$ as the active species. Reaction with tributylphosphine telluride gives rise to the 5-membered Si_2Te_3 heterocycle **37** (see Scheme 5.8).[40]

Si-containing chalcogen heterocycles have also been reported by West *et al.* in the course of his pioneering work in silylene and disilene chemistry. Reaction of the tetramesityldisilene with elemental chalcogens form the 3-membered Si_2Ch heterocycles **38**,[41,42] whereas 4-membered Si_2Ch_2 rings in **39** formed from a stable N-heterocyclic silylene, which in the case of the sulfur species **39a** could further be reacted to compound **40**, featuring a bicyclic Si_3S_4 core (see Scheme 5.9).[43]

A similar motif was reported when Jambor and co-workers studied the activation of tellurium by their $L^{N,C,N}$ pincer ligand stabilized dimeric Sn(I) reagent $(L^{N,C,N}Sn)_2$ ($L = 2,6\text{-}(Me_2NCH_2)_2C_6H_3$, *vide supra*), which first forms a tellurium-bridged species and undergoes formation of an unstable bis(organotin telluride) with additional tellurium and finally forms the Sn/Te heterocycle **41** upon irradiation with visible light (see Scheme 5.10).[44]

4-Membered Sn_2Ch_2 heterocycles and related compounds are of interest as precursors for tin-chalcogen materials such as films or nanoparticles. In 1999, Richeson and co-workers described that the reaction of the Sn(II) precursors $CyNC(R)NCy]Sn[N(SiMe_3)_2$ ($R = Me$, ^tBu) with elemental sulfur gave rise to the tetrasulfido species **42**, comprising a $R_2Sn\text{-}(\kappa^2\text{-}S_4)$ moiety. Compounds **42** in turn can be converted to the $[Sn(\mu\text{-}S)]_2$ heterocycle **43** *via* sulfur abstraction with triphenylphosphine. An alternative synthetic route to

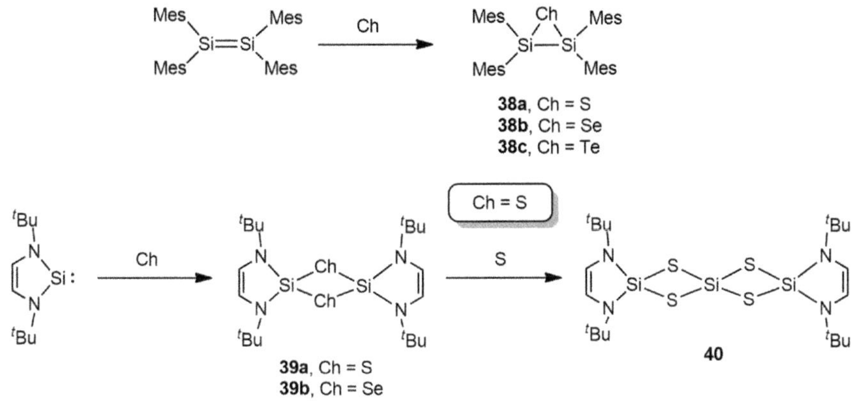

Scheme 5.9 Formation of Si/Ch heterocycles from disilenes and N-heterocyclic silylenes.

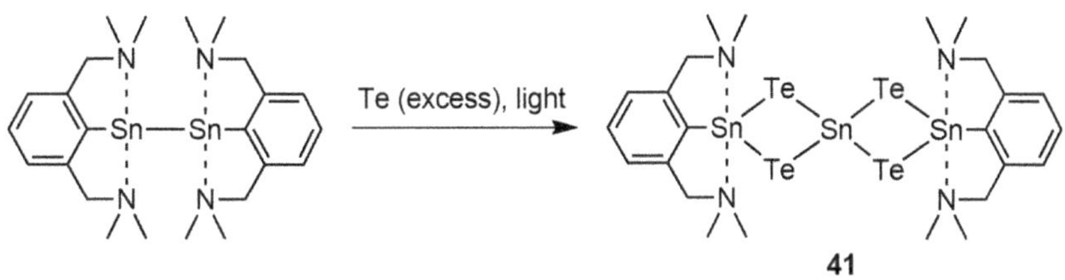

Scheme 5.10 N,C,N-Pincer ligand stabilized Sn/Te heterocycle **41**.

43 is the direct reaction of the Sn(II) precursors with sulfur atom sources such as styrene sulfide or propylene sulfide (see Scheme 5.11).[45] Compounds with a structurally similar $SnCh_4$ (Ch = S, Se) moiety as in **42** have also been reported by Okazaki and co-workers and served as precursors for the isolation of stable Sn=Ch double bond compounds.[46]

Upon reaction of the 2-(*N*,*N*-dimethylaminomethyl)phenyl stannylene with chalcogens, the 4-membered Sn_2Ch_2 cycles **44** were obtained. However, using CS_2 as the solvent, the same reaction with elemental sulfur was not observed, instead the sulfur acted as a catalyst in the formation of bicyclic species **45**, containing a $S_2C{=}CS_2$ fragment, as reported by Růžička *et al.* (see Scheme 5.12).[47]

Using the same 2-(*N*,*N*-dimethylaminomethyl)phenyl ligand system with chalcogens, Sarazin and co-workers prepared dimeric S and Se precursors of the type $[2\text{-}(Me_2NCH_2)C_6H_4]_2Ch_2$. Reaction of these dimers with $Ge[N(SiMe_3)_2]_2$, $Sn[N(SiMe_3)_2]_2$ and $Pb[N(SiMe_3)_2]_2$ yielded 6-membered cyclic germylenes **46** and **49**, stannylenes **47** and **50**, and plumbylenes **48** and **51** (see Scheme 5.13). Single crystal X-ray diffraction analysis of **48** and **51** revealed weak interactions between two plumbylene units in the solid state (see Figure 5.2). Employing the same *N*,*N*-dimethylamino pendant arm in sulfuryl or selenyl alkoxide ligands such as **52** did not lead to clean coordination of either nitrogen or the chalcogen, but mixtures containing

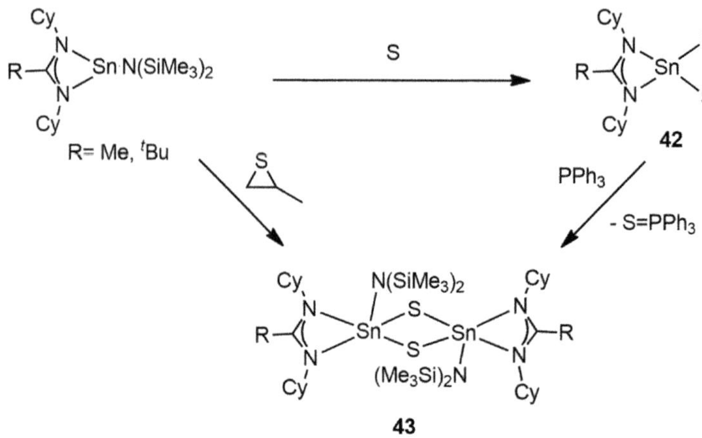

Scheme 5.11 Preparation and reactivity of amido/amidinato $[Sn(\mu\text{-}S)]_2$ complexes.

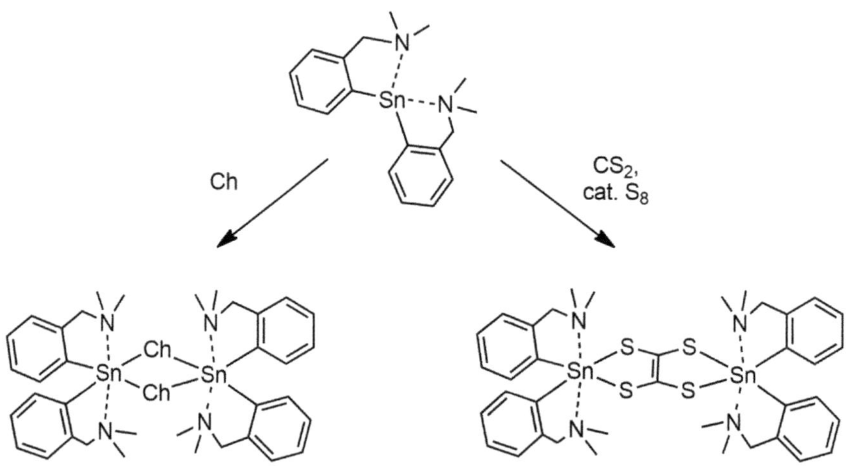

Scheme 5.12 Amino-stabilized stannylene reactivity with chalcogens and activation of carbon disulfide.

oxygen-bridged dimers were obtained in most cases. However, by pretreating **52** with Bochmann's acid, coordination of both N and Se was achieved leading to the cationic dimeric Pb complex **53** (See Scheme 5.13). Introducing electron-withdrawing $-CF_3$ substituents α to the alkoxide, as in Se-ligand **54**, enforced coordination of Se to Sn (**55**) and Pb (**56**), however a second equivalent of ligand was sequestered regardless of stoichiometry (see Scheme 5.13).[48]

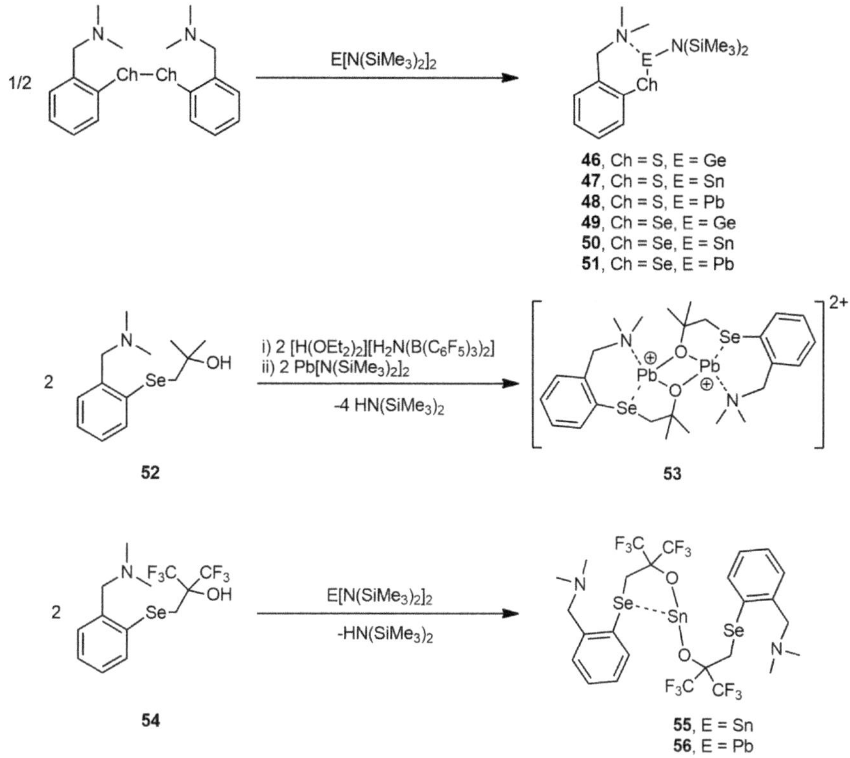

Scheme 5.13 Tetrel–chalcogen heterocycles supported by *N,N*-dimethylamino pendant arms.

A recent striking example of steric influence is reported by Johnson *et al.* by employing Sn(II) guanidinate precursors. *Iso*-propyl substituents led to isolation of the 4-membered Sn_2Se_2 heterocycle **57**, whereas cyclohexyl substituents gave rise to compounds containing a Sn=Ch moiety (from propylene sulfide and triethylphosphine selenide) and the $SnCh_4$ cycles **58**, when elemental chalcogen was employed (see Scheme 5.14).[49]

The group of Siemeling reported on N-heterocyclic germylenes and stannylenes, which both yielded the 4-membered E_2Ch_2 ring systems **59–61** upon reaction with sulfur or selenium (see Scheme 5.15).[50,51] The Sn/Ch compounds **62**,[52] **63**,[53] **64**[54] and **65**[55] illustrated in Scheme 5.16 present further examples of recently reported Sn/Ch heterocycles bearing the $[Sn(\mu\text{-}Ch)]_2$ motif.

In the mid-2000s, Chivers and co-workers investigated imidotin cage complexes and observed $[Sn(\mu\text{-}Ch)]_2$ species, such as **66** in equilibrium with the target compounds featuring a Sn=Ch bond. The $[Sn(\mu\text{-}Ch)]_2$ species **67** was obtained by the reaction of the neutral cubane $[(SnN(dipp))_4]$ (dipp = 2,6-di*iso*propylphenyl) with elemental selenium and catalytic amounts of the analogous anionic lithium cubane (see Scheme 5.17).[56,57]

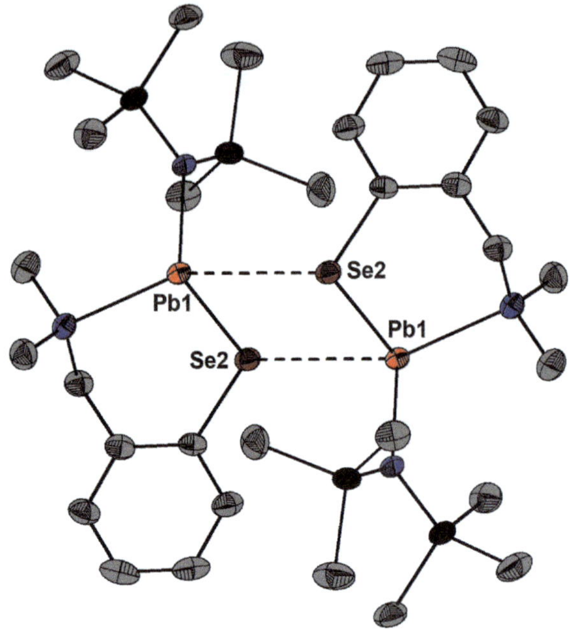

Figure 5.2 Intermolecular Pb···Se interactions within the molecular structure of **51** plotted with 50% thermal ellipsoids. H-atoms omitted for clarity.

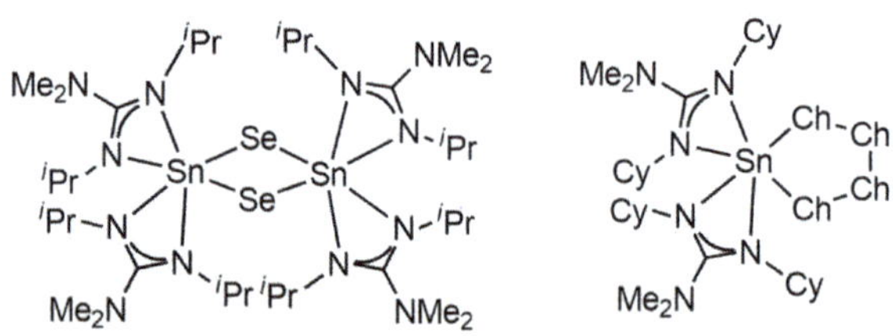

57

58a, Ch = S
58b, Ch = Se

Scheme 5.14 Johnson's bulky guanidinato-tin-sulfur and -selenium cycles.

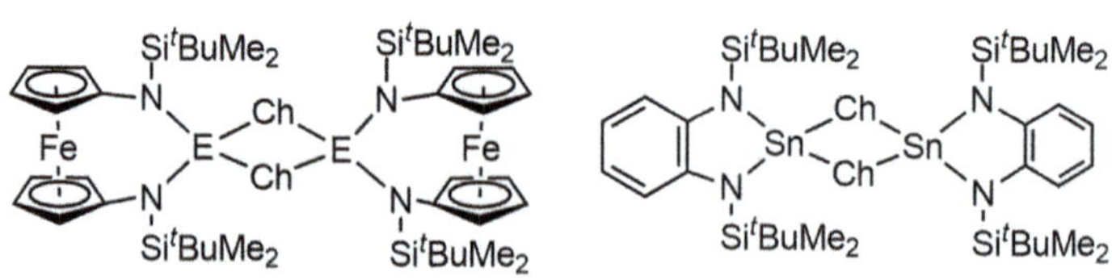

59a, E = Ge; Ch = S
59b, E = Ge; Ch = Se
60a, E = Sn; Ch = S
60b, E = Sn; Ch = Se

61a, Ch = S
61b, Ch = Se

Scheme 5.15 Siemeling's N-heterocyclic germanium and tin cycles.

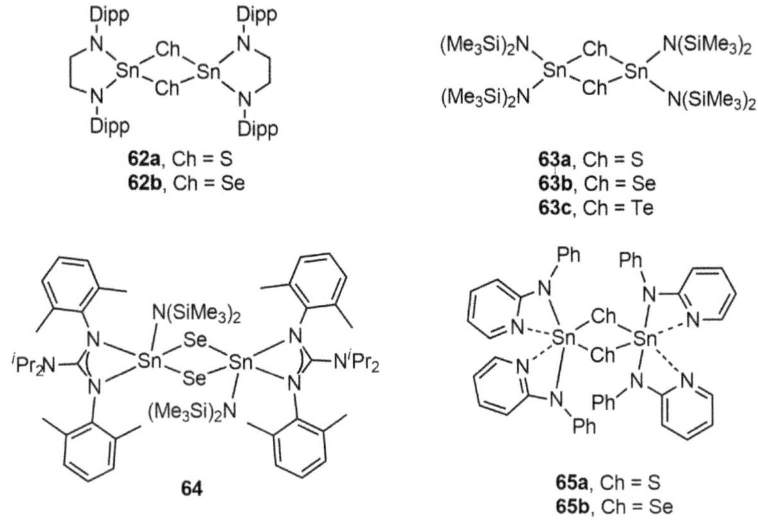

Scheme 5.16 Recent examples of $[Sn(\mu\text{-}Ch)]_2$-type complexes.

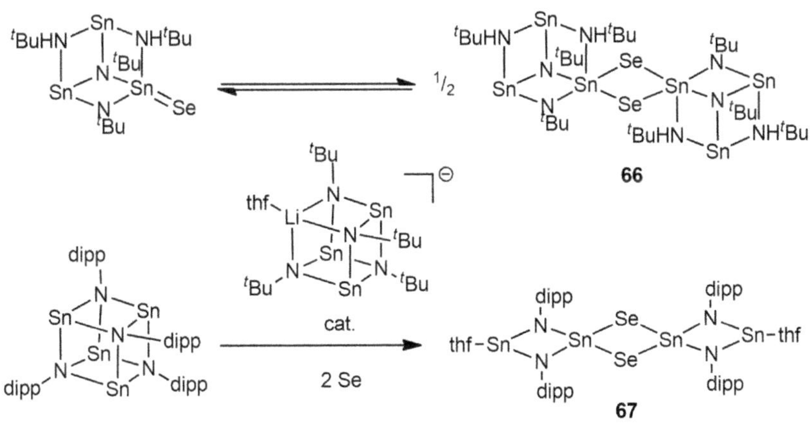

Scheme 5.17 Imidotin cage complexes and their reactivity with Se.

Inorganic cage compounds can be considered as polyheterocycles, and those consisting of group 14 and 16 elements, are well-studied. This class of compounds represents a fruitful field of investigation, which has been thoroughly reviewed elsewhere.[58–60] Several notable examples from recent years are highlighted here. Dehnen *et al.* prepared various cage compounds with organic substituents, which allows for further derivatization and construction of larger clusters.[61–64]

In one example, adamantane-like Sn/S and Sn/Se complexes **68** were prepared from $FcSnCl_3$ (Fc = ferrocenyl) with sodium or potassium salts of the chalcogens. In contrast, the reaction with K_2Te yielded the 5-membered Sn_3Te_2 ring **69**, with the ferrocene substituents redistributed over the tin atoms (see Scheme 5.18).[61,62] An analogues Sn/S adamantane type cluster to **68a** was reported by Dehnen and her team, namely the vinylbenzene substituted $(RSn)_4S_6$ (R = 4-CH_2CH–C_6H_4), which shows impressive white-light emission.[65] The same group has also reported the synthesis of germanium-chalcogen cages from nitrile-functionalized $RGeCl_3$ with chalcogen sodium salts. For sulfur and selenium, the adamantane-like compounds **70** were obtained, respectively, with tellurium showing different reactivity, yielding the noradamantane-like compound **71** (see Scheme 5.19).[63]

The group of Schnepf recently reported a different approach to germanium cages, reacting $GeCl_2$(dioxane) with lithium salts of the type $LiChC(SiMe_3)_3$. In the case of selenium, they obtained the dimeric Ge(IV) cage compound **72**. For tellurium, the results depended on the work-up procedure. Extraction with benzene allowed isolation of the $[Ge_2(\mu\text{-}Te)]_2$ compound **73**, whereas extraction with diethyl ether revealed C–H activation on the solvent THF, and the lithium-bridged dimeric Ge_3Te_3 cage **74** was isolated (see Scheme 5.20).[66]

Lips *et al.* reported the preparation of amido-substituted cage compounds from the bicyclic silicon compound **75**. Interestingly, they obtained different products depending on whether gray or red selenium was used. Reaction with gray selenium required higher temperatures and mainly led to the tricycloheptane-like structure **76**, whereas the reaction with red selenium proceeded at lower temperatures and yielded the nortricyclane-type cluster **77**. Both **76** and **77** could undergo thermal rearrangement to the new structural motif of a defect hetero-cubane **78**. Reactions of **75** with elemental tellurium were unsuccessful, but tributylphosphine telluride as the stoichiometric tellurium source yielded the adamantane-type cluster **79** (see Scheme 5.21).[67]

In 2020, the Scheschkewitz group reported the oxidation of their siliconoid **80** by chalcogens. Presumably, the ligated silylene reacts with the chalcogen, forming a Si=Ch species, which then rearranges to the chalcogen-expanded Si_7Ch siliconoids **81** (see Scheme 5.22).[68]

Unsaturated 3-membered silicon rings are also suitable precursors for mixed heterocyclic compounds. In 2008, the Sekiguchi group reacted (bis-*t*-butylmethyl)silyl-substituted cyclotrisilene with chalcogens, leading to the formation of bicyclobutanes **82**. In the case of the sulfur and selenium

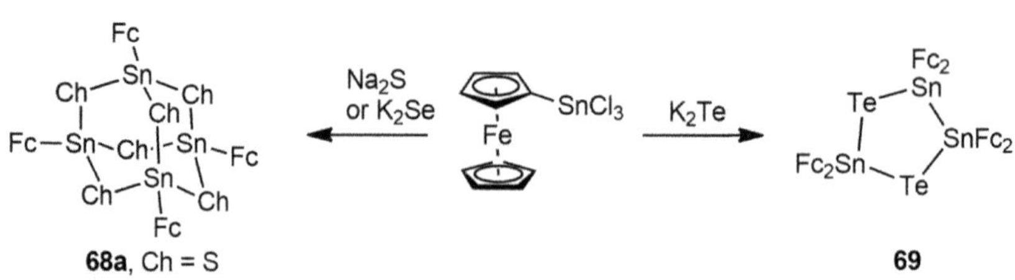

68a, Ch = S
68b, Ch = Se

69

Scheme 5.18 Ferrocenyl-substituted Sn/S and Sn/Se cages, and Sn/Te cycle.

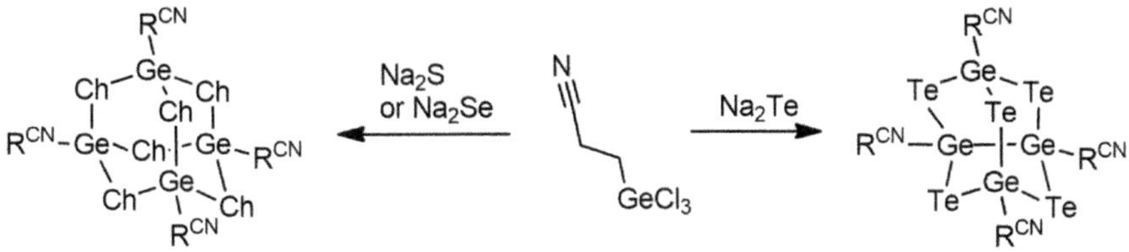

70a, Ch = S
70b, Ch = Se

71

Scheme 5.19 Nitrile-decorated germanium-chalcogen cages.

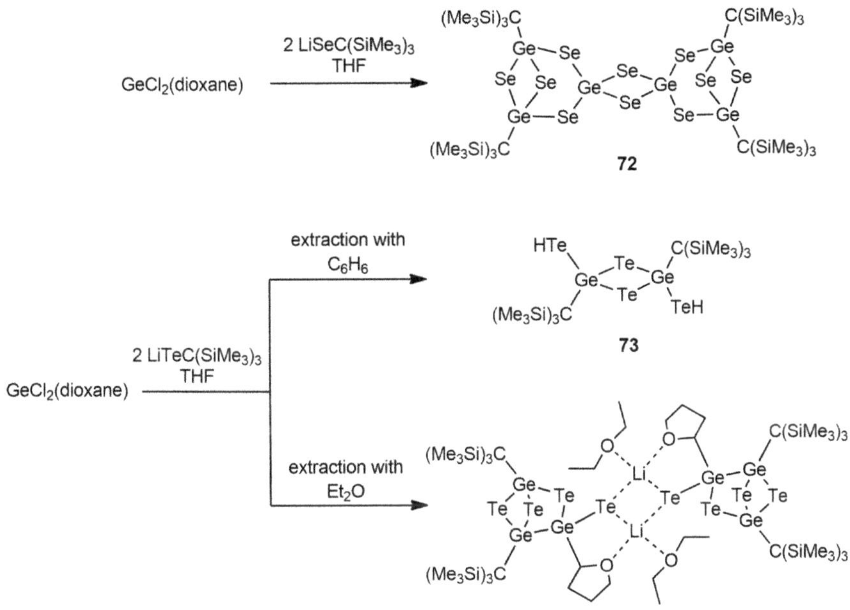

72

73

74

Scheme 5.20 Cyclic and polycyclic compounds of germanium with the heavier chalcogens.

species **82a/b**, these underwent photochemical rearrangement to form the cyclobutenes **83a** and **83b**, respectively (see Scheme 5.23).[69]

Scheschkewitz revisited this system a decade later, replacing the *t*-butylmethylsilyl groups with Tip substituents (Tip = 2,4,6-(^{i}Pr)$_2$C$_6$H$_3$). The bicyclopentanes **84** were obtained by the reaction with elemental sulfur or an excess of selenium. The use of propylene sulfide or stoichiometric amounts of selenium or tellurium gave rise to the Si$_3$Ch heterocycles **85**. Thermal rearrangement of **85** led to the formation of 4-membered rings **86**, which could not be isolated, but controlled hydrolysis of the Si=Si double bond in the Si/Te species **86c** gave the oxygen-bridged, bicyclic compound **87** (see Scheme 5.24).[70]

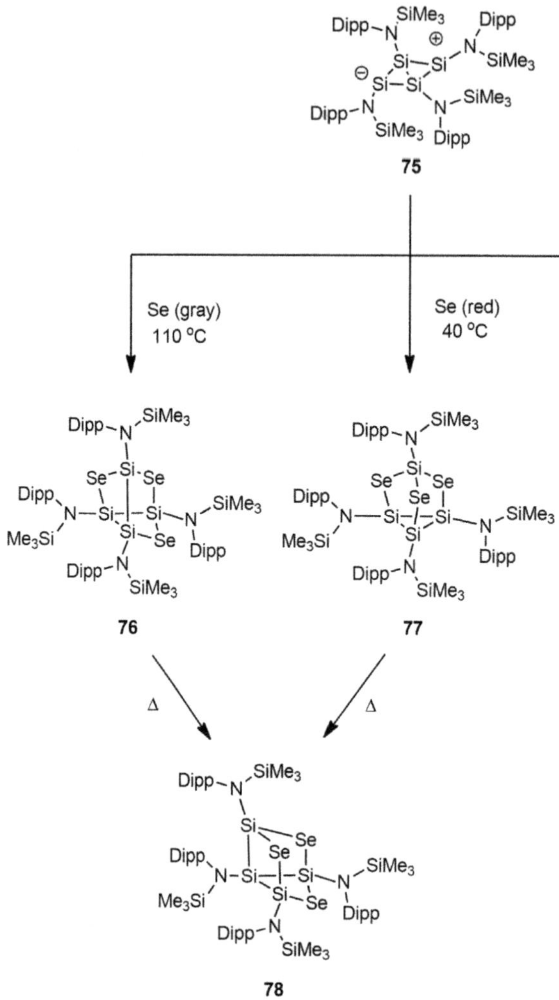

Scheme 5.21 Si/Se and Si/Te cages derived from a bicyclic Si_4 species.

As previously outlined for group 13 chalcogen heterocycles, Chivers and co-workers investigated mono-anionic dichalcogenido PCP-ligands and their interactions with main group elements including Sn and Te. The lithium salt of the ligand reacts directly with various metal dichlorides, *e.g.*, $SnCl_2$, giving rise to the Sn/Se heterocycle **88**. The C–H moiety in the ligand backbone displays interesting reactivity. In a bimolecular reaction, one hydrogen is transferred to a second equivalent of ligand, which surrenders one of its seleniums, leading to the quadricyclic **89** under elimination of elemental Sn and two equivalents of neutral $Ph_2PCH_2PPh_2{=}Se$ (see Scheme 5.25).[71] It was later shown that similar compounds, including mixed S/Se species, can be obtained starting from $SnBr_4$ and *in situ* generated trichalcogenido PCP-ligand.[72]

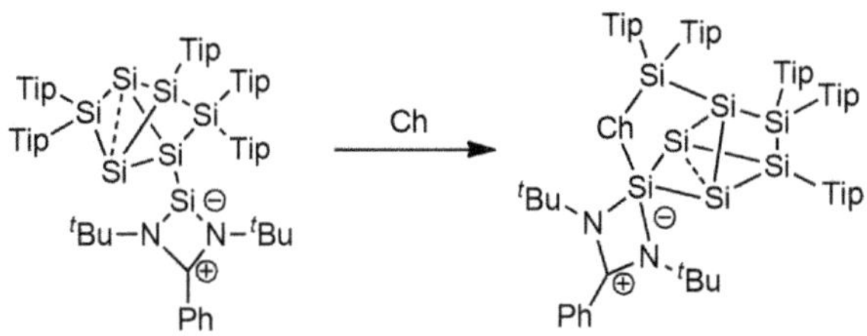

80

81a, Ch = S
81b, Ch = Se
81c, Ch = Te

Scheme 5.22 Scheschkewitz's siliconoid and its reactivity with elemental chalcogens, Tip = 2,4,6-(^{i}Pr)$_2$C$_6$H$_3$.

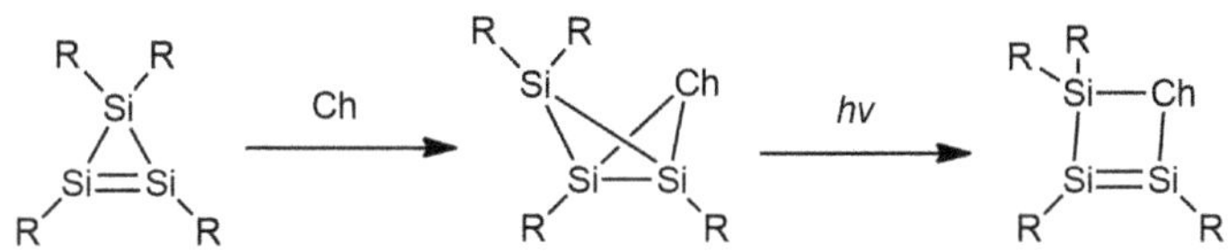

82a, Ch = S
82b, Ch = Se
82c, Ch = Te

83a, Ch = S
83b, Ch = Se

Scheme 5.23 The reactivity of bulky cyclotrisilenes as Si/Ch heterocycle precursors (R = ^{t}Bu$_2$MeSi).

Partially organic heterocycles containing tetrel and chalcogen elements are comparatively rare. In 2006, Amosova *et al.* reported the reactions of Se(II) as well as Se(IV) halides with various diorgano-diethynyl silanes. These reactions allowed the preparation of a variety of halide-substituted, unsaturated 5-membered rings, such as **90** and **91**, containing Si and Se heterocycles as outlined in Scheme 5.26. In the case of selenium dihalides, an exocyclic C=C double bond was retained, substituted with a single halide. For tetrahalide selenium sources, the additional two halides are found to substitute this double bond. Additionally, using a cyclic diethynylsilane precursor, the heterocyclic product contained a spirocycle at silicon.[73]

Another impressive collection of group 14–16 mixed heterocycles was established by the Block group in 2006.[74] They prepared over 40 new heterocyclic compounds from small α,ω-dihalides with one or two dimethyltetrel groups in the backbone. As summarized in Schemes 5.27 and 5.28, starting from Me$_2$E(CH$_2$I)$_2$ (E = Si, Sn), 4-membered C$_2$ECh heterocycles **92** and **93** have been prepared by the reaction with lithium or sodium salts of the chalcogen of choice. The Si/Se 4-membered ring **92b** undergoes catalytic oligomerization to a mixture of products which include the 8- and 12-membered rings **94** and **95**, respectively. Reaction of Me$_2$E(CH$_2$I)$_2$ (E = Si, Sn) with Na$_2$Se$_2$ instead led to the formation of the 5-membered rings **96** and **97**, which upon reaction with a further equivalent of Me$_2$E(CH$_2$I)$_2$ and subsequent reduction, yielded the 8-membered heterocycles **98b**/**99b**.

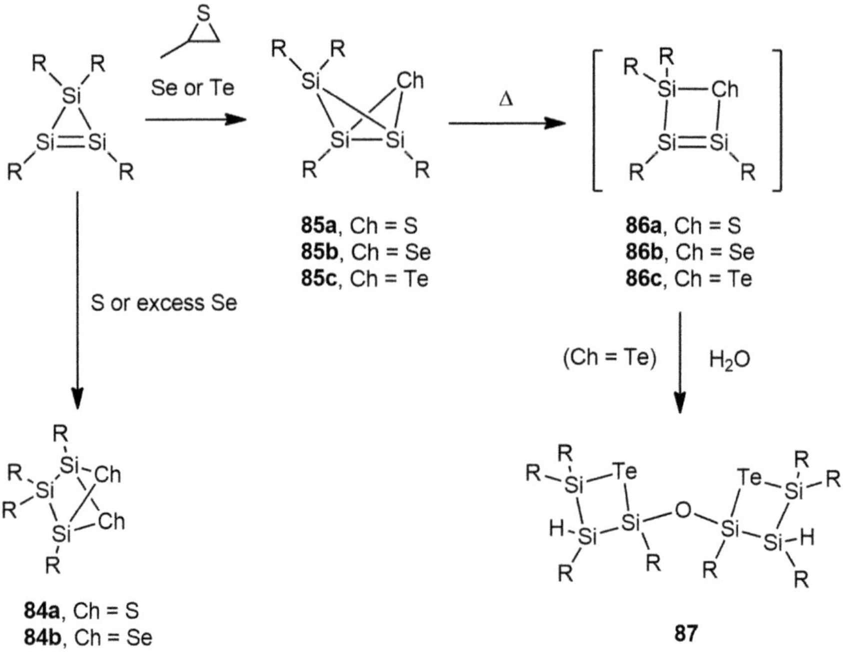

Scheme 5.24 Synthesis of Si/Ch heterocycles from bulky cyclotrisilenes (R = 2,4,6-(^{i}Pr)$_2$C$_6$H$_3$).

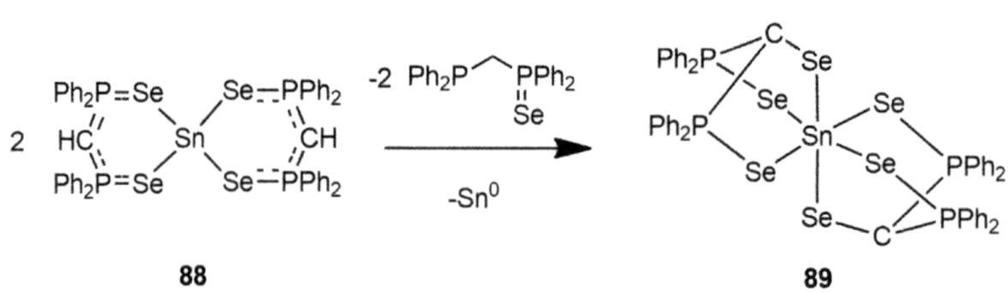

Scheme 5.25 Bicyclic and quadricyclic Sn/Ch compounds based on mono- and di-anionic PCP-type ligands.

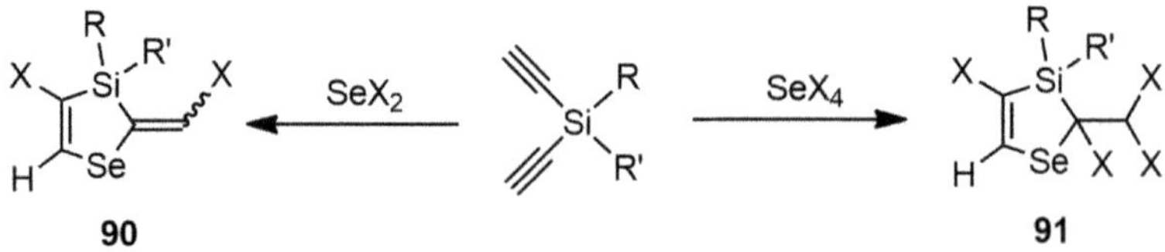

Scheme 5.26 Formation of a range of unsaturated Si/Se cyclopentanes from diethynylsilanes.

The equivalent sulfur and tellurium cycles **98a/99a** and **98c** required the derivatization and subsequent reaction with Me$_2$E(CH$_2$I)$_2$ followed by reduction. In the case of the Si/Ch heterocycles **98**, 1-electron oxidation of

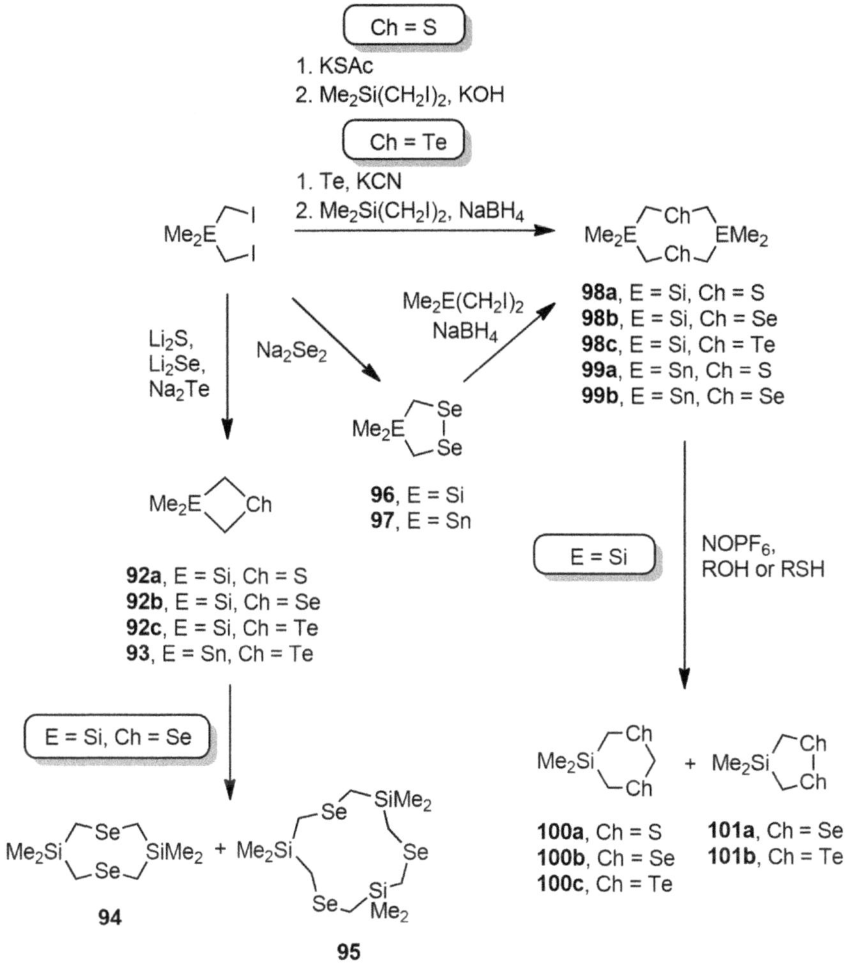

Scheme 5.27 Preparation and interplay of Si/Ch and Sn/Ch heterocycles as prepared by Block *et al.*

these 8-membered rings with $NOPF_6$ yielded the corresponding 5-membered and 6-membered rings **100** and **101** in various mixtures. The 6-membered rings **100** could also be prepared directly from **96** by reaction with diiodomethane and subsequent reduction with sodium borohydride. Similarly, introducing another $-Me_2Si$-functionality in the starting material, the 5-, 6- and 10-membered heterocycles **102–104** were obtained as depicted in Scheme 5.28.[74]

In a recent approach to cationic Si/Ch cycles, the Müller group made use of *peri*-substituted naphthalene or acenaphthene backbones, which are versatile ligands in the preparation of various E/Ch heterocycles, *vide infra*. Compounds of the type **105** were prepared by installing a dimethylsilane and an organochalcogen functionality in the *peri* positions and are forced in

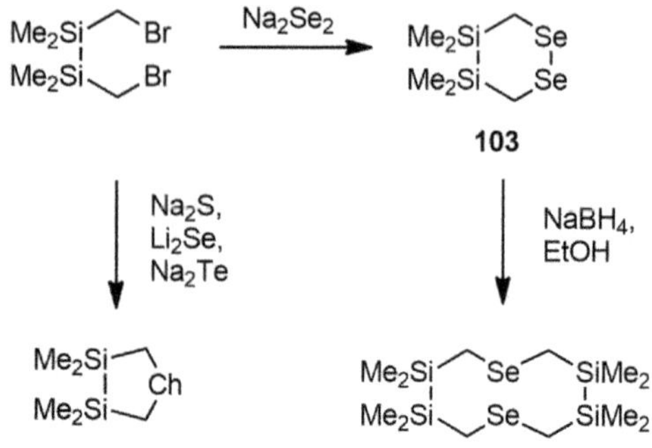

Scheme 5.28 Preparation and interplay of Si/Ch and Sn/Ch heterocycles as prepared by Block *et al.*

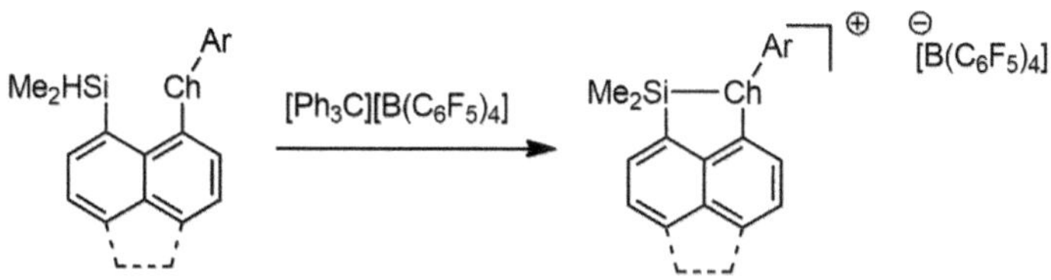

Scheme 5.29 Formation of cationic Si/Ch heterocycles by hydride abstraction from *peri*-substituted precursors (Ar = Ph, Mes).

close proximity. A hydride could be readily abstracted by addition of a trityl cation, forming a Si–Ch bond, thus leading to the formation of rigid 5-membered cycles for all the chalcogen elements (see Scheme 5.29).[75]

5.4 Group 15 Chalcogen Heterocycles

Prominent examples of group 15 chalcogen heterocycles include Lawesson's $((4\text{-}(MeO)C_6H_4)PS_2)_2$[76–79] and Woollins' reagent, $(PhPSe_2)_2$ (**106**),[80,81] consisting of 4-membered P_2S_2 and P_2Se_2 cores, respectively. The full synthetic potential of Woollins' reagent was revealed once a large-scale synthesis became available in 2005.[82] The synthesis was further improved by Hua and Woollins in 2021, illustrating the importance of this reagent.[83] The reactivity of Woollins' reagent stems from the monomeric form, $PhPSe_2$,[84,85] which could be experimentally isolated and spectroscopically analyzed in an argon matrix by Mardyukov and co-workers.[86]

The synthetic potential of Woollins' reagent include applications in inorganic and organic chemistry.[87] Schemes 5.30 and 5.31 illustrate selected

P/Se heterocycles **107** to **112**, which were synthesized from Woollins' reagent **106**. The reaction with the small molecule CS_2 leads to heterocycle **107** with the central P_2Se_2 core remaining intact and two CP_3Se heterocycles being attached at each of the P-sides (see Scheme 5.30).[88] The reaction with the dienes $H_2C{=}C(Bn){-}C(Bn){=}CH_2$ (Bn = benzyl) and $HC{=}C(Me)C_2H_4C(Me){=}CH$ led to the formation of P/Se and P/Se/P heterocycles **108** and **109**, respectively,[89] whereas heterocycles **110** and **111** containing a P/Se/Se fragment can be obtained *e.g.*, by the reaction with norbornene[90] or alkynes such as Ph–C≡C–Ph.[91] Interestingly, the heterocycle **112**, consisting of two fused C_3PSe units reminiscent of pentalene, was obtained by the reaction of Woollins' reagent with the diyne ^{t}Bu–C≡C–C≡C–^{t}Bu (see Scheme 5.31) and the molecular structure derived by single-crystal X-ray measurements are shown in Figure 5.3.[92] Other P/Se heterocycles with a C_2P_2Se core have been reported by Hey-Hawkins by using carboranes as the C_2-bridging ligand.[93]

P(III) analogues of Lawesson's and Woollins' reagents were reported by Ragogna and his team, namely $(TerPCh)_2$ (**113a**, Ch = S; **113b**, Ch = Se; Ter = 2,6-$Mes_2C_6H_3$), by the reaction of $TerPCl_2$ with $Ch(SiMe_3)_2$ (see Scheme 5.32). Notably, in case of **113a**, the monomeric form, P(S)Ter, could be base-stabilized by the use of the N-heterocyclic olefin $Im^iPr_2Me_2$ (1,3-di*iso*propyl-4,5-dimethylimidazol-2-ylidene) to form the adduct NHC → P(S)Ter. This synthetic route proved to be challenging in the case of selenium and in a different approach, reaction of **113b** with 2,3-dimethylbutadiene at elevated temperatures led to the formation of the 6-membered heterocycle **114**, which indicates the existence of a monomeric phosphinidene selenide, TerPSe, unit in the course of the reaction (see Scheme 5.32).[94]

Beside the 4-membered P/Se ring present in **106** and **113**, various $(RPCh)_n$ ring sizes and cages have been reported.[3,87,95] The chemistry evolved from sulfur over selenium[96] to the more recently investigated P/Te heterocycles. Notably, it was not until 2014, before the first structurally characterized P(III)/Te heterocycles were reported by Woollins and his team, although P/Te containing heterocycles have been investigated and analyzed before, *e.g.*, by Du Mont,[97,98] Karaghiosoff,[99,100] and Tokitoh.[101] Scheme 5.33 presents the 3-, 4-, 5-, and 6-membered P/Te heterocycles **115–118**, featuring a P_2Te (**115**), P_3Te (**116**), P_4Te (**117**), and P_3Te_3 (**118**) core, respectively, reported by Woollins in 2014.[102]

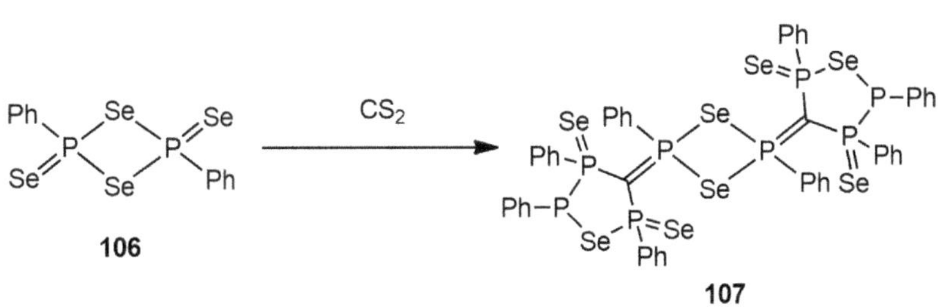

Scheme 5.30 P/Se heterocycle **107** prepared from Woollins' reagent with CS_2.

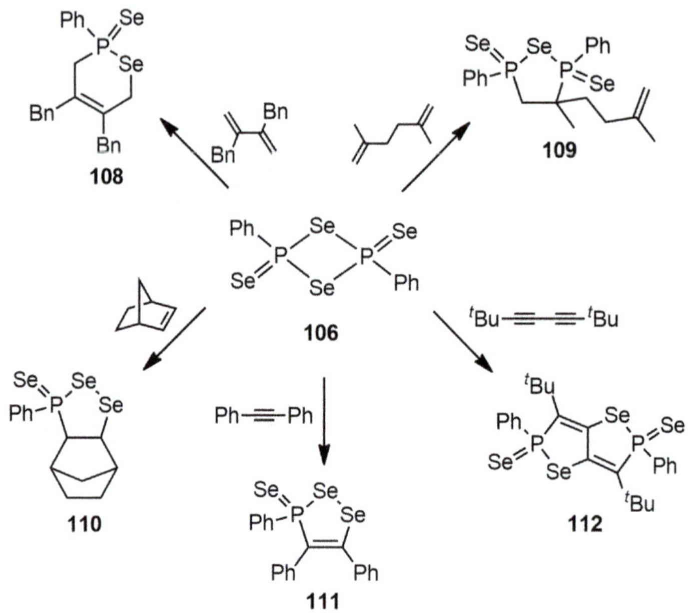

Scheme 5.31 Selected P/Se heterocycles **108** to **112** prepared from Woollins' reagent **106** with alkenes and alkynes.

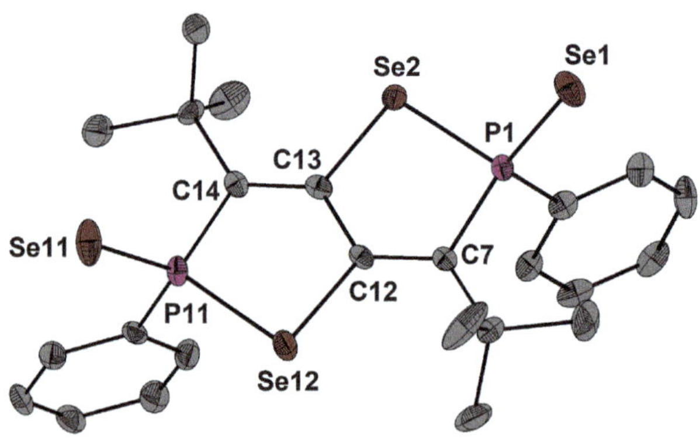

Figure 5.3 Molecular structure of **112** showing 50% thermal ellipsoids. H-atoms omitted for clarity.

Group 15 chalcogen heterocycles comprising a P_2NSe core have been reported by Weigand and co-workers by reacting the 8-membered $N_2P_4Se_2$ heterocycle **119** with elemental selenium leading to the 4-membered heterocycle **120** (see Scheme 5.34).[103] Another synthetic route to a P_2NSe heterocycle is the reaction of Woollins' reagent **106** with the Schiff base **121** leading to heterocycle **122**.[104]

Schulz and co-workers investigated the synthetic potential of phosphorous biradicaloids $[P(\mu\text{-NTer})]_2$ (Ter = 2,6-$Mes_2C_6H_3$) toward the activation of

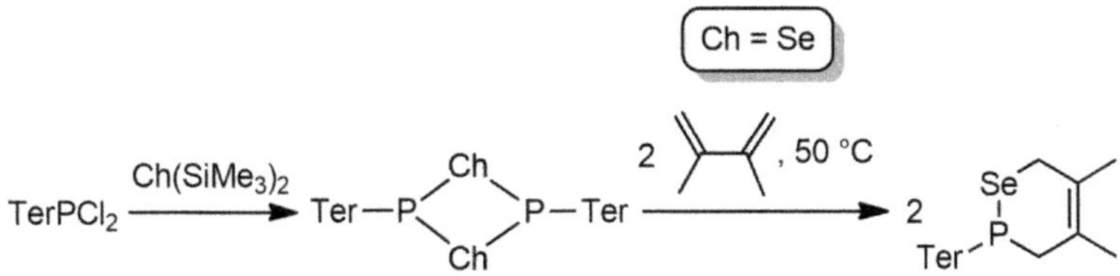

113a, Ch = S
113b, Ch = Se

114

Scheme 5.32 4-Membered (TerP(III)Ch)$_2$ heterocycles **113** and trapping reactions, Ter = 2,6-Mes$_2$C$_6$H$_3$.

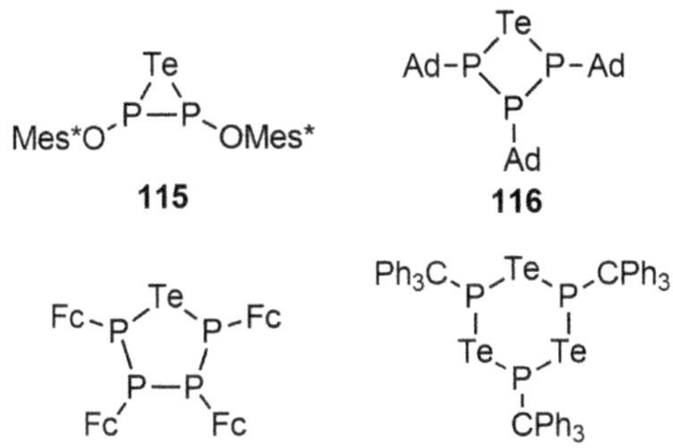

115 **116** **117** **118**

Scheme 5.33 3-, 4-, 5-, and 6-membered P/Te heterocycles **115** to **118**, Mes* = 2,4,6-^{t}Bu$_3$C$_6$H$_2$, Fc = ferrocenyl, Ad = adamantyl.

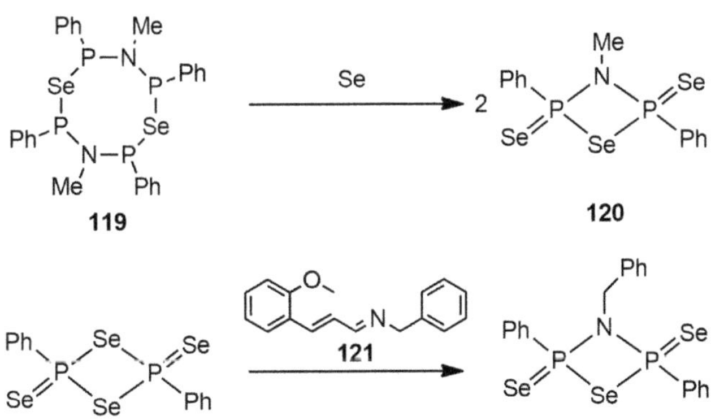

119 **120** **106** **122**

Scheme 5.34 P$_2$NSe heterocycles **119**, **120**, and **122** reported by Weigand and Woollins, respectively.

small molecules containing double and triple bonds and also reacted the biradicaloids with chalcogens leading to the bridging heterocycles **123** containing a P–Ch–Ch–P moiety (see Scheme 5.35). Interestingly, the lightest

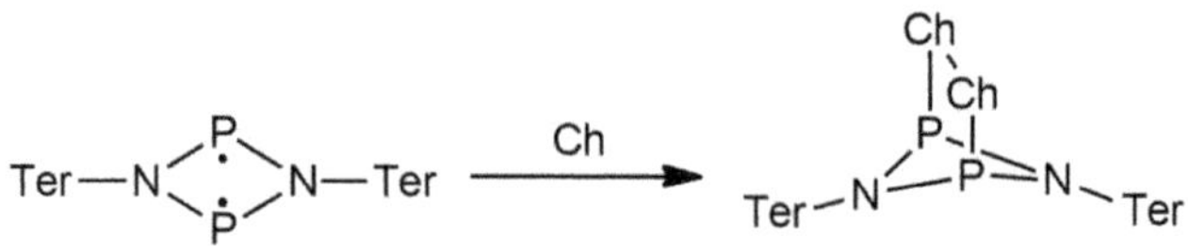

123a, Ch = S
123b, Ch = Se
123c, Ch = Te

Scheme 5.35 Selected Ch_2-bridging P_2N_2 heterocycles **123**, Ter = 2,6-$Mes_2C_6H_3$.

chalcogen congener oxygen, does not give the bridged compound, but bonding to only one P-atom occurs.[105]

Scheme 5.36 illustrates other P_2N_2 containing chalcogen heterocycles including compounds **124** to **127**, comprising bridging ChPCh (**124**),[102,106] Te_3 (**125**),[107] Se_4 (**126**),[108] and mixed $TeSe_nTe$ (**127**)[106] units of the P_2N_2 core.

Chalcogen(II) species have been investigated by Ragogna and his team and reported a versatile $ChCl_2$ source by base stabilizing the chalcogen dichlorides with bipy (bipy = 2,2′-bipyridine) forming $ChCl_2$·bipy compounds **128**.[109] These precursors do not only serve as the starting point for the generation of cationic N_2Ch heterocycles **129** (see Scheme 5.37), but can also be used as a Te transmetalation reagent for the generation of tellurophenes with interesting phosphorescent properties as shown by Rivard *et al.*[110] Other base stabilized chalcogen dichlorides include compounds **130**, which upon reaction with silver triflate give the dicationic species **131**.[111,112] Interestingly, ligand exchange of **131** with *e.g.*, bis(diphenylarsino)ethane gives compounds **132** comprising of As_2Ch heterocycles (see Scheme 5.37).[113] Analogue compounds of **132**, containing N_2Ch and P_2Ch rings have also been reported by Ragogna *et al.*[114] Notably, other N_2Ch-containing compounds include the compound classes of chalcogenadiazoles[115,116] and benzochalcogenadiazoles.[117]

A different approach to generate a p-block element containing chalcogen heterocycles is the use of a rigid organic backbone to force two elements into close proximity, see *e.g.*, Müller's Si/Ch (ace)naphthyl species **105**. The *peri*-substituted N/Te naphthyl tellurides **133** comprising a C_3NTe heterocycle have been reported as early as 1996 by Singh[118] and co-workers and later on by the group of Beckmann in 2011.[119] A natural progression of this chemistry involved the study of the heavier pnictogenes. Consequently, Beckmann and Woollins reported in three joint studies a series of *peri*-substituted P/Te acenaphthenes[120] including the P/Te species **134**,[121] the PETe compounds **135**,[122] as well as the tellurenyl triflates **136** (Pn = P) in addition to the heavier pnictogen analogues **137** (Pn = As) and **138** (Pn = Sb) (see Scheme 5.38).[120] More recently, ditellurides featuring a P/Te and P/O/Te core have been reported by Beckmann.[123] In addition, compounds comprising C_3NTe heterocycles have been realized by applying an *ortho* substituted imino donor leading to the tellurenyl cation [2-(^{t}BuNCH)C_6H_4Te][$CB_{11}H_{12}$], in which the Brønsted acidity of the carborane is dramatically increased.[124]

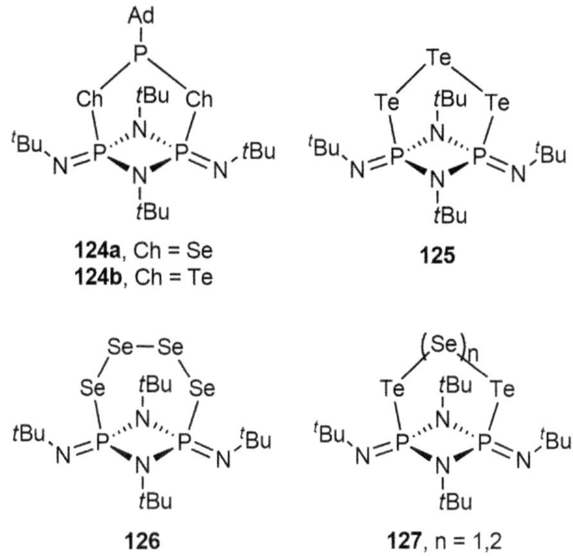

Scheme 5.36 Selected P_2N_2 selenium and tellurium heterocycles **124–127**, Ad = adamantyl.

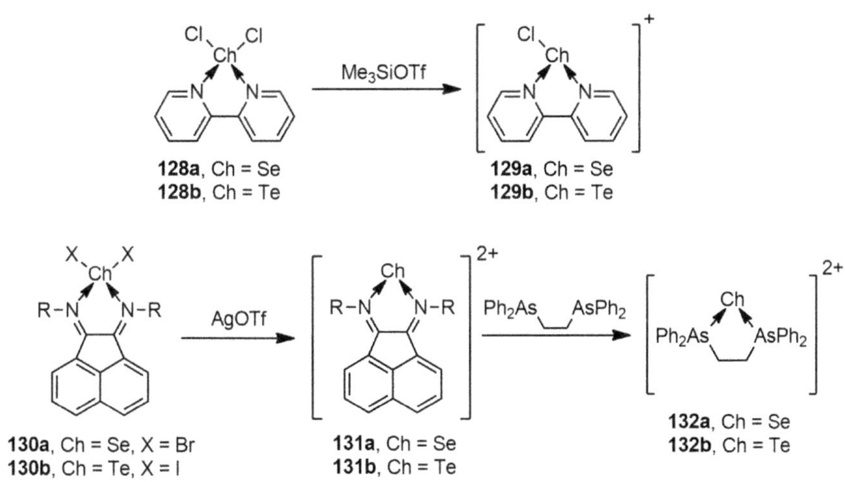

Scheme 5.37 Selected pnictogen containing selenium and tellurium heterocycles **128–132**, R = 2,6-$^iPr_2C_6H_3$.

Furthermore, Woollins and Ashbrook showed that *peri*-substituted 1,8-naphthodichalcogenoles serve as synthetic building blocks for Ch/P/Ch heterocycles 1,8-Nap[Ch(PR)Ch] (Ch = S, Se).[125]

In the course of their interest in dipnictogenes, Tokitoh and co-workers reported the formation of a 3-membered P_2Ch (Ch = S, Se)[126] heterocycle by the reaction of the respective diphosphene with elemental chalcogens. In 2005 Tokitoh *et al.* reported the heavier heterocycles Pn_2Te **139** (Pn = Sb) and

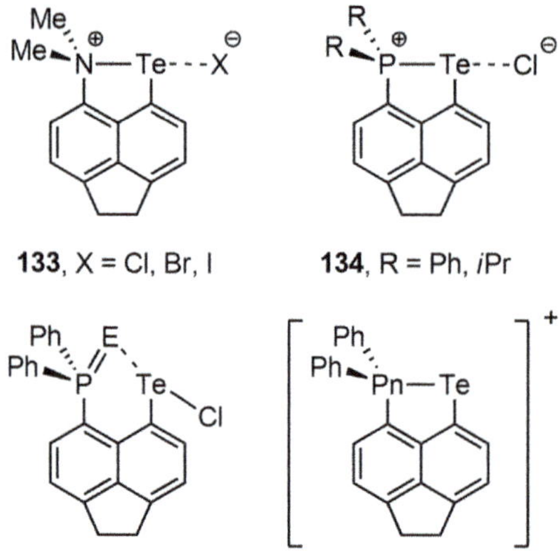

133, X = Cl, Br, I
134, R = Ph, *i*Pr
135a, E = O
135b, E = S
135c, E = Se
136, Pn = P
137, Pn = As
138, Pn = Sb

Scheme 5.38 Selected *peri*-substituted Pn/Te acenaphthenes **133** to **138**.

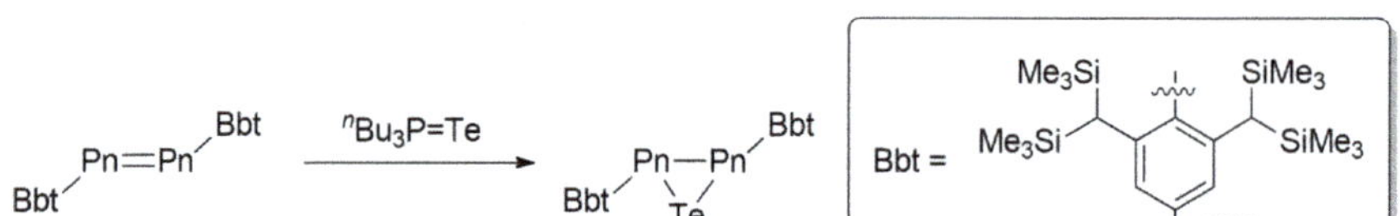

139, Pn = Sb
140, Pn = Bi

Scheme 5.39 3-Membered heavier Pn_2Te heterocycles **139** and **140**.

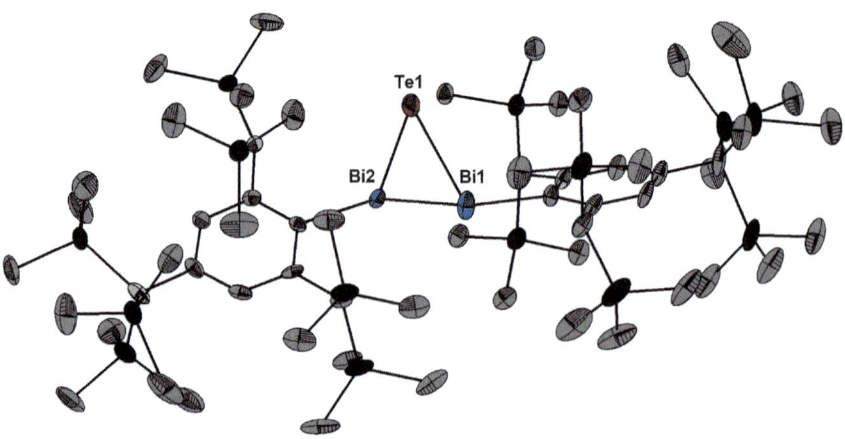

Figure 5.4 Molecular structure of **140** showing 50% thermal ellipsoids. H-atoms omitted for clarity.

140 (Pn = Bi), which were obtained by the reaction of the respective distibene and dibismuthene with the tellurium source $^{n}Bu_3P{=}Te$ (see Scheme 5.39).[127] Figure 5.4 displays a molecular structure derived by single-crystal X-ray measurements from the telluradibismirane **140**.

References

1. T. Kawashima, *Coord. Chem. Rev.*, 2003, **244**, 137.
2. T. Chivers and I. Manners, *Inorganic Rings and Polymers of the p-Block Elements: From Fundamentals to Applications*, Royal Society of Chemistry, 1st edn, 2009, vol. 1.
3. G. Hua and J. D. Woollins, *Angew. Chem., Int. Ed.*, 2009, **48**, 1368.
4. J. D. Woollins and R. S. Laitinen, *Selenium and Tellurium Chemistry: From Small Molecules to Biomolecules and Materials*, Springer, Berlin, Heidelberg, 2011.
5. F. A. Devillanova and W.-W. du Mont, *Handbook of Chalcogen Chemistry: New Perspective in Sulfurm Selenium and Tellurium*, Royal Society of Chemistry, 2nd edn, 2013, vol. 1.
6. G. He, O. Shynkaruk, M. W. Lui and E. Rivard, *Chem. Rev.*, 2014, **114**, 7815.
7. A. Nordheider, J. D. Woollins and T. Chivers, *Chem. Rev.*, 2015, **115**, 10378.
8. T. Chivers and R. S. Laitinen, *Dalton Trans.*, 2017, **46**, 1357.
9. R. S. Laitinen and R. Oilunkaniemi, *Selenium and Tellurium Reagents: In Chemistry and Materials Science*, De Gruyter, Berlin, Boston, 2019.
10. S. Schulz, H. W. Roesky, H. J. Koch, G. M. Sheldrick, D. Stalke and A. Kuhn, *Angew. Chem., Int. Ed. Engl.*, 1993, **105**, 1828.
11. S. Schulz, M. Andruh, T. Pape, T. Heinze, H. W. Roesky, L. Haeming, A. Kuhn and R. Herbst-Irmer, *Organometallics*, 1994, **13**, 4004.
12. M. G. Gardiner, C. L. Raston and V.-A. Tolhurst, *J. Chem. Soc., Chem. Commun.*, 1995, 2501.
13. D. Auerhammer, M. Arrowshmith, R. D. Dewhurst, T. Kupfer, J. Böhnke and H. Braunschweig, *Chem. Sci.*, 2018, **9**, 2252.
14. H. Braunschweig, P. Constantinidis, T. Dellermann, W. C. Ewing, I. Fischer, M. Hess, F. R. Knight, A. Rempel, C. Schneider, S. Ullrich, A. Vargas and J. D. Woollins, *Angew. Chem., Int. Ed.*, 2016, **55**, 5606.
15. S. R. Wang, M. Arrowsmith, J. Böhnke, H. Braunschweig, T. Dellermann, R. D. Dewhurst, H. Kelch, I. Krummenacher, J. D. Mattock, J. H. Müssig, T. Thiess, A. Vargas and J. Zhang, *Angew. Chem., Int. Ed.*, 2017, **56**, 8009.
16. S. Liu, M.-A. Légaré, A. Hofmann and H. Braunschweig, *J. Am. Chem. Soc.*, 2018, **140**, 11223.
17. S. Liu, M.-A. Légaré, A. Hofmann, A. Rempel, S. Hagspiel and H. Braunschweig, *Chem. Sci.*, 2019, **10**, 4662.
18. M. Dietz, M. Arrowsmith, A. Gärtner, K. Radacki, R. Bertermann and H. Braunschweig, *Chem. Commun.*, 2021, **57**, 13526.

19. G. K. H. Goh, Y. Li and R. Kinjo, *Dalton Trans.*, 2019, **48**, 7514.
20. F. A. Tsao, A. J. Lough and D. W. Stephan, *Chem. Commun.*, 2015, **51**, 4287.
21. F. A. Tsao and D. W. Stephan, *Chem. Commun.*, 2017, **53**, 6311.
22. F. A. Tsao, L. Cao, S. Grimme and D. W. Stephan, *J. Am. Chem. Soc.*, 2015, **137**, 13264.
23. P. A. Brown, C. D. Martin and K. L. Shuford, *Phys. Chem. Chem. Phys.*, 2019, **21**, 18458.
24. C. Cui, H. W. Roesky, M. Noltemeyer and H.-G. Schmidt, *Organometallics*, 1999, **18**, 5120.
25. K. S. Klimek, J. Prust, H. W. Roesky, M. Noltemeyer and H.-G. Schmidt, *Organometallics*, 2001, **20**, 2047.
26. V. Jancik, M. M. Moya Cabrera, H. W. Roesky, R. Herbst-Irmer, D. Neculai, A. M. Neculai, M. Noltemeyer and H.-G. Schmidt, *Eur. J. Inorg. Chem.*, 2004, 3508.
27. S. Gonzáles-Gallardo, A. S. Cruz-Zavala, V. Jancik, F. Cortés-Guzmán and M. Moya-Cabrera, *Inorg. Chem.*, 2013, **52**, 2793.
28. W. Zheng, N. C. Mösch-Zanetti, H. W. Roesky, M. Noltemeyer, M. Hewitt, H.-G. Schmidt and T. R. Schneider, *Angew. Chem., Int. Ed.*, 2000, **39**, 4276.
29. D. Franz, T. Szilvási, E. Irran and S. Inoue, *Nat. Commun.*, 2015, **6**, 10037.
30. M. D. Anker and M. P. Coles, *Angew. Chem., Int. Ed.*, 2019, **58**, 13452.
31. T. Řičica, L. Dostál, Z. Růžičková, L. Beneš, P. Němec, M. Bouška, J. M. Macak, P. Knotek, P. Ruleová and R. Jambor, *Chem. – Eur. J.*, 2018, **24**, 14470.
32. T. Řičica, Y. Milasheuskaya, Z. Růžičková, P. Němec, P. Švanda, Z. O. Zmrhalová, R. Jambor and M. Bouška, *Chem. – Asian J.*, 2019, **14**, 4229.
33. M. C. Copsey and T. Chivers, *Chem. Commun.*, 2005, 4938.
34. S. S. Garje, M. C. Copsey, M. Afzaal, P. O'Brien and T. Chivers, *J. Mater. Chem.*, 2006, **16**, 4542.
35. J. S. Ritch and T. Chivers, *Dalton Trans.*, 2010, **39**, 1745.
36. M. Risto, T. Chivers and J. Konu, *Dalton Trans.*, 2011, **40**, 8238.
37. H. Zhao and F. Gabbaï, *Nat. Chem.*, 2010, **2**, 984.
38. C. M. E. Graham, C. R. P. Millet, A. N. Price, J. Valjus, M. J. Cowley, H. M. Tuononen and P. J. Ragogna, *Chem. – Eur. J.*, 2018, **24**, 672.
39. T. Chivers and I. Manners, *Inorganic Rings and Polymers of the p-Block Elements*, RSC Publishing, Cambridge, United Kingdom, 2009, ch. 10, pp. 196–202.
40. P. Jutzi, A. Mohrke, A. Müller and H. Bögge, *Angew. Chem., Int. Ed.*, 1989, **28**, 1518.
41. R. West, D. J. De Young and K. J. Haller, *J. Am. Chem. Soc.*, 1985, **107**, 4942.
42. R. P.-K. Tan, G. R. Gillette, D. R. Powell and R. West, *Organometallics*, 1991, **10**, 546.

43. M. Haaf, A. Schmiedl, T. A. Schmedake, D. R. Powell, A. J. Millevolte, M. Denk and R. West, *J. Am. Chem. Soc.*, 1998, **120**, 12714.
44. M. Bouška, L. Dostál, Z. Padělková, A. Lyčka, S. Herres-Pawlis, K. Jurkschat and R. Jambor, *Angew. Chem., Int. Ed.*, 2012, **51**, 3478.
45. S. R. Foley, G. P. A. Yap and D. A. Richeson, *Organometallics*, 1999, **18**, 4700.
46. M. Saito, N. Tokitoh and R. Okazaki, *J. Am. Chem. Soc.*, 2004, **126**, 15572.
47. Z. Padělková, I. Císařová, M. S. Nechaev and A. Růžička, *J. Organomet. Chem.*, 2009, **694**, 2871.
48. A. Pop, L. Wang, V. Dorcet, T. Roisnel, J.-F. Carpentier, A. Silvestru and Y. Sarazin, *Dalton Trans.*, 2014, **43**, 16459.
49. I. Y. Ahmet, M. S. Hill, P. R. Raithby and A. L. Johnson, *Dalton Trans.*, 2018, **47**, 5031.
50. N. Weyer, R. Guthardt, B. A. Correia Bicho, J. Oetzel, C. Bruhn and U. Siemeling, *Z. Anorg. Allg. Chem.*, 2019, **645**, 188.
51. R. Guthardt, D. Bachmann, C. Bruhn and U. Siemeling, *Z. Anorg. Allg. Chem.*, 2020, **646**, 761.
52. S. M. Mansell, C. A. Russell and F. F. Wass, *Dalton Trans.*, 2015, **44**, 9756.
53. J. R. Thompson, I. Y. Ahmet, A. L Johnson and G. Kociok-Köhn, *Eur. J. Inorg. Chem.*, 2016, 4711.
54. M. K. Barman and S. Nembenna, *RSC Adv.*, 2016, **6**, 338.
55. N. Sen, S. Pal, V. V. Kade and S. Khan, *Eur. J. Inorg. Chem.*, 2019, 4450.
56. T. Chivers, D. J. Eisler and J. S. Ritch, *Z. Anorg. Allg. Chem.*, 2004, **630**, 1941.
57. D. J. Eisler and T. Chivers, *Chem. – Eur. J.*, 2006, **12**, 233.
58. S. Dehnen and M. Melullis, *Coord. Chem. Rev.*, 2007, **251**, 1259.
59. J. Heine and S. Dehnen, *Z. Anorg. Allg. Chem.*, 2012, **638**, 2425.
60. A. M. Polgar and J. F Korrigan, *Phys. Sci. Rev.*, 2019, **4**, 20170126.
61. C. Pöhlker, I. Schellenberg, R. Pöttgen and S. Dehnen, *Chem. Commun.*, 2010, **46**, 2605.
62. Z. You, J. Bergunde, B. Gerke, R. Pöttgen and S. Dehnen, *Inorg. Chem.*, 2014, **53**, 12512.
63. S. Heimann, G. Thiele and S. Dehnen, *J. Organomet. Chem.*, 2016, **821**, 214.
64. K. Hanau, N. Rinn, M. Argentari and S. Dehnen, *Chem. – Eur. J.*, 2018, **24**, 11711.
65. N. W. Rosemann, J. P. Eußner, A. Beyer, S. W. Koch, K. Volz, S. Dehnen and S. Chatterjee, *Science*, 2016, **352**, 1301.
66. T. Kunz, O. Şahin, C. Schrenk and A. Schnepf, *Dalton Trans.*, 2021, **50**, 2663.
67. K. Schwedtmann, A. Hepp, K. Schwedtmann, J. J. Weigand and F. Lips, *Eur. J. Inorg. Chem.*, 2019, 4719.
68. N. E. Poitiers, V. Huch, M. Zimmer and D. Scheschkewitz, *Chem. – Eur. J.*, 2020, **26**, 16599.

69. V. Y. Lee, S. Miyazaki, H. Yasuda and A. Sekiguchi, *J. Am. Chem. Soc.*, 2008, **130**, 2758.
70. H. Zhao, L. Klemmer, M. J. Cowley, V. Huch, M. Zimmer and D. Scheschkewitz, *Z. Anorg. Allg. Chem.*, 2018, **644**, 999.
71. J. Konu and T. Chivers, *Chem. Commun.*, 2010, **46**, 1431.
72. R. Thirumoorthi and T. Chivers, *Z. Anorg. Allg. Chem.*, 2014, **640**, 23.
73. S. V. Amosova, A. V. Martynov, N. A. Mahaeva, O. V. Belozerova, M. V. Penzik, A. I. Albanov, O. G. Yarosh and M. G. Voronkov, *J. Organomet. Chem.*, 2007, **692**, 946.
74. E. Block, E. V. Dikarev, R. S. Glass, J. Jin, B. Li, X. Li and S.-Z. Zhang, *J. Am. Chem. Soc.*, 2006, **128**, 14949.
75. N. Kordts, S. Kenzler, S. Rathjen, T. Sieling, H. Großekappenberg, M. Schmidtmann and T. Müller, *Chem. – Eur. J.*, 2017, **21**, 10068.
76. H. Z. Lecher, R. A. Greenwood, K. C. Whitehouse and T. H. Chao, *J. Am. Chem. Soc.*, 1956, **78**, 5018.
77. B. S. Pedersen, S. Scheibye, N. H. Nilsson and S.-O. Lawesson, *Bull. Soc. Chim. Belg.*, 1978, **87**, 223.
78. M. P. Cava and M. I. Levinson, *Tetrahedron*, 1985, **41**, 5061.
79. R. A. Cherkasov, G. A. Kutyrev and A. N. Pudovik, *Tetrahedron*, 1985, **41**, 2567.
80. P. T. Wood and J. D. Woollins, *J. Chem. Soc., Chem. Commun.*, 1988, 1190.
81. I. Baxter, A. F. Hill, J. M. Malget, A. J. P. White and D. J. Williams, *Chem. Commun.*, 1997, 2049.
82. I. P. Gray, P. Bhattacharyya, A. M. Z. Slawin and J. D. Woollins, *Chem. – Eur. J.*, 2005, **11**, 6221.
83. G. Hua and J. D. Woollins, *ACS Omega*, 2021, **6**, 32116.
84. G. Hua, Y. Li, A. M. Z. Slawin and J. D. Woollins, *Org. Lett.*, 2006, **8**, 5251.
85. L. Ascherl, A. Nordheider, K. S. Athukorala Arachchige, D. B. Cordes, K. Karaghiosoff, M. Bühl, A. M. Z. Slawin and J. D. Woollins, *Chem. Commun.*, 2014, **50**, 6214.
86. A. Mardyukov, F. Keul and P. R. Schreiner, *Eur. J. Org. Chem.*, 2019, 387.
87. G. Hua and J. D. Woollins, *Phys. Sci. Rev.*, 2019, **4**, 20170132.
88. J. C. Fitzmaurice, D. J. Williams, P. T. Wood and J. D. Woollins, *J. Chem. Soc., Chem. Commun.*, 1988, 741.
89. G. Hua, A. L. Fuller, A. M. Z. Slawin and J. D. Woollins, *Synlett*, 2012, **23**, 2453.
90. P. Bhattacharyya, A. M. Z. Slawin and J. D. Woollins, *J. Chem. Soc., Dalton Trans.*, 2001, 300.
91. G. Hua, Y. Li, A. M. Z. Slawin and J. D. Woollins, *Eur. J. Inorg. Chem.*, 2007, 891.
92. G. Hua, Y. Li, A. M. Z. Slawin and J. D. Woollins, *Chem. Commun.*, 2007, 1465.
93. P. Coburger, R. Aures, P. Schulz and E. Hey-Hawkins, *ChemPlusChem*, 2018, **83**, 1057.

94. C. M. E. Graham, T. E. Pritchard, P. D. Boyle, J. Valjus, H. M. Tuononen and P. J. Ragogna, *Angew. Chem., Int. Ed.*, 2017, **56**, 6236.
95. R. Davies and L. Patel, *Chalcogen-Phosphorus (and Heavier Congener) Compounds in Handbook of Chalcogen Chemistry: New Perspective in Sulfurm Selenium and Tellurium*, ed. F. A. Devillanova and W.-W. du Mont, Royal Society of Chemistry, 2nd edn, 2013, vol. 1, pp. 238–306.
96. P. Kilian, S. Parveen, A. L. Fuller, A. M. Z. Slawin and J. D. Woollins, *Dalton Trans.*, 2008, 1908.
97. W.-W. du Mont, T. Severengiz and B. Meyer, *Angew. Chem., Int. Ed.*, 1983, **22**, 983.
98. W.-W. du Mont and T. Severengiz, *Z. Anorg. Allg. Chem.*, 1993, **619**, 1083.
99. S. Grimm, K. Karaghiosoff, P. Mayer and D. Ross, *Phosphorus, Sulfur Silicon Relat. Elem.*, 2001, **169**, 51.
100. K. Lux, O. Schön and K. Karaghiosoff, *Z. Anorg. Allg. Chem.*, 2009, **635**, 2465.
101. N. Nagahora, T. Sasamori and N. Tokitoh, *Heteroat. Chem.*, 2008, **19**, 443.
102. A. Nordheider, T. Chivers, O. Schön, K. Karaghiosoff, K. S. Athukorala Arachchige, A. M. Z. Slawin and J. D. Woollins, *Chem. – Eur. J.*, 2014, **20**, 704.
103. S. Yogendra, F. Hennersdorf and J. J. Weigand, *Inorg. Chem.*, 2017, **56**, 8698.
104. G. Hua, D. B. Cordes, A. M. Z. Slawin and J. D. Woollins, *Eur. J. Inorg. Chem.*, 2019, 4682.
105. A. Hinz, R. Kuzora, U. Rosenthal, A. Schulz and A. Villinger, *Chem. – Eur. J.*, 2014, **20**, 14659.
106. A. Nordheider, K. Hüll, J. K. D. Prentis, K. S. Athukorala Arachchige, A. M. Z. Slawin, J. D. Woollins and T. Chivers, *Inorg. Chem.*, 2015, **54**, 3043.
107. A. Nordheider, T. Chivers, R. Thirumoorthi, K. S. Athukorala Arachchige, A. M. Z. Slawin, J. D. Woollins and I. Vargas-Baca, *Dalton Trans.*, 2013, **42**, 3291.
108. A. Nordheider, T. Chivers, R. Thirumoorthi, I. Vargas-Baca and J. D. Woollins, *Chem. Commun.*, 2012, **48**, 6346.
109. J. L. Dutton, G. J. Farrar, M. J. Sgro, T. L. Battista and P. J. Ragogna, *Chem. – Eur. J.*, 2009, **15**, 10263.
110. (a) W. Torres Delgado, C. A. Braun, M. P. Boone, O. Shynkaruk, Y. Qi, R. McDonald, M. J. Ferguson, P. Data, S. K. C. Almeida, I. de Aguiar, G. L. C. de Souza, A. Brown, G. He and E. Rivard, *ACS Appl. Mater. Interfaces*, 2018, **10**, 12124; (b) K. Takahashi, S. Shimo, E. Hupf, J. Ochiai, C. A. Braun, W. Torres Delgado, L. Xu, G. He, E. Rivard and N. Iwasawa, *Chem. – Eur. J.*, 2019, **25**, 8479; (c) C. A. Braun, N. Martinek, Y. Zhou, M. J. Ferguson and E. Rivard, *Dalton Trans.*, 2019, **48**, 10210; (d) E. Hupf, Y. Tsuchiya, W. Moffat, L. Xu, M. Hirai, Y. Zhou,

M. J. Ferguson, R. McDonald, T. Murai, G. He and E. Rivard, *Inorg. Chem.*, 2019, **58**, 13323; (e) B. T. Luppi, R. McDonald, M. J. Ferguson, L. Sang and E. Rivard, *Chem. Commun.*, 2019, **55**, 14218; (f) E. Rivard, *Chem. Rec.*, 2020, **20**, 640; (g) C. A. Braun, M. J. Ferguson and E. Rivard, *Inorg. Chem.*, 2021, **60**, 2672.

111. J. L. Dutton, H. M. Tuononen and P. J. Ragogna, *Angew. Chem., Int. Ed.*, 2009, **48**, 4409.
112. J. L. Dutton, T. L. Battista, M. J. Sgro and P. J. Ragogna, *Chem. Commun.*, 2010, **46**, 1041.
113. J. W. Dube, M. M. Hänninen, J. L. Dutton, H. M. Tuononen and P. J. Ragogna, *Inorg. Chem.*, 2012, **51**, 8897.
114. J. L. Dutton and P. J. Ragogna, *Chem. – Eur. J.*, 2010, **16**, 12454.
115. N. A. Semenov, A. V. Lonchakov, N. A. Pushkarevsky, E. A. Suturina, V. V. Korolev, E. Lork, V. G. Vasiliev, S. N. Konchenko, J. Beckmann, N. P. Gritsan and A. V. Zibarev, *Organometallics*, 2014, **33**, 4302.
116. N. A. Semenov, A. V. Lonchakov, N. P. Gritsan and A. V. Zibarev, *Russ. Chem. Bull.*, 2015, **64**, 499.
117. G. E. Garrett, G. L. Gibson, R. N. Straus, D. S. Seferos and M. S. Taylor, *J. Am. Chem. Soc.*, 2015, **137**, 4126.
118. S. C. Menon, H. B. Singh, J. M. Jasinski, J. P. Jasinski and R. J. Butcher, *Organometallics*, 1996, **15**, 1707.
119. J. Beckmann, J. Bolsinger and A. Duthie, *Chem. – Eur. J.*, 2011, **17**, 930.
120. A. Nordheider, E. Hupf, B. A. Chalmers, F. R. Knight, M. Bühl, S. Mebs, L. Chęcińska, E. Lork, P. Sanz Camacho, S. E. Ashbrook, K. S. Athukorala Arachchige, D. B. Cordes, A. M. Z. Slawin, J. Beckmann and J. D. Woollins, *Inorg. Chem.*, 2015, **54**, 2435.
121. T. G. Do, E. Hupf, A. Nordheider, E. Lork, A. M. Z. Slawin, S. G. Makarov, S. Y. Ketkov, S. Mebs, D. Woollins and J. Beckmann, *Organometallics*, 2015, **34**, 5341.
122. E. Hupf, T. G. Do, A. Nordheider, M. Wehrhahn, P. Sanz Camacho, S. E. Ashbrook, E. Lork, A. M. Z. Slawin, S. Mebs, J. D. Woollins and J. Beckmann, *Organometallics*, 2017, **36**, 1566.
123. T. G. Do, E. Lork and J. Beckmann, *Z. Anorg. Allg. Chem.*, 2018, **644**, 1190.
124. M. Hejda, D. Duvinage, E. Lork, A. Lyčka, Z. Černošek, J. Macháček, S. Makarov, S. Ketkov, S. Mebs, L. Dostál and J. Beckmann, *Chem. – Eur. J.*, 2021, **27**, 14577.
125. P. Sanz Camacho, M. W. Stanford, D. McKay, D. M. Dawson, K. S. Athukorala Arachchige, D. B. Cordes, A. M. Z. Slawin, J. D. Woollins and S. E. Ashbrook, *Chem. – Eur. J.*, 2018, **24**, 11067.
126. T. Sasamori, N. Takeda and N. Tokitoh, *J. Phys. Org. Chem.*, 2003, **16**, 450.
127. T. Sasamori, E. Mieda, N. Takeda and N. Tokitoh, *Angew. Chem., Int. Ed.*, 2005, **44**, 3717.

CHAPTER 6

Chalcogen-containing and Chalcogen-functionalized Heterocycles

FILIPE PENTEADO*[a] AND EDUARDO. E. ALBERTO*[b]

[a] Departamento de Química, Centro de Ciências Naturais e Exatas, Universidade Federal de Santa Maria - UFSM, 97105-340, Santa Maria, RS, Brazil; [b] Grupo de Síntese e Catálise Orgânica - GSCO, Departamento de Química, Universidade Federal de Minas Gerais - UFMG, 31.270-901, Belo Horizonte, MG, Brazil
*Emails: filipe.penteado@ufsm.br; albertoee@ufmg.br

6.1 Introduction

Heterocycles compose one of the most impressive classes of compounds, being widely found in an extensive range of organisms where they exhibit several biochemical functions.[1,2] Remarkable examples are: (1) purine- and pyrimidine-based nitrogenous bases, building blocks for the construction of the nucleic acids DNA and RNA; (2) the amino acids proline, histidine, and tryptophan which are building blocks of proteins. Additionally, histidine is the bioprecursor of serotonin, a neurotransmitter that regulates several mood functions in humans.[2–5] Heterocycles play a particularly important role in the context of world public health, and they are found in the core of a wide range of worldwide commercialized drugs. It is estimated that in nearly 70% of the drugs used for the treatment of several pathologies, at least one heterocyclic unit is found in their structure. They are also important in the world food supply chain, particularly considering that since the 90s, around

Chalcogen Chemistry: Fundamentals and Applications
Edited by Vito Lippolis, Claudio Santi, Eder J. Lenardão and Antonio L. Braga

Published by the Royal Society of Chemistry, www.rsc.org

70% of developed and commercialized agrochemicals are based on these compounds.[6] Additionally, heterocyclic compounds have a pivotal role in the material science industry, mainly due to their photoactive properties.[7]

Organochalcogen compounds are a widely known class of compounds, owing to their applicability in synthetic chemistry, playing an important role as building blocks to construct complex structures and as catalysts in several transformations.[8] The main example is the *syn*-elimination of selenoxides, which was disclosed in the 70s as an elegant and efficient strategy to access alkenes.[9] Furthermore, during the last few decades organochalcogen compounds have emerged as potent bioactive compounds, showing to be very efficient in the treatment of various pathologies.[10,11] In this context, the molecular hybridization of heterocyclic and organochalcogen compounds has become a trend in the prospection of bioactive molecules during recent years.

Accordingly, during recent years plenty of new synthetic technologies have emerged to construct chalcogen-decorated heterocycles, employing several ingenious strategies. Thus, in this chapter we aim to cover some of the recent important advances in the synthesis of chalcogen-functionalized and chalcogen-containing heterocycles, highlighting remarkable synthetic features.

6.2 Chalcogen-containing Heterocycles

Penicillin, which was isolated by Alexander Fleming in 1928,[12] and 2-phenylbenzo[*d*][1,2]selenazol-3(2*H*)-one, firstly prepared in 1924[13] and broadly known as Ebselen, are perhaps the most famous examples of chalcogen-containing heterocycles (Figure 6.1). Heterocycles embedded with heavier chalcogen atoms (S, Se, or Te) are recognized as important substances due to the myriad applications they exhibit in many science fields. They are commonly employed in medicinal chemistry owing to their cytotoxicity, pharmacology, and application as therapeutic agents[14–19] in material science

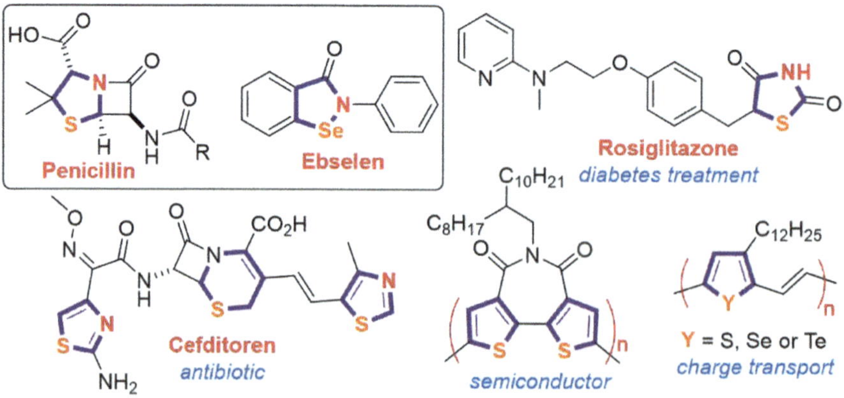

Figure 6.1 Illustrative examples of chalcogen-containing heterocycles.

to name a few, their applicability as semiconductors,[20] organic solar cells[21] and optoelectronic materials,[22–25] and in supramolecular chemistry.[26,27] These heterocycles can be found in many naturally-occurring compounds (especially sulfur-heterocycles) and in man-made substances. Figure 6.1 highlights the structures and applicability of some of these compounds.

Not surprisingly, the search to implement efficient methodologies to produce these heterocycles and their analogues has blossomed in recent decades.[28–39] Among a vast variety of naturally-occurring and synthetic heterocycle-containing chalcogens, the chalcogenophene family is recognized as one of the most prominent. The discovery of thiophene can be traced back to the nineteenth-century,[40] and shortly later, the preparations of thiophene[41] and selenophene[42] were independently described. In the early years the main strategy for their synthesis involved a one-pot intermolecular reaction using a suitable organic scaffold and an electrophilic chalcogen source (Scheme 6.1).

Nowadays this approach is still very attractive. For instance, a straightforward synthesis of thiophene derivatives has been described by the reaction of aryl methyl ketones with elemental sulfur catalyzed by imidazolium salts.[43] Benzo[*b*]tellurophenes **1** could be assembled by a sequential electrophilic telluration of alkenylmagnesium species with $TeCl_4$. It is noteworthy that a Mg-to-Zn transmetalation was required for product formation.[44] Reaction of diynes with elemental selenium or sulfur mediated by Ag was reported to synthesize cyclopenta-fused chalcogenophenes **2**.[45] Tellura(benzo)bithiophenes **3**, which exhibits orange-red phosphorescence at room temperature, was synthesized using diynes and Te(II) precursors.[46] Recently, the synthesis of highly functionalized thiophenes **4** and selenophenes **5** was described. The former was obtained by a modified version of the Gewald reaction[47] and the latter by the electrophilic cyclization of 1,3-diynes promoted by an electrophilic organoselenium species generated *in situ* from a mixture of BuSeSeBu and Oxone as the oxidant.[48]

Years after the synthesis of thiophene and selenophene, a practical manner for the preparation of tellurophene was reported. On occasion, the desired product could be prepared either by the one-pot intermolecular reaction using an electrophilic tellurium source, as discussed above, or by the nucleophilic attack of Li_2Te on a conjugated C–C π system.[49] This strategy

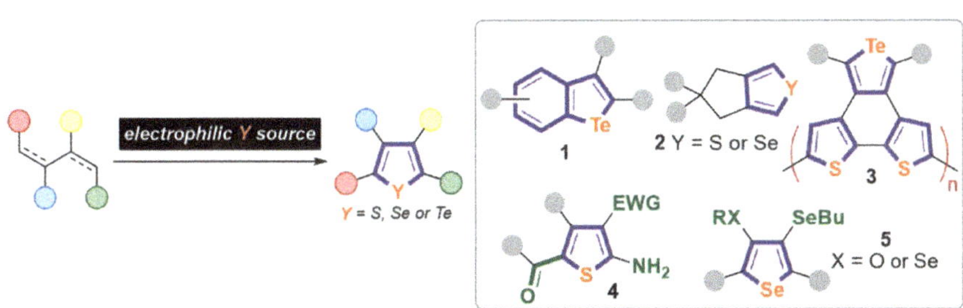

Scheme 6.1 Chalcogenophenes prepared using an electrophilic chalcogen species.

illustrates another classical prototype reaction for the preparation of chalcogenophenes. This approach works especially well for the addition of (Y^{2-}) species to conjugate C–C or C–heteroatom multiple bonds (Scheme 6.2).

Recently tellurophene derivatives **6** were prepared by the addition of a combination of elemental tellurium, $NaBH_4$, and the appropriate diyne. The thiol functionality (incorporated latter) enabled the formation of a monolayer of **6** on a gold electrode.[50] Likewise, di(selenophen)-diselenides **7** were synthesized from the *in situ* preparation of 1-aminobutadiynes, followed by the addition of a mixture of elemental Se and $NaBH_4$.[51] Alternative methods to produce (Y^{2-}) anions have also been described. For instance, the reaction between CS_2 and KOH delivers *in situ* a dithiocarbonate anion ($CS_3{}^{2-}$) which acts as a surrogate for the (S^{2-}) anion. Subsequent addition of 1,3-butadiynes led to the formation of 2,5-disubstituted thiophenes **8**.[52] Alternatively, under a superbasic condition (DMSO/KOH), elemental selenium undergoes an oxidation–reduction disproportionation giving the corresponding (Se^{2-}) anion. Upon treatment of this mixture with 1,3-butadiynes, selenophene derivatives **9** were promptly prepared.[53] Recently, a report on an expeditious synthesis of thiophene has been disclosed. It stems from the reaction between allylic alcohols, KBr, HBr, and DMSO under metal-free conditions. In a redox-divergent strategy, DMSO served as an oxidant and a nucleophilic sulfur source for the construction of thiophenes or bromothiophenes **10**. The product selectivity was conveniently manipulated by the appropriate DMSO/HBr ratio.[54]

Early in the 2000s an alternative method for the synthesis of furans[55] and thiophene analogues[56,57] using a pre-assembled organochalcogen substrate **A** was disclosed and rapidly became one of the most used strategies for the synthesis of chalcogenophenes and its derivatives. The simplified reaction mechanism, as outlined in Scheme 6.3, is initiated by the electrophilic activation of the C≡C triple bond. A well placed chalcogen atom then attacks the activated C–C π bond to form the cyclic intermediate **B**. At this stage, the allowed 5-*endo-dig* annulation step ensures the high regio- and stereoselectivity observed in the method. Finally, a Lewis base attacks the group attached to the chalcogenium salt on **B** to deliver the final product **C**.

The first examples of this methodology were based on the addition of halogen species as the electrophile source to induce the electrophilic cyclization.[58,59] This protocol was employed recently to construct highly

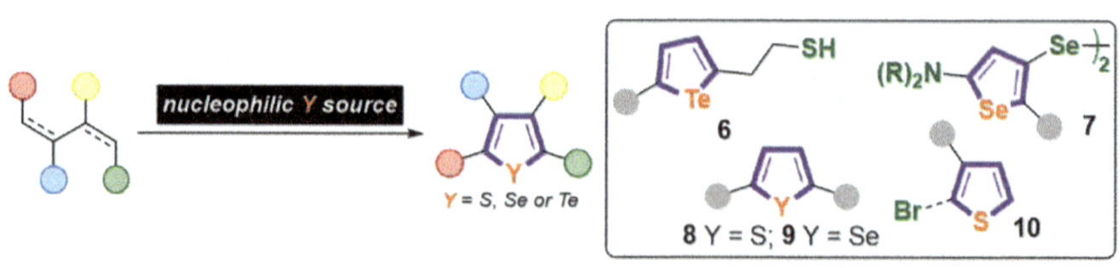

Scheme 6.2 Chalcogenophene motifs produced using nucleophilic chalcogen sources.

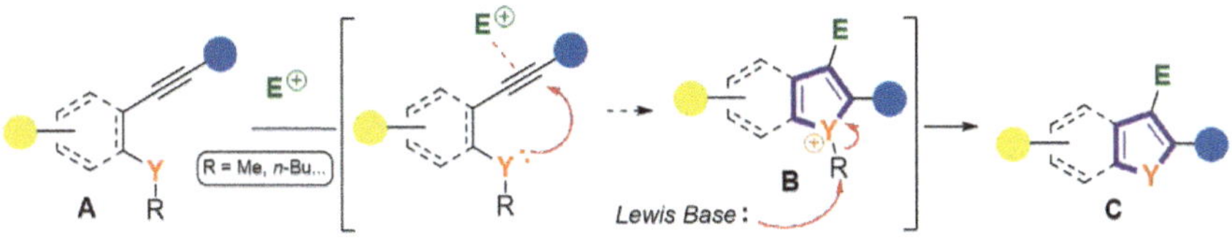

Scheme 6.3 The "Larock-type" synthesis of chalcogenophenes.

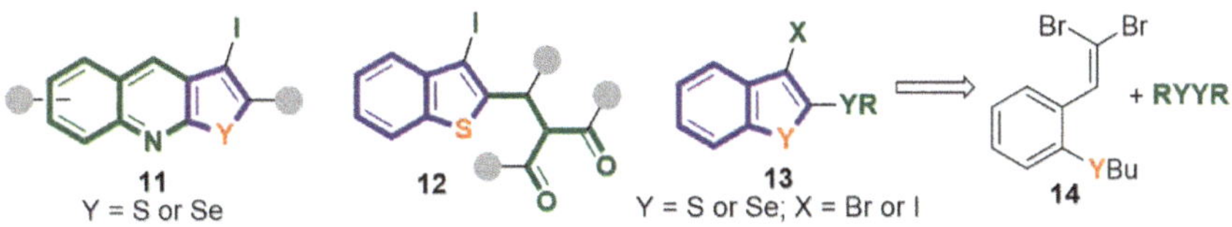

Figure 6.2 Examples of chalcogenophenes produced in a "Larock-type" fashion.

functionalized compounds such as 3-iodothieno- or 3-iodoselenopheno[2,3-*b*]quinoline derivatives **11**,[60] 3-iodobenzo[*b*]thiophene tethered to carbonyl compounds **12**,[61] and 3-halo-2-organochalcogenylbenzo[*b*]chalcogenophenes **13** (Figure 6.2).[62] It is worth mentioning that intermediate **A** used for the synthesis of **13** was prepared *in situ* from (2,2-dibromovinyl)-2-organochalcogenylbenzenes **14** and diorganyl dichalcogenides (RYYR).

One of the reasons for the electrophilic cyclization reaction becoming very popular is the fact that it allows the use of a broad range of electrophilic species (*e.g.*, halogens, transition metals, or chalcogen electrophiles) to smoothy induce the annulation process.[63] Some of the tailor-made products using variants of this methodology are depicted in Figure 6.3. For instance, 2-substituted benzo[*b*]thio- and benzo[*b*]selenophenes **15** can be achieved by the reaction of substrate **A** using Ag_2O as the catalyst.[64] Another interesting accomplishment was the preparation of borylated benzothiophenes **16** using BCl_3 as the electrophilic annulating agent.[65]

Furthermore, chalcogen-functionalized chalcogenophene derivatives can be prepared by the utilization of RYYR in combination with a catalytic amount of transition metals (*e.g.*, Pd, Ru, Cu, or Fe salts) to produce a "RY^+" species as the cyclization promoter. Synthesis of selenophene **17**,[66] chalcogenophene-fused to chromene **18**,[67] quinoxaline **19**,[68] or isochromene-based scaffolds **20**[69] exemplifies this protocol. Polyynes are also suitable substrates for this transformation, producing polyfused chalcogenophenes such as **21**[70] and the quinoline derivative **22**.[71] Noteworthy, for the synthesis of **21** MeYCl (Y = S, Se, or Te) was introduced as an ambiphilic reagent to trigger the polyelectrophilic cyclization. Another approach extensively explored to produce (RSe^+) electrophilic species is the combination of RSeSeR and an oxidizing agent such as Oxone. A broad range of products has been accomplished in this way.[72–75]

Conversely, numerous examples in the literature deal with the preparation of chalcogenophene and its congeners by cascade radical cyclization reactions (Figure 6.4). Recently, a protocol featuring the divergent synthesis of

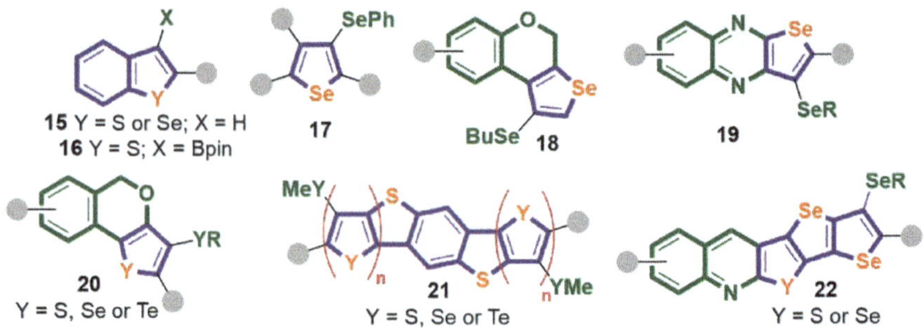

Figure 6.3 Diversification of products obtained by the electrophilic cyclization strategy.

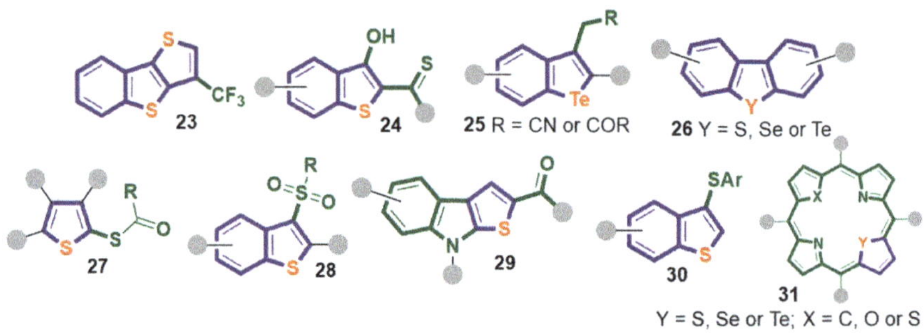

Figure 6.4 Featured chalcogenophene products synthesized by a cascade radical set-up.

polythiophenes (and dichalcogenides bearing thiophene) **23** from CF_3-containing 1,3-enynes, S_8, Cs_2CO_3, and Cu has been disclosed.[76] Multi-functionalized benzo[*b*]chalcogenophenes **24** and **25** were prepared by a radical cyclization strategy. The former from the reaction between 2-iodophenyl ketones and xanthate as the sulfur source catalyzed by Cu[77] and the latter by the C − H bond activation of ketones or acetonitrile with di-*tert*-butyl peroxide (DTBP) followed by reaction with an **A** substrate.[78] In the same manner dibenzo[*b*]tellurophenes **26** and π-conjugated materials were readily prepared by cascade radical cyclization steps from the corresponding diaryliodonium salts.[79] Using another approach, **26** derivatives were prepared from arylboronic acids with elemental Se or Te. The radical cyclization step necessary to produce them was achieved respectively by a silver-catalyzed reaction[80] or by using trimethylsilyl cyanide (TMSCN) as a catalyst.[81]

However, encouraged by the increasing demand from society for the development of new efficient and more environmentally benign processes, significant efforts have been devoted to finding alternative reaction conditions to synthesize these high-added value products. Under this concept, photochemical and electrochemical processes, among others, have been catching the attention of researchers for the preparation of chalcogenophenes and derivatives. As so, dibenzo[*b*]selenophenes **26** have been recently prepared *via* a mild, one-pot visible-light-mediated silver catalysis radical cyclization of aryl diazonium salts and KSeCN (Figure 6.5).[82] Similarly, a visible-light-induced [3 + 2] oxidative cyclization of various alkynes with ketene dithioacetals in the presence of a photosensitizer was employed to produce thiophene **27**.[83] On the other hand, the preparation of functionalized benzo[*b*]thiophene **28** was accomplished electrochemically from the sulfonylation/cyclization of 2-alkynylthioanisoles with sodium sulfinates.[84]

Finally, miscellaneous methods for the preparation or functionalization of chalcogenophenes will be briefly presented. A catalyst-free cascade deprotonation/intramolecular aldol reaction of α-carbonyl sulfonium ylides with 2-mercaptoindole-3-carbaldehydes was developed to smoothly deliver thienoindoles **29**.[85] A one-pot procedure for the synthesis of **30** was described upon a TfOH-promoted Friedel–Crafts cyclization of phenylthioacetic acid followed by the addition of a mixture of a thiophenol and a base.[86] Compounds bearing chalcogenophenes moieties such as porphyrin-like or tethered to macrocyclic boronic esters (not shown) **31** containing C, N, and Y donor atoms were synthesized under different conditions.[87–91] Collectively, these macrocycles can open new perspectives for coordination, hypervalent interactions, electronic properties, and supramolecular chemistry.

In another perspective, it is widely agreed that N, S-containing heterocycles constitute a major class of compounds, being broadly found in natural products. Not surprisingly, their preparation is extensively studied. Accordingly, recent reports on the synthesis of these compounds and their analogues will be succinctly disclosed here. Just as described for the synthesis of chalcogenophenes, the preparation of 1,3-thiazoles, isothiazoles (1,2-thiazoles), thiazolidines, or thiadiazoles derivatives will tentatively be categorized into three major mechanistic pictures: reaction of an organic scaffold (i) with an electrophilic "Y^+" species; (ii) with a nucleophilic "Y^-" species, or (iii) by cascade radical cyclization reactions.

For instance, the addition of an electrophilic chalcogen source was employed for the synthesis of 1,3-thiazole or benzo[*d*]thiazole **32** starting from elemental sulfur and phenylalanine derivatives (or by a mixture of anilines and arylacetic acids) using a mixture of NH_4I and K_3PO_4 as catalysts.[92] Alternatively, a redox cyclization strategy was employed to accomplish the synthesis of 2-substituted benzo[*d*]thiazoles **32** from 2,2′-dinitrodiphenyl disulfides and benzylamines under solvent and transition-metal free conditions.[93] Lewis acid-promoted activation of a mixture of acetophenones,

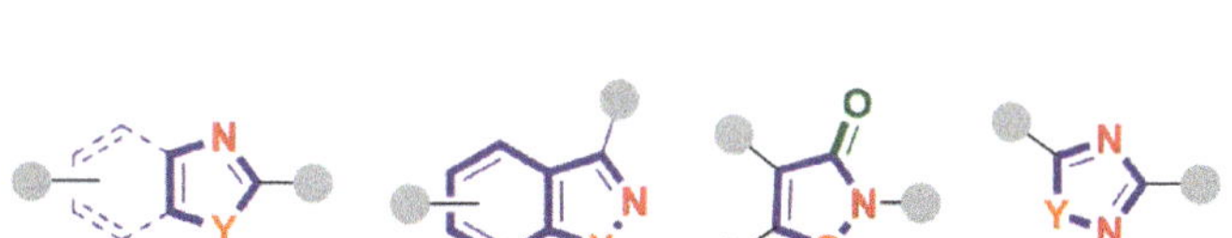

32 Y = S or Se **33** Y = S or Se **34** **35** Y = S or Se

Figure 6.5 N, Y-containing heterocycles prepared from an electrophilic Y source.

anilines, and SeO_2 was devised to deliver benzo[*d*]selenazoles **32**.[94] Formation of the benzo[*d*]isothiazole ring such as in **33** was possible due to a rhodium-catalyzed oxidative annulation of benzimidates with elemental sulfur.[95] Fully substituted isothiazol-3(2*H*)-ones (isothiazolones) **34** were prepared through Selectfluor-mediated intramolecular oxidative annulation of α-carbamoyl ketene dithioacetals under metal-free conditions in the presence of water.[96] 1,2,4-chalcogenodiazoles **35** were prepared from the corresponding benzoselenoamide or benzothioamide under aerobic conditions using organocatalysts[97] or by a NaI-promoted sequential double carbon–sulfur bond formation using imidazopyridines and bis(2-methoxyethyl) aminosulfur trifluoride (Deoxo-Fluor) as the sulfur source.[98]

Under the addition of a nucleophilic chalcogen species strategy to promote the synthesis of N, S-containing heterocycles, the preparation of benzo[*d*]thiazoles **32** was reported.[99] Similarly, a mixture of arylamines, S_8, and CO_2 was successfully employed to generate benzo[*d*]thiazol-2(3*H*)-one derivatives **36** under transition-metal-free reaction conditions (Figure 6.6).[100] Moreover, thiazole-2-thiones **37** were efficiently produced by a *tert*-butyl peroxybenzoate (TBPB)-promoted oxidative cascade cyclization from enaminones and S_8.[101] Using a different protocol, **37** was synthesized on a base-catalyzed three-component reaction between chalcones, isothiocyanates, and elemental sulfur.[102]

A carbonylative cyclization procedure of propargylic amines with elemental Se, an organic base and benzene-1,3,5-triyl triformate (TFBen) as the CO source was devised to produce 1,3-selenazolidin-2-ones **38**, as shown in Figure 6.7.[103,104] The parent 1,3-selenazolidin-2-imine **39** on the other hand, was stereospecifically obtained from an Al-catalyzed tandem C–N and C–Se bond formation using isoselenocyanates and chiral aziridines.[105] Isoselenocyanates were also used as the selenium source for the synthesis of 2-imino-1,3,4-selenadiazoles **40** *via* tributylphosphine-mediated annulation of *N*-aroyldiazenes.[106] A modular synthesis of highly functionalized 1,3-thiazolidin-4-ones **41** was described by a formal [3 + 2] cycloaddition of α-enolic dithioesters and α-halohydroxamates.[107] Demethoxylative chalcogenoannulation of alkynyl oxime ethers employing *in situ* prepared "Y^-" anions was reported for the preparation of isoseleneazole **42**[108] and bisisothiazole-4-disulfides **43**.[109]

Recent reports have appeared based on the synthesis N, S-containing heterocycles *via* radical-mediated reactions. A mixture of elemental sulfur and Cs_2CO_3 in DMSO produced a sulfur-centred radical which readily reacted with an imidazole-flanked cyclic iodonium substrate to deliver

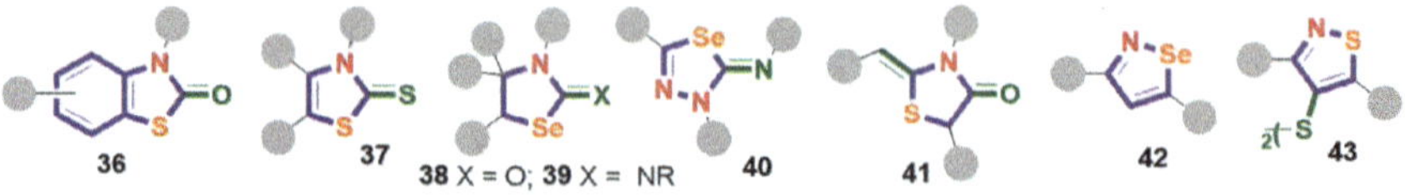

Figure 6.6 Illustrative chalcogen heterocycles assembled from nucleophilic Y sources.

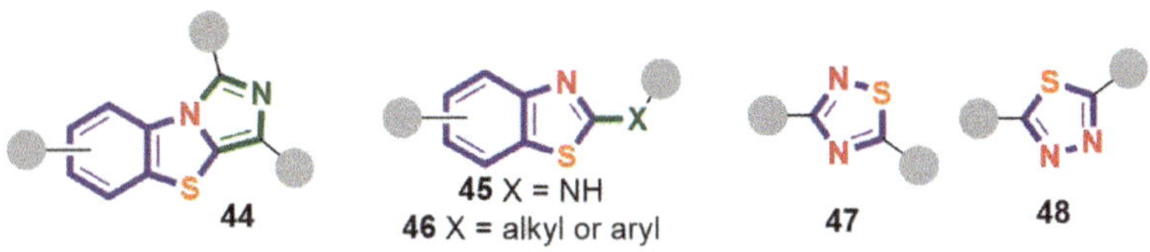

Figure 6.7 N, S-containing heterocycles recently prepared *via* radical-mediated reactions.

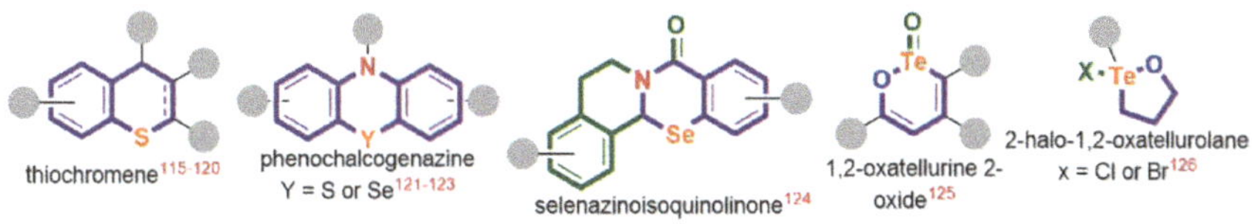

Figure 6.8 Structures of other chalcogen-based heterocycles recently synthesized.

benzo[*d*]imidazo[5,1-*b*]thiazoles **44.**[110] An expression-based *Vitreoscilla* haemoglobin (VHb) protein featuring a cobalt porphyrin active site, successfully promoted the conversion of 2-aminophenols and phenylisothiocyanate to the corresponding 2-aminobenzo[*d*]thiazoles **45** in water and air as the sole oxidant.[111] 2-aryl (alkyl) benzothiazoles **46** were conveniently synthesized upon a photocatalyst and transition-metal-free visible-light-induced cyclization of *ortho*-halothiobenzanilides assisted by Na_3PO_4 in DMSO.[112] Additionally, thioamides or a mixture of isothiocyanates and hydrazones were smoothly converted to 1,2,4-thiadiazoles **47** or 1,3,4-thiadiazoles **48.** While the former protocol was based on an aerobic visible-light induced intermolecular S–N bond construction without the addition of photosensitizer, metal, or base,[113] the latter relied on the electrochemical transformation of the starting material catalyzed by 2,3-dichloro-5,6-dicyano-1,4-benzoquinone (DDQ).[114]

Due to the countless possibilities to construct heterocycle-containing chalcogens and the interest that these compounds attract due to their outstanding properties, it is impossible to summarize all the aspects of their preparation in a single document. Accordingly, Figure 6.8 features the structures of miscellaneous chalcogen-based heterocycles in which their preparation has been reported throughout the current year. The details on the design, synthesis, and mechanistic aspects will not be covered here.[115–126]

6.3 Chalcogen-functionalized Heterocycles

The recent literature has been demonstrating the *in vitro* and *in vivo* biological potential of several chalcogen-functionalized heterocycle (CFH) analogues, against diverse types of pathologies. Some important examples can be highlighted to summarize this regard: (1) 3-sulfenylindole **49**,[127,128] which showed potent anticancer activity inhibiting tubulin polymerization in breast cancer cells, (2) analogue **50**, which inhibited the COX-2, with a similar efficiency in comparison to Etoricoxib.[129] Additionally, (3) 4-selanylquinoline **51**,[130] which showed a superior anti-inflammatory activity when compared to that of the worldwide marketed drug Meloxicam and (4) 3-selanylindole **52**,[131–133] which has been exhaustively investigated for mood and neurodegenerative diseases (Figure 6.9).

Based on the biological importance that CFH has been presenting over the last few decades, ingenious synthetic technologies to construct these compounds have been constantly emerging, establishing new horizons, by employing new oxidant species and alternative energy sources, including light, electricity, and ultrasound irradiation. In addition, advances have also been reached employing bench-stable chalcogen-based reagents. Thus, aiming to introduce the audience to some of the latest trends in organochalcogen chemistry to access CFH, a brief highlight will be delivered, covering some important recent advances.

During recent years, a couple of approaches to prepare CFH through electrophilic intramolecular annulation have emerged, promoted by the system Oxone®/RYYR (Y = Se and Te).[134–137] The innovation afforded by this method lies in the application of Oxone® as a green and environmentally safe oxidant agent, which is a very cheap, non-toxic, bench stable, and easy to handle reagent. The strategy core is based on the *in situ* formation of chalcogen-based electrophilic intermediates, under thermal or ultrasound-irradiation conditions (Figure 6.10).

New advances have also been reached employing halogen-based reagents, aiming to perform the *in situ* chalcogen – chalcogen bond oxidation. Two remarkable examples are the application of the system Selectfluor®/RSeSeR, generating *in situ* RSeF species (confirmed by ^{77}Se and ^{19}F NMR analysis), which were able to trigger the electrophilic annulation of alkynyl anisoles, as well as the direct selenylation of several heterocyclic compounds (Scheme 6.4).[138,139]

Additionally, important advances were also reached employing the system NFSI/RSeSeR to promote the annulation of 2-vinylanilines, affording 3-selanylindoles, through the *in situ* generation of electrophilic cationic species.[140] Recently, the electrophilic intramolecular annulation of 2-alkynyl benzoates, to construct chalcogen-functionalized isocoumarins was also reported. The protocol is based on the *in situ* formation of RYCl species, by a

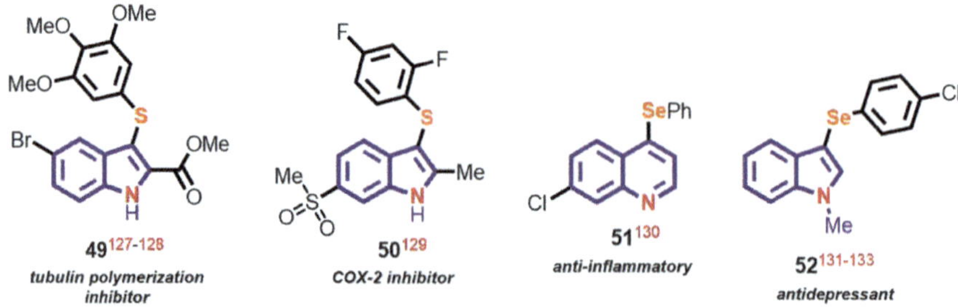

Figure 6.9 Bioactive chalcogen-functionalized heterocycles.

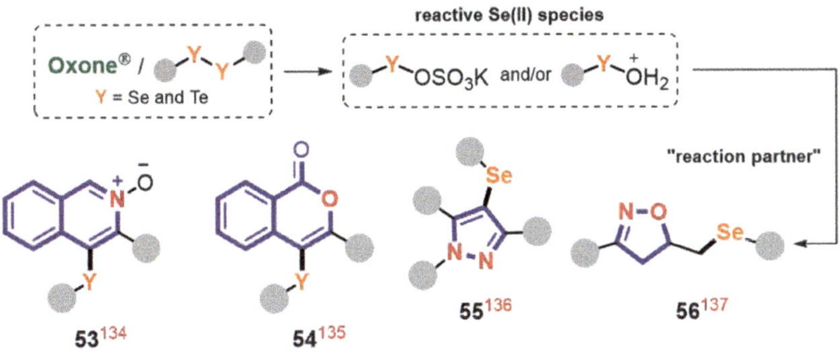

Figure 6.10 Oxone®/dichalcogenides-promoted synthesis of CFH.

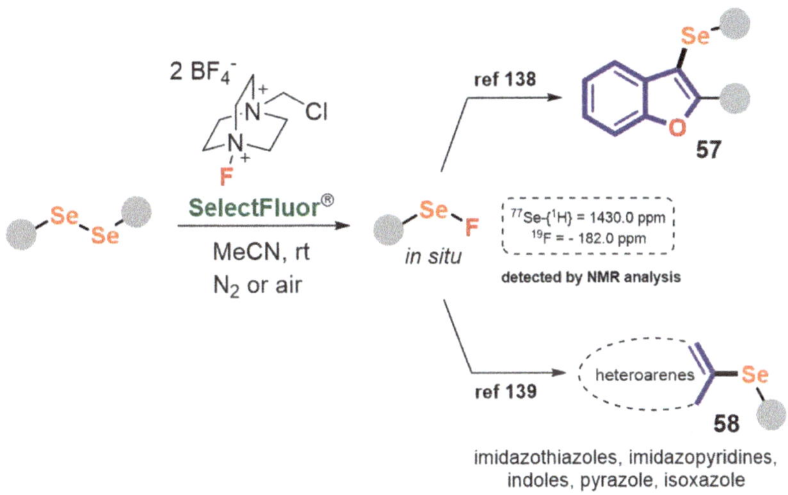

Scheme 6.4 RSeF-mediated synthesis of CHF.

mild reaction between diorganyl dichalcogenides and chlorine-containing hypervalent iodine species ($PhICl_2$) (Scheme 6.5).[141]

Sodium sulfinate (RSO_2Na) has been extensively employed as an odourless, bench-stable, and moisture-tolerant sulfonylating agent, emerging as an outstanding option to thiols in organic synthesis.[142] Inspired by sodium sulfinate (RSO_2Na) chemistry, recent advances have been reached employing sodium selenite ($RSeO_2Na$) and benzeneseleninic acids ($ArSeO_2H$), circumventing the use of air-moisture sensitive RSe-halogen species (RSeCl and RSeBr). In the presence of phosphorus-based reductant $(MeO)_2P(O)H$, sodium selenite is smoothly reduced to diorganyl diselenide, which in the presence of Cu(I) catalyst, is converted to a Se-based electrophilic species.[143] On the other hand, benzeneseleninic acids ($ArSeO_2H$) were able to be converted to 3-selanylindoles in the absence of catalyst or reductive species. ^{77}Se NMR studies revealed that, under thermal conditions, an initial conversion of benzeneseleninic acids ($ArSeO_2H$) to diselenide takes

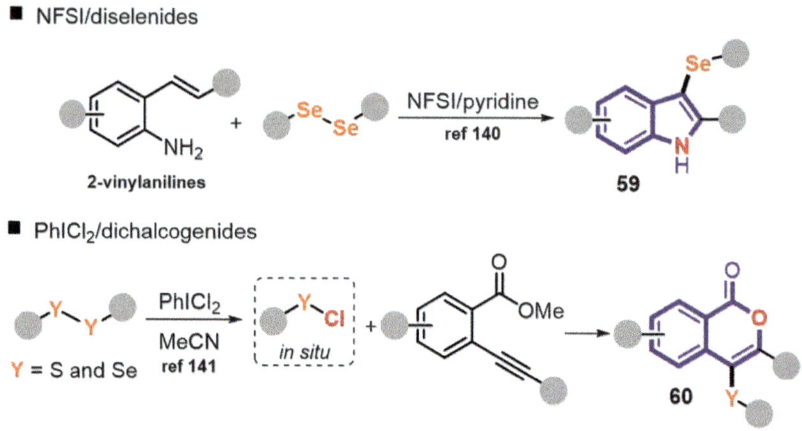

Scheme 6.5 Halogen-based reagents in the synthesis of CHF.

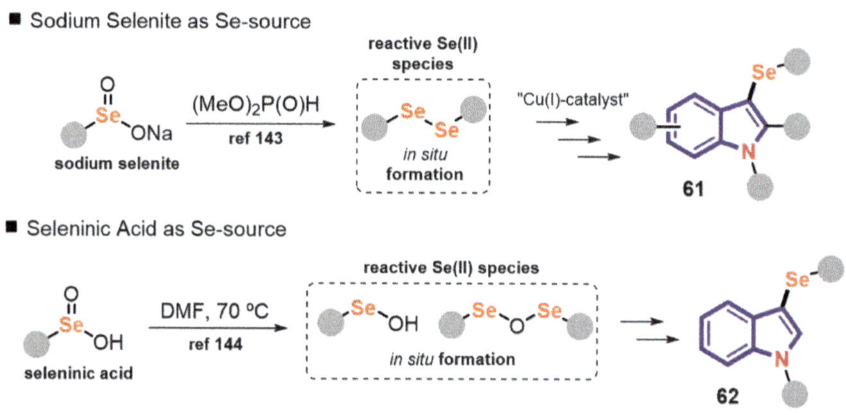

Scheme 6.6 Sodium selenite and seleninic acid as Se-source in the synthesis of CHF.

place, followed by a comproportionation process, to give the electrophilic Se(II) species (Scheme 6.6).[144]

Important advances were also delivered aiming to improve traditional catalytic models, increasing efficiency in terms of yield, reaction time, and selectivity. Impressive improvements were reached using Cu(I)- and I_2-catalysis employing SeO_2 as the co-oxidant, promoting the selective chalcogenylation of heteroarenes (indoles and imidazopyridines), under thermal and ultrasound-irradiation conditions.[145,146] Advances were also reached merging Cu(I)-catalysis and ultrasound irradiation as the energy source. In this context, a multicomponent process to afford mono- and di-substituted organylselanyl pyrroles, controlling the selectivity among the products using a catalytic amount of the Cu(I)-catalyst, was achieved.[147] Additionally, the duo Cu(I)/US was able to perform the selective mono- and bis-chalcogenylation of indolizines, on the positions C1 and C3, under mild reaction conditions.[148]

Furthermore, the system flavin/I_2 was shown to be efficient to give the C3-sulfenylation of indoles and C6-sulfenylation of quinolines, in the presence of thiols, providing good atom-efficiency and excellent functional compatibility.[149] Besides, the application of KIO_3 as a catalyst, in the presence of glycerol under thermal conditions, was shown to be an efficient technology to also promote the mild selenylation of heterocycles (Figure 6.11).[150]

Photochemical strategies have opened new horizons in organochalcogen chemistry. Among a wide range of recent advances, some remarkable examples can be highlighted. In this context, the dye-based photocatalyst-promoted SET process has been established as an elegant approach to construct CFH.[151–153] Besides, the Se – Se bond ability to absorb irradiation (blue and UV light) has been extensively explored to construct selenium-functionalized heterocycles, in the presence of O_2. The core of this green strategy is to trigger a radical addition pathway by the generation of selenium-centred radical species (Figure 6.12).[154–157]

Electrosynthesis has become an outstanding field in organic synthesis, allowing the construction of a diversity of compounds under mild reaction conditions, using electrolytic cells. In general, the main requirement for these strategies is the use of electrolytes in solution, to assist the electron flow from the anode to the cathode. Furthermore, an important aspect of the protocol is that the use of an external oxidant is not required. This is possible due to the control of the electric potential, which allows the manipulation of the electron rate transfer into the solution.[158]

In this context, mild protocols have been reported during the last years, describing the direct chalcogenylation of heteroarenes, including indoles and imidazopyridines. Such transformations could be conducted into undivided cells equipped with different electrodes pairs. In general, MeCN has been demonstrated to be a suitable solvent as the reaction medium, while different electrolytes can be satisfactorily employed (Figure 6.13).[159–164]

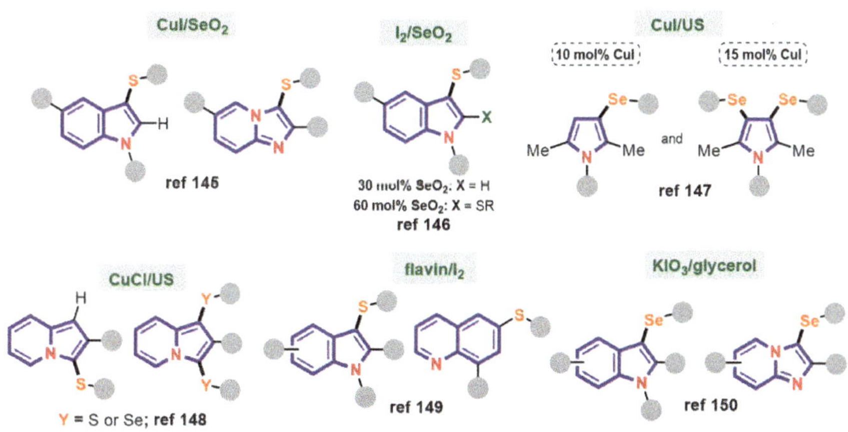

Figure 6.11 Recent advances in the traditional catalysis for the synthesis of CFH.

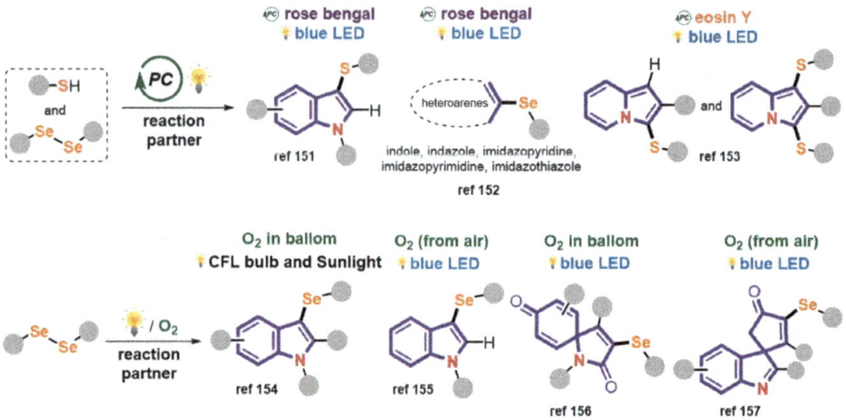

Figure 6.12 Light-meditated synthesis of CHF.

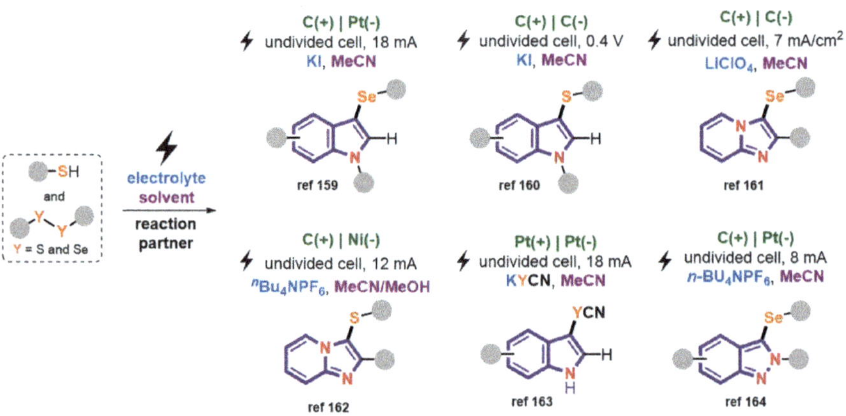

Figure 6.13 Electricity-mediated synthesis of CHF.

References

1. L. D. Quin and J. A. Tyrell, *Fundamentals of Heterocyclic Chemistry*, Wiley-VCH, Weinheim, 2010.
2. J. Alvarez-Builla, J. J. Vaquero and J. Barluenga, *Modern Heterocyclic Chemistry*, Wiley-VCH, Weinheim, 2010.
3. A. R. Katritzky and C. W. Rees, *Comprehensive Heterocyclic Chemistry*, Pergamon Press, Oxford, 1984.
4. A. R. Katritzky, C. W. Rees and E. F. V. Scriven, *Comprehensive Heterocyclic Chemistry II*, Pergamon Press, Oxford, 1996.
5. A. R. Katritsky, C. W. Rees and E. F. Scriven, *Comprehensive Heterocyclic Chemistry III*, Pergamon, Oxford, 1999.
6. R. Vardanyan and V. Hruby, *Synthesis of Best-Seller Drugs*, Academic Press, London, 2016.

7. D. Chen, S.-J. Su and Y. Cao, *J. Mater. Chem. C*, 2014, **2**, 9565.
8. E. J. Lenardão, C. Santi and L. Sancineto, *New Frontiers on Organoselenium Compounds*, Springer, Switzerland, 2018.
9. H. J. Reich, J. M. Renga and I. L. Reich, *J. Am. Chem. Soc.*, 1975, **97**, 5434.
10. C. W. Nogueira, G. Zeni and J. B. T. Rocha, *Chem. Rev.*, 2004, **104**, 6255.
11. D. de Souza, D. O. C. Mariano, F. Nedel, E. Schultze, V. F. Campos, F. Seixas, R. S. da Silva, T. S. Munchen, V. Ilha, L. Dornelles, A. L. Braga, A. B. T. Rocha, T. Collares and O. E. D. Rodrigues, *J. Med. Chem.*, 2015, **58**, 3329.
12. A. Fleming, *Br. J. Exp. Pathol.*, 1929, **10**, 226.
13. R. Lesser and R. Weiss, *Ber. Dtsch. Chem. Ges.*, 1924, **57**, 1077.
14. S. Pathania, R. K. Narang and R. K. Rawal, *Eur. J. Med. Chem.*, 2019, **180**, 486.
15. Archna, S. Pathania and P. A. Chawla, *Bioorg. Chem.*, 2020, **101**, 104026.
16. Z. Chen, H. Lai, L. Hou and T. Chen, *Chem. Commun.*, 2020, **56**, 179.
17. C. W. Nogueira, N. V. Barbosa and J. B. T. Rocha, *Arch. Toxicol.*, 2021, **95**, 1179.
18. P. Bhutani, G. Joshi, N. Raja, N. Bachhav, P. K. Rajanna, H. Bhutani, A. T. Paul and R. Kumar, *J. Med. Chem.*, 2021, **64**, 2339.
19. K. Laxmikeshav, P. Kumari and N. Shankaraiah, *Med. Res. Rev.*, 2021, 1.
20. K. Feng, H. Guo, H. Sun and X. Guo, *Acc. Chem. Res.*, 2021, **54**, 3804.
21. L. Yang, W. Gu, L. Lv, Y. Chen, Y. Yang, P. Ye, J. Wu, L. Hong, A. Peng and H. Huang, *Angew. Chem., Int. Ed.*, 2018, **57**, 1096.
22. S. M. Parke, M. P. Boone and E. Rivard, *Chem. Commun.*, 2016, **52**, 9485.
23. X. Wu, L. Lv, L. Hu, Q. Shi, A. Peng and H. Huang, *Chem. Phys. Chem.*, 2019, **20**, 2600.
24. P. Yadava and A. Patra, *Polym. Chem.*, 2020, **11**, 7275.
25. B. Fan, F. Lin, X. Wu, Z. Zhu and A. K.-Y. Jen, *Acc. Chem. Res.*, 2021, **54**, 3906.
26. S. Panda, A. Panda and S. S. Zade, *Coord. Chem. Rev.*, 2015, **300**, 86.
27. G. Haberhauer and R. Gleiter, *Angew. Chem., Int. Ed.*, 2020, **59**, 21236.
28. B. Godoi, R. F. Schumacher and G. Zeni, *Chem. Rev.*, 2011, **111**, 2937.
29. M. Elsherbini, W. S. Hamama and H. H. Zoorob, *Coord. Chem. Rev.*, 2017, **330**, 110.
30. B. Banerjee and M. Koketsu, *Coord. Chem. Rev.*, 2017, **339**, 104.
31. A. Alka, V. S. Shetti and M. Ravikanth, *Dalton Trans.*, 2019, **48**, 4444.
32. V. Jaiswal, B. Mondal and J. Saha, *Asian J. Org. Chem.*, 2020, **9**, 1466.
33. A. D. Sonawane, R. A. Sonawane, M. Ninomiya and M. Koketsu, *Adv. Synth. Catal.*, 2020, **362**, 3485.
34. F. Penteado, T. J. Peglow, M. S. Silva, G. Perin and E. J. Lenardão, *Curr. Opin. Green Sustain. Chem.*, 2020, **26**, 100372.
35. P. S. Hellwig, T. J. Peglow, F. Penteado, L. Bagnoli, G. Perin and E. J. Lenardão, *Molecules*, 2020, **25**, 5907.
36. G. Zeni, B. Godoi, C. K. Jurinic, A. L. Belladona and R. F. Schumacher, *Chem. Rec.*, 2021, **21**, 1342.
37. T. Chatterjee and B. C. Ranu, *J. Org. Chem.*, 2021, **86**, 13895.

38. C. K. Jurinic, A. L. Belladona, R. F. Schumacher and B. Godoi, *Synthesis*, 2021, **53**, 2545.
39. D. Sharma, V. Sharma, A. Sharma, R. Goyal, R. K. Tonk, V. K. Thakur and P. C. Sharma, *Sustainable Chem. Pharm*, 2021, **23**, 100496.
40. V. Meyer, *Ber. Dt. Chem. Ges.*, 1882, **15**, 2893.
41. V. Meyer and T. Sandmeyer, *Ber. Dt. Chem. Ges.*, 1883, **16**, 2176.
42. C. Paal, *Ber. Dtsch. Chem. Ges.*, 1885, **18**, 2255.
43. P. Gisbert and I. M. Pastor, *Eur. J. Org. Chem.*, 2020, 4319.
44. B. Wu, Melvina, X. Wu, E. K. L. Yeow and N. Yoshikai, *Chem. Sci.*, 2017, **8**, 4527.
45. Y. Wang, Y. Yuan, Z. Wang, Y. Gu, S. Fu, L. Kong and Y. Li, *Org. Lett.*, 2021, **23**, 5911.
46. C. A. Braun, M. J. Ferguson and E. Rivard, *Inorg. Chem.*, 2021, **60**, 2672.
47. T. B. Nguyen, D. H. Mac and P. Retailleau, *J. Org. Chem.*, 2021, **86**, 9418.
48. P. S. Hellwig, J. S. Guedes, A. M. Barcellos, G. Perin and E. J. Lenardão, *Molecules*, 2021, **26**, 2378.
49. E. H. Braye, W. Hübel and I. Caplier, *J. Am. Chem. Soc.*, 1961, **83**, 4406.
50. G. C. Hoover, J. Ham, C. Tang, E. I. Carrera and D. S. Seferos, *Can. J. Chem.*, 2018, **96**, 929.
51. P. Męcik, B. Pigulski and S. Szafert, *Org. Lett.*, 2021, **23**, 1066.
52. D. B. Paixão, D. S. Rampon, H. D. Salles, E. G. O. Soares, F. N. Bilheri, H. Paulo and Schneider, *J. Org. Chem.*, 2020, **85**, 12922.
53. H. D. Salles, F. L. Coelho, D. B. Paixão, C. A. Barboza, D. S. Rampon, F. S. Rodembusch and P. H. Schneider, *J. Org. Chem.*, 2021, **86**, 10140.
54. H. Liu, G.-C. He, C.-Y. Zhao, X.-X. Zhang, D.-W. Ji, Y.-C. Hu and Q.-A. Chen, *Angew. Chem., Int. Ed.*, 2021, **60**, 24284.
55. G. M. M. El-Taeb, A. B. Evans, S. Jones and D. W. Knight, *Tetrahedron Lett.*, 2001, **42**, 5945.
56. B. L. Flynn, G. P. Flynn, E. Hamel and M. K. Jung, *Bioorg. Med. Chem. Lett.*, 2001, **11**, 2341.
57. B. L. Flynn, P. Verdier-Pinard and E. Hamel, *Org. Lett.*, 2001, **3**, 651.
58. D. Yue and R. C. Larock, *J. Org. Chem.*, 2002, **67**, 1905.
59. K. O. Hessian and B. L. Flynn, *Org. Lett.*, 2003, **5**, 4377.
60. A. D. Sonawane, D. R. Garud, T. Udagawa and M. Koketsu, *Org. Biomol. Chem.*, 2018, **16**, 245.
61. C. Cunningham, M. Cloyd, A. Phillips, S. Khan, K. Whalen and T. Kesharwani, *Org. Biomol. Chem.*, 2021, **19**, 4107.
62. E. Q. Luz, G. L. Silvério, D. Seckler, D. B. Lima, F. S. Santana, R. V. Barbosa, C. R. M. D'Oca and D. S. Rampon, *Adv. Synth. Catal.*, 2021, **363**, 2610.
63. T. Kesharwani, S. A. Worlikar and R. C. Larock, *J. Org. Chem.*, 2006, **71**, 2307.
64. T. Cai, C. Feng, F. Shen, K. Bian, C. Wu, R. Shen and Y. Gao, *Eur. J. Org. Chem.*, 2021, 653.
65. A. J. Warner, A. Churn, J. S. McGough and M. J. Ingleson, *Angew. Chem., Int. Ed.*, 2017, **56**, 354.

66. K. K. Casola, M. R. Gomes, D. F. Back and G. Zeni, *J. Org. Chem.*, 2018, **83**, 6706.
67. G. Lutz, D. F. Back and G. Zeni, *Adv. Synth. Catal.*, 2020, **362**, 1096.
68. A. D. Sonawane, A. Shimozuma, T. Udagawa, M. Ninomiya and M. Koketsu, *Org. Biomol. Chem.*, 2020, **18**, 4063.
69. R. C. Pinheiro, D. F. Back and G. Zeni, *Adv. Synth. Catal.*, 2019, **361**, 1866.
70. A. S. Dillon and B. L. Flynn, *Org. Lett.*, 2020, **22**, 2987.
71. A. D. Sonawane, Y. Kubota and M. Koketsu, *J. Org. Chem.*, 2019, **84**, 8602.
72. H. A. Goulart, J. S. S. Neto, A. M. Barcellos, T. Barcellos, M. S. Silva, D. Alves, R. G. Jacob, E. J. Lenardão and G. Perin, *Adv. Synth. Catal.*, 2019, **361**, 3403.
73. G. Perin, L. K. Soares, P. S. Hellwig, M. S. Silva, J. S. S. Neto, J. A. Roehrs, T. Barcellos and E. J. Lenardão, *New J. Chem.*, 2019, **43**, 6323.
74. T. J. Peglow, R. H. Bartz, T. Barcellos, R. F. Schumacher, R. Cargnelutti and G. Perin, *Asian J. Org. Chem.*, 2021, **10**, 1198.
75. P. S. Hellwig, J. S. Guedes, A. M. Barcellos, R. G. Jacob, C. C. Silveira, E. J. Lenardão and G. Perin, *Org. Biomol. Chem.*, 2021, **19**, 596.
76. S. Jin, S.-J. Li, X. Ma, J. Su, H. Chen, Y. Lan and Q. Song, *Angew. Chem., Int. Ed.*, 2021, **60**, 881.
77. N. Sundaravelu, T. Singha, A. Nandy and G. Sekar, *Chem. Commun.*, 2021, **57**, 4512.
78. T. Cai, F. Shen, Y. Ni, H. Xu, R. Shen and Y. Gao, *J. Org. Chem.*, 2021, **86**, 1002.
79. M. Jiang, J. Guo, B. Liu, Q. Tan and B. Xu, *Org. Lett.*, 2019, **21**, 8328.
80. X. Zhang, X.-B. Huang, W.-X. Gao, Y.-B. Zhou, M.-C. Liu and H.-Y. Wu, *Adv. Synth. Catal.*, 2020, **362**, 5639.
81. X. Zhang, X.-B. Huang, Y.-B. Zhou, M.-C. Liu and H.-Y. Wu, *Chem. Eur. J.*, 2021, **27**, 944.
82. K. C. C. Aganda and A. Lee, *Adv. Synth. Catal.*, 2021, **363**, 5149.
83. B. Zheng, X. Li, Y. Song, S. Meng, Y. Li, Q. Liu and L. Pan, *Org. Lett.*, 2021, **23**, 3453.
84. M.-M. Zhang, Y. Sun, W.-W. Wang, K.-K. Chen, W.-C. Yang and L. Wang, *Org. Biomol. Chem.*, 2021, **19**, 3844.
85. L. Yang, S. Zhou, J.-Q Zhao, Y. You, Z.-H Wang, M.-Q. Zhou and W.-C. Yuan, *Org. Biomol. Chem.*, 2021, **19**, 3678.
86. K. Mitsudo, N. Habara, Y. Kobashi, Y. Kurimoto, H. Mandai and S. Suga, *Synlett*, 2020, **31**, 1947.
87. S. Ahmad, A. Singhal, K. Nisa and S. M. S. Chauhan, *Inorg. Chem.*, 2018, **57**, 11333.
88. S. Kumar, W.-Z. Lee and M. Ravikanth, *Org. Lett.*, 2018, **20**, 636.
89. K. Takahashi, S. Shimo, E. Hupf, J. Ochiai, C. A. Braun, W. T. Delgado, L. Xu, G. He, E. Rivard and N. Iwasawa, *Chem. – Eur. J.*, 2019, **25**, 8479.
90. T. D. Lash, D. I. AbuSalim and G. M. Ferrence, *Inorg. Chem.*, 2021, **60**, 9833.

91. A. Sinha and M. Ravikanth, *J. Org. Chem.*, 2021, **86**, 6100.
92. Y. Xia, H. Huang, W. Hu and G.-J. Deng, *Org. Biomol. Chem.*, 2021, **19**, 5108.
93. M. Teramoto, M. Imoto, M. Takeda, T. Mizuno, A. Nomoto and A. Ogawa, *Synthesis*, 2021, **53**, 2485.
94. M. Hu and Y. Ren, *Org. Biomol. Chem.*, 2021, **19**, 6692.
95. S. Moon, Y. Nishii and M. Miura, *Org. Lett.*, 2021, **23**, 49.
96. Z. Liu, Y. Wang, J. Huo, X.-J Li, S. Li and X. Song, *J. Org. Chem.*, 2021, **86**, 5506.
97. V. P. R. K. Putta, R. Gujjarappa, N. Vodnala, R. Gupta, P. P. Pujar and C. C. Malakar, *Tetrahedron Lett.*, 2018, **59**, 904.
98. S.-T. Ma, X.-X. Zhu, J.-Y. Xu, Y. Li, X.-M. Zhang, C.-T Feng and Y. Yan, *Chem. Commun.*, 2021, **57**, 5338.
99. X. Li, Y. Li, R. Liu, Z. Wang, X. Li and D. Shi, *Tetrahedron Lett.*, 2020, **61**, 152054.
100. C.-K. Ran, L. Song, Y.-N. Niu, M.-K. Wei, Z. Zhang, X.-Y. Zhou and D.-G. Yu, *Green Chem.*, 2021, **23**, 274.
101. B. Zhang, D. Liu, Y. Sun, Y. Zhang, J. Feng and F. Yu, *Org. Lett.*, 2021, **23**, 3076.
102. T. B. Nguyen and P. Retailleau, *Org. Lett.*, 2021, **23**, 5344.
103. H. Wang, J. Ying, M. Lai, X. Qi, J.-B Peng and X.-F. Wu, *Adv. Synth. Catal.*, 2018, **360**, 1693.
104. H. Wang, J. Ying, H.-J. Ai and X.-F Wu, *Eur. J. Org. Chem.*, 2019, 1553.
105. V. Satheesh, H. K. Srivastava, S. V. Kumar, M. Sengoden and T. Punniyamurthy, *Adv. Synth. Catal.*, 2019, **361**, 55.
106. Z. Huang, Q. Zhang, X. Yi, Z. Zhao, W. Yu and J. Chang, *Adv. Synth. Catal.*, 2021, **363**, 4894.
107. J. Sun, T. Wang, X. Zhu, W. Xu, B. Cheng and H. Zhai, *New J. Chem.*, 2021, **45**, 14508.
108. Z.-Z. Zhang, R. Chen, X.-H. Zhang and X.-G. Zhang, *J. Org. Chem.*, 2021, **86**, 632.
109. L. Li, Q. Chen, H.-H. Xu, X.-H. Zhang and X.-G. Zhang, *J. Org. Chem.*, 2020, **85**, 10083.
110. N. S. Antonkin, Y. A. Vlasenko, A. Yoshimura, V. I. Smirnov, T. N. Borodina, V. V. Zhdankin, M. S. Yusubov, A. Shafir and P. S. Postnikov, *J. Org. Chem.*, 2021, **86**, 7163.
111. Y. Xu, F. Li, N. Zhao, J. Su, C. Wang, C. Wang, Z. Li and L. Wang, *Green Chem.*, 2021, **23**, 8047.
112. H. Wang, Q. Wu, J.-D. Zhang, H.-Y. Li and H.-X. Li, *Org. Lett.*, 2021, **23**, 2078.
113. L. Zhuo, S. Xie, H. Wang and H. Zhu, *Eur. J. Org. Chem.*, 2021, 3398.
114. Z. Ma, X. Hu, Y. Li, D. Liang, Y. Dong, B. Wanga and W. Li, *Org. Chem. Front.*, 2021, **8**, 2208.
115. T. Shibata, A. Sekine, M. Akino and M. Ito, *Chem. Commun.*, 2021, **57**, 9048.
116. L. L. Rodrigues, A. S. Micallef, M. C. Pfrunder, V. X. Truong, J. C. McMurtrie, T. R. Dargaville, A. S. Goldmann, F. Feist and C. Barner-Kowollik, *J. Am. Chem. Soc.*, 2021, **143**, 7292.

117. N. Velasco, A. Suárez, F. Martínez-Lara, M. Á. Fernández-Rodríguez, R. Sanz and S. Suárez-Pantiga, *J. Org. Chem.*, 2021, **86**, 7078.
118. Y. Mo, X. Zhang, Y. Yao, C. Duan, L. Ye, Z. Shi, Z. Zhao and X. Li, *J. Org. Chem.*, 2021, **86**, 4448.
119. K. Ji, K. Lu, J. Huang, Z.-H. Li, H. Ke and Z.-M. Chen, *Org. Lett.*, 2021, **23**, 8028.
120. M. Shankar, A. Saha, S. Sau, A. Ghosh, V. Gandon and A. K. Sahoo, *Chem. Sci.*, 2021, **12**, 6393.
121. X. Rui, C. Wang, D. Si, X. Hui, K. Li, H. Wen, W. Li and J. Liu, *J. Org. Chem.*, 2021, **86**, 6622.
122. T. Matsuzawa, T. Hosoya and S. Yoshida, *Org. Lett.*, 2021, **23**, 2347.
123. C. Cremer, M. A. Eltester, H. Bourakhouadar, I. L. Atodiresei and F. W. Patureau, *Org. Lett.*, 2021, **23**, 3243.
124. X. Zhou, J. Hao, Y. Kong and R. Xu, *Chem. Commun.*, 2021, **57**, 5426.
125. J. P. A. Souza, L. R. A. Menezes, F. P. Garcia, D. B. Scariot, P. T. Bandeira, M. B. Bespalhok, S. O. K. Giese, D. L. Hughes, C. V. Nakamura, A. Barison, A. R. M. Oliveira, R. B. Campos and L. Piovan, *Chem. – Eur. J.*, 2021, **27**, 14427.
126. S. Wang, C. Yan, W. Zhao, X. Liu, C.-S. Yuan, H.-L. Zhang and X. Shao, *Chem. Sci.*, 2021, **12**, 5811.
127. G. De Martino, G. La Regina, A. Coluccia, M. C. Edler, M. C. Barbera, A. Brancale, E. Wilcox, E. Hamel, M. Artico and R. Silvestri, *J. Med. Chem.*, 2004, **47**, 6120.
128. G. De Martino, M. C. Edler, G. La Regina, A. Coluccia, M. C. Barbera, D. Barrow, R. I. Nicholson, G. Chiosis, A. Brancale, E. Hamel, M. Artico and R. Silvestri, *J. Med. Chem.*, 2006, **49**, 947.
129. J. A. Campbell, V. Bordunov, C. A. Broka, M. F. Browner, J. M. Kress, T. Mirzadegan, C. Ramesha, B. F. Sanpablo, R. Stabler, P. Tanakara, A. Villasenor, K. A. M. Walker, J.-H. Wang, M. Welch and P. Weller, *Bioorg. Med. Chem. Lett.*, 2004, **14**, 4741.
130. M. Pinz, A. S. Reis, V. Duarte, M. J. de Rocha, B. S. Goldani, D. Alves, L. Savegnago, C. Luchese and E. A. Wilhelm, *Eur. J. Pharma.*, 2016, **780**, 122.
131. A. M. Casaril, D. A. Lourenço, M. Domingues, T. A. Smaniotto, P. T. Birmann, B. Vieira, M. S. Sonego, F. K. Seixas, T. Collares, E. J. Lenardão and L. Savegnago, *Compr. Psychoneuroendocrinol.*, 2021, **6**, 1000054.
132. A. M. Casaril, M. Domingues, S. R. Bampi, D. A. Lourenço, T. A. Smaniotto, N. Segatto, B. Vieira, F. K. Seixas, T. Collares, E. J. Lenardão and L. Savegnago, *Brain, Behav., Immun.*, 2020, **84**, 229.
133. A. M. Casaril, M. Domingues, D. A. Lourenço, B. Vieira, K. Begnini, C. D. Corcini, R. T. França, A. S. V. Junior, F. K. Seixas, T. Collares, E. J. Lenardão and L. Savegnago, *Chem. – Biol. Interact.*, 2020, **331**, 109278.
134. D. R. Araujo, H. A. Goulart, A. M. Barcellos, R. Cargnelutti, E. J. Lenardão and G. Perin, *J. Org. Chem.*, 2021, **86**, 1721.

135. H. A. Goulart, J. S. S. Neto, A. M. Barcellos, K. B. Silva, M. C. de Moraes, R. G. Jacob, E. J. Lenardão, T. Barcellos and G. Perin, *J. Org. Chem.*, 2021, **86**, 14016.
136. D. R. Araujo, Y. R. Lima, A. M. Barcellos, M. S. Silva, R. G. Jacob, E. J. Lenardão, L. Bagnoli, C. Santi and G. Perin, *Eur. J. Org. Chem.*, 2020, 586.
137. G. Perin, P. C. Nobre, D. H. Mailahn, M. S. Silva, T. Barcellos, R. G. Jacob, E. J. Lenardão, C. Santi and J. A. Roehrs, *Synthesis*, 2019, **51**, 2293.
138. M. C. D. F. Xavier, E. M. A. Sandagorda, J. S. S. Neto, R. F. Schumacher and M. S. Silva, *RSC Adv.*, 2020, **10**, 13975.
139. A. L. Belladona, R. Cervo, D. Alves, T. Barcellos, R. Cargnelutti and R. F. Schumacher, *Tetrahedron Lett.*, 2020, **26**, 152035.
140. Z. Zhao, Y. Zhang, Y. Shao, W. Xiong, R. Li and J. Chen, *J. Org. Chem.*, 2020, **85**, 15015.
141. L. Xing, Y. Zhang, B. Li and Y. Du, *Org. Lett.*, 2019, **21**, 3620.
142. R. J. Reddy and A. H. Kumari, *RSC Adv.*, 2021, **11**, 9130.
143. Y. Cao, J. Liu, F. Liu, L. Jiang and W. Yi, *Org. Chem. Front.*, 2019, **6**, 825.
144. L. Abenante, N. B. Padilha, J. M. Anghinoni, F. Penteado, O. Rosati, C. Santi, M. S. Silva and E. J. Lenardão, *Org. Biomol. Chem.*, 2020, **18**, 5210.
145. B. M. Vieira, S. Thurow, M. da Costa, A. M. Casaril, M. Domingues, R. F. Schumacher, G. Perin, D. Alves, L. Savegnago and E. J. Lenardão, *Asian J. Org. Chem.*, 2017, **6**, 1635.
146. S. Thurow, F. Penteado, G. Perin, D. Alves, C. Santi, B. Monti, C. H. Schiesser, T. Barcellos and E. J. Lenardão, *Org. Chem. Front.*, 2018, **5**, 1983.
147. T. J. Peglow, G. P. da Costa, L. F. B. Duarte, M. S. Silva, T. Barcellos, G. Perin and D. Alves, *J. Org. Chem.*, 2019, **84**, 5471.
148. F. Penteado, L. Bettanin, K. Machado, G. Perin, D. Alves and E. J. Lenardão, *Asian J. Org. Chem.*, 2020, **9**, 1631.
149. X. Jiang, Z. Zhao, Z. Shen, K. Chen, L. Fang and C. Yu, *Eur. J. Org. Chem.*, 2020, 3889.
150. J. Rafique, S. Saba, M. S. Franco, L. Bettanin, A. R. Schneider, L. T. Silva and A. L. Braga, *Chem. – Eur. J.*, 2018, **24**, 4173.
151. W. Guo, W. Tan, M. Zhao, K. Tao, L. Y. Zheng, Y. Wu, D. Chen and X. L. Fan, *RSC Adv.*, 2017, **7**, 37739.
152. S. Saba, J. Rafique, M. S. Franco, A. R. Schneider, L. Espíndola, D. O. Silva and A. L. Braga, *Org. Biomol. Chem.*, 2018, **16**, 880.
153. F. Penteado, C. S. Gomes, L. I. Monzon, G. Perin, C. C. Silveira and E. J. Lenardão, *Eur. J. Org. Chem.*, 2020, 2110.
154. V. Rathore and S. Kumar, *Green Chem.*, 2019, **21**, 2670.
155. I. D. Lemir, W. D. Castor-Godoy, A. A. Heredia, L. C. Schmidt and K. E. Argüello, *RSC Adv.*, 2019, **9**, 22685.
156. H. Sahoo, A. Mandal, S. Dana and M. Baidya, *Adv. Synth. Catal.*, 2017, **360**, 1099.

157. X.-J. Zhou, H.-Y. Liu, Z.-Y. Mo, X.-L. Ma, Y.-Y. Chen, H.-T. Tang, Y.-M. Pan and Y.-L. Xu, *Chem. – Asian J.*, 2020, **15**, 1536.
158. C. Kingston, M. D. Palkowitz, Y. Takahira, J. C. Vantourout, B. K. Peters, Y. Kawamata and P. S. Baran, *Acc. Chem. Res.*, 2020, **53**, 72.
159. X. Zhang, C. Wang, H. Jiang and L. Sun, *Chem. Commun.*, 2018, **54**, 8781.
160. C. Chen, P. Niu, Z. Shen and M. Li, *J. Electrochem. Soc.*, 2018, **165**, G67.
161. Y. J. Kim and D. Y. Kim, *Tetrahedron Lett.*, 2019, **60**, 739.
162. Y. Yuan, Y. Cao, J. Qiao, Y. Lin, X. Jiang, Y. Weng, S. Tang and A. Lei, *Chin. J. Chem.*, 2019, **37**, 49.
163. X. Zhang, C. Wang, H. Jiang and L. Sun, *RSC Adv.*, 2018, **8**, 22042.
164. S. Lin, X. Cheng, B. Hasimujiang, Z. Xu, F. Li and Z. Ruan, *Org. Biomol. Chem.*, 2022, **20**, 117.

CHAPTER 7

Chalcogen–Nitrogen Heterocyclic Radicals

ANDREY V. ZIBAREV

Institute of Organic Chemistry, Siberian Branch, Russian Academy of Sciences, Novosibirsk 630090, Russia
Email: zibarev@nioch.nsc.ru

7.1 Introduction

Organic spin science is a well-developed discipline combining chemistry, physics, materials science, and biomedicine of open-shell compounds; of specific interest are $S \geq 1$ high-spin species.[1–15]

With chalcogens, organic π-cyclic derivatives of thiazyl $[SN]^{\bullet}$ and its putative heavier congeners are essentially important.[1–3,7,8,16–31] Some of them, especially 4-*R*-1,2,3,5-dichalcogenadiazolyls (chalcogen = S, Se), are comprehensively discussed elsewhere.[1–3,7,32–34] The present chapter is focused on neutral 1,2,3- and 1,3,2-benzodichalcogenazolyls (**1** and **2**; Herz and Wolmershäuser radicals, respectively), and anionic 2,1,3-benzochalcogenadiazolidyls and 1,2,3-benzodichalcogenazolidyls (**3** and **4**, respectively), involved in the design and synthesis of conductive and magnetic materials, including spin-crossover materials. The common feature of **1–4** is a 6–5 bicyclic scaffold and a delocalized antibonding π*-SOMO implying an intrinsic instability (for structural archetypes, see Scheme 7.1). Nevertheless, many of them can be isolated in the form of thermally-stable solids possessing highly-correlated electron systems with strong spin-lattice interactions.[1–3,7,14,18,19,22,25,26,32–37] Monocyclic analogues of **1–4** and other relevant ring systems are also

Chalcogen Chemistry: Fundamentals and Applications
Edited by Vito Lippolis, Claudio Santi, Eder J. Lenardão and Antonio L. Braga

Published by the Royal Society of Chemistry, www.rsc.org

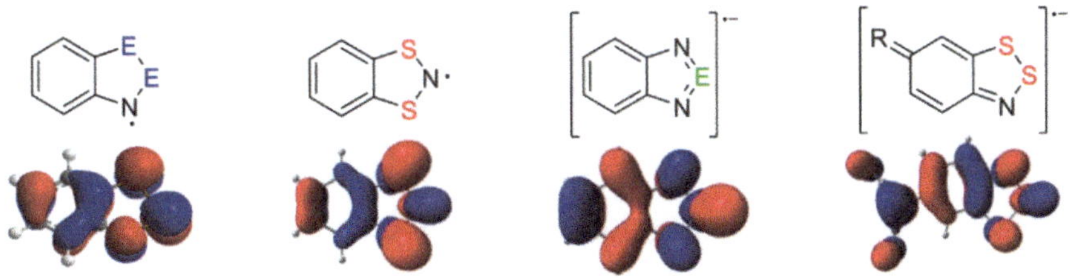

1 2 3 4, R = C(CN)$_2$

Scheme 7.1 Compounds **1–4** and their π*-SOMOs from DFT calculations; E = S, Se; E = S, Se, Te. Hereinafter, substituents, including fused rings, are unspecified or omitted where not significant; and π-delocalized radicals **1** and **2** are shown as *N*-centred to avoid resonance superpositions.

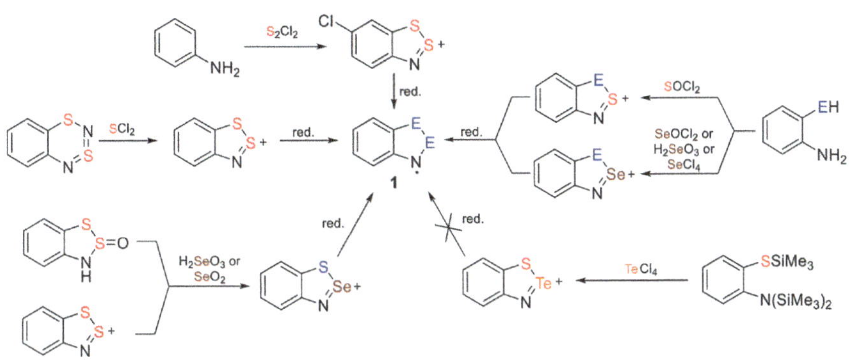

Scheme 7.2 Synthesis of HRs **1** *via* HCs, E = S, Se.

discussed where reasonable; open-shell derivatives of 2,2-dioxides of **3** (E = S) and congeners[23,38] are not covered.

7.2 1,2,3-Benzodichalcogenazolyls

The main approach to Herz radicals (HRs) is a chemical or electrochemical reduction of Herz cations (HCs). With S and Se, HCs are synthesized by ring-closure, ring-contraction, and chalcogen-exchange (*i.e.*, ring-modification) reactions; the most used is the Herz reaction between $ArNH_2$ and S_2Cl_2. With Te, a singular success is achieved with a ring-closure reaction. Except for the Te congener, the HCs are reduced into the target HRs by common reducing agents varying from Zn to $SbPh_3$ and $Fe(\eta^5\text{-}C_5Me_5)_2$[19,39–47] (see Scheme 7.2). For EPR studies, solution thermolysis or photolysis of various EN rings (E = S, Se) is synthetically useful; particularly, HRs unavailable *via* HCs, *e.g.*, polyfluorinated systems, can be generated following this stategy[19,40,48,49] (see Scheme 7.3).

Other Herz species synthesized by similar methods include dications **5**, radical cations **6**, resonance-stabilized radicals **7** and **8**, quinoids **9** and **10**, and bipolar ions **11** and **12**; compounds **9–12** can be considered as singlet

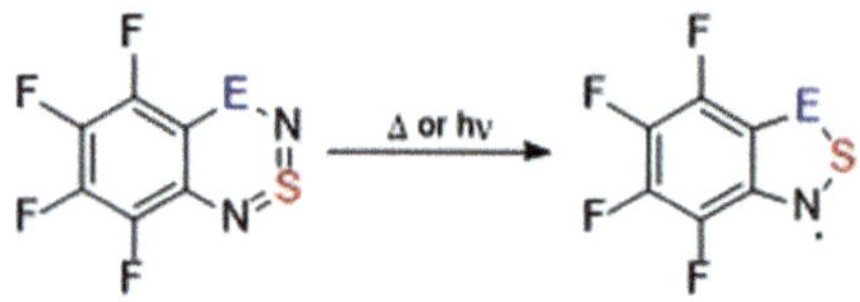

Scheme 7.3 Generation of fluorinated HRs; E = S, Se.

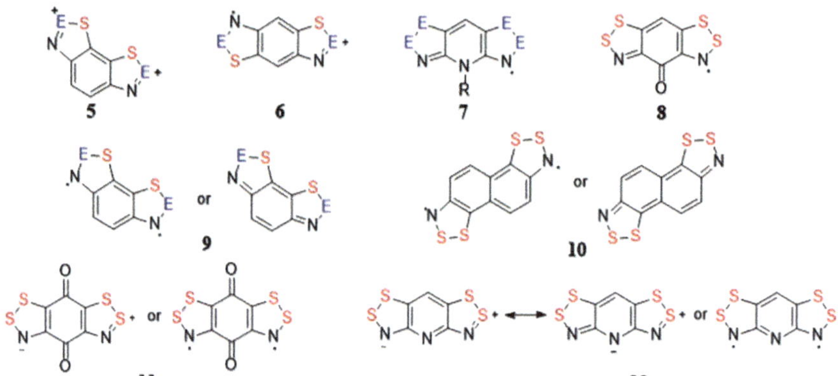

Scheme 7.4 Herz species **5–12**, E = S, Se.

diradicals[5,19,46,50,51] (see Scheme 7.4). The overall substitution/annulation pattern of HRs is very rich. Tri- and poly-HCs and HRs are unknown. Monocyclic analogues of HRs are reported only for S and reveal a high propensity toward C-centred recombination.[21,24,52]

HRs are π-delocalized (see Scheme 7.1); the replacement of S by Se induces minor perturbation of the electronic structures mainly involving neighbouring sites. Due to strong spin–orbit coupling (SOC; increases with the atomic number Z as Z^4) in Se atoms, the replacement, however, noticeably affects solution EPR spectra, in some cases transforming EPR signals into unresolved singlets.[19,45,53]

For putative diradicals **13** and **14** (see Scheme 7.5), the sign and value of the CASSCF-/DFT-calculated singlet–triplet splitting ΔE_{ST} depend on the annulation mode and the nature and positions of chalcogen atoms. For the all-S **13**, the ground state is a singlet. The replacement of S by Se reduces the absolute value of ΔE_{ST} and gives a triplet ground state for the all-Se derivative.[19] Donor substituents in the central ring stabilize the paramagnetic state of the all-S **13** and its aza-congeners (see Scheme 7.5); acceptor substituents stabilize the diamagnetic bipolar state.[54,55] For **14** independently of S and Se combinations, the ground state is a triplet with much larger ΔE_{ST} than for **13**. For putative triradicals **15**, CASSCF and DFT calculations suggest a quartet ground state. The replacement of S by Se affects the doublet-quartet splitting only slightly, with two doublet states being almost degenerate.[19]

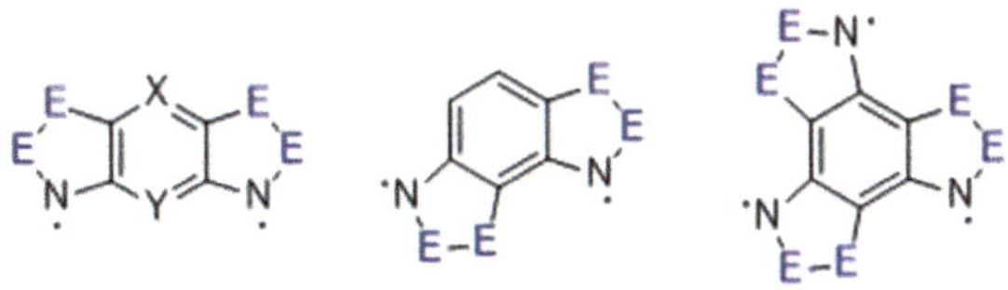

13 14 15

Scheme 7.5 HRs **13–15**, E = S, Se (all-S: X = C–H, C–R, N; Y = C–H, N).

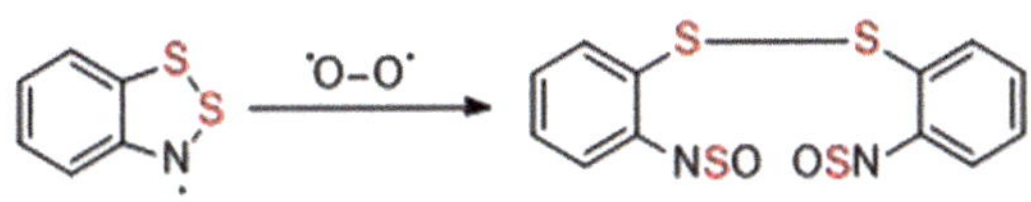

Scheme 7.6 Interaction of HRs with dioxygen.

HRs are sensitive to atmosphere and, probably, UV light. Interaction with O_2 combines two HRs and involves their nontrivial heteroatom rearrangement (see Scheme 7.6).[19,56]

In concentrated solutions, HRs undergo self-condensations affording, besides anticipated phenazines and phenothiazines, a variety of polycyclic quinoids hardly accessible otherwise, *e.g.*, **16–23**[19,40,46,57–59] (see Scheme 7.7). Compounds **22** and **23** are near-IR dyes, and **22** can afford $[22]^{2+}$, $[22]^{\bullet+}$, $[22]^{\bullet-}$, and $[22]^{2-}$ stable redox species.[59]

HRs **24** and **25** thermally transform into bipolar ions **26** and **11** (see Scheme 7.8), whereas analogues of **24** (R = H, Hal, Alk, Ar) and their Se congeners do not undergo this transformation.[60,61]

HRs are redox active. Chemically (with halogens) or electrochemically they oxidize into HCs.[19,46] Reduction of the 1,2-naphthoquinone analogue of **1** (E = S) gives an O-centred anion.[62]

HRs coordinate metal cations, *e.g.*, 1,4-naphthoquinone analogue of **1** (E = S) forms the complex $[Mn(hfac)_2]_3$ with the $S = 13/2$ ground state (hfac = hexafluoroacetylacetonate).[25]

In the solid state, HRs demonstrate polymorphism that can be driven by external stimuli. The solids can be paramagnetic or diamagnetic; monomeric or dimeric; and S and Se incarnations of the same scaffolds can be isomorphous or unisomorphous. For monomeric structures, various π-stacks are typical; and for dimeric structures, π-stacked pairs featuring pancake bonds can be observed; lateral σ-pairs are also disclosed. DFT calculations suggest instability of gas-phase π-dimers, whereas Se containing lateral σ-dimers can be stable. Pyridine-bridged HRs reveal the formation of σ-dimers with intermolecular S···E–E···S bonds instead of intramolecular S–E bonds (E = S, Se; see Scheme 7.9).[19,21,63–66]

Solid HRs are conductive and ferromagnetic (FM) or spin-canted antiferromagnetic (AF) materials. Specific crystal structures with inherent intermolecular interactions can modulate the electrical and magnetic properties. The replacement of S by Se increases electrical conductivity and magnetic exchange interactions; stronger SOC in Se atoms lead to magnetic anisotropy.[2,3,7,19,32]

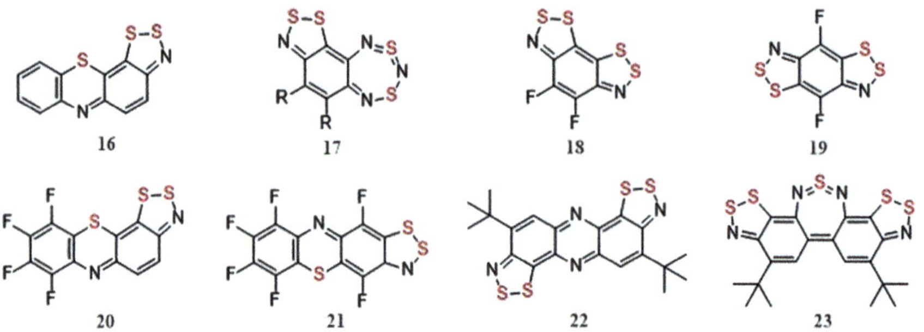

Scheme 7.7 Quinoids **16–23** from self-condensation of HRs.

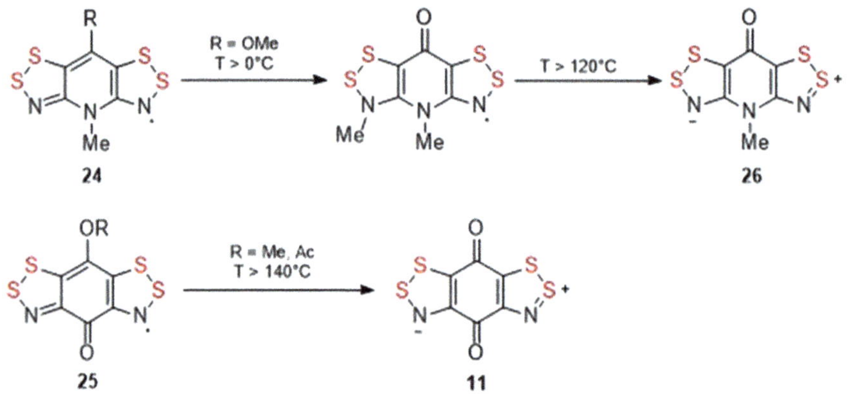

Scheme 7.8 Transformations of HRs into bipolar ions.

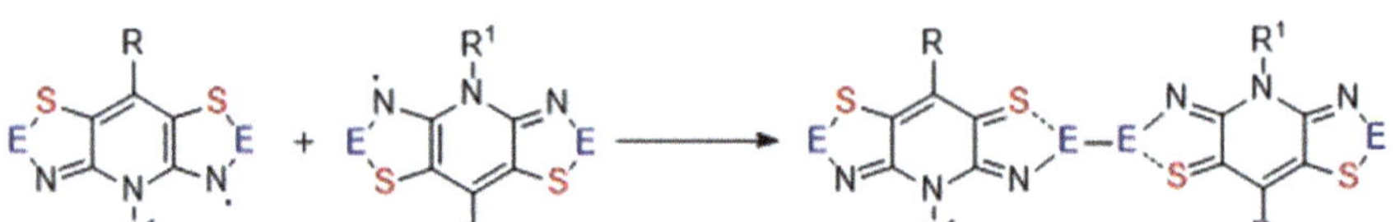

Scheme 7.9 Solid-state σ-dimers of HRs with an intermolecular S···E–E···S bond instead of intramolecular S–E bonds, E = S, Se.

Besides peculiarities of molecular and crystal structures, the electrical conductivity of HRs depends on the external pressure. For some HRs, pressure changes lead the activation energy E_a to approach the value zero which means band-gap closure and the formation of a metallic state. The highest room-temperature conductivity σ_{RT} ($S\,cm^{-1}$) belongs to the bipolar ion **11** (metallization at 8 GPa with $\sigma_{RT} \sim 10$; $\sigma_{RT} \sim 10^2$ at 12 GPa)[67] and to the salt $[\mathbf{10}_3]^{\bullet+}[GaCl_4]^-$ ($E_a \sim 0.18$ eV with $\sigma_{RT} \sim 0.5$ at normal pressure).[68,69] The best relevant molecular conductor $(SN)_x$, formally composed of $[SN]^\bullet$ units, has $\sigma_{RT} \sim 10^3$ along the macromolecules and becomes a BCS (*i.e.*, electron–phonon) superconductor at ∼0.3 K.[70]

Magnetic properties are frequently combined with electrical conductivity and depend on similar external factors.[19,32,71–73] Record temperatures of FM or AF ordering (Curee or Néel temperatures; T_C or T_N, respectively) belong to resonance-stabilized HRs **27–30**[34,74–79] (see Scheme 7.10). Amongst other relevant magnets, spin-canted AF (*i.e.*, weak FM) 4-R-1,2,3,5-dithiadiazolyl ($R = 4\text{-NC-}C_6F_4$) has $T_N = 36$ K at normal pressure and 70 K at 1.6 GPa.[2,32,80] The all-S analogue of **30** (R = F) reveals sharp hysteretic spin crossover ($T_{C\uparrow} = 380$, $T_{C\downarrow} = 375$ K) between π-monomeric and σ-dimeric species,[66] which can be driven by visible light.[72]

Amongst other solid-state properties of HRs, the spin frustration in the kagome-coupled chain structure of $[Cl_2\text{-}\mathbf{6}]^{\bullet+}[GaBr_4]^-$ (**6**, E = S),[81] and the nonlinear electrical transport and the low-field negative-resistance effect in $[\mathbf{10}_3]^{\bullet+}[GaCl_4]^-$,[37,68,69] can be mentioned.

7.3 1,3,2-Benzodichalcogenazolyls

Wolmershäuser radicals (WRs) are synthesized by chemical or electrochemical reduction of Wolmershäuser cations (WCs). The latter can be prepared by the interaction of $1,2\text{-}(SCl)_2$ derivatives with Me_3SiN_3 in the hydrocarbon and fluorocarbon series, or by the cycloaddition/oxidative dehydrogenation reaction of 1,4-haphthoquinone with $[SNS]^+$. Also, they can be afforded by cleavage of 2-R-1,3,2-dithiazoles ($R = H$, $ArSO_2$), and by ring-closure of $1,2\text{-}(SH)_2$ derivatives with $(NSCl)_3$.[82–106] The monocyclic analogues, EPR-detected for the first time by L. H. Sutcliffe *et al.* in the reaction of alkynes with $(SN)_4$,[107,108] are synthesized by the reduction of dithiazoliums from the cycloaddition of $[SNS]^+$ to alkynes[109–111] (see Schemes 7.11 and 7.12). Chemical reducing agents used are Ag, Zn/Cu, $SbPh_3$, and $[S_2O_4]^{2-}$.

Heavier chalcogen analogues of WRs were not isolated. For reduction of the archetypal all-Se WC, the corresponding WR was detected by EPR at 225 K. The final product, however, was tetraselenocine seemingly formed by self-condensation of the target WR.[82,83,112]

WRs are π-delocalized (for **2**, see Scheme 7.1); the spin density distribution depends on the nature of annulated rings (see Scheme 7.11).[88,93,102,107,109,113] EPR spectra in solution of **37**[85,93–95] display the exchange-coupled free diradical together with uncoupled associated species; unrestricted HF calculations estimate ΔE_{ST} as ~0.5 kcal mol^{-1}.[93]

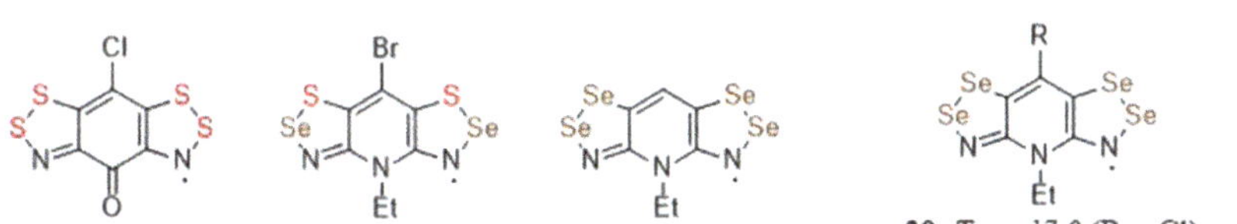

27, $T_N = 8$ **28**, $T_C = 14.1$ **29**, $T_N = 27$ **30**, $T_C = 17.0$ (R = Cl); $17.5 / 24$ (R = Br); $10.5 / 27.5$ (R = I)

Scheme 7.10 Magnetic ordering temperatures T_C or T_N (K) of HRs **27–30** at normal/enlarged pressure 100 kPa/2 GPa.

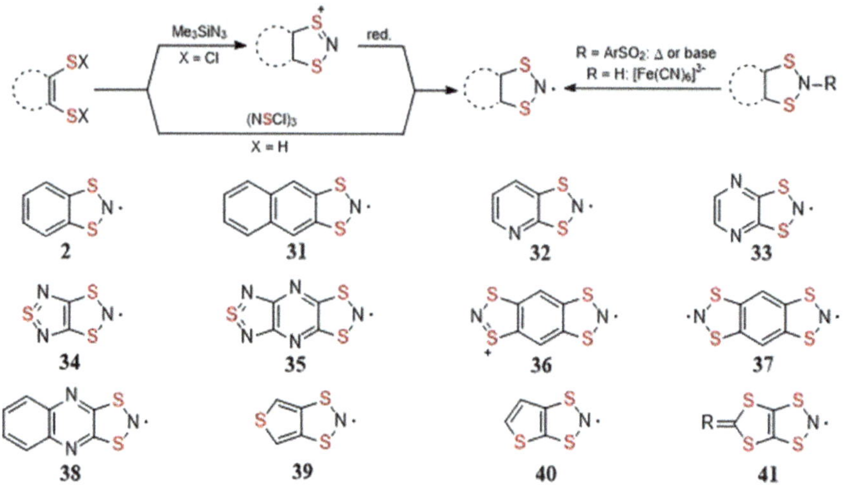

Scheme 7.11 Synthesis of WRs by ring-closuring and ring-modifying reactions.

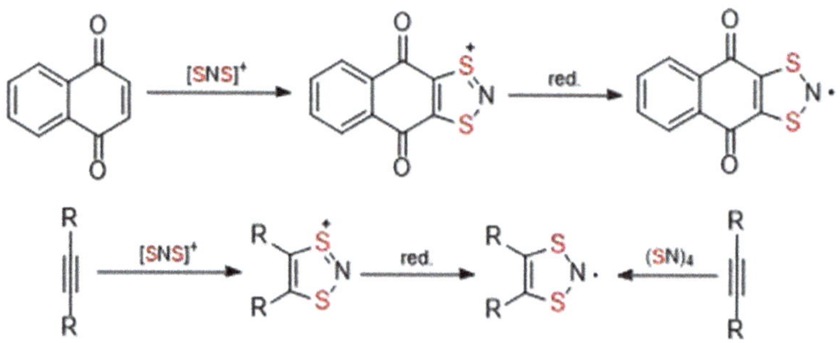

Scheme 7.12 Synthesis of WRs based on cycloaddition reactions.

WRs are air-sensitive; **39–41** are inherently unstable, and **41** reveals self-condensation with the loss of CO and rearrangement.[102–105] Quinone-fused WR undergoes thermal transformation into tetrathiine and thiadiazole[106] (see Scheme 7.13).

WRs form CT complexes with TCNQ and I_2,[82,83,114] and coordinate Cu, Zn, Co, and Mn cations as N-ligands[25,36,115–119] (WCs also form salts with Co, Ni, and Cu 1,2-dithiolenes).[114]

Similar to HRs, crystalline WRs reveal diamagnetic π-dimerization (π-stacks are typical), together with polymorphism (especially salts of radical cation **36**).[88,89,93,98,101,106,120–123] WR-based solids feature both AF (including spin-canted and metamagnetic) and FM properties (see Table 7.1 for salts of **36**). In complexes [M(hfac)$_2$][5-Me-**2**]$_2$ (M = Zn, Co, Mn), strong AF interactions occur between WRs and paramagnetic Mn^{2+} and Co^{2+} cations; in the case of diamagnetic Zn^{2+} cations, weak AF interactions between WRs are observed. Strong intermolecular exchange coupling in the Mn^{2+} complex

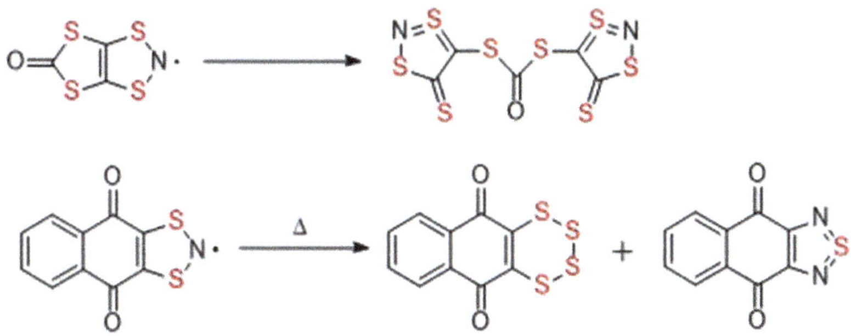

Scheme 7.13 Examples of chemical transformations of WRs.

Table 7.1 Magnetic ordering temperatures of salts [**36**][MX_4][122–127] and related.[a]

M	Tl		In	Ga		Fe
X	Br	I	I[b]	Br	Cl[c]	Cl[d]
T (K)	T_N, 11.6	T_N, 8.9	T_N, 11.3	T_N, 24	T_C, 7 (14.5, 1.6 GPa)	T_N, 6.6/6.3

[a][**36**][$Au(CN)_2$]: T_C, 8.2;[128] [**36**]$_2$[$ReBr_6$]: T_N, 12.2.[129]
[b]X = Cl, spin-Peierls transition at 108 K.[119]
[c]MeCN solvate.
[d]MeCN/Me_2CO solvates, desolvated salts are ferrimagnetics below 44 K.

Table 7.2 Magnetic bistability of WRs.[32,34,38,92,98,101,130–137]

T (K)	**34**	**35**	**31**[a]	**33**	4-CN-**2**[b]
$T_{C\uparrow}$	317	180	188	343	250
$T_{C\downarrow}$	225	120	128	297	250
ΔT_C[c]	92	60	60	46	0

[a]Diamagnetic phase consists of N–N σ-bonded dimers.[135]
[b]π-Dimeric **2** affords *via* melting and rapid cooling paramagnetic phase of unknown structure with $T_N = 11$ K.[120,121,131]
[c]Bistability region wherein two phases coexist.

comes from WRs' π-contacts.[36] The polymeric complex [$Cu(hfac)_2$][**34**] reveals FM interactions in coordination chains.[115,117]

WRs possess magnetic bistability associated with hysteretic spin crossover between $S = 0$ for π-dimeric and $S = 1/2$ for monomeric species, in which vibrational entropy plays the key role; the crossover can be driven by light and pressure. In contrast to HRs, the hysteretic loop is broad (see Table 7.2). Diradical **37** is diamagnetic in the range 5–400 K.[32,34,38,92,98,101,130–137]

CT complex [**2**][TCNQ] is a semiconductor with $E_a = 0.06$ eV ($\sigma_{RT} \sim 35$ S cm^{-1});[110] while the diradical **37** features an $E_a \sim 0.22$ eV.[38]

7.4 2,1,3-Benzochalcogenadiazolidyls

2,1,3-Benzochalcogenadiazoles, neutral precursors of radical anions (RAs) **3**, are easily available for S, Se, and Te in the hydrocarbon and fluorocarbon series.[30,138–143] Their DFT-calculated electron affinity (EA) is positive, *i.e.*, RAs are thermodynamically more preferable than the parent neutral molecules. In the isostructural series, EA increases with Z of the chalcogen, *i.e.*, with decreasing its electronegativity; which is most likely due to a better charge/spin delocalization in the diffuse π^*-SOMOs of RAs containing heavier chalcogens.[18,20,30]

RAs are synthesized by electrochemical or chemical reduction of neutral precursors. Chemical reducing agents are Na, K, KC_8, $[(Me_2N)_2C]_2$ (TDAE), $[PhE]^-$ (E = S, Se), $[S_2O_4]^{2-}$, $M(\eta^5\text{-}C_5R_5)_2$ (M = Co, Cr; R = H, Me), and $M(\eta^6\text{-Ar})_2$ (M = Cr, Mo; Ar = benzene, toluene, or mesitylene); $1,4\text{-}(RO)_2C_6H_4$ derivatives and RAs of C_6H_4COEt or 9,10-dihydroanthracene are also used. RAs **3** and **42–46** (See Scheme 7.14) are isolated in the form of thermally-stable salts.[18,20,30,138,144–159]

The RAs are π-delocalized (see Scheme 7.1 for **3**; with S, Se, and Te, the SOMOs are isolobal and the electron spin density distribution is quite uniform). Due to the π^* nature of SOMOs, the E–N bonds (E = S, Se, Te) elongate on going to RAs from neutral precursors.[18,20,30,138,144–146]

RA **3** (E = Te) is isolated in the form of a trimeric species formally composed of a neutral molecule and two RAs asymmetrically bound by shortened Te···N contacts (see Figure 7.1). Both DFT and CASSCF calculations suggest the singlet ground state caused by AF exchange interactions between unpaired electrons; according to CASSCF, this state contains 15% contribution of a diradical character. The negative charge is highly delocalized, which makes it impossible to distinguish neutral and charged units within the trimeric species. Related sterically-hindered RA **46** is isolated in the monomeric form (see Figure 7.1); calculations suggest that the unpaired electron is delocalized over the entire molecule. Ambient-temperature solution EPR spectra of RAs **3** (E = Te) and **46** feature broadened lines and are significantly *g*-shifted due to a strong SOC in the Te atoms.[144,145]

In air-protected solutions, the RAs are long-lived. This includes polyfluorinated derivatives of **3**; this is in contrast with the high instability of polyfluoro (het)arene RAs such as $[C_6F_6]^{\bullet-}$, $[C_6F_5N]^{\bullet-}$ and their analogues. The reason for this is the effective sharing of the negative charge by N, E, and F atoms reducing its on-site values and preventing F^- elimination.[148,155]

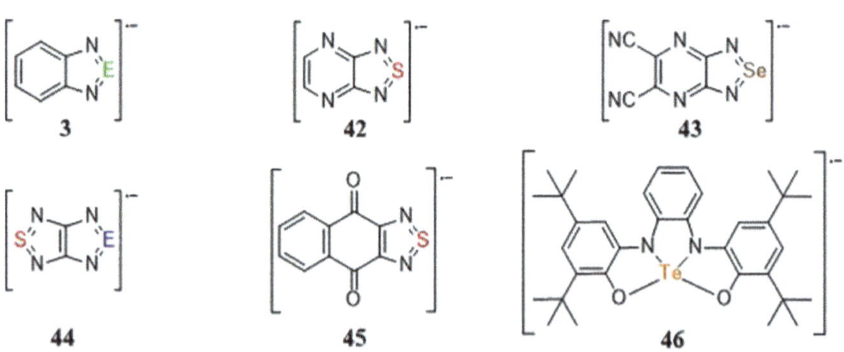

Scheme 7.14 RAs 3, 42–46; E = S, Se, Te; E = S, Se.

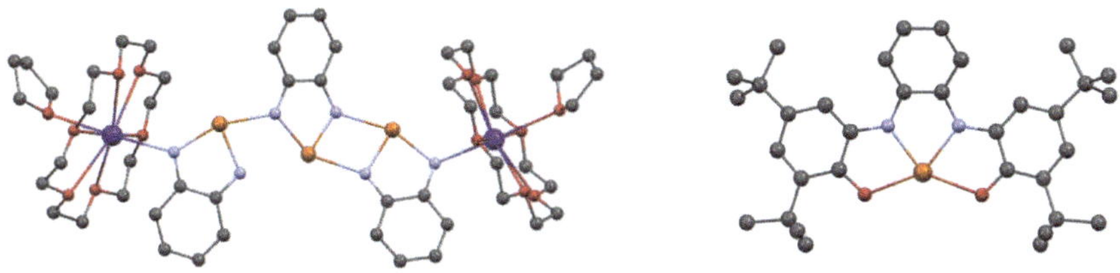

Figure 7.1 XRD structures of salts $[K(18\text{-crown-6})(THF)]^{+}{}_{2}[\mathbf{3}_{3}]^{2(\bullet-)}$ (**3**, E = Te) and $[Co(\eta^{5}\text{-}C_5H_5)_2]^{+}[\mathbf{46}]^{\bullet-}$ (cations omitted); colour code: C, grey; K, purple; N, blue; O, red; Te, orange; H, omitted.

Upon contact with air, RAs **3** decompose in solution into anions $[ECN]^{-}$ (E = S, Se), or dianions $[\mathbf{3}\text{–}E_2]^{2-}$ and $[(\mathbf{3}\text{–}E_2\text{–})_2]^{2-}$ (E = Te) featuring coordinate Te–Te bonds; the reaction pathways are unclear.[146] This instability can restrict practical applicability of **3** (E = S) and its carbocycle-substituted derivatives which were suggested as charge carriers for organic redox flow batteries.[156–159] Hydrolysis of RA **44** (E = S) gives $[S(SO_3)_2]^{2-}$.[160]

Similar to WRs, the RAs are N-ligands for coordination compounds, *e.g.*, $M(CO)_5(\mathbf{3})$ (M = W, Mo, Cr; E = S, Se).[27]

With RAs **3** and **42–46** (See Scheme 7.14), homo- (only the anion is paramagnetic) and heterospin (both ions are paramagnetic) salts are synthesized. Cations of the homospin salts are crown ether-encapsulated Li^{+}, Na^{+}, and K^{+}, $[(Me_2N)_3S]^{+}$, $[TDAE]^{2+}$, $[Co(\eta^{5}\text{-}C_5R_5)_2]^{+}$; and of the heterospin salts, $[Cr(\eta^{5}\text{-}C_5Me_5)_2]^{\bullet+}$ and $[M(\eta^{6}\text{-}Ar)_2]^{\bullet+}$ (M = Cr, Mo).[18,30,138,144–147,149,151–154,161] The most interesting structural features are RAs' π-dimers in [K(18-crown-6)][**43**][30] and $[TDAE][\mathbf{44}]_2$ (E = S)[161] (other known salts of **44** are monomeric), and co-packing of RA **45** with π-stacked units composed of **45** and its neutral precursor.[150] CASSCF and CASSCF/NEVPT2 calculations suggest that the RAs' π-dimers have the singlet ground state though with a noticeable singlet-diradical contribution.[30,161] Accordingly, the solid salts are EPR-silent, whereas their solutions are EPR-active due to dissociation of the dimers. For the salts of **43**, the replacement of K by Na in the cation affords a monomeric structure.[30]

Magnetic structures of monomeric RA salts are complex; the RAs are multidentate magnetic couplers participating in numerous AF and FM interactions. In both homo- and heterospin salts, the former dominate over the latter. Further progress is expected with heterospin salts. Contacts of unlike spin density of positively spin-polarized RAs and negatively spin-polarized cations (*e.g.*, metal sandwich cations) can originate FM ground state. Furthermore, strong SOC in salts containing atoms of heavier chalcogens and heavier transition metals can give a ferrimagnetic ground state *via* spin canting.[18,30,138,139,144–147,149,151,152,154]

7.5 1,2,3-Benzodichalcogenazolidyls

The initial materials for a two-step synthesis of RAs **4** are HCs (see Scheme 7.15); the reduction of intermediate quinoids can be chemical or electrochemical. Similar to **1–3**, these RAs are π-delocalized[28,29,162]

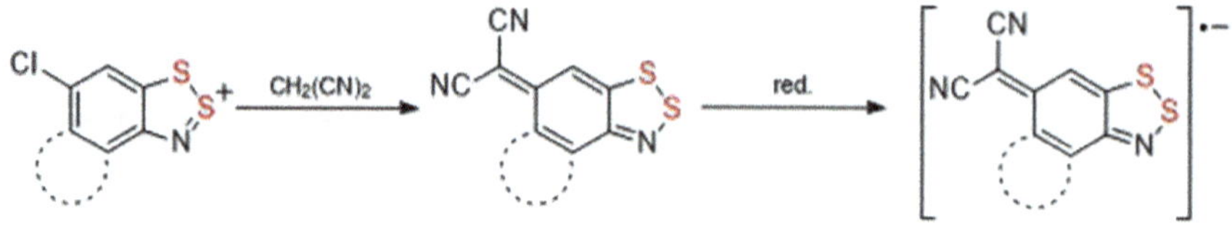

Scheme 7.15 Synthesis of RAs **4**.

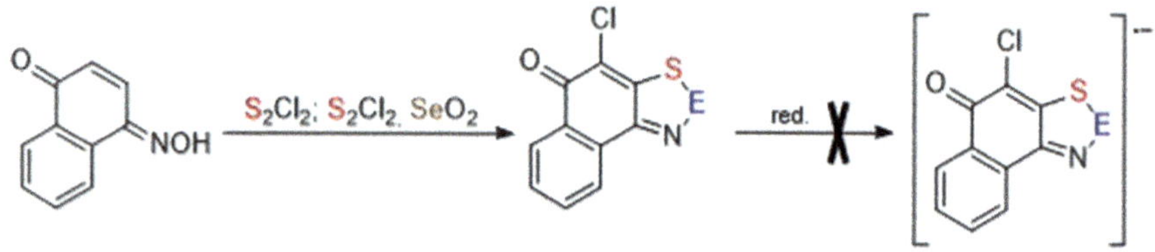

Scheme 7.16 Attempted synthesis of RAs **47**, E = S, Se.

Figure 7.2 Example of elongated S–S bond (Å) and the σ*-SOMO of 1,2,3-dithiazolidyl from DFT calculations; colour code: C, grey; H, light grey; Cl, green; N, blue; O, red; S, yellow.

(see Figure 7.1). Stable homo- and hetero-spin salts [TDAE][**4**]$_2$ and [Cr(η^6-Ar)$_2$][**4**] are isolated (**4**, R = C(CN)$_2$); magnetic properties of [Cr(η^6-Ar)$_2$][**4**] suggest π-dimerization of **4** in the solid state.[28] Other potential precursors did not give the target RAs (*e.g.*, **47**, see Scheme 7.16) under electrochemical conditions.[163,164] Unprecedented in the field, DFT calculations suggest the formation of σ-RAs (see Figure 7.2) featuring an abnormally elongated S–E bond (E = S, Se) predicting its weakening or breaking.[164]

7.6 Conclusions

Chalcogen–nitrogen heterocyclic radicals **1–4** contribute much in organic spin science. The S incarnations are better studied than the corresponding Se and, especially, Te ones, whereas the latter are of essential interest for fundamentals, advances, and applications due to stronger SOC, higher polarizability, and a larger propensity to secondary soft–soft bonding interactions. However, their synthesis and characterization are challenging and require further research.

Acknowledgements

The author is grateful to Prof. A. Yu. Makarov, Prof. L. A. Shundrin, E. A. Radiush, Dr. E. A. Chulanova, Dr. N. A. Semenov, and all other co-authors featuring the references below.

References

1. *World Scientific Reference on Spin in Organics*, ed. Z. V. Vardeny, M. Wohlgenannt and J. S. Miller, World Scientific, Singapore, 2018, vol. 4.
2. *Stable Radicals. Fundamentals and Applied Aspects of Odd-Electron Compounds*, ed. R. G. Hicks, Wiley, New York, 2010.
3. J. M. Rawson and J. J. Hayward, in *Spin Crossover Materials: Properties and Applications*, ed. M. A. Halcrow, Wiley, New York, 2013.
4. Z. X. Chen, Y. Li and F. Huang, *Chem.*, 2021, **7**, 288.
5. T. Stuyver, B. Chen, T. Zeng, P. Geerlings, F. De Proft and R. Hoffmann, *Chem. Rev.*, 2019, **119**, 11291.
6. M. Baumgarten, *Phys. Status Solidi B*, 2019, **256**, 1800642.
7. S. Kumar, Y. Kumar, S. K. Keshri and P. Mukhopadhyay, *Magnetochemistry*, 2016, **2**, 42.
8. G. E. Cutsail, *Dalton Trans.*, 2020, **49**, 12128.
9. N. Gallagher, H. Zhang, T. Junghoefer, E. Giangrisostomi, R. Ovsyannikov, M. Pink, S. Rajca, M. B. Casu and A. Rajca, *J. Am. Chem. Soc.*, 2019, **141**, 4764.
10. C. Shu, H. Zhang, A. Olankitwanit, S. Rajca and A. Rajca, *J. Am. Chem. Soc.*, 2019, **141**, 17287.
11. N. M. Gallagher, J. J. Bauer, M. Pink, S. Rajca and A. Rajca, *J. Am. Chem. Soc.*, 2016, **138**, 9377.
12. N. M. Gallagher, A. Olankitwanit and A. Rajca, *J. Org. Chem.*, 2015, **80**, 1291.
13. M. Abe, *Chem. Rev.*, 2013, **113**, 7011.
14. I. Ratera and J. Veciana, *Chem. Soc. Rev.*, 2012, **41**, 303.
15. A. Rajca, *Adv. Phys. Org. Chem.*, 2005, **40**, 153.
16. R. T. Boere, H. M. Tuononen, T. Chivers and T. L. Roemmele, *J. Organomet. Chem.*, 2007, **692**, 2683.
17. K. V. Shuvaev and J. Passmore, *Coord. Chem. Rev.*, 2013, **257**, 1067.
18. E. A. Chulanova, N. A. Semenov, N. A. Pushkarevsky, N. P. Gritsan and A. V. Zibarev, *Mendeleev Commun.*, 2018, **28**, 453.
19. Y. M. Volkova, A. Y. Makarov, E. A. Pritchina, N. P. Gritsan and A. V. Zibarev, *Mendeleev Commun.*, 2020, **30**, 385.
20. E. A. Pritchina, N. P. Gritsan, O. A. Rakitin and A. V. Zibarev, *Targets Heterocycl. Syst.*, 2019, **23**, 143.
21. K. E. Preuss, *Polyhedron*, 2014, **79**, 1.
22. W. Fujita, K. Awaga, H. Matsuzaki and H. Okamoto, *Phys. Rev. B: Condens. Matter Mater. Phys.*, 2002, **65**, 064434.

23. Y. Shuku and K. Awaga, *Molecules*, 2014, **19**, 609.
24. C. P. Constantinidis and P. A. Koutentis, *Adv. Heterocycl. Chem.*, 2016, **119**, 173.
25. K. E. Preuss, *Coord. Chem. Rev.*, 2015, **289–290**, 49.
26. K. E. Preuss, *Dalton Trans.*, 2007, 2357.
27. S. Plebst, M. Bubrin, D. Schweinfurth, S. Zalis and W. Kaim, *Z. Naturforsch., B: J. Chem. Sci.*, 2017, **72**, 839.
28. A. Y. Makarov, E. A. Chulanova, N. A. Semenov, N. A. Pushkarevsky, A. V. Lonchakov, A. S. Bogomyakov, I. G. Irtegova, N. V. Vasilieva, E. Lork, N. P. Gritsan, S. N. Konchenko, V. I. Ovcharenko and A. V. Zibarev, *Polyhedron*, 2014, **72**, 43.
29. E. A. Chulanova, I. G. Irtegova, N. V. Vasilieva, I. Y. Bagryanskaya, N. P. Gritsan and A. V. Zibarev, *Mendeleev Commun.*, 2015, **25**, 336.
30. N. A. Semenov, E. A. Radiush, E. A. Chulanova, A. M. Z. Slawin, J. D. Woollins, E. M. Kadilenko, I. Y. Bagryanskaya, I. G. Irtegova, A. S. Bogomyakov, L. A. Shundrin, N. P. Gritsan and A. V. Zibarev, *New J. Chem.*, 2019, **43**, 16331.
31. O. A. Rakitin, *Russ. Chem. Rev.*, 2011, **80**, 679.
32. J. M. Rawson, C. P. Constantinides, in *World Scientific Reference on Spin in Organics*, ed. Z. V. Vardeny, M. Wohlgenannt and J. S. Miller, World Scientific, Singapore, 2018, vol. 4, pp. 95–124.
33. D. A. Haynes, *CrystEngComm*, 2011, **13**, 4793.
34. J. M. Rawson, A. Alberola and A. Whalley, *J. Mater. Chem.*, 2006, **16**, 2560.
35. Y. Beldjoudi, M. A. Nascimento, Y. J. Cho, H. Yu, H. Aziz, D. Tonouchi, K. Eguchi, M. M. Matsushita, K. Awaga, I. Osorio-Roman, C. P. Constantinidis and J. M. Rawson, *J. Am. Chem. Soc.*, 2018, **140**, 6260.
36. D. Leckie, N. T. Stephaniuk, A. Arauzo, J. Campo and J. M. Rawson, *Chem. Commun.*, 2019, **55**, 9849.
37. K. Awaga, K. Namura, H. Kishida, W. Fujita, H. Yoshikawa, M. M. Matsushida, L. Ho, Y. Shuku and R. Suizu, *Bull. Chem. Soc. Jpn.*, 2014, **87**, 234.
38. P. Pakulski and D. Pinkowicz, *Molecules*, 2021, **26**, 4873.
39. R. T. Oakley, R. W. Reed, C. M. Robertson and J. F. Richardson, *Inorg. Chem.*, 2005, **44**, 1837.
40. F. Blockhuys, N. P. Gritsan, A. Y. Makarov, K. Tersago and A. V. Zibarev, *Eur. J. Inorg. Chem.*, 2008, **2008**, 655.
41. L. S. Konstantinova and O. A. Rakitin, *Mendeleev Commun.*, 2021, **31**, 433.
42. M. Risto, A. Assoud, S. M. Winter, R. Oilunkaniemi, R. S. Laitinen and R. T. Oakley, *Inorg. Chem.*, 2008, **47**, 10100.
43. A. Y. Makarov, I. Y. Bagryanskaya, Y. M. Volkova, M. M. Shakirov and A. V. Zibarev, *Eur. J. Inorg. Chem.*, 2018, **2018**, 1322.
44. A. Y. Makarov, F. Blockhuys, I. Y. Bagryanskaya, Y. V. Gatilov, M. M. Shakirov and A. V. Zibarev, *Inorg. Chem.*, 2013, **52**, 3699.

45. A. V. Pivtsov, L. V. Kulik, A. Y. Makarov and F. Blockhuys, *Phys. Chem. Chem. Phys.*, 2011, **13**, 3873.
46. A. J. Nicholls and I. R. Baxendale, *Reactions*, 2021, **2**, 175.
47. P. Vasco, J. Hurmalainen, A. Mansikkamäki, A. Peuronen, A. Mailman and H. M. Tuononen, *Dalton Trans.*, 2017, **46**, 16004.
48. N. P. Gritsan, A. Y. Makarov and A. V. Zibarev, *Appl. Magn. Reson.*, 2011, **41**, 449.
49. Y. M. Volkova, A. Y. Makarov, S. B. Zikirin, A. M. Genaev, I. Y. Bagryanskaya and A. V. Zibarev, *Mendeleev Commun.*, 2017, **27**, 19.
50. Z. Zeng, X. Shi, C. Chi, J. T. Lopez Navarrete, J. Casado and J. Wu, *Chem. Soc. Rev.*, 2015, **44**, 6578.
51. K. Takauji, R. Siuzu, K. Awaga, H. Kishida and A. Nakamura, *J. Phys. Chem. C*, 2014, **118**, 4303.
52. L. S. Konstantinova, O. A. Bolshakov, I. A. Baranovsky, A. M. Bogacheva, V. V. Strunyasheva and O. A. Rakitin, *Mendeleev Commun.*, 2015, **25**, 427.
53. R. T. Oakley, R. W. Reed, C. M. Robertson and J. F. Richardson, *Inorg. Chem.*, 2005, **44**, 1837.
54. L. Beer, R. T. Oakley, J. R. Mingie, K. E. Preuss and N. J. Taylor, *J. Am. Chem. Soc.*, 2000, **122**, 7602.
55. C. P. Constantinides, T. A. Ioannou and P. A. Koutentis, *Polyhedron*, 2013, **64**, 172.
56. N. P. Gritsan, S. N. Kim, A. Y. Makarov, E. N. Chesnokov and A. V. Zibarev, *Photochem. Photobiol. Sci.*, 2006, **5**, 95.
57. V. V. Zhivonitko, A. Y. Makarov, I. Y. Bagryanskaya, Y. V. Gatilov, M. M. Shakirov and A. V. Zibarev, *Eur. J. Inorg. Chem.*, 2005, **2005**, 4099.
58. A. Y. Makarov, V. V. Zhivonitko, A. G. Makarov, S. B. Zikirin, I. Y. Bagryanskaya, V. A. Bagryansky, Y. V. Gatilov, I. G. Irtegova, M. M. Shakirov and A. V. Zibarev, *Inorg. Chem.*, 2011, **50**, 3017.
59. A. Y. Makarov, Y. M. Volkova, L. A. Shundrin, A. A. Dmitriev, I. G. Irtegova, I. Y. Bagryanskaya, I. K. Shundrina, N. P. Gritsan, J. Beckmann and A. V. Zibarev, *Chem. Commun.*, 2020, **56**, 727.
60. S. M. Winter, R. J. Roberts, A. Mailman, K. Cvrkalj, A. Assoud and R. T. Oakley, *Chem. Commun.*, 2010, **46**, 4496.
61. A. Mailman, C. M. Robertson, S. M. Winter, P. A. Dube and R. T. Oakley, *Inorg. Chem.*, 2019, **58**, 6495.
62. S. Smithson, D. J. MacDonald, T. M. Letvenuk, C. E. Carello, M. Jennings, A. J. Lough, J. Britten, A. Decken and K. E. Preuss, *Dalton Trans.*, 2016, **45**, 9608.
63. M. Kertezs, *Chem. – Eur. J.*, 2019, **25**, 400.
64. D. Jose and A. Datta, *Cryst. Growth Des.*, 2011, **11**, 3137.
65. A. A. Leitch, X. Yu, C. M. Robertson, R. A. Secco, J. S. Tse and R. T. Oakley, *Inorg. Chem.*, 2009, **48**, 9874.
66. K. Lekin, S. M. Winter, L. E. Downie, X. Bao, J. S. Tse, S. Desgreniers, R. A. Secco, P. A. Dube and R. T. Oakley, *J. Am. Chem. Soc.*, 2010, **132**, 16212.

67. A. Mailman, A. A. Leitch, W. Yong, E. Steven, S. M. Winter, R. C. M. Claridge, A. Assoud, J. S. Tse, S. Desgreniers, R. A. Secco and R. T. Oakley, *J. Am. Chem. Soc.*, 2017, **139**, 2180.
68. K. Okamoto, T. Tanaka, W. Fujita, K. Awaga and T. Inabe, *Phys. Rev. B: Condens. Matter Mater. Phys.*, 2007, **76**, 075328.
69. K. Okamoto, T. Tanaka, W. Fujita, K. Awaga and T. Inabe, *Angew. Chem., Int. Ed.*, 2006, **45**, 4516.
70. A. J. Banister and I. B. Gorrell, *Adv. Mater.*, 1998, **10**, 1415.
71. S. M. Greer, R. T. Oakley, J. van Tol, M. Shatruk and S. Hill, *Polyhedron*, 2018, **153**, 99.
72. K. Lekin, H. Phan, S. M. Winter, J. W. L. Wong, A. A. Leitch, D. Laniel, W. Young, R. A. Secco, J. S. Tse, S. Desgreniers, P. A. Dube, M. Shatruk and R. T. Oakley, *J. Am. Chem. Soc.*, 2014, **136**, 8050.
73. H. Phan, K. Lekin, S. M. Winter, R. T. Oakley and M. Shatruk, *J. Am. Chem. Soc.*, 2013, **135**, 15674.
74. X. Yu, A. Mailman, K. Lekin, A. Assoud, C. M. Robertson, B. C. Noll, C. F. Campana, J. A. K. Howard, P. A. Dube and R. T. Oakley, *J. Am. Chem. Soc.*, 2012, **134**, 2264.
75. K. Irie, K. Shibayama, M. Mito, S. Takagi, M. Ishizuka, K. Lekin and R. T. Oakley, *Phys. Rev. B: Condens. Matter Mater. Phys.*, 2019, **99**, 014417.
76. K. Lekin, K. Ogata, A. Maclean, A. Mailman, S. M. Winter, A. Assoud, M. Mito, J. S. Tse, S. Desgreniers, N. Hirao, P. A. Dube and R. T. Oakley, *Chem. Commun.*, 2016, **52**, 13877.
77. A. A. Leitch, K. Lekin, S. M. Winter, L. E. Downie, H. Tsuruda, J. S. Tse, M. Mito, S. Desgreniers, P. A. Dube, S. Zhang, Q. Liu, C. Jin, Y. Ohishi and R. T. Oakley, *J. Am. Chem. Soc.*, 2011, **133**, 6051.
78. C. M. Robertson, A. A. Leitch, K. Cvrkalj, D. J. T. Myles, R. W. Reed, P. A. Dube and R. T. Oakley, *J. Am. Chem. Soc.*, 2008, **130**, 14791.
79. C. M. Robertson, D. J. T. Myles, A. A. Leitch, R. W. Reed, B. M. Dooley, N. L. Frank, P. A. Dube, L. K. Thompson and R. T. Oakley, *J. Am. Chem. Soc.*, 2007, **129**, 12688.
80. R. I. Thomson, C. M. Pask, G. O. Lloyd, M. Mito and J. M. Rawson, *Chem. - Eur. J.*, 2012, **18**, 8629.
81. L. Postulka, S. M. Winter, A. G. Mikhailov, A. Mailman, A. Assoud, C. M. Robertson, B. Wolf, M. Lang and R. T. Oakley, *J. Am. Chem. Soc.*, 2016, **138**, 10738.
82. G. Wolmershäuser, M. Schnauber and T. Wilhelm, *Chem. Commun.*, 1984, 573.
83. G. Heckmann, R. Johann, G. Kraft and G. Wolmershäuser, *Synth. Met.*, 1991, **43**, 3287.
84. G. Wolmershäuser and R. Johann, *Angew. Chem., Int. Ed.*, 1989, **28**, 920.
85. G. Wolmershäuser, M. Schnauber, T. Wilhelm and L. H. Sutcliffe, *Synth. Met.*, 1986, **14**, 239.
86. A. Alberola, G. D. McManus and J. M. Rawson, *Phosphorus, Sulfur Silicon Relat. Elem.*, 2004, **179**, 979.

87. A. Alberola, D. Eisler, R. J. Less, E. Navarro-Moratella and J. M. Rawson, *Chem. Commun.*, 2010, **46**, 6114.
88. A. Alberola, O. P. Clements, R. J. Collis, L. Cubbitt, C. M. Grant, R. J. Less, R. T. Oakley, J. M. Rawson, R. W. Reed and C. M. Robertson, *Cryst. Growth Des.*, 2008, **8**, 155.
89. A. Alberola, R. J. Collis, R. J. Less and J. M. Rawson, *J. Organomet. Chem.*, 2007, **692**, 2743.
90. A. Alberola, J. Burley, R. J. Collis, R. J. Less and J. M. Rawson, *J. Organomet. Chem.*, 2007, **692**, 2750.
91. R. J. Alberola, S. M. Collis, R. J. Humphrey, Less and J. M. Rawson, *Inorg. Chem.*, 2006, **45**, 1903.
92. T. M. Barclay, A. W. Cordes, N. A. George, R. C. Haddon, M. E. Itkis, M. S. Mashuta, R. T. Oakley, G. W. Patenaude, R. W. Reed, J. F. Richardson and H. Zhang, *J. Am. Chem. Soc.*, 1998, **120**, 352.
93. T. M. Barclay, A. W. Cordes, R. H. de Laat, J. D. Goddard, R. C. Haddon, D. Y. Jeter, R. C. Mawhinney, R. T. Oakley, T. T. M. Palstra, G. W. Patenaude, R. W. Reed and N. P. C. Westwood, *J. Am. Chem. Soc.*, 1997, **119**, 2633.
94. E. Dormann, M. J. Nowak, K. A. Williams, R. O. Angus and F. Wudl, *J. Am. Chem. Soc.*, 1987, **109**, 594.
95. K. A. Williams, M. J. Nowak, E. Dormann and F. Wudl, *Synth. Met.*, 1986, **14**, 233.
96. G. Wolmershäuser and G. Kraft, *Chem. Ber.*, 1990, **123**, 881.
97. G. Wolmershäuser and G. Kraft, *Chem. Ber.*, 1989, **122**, 385.
98. G. D. McManus, J. V. Rawson, N. Feeder, F. Palacio and P. Oliete, *J. Mater. Chem.*, 2000, **10**, 2001.
99. G. Wolmershäuser, G. Wortmann and M. Schnauber, *J. Chem. Res. S*, 1988, 358.
100. R. Johann and G. Wolmershäuser, *Phosphorus, Sulfur Silicon Relat. Elem.*, 1997, **124**, 233.
101. J. L. Brusso, O. W. Clements, R. C. Haddon, M. E. Itkis, A. A. Leitch, R. T. Oakley, R. W. Reed and J. F. Richardson, *J. Am. Chem. Soc.*, 2004, **126**, 8256.
102. A. Alberola, R. D. Farley, S. M. Humphrey, G. D. McManus, D. M. Murphy and J. M. Rawson, *Dalton Trans.*, 2005, 3838.
103. V. V. Popov, S. A. Amelichev, I. S. Bushmarinov, L. S. Konstantinova, K. A. Lyssenko, L. V. Mikhalchenko, V. V. Novikov, V. P. Gultyai and O. A. Rakitin, *Tetrahedron*, 2013, **69**, 8790.
104. L. S. Konstantinova, V. V. Popov, A. V. Lalov, L. V. Mikhalchenko, V. P. Gultyai and O. A. Rakitin, *Tetrahedron Lett.*, 2012, **53**, 3767.
105. S. A. Amelichev, V. V. Popov, L. S. Konstantinova, S. P. Golova, V. V. Novikov, E. D. Lubuzh, L. V. Mikhalchenko, V. P. Gultyai and O. A. Rakitin, *Mendeleev Commun.*, 2010, **20**, 82.
106. A. Decken, A. Mailman, J. Passmore, J. M. Rautiainen, W. Scherer and E. W. Scheidt, *Dalton Trans.*, 2011, **40**, 868.
107. E. G. Awere, N. Burford, C. Mailer, J. Passmore, M. J. Schriver, P. S. White, A. J. Banister, H. Oberhammer and L. S. Sutcliffe, *Chem. Commun.*, 1984, 66.

108. S. R. Harrison, R. G. Pilkington and L. H. Sutcliffe, *J. Chem. Soc., Faraday Trans. 1*, 1984, **80**, 669.
109. S. Brownridge, H. Du, S. A. Fairhurst, R. C. Haddon, H. Oberhammer, S. Parsons, J. Passmore, M. J. Schriver, L. H. Sutcliff and N. P. C. Westwood, *Dalton Trans.*, 2000, 3365.
110. G. K. MacLean, J. Passmore, M. N. S. Rao, M. J. Schriver and P. S. White, *Dalton Trans.*, 1985, 1405.
111. S. A. Fairhurst, R. G. Pilkington and L. H. Sutcliffe, *J. Chem. Soc., Faraday Trans. 1*, 1983, **79**, 925.
112. G. Wolmershäuser, W. Kaim, G. Heckmann and A. Lichtbau, *Z. Naturforsch., B: J. Chem. Sci.*, 1992, **47**, 675.
113. Y. L. Chung, S. A. Fairhurst, D. G. Gillies, G. Kraft, A. M. L. Krebber, K. F. Preston, L. H. Sutcliffe and G. Wolmershäuser, *Magn. Reson. Chem.*, 1992, **30**, 774.
114. K. Awaga, Y. Umezono, W. Fujita, H. Yoshikawa, H. Cui, H. Kobayashi, S. S. Staniland and N. Robertson, *Inorg. Chim. Acta*, 2008, **361**, 3761.
115. K. Awaga, T. Tanaka, T. Shirai, M. Fujimori, Y. Suzuki, H. Yoshikawa and W. Fujita, *Bull. Chem. Soc. Jpn.*, 2006, **79**, 25.
116. M. Kepenikian, B. Le Guennic, K. Awaga and V. Robert, *Phys. Chem. Chem. Phys.*, 2009, **11**, 6066.
117. W. Fujita and K. Awaga, *J. Am. Chem. Soc.*, 2001, **123**, 3601.
118. W. Fujita, K. Kukuchi and K. Awaga, *Angew. Chem., Int. Ed.*, 2008, **47**, 9480.
119. W. Fujita, K. Awaga, R. Kondo and S. Kagoshima, *J. Am. Chem. Soc.*, 2006, **128**, 6016.
120. E. G. Awere, N. Burford, R. C. Haddon, S. Parsons, J. Passmore, J. V. Waczczak and P. S. White, *Inorg. Chem.*, 1990, **29**, 4821.
121. W. Fujita, K. Awaga, K. Saito and M. Sorai, *Phys. Chem. Lett.*, 2002, **352**, 348.
122. W. Fujita and K. Awaga, *Chem. Phys. Lett.*, 2004, **388**, 186.
123. W. Fujita, K. Awaga, M. Takahashi, M. Takeda and T. Yamazaki, *Chem. Phys. Lett.*, 2002, **362**, 97.
124. W. Fujita and K. Awaga, *Chem. Phys. Lett.*, 2002, **357**, 385.
125. W. Fujita, T. Takahashi and H. Kobayashi, *Cryst. Growth Des.*, 2011, **11**, 575.
126. M. Mito, M. Fujino, Y. Komorida, H. Deguchi, S. Takagi, W. Fujita and K. Awaga, *J. Phys. Soc. Jpn.*, 2008, **77**, 124713.
127. W. Fujita and K. Kukuchi, *Chem. - Asian J.*, 2009, **4**, 400.
128. W. Fujita, *Dalton Trans.*, 2015, **44**, 903.
129. S. Suzuki, T. Kodama, K. Kukuchi and W. Fujita, *Chem. Lett.*, 2010, **39**, 1096.
130. A. Alberola, D. J. Eisler, L. Harvey and J. M. Rawson, *CrystEngComm*, 2011, **13**, 1794.
131. W. Fujita and K. Awaga, *Synth. Met.*, 2003, **137**, 1263.
132. W. Fujita and K. Awaga, *Science*, 1999, **286**, 261.
133. A. Alberola, R. C. Collis, S. M. Humphrey, R. J. Less and J. M. Rawson, *Inorg. Chem.*, 2006, **45**, 1903.

134. S. Vela, F. Mota, M. Deumal, R. Duizu, Y. Shuku, K. Awaga, M. Shiga, J. J. Novoa and J. Ribas-Arino, *Nat. Commun*, 2014, **5**, 4411.
135. D. Bates, C. M. Robertson, A. A. Leitch, P. A. Dube and R. T. Oakley, *J. Am. Chem. Soc.*, 2018, **140**, 3846.
136. J. L. Brusso, O. W. Clements, R. C. Haddon, M. E. Itkis, A. A. Leitch, R. T. Oakley, R. W. Reed and J. F. Richardson, *J. Am. Chem. Soc.*, 2004, **126**, 14692.
137. J. G. Richardson, A. Mizuno, Y. Shuku, K. Awaga, N. Robertson, C. A. Morrison, M. R. Warren, D. R. Allan and S. A. Moggach, *Cryst EngComm*, 2021, **23**, 4444.
138. A. V. Lonchakov, O. A. Rakitin, N. P. Gritsan and A. V. Zibarev, *Molecules*, 2013, **18**, 9850.
139. O. A. Rakitin, *Tetrahedron Lett.*, 2020, **61**, 152230.
140. O. A. Rakitin, *Synthesis*, 2019, **51**, 4338.
141. O. A. Rakitin and A. V. Zibarev, *Asian J. Org. Chem.*, 2018, **7**, 2397.
142. L. S. Konstantinova, E. A. Knyazeva and O. A. Rakitin, *Org. Prep. Proced. Int.*, 2014, **46**, 475.
143. B. A. D. Neto, A. A. M. Lapis, E. N. da Silva and J. Dupont, *Eur. J. Org. Chem.*, 2013, **2013**, 228.
144. N. A. Puskarevsky, A. I. Smolentsev, A. A. Dmitriev, I. Vargas-Baca, N. P. Gritsan, J. Beckmann and A. V. Zibarev, *Chem. Commun.*, 2020, **56**, 1113.
145. P. A. Petrov, E. M. Kadilenko, T. S. Sukhikh, I. V. Eltsov, A. L. Guschin, V. A. Nadolinny, M. N. Sokolov and N. P. Gritsan, *Chem. – Eur. J.*, 2020, **26**, 14688.
146. N. A. Pushkarevsky, E. A. Chulanova, L. A. Shundrin, A. I. Smolentsev, G. E. Salnikov, E. A. Pritchina, A. M. Genaev, I. G. Irtegova, I. Y. Bagryanskaya, S. N. Konchenko, N. P. Gritsan, J. Beckmann and A. V. Zibarev, *Chem. – Eur. J.*, 2019, **25**, 806.
147. Y. Shuku, Y. Hirai, N. A. Semenov, E. M. Kadilenko, N. P. Gritsan, A. V. Zibarev, O. A. Rakitin and K. Awaga, *Dalton Trans.*, 2018, **47**, 9897.
148. L. A. Shundrin, I. G. Irtegova, P. A. Avrorov, T. F. Mikhailovskaya, A. G. Makarov, A. Y. Makarov and A. V. Zibarev, *Arkivoc*, 2017, **iii**, 168.
149. N. A. Pushkarevsky, N. A. Semenov, A. A. Dmitriev, N. V. Kuratieva, A. S. Bogomyakov, I. G. Irtegova, N. V. Vasilieva, B. E. Bode, N. P. Gritsan, L. S. Konstantinova, J. D. Woollins, O. A. Rakitin, S. N. Konchenko, V. I. Ovcharenko and A. V. Zibarev, *Inorg. Chem.*, 2015, **54**, 7007.
150. L. S. Konstantinova, I. E. Bobkova, Y. V. Nelyubina, E. A. Chulanova, I. G. Irtegova, N. V. Vasilieva, P. S. Camacho, S. E. Ashbrook, G. Hua, A. M. Z. Slawin, J. D. Woollins, A. V. Zibarev and O. A. Rakitin, *Eur. J. Org. Chem.*, 2015, **2015**, 5585.
151. N. A. Semenov, N. A. Pushkarevsky, E. A. Suturina, E. A. Chulanova, N. V. Kuratieva, A. S. Bogomyakov, I. G. Irtegova, N. V. Vasilieva, L. S. Konstantinova, N. P. Gritsan, O. A. Rakitin, V. I. Ovcharenko, S. N. Konchenko and A. V. Zibarev, *Inorg. Chem.*, 2013, **52**, 6654.

152. N. P. Gritsan and A. V. Zibarev, *Russ. Chem. Bull.*, 2011, **60**, 2131.
153. S. N. Konchenko, N. P. Gritsan, A. V. Lonchakov, U. Radius and A. V. Zibarev, *Mendeleev Commun.*, 2009, **19**, 7.
154. J. S. Morgan, M. Jennings, A. Vindigni, R. Clerac and K. E. Preuss, *Cryst. Growth Des.*, 2011, **11**, 2520.
155. N. V. Vasilieva, I. G. Irtegova, N. P. Gritsan, A. V. Lonchakov, A. Y. Makarov, L. A. Shundrin and A. V. Zibarev, *J. Phys. Org. Chem.*, 2010, **23**, 536.
156. L. A. Robertson, I. A. Shkrob, G. Agarwal, Y. Zhao, Z. Yu, R. S. Assary, L. Cheng, J. S. Moore and L. Zhang, *ACS Energy Lett.*, 2020, **5**, 3062.
157. Z. Zhang, J. Huang, L. A. Robertson, R. S. Assary, I. A. Shkrob and L. Zhang, *J. Phys. Chem. C*, 2018, **122**, 8116.
158. J. Huang, W. Duan, J. Zhang, I. A. Shkrob, R. S. Assary, B. Pan, C. Liao, Z. Zhang, X. Wei and L. Zhang, *J. Mater. Chem. A*, 2018, **6**, 6251.
159. W. Duan, J. Huang, J. A. Kowalski, I. A. Shkrob, M. Vijayakumar, E. Walter, B. Pan, Z. Yang, J. D. Milstein, B. Li, C. Liao, Z. Zhang, W. Wang, J. Liu, J. S. Moore, F. R. Brushett, L. Zhang and X. Wei, *ACS Energy Lett.*, 2017, **2**, 1156.
160. E. Lork, R. Mews and A. V. Zibarev, *Mendeleev Commun.*, 2009, **19**, 147.
161. N. P. Gritsan, A. V. Lonchakov, E. Lork, R. Mews, E. A. Pritchina and A. V. Zibarev, *Eur. J. Inorg. Chem.*, 2008, **2008**, 1994.
162. P. A. Koutentis and C. W. Rees, *J. Chem. Soc., Perkin Trans. 1*, 2002, **3**, 315.
163. L. S. Konstantinova, I. V. Baranovsky, I. G. Irtegova, I. Y. Bagryanskaya, L. A. Shundrin, A. V. Zibarev and O. A. Rakitin, *Molecules*, 2016, **21**, 596.
164. L. S. Konstantinova, I. V. Baranovsky, E. A. Pritchina, M. S. Mikhailov, I. Y. Bagryanskaya, N. A. Semenov, I. G. Irtegova, G. E. Salnikov, K. A. Lyssenko, N. P. Gritsan, A. V. Zibarev and O. A. Rakitin, *Chem. Eur. J.*, 2017, **23**, 17037.

CHAPTER 8

Synthesis and Applications of Organophosphorus Compounds with a P=Se Group

YUUKI MAEKAWA AND TOSHIAKI MURAI*

Department of Chemistry and Biomolecular Science, Faculty of Engineering, Gifu University, Yanagido, Gifu 501-1193, Japan
*Email: mtoshi@gifu-u.ac.jp

8.1 Introduction

Increasing attention has been paid to four-coordinate pentavalent organophosphorus compounds with a P=Se group because of their wide applicability to a range of research fields.[1] The phosphorus atom of their P=Se group could have three different substituents, which provide numerous substitution patterns. For example, phosphine selenides have three carbon-containing substituents on the phosphorus atom (P–C bonds). In general, they are synthesized from trivalent phosphines and elemental selenium. Treating phosphines **1** with selenium powder in benzene gave the corresponding phosphine selenides **2** in good yields (Scheme 8.1).[2]

Phosphine selenides have been utilized for various purposes. For instance, tri(*n*-octyl)phosphine selenide provided the selenium atom to Cd oleate to synthesize CdSe quantum dots.[3,4] Likewise, the reaction of $[Ni(1,5\text{-}COD)_2]$ with $Et_3P{=}Se$ gave a $[Ni_{23}Se_{12}Cl_3(PEt_3)_{10}]$ cluster in which selenium atoms constructed the main cluster shell.[5] Cocrystallization of triphenylphosphine selenide (**3**) with iodobenzenes (1,3-F_4DIB, IF_5B) revealed their unique intermolecular interactions (Scheme 8.2).

Chalcogen Chemistry: Fundamentals and Applications
Edited by Vito Lippolis, Claudio Santi, Eder J. Lenardão and Antonio L. Braga

Published by the Royal Society of Chemistry, www.rsc.org

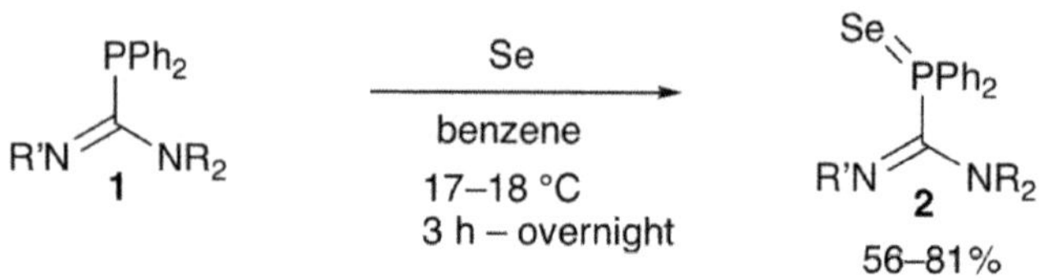

Scheme 8.1 Selenation of phosphines.

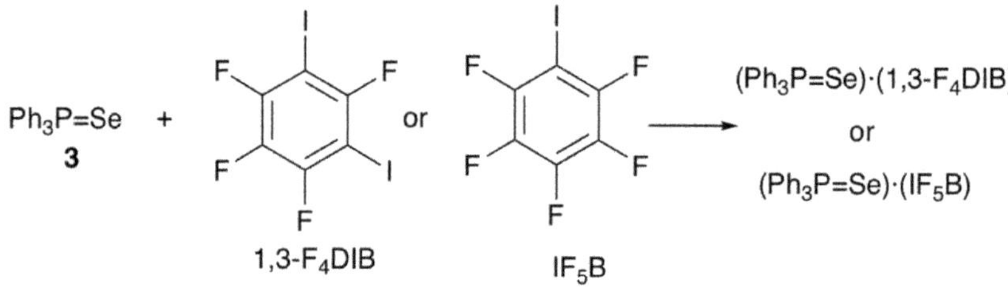

Scheme 8.2 Cocrystallization of phosphine selenide with iodofluorobenzenes.

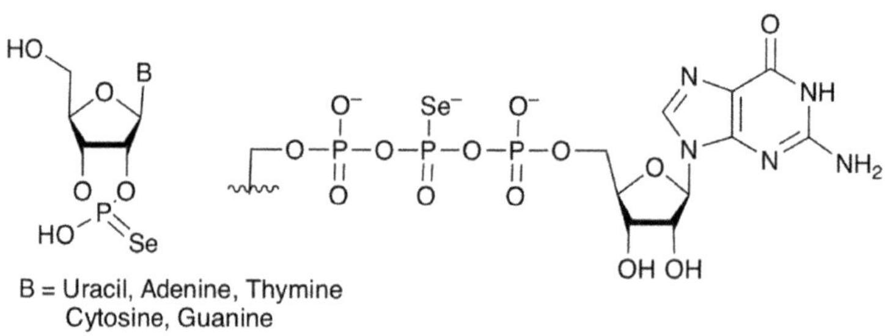

Scheme 8.3 Biologically-relevant compounds with $(RO)_3P{=}Se$.

The X-ray structure analyses of their cocrystals showed a chalcogen bond between the selenium atom and iodine atom providing a different packing arrangement from a cocrystal of the phosphine oxide with iodobenzenes.[6] Phosphine selenides have also been used as a tool to evaluate properties of trivalent phosphines. The spin–spin coupling constants between ^{31}P atom and ^{77}Se atom in phosphine selenides correlate with the donor ability of trivalent phosphines as well as their basicity.[7]

Selenophosphates, in which the P=Se group has three alkoxy groups (P–O bonds), have been well studied. This class of compounds appears as biologically relevant compounds as shown in Scheme 8.3.[8,9]

In this chapter, we will focus on the following four topics, *i.e.*, (i) reactions of secondary phosphine selenides, (ii) recent applications of 2,4-bis(phenyl)-1,3-diselenadiphosphetane-2,4-diselenide (Woollins' reagent), (iii) catalytic reactions with compounds with a P=Se group, and (iv) phosphoroselenoates with a binaphthyl group.

8.2 Reaction of Secondary Phosphine Selenides

Since secondary phosphine selenides **4** can exist in equilibrium with selenophosphinites **5**, the phosphorus atom in **4** can be a nucleophile. In fact, the reaction of **4** with methyl iodide in the presence of KOH gave phosphine selenide **6** in good yield (Scheme 8.4).[10]

Aldehydes can work as an electrophile in the reaction with secondary phosphine selenide **4a** (Scheme 8.5).[11]

The simple mixing of **4a** and aldehydes in THF gave α-hydroxyphosphine selenides **7** in good to high yields. This reaction tolerated various types of aldehydes such as trichloro, propargyl, aliphatic, aromatic, and heteroaromatic aldehydes. THF could be replaced by less polar toluene as a solvent.[12] Secondary phosphine selenides having two pyridylethyl groups that participate in the reaction with aldehydes to give the corresponding products.[13–15] These types of reactions with aldehydes also took place in the absence of a solvent, although the higher temperature (around 50 °C) was necessary.[16]

A P–H bond in secondary phosphine selenide **4a** could add to terminal alkenes upon irradiation of UV to give various phosphine selenides **8** (Scheme 8.6).[17]

Scheme 8.4 Tautomeric form and methylation of secondary phosphine selenides.

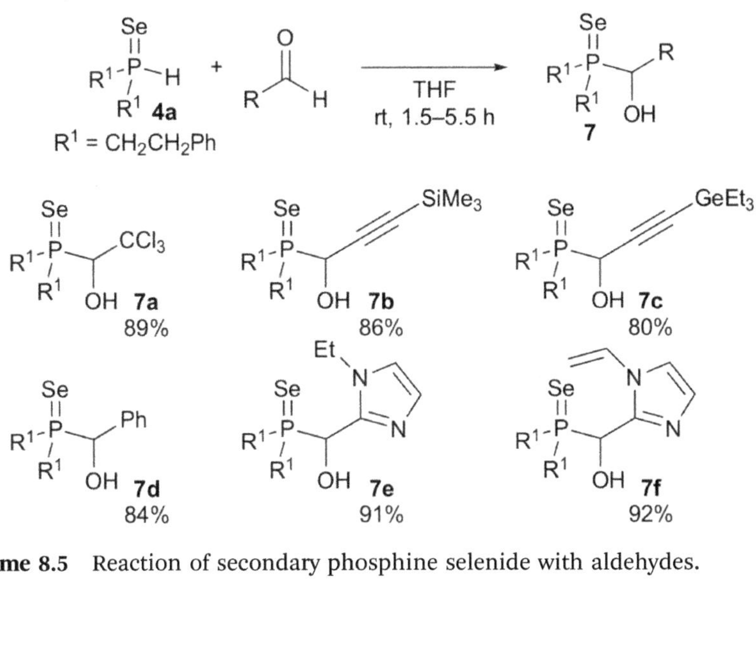

Scheme 8.5 Reaction of secondary phosphine selenide with aldehydes.

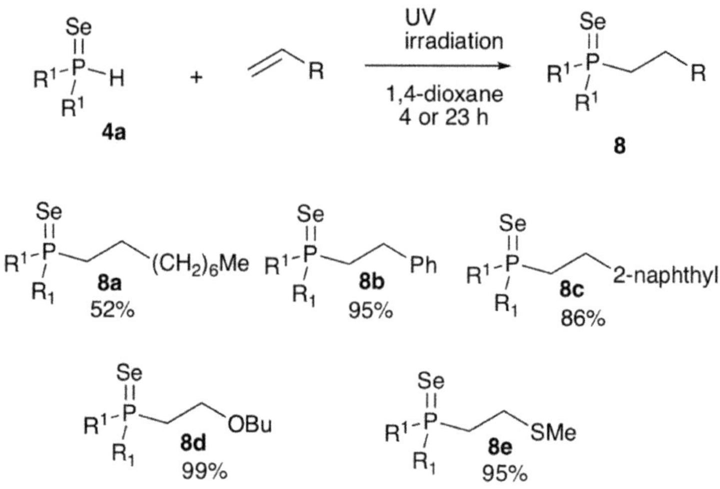

Scheme 8.6 Reaction of secondary phosphine selenide with alkenes.

The reaction with 1-nonene required 23 hours for the completion, while the reaction with aromatic alkenes, vinyl ether, and vinyl thioether completed within 4 hours to give the corresponding products in high yields. Vinyl selenides[18,19] and vinyl tellurides[20,21] could participate in this type of reaction as alkene substrates in the presence of AIBN at 65–70 °C. The addition reaction of secondary phosphine selenides was extended to several types of vinyl ethers,[22–24] allylic amines,[25,26] vinylbenzyl phosphine oxide,[27] *N*-vinyl imidazole, and *N*-vinyl benzimidazole.[28] Thermal conditions without UV irradiation or any additives could initiate an addition reaction of secondary phosphine selenides to terminal alkenes.[29] The treatment of these substrates at 80 °C without solvent produced the corresponding adducts in good yields. The reaction time to achieve full conversion depended on the substituents on the alkenes. Although the above-mentioned reports employed different reaction conditions and alkenes with a variety of functional groups, all the reactions gave products in an anti-MarKovnikov fashion.

In contrast, the use of ferrocenylmethyl vinyl ether **9** afforded Markovnikov adduct **10** and anti-Markovnikov adduct **11** in a ratio of ~3 : 1 (Scheme 8.7).[30]

This is in marked contrast to the reaction with secondary phosphine sulfide, wherein anti-Markovnikov adduct was selectively obtained. Attempts to isolate each adduct were not successful because of the low stability of **10** and **11**. However, the addition reaction of diphenylphosphine selenide **4b** to alkenes **9a** and **9b** proceeded smoothly without AIBN at room temperature to form Markovnikov adduct **10a** and **10b**, which underwent an extrusion reaction of acetaldehyde during the storage to give phosphine selenides **12a** and **12b** (Scheme 8.8).

The addition of phosphine selenide **4b** to the alkenyl group of β,γ-unsaturated ketones completed within 8 hours at 80 °C to give γ-ketophosphine selenides **14** in good yields (Scheme 8.9).[31]

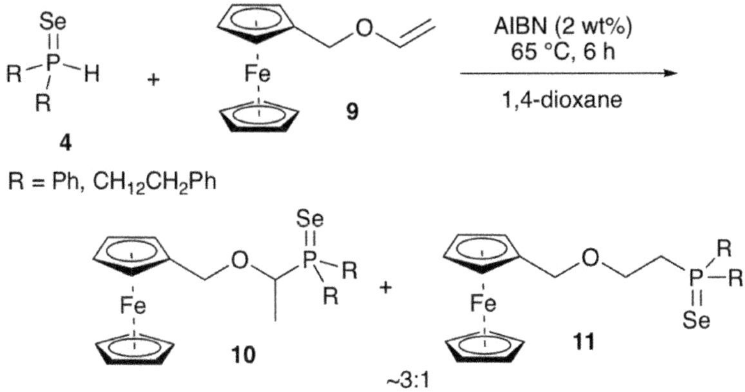

Scheme 8.7 Reaction with ferrocenylmethyl vinyl ether.

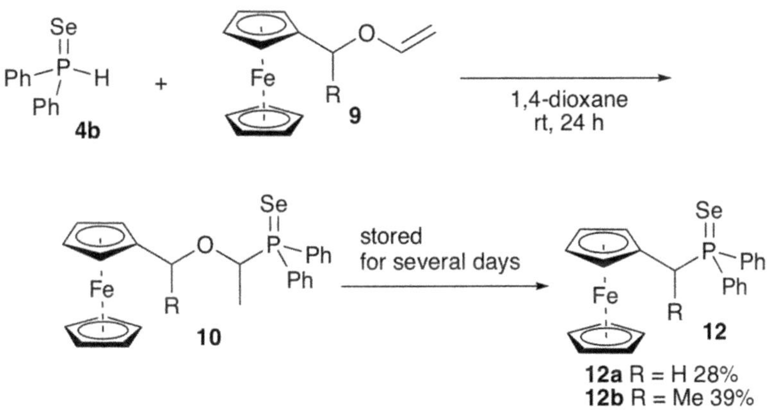

Scheme 8.8 Reaction of diphenylphosphine selenide **4b** with ferrocenylmethyl vinyl ether.

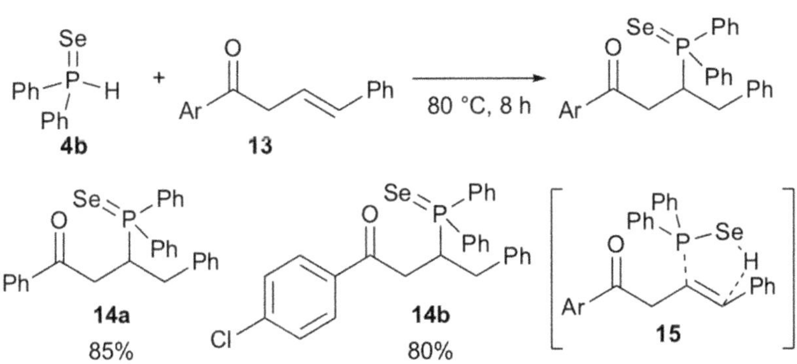

Scheme 8.9 Reaction of diphenylphosphine selenide **4b** with β,γ-unsaturated ketones **13**.

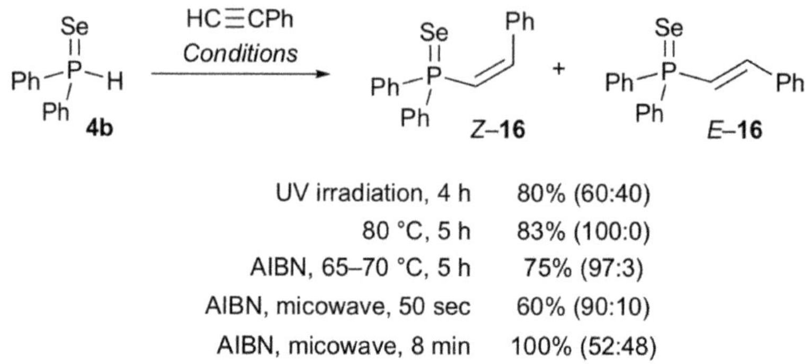

Scheme 8.10 Reaction of diphenylphosphine selenide **4b** with phenylacetylene.

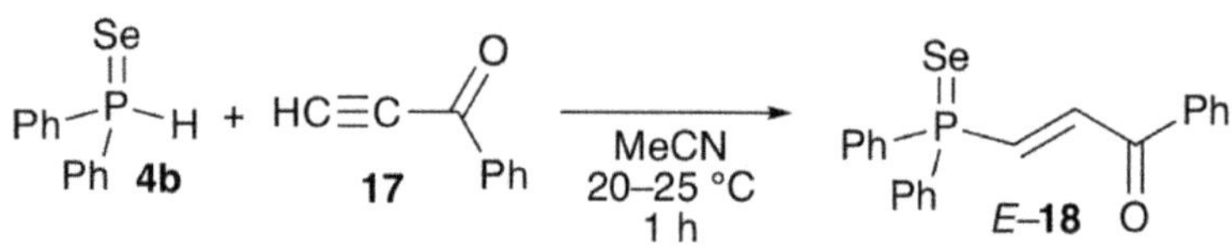

Scheme 8.11 Reaction of diphenylphosphine selenide **4b** with ethynyl ketone **17**.

The reaction would proceed through **15** to afford the products with high regioselectivity while the carbonyl group remained intact.

The addition of secondary phosphine selenides **4b** to alkynes was explored under several reaction conditions (Scheme 8.10).[32,33]

The reaction of **4b** with phenylacetylene under UV irradiation conditions gave anti-Markovnikov adduct **16** as a mixture of *Z*/*E* isomers in a ratio of 60 : 40. In contrast, the reaction at 80 °C gave only the *Z*-isomer in 83% yield. AIBN could also initiate the reaction instead of UV irradiation to afford the product in high *Z*-selectivity. Microwave irradiation accelerated the reaction, however, resulting in lower regioselectivity of the products over a prolonged reaction time. The reaction of **4b** with ethynyl ketone **17** selectively led to *E*-**18** (Scheme 8.11).[34]

Recently, a three component reaction of phosphine selenides **4**, pyridines, and alkynyl esters or ketones was found to afford coupling products **21**, wherein the phosphinoselenoyl group was introduced to the 4-position of the pyridyl ring, and the nitrogen atom was attached to the terminal carbon atom of alkynyl esters.[34–37] The reaction with ethynyl esters **20** proceeded smoothly in MeCN to give *E*-*N*-ethenyl-1,4-dihydropyridines **21** (Scheme 8.12).[36]

The reaction might begin with the nucleophilic attack of pyridine **19** with ethynyl ester **20** to generate twitter ion **22**. Then, **22** could deprotonate **4a** to generate phosphorus anion **23** and pyridinium ion **24**. Finally, anion **23** could attack the 4-position of the pyridine moiety in **24** to afford the corresponding product **21**. Instead of ethynyl ester, the use of alkynyl ketone **25** gave 4-phosphinoselenoylpyridine **26** along with chalcone (**27**) (Scheme 8.13).[38]

Firstly, the reaction would afford intermediate **28** and its isomer **29** similarly to Scheme 12, then, **27** might be eliminated from them to give **26**.

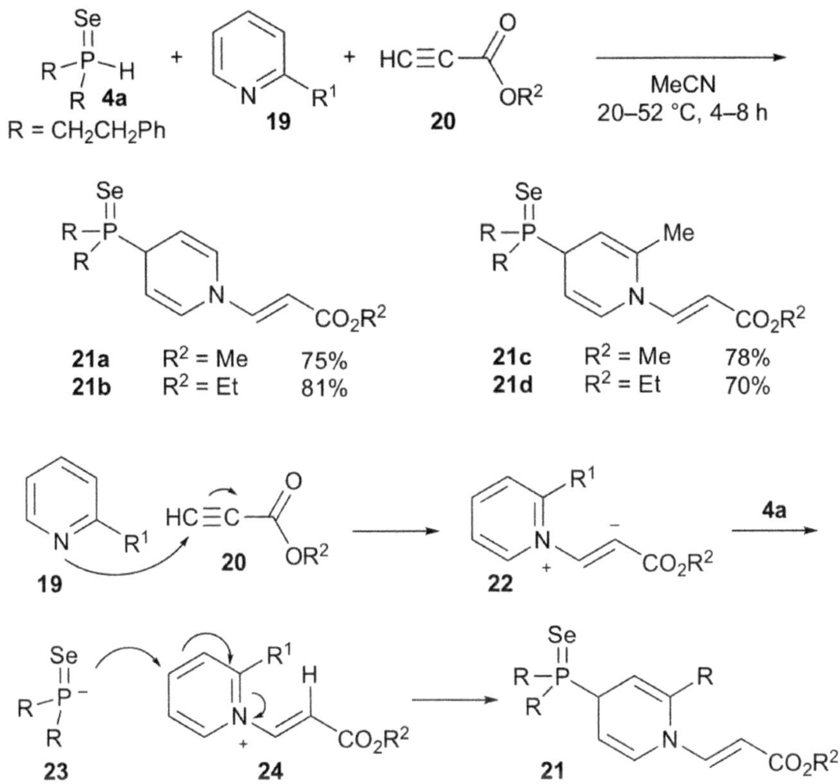

Scheme 8.12 Three-component reaction of **4a**, **19**, and **20**.

The reaction of **4a** with acridine (**30**) afforded 9-phosphinoselenoyl-9,10-dihydroacridine **31** without using alkynyl ketone **25** (Scheme 8.14).[39]

The P–O bond forming reaction was achieved by reacting secondary phosphine selenides **4** with 4-hydroxy-6-methyl-2-pyrone (**32**) in the presence of Et_3N with CCl_4 as a solvent (Scheme 8.15).[40]

Deprotonation from **4** with Et_3N might initially take place to generate phosphorus anions **23**. The chlorination of **23** with CCl_4 would give phosphinoselenoyl chlorides **34**, which then reacted with **32** to yield the products **33**. A range of aromatic and aliphatic alcohols were also tolerated under these conditions to yield the corresponding products in good to high yields (Scheme 8.16).[41–46]

The reaction was compatible with reactive functional groups such as azo (**35b**) and carbonyl groups (**35c**). The reaction of 5-nitro-5-aminophenol took place on the hydroxy group selectively instead of the amino group. Both primary and secondary alkyl alcohols also reacted with **4a** to yield the corresponding products **36** (Scheme 8.17). NEt_3 was not necessary for the formation of **36b** and **36c**.[47,48]

This method was extended to amine substrates to form a P–N bond.[49] As chloride-free conditions, the simple reaction of secondary phosphine

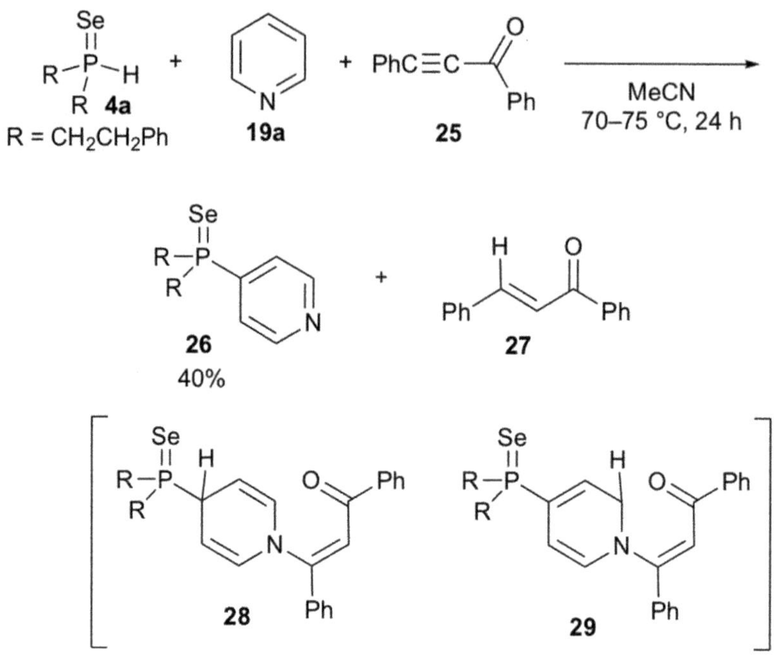

Scheme 8.13 Reaction of **4a** with pyridine in the presence of alkynyl ketone.

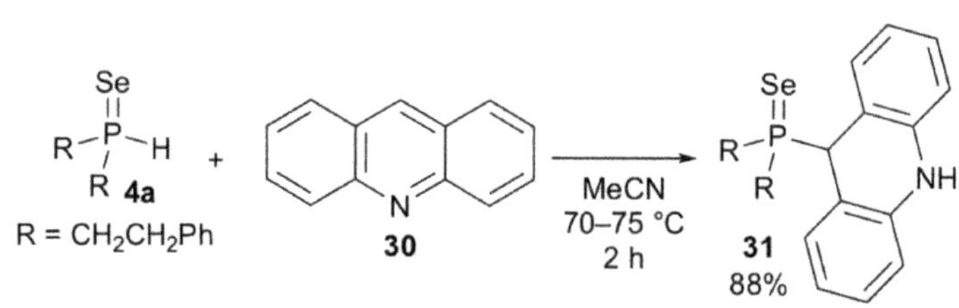

Scheme 8.14 Reaction of **4a** with acridine.

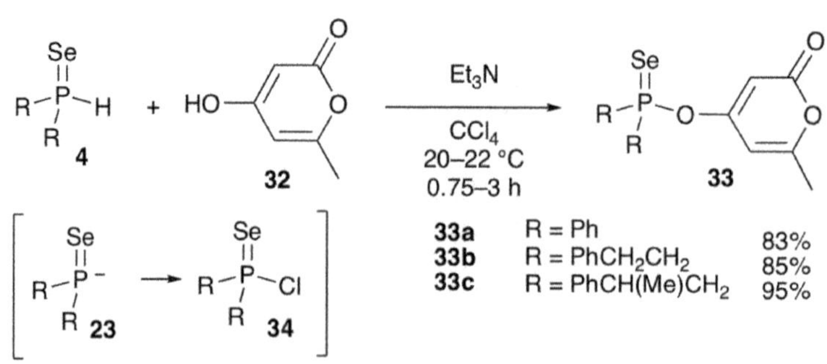

Scheme 8.15 Reaction of **4** with pyrone **32**.

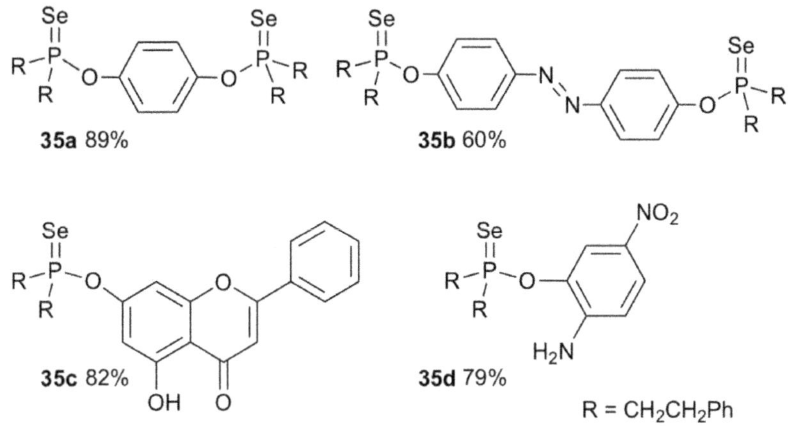

Scheme 8.16 Products derived from the reaction of **4a** with aromatic alcohols.

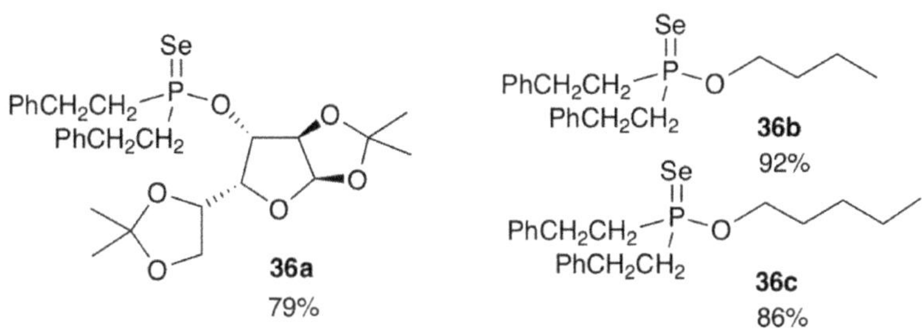

Scheme 8.17 Products derived from the reaction of **4a** with aliphatic alcohols.

selenides **4** with primary and secondary amines gave phosphinoselenoic amides as products, however, the yields of the products were moderate (Scheme 8.18).[50] The reaction at higher temperature did not improve the product yields.

The oxidative coupling between secondary phosphine selenides **4** and thiols took place with Et_3N as a base in CCl_4 to form phosphinoselenothioic acid esters **38** (Scheme 8.19).[51]

As a scope of thiols, primary and even sterically-hindered tertiary thiols could react with **4** to give the products **38a** and **38b**. Heteroaromatic thiols were also tolerated to yield the products **38c** and **38d**.

Secondary phosphine selenide **4a** reacted with sulfur powder and lupinine at 50–65 °C for 1 h to give phosphioselenothioic acid ammonium salt **39** in high yield (Scheme 8.20).[52] Lupinine worked as a base to deprotonate from **4a** followed by the reaction with elemental sulfur to form the corresponding salt **39**.

The reaction of secondary phosphine selenides **4** with elemental selenium in the presence of alkali metal hydroxide completed within 30 s to give thermally-stable phosphinodiselenoic acid metal salts **40** in high yields (Scheme 8.21).[53,54]

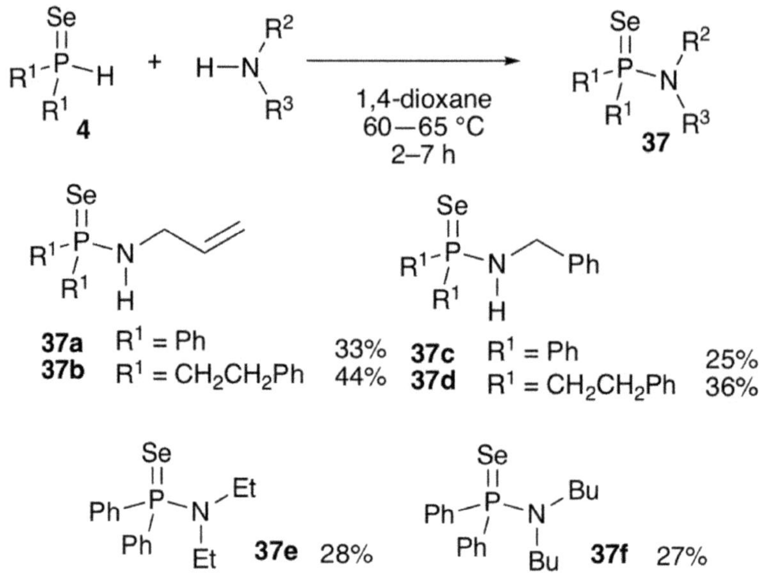

Scheme 8.18 Reaction of **4** with amines.

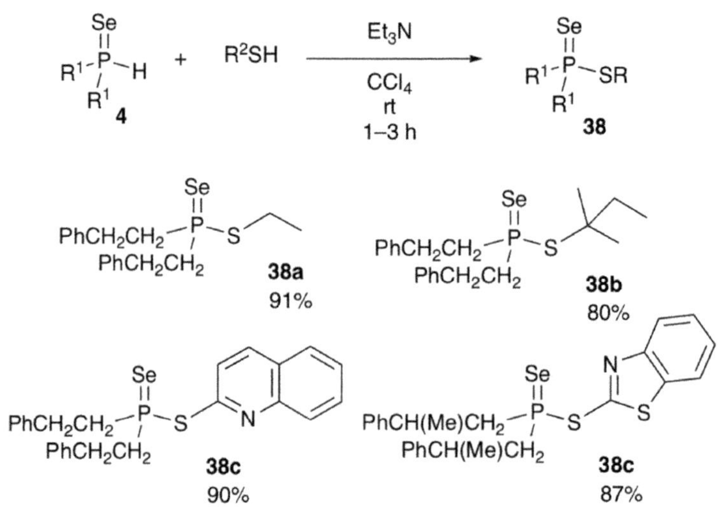

Scheme 8.19 Oxidative coupling reaction of **4** with thiols.

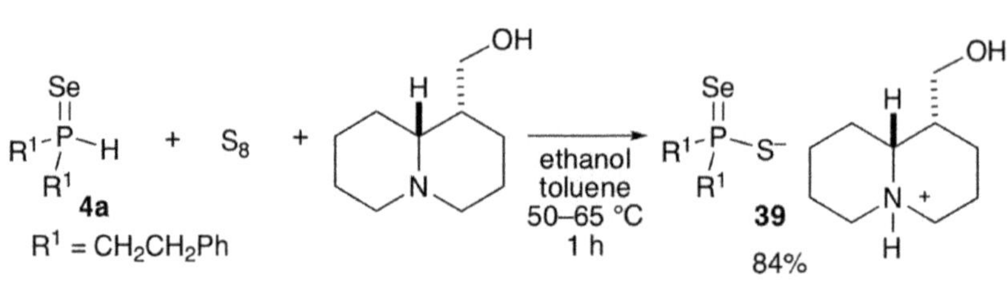

Scheme 8.20 Synthesis of phosphinoselenothioates.

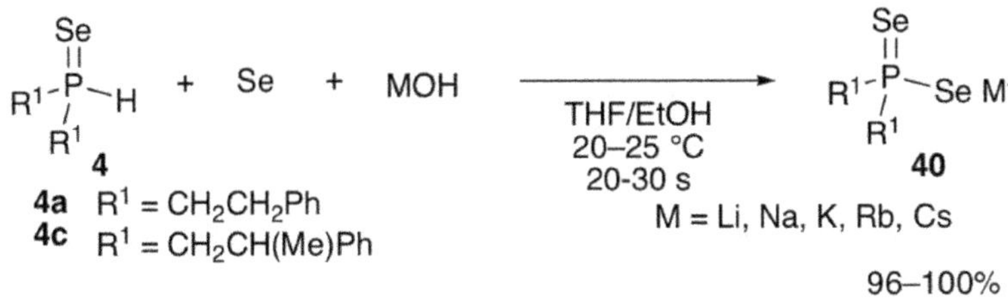

Scheme 8.21 Reaction of **4** with elemental selenium in the presence of alkali metal hydroxides.

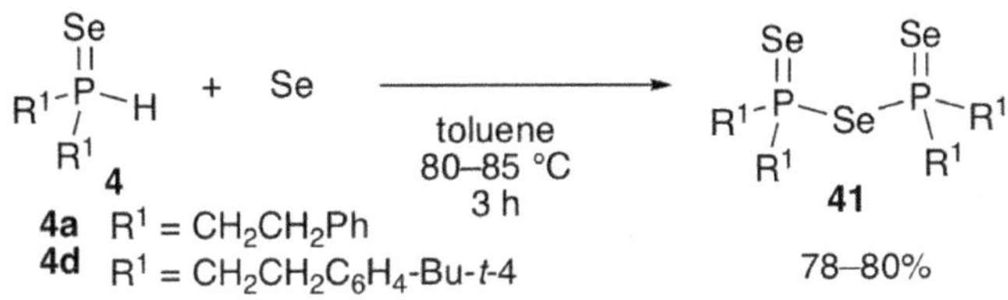

Scheme 8.22 Reaction of **4** with elemental selenium.

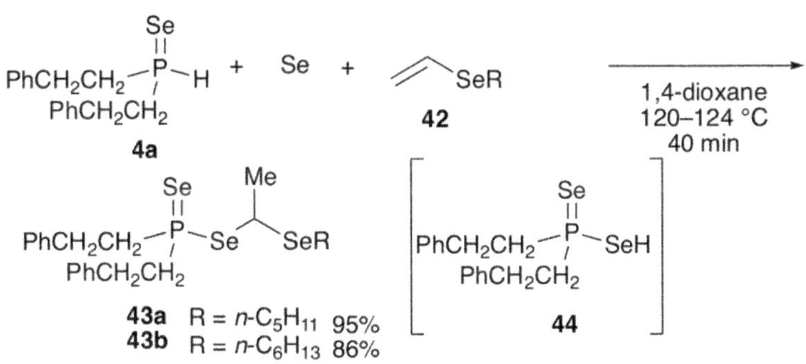

Scheme 8.23 Reaction of **4** with elemental selenium and vinyl selenides **42**.

Ammonium salts could be also prepared by using ammonia[55] and hydrazine[56] instead of alkali metal hydroxides. Simple heating of **4** and elemental selenium in toluene gave bis(diorganophosphinoselenoyl) selenides **41** (Scheme 8.22).[57]

A similar treatment of **4a** and elemental selenium in the presence of vinyl selenides **42** gave phosphinodiselenoic acid esters **43** as a product. Initially, diselenoic acid **44** might be generated and undergo the addition reaction to **42** in a Markovnikov fashion (Scheme 8.23).[58]

In this chapter, reactions of secondary phosphine selenides have been overviewed. A range of unsaturated compounds, alcohols, and amines can be used as coupling partners. Further development of their reactions, in particular asymmetric reactions leading to the formation of organophosphorus compounds with chiral carbon centers adjacent to the phosphorus atom, is expected.

8.3 2,4-Bis(phenyl)-1,3-diselenadiphosphetane-2,4-diselenide (Woollins' Reagent)

2,4-Bis(phenyl)-1,3-diselenadiphosphetane-2,4-diselenide was synthesized by Woollins and his co-workers in 1988.[59] It has been used as the reagent for the conversion of amides to selenoamides since 2001.[60] It is now called Woollins' reagent (**WR**) and is commercially available. It is also a key substrate for the preparation of a range of compounds containing phosphorus and selenium atoms. Therefore, its preparation and synthetic applications have been reviewed in several reports.[1,61,62] This chapter then focuses on its applications after 2017. In the first report, Woollins' reagent was synthesized from $(PhP)_5$ and selenium,[59] however, the modified method used sodium selenide, generated from sodium metal, selenium, and naphthalene, which then reacted with $PhPCl_2$ and selenium to give 82 g of **WR** in 96.4% (Scheme 8.24).[63] The resulting **WR** is a dimeric form of phenylphosphine diselenide (**45**) (Scheme 8.25).[64] The monomeric form **45** was generated through gas phase thermolysis of **WR** at 150 °C followed by trapping at 10 K under Ar. The matrix-isolated **45** was further converted to diselenophosphirane **46** upon irradiation of UV ($\lambda = 334$ nm), and the visible light irradiation ($\lambda = 465$ nm) of **46** regenerated **45**.

A range of functional groups were tolerated under various conditions in the conversion of C=O to C=Se with **WR** (Scheme 8.26). The carbonyl group in 6-*tert*-butyl-4-methylpyridazin-3(2*H*)-one was selectively converted to selenocarbonyl group to form **47**.[65] Selenomaltol **48** was also obtained in good yield.[66] A variety of amides including secondary and tertiary aliphatic and aromatic substrates such as **49** and **50** were converted into the corresponding selenoamides.[67] **WR** was also employed for the synthesis of selenosugar **52**.[68] Initially, the acetyl group on the nitrogen atom in **51** could be

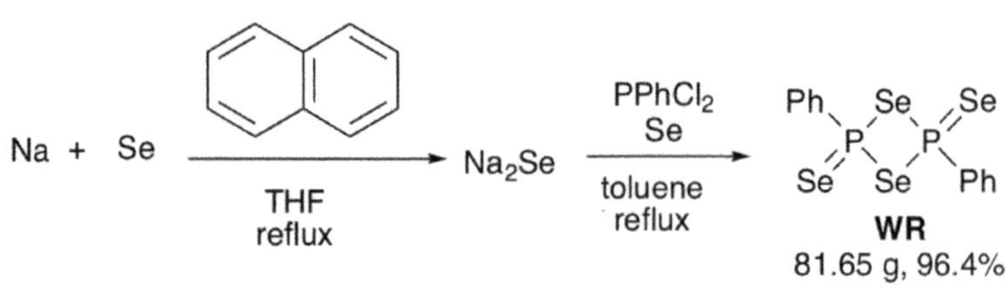

Scheme 8.24 Preparative method for Woollins' reagent **WR**.

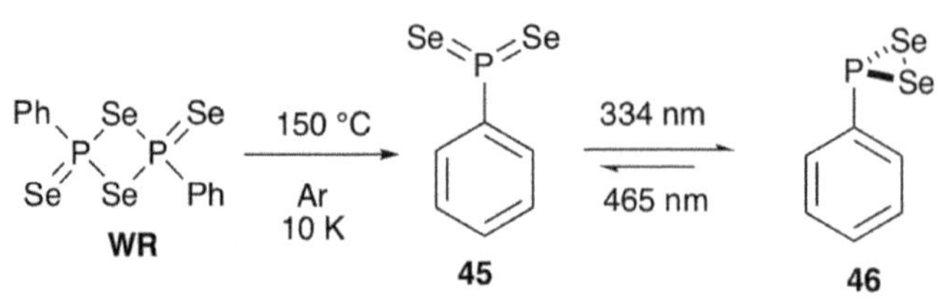

Scheme 8.25 Woollins' reagent **WR** and its monomeric form.

converted to the selenoacetyl group followed by the intramolecular cyclization to form **52**.

WR promotes reduction of alkenyl and carbonyl groups in some specific substrates (Scheme 27). The treatment of α,β-unsaturated thiazo and selenazolidinones **53** with **WR** gave the products **54**.[69] Instead of converting their C═O to C═Se groups, **WR** reduced the double bond between the amide and

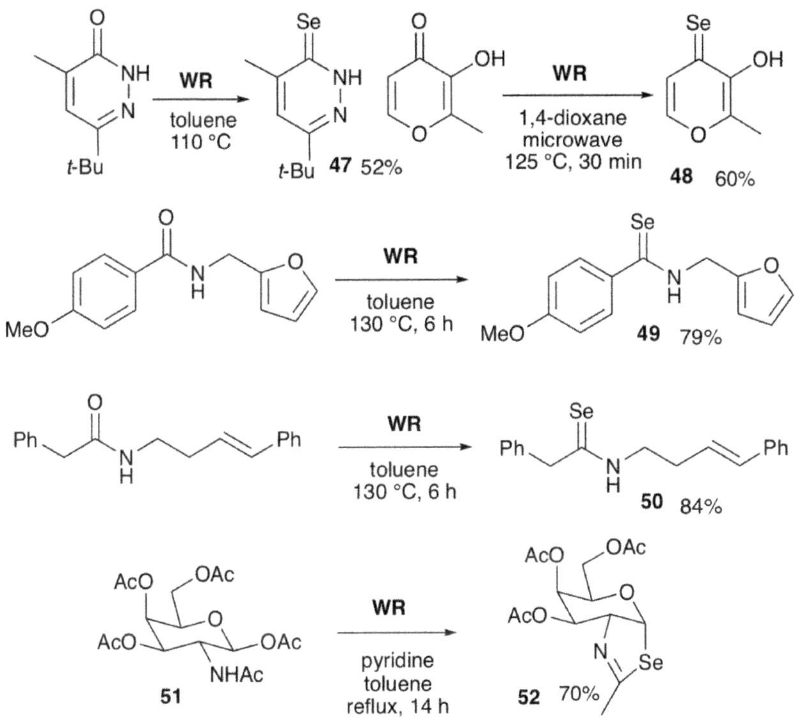

Scheme 8.26 Conversion of C═O to C═Se with Woollins' reagent **WR**.

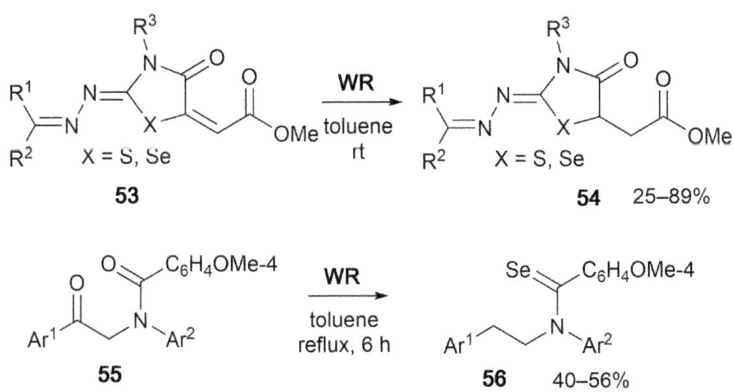

Scheme 8.27 Reduction with Woollins' reagent **WR**.

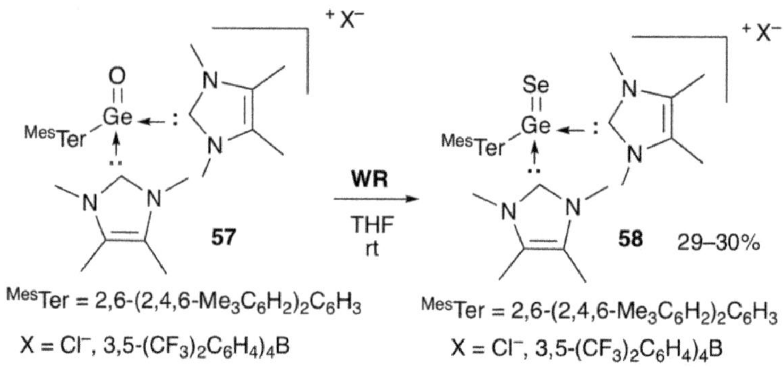

Scheme 8.28 Selenation of germa-acylium ions with Woollins' reagent **WR**.

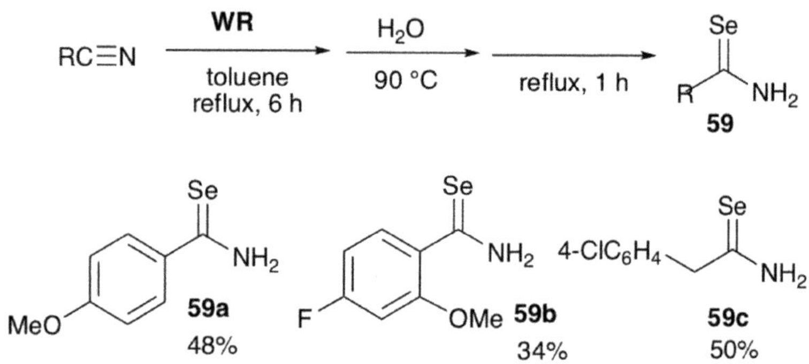

Scheme 8.29 Reaction of nitriles with Woollins' reagent **WR**.

ester moieties selectively. In the reaction of **55** with **WR**, not only selenation of the amide group but also reduction of the other carbonyl group took place in one-pot to give **56**.[70]

WR was further used for the selenation of germa-acylium ions **57** to form germa-selenoacylium ions **58**,[71] and its structure was successfully determined by X-ray analyses (Scheme 8.28).

WR could mediate transformation of cyano substrates to primary selenoamides **59**, which were generally labile compounds (Scheme 8.29).[67] The reaction was applicable for the formation of aromatic and aliphatic primary selenoamides **59**.

Several types of derivatization of **WR** have also been studied. The reaction of **WR** with diols followed by dialkylation proceeded smoothly to give cyclic products **60** as stable compounds in good yields (Scheme 8.30).[72] In this case, hydroxy groups in a diol attached to a different phosphorus atom in **WR** to generate diselenoate, which then reacted with dibromide.

In contrast, the reaction of **WR** with dithiols gave products **61** wherein two sulfenyl groups were attached to the identical phosphorus atom (Scheme 8.31).[73]

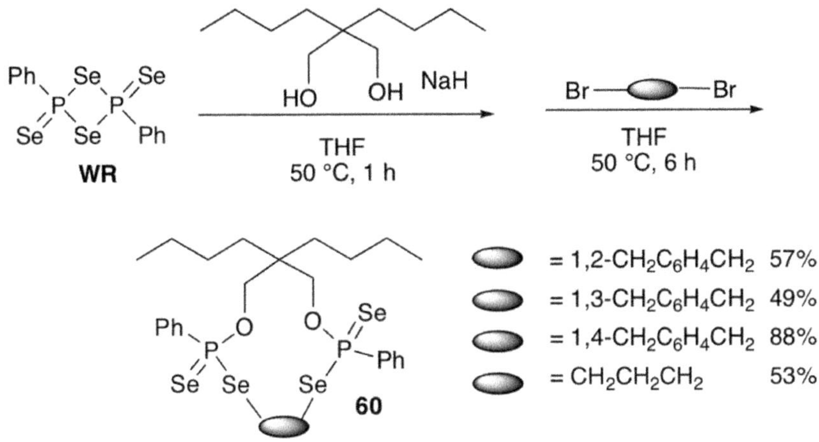

Scheme 8.30 Reaction of **WR** with diol and dibromides.

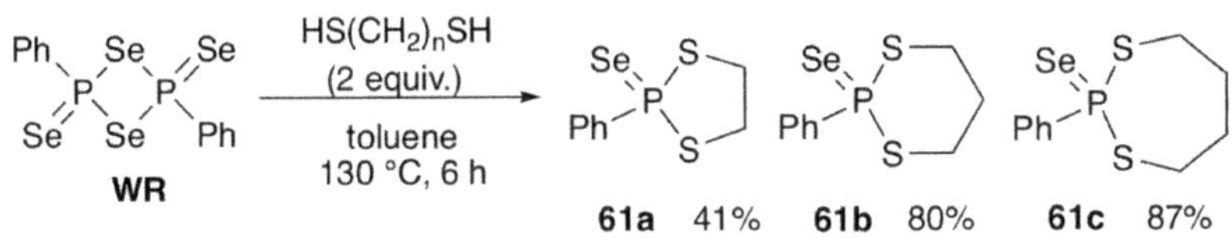

Scheme 8.31 Reaction of **WR** with dithiols.

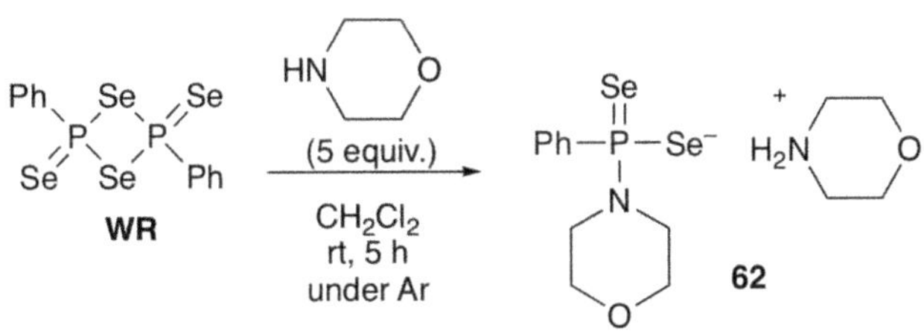

Scheme 8.32 Reaction of **WR** with morpholine.

Two phosphorus–selenium bonds in **WR** were cleaved with morpholine to give diselenophosphinoamidic acid salt **62** (Scheme 8.32).[74] The resulting salt **62** was employed as a hydrolysis-based H_2Se donor for biological application.

The reaction of **WR** with α,β-unsaturated ketone **63** gave cyclic products **65** and **66** (Scheme 8.33).[75] **WR** undergoes different types of cycloaddition reaction depending on the reaction partners to construct various ring sizes. [4 + 2]-cycloaddition of **63** with monomeric form **45** of **WR** took place to produce **65**, whereas **63** was converted to selenoketone **64**, which then underwent [4 + 2]-cycloaddition with **45** to form **66**.

WR could participate in the [2 + 2]-cycloaddition reaction with aldehyde **67** to form **68** (Scheme 8.34).[76]

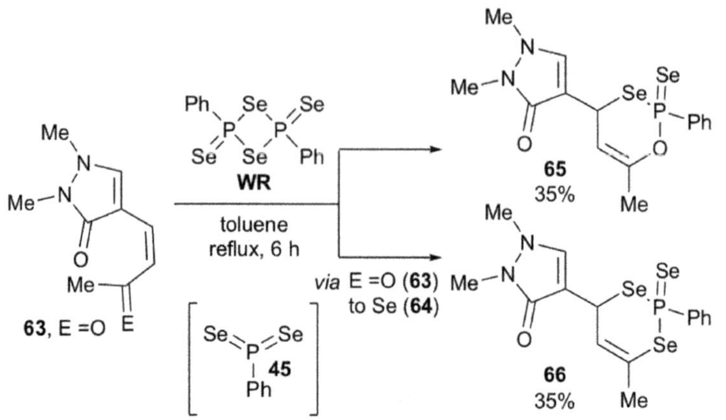

Scheme 8.33 Reaction of **WR** with α,β-unsaturated ketone.

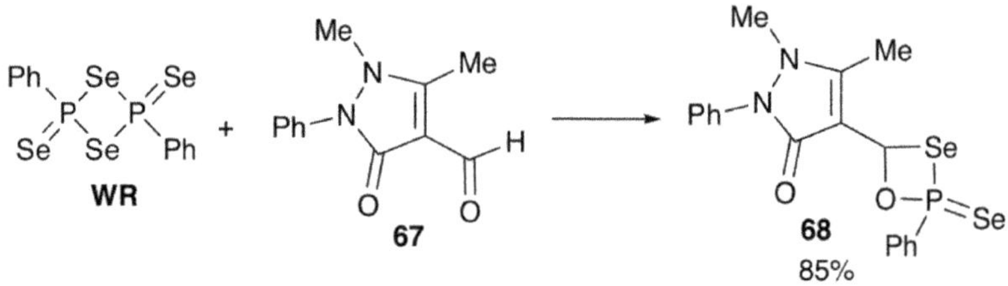

Scheme 8.34 Reaction of **WR** with α,β-unsaturated aldehyde.

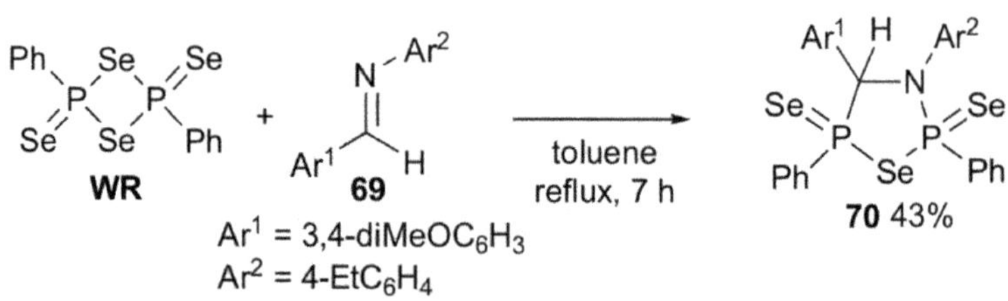

Scheme 8.35 Reaction of **WR** with imine.

However, reaction of **WR** with imine **69** formed cyclized product **70** *via* [3 + 2]-cycloaddition (Scheme 8.35).[77]

Reaction of **WR** with an amido-Schiff base has been studied. The product structure depended on the substitution patterns of substrates. The reaction with the substrates **71** led to the formation of ring-fused heterocycles **72** (Scheme 8.36).[78]

In the reaction of **WR** with cyanamides **73**, [3 + 2]-cycloaddition reaction between selenium-extruded species from **WR** and a cyano group took place to form phosphorus and selenium-containing five-membered heterocycles

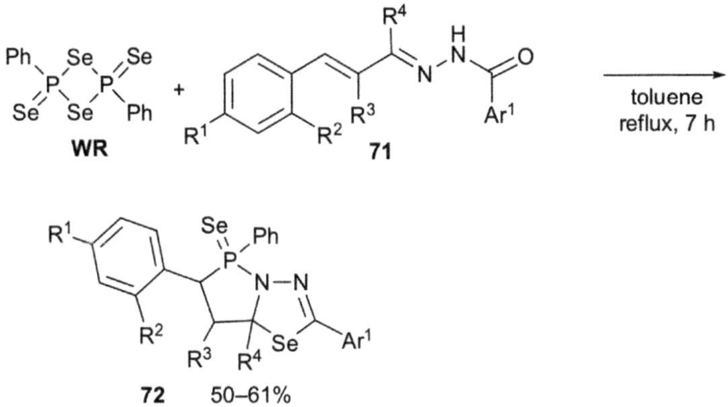

Scheme 8.36 Reaction of **WR** with amido-schiff base.

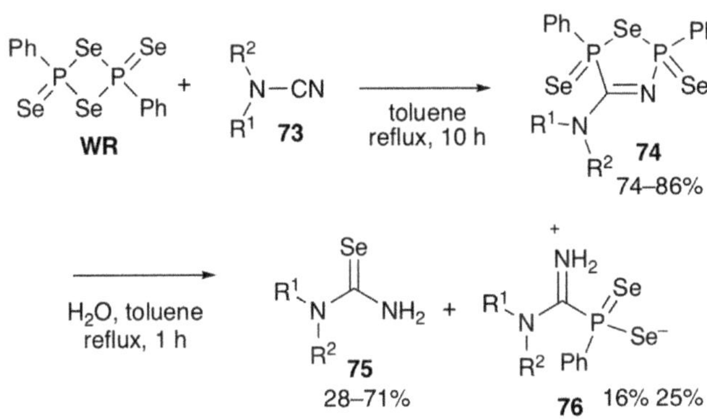

Scheme 8.37 Reaction of **WR** with cyanamides.

74 (Scheme 8.37).[79] The hydrolysis of the products **74** gave selenoureas **75** and phosphonodiselenoates **76**.

In this section, the wide applicability of Woollins' reagent was proved as a selenating agent of carbonyl, cyano, and Ge=O groups and of key substrates leading to various types of phosphorus and selenium-containing heterocycles. Further development of **WR** is highly expected.

8.4 P=Se Compounds as a Lewis Base Catalyst

Lewis base activation of Lewis acids is one of the powerful strategies used in catalytic, asymmetric reactions. Since P=Se compounds have a Lewis basic selenium atom, it can form adducts, namely cationic Lewis acid species, with Lewis acidic reactants, enhancing the electrophilicity of the reactants. In the case of alkene functionalization, the coordination of the selenium atom to Lewis acidic elements enhances the acidity of the elements, which facilitates electrophilic coordination of alkenes. The reaction between alkenes with an alcohol tether **77** and *N*-phenylsulfanyl phthalimide (**78a**) was

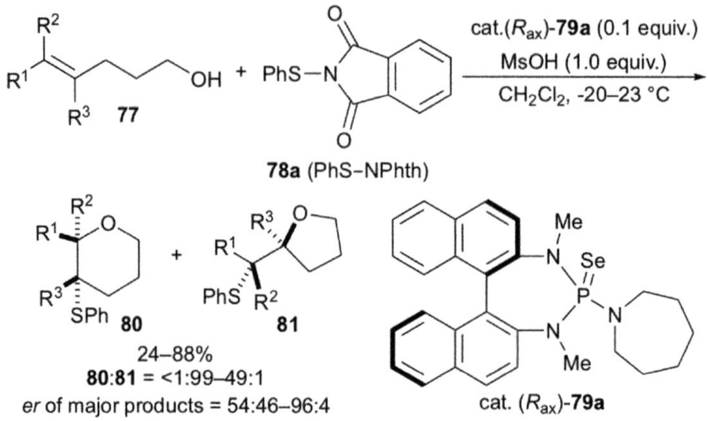

Scheme 8.38 Lewis base-catalyzed intramolecular asymmetric sulfanylation.

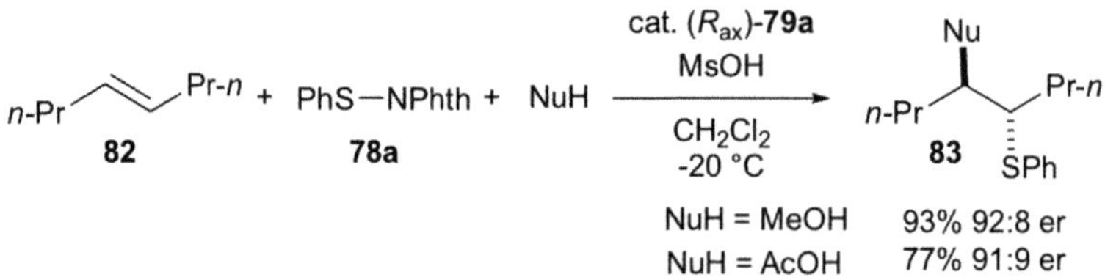

Scheme 8.39 Lewis base-catalyzed intermolecular asymmetric sulfanylation.

catalyzed by 0.1 equivalent of phosphoroselenoylamide (R_{ax})-**79a** at low temperatures in the presence of methanesulfonic acid (MsOH) to give tetrahydropyrans **80** and tetrahydrofurans **81** (Scheme 8.38).[80] The ratio of **80** and **81** as well as the enantioselectivity of the products depended on the substitution patterns of the double bond in **77**. Terminal alkenes and *Z*-alkenes generally gave five-membered ring **81** with high selectivity. Among various alkenes, *E*-alkene **77** ($R^1 = CH_2CH_2Ph$, R^2 and $R^3 = H$) underwent the reaction with the highest enantioselectivity. Introducing substituents on 2,6-positions of *N*-phenylsulfanyl phthalimide highly enhanced the enantiomeric ratio of the products.[81]

Intermolecular sulfanylation of unactivated alkene **82** with methanol or acetic acid also proceeded smoothly to afford the products **83** in high yields with a high enantiomeric ratio (Scheme 8.39). The details of a catalytic cycle are shown in Scheme 8.40. Protonation of phthalimide **78a** may give cationic species **84** followed by nucleophilic attack of the selenium atom in **79** to generate active catalytic species **85**. The phenylsulfanyl group in **85** may transfer to alkenes to form sulfonium ions **86**, which can be the enantio-determining step. Finally, nucleophiles stereoselectively attack the carbon atom attached to the sulfur atom in **86** to give the final products and regenerate the catalyst **79**. The reaction mechanism was studied by kinetic,

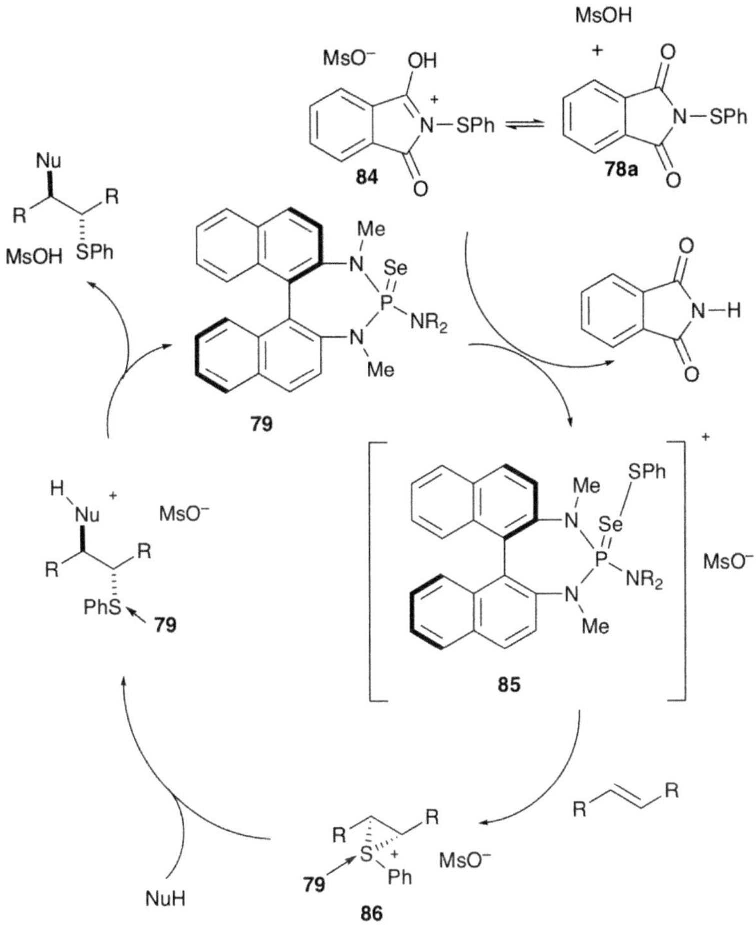

Scheme 8.40 Proposed catalytic cycle for asymmetric sulfenylation.

crystallographic, and computational investigations, leading to the understanding of how the catalyst controls enantioselectivity of the reaction.[81]

(*E*)-Cinnamyl phenols **87** could also participate in intramolecular asymmetric sulfanylation (Scheme 8.41).[82] The use of the azepane-substituted catalyst (R_{ax})-**79a** gave the corresponding products with a lower enantiomeric ratio (65.6 : 34.4). Instead, the catalyst (R_{ax})-**79b**, wherein a di(isopropyl) amino group is attached to the phosphorus atom, worked efficiently to give the corresponding products **88** with a high enantiomeric ratio.

In the reaction of *ortho*-geranylphenols **89** with *N*-arylsulfanyl phthalimide **78b** using catalyst **79b** in hexafluoroisopropyl alcohol (HFIP), a catalytic cascade cyclization took place to form tricyclic products **90** (Scheme 8.42).[83] While the reaction in Scheme 8.38 was observed to be first order in both catalyst and olefin and zeroth order in sulfenylating agent, this polyene cyclization showed a unique kinetic behaver that the reaction had a first-order

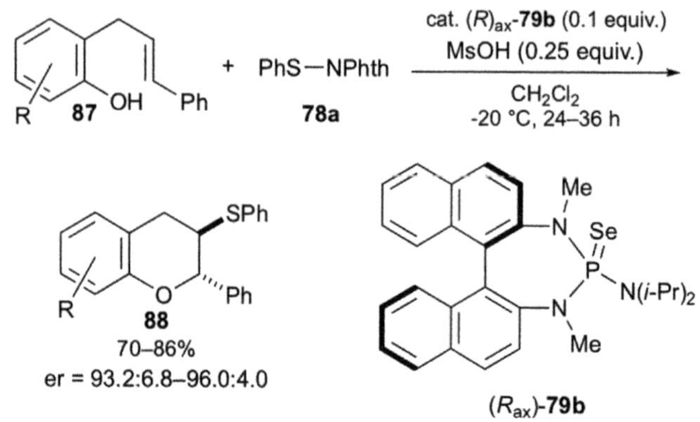

Scheme 8.41 Lewis base-catalyzed intramolecular asymmetric sulfanylation of (*E*)-cinnamylphenols.

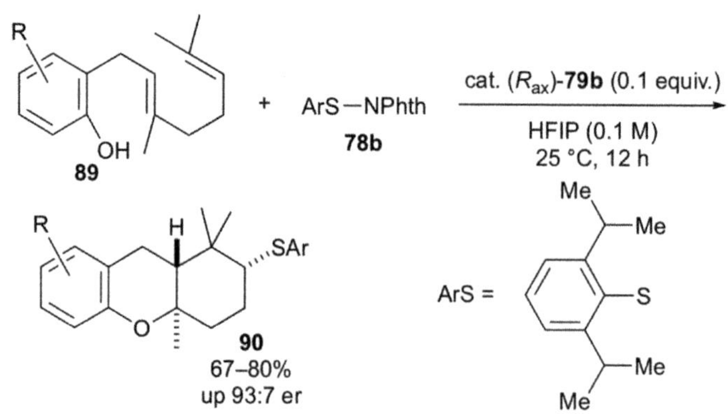

Scheme 8.42 Lewis base-catalyzed intramolecular asymmetric sulfanylation of *ortho*-geranylphenols.

concentration dependence on the catalyst and a fractional-order concentration dependence on olefin.

The catalytic reaction with phosphoroselenoylamide **79** was extended to carbosulfanylation of butenylarenes **91** with **78**. The reaction proceeded in the presence of ethanesulfonic acid and 0.1 equivalent of (S_{ax})-**79c** to give tetrahydronaphthalenes **92** with good to high enantiomeric ratio (Scheme 8.43).[84] The use of (S_{ax})-**79c** as a catalyst afforded *trans* products **92** with higher enantioselectivities than the use of (S_{ax})-**79a** and (S_{ax})-**79b**. Surprisingly, the reaction rate without the catalyst **79c** was nearly equal to that of the catalyzed reaction.[85] The mechanistic studies suggested that sulfonate and phthalimide byproducts might suppress the background reaction catalyzed by Brønsted acid. The detailed stereochemical aspects of the carbosulfanylation[86] and mechanistic investigations of sulfenofunctinalization of alkenes[87] were further reported.

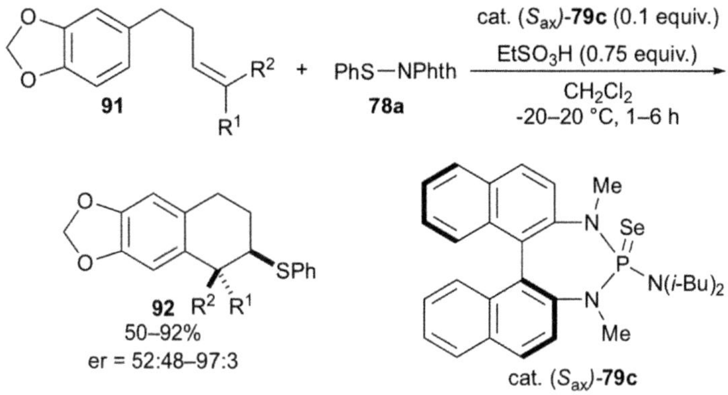

Scheme 8.43 Lewis base-catalyzed intramolecular carbosulfanylation.

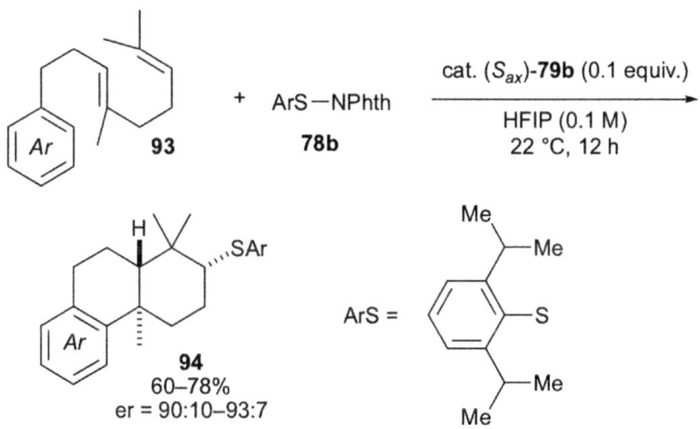

Scheme 8.44 Lewis base-catalyzed intramolecular asymmetric sulfanylation of homogeranylarenes.

Similarly to Scheme 8.42, the phosphoroselenoylamide (S_{ax})-**79b** could catalyze the cascade cyclization reaction of **93** with phthalimide **78b** to give polycyclic compounds **94** with a high enantiomeric ratio (Scheme 8.44).[88]

Up until this point, only achiral Brønsted acid has been employed in sulfanofunctionalization reactions. The use of chiral Brønsted acid (S_{ax})-**97** as a co-catalyst was found to assist enantioselective sulfanylative ring expansion reaction of cyclobutanols **95** with *N*-arylthiosaccharin **96a** in the presence of a Lewis base (R_{ax})-**79** to give cyclopentanones **98** with a chiral quaternary carbon (Scheme 8.45).[89] Mixing (S_{ax})-**97**, (R_{ax})-**79**, and **96a** afforded the species **99** confirmed by ^{31}P NMR spectra. Based on this observation, detailed mechanistic studies were conducted for the following ring expansion steps of substrates **95** involving the active species **99**.

6-*endo*-Cyclization of alkenyl carboxylic acids with *N*-phenylthiosaccharin **96b** was catalyzed by (S_{ax})-**79** to give δ-lactones **101** (Scheme 8.46).[90,91]

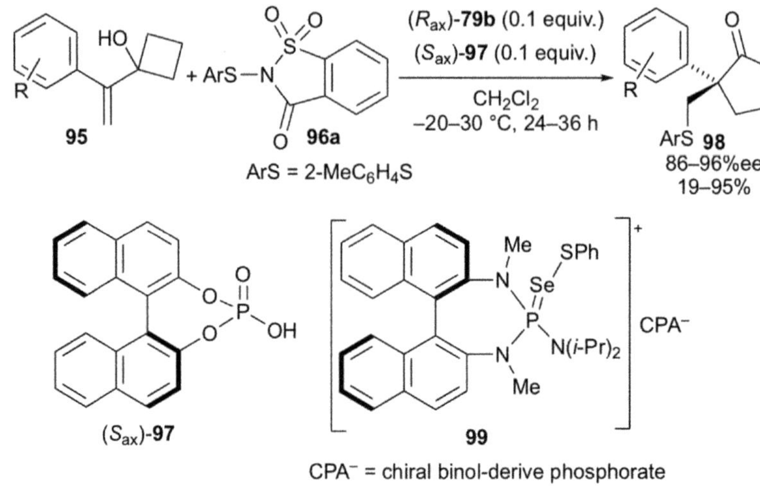

Scheme 8.45 Lewis base/Brønsted acid co-catalyzed sulfanylation.

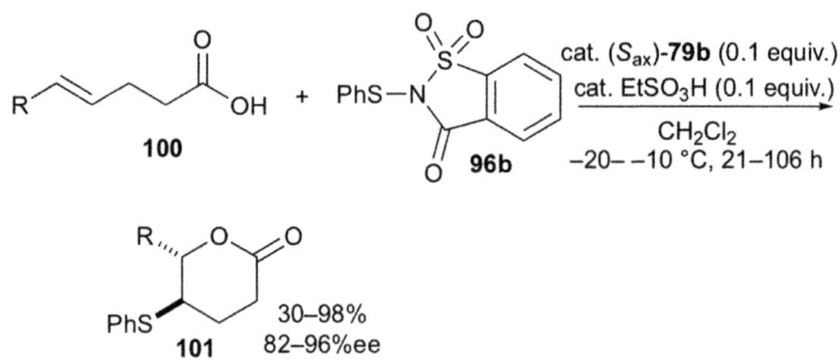

Scheme 8.46 Lewis base-catalyzed asymmetric 6-*endo*-cyclization.

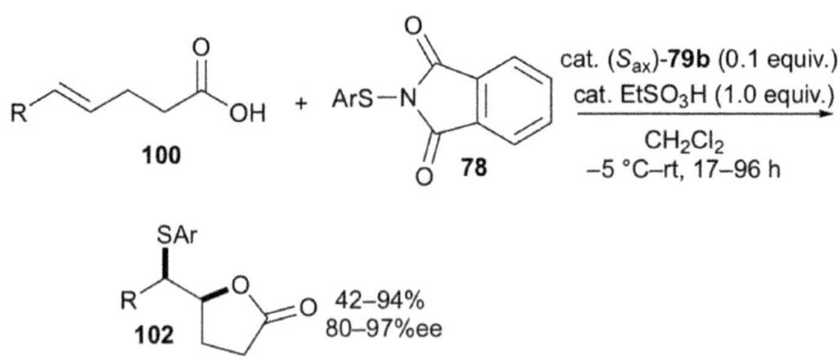

Scheme 8.47 Lewis base-catalyzed asymmetric 5-*exo*-cyclization.

Notably, the use of *N*-arylsulfanyl phthalimides **78**, instead of **96b**, facilitated 5-*exo*-cyclization of **100** to give γ-lactones **102** (Scheme 8.47).

Asymmetric sunfanylation could be applied to not only inactivate alkenes but also silyl enol ethers. Silyl enol ether **103** reacted with *N*-phenylthiosaccharin

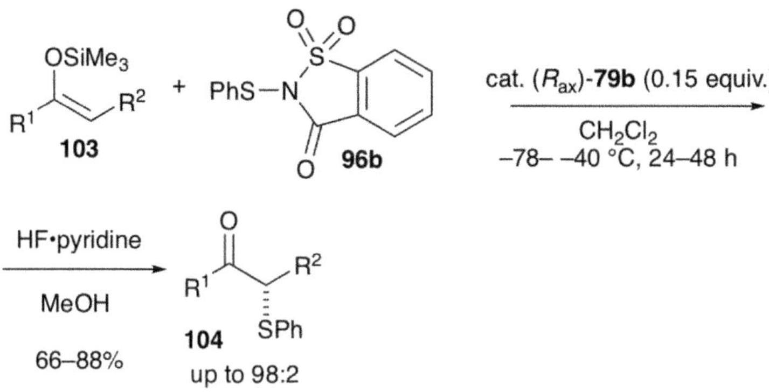

Scheme 8.48 Lewis base-catalyzed asymmetric sulfanylation of silyl enol ethers.

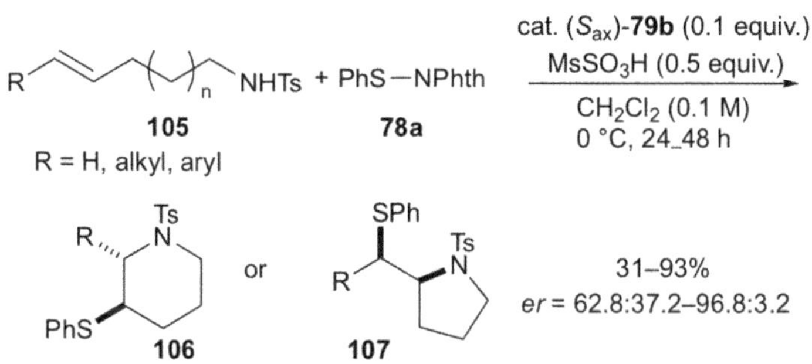

Scheme 8.49 Lewis base-catalyzed asymmetric cyclization of alkenyl *N*-tosyl amines.

96b and a catalytic amount of (R_{ax})-**79b** to afford α-sulfanyl ketones **104** (Scheme 8.48).[92]

The sulfanofunctionalization reaction was extended to amine substrates to synthesize cyclic amines. *N*-Tosyl alkenyl amines **105** underwent (S_{ax})-**79b**-catalyzed intramolecular sulfanoamination with *N*-phenylphthalimide (**78a**) to form *endo* products **106** or *exo* products **107**. The *end*/*exo* selectivity of the product depended on the alkene substituents R and the lengths of alkyl chains in **105** (Scheme 8.49).[93] In the case of electron-rich and neutral substrates, the products were formed in good to high yields, although the reaction of **105** with an electron-withdrawing group ($R = C_6H_4CF_3$-4) was not complete after 48 h and resulted in a low yield (39%) of the corresponding product **106**. *Z*-Alkenyl amine reacted slowly to give the product with a low enantiomeric ratio. The suggested catalytic cycle was similar to Scheme 8.40, but an amine tether captured thiiranium ion as a nucleophile in this case.

Intermolecular sulfanoamination of alkenes **108** was also developed in the reaction with **78b** using (S_{ax})-**79b** as a catalyst (Scheme 8.50).[94] The reaction showed a broad substrate scope with high regio- and stereo-selectivities. In the reaction of aromatic alkenes such as styrene, an amino group was selectively introduced to the carbon atom attached to the aromatic ring, and an arylsulfanyl group was introduced to the other carbon atom in the alkene

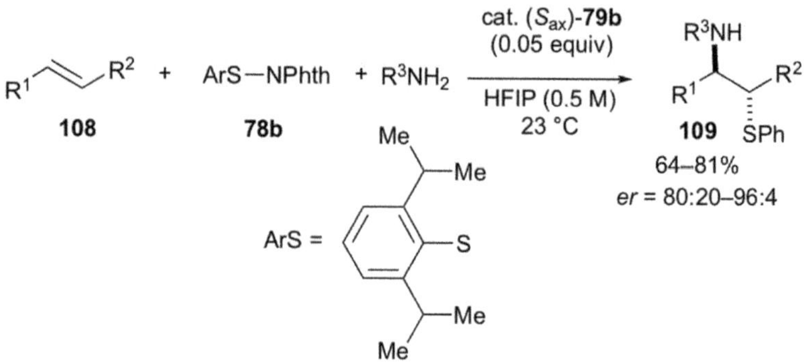

Scheme 8.50 Lewis base-catalyzed intermolecular asymmetric sulfanoamination.

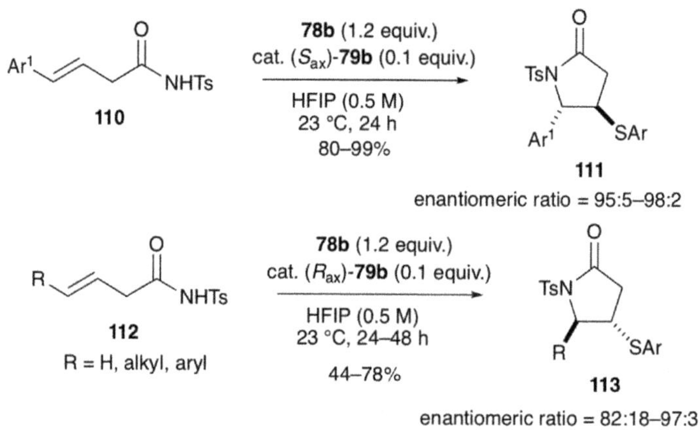

Scheme 8.51 Lewis base-catalyzed asymmetric 5-*endo*-cyclization.

moiety. In the scope of amine substrates, aromatic amines with a wide range of substituents such as halogen, methoxy, hydroxy, pivaloyloxy, acetoxy, ethoxycarbonyl, Boc-protected amino pinacolboryl, and 2-pyrdyl groups as well as benzylic amines are tolerated under optimized conditions. The use of HFIP as a solvent was key to facilitating sulfur transfer of **78** by the protonation.

β,γ-Unsaturated sulfonyl carboxyamides **110** and **112** were subjected to an intramolecular sulfanoamidation reaction leading to the formation of γ-lactams **111** and **113** (Scheme 8.51).[95] In both cases, *endo*-cyclization took place, and the nitrogen atom of the amide group in **110** and **112** was attached to the carbon atom at the γ-position of products **111** and **113**.

The three-component coupling reaction of alkenyl aldehydes **114** and **116** with **78b** was catalyzed by (S_{ax})-**79b** in HFIP solvent (Scheme 8.52).[96] The reaction proceeded through intramolecular cyclization to generate an oxocarbenium ion followed by capturing it with HFIP to afford tetrahydropyrans **115** and 1,3-disubstituted isochromans **117** in good yields with a good to excellent enantiomeric ratio.

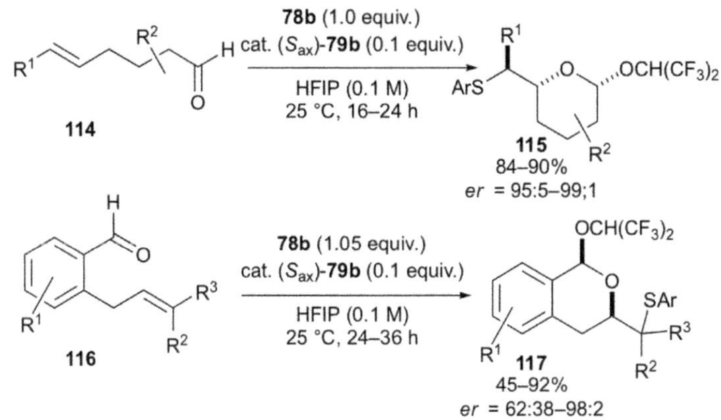

Scheme 8.52 Lewis base-catalyzed three-component coupling reaction.

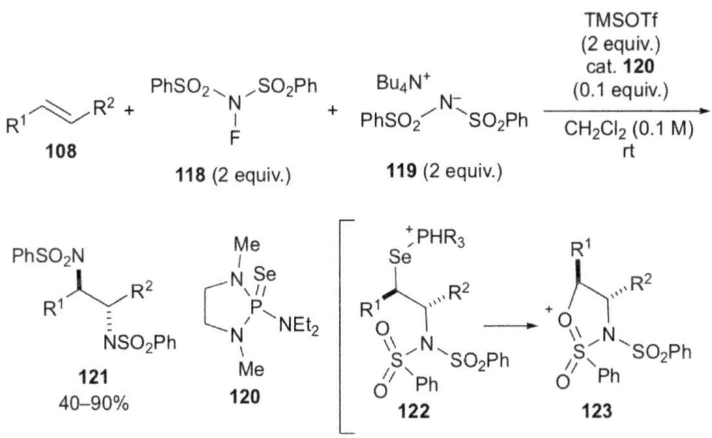

Scheme 8.53 Lewis base-catalyzed 1,2-diamination of alkenes.

$Ph_3P{=}Se$ and $(PhO)_3P{=}Se$ were also used as Lewis basic catalysts for intramolecular halocyclization.[97,98] Phosphoroselenoylamide **120** could catalyze the 1,2-diamination reaction of alkenes **108** with *N*-fluorobenzenesulfonimide **118** and ammonium benzenesulfomidate **119** to give 1,2-diaminoalkanes **121** in moderate to excellent yields (Scheme 8.53).[99] The use of (*E*)-alkenes gave *anti*-products **121** selectively. The *anti*-stereoselectivity may be derived from the initial *anti*-addition of the catalyst **120** and *N*-fluorobenzenesulfonimide **118** to an alkene to generate the adduct **122**. Then, **122** undergoes intramolecular S_N2 *via* the attack of the sulfonyl oxygen to the carbon atom attached to the selenium atom to form intermediate **123** and regenerate **120**. Finally, the intermolecular S_N2 reaction of **123** with **119** may give rise to the formation of *anti*-**121**. Functional groups such as halogen, alkoxycarbonyl, formyl, and azide groups remained intact during the reaction.

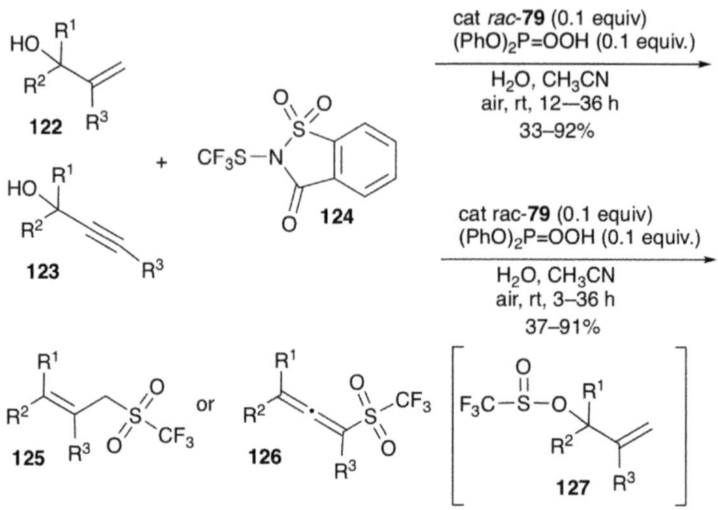

Scheme 8.54 Lewis base-catalyzed sulfonylation of allylic and propargylic alcohols.

Allylic alcohols **122** and propargylic alcohols **123** reacted with *N*-trifluoromethylsulfanylsaccharin (**124**) in the presence of water using a catalytic amount of racemic **79** and diphenyl phosphate (Scheme 8.54).[100] The reaction provided allylic and allenic triflones wherein one of the oxygen atoms in the triflyl group was derived from water and the triflyl group was formally introduced to the remote alkenyl and alkynyl carbon from the hydroxy group. One of the proposed catalytic cycles involved the generation of intermediates **127**, which then underwent [2,3]-sigmatropic rearrangement to form the final products **125**. The reaction was limited to substrates with aromatic substituents as the R^1 or R^2 group so that aliphatic allylic alcohol could not participate in the sulfonylation reaction. The reaction of the allylic alcohols with two different carbon substituents as R^1 and R^2 exhibited high stereoselectivity of the resulting alkenyl groups.

8.5 P=Se Compounds with a Binaphthyl Group

The binaphthol skeleton represents one of the widely used chiral sources in asymmetric synthesis. A binaphthyloxy group has been incorporated in organophosphorus compounds with a P=Se group such as **128–132** as shown in Scheme 8.55. Among them, compounds **130** and **131** can be prepared by reacting the corresponding phosphites[101–104] and phosphoramidites[105–108] with elemental selenium. These selenium adducts are used to evaluate the σ-donor ability of the phosphorus atom in their parent compounds based on the coupling constants between ^{31}P and ^{77}Se atoms.

Alternatively, phosphoroselenoyl chloride **128**, reported in 2006, can be a starting material for P=Se compounds with a binapthyloxy group (Scheme 8.56).[109] Treatment of phosphoryl chloride with (R_{ax}) or (S_{ax})-1,1′-bi-2-naphthol

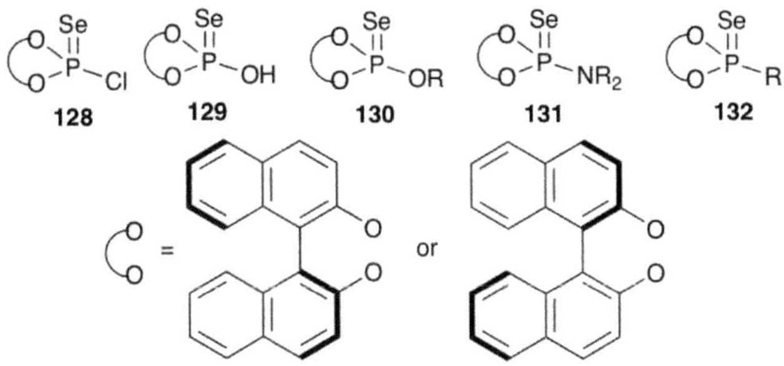

Scheme 8.55 Representative examples of P=Se compounds with a binaphthyloxy group.

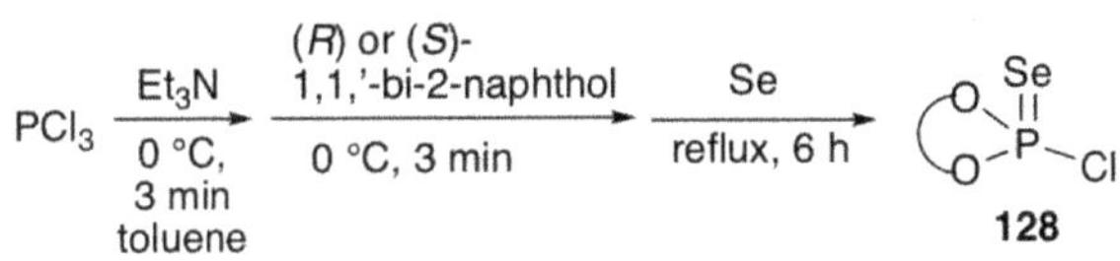

Scheme 8.56 Synthesis of phosphoroselenoyl chlorides with a binaphthyloxy group.

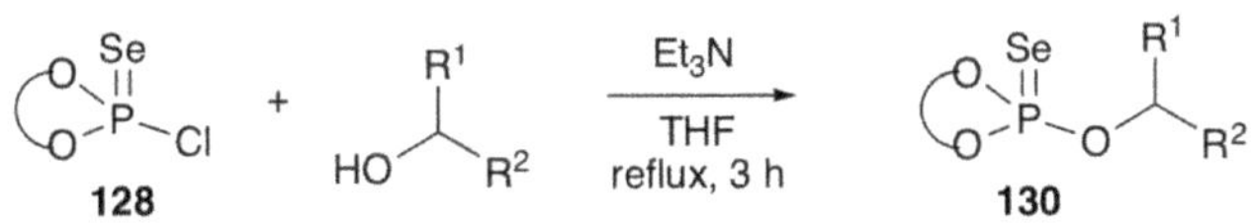

Scheme 8.57 Reaction of phosphoroselenoyl chloride with secondary alcohols.

and elemental selenium in the presence of NEt_3 gave **128** in high yield. 1,1′-Bi-2-naphthol with substituents at the 3,3′-positions as a starting material could be used to synthesize derivatives of **128**.[110] The chloride **128** is a bench-stable colorless solid, which readily reacts with a range of nucleophiles under basic reaction conditions.

The reaction of **128** with secondary alcohols in the presence of Et_3N under reflux in THF went to completion within 3 h to give diastereomeric mixtures of **130** (Scheme 8.57).[111] The diastereomers were readily distinguished by ^{31}P and ^{77}Se NMR spectra, and this discrimination could also be observed in the case of primary alcohols with a chiral carbon atom far from the hydroxy group.[112] The reaction of **128** with secondary 2-aryl-1-cycloalkanols led to their kinetic resolution to give enantioenriched unreacted 2-aryl-1-cycloalkanols and diastereomerically-enriched phosphoroselenoic acid esters **130**.[113]

Diastereoselective phosphorylation of *meso*-inositol **133** was carried with chloride **128** to give the product **134** with good diastereoselectivity (Scheme 8.58).[114] A similar reaction with oxygen and sulfur analogs of **128** was not as selective and efficient as that with **128**. Introducing a 3,3′-di-bromo-octahydrobinaphthyl group **135** instead of a simple binaphthyloxy group improved the diastereoselectivity to 99 : 1.[115]

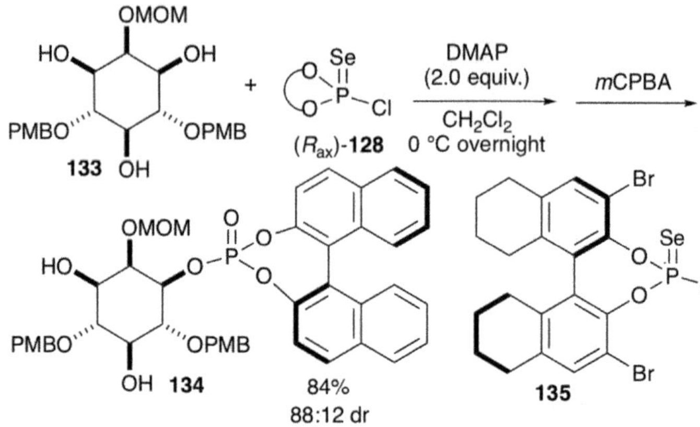

Scheme 8.58 Diastereoselective phosphorylation of *meso*-inositol.

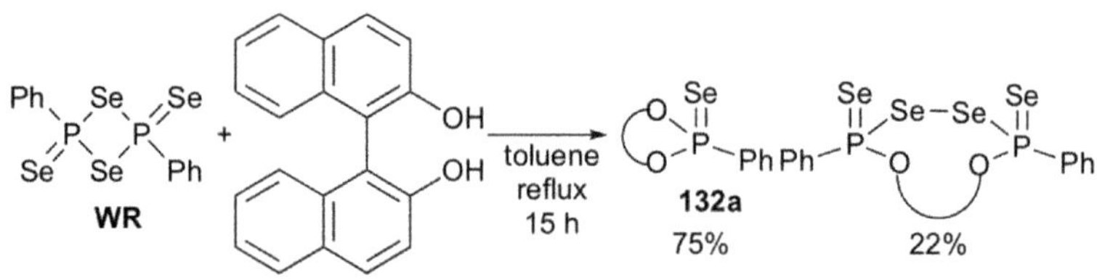

Scheme 8.59 Reaction of 1,1′-bi-2-naphthol with Woollins' reagent **WR**.

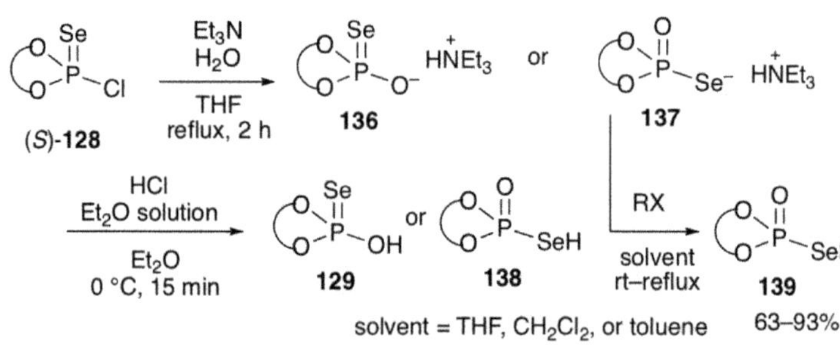

Scheme 8.60 Generation of phosphoroselenoic acids and ammonium salts and reaction with alkyl halides.

Woollins' reagent (**WR**) reacted with 1,1′-bi-2-naphthol under reflux in toluene to give phenyl phosphonoselenoate **132** along with the diselenide, wherein hydroxy groups were attached to the different phosphorus atom in **WR** (Scheme 8.59).[116]

Hydrolysis of the chlorides (S_{ax})-**128** in the presence of Et_3N generated ammonium salts **136** and/or **137** (Scheme 8.60).[117] Alkylation of the salts selectively took place at the selenium atom of **136** and/or **137** to give phosphoroselenoic acid Se-esters **139** in good yields. The treatment of the salts **136** and/or **137** with ether solution of HCl could form phosphoroselenoic acid **129** and/or **138**. The predominant structures (**136** *vs.* **137** and **129** *vs.* **138**) in solution were determined by J_1 coupling constants of ^{31}P and ^{77}Se atoms showing that **129** and **136** were major species in $CDCl_3$.[110] The acidity of acid **129** was calculated suggesting that the pK_a values of **129** were smaller than those of the corresponding phosphoric acid with the order of 10 to the 4th power.[118]

Primary and secondary amines reacted with chloride **128** to give phosphoroselenoylamides **140** (Scheme 8.61).[119] In the reaction of 2 equivalent of racemic amines, two diastereomers are formed in a different ratio and separated by fractional crystallization. The selenium atom of the obtained diastereomerically-enriched amides was extracted with tributylphosphine to form phosphoramidites, which were used as ligands for Ir-catalyzed hydrogenation of an imine. The lability of trivalent phosphoramidites depended on the substitution patterns on the nitrogen atom. Phosphoramidites derived from primary amines were generally labile and readily oxidized under air.

Phosphoroselenoylamides (R_{ax}) or (S_{ax})-**141**[120–123] and (S_{ax})-**142**[124] have been employed as Brønsted acids and Lewis base catalysts, respectively (Scheme 8.62).

To provide phosphonoselenoic acid esters **132** wherein the phosphorus–carbon bond is present, phosphorodiselenoic acid phenyl esters **143** reacted with alkenes **108** under radical reaction conditions to give products **144** (Scheme 8.63).[125] The reaction proceeds in an anti-Markovnikov fashion.

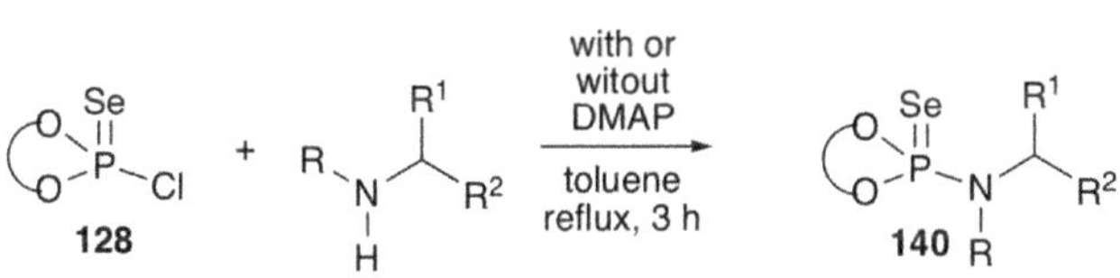

Scheme 8.61 Reaction of phosphoroselenoyl chloride with amines.

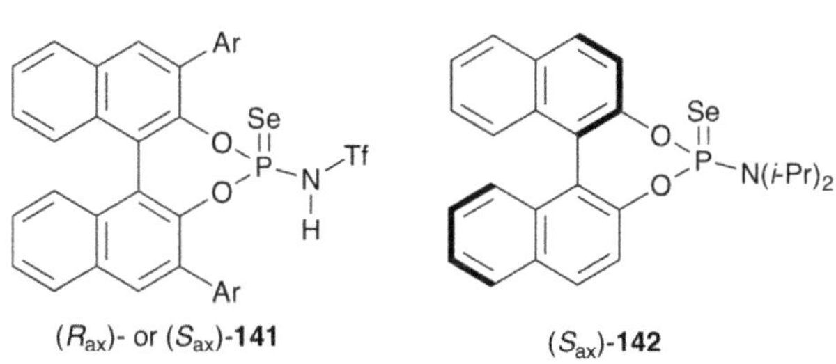

Scheme 8.62 Phosphoroselenoic amides used as a catalyst.

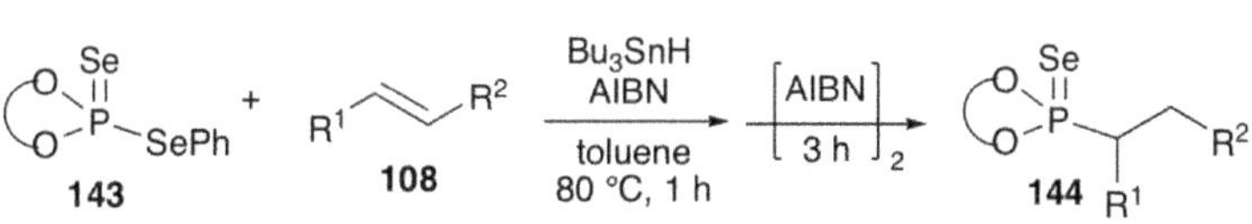

Scheme 8.63 Reaction of phosphorodiselenoic acid phenyl esters with alkenes.

The reaction of cyclic alkenes gave the products **144** with relatively higher diastereoselectivities than those of acyclic alkenes.

The substitution reaction at the phosphorus atom of chloride **128** with Grignard reagents efficiently took place to give the corresponding phosphonoselenoic acid esters **132** in moderate to high yields (Scheme 8.64).[126] A similar reaction to phosphoric acid chlorides gave the acid esters only in low yields along with the formation of phosphinoic acid esters. No reaction took place with phosphorothionyl chlorides under identical reaction conditions. The reaction conditions tolerated a variety of Grignard reagents such as primary to tertiary alkyl Grignard reagents and aromatic Grignard reagents.

The adjacent carbon to the P=Se group in **145** could be deprotonated by lithium di(isopropyl)amide (LDA) to generate carbanions **146** next to the phosphorus atom (Scheme 8.65).[127] The subsequent alkylation of **146** gave phosphonoselenoates **147** with a chiral trisubstituted carbon. The reaction generally showed good to high diastereoselectivity except for the reaction of **145** with an ethyl group on the phosphorus atom. The resulting phosphonoselenoates were converted to chiral tertiary and primary phosphines.

The reaction of **147a** with excess ^{n}BuLi gave diastereomerically-enriched tertiary phosphine **148** (Scheme 8.66). Reduction of **144a** with lithium

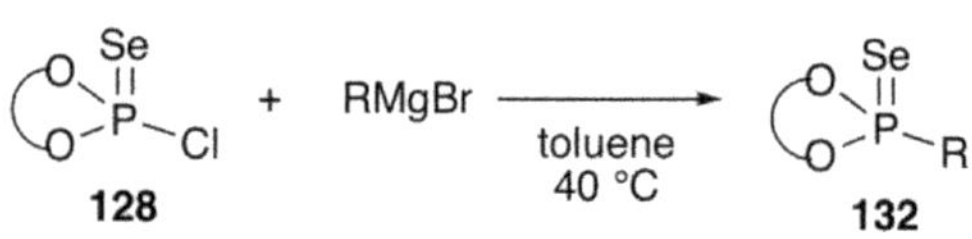

Scheme 8.64 Reaction of phosphoroselenoyl chloride with Grignard reagents.

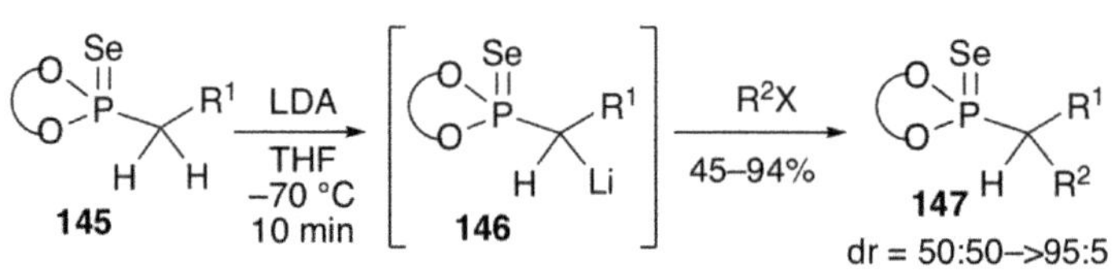

Scheme 8.65 Deprotonation of phosphonoselenoates and subsequent alkylation.

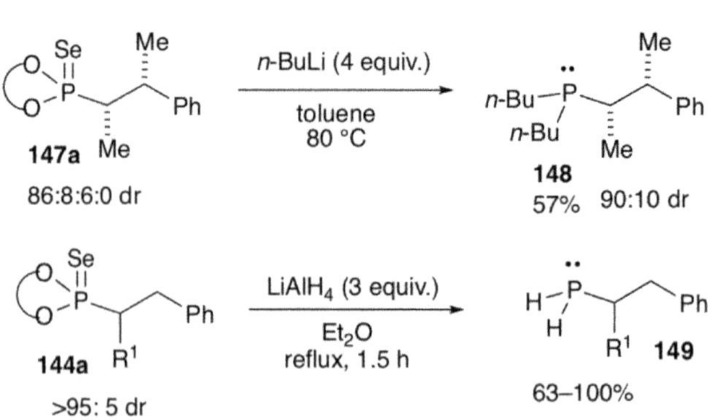

Scheme 8.66 Conversion of phosphonoselenoates to phosphines.

aluminum hydride formed primary phosphines **149**, although an enantiomeric ratio of **149** was not determined.[128]

In the transformations described above, phosphorus–oxygen (P–O) bonds stitching a binaphthyloxy group to a phosphorus atom was retained intact in the products. On the other hand, fluoride ion cleaved P–O bonds in **140** (Scheme 8.67).[129] Treatment of **140** with tetrabutylammonium fluoride in THF gave ammonium salts **150** and **151**. The ratio of these products depended on the substituents on the nitrogen atom of **140**. The reaction may begin with the nucleophilic attack of fluoride ion to the phosphorus atom of **140** followed by the cleavage of phosphorus–oxygen bonds to generate intermediate **152**. Further hydrolysis of **152** eliminates a binaphthyloxy group to give **150**, while the elimination of a fluoride ion forms **151**.

Hydrolysis of phosphonoselenoate (S_{ax})-**145a** under basic conditions readily takes place to give phosphonoselenoic acid **153** (Scheme 8.68).[130] The reaction proceeds *via* an axis-to-center chirality transfer to afford *P*-stereogenic acid **153** with high diastereoselectivity.

In this section, the synthesis and reactions of P=Se compounds with a binaphthyloxy group were overviewed. Since the first synthesis of 1,1′-binaphthyl-2,2′-diyl phosphoroselenoyl chloride (**128**),[109] it has been used as a precursor for a wide range of derivatives with phosphorus–carbon, oxygen, –nitrogen, and selenium bonds. Notably, most of them are stable and handled under air. The corresponding phosphoric acid derivatives with a P=O bond are often more reactive than the selenium isologue under various reaction conditions, which causes undesired side reactions. In contrast, sulfur isologues are less reactive than selenium isologues and require harsh reaction conditions leading to over-reaction of the desired products.

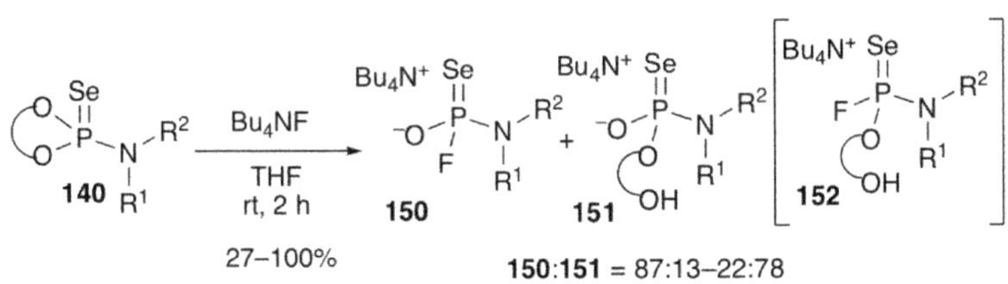

Scheme 8.67 Reaction of phosphoroselenoamides with fluoride ion.

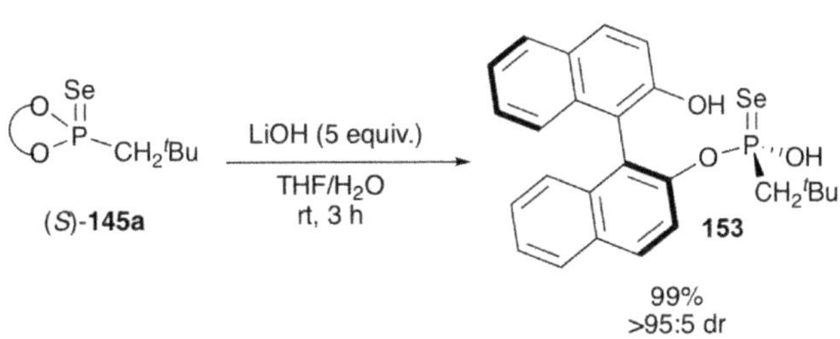

Scheme 8.68 Hydrolysis of phosphonoselenoates.

The reactivity trends toward the phosphorus center and can be depicted as follows, *i.e.*, P=O > P=Se > P=S.

8.6 Conclusions

In this chapter, we have described syntheses and synthetic applications of a range of derivatives with P=Se bonds in this century. These compounds show unique reactivity toward various transformations unlike their oxygen and sulfur isologue. In particular, phosphine selenides and Woollins' reagent have been recognized as powerful selenium donors in synthetic and material chemistry. Binaphthyl group-incorporated derivatives have also played an important role in asymmetric synthesis as a Lewis base catalyst and chirality transfer reagents. However, there are still many unexplored areas for these compounds considering how widely organophosphorus compounds with a P=O bond and trivalent phosphines have been studied in various chemistry fields. Organoselenium compounds are known as unstable molecules causing a distinctive odor which keeps researchers from being willing to study these compounds. But actually, P=Se compounds are often bench-stable and easy to handle. As these properties become known, we hope the chemistry of P=Se compounds will be developed to open up their novel utilities.

Acknowledgements

This research was partly supported by a Grant-in-Aid for Scientific Research (B) (19H02712) from MEXT.

References

1. *Best Synthetic Methods: Organophosphorus (V) Chemistry*, ed. C. M. Timperley, Elsevier, London, UK, 2015.
2. A. Marchenko, G. Koidan, A. Hurieva, V. V. Dyakonenko, S. V. Shishkina, E. B. Rusanov, A. A. Kyrylchuk and A. Kostyuk, *Eur. J. Inorg. Chem.*, 2021, 969.
3. I. Fedin, M. Goryca, D. Liu, S. Tretiak, V. I. Klimov and S. A. Crooker, *ACS Nano*, 2021, **15**, 14444.
4. W. Baek, M. S. Bootharaju, K. M. Walsh, S. Lee, D. R. Gamelin and T. Hyeon, *Nat. Mater.*, 2021, **20**, 650.
5. A. J. Touchton, G. Wu and T. W. Hayton, *Inorg. Chem.*, 2021, **60**, 17586.
6. A. J. Peloquin, C. D. McMillen, S. T. Iacono and W. T. Pennington, *ChemPlusChem*, 2021, **86**, 549.
7. T. Scattolin, V. A. Voloshkin, E. Martynova, S. M. P. V. Broeck, M. Beliš, C. S. J. Cazin and S. P. Nolan, *Dalton Trans.*, 2021, **50**, 9491.
8. M. Nassir, L. Balaomb and B. Fischer, *Chem. Commun.*, 2020, **56**, 11633.
9. M. Warminski, J. Kowalska, E. Nowak, D. Kubacka, R. Tibble, R. Kasprzyk, P. J. Sikorski, J. D. Gross, M. Nowotny and J. Jemielity, *ACS Chem. Biol.*, 2021, **16**, 334.

10. N. K. Gusarova, N. I. Ivanova, S. F. Malysheva, A. V. Artem'ev, B. V. Timokhin and B. A. Trofimov, *R. J. Gen. Chem.*, 2008, **78**, 1628.
11. N. K. Gusarova, N. I. Ivanova, N. A. Konovalova, B. G. Sukhov, L. V. Baikalova, L. M. Sinegovskaya, D. V. Pavlov and B. A. Trofimov, *Synthesis*, 2006, 4159.
12. S. F. Malysheva, V. A. Kuimov, N. A. Belogorlova, N. K. Gusarova, I. V. Taydakov, A. I. Albanov, I. L. Eremenkoc and B. A. Trofimov, *Mendeleev Commun.*, 2019, **29**, 683.
13. N. I. Ivanova, P. A. Volkov, L. V. Baikalova, N. K. Gusarova and B. A. Trofimov, *Chem. Heterocycl. Compds*, 2008, **44**, 1359.
14. N. I. Ivanova, N. K. Gusarova, N. A. Konovalova, P. A. Volkov, G. G. Levkovskaya, L. I. Larina and B. A. Trofimov, *Russ. J. Gen. Chem.*, 2009, **79**, 2102.
15. N. I. Ivanova, P. A. Volkov, N. K. Gusarova, L. I. Larina and B. A. Trofimov, *Russ. J. Gen. Chem.*, 2011, **81**, 315.
16. N. K. Gusarova, N. I. Ivanova, P. A. Volkov, K. O. Khrapova, L. I. Larina, V. I. Smirnov, T. N. Borodina and B. A. Trofimov, *Synthesis*, 2015, **47**, 1611.
17. N. K. Gusarova, S. F. Malysheva, N. A. Belogorlova, B. G. Sukhov and B. A. Trofimov, *Synthesis*, 2007, 2849.
18. N. K. Gusarova, N. A. Chernysheva, S. V. Yas'ko, T. I. Kazantseva, I. A. Ushakov and B. A. Trofimov, *Synthesis*, 2008, 2743.
19. N. K. Gusarova, P. A. Volkov, N. I. Ivanova, N. A. Chernysheva, S. V. Yas'ko, A. I. Albanov and B. A. Trofimov, *Russ. J. Gen. Chem.*, 2010, **80**, 1602.
20. N. A. Chernysheva, S. V. Yas'ko, N. K. Gusarova, A. A. Tatarinova, L. V. Klyba and B. A. Trofimov, *Mendeleev Commun.*, 2010, **20**, 346.
21. N. K. Gusarova, N. A. Chernysheva, S. V. Yas'ko, L. V. Klyba and B. A. Trofimov, *Russ. J. Gen. Chem.*, 2011, **81**, 2506.
22. B. A. Trofimov, S. F. Malysheva, N. A. Belogorlova, V. A. Kuimov, A. I. Albanov and N. K. Gusarova, *Eur. J. Org. Chem.*, 2009, 3427.
23. L. A. Oparina, S. F. Malysheva, N. K. Gusarova, N. A. Belogorlova, O. V. Vysotskaya, A. V. Stepanov, A. I. Albanov and B. A. Trofimov, *Synthesis*, 2009, 3427.
24. L. A. Oparina, N. K. Gusarova, O. V. Vysotskaya, N. A. Kolyvanov, A. V. Artem'ev and B. A. Trofimov, *Synthesis*, 2012, 2938.
25. N. K. Gusarova, S. I. Verkhoturova, T. I. Kazantseva, S. N. Arbuzova, A. I. Albanov, A. A. Tatarinova and B. A. Trofimov, *Russ. J. Gen. Chem.*, 2013, **83**, 1895.
26. S. I. Verkhoturova, T. I. Kazantseva, S. N. Arbuzova, A. I. Albanov, N. K. Gusarova and B. A. Trofimov, *Russ. J. Gen. Chem.*, 2014, **84**, 1742.
27. N. K. Gusarova, V. A. Kuimov, S. F. Malysheva, N. A. Belogorlova, A. I. Albanov and B. A. Trofimov, *Tetrahedron*, 2012, **68**, 9218.
28. N. K. Gusarova, S. F. Malysheva, N. A. Belogorlova, L. N. Parshina and B. A. Trofimov, *Synthesis*, 2011, 1777.

29. S. F. Malysheva, N. K. Gusarova, A. V. Artem'ev, N. A. Belogorlova, A. I. Albanov, T. N. Borodina, V. I. Smirnov and B. A. Trofimov, *Synthesis*, 2014, **46**, 2656.
30. A. V. Artem'ev, L. A. Oparina, N. K. Gusarova, O. V. Vysotskaya, O. A. Tarasova, Y. V. Gatilov, A. I. Albanov and B. A. Trofimov, *Tetrahedron*, 2015, **71**, 1998.
31. A. V. Artem'ev, S. F. Malysheva, N. A. Belogorlova, N. I. Protzuk, A. I. Albanov, I. Y. Bagryanskaya, E. Y. Schmidt, N. K. Gusarova and B. A. Trofimov, *Heteroat. Chem.*, 2015, 26, 455.
32. B. A. Trofimov, N. K. Gusarova, S. N. Arbuzova, N. I. Ivanova, A. V. Artem'ev, P. A. Volkov, I. A. Ushakov, S. F. Malysheva and V. A. Kuimov, *J. Organomet. Chem.*, 2009, **694**, 677.
33. A. V. Artem'ev, S. F. Malysheva, N. K. Gusarova, N. A. Belogorlova, V. A. Shagun, A. I. Albanov and B. A. Trofimov, *Synthesis*, 2015, **47**, 263.
34. P. A. Volkov, A. A. Telezhkin, K. O. Khrapova, N. I. Ivanova, A. I. Albanov, N. K. Gusarova and B. A. Trofimov, *New J. Chem.*, 2021, **45**, 6206.
35. P. A. Volkov, N. I. Ivanova, K. O. Khrapova, A. I. Albanov, S. N. Arbuzova, N. K. Gusarova and B. A. Trofimov, *Russ. J. Gen. Chem.*, 2015, **85**, 1978.
36. N. K. Gusarova, P. A. Volkov, N. I. Ivanova, S. N. Arbuzova, K. O. Khrapova, A. I. Albanov, V. I. Smirnov, T. N. Borodina and B. A. Trofimov, *Tetrahedron Lett.*, 2015, **56**, 4804.
37. P. A. Volkov, A. A. Telezhkin, N. I. Ivanova, K. O. Khrapova, A. I. Albanov, N. K. Gusarova and B. A. Trofimov, *Russ. J. Gen. Chem.*, 2018, **88**, 912.
38. B. A. Trofimov, P. A. Volkov, K. O. Khrapova, A. A. Telezhkin, N. I. Ivanova, A. I. Albanov, N. K. Gusarovaa and O. N. Chupakhin, *Chem. Commun.*, 2018, **54**, 3371.
39. P. A. Volkov, K. O. Khrapova, A. A. Telezhkin, N. I. Ivanova, A. I. Albanov, N. K. Gusarova and B. A. Trofimov, *Org. Lett.*, 2018, **20**, 7388.
40. N. K. Gusarova, P. A. Volkov, N. I. Ivanova, B. G. Sukhov, L. I. Larina, O. N. Kazheva, G. G. Alexandrov, O. A. D'yachenko and B. A. Trofimov, *Tetrahedron Lett.*, 2013, **54**, 6772.
41. N. K. Gusarova, P. A. Volkov, N. I. Ivanova, L. I. Larina and B. A. Trofimov, *Heteroat. Chem.*, 2012, **23**, 322.
42. P. A. Volkov, N. N. Pogodaeva, N. I. Ivanova, K. O. Khrapova, L. I. Larina, B. G. Sukhov, Y. V. Gatilov, N. K. Gusarova and B. A. Trofimov, *Tetrahedron Lett.*, 2014, **55**, 4927.
43. P. A. Volkov, K. B. Petrushenko, N. I. Ivanova, K. O. Khrapova, L. I. Larina, N. K. Gusarova and B. A. Trofimov, *Tetrahedron Lett.*, 2017, **58**, 1992.
44. K. O. Khrapova, P. A. Volkov, N. I. Ivanova, A. A. Telezhkin, N. K. Gusarova and B. A. Trofimov, *Russ. J. Gen. Chem.*, 2018, **88**, 2223.
45. P. A. Volkov, N. I. Ivanova, A. A. Telezhkin, K. O. Khrapova, L. I. Larina, N. K. Gusarova and B. A. Trofimov, *Russ. J. Gen. Chem.*, 2019, **89**, 59.

46. K. O. Khrapova, A. A. Telezhkin, P. A. Volkov, L. I. Larina, D. V. Pavlov, N. K. Gusarova and B. A. Trofimov, *Org. Biomol. Chem.*, 2021, **19**, 5098.
47. P. A. Volkov, N. I. Ivanova, N. K. Gusarova, B. G. Sukhov, K. O. Khrapova, L. E. Zelenkov, V. I. Smirnov, T. N. Borodina, T. I. Vakul'skaya, S. S. Khutsishvili and B. A. Trofimov, *Heteroat. Chem.*, 2015, **26**, 329.
48. B. A. Trofimov, N. K. Gusarova, P. A. Volkov, N. I. Ivanova and K. O. Khrapova, *Heteroat. Chem.*, 2016, **27**, 44.
49. N. I. Ivanova, P. A. Volkov, L. I. Larina, N. K. Gusarova and B. A. Trofimov, *Chem. Heterocycl. Compd.*, 2012, **47**, 1384.
50. N. K. Gusarova, S. I. Verkhoturova, T. I. Kazantseva, S. N. Arbuzova, A. I. Albanov and B. A. Trofimov, *Mendeleev Commun.*, 2013, **23**, 253.
51. N. K. Gusarova, P. A. Volkov, N. I. Ivanova, Y. V. Gatilov and B. A. Trofimov, *Tetrahedron Lett.*, 2013, **54**, 3543.
52. A. V. Artem'ev, N. K. Gusarova, S. F. Malysheva, Y. V. Gatilov and V. I. Mamatyuk, *Org. Prep. Procd. Int.*, 2012, **44**, 262.
53. B. A. Trofimov, A. V. Artem'ev, S. F. Malysheva and N. K. Gusarova, *Dok. Chem.*, 2009, **428**, 225.
54. B. A. Trofimov, A. V. Artem'ev, S. F. Malysheva and N. K. Gusarova, *J. Organomet. Chem.*, 2009, **694**, 4116.
55. A. V. Artem'ev, N. K. Gusarova, S. F. Malysheva, I. A. Ushakov and B. A. Trofimov, *Russ. J. Gen. Chem.*, 2010, **80**, 2063.
56. A. V. Artem'ev, N. K. Gusarova, S. F. Malysheva and B. A. Trofimov, *Russ. Chem. Bull. Int. Ed.*, 2010, **59**, 1671.
57. A. V. Artem'ev, N. K. Gusarova, S. F. Malysheva, A. A. Tatarinova and B. A. Trofimov, *Russ. J. Gen. Chem.*, 2010, **80**, 1383.
58. A. V. Artem'ev, N. A. Chernysheva, N. K. Gusarova, S. V. Yas'ko, J.-H. Liao, C. W. Liu, A. I. Albanov and B. A. Trofimov, *Heteroat. Chem.*, 2014, **25**, 135.
59. P. T. Wood and J. D. Woollins, *J. Chem. Soc., Chem. Commun.*, 1988, 1190.
60. P. Bhattacharyya and J. D. Woollins, *Tetrahedron Lett.*, 2001, **42**, 5949.
61. J. D. Woollins, *Synlett*, 2012, **23**, 1154.
62. G. Hua and J. D. Woollins, *Phys. Sci. Rev.*, 2019, **4**, 20170132.
63. G. Hua and J. D. Woollins, *ACS Omega*, 2021, **6**, 31226.
64. A. Mardyukov, F. Keul and P. R. Schreiner, *Eur. J. Org. Chem.*, 2019, 387.
65. S. Holler, M. Tüchler, M. C. Roschger, F. Belaj, L. F. Veiros, K. Kirchner and N. C. Mösch-Zanetti, *Inorg. Chem.*, 2017, **56**, 12670.
66. M. T. Spiegel, A. Hoogerbrugge, S. Truksa and A. G. Smith, *Dalton Trans.*, 2018, **47**, 9030.
67. S. Thurow, E. J. Lenardão, X. Just-Baringo and D. J. Procter, *Org. Lett.*, 2017, **56**, 12670.
68. I. Compañón, A. Guerreiro, V. Mangini, J. Castro-López, M. Escudero-Casao, A. Avenoza, J. H. Busto, S. Castillón, J. Jiménez-Barbero, J. L. Asensio, G. Jiménez-Osés, O. Boutureira, J. M. Peregrina, R. Hurtado-Guerrero, R. Fiammengo, G. J. L. Bernardes and F. Corzana, *J. Am. Chem. Soc.*, 2019, **141**, 4063.

69. C. Pizzo and G. Mahler, *Tetrahedron Lett.*, 2017, **58**, 1445.
70. G. Hua, C. L. Carpenter-Warren, D. B. Cordes, A. M. Z. Slawin and J. D Woollins, *Molecules*, 2021, **26**, 2367.
71. D. Sarkar, C. Weetman, S. Dutta, E. Schubert, C. Jandl, D. Koley and S. Inoue, *J. Am. Chem. Soc.*, 2020, **142**, 15403.
72. G. Hua, J. Du, A. M. Z. Slawin and J. D. Woollins, *Chem. – Eur. J.*, 2016, **22**, 7782.
73. G. Hua, D. B. Cordes, A. M. Z. Slawin and J. D. Woollins, *Arkivoc*, 2017, 9.
74. T. D. Newton and M. D. Pluth, *Chem. Sci.*, 2019, **10**, 10723.
75. M. A. Abd-El-Maksoud, M. El-Hussieny, H. M. Awad, A.-T. H. Mossa and F. M. Soliman, *Russ. J. Gen. Chem.*, 2020, **90**, 2356.
76. M. A. Abd-El-Maksoud, T. K. Khatab, S. S. Maigali, F. M. Soliman and A. A. Hamed, *J. Hecrocyclic Chem.*, 2018, **55**, 2883.
77. G. Hua, D. B. Cordes, A. M. Z. Slawin and J. D. Woollins, *Eur. J. Inorg. Chem.*, 2019, 4682.
78. G. Hua, D. B. Cordes, A. M. Z. Slawin and J. D. Woollins, *ACS Omega*, 2020, **5**, 11737.
79. G. Hua, J. Du, A. M. Z. Slawin and J. D. Woollins, *ChemistrySelect*, 2016, **1**, 6810.
80. S. E. Denmark, D. J. P. Kornfilt and T. Vogler, *J. Am. Chem. Soc.*, 2011, **133**, 15308.
81. S. E. Denmark, E. Hartmann, D. J. P. Kornfilt and H. Wang, *Nat. Chem.*, 2014, **6**, 1056.
82. S. E. Denmark and D. J. P. Kornfilt, *J. Org. Chem.*, 2017, **82**, 3192.
83. K. A. Robb, S. V. Athavale and S. E. Denmark, *Synlett*, 2019, **30**, 1656.
84. S. E. Denmark and A. Jaunet, *J. Am. Chem, Soc.*, 2013, **135**, 6419.
85. S. E. Denmark and H. M. Chi, *J. Am. Chem, Soc.*, 2014, **136**, 3655.
86. S. E. Denmark and A. Jaunet, *J. Org. Chem.*, 2014, **79**, 140.
87. E. Hartmann and S. E. Denmark, *Helv. Chim. Acta*, 2017, **100**, e1700158.
88. Z. Tao, K. A. Robb, K. Zhao and S. E. Denmark, *J. Am. Chem. Soc.*, 2018, **140**, 3569.
89. Y.-Y. Xie, Z.-M. Chen, H.-Y. Luo, H. Shao, Y.-Q. Tu, X. Bao, R.-F. Cao, S.-Y. Zhang and J.-M. Tian, *Angew. Chem. Int. Ed.*, 2019, **58**, 12491.
90. H.-Y. Luo, Y.-Y. Xie, X.-F. Song, J.-W. Dong, D. Zhu and Z.-M. Chen, *Chem. Commun.*, 2019, **55**, 9367.
91. H.-Y. Luo, J.-W. Dong, Y.-Y. Xie, X.-F. Song, D. Zhu, T. Ding, Y. Liu and Z.-M. Chen, *Chem. Eur. J.*, 2019, **25**, 15411.
92. S. E. Denmark, S. Rossi, M. P. Webster and H. Wang, *J. Am. Chem. Soc.*, 2014, **136**, 13016.
93. S. E. Denmark and H. M. Chi, *J. Am. Chem. Soc.*, 2014, **136**, 8915.
94. A. Roth and S. E. Denmark, *J. Am. Chem. Soc.*, 2019, **141**, 13767.
95. J. L. Panger and S. E. Denmark, *Org. Lett.*, 2020, **22**, 2501.
96. A. Matviitsuk and S. E. Denmark, *Angew. Chem., Int. Ed.*, 2019, **58**, 12486.
97. C. Grandclaudon, M. Ruamps, R. Duboc, V. Michelet and P. Y. Toullec, *Synlett*, 2018, **29**, 1181.

98. S. Guria, C. G. Daniliuc and U. Henneckea, *Adv. Synth. Catal.*, 2021, **363**, 3852.
99. J. R. Tabor, D. C. Obenschain and F. E. Michael, *Chem. Sci.*, 2020, **11**, 1677.
100. D. Zhu, H.-Y. Luo and Z.-M. Chen, *Org. Lett.*, 2021, **23**, 1044.
101. E. Balaraman and K. C. K. Swamy, *Tetrahedron: Asymmetry*, 2007, **18**, 2037.
102. D. Selent, W. Baumann, K.-D. Wieseb and A. Börner, *Chem. Commun.*, 2008, 6203.
103. G. Farkas, S. Balogh, Á. Szöllösy, L. Ürge, F. Darvas and J. Bakos, *Tetrahedron: Asymmetry*, 2011, **22**, 2104.
104. G. Farkas, S. Balogh, J. Madarász, Á. Szöllösy, F. Darvas, L. Ürge, M. Gouygoud and J. Bakos, *Dalton Trans.*, 2012, **41**, 9493.
105. G. Erre, K. Junge, S. Enthaler, D. Addis, D. Michalik, A. Spannenberg and M. Beller, *Chem. – Asian J.*, 2008, **3**, 887.
106. S. Enthaler, G. Erre, K. Junge, K. Schröder, D. Addis, D. Michalik, M. Hapke, D. Redkin and M. Beller, *Eur. J. Org. Chem.*, 2008, 3352.
107. S. Balogh, G. Farkas, Á. Szöllösy, F. Darvas, L. Ürge and J. Bakos, *Tetrahedron: Asymmetry*, 2013, **24**, 66.
108. S. Balogh, G. Farkas, I. Tóth and J. Bakos, *Tetrahedron: Asymmetry*, 2015, **26**, 666.
109. T. Murai, D. Matsuoka and K. Morishita, *J. Am. Chem. Soc.*, 2006, **128**, 4584.
110. T. Murai, M. Monzaki, T. Katoh, T. Suzuki and T. Akiyama, *Phosphorus, Sulfur, and Silicon*, 2010, **185**, 964.
111. T. Murai, H. Tsuji, S. Imaizumi and T. Maruyama, *Chem. Lett.*, 2010, **39**, 524.
112. T. Murai and H. Itoh, *Phosphorus, Sulfur, and Silicon*, 2016, **191**, 163.
113. T. Murai and T. Miwa, *Heteroat. Chem.*, 2014, **25**, 3.
114. T. Aiba, S. Suehara, S.-L. Choy, Y. Maekawa, H. Lotter, T. Murai, S. Inuki, K. Fukase and Y. Fujimoto, *Chem Eur. J.*, 2017, **23**, 8304.
115. Y. Arai, S. Torigoe, T. Matsumaru, S. Yamasaki and Y. Fujimoto, *Org. Biomol. Chem.*, 2020, **18**, 3659.
116. G. Hua, A. L. Fuller, A. M. Z. Slawin and J. D. Woollins, *Polyhedron*, 2011, **30**, 805.
117. T. Murai, M. Monzaki and F. Shibahara, *Chem. Lett.*, 2007, **36**, 852.
118. C. Yang, X.-S. Xue, J.-L. Jin, X. Li and J.-P. Cheng, *J. Org. Chem.*, 2013, **78**, 7076.
119. T. Murai, S. Inaji, K. Morishita, F. Shibahara, M. Tokunaga, Y. Obora and Y. Tsuji, *Chem. Lett.*, 2006, **35**, 1424.
120. C. H. Cheon and H. Yamamoto, *J. Am. Chem. Soc.*, 2008, **130**, 9246.
121. C. H. Cheon and H. Yamamoto, *Org. Lett.*, 2010, **12**, 2476.
122. C. Yang, X.-S. Xue, X. Li and J.-P. Cheng, *J. Org. Chem.*, 2014, **79**, 4340.
123. M. Sai and H. Yamamoto, *J. Am. Chem. Soc.*, 2015, **137**, 7091.
124. T. Menard, A. Laverny and S. E. Denmark, *J. Org. Chem.*, 2021, **86**, 14290.

125. T. Murai, Y. Maekawa, M. Monzaki, T. Ando and T. Maruyama, *Chem. Commun.*, 2013, **49**, 9675.
126. T. Murai, Y. Maekawa, Y. Hirai, K. Kuwabara and M. Minoura, *RSC Adv.*, 2016, **6**, 15180.
127. Y. Maekawa, T. Maruyama and T. Murai, *Org. Lett.*, 2016, **18**, 5264.
128. T. Murai, R. Wada, K. Iwata, Y. Maekawa, K. Kuwabara and M. Minoura, *Organics*, 2021, **2**, 395.
129. T. Murai, S. Inaji and T. Takenaka, *Heteroat. Chem.*, 2009, **20**, 255.
130. K. Kuwabara, Y. Maekawa, M. Minoura and T. Murai, *Org. Lett.*, 2018, **20**, 1375.

CHAPTER 9

Optically Active Derivatives with a Stereogenic Sulfur Center: Synthesis and Applications

DOROTA KRASOWSKA,*[a] AGATA. J. PACUŁA-MISZEWSKA,[b] JACEK ŚCIANOWSKI[b] AND JOZEF DRABOWICZ[a,c]

[a] Centre of Molecular and Macromolecular Studies, Polish Academy of Sciences, 90-363 Łódź, Sienkiewicza 112, Poland; [b] Department of Organic Chemistry, Faculty of Chemistry, Nicolaus Copernicus University in Torun, 87-100 Torun, Gagarina Street 7, Poland; [c] Jan Dlugosz University in Czestochowa, 42-200 Czestochowa, Armii Krajowej 13/15, Poland
*Email: dkras@cbmm.lodz.pl

9.1 Introduction

The preparation of optically active, especially enantiomerically pure, organic molecules is a challenging topic in chemistry. This issue has attracted widespread attention and has been studied for decades. Among these molecules there are a large number of sulfur-containing derivatives whose chirality results from the presence of a stereogenic sulfur atom with different valence and coordination numbers. Recently, they have gained great interest as biologically active compounds, components of new materials, as well as reagents, catalysts, and/or auxiliary agents that can be used effectively in asymmetric synthesis. This chapter is devoted to the presentation of the main groups of optically active sulfur derivatives and the most common approaches that have made it possible to isolate them as optically active forms and their selected synthetic applications.

Chalcogen Chemistry: Fundamentals and Applications
Edited by Vito Lippolis, Claudio Santi, Eder J. Lenardão and Antonio L. Braga

Published by the Royal Society of Chemistry, www.rsc.org

9.2 Optical Activity of Organosulfur Derivatives – Historical Overview

Optically active organic sulfur compounds constitute the second oldest group among the family of optically active organic derivatives. Their history goes back more than 120 years, as the first representatives of this group of organic compounds (Figure 9.1), optically active sulfonium salts **1**[1] and **2**[2] were isolated by optical resolution of their racemic mixtures in 1900.

This optical resolution of sulfonium ions **1** and **2** was of great importance in the development of the theory of optical activity, as it showed that in the tetrahedral model the role of a fourth substituent can be played by a lone pair of electrons. The first isolation of the enantiomeric form of *p*-aminophenyl–phenyl sulfoxide **3**[3] and 3-carboxyphenyl methyl sulfoxide **4**[3] and the first asymmetric synthesis of diastereomerically pure *O*-menthyl *p*-toluenesulfinate **5**,[4] (see their structures in Figure 9.2) described a quarter of a century later, had a similar importance.

After more than a 20 year hiatus, the isolation of several chiral sulfinyl derivatives (thiosulfinates and α-amino acids functionalized with a sulfinyl moiety) from natural sources around 1950 again sparked growing interest in research on new chiral sulfinyl derivatives. Over the next two decades, this interest was stimulated very much by research devoted to explaining the mechanism and steric course of nucleophilic substitution reaction at a stereogenic sulfur atom. Since 1970 optically active organosulfur derivatives have attracted much attention as reagents in asymmetric synthesis. The reviews and literature compilations dealing with optically active sulfur derivatives, which appeared before 1980 are collected in the chapter co-authored by one of us and included to vol. 13 of the series Topics in Stereochemistry.[5] The most important achievements in sulfur stereochemistry described over the years 1900–1996 have been summarized in a book published in 1997.[6] Since 2000, advances in research devoted to the stereochemical aspects of the chemistry of organic sulfur derivatives, especially chiral sulfoxides, have been reviewed frequently.[7–21] The above-mentioned books, book chapters, and reviews present more or less deeply the preparation, structural determination, and synthetic applications, as chiral auxiliaries in asymmetric synthesis, optically active organosulfur derivatives the chirality of which results mainly from the presence of a stereogenic sulfur atom in compounds with the general pyramidal (a distorted tetrahedron) structure **6a** or tetrahedral geometry **6b** (Figure 9.3). Among

Figure 9.1 Optically active sulfonium salts isolated as early as 1900.

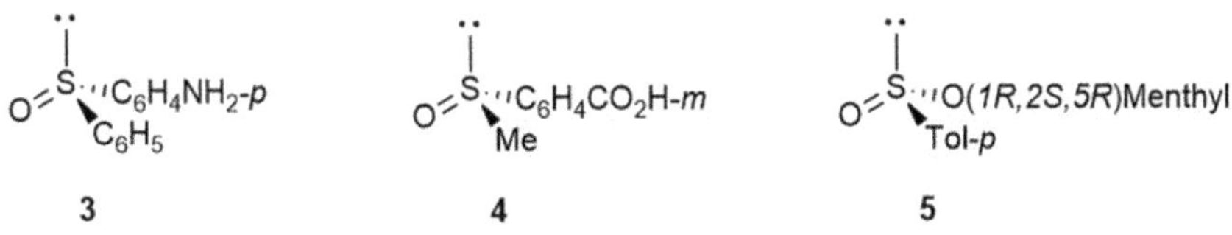

Figure 9.2 First isolated optically active sulfoxides **3** and **4** and *O*-menthyl *p*-toluenesulfinate **5**.

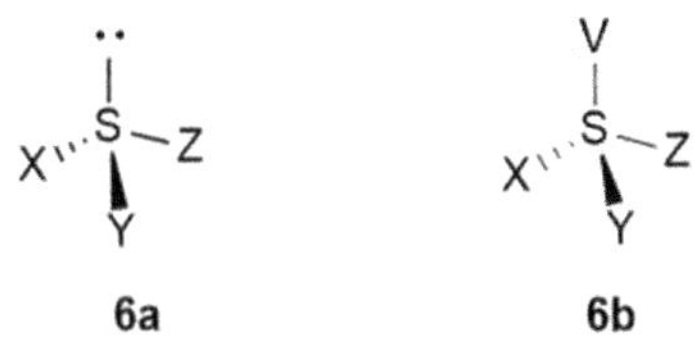

Figure 9.3 General structures of chiral organosulfur derivatives with a pyramidal (a distorted tetrahedron) **6a** or tetrahedral geometry **6b**.

them are various sulfonium salts **7** (X, Y, Z = R in **6a**) and its alkoxysulfonium analogs **8** (X, Y = R and Z = OR in **6a**), sulfoxides **9** (X, Y = R and Z = O in **6a**), sulfimides **10** (X, Y = R and Z = NR in **6a**), sulfinic acid derivatives such as sulfinates **11** (X = R, Y = OR and Z = O in **6a**), thiosulfinates **12** (X = R, Y = SR and Z = O in **6a**), sulfinamides **13** (X = R, Y = NR^1R^2 and Z = O in **6a**), sulfinic acid [^{16}O,^{18}O] **14** (X = R, Y = OH and Z = ^{18}OH in **6a**), thiosulfinic acid **15** (X = R, Y = OH and Z = S in **6a**), and sulfinyl chlorides **16** (X = R, Y = Cl and Z = O in **6a**), and chiral tri-coordinated sulfinyl compounds derived formally from achiral sulfurous acid, $S(O)(OH)_2$, such as sulfites **17a** (X = OR^1, Y = OR^2 and Z = O in **6a**), amidosulfites **17b** (X = OR^1, Y = NR^1 R^2 and Z = O in **6a**), and amidothiosulfites **17c** (X = SR^1, Y = NR^1 R^2 and Z = O in **6a**).

Among chiral sulfur compounds with four ligands lacking a nonbonding lone electron pair the most common are sulfoximines **18** (X = R^1, Y = R^2, Z = NR^1 and V = O in **6b**). A few chiral, unsymmetrical sulfodiimides **19** (X = R^1, Y = R^2, Z = NR^1 and V = NR^2 in **6b**) are known[5,6,22] and optically active sulfonimidoyl chloride **20** (X = R^1, Y = O, Z = NR^1 and V = Cl in **6b**) has also been described.[5,6,22] They were converted with a high stereoselectivity to the corresponding esters **21** (X = R^1, Y = O, Z = NR^1 and V = OR in **6b**) and amides **22** (X = R^1, Y = O, Z = NR^1, and V = NR^1R^2 in **6b**).[5,6,22] Furthermore, there are also optically active sulfonyl derivatives in which the chirality is generated by isotopic substitution such as benzyl (D_2benzyl) *p*-tolyl(1-naphthyl) [^{16}O, ^{18}O]-sulfones **23** (X = CH_2Ph, Y = CD_2Ph', Z = O and V = ^{18}O in **6b**),[23,24] diastereomerically pure (−)-menthyl [^{16}O, ^{18}O]-phenylmethane-sulfonate **24** (X = CH_2Ph, Y = *O*-menthyl, Z = O and V = ^{18}O in **6b**)[25] and the lithium salt of [^{16}O, ^{18}O]-*p*-toluenethiosulfonic acid **25** (X = *p*-Tol, Y = SLi, Z = O and V = ^{18}O in **6b**).[26] A single oxosulfonium salts **26**[27] (X = Me Y = Et, Z = Ph, V = O) and the oxosulfonium ylides **27** (X = Ar, Y = NR, Z = CHR, V = O),[28] which can be prepared from the corresponding optically active sulfoximides constitute other groups of tetrahedral tetra-coordinated sulfur compounds isolated in enantiomeric forms. The ylides **27** were used as nucleophilic alkylidene transfer reagents for the asymmetric synthesis of optically active cyclopropanes.[28] In this context it is worth noting that optically active sulfonium ylide **28** (X = 1-adamanthyl, Y = allyl, Z = CHMe in **6a**)[29] generated from the corresponding sulfonium adamantylethylallyl tetrafluoroborate, in spite of its lower optical stability,[30] afforded *via* [2,3]-sigmatropic rearrangement the optically active 1-adamantyl 2-pent-4-enyl sulfide **29** (X = 1-adamantyl, Y = pent-4-en-2-yl, Z = :, in **6a**) with a very high stereoselectivity (ee = 94%). Another group of organosulfur derivatives that exist and have been isolated as optically active species constitute tetra-coordinated compounds with a lone electron pair at the sulfur or their penta-coordinated oxidized analogs with a more or less distorted trigonal–bipyramidal structure. They are commonly known as sulfuranes or sulfurane oxides.[31] The first example, and to date only single compound with a monocyclic structure, of an optically active sulfurane was the dextrorotatory chlorosulfurane **30**[32] (Figure 9.4). The other optically active sulfuranes **31**,[33] **33**,[33] **35**,[34] **36**,[34] **37**,[35] and the sulfurane oxides **32**[33] and **34**[33] that have been prepared are spiro derivatives.

It is also interesting to note that in the sulfenamides **38** (Figure 9.5), due to their axial chirality resulting from the hindered rotation around the sulfur–nitrogen bond, the existence of two enantiomeric forms was observed.[36]

In recent years, research devoted to the preparation and application of a number of the above-mentioned compounds has been more or less widely continued. The most representative results taken from the original publications that appeared after 2010 are discussed below.

9.2.1 Sulfoxides

Since the results devoted to the preparation and synthetic utility of optically active sulfoxides as chiral auxiliaries that emerged up to 2019 have been discussed in several recent reviews,[13,15,16,18,37,38] only the results from 2020–2021 are presented below. Several procedures based on the enzymatic oxidation of prochiral sulfides published during these two years gave the

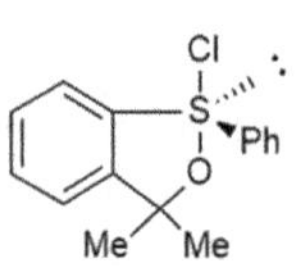

30

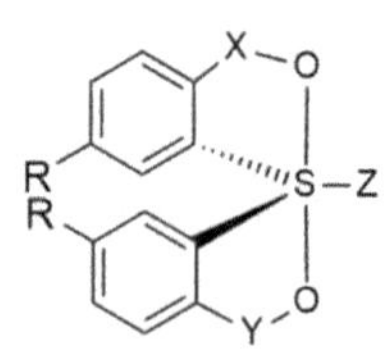

31, X = Y = CMe_2, R = H, Z = :
32, X = Y = CMe_2, R = H, Z = O
33, X = Y = $C(CF_3)_2$, R = *t*-Bu, Z = :
34, X = Y = $C(CF_3)_2$, R = *t*-Bu, Z = O
35, X = Y = C(O), R = H, Z = :
36, X = CH_2, Y = C(O), R = H, Z = :

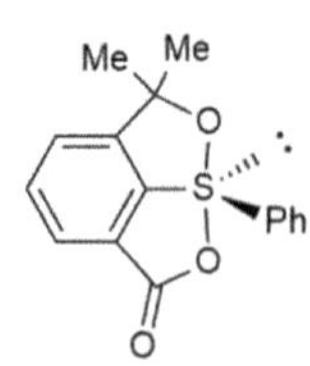

37

Figure 9.4 Tetra-coordinated or penta-coordinated hypervalent organosulfur compounds having a distorted trigonal-bipyramidal structure.

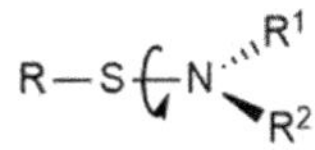

38

Figure 9.5 An axis of chirality in sulfenamides.

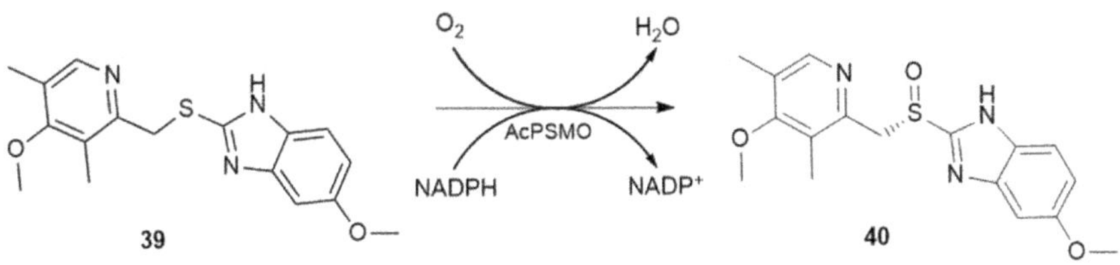

Scheme 9.1 Asymmetric sulfoxidation of omeprazole sulfide **39** catalyzed by prazole sulfide monooxygenase (AcPSMO).

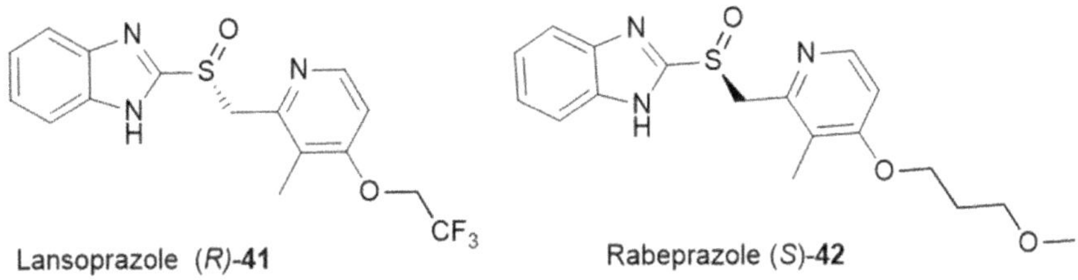

Figure 9.6 Bioactive enantiomeric sulfoxides **41** and **42**.

corresponding optically active sulfoxides with very high chemical yields, and in most cases almost enantioselectively. Thus, optimizing asymmetric sulfoxidation of omeprazole sulfide **39** catalyzed by prazole sulfide monooxygenase (AcPSMO) allowed isolation of (*S*)-omeprazole sulfoxide **40** [the commonly used drug for treating gastroesophageal reflux disease (acting as a proton pump inhibitor (PPI)] on the 0.39 kg scale with 57.8% overall yield and 99.1% purity and 99.9% enantiomeric excess (Scheme 9.1).[39]

In two other very recent reports response surface methodology (RSM) was applied to optimize the asymmetric oxidation of sulfide **39** by soybean pad peroxidase (SPP) in cetyltrimethylammonium bromide/isooctane/*n*-butanol/water in oil microemulsion containing hydrogen peroxide[40] and by immobilized cells mutant of *Rhodococcus rhodocrous* ATCC4276.[41] Both procedures afforded (*S*)-omeprazole sulfoxide **40** with ee above 95%. Very detailed studies on the sequence of BoBVMO (Baeyer–Villiger monooxygenase from *Bradyrhizobium oligotrophicum*) allowed the identification of BVMOs *via* genomic mining for the highly efficient synthesis of (*R*)-omeprazole sulfoxide **40** and other pharmaco-sulfoxides including (*R*)-lansoprazole sulfoxide **41** and (*S*)-rabeprazole sulfoxide **42** (Figure 9.6) by the asymmetric oxidation of the corresponding sulfides (ee 99%).[42]

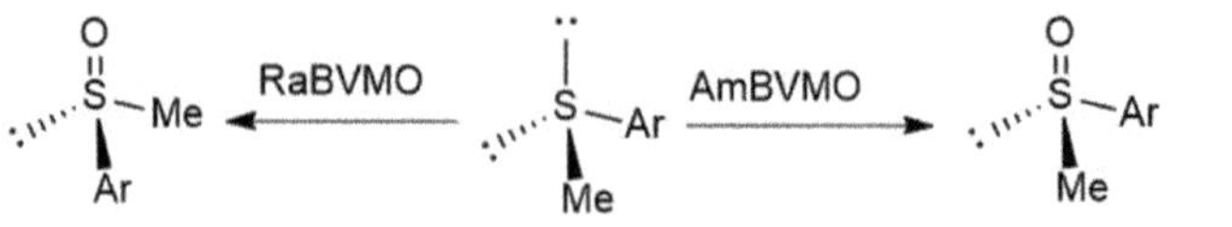

(*S*)-**44a-j** **43a-j** (*R*)-**44a-j**

ee= 6-99% *ee*= 95-99%

a, Ar=Ph; **b**, Ar=*o*-ClC$_6$H$_4$; **c**, Ar=*m*-ClC$_6$H$_4$; **d**, Ar=*p*-ClC$_6$H$_4$; **e**, Ar=*o*-BrC$_6$H$_4$;
f, Ar=*m*-BrC$_6$H$_4$; **g**, Ar=*p*-BrC$_6$H$_4$; **h**, Ar=*o*-MeOC$_6$H$_4$; **i**, Ar=*m*-MeOC$_6$H$_4$; **j**, Ar=*p*-MeOC$_6$H$_4$

Scheme 9.2 RaBVMO and AmBVMO catalyzed oxidation of thioanisoles **43a–j** to the corresponding sulfoxides **44a–j**.

Independently, in another research group, two enantiocomplementary Baeyer-Villiger monooxygenases RaBVMO and AmBVMO, identified by genome mining, were used to asymmetrically oxidize a series of functionalized thioanisoles **43a–j** to the corresponding optically active sulfoxides **44a–j** (Scheme 9.2).[43] It was found that RaBVMO gives mainly (*S*)-sulfoxides, while AmBVMO gives (*R*)-sulfoxides. In addition, it was observed that the substituents at different positions on the phenyl ring of sulfides **43a–j** had little effect on the enantioselectivity of RaBVMO (ee ranging from 95 to 99%), while a significant effect was noted for AmBVMO (ee ranging from 6 to 99%).

Methyl phenyl sulfide **43a** was used as a model for alkyl aryl sulfides in their asymmetric oxidation to the corresponding alkyl aryl sulfoxides with Streptomyces glaucescens GL.A.0. as a whole-cell biocatalyst. This soil actinobacterium was found to give methyl phenyl sulfoxide **44a** with (*R*) absolute configuration at the sulfinyl sulfur atom and ee value higher than 99%.[44]

Considering chemical oxidants it is worth noting that a relatively high enantioselectivity was observed very recently in asymmetric oxidation of a series of alkyl *ortho*-alkoxy aryl sulfides **45a–h** to the corresponding sulfoxides **46a–h** by an original Sharpless oxidation procedure (Scheme 9.3).[45]

A modification of this procedure was used for the asymmetric synthesis of methiin (*S*-methyl-(*R*)-cysteine sulfoxide) diastereomers **48** starting from (*R*)-*S*-methylcysteine **47** oxidized with *t*-butyl hydroperoxide in the presence of titanium tetra(isopropylate) and (*S*)- or (*R*)-BINOL (Scheme 9.4).[46]

Another efficient approach for the preparation of optically active sulfoxides relies on the kinetic resolution of racemic mixtures. This can be based on oxidative or non-oxidative procedures. Very recently, there have been reports of new catalytic protocols based on the use of organometallic complexes. Thus, the Ir(III)-catalyzed enantioselective C–H amidation of dibenzyl sulfoxides **49a–r** through its desymmetrization taking place with 3-phenyl-1,4,2-dioxazol-5-one **50** in the presence of an Ir(III) complex containing a *t*-butyl cyclopentadienyl ligand and modified chiral proline derivatives were found to afford, after optimizing the reaction conditions, a rich family of optically active sulfoxides **51a–r** with very high enantiomeric excesses (Scheme 9.5).[47] In addition, this family of optically active, unsymmetric

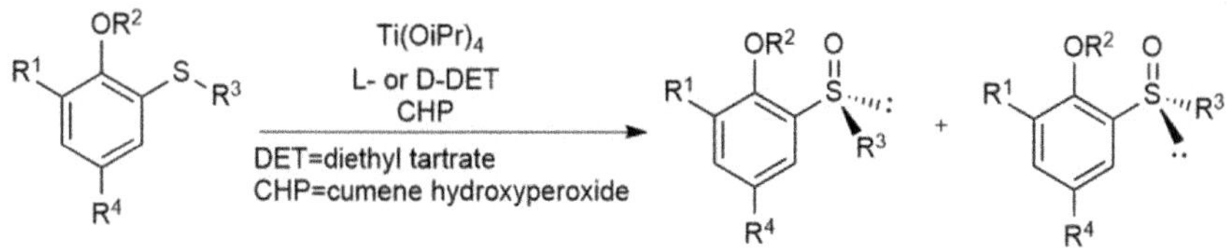

45a-h (*S*)-**46a-h** (*R*)-**46a-h**

		Abs. conf.	*ee* [%]
a, R^1=H; R^2= H; R^3= Me; R^4= H;	L-DET	(*R*)	4.3
b, R^1=H; R^2= Me; R^3= Me; R^4= H;	L-DET	(*R*)	91.0
c, R^1=H; R^2= MOM; R^3= Me; R^4= H;	L-DET	(*R*)	91.0
d, R^1=Me; R^2= MOM; R^3= Me; R^4= H;	L-DET	(*R*)	94.0
e, R^1=H; R^2= MOM; R^3= Et; R^4= H;	L-DET	(*R*)	68.0
f, R^1=H; R^2= MOM; R^3= iPr; R^4= H;	L-DET	(*R*)	26.0
g, R^1=Me; R^2= MOM; R^3= Tol; R^4= H;	L-DET	(*R*)	7.2
h, R^1=H; R^2= CH_2OMe; R^3= Me; R^4= CO_2Me;	L-DET	(*R*)	96.0

Scheme 9.3 Asymmetric oxidation of a series of alkyl(aryl) *ortho*-alkoxy aryl sulfides **45**.

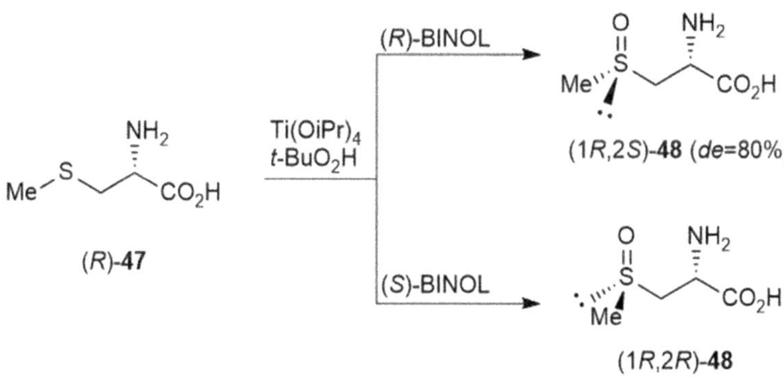

Scheme 9.4 Diastereoselective oxidation of *S*-methylcysteine (*R*)-**47**.

dibenzyl sulfoxides was very much extended by the use as amidation component dioxoazalones (**50** for example in Scheme 9.5) functionalized with a variety of the aryl substituents (ee values from 84 to 96%).[47]

The analogous Ir(III)-catalyzed C–H amidation of racemic unsymmetric dibenzyl sulfoxides **52a–f** was found to allow their efficient kinetic resolution by desymmetrization taking place with 3-phenyl-1,4,2-dioxazol-5-one **50** as the amination reagent leading to a mixture of two *ortho*-functionalized benzyl sulfoxides **53a–f** and **54a–f** (Scheme 9.6).[47]

Another very recent protocol is based on the Rh-mediated hydrogenative kinetic resolution of α,β-unsaturated sulfoxides. Detailed studies with racemic unsaturated *E* and *Z* sulfoxides **55** containing alkyl and aryl substituents and Rh-precatalyst $[Rh(nbd)(\mathbf{L})]BF_4$[48,49] (nbd = norbornandiene)

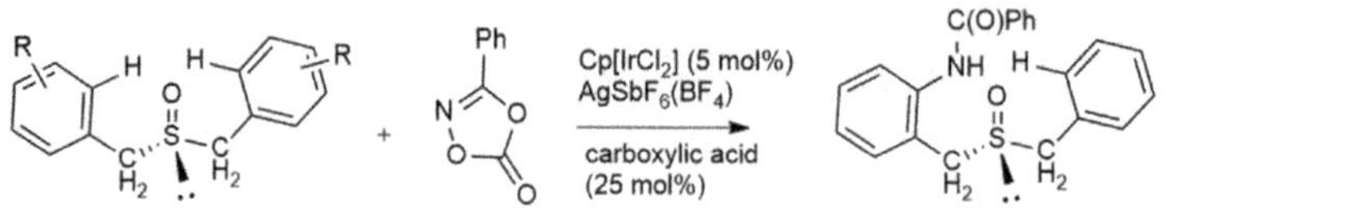

49a-r 50 (S)-51a-r

Yield: 23-79%
ee =19-79%

a, R= p-H; b, R= p-Me; c, R= p-t-Bu; d, R= p-OMe; e, R= p-CO_2Me; f, R= p-CF_3; g, R= p-OCF_3; h, R= p-SCF_3; i, R= p-trans-CH=CHPh; j, R= p-F; k, R= p-Cl; l, R= p-Br; m, R= p-F; n, R= p-I; o, R= o-Me; p, R= o-F; q, R= o-Br; r, R= m-Br

carboxylic acids:
N-Pht-Phe-OH (N-Phthaloyl-L-phenylalanine); N-Boc-Phe-OH (N-(tert-Butoxycarbonyl)-L-phenylalanine); N-Boc-Val-OH (Boc-L-valine); N-Boc-Leu-OH (Boc-L-leucine); N-Boc-Pro-OH (Boc-L-proline); N-Piv-Pro-OH (N-pivaloyl-L-proline); N-Piv-Me-Pro-OH (N-pivaloyl-α-methyl-L-proline).

Scheme 9.5 The Ir(III)-catalyzed enantioselective C–H amidation of dibenzyl sulfides **49a–r**.

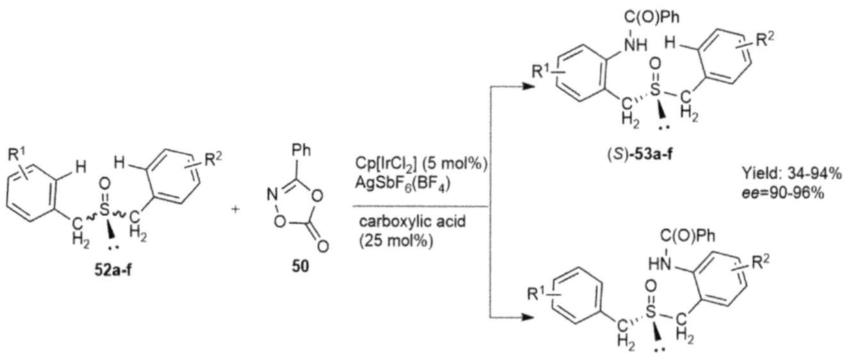

52a-f 50 (S)-53a-f (S)-54a-f

Yield: 34-94%
ee=90-96%

For 52, 53 and 54
a, R^1= Me, R^2 = p-t-Bu; b, R^1= p-OCF_3, R^2 = p-SCF_3; c, R^1= p-CF_3, R^2 = p-OCF_3; d, R^1= p-Br, R^2 = p-H; e, R^1= p-Br, R^2 = p-CO_2Me; f, R^1= p-I, R^2 = p- trans-CH=CHCO_2Me;

Scheme 9.6 The Ir(III)-catalyzed kinetic resolution of racemic dibenzyl sulfoxides **52a–f**.

allowed the isolation of highly enantioenriched unsaturated starting sulfoxides **55a–d** and their saturated analogs **56a–d** (Scheme 9.7).[50]

The isolation of optically active sulfoxides is also possible in procedures based on the kinetic resolution of racemic mixtures during their partial reduction to the corresponding sulfides. The application of enantioselective sulfoxide reductases for this purpose has very recently been reported.[51–53] It should be added here that this topic is a subject of a very recent mini-review.[54]

Another efficient approach for the preparation of optically active sulfoxides **59** is based on stereoselective alkylation of sulfenic acid anions **58,** which are generally generated *in situ* from the appropriate designed sulfoxides **57** (Scheme 9.8). A very recent mini-review[55] summarized the obtained results, along with the results devoted to the use of catalytic

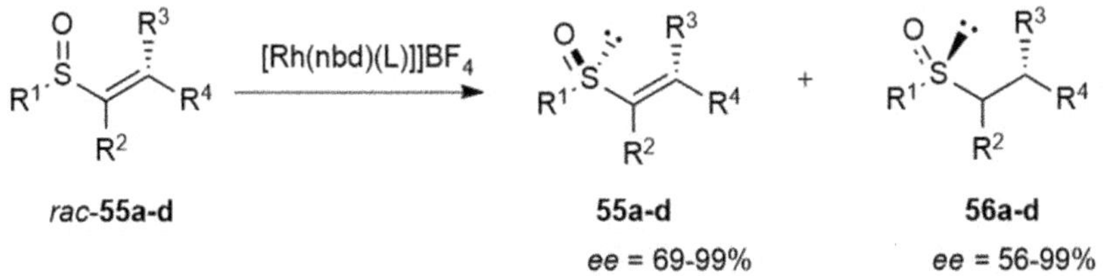

a, R^1= Me, R^2= R^3= H, R^4= $(CH_2)_2OBz$; **b**, R^1= *p*-ClC_6H_4, R^2= R^3= H, R^4= $(CH_2)_3CO_2Me$;
c, R^1= Bn, R^2= R^3= R^4= H; **d**, R^1= *o*-FC_6H_4, R^2= R^3= R^4= H
nbd = bicyclo[2.2.1]hepta-2,5-diene(norbornadiene)

Scheme 9.7 The Rh-mediated hydrogenative kinetic resolution of α,β-unsaturated sulfoxides **55**.

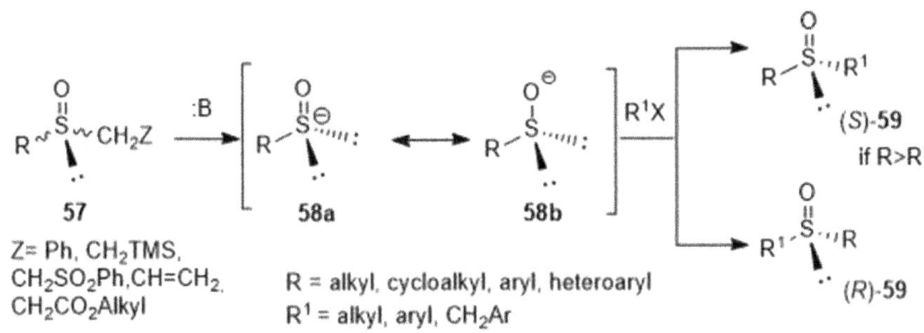

Scheme 9.8 Stereoselective alkylation of sulfenic acid anions **58**.

procedures leading to the isolation of optically active sulfoxides **59** in very high chemical yields and with very high enantiomeric excesses.

Above it has been noted that the results on synthetic utility of optically active sulfoxides as chiral auxiliaries that appeared in the chemical literature up to 2019 have been discussed extensively in several recent reviews.[13,15,16,18,37,38] Some of them are also presented in another 3 very recent reviews.[56–58] The search of the literature on this topic from 2020–2021 ended with the recognition of only one article in which very interesting results regarding the use of metal complexes with Schiff base ligands containing a sulfinyl moiety for Pd(II)-catalyzed asymmetric intramolecular allylic C–H amination reactions. Detailed studies led to the discovery that a Schiff base-Cu(II)/sulfoxide complex **60** in combination with $Pd(OAc)_2$ gives the best results with internal and terminal alkenes. They were converted to the corresponding C–H amination products (cyclic aminocarbonates) **62**, according to the general process, shown in Scheme 9.9, with a very high selectivity (ee up to 82%).[59]

9.2.2 Sulfimides

Sulfimides **10** containing two carbon–sulfur single bonds and a sulfur double bond to a nitrogen substituent (ylide-type) can be considered as

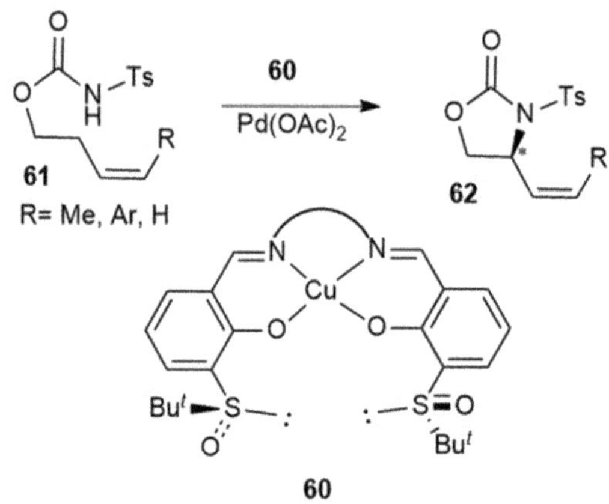

Scheme 9.9 The allylic C–H amination of Ts-protected aminocarbonates **61**.

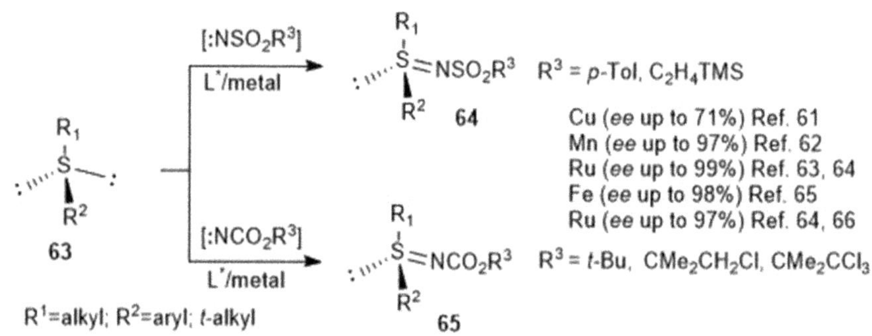

Scheme 9.10 Enantioselective sulfimidation reactions of sulfides **63** by metal catalysis.

nitrogen analogs of sulfoxides **9**. And like sulfoxides, they can exist in enantiomeric forms. Enantioselective approaches toward optically active sulfimides **64** and **65** most frequently rely on enantioselective sulfimidation reactions obtained by metal catalysis on metal-catalyzed nitrene transfer reaction to unsymmetrical sulfides **63** taking place in the presence of a chiral ligand.[60] Thus, over 2 decades (1996–2007) sulfimidation reactions were mainly carried out with *N*-carboxylated and *N*-sulfonylated nitrene precursors using copper,[61] manganese,[62] and ruthenium,[63,64] as catalytically-active metal centers and various optically active ligand L*. More recently (2013–2016) a few protocols based on the use of iron was reported[65] and very recently another protocol based on ruthenium was published[66] (Scheme 9.10).

It has very recently been found that 3-thiosubstituted 2-quinolones **66** and 2-pyridones **67** can be converted to the corresponding sulfimides **68** and **69** by the sulfimidation procedure using *N*-(*p*-nitrophenylsulfonyl)iminophenyliodinane (PhI = NNs **70**) as the nitrene precursor and a silver-based catalyst system containing of a chiral phenanthroline-based ligand **71** with a hydrogen bonding site (Scheme 9.11).[67] Optically active sulfimides were obtained in high chemical yields (44–99%) and with significant enantiomeric excess (70–99% ee).

The participation of a heteroleptic silver complex, in which the silver atom is bound to one molecule of the chiral ligand **71** and one molecule of an achiral 1,10-phenanthroline, in the reaction course was supported by ESI mass spectrometry, DFT calculations, and Hammett analysis. It is worth noting that this procedure can be applied for the kinetic resolution of racemic sulfoxides.

9.2.3 Sulfonium Ylides

Literature reports on the synthesis and application of optically active sulfonium ylides (a single structure from this group is represented by the sulfonium ylide **27**), which appeared up to 2017, are mentioned in two recent reviews.[68,69] Among several very recent original publications on this issue, an article which describes the multigram in situ preparation of sulfonium ylid **74** can be listed as one of the most interesting. This ylide can be derived from the corresponding benzylsulfonium salt **73** (prepared on >50 g scale by alkylation of the chiral sulfide isothiocineole **72** obtained from limonene and elemental sulfur). Using this sulfonium salt the sulfur ylide-mediated asymmetric epoxidation of benzaldehyde and the asymmetric aziridination of imine **75** afforded the corresponding epoxide **76** or aziridine **77** on a 1–2 g scale (Scheme 9.12).[70]

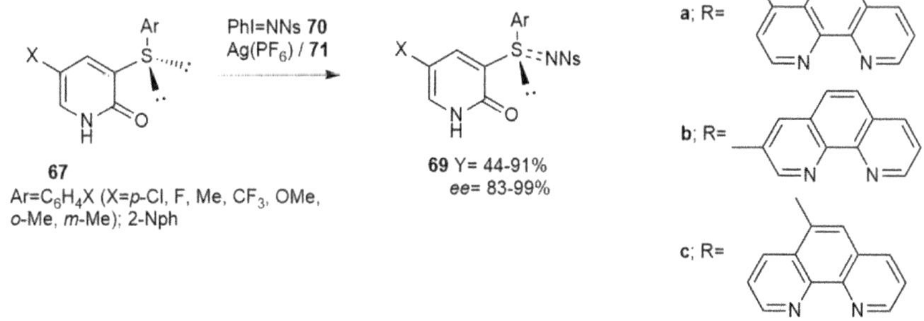

Scheme 9.11 Enantioselective sulfimidation reactions of 3-thiosubstituted 2-quinolones **66** and 2-pyridones **67**.

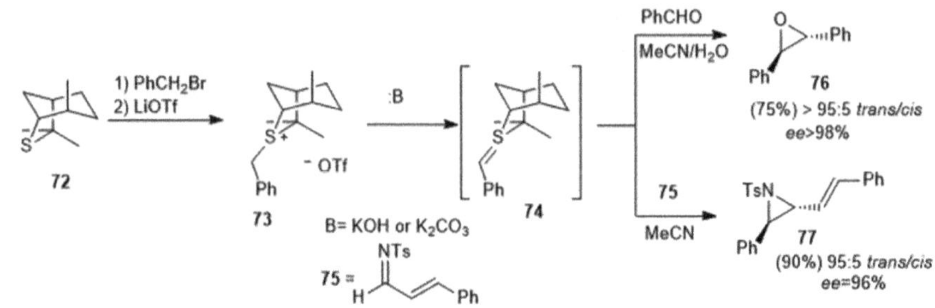

Scheme 9.12 Enantioselective reactions of *in situ* generated ylide **74** with benzaldehyde and imine **75**.

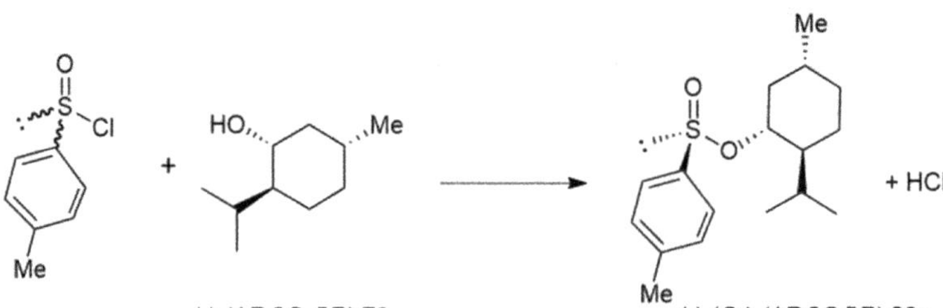

78 (-)-(1*R*,2*S*, 5*R*)-**79** (-)-(S_s)-(1*R*,2*S*,5*R*)-**80a**

Scheme 9.13 Fully diastereoselective synthesis of (−)-(S_S)-(1*R*,2*S*,5*R*)-*O*-menthyl *p*-toluenesulfinate **80a**.

9.2.4 Sulfinic Esters

Literature reports on the synthesis and application of optically active sulfinic esters **11**, which appeared up to 2017, are listed in a previous review already mentioned in this chapter.[37] Only two new protocols were reported after 2017 for the preparation of diasteroisomeric sulfinic esters derived from menthol **79** enantiomers. The first protocol comes from the Polish patent pending, according to which the crystalline (−)-(S_S)-(1*R*,2*S*,5*R*)-*O*-menthyl *p*-toluenesulfinate **80a** and its enantiomer (+)-(R_S)-(1*S*,2*R*,5*S*) **80b** are formed with a yield nearly quantitative, maintaining an equimolar mixture of menthol **79** enantiomers and *p*-toluenesulfinyl chloride **78** for 14 days at room temperature (Scheme 9.13).[71]

The second report describes the diastereoselective oxidation of (−)-(1*R*,2*S*,5*R*)-*O*-menthyl arenesulfenates **81a–c** to the corresponding diastereoisomeric sulfinic esters **82a–c** with diastereomeric excesses ranging from 10 to 48% (Scheme 9.14).[72]

9.2.5 Sulfoximines

Literature reports on the synthesis and use of optically active sulfoximines **18** that have appeared up to 2019 in the reviews already mentioned in this chapter.[7,11,20,37,60] In a very recent account published in 2021[73] most of these

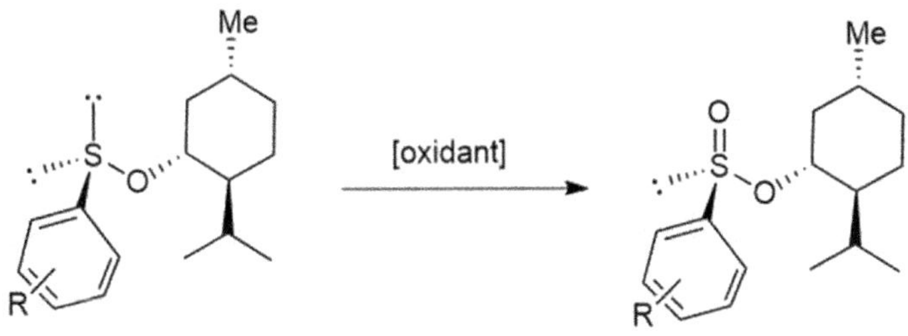

(-)-(1*R*,2*S*,5*R*)-**81a-c** (-)-(1*R*,2*S*,5*R*)-**82a-c**

a, R = *o*-Me; **b** , R = *o*-NO_2; **c**, R = *p*-NO_2

Scheme 9.14 Diastereoselective oxidations of (−)-(1*R*,2*S*,5*R*)-*O*-menthyl arenesulfenates **81a–c**.

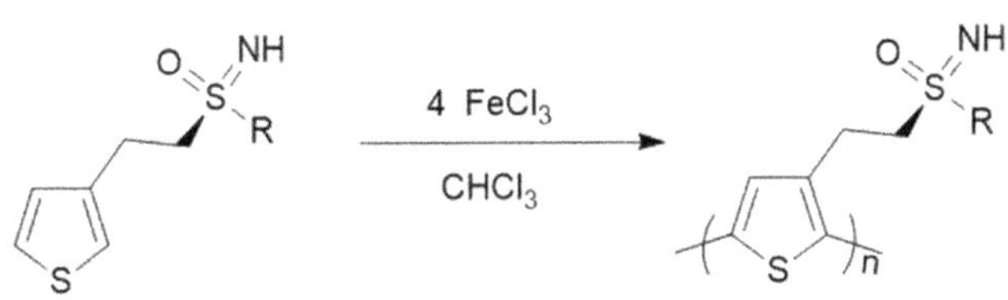

83a-b **a**, R = *p*-Tol **b**, R = *n*-$C_{16}H_{33}$ **84a-b**

Scheme 9.15 Polymerization of 2-(3′-thienyl)ethyl alkyl (aryl) sulfoximines **83a–b** *via* chemical oxidative coupling using $FeCl_3$.

reports have been recalled, and a large number of articles describing NH–sulfoximine transformations are comprehensively analyzed. A search on this topic for 2020 and 2021 shows only a single paper[74] on the synthesis and structural characterization of monomeric **83a–b** and polymeric **84a–b** thiophenes functionalized at the 3-position with a pendant containing a stereogenic sulfoximine sulfur atom (Scheme 9.15).

References

1. W. J. Pope and S. J. Paechey, *J. Chem. Soc.*, 1900, **77**, 1072.
2. S. J. Smiles, *J. Chem. Soc.*, 1900, **77**, 1174.
3. P. W. B. Harrison, J. Kenyon and H. Phillips, *J. Chem. Soc.*, 1926, **128**, 2079.
4. H. Phillips, *J. Chem. Soc.*, 1925, **127**, 2552.
5. M. Mikolajczyk and J. Drabowicz, *Top. Stereochem.*, 1982, **13**, 333.
6. M. Mikolajczyk, J. Drabowicz and P. Kielbasinski, *Chiral Sulfur Reagents: Applications in Asymmetric and Stereoselective Synthesis*, CRC Press, Boca Raton, New York, 1997.
7. R. Bentley, *Chem. Soc. Rev.*, 2005, **34**, 609–624.
8. M. Mellah, A. Voituriez and E. Schulz, *Chem. Rev.*, 2007, **107**, 5133–5209.

9. H. Pellissier, *Tetrahedron*, 2007, **63**, 1297–1330.
10. J. Drabowicz, P. Kiełbasinski, D. Krasowska and M. Mikołajczyk, Asymmetric Synthesis of Optically Active Sulfinic Acid Esters, in *Organosulfur Chemistry in Asymmetric Synthesis*, ed. T. Toru and C. Bolm, Wiley-VCH, Wienheim, 2008, ch. 2, pp. 31–54.
11. C. Worch, A. C. Mayer and C. Bolm, Synthesis and Use of Chiral Sulfoximines, in *Organosulfur Chemistry in Asymmetric Synthesis*, ed. T. Toru and C. Bolm, Wiley-VCH, Wienheim, 2008, ch. 6, pp. 209–232.
12. C. H. Senanayake, Z. Han and D. Krishnamurthy, Synthesis and Use of Chiral Sulfinamides, in *Organosulfur Chemistry in Asymmetric Synthesis*, ed. T. Toru and C. Bolm, Wiley-VCH, Wienheim, 2008, ch. 7, pp. 233–264.
13. E. Wojaczyńska and J. Wojaczyński, *Chem. Rev.*, 2010, **110**, 4303.
14. H. C. Xu, S. Chowdhury and J. A. Ellman, *Nat. Protoc.*, 2013, **8**, 2271.
15. B. M. Trost and M. Rao, *Angew. Chem., Int. Ed.*, 2015, **54**, 5026.
16. G. Sipos, E. E. Drinkel and R. Dorta, *Chem. Soc. Rev.*, 2015, **44**, 3834.
17. I. Colomer, M. Velado, R. Fernandez de la Pradilla and A. Viso, *Chem. Rev.*, 2017, **117**, 14201.
18. S. Otocka, M. Kwiatkowska, L. Madalinska and P. Kiełbasinski, *Chem. Rev.*, 2017, **117**, 4147.
19. J. Drabowicz, *Phosphorus, Sulfur Silicon Relat. Elem.*, 2017, **192**, 145.
20. D. Kaiser, I. Klose, R. Oost, J. Neuhaus and N. Maulide, *Chem. Rev.*, 2019, **119**, 8701.
21. U. Lücking, *Org. Chem. Front.*, 2019, **6**, 1319.
22. A. Nudelman, *The Chemistry of Optically Active Sulfur Compounds*, Gordon and Breach, New York, 1984.
23. C. J. M. Stirling, *J. Chem. Soc.*, 1963, 5741.
24. R. Annunziata, M. Cinquini and S. Colonna, *J. Chem. Soc., Perkin Trans. 1*, 1972, 2057.
25. M. Sabol and K. K. Andersen, *J. Am. Chem. Soc.*, 1969, **91**, 3603.
26. J. Drabowicz, P. Łyżwa, B. Bujnicki and M. Mikolajczyk, *Phosphorus, Sulfur Silicon Relat. Elem.*, 1994, **95–96**, 293.
27. K. Kamiyama, H. Minato and M. Kobayashi, *Bull. Chem. Soc. Jpn.*, 1973, **46**, 3895.
28. (a) C. R. Johnson, *Acc. Chem. Res.*, 1973, **6**, 341; (b) C. R. Johnson and C. W. Schroeck, *J. Am. Chem. Soc.*, 1973, **93**, 7418.
29. B. S. Trost and R. F. Hammen, *J. Am. Chem. Soc.*, 1973, **95**, 962.
30. (a) D. Darwish and G. Tourigny, *J. Am. Chem. Soc.*, 1966, **88**, 4303; (b) D. Darwish, S. Hui and R. Tomilson, *J. Am. Chem. Soc.*, 1968, **90**, 5631.
31. (a) J. Drabowicz, *Chemistry of Hypervalent Compounds*, ed. K. Akiba, Wiley-VCH, New York, 1999, ch. 7, pp. 211–240; (b) J. Drabowicz, P. Łyżwa and M. Mikolajczyk, *Supplement S: the Chemistry of Sulfur-Containing Groups*, ed. S. Patai and Z. Rappoport, John Wiley and Sons, Chichester, 1993, pp. 799–956; (c) R. A. Hayes and J. C. Martin, *Organic Sulfur Chemistry: Theoretical and Experimental Advances*, ed. F. Bernardi, J. G. Csizmadia and A. Mangini, Elsevier, Amsterdam, 1985, ch. 8, pp. 408–458.

32. J. C. Martin and T. M. Balthazor, *J. Am. Chem. Soc.*, 1977, **99**, 152.
33. (a) J. Drabowicz and J. C. Martin, *Phosphorus, Sulfur Silicon Relat. Elem.*, 1993, **74**, 439; (b) J. Drabowicz and J. C. Martin, *Pure Appl. Chem.*, 1996, **68**, 951.
34. I. Kapovits, *Phosphorus, Sulfur Silicon Relat. Elem.*, 1991, **58**, 39.
35. J. Drabowicz and J. C. Martin, *Tetrahedron: Asymmetry*, 1993, **4**, 297–300.
36. D. Kost and M. Raban, *The Chemistry of Sulfenic Acids and Their Derivatives*, ed. S. Patai and Z. Rappoport, John Wiley and Sons, Chichester, 1990, pp. 23–82.
37. E. Wojaczyńska and J. Wojaczyński, *Chem. Rev.*, 2020, **120**, 4578.
38. (a) J. Han, V. A. Soloshonok, K. Klika, J. Drabowicz and A. Wzorek, *Chem. Soc. Rev.*, 2018, **47**, 1307; (b) W. Maczka, K. Wińska and M. Grabarczyk, *Catalysts*, 2018, **8**, 624; (c) F. Garzón-Posse, L. Becerra-Figueroa, J. Hernández-Arias and D. Gamba-Sánchez, *Molecules*, 2018, **23**, 1265.
39. N. Xu, J. Zhu, Y.-Q. Wu, Y. Zhang, J.-Y. Xia, Q. Zhao, G.-Q. Lin and J.-H. Xu, *Org. Process Res. Dev.*, 2020, **24**, 1124–1130.
40. H. Tang, Y. Zhang, Y. Deng, S. Du, D. Li, Z. Wang, H. Li, X. Gao and F. Wang, *Catal. Lett.*, 2021, **151**, 2928.
41. Y. Zhang, Q. Zhao, H. Tang, H. Li, D. Li, Z. Wang, X. Gao and F. Wang, *Korean J. Chem. Eng.*, 2021, **38**, 989.
42. F. Liu, C. Shou, Q. Geng, C. Zhao, J. Xu and H. Yu, *Appl. Microbiol. Biotechnol.*, 2021, **105**, 3169.
43. S. Wei, Y. Liu, J. Zhou, G. Xu and Y. Ni, *Mol. Catal.*, 2021, **543**, 111784.
44. S. Salama, T. Dishisha, M. H. Habib, A. Z. Abdelazem, W. Bakeer, M. Abdel-Latif and Y. Gaber, *RSC Adv.*, 2020, **10**, 32335.
45. T. Takei, J. Takayama, M. Xuan, M. Tomoda, H. Miyamae and T. Sakamoto, *J. Chem. Sci.*, 2021, **133**, 28.
46. A. O. Kolodiazhna, I. Skliarov, A. A. Slastennikova and O. I. Kolodiazhnyi, *Phosphorus, Sulfur Silicon Relat. Elem.*, 2020, **195**, 713.
47. W. Liu, W. Yang, J. Zhu, Y. Guo, N. Wang, J. Ke, P. Yu and C. He, *ACS Catal.*, 2020, **10**, 7207.
48. J. R. Lao, H. Fernández-Pérez and A. Vidal-Ferran, *Org. Lett.*, 2015, **17**, 4114.
49. B. Balakrishna and A. Vidal-Ferran, *Synthesis*, 2016, **48**, 997.
50. H. Fernández-Pérez, J. R. Lao, A. Grabulosa and A. Vidal-Ferran, *Eur. J. Org. Chem.*, 2020, 4331.
51. Y. Wen, L. Peng, Y. Zhou, T. Peng, Y. Chen and X. Cheng, *et al.*, *Catal. Commun.*, 2020, **136**, 105908.
52. V. Tarallo, K. Sudarshan, V. Nosek and J. Misek, *Chem. Commun.*, 2020, **56**, 5386.
53. S. Kim, K. Lee, S. H. Park, G. H. Kwak, M. S. Kim and H. Y Kim, *et al.*, *Antioxidants*, 2021, **10**, 389.
54. T. Peng, X. Cheng, Y. Chen and J. I. Yang, *Front. Chem.*, 2021, **7**, 714899.
55. L. Yang, B. Wang, X. Yin and Q. Zeng, *Chem. Rec.*, 2021, **21**, 1.
56. Q. Tang, X. Yin, R. R. Kuchukulla, Q. Zeng and R. R. Kuchukulla, *Chem. Rec.*, 2021, **21**, 893–905.

57. J. Lou, Q. Wang, P. Wu, H. Wang, Y.-G. Zhou and Z. Yu, *Chem. Soc. Rev.*, 2020, **49**, 4307.
58. N. Sundaravelu, S. Sangeetha and G. Sekar, *Org. Biomol. Chem.*, 2021, **19**, 1459.
59. Y. Bunno, Y. Tsukimawashi, M. Kojima and S. Matsunaga, *ACS Catal.*, 2021, **11**, 2663.
60. V. Bizet, C. M. M. Hendriks and C. Bolm, *Chem. Soc. Rev.*, 2015, **44**, 3378.
61. (a) H. Takada, Y. Nishibayashi, K. Ohe and S. Uemura, *Chem. Commun.*, 1996, 931; (b) Y. Miyake, H. Takada, K. Ohe and S. Uemura, *J. Chem. Soc., Perkin Trans. 1*, 1998, 2373.
62. (a) H. Nishikori and T. Katsuki, *Appl. Catal., A*, 2000, **194**, 475; (b) C. Ohta and T. Katsuki, *Tetrahedron Lett.*, 2001, **42**, 3885.
63. (a) M. Murakami, T. Uchida and T. Katsuki, *Tetrahedron Lett.*, 2001, **42**, 7071; (b) H. Fujita, T. Uchida, R. Irie and T. Katsuki, *Chem. Lett.*, 2007, **36**, 1092.
64. T. Uchida, Y. Tamura, M. Ohba and T. Katsuki, *Tetrahedron Lett.*, 2003, **44**, 7965.
65. (a) J. Wang, M. Frings and C. Bolm, *Angew. Chem., Int. Ed.*, 2013, **52**, 8661; (b) C. C. Farwell, J. A. McIntosh, T. K. Hyster, Z. J. Wang and F. H. Arnold, *J. Am. Chem. Soc.*, 2014, **136**, 8766; (c) C. K. Prier, T. K. Hyster, C. C. Farwell, A. Huang and F. H. Arnold, *Angew. Chem., Int. Ed.*, 2016, **55**, 4711.
66. M. Yoshitake, H. Hayashi and T. Uchida, *Org. Lett.*, 2020, **22**, 4021.
67. R. R. Annapureddy, F. Burg, J. Gramüller, T. P. Golub, C Merten, S. M. Huber and T. Bach, *Angew. Chem., Int. Ed.*, 2021, **60**, 7920.
68. J. D. Neuhaus, R. Oost, J. Merad and N. Maulide, *Top. Curr. Chem.*, 2018, **376**, 15.
69. P. Li, *Synlett*, 2021, **32**, 1275.
70. M. P. Ó Fearraigh, J. V. Matlockb, O. Illa, E. M. McGarrigle and V. K. Aggarwal, *Synthesis*, 2018, **50**, 3337.
71. J. Drabowicz, E. Krawczyk-Sójka, A. Jasiak, D. Krasowska, G. Mielniczak, M. Koprowski and K. Owsianik, *Pat.*, P.433225, 2021.
72. M. Hamel, M. Vazeux and M. Gulea, *Chem. Pap.*, 2021, **75**, 6137.
73. W. Zheng, X. Chen, F. Chen, Z. He and Q. Zeng, *Chem. Rec.*, 2021, **21**, 396.
74. D. Krasowska, S. Wysocki, M. Janicka, T. Girek, A. Mielniczak and J. Drabowicz, *Polym. Chem.*, 2021, **12**, 1707.

CHAPTER 10

Chiral Selenium and Tellurium-containing Derivatives: Synthesis and Applications

AGATA. J. PACUŁA-MISZEWSKA,[a] JACEK ŚCIANOWSKI,*[a] DOROTA KRASOWSKA[b] AND JOZEF DRABOWICZ[b,c]

[a] Department of Organic Chemistry, Faculty of Chemistry, Nicolaus Copernicus University in Torun, Gagarina Street 7, 87-100 Torun, Poland; [b] Centre of Molecular and Macromolecular Studies, Polish Academy of Sciences, Sienkiewicza 112, 90-363 Łódź, Poland; [c] Jan Dlugosz University in Czestochowa, Armii Krajowej 13/15, 42-200 Czestochowa, Poland
*Email: jsch@umk.pl

10.1 Introduction

The presence of heteroatoms in organic molecules is the key feature that makes them reactive and susceptible to transformation. The chemistry of heterorganic compounds is a very broad field of research covering almost all developing branches of organic chemistry among which the group of organochalcogens plays an important role. Their applicability in organic synthesis as reagents and catalyst, in material science, and as bio-active compounds has been studied for decades and is still evolving. Herein, the synthesis and applicability of chiral organoselenium and organotellurium compounds will be presented, focusing on the most recent reports with a brief historical background.

Chalcogen Chemistry: Fundamentals and Applications
Edited by Vito Lippolis, Claudio Santi, Eder J. Lenardão and Antonio L. Braga

Published by the Royal Society of Chemistry, www.rsc.org

10.2 Chiral Organoselenium Derivatives

In 1973, when Sharpless reported that diphenyl diselenide could efficiently convert epoxides to allylic alcohols through selenoxide elimination,[1] a new scientific field emerged – the development of reagents possessing a highly reactive selenium center. Rapid evolution of research expertise revealed that organoselenium compounds could be easily used as electrophilic, nucleophilic, and radical reagents to selectively functionalize more complex molecules. Facile manipulation of the incorporated Se-moiety, including its elimination from the final product's structure, along with mild reaction conditions, increased the popularity of Se-mediated protocols, *e.g.*, selenenylations, selenocyclizations, SeO_2-eliminations, and 2,3-sigmatropic rearrangements. Many current chapters and reviews summarize these Se-mediated transformations.[2]

The application of organoselenium reagents in catalysis begun in 1975 from the work of Sonoda *et al.* that presented the formation of ureas and carbamates by the reaction of amine or alcohol with carbon monoxide in the presence of elemental selenium. Carbonyl selenide SeCO was the catalytically-active species that initiated the reaction.[3] To date, selenium-based catalysts have been used in many transformations.[4] As organoselenium compounds are well-known for their substantial oxidative properties, a significant part of the field concerns their application in oxidation reactions.[5] A catalytic amount of Se-species combined with various oxidants like hydrogen peroxide, *tert*-butyl hydroperoxide, iodoxybenzene, and ammonium persulfate enables the oxidation of alcohols, carbonyl compounds, and unsaturated bonds (C=C, C=NOH). Furthermore, positively-charged selenium ions RSe^+ can activate π-bonds which results in Se-catalyzed halogenations, aminations, lactonizations, and aminocyclizations. Other Se-catalyzed transformations include trifluoromethylthiolations, esterifications, and oxidative ring expansions.[6]

As has been presented, the utility of organoselenium compounds in modern organic synthesis is widely explored and quite common. However, the asymmetric types of Se-mediated reactions, especially in catalytic versions, are much less examined. The attachment of a chiral auxiliary directly to the Se atom, forming a chiral reagent, broadens the perspective of potential applications in *e.g.*, the synthesis of natural products, chiral building blocks and drug design.

The synthesis of chiral organoselenium reagents, catalyst, or ligands is based on the incorporation of the Se-functionality in the structure of the optically-active precursor.[7] A general synthetic methodology to obtain chiral selenides **1**, diselenides **2**, and selenols **3** is presented in Scheme 10.1. The enantiomerically-pure substrate in the form of halide R*X can be converted to compound **1** or **2** through nucleophilic substitution with metal selenide/diselenide. Optically-active selenols **3**, obtained by the reaction of elemental selenium with metalated chiral substrates, can also be oxidized to diselenides **2** and reversibly, compounds **2** can be easily reduced to R*SeH.

Scheme 10.1 General synthesis of selenium-based reagents.

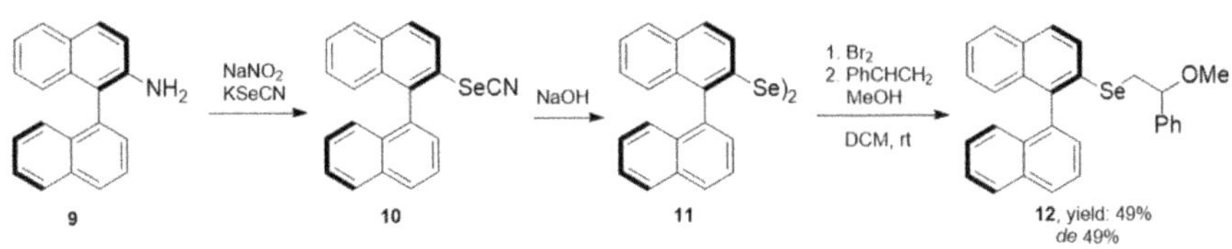

Scheme 10.2 Tomoda's chiral diselenide **11** and its application in methoxyselenenylation.

The Se–Se bond can be readily cleaved and was frequently converted to electrophilic **4** and nucleophilic-type functionality **5** efficiently used in selenium-mediated asymmetric transformations. Selenides on the other hand can be oxidized to selenoxides **6**, converted to selenonium salts **7** and ylides **8**, forming a different class of chiral organoselenium reagents (Scheme 10.1).

The chapter is divided into sections according to the type and applicability of the described organoselenium derivatives.

10.2.1 Chiral Electrophilic Organoselenium Reagents

Conversion of diselenides to chiral electrophiles and their utility as selenenylating agents was the first example of a stereoselective reaction conducted by non-racemic Se-reagent. In 1988 Tomoda and co-workers presented the synthesis of optically-active binaphtyl diselenides **11** by the reaction of corresponding amines **9** with potassium selenocyanate and an aqueous solution of NaOH.[8] The compounds were applied as electrophiles in the methoxyselenenylation of alkenes (Scheme 10.2).

The work of Tomoda inspired others and in the 90s several chiral diselenides were obtained and in most cases also evaluated as reagents in the methoxyselenenylation of styrene. All reported derivatives resembled a structural similarity: a diaryl diselenide core with an attached chiral scaffold in the *ortho*-position. Diezel and co-workers synthetized disubstituted compounds with an ethoxyethyl **13** and tetrahydrofuranyl **14** moieties revealing that the C_2-symmetrical diselenides can provide promising asymmetric induction.[9] The influence of C_2-symmetry was also explored by Tomoda.[10] However, the attachment of a pyrrolidine-based moiety, to construct derivative **15**, decreased the diastereomeric excess. In 1994 Uemura *et al.* exchanged the aryl ring to a chiral ferrocenyl core directly connected to the selenium atom.[11] Compound **16**, converted to a hexafluorophosphate $RSePF_6$, significantly improved the stereoselectivity of the selenenylation reaction (de 97%). Later on, a series of simple chiral scaffolds was used by Wirth to obtain various selenium electrophiles.[12] It was observed that the presence of different heteroatoms influences the chirality transfer, compounds **17** and **18** (Scheme 10.3).

Similar compounds were later synthetized by Tiecco *et al.*[13] The authors proved that the presence of a Se-heteroatom non-bonding interaction between the antibonding orbital of selenium with the lone pair of oxygen **19**, nitrogen **20**, or sulfur **21**, shortens the distance between the catalytically-active center and the chiral carbon making the structure more rigid and thus enhancing the asymmetric induction. Compound **22** without an additional heteroatom furnished no asymmetric induction (Figure 10.1).

More recently, Wirth and co-workers presented aromatic sulfoxide-containing diselenides **23** and **24** chiral due to the tetrahedral geometry of the sulfur center.[14] Methoxyselenenylation of 2-chlorostyrene revealed that the incorporation of an additional methoxy group decreased the diastereomeric excess of the process (Scheme 10.4).

Generally, the majority of chiral diselenides tested as selenenylating agents have been further evaluated in asymmetric selenocyclizations. These

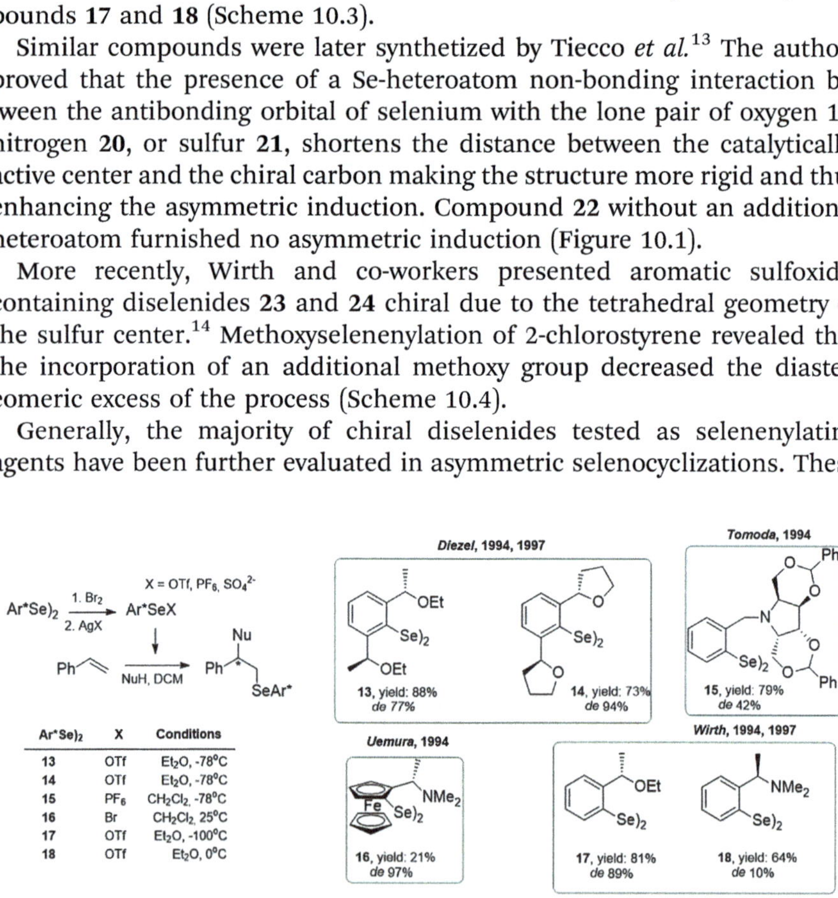

Ar*Se)2	X	Conditions
13	OTf	Et2O, -78°C
14	OTf	Et2O, -78°C
15	PF6	CH2Cl2, -78°C
16	Br	CH2Cl2, 25°C
17	OTf	Et2O, -100°C
18	OTf	Et2O, 0°C

Scheme 10.3 Chiral diaryl diselenides used as selenenylating agents.

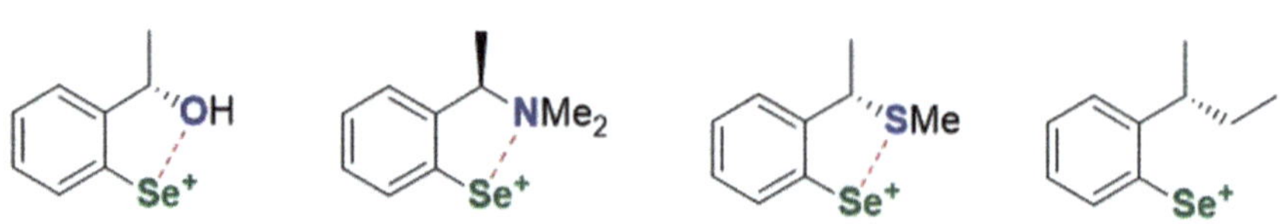

19 de 83% — **20** de 35% — **21** de 92% — **22** de 0%

Figure 10.1 Selenium-heteroatom non-bonding interactions.

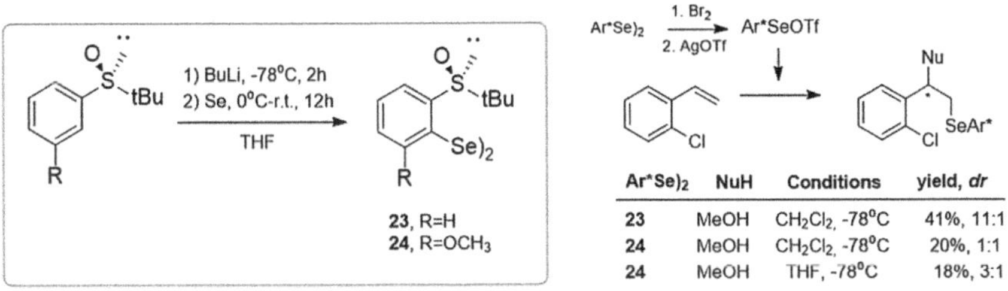

Ar*Se)$_2$	NuH	Conditions	yield, *dr*
23	MeOH	CH_2Cl_2, -78°C	41%, 11:1
24	MeOH	CH_2Cl_2, -78°C	20%, 1:1
24	MeOH	THF, -78°C	18%, 3:1

Scheme 10.4 Sulfoxide-containing diselenides in the methoxyselenenylation of 2-chlorostyrene.

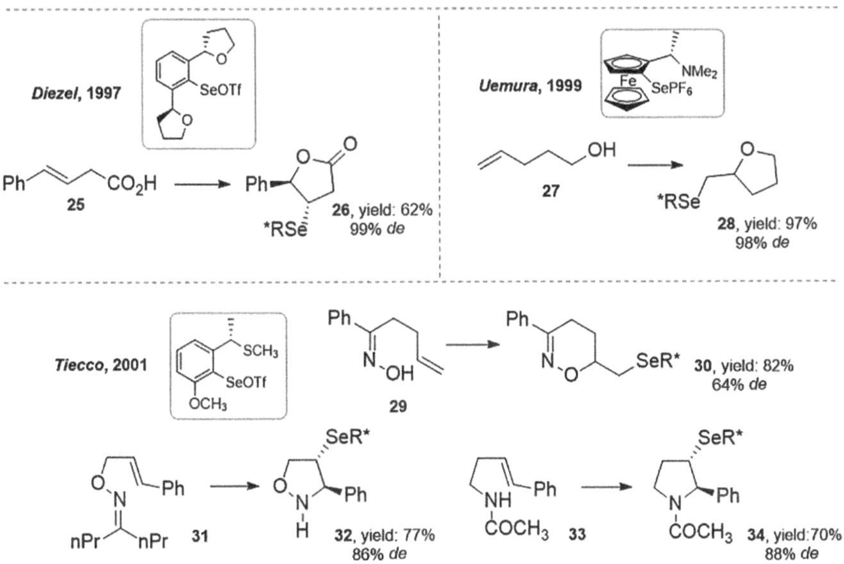

Scheme 10.5 Selenium-driven cyclizations.

types of transformations can occur when the nucleophile is an internal part of the unsaturated substrate. After the selenenylation step and the Nu-attack on the seleniranium ion, a cyclic product is obtained. The enantiomeric enrichment of the final products, achieved in these Se-promoted cyclizations, made them very useful tools in modern asymmetric syntheses.

To date, numerous research groups efficiently converted structurally-diversified substrates to various heterocycles – unsaturated carboxylic acids **25** to lactones **26**,[15] allylic alcohols **27** to cyclic ethers **28**[16] and N-heterocycles like 1,2-oxazines **30** from alkenyl oximes **29**,[17] isoxazolines **32** from allyl oximes **31**,[18] and *N*-acetylpyrrolidines **34** from *N*-alkenyl acetamides **33**[19] (Scheme 10.5).

As a natural and commercially-available source of chirality terpenes have been broadly used to synthetize structurally divers asymmetrical diselenides as precursors of selenenylating agents. The most exploited in the literature was the C^3-substituted camphor diselenide **35** obtained by Back and co-workers through the formation of a monoterpene selenolate at the α-carbon and subsequent acidification and air oxidation. Compound **35** was used as a substrate for the synthesis of other diselenides, transforming the carbonyl to an oxime moiety **36**, **37**, or a spiro-oxazolidinone **38**. As electrophiles, in the form of chlorides **35–38a** the reagents were tested in cyclization of unsaturated alcohols, furnishing promising results for compound **38a**. Transformed to triflate electrophiles **35–38b** were used to promote the methoxyselenenylation of *trans*-dec-5-ene. In contrast to the selenocyclization reaction, compound **38a** yielded the lowest diastereomeric ratio compared to other tested diselenides (Scheme 10.6).[20–30]

Tiecco *et al.* transformed diselenide **35** to a selenenyl sulfate **39** and evaluated it as an electrophile in a series of selenenylatio-elimination protocols. First, allylic ethers were obtained by alkoxyselenenylation of β,γ-unsaturated esters **40**. The elimination of the Se-moiety was achieved through the treatment with excess of ammonium persulfate. Next, oxazolines **44** were obtained by the amidoselenenylation of alkenes **42** and deselenylation in the presence of PhSeOTf. Conversion of amides **43** to thioamides **45** with Lawesson's reagent and treatment with PhSeCl correspondingly led to enantiomerically-pure cyclic thiazolines **46** (Scheme 10.7).[31–34]

Later on, numerous examples of various Se-functionalizations of monoterpenes were presented by Ścianowski and co-workers leading to a series of structurally diverse chiral electrophilic reagents. Monoterpene alcohols **47** from *p*-menthane, carane, and pinane systems were used as starting materials and converted to the corresponding chlorides **48** and tosylates **49** through the reaction with carbon tetrachloride/triphenylphosphine or tosyl chloride in pyridine, respectively. Reaction of **48** or **49** with sodium diselenide led to the final products **50**. All synthetized compounds were evaluated as reagents in the methoxyselenenylation of styrene and cyclization of *o*-allylphenol. The bicyclic structures derived from pinane **51**, hydroxycarane **52**, and myrtenol **53** were the most efficient ones (Scheme 10.8).[35–40]

10.2.2 Chiral Nucleophilic Organoselenium Reagents

The first application of organoselenium nucleophiles was presented by Sharpless in 1973 in the ring opening of epoxides,[1] and the use of chiral Se-nucleophiles in stereoselective synthesis began in 1988 in the work of Tomoda *et al.* in the cleavage of *meso*-oxiranes[41] and later on also by Uemura.[42] Nucleophiles generated from binaphthyl **54** and ferrocenyl diselenides **55** exhibited moderate de values (Scheme 10.9).

In 2016, Ludke and Wessjohan presented the utility of chiral carbohydrate diselenides **56** for the ring opening of steroidal epoxides **57**.[43] These interesting natural product conjugates **58** were obtained through the $NaBH_4$-induced generation of Se-pyranoside and furanoside nucleophiles (Scheme 10.10).

Other sugar-based nucleophilic reagents were obtained by Ishihara and co-workers.[44] β-*p*-Methylbenzoyl selenoglycosides were synthetized and converted to nucleophiles using Cs_2CO_3. Through the reaction with diverse

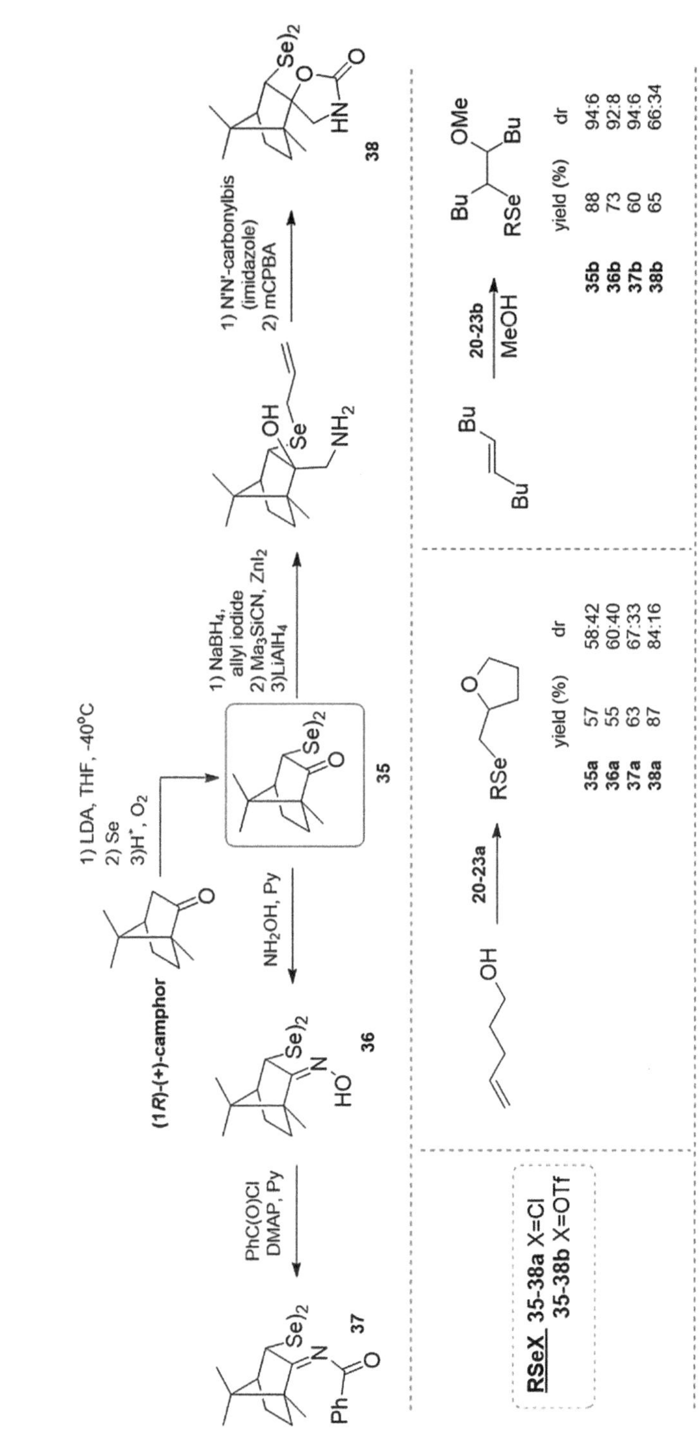

Scheme 10.6 Results of cyclization and methoxyselenenylation with the use of selenenyl chlorides and triflates.

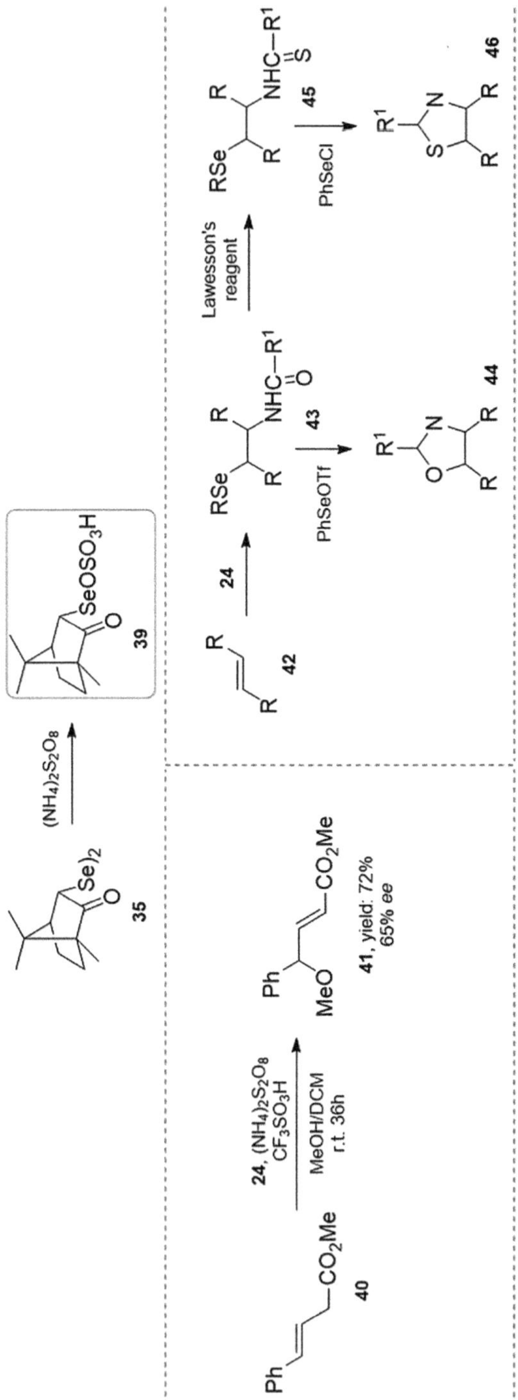

Scheme 10.7 Selenylation–deselenylation performed with selenyl sulfate **39**.

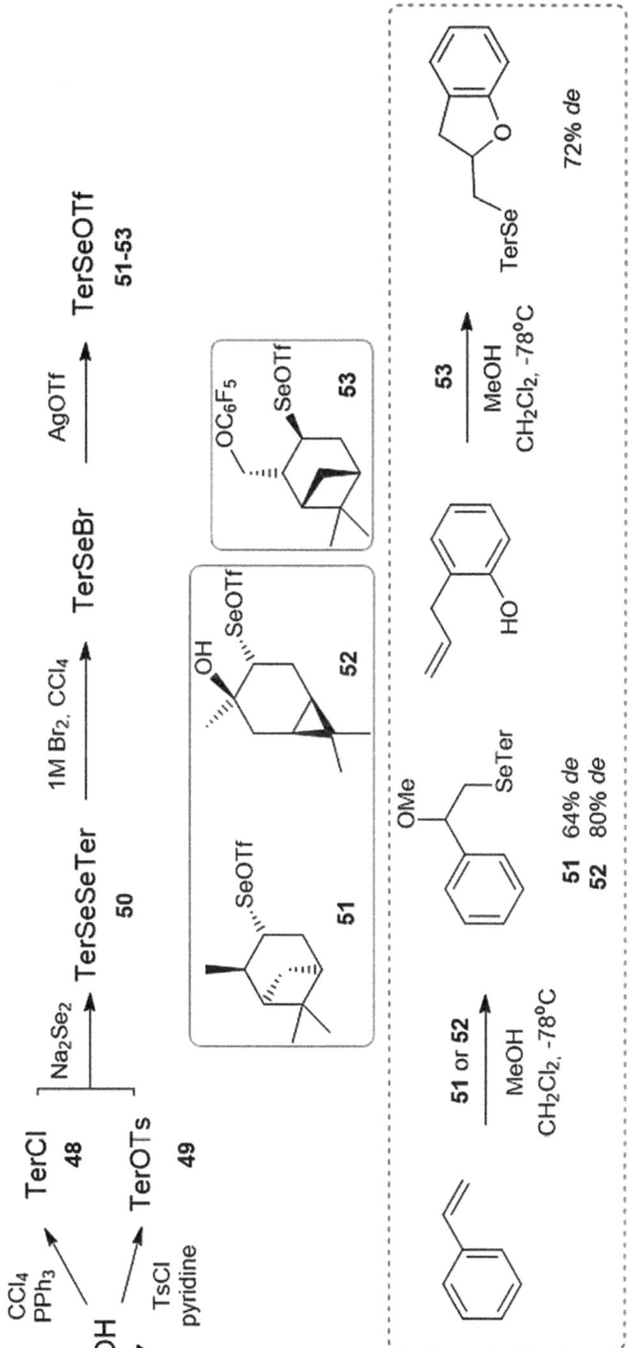

Scheme 10.8 Methoxyselenylation and cyclization of *o*-allylphenol induced by Se–terpene electrophiles.

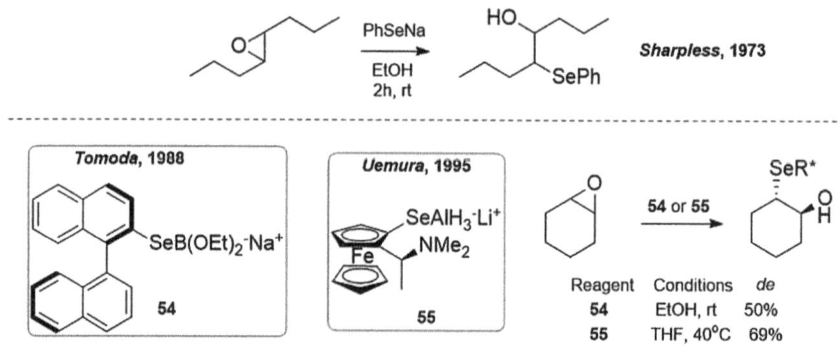

Scheme 10.9 Epoxide-ring opening with reagents **54** and **55**.

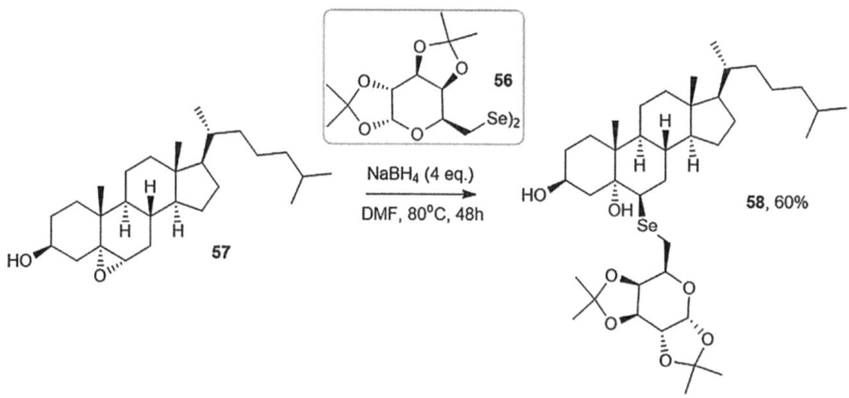

Scheme 10.10 Opening of steroidal epoxides with sugar diselenides.

electrophiles, *e.g.*, alkyl, aryl, sugar, and amino acid-derived, a series of new selenoglycosides was obtained.

Chiral selenols have been also used as nucleophiles to construct other chiral derivatives. In 2008, Santi *et al.* presented a procedure to generate selenols **59–61** from diselenides by the treatment with zinc under acidic conditions.[45] Reaction of nucleophilic selenols with iodides enabled selenides to be obtained in moderate yields (Scheme 10.11).

Another application of optically-active selenol as the nucleophile is the peptide conjugation mediated by selenocysteine **62**. The –SeH group of the N-terminal peptide attacks the thioester carbon of the C-terminal **63** and elongated selenopeptides **65** can be formed (Scheme 10.12).[46]

Organoselenium nucleophiles have also been explored in stereoselective aldol condensations. After the work of Toru *et al.* in which the α-selenocarbonyl compounds **66** were transformed to α-seleno titanium enolates, they were efficiently tested in the reaction with aldehydes and ketones to afford, after Se-elimination, α,β-unsaturated carbonyl compounds **68**.[47,48] Later, Tiecco and co-workers explored the reaction further and attached a

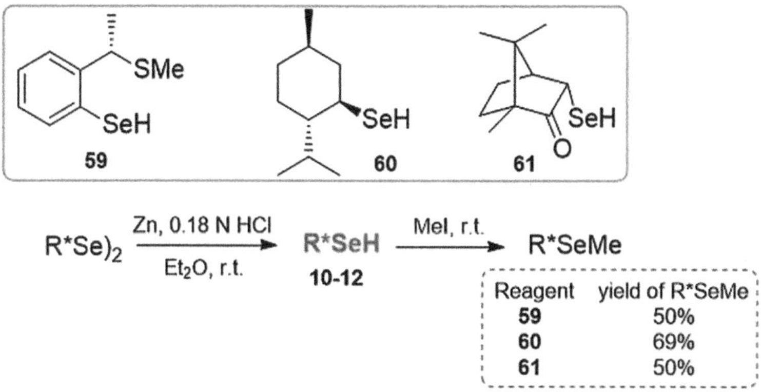

Scheme 10.11 Generation of selenols with zinc in acidic conditions.

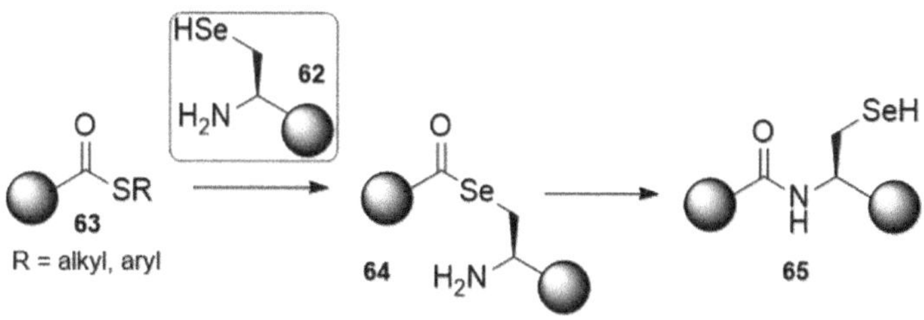

Scheme 10.12 Peptide elongation with nucleophilic selenocysteine.

chiral camphor auxiliary to the Se-nucleophile. The titanium enolates formed from (*R*)-camphorselenoacetone **70** and methyl (*R*)-camphorselenoacetate **69** afforded after separation enantiomerically-pure *syn-* and *anti*-aldols **71** and **72**.[49,50] Later, Franck *et al.* presented the synthesis of *N*-phenylselanylacetyloxazolidin-2-thione **73** and thiazolidine-2-thione **74** which after the generation of titanium enolates could react with aldehydes to obtain aldols **75** and **76** with good yields and diastereoselectivities[51] (Scheme 10.13).

Furthermore, Silveira and co-workers explored the reactivity of α-phenylseleno chlorotitanium enolates. In this case, a Mannich-type reaction with aromatic aldimines was described. A stereoselective α-phenylseleno-β-amino ester preparation was presented.[52]

10.2.3 Chiral Selenide-based Reagents

The presented section of chiral selenide-based reagents covers derivatives that were designed on the basis of single selenium functionality like selenoxides, selenonium salts, and selenonium ylides. Taking into account that the chemistry of these compounds was thoroughly discussed in recent reviews,[53,54] herein we will present only those examples where a chiral auxiliary is connected to the selenium atom, without the considerable number of chiral derivatives due to the presence of the quaternary selenium.

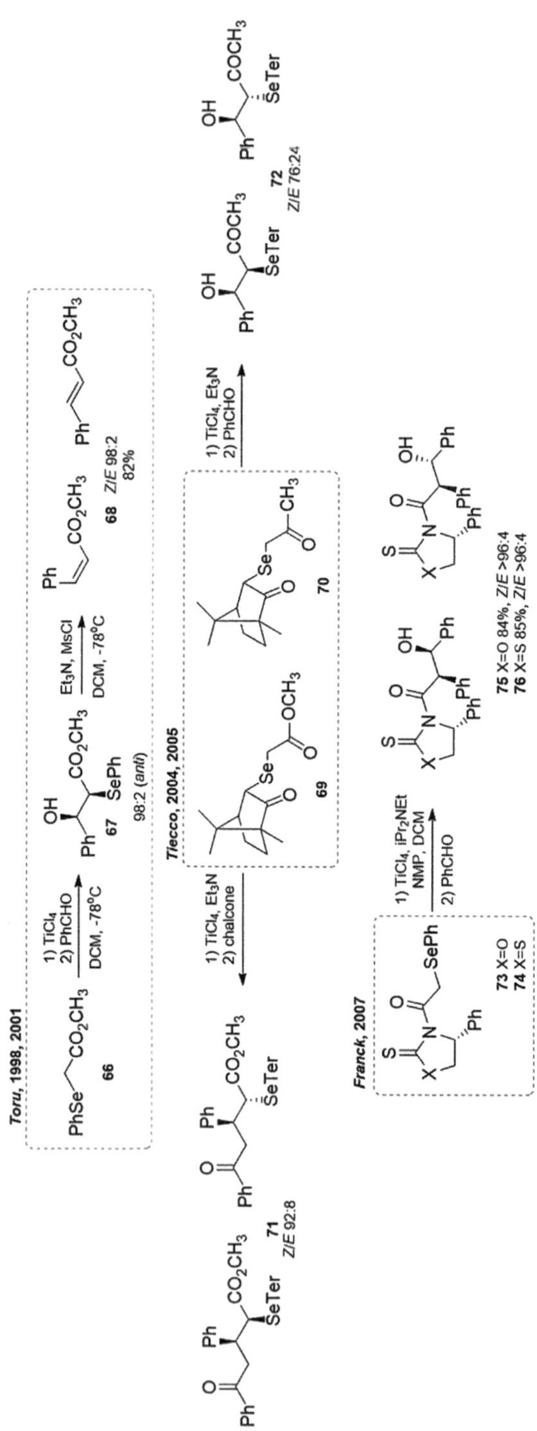

Scheme 10.13 Se-nucleophiles in stereoselective aldol condensations.

Chiral selenoxides can be formed by oxidation of enantiomerically-pure substrates or through the asymmetric oxidation of prochiral selenides. This topic has been recently reviewed by Drabowicz and co-workers.[54] A facile *syn*-elimination of RSe(O)OH enables alkenes to be obtained, valuable intermediates in organic syntheses. In the early 60s a series of steroidal selenoxides was presented and after oxidation vast elimination of the Se-moiety also observed leading to double bond formation in the steroidal core.[55,56] This discovery was later explored and characterized by Sharpless showing that alkyl phenyl selenoxides readily decompose to olefins at room temperature. At the same time, Sharpless also presented the first [2,3]-sigmatropic rearrangement of Se-terpenyl derivatives. Geranyl selenol **77**, selenide **78**, and diselenide **79** after oxidation with H_2O_2 easily converted to linalool **83**.[57–59] The ability of allylic selenoxides to form allylic alcohols was also evaluated in the asymmetric version. Reich *et al.* were the first to prepare optically-active alcohols using this methodology. A geranyl [2.2]paracyclophanyl selenide **80** was oxidized with mCPBA and after [2,3]-sigmatropic rearrangement forming selenenic ester **82** the final alcohol was obtained with 65% ee.[60] A similar procedure was performed with ferrocenyl selenide **81** improving the enantioselectivity to 83% ee[61] (Scheme 10.14).

Also the work of Koizumi *et al.* described the vast conversion of chiral selenoxides to allylic alcohols. First, a series of haloselenuranes **83** were obtained from camphor bromide *via* selenide **82** formation and treatment with *tert*-butylhypochloride. The allylic chloroselenuranes **84** were further converted to selenoxides **85**, selenonium ylides **86**, and allylic selenimines **87** by the treatment with an aqueous solution of $NaHCO_3$, carbon nucleophiles, and lithium *N*-protected amides, respectively. The subsequent [2,3]-sigmatropic rearrangement yielded allylic alcohols **88**, allylic selenides **89**, and allylic amines **90** (Scheme 10.15).[62–66]

In this diastereoselective synthesis of selenoxides **85** the racemization of the compounds is inhibited due to several factors such as the intramolecular hydrogen bonding between the seleninyl oxygen and the hydroxy group of the bornane auxiliary and the sterical hindrance of the terpene skeleton. The stability of these selenoxides was tested in the enantioselective protonation of metal enolates.[67] Achiral lithium or zinc enolate **92** was converted to ketone **93** with an ee up to 89%. The bornane –OH group acted as the proton source while the selenium moiety **91** chelated the metal cation (Scheme 10.16).

Later, Ścianowski *et al.* transformed other terpene phenyl selenides to [2,3]-sigmatropic rearrangement precursors.[68–72] The series included acyclic (geranyl, neryl), monocyclic ((+)-carvotanacetyl, (–)-carvyl, perillyl) and bicyclic (caranyl, pinanyl) skeletons. As an example (+)-10-phenylseleno-2-pinene **96** was efficiently converted to selenoxide **97** *via* oxidation with 30% H_2O_2, to methylcarbamate **98** by the reaction with methylcarbamate and NCS and toluenosulfonamide **99** through the treatment with chloramine T. Compounds underwent [2,3]-sigmatropic rearrangement and hydrolysis affording the corresponding allyl alcohol **100**, methylcarbamate **101**, and amine **102** (Scheme 10.17).

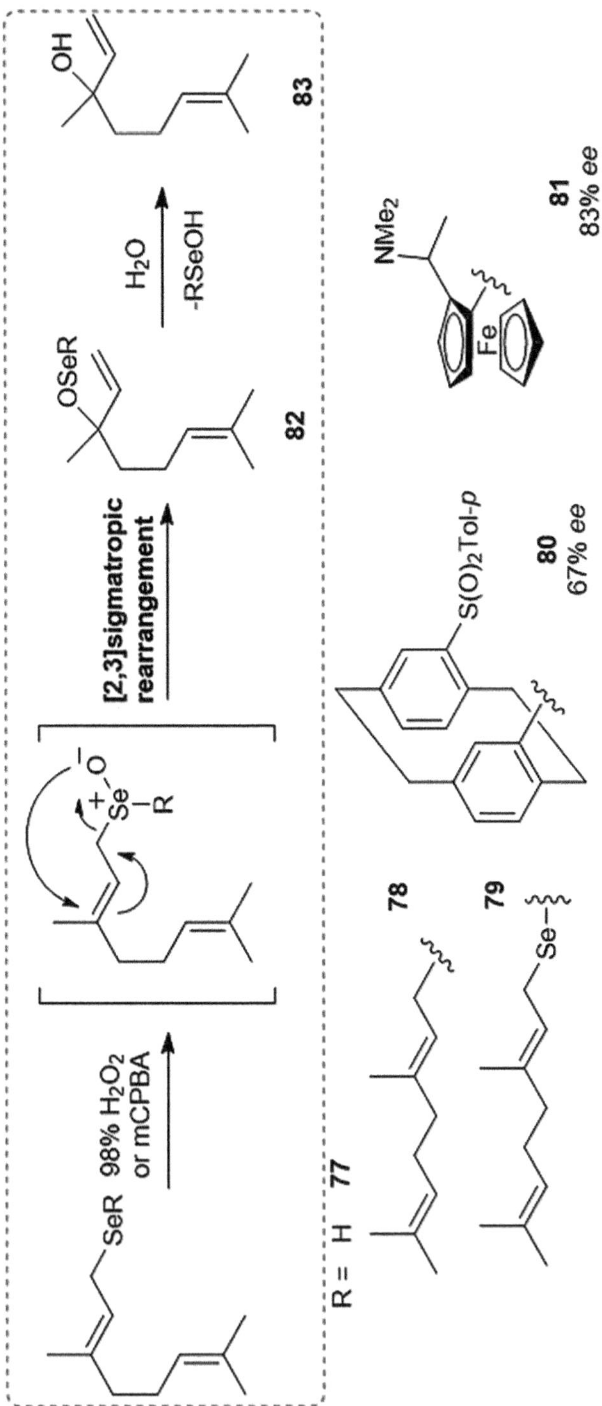

Scheme 10.14 Allylic alcohol synthesis *via* [2,3]-sigmatropic rearrangement of chiral selenoxides.

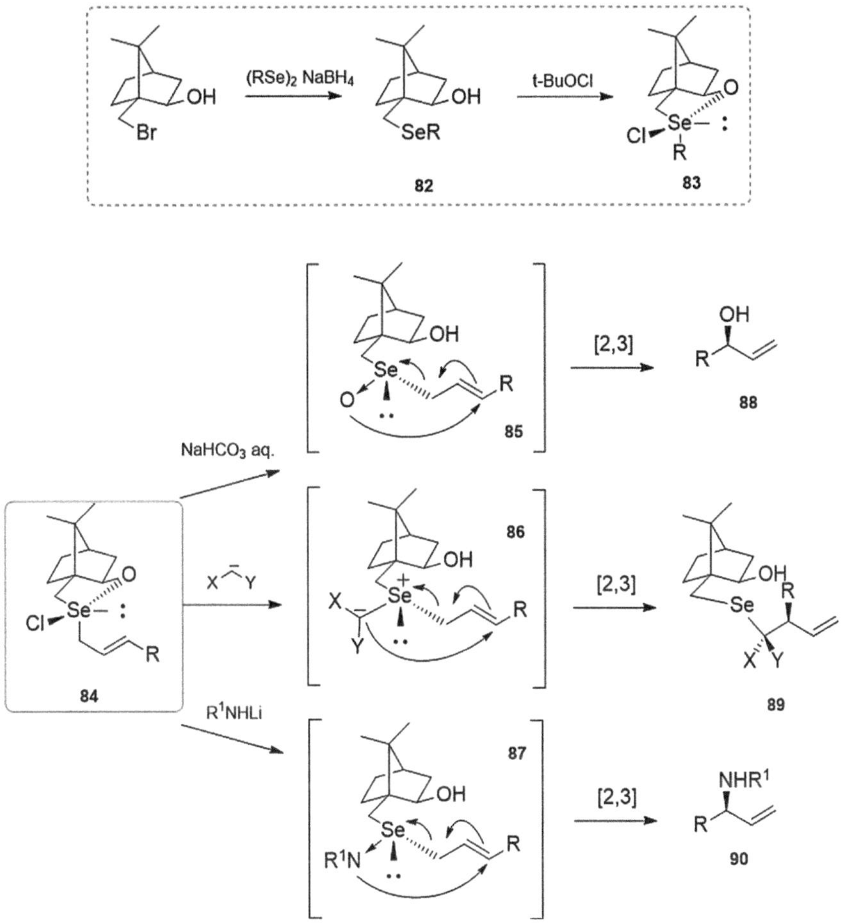

Scheme 10.15 Allylic haloselenuranes as substrates for the synthesis of selenide-based reagents.

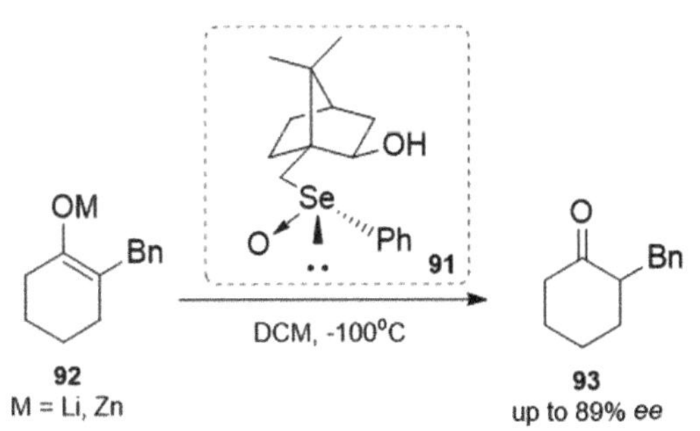

Scheme 10.16 Allylic haloselenuranes as reagents in the protonation of metal enolates.

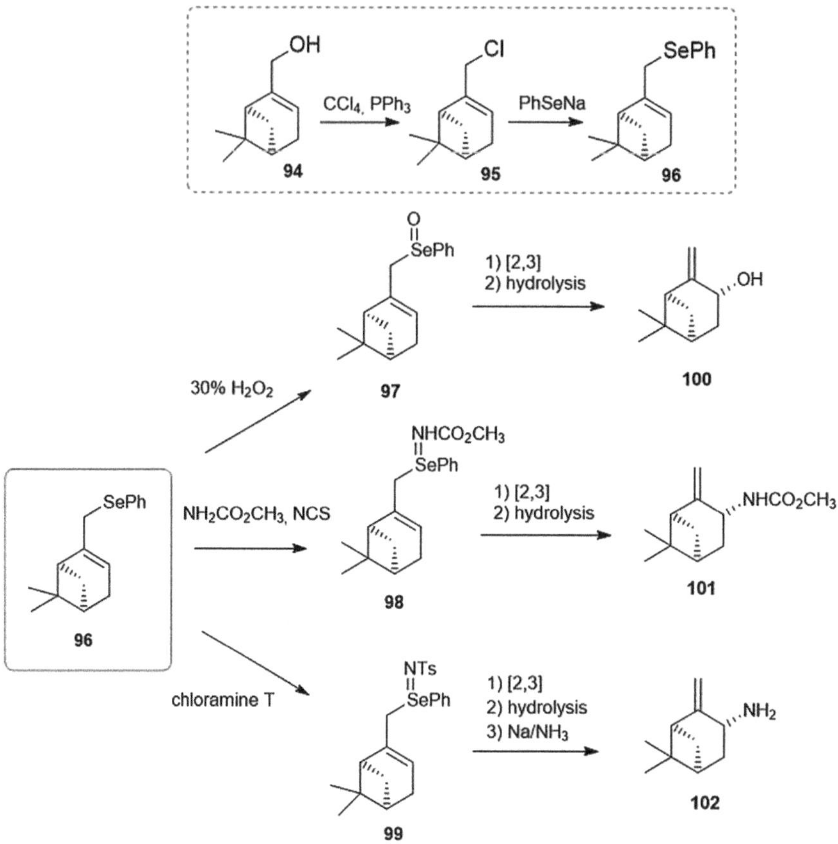

Scheme 10.17 (+)-10-Phenylseleno-2-pinene as a substrate for the synthesis of [2,3]-sigmatropic rearrangement precursors.

Next, in 2013 the same group transformed *p*-menthane, carane, and pinane selenols to a series of allylic selenides and also evaluated their utility in the synthesis of chiral allylic alcohols.[73] The rearrangement was preceded with mCPBA oxidation of allylic selenide **105.** Caranylselenol **104** afforded the highest enantiomeric excess with 34% ee (Scheme 10.18).

Optically-active selenonium ylides and salts form a fairly new group of reagents to use in asymmetric synthesis. Se-ylides have already been presented in Scheme 10.14, where Koizumi used a 2-*exo*-hydroxy-10-bornyl group as the chiral auxiliary to synthetize compound **86.** These Se-ylides underwent rearrangement and enabled enantiomerically enriched allylic selenides **89** to be obtained. The first example of optically-active selenonium ylide was described in 1991 by Kamigata *et al.*[74,75] Commercially available L-menthol **107** was first transformed to methyl selenides **109** by DCC-coupling. Next, ^{t}BuOCl oxidation led to selenoxide **110** which was further treated with dimedone to afford a diastereomeric mixture of ylides **111** in 69% yield (Scheme 10.19).

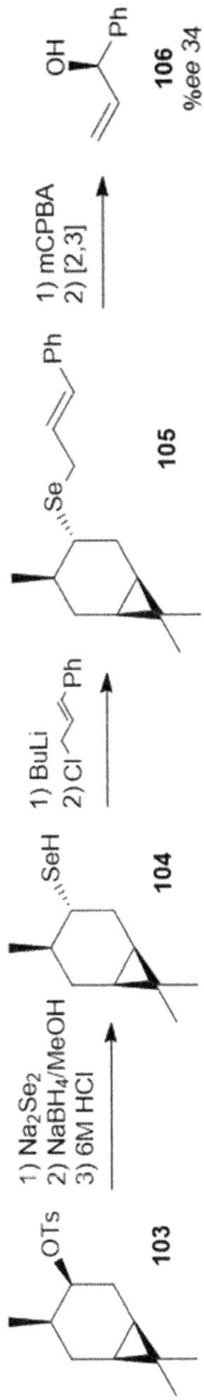

Scheme 10.18 Synthesis of allylic alcohols from allylic selenides.

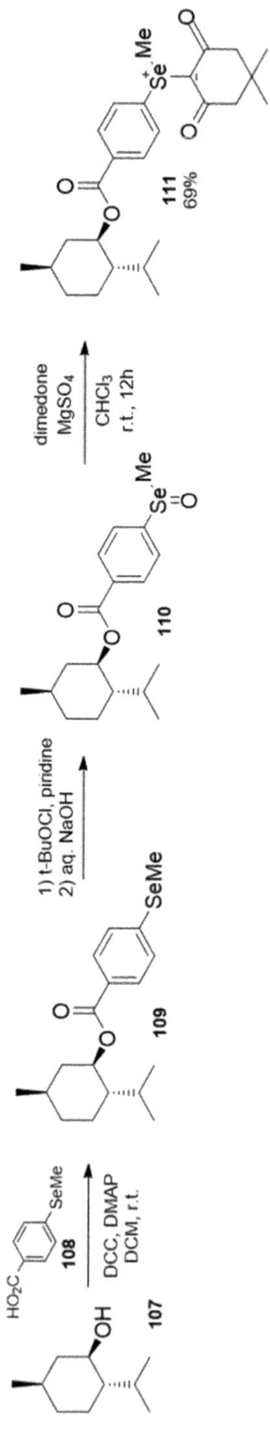

Scheme 10.19 First synthesis of optically-active ylide 111.

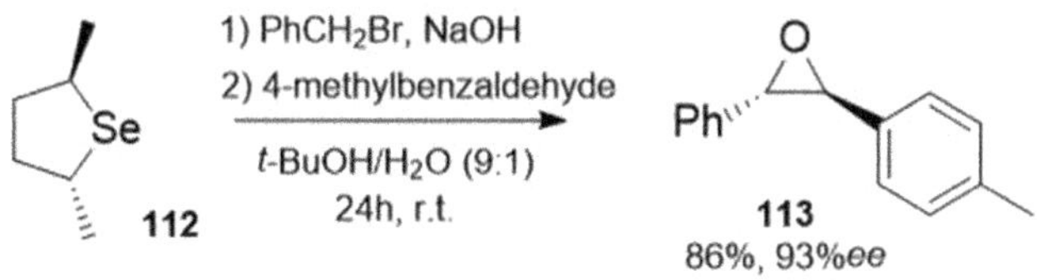

Scheme 10.20 First asymmetric epoxidation using chiral selenonium ylide.

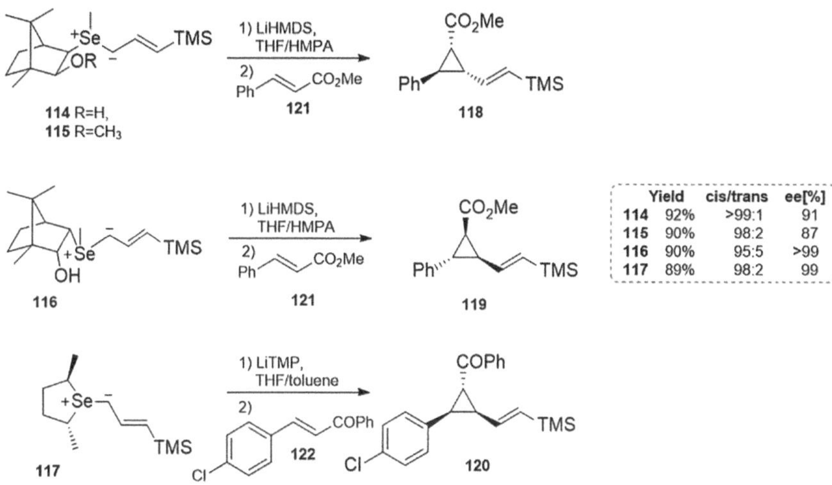

Scheme 10.21 Synthesis of cyclopropanes from camphor *exo*- and *endo*-ylides.

Chiral selenonium ylides and salts can be efficiently used as precursors of chiral epoxides and cyclopropanes. The first application of optically-active selenonium ylide in asymmetric epoxidation was reported in 2001 by Metzner and co-workers.[76] Reaction of 2,5-dimethylselenolane **112** with benzyl bromide and NaOH furnished the selenonium ylide which after the reaction with various aldehydes afforded epoxides with high enantioselectivities (92–93%). An example is presented in Scheme 10.20.

In 2009 Huang and co-workers used optically-active organoselenium substrates for the first time in asymmetric cyclopropanation.[77] The reaction was conducted between *exo*-**114**, **115**, and *endo*-camphor **116** selenonium ylides and α,β-unsaturated carbonyl compounds. In both cases the diastereoselectivities and enantioselectivities of the obtained *cis*-(1*R*,2*R*,3*R*)- **118** and *cis*-(1*S*,2*S*,3*S*)-trisubstituted cyclopropanes **119** were excellent. The same authors also evaluated the utility of C_2-symmetrical ylide **117** obtaining good yields and stereoselectivities for a variety of unsaturated compounds including ester, ketones, amides, and nitriles. Examples utilizing unsaturated methyl ester **121** and ketone **121** are presented in Scheme 10.21.

Huang *et al.* also tested the C_3-camphor selenonium salts in the synthesis of oxiranes.[78] The highest enantio- and diastereo-selectivity was observed for

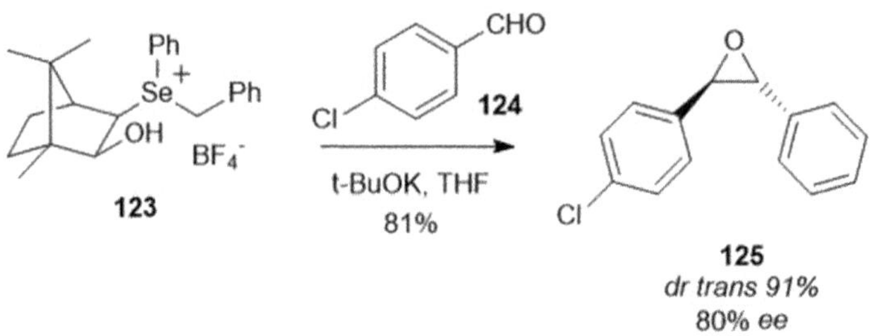

Scheme 10.22 Epoxidation with the use of C_3-camphor selenonium salt **123**.

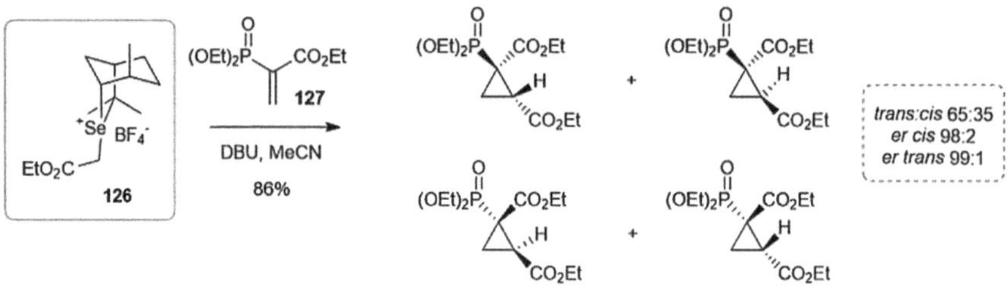

Scheme 10.23 Cyclopropanation of vinylphosphonate.

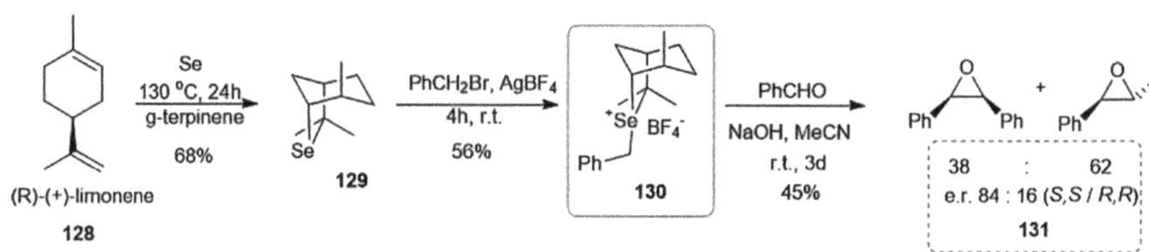

Scheme 10.24 Terpenyl selenonium salt **130** in asymmetric epoxidation.

the epoxidation performed with salt **123** and 4-chlorobenzaldehyde **124** (Scheme 10.22).

Quite recently, Ścianowski and co-workers obtained a series of terpene-derived selenonium salts applicable in asymmetric cyclopropanation and epoxidation. Compounds derived from (+)-limonene and (−)-menthol, previously converted to ylides, were tested in the cyclopropanation of vinylphosphonate **127**.[79] The best results were observed for salt **126** with enantiomeric ratios 98 : 2 (*cis*) and 99 : 1 (*trans*) (Scheme 10.23).

Salts based on *p*-menthane, carane, pinane, and (+)-limonene skeletons were evaluated in asymmetric epoxidation.[80] Derivative **130**, obtained from terpene **128** gave the best enantioselectivities with an enantiomeric ratio of 84 : 16 (Scheme 10.24).

10.2.4 Chiral Organoselenium Catalysts

The first application of a chiral organoselenium compound as a catalyst was presented by Uemura *et al.* in 1994.[81] Diferrocenyl diselenides were used as ligands in the Rh(I)-catalyzed asymmetric hydrosilylation of ketones. Since that time, a lot has been accomplished in the field, a variety of Se-catalysts and ligands have been efficiently used in asymmetric synthesis, and several reviews have been published to summarize the topic.[82]

In 1995 Wirth described the utility of chiral organoselenium derivatives as ligands in the well-known addition of diethylzinc to aldehydes.[83] A series of chiral diaryl diselenides was evaluated in the Et_2Zn addition to benzaldehyde, and selected examples **132** and **133** are presented in Scheme 10.25. The used catalysts yielded the final alcohol with excellent enantiomeric excess. Different bidendate N, Se-ligands were synthetized by Braga *et al.* Chiral oxazolidine **134**[84] and aliphatic amino diselenides **135**,[85] obtained from chiral aminoalcohols, enabled the conversion of benzaldehyde to alcohol with good yield and enantioselectivity.

Interesting diferrocenyl ligands possessing an oxazilidine **136** functionality were synthetized by Bolm *et al.* and tested in the zinc-promoted aryl transfer to aldehydes.[86] It was reported that better enantiomeric ratios were observed when the reaction was performed with aromatic aldehydes as substrates (Scheme 10.26).

The Pd-catalyzed Tsuji–Trost allylic substitution is another example of a well-known metal-mediated process were Se-ligands were proven to be efficient. The first example of chiral Se-derivative applied in this reaction was presented by Helmchen and co-workers in 1994.[87] Aminoalcohol **137** was

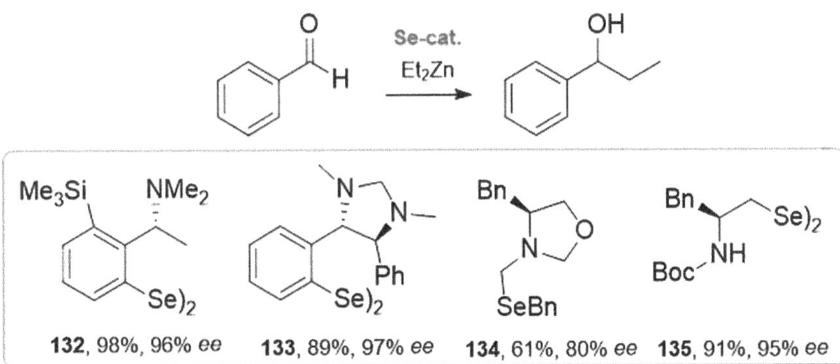

Scheme 10.25 Se-ligands in diethylzinc addition to aldehydes.

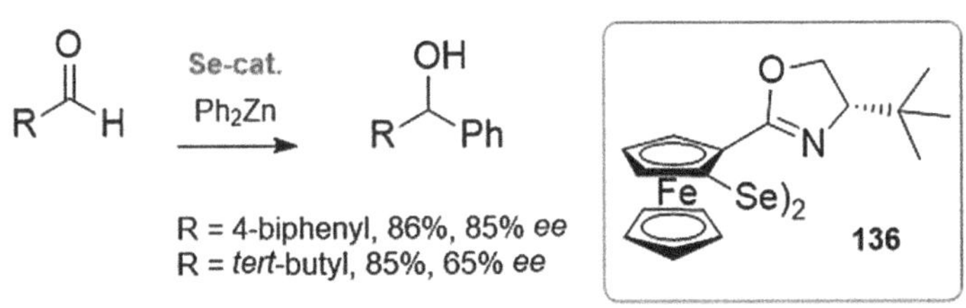

Scheme 10.26 Ferrocenyl Se-ligand **136** in diphenylzinc addition to aldehydes.

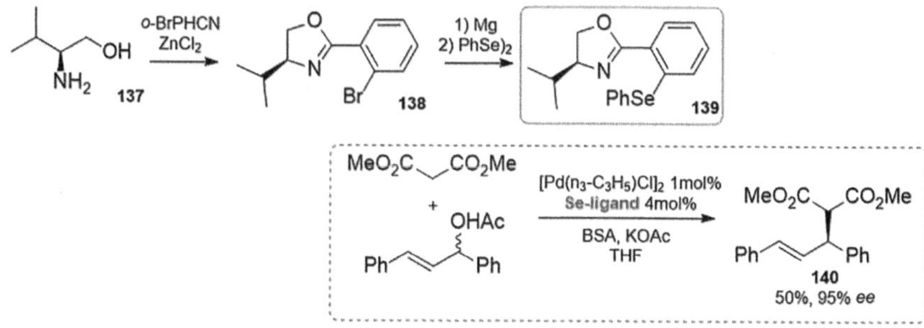

Scheme 10.27 First example of Pd-catalyzed allylic alkylation using Se-ligand **139**.

first converted to bromide **138** and then a further Grignard procedure and diphenyl diselenide substitution gave oxazoline-based phenylselenide **139**. The asymmetric allylic alkylation of *rac*-1,3-diphenyl-2-propenyl acetate using 4 mol% of ligand **139** gave excellent enantiomeric excess of the final product (Scheme 10.27).

As a result of these findings numerous research groups have made attempts to design Se-based protocols for efficient Pd-catalyzed Tsuji–Trost reactions. Examples of diverse organoselenium ligands **141–146** are presented in Table 10.1.[88–92]

Additionally, in 2016 Ścianowski *et al.* synthetized terpene-derived diselenides as procatalysts to be used in the Tsuji–Trost reaction.[93] The diselenides bearing *o*-(*S*-caranyl)- **147** and *o*-(*S*-isocaranyl)- **148** linkages were transformed to methyl selenides **149** and **150**, further tested as chiral ligands. The allylic alkylation furnished only moderate enantioselectivities. Next, the compounds were also evaluated in the nitroaldol reaction with observed enantiomeric excess in the case of *o*-(*S*-caranyl) derivative **109** as the first example of an organoselenium ligand applicable in the Henry reaction (Scheme 10.28).

Recently, Skarżewski and co-workers presented new chiral N, Se-ligands for the Tsuji–Trost alkylation.[94] Phenylselenides **153** were obtained through the nucleophilic substitution of sulfonamides **152**, previously obtained from amino alcohols **151**. The allylic alkylation product was obtained with 50% ee (Scheme 10.29).

Chiral organoselenium catalysts can also be used to promote heterocycle-forming reactions. Indane-derived selenide **154** through the selenenylation–deselenylation sequence, in the presence of NFSI as the terminal oxidant, enabled the lactonization of unsaturated carboxylic acids **155** to butenolides **156**.[95] Recently, a structurally similar diselenide **157** was efficiently used in the iminolactonization of unsaturated amides **158** yielding iminolactones **159**.[96] Another organoselenium indane-based catalyst **160** was used to obtain the trifluoromethylthiolated lactone **162** starting from phenylbutanoic acid **161** and *N*-CF_3S-saccharine as the source of –SCF_3.[97] Monoterpene-based diselenide **163** was also used to catalyze lactonization of unsaturated carboxylic acids **164**. After the selenenylation step, the deselenylation was achieved

Table 10.1 Examples of organoselenium ligands used for the Pd-catalyzed Tsuji–Trost reaction.

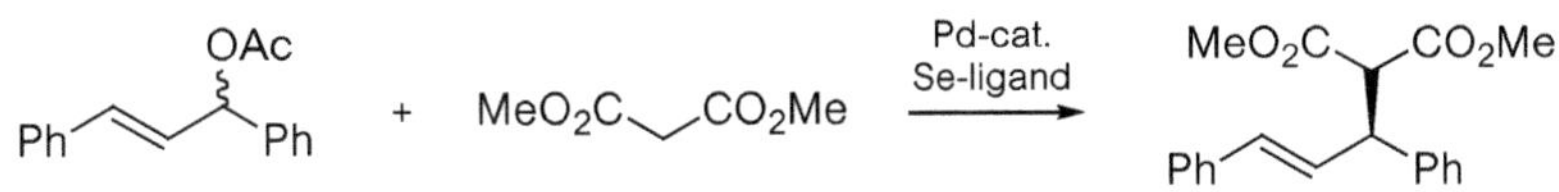

Se-ligand	Conditions	Yield	ee
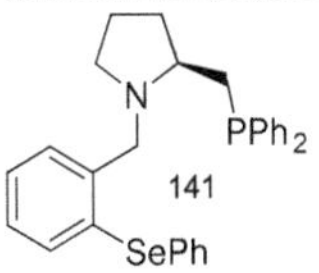 141	Se-ligand 6 mol%, Pd-cat. 3 mol%, r.t., 60 h	75%	79%
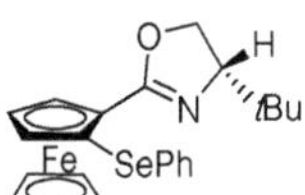 142	Se-ligand 6 mol%, Pd-cat. 2 mol%, r.t., 4 d	71%	99%
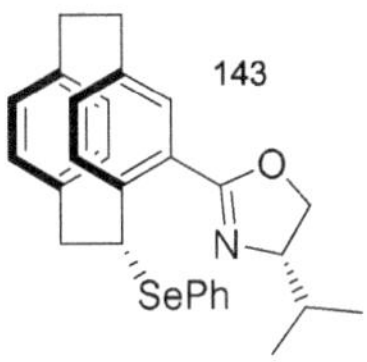 143	Se-ligand 6 mol%, Pd-cat. 2 mol%, r.t., 2 h	98%	93%
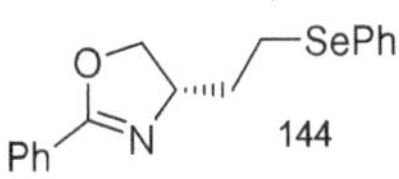 144	Se-ligand 10 mol%, Pd-cat. 2.5 mol%, r.t., 24 h	99%	91%
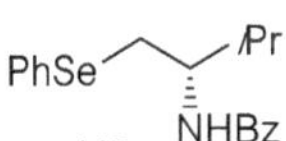 145	Se-ligand 5 mol%, Pd-cat. 2.5 mol%, r.t., 24 h	97%	98%
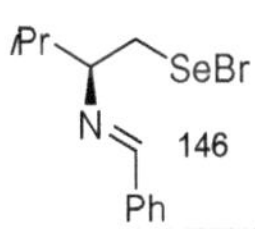 146	Se-ligand 5 mol%, Pd-cat. 2.5 mol%, r.t., 48 h	93%	97%

through the H_2O_2-oxidation of the incorporated Se-moiety.[98] Several examples of Se-catalyzed aminocyclizations are also known. A cyclic selenide **166**, synthetized by four reaction steps from mannitol, can catalyze the bromoaminocyclization of amides **167** to pyrrolidines **168** in the presence of NBP as the bromine source.[99] A procedure to synthetize trifluoromethylthiolated pyrrolidines **171** was also designed using indane-derived catalyst **169**.[100] Recently, chiral diferrocenyl phenylselenides bearing a urea group **172** were obtained by Schirakawa and successfully tested in asymmetric bromoaminocyclization of 2-allyl aniline substrates **173**[101] (Scheme 10.30).

A different type of reaction catalyzed by the chiral organoselenium derivative was presented by Santi *et al.*[102] Selenium dioxide oxidation of β-pinene **175** resulted in the synthesis of selenide **176** which was further used to promote the allylic chlorination of olefins, including carvone **177**. The reactions could be conducted in mild conditions giving excellent yields of the final products **178** (Scheme 10.31).

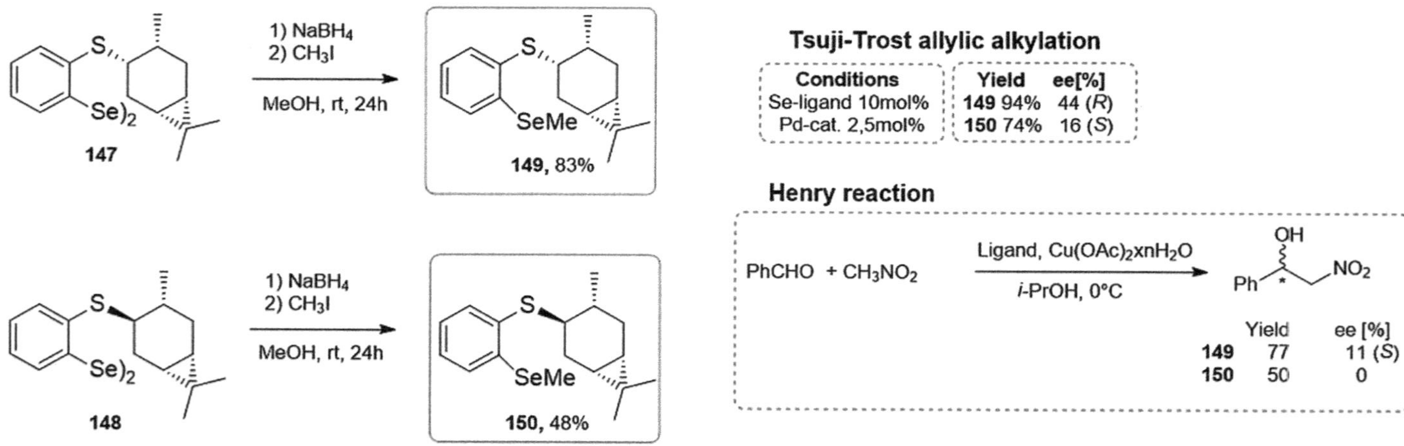

Scheme 10.28 Application of terpene-based methylselenide **149** and **150** as chiral ligands.

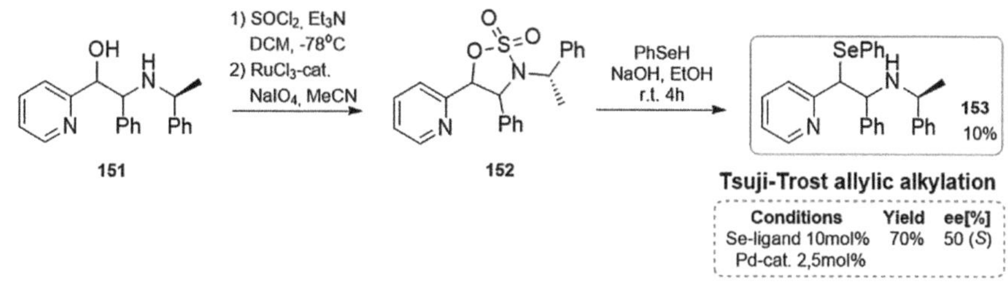

Scheme 10.29 Evaluation of N, Se-ligand **153** in the Tsuji–Trost reaction.

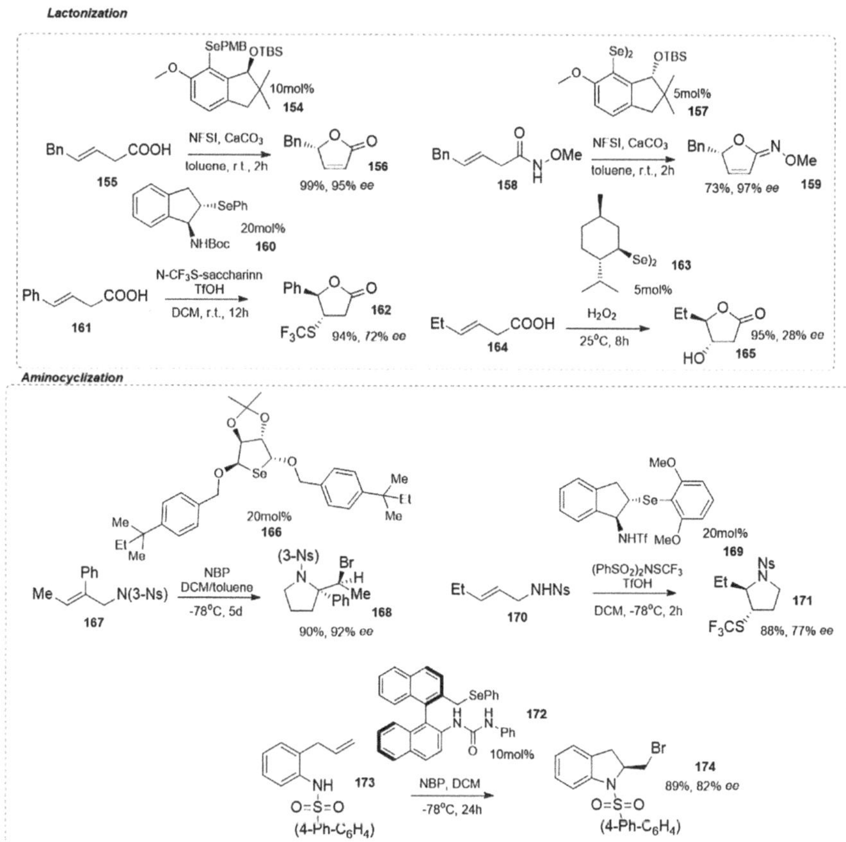

Scheme 10.30 Se-catalyzed asymmetric lactonizations and aminocyclizations.

Another example of bis(terpene) selenide was synthetized by Katanoka.[103] The camphor selenide **179** efficiently catalyzed the formation of epoxyketones **180** in the Darzens reaction. The highest ee values were observed when benzaldehyde was used as the substrate. In the case of aliphatic aldehydes, the enantioselectivity significantly decreased (Scheme 10.32).

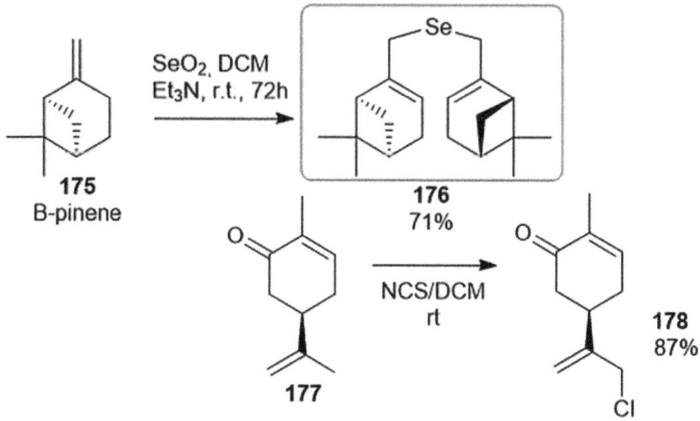

Scheme 10.31 Allylic chlorination of olefins catalyzed by pinene selenide **176**.

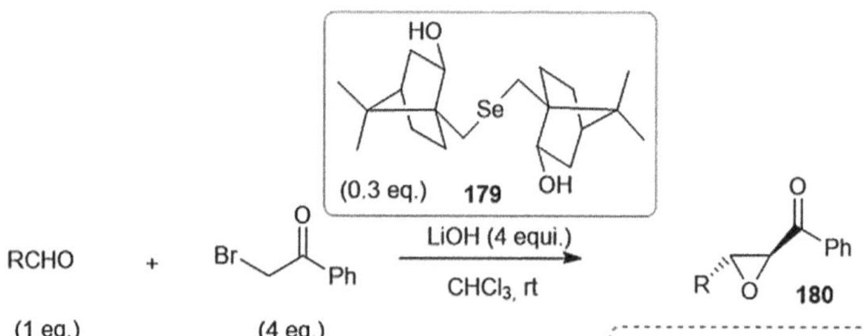

R	Yield	ee[%]
Ph	71	62
4-$NO_2C_6H_4$	86	46
4-BrC_6H_4	65	56
Et	49	4

Scheme 10.32 Utility of selenide **179** in the Darzens reaction.

10.3 Chiral Organotellurium Derivatives

Although the first organotellurium compound, diethyl ditelluride prepared by Wohler in 1840, was discovered earlier then the organoselenium one (Wohler and Siemens, 1847), organotellurium compounds are far less explored than their Se-analogs. Their high reactivity, which is an advantage, also brings certain drawbacks such as toxicity and handling difficulties with the possibility of rapid decomposition. However, along with the vast development of organoselenium chemistry, increasing attention has been captured by the still undeveloped and synthetically promising organochalcogens that bear an even more reactive tellurium atom. Since the 1980s, when considerable progress in the field was observed, various Te-derivatives have been successfully applied as efficient reagents and bio-active molecules, with a broad summary in several review articles.[104–107]

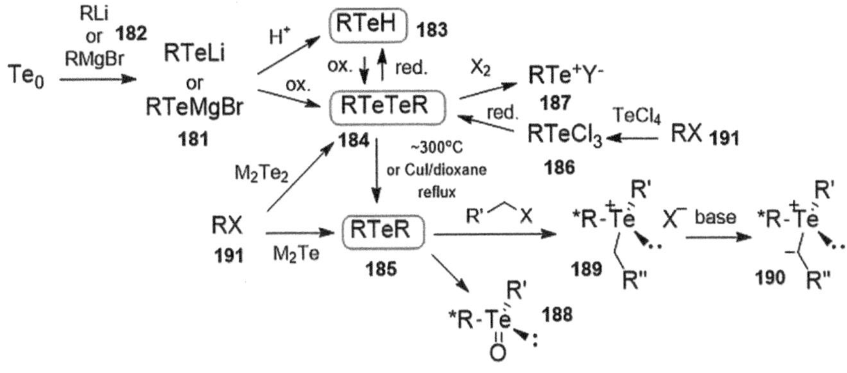

Scheme 10.33 General synthesis of tellurium-based reagents.

As reagents organotellurium compounds act similarly to the Se-analogs. Starting from elemental tellurium, nucleophiles like **181** can be generated by the reaction of Te^0 with organolithium or Grignard reagents **182**. Next, **181** can be acidified to obtain organotellurols **183** or oxidized to ditellurides **184**. Compounds **184** can be also generated from halogenated substrates **191** by the treatment with metal ditelluride M_2Te_2 or by the reduction of electrophilic species **186**, previously generated from halides **191** and tellurium tetrachloride. Other organotellurium electrophiles **187** can be synthetized from ditellurides **184** by halogenolysis. Tellurides **185** are obtained through the reaction of halogens **191** with metal monotelluride or by the elimination of one Te atom from ditelluride **184**, in the presence of CuI/dioxane or by heating up to 300 °C. Derivatives **185** can be further oxidized to telluroxides **188** or transformed to telluronium salts **189** and ylides **190** (Scheme 10.33).

Over the years, organotellurium compounds have been efficiently used in several reactions like epoxidation, cyclopropanation and aziridination (telluronium salts and ylides),[108] oxidation (telluroxides), as [2,3]-sigmatropic rearrangement precursors to form new C–C, C–O and C–N bonds,[109] and metal-catalyzed cross-coupling reactions (vinylic tellurides).[110] Similarly to organoselenium compounds, the synthesis and applicability of chiral organotellurium derivatives are limited in comparison to the racemic ones. There are several examples of chiral Te-compounds derived from amino acids **191**, **192**,[111,112] terpenes **193–197**,[113–116] aromatic/alkyl ditellurides **198–200**[117–119] or ylides **201**[120] however not utilized in asymmetric synthesis, frequently due to light sensitivity and rapid decomposition when exposed to air (Figure 10.2). In this part of the chapter the known non-racemic examples of organotellurium reagents and catalysts will be discussed.

In 1993 Ogura and co-workers obtained binaphthyl ditellurides,[121] corresponding to the binaphthyl diselenides synthetized by Tomoda in 1988.[8] Compound **202**, after treatment with diisobutylaluminium hydride, was efficiently used as a nucleophile in the 1,4-addition reaction to α,β-unsaturated carbonyls yielding β-aryltelluro ketone **203** with 57% yield and 70% diastereomeric excess (Scheme 10.34).

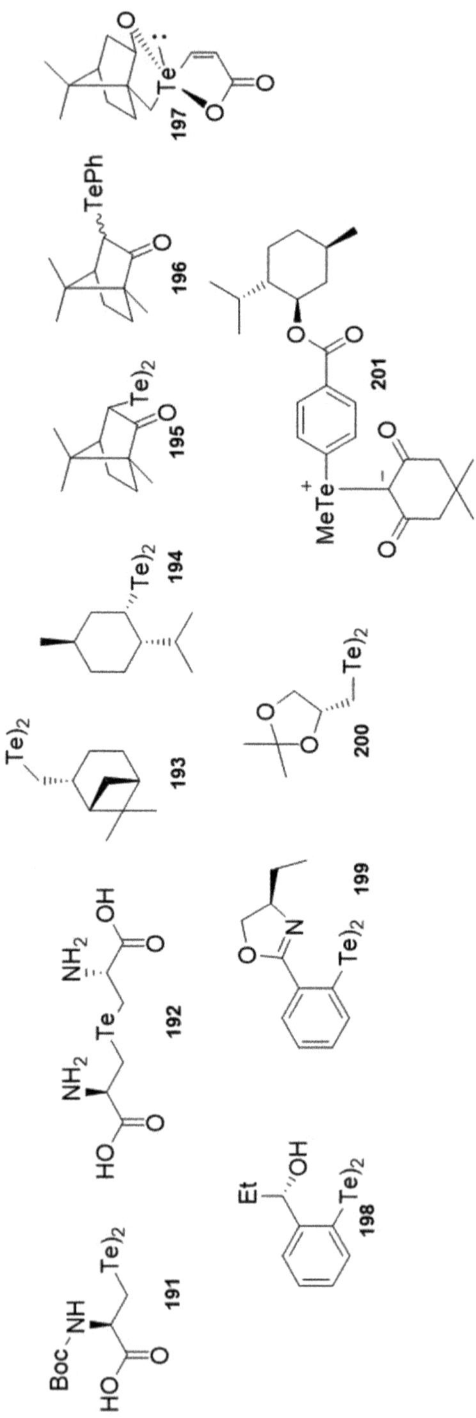

Figure 10.2 Known chiral organotellurium compounds.

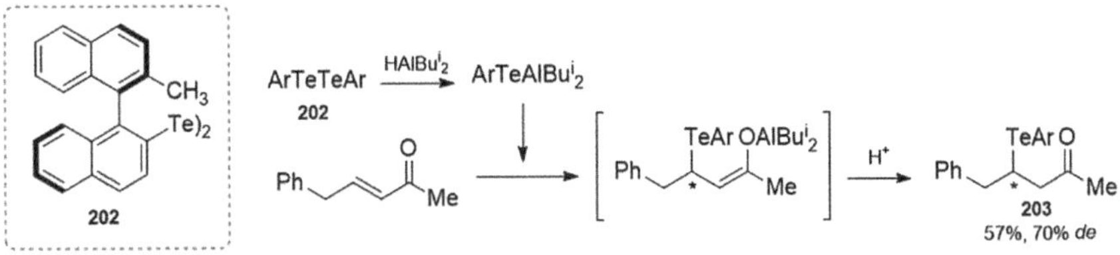

Scheme 10.34 Binaphthyl ditelluride **202** in 1,4-addition to unsaturated ketones.

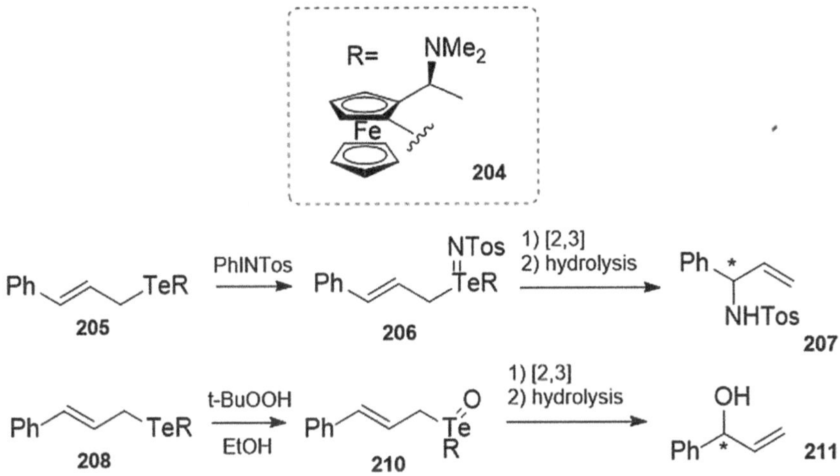

Scheme 10.35 Allylic ferrocenyl tellurides in the synthesis of allylic amines **207** and alcohols **211**.

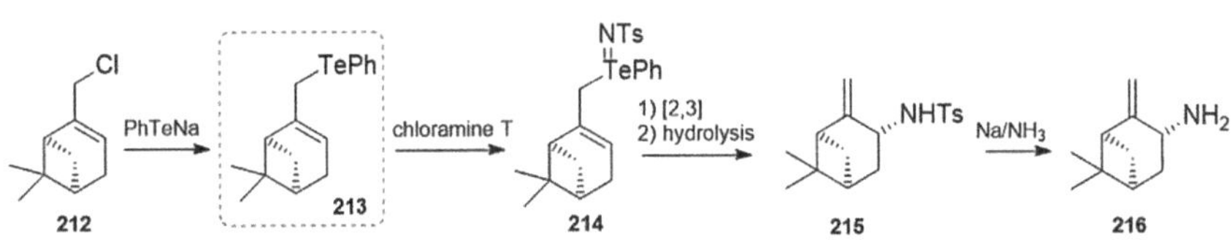

Scheme 10.36 [2,3]-Sigmatropic rearrangement of phenyltellurides to allylic amines.

Next, in 1995 Uemura *et al.* presented the utility of allylic ferrocenyl tellurides **205** and **208** in the synthesis of allylic amines **207**, through imination to **206** and further [2,3]-sigmatropic rearrangement, and allylic alcohols **211** by oxidation to **210** and then rearrangement and hydrolysis (Scheme 10.35). Compounds **205** and **208** were obtained by the reaction of allylic bromides with ferrocenyl sodium telluride.[122,123]

The same rearrangement phenomena were observed by Ścianowski and co-workers. A series of bicyclic, cyclic, and acyclic phenyltellurides was obtained from the corresponding chlorides by treatment with PhTeNa (exemplary reaction starting from (+)-10-chloro-pinene **212** is presented in Scheme 10.36). Phenyltelluride **213** underwent [2,3]-sigmatropic rearrangement

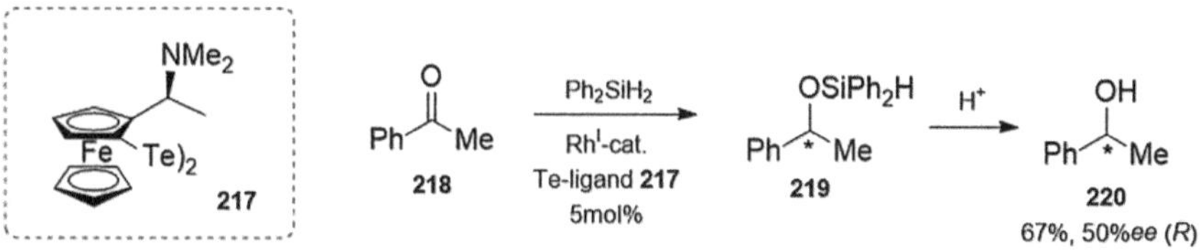

Scheme 10.37 Hydrosilylation of ketones using chiral ditelluride as the ligand.

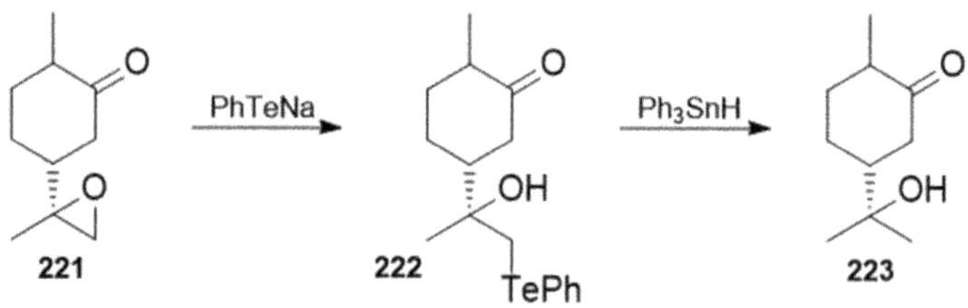

Scheme 10.38 Triphenyltin hydride reduction of phenyl telluride **222** to chiral alcohol **223**.

to toluenosulfonamide **214**, and further treatment with sodium in liquid ammonia enabled the synthesis of chiral allylic amine **216**.[124–127]

Uemura used the previously presented ferrocenyl diselenide **217** as a chiral ligand in the rhodium(I)-catalyzed hydrosilylation of ketones. The further hydrolysis of silylated product **219** afforded the chiral alcohol **220** with 50% enantiomeric excess (Scheme 10.37).[128]

Clive *et al.* showed that chiral alcohols could also be obtained by the reduction of organotellurium substrates. The ring opening of carvone 7,8-epoxide **221** with sodium phenyl telluride yielded the hydroxy phenyl telluride **222**. A selective reduction using triphenyltin hydride furnished compound **223** (Scheme 10.38).[129]

In 1989 Michalska and co-workers obtained the first chiral organotellurium compounds derived from carbohydrates. α-Acetobromoglucose **224** was transformed to phosphorotelluroate **225**, hydrolized in methanol to tellurol **226**, and oxidized to bis(β-D-glucosyl) ditelluride **227**.[130] In 1999 Yoshida *et al.* used the corresponding telluroglycosides **228** in a radical addition to alkynes. Compound **228** under UV-irradiation cleaved to a glycosyl radical, which further formed the vinylic glycoside **229** (Scheme 10.39).[131]

In 2003–2005 a simple C_2-symmetrical 2,5-dimethyltellurolane **230** was found to be a perfect precursor of telluronium salts and ylides efficiently used in cyclopropanation and epoxidation reactions. First, Tang and co-workers obtained Te-salt **232** which after *in situ* deprotonation using LiTMP/HMPA gave the Te-ylide that reacted with unsaturated carbonyls like **233** to form vinylic cyclopropanes (example **234**) with high yield and excellent enantioselectivity (Scheme 10.40).[132]

Next, Huang and Metzner synthetized allylic **235**[133] and benzyl **238**[134] telluronium salts derived from 2,5-dimethyltellurolane **230** and utilized

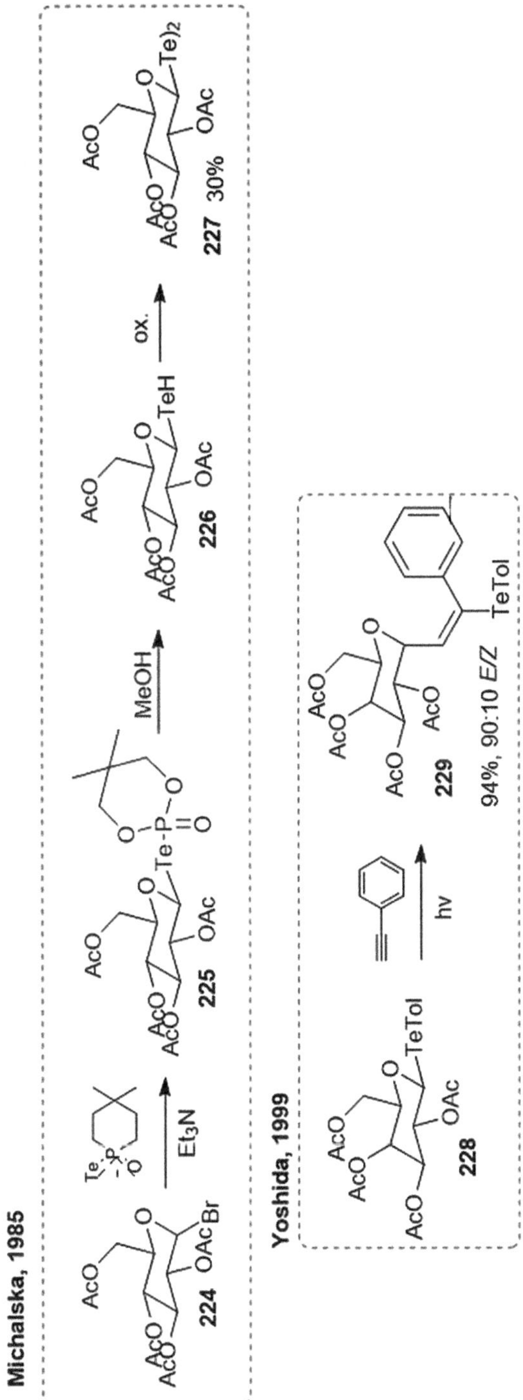

Scheme 10.39 Telluroglycosides **227** and **228** with the application as radical reagents.

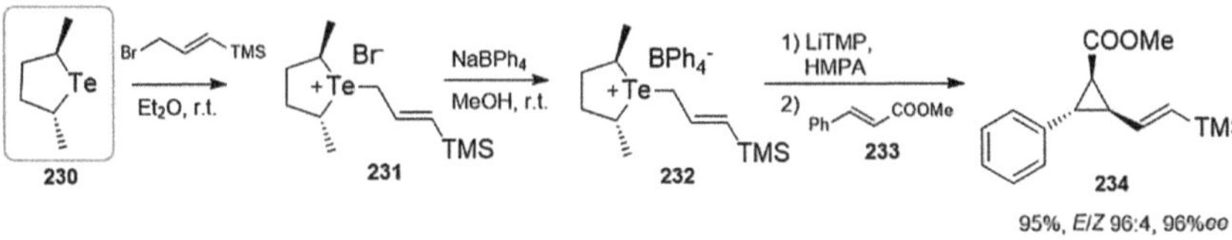

Scheme 10.40 2,5-dimethyltellurolane **230** in asymmetric cyclopropanation.

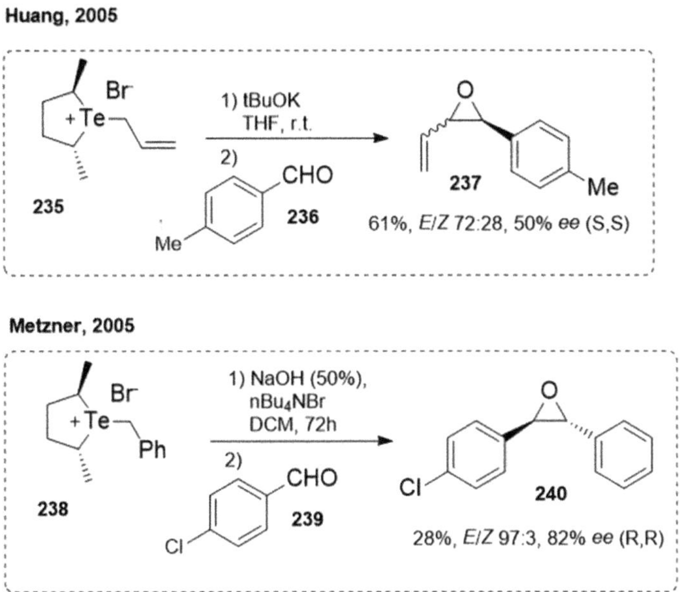

Scheme 10.41 2,5-dimethyltellurolane salts **235** and **238** in asymmetric epoxidation.

these reagents in the epoxidation of substituted aromatic aldehydes *e.g.*, **236** and **239.** The ylide formed from benzyl salt **238** furnished better enantiomeric excess, up to 82% ee (Scheme 10.41).

In 2015 Ścianowski *et al.* presented the application of methyl **241, 243** and phenyl **242,** and **244** terpenyl tellurides in asymmetric epoxidation. The compounds gave moderate results with the best *cis*/*trans* ratio observed for bicyclic derivative **244** (Scheme 10.42).[135]

In 2007 dos Santos published the synthesis of lactones and spiroketals from chiral alkyl tellurides **245.** Acetate **245** hydrolyzed to alcohol **246** and after treatment with butyllithium furnished a 1,4-*C*,*O*-dianion **247** that after reaction with CO_2 and acidification yielded lactone **248.** Mixing telluride **246** with cerium trichloride and subsequent treatment with *n*BuLi afforded a dicerium salt **249.** Reaction of **249** with lactones **250** or anhydrides **251** enabled spiroketals **252** and **253** to be obtained (Scheme 10.43).[136]

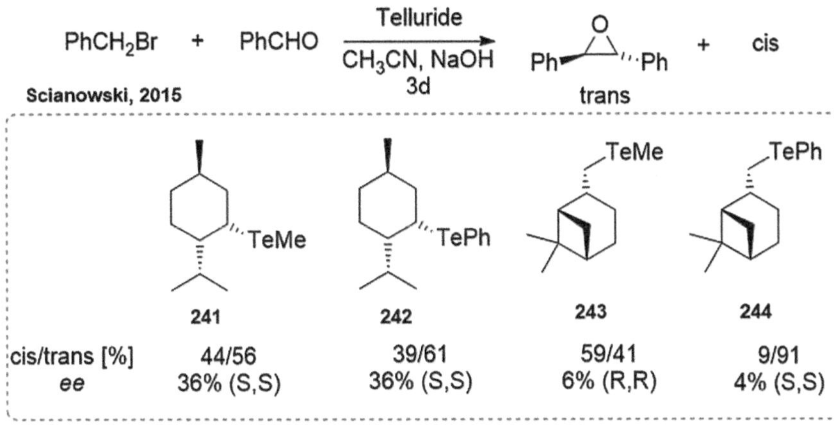

Scheme 10.42 Terpenyl tellurides **241–244** utilized in asymmetric epoxidation.

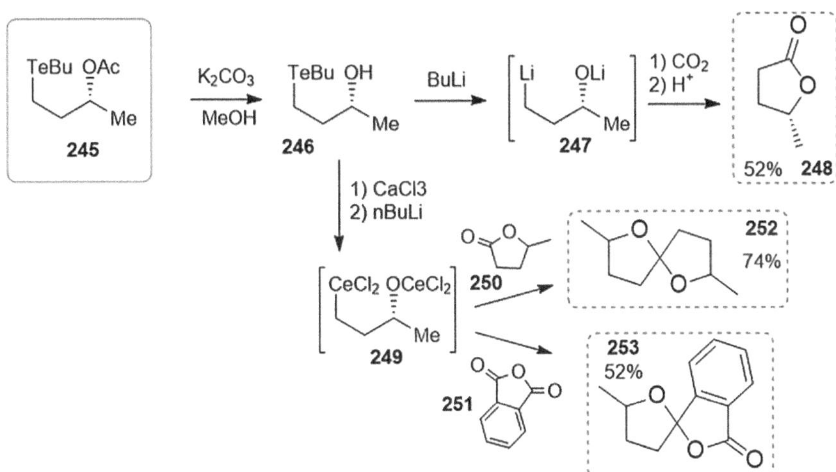

Scheme 10.43 Telluride **245** as a substrate for the synthesis of lactones **248** and spiroketals **252** and **253**.

10.4 Conclusions

This presented review covers the broad field of chiral organoselenium and organotellurium compounds used as reagents and catalysts in various synthetic transformations. A summary of the known procedures to obtain Se- and Te-derivatives has been presented. The utility of organoselenium reagents as electrophiles (methoxyselenenylation and selenocyclizations reactions), nucleophiles (epoxide ring opening and stereoselective aldol condensations), selenide-based reagents (undergoing [2,3]-sigmatropic rearrangement to afford allylic amines, selenides, and alcohols, epoxidation and cyclopropanation promotors) and catalyst (diethyl zinc addition,

Tsuji–Trost Pd-catalyzed allylic alkylation, Henry reaction, and promotion of various heterocycle forming reactions) has been presented. The application of organotellurium reagents and catalysts included the utility in 1,4-addition to unsaturated ketones, hydrosilylation of ketones, and asymmetric cyclopropanation and epoxidation.

References

1. K. B. Sharpless and R. F. J. Lauer, *J. Am. Chem. Soc.*, 1973, **95**, 2697.
2. (a) C. Santi, *Organoselenium Chemistry: Between Synthesis and Biochemistry*, Bentham Science Publishers, United Arab Emirates, 2014; (b) G. Mugesh, *Organoselenium Chemistry: Synthesis and Reactions*, ed. T. Wirth, Wiley-VCH, Weinheim, 2012.
3. N. Sonoda, G. Yamamoto, K. Natsukawa, K. Kondo and S. Murai, *Tetrahedron Lett.*, 1975, **24**, 1969.
4. C. Santi, S. Santoro and B. Battistelli, *Curr. Org. Chem.*, 2010, **14**, 2442.
5. (a) T. G. Back, *Curr. Green Chem.*, 2016, **3**, 76; (b) J. Młochowski and H. Wójtowicz-Młochowska, *Molecules*, 2015, **20**, 10205.
6. (a) F. V. Singh and T. Wirth, *Catal. Sci. Technol.*, 2019, **9**, 1073; (b) E. M. McGarrigle, E. L. Myers, O. Illa, M. A. Shaw, S. L. Riches and V. K. Aggarwal, *Chem. Rev.*, 2007, **107**, 5841.
7. T. Wirth, *Tetrahedron*, 1999, **55**, 1.
8. S. Tomoda and M. Iwaoka, *Chem. Lett.*, 1988, 1895.
9. (a) R. Déziel and E. J. Malenfant, *Org. Chem.*, 1995, **60**, 4660; (b) R. Déziel, E. Malenfant, C. Thibault, S. Fréchette and M. Gravel, *Tetrahedron Lett.*, 1997, **38**, 4753.
10. K. Fujita, M. Iwaoka and S. Tomoda, *Chem. Lett.*, 1994, 923.
11. Y. Nishibayashi, J. D. Singh, K. Segawa, S.-I. Fukuzawa and S. Uemura, *J. Chem. Soc., Chem. Commun.*, 1994, 1375.
12. (a) T. Wirth, *Angew. Chem.*, 1995, **107**, 1872; (b) T. Wirth, *Angew. Chem. Int. Ed. Engl.*, 1995, **34**, 1726; (c) T. Wirth and G. Fragale, *Chem. Eur. J.*, 1997, **3**, 1894.
13. M. Tiecco, L. Testaferri, C. Santi, C. Tomasini, S. Santoro, F. Marini, L. Bagnoli, A. Temperini and F. Costantino, *Eur. J. Org. Chem.*, 2006, 4867.
14. D. M. Freudendahl, M. Iwaoka and T. Wirth, *Eur. J. Org. Chem.*, 2010, **20**, 3934.
15. R. Déziel, E. Malenfant, C. Thibault, S. Fréchette and M. Gravel, *Tetrahedron Lett.*, 1997, **38**, 4753.
16. H. Takada, Y. Nishibayashi and Sakae Uemura, *J. Chem. Soc., Perkin Trans. 1*, 1999, 1511.
17. M. Tiecco, L. Testaferri, L. Bagnoli, V. Purgatorio, A. Temperini, F. Marini and C. Santi, *Tetrahedron: Asymmetry*, 2001, **12**, 3297.
18. M. Tiecco, L. Testaferri, F. Marini, S. Sternativo, C. Santi, L. Bagnoli and A. Temperini, *Tetrahedron: Asymmetry*, 2001, **12**, 3053.
19. M. Tiecco, L. Testaferri, F. Marini, S. Sternativo, L. Bagnoli, C. Santi and A. Temperini, *Tetrahedron: Asymmetry*, 2001, **12**, 1493.

20. T. G. Back, B. P. Dyck and M. Parvez, *J. Chem. Soc., Chem. Commun.*, 1994, 515.
21. T. G. Back, B. P. Dyck and M. Parvez, *J. Org. Chem.*, 1995, **60**, 703.
22. D. Schley and J. Liebscher, *Eur. J. Org. Chem.*, 2007, 2945.
23. T. G. Back and B. P. Dyck, *Chem. Commun.*, 1996, 2567.
24. T. G. Back and S. Nan, *J. Chem. Soc., Perkin Trans. 1*, 1998, 3123.
25. T. G. Back, B. P. Dyck and S. Nan, *Tetrahedron*, 1999, **55**, 3191.
26. T. G. Back and Z. Moussa, *Org. Lett.*, 2000, **2**, 3007.
27. T. G. Back, Z. Moussa and M. Parvez, *J. Org. Chem.*, 2002, **67**, 499.
28. T. G. Back, Z. Moussa and M. Parvez, *Phosphorus, Sulfur Silicon*, 2004, **179**, 2569.
29. T. G. Back and B. P. Dyck, *J. Am. Chem. Soc.*, 1997, **119**, 2079.
30. T. G. Back and Z. Moussa, *J. Am. Chem. Soc.*, 2002, **124**, 12104.
31. M. Tiecco, L. Testaferri, C. Santi, F. Marini, L. Bagnoli and A. Temperini, *Tetrahedron Lett.*, 1998, **39**, 2809.
32. M. Tiecco, L. Testaferri, F. Marini, C. Santi, L. Bagnoli and A. Temperini, *Tetrahedron: Asymmetry*, 1999, **10**, 747.
33. M. Tiecco, L. Testaferri, C. Santi, C. Tomassini, F. Marini, L. Bagnoli and A. Temperini, *Eur. J. Org. Chem.*, 2000, 3451.
34. M. Tiecco, L. Testaferri, C. Santi, C. Tomassini, F. Marini, L. Bagnoli and A. Temperini, *Tetrahedron: Asymmetry*, 2002, **13**, 429.
35. J. Ścianowski, Z. Rafiński and A. Wojtczak, *Eur. J. Org. Chem.*, 2006, 3216.
36. J. Ścianowski, *Tetrahedron Lett.*, 2005, **46**, 3331.
37. Z. Rafiński and J. Ścianowski, *Tetrahedron: Asymmetry*, 2008, **19**, 1237.
38. J. Ścianowski, Z. Rafiński, A. Wojtczak and K. Burczyński, *Tetrahedron: Asymmetry*, 2009, **20**, 2871.
39. Z. Rafiński, J. Ścianowski and A. Wojtczak, *Lett. Org. Chem.*, 2009, **6**, 321.
40. J. Ścianowski, Z. Rafiński, A. Szuniewicz and A. Wojtczak, *Tertrahedron*, 2009, **65**, 10162.
41. S. Tomoda and M. lwaoka, *J. Chem. Soc., Chem. Commun.*, 1988, 1283.
42. Y. Nishibayashi, J. D. Singh, S.-I. Fukuzawa and S. Uemura, *J. Chem. Soc., Perkin Trans. 1*, 1995, 2871.
43. R. Affeldt, F. P. Santos, R. S. da Silva, O. E. D. Rodrigues, L. A. Wessjohann and D. S. Ludtke, *RSC Adv.*, 2016, **6**, 93905.
44. Y. Kawai, H. Ando, H. Ozeki, M. Koketsu and H. Ishihara, *Org. Lett.*, 2005, **7**, 4653.
45. C. Santi, S. Santoro, L. Testaferri and M. Tiecco, *Synlett*, 2008, **10**, 1471.
46. R. J. Hondal, B. L. Nilsson and R. T. Raines, *J. Am. Chem. Soc.*, 2001, **123**, 5140.
47. T. Toru, T. Hayakawa, T. Nishi, Y. Watanabe and Y. Ueno, *Phosphorus Sulfur Silicon*, 1998, **136**, 653.
48. T. Toru, *Tetrahedron*, 2001, **57**, 6703.
49. M. Tiecco, L. Testaferri, F. Marini, S. Sternativo, C. Santi, L. Bagnoli and A. Temperini, *Tetrahedron: Asymmetry*, 2004, **15**, 783.

50. M. Tiecco, L. Testaferri, F. Marini, S. Sternativo, C. Santi, L. Bagnoli and Andrea Temperini, *Eur. J. Org. Chem.*, 2005, 543.
51. X. Franck, E. Langlois and F. Outurquin, *Synthesis*, 2007, **5**, 719.
52. C. C. Silveira, A. S. Vieira, A. L. Bragaa and D. Russowsky, *Tetrahedron*, 2005, **61**, 9312.
53. D. Krasowska, L. Sancineto, M. Deska and Józef Drabowicz, *Symmetry*, 2020, **12**, 349.
54. J. Drabowicz, A. Rzewnicka and R. Żurawiński, *Molecules*, 2020, **25**, 2420.
55. G. Ayrey, D. Barnard and D. T. Woodbridge, *J. Chem. Soc.*, 1962, 2089.
56. W. Salmond, M. Barta, A. Cain and M. Sobala, *Tetrahedron Lett.*, 1977, **18**, 1683.
57. K. B. Sharpless and R. F. Lauer, *J. Am. Chem. Soc.*, 1972, **94**, 7154.
58. K. B. Sharpless and R. F. Lauer, *J. Org. Chem.*, 1972, **37**, 3973.
59. K. B. Sharpless, M. W. Young and R. F. Lauer, *Tetrahedron Lett.*, 1973, **22**, 1979.
60. H. J. Reich and K. E. Yelm, *J. Org. Chem.*, 1991, **56**, 5672.
61. R. Kawakami, J. D. Singh, S.-I. Fukuzawa and S. Uemura, *J. Org. Chem.*, 1995, **60**, 4114.
62. T. Takahashi, N. Kurose, S. Kawanami, Y. Arai and T. Koizumi, *J. Org. Chem.*, 1994, **59**, 3262.
63. Sh. Saito, J. Zhang and T. Koizumi, *J. Org. Chem.*, 1998, **63**, 6029.
64. N. Kurose, T. Takahashi and T. Koizumi, *J. Org. Chem.*, 1996, **61**, 2932.
65. N. Kurose, T. Takahashi and T. Koizumi, *J. Org. Chem.*, 1997, **62**, 4562.
66. N. Kurose, T. Takahashi and T. Koizumi, *Tetrahedron: Asymmetry*, 1997, **53**, 12115.
67. T. Takahashi, N. Nakao and T. Koizumi, *Tetrahedron: Asymmetry*, 1997, **8**, 3293.
68. A. Uzarewicz, J. Ścianowski and J. Bąkowska-Janiszewska, *Pol. J. Chem.*, 1999, **73**, 1791.
69. A. Uzarewicz, J. Ścianowski and J. Bąkowska-Janiszewska, *Pol. J. Chem.*, 2000, **74**, 777.
70. A. Uzarewicz, J. Ścianowski and J. Bąkowska-Janiszewska, *Pol. J. Chem.*, 2000, **74**, 1079.
71. A. Uzarewicz, J. Ścianowski and J. Bąkowska-Janiszewska, *Pol. J. Chem.*, 2001, **74**, 649.
72. J. Ścianowski and A. Tafelska-Kaczmarek, *Pol. J. Chem.*, 2007, **81**, 529.
73. J. Ścianowski, J. Rafalski, A. Banach, J. Czaplewska and A. Komoszyńska, *Tetrahedron: Asymmetry*, 2013, **24**, 1089.
74. N. Kamigata, Y. Nakamura, K. Kikuchi, I. Ikemoto, T. Shimizu and H. Matsuyama, *J. Chem. Soc., Perkin Trans. 1*, 1992, **13**, 1721.
75. N. Kamigata, Y. Nakamura, H. Matsuyama and T. Shimizu, *Chem. Lett.*, 1991, **20**, 249.
76. H. Takada, P. Metzner and C. Philouze, *Chem. Commun.*, 2001, 2350.
77. H.-Y. Wang, F. Li, X.-L. Yang, X.-M. Yan and Z.-Z. Huang, *Chem. – Eur. J.*, 2009, **15**, 3784.

78. X.-L. Li, H.-Y. Wang and Z.-Z. Huang, *Aust. J. Chem.*, 2005, **58**, 749.
79. W. H. Midura, J. Ścianowski, A. Banach and A. Zając, *Tetrahedron: Asymmetry*, 2014, **25**, 1488.
80. A. Banach, J. Ścianowski, M. Uzarewicz-Baig and A. Wojtczak, *Eur. J. Org. Chem.*, 2015, **16**, 3477.
81. Y. Nishibayashi, J. D. Singh, K. Segawa, S.-I. Fukuzawa and S. Uemura, *J. Chem. Soc., Chem. Commun.*, 1994, 1375.
82. A. L. Braga, D. S. Lüdtke, F. Vargas and R. C. Braga, *Synlett*, 2006, 1453.
83. T. Wirth, *Tetrahedron Lett.*, 1995, **36**, 7849.
84. A. L. Braga, O. E. D. Rodrigues, M. W. Paixão, H. R. Appelt, C. C. Silveira and D. P. Bottega, *Synthesis*, 2002, 2338.
85. A. L. Braga, M. W. Paixão, D. S. Lüdtke, C. C. Silveira and O. E. D. Rodrigues, *Org. Lett.*, 2003, **5**, 2635.
86. C. Bolm, M. Kesselgruber, A. Grenz, N. Hermanns and J. P. Hildebrand, *New J. Chem.*, 2001, **25**, 13.
87. J. Sprinz, M. Kiefer, G. Helmchen, M. Reggelin, G. Huttner, O. Walter and L. Zsolnai, *Tetrahedron Lett.*, 1994, **35**, 1523.
88. K. Hiroi, Y. Suzuki and I. Abe, *Tetrahedron: Asymmetry*, 1999, **10**, 1173.
89. S.-L. You, X.-L. Hou and L.-X. Dai, *Tetrahedron: Asymmetry*, 2000, **11**, 1495.
90. G. M. Ksander, R. de Jesus, A. Yuan, R. D. Ghai, A. Trapani, C. McMartin and R. J. Bohacek, *Med. Chem.*, 1997, **40**, 495.
91. A. L. Braga, M. W. Paixão and G. Marin, *Synlett*, 2005, 1975.
92. A. L. Braga, F. Vargas, J. A. Sehnem and R. C. Braga, *J. Org. Chem.*, 2005, **70**, 9021.
93. J. Ścianowski, A. J. Pacuła, M. Zielińska-Błajet and A. Wojtczak, *New J. Chem.*, 2016, **40**, 6697.
94. M. Wosińska-Hrydczuk and J. Skarżewski, *Molecules*, 2021, **26**, 3493.
95. Y. Kawamata, T. Hashimoto and K. Maruoka, *J. Am. Chem. Soc.*, 2016, **138**, 5206.
96. Y. Otsuka, Y. Shimazaki, H. Nagaoka, K. Maruoka and T. Hashimoto, *Synlett*, 2019, **30**, 1679.
97. X. Liu, R. An, X. Zhang, J. Luo and X. Zhao, *Angew. Chem., Int. Ed.*, 2016, **128**, 5940.
98. L. Sancineto, F. Mangiavacchi, C. Tidei, L. Bagnoli, F. Marini, A. Gioiello, J. Ścianowski and C. Santi, *Asian J. Org. Chem.*, 2017, **6**, 988.
99. F. Chen, C. K. Tan and Y.-Y. Yeung, *J. Am. Chem. Soc.*, 2013, **135**, 1232.
100. J. Luo, Q. Cao, X. Cao and X. Zhao, *Nat. Commun.*, 2018, **9**, 527.
101. K. Okuno, T. Nakamura and S. Shirakawa, *Asian J. Org. Chem.*, 2021, **10**, 655.
102. B. Boualy, S. El Houssame, L. Sancineto, C. Santi, M. Ait Ali, H. Stoeckli-Evans and L. El Firdoussi, *New J. Chem.*, 2016, **40**, 3395.
103. S. Watanabe, R. Hasebe, J. Ouchi, H. Nagasawa and T. Kataoka, *Tetrahedron Lett.*, 2010, **51**, 5778.
104. S. Patai, *The Chemistry of Organic Selenium and Tellurium Compounds*, Wiley, New York, 1987, vol. 2.

105. N. Petragnani, *Tellurium in Organic Synthesis*, Academic Press, London, 1994.
106. J. V. Comasseto, L. W. Ling, N. Petragnani and H. A. Stefani, *Synthesis*, 1997, 373.
107. C. W. Nogueira, G. Zeni and J. B. T. Rocha, *Chem. Rev.*, 2004, **104**, 6255.
108. X.-L. Sun and Y. Tang, *Acc. Chem. Res.*, 2008, **41**, 937; Y. Tang, S. Ye and X.-L. Sun, *Synlett*, 2005, **18**, 2720.
109. W. H. Midura, P. Kiełbasiński and J. Drabowicz, *PATAI'S Chemistry of Functional Groups*, John Wiley & Sons, 2011.
110. G. Zeni, D. S. Ludtke, R. B. Panatieri and A. L. Braga, *Chem. Rev.*, 2006, **106**, 1032.
111. A. Schneider, O. E. D. Rodrigues, M. W. Paixao, H. R. Appelt, A. L. Braga and L. A. Wessjohann, *Tetrahedron Lett.*, 2006, **47**, 1019.
112. P. Singh, H. B. Singh and R. J. Butcher, *J. Organomet. Chem.*, 2018, **876**, 1.
113. J. Ścianowski, A. J. Pacuła and A. Wojtczak, *Tetrahedron: Asymmetry*, 2015, **26**, 400.
114. T. Takahashi, J. Zhang, N. Kurose, S. Takahashi and T. Koizumi, *Tetrahedron: Asymmetry*, 1996, **7**, 2797.
115. J. Zhang and T. Koizumi, *Tetrahedron*, 2000, **11**, 3323.
116. T. G. Back, P. B. Dyck and M. Parvez, *J. Org. Chem.*, 1995, **60**, 4657.
117. P. Rakesh, H. B. Singh and R. J. Butcher, *Dalton Trans.*, 2012, **41**, 10707.
118. E. L. Borges, T. J. Peglow, M. S. Silva, C. G. Jacoby, P. H. Schneider, E. J. Lenardão, R. G. Jacob and G. Perin, *New J. Chem.*, 2016, **40**, 2321.
119. F. W. Bürgler, G. Fragale and T. Wirth, *ARKIVOC*, 2007, 21.
120. N. Kamigata, A. Matsuhisa, H. Taka and T. Shimizu, *J. Chem. Soc. Perkin Trans. 1*, 1995, 821.
121. M. Irie, Y. Doi, M. Ohsuka, Y. Aso, T. Otsubo and F. Ogura, *Tetrahedron: Asymmetry*, 1993, **4**, 2127.
122. T. Chiba, Y. Nishibayashi, J. D. Singh, K. Ohe and S. Uemura, *Tetrahedron Lett.*, 1995, **36**, 1519.
123. Y. Nishibayashi, S. K. Srivastava, K. Ohe and S. Uemura, *Tetrahedron Lett.*, 1995, **36**, 6725.
124. A. Uzarewicz, J. Ścianowski and J. Bąkowska-Janiszewska, *Polish J. Chem.*, 1999, **73**, 1791.
125. A. Uzarewicz, J. Ścianowski and J. Bąkowska-Janiszewska, *Polish J. Chem.*, 2000, **74**, 777.
126. A. Uzarewicz, J. Ścianowski and J. Bąkowska-Janiszewska, *Polish J. Chem.*, 2000, **74**, 1079.
127. A. Uzarewicz, J. Ścianowski and J. Bąkowska-Janiszewska, *Polish J. Chem.*, 2001, **75**, 649.
128. Y. Nishibayashi, J. D. Singh, K. Segawa, S.-I. Fukuzawa and S. Uemura, *J. Chem. Soc., Chem. Commun.*, 1994, 1375.
129. D. L. J. Clive, G. J. Chittattu, V. Farina, W. A. Kiel, S. M. Menchen, C. G. Russel, A. Singh, C. K. Wong and N. J. Curtis, *J. Am. Chem. Soc.*, 1980, **102**, 4438.

130. J. Czytewska-Chlebny and M. Michalska, *J. Chem. Soc., Chem. Commun.*, 1985, 693.
131. S. Yamago, H. Miyazoe and J. Yoshida, *Tetrahedron Lett.*, 1999, **40**, 2343.
132. W.-W. Liao, K. Li and Y. Tang, *J. Am. Chem. Soc.*, 2003, **125**, 13030.
133. W.-H. Ou and Z.-Z. Huang, *Synthesis*, 2005, **17**, 2857.
134. J.-F. Briere, H. Takada and P. Metzner, *Phosphorus, Sulfur, and Silicon*, 2005, **180**, 965.
135. J. Ścianowski, A. Banach and A. J. Pacuła, *Phosphorus, Sulfur, Silicon*, 2016, **191**, 263.
136. A. A. Dos Santos, J. L. Princival, J. V. Comasseto, S. M. G. de Barrosc and J. E. B. Neto, *Tetrahedron*, 2007, **63**, 5167.

CHAPTER 11

Electrophilic Chalcogen-containing Reagents†

JAMAL RAFIQUE,*[a,b] JULIANO B. AZEREDO[c] AND SUMBAL SABA[b]

[a] Instituto de Química (INQUI), Universidade Federal de Mato Grosso do Sul – UFMS, Campo Grande 79074-460MS, Brazil; [b] Instituto de Química (IQ), Universidade Federal de Goiás – UFG, Goiânia 74690-900GO, Brazil; [c] Departamento de Farmácia, Universidade Federal do Pampa, Uruguaiana 97500-970RS, Brazil

*Emails: jamal.chm@gmail.com; jamal.rafique@ufms.br

11.1 Introduction

The construction of C–Chalcogen (S/Se/Te) bonds are becoming increasingly appreciated, due to the biological properties associated with the presence of these heteroatoms.[1] A large number of naturally existing biologically-active compounds have sulfur present in them.[2] Historically, organoselenium/tellurium compounds were considered to be toxic toward human health and an environmental hazard. However, in the 1970s,[3,4] the discovery of the presence of the selenium atom in the structure of some proteins, caused increasing interest in the synthesis of organoselenium compounds. This also resulted in increased interest in the synthesis of organotellurium compounds and their biological application.[4] Numerous reports demonstrate that several therapeutic activities are attributed to organochalcogen

†Dedication: The authors dedicate this chapter with gratitude to Prof. Dr. Antonio Luiz Braga for his notable contributions in organochalcogen chemistry over the past 40 years.

Chalcogen Chemistry: Fundamentals and Applications
Edited by Vito Lippolis, Claudio Santi, Eder J. Lenardão and Antonio L. Braga

Published by the Royal Society of Chemistry, www.rsc.org

compounds such as their function as antioxidants, in several organs including the brain, prostate, stomach, colon, liver, lungs, kidneys, and skin, as apoptosis inducers, and chemopreventors.[4,5] Considering this, in the last decade, synthetic organochalcogen compounds have been vastly studied for numerous biological activities, such as an enzyme mimetic,[4,6] enzyme inhibition, anti-HIV,[4,5] antihypertensive activities, immune response, anticancer, anti-Alzheimer, and thyroid hormone production,[7] among others (Figure 11.1).[4,5,8] Furthermore, a huge variety of organochalcogens have been successfully synthesized and screened as antimicrobial substances.[8f,9]

Besides biological applications, organochalcogen compounds have also achieved a prominent place in modern organic synthesis. They are renowned for their use as synthetic intermediates in total synthesis, ligands, organocatalysts, and ionic liquids.[10,11] There are also reports of their use as fluorescent probes in material science.[12] Therefore, the design and synthesis of chalcogen containing molecules is a hot topic both in chemical/pharmaceutical industries as well as in academic laboratories.

On the other hand, in the last three decades concern for the preservation of the environment has been growing and the use of greener methodologies to access organochalcogen compounds has revolutionized the approach to modern organic synthesis.[13] The use of sustainable routes has seen a significant increase in the number of alternative protocols for chalcogenofunctionalization of organic moieties. This can be observed, for instance, in terms of the use of alternative energy sources, renewable and non-toxic solvents, and solvent-free reactions and catalysis.[14]

Considering the importance of organochalcogen compounds, a number of strategies are used to access the C – Chalcogen bond, involving transition metal-free approaches or the use of transition metals. In all these methods, the chalcogen moiety is introduced into the organic structure using electrophilic, nucleophilic, or radical chalcogen species. Among all these approaches, electrophilic chalcogen species (either generated *in situ* or presynthesized) is the method of choice among the scientific community due to the ease in handling, chemo-, region-, and stereo-selectivity with mild

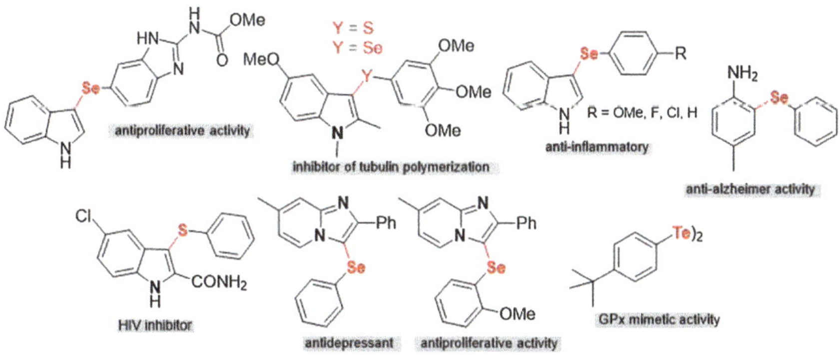

Figure 11.1 Biologically-active synthetic organochalcogen compounds.

experimental conditions.[15] This can be evidenced by the significant increase in the number of alternative and greener procedures for the chalcogeno-functionalization of organic compounds, *via* electrophilic chalcogen species.

In this context, this chapter highlights recent scientific literature (excluding the patent literature) involving the use of different electrophilic chalcogen species in the construction of the C – Chalcogen bond.

11.2 Electrophilic Chalcogen Species

In comparison to the nucleophilic nature of organochalcogen compounds, they are renowned for their use as electrophiles, mainly due to the ease of their formation. In general, a variety of electrophilic chalcogen species can be generated simply by the cleavage of the chalcogen – chalcogen bond in dichalcogenides. These electrophiles are very reactive and react easily with electron rich arenes and with unsaturated carbon (sp^2 and sp). Besides, these species are also used in the chalcogenation of aliphatic structures.

In the following topics, we will briefly discuss the recent advances from the last decade, where different types of electrophilic chalcogen species (generated *in situ* or pre-synthesized) were used in the chalcogeno-functionalization of organic compounds.

11.2.1 Halogenated Electrophilic Chalcogen Species

11.2.1.1 Monovalent Halogen Source

11.2.1.1.1 Fluorine-containing Species. The fluorinated selenium species was generated using diorganoyl diselenides **2** in the presence of select-Fluor in the work reported by Schumacher and co-workers. In the related protocol, a series of selenyl-benzo[*b*]furans **3** was prepared in up to 97% yield, through the seleno-cyclization of 2-(phenylethynyl)anisoles **1**, using acetonitrile as the solvent and avoiding the use of metals in the reaction system (Scheme 11.1).[16]

According to the authors, one of the routes for the cyclization occurs through the interaction of substrate **1** with the fluorinated selenium species generated *in situ*, affording the seleniranium intermediate **4**. An intramolecular cyclization of **4** leads to the formation of the selenylated benzofurans **3** (Scheme 11.1).

In another approach, SelectFluor was used as the oxidant in the direct chalcogenylation of quinoxalin-2(1*H*)-ones **5** with diaryl diselenides and disulfides **6**. The developed protocol enables the synthesis of 30 chalcogenated product **7** in good yields (Scheme 11.2).[17]

11.2.1.1.2 Bromine-containing Species. The generation of brominated electrophilic chalcogen species constitutes an important strategy for the selenylation of organic substrates. In this context, Yan and co-workers described a KBr-catalyzed method for the acetoxyselenylation of alkenes **8** in

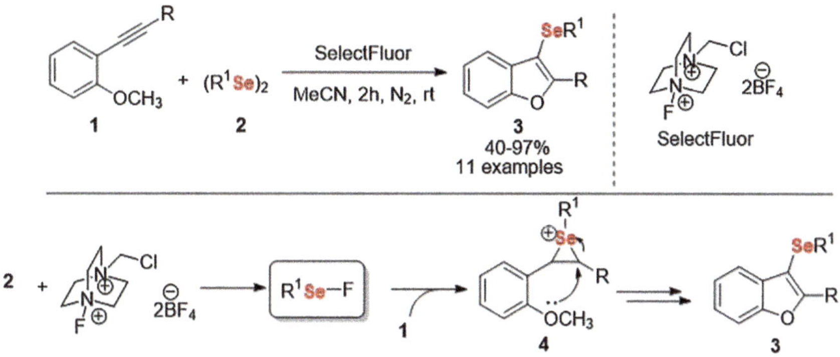

Scheme 11.1 SelectFluor mediated synthesis of selenyl-benzo[*b*]furans.

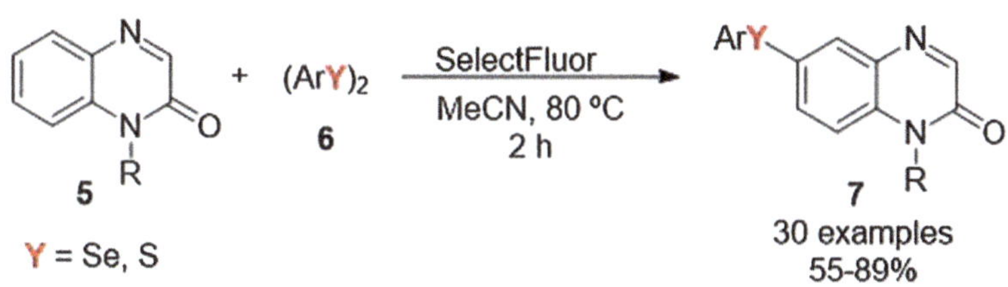

Scheme 11.2 SelectFluor mediated direct chalcogenylation of quinoxalin-2(1*H*)-ones.

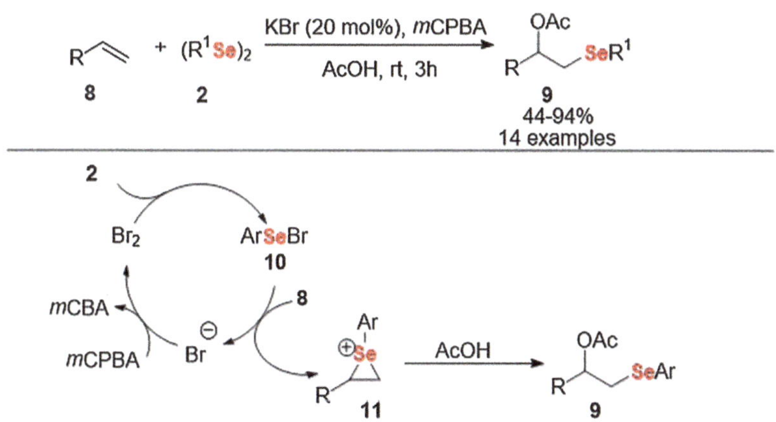

Scheme 11.3 KBr-catalyzed acetoxyselenylation of alkenes.

the presence of *m*CPBA and acetic acid. From this methodology, the desired selenylated products **9** were obtained in high yields at room temperature in 3 hours (Scheme 11.3).[18]

In the proposed mechanism, the role of *m*CPBA is to oxidize the bromine anion to Br_2, which in the presence of diselenides, generates the ArSeBr specie **10**. After an electrophilic addition in alkene **8**, the seleniranium intermediate **11** is formed and bromine anion is regenerated in the system. The acetic acid promotes the ring opening of seleniranium, providing the formation of the acetoxyselenylated product **9** (Scheme 11.3).

The same group employed a similar protocol in the selenylation of indoles, which are important scaffolds from a biological point of view. The reaction involves the use of indoles **12**, diaryl or dibezyl diselenides **2**, in the presence of 0.5 equiv. of NaBr and 1.2 equiv. of *m*CPBA in acetonitrile as the solvent. The protocol allowed the preparation of 15 examples of selenylindoles **13** at room temperature, after 24 hours in yields that ranged from 23 to 87% (Scheme 11.4).[19]

In 2015, Barman's group reported an alternative brominating agent for the sulfenylation of active methylene compounds. They used tetrabutylammonium tribromide (TBATB), which upon reaction with diorganyl disulfides **15** generates *in situ* a sulfenyl bromide intermediate. This intermediate reacts with the active methylene of **14** resulting in the formation of sulfenylated products **16** (Scheme 11.5).[20]

11.2.1.1.3 Iodine-containing Species. In 2015, Braga's group developed a sustainable method for the alkoxyselenylation of alkenes *via in situ* generation of the iodate electrophilic selenium specie RSeI. The methodology allowed the synthesis of several selenylated products **19** in up to 96% yield, obtained from the reaction of styrenes **17**, diorganoyl diselenides **2**, and alcohols **18** in the presence of the catalytic system I_2/equivalent amount of DMSO, under microwave irradiation (Scheme 11.6). Sulfenylated and tellurated compounds can also be prepared by this method.[21]

The proposed mechanism involves the formation of the electrophilic iodinated selenium specie, through the combination of the diselenide with the iodine. This specie reacts with the styrene in order to form both the

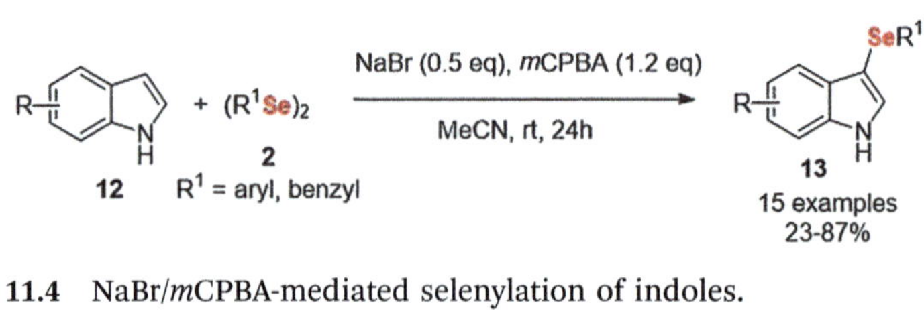

Scheme 11.4 NaBr/*m*CPBA-mediated selenylation of indoles.

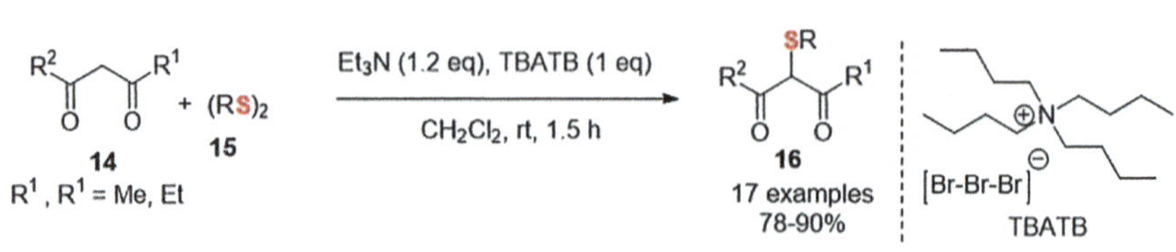

Scheme 11.5 Tetrabutylammonium tribromide-mediated sulfenylation of active methylene compounds.

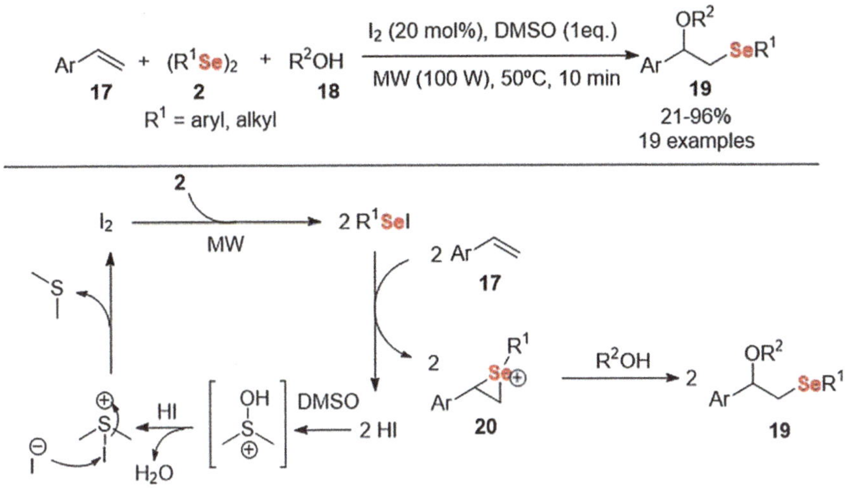

Scheme 11.6 I_2/DMSO-catalyzed alkoxy-selenylation of alkenes.

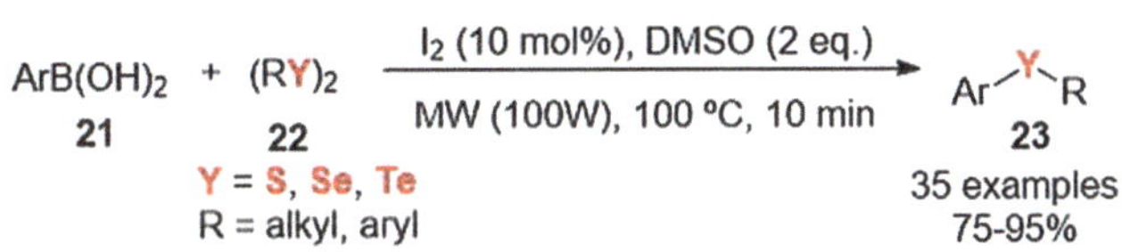

Scheme 11.7 I_2/DMSO-catalyzed synthesis of unsymmetrical diorganoyl chalcogenides.

seleniranium intermediate **20**, which undergoes nucleophilic attack by the alcohol affording the corresponding product **19**, and hydrogen iodide. The dimethyl sulfoxide promotes the oxidation of iodine in two steps, regenerating I_2 in the catalytic cycle and eliminating H_2O and dimethyl sulfide (Scheme 11.6).

Later on, the same group employed the catalytic system I_2/DMSO in the synthesis of unsymmetrical dioganoyl chalcogenides. In this approach, the desired products **23** were obtained from aryl boronic acids **21**, diorganoyl dichalcogenides **22**, 10 mol% of I_2 and 2 equiv. of DMSO in only 10 minutes under microwave irradiation (Scheme 11.7). The methodology allowed the preparation of sulfides, selenides, and tellurides in high yields, in the absence of solvents and metals (Scheme 11.7).[22]

The I_2/DMSO catalytic system was also successfully employed in the chalcogeno-functionalization of monocyclic activated arenes **24**. For this purpose, **24** in the presence of dichalcogenides **25**, 20 mol% iodine, and 3 equiv. of DMSO allow the synthesis of a series of selenylated arenes **26** in up to 97% yield, under microwave irradiation (Scheme 11.8A).[23]

The same group extended the study to bicyclic arenes, by using **27** and dichalcogenides **25**, affording the corresponding products **28**, in up to 99% yields. Comparative studies have demonstrated that both microwave

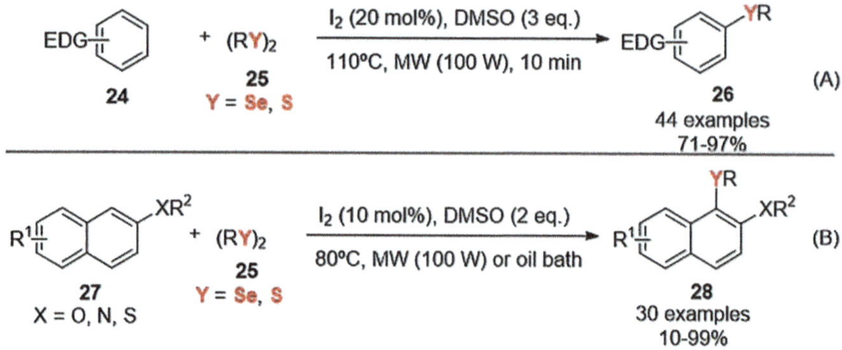

Scheme 11.8 I_2/DMSO catalyzed chalcogeno-functionalization of activated (hetero)-arenes: (A) monocyclic and (B) bicyclic.

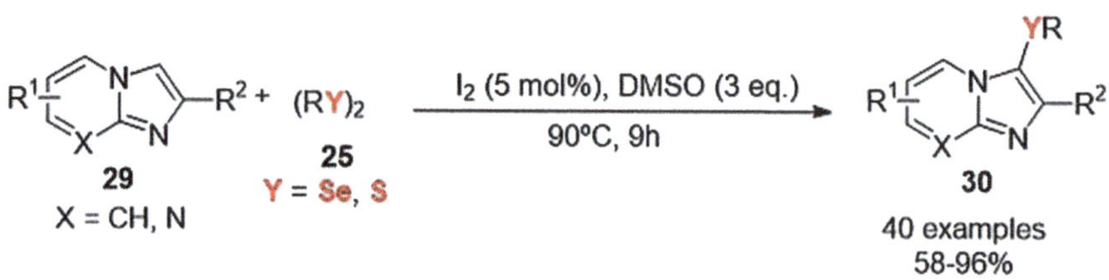

Scheme 11.9 I_2/DMSO catalyzed chalcogeno-functionalization of imidazo[1,2-*a*]-pyridines.

irradiation and conventional heating are efficient for this methodology (Scheme 11.8B).[24]

The I_2/DMSO catalytic system is also effective for the chalcogenation of heteroarenes. In 2016, Braga's group reported this catalytic system in the chalcogenation of imidazo[1,2-*a*]pyridines **29**, under conventional heating. Optimized conditions were found by using 5 mol% of I_2 as the catalyst, 3 molar equiv. of DMSO, and 0.5 molar equiv. of dichalcogenides **25** resulting in the chalcogenation product **30** in up to 96% yields (Scheme 11.9).[25]

Another synthetic utility of the I_2/DMSO catalytic system in the oxidation of thiols **31** to disulfide **32** was demonstrated by the same group. The use of 10% of I_2 and 1 molar equiv. of DMSO as the oxidant resulted in the symmetrical diorganyl disulfides **32** in up to 98% yields (Scheme 11.10), under complete solvent-free conditions.[26]

In 2018, Braga's group reported an alternative approach for the electrophilic iodinated chalcogen specie, by using a catalytic amount of NH_4I in acidic medium.[27a] The reaction involves 10 mol % of NH_4I in 1 equiv. of acetic acid and in the presence of 5 equiv. of DMSO/water (50% solution) as the oxidant source. A number of heteroarenes **33**, *e.g.*, imidazo[1,2-*a*]-pyridines, indoles, indazole, *etc.* were successfully chalcogenated by using dichalcogenides **25** affording the corresponding products **34** in 37 – 98%

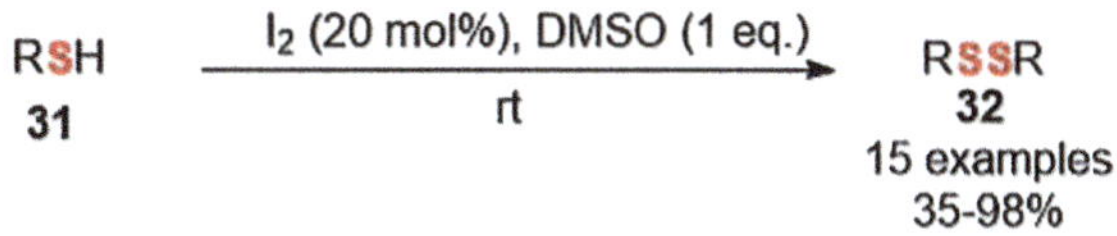

Scheme 11.10 I_2/DMSO catalyzed oxidation of thiols to disulfides.

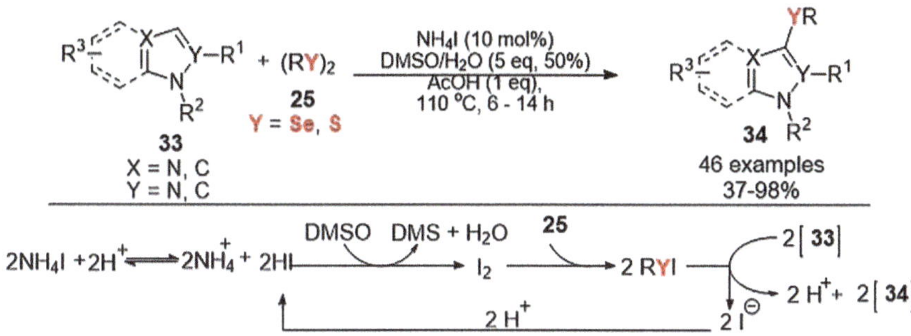

Scheme 11.11 NH_4I-catalyzed chalcogenation of *N*-heteroarens.

yields (Scheme 11.11).[27a] These heteroarenes are well-known for their diverse application in organic syntesis.[27g]

In the proposed mechanism, under acidic medium NH_4I generates HI, which is subsequently oxidized to I_2 in the presence of DMSO. In the next step, electrophilic iodinated chalcogen specie was formed by the reaction of the dichalcogenide **25** with I_2., which undergoes attack of the nucleophile **33** to result in the chalcogenated product **34** (Scheme 11.11).

11.2.1.2 Selenium Dihalide

The use of $SeCl_2$ as a source of electrophilic selenium in 3-(arylalkynyl)indoles **35** cyclization reactions was described by Silveira and co-workers. From the reaction of thionyl chloride with Se^0, it was possible to generate $SeCl_2$ *in situ*, which in the presence of substrate **35**, allowed the synthesis of the tricyclic compounds **36** in up to 82% yield (Scheme 11.12).[28] According to the authors, the triple bond of **35** undergoes an electrophilic addition in the presence of $SeCl_2$, previously generated *in situ* by the elemental selenium and SO_2Cl_2, resulting in the seleniranium cation **37.** A chloride anion promotes the ring-opening of the selenium ring leading to the formation of intermediate **38,** where the selenium atom bonded to chlorine undergoes an intramolecular attack of the carbon-2 of the indole ring, forming the tricyclic intermediate **39.** Deprotonation of **39** results in the formation of the product **36** (Scheme 11.12).

In the same year, Potapov's group reported the use of $SeCl_2$ and $SeBr_2$ in the synthesis of functionalized diorganyl selenides **41a** and **42b** from allylic aromatics **40** (Scheme 11.13).[29]

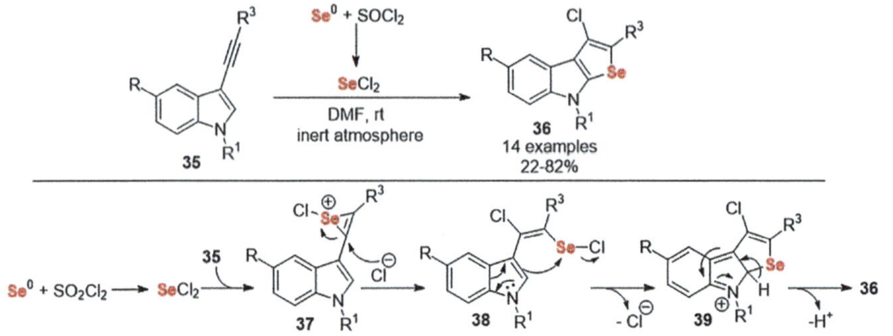

Scheme 11.12 $SeCl_2$ as a source of electrophilic selenium in 3-(arylalkynyl)indole cyclization reactions.

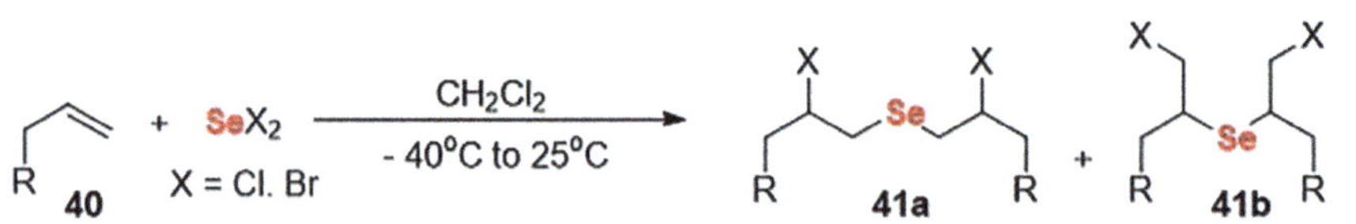

Scheme 11.13 SeX_2 (X = Cl, Br) in the synthesis of functionalized diorganoyl selenides.

11.2.1.3 Hypervalent Iodine Source

A variety of iodo-hypervalent compounds can be used to generate electrophilic chalcogen species. Generally, such hypervalent compounds are applied in stochiometric amounts. Recently, Opai *et al.* reported the use of 2 equiv. of (diacetoxyiodo)benzene (PIDA) in the selenylation of imidazoheteroarenes **42** by using diphenyl diselenide **43**, affording the selenylated products **44** (Scheme 11.14).[30] In the proposed mechanism, the authors demonstrated that the reaction of diselenide **43** with PIDA results in the formation of 2 equiv. of electrophilic benzeneselenenyl acetate (*via* intermediate **45**). These electrophilic species react with **42**, forming σ-complex **46** which undergoes aromatization to furnish **44** (Scheme 11.14).

Instead of using PIDA directly, Yan's group demonstrated an alternative approach, involving it's *in situ* formation from the reaction of a catalytic amount of PhI (10 mol%), *m*CPBA (0.8 equiv.) as the oxidant and AcOH as the solvent. After the generation of PIDA, it reacts with diselenide to form electrophilic benzeneselenenyl acetate, which was used in the acetoxyselenylation of alkenes. Under the optimized conditions, they used two types of diselenides **48** and different alkenes **47**, resulting in the selenylated products **49** in up to 97% yield (Scheme 11.15).[31] Formylated products were obtained, when formic acid was used instead of AcOH.

In another report, Tsai *et al.* reported the use of [bis(trifluoroacetoxy)iodo]benzene (PIFA) in the synthesis of acylsulfenic acid alkyl esters and benzoyl alkyl disulfides **52** from thiobenzoic acids **50** and alkyl

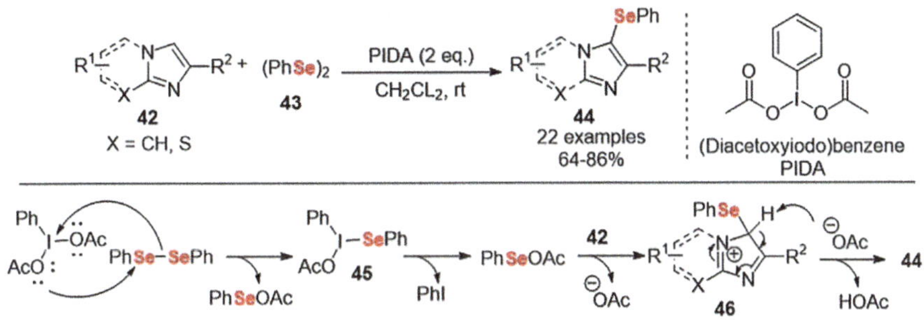

Scheme 11.14 PIDA-mediated seleno-functionalization of imidazo-heteroarenes.

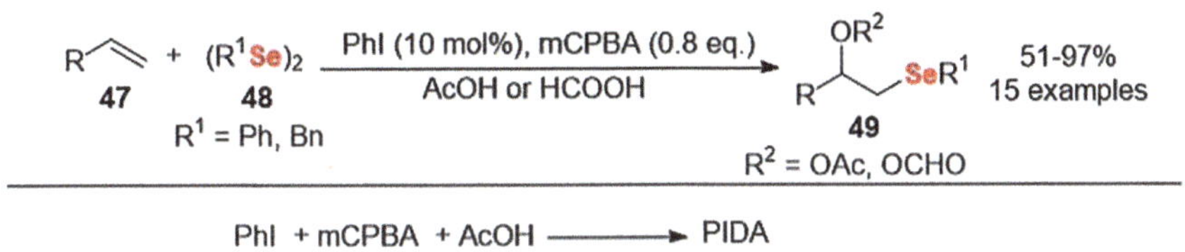

Scheme 11.15 PIDA-mediated acetoxy-selenylation of alkenes.

alcohols/thiols **51**. The developed protocol allowed the synthesis of the desired products **52** in up to 92% yield (Scheme 11.16).[32]

Recently Zhao's group reported the iodosobenzene (PhIO) mediated intramolecular selenocyclization of olefins **53** with diorganyl diselenides **2**. Under the optimized conditions, the use of 1 equiv. of **53** and **2** in the presence of 1 equiv. of the promoter PhIO, resulted in the cyclized selenylated products **54** in 60 – 95% yields (Scheme 11.17).[33] In this selenocyclization transformation, the authors utilized a diverse range of terminal olefines with *N*-alkylsulfonamides, amides, carboxylic acids, alcohols, and diketones functionalities. According to the proposed mechanism, diselenides **2** are oxidized by PhIO to generate RSe^+, which serves as a reactive electrophilic intermediate to induce the selenocyclization.

11.2.1.4 N-Chalcogeno-amides

Among bench stable electrophilic chalcogen species, *N*-chalcogenoimide are the most common ones. They can be prepared easily from the reaction of chalcogenols or dichalcogenides with a different halo-amides, *e.g.*, *N*-halosuccinimide (Cl, Br, I), *N*-halophthalimide. Besides, these halo-amides and other related substrates *e.g.*, 1,3-dichloro-5,5-dimethylhydantoin (DCDMH), saccarine, trichloroisocyanuric acid (TCCA), *N*-fluorobenzenesulfonimide (NFSI) *etc.*, can be used in the *in situ* generation of the electrophilic organochalcogens.

Chen and co-workers reported the use of *N*-(arylthio) succinimides/phthalimides **56** in the enantioselective sulfenylation of *β*-naphthols **55**.

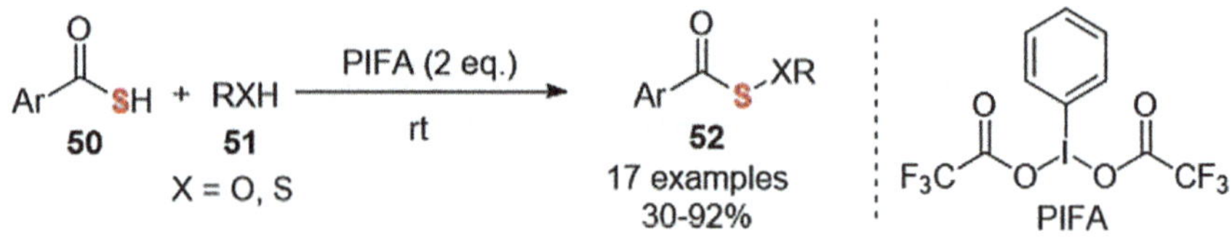

Scheme 11.16 PIFA-mediated synthesis of acylsulfenic acid alkyl esters and benzoyl alkyl disulfides.

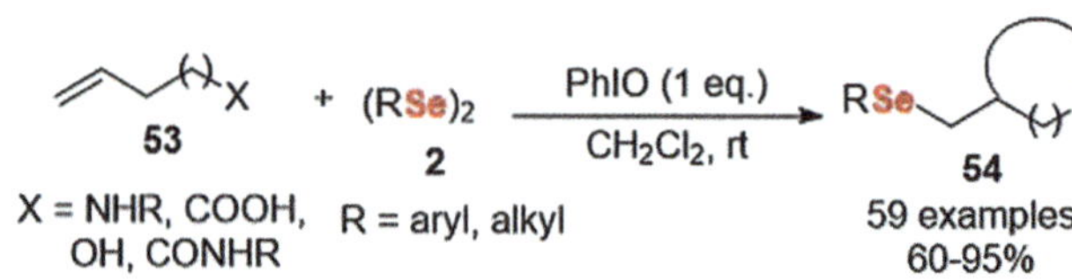

Scheme 11.17 PhIO-mediated intramolecular selenocyclization of olefins.

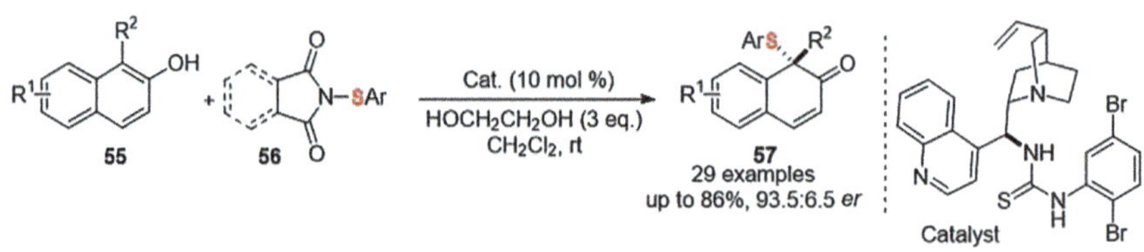

Scheme 11.18 Enantioselective sulfenylation of β-naphthols by *N*-(arylthio) succinimides/phthalimides.

A series of enantioenriched naphthalenones **57** with an S-containing all-substituted stereocenter were synthesized *via* a dearomatization approach in the presence of cinchona-derived thiourea as the catalyst (Scheme 11.18).[34]

In 2019, Sun's group reported the use of *N*-thiosuccinimide **58a**, generated *in situ* from the reaction of diorganodisulfides **58** and *N*-halosuccinimide (Cl, Br) in the presence of TEMPO, in the multicomponent reactions with isocyanides **59** and a range of nucleophiles **60** in the presence of *N*-thiosuccinimide **58a.** The reaction allows access to isothioureas and related molecular moieties **61** in up to 92% yields (Scheme 11.19).[35]

Xiao and co-workers developed a method for the sulfenylation of *H*-phosphonates **62**, employing DCDMH (1,3-dichloro-5,5-dimethylhydantoin) as a mediator in the generation of the electrophilic sulfur specie RSCl. The thiophosphonates **64** were obtained in 30 to 95% yields, using **62**, arylic, and alkylic thiols **63** (Scheme 11.20).[36] The electrophilic sulfur specie would result from the attack of thiols on the highly polarized chlorine atoms on the DCDMH molecule.

In another approach, trichloroisocyanuric acid (TCCA) was also used with the objective to generate the RSCl. In the methodology described by Chen, the coupling of thiols **63** with phosphites **65** was promoted by TCCA leading to the achievement of several thiophosphates **66** at room temperature and

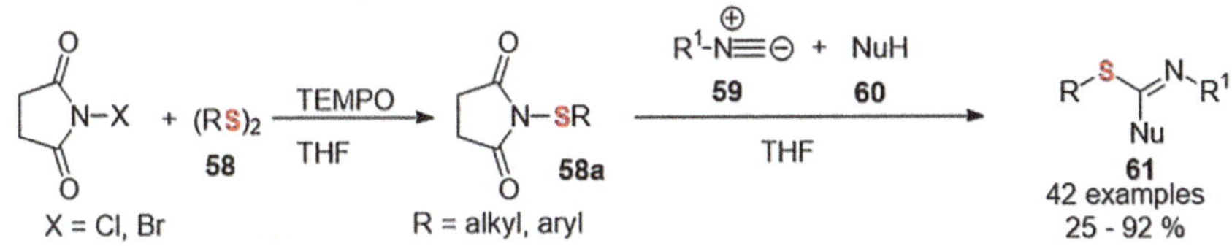

Scheme 11.19 Reactivity of *N*-thiosuccinimides with isocyanides and nucleophiles.

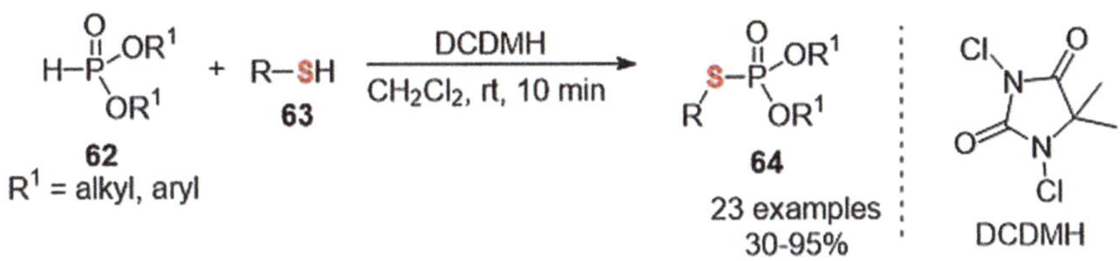

Scheme 11.20 DCDMH-mediated sulfenylation of *H*-phosphonates.

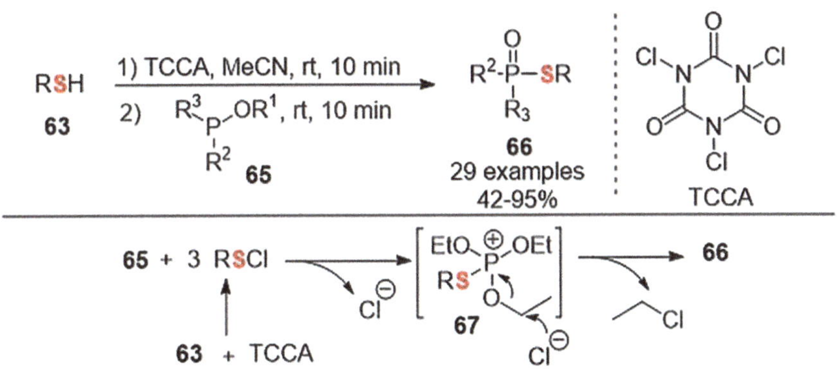

Scheme 11.21 TCCA-mediated synthesis of thiophosphonate.

yields up to 95% (Scheme 11.21).[37] The authors reported the formation *in situ* of the electrophilic RSCl specie, which can undergo nucleophilic attack from the lone electron pair of the phosphorous atom from the phosphite, generating intermediate **67**, which subsequently produces thiophosphonate **66** (Scheme 11.21).

The sulfimide group can also be utilized in the chalcogenation reactions. One example is saccharin, which can form a bench stable electrophilic chalcogen source. In this regard, Shen and co-workers have reported the use of trifluoromethylthiosaccharin **69** in lactonization/lactamization reactions, mediated by a Lewis acid. For this transformation, the authors used a range of vinyl benzoic acid and vinyl benzamides **68** and reacted them with **69**, in the presence of Me_3SeCl as a Lewis acid, resulting in the corresponding trifluoromethylthiolated lactone/lactam **70** in up to 98% yields (Scheme 11.22).[38] It was proposed that a thiiranium ion intermediate was

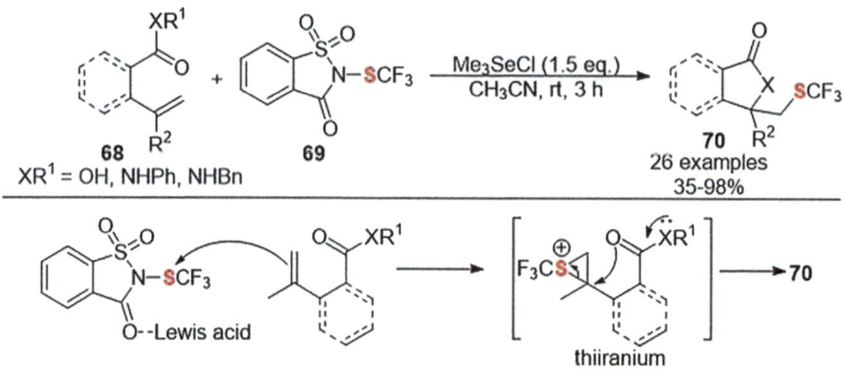

Scheme 11.22 Me_3SeCl-mediated lactonization/lactamization reactions.

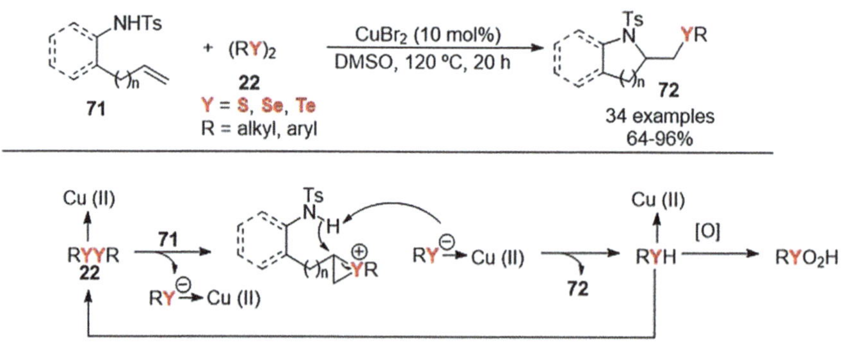

Scheme 11.23 $CuBr_2$-catalyzed electrophilic chalcogeno-cyclization of olefins.

initially formed, which was further attacked by the carboxylic acid or amide to generate the corresponding product.

11.2.2 Lewis Acids Containing Electrophilic Sources

Another approach in the formation of electrophilic chalcogens is Lewis acid activation of the dichalcogenide bond. For this, a number of transition metal-free Lewis acids (*e.g.*, $BF_3.OEt_2$) and transition metal-based Lewis acids (*e.g.*, $CuBr_2$, $ZnCl_2$, $Sc(OTf)_3$, $SbCl_5$, *etc.*) are used. Zhong's group reported copper catalyzed intramolecular electrophilic chalcogenation of *N*-tosylated olefins with dichalcogenides **22**. $CuBr_2$ (10 mol%) generates *in situ* electrophilic chalcogen species, which trigger electrophilic cyclization of olefins **71**, affording chalcogenyl N-heterocycles **72** under aerobic conditions (Scheme 11.23).[39] Mechanistically, complexation of copper with chalcogen atom polarized the chalcogen – chalcogen bond of **22**, making another chalcogen atom electrophilic, which undergoes electrophilic addition to **71** to generate a chalcogen cation. Subsequently, the cation suffers regioselective ring-opening transformation, affording product **72**.

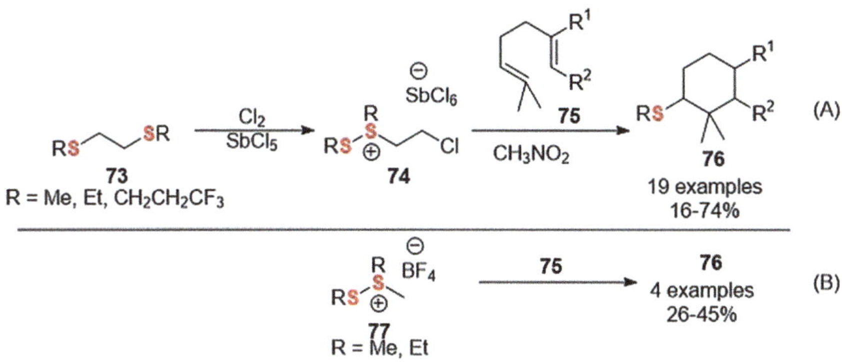

Scheme 11.24 Alkyldisulfanium salts as alkyl-based S-electrophiles for thiolation processes.

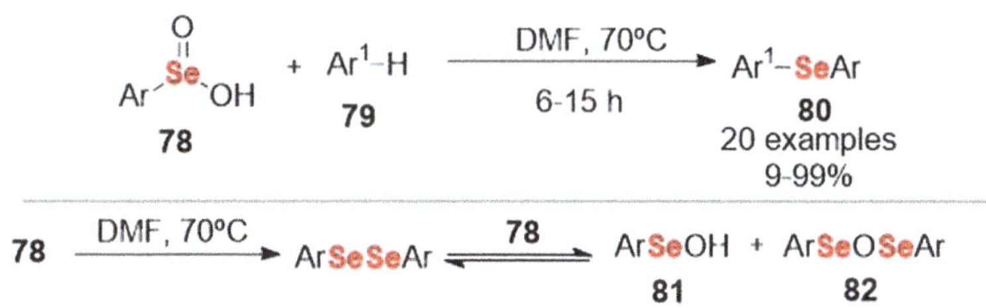

Scheme 11.25 Use of aryl seleninic acid in the electrophilic substitution reactions of activated arenes.

In 2017, Snyder and co-workers reported the reaction of 1,2-dithioethers **73** with Cl_2 and $SbCl_5$ to generate isolable and storable alkyldisulfanium salts **74** and used them as alkyl-based S-electrophiles to install –SR groups on a diverse range of cyclization in racemic forms, when reacted with dienes **75** (Scheme 11.24A).[40] The authors also demonstrated that dimethyl(methylthio)-sulfonium tetrafluoroborate (DMTSF) and diethyl(methylthio)sulfonium tetrafluoroborate **77** can accomplish a similar transformation (Scheme 11.24B).

11.2.3 Miscellaneous Sources

11.2.3.1 Organoseleninic Acid ($RSeO_2H$)

The use of aryl seleninic acid **78** as a precursor for the generation of electrophilic selenium species was reported by Lenardão and co-workers. Through the developed method, it was possible to perform electrophilic substitution in activated arenes **79** *e.g.*, anilines and indoles, in the presence of DMF as the solvent. The corresponding products **80** were obtained in up to 99% yield in reaction times that ranged from 6 to 15 hours (Scheme 11.25).[41] According to the proposed mechanism, arylseleninic acid **78** in DMF at 70 °C result in the formation of diselenides, which

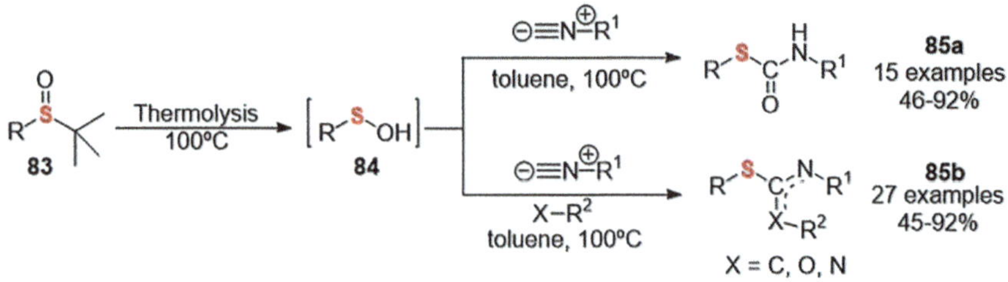

Scheme 11.26 Reactivity of orgnosulfenic acid toward isocyanides.

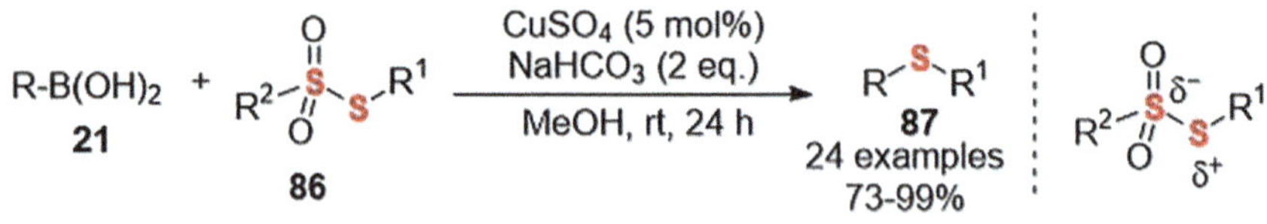

Scheme 11.27 Catalytic thiolation of organoboronic acids by thiosulfonates.

subsequently react with **78** to generate species **81** and **82**. Both species would undergo protonation, generating the selenylated electrophilic intermediates, which reacted with arenes **79** to afford **80**.

11.2.3.2 Organosulfenic Acid (RSOH)

Sun and co-workers reported *in situ* formation of organosulfenic acid **84**, an electrophilic *S*-source, thermolysis for organo-*tert*-butyl sulfoxide **83**. These RSOH species **84** on direct reaction with isocyanides led to the formation of thiocarbamic acid *S*-esters **85a** in good to high yields (Scheme 11.26).[42] Furthermore, they also applied this approach in a multicomponent reaction involving isocyanide, sulfoxide **83**, and a suitable nucleophile to access a diverse range of S-containing compounds **85b**, including isothioureas, carbonimidothioic acid esters, and carboximidothioic acid esters.

11.2.3.3 Thiosulfonates ($RSSO_2R$)

Thiosulfonates **86** are known to be readily available, easy to synthesize, and bench stable electrophilic *S*-source to provide sulfides by reaction with carbon nucleophiles. Considering the good leaving ability of the sulfonyl group of **86**, Yoshida and co-workers reported the thiolation of organoboronic acids **21** by using a catalytic amount of $CuSO_4$ under basic conditions. Under the optimized conditions, they obtained the thiolated products **87** in good to excellent yields (Scheme 11.27).[43]

11.3 Conclusions and Perspectives

In this chapter, recent advances in the formation of a C – Chalcogen bond *via* electrophilic organochalcogen species are described. The chalcogen moiety can be introduced into an organic structure easily by using electrophilic S/Se/Te species, which can be readily generated *in situ* or pre-synthesized as bench stable reagents and reacted with appropriate carbanion/nucleophilic centers through diverse methodologies. In general, the choice of suitable electrophilic chalcogen is determined by the structure of the organic substrate that reacts with the chalcogen source.

Due to the great potential of organochalcogen compounds, the content of this chapter will help future research on the use of electrophilic chalcogen species in the C–Chalcogen bond formation, providing new synthetic routes and allowing improvements to the previous methodologies. This approach could also encourage the development of hybrid systems, with the use of flow or electrochemical or mechanochemical processes, for the chalcogeno-functionalization. We believe that the progress and perspective highlighted in this chapter will stimulate further efforts from researchers all across the organochalcogen community and related areas.

Acknowledgements

We gratefully acknowledge CAPES (Finance Code 001), CNPq, J. R. and S. S. would like to acknowledge CNPq (315399/2020-1; 422645/2021-4; and 403210/2021-6) and FUNAPE (Call No. 01/2022, No.: 210, and Call No. 01/2022, No.: 223).

References

1. (a) D. Manna, G. Roy and G. Mugesh, *Acc. Chem. Res.*, 2013, **46**, 2706; (b) S. Kumar, J. Yan, J.-F. Poon, V. P. Singh, X. Lu, M. K. Ott, L. Engman and S. Kumar, *Angew. Chem., Int. Ed.*, 2016, **55**, 3729; (c) D. J. Press and T. G. Back, *Org. Lett.*, 2011, **13**, 4104; (d) S. T. Manjare, Y. Kim and D. G. Churchill, *Acc. Chem. Res.*, 2014, **47**, 2985; (e) J. Rafique, S. Saba, R. F. S. Canto, T. E. A. Frizon, W. Hassan, E. P. Waczuk, M. Jan, D. F. Back, J. B. T. Rocha and A. L. Braga, *Molecules*, 2015, **20**, 10095; (f) J. Rafique, R. F. S. Canto, S. Saba, F. A. R. Barbosa and A. L. Braga, *Curr. Org. Chem.*, 2016, **20**, 166.
2. (a) A. Francioso, A. B. Conrado, L. Mosca and M. Fontana, *Oxid. Med. Cell. Longevity*, 2020, 8294158; (b) Y. Hai, M. Wei, C. Wang, Y. Gu and C. Shao, *Mar. Life Sci. Technol.*, 2021, **3**, 488.
3. (a) D. L. Klayman and W. H. H. Günther, *Organic Selenium Compounds: Their Chemistry and Biology (Chemistry of Organometallic Compounds)*, John Wiley & Sons Inc., New York, 1973; (b) A. L. Braga and J. Rafique, Synthesis of biologically relevant small molecules containing selenium. Part A. Antioxidant compounds, in *The Chemistry of Organic Selenium and Tellurium Compounds*, ed. Z. Rappoport, Wiley & Sons, Ltd, Chichester, 2014, ch. 13, vol. 4, p. 989; (c) H. J. Reich and R. J. Hondal, *ACS Chem. Biol.*, 2016, 11, 821.
4. C. W. Nogueira, G. Zeni and J. B. T. Rocha, *Chem. Rev.*, 2004, **104**, 6255.

5. (a) A. L. Braga and J. Rafique, Synthesis of biologically relevant small molecules containing selenium. Part B. Anti-infective and anticancer compounds, in *The Chemistry of Organic Selenium and Tellurium Compounds*, ed. Z. Rappoport, Wiley & Sons, Ltd, Chichester, UK, 2013, vol. 4, p. 1053; (b) H. C. Braga, H. A. Stefani, M. W. Paixão, F. W. Santos and D. S. Lüdtke, *Tetrahedron*, 2010, **66**, 3441; (c) R. L. O. R. Cunha, I. E. Gouvea and L. Juliano, *An. Acad. Bras. Cienc.*, 2009, **81**, 393; (d) D. C. dos Santos, J. Rafique, S. Saba, G. M. Almeida, T. Siminski, C. Padua, D. W. Filho, A. Zamoner, A. L. Braga, R. C. Pedrosa and F. Ourique, *J. Biochem. Mol. Toxicol.*, 2021, **35**, e22663.
6. (a) A. L. Braga and J. Rafique, Synthesis of biologically relevant small molecules containing selenium. Part C. Miscellaneous biological activities, in *The Chemistry of Organic Selenium and Tellurium Compounds*, ed. Z. Rappoport, Wiley & Sons, Ltd, Chichester, 2014, ch. 15, vol. 4, p. 1119; (b) G. V. Botteselle, W. C. Elias, L. Bettanin, R. F. S. Canto, D. N. O. Salin, F. A. R. Barbosa, S. Saba, H. Gallardo, G. Ciancaleoni, J. B. Domingos, J. Rafique and A. L. Braga, *Molecules*, 2021, **26**, 4446.
7. S. W. May, L. Wang, M. M. Gill-Woznichak, R. F. Browner, A. A. Ogonowski, J. B. Smith and S. H. Pollock, *J. Pharmacol. Exp. Ther.*, 1997, **283**, 470.
8. (a) C. Jacob, G. I. Giles, N. M. Giles and H. Sies, *Angew. Chem., Int. Ed.*, 2003, **42**, 4742; (b) F. Martinez-Ramos, H. Salgado-Zamora, M. E. Campos-Aldrete, Y. Melendez-Camargo, Y. Márquez-Flores and M. Soriano-García, *Eur. J. Med. Chem.*, 2008, **43**, 1432; (c) J. Rafique, G. Farias, S. Saba, E. Zapp, I. B. Casagrande, C. A. M. Salla, I. H. Bechtold, M. R. Scheide, J. S. S. Neto, D. M. Souza Jr, H. C. Braga, L. F. B. Ribeiro, F. Gastaldon, C. T. Pich and T. E. A. Frizon, *Dyes Pigm.*, 2020, **180**, 108519; (d) L. S. Galant, J. Rafique, A. L. Braga, F. C. Braga, S. Saba, R. Radi, J. B. T. Rocha, C. Santi, M. Monsalve, M. Farina and A. F. Bem, *Neurochem. Res.*, 2021, **46**, 120; (e) T. E. A. Frizon, J. H. Cararo, S. Saba, G. Dal-Pont, M. Michels, H. C. Braga, T. Pimentel, F. Dal-Pizzol, S. S. Valvassori and J. Rafique, *Oxid. Med. Cell. Longevity*, 2020, **2020**, 5417024; (f) G. M. Almeida, J. Rafique, S. Saba, T. Siminski, N. S. R. S. Mota, D. W. Gilho, A. L. Braga, R. C. Pedrosa and F. Ourique, *Biochem. Biophys. Res. Commun.*, 2018, **503**, 1291; (g) I. C. Veloso, E. Delanogare, A. E. Machado, S. P. Braga, G. K. Rosa, A. F. Bem, J. Rafique, S. Saba, R. N. Trindade, F. Z. Galetto and E. L. G. Moreira, *J. Pharm. Pharmacol.*, 2021, **73**, 673; (h) J. Rodrigues, S. Saba, A. C. Joussef, J. Rafique and A. L. Braga, *Asian J. Org. Chem.*, 2018, **7**, 1819; (i) R. T. Ruhee, L. A. Roberts, S. Ma and K. Suzuki, *Front. Nutr.*, 2020, **7**, 1.
9. (a) D. Deidda, G. Lampis, C. Maullu, R. Pompei, F. Isaia, V. Lippolis and G. Verani, *Pharm. Res.*, 1997, **36**, 193; (b) S. H. A. Hafez, *Russ. J. Org. Chem.*, 2005, **41**, 396; (c) J. W. Foley, X. Song, T. N. Demidova, F. Jilal and M. R. Hamblin, *J. Med. Chem.*, 2006, **49**, 5291; (d) S. H. A. Hafez, *Eur. J.*

Med. Chem., 2008, **43**, 1971; (e) A. B. Sabti, A. A. Al-Fregi and M. Y. Yousif, *Molecules*, 2020, **25**, 3439.

10. (a) L. Shao, Y. Li, J. Lu and X. Jiang, *Org. Chem. Front.*, 2019, **6**, 2999; (b) J. E. Tungen, R. Kristianlund, A. Vik and T. V. Hansen, *J. Org. Chem.*, 2019, **84**, 11373; (c) H. Cao, R. Qian and L. Yu, *Catal. Sci. Technol.*, 2020, **10**, 3113; (d) S. Panda, A. Panda and S. S. Zade, *Coord. Chem. Rev.*, 2015, **300**, 86.
11. (a) S. Otocka, M. Kwiatkowska, L. Madalińska and P. Kiełbasiński, *Chem. Rev.*, 2017, **117**, 4147; (b) A. R. Santos, M. W. D. Hanson-Heine, N. A. Besley and P. Licence, *Chem. Commun.*, 2018, **54**, 11403; (c) S. Wolf and C. Feldmann, *Dalton Trans.*, 2019, **48**, 15521; (d) A. Mihoya, S. Koguchi, Y. Shibuya, M. Mimura and M. Oba, *Catalysts*, 2020, **10**, 398.
12. (a) S. T. Manjare, Y. Kim and D. G. Churchill, *Acc. Chem. Res.*, 2014, **47**, 2985; (b) T. E. Frizon, J. Rafique, S. Saba, I. H. Bechtold, H. Gallardo and A. L. Braga, *Eur. J. Org. Chem.*, 2015, 3470; (c) G. C. Hoover and D. S. Seferos, *Chem. Sci.*, 2019, **10**, 9182; (d) P. C. Ho, J. Rafique, J. Lee, L. M. Lee, H. A. Jenkins, J. F. Britten, A. L. Braga and I. Vargas-Baca, *Dalton Trans.*, 2017, **46**, 6570; (e) W. Chen, C. Liu, B. Peng, Y. Zhao, A. Pacheco and M. Xian, *Chem. Sci.*, 2013, **4**, 2892.
13. (a) Z. Guan, Y. Wang, Y. Huang, S. Wang, H. Tang, H. Zhang and A. Lei, *Green Chem.*, 2019, **21**, 4976; (b) M. R. Scheide, M. M. Peterle, S. Saba, J. S. S. Neto, G. F. Lenz, R. D. Cezar, J. F. Felix, G. V. Botteselle, A. R. Schneider, J. Rafique and A. L. Braga, *Sci. Rep.*, 2020, **10**, 15233; (c) M. M. Peterle, M. R. Scheide, L. T. Silva, S. Saba, J. Rafique and A. L. Braga, *ChemistrySelect*, 2018, **3**, 13191; (d) G. Wu, L. Min, H. Li, W. Gao, J. Ding, X. Huang, M. Liu and H. Wu, *Green Chem.*, 2018, **20**, 1560; (e) A. G. Meirinho, V. F. Pereira, G. M. Martins, S. Saba, J. Rafique, A. L. Braga and S. R. Mendes, *Eur. J. Org. Chem.*, 2019, 6465; (f) J. Rafique, S. Saba, M. S. Franco, L. Bettanin, A. R. Schneider, L. T. Silva and A. L. Braga, *Chem. – Eur. J.*, 2018, **24**, 4173; (g) M. R. Scheide, A. R. Schneider, G. A. M. Jardim, G. M. Martins, D. C. Durigon, S. Saba, J. Rafique and A. L. Braga, *Org. Biomol. Chem.*, 2020, **18**, 4916; (h) S. Saba, J. Rafique, M. S. Franco, A. R. Schneider, L. Espindola, D. O. Silva and A. L. Braga, *Org. Biomol. Chem.*, 2018, **16**, 880.
14. (a) J. Rafique, D. S. Rampom, J. B. Azeredo, F. L. Coelho, P. H. Schneider and A. L. Braga, *Chem. Rec.*, 2021, **21**, 2739; (b) G. Zeni, B. Godoi, C. K. Jurinic, A. L. Belladona and R. F. Schumacher, *Chem. Rec.*, 2021, **21**, 2880.
15. (a) G. Pandey and B. B. V. S. Sekhar, *J. Org. Chem.*, 1992, **57**, 4019; (b) R. Guo, L. Liao and X. Zhao, *Molecules*, 2017, **22**, 835; (c) F. V. Singh and T. Wirth, *Selenium and Tellurium Reagents*, De Gruyter, Berlin, 2019; (d) S. Saba, G. V. Botteselle, M. Godoi, T. E. A. Frizon, F. Z. Galetto, J. Rafique and A. L. Braga, *Molecules*, 2017, **22**, 1367; (e) M. S. T. Rocha, J. Rafique, S. Saba, J. B. Azeredo, D. Back, M. Godoi and A. L. Braga, *Synth. Commun.*, 2017, **47**, 219; (f) M. Godoi, A. Leitemberger,

L. M. C. Bohs, M. V. Silveira, J. Rafique and M. G. M. D'Oca, *Environ. Chem. Lett.*, 2019, **17**, 1441.
16. M. C. D. F. Xavier, E. M. A. Sandagorda, J. S. S. Neto, R. F. Schumacher and M. S. Silva, *RSC Adv.*, 2020, **10**, 13975.
17. J.-W. Yuan, Y. Zhang, G.-C Huang, M. Y. Ma, T.-Y Yang, L.-R Yang, S.-R. Zhang, P. Mao and L.-B. Qu, *Org. Chem. Front.*, 2021, **8**, 6937.
18. H.-W. Shi, C. Yu and J. Yan, *Chin. Chem. Lett.*, 2015, **26**, 1117.
19. H. Li, J. Wang, X. Wang and J. Yan, *Phosphorus, Sulfur Silicon Relat. Elem.*, 2018, **193**, 394.
20. R. Rahaman, N. Devi and P. Barman, *Tetrahedron Lett.*, 2015, **56**, 4224.
21. A. A. Vieira, J. B. Azeredo, M. Godoi, C. Santi, E. N. da Silva Junior and A. L. Braga, *J. Org. Chem.*, 2015, **80**, 2120.
22. S. Saba, J. Rafique and A. L. Braga, *Adv. Synth. Catal.*, 2015, **357**, 1446.
23. S. Saba, J. Rafique and A. L. Braga, *Catal. Sci. Technol.*, 2016, **6**, 3087.
24. L. T. Silva, J. B. Azeredo, S. Saba, J. Rafique, A. J. Bortoluzzi and A. L. Braga, *Eur. J. Org. Chem.*, 2017, 4740.
25. J. Rafique, S. Saba, A. R. Rosário and A. L. Braga, *Chem. – Eur. J.*, 2016, **22**, 11854.
26. L. Bettanin, S. Saba, F. Z. Galetto, G. A. Mike, J. Rafique and A. L. Braga, *Tetrahedron Lett.*, 2017, **58**, 4713.
27. (a) L. Bettanin, S. Saba, C. V. Doerner, M. S. Franco, M. Godoi, J. Rafique and A. L. Braga, *Tetrahedron*, 2018, **74**, 3971; (b) M. S. Franco, S. Saba, J. Rafique and A. L. Braga, *Angew. Chem., Int. Ed.*, 2021, **60**, 18154; (c) S. Saba, C. R. Santos, B. R. Zavarise, A. A. S. Naujorks, M. S. Franco, A. R. Schneider, M. R. Scheide, R. F. Affeldt, J. Rafique and A. L. Braga, *Chem. – Eur. J.*, 2020, **26**, 4461; (d) B. L. Tornquist, G. P. Bueno, J. C. M. Willing, I. M. Oliveira, H. A. Stefani, J. Rafique, S. Saba, B. A. Iglesias, G. V. Botteselle and F. Manain, *ChemistrySelect*, 2018, **3**, 6358; (e) J. S. S. Neto, R. A. Balaguez, M. S. Franco, V. C. S. Machado, S. Saba, J. Rafique, F. Z. Galetto and A. L. Braga, *Green Chem.*, 2020, **22**, 3410; (f) Y. H. Matzkeit, B. L. Tornquist, F. Manarin, G. V. Botteselle, J. Rafique, S. Saba, A. L. Braga, J. F. Felix and R. Schneider, *J. Non-Cryst. Solids*, 2018, **498**, 153.
28. G. M. Martins, G. do Carmo, A. F. Morel, T. S. Kaufman and C. C. Silveira, *Asian J. Org. Chem.*, 2019, **8**, 369.
29. M. V. Musalov, V. A. Yakimov, V. A. Potapov, S. V. Amosova and S. V. Zinchenko, *Russ. J. Org. Chem.*, 2019, **55**, 1153.
30. I. A. Opai, P. D. Shirsath, A. H. Kalbandhe and N. N. Karade, *ARKIVOC*, 2021, 101.
31. Y. Zhang, S. Wu and J. Yan, *Helv. Chim. Acta*, 2017, **100**, 1600306.
32. C.-H. Tsai, D. M. Reddy, P.-A. Hsieh, Y.-C Liu, M. Kandasamy, W.-Y. Lin and C.-F. Lee, *Synthesis*, 2016, **48**, 4459.
33. P.-F. Wang, W. Yi, Y. Ling, L. Ming, G.-Q. Liu and Y. Zhao, *Chin. Chem. Lett.*, 2021, **8**, 2587.
34. J.-J. Wang, H. Yang, B.-B. Gou, L. Zhou and J. Chen, *J. Org. Chem.*, 2018, **83**, 4730.

35. X. Lei, Y. Wang, E. Fan and Z. Sun, *Org. Lett.*, 2019, **21**, 1484.
36. X. Bi, J. Li, F. Meng, H. Wang and J. Xiao, *Tetrahedron*, 2016, **72**, 706.
37. Y. Chen, M. Li, Z. Gong and Z. Shen, *Phosphorus, Sulfur Silicon Relat. Elem.*, 2020, **196**, 19.
38. C. Xu and Q. Shen, *Org. Lett.*, 2015, **17**, 4561.
39. Y. Ni, H. Zuo, Y. Li, Y. Wu and F. Zhong, *Org. Lett.*, 2018, **20**, 4350.
40. F. T. Schevenels, M. Shen and S. A. Snyder, *Org. Lett.*, 2017, **19**, 2.
41. L. Abenante, N. B. Padilha, J. M. Anghinoni, F. Penteado, O. Rosati, C. Santi, M. S. Silva and E. J. Lenardão, *Org. Biomol. Chem.*, 2020, **18**, 5210.
42. S. Wu, X. Lei, E. Fan and Z. Sun, *Org. Lett.*, 2018, **20**, 522.
43. S. Yoshida, Y. Sugimura, Y. Hazama, Y. Nishiyama, T. Yano, S. Shimizu and T. Hosoya, *Chem. Commun.*, 2015, **51**, 16613.

CHAPTER 12

Nucleophilic Chalcogen-containing Reagents

DAMIANO TANINI,* ANTONELLA CAPPERUCCI AND STEFANO MENICHETTI

University of Florence, Department of Chemistry "Ugo Schiff", Via della Lastruccia 3-13, I-50019 Sesto Fiorentino, FI, Italy
*Email: damiano.tanini@unifi.it

12.1 Introduction

The use of sulfur and selenium nucleophiles in organic and inorganic synthesis has been widely investigated. A broad range of functional group transformations can be accomplished by using nucleophilic chalcogen reagents. Additionally, biologically and synthetically valuable sulfur- and selenium-containing compounds have been prepared by using such nucleophilic species. The synthesis and the reactivity of nucleophilic sulfur and selenium reagents have already been reviewed in books and book chapters.[1,2] Metal thiolates and selenolates, which represent the commonly employed chalcogen nucleophilic reagents, have been comprehensively discussed therein. This chapter will focus on "unconventional" nucleophilic sulfur and selenium reagents containing chalcogen–boron, chalcogen–aluminium, chalcogen–silicon, and chalcogen–tin bonds. Features of the synthesis and reactivity of such species will be presented with particular emphasis on recent advances in the field, surveying the literature between 2000 and 2021.

Chalcogen Chemistry: Fundamentals and Applications
Edited by Vito Lippolis, Claudio Santi, Eder J. Lenardão and Antonio L. Braga

Published by the Royal Society of Chemistry, www.rsc.org

12.2 Nucleophilic Species Containing the Chalcogen–Boron Bond

Sulfur and selenium boron species have emerged as versatile reagents for the formation of C–S and C–Se bonds. The reduction of diphenyl diselenide with sodium borohydride, which is one of the easiest and most widely used procedures for the generation of nucleophilic selenium species, affords the sodium selanylborate reagent **1** (Scheme 12.1) as an exquisite nucleophile in a wide array of transformations, such as the conversion of epoxy ketones and esters into the corresponding β-hydroxy ketones and esters.[3] This Se-mediated functional group interconversion has also been exploited for the construction of the WXYZ ring of maitotoxin and the HIJK ring of brevisulcenal-F.[4]

Sulfanylboranes and selanylboranes are used in a variety of organic reactions including additions, substitutions, and nucleophilic ring-opening reactions. For example, the Pd-catalyzed thioboration of terminal alkynes with 9-(organylsulfanyl)-9-borabicyclo[3.3.1]nonane derivatives provide the corresponding addition products with high regio- and stereo-selectivity.[5] Tris(phenylseleno)borane and tris(methylseleno)borane [$(PhSe)_3B$ and $(MeSe)_3B$, respectively] react with alkynes through a radical mechanism affording the corresponding (*Z*)-vinyl selenides with high stereoselectivity. The reaction is initiated by oxygen and can be employed for the synthesis of functionalized heterocycles and carbocycles *via* radical cyclization of en-ynes.[6] The use of selanylboranes in nucleophilic ring-opening reactions (NROR) of epoxides[7] and in the conversion of carbonyl compounds into the corresponding selenoacetals[8] has also been described.

Sulfanylboranes are commonly prepared from thiols or metal thiolates and haloboranes. Disulfides and aryldiiodoboranes can be employed for the synthesis of sulfanylboranes. Similarly, the preparation of selanylboranes generally relies on the reactivity of selenols, selenolates, and diselenides with haloboranes. The synthesis and the reactivity of sulfanylboranes and selanylboranes were reviewed in 2005 by Habben and Kaufmann.[9] Therefore, this chapter will not focus on topics covered therein.

Besides sulfanylboranes and selanylboranes, due to their peculiar reactivity and significant synthetic potential, thioborates [$R^1S–B(OR^2)_2$] and selenoborates [$R^1Se–B(OR^2)_2$] are attracting considerable interest among organic chemists. Thioborates are used as precursors of a wide array of sulfur-containing biologically-active compounds.[10] Thioborates are commonly prepared through the dehydrocoupling of thiols with boranes. Both

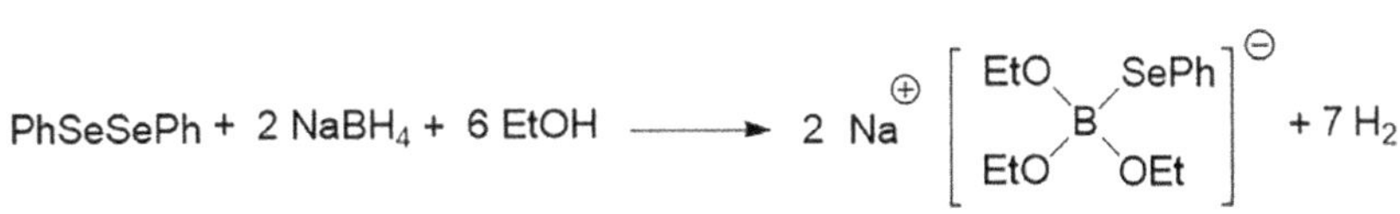

1

Scheme 12.1 Formation of sodium selanylborate **1**.

metal-catalyzed and metal-free approaches have been reported for the formation of S–B bonds (Scheme 12.2). For example, the Nolan's ruthenium-catalyzed methodology enables the synthesis of a broad range of thioborates in excellent yield.[11] A number of thioborates can also be prepared by using an aluminium dihydride catalyst.[12] An alternative approach towards thioborates relies on the use of the frustrated Lewis pair (FLP) $[NMe_2–C_6H_4–BH_2]_2$ as the catalyst for the dehydrogenative coupling of thiols with pinacolborane (HBpin).[13] A mild catalyst-free procedure for the formation of S–B bonds, proceeding *via* Lewis acid–base adducts featuring a hydridic B–H group and an acidic S–H moiety in close proximity, has also been reported.[14]

Thioborates are employed in the thioboration of α,β-unsaturated aldehydes and ketones. Owing to its Lewis acid properties, the boryl moiety of RS–Bpin interacts with the carbonyl oxygen of the substrate promoting the nucleophilic attack of the RS unit onto the α,β-unsaturated compound, which leads to a 1,4- or 1,2-addition product depending on the nature of the substrate. Both 1,4- and 1,2-thioboration products are converted into β-sulfido carbonyl derivatives **2**, **3** upon treatment with methanol; the 1,2-thioborane intermediate plausibly undergoes rearrangement during the *in situ* protic work up.[15] PhSeBpin, prepared by Rh-catalyzed dehydrogenative borylation of benzeneselenol (PhSeH) with pinacolborane, reacts efficiently with α,β-unsaturated carbonyl derivatives under catalyst-free conditions to afford the corresponding conjugate addition products (Scheme 12.3, eqn (I)). Interestingly, additions of PhSeBpin to alkenes do not require the use of co-solvents, such as MeOH.[16]

The "pull–push" effect of the boryl moiety of chalcogenoborates R^3X–Bpin (X = S, Se) also enables the delivery of chalcogen-containing groups onto electron-poor alkynes. Ynones react with thioborates and selenoborates under catalyst-free conditions providing a good yield of the corresponding vinyl sulfides and selenides. The conjugate addition occurs with good stereoselectivity leading preferentially to (*Z*)-alkenyl chalcogenides **4** and **5** (Scheme 12.3, eqn (II)). Reactions using selenoboranes generally exhibit a higher *Z*/*E* ratio compared to the related additions with thioboranes.[17]

PhSe–Bpin also reacts with α,β-acetylenic esters and amides to afford the corresponding α- and β-vinyl selenides as a function of the reaction conditions (Scheme 12.4). Products **6**, arising from the *anti*-3,4-selenoboration path, are formed upon treatment of α,β-acetylenic esters and amides with PhSe–Bpin in the presence of PCy_3. Such products can be easily converted into the corresponding α-hydroselenated derivatives *via* protodeboronation using K_2CO_3. On the other hand, when PhSe–Bpin is reacted with

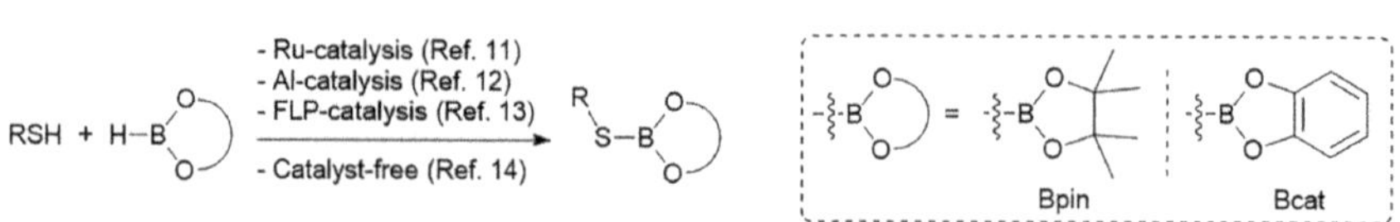

Scheme 12.2 Synthesis of thioborates.

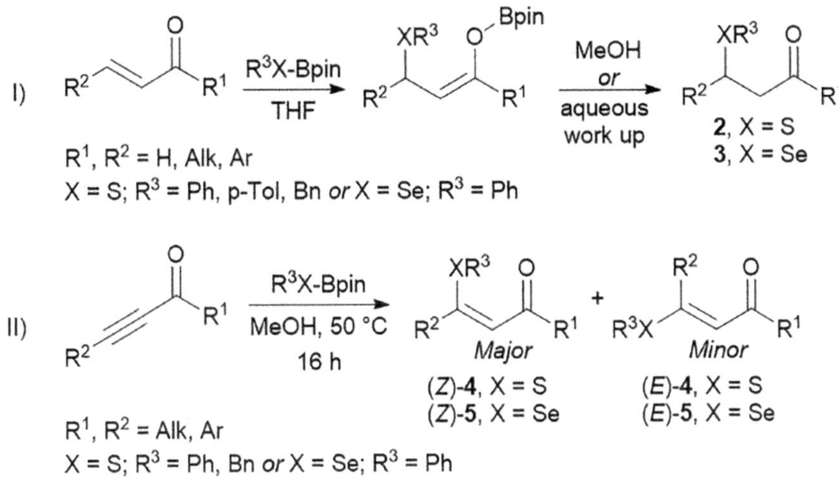

Scheme 12.3 Addition of chalcogenoboranes to α,β-unsaturated carbonyl compounds.

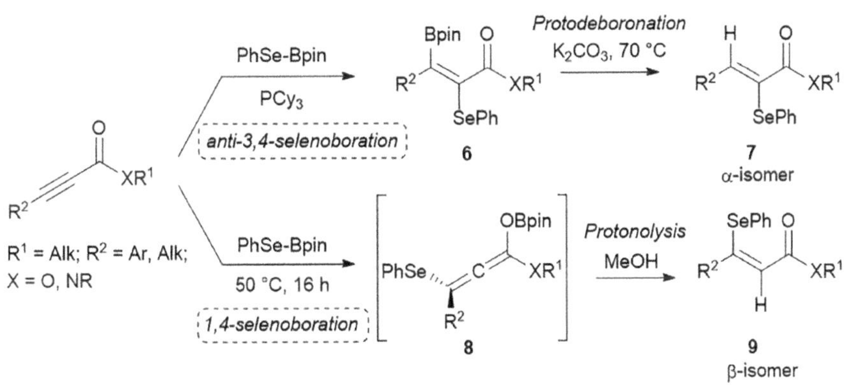

Scheme 12.4 Anti-selenoboration of alkynes.

α,β-acetylenic derivatives at 50 °C under catalyst-free conditions the reaction proceeds through 1,4-selenoboration to afford intermediates **8**, which undergo protonolysis providing β-seleno-α,β-unsaturated derivatives **9** (Scheme 12.4).[18]

Thioborates have also been used in reactions with trimethylsilyldiazomethane (Me_3SiCHN_2) to afford polyfunctionalized derivatives **10**. The insertion of the diazo compound into the B–S σ bond of thioborates provides access to compounds bearing a polysubstituted sp^3 carbon [H–C(SR)(Bpin)($SiMe_3$)], which can be further subjected to a range of base-assisted transformations, including deborylative and desilylative C–C bond forming strategies (Scheme 12.5).[19]

Thioborates can be used as thiolation reagents in the Ni-catalyzed hydrothiolation of alkenes and alkynes. The reaction enables the remote C–H

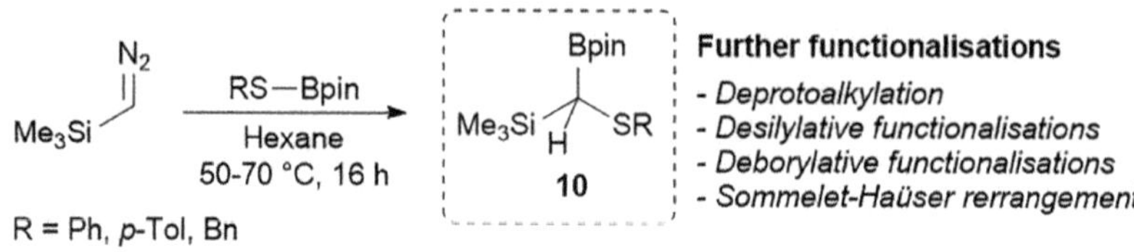

Scheme 12.5 Synthesis of poly-substituted functionalizable one-carbon fragments **10**.

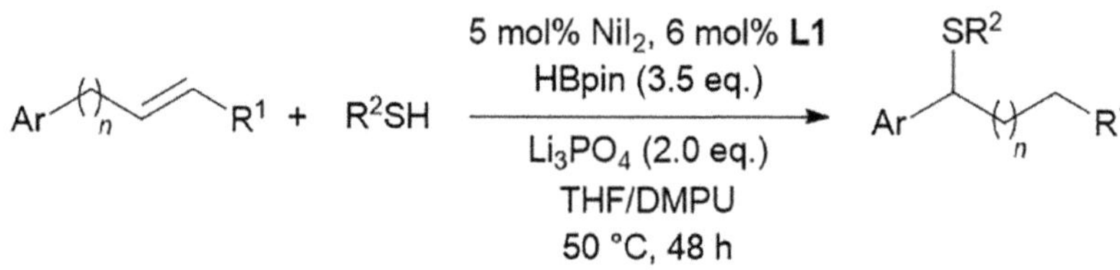

Scheme 12.6 Ni-catalyzed selective hydrothiolation of alkenes with thiols. Key R^2S–Bpin species are formed *in situ*. **L1**, 2,9-dimethyl-4,7-diphenyl-1,10-phenanthroline.

thiolation of different alkenes bearing a variety of remote aryl rings *via* a NiH-catalyzed migratory hydrothiolation mechanism. This reaction can be efficiently performed using thiols as thiolation regents in the presence of pinacolborane (HBpin), as reported in Scheme 12.6. This mechanism is demonstrated to proceed *via* RS–Bpin species, generated *in situ* from the thiol and pinacolborane through a Ni-catalyzed oxidative addition–reductive elimination sequence.[20]

TolSBpin and PhSeBpin were also efficiently employed in reactions with acyl chlorides affording the corresponding thiol- and selenol-esters along with ClBpin.[21]

12.3 Nucleophilic Species Containing the Chalcogen–Aluminium Bond

Aluminium thiolates and selenolates are used as versatile reagents for the selective formation of new C–S and C–Se bonds. Aluminium thiolates are generally prepared upon reaction of alkylaluminium compounds with thiols or by treatment of aluminium hydrides with thiols or disulfides.[22] Analogously, aluminium selenolates are commonly prepared from alkylaluminium compounds and elemental selenium or from aluminium hydride and selenols or diselenides.[23]

Aluminium thiolates react with a broad range of functional groups enabling the delivery of sulfurated moieties. For example, the reaction of Et_2AlSPh with vinyl epoxides affords the corresponding 1,4-addition products with high stereoselectivity (Scheme 12.7).[24]

Aluminium thiolates also react with esters providing simple access to thiolesters. This route enables the straightforward conversion of methyl and

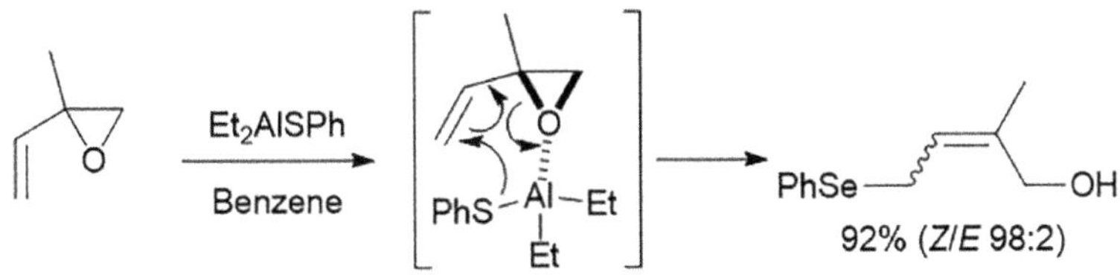

Scheme 12.7 1,4-Addition of Et_2AlSPh to vinyl epoxides.

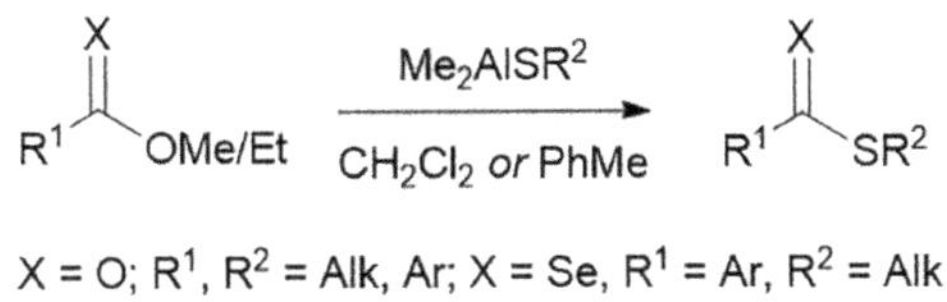

Scheme 12.8 Reaction of aluminium thiolates with esters and selenoic acid *O*-methyl esters.

ethyl esters into the corresponding synthetically valuable thiolesters (Scheme 12.8).[25] The reaction of aluminium thiolates with α,β-unsaturated esters generally yields the conjugate addition products. A related procedure, relying on the reaction of selenoic acid *O*-methyl esters with aluminium thiolates can be employed for the synthesis of *S*-alkyl arenecarboselenothioates.[26] Aluminium thiolates and selenolates also react with aldehydes to afford the corresponding chalcogenoesters *via* a Tishchenko-type hydride transfer process.[27,28]

Alkyl fluorides can be efficiently reacted with dialkylaluminium thiolates and selenolates to afford the corresponding chalcogenides. Primary alkyl fluorides smoothly react with $^iBu_2AlXPh$ (X = S, Se, Te) through a S_N2 mechanism providing the corresponding alkyl–phenyl chalcogenides in good yields.[29]

The reaction of aluminium thiolates and selenolates with α,β-unsaturated ketones provides the corresponding conjugate addition products. The resulting aluminium enolate intermediates can be efficiently reacted with aldehydes to give the aldol products which, after formal thiol or selenol elimination, lead to α-substituted-α,β-unsaturated ketones (Scheme 12.9).[30] α,β-Unsaturated nitriles also undergo conjugate addition upon treatment with aluminium thiolates.[31]

The reactivity of acetals and hemiacetals with aluminium thiolates and selenolates has also been investigated. Acetals can be converted into monothioacetals and monoselenoacetals upon treatment with Et_2AlSPh or $^iBu_2AlSePh$.[32,33] An excess of aluminium reagent can be used to promote the conversion of acetals into diselenium acetals. Soft selenium-centred nucleophiles such as $^iBu_2AlSePh$ and $^iBu_2AlSe^nBu$ react with perfluoro hemiacetals leading to the corresponding 1-(organoselanyl)perfluoroalkanols in quantitative yield (Scheme 12.10).[34]

The conversion of an *O,O*-acetal into the corresponding *O,Se*-acetal with $^iBu_2AlSePh$ has also been exploited to access a key intermediate used in one

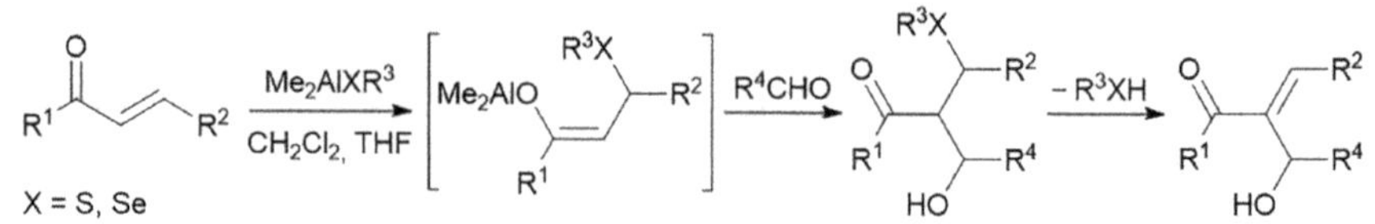

Scheme 12.9 Functionalization of α,β-unsaturated ketones through aluminium thiolates and selenolates.

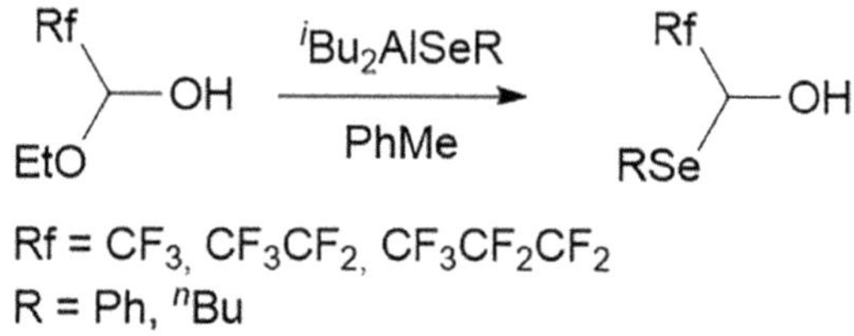

Scheme 12.10 Reaction of perfluoro hemiacetals with aluminium selenolates.

of the available routes for the synthesis of the left wing of the ciguatoxin CTX1B.[35] The intramolecular radical cyclization of *O,S*- and *O,Se*-acetals, prepared *via* the reactivity of aluminium chalcogenolates, also enabled the construction of the oxepane ring G of ciguatoxin CTX3C.[36,37]

Bis(dimethylaluminium)selenide **11**, which can be prepared either by reaction of dimethylaluminium chloride with bis(trimethylsilyl)selenide or upon treatment of trimethylaluminium with bis(tributylstannyl)selenide in toluene,[23,38] is an effective selenylating reagent towards carbonyl compounds. Ketones and aldehydes are easily converted into the corresponding selenocarbonyl derivatives, which can be trapped with dienes to afford the Diels–Alder adducts. Selenocarbonyl compounds are unstable and generally not isolable; however, sterically-hindered derivatives such as selenofenchone can be isolated.[38] The reaction of **11** with α,β-unsaturated selenoaldehydes and selenoketones also provides the corresponding α,β-unsaturated selenoaldehydes and selenoketones. Although the monomeric form of these derivatives is too unstable to be isolated, they undergo "head-to-head" [4 + 2] dimerization to afford diselenin derivatives **12**. α,β-Unsaturated selenocarbonyl compounds also serve as 2π dienophiles in reactions with cyclopentadiene and as 4π heterodienes in Diels–Alder cycloadditions with norbornadiene, enabling the rapid synthesis of selena-heterocycles **13** and **14** (Scheme 12.11).[39]

The [4 + 2] cycloadducts **15**, obtained upon treatment of aromatic or aliphatic aldehydes with **11** in the presence of anthracene, are stable at room temperature but easily undergo thermal retro Diels–Alder reactions to provide the corresponding selenoaldehydes (or selenals), which serve as versatile precursors of a number of selenium-containing derivatives (Scheme 12.12).[40,41]

Acetals can also be employed as the starting material to generate selenoaldehydes, and their *in situ* trapping with dienes provides the

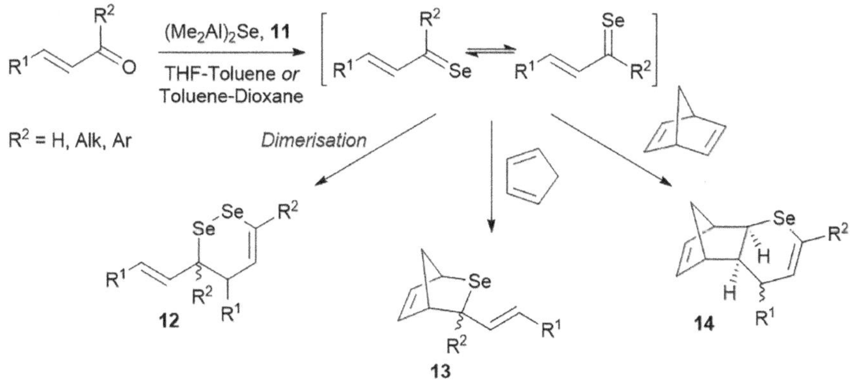

Scheme 12.11 Synthesis and reactivity of α,β-unsaturated selenocarbonyl derivatives.

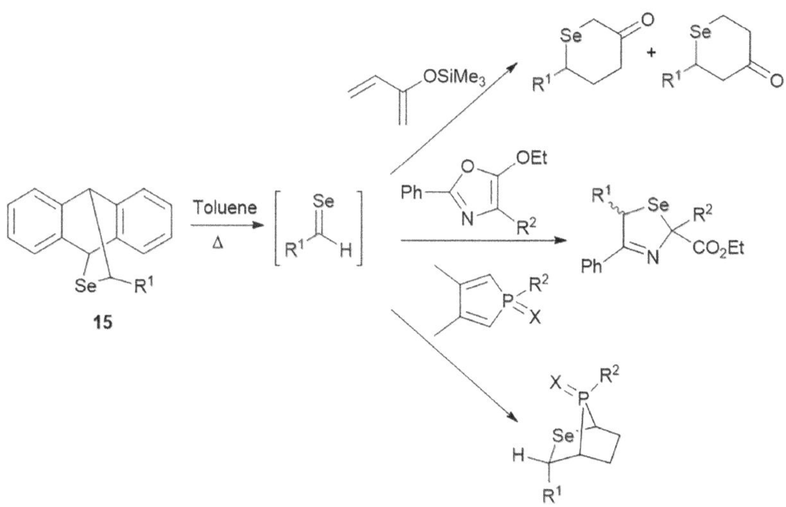

Scheme 12.12 Generation and functionalization of selenoaldehydes.

corresponding Diels–Alder adducts. Cyclic acetals with one exocyclic acetal oxygen atom afford systems bearing a hydroxy group at the terminal position. Dihydropyranyl methyl ethers give functionalized derivatives with a terminal formyl group.[38]

12.4 Nucleophilic Species Containing the Chalcogen–Silicon Bond

The sulfide moiety is present in many important structures, as pharmaceutical and agrichemical compounds and natural products and finds applications as a building block in organic chemistry and organocatalysis.[42–45] In this context, organothiosilanes and organoselenosilanes[46] are used as more efficient nucleophiles to transfer a sulfurated or selenated moiety with respect to the corresponding thiols or selenols for milder functionalization with electrophilic species.[47]

12.4.1 Reaction with Acetates and Ethers

Aryl- and alkyl-thiosilanes are suitable reagents in the nucleophilic substitution on substrates containing different functional groups, such as acetyloxy, ethers and silyl ethers, and halides, to provide a variety of unsymmetrical sulfides under suitable catalytic conditions.

InI_3 promoted substitution of acetoxy and siloxy groups by aryl- and alkyl-thiosilanes in alkyl acetates, including carbohydrate substrates, enabled the synthesis of a variety of thioethers, avoiding the use of metal thiolates (RSMet), which often led to several byproducts (Scheme 12.13).[48,49]

The substitution of an acetoxy group by a sulfurated moiety is extensively used in carbohydrate chemistry as thioglycosides represent versatile and widely used intermediates able to activate the anomeric position to further elaborations to obtain a plethora of more complex oligosaccharides. The synthesis of thioglycosides by treatment of glycosyl acetates with thiols is frequently employed, but often leads to undesired side reactions. The lower nucleophilicity of the silyl sulfide allowed a higher stereoselective attack on the anomeric carbon with respect to the related thiols. A huge number of examples are reported on the stereo-controlled replacement of the anomeric acetoxy group in reagents **16** by treatment with TMSSMe, usually promoted by Lewis acids, as the general example depicted in Scheme 12.14. The obtained methylthio derivatives **17** are widely used as intermediates to access oligosaccharides **18**.

An example of the solid-phase synthesis of oligosaccharides using thioglycosides has also been reported.[50] Numerous reactions of thiosilanes with carbohydrates were catalyzed by trimethylsilyl trifluoro-methanesulfonate (TMSOTf). Some examples are the preparation of *N*-acetylneuraminic acid thiomethylesters,[51–53] galactosides,[54,55] and methylthio-L-*glycero*-α-D-*manno*-heptopyranoside derivatives.[56] Likewise, fluorine-containing thioglycosides,[57] methylthio-substituted rhamnose derivatives,[58,59] and *S*-glucofuranosyl compounds were prepared with TMSSMe under TMSOTf catalysis.[60] Introduction of the trichloroacetimidate group on the anomeric position of a D-xylopyranosyl-derivative was favored by exploiting the synthetic flexibility of thioglicoside intermediates, obtained from TMSSMe/TMSOTf.[61] Catalysis of boron trifluoride etherate ($BF_3{\cdot}Et_2O$) was also efficient to transfer a –SMe group on β-D-xylopyranoside derivatives.[62]

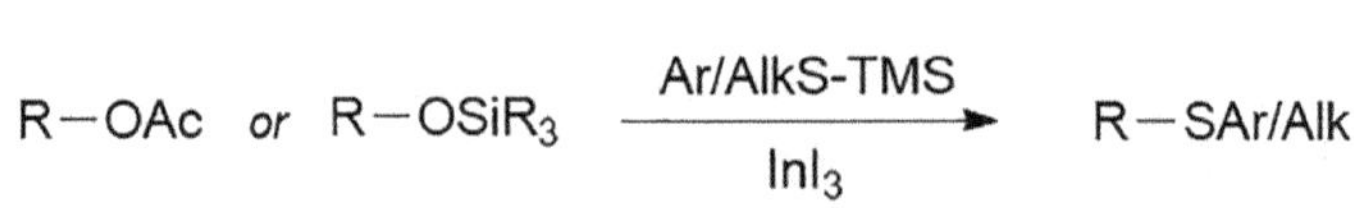

Scheme 12.13 In_3-promoted thiolation of acetoxy and siloxy groups.

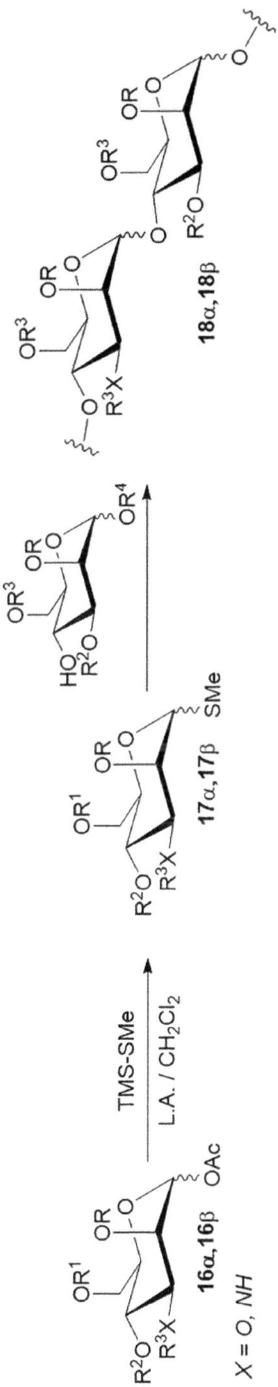

Scheme 12.14 Synthesis of thioglycosides as oligosaccharide precursors.

In this context selenoglycosides also represent important intermediates to provide carbohydrate derivatives. Selenoglycosidation was directly achieved by treatment of an amino sugar (galacto-, gluco-, and manno-derivatives) with phenylseleno(trimethylsilane) (PhSeTMS) and TMSOTf, leading to the corresponding selenides as precursors of α-*C*-glycosides (Scheme 12.15). When benzenselenol was reacted instead of the selenosilane, a two-step procedure was required (*via* oxazoline).[63]

Not only acetoxy, but also other groups were suitably substituted by thiosilanes. Methylthio-substituted thioglycosides from sialic acid were accomplished by selective nucleophilic substitution of a silyl-containing alkyl ether [$O(CH_2)_2TMS$, OSE] by TMSSMe/TfOTMS.[64,65] Similarly, the replacement of the anomeric –OSE group by reaction with TMSSMe was applied to prepare the thioglycoside of a 3-*O*-levulinoylated derivative.[66] The replacement of an *O*-allyl moiety by treatment with methylthiosilane allowed the introduction of a smaller methylthio group on α-L-rhamnopyranosyl derivatives, to examine their potential as acceptors for *Np*AS (*Neisseria polysaccharea*).[67]

In the search for the construction of the ABCD ring system of the more complex azaspiracid (a novel shellfish poison), a central step was the use of the thioacetal intermediate (Scheme 12.16), obtained through reaction of methyl acetal with TMSSMe in the presence of $Zn(OTf)_2$, followed by acidic treatment.[68]

Moreover, for the preparation of amino sugars starting from enantiopure 1,2-oxazines, PhSTMS was used to introduce a phenylthio group on a dioxolane ring to afford *syn*- and *anti*-isomers (Scheme 12.17). Likewise, the corresponding phenylseleno(trimethylsilane) led to the 2-phenylseleno-substituted dioxolane. Further elaboration of the substituted dioxolanes provided a stereodivergent synthesis of amino sugar derivatives, exploiting the suitable electronic properties of the chalcogenated groups (Scheme 12.17).[69]

High stereoselective ring-opening was applied to a α(1,4)-galactoside dioxolane by PhSTMS, under ZnI_2 catalysis, to provide a β-thioglycoside as the intermediate of Terpioside B.[70]

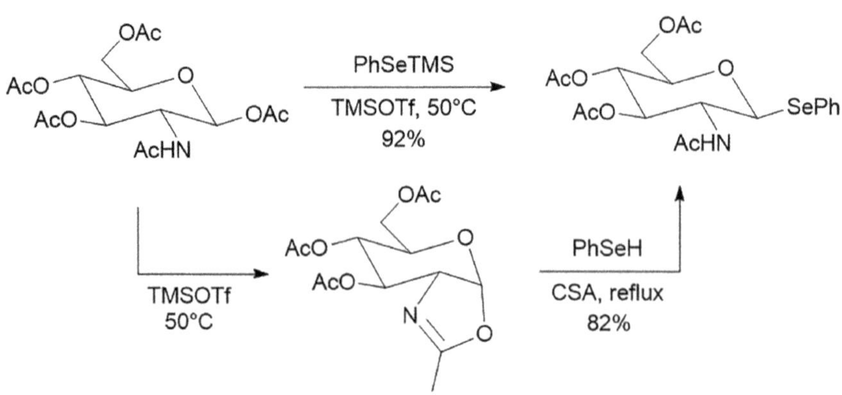

Scheme 12.15 Synthesis of phenylselenoglycosides.

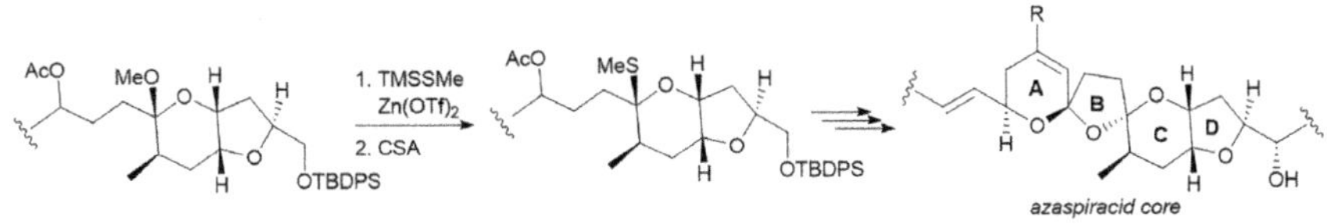

Scheme 12.16 Methylthioacetal as an intermediate in the synthesis of the azaspiracid ring system.

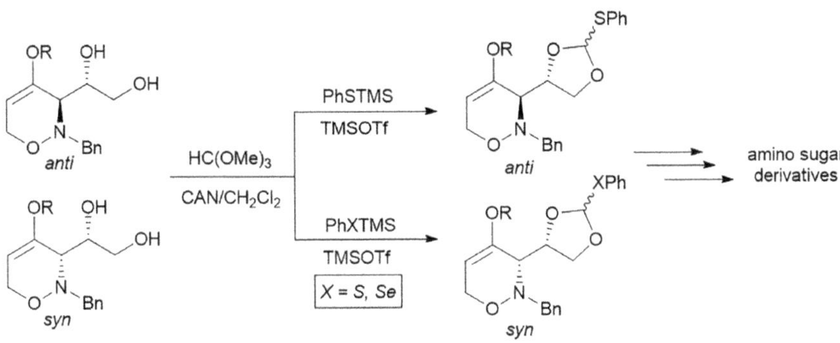

Scheme 12.17 Synthesis of 2-phenylthio- and phenyl-seleno substituted dioxolanes.

Phenylseleno(trimethylsilane) in combination with zinc chloride and cyclic ethers,[71] or aluminium tribromide and allylic alcohols,[72] was employed for the synthesis of exocyclic alkenes and selenochroman derivatives, respectively (Scheme 12.18).

12.4.2 Reaction with Acyl Chlorides

Selenosilanes were able to transfer the organo-selenyl moiety onto aryl– and alkyl–acyl chlorides. Depending on the type of silyl selenide, PhSeTMS or bis(trimethylsilyl)selenide [$(TMS)_2Se$, HMDSS] and on the stoichiometric ratio, selenolesters, selenoanhydrides, and diacyl diselenides were synthesized.[73] Acyl chlorides were also reacted at room temperature both with PhSeTMS and with different organoselenosilanes containing two –SeTMS groups, such as 1,4-TMSSe–C_6H_4–SeTMS, 4,4′-TMSSe–$(C_6H_4)_2$–SeTMS, and 1,1′-Fe(η^5-C_5H_4SeTMS)$_2$ to provide aromatic mono- and diselenoesters, showing that silylselenides behave as a convenient and reactive source of $ArSe^-$. With more hindered acyl chlorides higher temperature was necessary.[74]

12.4.3 Palladium-promoted Carbothiolations

Phenylthio- and alkylthio triisopropylsilanes (TIPS-SR) were successfully employed in the synthesis of alkyl–aryl and dialkyl sulfides *via* a palladium-catalyzed carbothiolation on a σ-alkyl Pd-intermediate (Scheme 12.19, left).[75] A palladium-promoted methylthiolation with TMSSMe was obtained

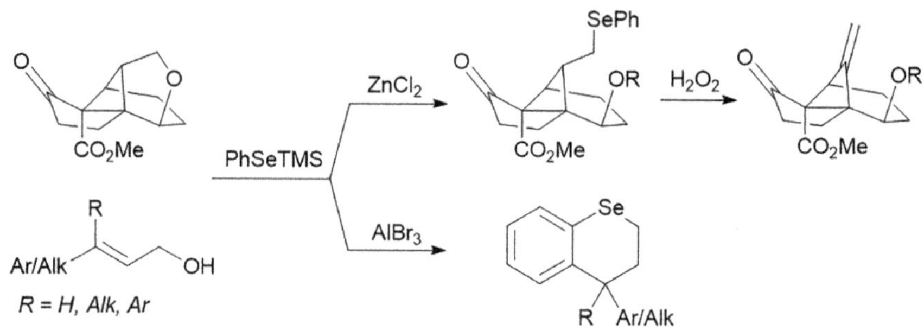

Scheme 12.18 PhSeTMS/L.A. combination for the synthesis of bridged bicyclic alkenes and selenochromans.

on the allylic carbonate in Scheme 12.19 (right) to prepare the corresponding thioethers, precursors of mannostatin A analogues.[76] The reaction with thiophenol failed, probably due to the high affinity of sulfur compounds with Pd, with subsequent poisoning of the catalyst. The use of a thio-derivative bearing a bulky protecting group (PhS-TIPS), unable to coordinate the Pd, was capable of overcoming this drawback.

12.4.4 Ring-opening of Heterocycles by Chalcogenosilanes

Thiosilanes were usefully employed in the ring-opening of strained heterocycles, as epoxides and aziridines, to access a variety of versatile building blocks to prepare more complex molecules. In the study for the synthesis of the amino sugar fragment of the lincosamide antibiotics, the epoxide of a glycal was reacted with TMSSMe/TMSOTf to afford the corresponding β-OTMS methyl sulfide with high regio- and stereo-control (Scheme 12.20, left).[77]

Phenylthio(trimethyl)silane was reacted with epoxides also under tetrabutylammonium phenoxide ($PhON^nBu_4$) catalysis to afford β-hydroxyalkyl phenyl sulfides in good yields.[78] A bifunctionalized silyl sulfide, namely bis(trimethylsilyl)sulfide (TMS–S–TMS, HMDST), found a broad application in the TBAF or $PhON^nBu_4$ catalyzed opening of differently substituted epoxides, allowing direct access to a wide range of β-mercaptoalcohols in a highly regio- and stereo-selective way.[78,79] Thiiranes also enabled access to 1,2-dithiols after treatment with $(TMS)_2S$ and TBAF.[80,81] Bis(trimethylsilyl)sulfide and (phenylthio)trimethysilane were also reacted with aziridines to achieve chiral enantioenriched 1,2-mercaptoamines and β-amino phenylsulfides through a regio- and stereo-selective nucleophilic substitution. The procedure was applied to *N*-Ts and *N*-Boc protected aziridines, as well as to unsubstituted derivatives.[82]

The nucleophilic substitution onto three membered heterocycles was also performed using selenosilanes.[83] Regio- and enantio-selective ring-opening of epoxides, thiiranes, and aziridines with bis(trimethylsilyl)selenide – $(TMS)_2Se$, HMDSS – afforded an entry to β-hydroxy, β-mercapto, and β-amino

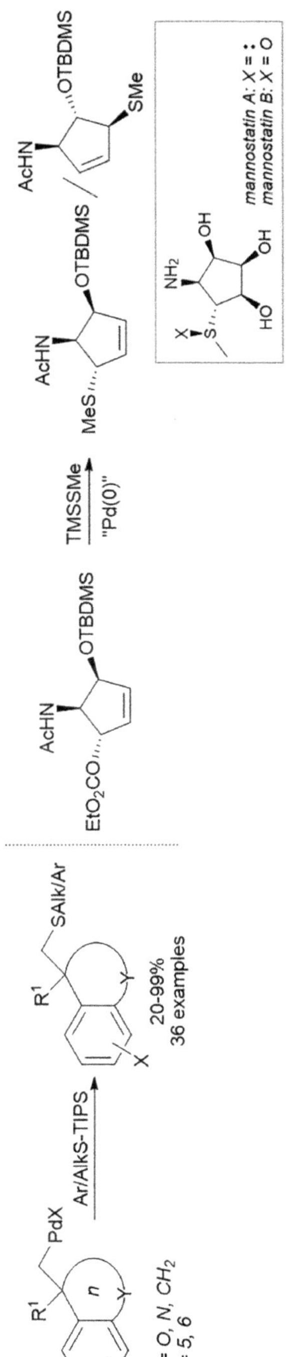

Scheme 12.19 Palladium-promoted methylthiolation.

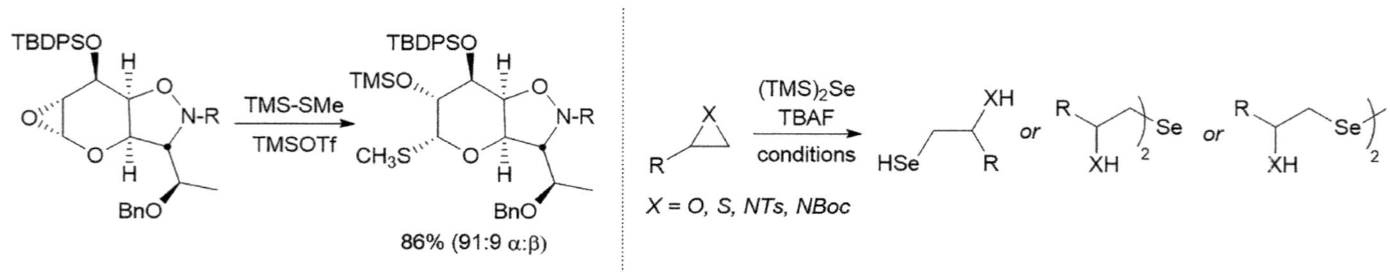

Scheme 12.20 Ring-opening of three-membered heterocycles.

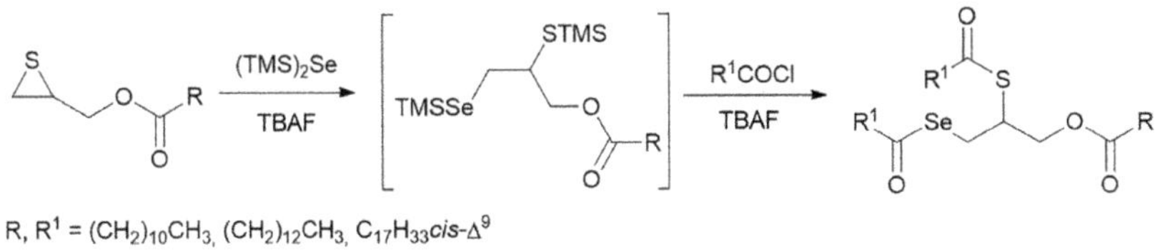

Scheme 12.21 Synthesis of sulfur- and selenium-isosters of triacyl glycerols.

diselenides and selenides, through tuning of the reaction conditions (Scheme 12.20, right).[84] The reaction carried out under strictly-controlled conditions (eq. of TBAF, time, *T*) allowed direct access to β-substituted alkyl selenols.[85] Ring-opening of suitable substituted thiiranes of glycidol by HMDSS led to sulfur- and selenium isosters of triacyl glycerols, exploiting the activation of the Si–chalcogen bonds of the chalcogenosilane intermediates for the following *in situ* reaction with fatty acid acyl chlorides (Scheme 12.21).[86]

Asymmetric ring-opening of *meso*-epoxides by phenyl(selenosilanes) under salen(Cr)complexes led to acyclic and cyclic β-hydroxy selenides (Scheme 12.22).[87] Likewise, desymmetrization of *meso*-azyridines was achieved with substituted silyl sulfides RSTMS (R = Me, Bn, CN) or silyl selenides ($PhSeSiR_2R^1$) under VAPOL-phosphate salt catalysis, leading to chiral β-amino chalcogenides (Scheme 12.22).[88,89]

12.4.5 Nucleophilic Substitution on Halogenated Compounds

Halogenated substrates also conveniently reacted with chalcogenosilanes. Echinomycin is a bicyclic octadecadepsipeptide bridged with a methylthioacetal moiety, with relevant biological properties (Scheme 12.23). A total synthesis was recently reported, and in the last step the S_N2 reaction of the chlorine to introduce the methylthio group was unsuccessful with NaSMe, while using TMSSMe, and $ZnCl_2$ as a promoter, allowed isolation of echinomycin.[90]

Thiosilanes also behaved as efficient nucleophiles to access thio-derivatives of tribenzotriquinacenes, useful rigid building blocks in molecular architecture with C_3 symmetry. The substitution of the bromine was achieved by reacting alkyl(silyl) sulfides instead of the corresponding thiols.[91] Selenosilanes [RSeTMS, $(TMS)_2Se$] were also used to synthesize cyclohepta-2,4,6-trienyl selenides by nucleophilic substitution from the corresponding tropylium bromide.[92]

12.4.6 Aromatic Nucleophilic Substitution with Organothiosilanes

TMSSMe and TMSSPh were involved in aromatic nucleophilic substitutions on properly activated aromatic substrates, as arenediazonium tetrafluoroborates to

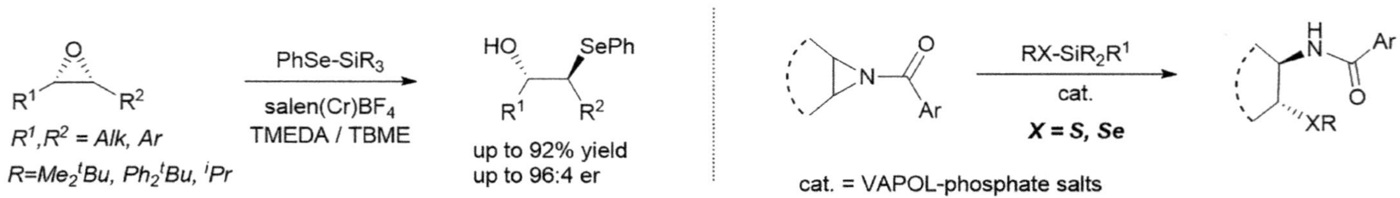

Scheme 12.22 Asymmetric ring-opening of meso-epoxides and -aziridines.

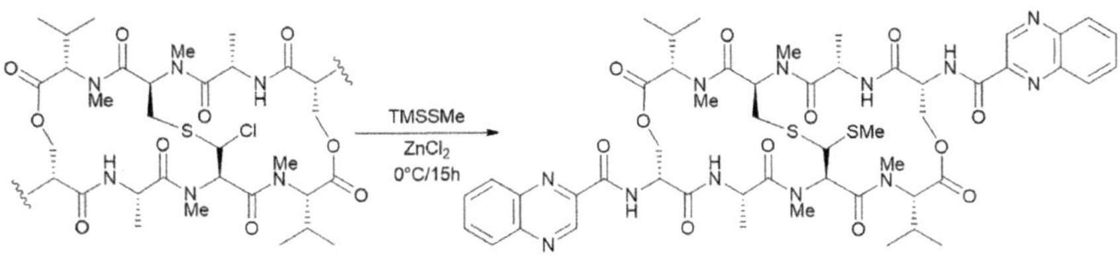

Scheme 12.23 Methylthiolation with TMSSMe in the total synthesis of echinomycin.

prepare aryl methyl- and aryl phenyl sulfides.[93] A combination of methylthio(trimethylsilane) and cesium carbonate was found to be a more efficient methylthiolation system in the reaction with nitroarenes than methanethiol or sodium methoxide (NaSMe) (Scheme 12.24, left).[94]

Silylthiolation of iodoaryl triflates with thiosilanes led to a variety of polyfunctionalized aromatic compounds (Scheme 12.24, right).[95,96] Likewise, activated aryl- and hetero-aryl halides were reacted with thiosilanes through functionalization of S–Si bonds under Cl/Br ion catalysis.[97] Nucleophilic substitution of an amino group was also achieved with TMSSMe/isoamyl nitrite on 2-amino thiazoles.[98]

12.4.7 Reaction with Carbonyl-containing Compounds

Addition reactions towards carbonyl substrates of alkyl- and aryl-thiosilanes is a well-known procedure for selective carbonyl protection under appropriate acid catalysis.[99] In this context, thiosilanes were involved in the dienone–phenol type rearrangement to synthesize aryl thioethers under acidic conditions (Scheme 12.25).[100] Conversely, the treatment of cyclohexadienone with thiols and Lewis acids gave the unsymmentrical sulfides in very poor yields, showing the advantages of using silyl sulfides.

A proposed reaction mechanism is illustrated in Scheme 12.26. Intermediate **19** would be formed by an unusual 1,2 addition of the thiosilane onto the C=O group. After elimination of silanol, the thionium ion intermediate **20** undergoes 1,2-migration of the alkyl group and aromatization to aryl thioether.

Addition of TMSSMe to a keto sugar was employed to prepare the corresponding thioketal in high yield, which was an intermediate for the total synthesis of shishijimicin A and its analogues.[101]

The treatment of different carbonyl compounds, as aldehydes, ketones, and acylsilanes with bis(trimethylsilyl)sulfide (HMDST) under $CoCl_2{\cdot}6H_2O$ or TMSOTf catalysis, allowed direct access to the corresponding thiocarbonyls, which can be trapped as Diels–Alder cycloadducts.[102] Alternative access to thioacylsilanes was obtained through the benzotriazole-mediated methodology.[102] Thionation of bis(acylsilanes) with HMDST provided sulfurated heterocycles with different ring sizes, depending on the spacers length between the C(=O)Si moieties.[102] The corresponding bis(trimethylsilyl)selenide

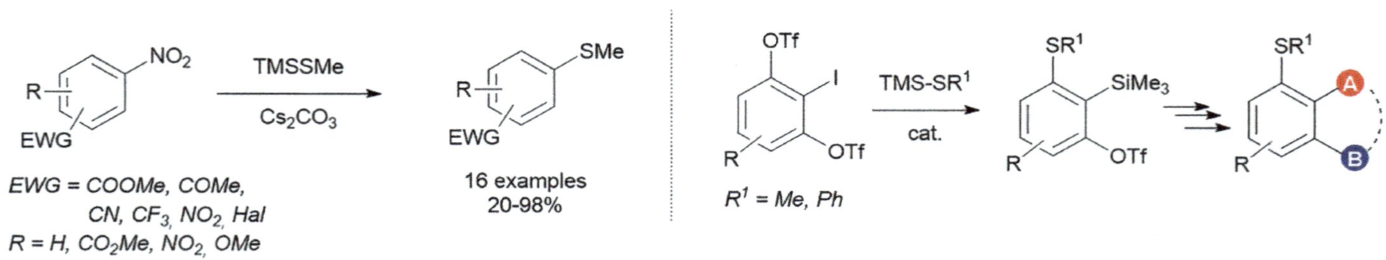

Scheme 12.24 Aromatic nucleophilic substitution with silyl sulfides.

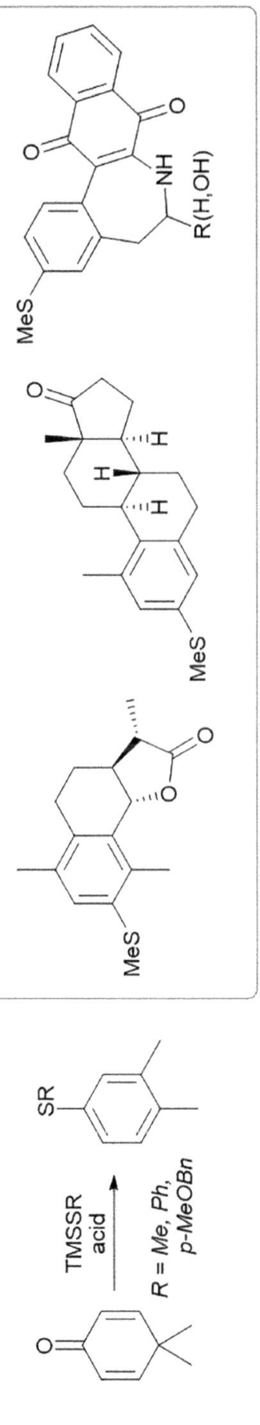

Scheme 12.25 Dienone–phenol type rearrangement *via* thiosilanes.

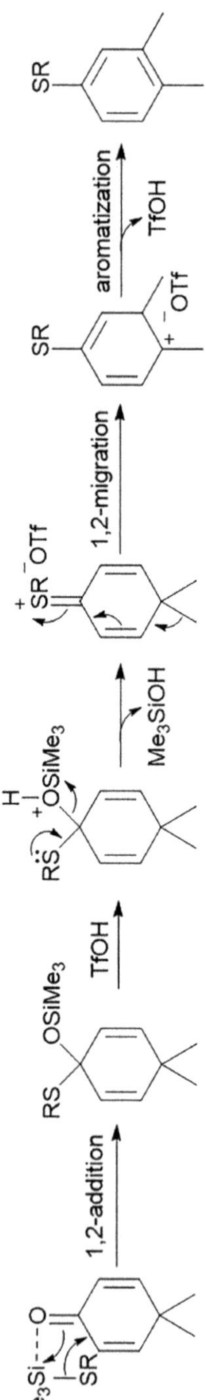

Scheme 12.26 Proposed reaction mechanism.

enabled the synthesis of alkyl and aryl selenoaldehydes, trapped as Diels–Alder adducts.[73]

Reaction of α,β-unsaturated carbonyl compounds with thiosilanes enabled synthesis of β-phenylthio silyl ethers as intermediates to access 3-substituted cyclohexenones using an umpolung strategy (Scheme 12.27, left).[103] Thia-Michael addition of the farnesylthio(triisopropylsilane) (Far-STIPS) onto an α,β-unsaturated carbonyl derivative of a peptide was employed in a convergent synthesis of peptide conjugates (Scheme 12.27, right).[104]

12.4.8 Reaction with Allenes and Alkynes

The reaction of bis(trimethylsilyl)sulfide with differently silylated and stannylated allenes promoted by $CoCl_2{\cdot}6H_2O$ provided, respectively, α,β-unsaturated thioacylsilanes and α,β-unsaturated thioacylstannanes as heterodienes/heterodienophiles (Scheme 12.28).[102]

Hydrothiolation of alkynes with alkyl thiosilanes led to stereoselective access to alkenyl sulfides. The use of thiosilanes instead of thiols allowed for further functionalization upon reaction with aldehydes to form new carbon–carbon bonds (Scheme 12.29).[105]

12.5 Species Containing the Chalcogen–Tin Bond

Trialkyl- and triaryl-stannyl sulfides and selenides (R_3SnSR', R_3SnSeR') are useful chalcogen nucleophiles that have found applications in organic synthesis, medicinal chemistry, and material science. These species can be directly prepared by reacting thiols and selenols with stannyl halides. However, the reactions of stannyl hydrides or bis-stannyl derivatives with disulfides and diselenides are valuable alternatives.[106–109] These latter opportunities represent a crucial advantage offered by R_3SnSR' and R_3SnSeR' reagents as often thiols, and above all selenols, are difficult to prepare and handle.

In this chapter we describe the principal applications based on the nucleophilic behaviour of triorganylstannyl sulfides and selenides.

12.5.1 Stannyl Sulfides and Selenides in Stille-like and Related Metal-catalyzed Cross-coupling Reactions

Trialkyl and tryarylstannyl sulfides and selenides have found several applications as chalcogen nucleophiles in Stille-like reactions with aryl and vinyl halides. After the very first report by Terada and others,[110] several modifications and upgrades have appeared. The insertion of a sulfur residue on an aromatic ring has been reported to be operative from aryl and heteroaryl iodides and bromides in the presence of a Pd(0) catalyst, like $Pd(PPh_3)_4$, in boiling toluene or xylenes.[111–123] The procedure was revealed to be also successful for the insertion of selenium aryl residues operating under pretty

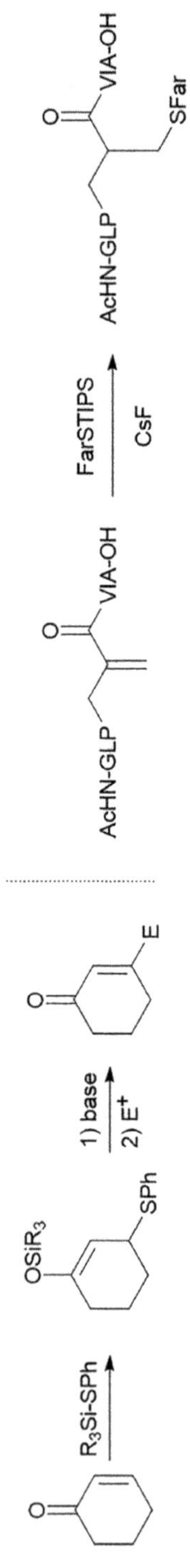

Scheme 12.27 Thia-Michael addition.

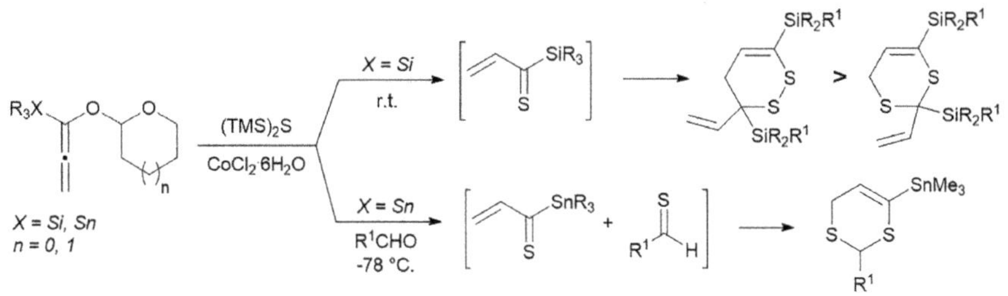

Scheme 12.28 Reaction of chalcogenosilanes with allenes.

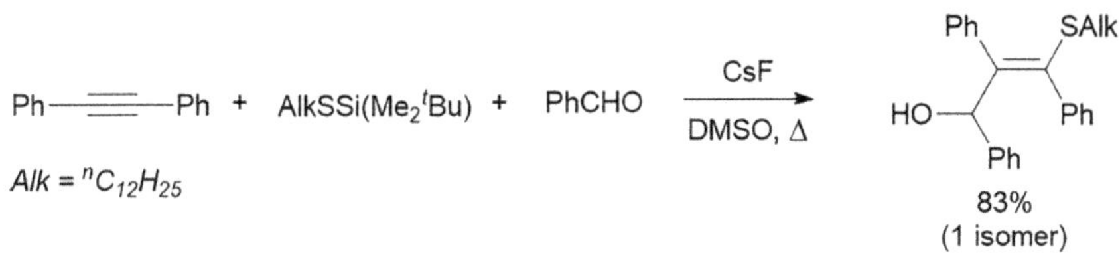

Scheme 12.29 Hydrothiolation of alkynes with silyl sulfides.

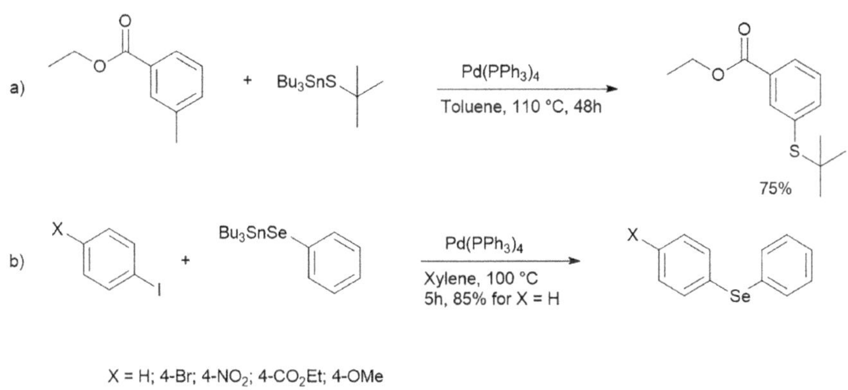

Scheme 12.30 Stille-like reactions of stannyl sulfides and selenides on aryl electrophiles.

similar conditions.[107,120,124,125] As an example, in Scheme 12.30 path a, is shown the reaction of 3-iodo ethylbenzoate with tributyl(*tert*-butylthio)stannane to give a sulfide used for the preparation of chemiluminescent probes useful for the detection of cholinesterase activity.[112] Under very similar conditions aryl iodides and tributyl(phenylseleno)stannane react to allow the preparation of unsymmetrical diaryl selenides [Scheme 12.30, path b].[124]

The methodology tolerates a variety of substituents on the aromatic ring and, at least in the sulfur series for the first decade of this century, has been used for the introduction of *S*-alkyl and *S*-aryl moieties on derivatives with applications in medicinal chemistry.[111–115] More recently, probably due to the common concerns about toxicity of stannyl derivatives, this reaction has found many more applications for the construction of sulfides and selenides exploited in material science.[116–125] As a matter of fact, many different variations and upgrades have been settled since to facilitate and increase the range of this chalcogen version of the Stille reaction. Thus, it has been reported that thio-functionalization can be promoted working under microwave irradiation,[126] while insertion of arylselenyl residues has been carried out using copper (CuI) and nickel ($NiBr_2$) catalysts in the presence of suitable ligands, either under classical heating[127] or microwave irradiation.[128]

Moreover, it has been demonstrated that the Pd-catalyzed insertion of selenyl residues can be carried out using ionic liquids as solvents,[129] using polyfluoroarenes as substrates,[130] and aryl triflates as electrophiles.[131] Thus, the synthesis of aryl and heteroaryl sulfides and selenides using stannyl chalcogenides as nucleophiles is a valuable and flexible procedure that allows the use of thiols/thiolates and selenols/selenolates to be avoided. Together with aryl halides and triflates, metal catalyzed cross-couplings allowed the insertion of alkyl- and arylthiol-residues on vinyl halides.[132–136] Thus, for example, the insertion of different *S*-alky and *S*-aryl groups on the six-membered ring of a cefalosporin skeleton, showing activity as a β-lactamase inhibitor, has been carried out with *S*-arylstannanes under $Pd_2(dba)_3$ catalysis as depicted in Scheme 12.31.[134,135]

Trialkylthiostannanes have been used under Pt catalysis in a three component reaction for the *ortho*-vinylation of aryl halides in the presence of terminal alkynes.[137,138] Different substituted aryl and heteroaryl iodides can be used as electrophiles to give the corresponding trisubstituted olefin as depicted in Scheme 12.32.

Due to the difficulty to prepare and store selenols and selenolates, the use of selenostannanes as suitable nucleophiles in metal-catalyzed cross-couplings has been also applied for the preparation of selenolesters.[139–142] These derivatives can be obtained *via* a three component process by reacting an aryl halide with a trialkylseleno stananne under a positive (5 atm) CO pressure in refluxing toluene using $Pd(PPh_3)_4$ as the catalyst (Scheme 12.33).[139] Selenolesters can be, even more directly, obtained reacting a selenostannane with an acyl chloride and $Pd(PPh_3)_4$ under conditions similar to those reported in

Scheme 12.31 Stille-like reactions of stannyl sulfides on vinyl electrophiles.

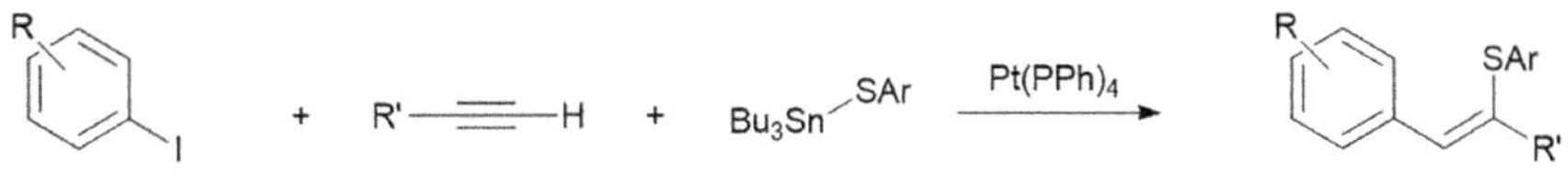

Scheme 12.32 Three component vinylation with stannyl sulfides and terminal alkynes.

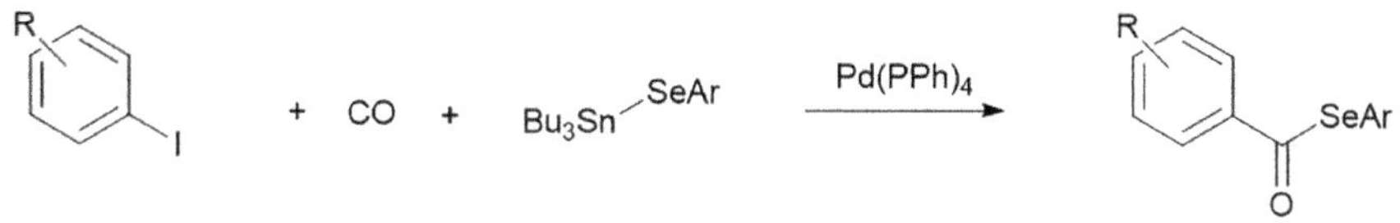

Scheme 12.33 Three component acylation of stannyl selenide with CO.

Scheme 12.31. Actually, literature data available for this transformation are a bit controversial, since it has been reported that metal catalysis is not necessary with acyl chlorides while it is required using anhydrides as electrophiles.[141,142]

The conditions revealed to be suitable for the introduction of seleny residues onto electrophilic sp^2 carbons are also suitable for the selenofunctionalization of electrophilic sp^3 carbons. The reactions of trialkylselenyl stannanes with α-bromo carbonyl compounds, as well as with propargylic, allylic, and benzylic bromides occur quite well affording the corresponding sulfides.[140,143] Using propargylic, allylic, and benzylic bromides as electrophiles, palladium pincer-complexes have often been the catalyst of choice since to improve the final yield of the reactions.[144]

12.5.2 Other Nucleophilic Processes Involving Stannyl Sulfides and Selenides

When making thioglycosides, the high nucleophilicity of alkanethiolates in the presence of Lewis acids can lead to undesirable side reactions. Consequently, tributylstannyl (and trimethylsilyl, *vide* **12.4.1**) derivatives of thiols are employed to reduce the nucleophilicity of the thiol reagent. Indeed, the insertion of an *S*-alkyl or *S*-aryl group on a sugar anomeric carbon can be carried out with the corresponding tribultylstannyl sulfide as the glycosyl acceptor, an anomeric bromide or fluoride as the glycosyl donor in the presence of tin(IV) chloride,[145] or $BF_3{\cdot}OEt_2$ as promoters.[146] The control of the stereochemistry at anomeric carbon depends upon the reaction conditions used that can allow the complete inversion during glycosylation as depicted in Scheme 12.34.[146]

A peculiar example of the peculiar nucleophilic character of stannyl sulfides is illustrated with the behaviour of the S–Sn Lewis pair in the ring-opening polymerization of α-amino acid *N*-carboxyanhydrides that allows rapid and controlled formation of polypeptides with high molecular weight (Scheme 12.35).[109] Remarkably, in this process, the promoter phenylthiotrimethyl stannane is photochemically generated *in situ* from

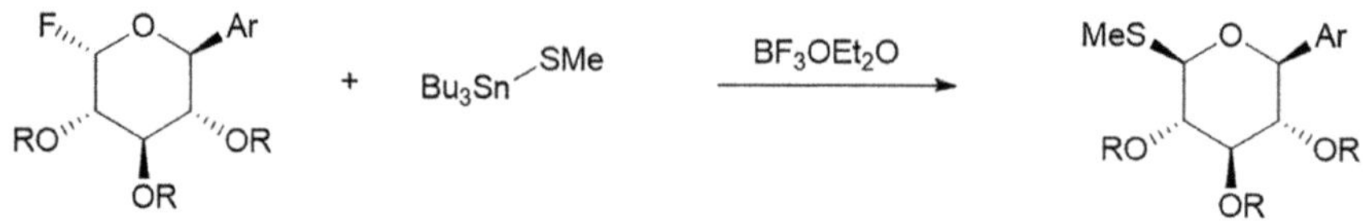

Scheme 12.34 Thioglycosylation using stannyl sulfides.

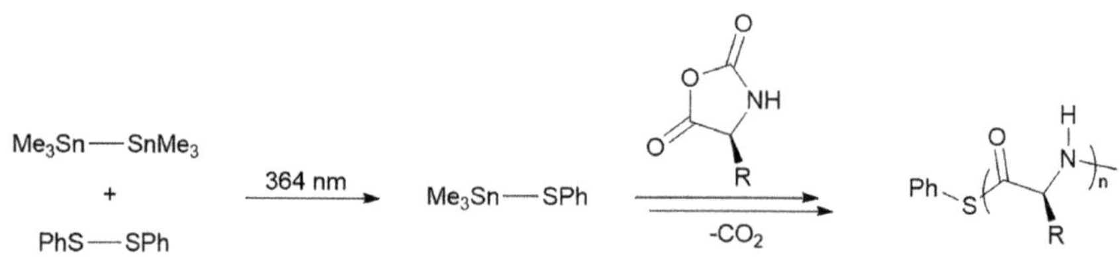

Scheme 12.35 Trimethylstannyl sulfides as polymerization promoters.

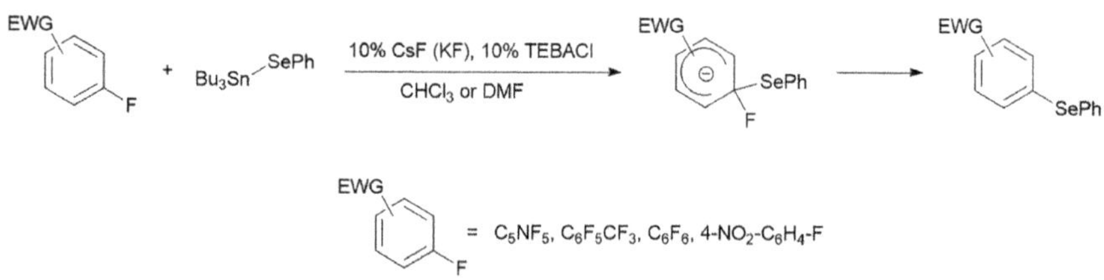

Scheme 12.36 S_NAr reactions using stannyl selenides as nucleophiles.

diphenyl disulfide and hexadimethyldistannane under polymerization conditions.

Another remarkable opportunity of using arylselenylstannane as synthetic equivalents of the corresponding selenols and selenolates is the preparation of unsymmetrical diaryl selenides of very electron-poor arenes *via* an aromatic nucleophilic substitution process.[147] Depending upon the substitution pattern of the aryl fluoride used as the electrophile, the reaction requires the use of sub-stoichiometry amounts of alkaline fluorides (KF or CsF) and/or triethyl benzyl ammonium chloride (TEBACl) (Scheme 12.36).[147]

Eventually, either tributylstannyl sulfides[148] or selenides,[149] react with arynes, generated from *ortho*-trimethylsilyl triflates, KF and 18-crown-6, to give the corresponding *ortho*-tributylstannyl aryl sulfides and selenides (Scheme 12.37). The proposed mechanism foresees the nucleophilic attack of the stannyl chalcogen to the aryne with formation of a zwitterionic sulfonium

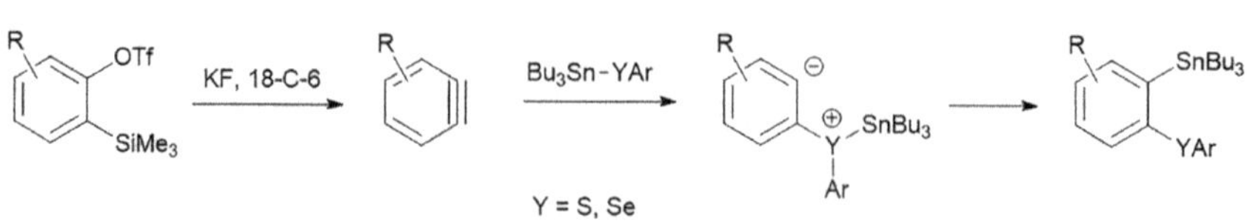

Scheme 12.37 Tributylstannyl sulfide and selenide reaction with arynes.

(or selenonium) ion that evolves to the final compound, reasonably, by an intramolecular attack of the aryl nucleophilic carbon to the tryalkyltin moiety.

Acknowledgements

We thank MIUR-Italy ("Progetto Dipartimenti di Eccellenza 2018–2022" allocated to the Department of Chemistry "Ugo Schiff").

References

1. P. Metzner and A. Thuillier, *Sulfur reagents in Organic Synthesis*, Academic Press Limited, London, U.K., 1994, and references cited therein.
2. M. Iwaoka, in *Organoselenium Chemistry*, ed. T. Wirth, Wiley-WHC Verlag & Co, Weinheim, Germany, 2012, and references cited therein.
3. M. Miyashita, M. Hoshino, T. Suzuki and A. Yoshikoshi, *Chem. Lett.*, 1988, 507.
4. N. Osato, H. Onoue, Y. Toma, K. Torikai, M. Ebine, M. Satake and T. Oishi, *Chem. Lett.*, 2018, **47**, 265.
5. T. Ishiyama, K. Nishijima, N. Miyaura and A. Suzuki, *J. Am. Chem. Soc.*, 1993, **115**, 7219.
6. T. Kataoka, M. Yoshimatsu, Y. Noda, T. Sato, H. Shimizu and M. Hori, *J. Chem. Soc. Perkin Trans. 1*, 1993, 121.
7. A. Cravador and A. Krief, *Tetrahedron Lett.*, 1981, **22**, 2491.
8. D. L. J. Clive and S. M. Menchen, *J. Chem. Soc., Chem. Commun.*, 1978, 356.
9. C. Habben and D. E. Kaufmann, in *Science of Synthesis*, ed. D. E. Kaufmann and D. S. Matteson, George Thieme Verlag, 2005, vol. 6, p. 443.
10. M. Feng, B. Tang, S. H. Liang and X. Jiang, *Curr. Top. Med. Chem.*, 2016, **16**, 1200.
11. J. A. Fernández-Salas, S. Manzini and S. P. Nolan, *Chem. Commun.*, 2013, **49**, 5829.
12. Z. Yang, M. Zhong, X. Ma, K. Nijesh, S. De, P. Parameswaran and H. W. Roesky, *J. Am. Chem. Soc.*, 2016, **138**, 2548.
13. É. Rochette, H. Boutin and F.-G. Fontaine, *Organometallics*, 2017, **36**, 2870.
14. F. A. Romero, J. L. Peltier, R. Jazzar and G. Bertrand, *Chem. Commun.*, 2016, **52**, 10563.
15. M. G. Civit, X. Sanz, C. M. Vogels, J. D. Webb, S. J. Geier, A. Decken, C. Bo, S. A. Westcott and E. Fernández, *J. Org. Chem.*, 2015, **80**, 2148.
16. X. Sanz, C. M. Vogels, A. Decken, C. Bo, S. A. Westcott and E. Fernández, *Chem. Commun.*, 2014, **50**, 8420.
17. M. G. Civit, X. Sanz, C. M. Vogels, C. Bo, S. A. Westcott and E. Fernández, *Adv. Synth. Catal.*, 2015, **357**, 3098.
18. D. García-López, M. G. Civit, C. M. Vogels, J. M. Ricart, S. A. Westcott, E. Fernández and J. J. Carbó, *Catal. Sci. Technol.*, 2018, **8**, 3617.

19. M. G. Civit, J. Royes, C. M. Vogels, S. A. Westcott, A. B. Cuenca and E. Fernández, *Org. Lett.*, 2016, **18**, 3830.
20. Y. Zhang, X. Xu and S. Zhu, *Nat. Commun.*, 2019, **10**, 1752.
21. M. C. Murphy, A. Trofimova, J. H. W. LaFortune, C. M. Vogels, S. J. Geier, J. F. Binder, C. L. B. Macdonald, D. W. Stephan and S. A. Westcott, *Dalton Trans.*, 2020, **49**, 5092.
22. T. Ooi and K. Maruoka, in *Science of Synthesis*, ed. H. Yamamoto, George Thieme Verlag, 2004, vol. 7, p. 197.
23. T. Ooi and K. Maruoka, in *Science of Synthesis*, ed. H. Yamamoto, George Thieme Verlag, 2004, vol. 7, p. 215.
24. A. Yasuda, M. Takahashi and H. Takaya, *Tetrahedron Lett.*, 1981, **22**, 2413.
25. D. P. Martin, R. T. Bibart and D. G. Drueckhammer, *J. Am. Chem. Soc.*, 1994, **116**, 4660.
26. T. Murai, Y. Ogino, T. Mizutani, T. Kanda and S. Kato, *J. Org. Chem.*, 1995, **60**, 2942.
27. T. Inoue, T. Takeda, N. Kambe, A. Ogawa, I. Ryu and N. Sonoda, *J. Org. Chem.*, 1994, **59**, 5824.
28. Y. Nishiyama, T. Asano, Y. Kishimoto, K. Itoh and Y. Ishii, *Tetrahedron Lett.*, 1998, **39**, 8685.
29. J. Terao, S. A. Begum, Y. Shinohara, M. Tomita, Y. Naitoh and N. Kambe, *Chem. Commun.*, 2007, 855.
30. A. Itoh, S. Ozawa, K. Oshima and H. Nozaki, *Tetrahedron Lett.*, 1980, **21**, 361.
31. A. E. A. Hassan, N. Nishizono, N. Minakawa, S. Shuto and A. Matsuda, *J. Org. Chem.*, 1996, **61**, 6261.
32. Y. Masaki, Y. Serizawa, K. Nagata and H. Oda, *Tetrahedron Lett.*, 1986, **27**, 231.
33. Y. Nishiyama, S. Nakaya and S. Hamanaka, *Chem. Lett.*, 1991, 1775.
34. T. Yamamoto, E. Moriura, A. Sawa and M. Yoshimatsu, *Chem. Lett.*, 2008, **37**, 1046.
35. S. Yamashita, K. Takeuchi, T. Koyama, M. Inoue, Y. Hayashi and M. Hirama, *Chem. – Eur. J.*, 2015, **21**, 2621.
36. M. Sasaki, T. Noguchi and K. Tachibana, *J. Org. Chem.*, 2002, **67**, 3301.
37. H. Imai, H. Uehara, M. Inoue, H. Oguri, T. Oishi and M. Hirama, *Tetrahedron Lett.*, 2001, **42**, 6219.
38. M. Segi, T. Takahashi, H. Ichinose, M. G. Li and T. Nakajima, *Tetrahedron Lett.*, 1992, **33**, 7865.
39. G. M. Li, S. Niu, M. Segi, R. A. Zingaro, H. Yamamoto, K. Watanabe, T. Nakajima and M. B. Hall, *J. Org. Chem.*, 1999, **64**, 1565.
40. M. Segi, A. Zhou and M. Honda, *Phosphorus, Sulfur Silicon Relat. Elem.*, 2005, **180**, 1045.
41. M. Segi, K. Kawaai, M. Honda and S. Fujinami, *Tetrahedron Lett.*, 2007, **48**, 3349.
42. H. Chen, W. Jiang and Q. Zeng, *Chem. Rec.*, 2020, **20**, 1269.
43. C. T. Barce Ferro, B. F. dos Santos, C. D. G. da Silva, G. Brand, B. A. L. da Silva and N. L. de Campos Domingues, *Curr. Org. Synth.*, 2020, **17**, 192.

44. P. Devendar and G.-F. Yang, *Top. Curr. Chem.*, 2017, **375**, 82.
45. E. M. McGarrigle, E. L. Myers, O. Illa, M. A. Shaw, S. L. Riches and V. K. Aggarwal, *Chem. Rev.*, 2007, **107**, 5841.
46. D. A. Armitage, *The Chemistry of Organic Silicon Compounds*, John Wiley & Sons, Hoboken (USA), 1998, vol. 2.
47. J. Gawronski, N. Wascinska and J. Gajewy, *Chem. Rev.*, 2008, **108**, 5227, and references therein.
48. Y. Nishimoto, A. Okita, M. Yasuda and A. Baba, *Org. Lett.*, 2012, **14**, 1846.
49. Y. Nishimoto, A. Okita, A. Baba and M. Yasuda, *Molecules*, 2016, **21**, 1330.
50. T. Kanemitsu, C.-H. Wong and O. Kanie, *J. Am. Chem. Soc.*, 2002, **124**, 3591.
51. M. Tsuji, N. Sriwilaijaroen, H. Inoue, K. Miki, K. Kinoshita, K. Koyama, K. Furuhata, Y. Suzuki and K. Takahashi, *Bioorg. Med. Chem.*, 2018, **26**, 17.
52. S. V. Shelke, G.-P. Gao, S. Mesch, H. Gäthje, S. Kelm, O. Schwardt and B. Ernst, *Bioorg. Med. Chem.*, 2007, **15**, 4951.
53. L. O. Kononov, A. M. Shpirt, Y. Ito and T. Ogawa, *Russ. Chem. Bull.*, 2003, **52**, 1442.
54. S. Bhattacharyya, B. G. Magnusson, U. Wellmar and U. J. Nilsson, *J. Chem. Soc., Perkin Trans. 1*, 2001, 886.
55. H. van Hattum, H. M. Branderhorst, E. E. Moret, U. J. Nilsson, H. Leffler and R. J. Pieters, *J. Med. Chem.*, 2013, **56**, 1350.
56. K. Mannerstedt, E. Segerstedt, J. Olsson and S. Oscarson, *Org. Biomol. Chem.*, 2008, **6**, 1087.
57. A. Tatami, Y.-S. Hon, I. Matsuo, M. Takatani, H. Koshino and Y. Ito, *Biochem. Biophys. Res. Commun.*, 2007, **364**, 332.
58. R. Suhr, P. Pfefferkorn, S. Weingarten and J. Thiem, *Org. Biomol. Chem.*, 2003, **1**, 4373.
59. B. Nolting, H. Boye and C. Vogel, *J. Carbohydr. Chem.*, 2001, **20**, 585.
60. R. H. Furneaux, B. Martin, P. M. Rendle and C. M. Taylor, *Carbohydr. Res.*, 2002, **337**, 1999.
61. R. Suhr and J. Thiem, *J. Carbohydr. Chem.*, 2004, **23**, 261.
62. M. Kishida, M. Nishiuchi, K. Kato and H. Akita, *Chem. Pharm. Bull.*, 2004, **52**, 1105.
63. L. Grant, Y. Liu, K. E. Walsh, D. S. Walter and T. Gallagher, *Org. Lett.*, 2002, **4**, 4623.
64. M. Yamaguchi, H. Ishida and M. Kiso, *Heterocycles*, 2009, **77**, 333.
65. A. V. Demchenko and G.-J. Boons, *J. Org. Chem.*, 2001, **66**, 2547.
66. K. Hori, N. Sawada, H. Ando, H. Ishida and M. Kiso, *Eur. J. Org. Chem.*, 2003, 3752.
67. S. Salamone, C. Guerreiro, E. Cambon, J. Hargreaves, N. Tarrat, M. Remaud-Siméon, I. André and L. A. Mulard, *J. Org. Chem.*, 2015, **80**, 11237.
68. Y. Ishikawa and S. Nishiyama, *Heterocycles*, 2004, **63**, 539.

69. F. Pfrengle and H.-U. Reissig, *Chem. – Eur. J.*, 2010, **16**, 11915.
70. K. Inaba, M. Endo, N. Iibuchi, D. Takahashi and K. Toshima, *Chem. – Eur. J.*, 2020, **26**, 10222.
71. M. T. Crimmins and E. B. Hauser, *Org. Lett.*, 2000, **2**, 281.
72. H. Abe, A. Yamasaki, N. Koshiba, Y. Takeuchi and T. Harayama, *Chem. Pharm. Bull.*, 2001, **49**, 1223.
73. A. Capperucci, A. Degl'Innocenti and C. Tiberi, *Synlett*, 2011, 2248, and references therein.
74. D. Taher and J. F. Corrigan, *Organometallics*, 2011, **30**, 5943.
75. Y. Hosoya, I. Kobayashi, K. Mizoguchi and M. Nakada, *Org. Lett.*, 2019, **21**, 8280.
76. S. J. Cho, R. Ling, A. Kim and P. S. Mariano, *J. Org. Chem.*, 2000, **65**, 1574.
77. M. J. Mitcheltree, J. W. Stevenson, A. Pisipati and A. G. Myers, *J. Am. Chem. Soc.*, 2021, **143**, 6829.
78. A. Capperucci, C. Tiberi, S. Pollicino and A. Degl'Innocenti, *Tetrahedron Lett.*, 2009, **50**, 2808.
79. A. Degl'Innocenti, A. Capperucci, A. Cerreti, S. Pollicino, S. Scapecchi, I. Malesci and G. Castagnoli, *Synlett*, 2005, 3063.
80. A. Degl'Innocenti, S. Pollicino and A. Capperucci, *Chem. Commun.*, 2006, 4881.
81. A. Capperucci, D. Tanini, C. Borgogni and A. Degl'Innocenti, *Heteroat. Chem.*, 2014, **25**, 678.
82. D. Tanini, C. Borgogni and A. Capperucci, *New J. Chem.*, 2019, **43**, 6388.
83. D. Tanini and A. Capperucci, *New J. Chem.*, 2019, **43**, 11451.
84. D. Tanini, A. Degl'Innocenti and A. Capperucci, *Eur. J. Org. Chem.*, 2015, 357.
85. D. Tanini, C. Tiberi, C. Gellini, P. R. Salvi and A. Capperucci, *Adv. Synth. Catal.*, 2018, **360**, 3367.
86. D. Tanini, V. D'Esopo, D. Tatini, M. Ambrosi, P. Lo Nostro and A. Capperucci, *Chem. – Eur. J.*, 2020, **26**, 2719.
87. M. Tiecco, L. Testaferri, F. Marini, S. Sternativo, F. Del Verme, C. Santi, L. Bagnoli and A. Temperini, *Tetrahedron*, 2008, **64**, 3337.
88. M. Senatore, A. Lattanzi, S. Santoro, C. Santi and G. Della Sala, *Org. Biomol. Chem.*, 2011, **9**, 6205.
89. G. Della Sala, *Tetrahedron*, 2013, **69**, 50.
90. K. Kojima, F. Yakushiji, A. Katsuyama and S. Ichikawa, *Org. Lett.*, 2020, **22**, 4217.
91. D. Kuck, A. Schuster, R. A. Krause, J. Tallenbröker, C. P. Exner, M. Penk, H. Bögge and A. Müller, *Tetrahedron*, 2001, **57**, 3587.
92. B. Wrackmeyer, Z. Garcìa Hernàndez and M. Herberhold, *Magn. Reson. Chem.*, 2007, **45**, 187.
93. G. K. Surya Prakash, D. Hoole, D. S. Ha, J. Wilkinson and G. A. Olah, *Arkivoc*, 2002, **xiii**, 50.
94. Q. Qiao, R. Dominique, A. Sidduri, J. Lou and R. A. Goodnow Jr, *Synth. Commun.*, 2010, **40**, 3691.

95. Y. Nakamura, Y. Miyata, K. Uchida, S. Yoshida and T. Hosoya, *Org. Lett.*, 2019, **21**, 5252.
96. Y. Nakamura, Y. Sakata, T. Hosoya and S. Yoshida, *Org. Lett.*, 2020, **22**, 8505.
97. X. Jia, L. Yu, J. Liu, Q. Xu, M. Sickert, L. Chena and M. Lautens, *Green Chem.*, 2014, **16**, 3444.
98. Q. Dang, Y. Liu, D. K. Cashion, S. R. Kasibhatla, T. Jiang, F. Taplin, J. D. Jacintho, H. Li, Z. Sun, Y. Fan, J. Da Re, F. Tian, W. Li, T. Gibson, R. Lemus, P. D. van Poelje, S. C. Potter and M. D. Erion, *J. Med. Chem.*, 2011, **54**, 153.
99. D. A. Evans, L. K. Truesdale, K. G. Grimm and S. L. Nesbitt, *J. Am. Chem. Soc.*, 1977, **99**, 5009.
100. Y. Wada, K. Otani, N. Endo, Y. Kita and H. Fujioka, *Chem. Commun.*, 2010, **46**, 797.
101. K. C. Nicolaou, R. Li, Z. Lu, E. N. Pitsinos, L. B. Alemany, M. Aujay, C. Lee, J. Sandoval and J. Gavrilyuk, *J. Am. Chem. Soc.*, 2018, **140**, 12120.
102. A. Degl'Innocenti, A. Capperucci, G. Castagnoli and I. Malesci, *Synlett*, 2005, 1965, and references cited therein.
103. H. Lechuga-Eduardo, E. Zarza-Acuña and M. Romero-Ortega, *Tetrahedron Lett.*, 2017, **58**, 3234.
104. Y. Zhu and W. A. van der Donk, *Org. Lett.*, 2001, **3**, 1189.
105. A. Kondoh, K. Takami, H. Yorimitsu and K. Oshima, *J. Org. Chem.*, 2005, **70**, 6468.
106. L. Ford, U. Wille and C. H. Schiesser, *Helv. Chim. Acta*, 2006, **89**, 2306.
107. M. Bonaterra, S. E. Martin and R. A. Rossi, *Tetrahedron Lett.*, 2006, **47**, 3511.
108. M. Maheswara, M. Kim, S.-J. Yun, J. J. Ju and J. Y. Do, *Tetrahedron Lett.*, 2009, **50**, 480.
109. J. Yuan, Y. Zhang, Z. Li, Y. Wang and H. Lu, *ACS Macro Lett.*, 2018, **7**, 892.
110. M. Terada, T. Migata, H. Sano, T. Ogata and M. Kosagi, *Bull. Chem. Soc. Jpn.*, 1985, **58**, 3657.
111. M. J. Brown, P. S. Carter, A. E. Fenwick, A. P. Fosberry, D. W. Hamprecht, M. J. Hibbs, R. L. Jarvest, L. Mensah, P. H. Milner, P. J. O'Hanlon, A. J. Pope, C. M. Richardson, A. West and D. R. Witty, *Bioorg. Med. Chem. Lett.*, 2002, **12**, 3171.
112. S. Sabelle, P.-Y. Renard, K. Pecorella, S. de Suzzoni-Dezard, C. Creminon, J. Grassi and C. Mioskowski, *J. Am. Chem. Soc.*, 2002, **124**, 4874.
113. S. Sabelle, J. Hydrio, E. Leclerc, C. Mioskowski and P.-Y. Renard, *Tetrahedron Lett.*, 2002, **43**, 3645.
114. A. Benjahad, S. Oumouch, J. Guillemont, E. Pasquier, D. Mabire, K. Andries, C. H. Nguyena and D. S. Grierson, *Bioorg. Med. Chem. Lett.*, 2007, **17**, 712.
115. C. Lunniss, C. Eldred, N. Aston, A. Craven, K. Gohil, B. Judkins, S. Keeling, L. Ranshaw, E. Robinson, T. Shipley and N. Trivedi, *Bioorg. Med. Chem. Lett.*, 2010, **20**, 137.

116. B. Chesneau, A. Passelande and P. Hudhomme, *Org. Lett.*, 2009, **11**, 649.
117. L. Preti, R. Cagnoli, F. Parenti, M. Lanzi, E. Libertini, D. Pinetti, L. Schenetti and A. Mucci, *Synthesis*, 2010, 1659.
118. H. Kamo, A. Ueda, T. Isono, K. Takahashi and H. Mori, *Tetrahedron Lett.*, 2012, **53**, 4385.
119. T. Biet, A. Fihey, T. Cauchy, N. Vanthuyne, C. Roussel, J. Crassous and N. Avarvari, *Chem. – Eur. J.*, 2013, **19**, 13160.
120. E. Palao, T. Slanina and P. Klan, *Chem. Commun.*, 2016, **52**, 11951.
121. M. Murata, S. Kaji, H. Nishimura, A. Wakamiya and Y. Murata, *Eur. J. Inorg. Chem.*, 2016, 3228.
122. A. Ueda and H. Mori, *Mater. Chem. Front*, 2018, **2**, 566.
123. J. Grolleau, S. Legoupy, P. Frère and F. Gohier, *Polymer*, 2019, **182**, 121811.
124. I. P. Beletskaya, A. S. Sigeev, A. S. Peregudov and P. V. Petrovskii, *J. Organomet. Chem.*, 2000, **605**, 96.
125. A. Ueda, H. Kamo and H. Mori, *Chem. Lett.*, 2015, **44**, 1538.
126. L. Cai, J. Cuevas, Y.-Y. Peng and V. W. Pike, *Tetrahedron Lett.*, 2006, **47**, 4449.
127. I. P. Beletskaya, A. S. Sigeev, A. S. Peregudov and P. V. Petrovskii, *Tetrahedron Lett.*, 2003, **44**, 7039.
128. I. P. Beletskaya, A. S. Sigeev, A. S. Peregudov, P. V. Petrovskii and V. N. Khrustalev, *Chem. Lett.*, 2010, **39**, 720.
129. H. Zhao, W. Hao, Z. Xia and M. Cai, *New J. Chem.*, 2011, **35**, 2661.
130. M. N. Lanteri, R. A. Rossi and S. E. Martín, *J. Organomet. Chem.*, 2009, **694**, 3425.
131. M. Bonaterra, R. A. Rossi and S. E. Martin, *Organometallics*, 2009, **28**, 933.
132. A. Carpita, R. Rossi and B. Scaramuzzi, *Tetrahedron Lett.*, 1989, **30**, 2699.
133. L. S. Liebeskind and J. Wang, *Tetrahedron Lett.*, 1990, **31**, 4293.
134. J. D. Buynak, L. Vogeti and H. Chen, *Org. Lett.*, 2001, **3**, 2953.
135. J. D. Buynak, L. Vogeti, V. R. Doppalapudi, G. M. Solomon and H. Chen, *Bioorg. Med. Chem. Lett.*, 2002, **12**, 1663.
136. D. R. Boyd, N. D. Sharma, B. E. Byrne, S. A. Haughey, M. A. Kennedy and C. C. R. Allenc, *Org. Biomol. Chem.*, 2004, **2**, 2530.
137. T. Hirai, H. Kuniyasu and N. Kambe, *Tetrahedron Lett.*, 2005, **46**, 117.
138. H. Kuniyasu, F. Yamashita, T. Hirai, J.-H. Ye, S. Fujiwara and N. Kambe, *Organometallics*, 2006, **25**, 566.
139. Y. Nishiyama, K. Tokunaga, H. Kawamatsu and N. Sonoda, *Tetrahedron Lett.*, 2002, **43**, 1507.
140. Y. Nishiyama, H. Kawamatsu, S. Funato, K. Tokunaga and N. Sonoda, *J. Org. Chem.*, 2003, **68**, 3599.
141. I. P. Beletskaya, A. S. Sigeev, A. S. Peregudov and P. V. Petrovskii, *Russ. J. Org. Chem.*, 2001, **37**, 1703.
142. I. P. Beletskaya, A. S. Sigeev, A. S. Peregudov and P. V. Petrovskii, *Mendeleev Commun.*, 2000, **10**, 127.

143. I. P. Beletskaya, A. S. Sigeev, A. S. Peregudov and P. V. Petrovskii, *Chem. Lett.*, 2005, **34**, 1348.
144. O. A. Wallner and K. J. Szabo, *J. Org. Chem.*, 2005, **70**, 9215.
145. G. Marnera and M. d'Alarcao, *Carbohydrate Res.*, 2006, **341**, 1105.
146. X. Zhou, H. Ding, P. Chen, L. Liu, Q. Sun, X. Wang, P. Wang, Z. Lv and M. Li, *Angew. Chem., Int. Ed.*, 2020, **59**, 4138.
147. I. P. Beletskaya, A. S. Sigeev, A. S. Peregudov and P. V. Petrovskii, *Mendeleev Commun.*, 2000, **10**, 213.
148. H. Yoshida, T. Terayama, J. Ohshita and A. Kunai, *Chem. Commun.*, 2004, 1980.
149. F. T. Toledo, J. V. Comasseto and C. Raminelli, *J. Braz. Chem. Soc.*, 2010, **21**, 2164.

CHAPTER 13

The Chemistry of Selenium-catalyzed Redox Processes

THOMAS G. BACK

Department of Chemistry, University of Calgary, Calgary, Alberta T2N 1N4, Canada
Email: tgback@ucalgary.ca

13.1 Introduction

The element selenium has provided a fertile field of study, stemming from its many roles and applications in biology, materials chemistry, and synthesis.[1] The organic chemistry of selenium is quite distinct from that of sulfur and tellurium, and often leads to unexpected and valuable types of reactivity,[2] of which redox reactions account for much of its behaviour. The range of oxidation states of selenium, from −2 to +6, and their facile interchange has enabled a variety of useful oxidation and reduction processes. The deployment of selenium and its compounds in a catalytic fashion mitigates some of the concerns related to its potential toxicity, lowers the environmental impact of such processes, and decreases the cost of using these materials, particularly for large-scale applications.[3,4] The biological and materials aspects of the element, including their related redox processes, are covered in other chapters of this book, while the present chapter describes catalytic transformations that are of particular interest in green organic synthesis.

13.2 Oxidative Additions and Cyclizations of Alkenes

Selenium electrophiles such as selenenyl halides and pseudohalides are highly polarizable Lewis acids that react readily with alkenes. While electrophilic

Chalcogen Chemistry: Fundamentals and Applications
Edited by Vito Lippolis, Claudio Santi, Eder J. Lenardão and Antonio L. Braga

Published by the Royal Society of Chemistry, www.rsc.org

reactions are covered elsewhere in this book, this chapter will focus on recent catalytic variations where an addition or cyclization is accompanied by an oxidation step. Several reviews of alkene functionalization, including oxidative processes, with selenium compounds by Ortgies and Breder[5] describe additional examples and mechanistic insights into these processes.

13.2.1 Oxidative Halogenations

Several oxidative variations employ catalytic amounts of selenium(II) compounds, along with electrophilic halogenating agents, to afford the corresponding halide-substituted products. Pioneering work by Hori and Sharpless[6] demonstrated that benzeneselenenyl chloride (PhSeCl) could be employed catalytically with stoichiometric *N*-chlorosuccinimide (NCS) in the overall nonradical allylic chlorination of alkenes. PhSeCl could also be produced *in situ* from the corresponding diselenide and NCS, together with *N*-(phenylseleno)succinimide (NPSS). Mechanistic studies indicated a complex process, involving possible equilibration at several points that determines regioselectivity. Dichloroselenurane **1** and adduct **2** were postulated as possible intermediates (Scheme 13.1).

Tunge *et al.*[7] indicated that the addition of benzeneselenenyl chloride (PhSeCl) to alkenes generated a mixture of equilibrating 1,2-adducts, which underwent further chlorination of selenium with NCS, followed by elimination to afford the corresponding allylic and vinylic chlorides. The 1,2-addition regioisomer that reacted faster with NCS then led to the dominant product (Scheme 13.2). These results and the catalytic mechanism proposed by the authors, based in part on the previous studies by Sharpless depicted in Scheme 13.1, are shown in Scheme 13.3. Tunge *et al.* also reported that β,γ-unsaturated carboxylic acids underwent primarily *endo* cyclizations when treated with NCS or *N*-bromosuccinimide (NBS) in the presence of 5 mol% of diphenyl diselenide to afford γ-lactones instead of more strained β-lactones from *exo* cyclization.[8]

Kumar *et al.* discovered that oxidative cyclizations of terminal unsaturated carboxylic acids induced with 10 mol% of di(*p*-methoxyphenyl) diselenide in the presence of 4-(*N,N*-dimethylamino)pyridine (DMAP), NBS, and sodium bicarbonate afforded 8–11-membered lactones, mainly by *exo* addition, in 30–69% yield, while similar cyclizations of alcohols produced oxepanes in 58–66% yield.[9] Kumar, Detty, and co-workers reported that isoselenazolones such as **3** catalyzed the vicinal addition of bromine and carboxylic acid groups to alkenes or the bromolactonization of unsaturated carboxylic acids with NBS.[10] The reactions occur *via* oxidative addition of NBS to **3**, followed by bromination of the alkene by **4**, accompanied by nucleophilic addition by carboxylate (Scheme 13.4). Similar acyclic 1,2-additions have been reported with *N*-halosuccinimides by other researchers, including with NCS and catalytic diphenyl selenide,[11] or with *N*-iodosuccinimide (NIS) and diphenyl diselenide.[12] Nucleophiles in the latter two examples included acetonitrile to afford amides after hydrolysis,

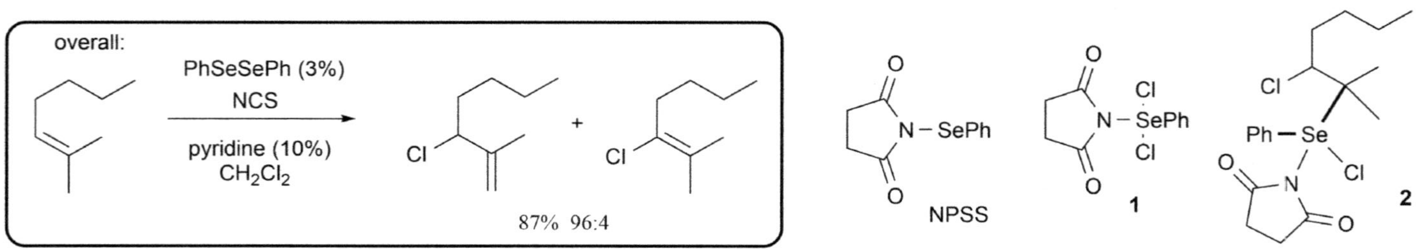

Scheme 13.1 Catalytic oxidative halogenation of alkenes with PhSeSePh and NCS.

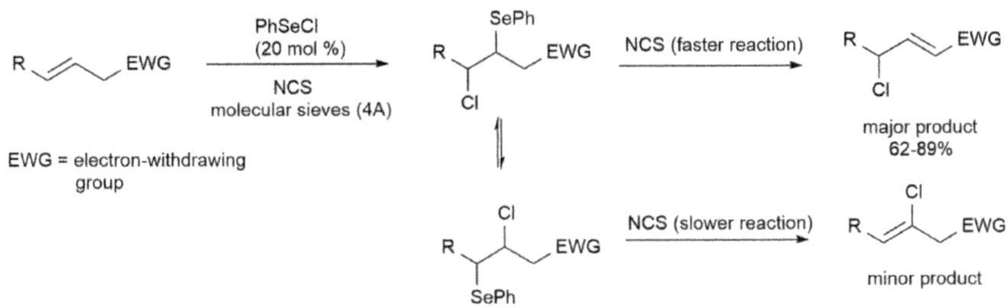

Scheme 13.2 Equilibration of 1,2-adducts of PhSeCl with alkenes in the presence of NCS.

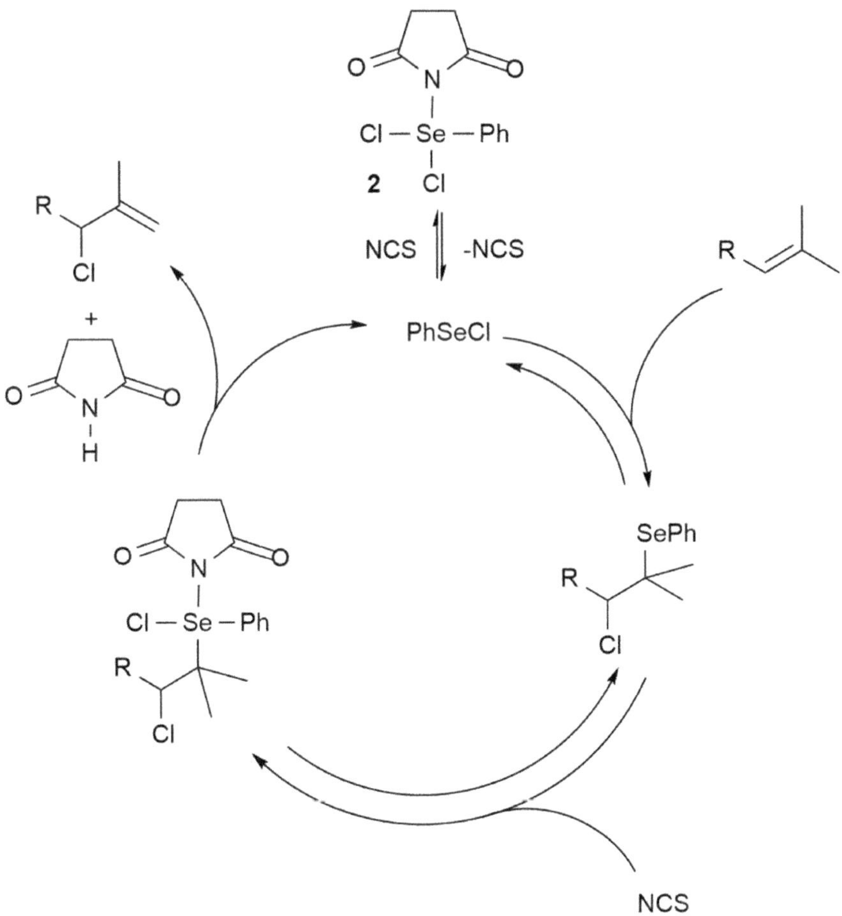

Scheme 13.3 Catalytic cycle of oxidative halogenation with PhSeCl catalyst and NCS.

and water to produce vicinal iodohydrins, respectively. The Detty group also reported other selenium-mediated methods for generating electrophilic bromine ("Br^+") *in situ*, with similar results. For example, the

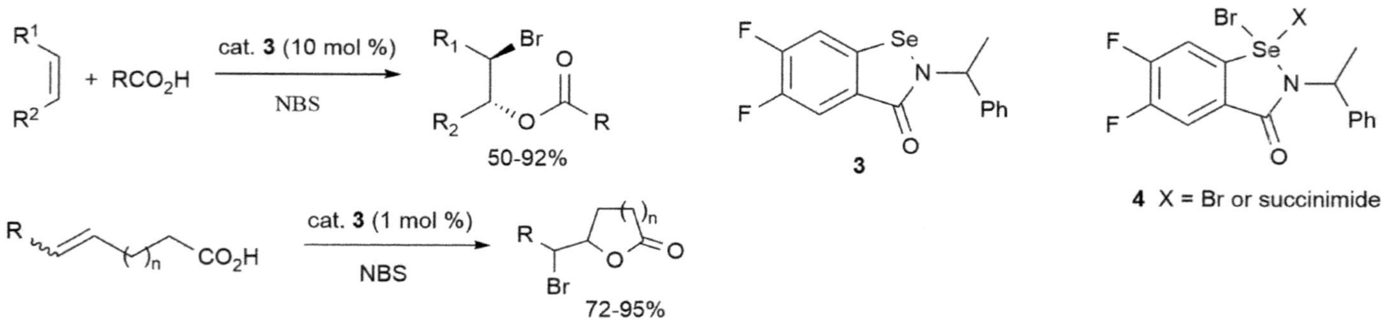

Scheme 13.4 Oxidative halogenations and cyclizations catalyzed by isoselenazolones with NBS.

oxidation of NaBr with hydrogen peroxide and a catalytic amount of selenoxides in the presence of 4-pentenoic acid afforded the lactone **5** in high yield. Kinetic investigations indicated that the reaction is facilitated by electron-donating substituents on the selenoxide, with the best results obtained with **7**, which contains a coordinating dimethylamino group.[13] Benzeneseleninic acid ($PhSeO_2H$) could be employed instead of a selenoxide with comparable results, except that vicinal dibromides such as **6** were formed as byproducts.[14] Again, seleninic acids with electron-donating substituents, or with no substituent, were more effective than ones containing electron-withdrawing groups. Both methods also permitted the bromination of electron-rich arenes, as in the case of **9**. Examples are provided in Scheme 13.5. A similar cyclization of 4-penten-1-ol produced the corresponding tetrahydrofuran derivative in only 17% yield, along with the 1,2-dibromo adduct in 48% yield. These reactions were also catalyzed by the water-soluble diselenides **8**, which permit the use of aqueous conditions and are readily oxidized by hydrogen peroxide to the corresponding seleninic acids in an environmentally compatible manner.[15]

In related work, Detty *et al.*[16] prepared dendritic selenides containing either 6 or 12 phenylseleno groups (*e.g.* **10**) for the vicinal dibromination of alkenes with hydrogen peroxide and NaBr (*e.g.* Scheme 13.6). These compounds exhibited a "dendritic effect" with rates greater than the expected statistical increase of 6 or 12 times that of the corresponding monoselenide. Kinetic experiments revealed that initial slow oxidation of the selenide groups to their selenoxides was followed by the selenoxide-catalyzed oxidation of bromide ion to "Br^+" with hydrogen peroxide, which in turn promoted the oxidation of the next selenide. Thus, selenoxide formation was autocatalytic under these conditions. The corresponding tellurides did not exhibit the dendritic effect because direct oxidation of tellurium by hydrogen peroxide was extremely fast on its own. These authors also prepared selenides sequestered in xerogels,[17] which proved effective in catalyzing cyclization of unsaturated carboxylic acids to lactones and bromination of arenes with NaBr and hydrogen peroxide. Such catalysts again promoted the oxidation of bromide ion by hydrogen peroxide to produce "Br^+" species like hypobromous acid, with potential use as antifouling agents for marine vessels.[17b]

Di(*endo*-3-camphoryl) diselenide, which had been previously employed for stoichiometric asymmetric additions and cyclofunctionalizations,[18] later proved effective at 0.1 mol% concentrations in oxidative bromolactonizations conducted with NBS.[19] However, the enantioselectivities of these reactions were not reported. On the other hand, several examples of asymmetric cyclizations achieved with NBS or other *N*-haloimides and chiral selenium catalysts,[20] such as **11**[20a] and **12**[20b] are shown in Scheme 13.7. A polymer-supported selenenyl bromide was used in conjunction with NCS to provide a highly selective allylic chlorination at the terminal isopropylidene units of a variety of different acyclic polyolefinic polyprenoids.[21]

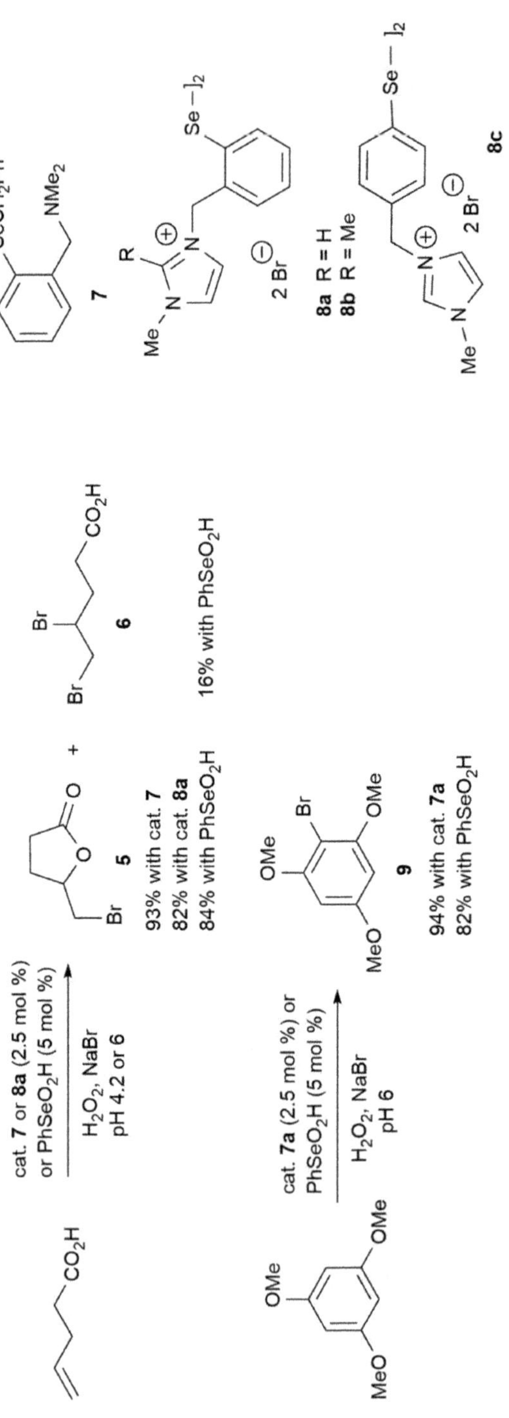

Scheme 13.5 Bromination of terminal unsaturated carboxylic acids and electron-rich arenes with $PhSeO_2H$ or water-soluble diselenide catalysts and $NaBr/H_2O_2$.

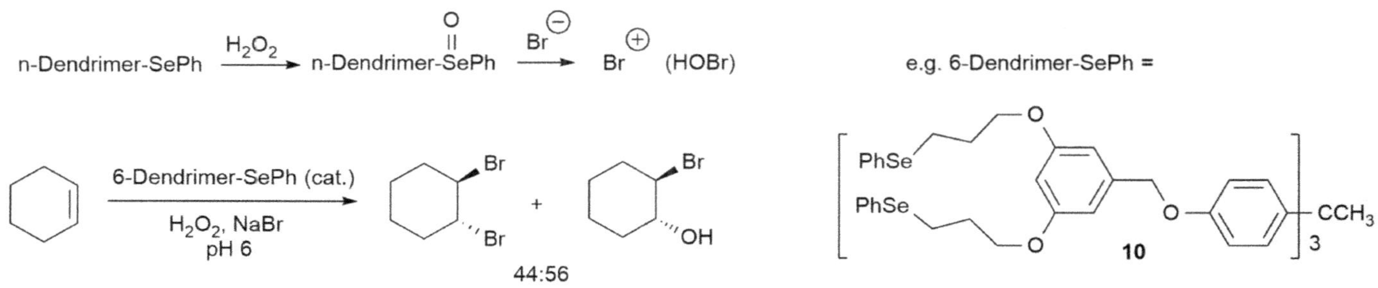

Scheme 13.6 Dendritic selenide catalysts for vicinal dibromination of alkenes.

13.2.2 Oxidative Additions and Cyclizations Mediated by Electrophilic Fluorinating Agents

While NCS and NBS were typically employed with selenium catalysts to promote oxidative halogenations, electrophilic *N*-fluoropyridinium salts such as **13** provide an alternative approach to both halogenated and nonhalogenated products. Thus, Denmark and co-workers[22] reported a remarkable protocol for the preparation of vicinal *syn*-dichlorides, instead of the *anti*-isomers available from classical alkene chlorination methods. The general method and the authors' proposed mechanism are shown in Scheme 13.8. Diphenyl diselenide served as a pre-catalyst that underwent oxidative chlorination by *N*-fluoropyridinium tetrafluoroborate (**13**) in the presence of a chloride ion. The resulting $PhSeCl_3$ added to the alkene and the bridged intermediate **14** was subjected to S_N2 ring-opening by chloride ion to produce **15**. In a departure from conventional methods, a second inversion of configuration by S_N2 displacement and reductive elimination of **15** by a second chloride ion then afforded the *syn*-dichloro product. The byproduct PhSeCl was similarly oxidized by **13** to continue the cycle. The inclusion of Me_3SiCl served to trap fluoride ions that otherwise interfered by displacing chloride from **15** to produce a less reactive electrophile. Competing β-elimination from **15** generated vinylic and allylic chloride byproducts. Furthermore, the authors developed an enantioselective variation of this method by employing chiral diselenides instead of the diphenyl derivative.[23] While 23 catalysts were screened, diselenide **16** proved most generally effective. A limitation to regio- and stereo-selectivity stems from reversibility in the formation of the seleniranium ion intermediate and alkene to alkene crossover,[24] which can be accompanied by racemization.

Several other groups have reported cyclizations *via* oxidative C–H activation promoted by electrophilic fluorinating agents like **13** or

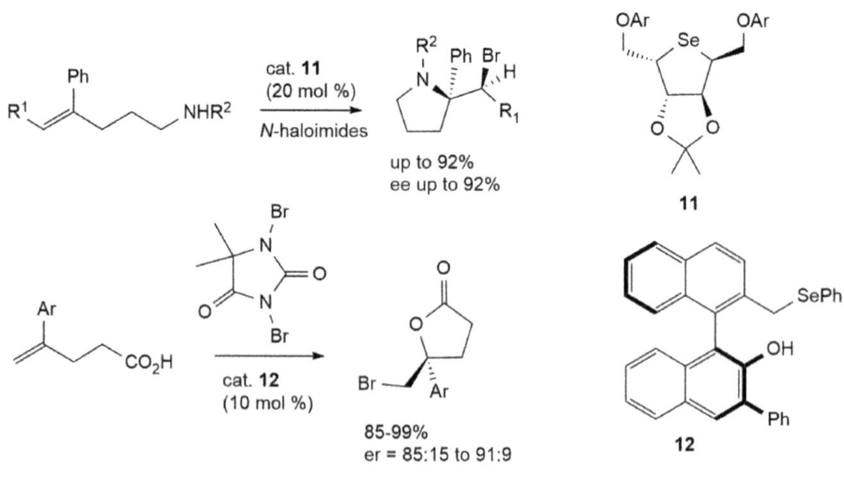

Scheme 13.7 Asymmetric cyclizations with *N*-haloimides and chiral selenide catalysts.

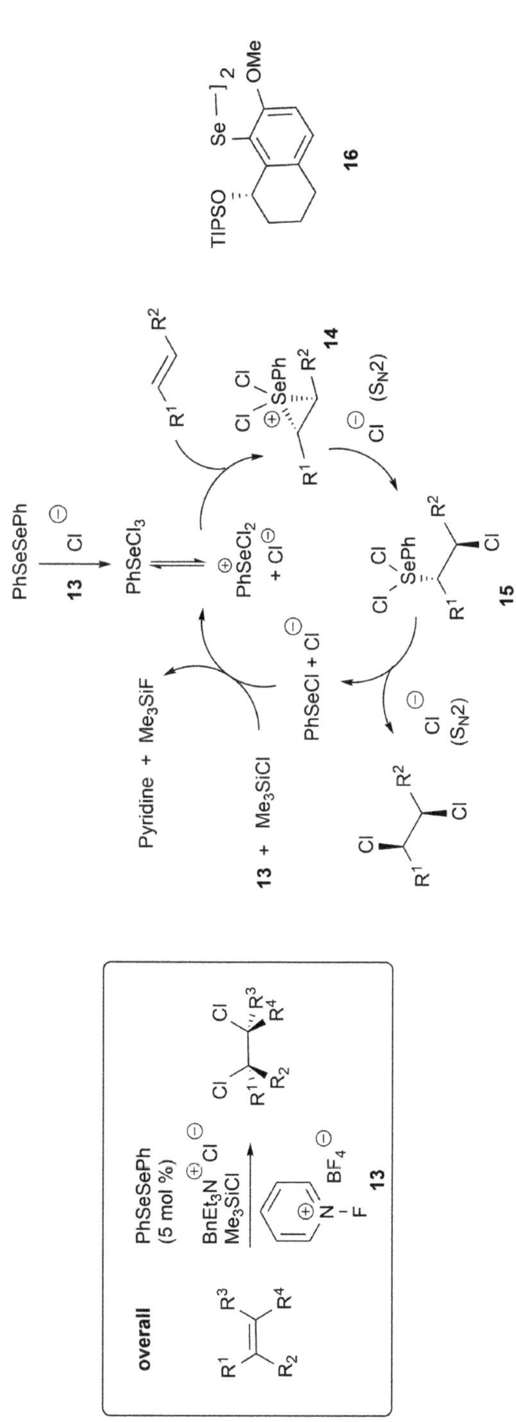

Scheme 13.8 Catalytic cycle for the preparation of vicinal *syn*-dichlorides from alkenes with a PhSeSePh catalyst and the electrophilic fluorinating agent 13.

N-fluorobenzenesulfonimide (NFSI, **17**) and selenium-based catalysts. Thus, Breder *et al.* prepared lactones **18** from *o*-allylbenzoic acids,[25] as well as indoles **19** and azaindoles from *o*-vinyl anilines and aminopyridines,[26] respectively, in this fashion, using catalytic diphenyl diselenide and NFSI as the oxidant. Maruoka *et al.* obtained butenolides **20**[27] and their oxime derivatives **21**[28] enantioselectively by means of chiral catalysts **22** and **23**, respectively, again with NFSI as the oxidant (for examples, see Scheme 13.9).

Zhao and co-workers reported a similar indole preparation,[29] along with other oxidative cyclizations of unsaturated amines with oxidant **24** leading to α-quaternary pyrrolidines **25** or β,γ-unsaturated azepines **26**, depending on the chain length.[30] Unsaturated alcohols similarly afforded tetrahydrofurans or pyrans. Breder *et al.*[31] demonstrated oxidative C–H aminations of alkenes and cycloalkenes, where NFSI (**17**) served as both the oxidant and the aminating reagent, as exemplified by products **27** and **28.** Zhao *et al.* extended this process to the amination of allylic alcohols,[32] and conjugated dienes.[33] They also reported a variation of the method where electrophilic fluorinating agents such as **24**, in the presence of TEMPO (2,2,6,6-tetramethylpiperidine-1-oxyl), afforded allylic fluorides **29** from alkene starting materials.[34] Furthermore, Breder *et al.* investigated a series of chiral diselenides for C–H amination of benzyl 3-pentenoate with NFSI, but obtained generally modest yields and enantioselectivities.[35] A new synthesis of 1,2-diarylpyrazolo[5,1-*b*]quinazolin-9(1*H*)-ones

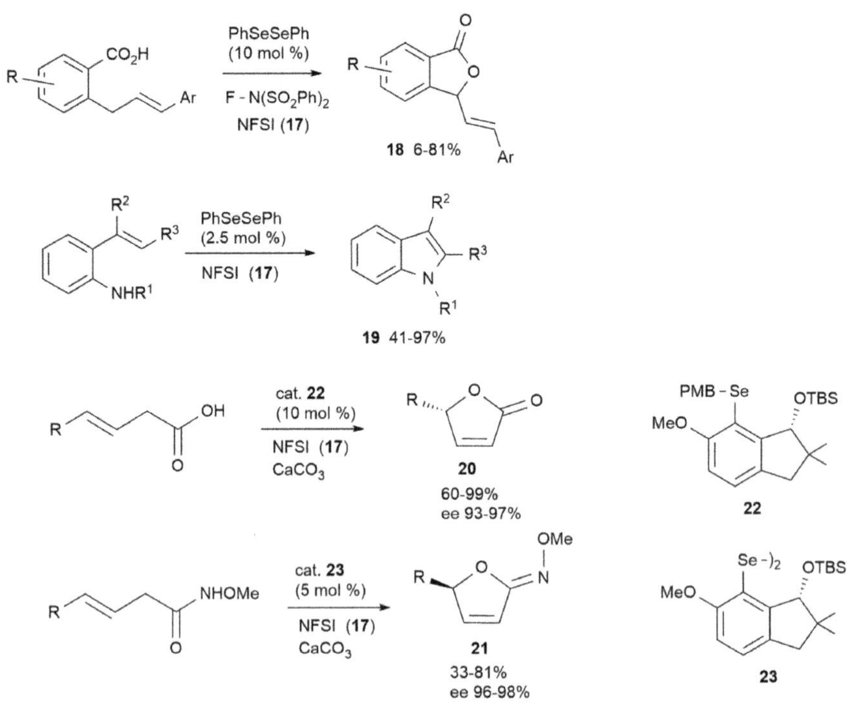

Scheme 13.9 Oxidative cyclizations promoted by the electrophilic fluorinating agent NFSI and selenium-based catalysts.

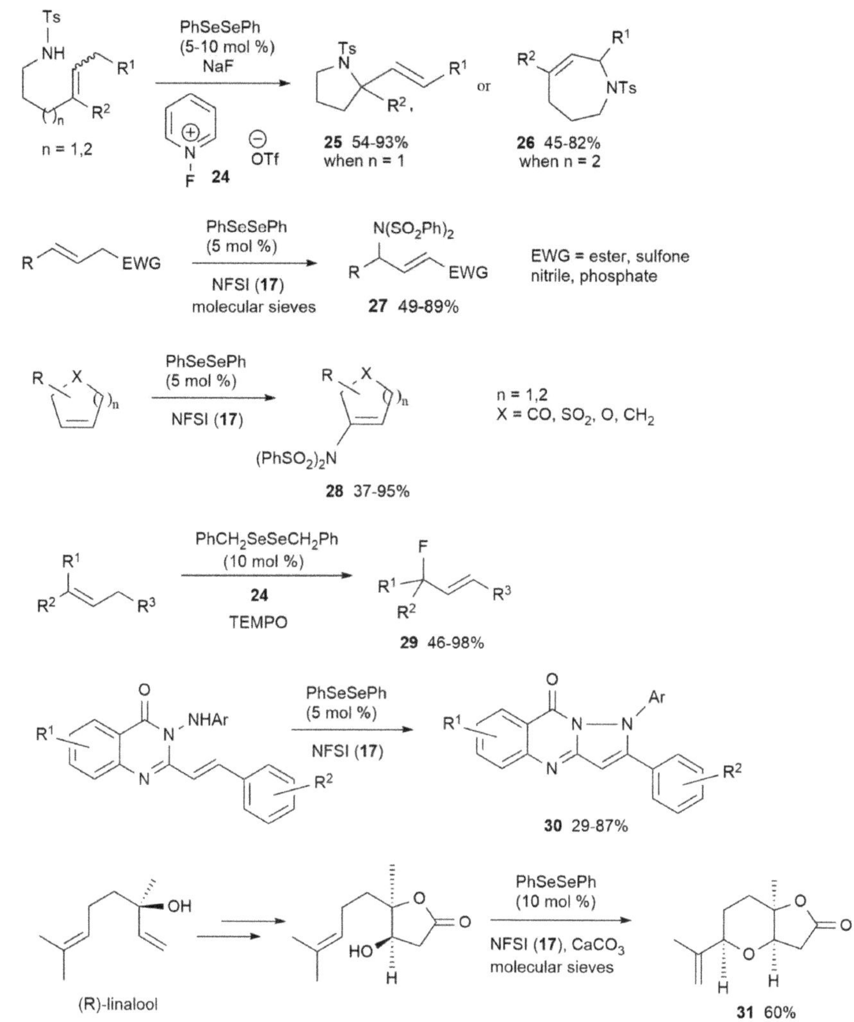

Scheme 13.10 Oxidative cyclizations of various functionalized alkenes with electrophilic fluorinating agents and diselenide catalysts.

30 was reported by Chen *et al.* by oxidative amination,[36] while Christmann *et al.* synthesized the (+)-enantiomer of the Greek tobacco lactone **31** by oxidative cyclic ether formation[37] (Scheme 13.10).

13.2.3 Oxidative Additions and Cyclizations Mediated by Hypervalent Iodine Compounds

Wirth and co-workers[38] employed [bis(trifluoroacetoxy)iodo]benzene (**32**) in a dual role in the formation of butenolides **33** from the corresponding 3-butenoic acids (Scheme 13.11). The hypervalent iodine reagent first

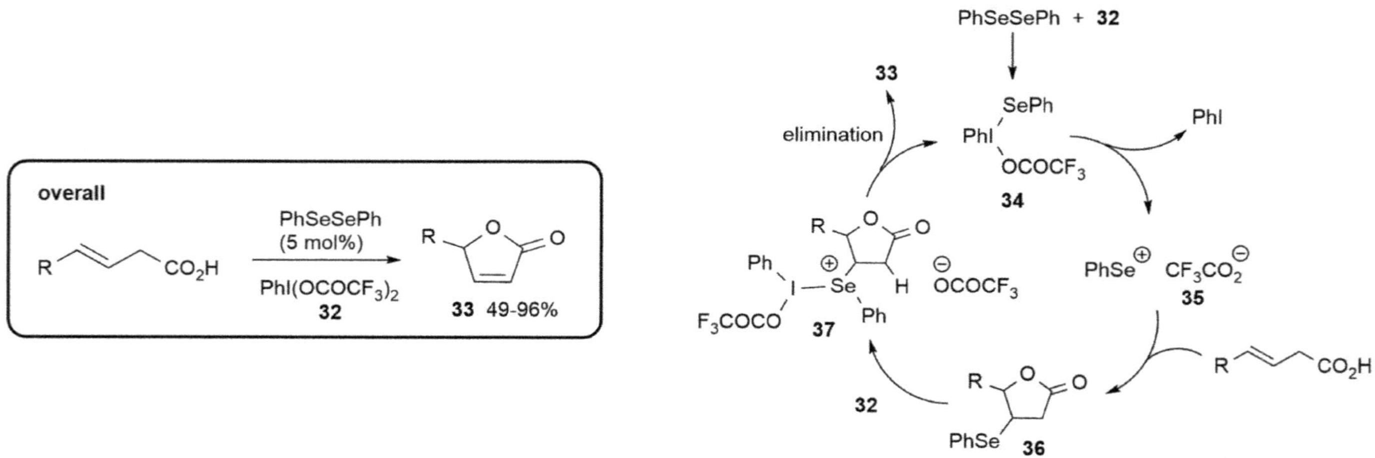

Scheme 13.11 Formation of butenolides **33** by oxidative cyclizations of 3-butenoic acids mediated by the $PhI(OCOCF_3)_2$ and PhSeSePh catalyst.

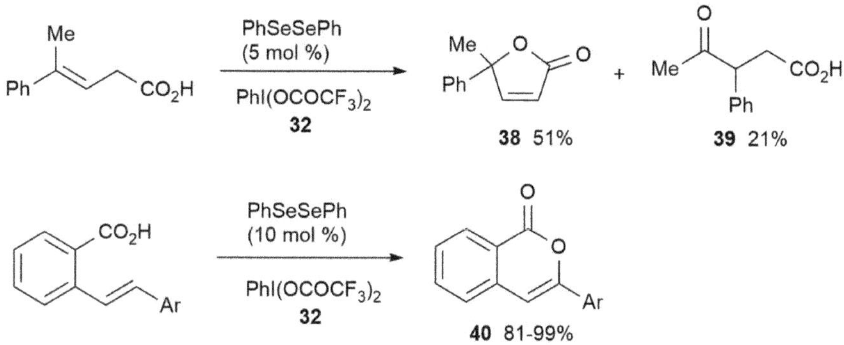

Scheme 13.12 Further examples of PhSeSePh-catalyzed oxidative cycloadditions mediated by $PhI(OCOCF_3)_2$.

converted diphenyl diselenide to the selenenyl trifluoroacetate derivative **35** *via* **34**. The electrophile **35** effected the cyclization step to form lactone **36**, followed by a second oxidation with **32** and elimination from **37** to regenerate catalyst **34** and afford the butenolide product **33**.[38a] Several chiral diselenides were investigated, but gave poor enantiomeric ratios, except at −100 °C. The formation of butenolide **38** was accompanied by that of keto acid **39**, the result of a phenonium ion shift.[38b] The latter was the sole product in the absence of diphenyl diselenide. These authors also reported that *o*-stilbenecarboxylic acids afforded isocoumarins **40** (Scheme 13.12),[39] while substituted 4-pentenoic and 5-hexenoic acid derivatives produced 2-pyranones with *endo*- and *exo*-cyclic double bonds, respectively.[40]

13.2.4 Other Oxidative Additions and Cyclizations

In early work, Iwaoka and Tomoda[41] reported the oxidation of alkenes with catalytic amounts of diselenides **41** or **42** in the presence of sodium persulfate and various alcohols, affording rearranged allylic ethers in yields of 37–62%. In a similar approach, Wirth and co-workers[42] reported that potassium or ammonium persulfate oxidized diselenides to the corresponding selenenyl sulfates ($ArSeOSO_3^-$), which performed 1,2-additions to alkenes with methanol as the nucleophile, followed by further oxidation of the selenide moiety and elimination. Chiral diselenide **43** afforded products with ee (enantiomeric excess) up to 75%. This group also reported an electrochemical process where bromide anion was converted into bromine by anodic oxidation, thereby promoting the bromination and elimination of an arylseleno moiety from an initial 1,2-adduct generated from ArSeBr, an oxygen-based nucleophile and the substrate alkene.[43]

Tiecco *et al.*[44] also reported several oxidative additions and cyclizations with ammonium persulfate as the stoichiometric oxidant in conjunction with chiral diselenide catalyst **44** to produce allylic alcohols and ethers, as well as the lactone **45**.[44a] Oxidation of the selenide moiety of the initial

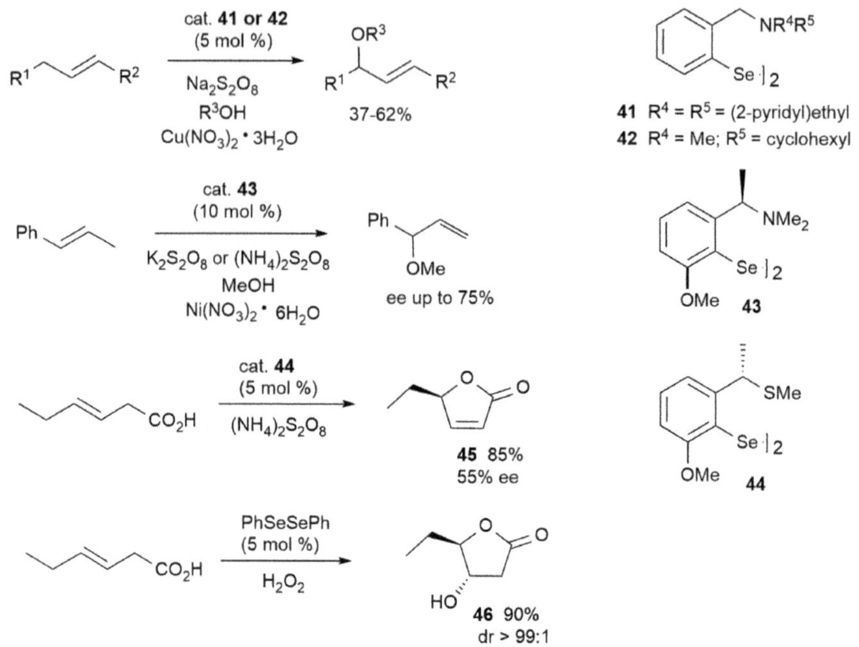

Scheme 13.13 Preparation of allylic ethers and lactones with diselenide catalysts and various oxidants.

β-hydroxy selenide adduct followed by hydrolysis also afforded mixtures of *cis* and *trans* diols with variable, substrate-dependent stereoselectivity.[44b] The use of hydrogen peroxide and diphenyl diselenide by Santi and co-workers[45] provided a particularly simple and ecologically friendly approach to lactones such as **46** (Scheme 13.13), although more highly substituted analogues were generally formed as mixtures of diastereomers.

Several groups have investigated photoredox approaches to selenium-catalyzed oxidative additions.[46] Breder *et al.*[35,47] employed diphenyl diselenide with photocatalysts like 2,4,6-tris(4-anisyl)pyrylium tetrafluoroborate (TAPT, **47**) for the preparation of a series of lactones (*e.g.*, **48**) by irradiation with blue light (465 nm) in the presence of air. A variation of this method was performed with an assortment of chiral catalysts such as **49**, but enantioselectivities were generally modest, as in the case of lactone **50**. Scheme 13.14 contains representative examples. The cyclization leading to the Greek tobacco lactone **31** in Scheme 13.10 was also achieved *via* a photoredox method employing diphenyl diselenide (10 mol%) and TAPT as the photocatalyst (5 mol%) under aerobic conditions, affording 83% of **31** with a diastereomeric ratio (dr) of 84:16.[37]

A method that incorporates trifluoromethylthio groups into alkenes to afford products such as **52** from the saccharin-derived reagent **51** was developed by Zhao *et al.* (Scheme 13.15),[48] using catalytic di-(*p*-methoxyphenyl) selenide. Enantioselective trifluoromethylthiolations of unsaturated acids,

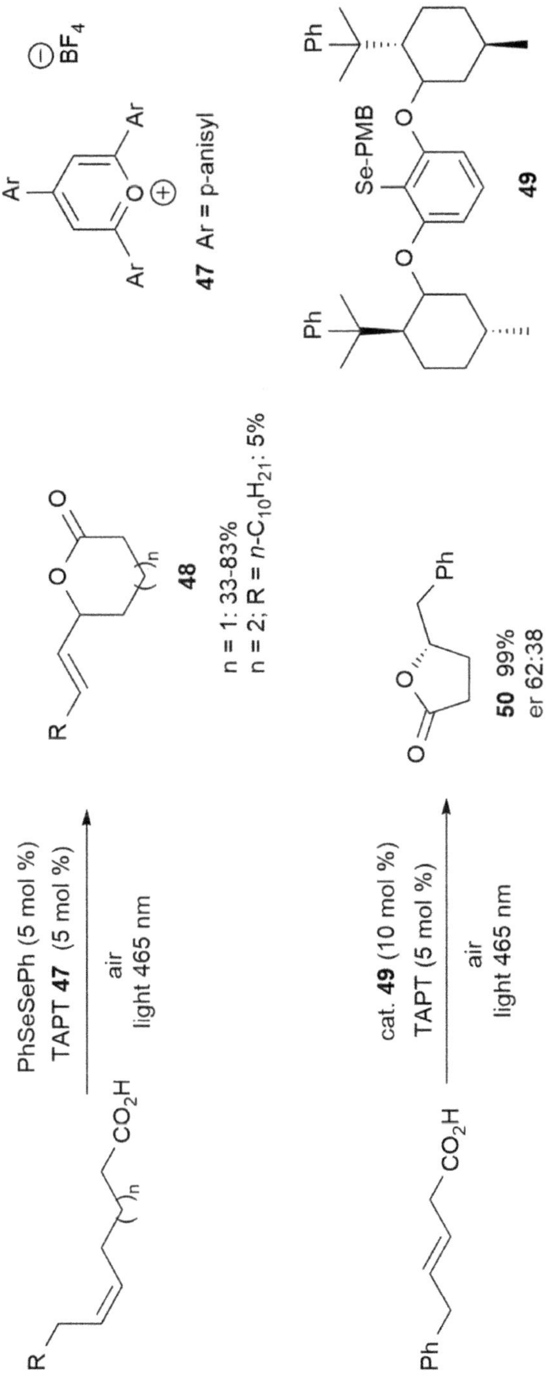

Scheme 13.14 Photoredox cyclizations in the presence of selenium-based catalysts.

affording the corresponding lactones with chiral sulfur- or selenium-based catalysts, were also achieved.[49] A ring-expansion of cyclopropylidenes **53** to cyclobutanones **54** was reported by Yu and co-workers[50] *via* oxidation with hydrogen peroxide in the presence of di-(*o*-fluorophenyl) and other diselenides (Scheme 13.15).

13.3 Oxidations of Alkenes and Alkynes to Allylic or Propargylic Alcohols and Carbonyl Compounds

13.3.1 Oxidation of Alkenes

Oxidations of activated methylene groups by selenium dioxide were among the first widespread applications of selenium chemistry in synthesis,[51] but earlier work typically employed stoichiometric amounts of the reagent. However, in 1977 Umbreit and Sharpless[52] reported that it could be used catalytically for allylic oxidations in small amounts (1.5–2.0%) in the presence of *t*-butyl hydroperoxide. This approach avoided the formation of byproducts such as epoxides and diols, as well as of reduced selenium species. Since allylic oxidations with selenium dioxide have been widely studied and thoroughly reviewed,[4c,53] this section will focus on more recent methods.

The seleninic acids [RSe(=O)OH] and their congeners[4a] have proven to be versatile oxidants, largely through the pioneering efforts of Barton and coworkers. Thus, Barton and Crich[54] reported that 2-pyridineseleninic anhydride (**55**) was generated *in situ* from its diselenide with iodoxybenzene ($PhIO_2$) or other hypervalent iodine compounds. It then served as the catalyst for the oxidation of a variety of unsaturated products, such as the formation of the ketone derivative **56** of β-pinene (Scheme 13.16). The authors noted that **55** was more reactive and gave cleaner products than benzeneseleninic anhydride and afforded regioselectivity as predicted by the Guillemonat rules for selenium dioxide.[55] They invoked a similar mechanism to that for allylic oxidations by selenium dioxide, involving an ene reaction followed by a [2,3]sigmatropic rearrangement of the resulting selenoxide intermediate.[52b] Examples of an enamide oxidation used by Magnus *et al.*[56] to produce the aspidosperma alkaloid derivatives **57** and **58**, along with the use of a fluorinated seleninic acid **59** by Crich and Zou[57] in the allylic oxidation of alkenes to ketones such as **60**, are also provided in Scheme 13.16. Fluorous extraction of the corresponding diselenide permits its efficient recovery and reuse.

Apart from allylic oxidations, Barton and Wang[58] also reported several benzylic oxidations catalyzed by pentafluorobenzeneseleninic acid, as in the preparation of xanthone **61**. Among other examples, Crich and Zou[59] obtained *o*-quinone **62** from tetrahydronaphthalene under fluorous conditions that again facilitated recovery of the catalyst (Scheme 13.17). Related oxidations catalyzed by various selenium compounds include the conversion of anthracene to 9,10-anthraquinone by means of a selenazolium ion catalyst,[60] the benzylic oxidation of substituted pyridines with catalytic PhSeBr in the

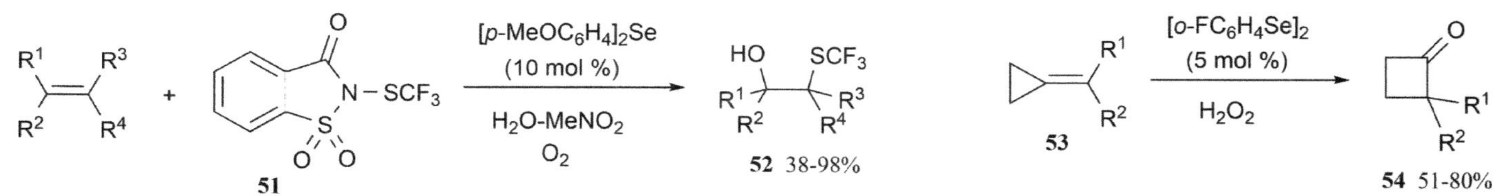

Scheme 13.15 Oxidative trifluoromethylthiolation of alkenes and ring expansion of cyclopropylidenes with diselenide catalysts.

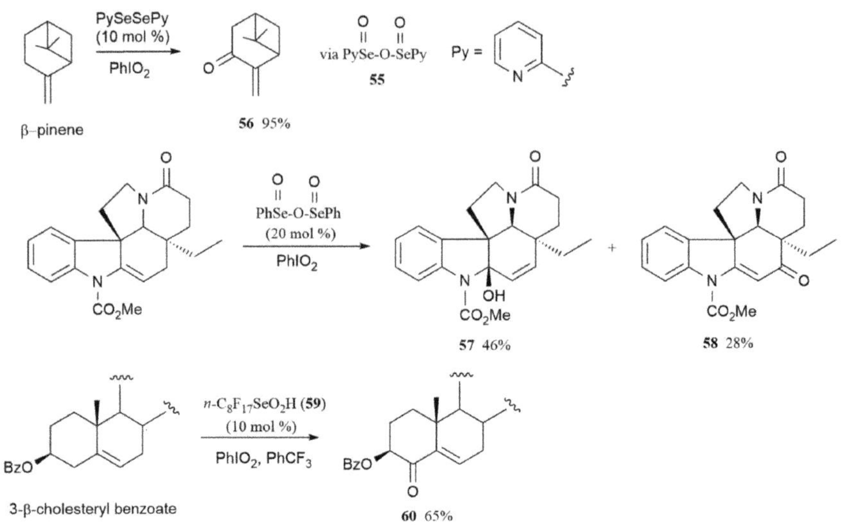

Scheme 13.16 Allylic oxidations catalyzed by seleninic acids and anhydrides.

presence of oxygen,[61] and the cleavage of substituted naphthalenes to *trans*-2-carboxycinnamic or 2-naphthoic acids with hydrogen peroxide and a polymeric diselenide catalyst.[62]

13.3.2 Oxidation of Acetylenes

In 1979, Sharpless *et al.*[63] investigated the oxidation of a variety of acetylenes with catalytic selenium dioxide and *t*-butyl hydroperoxide, affording mixtures of propargylic alcohols and diols as principal products, with minor amounts of ketones. More recently, Santi *et al.* employed diphenyl diselenide and ammonium persulfate under aqueous conditions to obtain α-diketones such as **63** from internal acetylenes and α-keto hemiacetals **64** from terminal ones in the presence of alcohols.[64] Zhao and co-workers[65,66] used electrophilic fluorine species like **24** as the oxidant with diphenyl diselenide catalyst to obtain rearranged ynone phosphonates from propargylic phosphonates, while an electrochemical process devised by Pan *et al.*[67] with *o*-nitrophenyl acetylenes produced benzoxazoles **65**. Examples are provided in Scheme 13.18.

13.4 Epoxidations and Dihydroxylations Catalyzed by Selenium Compounds

In 1977, Grieco[68] and Kametani[69] reported that seleninic acids and selenides, respectively, could effect the epoxidation of alkenes under relatively mild conditions in the presence of hydrogen peroxide. The following year, Reich[70] and Sharpless[71] independently reported catalytic versions of this

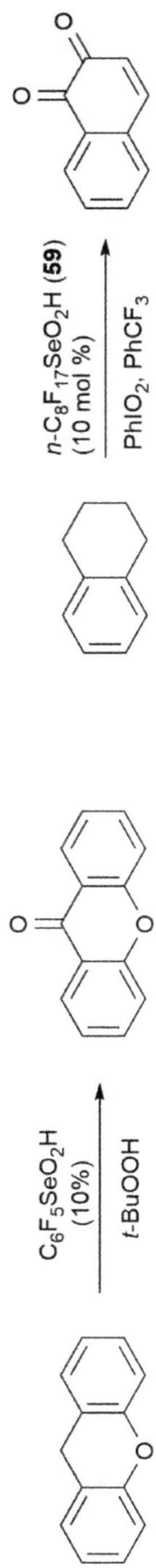

61 quantitative

62 50%

Scheme 13.17 Benzylic oxidations catalyzed by fluorinated seleninic acids.

process, using benzeneseleninic acid or its 2-nitro or 2,4-dinitro derivatives as catalysts with hydrogen peroxide. The required seleninic acids could also be generated *in situ* from the oxidation of diselenides or selenenic acids (RSeOH). Grieco proposed that the corresponding peroxyseleninic acid [PhSe(=O)OOH] (**66**) was formed by further oxidation of its seleninic acid counterpart and comprised the active oxidant, a conclusion endorsed by Sharpless, Reich, and later by numerous other researchers. More recently, it was demonstrated by Back *et al.*[72] that **66** is indeed formed in such processes, but is actually a poor epoxidizing agent because in solution its conversion to the mixed selenonium selenonate salt **67** *via* spontaneous loss of oxygen occurs more rapidly than the epoxidation itself. The salt then undergoes further oxidation with hydrogen peroxide to form the inert selenonic acid **68**, which is in turn converted to the postulated peroxyselenonic acid **69**, resulting in rapid epoxidation (Scheme 13.19). Thus, epoxidation

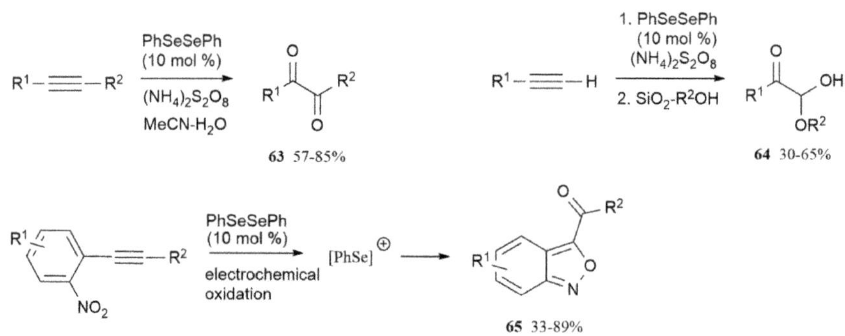

Scheme 13.18 Electrochemical and persulfate oxidations of acetylenes with diselenide catalysts.

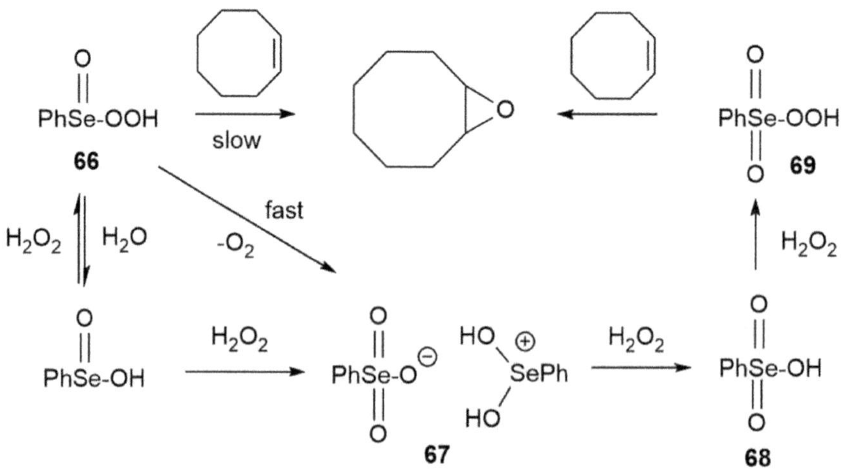

Scheme 13.19 Peroxyselenonic acid **69** as a potent epoxidation catalyst.

proceeds mainly *via* the Se(VI) oxidation state and not Se(IV) as originally believed.

Other researchers have developed alternative catalysts for epoxidation, or *trans*-diol formation from further hydrolysis, with hydrogen peroxide. These include a polymer-supported seleninic acid reported by Taylor and Flood[73] and a matrix prepared by Yu *et al.* from heating selenium with glucose at 500 °C, for the epoxidation of β-ionone.[74] Knochel[75] employed the seleninic acid generated *in situ* from the selenoxide elimination of the corresponding *n*-butyl selenide **70** in *n*-$C_8F_{17}Br$ as the solvent, while Sheldon[76] performed epoxidations with seleninic acid **71** in trifluoroethanol. The fluorous conditions permitted facile recovery of the catalyst. Catalyst **71** was found by Arends *et al.* to be particularly effective in glycerol-based solvents,[77] and was used in a green epoxidation of unsaturated fatty esters and biodiesel fuel.[78] It proved more effective than elemental selenium, selenium dioxide, and various selenides and diselenides in the selective epoxidation of the endocyclic double bond of β-ionone, while other catalysts gave significant amounts of Baeyer–Villiger products.[79] The selenoxide **72**[80] and cyclic seleninate ester **73**[81] also served as effective epoxidation catalysts with hydrogen peroxide as the co-oxidant. Relay epoxidations based on the selenium-catalyzed formation of oxaziridine **75** by DuBois *et al.*[82] or chiral oxaziridinium species **77** by Bulman Page *et al.*[83] from the corresponding imine **74** or iminium salts such as **76**, respectively, have been reported. The products **75** and **77** were in turn employed in the epoxidation of alkenes, with good enantioselectivity observed with **77** (Scheme 13.20). Oxaziridine **75** was also found to oxidize tertiary C–H bonds to alcohols by this method. In view of the results shown in Scheme 13.19, it is likely that other selenium-catalyzed epoxidations also proceed *via* Se(VI) instead of Se(IV) intermediates.

While hydrolysis of epoxides to *trans*-diols is well-known, these products can be obtained in one step from alkenes using either selenium dioxide[84] or polymer-based seleninic/selenonic acid catalysts.[73,85] Santi, Wirth, and their co-workers[86] developed an asymmetric approach to diol formation by means of catalysts **78–80** (*e.g.*, see Scheme 13.21). The dihydroxylation of cyclohexene catalyzed by the diselenide precursor of seleninic acid **71** in 96% yield has also been reported by Yu *et al.*[87] Hydrogen peroxide was typically employed as the co-oxidant. Side reactions of selenium-catalyzed epoxidations can include competing Baeyer–Villiger oxidations (see Section 13.6.3), which are, however, typically slower than epoxidations. Cleavage of alkenes to carbonyl compounds has also been observed in some cases.[81,88]

13.5 Oxidations of Alcohols and Phenols

13.5.1 Alcohols

The first oxidations of primary and secondary alcohols to aldehydes and ketones with benzeneseleninic anhydride were reported by Barton *et al.*[89] and Back,[90] using stoichiometric amounts of the reagent. This was followed

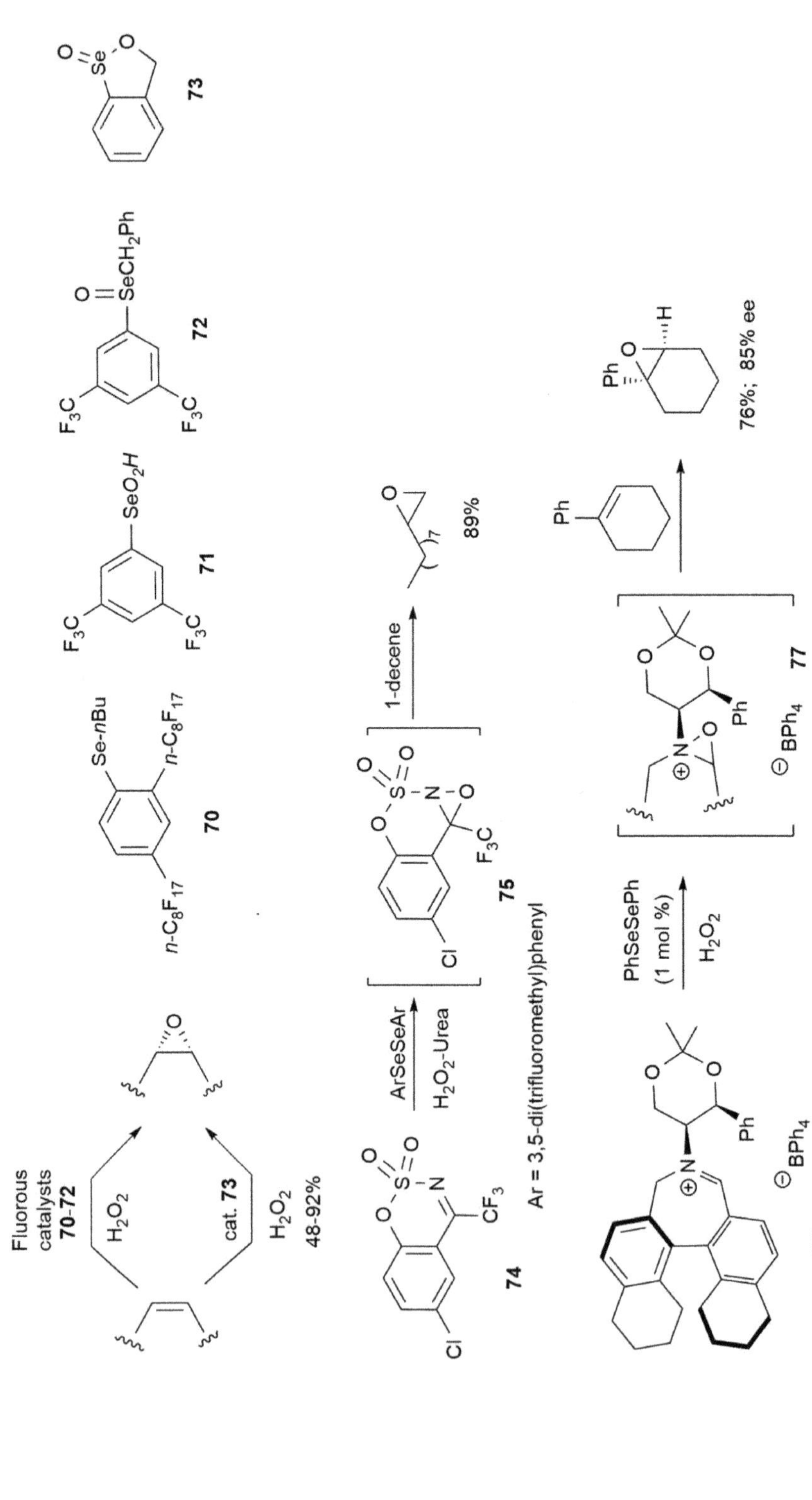

Scheme 13.20 Examples of selenium-catalyzed epoxidations.

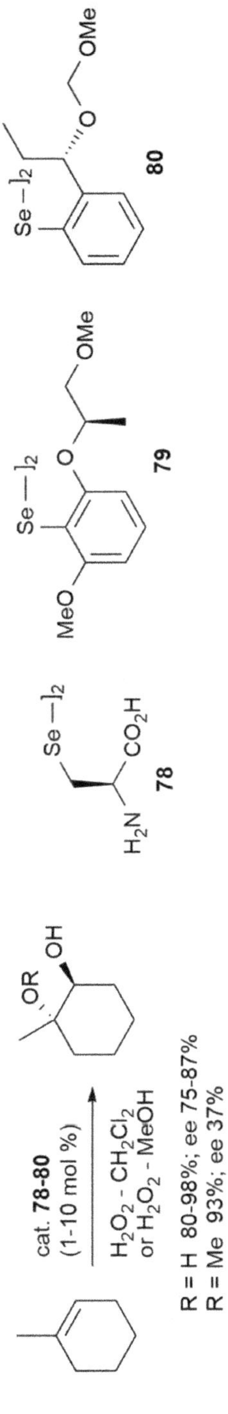

Scheme 13.21 Asymmetric dihydroxylations of alkenes with diselenide-based catalysts.

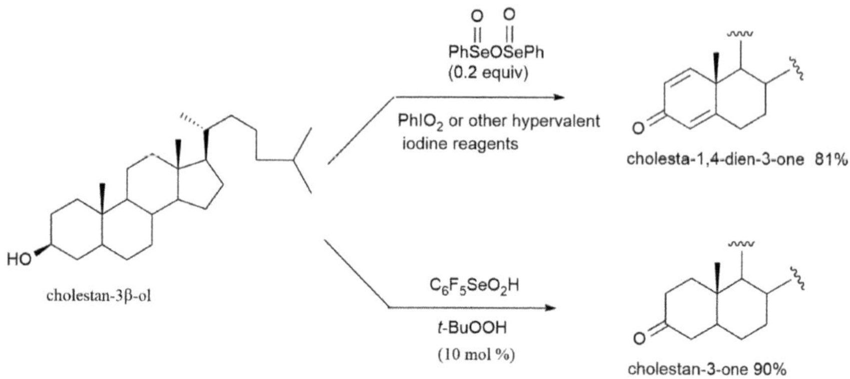

Scheme 13.22 Oxidation and dehydrogenation of cholestan-3β-ol with benzeneseleninic anhydride and pentafluorobenzeneseleninic acid catalysts.

in the Barton group by a catalytic process employing benzeneseleninic anhydride and iodoxybenzene ($PhIO_2$) or other hypervalent iodine reagents as the stoichiometric oxidant.[91,92] It was thus possible to oxidize cholestan-3β-ol to the 3-ketone, followed by dehydrogenation to afford cholesta-1,4-dien-3-one in one pot. The combination of pentafluorobenzeneseleninic acid (10%) and *t*-butyl hydroperoxide afforded cholestan-3-one in 90% yield (Scheme 13.22).[58a]

Oxidations of alcohols with diselenide catalysts likely proceed *via* oxidation to seleninic acids or anhydrides, then selenoxide elimination of intermediate seleninate esters **81**, followed by reoxidation of the byproduct selenenic acid (RSeOH). Several other selenium-based catalyst-co-oxidant pairs have been used in similar oxidations of alcohols: isoselenazolone **3** with bromine,[10] diselenides **82** with *N*-chloro-4-chlorobenzenesulfonamide salts,[93] di-*p*-chlorophenyl, diphenyl, or dimesityl diselenide with *t*-butyl hydroperoxide,[94,95] and an imidazolium seleninic acid **83** with hydrogen peroxide.[96] Catalytic selenium dioxide in the presence of di-(*p*-methoxyphenyl) selenoxide,[97] as well as the polymer-supported seleninic acid that was used in epoxidations by Taylor and Flood (Section 13.4) also served to oxidize alcohols to ketones or aldehydes.[73] Singh *et al.* reported that a series of ruthenium chalcogenide salts such as **84** and **85** performed such oxidations in up to 98% yield, with a turnover number of up to 9.8×10^4, thus enabling as little as 0.001 mol% to be used. These catalysts could be employed with *N*-methylmorpholine *N*-oxide, *t*-butyl hydroperoxide, sodium hypochlorite, or sodium periodate as the co-oxidant. Interestingly, these catalysts could also effect the reverse process, reduction of ketones to alcohols, by transfer hydrogenation in alkaline isopropanol[98] (Scheme 13.23).

13.5.2 Phenols

The oxidation of phenols to hydroxydienones or quinones with equivalent amounts of benzeneseleninic acid or anhydride was first discovered by

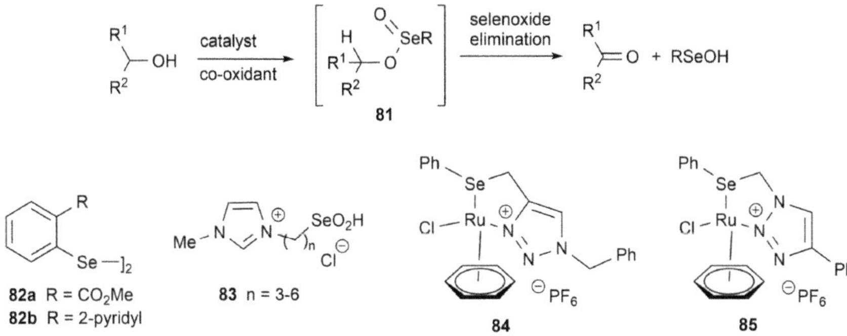

Scheme 13.23 Oxidations of alcohols to ketones with selenium-based catalysts.

Barton *et al.*[99] More recently, muconolactones like **87** were prepared by Giurg, Młochowski, and co-workers by means of various catalytic diselenides and hydrogen peroxide.[100] The electron-deficient diselenides **86** proved particularly effective. Both 1- and 2-naphthols were cleaved under similar conditions to afford dicarboxylic acids **88**, possibly *via* the corresponding *o*-quinones.[101] These authors also reported that, interestingly, *o*-aminophenols dimerized to products **90** when treated with ebselen (**89**) or various diselenides and hydrogen peroxide or *t*-butyl hydroperoxide. The enzymes laccase and horseradish peroxidase were also investigated in this process.[102] The products **90** included questiomycin A ($R^1 = R^2 = H$), cinnabarinic acid ($R^1 = CO_2H$; $R^2 = H$), and actinocin ($R^1 = CO_2H$; $R^2 = CH_3$). The oxidation of the antioxidant Trolox C with selenazolium catalysts such as **91** afforded *p*-quinone **92**, while 2-aminophenol generated questiomycin A similarly (Scheme 13.24).[60]

13.6 Oxidations of Aldehydes, Ketones, and Ketals

13.6.1 Oxidation of Aldehydes

In 1988, Choi *et al.* reported that both alkyl and aryl aldehydes can be oxidized to carboxylic acids upon treatment with benzeneseleninic acid (5 mol%) and hydrogen peroxide.[103] Mlochowski and co-workers[104] employed selenium dioxide or silica-supported selenenamides with either hydrogen peroxide or *t*-butyl hydroperoxide, with the latter providing slower reaction rates. They obtained particularly successful outcomes with aromatic aldehydes by using ebselen (**89**, 5 mol%) as the catalyst and *t*-butyl hydroperoxide as the co-oxidant, often obtaining yields greater than 90%. More recently, Santi *et al.*[105] reported that both aryl and alkyl aldehydes can be oxidized to carboxylic acids with diphenyl diselenide (2%) and hydrogen peroxide in aqueous media in yields of 75 to >99%. The catalyst was easily recycled and remained active for at least five reactions. When the process was conducted in the presence of alcohols, esters were obtained in good yield.

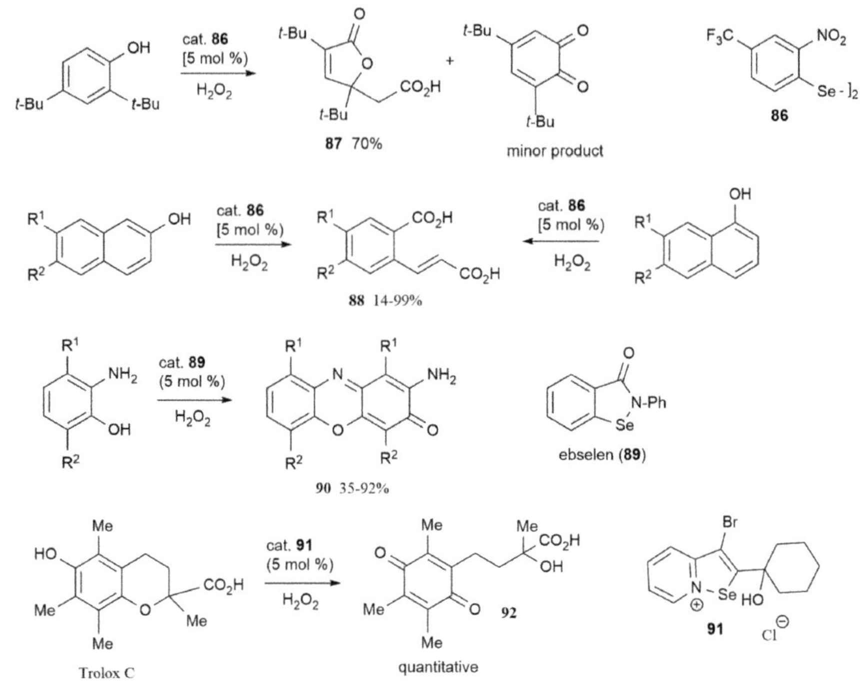

Scheme 13.24 Oxidations of phenols with selenium-based catalysts.

13.6.2 Oxidation of Ketones

As shown in Scheme 13.22, selenium-mediated catalytic α,β-dehydrogenations of ketones, can be performed in conjunction with prior oxidation of the corresponding alcohols. Such processes are typically conducted *via* the reaction of enols or enolates with seleninic acids or anhydrides, followed by selenoxide elimination. In other examples of ketone oxidations, the bicyclic enone **93** was similarly dehydrogenated, as shown in Scheme 13.25.[106] The catalytic dehydrogenation of two sapogenin ketals was reported by Morzycki *et al.*, albeit in low yield, presumably arising *via* the corresponding ketone. Other oxidation products were observed in higher yield, but required stoichiometric amounts of benzeneseleninic anhydride.[107] Aryl methyl ketones afforded keto carboxylic acids **94** by oxidation with seleninic acid **59** under fluorous conditions.[59] When the oxidation was carried out in alcohol solvents with diphenyl diselenide in the presence of CuBr and oxygen to generate redox-active species, the corresponding acetals **95** were obtained.[108] Cycloalkanones afforded ring-contracted products such as **98** by oxidation with di(2-quinolinyl) or polymeric anthracenyl diselenides (**96** and **97**, respectively).[109] α-Halogenations, instead of selenenylations, of ketones and β-keto esters with NCS or NBS were catalyzed by benzeneselenenyl chloride or bromide to give products such as **99**.[8c] α-Hydroxylations of ketones were achieved *via* oxidation of their enamines **100** with the cyclic seleninate ester

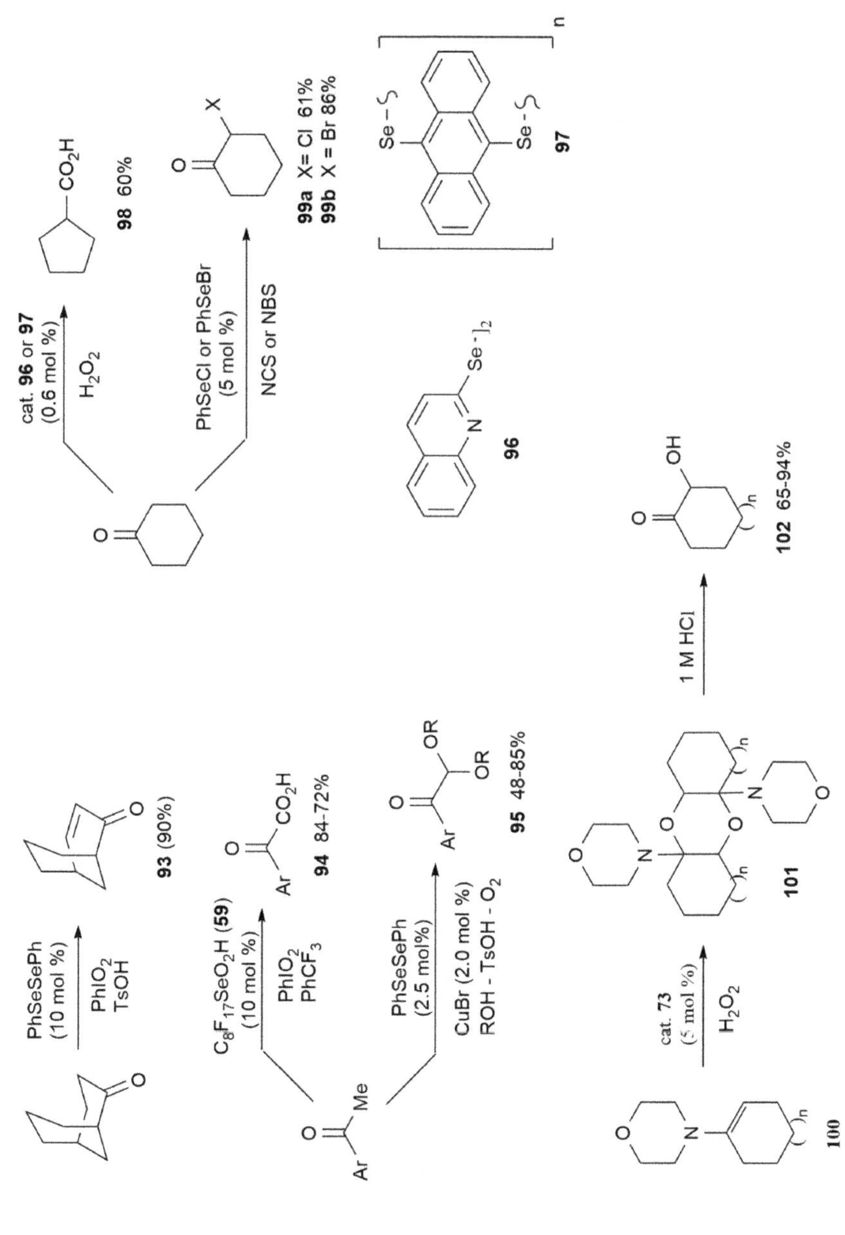

Scheme 13.25 Oxidations of ketones with selenium-based catalysts.

73 and hydrogen peroxide. The latter process afforded isolable dimers **101**, which were hydrolyzed quantitatively to the target compounds **102**.[81] Examples are shown in Scheme 13.25.

13.6.3 Baeyer–Villiger Oxidations

The combination of benzeneseleninic acid and excess hydrogen peroxide was first employed for Baeyer–Villiger oxidations by Grieco and co-workers in 1977.[110] They, and many subsequent researchers, attributed this process to the formation of the corresponding peroxyseleninic acid **66**. Syper and Młochowski[111] reported examples of Baeyer–Villiger reactions conducted with benzeneseleninic, as well as with its 2- and 2,4-dinitro derivatives, and hydrogen peroxide. They indicated that the intermediate peroxy acid **66** isomerized to the corresponding selenonic acid [PhSe(=O)$_2$OH, **68**] upon heating in acetonitrile. However, it was subsequently discovered that the latter product was actually the mixed salt **67** (See Scheme 13.19), which formed rapidly even at room temperature and was unequivocally identified by X-ray crystallography.[72] Thus, as in the case of epoxidation, the assignment of structure **66**, or its nitro derivatives, as the active oxidant in Baeyer–Villiger oxidations must also be treated with reservation.

Although Baeyer–Villiger reactions are generally slower than epoxidations, Syper[112] reported that acrolein derivatives produced formate esters as the major products when treated with various diselenide catalysts and hydrogen peroxide. Other researchers[79,113,114] have also observed that α,β-unsaturated ketones afforded lactones (*e.g.* **103**[113]) or esters (*e.g.* **104**[114a]) as principal products in such oxidations. This can be attributed to the electron-deficient C=C bond conjugated to the carbonyl moiety that suppresses competing electrophilic epoxidation, as well as relief of angle strain in the formation of **103**. Isatins afforded the corresponding isatoic acid anhydrides **105** with diphenyl diselenide and hydrogen peroxide.[115] The first asymmetric Baeyer–Villiger oxidations were conducted by Uemura *et al.*,[116] albeit in low ee, with symmetrically substituted cyclic ketones **106** by means of the chiral catalyst **107**. Several other catalysts have proven successful in Baeyer–Villiger oxidations (Scheme 13.26). These include seleninic acid **108**,[117] fluorinated compounds like selenoxide **72**,[80] seleninic acids **71**,[82,118,119] and **109**[119] and its precursor selenide **110** *via* selenoxide elimination and further oxidation of the selenenic acid, as in the case of **70**. Fluorinated solvents permitted facile recovery of the catalysts under mono-, bi-, or triphasic conditions.[119] Furthermore, the polymer-supported seleninic acid employed by Taylor and Flood[73] for epoxidation and oxidation of alcohols also proved successful in Baeyer–Villiger reactions.

13.7 Oxidation of Nitrogen Compounds

13.7.1 Amines and Anilines

In the 1970s and 1980s, several groups reported that oxidations of primary or secondary amines with equivalent amounts of benzeneseleninic anhydride

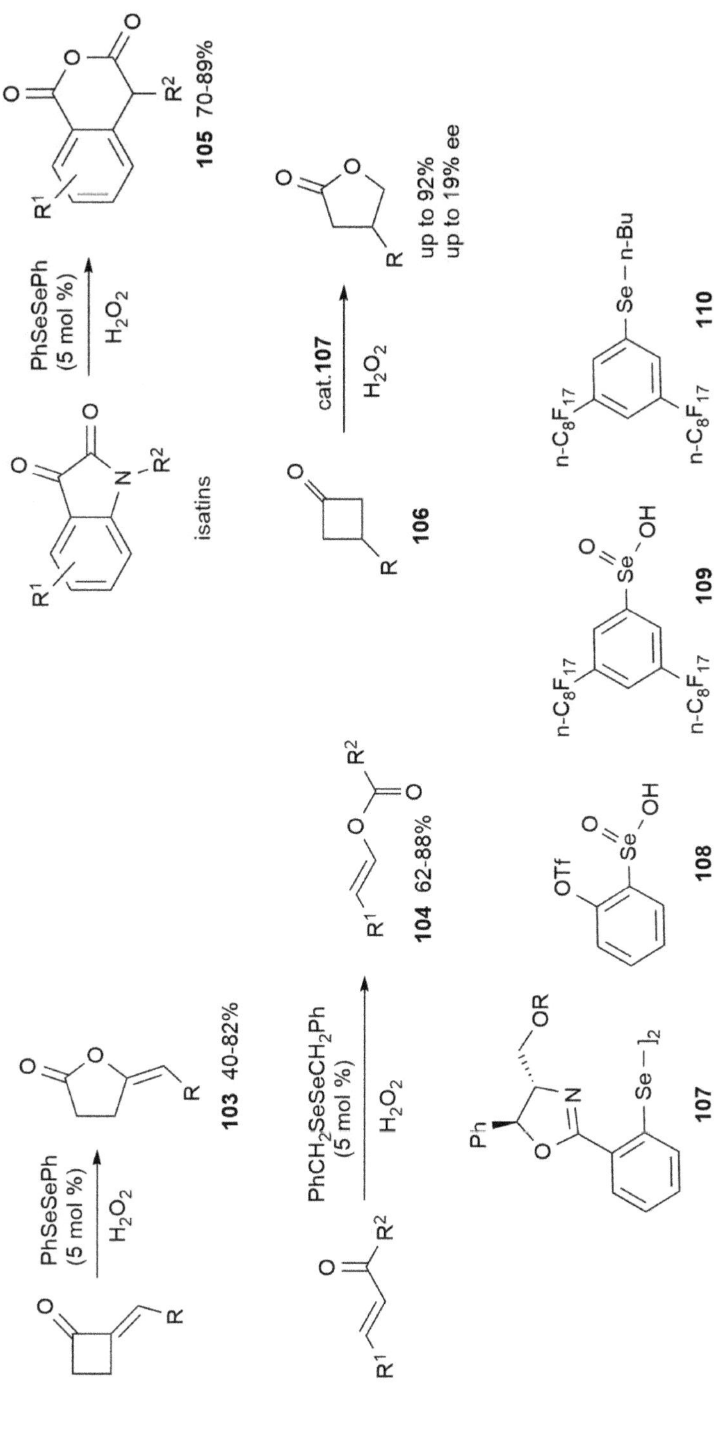

Scheme 13.26 Baeyer–Villiger oxidations with selenium-based catalysts.

or similar reagents produced imines in a manner analogous to the oxidation of alcohols in Section 13.5.1.[120–122] The imines were then hydrolyzed to aldehydes or ketones, oxidized further to nitriles or converted into α-cyanoamines by the addition of cyanide anion. Catalytic variations of these processes followed. Młochowski *et al.*[123] employed various diselenides, ebselen and a silica-supported selenenamide to convert primary amines to nitriles, secondary amines to nitrones, and tetrahydroisoquinoline to isoquinoline.

Under similar conditions, Bäckvall[124] reported that anilines were oxidized with diphenyl diselenide and hydrogen peroxide to mixtures of nitroso compounds **111** and azo derivatives **112** as the major and minor products, respectively. The nitroso products were then used in hetero-Diels–Alder reactions to afford oxazines. Priewisch and Rück-Braun[125] employed selenium dioxide as the catalyst and found that the outcome of the reaction was dependent on the substituents and conditions, as exemplified by the formation of products **113–115**, along with byproduct **116**, obtained from heptane. Luo *et al.*[126] demonstrated that nitroso compounds prepared in this manner can be trapped by enolates of β-keto esters *via* nitroso aldol reactions in the presence of a chiral base to afford products **117** (Scheme 13.27).

13.7.2 Hydrazones, Oximes, and Hydroxylamines

The use of seleninic anhydrides for the liberation of ketones from their hydrazone, oxime, or related derivatives dates back to 1980.[127] More recently, Młochowski *et al.*[104c,128] devised a catalytic process wherein aldehyde derivatives of structure **118** produced carboxylic acids when oxidized with hydrogen peroxide and polymeric diselenide catalysts **119**. Dimethylhydrazones afforded nitriles or amides when oxidized under similar conditions in acetonitrile.[128b] Yu and co-workers reported that aldoximes afforded nitriles with a di-(3-fluorophenyl) diselenide catalyst and hydrogen peroxide as the co-oxidant,[129] while such oxidations conducted with benzeneseleninic acid do not require a peroxide co-oxidant, as the byproduct selenenic acid (PhSeOH) is reoxidized by air.[130] The process was scaled up to produce 1 kg of anisonitrile without the use of a solvent.[131] Ketoximes typically produced ketones **120** when catalyzed with dibenzyl diselenide and hydrogen peroxide, while aldoximes afforded mixtures of aldehydes and nitriles with this catalyst.[114b,132] The oxidation of aldoximes with a polystyrene-supported seleninic acid produced esters in the presence of alcohols.[133] Lactones from the Baeyer–Villiger oxidation of initially-formed ketones, as well as ring-contracted and oxidative cleavage products were obtained when a taraxerone oxime was treated with catalytic selenium dioxide and hydrogen peroxide.[134] A remarkable rearrangement of hydroxylamine derivatives **121** with catalytic benzeneselenenyl bromide was reported by Zhao *et al.* to proceed *via* a double [2,3]sigmatropic rearrangement of intermediates **122** to afford phenols **123** or dienones **124**.[135] Examples are provided in Scheme 13.28.

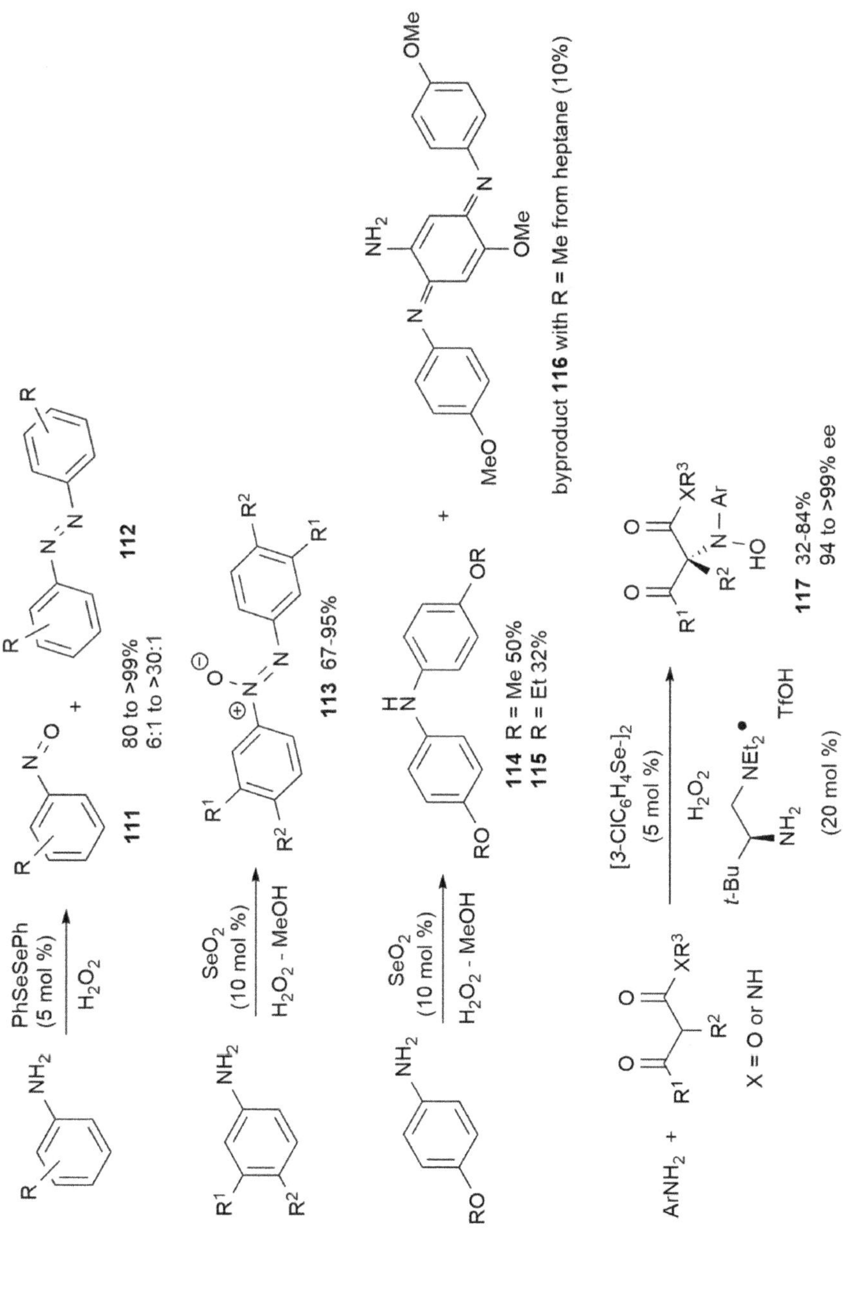

Scheme 13.27 Oxidations of anilines with selenium-based catalysts.

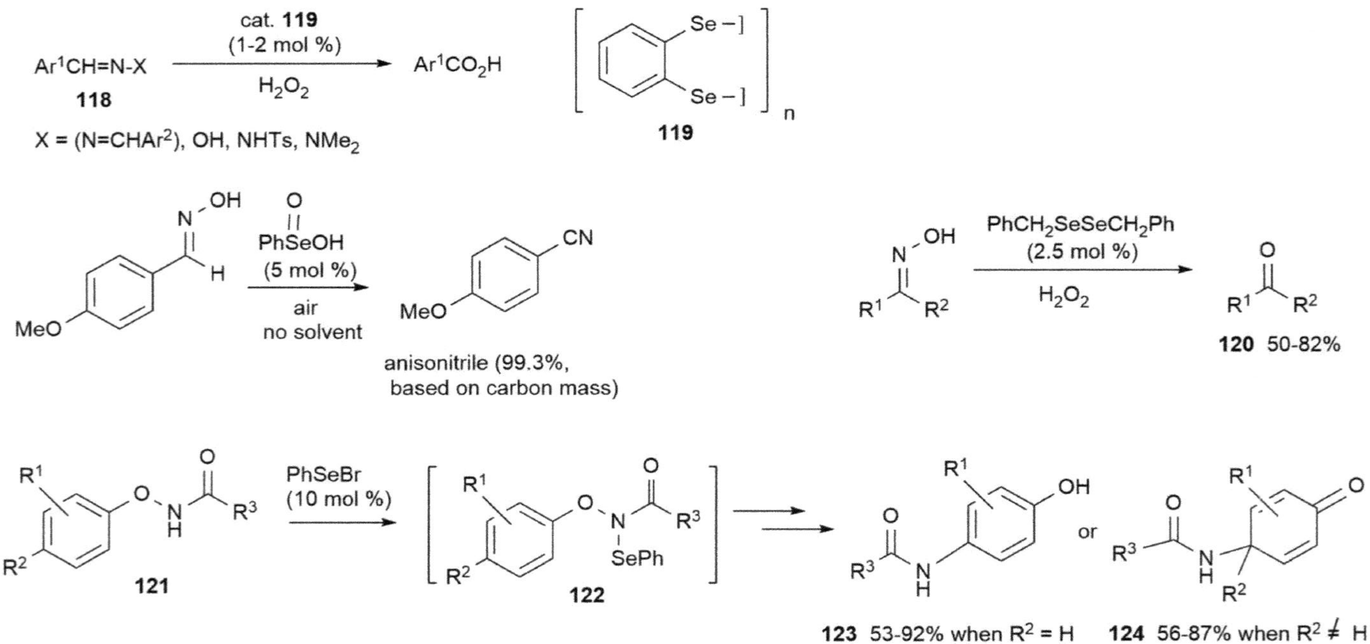

Scheme 13.28 Oxidations of hydrazones, oximes, and related derivatives with selenium-based catalysts.

13.8 Oxidations of Sulfur Compounds

Sulfides are readily oxidized by hydrogen peroxide in the presence of selenium-based catalysts, but the initially produced sulfoxides are easily oxidized further to the corresponding sulfones. In 1978, Reich and co-workers[70] reported that methyl phenyl sulfide can be oxidized selectively to either sulfoxide **125** or sulfone **126** with *o*-nitrobenzeneseleninic acid as the catalyst, depending on the conditions. Other researchers have employed benzeneseleninic acid together with iodosobenzene (PhI=O),[136] or diphenyl diselenide with the urea–hydrogen peroxide complex,[137] for the selective preparation of sulfoxides. Recently, biphasic flow conditions were used by Santi *et al.*[138] in the oxidation of sulfides such as ethyl phenyl sulfide with hydrogen peroxide and selenium dioxide catalyst. It was possible to prepare either the sulfoxide or the sulfone by adjustment of the flow conditions. The oxidation of disulfides to thiolsulfinates **127** is similarly challenging because of the possible overoxidation to thiolsulfonates **128**. However, good selectivity for **127** was achieved by oxidation with the cyclic seleninate ester **73** and hydrogen peroxide.[139] Interestingly, unsymmetrical alkyl aryl disulfides were oxidized at the sulfur centre proximal to the alkyl substituent even in the case of a *t*-butyl derivative. The reaction was accelerated by acids like trifluoroacetic acid (TFA) and arrested by bases such as Cs_2CO_3. Certain thiolsulfinates (*e.g.*, the diallyl derivative allicin) are flavour components of allium species that also display mild antibiotic activity.[140] The conversion of thiols to disulfides was effected with the zinc reagent **129** and afforded high yields with a stoichiometric amount of **129**, while slower reactions were reported with catalytic amounts.[141] For examples, see Scheme 13.29.

13.9 Catalytic Reduction Reactions

While most of the synthetically useful redox chemistry of selenium has involved oxidation reactions conducted with selenium(IV) or (VI) species, lower oxidation states of the element lend themselves to a variety of useful reductions.[142]

13.9.1 Reduction of Nitro Groups with Carbonyl Selenide

Carbonyl selenide (Se=C=O, **130**) is a gas at room temperature and is awkward to handle. While an effective and detailed method for its preparation has been reported,[143] it is more conveniently generated *in situ* from elemental selenium and carbon monoxide at atmospheric or higher pressures. Carbonyl selenide is hydrolyzed to hydrogen selenide, which can be harnessed *in situ* as a reducing agent, while recycling elemental selenium in the process. The reagent can also serve as a carbonylating agent and surrogate for phosgene, again releasing hydrogen selenide that is recycled with air or oxygen. Sonoda *et al.*[144] reported that aryl nitro compounds are catalytically reduced to anilines **131** by **130** in the presence of amines under such

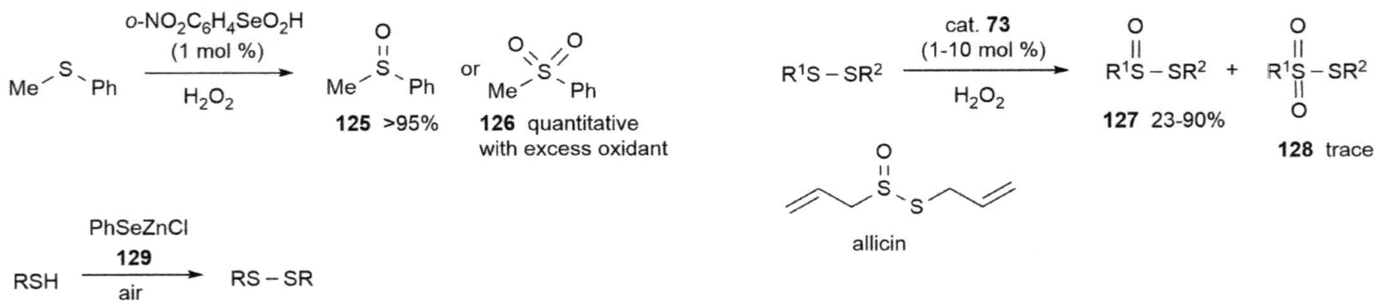

Scheme 13.29 Oxidations of sulfides, disulfides, and thiols with selenium-based catalysts.

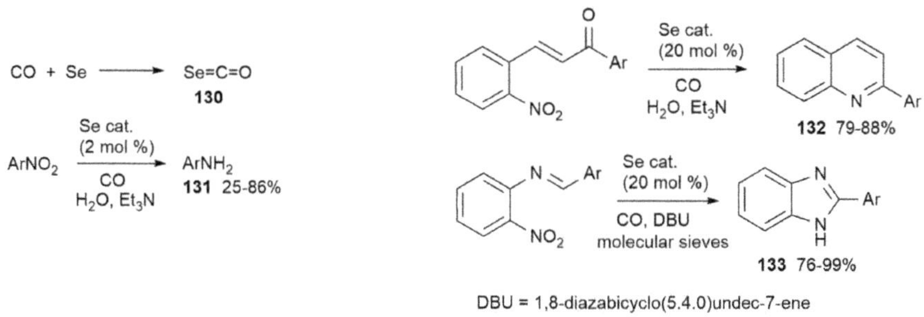

Scheme 13.30 Catalytic reductions of nitro groups and the formation of heterocycles with carbonyl selenide.

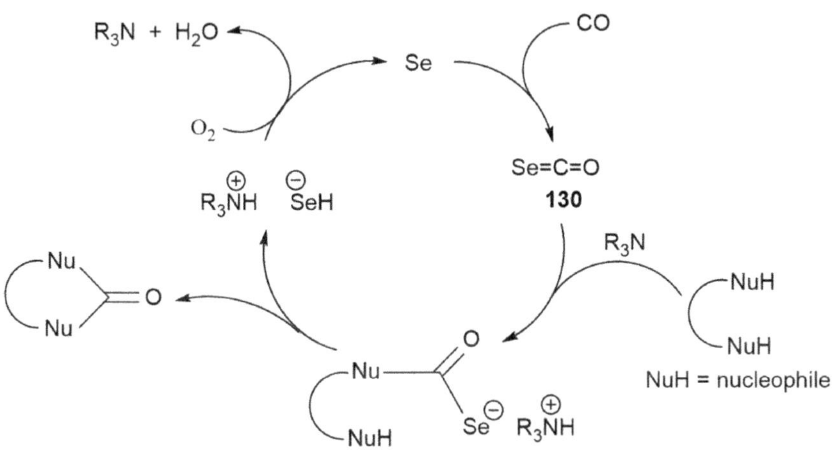

Scheme 13.31 Catalytic cycle for carbonylative coupling of nucleophiles with carbonyl selenide.

conditions. Later, Lu *et al.* reported similar results.[145] The resulting aniline can then undergo condensation and cyclization with other functionalities, thus leading to convenient syntheses of quinolines **132**[146] (Scheme 13.30), as well as indoles,[146,147] pyrroles,[148] and 3,4-dihydroquinazolin-3-ones.[149] Nitrobenzenes containing *o*-imino substituents afforded benzimidazoles **133** when reduced by **130** (Scheme 13.30),[150] while other nitrobenzenes generated anilines that condense with aldehydes to produce the corresponding imines.[151] Benzamide and nicotinamide underwent *N*-arylation by nitroarene or aniline derivatives under similar conditions.[152]

13.9.2 Carbonylations with Carbonyl Selenide

When carbonyl selenide (**130**) is generated in the presence of nucleophiles, carbonylative coupling ensues, as shown in Scheme 13.31. In the case of a primary or secondary amine and a nitro compound subjected to these conditions, the latter is reduced to the corresponding aniline and the two amines are coupled *in situ* by **130** to afford the corresponding unsymmetrical cyclic or acyclic ureas,[153] as well as various quinazoline-2,4-diones[154] and indoloquinazolinones.[155] Alternatively, two different amines can be coupled directly with **130**, without the need for a nitro precursor.[156] Similar carbonylations with selenium-containing ionic liquids have been reported.[157,158] The conversion of amines and alcohols, or amino alcohols, to acyclic and cyclic carbamates, respectively, can also be achieved with selenium and carbon monoxide (*i.e. via* **130**).[153f,159,160] In many examples the amine functionality was prepared *in situ* from the corresponding nitro compound in the usual manner. Alternative catalysts for carbamate formation include $CuSe_2/CeO_2$ with cetyl triethylammonium bromide[161] and immobilized selenium-functionalized mesoporous silica-supported ionic liquids.[162] Similarly, vicinal diols afford cyclic carbonates with **130**,[163] while nitroarenes or anilines react with thiols to produce thiocarbamates[159a,164] and disulfides provide thiocarbonates when treated with alcohols and catalyst **130**.[165] Illustrative examples of **134**,[153g] **135**,[159a] **136**,[163] and **137**[165] are provided in Scheme 13.32.

13.9.3 Other Selenium-catalyzed Reductions

A number of functional group reductions with carbonyl selenide (**130**) were reported by Sonoda, Nishiyama, and their co-workers, as well as by other researchers utilizing different selenium catalysts. Vicinal dibromides and α-halo ketones underwent reductive dehalogenation to alkenes and ketones, respectively, with hydrogen selenide produced from **130** in the presence of water and *N*-methylpyrrolidine.[166] Crich[167] employed sodium borohydride and diselenide catalyst **139** under fluorous conditions to effect the elimination of the vicinal dimesylate moiety of nucleoside **138**. α,β-Unsaturated ketones were reduced to alkyl ketones with **130** generated from catalytic selenium and carbon monoxide in the presence of water,[168] or with the reagents NaHSe and LiHSe.[169] Deoxygenations of α- and β-diketones to monoketones[170] and of aryl ketones to benzylic products[171] with *in situ* generated **130** under aqueous conditions were also reported by Nishiyama *et al.* A more elaborate system devised by Mamane *et al.*[172] was based on the activation of Hantzsch ester **141** by the electron-deficient bipyridyl selenide catalyst **140** through a sigma hole interaction, which enabled the reduction of 2-phenylquinoline to its tetrahydro derivative, although enantioselectivity could not be induced with the pure (*P*)-enantiomer of **140**. Uemura *et al.*[173] reported the asymmetric hydrosilylation and reduction of ketones to afford alcohols by means of Rh(I) or Ir(I) catalysts in conjunction with chiral ferrocenyl selenide or diselenide species such as **142**. The latter process was extended to the enantioselective reduction of imines and dehydroamino acids (Scheme 13.33).[173c]

The selenium-catalyzed asymmetric addition of diethylzinc to aldehydes was reported by Wirth *et al.*,[174] who proposed the intermediate **144** in aggregate form as the active catalytic species, derived from diselenides such as

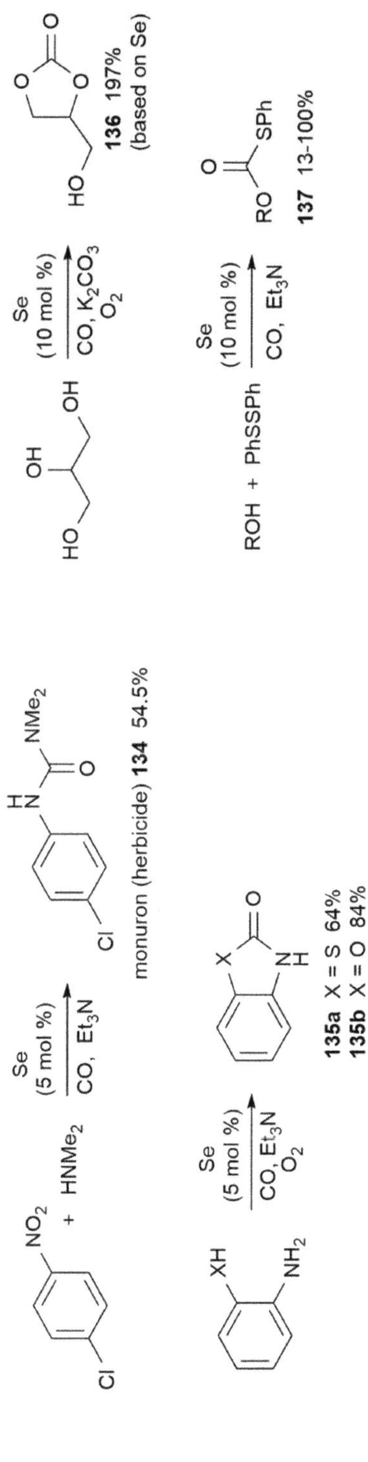

Scheme 13.32 Carbonylations of amines and alcohols with carbonyl selenide.

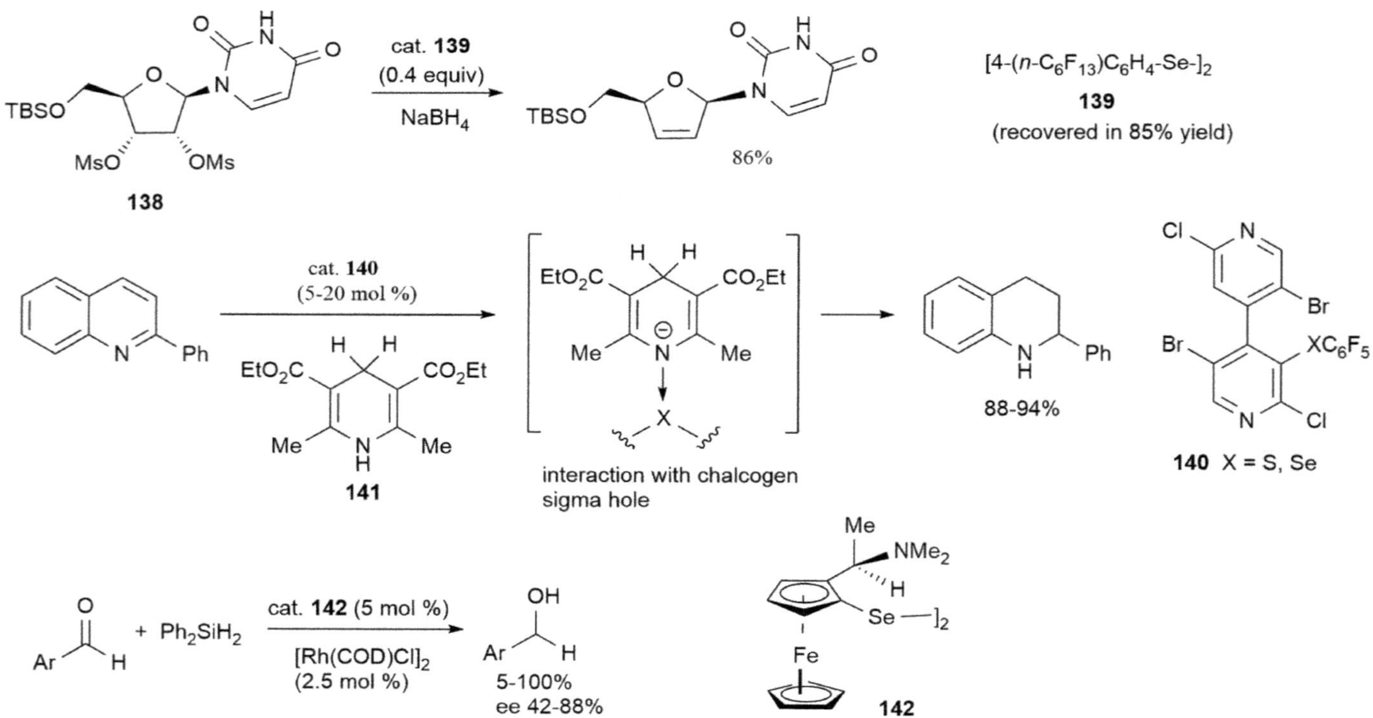

Scheme 13.33 Miscellaneous selenium-catalyzed reductions.

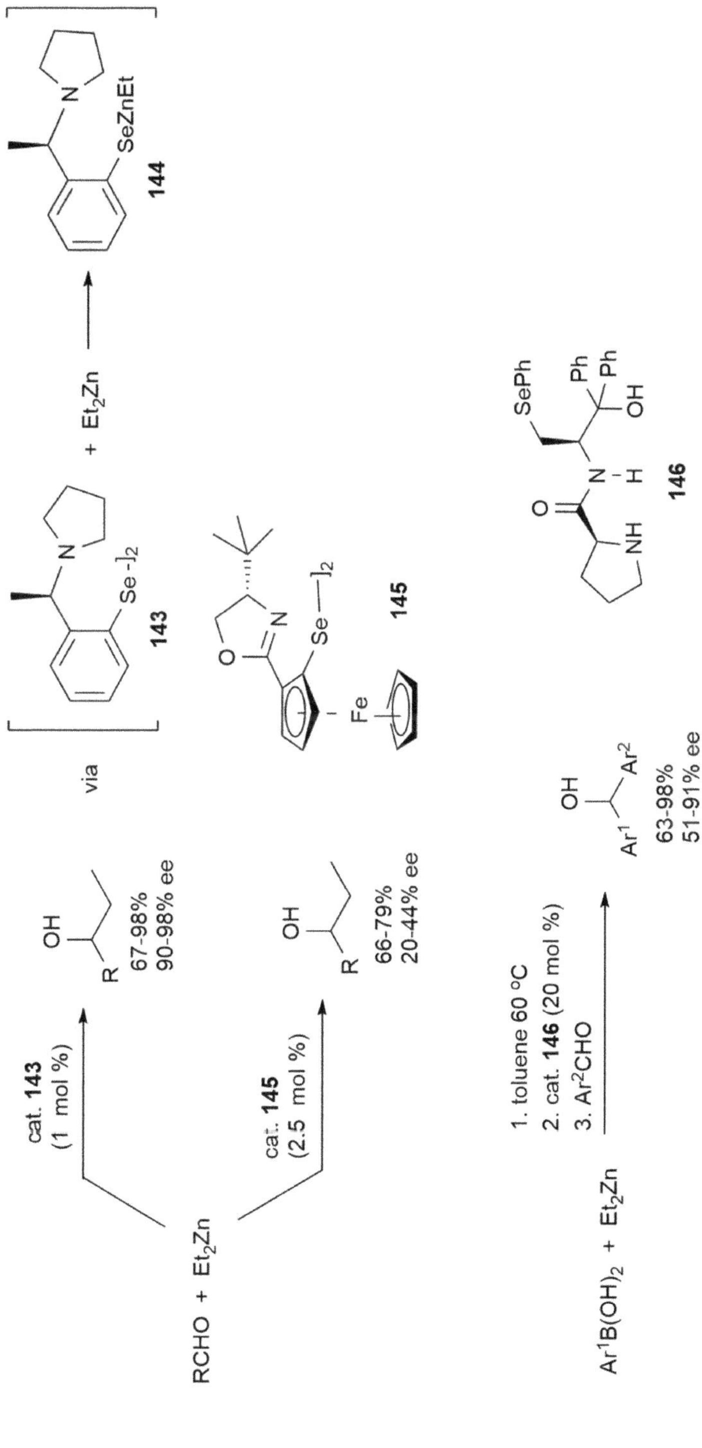

Scheme 13.34 Selenium-catalyzed asymmetric addition of diethylzinc to aldehydes and boronic acids.

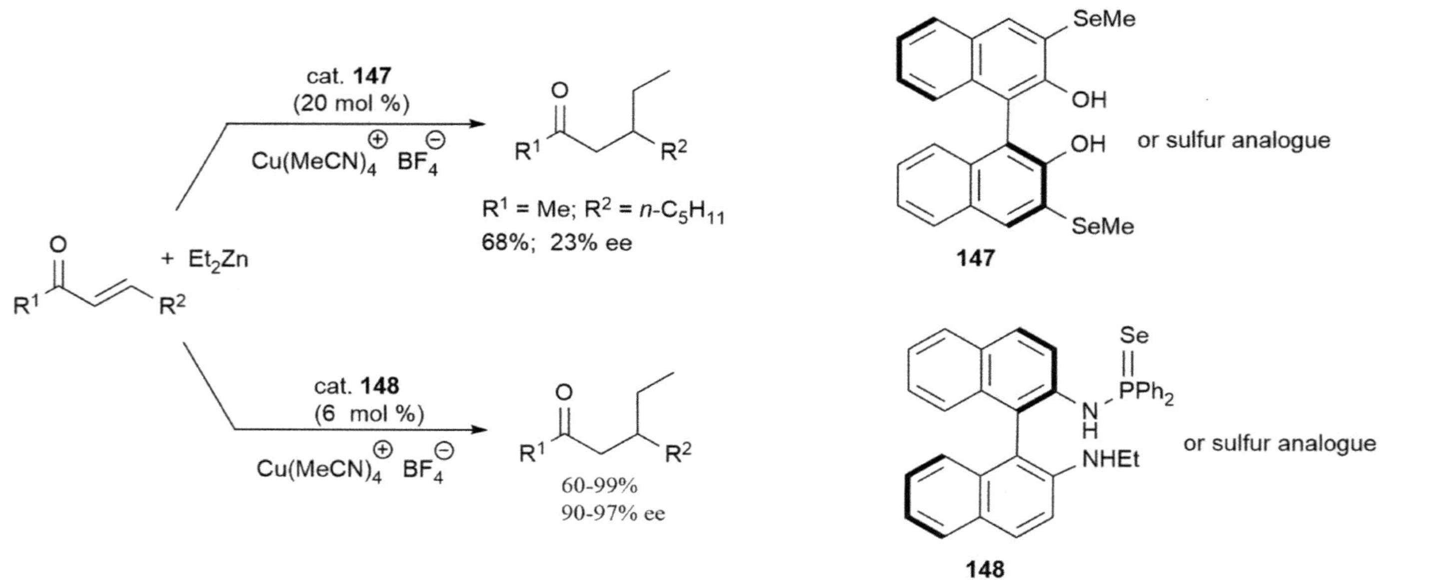

Scheme 13.35 Conjugate addition of diethylzinc to enones catalyzed by chiral binaphthyl selenium derivatives with Cu(II) as the cocatalyst.

143. An interesting asymmetric amplification was observed during the process. Other chiral diselenide catalysts for this process were later reported by Santi and Wirth,[175] and by Braga *et al.*,[176] while Bolm and co-workers[177] employed the ferrocene diselenide **145** and extended the process to include diphenylzinc, which provided higher yields and enantiomeric excesses. The Braga group also reported a reductive coupling of diethylzinc with boronic acids in the presence of chiral selenide **146** (Scheme 13.34).[178]

The conjugate addition of diethylzinc to enones like non-3-en-2-one in the presence of a Cu(I) cocatalyst and the chiral diselenide **147** was first reported by Woodward *et al.*[179] and was later extended by Shi and co-workers[180] to a broad range of cyclic and acyclic enones with similar binaphthyl-based phosphine sulfide and selenide catalysts **148**, affording improved ee of >90% (Scheme 13.35). Copper-catalyzed asymmetric conjugate additions of Grignard reagents to enones with chiral oxazoline-based diselenides were reported by Braga *et al.*[181]

The deoxygenation of sulfoxides to sulfides and reduction of azobenzene to the corresponding hydrazine was reported in 1966 by Gunther,[182] who reduced diselenides with hypophosphorous acid to produce selenols, which presumably served as the active catalysts. This method was also used to reduce disulfides to thiols. Moreover, **130** generated *in situ* in the presence of water was also reported to catalyze the reduction of disulfides, as well as diselenides, to thiols and selenols, respectively.[183]

References

1. For recent books on organoselenium chemistry, see: (a) *Organoselenium Chemistry – A Practical Approach*, ed. T. G. Back, Oxford University Press, Oxford, 1999; (b) *Organoselenium Chemisry – Modern Developments in Organic Synthesis*, ed. T. Wirth, Springer, Berlin, 2000; (c) *Organoselenium Chemistry – Synthesis and Reactions*, ed. T. Wirth, Wiley-VCH, Weinheim, 2012; (d) *The Chemistry of Organic Selenium and Tellurium Compounds*, ed. Z. Rappoport, J. Wiley and Sons, Chichester, 1986, vol 1; 1987, vol. 2; 2012, vol. 3; 2014, vol. 4; (e) *Organoselenium Chemistry: Between Synthesis and Biochemistry*, ed. C. Santi, Bentham, Sharjah, 2014; (f) E. J. Lenardão, C. Santi and L. Sancineto, *New Frontiers in Organoselenium Compounds*, Springer, Heidelberg, 2018; (g) V. K. Jain and K. I. Priyadarsini, *Organoselenium Compounds in Biology and Medicine: Synthesis, Biological and Therapeutic Treatments*, Royal Society of Chemistry, London, 2018.
2. H. J. Reich and R. J. Hondal, *ACS Chem. Biol.*, 2016, **11**, 821.
3. For reviews of catalytic and green selenium chemistry, see (a) A. L. Braga, D. S. Lüdtke, F. Vargas and R. C. Braga, *Synlett*, 2006, 1453; (b) C. Santi, S. Santoro and B. Battistelli, *Curr. Org. Chem.*, 2010, **14**, 2442; (c) C. Santi, *Curr. Green Chem.*, 2019, **6**, 9; (d) S. Santoro, J. B. Azeredo, V. Nascimento, L. Sancineto, A. L. Braga and C. Santi, *RSC Adv.*, 2014, **4**, 31521; (e) D. M. Freudendahl, S. Santoro, S. A. Shahzad,

C. Santi and T. Wirth, *Angew. Chem., Int. Ed.*, 2009, **48**, 8409; (f) F. V. Singh and T. Wirth, *Catal. Sci. Technol.*, 2019, **9**, 1073.

4. For reviews of selenium-catalyzed oxidations, see: (a) T. G. Back, *Curr. Green Chem.*, 2016, **3**, 76; (b) J. Młochowski and H. Wójtowicz-Młochowska, *Mini-Rev. Org. Chem.*, 2016, **13**, 473; (c) J. Młochowski and H. Wójtowicz-Młochowska, *Molecules*, 2015, **20**, 10205; (d) J. Młochowski, M. Brząszcz, M. Giurg, J. Palus and H. Wójtowicz, *Eur. J. Org. Chem.*, 2003, 4329.
5. (a) S. Ortgies and A. Breder, *ACS Catal.*, 2017, **7**, 5828; (b) A. Breder and S. Ortgies, *Tetrahedron Lett.*, 2015, **56**, 2843; (c) A. Breder, *Synlett*, 2014, **25**, 899.
6. (a) T. Hori and K. B. Sharpless, *J. Org. Chem.*, 1979, **44**, 4204; (b) T. Hori and K. B. Sharpless, *J. Org. Chem.*, 1979, **44**, 4208.
7. J. A. Tunge and S. R. Mellegaard, *Org. Lett.*, 2004, **6**, 1205.
8. (a) S. R. Mellegaard and J. A. Tunge, *J. Org. Chem.*, 2004, **69**, 8979; (b) S. R. Mellegaard-Waetzig, C. Wang and J. A. Tunge, *Tetrahedron*, 2006, **62**, 7191; (c) C. Wang and J. Tunge, *Chem. Commun.*, 2004, 2694.
9. A. Verma, S. Jana, C. D. Prasad, A. Yadav and S. Kumar, *Chem. Commun.*, 2016, **52**, 4179.
10. S. J. Balkrishna, C. D. Prasad, P. Panini, M. R. Detty, D. Chopra and S. Kumar, *J. Org. Chem.*, 2012, **77**, 9541.
11. D. W. Tay, I. T. Tsoi, J. C. Er, G. Y. C. Leung and Y.-Y. Yeung, *Org. Lett.*, 2013, **15**, 1310.
12. I. Carrera, M. C. Brovetto and G. A. Seoane, *Tetrahedron Lett.*, 2006, **47**, 7849.
13. M. A. Goodman and M. R. Detty, *Organomet.*, 2004, **23**, 3016.
14. M. D. Drake, M. A. Bateman and M. R. Detty, *Organomet.*, 2003, **22**, 4158.
15. E. E. Alberto, A. L. Braga and M. R. Detty, *Tetrahedron*, 2012, **68**, 10476.
16. (a) C. Francavilla, F. V. Bright and M. R. Detty, *Org. Lett.*, 1999, **1**, 1043; (b) C. Francavilla, M. D. Drake, F. V. Bright and M. R. Detty, *J. Am. Chem. Soc.*, 2001, **123**, 57; (c) M. D. Drake, F. V. Bright and M. R. Detty, *J. Am. Chem. Soc*, 2003, **125**, 12558.
17. (a) S. M. Bennett, Y. Tang, D. McMaster, F. V Bright and M. R. Detty, *J. Org. Chem.*, 2008, **73**, 6849; (b) C. M. Gatley, L. M. Muller, M. A. Lang, E. E. Alberto and M. R. Detty, *Molecules*, 2015, **20**, 9616.
18. (a) T. G. Back, B. P. Dyck and S. Nan, *Tetrahedron*, 1999, **55**, 3191; (b) The term "cyclofunctionalization" was first used by Clive *et al.*: D. L. J. Clive, G. Chittattu, N. J. Curtis, W. A. Kiel and C. K. Wong, *J. Chem. Soc., Chem. Commun.*, 1977, 725
19. J. E. Tungen, R. Kristianslund, A. Vik and T. V. Hansen, *J. Org. Chem.*, 2019, **84**, 11373.
20. (a) F. Chen, C. K. Tan and Y.-Y. Yeung, *J. Am. Chem. Soc.*, 2013, **135**, 1232; (b) R. Nishiyori, J. R. J. Maynard and S. Shirakawa, *Asian J. Org. Chem.*, 2020, **9**, 192.
21. A. F. Barrero, J. F. Quilez del Moral, M. M. Herrador, M. Cortés, P. Arteaga, J. V. Catalán, E. M. Sánchez and J. F. Arteaga, *J. Org. Chem.*, 2006, **71**, 5811.

22. A. J. Cresswell, S. T.-C. Eey and S. E. Denmark, *Nat. Chem.*, 2015, **7**, 146.
23. B. B. Gilbert, S. T.-C. Eey, P. Ryabchuk, O. Garry and S. E. Denmark, *Tetrahedron*, 2019, **75**, 4086.
24. (a) T. Wirth, G. Fragale and M. Spichty, *J. Am. Chem. Soc.*, 1998, **120**, 3376; (b) S. E. Denmark and M. G. Edwards, *J. Org. Chem.*, 2006, **71**, 7293; (c) S. E. Denmark, W. R. Collins and M. D. Cullen, *J. Am. Chem. Soc.*, 2009, **131**, 3490.
25. F. Krätzschmar, M. Kaßel, D. Delony and A. Breder, *Chem. – Eur. J.*, 2015, **21**, 7030.
26. S. Ortgies and A. Breder, *Org. Lett.*, 2015, **17**, 2748.
27. Y. Kawamata, T. Hashimoto and K. Maruoka, *J. Am. Chem. Soc.*, 2016, **138**, 5206.
28. Y. Otsuka, Y. Shimazaki, H. Nagaoka, K. Maruoka and T. Hashimoto, *Synlett*, 2019, **30**, 1679.
29. X. Zhang, R. Guo and X. Zhao, *Org. Chem. Front.*, 2015, **2**, 1334.
30. R. Guo, J. Huang, H. Huang and X. Zhao, *Org. Lett.*, 2016, **18**, 504.
31. J. Trenner, C. Depken, T. Weber and A. Breder, *Angew. Chem., Int. Ed.*, 2013, **52**, 8952.
32. Z. Deng, J. Wei, L. Liao, H. Huang and X. Zhao, *Org. Lett.*, 2015, **17**, 1834.
33. L. Liao, R. Guo and X. Zhao, *Angew. Chem., Int. Ed.*, 2017, **56**, 3201.
34. R. Guo, J. Huang and X. Zhao, *ACS Catal.*, 2018, **8**, 926.
35. F. Krätzschmar, S. Ortgies, R. Y. N. Willing and A. Breder, *Catalysis*, 2019, **9**, 153.
36. Y. Zhang, Y. Shao, J. Gong, J. Zhu, T. Cheng and J. Chen, *J. Org. Chem.*, 2019, **84**, 2798.
37. S. Leisering, I. Riaño, C. Depken, L. J. Gross, M. Weber, D. Lentz, R. Zimmer, C. B. W. Stark, A. Breder and M. Christmann, *Org. Lett.*, 2017, **19**, 1478.
38. (a) D. M. Browne, O. Niyomura and T. Wirth, *Org. Lett.*, 2007, **9**, 3169; (b) D. M. Browne, O. Niyomura and T. Wirth, *Phosphorus, Sulfur Silicon Relat. Elem.*, 2008, **183**, 1026.
39. S. A. Shahzad, C. Venin and T. Wirth, *Eur. J. Org. Chem.*, 2010, 3465.
40. F. V. Singh and T. Wirth, *Org. Lett.*, 2011, **13**, 6504.
41. M. Iwaoka and S. Tomoda, *J. Chem. Soc., Chem. Commun.*, 1992, 1165.
42. T. Wirth, S. Häuptli and M. Leuenberger, *Tetrahedron: Asymmetry*, 1998, **9**, 547.
43. O. Niyomura, M. Cox and T. Wirth, *Synlett*, 2006, 251.
44. (a) M. Tiecco, L. Testaferri, C. Santi, C. Tomassini, F. Marini, L. Bagnoli and A. Temperini, *Chem. – Eur. J.*, 2002, **8**, 1118; (b) C. Santi, M. Tiecco, L. Testaferri, C. Tomassini, S. Santoro and G. Bizzoca, *Phosphorus, Sulfur Silicon Relat. Elem.*, 2008, **183**, 956; (c) For an earlier racemic variation, see: M. Tiecco, L. Testaferri, M. Tingoli, L. Bagnoli and C. Santi, *Synlett*, 1993, 798
45. L. Sancineto, F. Mangiavacchi, C. Tidei, L. Bagnoli, F. Marini, A. Gioiello, J. Scianowski and C. Santi, *Asian J. Org. Chem.*, 2017, **6**, 988.

46. For a review of other single electron transfer reactions of organoselenium compounds, see: T. G. Back, in *The Chemistry of Organic Selenium and Tellurium Compounds*, ed. Z. Rappoport, J. Wiley and Sons, Chichester, 2012, ch. 8, vol. 3, Part 1, pp. 486–489.
47. (a) S. Ortgies, R. Rieger, K. Rode, K. Koszinowski, J. Kind, C. M. Thiele, J. Rehbein and A. Breder, *ACS Catal.*, 2017, **7**, 7578; (b) S. Ortgies, C. Depken and A. Breder, *Org. Lett.*, 2016, **18**, 2856.
48. Z. Zhu, J. Luo and X. Zhao, *Org. Lett.*, 2017, **19**, 4940.
49. X. Liu, R. An, X. Zhang, J. Luo and X. Zhao, *Angew. Chem., Int. Ed.*, 2016, **55**, 5846.
50. L. Yu, F. Chen and Y. Ding, *ChemCatChem*, 2016, **8**, 1033.
51. H. L. Riley, J. F. Morley and N. A. C. Friend, *J. Chem. Soc.*, 1932, 1875.
52. (a) M. A. Umbreit and K. B. Sharpless, *J. Am. Chem. Soc.*, 1977, **99**, 5526; (b) For an earlier study of the mechanism of these processes, see: K. B. Sharpless and R. F. Lauer, *J. Am. Chem. Soc.*, 1972, **94**, 7154.
53. (a) N. Rabjohn, *Org. React.*, 1949, **5**, 331; (b) N. Rabjohn, *Org. React.*, 1976, **24**, 261; (c) G. R. Waitkins and C. W. Clark, *Chem. Rev.*, 1945, **36**, 235; (d) R. A. Jerussi, *Selective Organic Transformations*, 1970, vol. 1, pp. 301–326; (e) E. N. Trachtenberg, in *Oxidation*, ed. R. L. Augustine, Marcel Dekker, New York, 1969, vol. 1, pp. 119–187; (f) S. E. Dibrell, Y. Tao and S. E. Reisman, *Acc. Chem. Res.*, 2021, **54**, 1360.
54. D. H. R. Barton and D. Crich, *Tetrahedron*, 1985, **41**, 4359.
55. A. Guillemonat, *Ann. Chim.*, 1939, **11**, 143.
56. P. Magnus, M. Ladlow and P. M. Cairns, *Tetrahedron Lett.*, 1987, **28**, 3307.
57. D. Crich and Y. Zou, *Org. Lett.*, 2004, **6**, 775.
58. (a) D. H. R. Barton and T.-L. Wang, *Tetrahedron Lett.*, 1994, **35**, 5149; (b) For earlier investigations of benzylic oxidations with stoichiometric amounts of benzeneseleninic anhydride, see D. H. R. Barton, R. A. H. F. Hui and S. V. Ley, *J. Chem. Soc., Perkin Trans. 1*, 1982, 2179
59. D. Crich and Y. Zou, *J. Org. Chem.*, 2005, **70**, 3309.
60. P. Arsenyan and J. Vasiljeva, *Mendeleev Commun.*, 2017, **27**, 621.
61. W. Jin, P. Zheng, W.-T. Wong and G.-L. Law, *Adv. Synth. Catal.*, 2017, **359**, 1588.
62. M. Giurg, L. Syper and J. Młochowski, *Polish J. Chem.*, 2004, **78**, 231.
63. B. Chabaud and K. B. Sharpless, *J. Org. Chem.*, 1979, **44**, 4202.
64. S. Santoro, B. Battistelli, B. Gjoka, C.-W. S. Si, L. Testaferri, M. Tiecco and C. Santi, *Synlett*, 2010, 1402.
65. L. Liao, H. Zhang and X. Zhao, *ACS Catal.*, 2018, **8**, 6745.
66. For a review of selenium-catalyzed functionalization of alkynes, see: L. Liao and X. Zhao, *Chem. Lett.*, 2021, **50**, 1104.
67. L.-W. Wang, Y.-F. Feng, H. M. Lin, H.-T. Tang and Y.-M. Pan, *J. Org. Chem.*, 2021, **86**, 16121.
68. P. A. Grieco, Y. Yokoyama, S. Gilman and M. Nishizawa, *J. Org. Chem.*, 1977, **42**, 2034.
69. T. Kametani, H. Nemoto and K. Fukumoto, *Heterocycles*, 1977, **6**, 1365.

70. H. J. Reich, F. Chow and S. L. Peake, *Synthesis*, 1978, 299.
71. T. Hori and K. B. Sharpless, *J. Org. Chem.*, 1978, **43**, 1689.
72. K. N. Sands, E. Mendoza Rengifo, G. N. George, I. J. Pickering, B. S. Gelfand and T. G. Back, *Angew. Chem., Int. Ed.*, 2020, **59**, 4283.
73. R. T. Taylor and L. A. Flood, *J. Org. Chem.*, 1983, **48**, 5160.
74. Y. Yang, X. Fan, H. Cao, S. Chu, X. Zhang, Q. Xu and L. Yu, *Catal. Sci. Technol.*, 2018, **8**, 5017.
75. B. Betzemeier, F. Lhermitte and P. Knochel, *Synlett*, 1999, 489.
76. G.-J. ten Brink, B. C. M. Fernandes, M. C. A. van Vliet, I. W. C. E. Arends and R. A. Sheldon, *J. Chem. Soc., Perkin Trans. 1*, 2001, 224.
77. H. García-Marín, J. C. van der Toorn, J. A. Mayoral, J. I. García and I. W. C. E. Arends, *Green Chem.*, 2009, **11**, 1605.
78. M. De Torres, I. W. C. E. Arends, J. A. Mayoral, E. Pires and G. Jiménez-Osés, *Appl. Catal., A*, 2012, **425–426**, 91.
79. L. Yu, Z. Bai, X. Zhang, X. Zhang, Y. Ding and Q. Xu, *Catal. Sci. Technol.*, 2016, **6**, 1804.
80. M. A. Goodman and M. R. Detty, *Synlett*, 2006, 1100.
81. E. A. Mercier, C. D. Smith, M. Parvez and T. G. Back, *J. Org. Chem.*, 2012, **77**, 3508.
82. B. H. Brodsky and J. Du Bois, *J. Am. Chem. Soc.*, 2005, **127**, 15391.
83. B. R. Buckley, C. E. Elliott, Y. Chan, N. Dreyfus and P. C. Bulman Page, *Synlett*, 2013, **24**, 2266.
84. P. Gogoi, S. Das Sharma and D. Konwar, *Lett. Org. Chem.*, 2007, **4**, 249.
85. Y. Wang, L. Yu, B. Zhu and L. Yu, *J. Mater. Chem. A*, 2016, **4**, 10828.
86. (a) C. Santi, R. Di Lorenzo, C. Tidei, L. Bagnoli and T. Wirth, *Tetrahedron*, 2012, **68**, 10530; (b) For a racemic variation using PhSeSePh, see: S. Santoro, C. Santi, M. Sabatini, L. Testaferri and M. Tiecco, *Adv. Synth. Catal.*, 2008, **350**, 2881
87. L. Yu, J. Wang, T. Chen, Y. Wang and Q. Xu, *Appl. Organomet. Chem.*, 2014, **28**, 652.
88. T. Wang, X. Jing, C. Chen and L. Yu, *J. Org. Chem.*, 2017, **82**, 9342.
89. D. H. R. Barton, A. G. Brewster, R. A. H. F. Hui, D. J. Lester, S. V. Ley and T. G. Back, *J. Chem. Soc., Chem. Commun.*, 1978, 952.
90. T. G. Back, *J. Org. Chem.*, 1981, **46**, 1442.
91. D. H. R. Barton, J. W. Morzycki, W. B. Motherwell and S. V. Ley, *J. Chem. Soc., Chem. Commun.*, 1981, 1044.
92. D. H. R. Barton, C. R. A. Godfrey, J. W. Morzycki, W. B. Motherwell and S. V. Ley, *J. Chem. Soc., Perkin Trans. 1*, 1982, 1947.
93. (a) T. Onami, M. Ikeda and S. S. Woodard, *Bull. Chem. Soc. Jpn.*, 1996, **69**, 3601; (b) H. Ehara, M. Noguchi, S. Sayama and T. Onami, *J. Chem. Soc., Perkin Trans. 1*, 2000, 1429.
94. I. Kuwajima, M. Shimizu and H. Urabe, *J. Org. Chem.*, 1982, **47**, 837.
95. (a) J. C. van der Toorn, G. Kemperman, R. A. Sheldon and I. W. C. E. Arends, *J. Org. Chem.*, 2009, **74**, 3085; (b) J. C. van der Toorn, G. Kemperman, R. A. Sheldon and I. W. C. E. Arends, *Eur. J. Org. Chem.*, 2011, 4345.

96. X. Y. Cheng, K. F. Li, Q. J. Wang, C. Y. Wang and T. Kai Ying, *Chin. Chem. Lett.*, 2012, **23**, 801.
97. F. Ogura, T. Otsubo, K. Ariyoshi and H. Yamaguchi, *Chem. Lett.*, 1983, 1833.
98. (a) P. Singh and A. K. Singh, *Eur. J. Inorg. Chem.*, 2010, 4187; (b) F. Saleem, G. K. Rao, A. Kumar, G. Mukherjee and A. K. Singh, *Organometallics*, 2013, **32**, 3595.
99. (a) D. H. R. Barton, S. V. Ley and M. N. Rosenfeld, *J. Chem. Soc., Perkin Trans. 1*, 1977, 567; (b) D. H. R. Barton, A. G. Brewster, S. V. Ley, C. M. Read and M. N. Rosenfeld, *J. Chem. Soc., Perkin Trans. 1*, 1981, 1473.
100. M. Giurg, E. Kowal, H. Muchalski, L. Syper and J. Młochowski, *Synth. Commun.*, 2009, **39**, 251.
101. M. Giurg, H. Muchalski and E. Kowal, *Synth. Commun.*, 2012, **42**, 2526.
102. (a) M. Giurg, E. Wiech, K. Piekielska, M. Gębala, J. Młochowski, M. Wolański, B. Ditkowski and W. Peczyńska-Czoch, *Pol. J. Chem.*, 2006, **80**, 297; (b) M. Giurg, K. Piekielska, M. Gębala, B. Ditkowski, M. Wolański, W. Peczyńska-Czoch and J. Młochowski, *Synth. Commun.*, 2007, **37**, 1779.
103. J.-K. Choi, Y.-K. Chang and S. Y. Hong, *Tetrahedron Lett.*, 1988, **29**, 1967.
104. (a) M. Brząszcz, K. Kloc, M. Maposah and J. Młochowski, *Synth. Commun.*, 2000, **30**, 4425; (b) H. Wójtowicz, M. Brząszcz, K. Kloc and J. Młochowski, *Tetrahedron*, 2001, **57**, 9743; (c) M. Giurg, M. Brząszcz and J. Młochowski, *Pol. J. Chem.*, 2006, **80**, 417.
105. L. Sancineto, C. Tidei, L. Bagnoli, F. Marini, E. J. Lenardão and C. Santi, *Molecules*, 2015, **20**, 10496.
106. E. Orentas, G. Bagdžiūnas, U. Berg, A. Žilinskas and E. Butkus, *Eur. J. Org. Chem.*, 2007, 4251.
107. I. Jastrzębska, A. Dobrogowska, E. Lutostańska and J. W. Morzycki, *Tetrahedron*, 2010, **66**, 5024.
108. C. Chen, Z. Cao, X. Zhang, Y. Li, L. Yu and X. Jiang, *Chin. J. Chem.*, 2020, **38**, 1045.
109. M. Giurg and J. Młochowski, *Synth. Commun.*, 1999, **29**, 2281.
110. P. A. Grieco, Y. Yokoyama, S. Gilman and Y. Ohfune, *J. Chem. Soc., Chem. Commun.*, 1977, 870.
111. L. Syper and J. Młochowski, *Tetrahedron*, 1987, **43**, 207.
112. L. Syper, *Tetrahedron*, 1987, **43**, 2853.
113. L. Yu, Y. Wu, H. Cao, X. Zhang, X. Shi, J. Luan, T. Chen, Y. Pan and Q. Xu, *Green Chem.*, 2014, **16**, 287.
114. (a) X. Zhang, J. Ye, L. Yu, X. Shi, M. Zhang, Q. Xu and M. Lautens, *Adv. Synth. Catal.*, 2015, **357**, 955; (b) H. Cao, M. Liu, R. Qian, X. Zhang and L. Yu, *Appl. Organomet. Chem.*, 2019, **33**, e4599.
115. L. Yu, J. Ye, X. Zhang, Y. Ding and Q. Xu, *Catal. Sci. Technol.*, 2015, **5**, 4830.
116. Y. Miyake, Y. Nishibayashi and S. Uemura, *Bull. Chem. Soc. Jpn.*, 2002, **75**, 2233.

117. H. Ichikawa, Y. Usami and M. Arimoto, *Tetrahedron Lett.*, 2005, **46**, 8665.
118. G.-J. ten Brink, J.-M. Vis, I. W. C. E. Arends and R. A. Sheldon, *J. Org. Chem.*, 2001, **66**, 2429.
119. G.-J. ten Brink, J.-M. Vis, I. W. C. E. Arends and R. A. Sheldon, *Tetrahedron*, 2002, **58**, 3977.
120. (a) M. R. Czarny, *Synth. Commun.*, 1976, **6**, 285; (b) M. R. Czarny, *J. Chem. Soc, Chem. Commun*, 1976, 81.
121. (a) D. H. R. Barton, A. Billion and J. Boivin, *Tetrahedron Lett.*, 1985, **26**, 1229; (b) D. H. R. Barton, X. Lusinchi and P. Milliet, *Tetrahedron*, 1985, **41**, 4727.
122. J. P. Marino and R. D. Larsen Jr, *J. Am. Chem. Soc.*, 1981, **103**, 4642.
123. (a) M. Brząszcz, K. Kloc and J. Młochowski, *Pol. J. Chem.*, 2003, **77**, 1579; (b) M. Giurg, M. Brzaszcz and J. Młochowski, *Pol. J. Chem.*, 2006, **80**, 417.
124. D. Zhao, M. Johansson and J.-E. Bäckvall, *Eur. J. Org. Chem.*, 2007, 4431.
125. (a) C. Gebhardt, B. Priewisch, E. Irran and K. Rück-Braun, *Synthesis*, 2008, 1889; (b) B. Priewisch and K. Rück-Braun, *J. Org. Chem.*, 2005, **70**, 2350.
126. W. Chen, Y. Wang, X. Mi and S. Luo, *Org. Lett.*, 2019, **21**, 8178.
127. D. H. R. Barton, D. J. Lester and S. V. Ley, *J. Chem. Soc., Perkin Trans. 1*, 1980, 1212.
128. (a) M. Giurg, S. B. Said, L. Syper and J. Młochowski, *Synth. Commun.*, 2001, **31**, 3151; (b) M. Giurg, J. Młochowski and A. Ambrozak, *Pol. J. Chem.*, 2002, **76**, 1713.
129. L. Yu, H. Li, X. Zhang, J. Ye, J. Liu, Q. Xu and M. Lautens, *Org. Lett.*, 2014, **16**, 1346.
130. X. Zhang, J. Sun, Y. Ding and L. Yu, *Org. Lett.*, 2015, **17**, 5840.
131. X. Jing, T. Wang, Y. Ding and L. Yu, *Appl. Catal., A*, 2017, **541**, 107.
132. X. Jing, D. Yuan and L. Yu, *Adv. Synth. Catal.*, 2017, **359**, 1194.
133. S.-R. Sheng and X. Huang, *J. Chem. Res.*, 2003, 491.
134. R. Singha and P. Ghosh, *J. Med. Chem. Sci.*, 2020, **3**, 95.
135. D. Yan, G. Wang, F. Xiong, W.-Y. Sun, Z. Shi, Y. Lu, S. Li and J. Zhao, *Nat. Commun.*, 2018, **9**, 4293.
136. K. R. Roh, K. S. Kim and Y. H. Kim, *Tetrahedron Lett.*, 1991, **32**, 793.
137. P. C. Bulman Page, B. R. Buckley, C. Elliott, Y. Chan, N. Dreyfus and F. Marken, *Synlett*, 2016, **27**, 80.
138. F. Mangiavacchi, L. Crociani, L. Sancineto, F. Marini and C. Santi, *Molecules*, 2020, **25**, 2711.
139. N. M. R. McNeil, C. McDonnell, M. Hambrook and T. G. Back, *Molecules*, 2015, **20**, 10748.
140. (a) E. Block, *Garlic and Other Alliums: The Lore and the Science*, Royal Society of Chemistry, Cambridge, 2010; (b) C. J. Cavallito and J. H. Bailey, *J. Am. Chem. Soc.*, 1944, **66**, 1950.
141. C. Tidei, M. Piroddi, F. Galli and C. Santi, *Tetrahedron Lett.*, 2012, **53**, 232.

142. Y. Nishiyama and N. Sonoda, in *Organoselenium Chemistry – A Practical Approach*, ed. T.G. Back, Oxford University Press, Oxford, 1999, ch. 6, pp. 113–129.
143. S.-I. Fujiwara, N. Kambe and N. Sonoda, in *Organoselenium Chemistry – A Practical Approach*, ed. T. G. Back, Oxford University Press, Oxford, 1999, ch. 12, pp. 223–240.
144. T. Miyata, K. Kondo, S. Murai, T. Hirashima and N. Sonoda, *Angew. Chem., Int. Ed.*, 1980, **19**, 1008.
145. (a) X. Liu and S. Lu, *J. Mol. Catal. A: Chem.*, 2004, **212**, 127; (b) X.-Z. Liu and S.-W. Lu, *J. Mol. Catal. A: Chem.*, 2009, **300**, 36.
146. R. Umeda, H. Kouno, T. Kitagawa, T. Okamoto, K. Kawashima, T. Mashino and Y. Nishiyama, *Heteroat. Chem.*, 2014, **25**, 698.
147. Y. Nishiyama, R. Maema, K. Ohno, M. Hirose and N. Sonoda, *Tetrahedron Lett.*, 1999, **40**, 5717.
148. R. Umeda, T. Mashino and Y. Nishiyama, *Tetrahedron*, 2014, **70**, 4395.
149. Y. Nishiyama, M. Hirose, W. Kitagaito and N. Sonoda, *Tetrahedron Lett.*, 2002, **43**, 1855.
150. Y. Nishiyama, M. Fujimoto and N. Sonoda, *Synlett*, 2006, 109.
151. X. Liu, S. Lan, C. Li, Y. Gao, J. Li and X. Zhang, *J. Chem. Res.*, 2014, 16.
152. J. Chen, G. Ling, Z. Yu, S. Wu, X. Zhao, X. Wu and S. Lu, *Adv. Synth. Catal.*, 2004, **346**, 1267.
153. (a) Y. Yang and S. Lu, *Tetrahedron Lett.*, 1999, **40**, 4845; (b) G. Ling, J. Chen and S. Lu, *J. Chem. Res.*, 2003, 442; (c) G. Ling, J. Chen and S. Lu, *J. Mol. Catal. A: Chem.*, 2003, **202**, 23; (d) J. Chen, G. Ling and S. Lu, *Eur. J. Org. Chem.*, 2003, 3446; (e) J. Chen, G. Ling and S. Lu, *Tetrahedron*, 2003, **59**, 8251; (f) X. Wang, G. Ling, Y. Xue and S. Lu, *Eur. J. Org. Chem.*, 2005, 1675; (g) J. Mei, Y. Yang, Y. Xue and S. Lu, *J. Mol. Catal. A: Chem.*, 2003, **191**, 135.
154. X. Wu and Z. Yu, *Tetrahedron Lett.*, 2010, **51**, 1500.
155. R. Umeda, S. Morishita and Y. Nishiyama, *Heteroat. Chem.*, 2011, **22**, 571.
156. (a) X. Zhang, D. Li, X. Ma, Y. Wang and G. Zhang, *Synthesis*, 2013, **45**, 1357; (b) X. Zhang, Z. Li, Y. Wang, S. Dong, X. Niu and G. Zhang, *Arkivoc*, 2016, 197; (c) X. Zhang, S. Dong, Q. Ding, X. Fan and G. Zhang, *Chin. Chem. Lett.*, 2019, **30**, 375.
157. H. S. Kim, Y. J. Kim, H. Lee, K. Y. Park, C. Lee and C. S. Chin, *Angew. Chem., Int. Ed.*, 2002, **41**, 4300.
158. F. Tian, Y. Chen, X. Wang, P. Li and S. Lu, *J. Chem.*, 2015, 210806.
159. (a) N. Sonoda, G. Yamamoto, K. Natsukawa, K. Kondo and S. Murai, *Tetrahedron Lett.*, 1975, 1969; (b) Y. Nishiyama, Y. Naitoh and N. Sonoda, *Synlett*, 2004, 886.
160. (a) X. Zhang and H. Jing, *J. Mol. Catal. A: Chem.*, 2009, **302**, 137; (b) X. Zhang, H. Jing and G. Zhang, *Synth. Commun.*, 2010, **40**, 1614; (c) X. Zhang, P. Wang, X. Niu, Z. Li, X. Fan and G. Zhang, *Chin. J. Catal.*, 2016, **37**, 2034.
161. A. V. Tran, T. T. Nguyen, H. J. Lee, S. W. Bae, J. Baek, H. S. Kim and Y. J. Kim, *Appl. Catal., A*, 2019, **587**, 117245.

162. Y. Ma, F. Shi and Y. Deng, *J. Chem. Res.*, 2010, 344.
163. T. Mizuno, T. Nakai and M. Mihara, *Heteroat. Chem.*, 2010, **21**, 541.
164. (a) X. Zhang and S. Lu, *Synlett*, 2005, 1535; (b) X. Zhang and S. Lu, *Synth. Commun.*, 2007, **37**, 3291.
165. Y. Nishiyama, K. Maehira, J. Nakase and N. Sonoda, *Tetrahedron Lett.*, 2005, **46**, 7415.
166. Y. Nishiyama, S. Katsuen, H. Jounen, S. Hamanaka, A. Ogawa and N. Sonoda, *Heteroat. Chem.*, 1990, **1**, 467.
167. D. Crich, S. Neelamkavil and F. Sartillo-Piscil, *Org. Lett.*, 2000, **2**, 4029.
168. F. Tian and S. Lu, *Synlett*, 2004, 1953.
169. Y. Nishiyama, M. Yoshida, S. Ohkawa and S. Hamanaka, *J. Org. Chem.*, 1991, **56**, 6720.
170. Y. Nishiyama, J. lnoue, K. Teranishi, M. Moriwaki and S. Hamanaka, *Tetrahedron Lett.*, 1992, **33**, 6347.
171. Y. Nishiyama, S. Hamanaka, A. Ogawa, N. Kambe and N. Sonoda, *J. Org. Chem.*, 1988, **53**, 1326.
172. R. Weiss, E. Aubert, P. Peluso, S. Cossu, P. Pale and V. Mamane, *Molecules*, 2019, **24**, 4484.
173. (a) Y. Nishibayashi, J. D. Singh, K. Segawa, S. Fukuzawa and S. Uemura, *J. Chem. Soc., Chem. Commun.*, 1994, 1375; (b) Y. Nishibayashi, J. D. Singh, Y. Arikawa, S. Uemura and M. Hidai, *J. Organomet. Chem.*, 1997, **531**, 13; (c) Y. Nishibayashi, K. Segawa, J. D. Singh, S. Fukuzawa, K. Ohe and S. Uemura, *Organometallics*, 1996, **15**, 370.
174. (a) T. Wirth, *Tetrahedron Lett.*, 1995, **36**, 7849; (b) T. Wirth, K. J. Kulicke and G. Fragale, *Helv. Chim. Acta*, 1996, **79**, 1957.
175. C. Santi and T. Wirth, *Tetrahedron: Asymmetry*, 1999, **10**, 1019.
176. (a) A. L. Braga, O. E. D. Rodrigues, M. W. Paixão, H. R. Appelt, C. C. Silveira and D. P. Bottega, *Synthesis*, 2002, 2338; (b) A. L. Braga, M. W. Paixão, D. S. Lüdtke, C. C. Silveira and O. E. D. Rodrigues, *Org. Lett.*, 2003, **5**, 2635.
177. C. Bolm, M. Kesselgruber, A. Grenz, N. Hermanns and J. P. Hildebrand, *New J. Chem.*, 2001, **25**, 13.
178. (a) A. L. Braga, D. S. Lüdtke, F. Vargas and M. W. Paixão, *Chem. Commun.*, 2005, 2512; (b) R. S. Schwab, L. C. Soares, L. Dornelles, O. E. D. Rodrigues, M. W. Paixão, M. Godoi and A. L. Braga, *Eur. J. Org. Chem*, 2010, 3574.
179. C. Börner, M. R. Dennis, E. Sinn and S. Woodward, *Eur. J. Org. Chem.*, 2001, 2435.
180. M. Shi, C.-J. Wang and W. Zhang, *Chem. Eur. J.*, 2004, **10**, 5507.
181. A. L. Braga, S. J. N. Silva, D. S. Lüdtke, R. L. Drekener, C. C. Silveira, J. B. T. Rocha and L. A. Wessjohann, *Tetrahedron Lett.*, 2002, **43**, 7329.
182. W. H. H. Günther, *J. Org. Chem.*, 1966, **31**, 1202.
183. A. Ogawa, Y. Nishiyama, N. Kambe, S. Murai and N. Sonoda, *Tetrahedron Lett.*, 1987, **28**, 3271.

CHAPTER 14

Chalcogen-based Probes

GAURI S. MALANKAR,[a] BEATRIZ S. CUGNASCA,[b] FELIPE WODTKE,[b] JOÃO L. PETRARCA DE ALBUQUERQUE,[b] PRATIKSHA P. DESHMUKH,[a] DIVYESH S. SHELAR,[a] ALCINDO A. DOS SANTOS*[b] AND SUDESH T. MANJARE*[a]

[a] Department of Chemistry, University of Mumbai, Mumbai 400098, India; [b] Institute of Chemistry, University of São Paulo, 05508-900 São Paulo, SP, Brazil

*Emails: sudeshmanjare@chemistry.mu.ac.in; alcindo@iq.usp.br

14.1 Introduction

Fluorescent sensors based on small organic molecules have been most widely developed as sensors for monitoring *in vitro* and *in vivo* biological targets because of their simplicity and high sensitivity.[1,2] Fluorescence spectroscopy works as a prominent technique to identify interactions between the analyte and the probe by observing changes occurring in the spectroscopic characteristics of the probe. These changes can be either based on reversible supramolecular (non-covalent) interactions, such as hydrogen bonding, π–π donor–acceptor, electrostatic, hydrophobic, hydrophilic, co-ordination based, or irreversible covalent interactions. The fluorescence technique is effective due to its high sensitivity, excellent temporal–spatial resolution, and fairly simple implementation.[3] Among the various imaging methods such as nuclear, ultrasound, and magnetic resonance imaging, fluorescence bioimaging gives a spatial approach for visualizing morphological details in tissue.[4–6] Thus, fluorescence spectroscopy is an essential technique to investigate biologically important analytes by using synthetic fluorescent probes.[7,8] Organochalcogen-based fluorescent sensors

Chalcogen Chemistry: Fundamentals and Applications
Edited by Vito Lippolis, Claudio Santi, Eder J. Lenardão and Antonio L. Braga

Published by the Royal Society of Chemistry, www.rsc.org

have received more attention due to their fast response and reversibility.[9–11] In organochalcogen-based probes, Se and Te act as active reaction sites for the sensitive and selective detection of reactive oxygen species (ROS), nitrogen species (RNS), biothiols, and metal ions.[12] These reactions include oxidation of selenium and tellurium to the corresponding chalcogenoxides, deselenation, binding of metal ions according to their chalcogenophilicity, substitution of Se by S, and so on.[12–14] In general, the absorption and emission maxima shift towards a longer wavelength on going from Se to Te due to a heavy atom effect.[13] Additionally, organotellurium compounds display a lower emission intensity (low quantum yield) compared to organoselenium compounds.[14]

The organochalcogen compounds have a high reducing power hence the chalcogens are oxidized easily when they interact with ROS/RNS.[15] Moreover, organoselenium compounds can be used as a GPx (Glutathione Peroxidase) mimic because of their antioxidant properties. Additionally, they possess other biological properties such as antiviral, antibacterial, *etc.*[16] The oxidation–reduction events involving the chalcogen atom can lead to changes in the fluorescence of chalcogen-based probes.[17–19] The organochalcogen-based fluorescence sensors can be developed by embedding a chalcogen in the skeleton of the fluorophores which offers modulation of the fluorescence when it reacts with ROS/RNS and biomolecules.[20]

Some research groups in the field of chemosensing have covered the organochalcogen-based probing systems through book chapters and reviews.[9–11,15,17,21] Moreover, the probes based on a nanomaterial are receiving more attention due to their short response time, high sensitivity, and better biocompatibility; furthermore, the nano-probes can detect the analyte up to picomolar levels.[22–24] Thus, the next phase in the probing system will be nano-sensors.

The present chapter deals with all the organoselenium and organotellurium-based sensors for the detection of biologically important analytes such as ROS, RNS, biothiols, and metal ions. The chapter unboxes the developments in the design of probes to target specific analytes as well as their particular location in the cell during bioimaging analysis.

14.2 Detection of Reactive Oxygen Species

The biology of ROS/RNS is linked to the involvement of intracellular species belonging to these classes with respect to various biological processes.[25] ROS/RNS play crucial roles in various biological processes like antitumour (tumourigenesis), inflammation, pain, fever, antifungal (proliferation), signal transduction, redox events mediation, and also in neurodegenerative diseases and nitrosative stresses.[26–29] Aerobic metabolism produces ROS,[30] and these can be classified into two types, those based on free radicals and those on non-radicals. Superoxide ($O_2^{\bullet -}$), hydroxyl ($HO^{\bullet}$), peroxyl ($RO_2^{\bullet}$), and hydroperoxyl ($HO_2^{\bullet}$) are free radicals and hydrogen peroxide (H_2O_2), hypochlorous acid (HOCl), and ozone (O_3) are non-radical-based ROS.[31]

There are several reports on the selective detection of these analytes by chalcogen-containing sensors. In ROS-sensible chalcogen-based probes, the chalcogen acts as an active reaction centre, triggering instantly the response to the analyte, and the chalcogen atom oxidation to the corresponding chalcogen oxide is accompanied by fluorescence emission switching (*on–off* or *off–on*). The oxidized probe may show reversibility with biothiol-assisted reduction reaction, also accompanied by fluorescence switching.

In this context, Nagano *et al.*[32] designed and synthesized the tellurium–rhodamine fluorescent reversible probe **1** for the detection of ROS. The authors demonstrated that this compound can be quickly oxidized by different ROS ($HO^{\bullet}$, $ONOO^-$, ClO^-) to the corresponding telluroxide **2**, and the process is accompanied by a large increase of the fluorescence emission after interaction with the analyte. The reversibility of the process was demonstrated by sequential treatment with ClO^- (oxidation) and the endogenous biothiol, GSH (reduction), in three cycles. Considering the specificity of the analytes and studies with HL-60 cells, the authors claimed monitoring of ROS as well as GSH, *in vivo* (Scheme 14.1).

Churchill and co-workers introduced the diselenide BODIPY-based sensor **3** for the detection of superoxide [$\lambda_{ex}/\lambda_{em} = 504/514$ nm; LOD (Limit of Detection) = 12.9 μM] (Figure 14.1).[33] A fluorescence turn-on response is observed by oxidation of the diselenide with 29-fold increment in fluorescence intensity. In 2018, Manjare *et al.* developed the phenylselenide-based probe **4** containing two BODIPY groups attached at the *ortho*-positions of a phenylselenide residue.[34] Upon reaction with superoxide, the selenide is converted to the corresponding selenoxide with 11-fold increment in the fluorescence intensity and a 4.42 μM LOD ($\lambda_{ex}/\lambda_{em} = 505/526$ nm).

Recently, the same group reported the cyclic diselenide–BODIPY probe **5** for the detection of superoxide ($\lambda_{ex}/\lambda_{em} = 506/521$ nm).[35] A strong green fluorescence emission is observed through oxidation of selenium to selenoxide with a 9-fold increment in fluorescence intensity and 0.924 μM LOD. The biological potential of both probes (**4** and **5**) to detect superoxide in living cells was demonstrated in MCF-7 cell line experiments.

The hemicyanine–selenide **6** (Figure 14.1) developed by Churchill *et al.* in 2018,[36] which presents no emission due to a photoinduced electron transfer

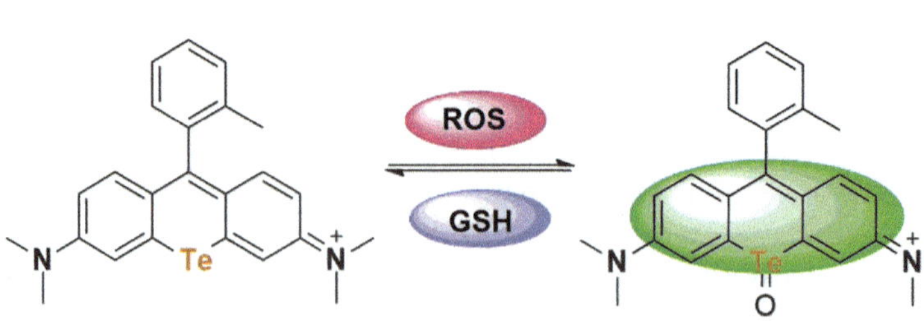

1 2

Scheme 14.1 ROS and GSH redox-triggered fluorescence emission by the tellurium–rhodamine probe **1**.

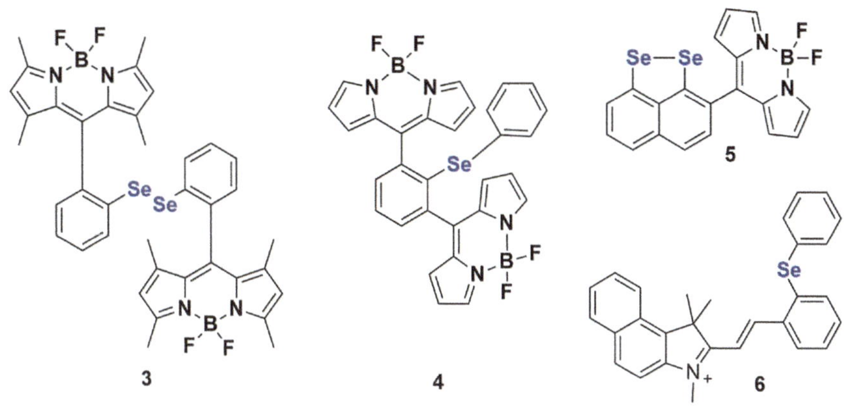

Figure 14.1 Selenium-based fluorophores used for the selective detection of superoxide.

(PET) effect from selenide to the hemicyanine unit. In the presence of the superoxide anion the compound becomes fluorescent (20-fold increase). Surprisingly, the phenomena were not attributed to the oxidation of the selenium atom, but to the superoxide reaction with the exocyclic double bond of **6**, eliciting a blue emission ($\lambda_{ex}/\lambda_{em} = 360/439$ nm) with a LOD of 11.9 nM. The applicability of **6** was also demonstrated as a turn-on probe for the detection of the superoxide anion in RAW264.7 cells.

In 2018, Chen and co-workers developed a probe for the combined detection of the superoxide anion ($O_2^{\bullet -}$) and Hg(II).[37] $O_2^{\bullet -}$ reacts with cyanine **7** to give **8** (Scheme 14.2), which is in equilibrium with **9** (**8** $\lambda_{ex}/\lambda_{em} = 755/800$ nm and **9** $\lambda_{ex}/\lambda_{em} = 405/567$ nm). The selenol group reacts with Hg(II) providing compound **10** ($\lambda_{ex}/\lambda_{em} = 510/607$ nm). Therefore, **7** acts as a three-channel ratiometric fluorescent probe and is successfully applied to study chronic mercurialism in cell and mice models.

Tang *et al.*[38] reported a tellurium–fluorenone-based molecular probe **11** ($\lambda_{ex}/\lambda_{em} = 380/440$ nm; LOD = 0.15 μM) for monitoring superoxide *via* oxidation of tellurium to afford **12** (Scheme 14.3). Additionally, the research group prepared carbon dots of the same compound ($\lambda_{ex}/\lambda_{em} = 380/440$ nm), which allowed the $O_2^{\bullet -}$ detection limit to be lowered up to 8 pM. The utility of this probe was also checked in normal abdomen tissue and breast tumour tissue of mice.

Hypochlorite (HClO) has been gaining attention due to its biological importance. It is also a strong biological oxidant and can oxidize proteins, lipids, and nucleic acids very rapidly and show a substantial effect on physiological processes.[39,40] Figure 14.2 presents the structures of several organoselenium-based probes (**13**,[41] **14**,[42] **15**,[43] **16**,[44] **17**,[45] **18**,[46] **19**,[47] **20**,[48] **21**,[49] **22**,[50] **23**,[51] **24**,[52] and **25**[53]) which are selective and sensitive for the detection of hypochlorite due to the reaction centred on the selenium atom. Wu *et al.* presented probe **13** with selectivity for hypochlorite with a 138-fold increase in the fluorescence intensity emission ($\lambda_{ex}/\lambda_{em} = 510/526$ nm;

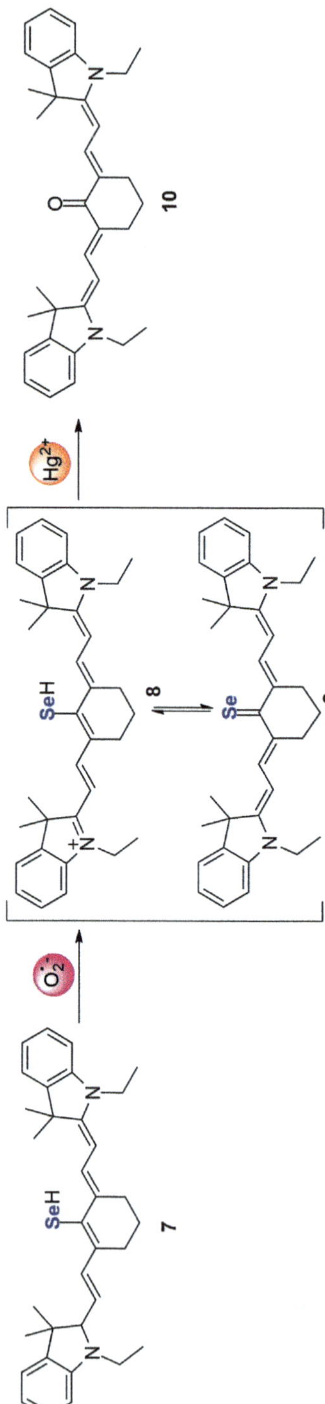

Scheme 14.2 Selenium-based probe for the combined detection of $O_2^{\bullet-}$ and Hg(II).

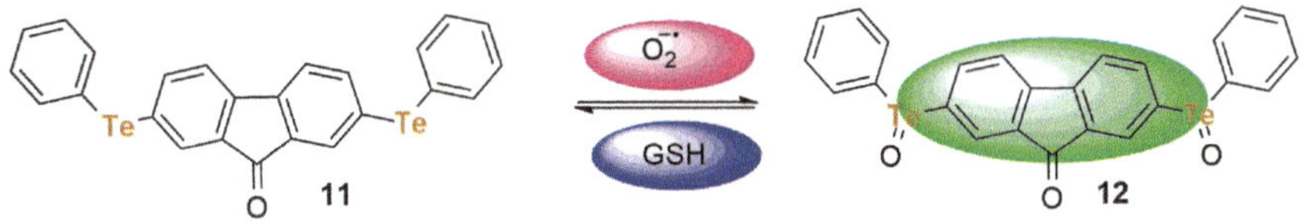

Scheme 14.3 Reversible superoxide-promoted fluorescence mediation in probe **11**.

LOD = 7.98 nM). In 2016, Churchill and co-workers reported probes **14a** ($\lambda_{ex}/\lambda_{em}$ = 492/507 nm; LOD = 30.9 nM) and **14b** ($\lambda_{ex}/\lambda_{em}$ = 511/526 nm; LOD = 4.5 nM) which selectively detect hypochlorite *via* a turn-on response in living cells. Furthermore, the same research group described the preparation and use of probe **15** ($\lambda_{ex}/\lambda_{em}$ = 500/525 nm) for hypochlorite detection. The derivative **16** ($\lambda_{ex}/\lambda_{em}$ = 512/526 nm; LOD = 19.6 nM) containing the mesitylene group at the *meso* position of the BODIPY fluorophore, selectively detects HOCl with a turn-on response. The turn-on response can be attributed to the blocking of the PET process due to oxidation of selenium. Yao *et al.* synthesized the selenium-containing benzooxadiazole-based probe **17** ($\lambda_{ex}/\lambda_{em}$ = 468/544 nm) selective for hypochlorite with a 3.3 nM LOD. In 2016, Churchill and co-workers reported a diselenide-containing dipyrazolopyridine-based probe **18** ($\lambda_{ex}/\lambda_{em}$ = 311/436 nm; LOD = 0.36 μM) for HClO detection with 180-fold increase in the fluorescence intensity emission, and the utility of the probe was shown in the MCF-7 cell line. Song *et al.* developed probes **19a** – **19c** ($\lambda_{ex}/\lambda_{em}$ = ~530/~550 nm) able to turn-on, selectively in the presence of HOCl, with green fluorescence (**19a** LOD = 0.3 nM; **19b** LOD = 0.8 nM; **19c** LOD = 9.2 nM). Thus, the selenide group attached at the *meso* position of the BODIPY fluorophore exhibits a very low detection limit as in probe **19a** and **19b** as compared to other organoselenide-based probes. In 2021 Manjare and co-workers reported the first selenium containing BOPHY-based probes **20a** ($\lambda_{ex}/\lambda_{em}$ = 452/482 nm; LOD = 0.63 μM) and **20b** ($\lambda_{ex}/\lambda_{em}$ = 452/484 nm; LOD = 1.08 μM) with selectivity for hypochlorite in the MCF-7 cell line. Benzofurane-1,3-dione-based selenium-containing probes **21a** ($\lambda_{ex}/\lambda_{em}$ = 404/502 nm) and **21b** ($\lambda_{ex}/\lambda_{em}$ = 416/523 nm; LOD = 90 nM) for hypochlorite were developed by Churchill *et al.* Zhang and co-workers presented a napthalamide-based probe **22** ($\lambda_{ex}/\lambda_{em}$ = 410/506 nm; LOD = 0.85 μM) for the detection of exogenous and endogenous HOCl in endoplasmic reticulum. Li *et al.* developed probe **23**, which presents selectivity to detect HOCl in lysosomes, due to the presence of the morpholine group. Lysosomes have a distinct stability pH range of 4.5–5.0, while morpholine has a pK_a value in the range 5.0–6.0, therefore, morpholine can be protonated in lysosomes.[54,55] The morpholine group of probe **23** can target the lysosomes. In 2013, Han and co-workers reported the phenylselenide containing napthalamide **24** ($\lambda_{ex}/\lambda_{em}$ = 450/523 nm; LOD = 0.586 μM) for visualizing HOCl in mice. Furthermore, probe **25** developed by Guo *et al.* can selectively sense HOCl in mitochondria (LOD = 4.8 nM) due to the presence of the triphenylphosphonium group (triphenylphosphonium is a cationic lipid that readily

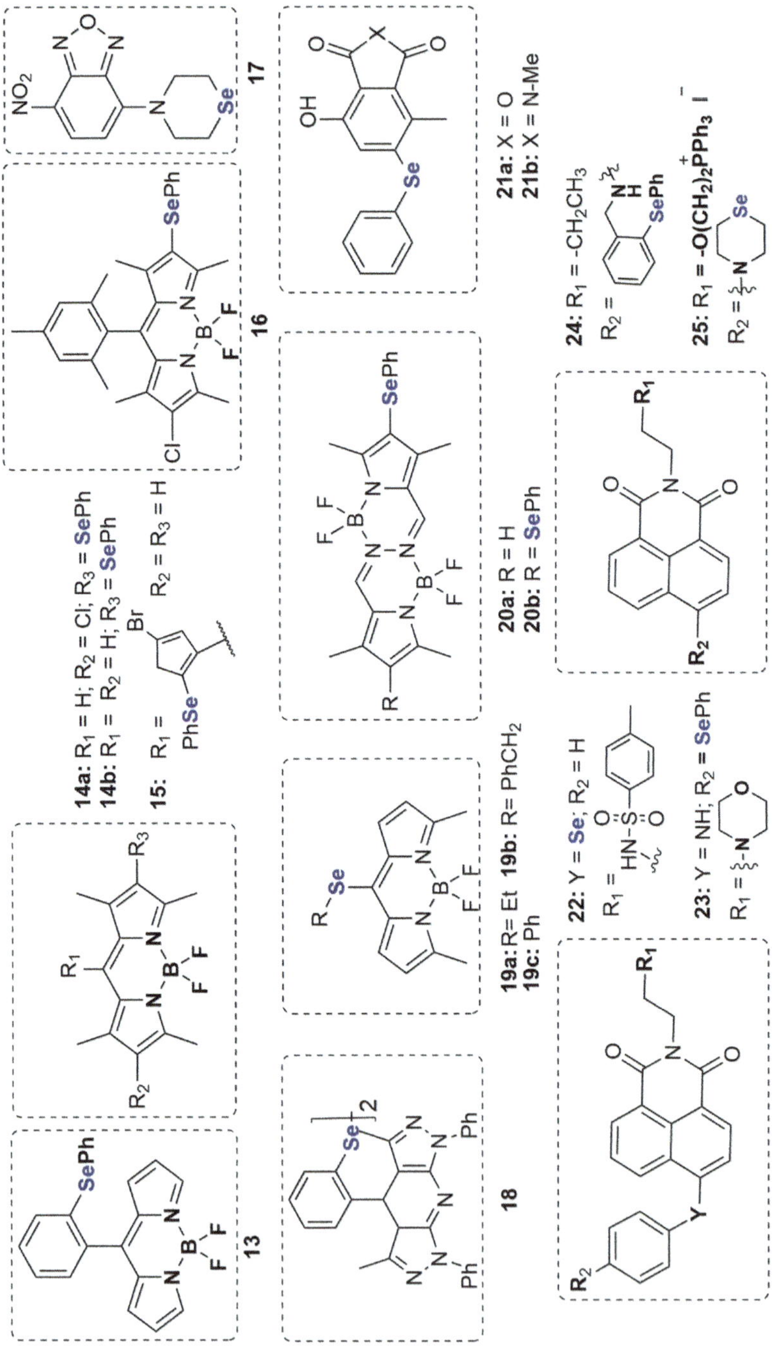

Figure 14.2 Structurally diverse selenium-containing fluorophores designed for the selective detection of hypochlorite (HClO).

accumulates and penetrates through the mitochondrial membrane due to the highly negative mitochondrial membrane potential).[56]

In 2013, Jiang *et al.* reported probes **26a** ($\lambda_{ex}/\lambda_{em}$ = 405/480 nm) and **26b** ($\lambda_{ex}/\lambda_{em}$ = 405/468 nm; LOD = 10 nM) for the detection of hypochlorite through selenoxide elimination, which restores the blue fluorescence of the fluorophore with visualization in progranulocytes and macrophages.[57] A similar strategy (selenoxide elimination) was used by Tang with probe **27** ($\lambda_{ex}/\lambda_{em}$ = 475/495 nm; LOD = 4.6 nM). The elimination upon reaction with hypochlorite extends the π-conjugation within the probe structure, thus determining a bathochromic (red) shift emission. Additionally, the biocompatibility of the probe was demonstrated in RAW264.7 macrophages and further extended for sensing HOCl in liver and kidney tissues.[58] Han *et al.* developed probe **28** for the detection of HBrO characterized by strong fluorescence emission (Figure 14.3, $\lambda_{ex}/\lambda_{em}$ = 610/635 nm; LOD = 0.97 μM). The connection of the fluorescent core to the Se-containing moiety through styrene contributes to extend the π-conjugation, which is responsible for the absorption of the probe at a longer wavelength. The author also reported the redox cycle of the probe in macrophage cells based on the oxidation by HOBr and back-reduction by H_2S.[59]

Peng *et al.*, in 2014, developed the Se–heptamethine–cyanine probe **29** selective for the detection of HClO.[60] In PBS buffer, **29** presents very weak fluorescence, due to aggregation effects. In the presence of NaOCl, the selenium atom is oxidized to selenoxide and as a result, the absorption peaks decrease drastically and an emission peak at λ_{em} = 786 nm (λ_{ex} = 690 nm; LOD = 0.31 μM) gradually increases, indicating aggregation suppression. The authors also showed the possibility of detecting endogenous HClO in living mice with **29**. Zeng *et al.*[61] reported probe **30** for monitoring the HOCl/H_2S reversibly ($\lambda_{ex}/\lambda_{em}$ = 500/600 nm; LOD = 1.3 nM). A fluorescence turn-on response was achieved through a selenide to selenoxide transformation. The selective detection of HOCl was also reported by Han and co-workers with the selenium-containing BODIPY-based probe **31** ($\lambda_{ex}/\lambda_{em}$ = 460/510 nm).[62] The probe was able to visualize a HOCl/H_2S-based redox cycle in the living RAW264.7 cell line. Recently, the phenylseleno-functionalized fluorescein **32** was reported to detect HClO as a consequence of the selenium atoms oxidation to the corresponding selenoxides and quenching of the PET effect,[63] which triggers the fluorescence emission ($\lambda_{ex}/\lambda_{em}$ = 450/528 nm; LOD = 1.1 nM), with high selectivity towards HClO against other oxidants. The process is reversible following a reduction reaction selectively assisted by cysteine (LOD = 3.0 nM), among other reducing agents. The probe shows low toxicity towards living cells (L929), with good applicability in *in vitro* studies. Chang and co-workers reported the hypochlorite selective phenylselenyl-fluorescein-based probe **33** ($\lambda_{ex}/\lambda_{em}$ = 500/519 nm; LOD = 39 nM) for monitoring the concentration of hypochlorite in tap water samples with a turn-on response *via* selenoxide formation.[64] In 2015, Tang *et al.*[65] reported fluorenone **34** ($\lambda_{ex}/\lambda_{em}$ = 415/520 nm) and its ability to detect HOCl, through the oxidation of selenium atoms to the respective selenoxides with a LOD of 0.35 μM. The oxidation reaction proved to

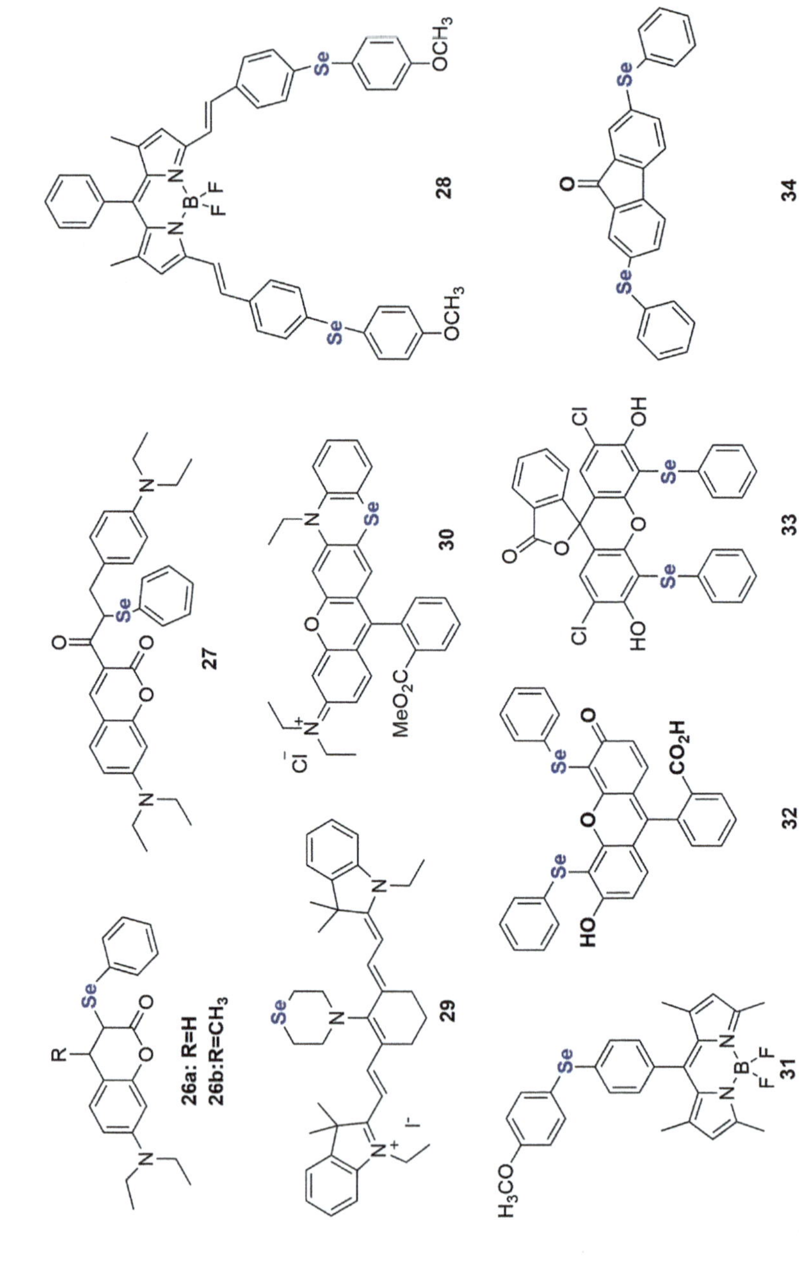

Figure 14.3 Selenium fluorophores designed for hypohalous acid detection.

be reversible through GSH causing a sharp drop in fluorescence intensity, due to the reestablishment of the PET effect. The probe was able to detect HOCl in live cells, zebrafish, and even in mice abdomen.

Huang and co-workers synthesized a series of fluorescent compounds (**35–38**) containing the N–P=Se residue, demonstrating their use for the rapid detection of hypochlorous acid at a nano-molar level of concentration. The reaction of hypochlorous acid oxidizes the N–P=Se moiety to N–P=Se=O (**36** and **38** $\lambda_{ex}/\lambda_{em} = 295/345$ nm) and, therefore, the bond length between P and Se atoms (Figure 14.4) changes from 2.27 to 2.71. As a result, the quenching effect imposed by selenium is suppressed and the fluorescence emission of the probe is recovered.[66]

Churchill *et al.* reported the first telluride-containing annulated BODIPY probe **39**[67] ($\lambda_{ex}/\lambda_{em} = 572/597$ nm; LOD = 3.7 μM) for the selective detection of hypochlorite through the oxidation of telluride. Also, the other organotelluride-containing sensors **40**[68] ($\lambda_{ex}/\lambda_{em} = 480/531$ nm; LOD = 41.3 nM), **41**[69] ($\lambda_{ex}/\lambda_{em} = 499/515$ nm; LOD = 0.12 μM) and **42**[70] ($\lambda_{ex}/\lambda_{em} = 630/738$ nm; LOD = 0.09 μM) detect hypochlorite in living cells *via* conversion of telluride to telluroxide (Figure 14.5). Generally, due to the heavy atom effect, absorption and emission of the chalcogen-containing probes shift towards a longer wavelength, moving from selenium to tellurium.

Two fluorescent probes, **43** and **49** (Scheme 14.4), containing a pyrene fluorophore core, connected to an aryl-selenium moiety[71] were developed and applied to detect hypochlorite and hydrogen peroxide, by distinct oxidative mechanisms. In the reduced form, both probes showed low fluorescence emission due to the PET effect imposed by the selenium atom. Following the reaction of **43** with H_2O_2, selenoxide **46** is formed, with consequent PET suppression and fluorescence emission increase; the reaction with a second equiv. of H_2O_2, leads to the formation of seleninic acid **48** with cleavage of the Se – C bond and increase of the fluorescence emission from the pyrene unit (**47** $\lambda_{em} = 470$ nm). Equimolar ClO^-, in aqueous solution, also promotes the oxidation of the selenium atom of **43**, with consequent Se – C cleavage resulting in the formation of the fluorescent pyrene-derivatives **44** and **45** with emission wavelengths at 392 and 380 nm, respectively. Probe **49**, on the other hand, reacts indistinctly with ClO^- or H_2O_2 giving rise to the same fluorescent product, the vinyl-pyrene **51** ($\lambda_{em} = 392$ nm), passing through the respective selenoxide **50**, which suffers intra-molecular-elimination (Scheme 14.4).

Yu *et al.* developed the selenium-containing AIE (Aggregation Induced Emission) turn-on fluorescent probes **52a** and **52b** for the selective detection of H_2O_2 taking advantage of a reaction strategy based on ebselen-like activity. Compound **52b** showed a selective response for H_2O_2 whereas **52a** did not show any notable fluorescence change in the presence of H_2O_2 or other ROS. Due to the lower hydrophobicity of **52a** ($\lambda_{ex}/\lambda_{em} = 350/436$ nm), the AIE process, presumably, was hampered with consequent florescence emission suppression. **52b** was functionalized with a dodecyl chain ($\lambda_{ex}/\lambda_{em} = 330/476$ nm) to increase the hydrophobic nature of the probe and to block the

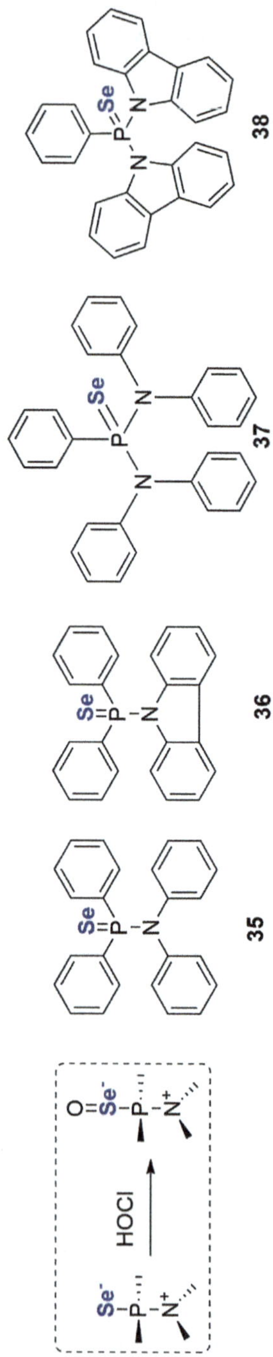

Figure 14.4 Selenium-phosphorylated fluorescent derivatives for HOCl detection in nanomolar concentration.

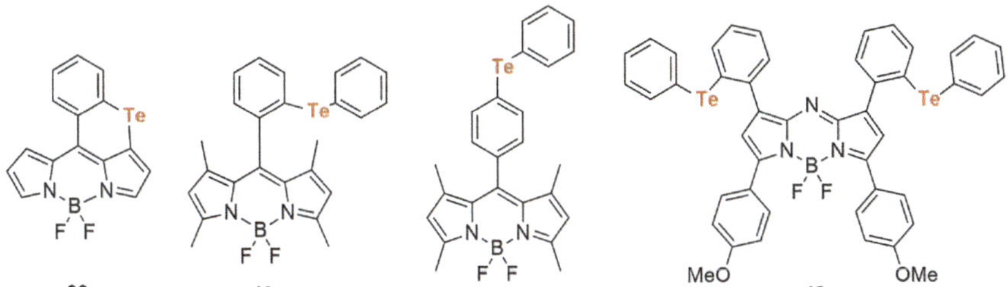

39 40 41 42

Figure 14.5 Tellurium-BODIPY/aza-BODIPY derivatives for the selective detection of HOCl.

π–π interactions between neighbouring molecules of the probe in aggregate components. The PET block process occurs due to oxidation of the probe and then AIE resulted in fluorescence turn-on (Figure 14.6).[72] Qian *et al.* reported the selenium-containing BODIPY-based compound **53**[73] ($\lambda_{ex}/\lambda_{em}$ = 460/504 nm; LOD = 0.13 μM) for the selective detection of hydrogen peroxide in cancer cells and zebrafish. Moreover, the protonation of probe **53** at the tertiary amino position allowed the H_2O_2 in a lysosome environment to be targeted. Mugesh and co-workers developed the naphthalimide-based probe **54** ($\lambda_{ex}/\lambda_{em}$ = 350/475 nm), containing the ebselen substituent, strategically positioned to detect hydrogen peroxide. H_2O_2 oxidizes the selenium atom, triggering cleavage of the Se–N bond and the formation of the corresponding fluorescent selenic acid in which the oxygen atom from the amide group forms an intramolecular interaction with the selenium atom (Se···O), thus leading to the interruption of the PET process (Figure 14.6).[74] Tellurium-based probe **55**[75] ($\lambda_{ex}/\lambda_{em}$ = 500/565 nm, LOD = 6 μM) was developed for selective detection of hydrogen peroxide *via* oxidation of tellurium to telluroxide.

14.3 Probes Selective for Reactive Nitrogen Species

Reactive nitrogen species (RNS) refers to peroxynitrite anion ($ONOO^-$), nitrogen dioxide (NO_2), nitric oxide (NO), *S*-nitrosothiols (RSNO), and nitroxyl (HNO).[76] Each form of RNS has its own distinctive properties based on reactivity, biological activity, solubility, and so forth.[77] RNS play a significant role in the physiological process of living systems which includes cardiomyocytes, smooth muscle cells, platelets, and nervous and juxtaglomerular cells. RNS react with a wide range of biomolecules such as proteins, nucleic acids, and lipids and act as NO transporters in cell signalling events.[78,79] Regularly produced RNS in normal cells are neutralized or eliminated by natural defence mechanisms which include glutathione peroxidase, superoxide dismutase, selenium, vitamin C, and glutathione. However, as the production of RNS exceeds the natural ability of the defence mechanism, its natural balance is disturbed and "oxidative stress" is developed.[80] Furthermore, this oxidative stress is responsible for DNA, protein, and lipid damages.[81,82] This may cause

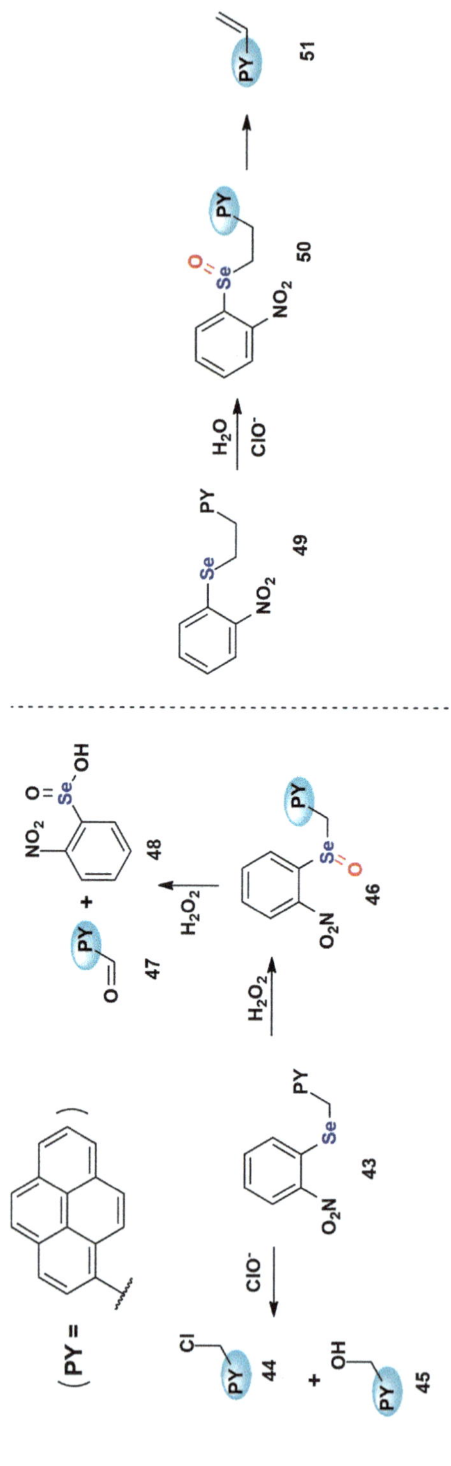

Scheme 14.4 Oxidative deselenilation in the reaction of **43** and **49** with H_2O_2 and HOCl.

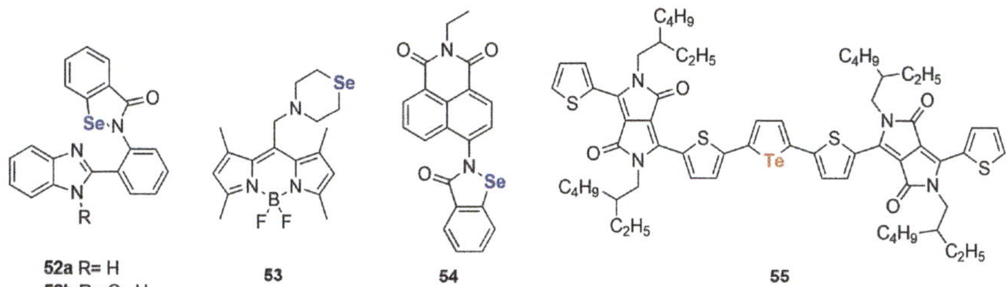

Figure 14.6 Hydrogen peroxide detection based on selenium and tellurium oxidation.

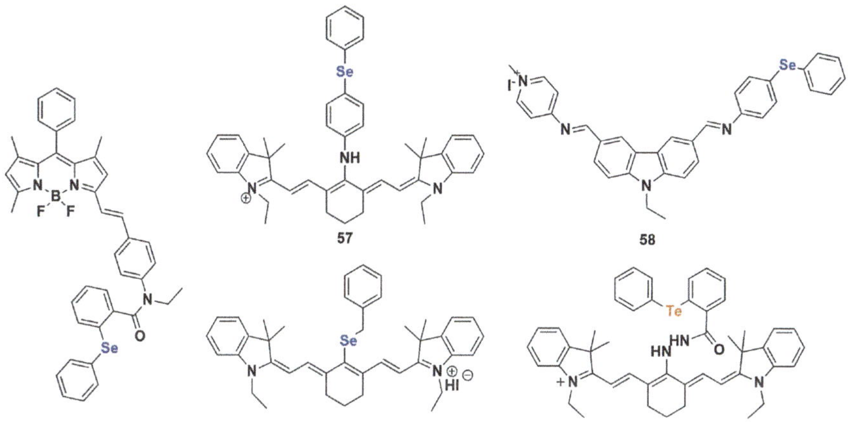

Figure 14.7 Probes for peroxynitrite detection mediated by oxidation of the chalcogen atom.

human diseases such as inflammatory, hepatic insomnia, atherosclerosis, neurodegenerative diseases, and cancer, *etc.*[83] Therefore, it is necessary to detect RNS in living systems.

Figure 14.7 comprises the structures of organoselenium- and organotellurium-based probes which were applied for the detection of RNS through oxidation of the chalcogen to the corresponding chalcogenoxide. Probes **56**,[84] **57**,[85] **58**,[86] and **59**,[87] selectively detect peroxynitrite over other RNS.

Han *et al.* developed probe **56** for the selective detection of peroxynitrite *via* a selenoxide spirocyclization reaction ($\lambda_{ex}/\lambda_{em} = 567/680$ nm; LOD = 5 μM). The probe was able to visualize intracellular peroxynitrite levels in a mouse macrophage cell line. The same research group synthesized probe **57** ($\lambda_{ex}/\lambda_{em} = 750/775$ nm) for the detection of peroxynitrite in RAW264.7 cells. Xie and co-workers reported probe **58** ($\lambda_{ex}/\lambda_{em} = 445/565$ nm; LOD = 31 nM) to detect peroxynitrite levels in RAW264.7 and HeLa cell lines. Tang *et al.* presented probe **59** ($\lambda_{ex}/\lambda_{em} = 770/800$ nm) for monitoring peroxynitrite in the RAW264.7 cell line *via* oxidation of selenide. The cyanine-based telluride **60** (Figure 14.7, $\lambda_{ex}/\lambda_{em} = 793/820$ nm; LOD = 0.917 μM) was synthesized by Han and submitted to the interaction with oxidants, reacting selectively with $ONOO^-$, through the oxidation of tellurium to telluroxide, which determines the suppression of the PET effect.[88] The process is reversible in the presence of GSH or cysteine. The probe was studied for redox homoeostasis in cells (RAW264.7) and mice.

Huimin Ma and co-workers[89] reported a new method for nitric oxide detection based on the seleno-rhodamine derivative **61** (Scheme 14.5). In the presence of NO, the colour of **61** changes from colourless to pink, and it is accompanied by an intense fluorescence emission ($\lambda_{ex}/\lambda_{em} = 520/580$ nm; LOD = 38 nM). Based on the absorption and emission spectra of the product, it was possible to observe a similarity with the spectral profile of rhodamine B. Thus, to elucidate the reaction mechanism, a study based on high-resolution mass spectrometry was carried out. The authors suggested that the reaction with NO causes the ring-opening of compound **61** leading to the formation of unstable compound **62**, which subsequently undergoes hydrolysis resulting in the formation of Rhodamine B (**63**). The formation of compound **64** was also detected by mass experiments. The applicability was demonstrated in HeLa cells experiments.

14.4 Probes Selective for Biothiols

Biothiols are species containing the –SH group, in cells these include glutathione (GSH), cysteine (cys), and homocysteine (hcy) which play irreplaceable roles in human physiological systems.[90] Normal levels of biothiols in the human body are directly related to human health as these biothiols play a crucial role in human systems such as signal transduction, regulation of oxidative stress, and chelation of metal ions. Specifically, the imbalance of biothiols is directly linked to several diseases including acquired immune deficiency syndrome (AIDS), cancer, cardiovascular diseases, and so forth.[91] Considering all these factors it is essential to develop rapid and selective sensors for the detection of thiols.

The rhodamine-based probe **65** (weakly fluorescent) was considered for the detection of thiols (Scheme 14.6).[92] Due to the thiophilicity of Se, the Se atom is attacked by the S atom in the presence of thiols (RSH), leading to Se – N bond cleavage and formation of rhodamine 6G (**66**) (strongly fluorescent) and selenenyl sulfide (**67**). Spectral studies using glutathione (GSH) were carried out under physiological conditions (PBS 20 mM, pH 7.4) and a good correlation between the concentration of GSH and fluorescence intensity emission ($\lambda_{ex}/\lambda_{em} = 525/550$ nm) was observed with a LOD of 1.4 nM. The probe presented high sensitivity and selectivity towards GSH in HL-7702 and HepG2 cells.

Tang *et al.*[93] reported a new rhodamine-based selenide **68** ($\lambda_{ex}/\lambda_{em} = 499/522$ nm; LOD = 144 pM) able to act as a fluorescent probe for thiols (Figure 14.8). The probe presents high sensitivity to GSH, and confocal microscopy experiments were performed with HL-7720 and HepG2cells, allowing the fast imaging of thiols by fluorescence. Compound **68** presented

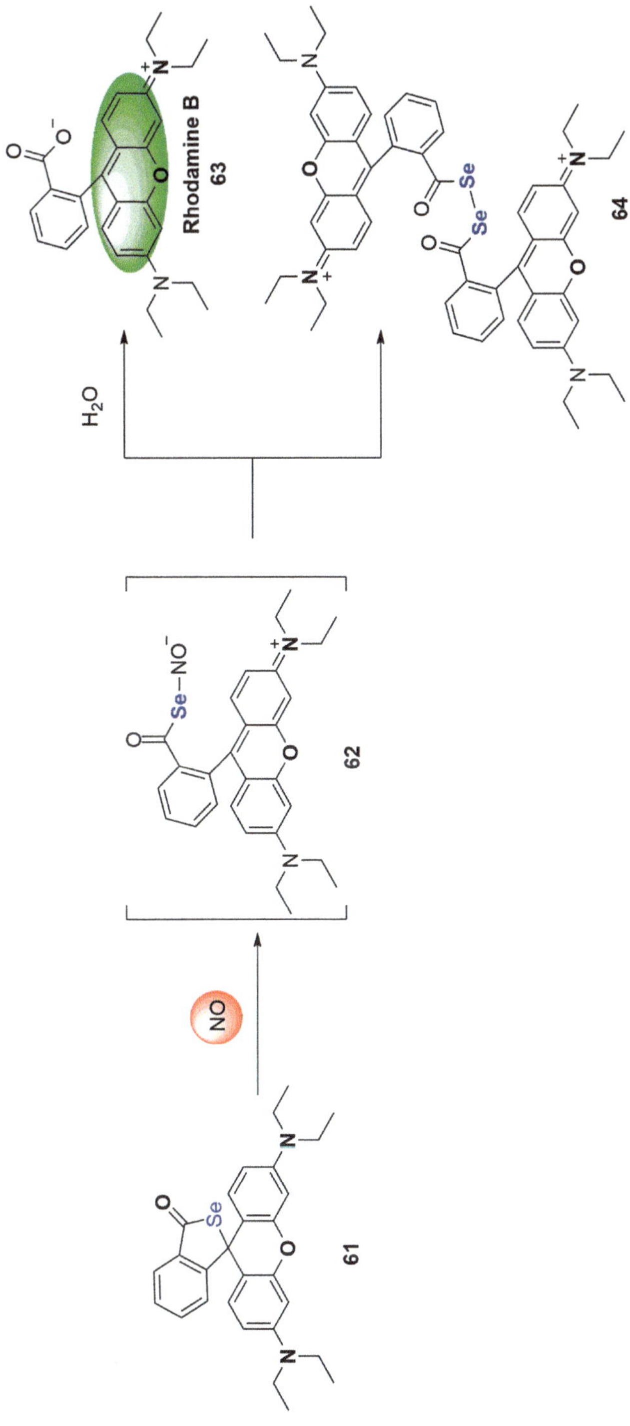

Scheme 14.5 NO-assisted selenium oxidation resulting in fluorescence emission of the Rhodamine B derivative **63**.

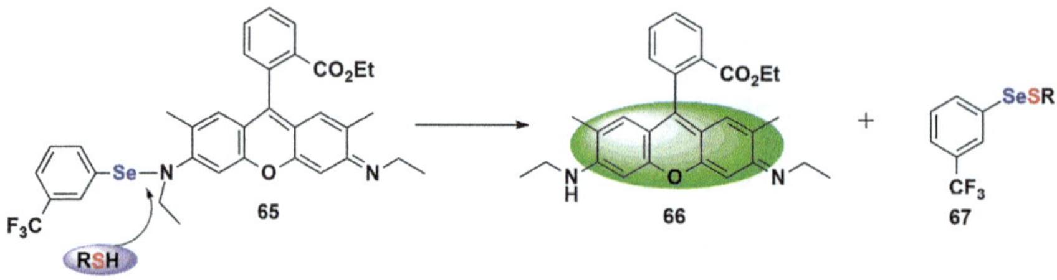

Scheme 14.6 Selenium–sulfur affinity driving Se–N bond cleavage for the detection of GSH in HL-7702 and HepG2 cells by probe **65**.

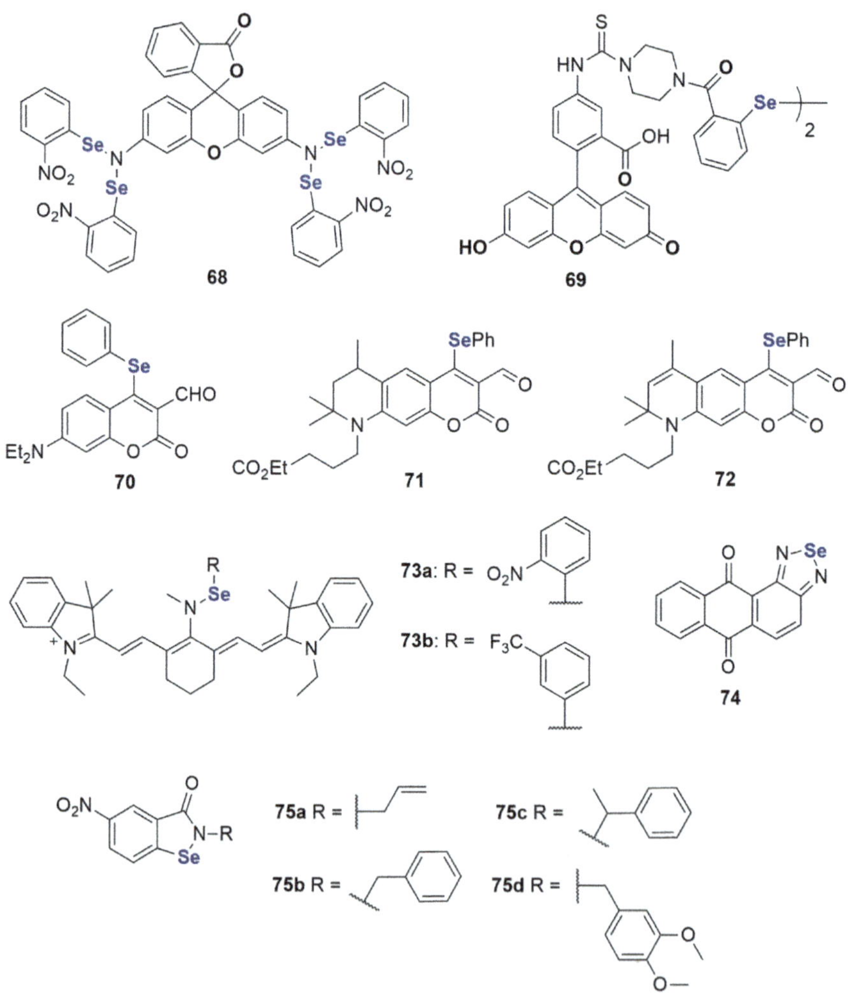

Figure 14.8 Structural collection of seleno-flourophores for biothiols detection.

the ability to detect mitochondrial GSH in apoptotic cells, using microchip electrophoresis with laser-induced fluorescence (MCE-LIF).[94] Diselenide **69** ($\lambda_{ex}/\lambda_{em}$ = 488/514 nm; LOD = 0.225 μM) was synthesized by Han and co-workers and applied for the detection of thiols such as glutathione (GSH) in solution as well as in living cells.[95] The probe presents weak fluorescence, attributed to the vicinity of the two fluorescent cores, and to the PET effect caused by the chalcogen atoms. The sensing mechanism is based on the cleavage of the Se – Se bond by reaction with GSH, resulting in two fluorescent derivatives (a selenol and a GSH-bonded adduct). In the presence of H_2O_2, the obtained products are re-oxidized to **69**, recovering the initial weak fluorescence emission. The sensor presented great fulfillment in living cells, recognizing GSH-enriched HeLa cells and showing selectivity for thiols, and consequently ROS such as H_2O_2. In 2015, Churchill *et al.* developed the coumarin-based phenylselenyl compound **70** ($\lambda_{ex}/\lambda_{em}$ = 518/550 nm; LOD 0.27 μM) for the selective detection of GSH over cysteine and homocysteine.[96] A turn-on response of **70** with GSH was observed with 100-fold increment in the fluorescence intensity emission. The transformation was accomplished by a cyclization process involving the PhSe-substitution of the coumarin derivative by the thiol-residue of GSH and the formation of an iminium group from the reaction of the adjacent aldehydic carbonyl group with the alfa-amino group of the cysteine residue of GSH. Furthermore, bioimaging was possible in Hep3B cells. The cysteine/homocysteine was also detected by independent reaction of probe **70** with cysteine (λ_{abs} = 380 nm), which was confirmed by mass spectrometry. This confirmed that the phenylselenide group undergoes the substitution reaction followed by S,N-displacement, which converts the aldehydic group into a thiazolidine moiety. However, the S,N-displacement was not possible with GSH due to the unfavourable macrocyclic transition state. Thus, probe **70** can efficiently discriminate GSH over Cys/Hcy.

In 2018, the same research group modified probe **70** and developed two other probes (**71** and **72**) for the selective detection of GSH (**71** $\lambda_{ex}/\lambda_{em}$ = 526/555 nm; LOD = 10.1 nM, **72** $\lambda_{ex}/\lambda_{em}$ = 550/590 nm; LOD = 17.0 nM) and Cys/Hcy (**71** $\lambda_{ex}/\lambda_{em}$ = 390/490 nm; LOD = 0.31 μM for cys, **72** $\lambda_{ex}/\lambda_{em}$ = 410/510 nm; LOD = 1.27 μM for cys). In both cases, the probe's water solubility was increased by introducing the ethyl butanoate group at the amino position. A red-shifted emission occurred due to the blockage of free rotation of the alkyl amino group of **70** into N-heterocyclic derivatives **71** and **72**. Probe **72** presented good cell permeability, attributed to the π-conjugation. A bioimaging study was performed with A549 lung cancer cells. Moreover, the detection of cysteine was reported in human fibrosis cells (Figure 14.8).[97] In 2012, Wang *et al.* developed NIR probes **73a–73b** ($\lambda_{ex}/\lambda_{em}$ = 635/750 nm) containing cyanine as a fluorophore for the detection of biothiols.[98] Thiols were detected by the cleavage of the Se–N bond of the probes. Bioimaging studies of both probes for the detection of GSH were conducted in living RAW 264.7 cells and rat liver tissues. Zhang and co-workers presented probe **74** for the detection of thiols. Probe **74** reacts with 1 equiv. of GSH

($\lambda_{ex}/\lambda_{em} = 470$ nm/585 nm; LOD = 16 nM) and Hcy ($\lambda_{ex}/\lambda_{em} = 488$ nm/596 nm; LOD = 9 nM) to give stable intermediates *via* Se – N bond cleavage and Se – S(GSH/Hcy) bond formation. However, the reaction with cys ends with cleavage of both Se – N bonds and the formation of the corresponding diamino derivative ($\lambda_{ex}/\lambda_{em} = 528$ nm/605 nm; LOD = 45 nM).[99] Sangit Kumar and co-workers developed a series of probes (**75a–75d**) based on a similar approach for the selective colorimetric detection of thiols (Figure 14.8).[100]

Tang and co-workers designed ebselen-based probe **76** using cyanine as the signalling platform for the reversible detection of GSH/H_2O_2 ($\lambda_{ex}/\lambda_{em} = 768/794$ nm).[101] The intracellular redox status during the apoptosis process was imaged by **76** using HepG2 and HL7702 cells. Additionally, the probe was able to monitor H_2O_2 at the wound margin in zebrafish larvae (Figure 14.9). Probe **77** was developed by Xue *et al.*[102] for monitoring GSH with fluorescence enhancement due to cleavage of the Se–N bond ($\lambda_{ex}/\lambda_{em} = 380/615$ nm; LOD = 0.178 mM).

Tang *et al.* synthesized probe **78**, which also detects selenocysteine through cleavage of the Se–N bond ($\lambda_{ex}/\lambda_{em} = 460/580$ nm; LOD = 7.0 nM).[103] Furthermore, the probe was able to monitor selenol in living HepG2 cells and *in vivo.*

In 2016, Chen and co-workers disclosed a ratiometric NIR-emissive cyanine probe **79** able to detect cysteine hydropersulfide.[104] The authors highlighted the short time necessary for detection, which is pivotal to accurately determine this analyte, since it is very reactive, being quickly metabolized in biological systems (Figure 14.9). The mechanism relied on the reduction of the diselenide linkage, followed by cyclic selenocarbonate elimination. The resulting methyl–amine dye showed a blue-shift in the emission spectrum, allowing its usage as a ratiometric probe for cysteine hydropersulfide, with a LOD of 12 μM. Cell imaging experiments were also accomplished allowing the detection of endogenous hydropersulfide produced by HL-7702 and HepG2 cells.

In 2015, Xian *et al.* reported the diselenide-based probes **80** ($\lambda_{ex}/\lambda_{em} = 340$ nm/455 nm) and **81** (Figure 14.10) for the selective and sensitive detection of H_2S over other biothiols. The diselenide bond cleavage by reaction with H_2S afforded reactive intermediates, which rapidly cyclized to 1,2-benzothiaseleno-3-one with concomitant release of the fluorescent coumarin and selenenic acid derivative, and these can be oxidized back to the original probe **80** in the presence of endogenous ROS. Probe **81** ($\lambda_{ex}/\lambda_{em} = 498/521$ nm; LOD = 0.06 μM), containing the Tokyo green fluorophore-residue, showed behaviours and responses similar to those observed for **80**, which was also confirmed in the HeLa cell line.[105] In 2019, Xian *et. al.* developed a series of nine fluorescein derivatives containing an R – Se – S – R′ substructure.[106] Seleno-sulfide derivative **82** (Figure 14.10) presented high selectivity for H_2S ($\lambda_{ex} = 498$ nm; LOD = 6.7 nM). The reaction of H_2S with **82** produces cyclic selenothioester along with high emissive fluorescein. Studies with HeLa cells were carried out showing a sensing ability towards Na_2S.

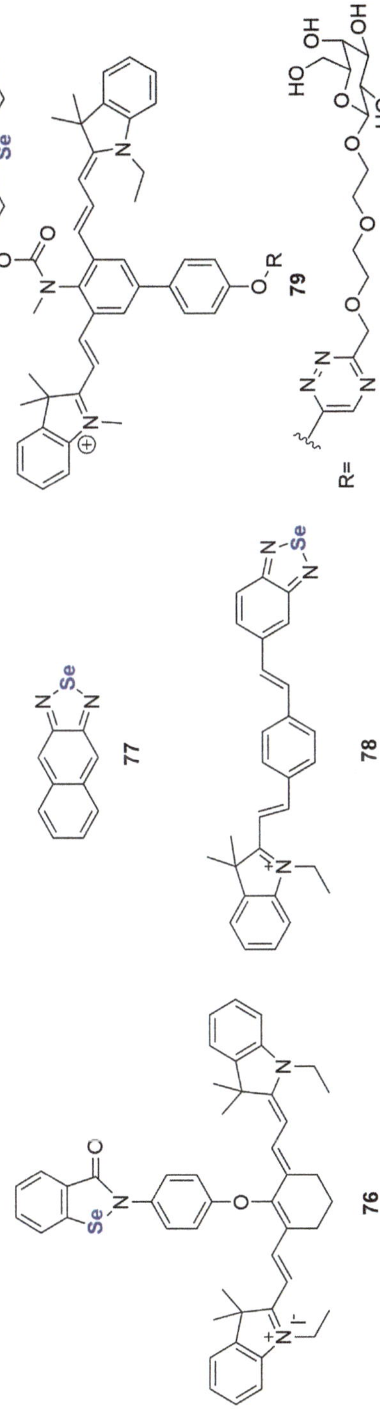

Figure 14.9 Designed selenides for GSH, selenocysteine, and selenol detection.

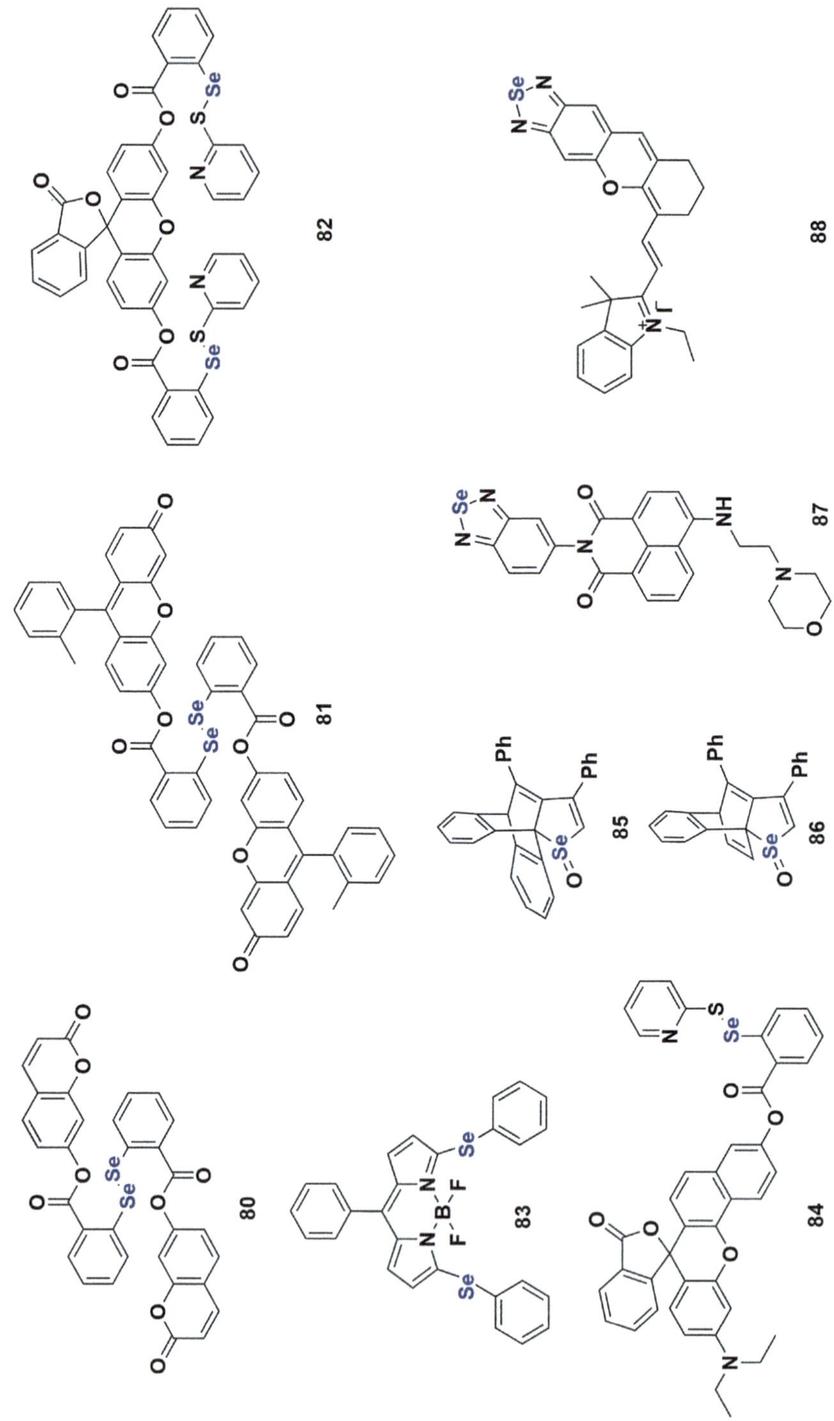

Figure 14.10 Organoselenides for H_2S and H_2Se detection in biologically important environments.

Furthermore, Guo *et al.* reported the phenylselenide substituted BODIPY-based sensor **83** ($\lambda_{ex}/\lambda_{em}$ = 582/610 nm) for the selective and sensitive detection of H_2S in BHK cells.[107] Probe **83** presents a lower detection limit (LOD = 2.5 nM) for H_2S by reaction with selenide as compared to other probes. Recently, in 2020 Yu and co-workers developed probe **84** ($\lambda_{ex}/\lambda_{em}$ = 580/635 nm; LOD = 34 nM) which selectively and rapidly detects hydrogen sulfide over other biothiols.[108] H_2S triggers the nucleophilic reaction and release of the fluorophore, which offers a turn-on fluorescence signal. Detection of exogenous and endogenous hydrogen sulfide was visualized in A549 living cells and zebrafish larvae. Ishii *et al.* developed the reversible probes **85** ($\lambda_{ex}/\lambda_{em}$ = 383/480 nm) and **86** ($\lambda_{ex}/\lambda_{em}$ = 389/463 nm) to detect H_2S *via* reduction of selenoxide to selenide. The corresponding selenide of the benzobarrelene skeleton was reverted back to selenoxide by reversible reaction with hypochlorite.[109]

Zhang *et al.* demonstrated selectivity for H_2Se detection using probe **87** ($\lambda_{ex}/\lambda_{em}$ = 435/535 nm; LOD = 1.2 μM) (Figure 14.10).[110] The probe was applied to monitor lysosomal H_2Se content in HepG2 cells. The same research group developed a NIR fluorescent probe **88**, for the detection of H_2Se *via* cleavage of the Se–N bond of the probe ($\lambda_{ex}/\lambda_{em}$ = 688/735 nm; LOD = 7.0 nM).[111] The probe was able to image endogenous H_2Se in HepG2 cells.

Chen *et al.* developed the NIR aza-BODIPY probe **89** ($\lambda_{ex}/\lambda_{em}$ = 707/737 nm; LOD = 2.3 nM) for the detection of sulfane sulfur *via* a turn-on mechanism. The relationship between sulfane sulfur and hypoxia stress was also examined with probe **89** in the hippocampus, mice, and zebrafish.[112] The same research group designed probe **90**, attributing to the triphenylphosphonium unit, the sulfane sulfur centre of targeting, in mitochondria ($\lambda_{ex}/\lambda_{em}$ = 680/720 nm; LOD = 3.1 nM) (Figure 14.11).[113] Probe **91**[114] ($\lambda_{ex}/\lambda_{em}$ = 410/527 nm; LOD = 0.178 mM) was developed by Tan and co-workers for the detection of 1,4-dithiothreitol (DTT: synthetic thiol widely used in cell biology, biochemistry, and biomedical application) in HeLa cells accompanied by fluorescence enhancement due to cleavage of the Se–N bond (Figure 14.11).

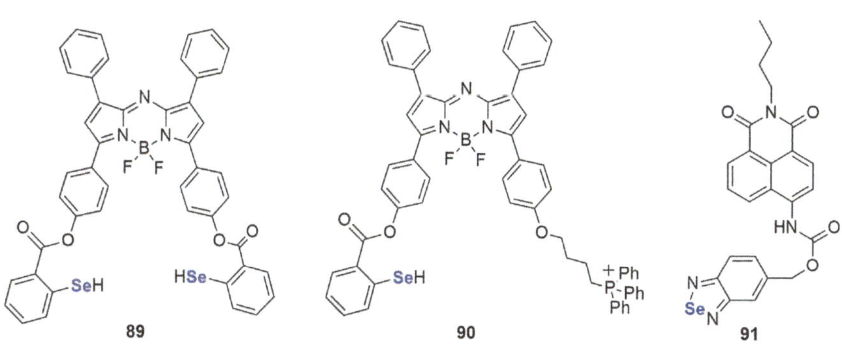

Figure 14.11 Selenium-governed sulfane, sulfur and DTT detection.

14.5 Probes for the Selective Detection of Metal Ions

Generally, a metal ion detection strategy is based on complexation by strategically positioned receptor units. Metal ions can be trapped in cavities featuring chalcogens and other donor heteroatoms. Moreover, heavy metal ions possess chalcogenophilicity and are prone to establish soft–soft interactions with chalcogen atoms, contributing to a fast and sensitive fluorescent detection.

In 2009, Tang *et al.* developed the selenium-containing fluorescein-based fluorescent probe **92** for the detection of Hg(II) ($\lambda_{ex}/\lambda_{em} = 492/513$ nm; LOD = 1.0 nM).[115] A fluorescence turn-on response of the probe upon the addition of Hg(II) was achieved *via* a deselenation process (Figure 14.12). The bio applicability of **92** to detect Hg(II) was confirmed in the living macrophage cell line, RAW264.7. Zeng *et al.* reported the hemicyanine-based colorimetric fluorophore **93**,[116] which features hemicyanine as a signalling unit and a NO_2Se_2 donating receptor unit for metal coordination, and the system is selective and sensitive for Hg(II) in both ethanol/water ($\lambda_{ex}/\lambda_{em} = 540/596$ nm) and aqueous medium ($\lambda_{ex}/\lambda_{em} = 530/590$ nm). The probe showed a colour change from red to colourless after reaction with Hg(II) which is due to an intramolecular charge transfer (ICT) mechanism. The detection limit was estimated to be 50 nM. Leray *et al.* designed and synthesized the chemodosimeter **94** containing a phosphorous–selenium P=Se moiety. The phosphane-selenide group integrated in a push–pull chromophore selectively detected Hg(II). The complexation of the P=Se moiety by Hg(II) is accompanied by the transformation of phosphane selenide to phosphane oxide leading to a fluorescence turn-on response ($\lambda_{ex}/\lambda_{em} = 330/406$ nm; LOD = 0.9 nM).[117] The tripodal NSe_3 donating probe **95**[118] was synthesized by Singh and co-workers without incorporation of a fluorophore. The probe was found to be selective for Hg(II) detection ($\lambda_{ex}/\lambda_{em} = 270/285$ nm; LOD = 0.1 nM) with turn-on fluorescence *via* Hg – Se bond formation.

Rhodamine-based probes containing chalcogen atoms have been gaining more attention in the recent years for the detection of various analytes, including metal ions. Generally, rhodamine-based compounds containing a spiro ring system are colourless and non-fluorescent derivatives. Nonetheless, ring opening, triggered by external stimulus (analyte interaction) may be accompanied by strong fluorescence emission and a colour change.[10,119,120] Wu *et al.* reported a very concise selenium-functionalization strategy of the rhodamine B core that was used as the key synthetic tool in the construction of the naked-eye and fluorescent highly selective molecular sensor **96** for *in vitro* and *in vivo* Hg(II) detection ($\lambda_{ex}/\lambda_{em} = 510/584$ nm; LOD = 12 nM). The compound presents high selectivity for Hg(II) in a large pH range (4.0–10.0) *via* metal complexation. The authors demonstrated the efficiency of the sensor for colorimetric naked-eye analytical and biochemical (zebrafish) application on the Hg(II) detection and also, the reversibility of the complexation upon the addition of Na_2S. Additionally, the authors prepared paper strips impregnated with the chemosensor to visually detect Hg(II) in solution *via* a colorimetric responsive differentiation in the

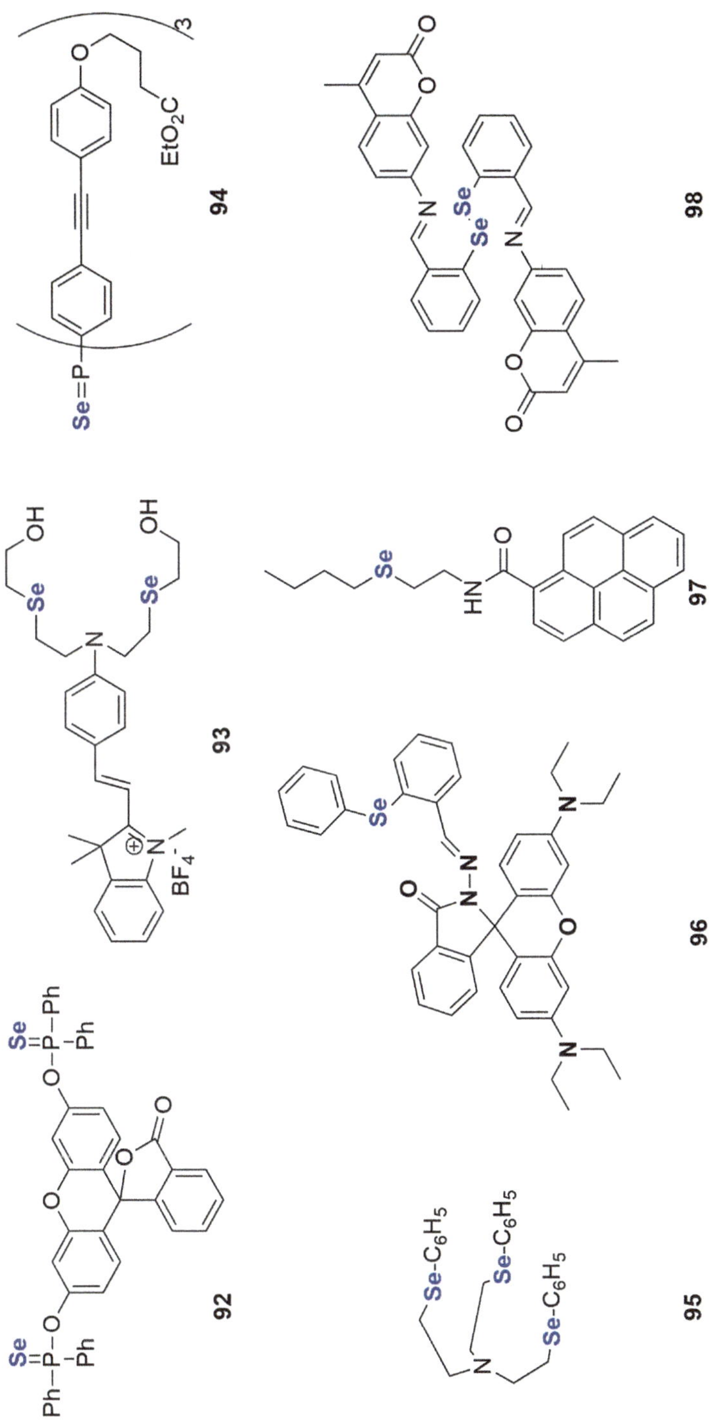

Figure 14.12 Selenide probes for Hg(II) detection in different analytical matrices.

concentration range of $10^{-7}-10^{-3}$ M and LOD = 1 μM. In a cytotoxicity test with HeLa cells, compound **96** (Figure 14.12) presented viability greater than 80%, allowing its use for *in vivo* imaging of Hg(II) using zebrafish.[121] In 2017, Dos Santos and co-workers[122] prepared a seleno-pyrene derivative **97** capable of detecting Hg(II) ions in an aqueous or aprotic system through metal complexation involving the chalcogen atom ($\lambda_{ex}/\lambda_{em}$ = 340/379 nm; LOD = 2 μM).

Recently, in 2022, Manjare *et al.* developed the coumarin-containing diselenide probe **98** for the selective detection of Hg(II) ($\lambda_{ex}/\lambda_{em}$ = 347/438 nm; LOD = 1.32 μM). A fluorescence turn-on signalling response of the probe was achieved through conversion of the diselenide moiety to monoselenide (*via* oxidation of the diselenide group), which was confirmed by single crystal X-ray diffraction (Figure 14.12). Utility of the probe to detect Hg(II) was demonstrated in the MCF-7 cell line.[123]

Organotellurium-based metal ion sensors are presented in Figure 14.13. Wu *et al.* reported tellurium-containing probes **99**[124] and **100**[125] for the selective detection of Hg(II) (**99** $\lambda_{ex}/\lambda_{em}$ = 500/560 nm; LOD = 19.2 μM, **100** $\lambda_{ex}/\lambda_{em}$ = 463/532 nm; LOD = 4.2 μM). Both probes exhibited a fluorescence turn-on response through chelation of Hg(II) in the RAW264.7 cell line. Dos Santos and co-workers[126] designed tellurium–rhodamine B derivative **101** in only two steps and demonstrated Hg(II) detection in aqueous solutions ($MeCN/H_2O$, 75 : 25 v/v) ($\lambda_{ex}/\lambda_{em}$ = 515/582 nm; LOD = 1.71 μM). The authors elucidated the involved mechanism, revealing a highly selective oxidation process, ratiometrically and selectively promoted by Hg(II). The compound was applied as a fluorescence sensor allowing fast detection of Hg(II), comparatively to other metal ions or conventional ROS. In 2019, the same Brazilian group reported the usage of the tellurium–rhodamine derivative **102** ($\lambda_{ex}/\lambda_{em}$ = 515/580 nm; LOD = 2.95 μM) to detect Hg(II) even in the presence of other metal ions, in aqueous media, based on the same mechanism (Figure 14.13).[127]

In 2009 a Chinese group[128] designed and synthesized the selenolactone–rhodamine B derivative **103** for imaging Hg(II) (LOD = 23 nM) and Ag(I) (LOD = 52 nM) in living cells and for investigating the toxicity of these cations (Figure 14.14) ($\lambda_{ex}/\lambda_{em}$ = 520/580 nm). Confocal fluorescence images of HeLa cells showed that the reaction of the probe with Ag(I) is slower than

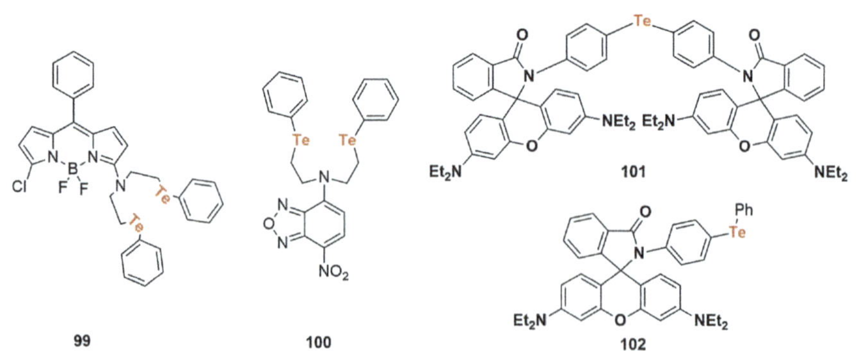

Figure 14.13 Tellurium-based probes for Hg(II) sensing.

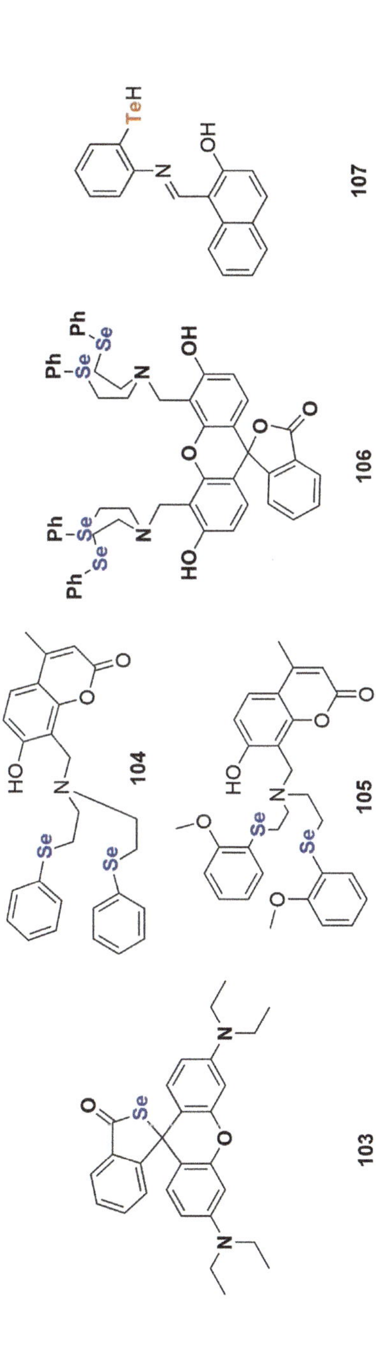

Figure 14.14 Chalcogen-based derivatives for Hg(II) and Ag(I) detection.

with Hg(II) due to the high concentration of Cl^- in the cellular medium. Yoon *et al.* also synthesized **103** ($\lambda_{ex}/\lambda_{em} = 550/580$ nm; LOD = 20 nM), but they demonstrated its application for the detection of methylmercury and Hg(II) species.[129] The probe showed high selectivity and sensitivity for these analytes. The probe was also successful in experiments using HeLa cells or zebrafish.

Zeng *et al.* reported the coumarin-based probes **104**[130] ($\lambda_{ex}/\lambda_{em} = 360/445$ nm) and **105**[131] ($\lambda_{ex}/\lambda_{em} = 373/445$ nm) containing a NSe_2 donor set for the coordination and selective detection of Ag(I). In both probes a fluorescence turn-on response was observed after reaction with Ag(I) due to complex formation (LOD = 52 nM for both the probes). The same research group developed the fluorescein derivative **106** and demonstrated its effectiveness for detecting Ag(I) selectively (Figure 14.14).[132] The Ag(I) coordination by the pendant arms located on the 4,5-positions of the fluorescein moiety, induces the ring-opening of the spirolactone portion, triggering maximum fluorescence emission ($\lambda_{ex}/\lambda_{em} = 463/538$ nm; LOD = 30 nM). An Indian group reported the naphthaline-based probe **107**[133] for the selective recognition of Ag(I) ($\lambda_{ex}/\lambda_{em} = 328/499$ nm; LOD = 2.34 nM). The fluorescence turn-on response was observed due to complex formation *via* interaction of Ag(I) with both the Schiff base and –TeH group. The probe was applied to a real water sample for Ag(I) monitoring.

Copper is one of most important trace elements after zinc and iron in biological systems. The unregulated concentration of copper is responsible for oxidative stress through a number of reactions, therefore, it is essential to monitor the concentration of copper metal ions. Wu *et al.* reported the BODIPY-based probe **108**[134] containing a NSe_2 coordinating cavity for the selective detection of Cu(II) ($\lambda_{ex}/\lambda_{em} = 500/516$ nm; LOD = 0.87 μM) *via* complexation. The selectivity towards Cu(II) was also confirmed in the RAW264.7 cell line. Selenium-containing coumarin-based compound **109**[135] was reported by Dos Santos for the selective detection of Cu(II) ($\lambda_{ex}/\lambda_{em} = 431/478$ nm; LOD = 0.2 μM) through complexation over various metal ions. In 2016, Yin *et al.* presented the BODIPY-based probe **110** for the selective detection of Cu(II) ($\lambda_{ex}/\lambda_{em} = 530/617$ nm; LOD = 0.92 μM) and Hg(II) ($\lambda_{ex}/\lambda_{em} = 530/590$ nm; LOD = 1.84 μM) at different emission wavelengths (Figure 14.15).[136] Significant changes in the emission spectra with Cu(II) and Hg(II) helped to construct OR, INHIBIT, and IMPLICATION logic gates at the molecular level. Sykes and co-workers, in 2013, developed the anthraquinone-based probe **111** bearing a selenoether cavity for *in vitro* detection of Cu(II) and Fe(III) ($\lambda_{ex}/\lambda_{em} = 400/593$ nm; LOD = 0.1 mM).[137] Probe **111** presents no fluorescence in solution, but in the presence of $Cu(ClO_4)_2$ or $Fe(ClO_4)_3$, cleavage of one selenide moiety occurs releasing an orange emissive residue. This behaviour was not observed when oxygen or sulfur analogues were exposed to the same solutions containing metal ions, demonstrating the importance of the selenium atom in the detection of Cu(II)/Fe(III) ions.

Chalcogen-containing probes for the selective detection of Fe(III), Ni(II), Pd(II), and Pb(II) metal ions have also been reported (Figure 14.16). The

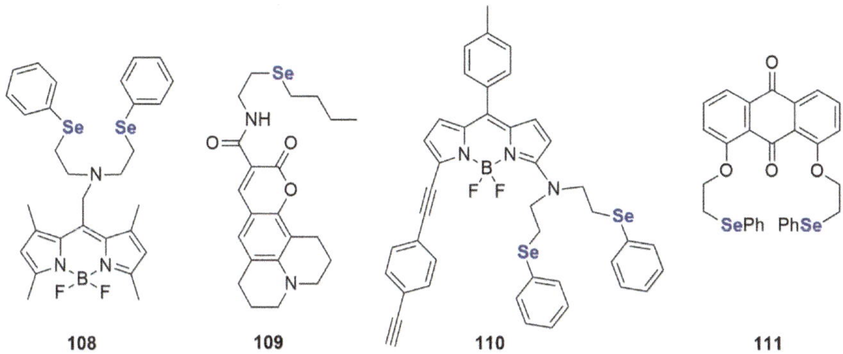

108 109 110 111

Figure 14.15 Selenide-governed probes for Cu(II), Hg(II), and Fe(III) detection.

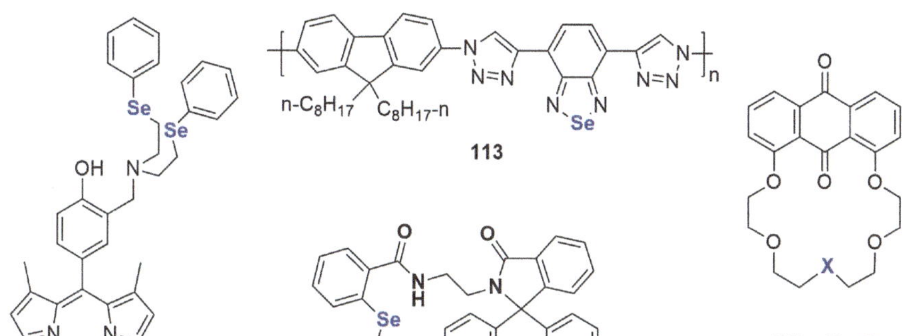

112 113 114

115a: X = S
115b: X = Se
115c: X = Te

Figure 14.16 Chalcogenides designed for the detection of less frequently considered metal ions like Fe(III), Ni(II), Pd(II), and Pb(II) by fluorescence monitoring.

seleno-BODIPY derivative **112** is selective for Fe(III) ($\lambda_{ex}/\lambda_{em} = 500/510$ nm; LOD = 96.3 μM).[138] The probe can distinguish Fe(II) and Fe(III) *via* Fenton chemistry. In 2014, Wu *et al.* developed the selenium-containing conjugated polymer fluorescent sensor **113** featuring benzoselenadiazole and triazole units. The sensor is selective and sensitive for the detection of Ni(II) over other metal ions ($\lambda_{ex}/\lambda_{em} = 425/535$; LOD = 2.4 nM). Ni(II) shows quenching of the fluorescence emission possibly due to an electron transfer process between the metal complex and the polymer backbone.[139] Another metallic ion that deserves mention is palladium. Recently, Dos Santos and co-workers published the highly selective, sensitive, and reversible seleno-rhodamine-based colorimetric and fluorescent probe **114** for nano-molar detection of Pd(II) ions in aqueous solution ($\lambda_{ex}/\lambda_{em} = 515/588$ nm; LOD = 32 nM) (Figure 14.16).[140] In the presence of various metal ions, the seleno-fluorophore is able to discriminate Pd(II), leading to an orange emissive complex in aqueous solutions. The potential use to detect Pd(II) in real synthetic API (Active Pharmaceutical Ingredient) samples was demonstrated.

In 2015, Sykes and co-workers reported a series of macrocycles containing an anthraquinone moiety and a polyheteroether (S, Se and Te) aliphatic chain.[141] The S and Se analogues were capable of detecting Pb(II) ions in acetonitrile (**115a** and **115b**) with the selenium-containing system (**115b**) presenting $\lambda_{ex}/\lambda_{em} = 390/520$ nm with LOD = 20 μM (Figure 14.16). The Te analogue **115c** did not present an optical response for any metal cation.

14.6 Conclusions

In summary, this chapter covers all organochalcogen-based probing systems to date for the detection of biologically important analytes. The chalcogen atoms in these compounds possesses a lone pair of electrons which can be transferred, resulting in the modification of the photophysical properties of the fluorophores. It was observed that most of the probes used for the detection of ROS/RNS have been developed by employing the advantage of redox properties of the chalcogens. Thus, the oxidation of the chalcogen to chalcogenoxide by reaction with ROS/RNS affects the PET/ICT/TICT mechanisms, resulting in fluorescence turn-on or turn-off responses. In contrast, when biothiols react with chalcogen-containing probes, nucleophilic attack of the thiol–sulfur on chalcogen (Se or Te) leads to the formation of a chalcogen–sufide bond, which alters the photophysical properties of the probing systems. In the case of the metal ion detection, frequently, chalcogen-containing receptor units have been employed for coordinating the metal ion selectively, which again changes the fluorescence properties of the conjugated fluorescent units. Additionally, incorporation of the chalcogen atoms in the probe exerts the heavy atom effect which offers a red shift in the emission of the probe. Furthermore, the lower cytotoxicity of the chalcogen-containing probes and detection of various biological analytes *in vitro* and *in vivo* confirmed the biocompatibility of the organochalcogen compounds. Thus, we feel that the development in the area of chalcogen-containing probes is one of the new feathers in the crown of chalcogen chemistry.

Acknowledgements

Sudesh T. Manjare, Gauri S. Malankar, Pratiksha P. Deshmukh, and Divyesh S. Shelar acknowledge SERB, New Delhi, India for the financial support (YSS/2014/000726) and the Department of Chemistry, University of Mumbai for the research platform and support.

Alcindo A. Dos Santos, Beatriz S. Cugnasca, Felipe Wodtke, and João L. Petrarca de Albuquerque thank FAPESP (2018/24434-3) for the financial support in recent years and the Institute of Chemistry from University of São Paulo for the support and research structure. Beatriz thanks FAPESP (2019/07634-1) for the PhD scholarship. Felipe thanks CAPES (88887.343036/2019-00) for the PhD scholarship. João thanks FAPESP (2020/04782-7) for the PhD scholarship.

References

1. J. Yoon, Y. Hua and J. Yin, *Chem. Soc. Rev.*, 2015, **44**, 4619.
2. M. Schferling, *Angew. Chem., Int. Ed.*, 2012, **51**, 3532.
3. Q.-Z. Yang, C.-H. Tung, L.-Z. Wu, H.-R. Zheng, Y.-Z. Chen and L.-Y. Niu, *Chem. Soc. Rev.*, 2015, **44**, 6143.
4. L. Xiong, Z. Chen, Q. Tian, Y. Cao, C. Xu and F. Li, *Anal. Chem.*, 2009, **81**, 8687.
5. Y. Yang, Q. Zhao, W. Feng and F. Li, *Chem. Rev.*, 2013, **113**, 192.
6. M. Yu, F. Li, Z. Chen, H. Hu, C. Zhan, H. Yang and C. Huang, *Anal. Chem.*, 2009, **81**, 930.
7. A. P. de Silva, H. Q. Nimal Gunaratne, T. Gunnlaugsson, A. J. M. Huxley, C. P. McCoy, J. T. Rademacher and T. E. Rice, *Chem. Rev.*, 1997, **97**, 1515.
8. Y. Wen, Z. Long, F. Huo and C. Yin, *Org. Chem. Front.*, 2021, **8**, 1302.
9. S. Panda, A. Panda and S. S. Zade, *Coord. Chem. Rev.*, 2015, **300**, 86.
10. S. T. Manjare, Y. Kim and D. G. Churchill, *Acc. Chem. Res.*, 2014, **47**, 2985.
11. Z. Lou, P. Li and K. Han, *Acc. Chem. Res.*, 2015, **48**, 1358.
12. W. Lee, S. V. Mulay, S. Shimodaira, A. Abdillaha, J. Palmaa, Y. Kim, T. Yudhistira and D. G. Churchill, *Methods Enzymol.*, 2020, **640**, 267.
13. D. Wu, L. Chen, N. Kwon and J. Yoon, *Chem.*, 2016, **1**, 674.
14. E. Fron, E. C. Gonzalez, L. Pandey, M. Sliva, M. V. Auweraer, F. D. De Schryver, J. Thomas, Z. Dong, V. Leen, M. Smet, W. Dehae and T. Vosch, *New J. Chem.*, 2009, **33**, 1490.
15. S. V. Mulay, Y. Kim, L. W. Hyun, S. Singh and D. G. Churchill, in *Organoselenium Compounds in Biology and Medicine Synthesis, Biological and Therapeutic Treatments*, ed. V. K. Jain and K. I. Priyadarsini, The Royal Society of Chemistry, United Kingdom, Croydon, 2018, pp. 178–195.
16. G. Mugesh, W. W. Du Mont and H. Sies, *Chem. Rev.*, 2001, **101**, 2125.
17. J. Fernandez-Lodeiro, M. F. Pinatto-Botelho, A. A. Soares-Paulino, A. C. Gonçalves, B. A. Sousa, C. Princival and A. A. Dos Santos, *Dyes Pigm.*, 2014, **110**, 28.
18. T. Yudhistira, W. Lee, J. Choi, M. Jang, M. B. Halle, P. M. Sonawane, D. Kim and D. G. Churchill, *SMC Bull.*, 2017, **8**, 10.
19. N. Soh, *Anal. Bioanal. Chem.*, 2006, **386**, 532.
20. T. Yudhistira, S. V. Mulay, Y. Kim, M. B. Halle and D. G. Churchill, *Chem. – Asian J.*, 2019, **18**, 3048.
21. A. Abdillah, P. M. Sonawane, D. Kim, D. Mametov, S. Shimodaira, Y. Park and D. G. Churchill, *Molecules*, 2021, **26**, 692.
22. P. Li, L. Liu, H. Xiao, W. Zhang, L. Wang and B. Tang, *J. Am. Chem. Soc.*, 2016, **138**, 2893.
23. W.-H. Zhang, D. Zhang, R.-J. Zhang, F. Xia and Y.-F. Liu, *Anal. Bioanal. Chem.*, 2012, **402**, 895.
24. J. Wan, X. Zhang, K. Zhang and Z. Su, *Anal. Chem.*, 2020, **39**, 209.

25. B. C. Dickinson and C. J. Chang, *Nat. Chem. Biol.*, 2011, **7**, 504.
26. H. Singh, K. Tiwari, R. Tiwari, S. K. Pramanik and A. Das, *Chem. Rev.*, 2019, **119**, 11718.
27. P. Arsenyan, K. Rubina, I. Shestakova and I. Domracheva, *Eur. J. Med. Chem.*, 2007, **42**, 635.
28. D. Bartolini, L. Sancineto, A. Fabro de Bem, K. D. Tew, C. Santi, R. Radi, P. Toquato and F. Galli, *Adv. Cancer Res.*, 2017, **136**, 259.
29. A. H. Ashoka, F. Ali, R. Tiwari, R. Kumari, S. K. Pramanik and A. Das, *ACS Omega*, 2020, **5**, 1730.
30. M. P. Murphy, *Biochem. J.*, 2009, **417**, 1.
31. H. Bayr, *Crit. Care Med.*, 2005, **33**, S498.
32. Y. Koide, M. Kawaguchi, Y. Urano, K. Hanaoka, T. Komatsu, M. Abo, T. Terai and T. Nagano, *Chem. Commun.*, 2012, **48**, 3091.
33. S. T. Manjare, S. Kim, W. D. Heo and D. G. Churchill, *Org. Lett.*, 2014, **16**, 410.
34. P. P. Deshmukh, A. Navalkar, S. K. Maji and S. T. Manjare, *Sens. Actuators, B*, 2019, **281**, 83.
35. K. S. Madibone, P. P. Deshmukh, A. Navalkar, S. K. Maji, P. M. Badani and S. T. Manjare, *ACS Omega*, 2020, **5**, 14186.
36. M. B. Halle, K. J. Lee, T. Yudhistira, J. H. Choi, H. S. Park and D. G. Churchill, *Chem. – Asian J.*, 2018, **13**, 3895.
37. Y. Wang, M. Gao, Q. Chen, F. Yu, G. Jiang and L. Chen, *Anal. Chem.*, 2018, **90**, 9769.
38. W. Zhang, R. Wang, W. Liu, X. Wang, P. Li, W. Zhang, H. Wang and B. Tang, *Chem. Sci.*, 2018, **9**, 721.
39. B. C. Dickinson and C. J. Chang, *Nat. Chem. Biol.*, 2011, **7**, 504.
40. J. M. Pullar, M. C. M. Vissers and C. C. Winterbourn, *IUBMB Life*, 2000, **50**, 259.
41. S. Liu and S. Wu, *Org. Lett.*, 2013, **15**, 878.
42. S. V. Mulay, M. Choi, Y. J. Jang, Y. Kim, S. Jon and D. G. Churchill, *Chem. – Eur. J.*, 2016, **22**, 9642.
43. Y. Kim, T. Jun, S. V. Mulay, S. T. Manjare, J. Kwak, Y. Lee and D. G. Churchill, *Dalton Trans.*, 2017, **46**, 4111.
44. S. V. Mulay, T. Yudhistira, M. Choi, Y. Kim, J. Kim, Y. J. Jang, S. Jon and D. G. Churchill, *Asian J. Chem.*, 2016, **11**, 3598.
45. Y. Shen, X. Zhang, X. Huang, S. Wen, M. Liu, Y. Deng, Y. Zhang, C. Zhang, J. Jin, H. Li and S. Yao, *RSC Adv.*, 2015, **5**, 79519.
46. Y. Kim, M. Choi, S. T. Manjare, S. Jon and D. G. Churchill, *RSC Adv.*, 2016, **6**, 32013.
47. X.-H. Xu, C. Liu, Y. Mei and Q.-H. Song, *J. Mater. Chem. B*, 2019, **7**, 6861.
48. G. S. Malankar, A. Sakunthala, A. Navalkar, S. K. Maji, S. Raju and S. T. Manjare, *Anal. Chim. Acta*, 2021, **1150**, 338205.
49. M. B. Halle, T. Yudhistira, K. J. Lee, J. H. Choi, Y. Kim, H. Park and D. G. Churchill, *ACS Omega*, 2018, **3**, 13474.
50. S. Zang, X. Kong, J. Cui, S. Su, W. Shu, X. Zhang and X. J. Zhang, *J. Mater. Chem. B*, 2020, **8**, 2660.

51. Z. Qu, J. Ding, M. Zhao and P. Li, *J. Photochem. Photobiol., A*, 2015, **299**, 1.
52. Z. Lou, P. Li, Q. Pan and K. Han, *Chem. Commun.*, 2013, **49**, 2445.
53. X. Lv, X. Yuan, Y. Wang and W. Guo, *New J. Chem.*, 2018, **42**, 15105.
54. X. Kong, L. Di, Y. Fan, Z. Zhou, X. Feng, L. Gai, J. Tian and H. Lu, *Chem. Commun.*, 2019, **55**, 11567.
55. J. R. Casey, S. Grinstein and J. Orlowski, *Nat. Rev. Mol. Cell Biol.*, 2010, **11**, 50.
56. G. Battogtokh, Y. S. Choi, D. S. Kang, S. J. Park, M. S. Shim, K. M. Huh, Y.-Y. Cho, J. Y. Lee, H. S. Lee and H. C. Kang, *Acta Pharm. Sin. B*, 2018, **8**, 862.
57. G. Li, D. Zhu, Q. Liu, L. Xue and H. Jiang, *Org. Lett.*, 2013, **15**, 2002.
58. X. Xie, T. Wu, X. Wang, Y. Li, K. Wang, Z. Zhao, X. Jiao and B. Tang, *Chem. Commun.*, 2018, **54**, 11965.
59. B. Wang, P. Li, F. Yu, J. Chen, Z. Qu and K. Han, *Chem. Commun.*, 2013, **49**, 5790.
60. G. Cheng, J. Fan, W. Sun, J. Cao, C. Hu and X. Peng, *Chem. Commun.*, 2014, **50**, 1018.
61. C. Liu, X. Jiao, S. Cai, S. He, L. Zhaob and X. Zeng, *Sens. Actuators, B*, 2021, **329**, 129213.
62. B. Wang, P. Li, F. Yu, P. Song, X. Sun, S. Yang, Z. Lou and K. Han, *Chem. Commun.*, 2013, **49**, 1014.
63. C. Liu, Y. Shang, T. Zhao, L. Liang, S. He, L. Zhao, X. Zeng and T. Wang, *Sens. Actuators, B*, 2021, **348**, 130632.
64. M. G. Choi, H. Ryu, M. J. Cho, S. K. Lee and S. Chang, *Sens. Actuators, B*, 2017, **244**, 307.
65. W. Zhang, W. Liu, P. Li, J. Kang, J. Wang, H. Wang and B. Tang, *Chem. Commun.*, 2015, **51**, 10150.
66. H. Jiang, L. Jia, Y. Li, S. Liu, R. Chen, L. Jin, J. Jin, C. Zheng, Q. Fan and W. Huang, *Chem. Commun.*, 2018, **54**, 2926.
67. S. T. Manjare, J. Kim, Y. Lee and D. G. Churchill, *Org. Lett.*, 2014, **16**, 520.
68. P. Venkatesan and S. Wu, *Analyst*, 2015, **140**, 1349.
69. W.-J. Shi, Y. Huang, W. Liu, D. Xu, S.-T. Chen, F. Liu, J. Hu, L. Zheng and K. Chen, *Dyes Pigm.*, 2019, **170**, 10756.
70. W.-J. Shi, L.-X. Feng, X. Wang, Y. Huang, Y.-F. Wei, Y.-Y. Huang, H.-J. Ma, W. Wang, M. Xiang and L. Gao, *Talanta*, 2021, **233**, 122581.
71. W. Chen, W. P. Bay, M. W. Wong and D. Huang, *Tetrahedron Lett.*, 2012, **53**, 3843.
72. Y. X. Liao, K. Li, M. Wu, T. Wu and X. Yu, *Org. Biomol. Chem.*, 2014, **12**, 3004.
73. C. Xu and Y. Qian, *J. Mater. Chem. B*, 2019, **7**, 2714.
74. H. Ungati, V. Govindaraj, M. Narayanan and G. Mugesh, *Angew. Chem.*, 2019, **131**, 8240.
75. M. Kaur, D. S. Yang, K. Choi, M. J. Cho and D. H. Choi, *Dyes Pigm.*, 2014, **100**, 118.

76. D. Procházková, N. Wilhelmová and M. Pavlík, Reactive Nitrogen Species and Nitric Oxide, in *Nitric Oxide Action in Abiotic Stress Responses in Plants*, ed. M. Khan, M. Mobin, F. Mohammad and F. Corpas, Springer, Chem, 2015.
77. M. C. Martínez and R. Andriantsitohaina, *Antioxid. Redox Signaling*, 2009, **11**, 3.
78. R. Radi, *J. Biol. Chem.*, 2013, **288**, 26464.
79. M. P. Gordge and F. Xiao, *Br. J. Pharmacol.*, 2010, **159**, 1572.
80. B. C. Dickinson, D. Srikun and C. J. Chang, *Curr. Opin. Chem. Biol.*, 2010, **14**, 50.
81. C. Szabo, H. Ischiropoulos and R. Radi, *Nat. Rev. Drug Discovery*, 2007, **6**, 662.
82. X. Chen, F. Wang, J. Y. Hyun, T. Wei, J. Qiang, X. Ren, I. Shin and J. Yoon, *Chem. Soc. Rev.*, 2016, **45**, 2976.
83. S. Toyokuni, K. Okamoto, J. Yodoi and H. Hiai, *FEBS Lett.*, 1995, **358**, 1.
84. B. Wang, F. Yu, P. Li, X. Sun and K. Han, *Dyes Pigm.*, 2013, **96**, 383.
85. F. Yu, P. Li, G. Li, G. Zhao, T. Chu and K. Han, *J. Am. Chem. Soc.*, 2011, **133**, 11030.
86. C. Sun, W. Du, P. Wang, Y. Wu, B. Wang, J. Wang and W. Xie, *Biochem. Biophys. Res. Commun.*, 2017, **494**, 518.
87. K. Xu, H. Chen, J. Tian, B. Ding, Y. Xie, M. Qiang and B. Tang, *Chem. Commun.*, 2011, **47**, 9468.
88. F. Yu, P. Li, B. Wang and K. Han, *J. Am. Chem. Soc.*, 2013, **135**, 7674.
89. C. Sun, W. Shi, Y. Song, W. Chen and H. Ma, *Chem. Commun.*, 2011, **47**, 8638.
90. J. Dai, C. Ma, P. Zhang, Y. Fu and B. Shen, *Dyes Pigm.*, 2020, **177**, 108321.
91. R. O. Ball, G. Courtney-Martin and P. B. Pencharz, *J. Nutr.*, 2006, **136**, 1682.
92. B. Tang, Y. Xing, P. Li, N. Zhang, F. Yu and G. Yang, *J. Am. Chem. Soc.*, 2007, **129**(38), 11666.
93. B. Tang, L. Yin, X. Wang, Z. Chen, L. Tong and K. Xu, *Chem. Commun.*, 2009, 5293.
94. Z. Chen, Q. Li, X. Wang, Z. Wang, R. Zhang, M. Yin, L. Yin, K. Xu and B. Tang, *Anal. Chem.*, 2010, **82**, 2006.
95. Z. Lou, P. Li, X. Sun, S. Yang, B. Wang and K. Han, *Chem. Commun.*, 2013, **49**, 391.
96. Y. Kim, S. V. Mulay, M. Choi, S. B. Yu, S. Jon and D. G. Churchill, *Chem. Sci.*, 2015, **6**, 5435.
97. S. V. Mulay, Y. Kim, M. Choi, D. Y. Lee, J. Choi, Y. Lee, S. Jon and D. G. Churchill, *Anal. Chem.*, 2018, **90**, 2648.
98. R. Wang, L. Chen, P. Liu, Q. Zhang and Y. Wang, *Chem. – Eur. J.*, 2012, **18**, 11343.
99. Y. Tian, B. Zhu, W. Yang, J. Jing and X. Zhang, *Sens. Actuators, B*, 2018, **262**, 345.
100. S. J. Balkrishna, A. S. Hodage, S. Kumar, P. Panini and S. Kumar, *RSC. Adv.*, 2014, **4**, 11535.

101. K. Xu, M. Qiang, W. Gao, R. Su, N. Li, Y. Gao, Y. Xie, F. Kong and B. Tang, *Chem. Sci.*, 2013, **4**, 1079.
102. X. Zeng, X. Zhang, B. Zhu, H. Jia, W. Yang, Y. Li and J. Xue, *Sens. Actuators, B*, 2011, **159**, 142.
103. F. Kong, B. Hu, Y. Gao, K. Xu, X. Pan, F. Huang, Q. Zheng, H. Chen and B. Tang, *Chem. Commun.*, 2015, **51**, 3102.
104. X. Han, F. Yu, X. Song and L. Chen, *Chem. Sci.*, 2016, **7**, 5098.
105. B. Peng, C. Zhang, E. Marutani, A. Pacheco, W. Chen, F. Ichinose and M. Xian, *Org. Lett.*, 2015, **17**, 1541.
106. Y. Wang, C. Yang, S. Xu, W. Chen and M. Xian, *Org. Lett.*, 2019, **21**, 7573.
107. D. Gong, X. Zhu, Y. Tian, S. Han, M. Deng, A. Iqbal, W. Liu, W. Qin and H. Guo, *Anal. Chem.*, 2017, **89**, 1801.
108. X. Zhang, W. Qu, H. Liu, Y. Ma, L. Wang, Q. Sun and F. Yu, *Anal. Chim. Acta*, 2020, **1109**, 37.
109. T. Annaka, N. Nakata and A. Ishii, *New J. Chem.*, 2019, **43**, 11643.
110. Y. Tian, F. Xin, X. Zhang and X. Zhang, *J. Mater. Chem. B*, 2019, **7**, 2829.
111. F. Kong, L. Ge, X. Pan, K. Xu, X. Liu and B. Tang, *Chem. Sci.*, 2016, **7**, 1051.
112. M. Gao, R. Wang, F. Yu, B. Li and L. Chen, *J. Mater. Chem. B*, 2018, **6**, 6637.
113. M. Gao, R. Wang, F. Yu and L. Chen, *Biomaterials*, 2018, **160**, 1.
114. B. Zhu, X. Zhang, H. Jia, Y. Li, H. Liu and H. Tan, *Org. Biomol. Chem.*, 2010, **8**, 1650.
115. B. Tang, B. Ding, K. Xu and L. Tong, *Chem. – Eur. J.*, 2009, **15**, 3147.
116. Y. Li, S. He, Y. Lu and X. Zeng, *Org. Biomol. Chem.*, 2011, **9**, 2606.
117. I. Samb, J. Bell, P. Y. Toullec, V. Michelet and I. Leray, *Org. Lett.*, 2011, **13**, 1182.
118. A. Kumar and J. D. Singh, *Inorg. Chem.*, 2012, **51**, 772.
119. H. Zheng, X. Zhan, Q. Bian and X. Zhang, *Chem. Commun.*, 2013, **49**, 429.
120. L. Chang, Q. Gao, S. Liu, C. Hu, W. Zhou and M. Zheng, *Dyes Pigm.*, 2018, **153**, 117.
121. P. Venkatesan, N. Thirumalisavan and S.-P. Wu, *RSC Adv.*, 2017, **7**, 21733.
122. C. A. Gonçalves, J. L. Capelo, C. Lodeiro and A. A. Dos Santos, *Sens. Actuators, B*, 2017, **239**, 311.
123. P. P. Deshmukh, G. S. Malankar, A. Sakunthala, A. Navalkar, S. K. Maji, D. P. Murale, S. Raju and S. T. Manjare, *Dalton Trans.*, 2022, **51**, 2269.
124. S. Kao and S. Wu, *Sens. Actuators, B*, 2015, **212**, 382.
125. S.-L. Kao, P. Venkatesan and S.-P. Wu, *New J. Chem.*, 2015, **39**, 3551.
126. A. A. Soares-Paulino, L. Giroldo, G. Celante, C. Lodeiro and A. A. Dos Santos, *Dyes Pigm.*, 2018, **159**, 121.
127. A. A. Soares-Paulino, L. Giroldo, G. Celante, E. Oliveita, S. M. Santos, R. F. Mendes, F. A. A. Paz, A. M. Fioroto, P. V. Oliveira, S. H. P. Serrano, C. Lodeiro and A. A. Dos Santos, *Dyes Pigm.*, 2019, **160**, 208.
128. W. Shi, S. Sun, X. Li and H. Ma, *Inorg. Chem.*, 2010, **49**, 1206.
129. X. Chen, K. Baek, Y. Kim, S. Kim, I. Shin and J. Yoon, *Tetrahedron*, 2010, **66**, 4016.

130. S. Huang, S. He, Y. Lu, F. Wei, X. Zeng and L. Zhao, *Chem. Commun.*, 2011, **47**, 2408.
131. J. Su, S. Huang, S. He and X. Zeng, *Chem. Res. Chin. Univ.*, 2016, **32**, 20.
132. C. Liu, S. Huang, H. Yao, H. Yao, S. He, Y. Lu, L. Zhao and X. Zeng, *RSC Adv.*, 2014, **4**, 16109.
133. A. K. K. Bhasin, J. Singh, H. Singh, P. Raj, N. Singh, N. Kaur and K. K. Bhasin, *Polyhedron*, 2017, **125**, 238.
134. C. Chou, S. Liu and S. Wu, *Analyst*, 2013, **138**, 3264.
135. A. C. Gonclaves, J. L. Capelo, C. Lodeiro and A. A. Dos Santos, *Photochem. Photobiol. Sci.*, 2017, **16**, 1174.
136. T. He, C. Lin, Z. Gu, L. Xu, A. Yang, Y. Liu, H. Fang, H. Qiu, J. Zhang and S. Yin, *Spectrochim. Acta, Part A*, 2016, **167**, 66.
137. K. Mariappan, P. N. Basa, V. Balasubramanian, S. Fuoss and A. G. Sykes, *Polyhedron*, 2013, **55**, 144.
138. D. P. Murale, S. T. Manjare, Y. Lee and D. G. Churchill, *Chem. Commun.*, 2014, **50**, 359.
139. Y. Lei, H. Li, W. Gao, M. Liu, J. Chen, J. Ding, X. Huang and H. Wu, *J. Mater. Chem. C*, 2014, **2**, 7402.
140. A. A. Soares-Paulino, L. Giroldo, N. A. Pradie, J. S. Reis, D. F. Back, A. A. C. Braga, H. A. Stefani, C. Lodeiro and A. A. Dos Santos, *Dyes Pigm.*, 2020, **179**, 108355.
141. K. Mariappan, M. Alaparthi, M. Hoffan, M. A. Rama, V. Balasubramanian, D. M. John and A. G. Sykes, *Dalton Trans.*, 2015, **44**, 11774.

CHAPTER 15

^{77}*Se Nuclear Magnetic Resonance: Experiments and* In Silico *Approaches*

MÁRCIO SANTOS SILVA*[a] AND LAURA ORIAN*[b]

[a] Laboratório de Síntese Orgânica Limpa – LASOL – Universidade Federal de Pelotas (UFPel), Campus Capão do Leão, RS, Brazil; [b] Dipartimento di Scienze Chimiche, Università degli Studi di Padova, Via Marzolo 1, 35131, Padova, Italy
*Emails: silva.ms@ufpel.edu.br; laura.orian@unipd.it

15.1 Introduction

The importance of ^{77}Se Nuclear Magnetic Resonance (NMR) is directly connected to the eminent role of this chalcogen atom in different fields, from inorganic chemistry[1] and materials science[2] to organic catalysis,[3] bioorganic and pharmaceutical chemistry,[4] to biology[5] and the biomedical field.[6] The first experiments, to the best of our knowledge, date back to the 50s, when Dharmatti and Weaver[7] reported the sign and magnitude of the selenium magnetic moment using a concentrated sample of selenous acid referring to ^{23}Na, and Walchli measured the chemical shifts of H_2Se, H_2SeO_2, and H_2SeO_4 referring to D_2O.[8] ^{77}Se NMR is considered a powerful tool to gain direct information on selenium states in reactions and in mechanistic studies.[9] In fact, its wide chemical shift range (over 6000 ppm) allows a fine resolution of different chemical sites. The major drawbacks are the low natural abundance (approximately 7.6%), and the low NMR sensitivity.

Chalcogen Chemistry: Fundamentals and Applications
Edited by Vito Lippolis, Claudio Santi, Eder J. Lenardão and Antonio L. Braga

Published by the Royal Society of Chemistry, www.rsc.org

In recent decades, powerful spectrometers were accessed using high-field magnets, computer hardware, tailored software and newer NMR pulse sequences, minimizing sensitivity and resolution problems in many areas of science.[10] Additionally, complex structures were elucidated employing advanced NMR techniques, especially 2D methods, overcoming the limitations of X-ray analysis with liquid and amorphous solid samples, and of mass spectrometry with volatile and unstable organic compounds. In the 90s, ^{77}Se NMR data of diverse organic compounds were already available in the literature,[11] demonstrating the simplicity and utility of ^{77}Se NMR spectroscopy. These dramatic advances have contributed to the application of ^{77}Se NMR spectroscopy in complex fields, from structural macromolecules[12] to sensor probes.[13]

Among the distinct NMR parameters, a ^{77}Se chemical shift is considered the most useful, because it can provide information about structure, oxidation state, stereochemistry, electronic neighbourhood, and stereoelectronic effects of a wide series of selenium-containing organic compounds.[14] Connected to this parameter, a quantitative evaluation of the integral values can be carried out in the ^{77}Se NMR spectra, demonstrating the importance to describe other NMR features, *i.e.*, the nuclear Overhauser effect (nOe) and relaxation delay. Finally, the scalar coupling offers valuable information to gain insight into the selenium neighbourhood,[15] mainly when a hydrogen atom is bound to selenium one.

The potential of ^{77}Se NMR can be even expanded by combining the experiment with computational protocols. In fact, *in silico* NMR is a flexible and powerful approach to assess with confidence the molecular structure as well as environment information. Pioneering computational studies appeared in the 90s; it is worth citing the non-relativistic GIAO calculations of selenium NMR shielding in SeH_2, $SeHCH_3$, and $Se(CH_3)_2$, developed by Magyarfalvi and Pulay,[16] the IGLO/GIAO-SCF and GIAO-MP2 calculations, by Bühl and co-workers,[17] the GIAO-CCSD chemical shifts, by Bühl *et al.*,[18] and the DFT-GIAO approach, by Ziegler and Schreckenbach.[19]

The aim of this chapter is to illustrate how this synergic approach is accurate and insightful, but not free from pitfalls, and to offer a comprehensive view of experimental and computational reliable protocols to investigate the structure and reactivity of organoselenium compounds by ^{77}Se NMR. Due to the vast scope of application of selenium, as mentioned above, and considering the simplicity in the measurements and interpretations, our focus is on liquid-phase experimental and computational ^{77}Se NMR applied to organic and bioorganic molecules. Thus, materials, inorganic, and biological compounds are not included in our discussion.

15.2 Experimental Background and Techniques

The selenium atom has proven to be a suitable nuclide in NMR spectroscopy. The great potential of ^{77}Se NMR experiments can be attributed to the ^{77}Se isotope properties (Table 15.1), which have no quadrupole moment

Table 15.1 Magnetogyric ratios, NMR frequencies (in a 9.4 T field), natural abundance, and receptivity of selected nuclides.[22]

Nuclide	Spin	$\gamma/10^7$ $T^{-1} s^{-1}$	ν_{NMR}/ MHz	Natural abundance (%)	Receptivity rel. ^{13}C
^{1}H	1/2	26.752	400.0	99.985	5.87×10^3
^{13}C	1/2	6.728	100.6	1.108	1.00
^{15}N	1/2	− 2.713	40.5	0.37	2.23×10^{-2}
^{17}O	5/2	− 3.628	54.3	0.037	6.50×10^{-2}
^{77}Se	1/2	5.12	76.3	7.63	3.15
^{125}Te	1/2	− 8.45	126.24	7.07	1.34×10^1

and negligible nOe effect.[20] The major chemical shift profile can be visualized in the spectral window of the ^{77}Se isotope between $\delta_{Se} = -700$ and +2200 ppm, in which organoselenium compounds can be accurately identified and quantified. This large shift dispersion overcomes the limitations of ^{1}H nuclide, which shows overlapping signals and high multiplicity standards, and of ^{13}C nuclide, due to its high nOe influence and the presence of a great number of carbon atoms in organic compounds, since typically organoselenium compounds present one or a few selenium atoms in a target molecule.[21]

In this context, we discuss the NMR parameters, particularly the chemical shift and spin-spin coupling, aiming at exploring the ^{77}Se nuclide characteristics, and at demonstrating some ^{77}Se NMR experiments. The discussions will benefit the current literature to clearly emphasize the advantages of ^{77}Se NMR spectroscopy and to highlight the strategies employed to achieve insights into distinct organoselenium reactions and applications.

First, it is worth recalling that the NMR frequency of a nucleus in a molecule emerges from the nuclear magnetic moment (μ), which is related to the magnetogyric ratio (γ) and to the spin angular moment (I) of each active nuclide, and to the strength of the applied external magnetic field (B^0). The magnitude and orientation of these parameters (γ and I) are connected to the sensitivity of the NMR experiments.

Referring to the data shown in Table 15.1, one can observe the highest sensitivity of the hydrogen nuclide, due to its greatest magnetogyric constant and high natural abundance. When other nuclides are compared to hydrogen, both NMR parameters decrease significantly, and thus a comparison with ^{13}C nuclide (receptivity) is used to predict the sensitivity. Comparatively, it is easier to acquire a ^{77}Se NMR spectrum than a ^{13}C NMR one, provided the receptivity of ^{77}Se related to ^{13}C is 3.15. ^{125}Te nuclide has the same advantages as selenium, while ^{17}O and ^{15}N nuclides are more difficult to record in a heteronuclear NMR experiment (Table 15.1).

To exemplify the ^{77}Se NMR sensitivity, very recently, Rozovsky and co-workers explored the use of selenomethionine amino acid residue as an active probe to provide data about the dynamics and structural changes at the protein interface and domain junctions.[23] The production of proteins with isotopic enrichment is a strategy performed to access the NMR spectra, and in this work, ^{77}Se incorporation as selenomethionine was higher than 85% (the remaining fraction was methionine). The natural abundancy of the ^{77}Se nuclide of the biological samples in a magnetic field of 11.74 T (500 MHz for hydrogen) was sensitive to protein variants. Although the high molecular weight of proteins implies lower concentrations and broadening of signals, which decrease the sensitivity of the NMR experiments, these results afforded new perspectives for ^{77}Se NMR spectroscopy. Finally, the interpretation is complex, especially compared to experiments on small organic molecules. A computational density functional theory (DFT) study was carried out to aid the discernment of the ^{77}Se NMR anisotropic effect.

The active nuclides in NMR spectroscopy do not have identical resonance frequencies, which depend on the position of the nucleus in the molecule, or, to be more precise, on the local electron distribution. This distinction, *i.e.*, the chemical shift, is one of the factors that make ^{77}Se NMR so attractive to organic chemists. The NMR parameter used to understand the chemical shift variation is termed nuclear shielding. It results from the external applied magnetic field, similarly to an electric current passing through a coil of wire, whose electronic density around the atoms generates a small magnetic field (B') in the opposite direction to B^0. Due to the electronic configuration of selenium and its distinct atomic orbitals, this active nuclide is very sensitive to electronic disparities.

Many factors play a role in determining the chemical shift values, like aromaticity, solvation, and acid-base interaction; they are often difficult to predict, justifying the use of computational data. Nevertheless, valuable information can often be extracted from ^{77}Se NMR spectra simply by noting the number of resonances and/or their relative intensities. A useful example is the ^{77}Se and ^{1}H NMR study of ebselen to identify the chalcogen bond (ChB) and hydrogen bond (HB) that are formed in solution with molecular targets.[24] Resnati and co-workers have evaluated the ability of the ebselen drug to act as a ChB donor in solution, testing distinct solvents in the presence of hexamethylphosphoramide (HMPA) as the chelating agent (Scheme 15.1). The increased shielding of ^{77}Se in the presence of incremental amounts of HMPA indicates that the ChB-driven formation of an ebselen/HMPA complex is due to the more electronegative oxygen attached to the selenium atom. A single peak was observed, showing that the

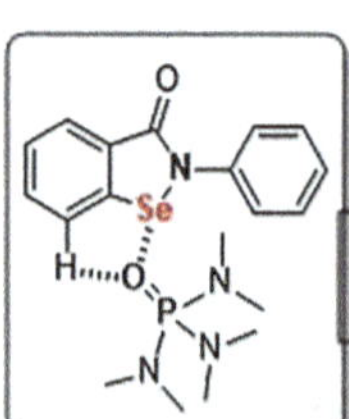

[HMPA]/[Ebselen] = 0.00	0.28	0.72	1.26	2.26	4.81	7.76
^{77}Se Chem. Shift = 940	939	937	934	930	925	922

Scheme 15.1 Ebselen/HMPA intermolecular interaction.

formation of the ebselen/HMPA complex is rapid in the NMR time scale, and this intermolecular interaction is dependent on the medium once there is competition between HMPA and solvent molecules to form the complex itself. For instance, the ^{77}Se $\Delta\delta$ (chemical shift variation – ppm) is significant in deuterated tetrahydrofuran (THF-d_8) solvent, and negligible in deuterated dimethylsulfoxide (DMSO-d_6) solvent.

A magnetic field can induce two kinds of electronic currents in a molecule, from which the diamagnetic and paramagnetic shielding constants originate. The diamagnetic effect arises from the circulation of electrons within atomic or molecular orbitals around the direction of the external field (B^0), generating a small local field opposed to B^0. On the other hand, the paramagnetic effect arises from the movement of electrons within molecules, but along a more circuitous route. This last current is clearly linked to the energy gaps involved in the electron circulation.

The contribution to the chemical shift from the local diamagnetic currents is strongly dependent on the electron density around the ^{77}Se nuclide. When a drop in the electron density around selenium element occurs, usually due to the presence of electronegative elements, there is a deshielding, leading to a higher chemical shift δ being observed. For instance, Andrade and Silva have observed the shielding effect of organoselenuranes, *i.e.*, compounds with hypervalent selenium(IV), when these compounds were reacted with L-cysteine in DMSO-d_6 deuterated solvent (Scheme 15.2).[25] The authors studied the stability and reactivity of distinct organoselenuranes, due to their pharmacological properties. Sulfur has a smaller electronegativity compared to chlorine and bromine, inducing a significant decrease in the chemical shift, and the selenium oxidation state remains unaffected due to the onset of bonding with the methoxy oxygen atom (Scheme 15.2). This variation can be easily detected by ^{77}Se NMR spectroscopy due to the higher local diamagnetic shift sensitivity. Interestingly, the reaction was effective only with L-cysteine amino acid, confirming the favourable interaction between chalcogen elements. This neighbourhood group effect was also evidenced at higher temperatures in a stability study.

A good method to explore the chemical shift dispersion of organoselenium compounds is to use their oxidized derivatives,[26] which present diamagnetic anisotropic effects because of the electronic situation created in

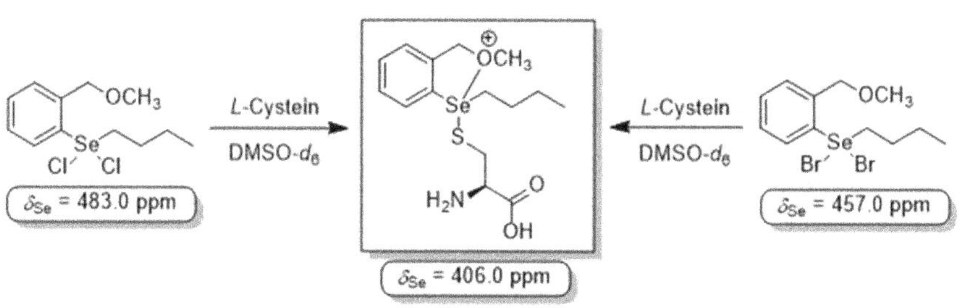

Scheme 15.2 Reaction between selenolluranes with L-cystein amino acid.

the selenium bonds. A green alternative to produce oxidized forms of selenium was developed by Perin and co-workers.[27] In 2019, the authors reported the use of Oxone® reagent as an oxidant in the synthesis of 4-(organoselanyl)-1*H*-pyrazoles. The reaction is based on the selenocyclization of α,β-alkynyl hydrazones using electrophilic selenium species generated *in situ* (Scheme 15.3). To gain some understanding of the role of the Oxone® in the reaction mechanism, a mixture of diphenyl diselenide (1.0 equiv.) and Oxone® (1.0 equiv.) was stirred at 70 °C in MeOD-d_4 and in DMSO-d_6, and after a few minutes, ^{77}Se-{^{1}H} NMR spectra were recorded at 25 °C. The spectrum in MeOD-d_4 presented a new resonance at $\delta_{Se} = 1218.0$ ppm, which is related to benzeneseleninate methyl ester, while the spectrum in DMSO-d_6 contained a new signal at $\delta_{Se} = 1160.0$ ppm, due to benzeneseleninic acid. ^{77}Se NMR spectroscopy revealed that the solvent is an important parameter in the chemical equilibrium of the oxidized forms of the electrophilic selenium species (Scheme 15.3).

The magnetic field of the spectrometer (B^0) induces electronic currents in the π electrons, which generate a local magnetic field. This induced field has the same general form as the magnetic field that would be produced by a magnetic dipole (bar magnet) sitting at the centre of the π electrons. The dipolar field lines for the aromatic ring are a contribution of π electron clouds, occasioned by the movement of these electrons, and producing large electronic currents (*e.g.*, aromaticity). This ring current shift is useful for distinct molecular recognitions, and it was used by Silva and co-workers for the chiral discrimination of selenium-containing secondary alcohols employing the (*S*)-Naproxen drug as a chiral derivatizing agent (CDA).[28] Due to the proximity of selenium to the aromatic ring, a paramagnetic anisotropic effect can be detected by ^{77}Se chemical shift variation (Scheme 15.4). Based on the Mosher acid model,[29] which involves the derivatization of secondary alcohols to their corresponding esters with an enantiopure aryl containing carboxylic acid, the key factor is that the ester has a conformational preference in which the two carbon stereocentres and two oxygen atoms are in

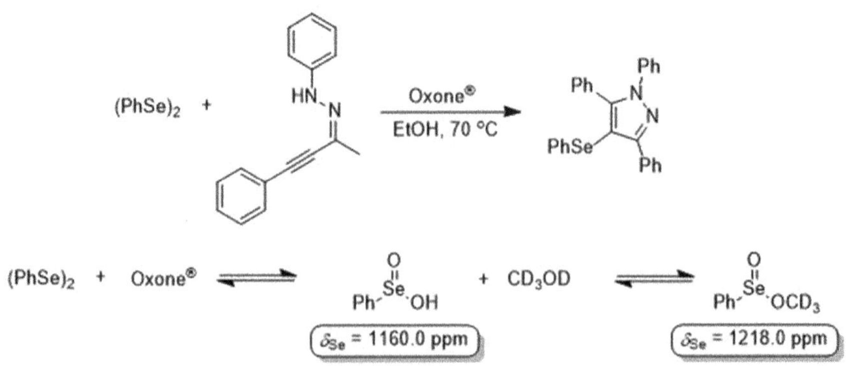

Scheme 15.3 Synthesis of 1,3,5-triphenyl-4-(phenylselanyl)-1*H*-pyrazole promoted by Oxone®.

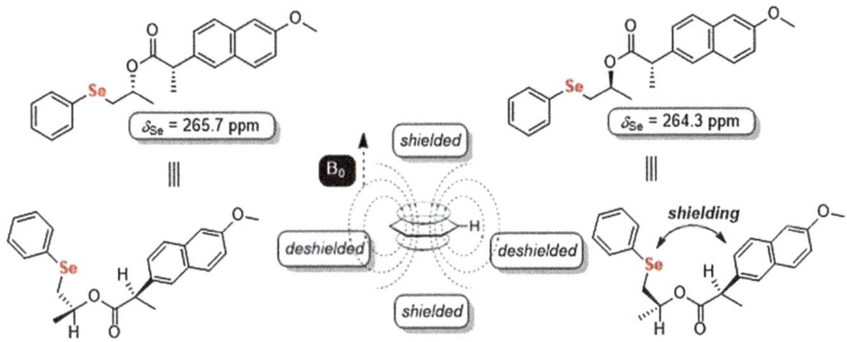

Scheme 15.4 Chiral recognition of selenium containing secondary alcohols by (*S*)-Naproxen.

the plane. This preference causes a shielding of the nuclei closer to the aromatic ring due to the paramagnetic effect of the ring current.

The use of aromatic rings into chiral auxiliaries to change the nuclide magnetic environment of the chiral target is another strategy to access the diastereoisomer signals, and provides a good example of the dependence of local paramagnetic effects derived from the double π bonds. Thus, Silva and co-workers developed a bidirectional approach exploring the aromatic ring effect (paramagnetic anisotropy) with the chemical shift sensitivity of the ^{77}Se nuclide.[30]

There is another valuable source of information in most NMR spectra, related to the interaction between nuclei, known as spin–spin coupling or scalar coupling, causing the NMR resonances to split into a small number of components (lines) with characteristic relative intensities and spacing. This interaction between the active nuclides is a result of the magnetic moment of each involved nuclide, since it generates a magnetic field that can be at the same or the opposite direction, *i.e.*, two situations with distinct energy levels. The scalar coupling depends on the spatial orientation and the proximity of the nuclides, so the splitting of signals usually occurs between the nuclei at the same neighbourhood.

The splitting diagram is derived from the spin angular moment, for which the spin $\frac{1}{2}$ provides the simplest components, while higher values produce a more complex multiplicity pattern. According to Table 15.1, ^{77}Se and ^{125}Te nuclides have spins $\frac{1}{2}$, similarly to ^{1}H and ^{13}C nuclides, facilitating the interpretation of the NMR spectra. To access the neighbourhood information from the ^{77}Se scalar coupling, a change in the NMR pulse sequence must be carried out. As observed in the mentioned examples, a proton-decoupled ^{77}Se NMR experiment (^{77}Se{^{1}H} NMR) is routinely performed to improve the NMR experiment sensitivity. However, it is possible to carry out a proton-coupled ^{77}Se NMR experiment (^{77}Se NMR), in which the scalar coupling with hydrogen atoms can be visualized.

Regularly, the ^{77}Se chemical shift and scalar coupling are evaluated together in various studies, and both these parameters should be considered to understand all changes in the ^{77}Se NMR spectrum. For example, it is worth mentioning the reaction mechanism studied by Perin and co-workers, which consists of the synthesis of chalcogenochromones employing a nucleophilic selenide generated *in situ*, reacting elemental selenium and $NaBH_4$ as the reducing agent (Scheme 15.5).[31] The structural unity chromone is present in a great number of natural products, amenable to pharmacological applications. At the beginning, to confirm the absolute structure of the synthesized chalcogenochromones, the ^{1}H–^{77}Se HMBC NMR experiment was an efficient analysis tool to collect information about the ^{77}Se chemical shift ($\delta_{Se} = 400.0$ ppm) and the exact position of the Se atom, due to the hydrogen correlations into the molecule derived from the scalar coupling (Figure 15.1). It is important to highlight that two- and three-bond (vicinal) coupling constants show some stereochemical control. Additionally, the yield of the reaction was satisfactory only when PEG-400 was used as the solvent. To rationalize this result, the authors performed a ^{1}H-coupled ^{77}Se NMR experiment to access the intermediate formed *in situ*. According to the ^{77}Se NMR spectrum, it is possible to confirm the formation of NaHSe (Scheme 15.5; $\delta_{Se} = -458.0$ ppm), through the hydrogen-selenium coupling constant value ($^{1}J_{H\text{–}Se} = 22.1$ Hz). It is remarkable that the visualization of the scalar coupling depends on the natural abundance of the active nuclide. If there is a low natural abundance, the splitting of signals in satellite peaks will occur, like a tetramethylsilane (TMS) reference, which has satellites peaks (4.68%) related to the scalar coupling with ^{29}Si.

Another advantage of NMR spectroscopy is the capability to detect dynamic equilibria. For simplicity, we consider a simple equilibrium between two species with equal forward and backward rate constants. Basically, when very slow exchange occurs, comparatively to the Larmor frequency of the target nuclide, one can detect two equally intense peaks. As the kinetic constant is increased, the two lines first broaden, then get progressively closer, broadening further, until they merge into a single peak.

To highlight this dynamic effect in ^{77}Se NMR spectroscopy, Menezes and co-workers developed a chiral discrimination protocol for selenium-containing carboxylic acids, employing (+)-methylbenzylamine (MBA) as the chiral solvating agent, based on an acid–base dynamic equilibrium.[32] In this case, the dynamic equilibrium is faster than the ^{77}Se NMR frequency and a single peak is recorded. Due to the formation of the diastereomeric complexes, two anisochronous resonances can be detected by ^{77}Se{^{1}H} NMR

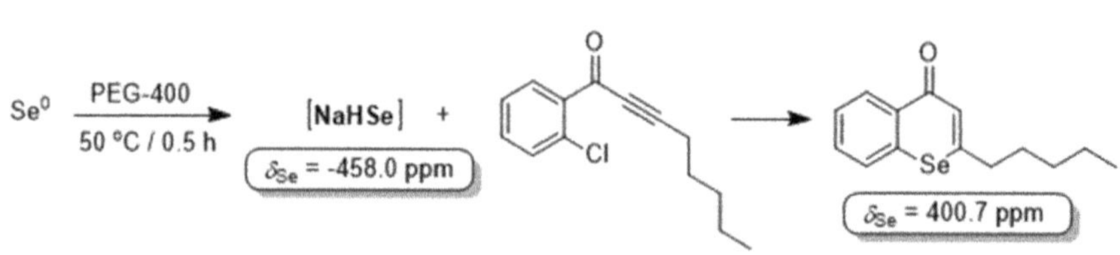

Scheme 15.5 Synthesis of chalcogenochromones employing a nucleophilic selenide generated *in situ*.

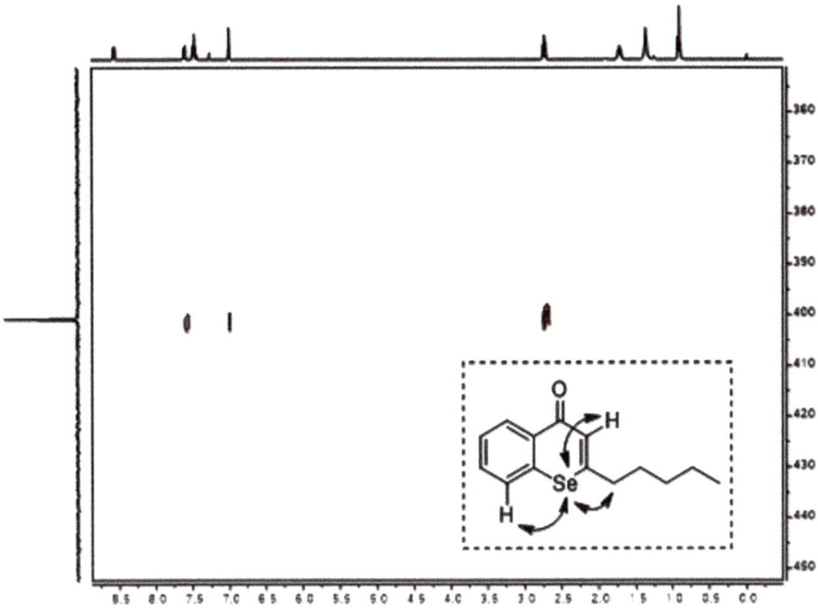

Figure 15.1 ^{1}H–^{77}Se HMBC NMR experiment of chalcogenochromones.

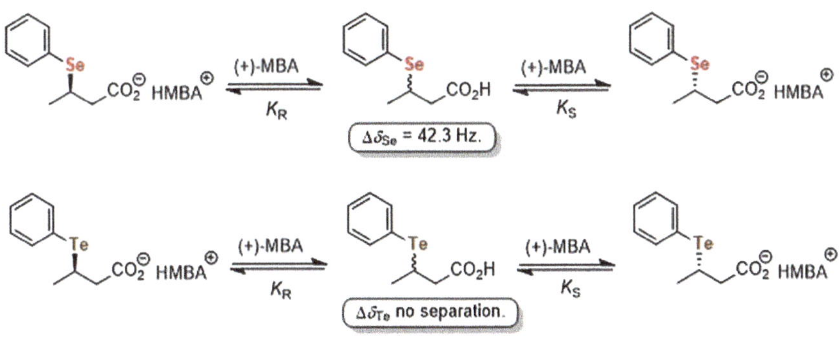

Scheme 15.6 Chiral discrimination of carboxylic acids by ^{77}Se and ^{125}Te NMR spectroscopies.

experiment (Scheme 15.6), for which the resonance separation is dependent on the size of the diastereomeric salt constants, K_R and K_S. When Silva and Pietrobom performed the same chiral discrimination protocol for carboxylic acids containing tellurium, the ^{125}Te{^{1}H} NMR experiment detected only one broad peak in this acid–base dynamic equilibrium (Scheme 15.6).[33] Although the Larmor frequency of ^{125}Te nuclide is higher than that of ^{77}Se, which makes the dynamic process slower, K_R and K_S diastereomeric constants had similar size, and the chiral discrimination was not possible because of this sensitivity toward the dynamic effect in NMR spectroscopy. To solve this issue, the authors used (+)-methylbenzylamine as the chiral derivatizing agent, using dicyclodicarbodiimide (DCC) to activate the carboxyl group and, consequently, eliminate the dynamic process.

Spin–lattice relaxation is a very weak interaction (coupling) with spins of other molecules, providing a pathway for the exchange of energy between a spin system and its surroundings. As the T_1 relaxation brings the spins to their energy equilibrium, it is important for the nuclear Overhauser effect (nOe) and for quantification protocols by NMR spectroscopy (qNMR). To illustrate this parameter in the ^{77}Se NMR context, an interesting study was performed by Silva and co-workers, who developed a synthetic method to prepare selenium-containing α-amino-phosphonates.[34] The synthetic strategy of this class of compounds, which is amply used in agrochemical and pharmaceutical fields, was based on the Kabachnik–Fields reaction. To access the diastereoisomers quantification by ^{77}Se{^{1}H} NMR experiments, the authors used the ^{31}P{^{1}H} NMR as a comparative tool. Additionally, the T_1, signal/noise ratio (*S*/*N*) and offset (O_1) of the ^{77}Se{^{1}H} NMR experiment were changed to evaluate the modification in the results, since only the *S*/*N* parameter has produced meaningful changes. Although the ^{77}Se T_1 of selenides can vary significantly,[35] the results demonstrated the robustness of this nuclide. Since no hydrogen is usually bonded to the selenium atom, no concerns are necessary for the nOe effect due to the increasing of the NMR resonance intensity.

In liquids, molecular collisions occur frequently, contacts are constant from all sides, modifying the rotational motion of molecules. The intramolecular magnetic interaction is modulated by this motion, *i.e.*, molecular tumbling, which affects the T_2 relaxation time. Spin–spin relaxation (T_2) increases as the tumbling gets faster and more effectively averages the dipolar coupling. In contrast, slow rotational motion reduces the spin–spin relaxation time, causing a line broadening in the signals of the order of $1/\pi T_2$. This effect can be expected for ^{77}Se nuclide. Ando and co-workers[36] prepared selenium-substituted carbohydrates and evaluated the interaction with the protein lectin by ^{77}Se NMR spectroscopy (Scheme 15.7). In the absence of lectin, the three Se atoms at 40.0, 256.0, and 413.0 ppm showed sharp signals (NMR linewidths) of 8.5, 9.1, and 9.7 Hz, respectively. Upon addition of lectin, changes in the ^{77}Se NMR linewidths were perceived, *i.e.*, a broadening of the resonances at 413.0 and 256.0 ppm to 28.5 and 18.2 Hz, respectively. According to Scheme 15.7, it is possible to observe the linewidth

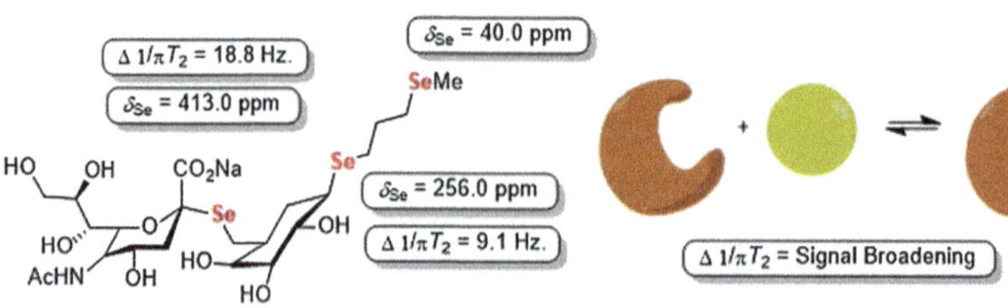

Scheme 15.7 ^{77}Se NMR analysis of the interaction between triseleno-galactose and lectin.

variation which establishes the epitope map for the ligand–protein interaction, identifying the main moiety that affects the binding affinity to sugar-binding lectins.

Chemical shift, scalar coupling, and relaxation times can distinguish nuclei in different environments of the molecule, and provide information on molecular identity, structure, and interactions. Although there is an extensive library of ^{77}Se NMR data, additional information could be useful to confirm the identities of organoselenium compounds. The Nuclear Overhauser effect is not discussed here, because of the ineffectiveness in the ^{77}Se nuclide. Anyway, the nOe is an important experimental procedure to access the conformation and the configuration of distinct organic structures. This effect has an impact on NMR sensitivity and is routinely minimized not to affect the integral values and, consequently, the concentration of the analyte. Actually, the nOe is derived from the dipolar interaction between the active nuclides in NMR spectroscopy, and it is possible to measure this dipolar coupling through the 1D NMR experiment, for instance the NOE-diff, and by 2D NMR experiment, such as NOESY.

15.3 Theoretical and Computational Aspects

Very recently, extensive and accurate reviews have been published on the fundamentals and state-of-the-art applications of the theoretical prediction of ^{77}Se chemical shift.[37] The described methods rely on accurate quantum chemistry methods, mainly Density Functional Theory (DFT) based approaches. In quantum mechanical calculations, the magnetic field is introduced into the total Hamiltonian of the system as a vector potential. This strategy, known as minimal coupling scheme, has the disadvantage that the potential depends on the choice of the origin of the coordinates. Since the solutions to the Hamiltonian problem are approximate, the gauge problem cannot be eliminated. Typically, two alternative solutions are used to overcome this issue, *i.e.*, the Individual Gauge for Localized Orbitals (IGLO) or the use of Gauge Including Atomic Orbitals (GIAO).

Shielding constants (isotropic averages of the shielding tensors) are computed and the chemical shifts are then obtained by subtracting the isotropic shielding constant of the compound of interest to the isotropic shielding constant of the reference compound:

$$\delta_i = \sigma_{\text{ref}} - \sigma_i \tag{15.1}$$

Within the GIAO formalism, the shielding constant can be decomposed in a diamagnetic and a paramagnetic contribution:

$$\sigma_i = \sigma_{\text{dia}} + \sigma_{\text{para}} \tag{15.2}$$

The diamagnetic contribution depends on the unperturbed electron density, while the paramagnetic one can be expressed as the sum of two terms:

$$\sigma_{i,para} = \sigma_{i,para}^{(\text{occ-occ})} + \sigma_{i,para}^{(\text{occ-virt})} \tag{15.3}$$

Based on eqn (15.3), the paramagnetic component of the isotropic shielding constant can be related to pairs of molecular orbitals (MOs) of the system of interest, thus allowing a chemically clear explanation of the computed chemical shift. Particularly, the contribution of MO pairs is proportional to the magnetic orbital coupling $\langle \varphi_i | \hat{B} | \varphi_j \rangle$. The effect of the magnetic moment operator is analogous to that of the angular momentum operator, *i.e.*, its effect is a rotation of the orbital on which it acts. In addition, each contribution to the paramagnetic shielding is divided by the energy separation between the two involved MOs. Thus, symmetry and energy distance of the couples of MOs determine the extent of the contribution to the paramagnetic part of the shielding constant, allowing a nice interpretation of the effect of chemical modifications on the chemical shift.

Like all quantum chemistry-based calculations of shielding constants, the theoretical prediction of ^{77}Se NMR chemical shift strongly depends on the molecular geometry. A nice computational study was performed by Bayse[38] combining DFT calculations with experimental selenium NMR spectroscopy to study chemical motifs taken from selenoproteins, highlighting the importance of combining the theoretical and experimental approach to gain structural information. The conformational effect emerged as the largest source of deviation. For example, the isotropic selenium shift of selenocystine when varying the dihedral angle C–Se–Se–C between $-180°$ and $+180°$ changes by more than 600 ppm. The effect of correlation on the ^{77}Se chemical shift was systematically analyzed by Bühl and co-workers[39] and by Bayse[40] who studied sets of small selenides using the GIAO-MP2 approach obtaining MAPE of approximately 10%. Good results for challenging compounds like tetra atomic chalcogen rings required higher level correlated calculations like the computationally expensive GIAO CASSCF and GIAO CASPT2.[41] A significant improvement was observed with the inclusion of relativistic effects. The breakthrough was undoubtedly the expressions of NMR parameters within the ZORA (Zeroth Order Regular Approximation) formalism by Ziegler and Autschbach[42] within the framework of DFT. ZORA is known to be a good approximation for the molecular properties that are "chemical" in the sense that they are determined by the valence orbitals. Autschbach applied a ZORA scalar DFT protocol to compute the isotropic average and anisotropy of the ^{13}C–^{77}Se and ^{77}Se–^{77}Se spin–spin coupling tensor of CSe_2.[43] A ZORA-BLYP approach was used by Nakanishi, Hada, and co-workers to evaluate the magnetic shielding tensor of forty Se species drawing interesting conclusions about the structural dependence of the spectroscopic values.[44] A scalar ZORA protocol was recently used to systematically investigate the mechanism of H_2O_2 reduction by model organic mono- and di-selenides.[45] The combination of measured and computed ^{77}Se chemical shift and the analysis of the paramagnetic contribution to the shielding constant allowed straight identification of the reaction intermediates and products and interpretation of the oxidation state of the chalcogen centre, with discrepancies in the range 15–40 ppm. Notably, the calculations were performed in gas-phase. The inclusion of solvent, *i.e.*, using a continuum dielectric model, which can bring variation up to 50 ppm to the result. An accurate computational work,[46] which can be characterized by the MAE of only 20 ppm and the MAPE of about 4%, was achieved within the non-hybrid four-component GIAO-DFT(KT2) scheme, with taking into account solvent, vibrational, and "relativistic geometry factor" corrections. The choice of the functional is important and it is recommended to run some tests with different potentials for the shielding constant calculation as well as for the geometry optimization. Keal and Tozer proposed KT3 which provides accurate structural and spectroscopic results without the inclusion of exact exchange.[47] The chemical shift is, as expected, strongly influenced by the conformational dynamics in solution.[48] When dealing with molecular systems characterized by large conformational freedom or complex environments like coordinating solvents, a good strategy is the combination of molecular dynamics (MD) and ZORA-DFT. Snapshots can be sampled along the simulation and the chemical shift computed for each of them is then averaged. This approach was successfully adopted to compute the ^{77}Se chemical shift of diphenyl diselenide in chloroform and in water, which is characterized by a bimodal distribution of the dihedral C – Se – Se – C values, associated to two opposite signed values.[49] This approach allowed us to take into account the rotation of the phenyl rings which is free in solution at the temperature of the experiment. The discrepancy in this example ranges from 30 to 70 ppm, but importantly no attempt of refining the non-relativistic quantum chemistry level of theory was done and no explicit solvent molecules were included in the calculations of the shielding constant, because the focus was to recover the essentials of the dynamics in solution which are encoded in the NMR signal.

From the concise overview of the state of the art of the ^{77}Se NMR chemical shift computation, it emerges that relativistic, vibrational, correlation, and solvent effects must be considered to reduce the error. Nevertheless, a compromise between accuracy and computational cost must be considered especially when tackling large systems or systems in a complex environment.

15.4 Conclusions and Prospective

Organoselenium compounds have revealed a unique reactivity, which has been considered and investigated by a large community of chemists. ^{77}Se NMR spectroscopy has brought details about structure elucidation and, but not limited to, insights for reaction mechanisms. For instance, the oxidized forms of selenium play a key role in biochemical processes mitigating the oxidative stress by reduction of H_2O_2 or hydroperoxides, typically using glutathione as the cofactor: these discernments were possible only using ^{77}Se NMR spectroscopy.[50] The ^{77}Se NMR spectroscopy also produced important insights in industrial processes, such as the synthesis of isatoic anhydrides catalyzed by the $(PhSe)_2/H_2O_2$ system, which is recyclable in a multi-gram scale.[51]

The development of ^{77}Se NMR spectroscopy has connected the synthesis of organoselenium compounds to the discovery of new applications, *i.e.*, pharmacologically active molecules, supramolecular chemistry, and chemical/biochemical sensors. The progress of NMR spectroscopy derived from electronic devices as well as from complex pulse sequences,[52] providing an improvement in the NMR sensitivity and resolution alongside a large diffusion of ^{77}Se NMR measurements. Indeed, the ^{77}Se NMR experiments have become a common routine in the characterization of organoselenium compounds and in pharmacological, biochemical, and catalysis investigations, as a direct consequence of the multifaceted NMR technology.

Although we have significantly improved the theoretical approach to NMR calculations, including ^{77}Se NMR, and enormously advanced their implementations in the last two decades, the *virtual* spectrometer is not of straightforward usage. Good correlation and accuracy are obtained by tuning the methodology and including solvation, vibrational, and relativistic effects, in some cases explicit electronic correlation effects too. As stressed by Rusakova,[37a] currently available basis sets are rather inefficient and the development of optimized and flexible basis sets for selenium is recommended for further improving the performance of the ^{77}Se NMR virtual spectrometer. It is also worth stressing that the combination of experimental and computational approaches, as has been emerging in many other fields in the last two decades, is optimal because the value of the theoretical contribution goes beyond the mere simulation or prediction, but rather allows interpretation of the measured chemical shift inferring precious information about the structure and the dynamics of the molecules in their environment. Due to the increasing contributions in the field, an ambitious goal is to define a general or even standard combined approach of broad applicability.

References

1. (a) H. Laasonen, J. Ikäheimonen, M. Suomela, J. M. Rautiainen and R. S. Laitinen, *Molecules*, 2019, **24**, 319; (b) A. Pop, L. Wang, V. Dorcet, T. Roisnel, J.-F. Carpentier, A. Silvestru and Y. Sarazin, *Dalton Trans.*, 2014, **43**, 16459.
2. Y. Wang, L. Yu, B. Zhu and L. Yu, *J. Mater. Chem. A*, 2016, **4**, 1082.
3. L. Shao, Y. Li, J. Lu and X. Jiang, *Org. Chem. Front.*, 2019, **6**, 2999.
4. (a) E. J. Lenardão, C. Santi and L. Sancineto, *New Frontiers in Organoselenium Compounds*, Springer, 1st edn, 2018; (b) D. H. Mailahn, L. E. B. Iarocz, P. C. Nobre, G. Perin, A. Sinott, A. P. Pesarico, P. T. Birmann, L. Savegnago and M. S. Silva, *Eur. J. Med. Chem.*, 2021, **231**, 113052.
5. (a) W. A. Hendrickson, J. R. Horton and D. M. LeMaster, *EMBO J.*, 1990, **9**, 1665; (b) G. Leszczynska, M. Cypryk, B. Gostynski, K. Sadowska, P. Herman, G. Bujacz, E. Lodyga-Chruscinska, E. Sochacka and B. Nawrot, *Int. J. Mol. Sci.*, 2020, **21**, 2882.
6. (a) P. Gettins and S. A. Wardlaw, *J. Biol. Chem.*, 1991, **266**, 3422; (b) Q. Chen, S. Xu, X. Lu, M. V. Boeri, Y. Pepelyayeva, E. L. Diaz,

S.-D. Soni, M. Allaire, M. B. Forstner, B. J. Bahnson and S. Rozovsky, *J. Phys. Chem. B*, 2020, **124**, 60; (c) S. A. Schaefer, M. Dong, R. P. Rubenstein, W. A. Wilkie, B. J. Bahnson, C. Thorpe and S. Rozovsky, *J. Mol. Biol.*, 2013, **425**, 222; (d) R. J. Hondal, *Protein Pept. Lett.*, 2005, **12**, 757.
7. S. S. Dharmatti and H. E. Weaver, *Phys. Rev.*, 1952, **86**, 259.
8. H. E. Walchli, *Phys. Rev.*, 1953, **90**, 331.
9. M. S. Silva, D. Alves, D. Hartwig, R. G. Jacob, G. Perin and E. J. Lenardão, *Asian J. Org. Chem.*, 2021, **10**, 91.
10. E. D. Becker, *Anal. Chem.*, 1993, **65**, 295.
11. H. Duddeck, *Prog. Nucl. Magn. Reson. Spectrosc.*, 1995, **27**, 1.
12. K. Fehér, I. Timári, K. Rákosi, J. Szolomájer, T. Z. Illyés, A. Bartok, Z. Varga, G. Panyi, G. K. Tóth and K. E. Kövér, *Chem. Sci.*, 2016, **7**, 2666.
13. A. Casula, A. Liopis-Lorente, A. Garau, F. Isaia, M. Kubicki, V. Lippolis, F. Sancenón, R. Martínez-Mánez, A. Owczarzak, C. Santi, A. Scorciapino and C. Caltagirone, *Chem. Commun.*, 2017, **53**, 3729.
14. D. Bhowwmick and G. Mugesh, *Org. Biomol. Chem.*, 2015, **13**, 10262.
15. A. R. M. Oliveira, L. Piovan, F. Simonelli, A. Barison, M. F. C. Santos and M. B. M. Mello, *J. Organomet. Chem.*, 2016, **806**, 54.
16. G. Magyarfalvi and P. Pulay, *Chem. Phys. Lett.*, 1994, **225**, 280.
17. M. Buhl, W. Thiel, U. Fleischer and W. Kutzelnigg, *J. Phys. Chem.*, 1995, **99**, 4000.
18. M. Buhl, J. Gauss and H. J. F. Stanton, *Chem. Phys. Lett.*, 1995, **241**, 248.
19. G. Schreckenbach, Y. Ruiz-Morales and T. Ziegler, *J. Chem. Phys.*, 1996, **104**, 8605.
20. G. A. Webb, Annual Reports on NMR Spectroscopy, ^{77}*Se NMR Spectroscopy and Its Applications in Chemistry*, 2004, pp. 105–166.
21. Y. Kim, M. Choi, S. T. Manjare, S. Jon and D. G. Churchill, *RSC Adv.*, 2016, **6**, 32013.
22. P. J. Hore, *Nuclear Magnetic Resonance*, Oxford University Press, 2nd edn, 2014.
23. Q. Chen, S. Xu, X. Lu, M. V. Boeri, Y. Pepelyayeva, E. L. Diaz, S.-D. Soni, M. Allaire, M. B. Forstner, B. J. Bahnson and S. Rozovsky, *J. Phys. Chem. B*, 2020, **124**, 601.
24. A. Daolio, P. Scilabra, M. E. D. Pietro, C. Resnati, K. Rissanen and G. Resnati, *New J. Chem.*, 2020, **44**, 20697.
25. (a) L. Piovan, P. Milani, M. S. Silva, P. G. Moraes, M. Demasi and L. H. Andrade, *Eur. J. Med. Chem.*, 2014, **73**, 280; (b) M. S. Silva and L. H. Andrade, *Org. Biomol. Chem.*, 2015, **13**, 5924.
26. L. Abenante, J. M. Anghinoni, N. B. Padilha, F. Penteado, O. Rosati, C. Santi, M. S. Silva and E. J. Lenardão, *Org. Biomol. Chem.*, 2020, **9**, 1631.
27. G. Perin, P. C. Nobre, D. H. Mallahn, M. S. Silva, T. Barcellos, R. G. Jacob, E. J. Lenardão, C. Santi and J. A. Roehrs, *Synthesis*, 2019, **51**, 2293.
28. N. B. G. Marques, R. G. Jacob, G. Perin, E. J. Lenardão, D. Alves and M. S. Silva, *Chirality*, 2019, **31**, 41.
29. T. J. Wenzel, *Tetrahedron: Asymmetry*, 2017, **28**, 1212.

30. (a) S. S. Oliveira, R. L. O. R. Cunha and M. S. Silva, *Tetrahedron Lett.*, 2016, **57**, 4556; (b) Y. R. Lima, T. J. Peglow, P. C. Nobre, P. T. Campos, G. Perin, E. J. Lenardão and M. S. Silva, *ChemistrySelect*, 2019, **4**, 4797.
31. P. C. Nobre, T. J. Peglow, R. H. Bartz, A. M. Barcellos, R. G. Jacob, M. S. Silva, T. Barcellos and G. Perin, *J. Braz. Chem. Soc.*, 2021, **32**, 1541.
32. P. H. Menezes, S. M. C. Gonçalves, F. Hallwass, R. O. Silva, L. W. Bieber and A. M. Simas, *Org. Lett.*, 2003, **5**, 1601.
33. M. S. Silva and D. Pietrobom, *New J. Chem.*, 2015, **39**, 8240.
34. A. G. Vogt, G. Perin, C. Luchese, P. C. Silva, E. A. Wilhelm and M. S. Silva, *Eur. J. Org. Chem.*, 2018, **7**, 627.
35. J. D. Odom, W. H. Dawson and P. D. Ellis, *J. Am. Chem. Soc.*, 1979, **101**, 5815.
36. T. Suzuki, C. Hayashi, N. Komura, R. Tamai, J. Uzawa, J. Ogawa, H.-N. Tanaka, A. Imamura, H. Ishida, M. Kiso, Y. Yamaguchi and H. Ando, *Org. Lett.*, 2019, **21**, 6393.
37. (a) I. L. Rusakova and Y. Y. Rusakov, *Magn. Reson. Chem.*, 2021, **59**, 359; (b) L. B. Krivdin, *Russ. Chem. Rev.*, 2021, **90**, 265.
38. F. Li, P. B. Lutz, Y. Pepelyayeva, E. S. J. Arner, C. A. Bayse and S. Rozovsky, *Proc. Natl. Acad. Sci.*, 2014, **111**, 6976.
39. (a) M. Bühl, W. Thiel, U. Fleischer and W. Kutzelnigg, *J. Phys. Chem.*, 1995, **99**, 4000; (b) M. Bühl, J. Gauss and J. F. Stanton, *Chem. Phys. Lett.*, 1995, **241**, 248.
40. (a) C. A. Bayse, *J. Chem. Theory Comput.*, 2005, **1**, 1119; (b) C. A. Bayse, *Inorg. Chem.*, 2004, **43**, 1208.
41. B. O. Roos, K. Andersson, M. P. Fülscher, P. A. Malmqvist, K. Serrano-Anders, K. Pierloot and M. Merchan, *Adv. Chem. Phys.*, 1996, **93**, 219.
42. J. Autschbach and T. Ziegler, *J. Chem. Phys.*, 2000, **113**, 936; J. Autschbach and T. Ziegler, *J. Chem. Phys.*, 2000, **113**, 9410.
43. J. Jokisaari and J. Autschbach, *Phys. Chem. Chem. Phys.*, 2003, **5**, 4551.
44. W. Nakanishi, S. Hayashi, Y. Katsura and M. Hada, *J. Phys. Chem. A*, 2011, **115**, 8721.
45. G. Ribaudo, M. Bellanda, I. Menegazzo, L. P. Wolters, M. Bortoli, G. Zagotto and L. Orian, *Chem. Eur. J.*, 2016, **23**, 2405.
46. Y. Y. Rusakov and L. B. Krivdin, *J. Comput. Chem.*, 2015, **36**, 1756.
47. T. W Keal and D. J. Tozer, *Mol. Phys.*, 2005, **103**, 1007.
48. M. Zielińska-Błajet, P. J. Boratyński, J. Palus and J. Skaeżewski, *Tetrahedron*, 2013, **69**, 10223.
49. M. Torsello, A. C. Pimenta, L. P. Wolters, I. Moreira, L. Orian and A. Polimeno, *J. Phys. Chem. A*, 2016, **120**, 4389.
50. H. J. Reich and R. J. Hondal, *ACS Chem. Biol.*, 2016, **11**, 821.
51. L. Yu, J. Ye, X. Zhang, Y. Ding and Q. Xu, *Catal. Sci. Technol.*, 2015, **5**, 4830.
52. Examples of NMR progress see: (a) L. Castañar and T. Parella, *Magn. Reson. Chem.*, 2015, **53**, 399; (b) J. L. Markley, R. Brüschweiler, A. S. Edison, H. R. Eghbalnia, R. Powers, D. Raftery and D. S. Wishart, *Curr. Opin. Biotechnol.*, 2017, **43**, 34; (c) D. A. Foley, A. L. Dunn and M. T. Zell, *Magn. Reson. Chem.*, 2016, **54**, 451; (d) B. Reif, S. E. Ashbrook, L. Emsley and M. Hong, *Nat. Rev.*, 2021, **1**, 2.

CHAPTER 16

Structural Features of Chalcogen Bonds and Weak Interactions Involving Chalcogens

M. CARLA ARAGONI*[a] AND YURY TORUBAEV[b]

[a] Dipartimento di Scienze Chimiche e Geologiche, Cittadella Universitaria Monserrato, SS. 554 – bivio Sestu, 09042 Monserrato, Cagliari, Italy;
[b] Department of Molecular Chemistry and Materials Science, Weizmann Institute of Science, Rehovot 7610001, Israel
*Email: aragoni@unica.it

16.1 Introduction

Chalcogen (Ch) elements of group 16 of the periodic table typically form neutral organic compounds such as chalcogenone $R_1R_2C{=}Ch$, chalcogenoethers R_1–Ch–R_2, and chalcogen oxides R_1–(Ch=O)–R_2, with anisotropic distribution of the electron density that determines the presence, on the Ch atom's surface, of areas of higher and lower electron density where the electrostatic potential can be negative and positive, respectively.[1] Regions of positive, or less negative, electrostatic potential, commonly named σ-holes, are frequently disposed opposite to these R_1–Ch bonds and can attractively interact with electron-rich sites such as Nu atoms and anions possessing a lone pair (Lp) of electrons, in the formation of the so-called chalcogen bonds (ChB, Scheme 16.1).

These attractive interactions involving chalcogen atoms as electrophiles can be identified by considering the close contacts between group 16 elements and Lp-possessing atoms in crystalline solids. The earliest

Chalcogen Chemistry: Fundamentals and Applications
Edited by Vito Lippolis, Claudio Santi, Eder J. Lenardão and Antonio L. Braga

Published by the Royal Society of Chemistry, www.rsc.org

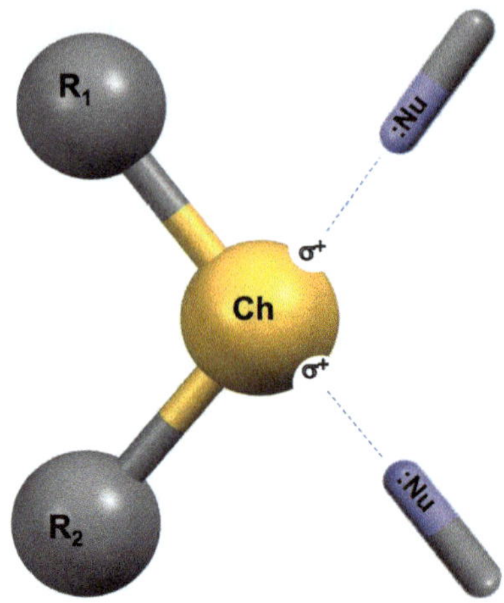

Scheme 16.1 Schematic representation of ChB formation.

crystallographic contribution to the understanding of these kinds of interactions involving chalcogen atoms was reported in 1977,[2] and describes the electrophilic attack to divalent sulfur in R_1–S–R_2 moieties roughly perpendicular to the R_1–S–R_2 plane and the nucleophilic attack along the extension of the S–R_1 or S–R_2 covalent bonds. However, attention to interactions driven by electrophilic chalcogen atoms started to increase only after 2007, when a model of the electronic basis of ChB was proposed.[3] Nowadays, ChBs are considered highly-directional noncovalent interactions between a covalently bonded Ch atom (O, S, Se, or Te) acting as an electrophilic species and a nucleophilic electron-rich donor centre Nu (Nu = electron donor atom, anion, π-system, radical, *etc.*) along the axes of the covalent bonds (Scheme 16.1) in agreement with the recent definition provided by the IUPAC.[4] Chalcogen-bonded systems typically display interatomic distances that are intermediate between those of typical single bonds and the sum of van der Waals radii.[5] In general, this interaction is strengthened by the presence of an electron-withdrawing group on the electron-acceptor Ch atom and it increases with the atomic number of the Ch species, from O to Te.[6] Following the surge of interest in halogen bonding (HaB), research on ChB is undergoing very fast expansion. Several recent reviews have covered fundamental aspects and applications of this type of interaction and highlighted examples from the literature.[7–9]

In this chapter the structural features of ChB will be mainly discussed highlighting the differences depending on the chalcogen element.

16.2 A Structural Approach to Chalcogen Bonds

ChB interactions involve electrostatic, covalent, and dispersion contributions (see Chapter 17 of this book). The electrostatic contribution comes from a Coulomb attraction between the negative electrostatic potential of the

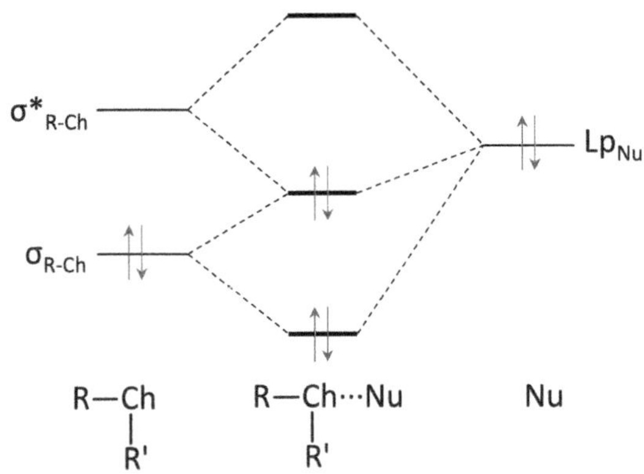

Figure 16.1 Perturbation molecular orbital scheme showing the 2e-delocalization of an electron lone pair of the nucleophile (Lp_{Nu}) into the σ^*_{R-Ch} orbital of the chalcogen.

Lewis base that behaves as the nucleophile (Nu), located at the lone pairs Lp_{Nu} or π-bond of Nu, and the positive electrostatic potential at the σ-hole region collinear to the covalent bond formed between the Ch atom and its most electronegative substituent.[10,11] The covalent contribution derives from the charge-transfer (CT) from the Lp_{Nu} of electrons localized on an occupied molecular orbital (MO) of the donor to the σ*(R–Ch) unoccupied MO (Figure 16.1) leading to a partial electron delocalization proportional to the Lp_{Nu}–σ* orbital overlap, which in turn is inversely proportional to the energy gap between the Lp_{Nu} and the σ*(R–Ch) orbitals.

The dominant contributions to ChB are strongly dependent on the nature of the R–Ch···Nu system and the ChB becomes stronger with increasing polarizability of the chalcogen atom (Te > Se > S) and polarizing effect of the R substituent, and higher donor ability of Lp_{Nu}.

This observation recalls the analogous finding for halogen bonding in crystalline halides RHa, where Ha = Cl, Br, or I. In these systems, close intermolecular contacts with nucleophiles tend to be along the extensions of the R–Ha bonds. These interactions known as halogen bonding HaB typically involves Ha and a Lewis base, and an increase in likelihood and in strength in the order Ha = Cl < Br < I.[12]

Information on the covalent or electrostatic nature of the ChBs can be obtained from the theoretical analysis of the energy density distribution at the electron density critical bond point according to the Quantum Theory of Atoms in Molecule (QTAIM).[11,13]

As previously recalled, compounds containing ChB show a marked directionality, the ChB being oriented approximately along the extensions of the R_1–Ch or R_2–Ch bonds (usually within a ±20° range). A "close" contact occurs when the internuclear distance is shorter than the sum of the van der Waals (vdW) radii of the relevant atomic species and implies a noncovalent interaction.

In the following, a systematic, although necessary concise, overview is presented about the structural features of the linear fragments R–Ch···Nu

(Ch = S, Se, Te; R–Ch···Nu angle in the range 160–180°) found in the Cambridge Structural Database (CSD) were a chalcogen bond is present. For each chalcogen, the analyses of the most frequently encountered fragments are presented sorted as a function of the nature of the atom directly bound. In order to better evaluate the strength of the ChBs involved and to compare the structural data of fragments R–Ch···Nu involving different atoms, the results were standardized for the relevant covalent and van der Waals radii (r_{cov} and r_{vdW}, respectively) by introducing the functions $\delta_{R\text{–}Ch}$ and $\delta_{Ch\cdots Nu}$ calculated by normalizing the differences between the involved R–Ch (Ch···Nu) distances $d_{R\text{–}Ch}$ ($d_{Ch\cdots Nu}$) and the sum of the relevant covalent (vdW) radii (eqn (16.1) and (16.2)):[14]

$$\delta_{R-Ch} = \frac{d_{R-Ch} - (r_{cov}R + r_{cov}Ch)}{r_{cov}R + r_{cov}Ch} \times 100 \quad (16.1)$$

$$\delta_{Ch\cdots Nu} = \frac{d_{Ch\cdots Nu} - (r_{vdW}Ch + r_{vdW}Nu)}{r_{vdW}Ch + r_{vdW}Nu} \times 100 \quad (16.2)$$

The values of $\delta_{R\text{–}Ch}$ and $\delta_{Ch\cdots Nu}$, defined above, represent the relative variations of R–Ch and Ch···Nu distances independently on the atomic species involved in the bonds and interactions, thus allowing for a direct comparison between strictly related systems.

Specific single/double bond covalent radii for the atoms engaged in R–Ch bonds were used depending on the nature of the bond in the considered fragments.[5]

A list of typical ChB donors and acceptors,[4] shows that R–Ch molecular entities commonly involved in ChB are chalcogen-containing heterocycles and azoles, and their fused polycyclic derivatives; chalcogen-anhydrides, and their analogues with sulfonyl groups in place of carbonyl ones; chalcogen-cyanates and chalcogen oxides, where the R atom directly bonded to the chalcogen is in most cases either a carbon, or a nitrogen, or another chalcogen atom. Entities behaving as ChB acceptors are nucleophiles with Lp-possessing atoms and anions where the atom Nu directly involved in ChB is essentially either a pnictogen (usually nitrogen), another chalcogen, or a halogen atom. Unsaturated systems where the acceptor is a multiple bond or an aromatic entity will not be examined here, and the reader is directed to specific publications to get a side view of this subject.[15–18] In the following sections a systematic survey of the structural features of the chalcogen bonds in the linear systems R–Ch···Nu (R = C, N, O, S, Se, Te and Nu = N, O, S, Se, Te, F, Cl, Br, I) is reported, categorized on the nature of the involved chalcogen atom (Ch = S, Se, Te). Among the about 34 000 fragments found, more than 80% are compounds where the Ch atom involved in the ChB is sulfur.

16.2.1 R–S···Nu fragments

The results of the search on the CSD,[19] for linear fragments R–S···Nu with R = B, C, N, P, As, O, S, Se, Te and Nu = N, P, As, Sb, O, S, Se, Te, F, Cl, Br,

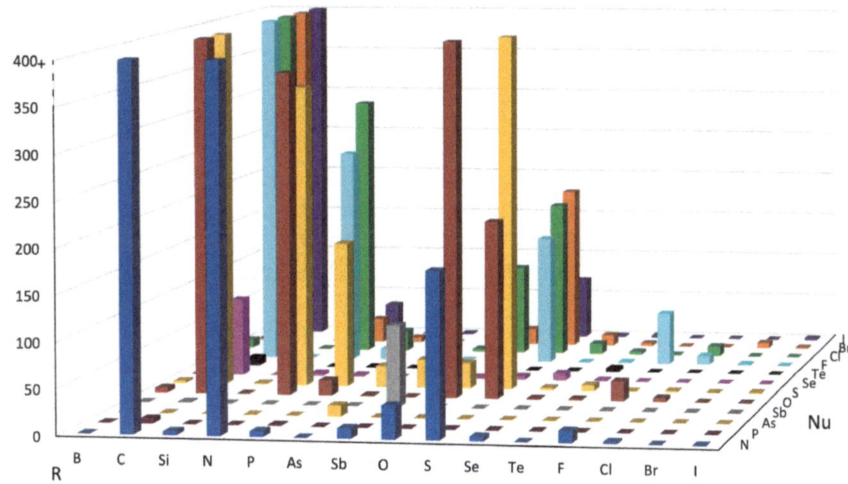

Figure 16.2 Linear R–S···Nu fragment occurrence (R = B, C, N, P, As, Sb, O, S, Se, Te, F, Cl, Br, I; Nu = N, P, As, Sb, O, S, Se, Te, F, Cl, Br, I) deposited at the CSD.

I are shown in Figure 16.2, where the numeric prevalence of R–S···Nu systems with R = C, N, O, and S is evidenced. Among the almost thirty thousand fragments (27 763 entries) found in the database, about 80% (21 993 items) belong to C–S···Nu systems with a carbon atom directly bound to the sulfur atom. The remaining 20% is mainly constituted of N–S···Nu (9%) and Ch–S···Nu (10%) fragments with a very minor contribute coming from all the other systems (1%).

Therefore, the survey here presented will be limited to the most frequently encountered systems belonging to the fragments R–S···Nu, with R = C, N, O, and S and Nu = N, Ch, and Ha.

16.2.1.1 C–S···Nu

A total of 21 993 C–S···Nu linear fragments showing structural features fitting the criteria described above and pointing to the presence of ChB interactions can be found to date in the CSD.[19] About 70% of these fragments involve ChB acceptors with Nu = chalcogen as the donor atoms. Following in abundance are C–S···Ha fragments (Ha = halogen, 21%), followed by C–S···Pn (Pn = pnictogen, 9%), with a net prevalence of nitrogen as the nucleophile donor atom. The compounds included in this survey are various and belong to the categories of sulfur-containing heterocycles such as thiophene and thiazole, and tetrathiafulvalene derivatives, thio- and dithio-carbamates, thioureas, dialkyl sulfoxides, thiocyanates, thioethers, and sulfur-containing macrocycles.

The CSD lists 15 417 linear fragments C–S···Ch, with a net prevalence for ChB acceptors featuring Ch = S (C–S···S 57%) and O (C–S···O 42%). Fragments involving halogen acceptors C–S···Ha are lesser, but still abundant with 4595 items (C–S···F 32.9%, C–S···Cl 40.7%, C–S···Br 14.4%, and

C–S···I 12.0%). Less numerous are the fragments involving pnictogen acceptors C–S···Pn, with 1981 items almost entirely concerning nitrogen (C–S···N 99.7%, C–S···P 0.25%, and C–S···Sb 0.05%). The mean distances calculated within the fragments C–S···Nu in the sub-categories are quite similar to each other, with values $d_{C\text{–}S} = 1.734$ (0.03), 1.746 (0.03), and 1.747 (0.05) Å, and $d_{S\cdots Nu} = 3.20$ (0.12), 3.33 (0.19), and 3.36 (0.22) Å for Nu = Pn, Ch, and Ha, respectively, (standard deviations s in parentheses). This notwithstanding, the scatterplots of $d_{S\cdots Nu}$ *vs.* $d_{C\text{–}S}$ for Nu = Pn (a), Ch (b), and Ha (c) show different ranges of existence depending on the nature of the interacting Nu atoms (evidenced by using different colours in Figure 16.3).

The mean values calculated for the C–S and S···Nu distances in the fragments C–S···Nu are summarized in Table 16.1, along with the mean of the delta $\delta_{C\text{–}S}$ and $\delta_{S\cdots Nu}$ values calculated, as explained above (eqn (16.1)

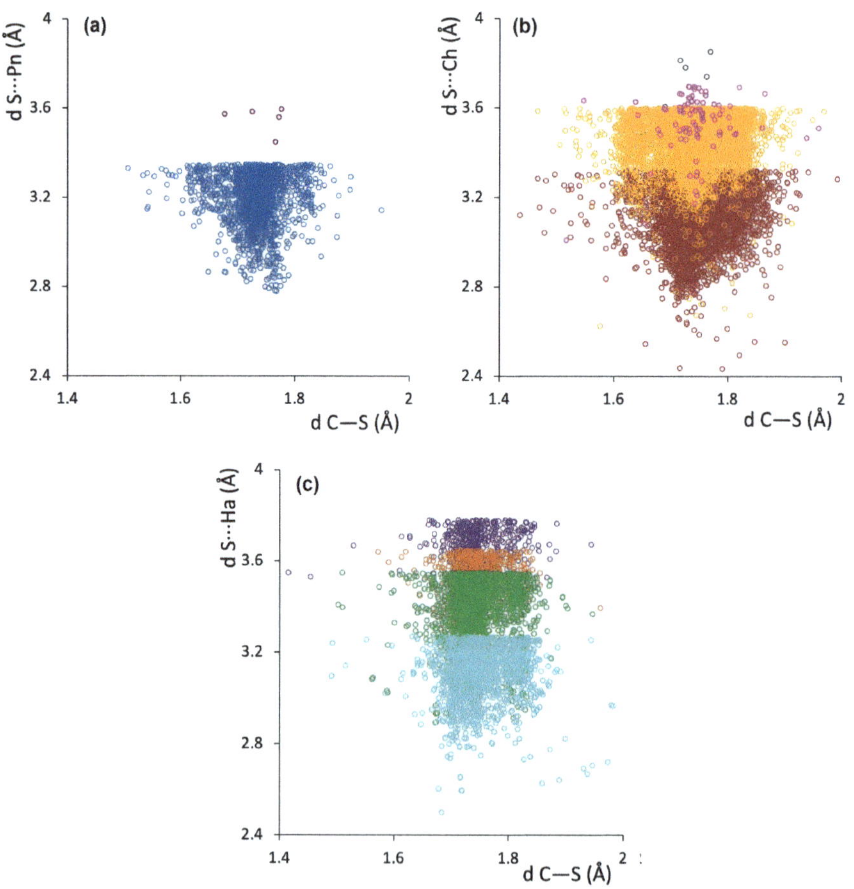

Figure 16.3 Scatterplots of $d_{S\cdots Nu}$ *vs.* $d_{C\text{–}S}$ distances in the C–S···Nu fragments for Nu = Pn (a); Ch (b); Ha (c). Nu colour codes: N = blue, P = purple in (a); O = red, S = yellow, Se = magenta, Te = black in (b); F = cyan, Cl = green, Br = orange, I = violet in (c).

Table 16.1 Occurrence, mean distances d [Å] (standard deviation, s), and δ values for C–S···Nu fragments.

Nu	Occurrence	$d_{C\text{–}S}$ (s)	$d_{S\cdots Nu}$ (s)	$\delta_{C\text{–}S}$ (%)	$\delta_{S\cdots Nu}$ (%)
O	6537	1.751 (0.05)	3.149 (0.10)	5.07	−5.15
S	8784	1.742 (0.05)	3.461 (0.11)	5.49	−3.87
Se	87	1.743 (0.06)	3.526 (0.10)	5.85	−4.69
Te	9	1.709 (0.05)	3.639 (0.16)	3.29	−5.71
N	1975	1.734 (0.05)	3.197 (0.11)	6.13	−4.57
P	5	1.744 (0.04)	3.552 (0.06)	5.17	−1.32
F	1511	1.753 (0.053)	3.112 (0.12)	5.22	−4.83
Cl	1871	1.746 (0.047)	3.402 (0.12)	5.50	−4.18
Br	661	1.743 (0.04)	3.519 (0.10)	5.58	−3.58
I	552	1.741 (0.054)	3.662 (0.08)	5.70	−3.13

and (16.2)), in order to compare systems involving atoms of different radii. Figure 16.4 shows the scatterplots of $\delta_{C\text{–}S}$ *vs.* $\delta_{S\cdots Nu}$.

The scatterplots clearly show that the relative C–S elongation, expressed as $\delta_{C\text{–}S}$, varies in a range between 0% and 9%. Depending on the nature of the molecule containing the sulfur atom involved in the ChB, three main groups are roughly defined in the scatterplots. Thiocyanates, tetrathiafulvalene derivatives, sulfonyl anions and compounds featuring the sulfur atom in planar and/or highly delocalized systems show marked elongations. On the contrary, systems with the sulfur atom bound to poorly electron-withdrawing or even electron-donating substituents show small elongations with some negative values related to systems such as dimethyl sulfoxide (DMSO) where the sulfur atoms are involved in very weak ChBs.

Without taking into account the Nu with limited occurrence, $\delta_{S\cdots Nu}$ values show the dependence of the interaction strength on σ-hole bond acceptors with a $\delta_{S\cdots Nu}$ shortening decreasing in the order O < F < Se < N < Cl < S < Br < I, according to the previously reported study by Wood on σ-hole interactions in small-molecule compounds containing divalent sulfur groups R_1–S–R_2.[20]

16.2.1.2 N–S···Nu

A number of 2845 linear fragments bearing a Pn–S···Nu ChB, with a net prevalence for the systems bearing a N–S covalent bond (2426 items, 85%) can be found in the CSD.[19] The P–S···Nu systems represent only a minority of these data (8%) and will not be dealt with in this chapter; the reader is directed to the recent work of W. T. Pennington.[21] Among the N–S···Nu fragments, 1116 (46%) N–S···N, 707 (29%) N–S···Ch, and 603 (25%) N–S···Ha items can be identified. The compounds comprised in this category are various and mainly belong to the class of sulfanilamide derivatives, or sulfur and nitrogen containing heterocycles such as thiazoles, and a fused polycyclic derivative thereof.

The number of linear fragments bearing a N–S···N ChB amounts to 1116, those involving chalcogen acceptors C–S···Ch amounts to 707 (N–S···O 51.3%, N–S···S 48.2%, and N–S···Se 0.5%), and those involving halogen

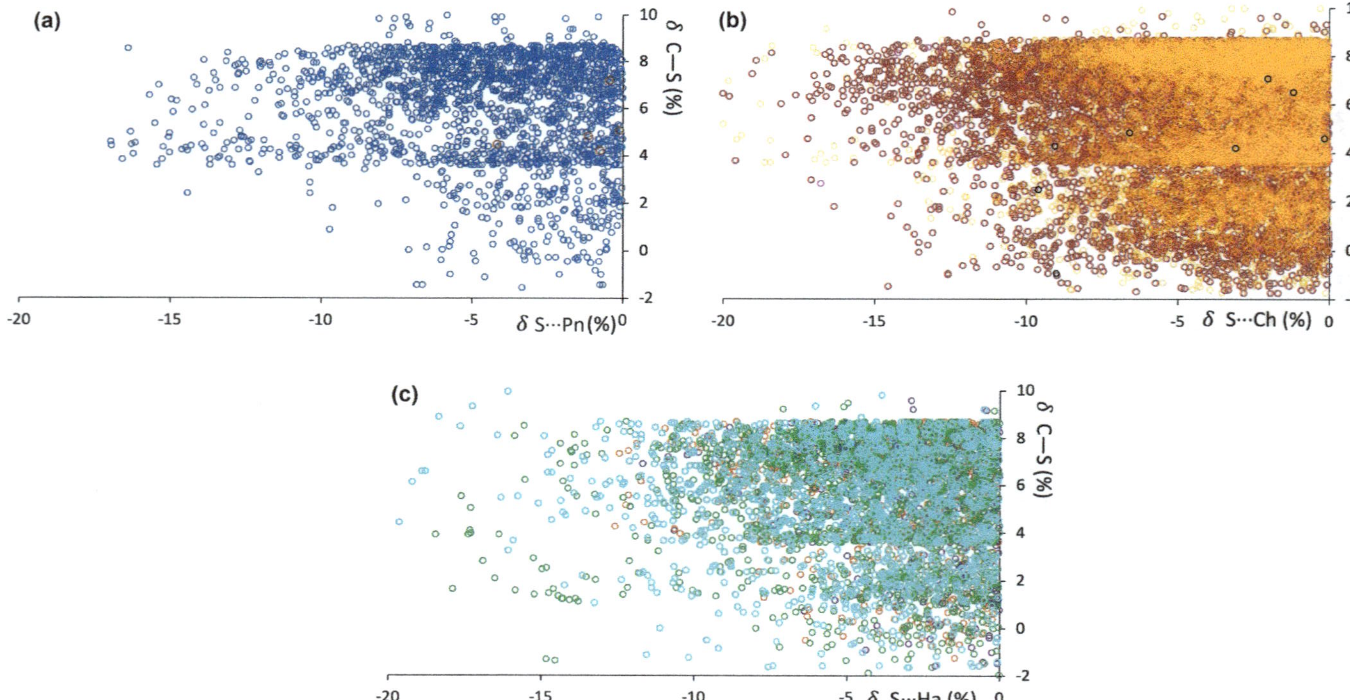

Figure 16.4 Scatterplots of $\delta_{C\text{–}S}$ *vs.* $\delta_{S\cdots Nu}$ calculated for the C–S···Nu fragments for Nu = Pn (a); Ch (b); Ha (c). Nu colour codes: N = blue, P = purple in (a); O = red, S = yellow, Se = magenta, Te = black in (b); F = cyan, Cl = green, Br = orange, I = violet in (c).

nucleophiles C–S···Ha amounts to 603 (C–S···F 40.7, C–S···F 32.9%, C–S···Br 14.4%, and C–S···I 12.0%). The mean distances d_{N-S} and $d_{S\cdots Nu}$ calculated within the fragments N–S···Nu in the three sub-categories with Nu = N, Ch, and Ha present similar values [d_{N-S} = 1.629 (0.03), 1.641 (0.04), and 1.615 (0.04) Å; $d_{S\cdots Nu}$ = 3.10 (0.13), 3.19 (0.28), and 3.16 (0.26) Å for Nu = N, Ch, and Ha, respectively]. Both covalent N–S and non-covalent S···Nu distances within N–S···Nu moieties are notably shorter than those found in the analogous C–S···Nu fragments, notwithstanding the similar values of the covalent radii of C and N atoms directly bonded to sulfur, the similarity of the involved substrates, and the same van der Waals radii of the interacting species. This experimental evidence confirms the expected ChB interaction increasing on passing from C–S···Nu to N–S···Nu fragments due to the larger electronegativity of N as compared to C. The distribution of the involved distances around the mean values is broader than that of the carbon analogues as showed by the scatterplots of S···Nu *vs.* N–S distances represented in Figure 16.5 for Nu = N (a), Ch (b), and Ha (c).

The mean values calculated for the N–S and S···Nu distances in the fragments N–S···Nu are summarized in Table 16.2, along with the mean of δ_{N-S} and $\delta_{S\cdots Nu}$ values. Figure 16.6 shows the scatterplots of δ_{N-S} *vs.* $\delta_{S\cdots Nu}$.

The scatterplots clearly show that the relative N–S elongation values, expressed as δ_{N-S}, mainly fall in a range between 2% and 10%, with a prevalence of the fragments C–S···O in the upper 8–10% range.

The $\delta_{S\cdots Nu}$ values decrease in the order Cl $\approx$ O < Br < F < N < S < Se < I. The remarkable shortenings with respect to the sum of relevant vdW radii observed in those systems where the ChB donor interacts with Cl, Br, and F acceptors is strongly dependent on the presence of numerous systems featuring anionic halides as Nu.

16.2.1.3 O–S···Nu

A number of 1101 linear O–S···Nu fragments can be found in the CSD with a prevalence of O–S···O systems (836, 76%), followed in occurrence by O–S···Cl (101, 9.2%) and O–S···F (71, 6.4%) moieties.[19] Less frequent are fragments of the type O–S···N (36, 3.3%) and O–S···S (29, 2.6%), with only very few examples for Nu = Se, Te, and I. The compounds comprised in this category mainly belong to the class of sulfoxides and alkyl derivatives thereof, thiophene, sulfate, sulfite, and dithionite anions.

The mean distances calculated within the fragments O–S···Nu in the three sub-categories with Nu = N, Ch, and Ha display similar values [d_{O-S} = 1.513 (0.04), 1.519 (0.05), and 1.525 (0.05) Å, $d_{S\cdots Nu}$ = 3.27 (0.07), 3.20 (0.17), and 3.32 (0.30) Å for Nu = N, Ch, and Ha, respectively]. The O–S···Nu fragments show the covalent O–S distances falling in ranges tighter than those found in the analogous C–S···Nu and N–S···Nu fragments because of the very large number of ChB donors sporting a double bond S=O. The non-covalent S···Nu distances fall in ranges similar to those previously described and, similarly to the C–S···Nu scatterplots (Figure 16.3) well defined areas are

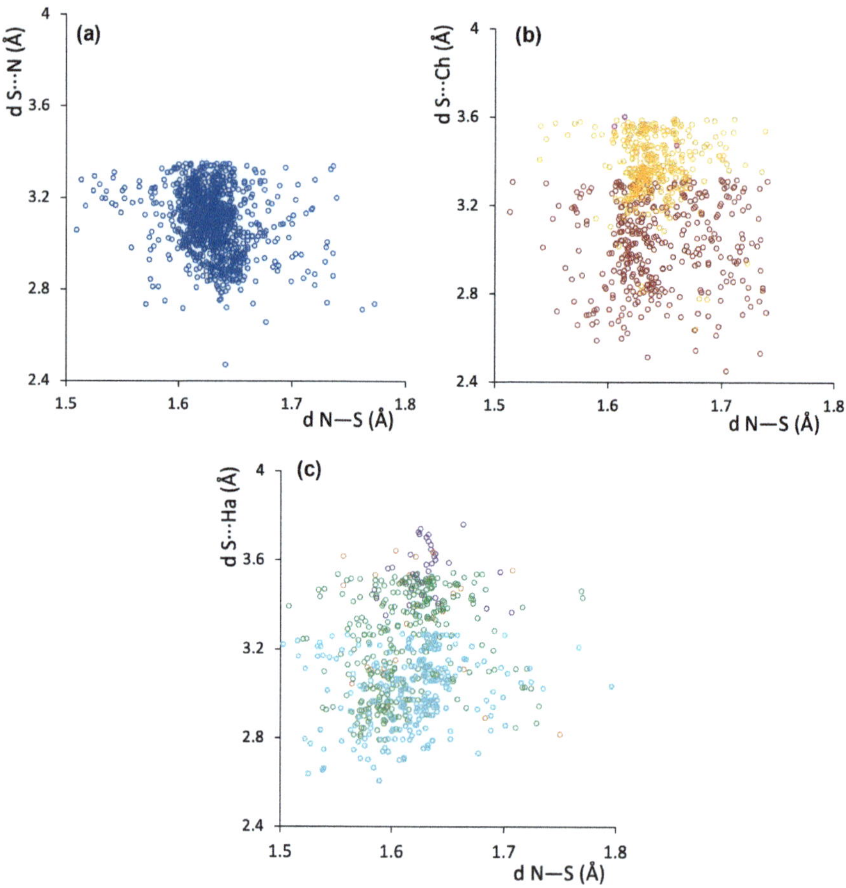

Figure 16.5 Scatterplots of $d_{S\cdots Nu}$ *vs.* $d_{N\text{-}S}$ distances in the N–S···Nu fragments for Nu = N (a); Ch (b); Ha (c). Nu colour codes: N = blue in (a); O = red, S = yellow, Se = magenta in (b); F = cyan, Cl = green, Br = orange, I = violet in (c).

Table 16.2 Occurrence, mean distances d [Å] (standard deviation, s), and δ values for N–S···Nu fragments.

Nu	Occurrence	$d_{N\text{-}S}$ (s)	$d_{S\cdots Nu}$ (s)	$\delta_{N\text{-}S}$ (%)	$\delta_{S\cdots Nu}$ (%)
O	363	1.644 (0.04)	3.017 (0.21)	3.16	−9.13
S	341	1.639 (0.03)	3.375 (0.21)	4.42	−6.25
Se	3	1.626 (0.03)	3.545 (0.07)	3.83	−4.18
N	1116	1.629 (0.03)	3.099 (0.13)	4.70	−7.48
Br	27	1.622 (0.05)	3.329 (0.25)	2.60	−8.81
Cl	297	1.613 (0.04)	3.222 (0.22)	3.34	−9.24
F	244	1.615 (0.05)	3.001 (0.21)	3.03	−8.22
I	35	1.630 (0.03)	3.559 (0.12)	4.33	−5.85

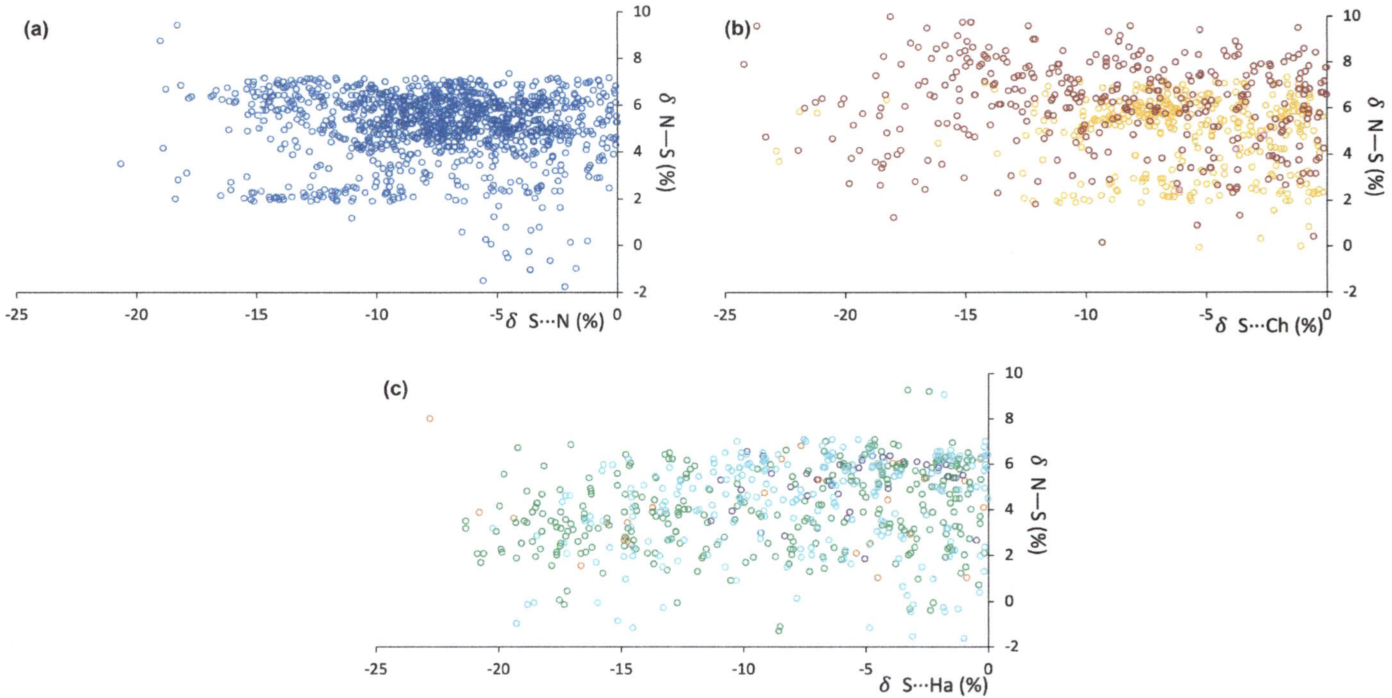

Figure 16.6 Scatterplots of $\delta_{N\text{–}S}$ *vs.* $\delta_{S\cdots Nu}$ calculated for the N–S···Nu fragments for Nu = N (a); Ch (b); Ha (c). Nu colour codes: N = blue in (a); O = red, S = yellow, Se = magenta in (b); F = cyan, Cl = green, Br = orange, I = violet in (c).

recognizable for the different acceptor atoms Nu involved in the interaction (Figure 16.7).

The mean values calculated for the O–S and S···Nu distances in the fragments O–S···Nu are summarized in Table 16.3, along with the mean of $\delta_{O\text{-}S}$ and $\delta_{S\cdots Nu}$ calculated values. Figure 16.8 shows the scatterplots of $\delta_{O\text{-}S}$ *vs.* $\delta_{S\cdots Nu}$.

The scatterplot in Figure 16.8 clearly shows a small range of values covered by the relative O–S elongations, expressed as $\delta_{O\text{-}S}$, mainly falling in a range between −2% and 2% for all the fragments, indicating a very weak nature for these ChB interactions. Fragments O–S···O and O–S···F (red and cyan circles in Figure 16.8) show values in the range from −7% and 7%, and stronger CHBs, as confirmed by the decreasing of $\delta_{S\cdots Nu}$ values in the order F < O < Cl < S < N < Br < I.

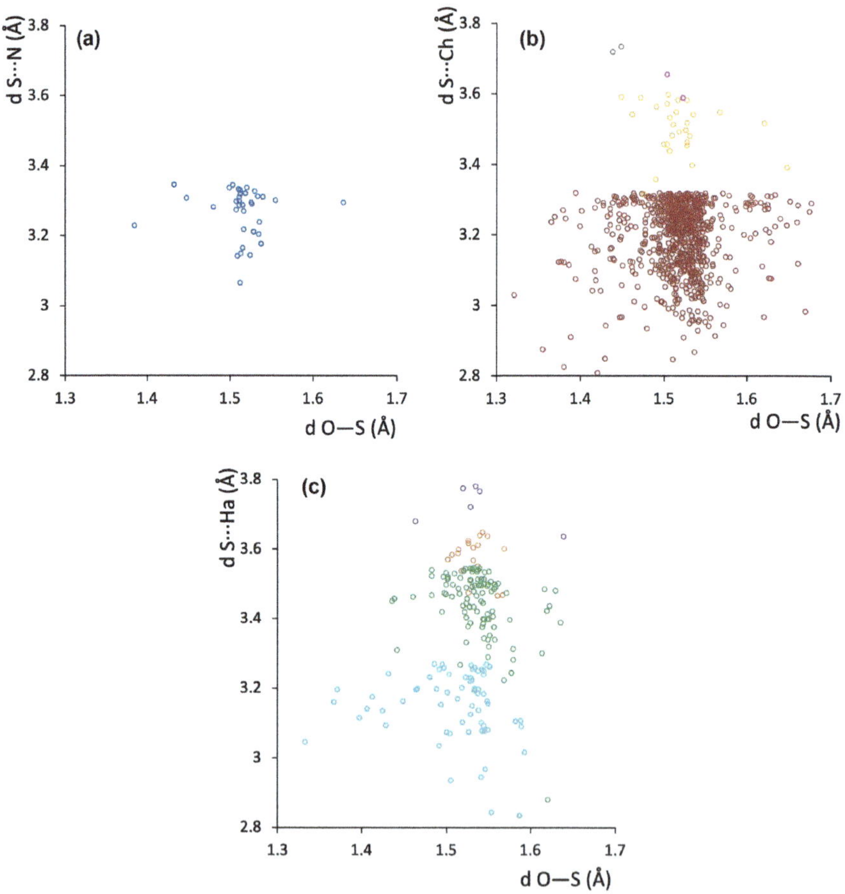

Figure 16.7 Scatterplots of $d_{S\cdots Nu}$ *vs.* $d_{O\text{-}S}$ distances in the O–S···Nu fragments for Nu = N (a); Ch (b); Ha (c). Nu colour codes: N = blue in (a); O = red, S = yellow, Se = magenta, Te = black in (b); F = cyan, Cl = green, Br = orange, I = violet in (c).

Table 16.3 Occurrence, mean distances d [Å] (standard deviation, s), and δ values for O–S···Nu fragments.

Nu	Occurrence	$d_{O\text{–}S}$ (s)	$d_{S\cdots Nu}$ (s)	$\delta_{O\text{–}S}$ (%)	$\delta_{S\cdots Nu}$ (%)
O	836	1.519 (0.05)	3.183 (0.16)	0.311	− 4.12
S	29	1.518 (0.04)	3.500 (0.07)	0.129	− 2.77
Se	2	1.512 (0.01)	3.622 (0.05)	0.166	− 2.12
Te	2	1.443 (0.007)	3.727 (0.01)	− 4.44	− 3.45
N	36	1.513 (0.04)	3.268 (0.07)	0.212	− 2.46
Br	18	1.532 (0.02)	3.578 (0.06)	1.49	− 1.97
Cl	101	1.537 (0.04)	3.429 (0.16)	1.38	− 3.41
F	71	1.505 (0.06)	3.075 (0.3)	− 0.041	− 5.95
I	6	1.537 (0.06)	3.727 (0.06)	0.87	− 1.41

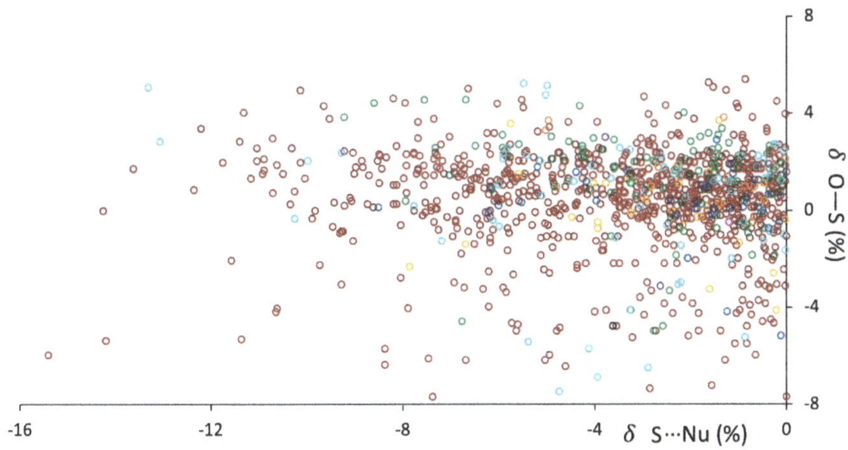

Figure 16.8 Scatterplot of $\delta_{O\text{–}S}$ *vs.* $\delta_{S\cdots Nu}$ calculated for the O–S···Nu fragments. Nu colour codes: N = blue, O = red, S = yellow, Se = magenta, Te = black, F = cyan, Cl = green, Br = orange, I = violet.

16.2.1.4 S–S···Nu

Deposited at the CSD are 1602 linear S–S···Nu fragments, including S–S···N (180, 11.3%), S–S···Ch (845, 52.7%), and S–S···Ha (577, 36.0%) moieties.[19] Belonging to this category are ChB donors such as disulfides and polysulfides, tetrathiocines, thiosulfate anions, and dithia-cyclic and polycyclic derivatives.

The mean distances calculated within the fragments S–S···Nu in the three sub-categories with Nu = N, Ch, and Ha display the covalent S–S distances varying in a very narrow range with values of the non-covalent S···Nu interactions falling in ranges close to those previously described for the analogous C–S···Nu ones [$d_{S\text{–}S}$ = 2.086 (0.04), 2.060 (0.05), and 2.051 (0.04) Å; $d_{S\cdots Nu}$ = 3.14 (0.12), 3.33 (0.20), and 3.20 (0.22) Å for Nu = N, Ch, and Ha, respectively]. Figure 16.9 reports the scatterplots of $d_{S\cdots Nu}$ *vs.* $d_{S\text{–}S}$ distances for Nu = N (a), Ch (b), and Ha (c) displaying similar values for the different ChB acceptor sub-categories.

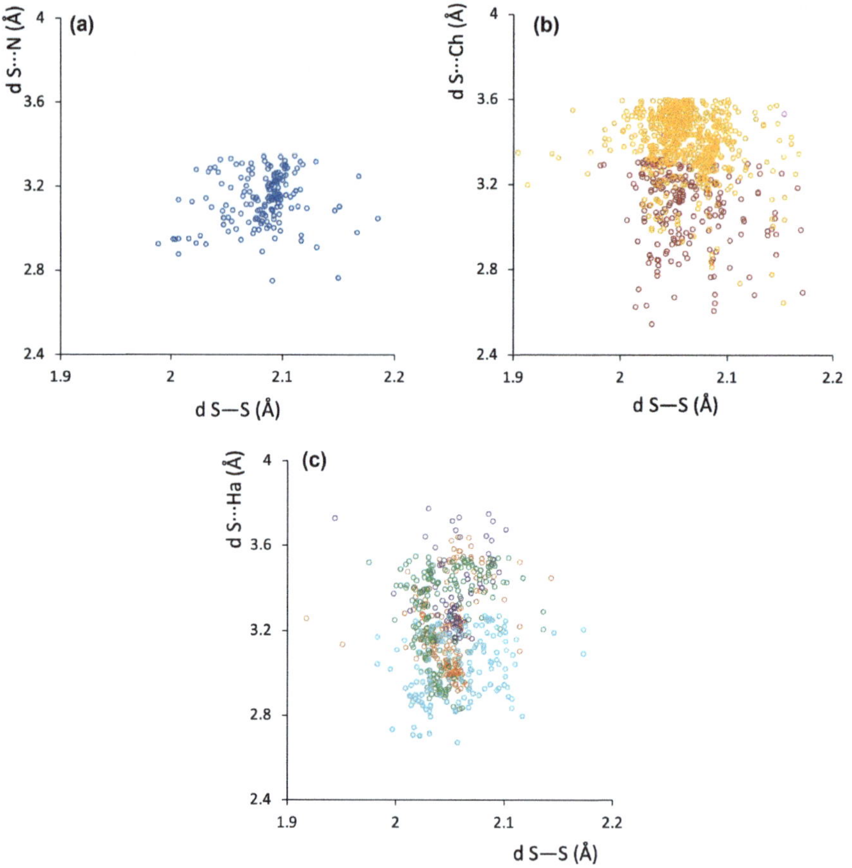

Figure 16.9 Scatterplots of $d_{S\cdots Nu}$ *vs.* $d_{S\text{-}S}$ distances in the S–S···Nu fragments for Nu = N (a), Ch (b); Ha (c). Nu colour codes: N = blue in (a); O = red, S = yellow, Se = magenta in (b); F = cyan, Cl = green, Br = orange, I = violet in (c).

The mean values calculated for the S–S and S···Nu distances in the fragments S–S···Nu are summarized in Table 16.4, along with the mean of $\delta_{S\text{-}S}$ and $\delta_{S\cdots Nu}$ values calculated as described above (eqn (16.1) and (16.2)); Figure 16.10 shows the scatterplots of $\delta_{S\text{-}S}$ *vs.* $\delta_{S\cdots Nu}$.

The scatterplots clearly confirm the small range of values covered by the relative S–S elongations, expressed as $\delta_{S\text{-}S}$, mainly falling in a range between −2% and 4%.

The $\delta_{S\cdots Nu}$ values show a decrease in the order Br < I < Cl < F < < O < N < S, showing that the stronger ChBs involve Nu = Ha. This trend can be probably explained taking into account that 72% of the compounds involved in S–S···Ha ChB interactions are salts where cations featuring the S–S moieties interact with the halide counterions.

Table 16.4 Occurrence, mean distances d [Å] (standard deviations s), and δ values for S–S···Nu fragments.

Nu	Occurrence	d_{S-S} (s)	$d_{S\cdots Nu}$ (s)	δ_{S-S} (%)	$\delta_{S\cdots Nu}$ (%)
O	199	2.062 (0.04)	3.086 (0.18)	1.09	−7.05
S	642	2.059 (0.05)	3.401 (0.15)	0.918	−5.54
Se	4	2.066 (0.06)	3.372 (0.11)	1.29	−8.87
N	180	2.086 (0.04)	3.144 (0.12)	2.27	−6.14
Br	186	2.049 (0.03)	3.203 (0.16)	0.45	−12.25
Cl	177	2.046 (0.04)	3.269 (0.21)	0.27	−7.92
F	145	2.060 (0.04)	3.027 (0.13)	0.966	−7.42
I	69	2.054 (0.02)	3.384 (0.15)	0.691	−10.47

16.2.2 R–Se···Nu fragments

A total number of 4109 linear fragments R–Se···Nu (R = B, C, N, P, As, O, S, Se, Te and Nu = N, P, As, Sb, O, S, Se, Te, F, Cl, Br, I) can be found to date in the CSD, notably lower than that recorded for the fragments containing the lighter congener S.[19] The following survey is limited to the same categories of fragments considered in the previous section (R = C, N, O, and S) with the addition of ChB donors featuring a Se–Se moiety, isologue to the S–S one discussed in the previous section. The histogram plot in Figure 16.11 shows the occurrence of fragments R–Se···Nu sorted by R and Nu, with a net prevalence of C–Se···Nu systems (2363, 57.5%), followed by N–Se···Nu (573, 13.9%), Se–Se···Nu (504, 12.3%), O–Se···Nu (220, 5.3%), and S–Se···Nu (97, 2.4%) fragments. The contribution coming from different systems that globally contributes to the 8.6% of the total fragments is very heterogeneous and remains untapped by this review.

16.2.2.1 C–Se···Nu

A number of 2363 C–Se···Nu linear fragments were found in the CSD; among them, 56.2% are C–Se···Ch systems involving ChB acceptors featuring nucleophile ChB-acceptors featuring a chalcogen Lewis donor atom;[19] the remaining fragments are of the type C–Se···Ha (32.2%) and C–Se···N (11.6%). No fragments with Nu = Pn different from N were found with the only exception of 1 fragment coming from a P–Se heterocycle based on peri-substituted naphthalene.[22]

The compounds comprised in this survey are isologues to those considered in Section 16.2.1.1 and belong to the classes of selenium-containing heterocycles such as selenophene and selenaazole, selenafulvalene derivatives, selenocarbamates, selenoureas, selenoxides, selenocyanates, selenoethers, and selenium-containing macrocycles.

The number of linear fragments C–Se···Ch amounts to 1329, with a net prevalence for ChB-acceptors featuring selenium (C–Se···Se 54%) and oxygen (C–Se···O 32%) as Nu interacting atoms, in analogy with what was found for the C–S···Ch items. The fragments C–Se···Ha involving ChB-acceptors interacting through halogen atoms are 761 (C–Se···F, 22.5%; C–Se···Cl,

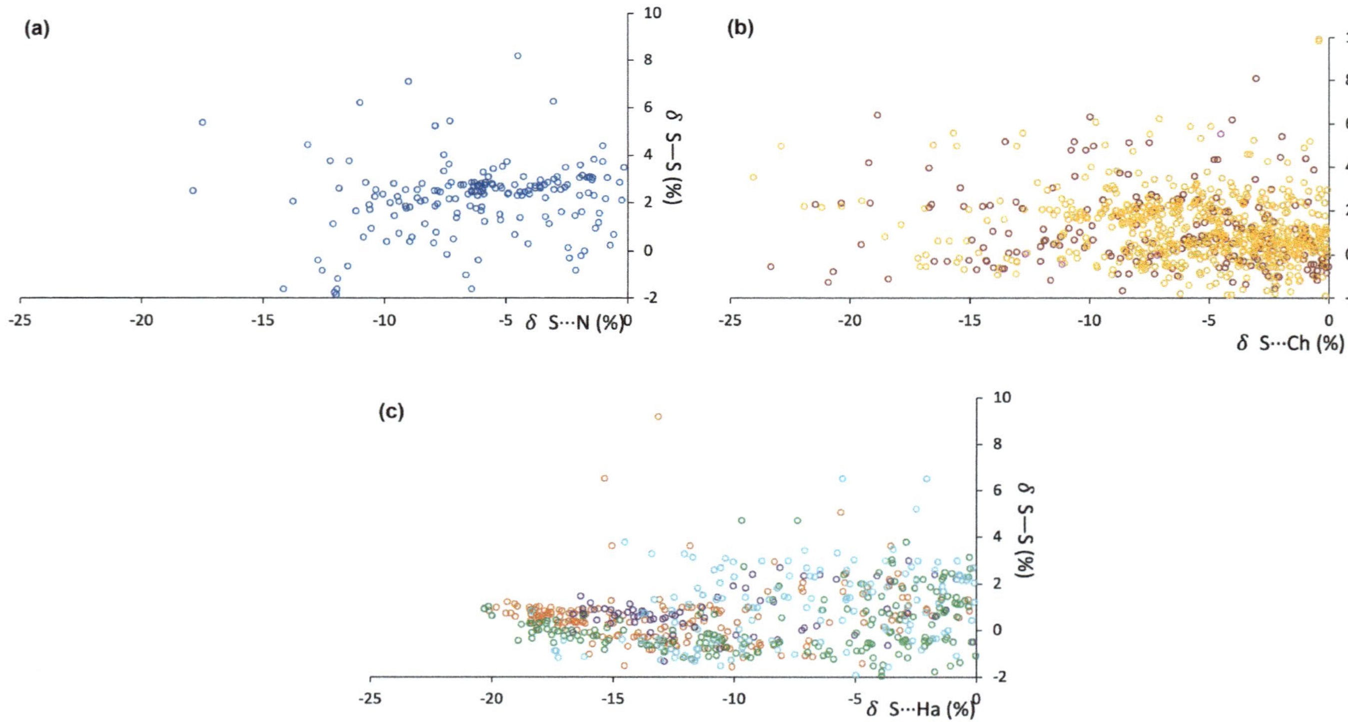

Figure 16.10 Scatterplots of $\delta_{S\text{–}S}$ *vs.* $\delta_{S\cdots Nu}$ calculated for the S–S···Nu fragments for Nu = N (a); Ch (b); Ha (c). Nu colour codes: N = blue in (a); O = red, S = yellow, Se = magenta in (b), F = cyan, Cl = green, Br = orange, I = violet in (c).

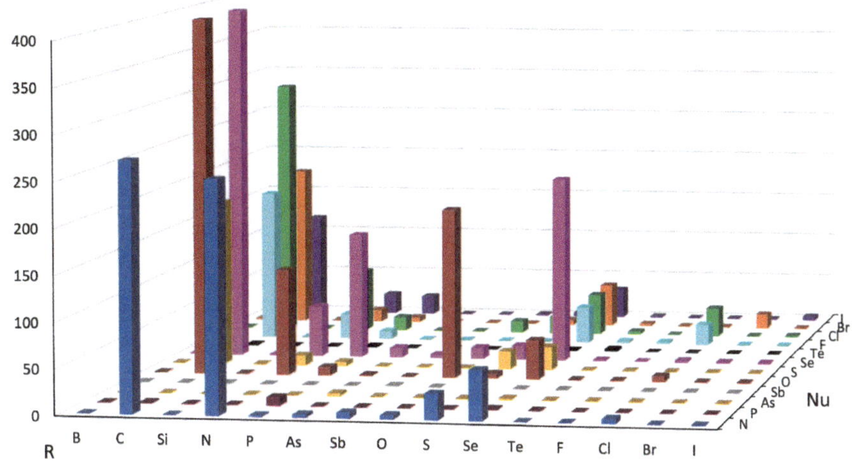

Figure 16.11 Linear R-Se···Nu fragment occurrence (R = B, C, Si, N, P, As, Sb, O, S, Se, Te, F, Cl, Br, I; Nu = N, P, As, Sb, O, S, Se, Te, F, Cl, Br, I) deposited at the CSD.

38.4%; C–Se···Br, 23.9%; C–Se···I, 15.2%), while those involving N-acceptors (C–Se···N) are 272. The mean covalent and non-covalent bond distances calculated within the fragments C–Se···Nu are roughly independent on the nature of Nu [$d_{C\text{–}Se}$ = 1.883 (0.05), 1.899 (0.05), and 1.910 (0.04) Å, and $d_{S\cdots Nu}$ = 3.19 (0.16), 3.46 (0.26), and 3.43 (0.21) Å, for Nu = N, Ch, and Ha, respectively]. Taking into account the different values of S and Se covalent and vdW radii, the mean distances found in C–Se···Nu fragments are comparable with those found in the analogous sulfur fragments as confirmed by the range of existence showed in scatterplots of Se···Nu *vs.* C–Se distances for Nu = N (a), Ch (b), and Ha (c) reported in Figure 16.12. This notwithstanding, probably due to the larger polarizability of Se with respect to S, wide distributions of the values can be noticed, especially for the C–Se···Ha systems (Figure 16.12c).

The mean values calculated for the C–Se and Se···Nu distances in the fragments C–Se···Nu are summarized in Table 16.5, along with the mean of $\delta_{C\text{–}Se}$ and $\delta_{Se\cdots Nu}$ values, calculated as explained above (eqn (16.1) and (16.2)). Figure 16.13 shows the scatterplots of $\delta_{C\text{–}Se}$ *vs.* $\delta_{Se\cdots Nu}$.

The scatterplots show relative C–Se elongations, expressed as $\delta_{C\text{–}Se}$, varying in a range between 0% and 13%. Akin to what was previously observed for the C–S···Nu fragments (Figure 16.4b), three main bands of data can be identified in the scatterplots depending on the nature of the molecule engaged in the ChB: selenocyanates, selenafulvalene derivatives, and selenium-heterocyclic compounds show evident elongations $\delta_{C\text{–}Se}$, whilst systems with the selenium atom directly bound to different electron-donating substituents show shorter elongations with some negative values corresponding to systems involved in very weak ChBs, such as in the case of dialkyl selenoxides.

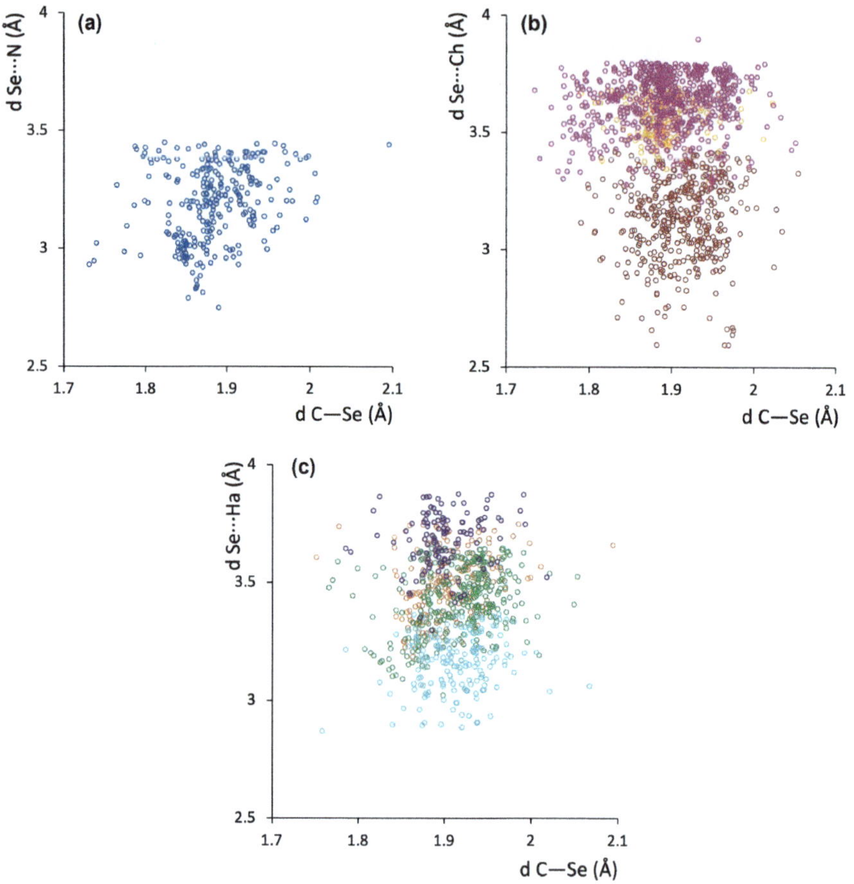

Figure 16.12 Scatterplots of $d_{\mathrm{Se\cdots Nu}}$ *vs.* $d_{\mathrm{C\text{–}Se}}$ distances in the C–Se···Nu fragments for Nu = N (a); Ch (b); Ha (c). Nu colour codes: N = blue in (a); O = red, S = yellow, Se = magenta in (b); F = cyan, Cl = green, Br = orange, I = violet in (c).

Table 16.5 Occurrence, mean distances d [Å] (standard deviation, s), and δ values for C–Se···Nu fragments.

Nu	Occurrence	$d_{\mathrm{C\text{–}Se}}$ (s)	$d_{\mathrm{Se\cdots Nu}}$ (s)	$\delta_{\mathrm{C\text{–}Se}}$ (%)	$\delta_{\mathrm{Se\cdots Nu}}$ (%)
O	424	1.912 (0.05)	3.137 (0.18)	6.57	− 8.19
S	186	1.891 (0.03)	3.560 (0.08)	7.07	− 4.20
Se	718	1.893 (0.06)	3.629 (0.12)	4.34	− 4.43
N	273	1.883 (0.05)	3.189 (0.16)	6.79	− 7.61
Br	182	1.905 (0.05)	3.498 (0.13)	6.12	− 6.71
Cl	292	2.053 (0.05)	3.427 (0.14)	4.26	− 6.10
F	171	1.913 (0.04)	3.167 (0.12)	4.07	− 6.03
I	116	1.900 (0.04)	3.682 (0.12)	6.36	− 5.11

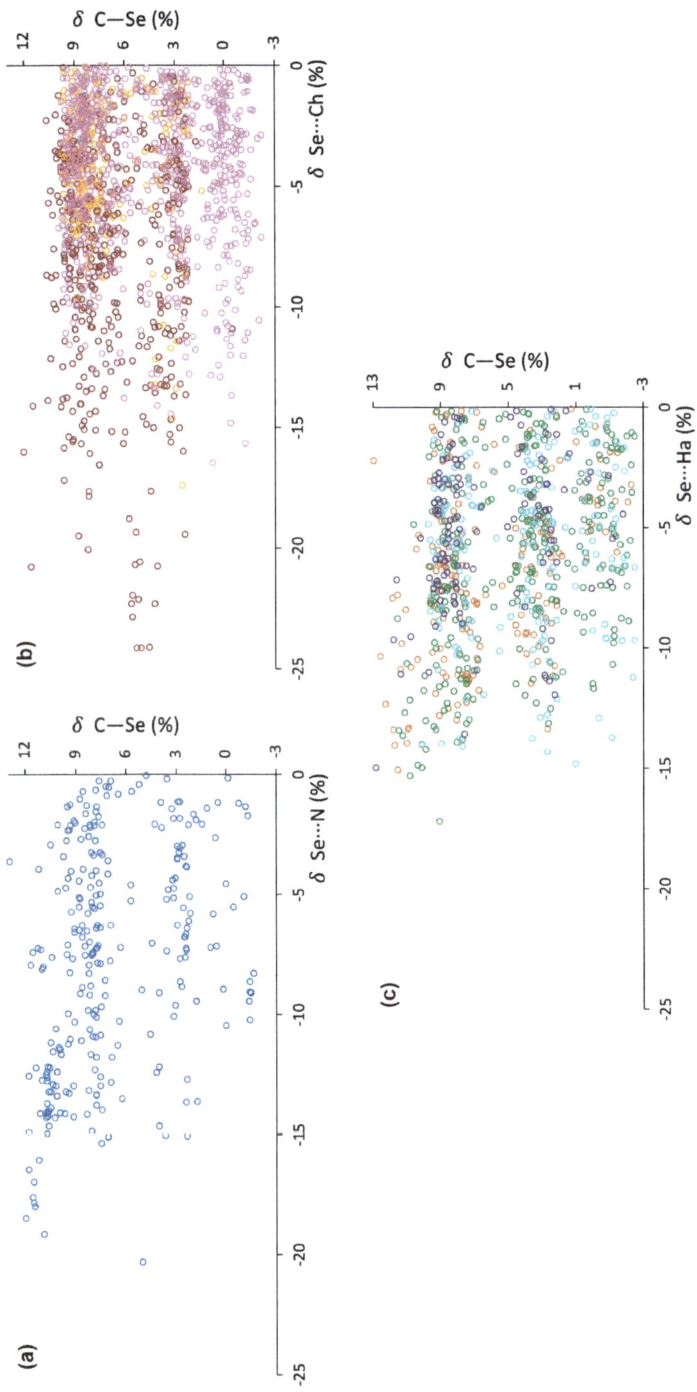

Figure 16.13 Scatterplots of $\delta_{\text{C-Se}}$ *vs.* $\delta_{\text{Se}\cdots\text{Nu}}$ calculated for the C–Se···Nu fragments for Nu = N (a); Ch (b); Ha (c). Nu colour codes: N = blue in (a); O = red, S = yellow, Se = magenta in (b); F = cyan, Cl = green, Br = orange, I = violet in (c).

The strength of the ChB, defined by the $\delta_{Se\cdots Nu}$ value follows the order O<N<Br<Cl<F<I<Se with increased shortenings with respect to the C–S···Nu isologues, confirming the expected strengthening of the ChB on passing from S to Se. The trend cannot be clearly related to any simple property of the nucleophile Lewis donor atom Nu, thus suggesting that the nature of the interaction, and hence the relative contributions of the electrostatic and CT terms, vary along the series.

16.2.2.2 N–Se···Nu

The number of linear Pn–Se···Nu fragments identified in the CSD amounts to 817,[19] but only those with a N–Se covalent bond (573 items, 70%) will be analyzed here in analogy to the choice adopted for the sulfur isologues reported in Section 16.2.1.2. In fact, the compounds encompassed are usually the corresponding Se-analogous of the previously described S-isologues, *i.e.*, selenenyl amide derivatives and Se-containing heterocycles such as selenophene and selenazoles and the fused polycyclic derivative thereof. Among the linear N–Se···Nu fragments can be identified 253 (44.1%) N–Se···N, 186 (32.5%) N–Se···Ch, and 134 (23.4%) N–S···Ha moieties. The mean N–Se and Se···Nu distances calculated for the three Nu sub-categories display similar values [d_{N-Se} = 1.800 (0.04), 1.827 (0.04), and 1.798 (0.05) Å; $d_{Se\cdots Nu}$ = 3.19 (0.16), 3.46 (0.26), and 3.42 (0.21) Å, for Nu = N, Ch, and Ha, respectively]. As observed for the isologues with sulfur, N–Se···Nu fragments show covalent N–Se and non-covalent Se···Nu distances notably shorter than those found in the analogous C–Se···Nu fragments, notwithstanding the similar values of the covalent and vdW radii of the involved atoms, confirming the expected strengthening of the ChB on passing from C–Se···Nu to N–Se···Nu fragments. The distribution of the involved distances around the mean values is broad as showed by the scatterplots of Se···Nu *vs.* N–Se distances presented in Figure 16.14 for Nu = N (a), Ch (b), and Ha (c).

In fact, the Se···Nu mean distances are shorter than the corresponding S···Nu ones, notwithstanding the larger Se vdW radium, and the means of the normalized shortening $\delta_{Se\cdots Nu}$ are almost doubled passing from −9.13% to −16.47% and from −7.48% to −14.65% for Nu = O and N, just to make the examples for the high-frequently occurring fragments (Tables 16.2 and 16.6).

The scatterplots of δ_{N-Se} *vs.* $\delta_{Se\cdots Nu}$ (Figure 16.15) confirm an amplified shortening with numerous fragments showing $\delta_{Se\cdots Nu}$ distances falling in the range from −15% to −30%.

16.2.2.3 O–Se···Nu

A search in the CSD reveals 220 linear O–Se···Nu fragments with a large prevalence of systems featuring a chalcogen as the ChB acceptor (O–Se···O, 86%; O–Se···Se, 6%, O–Se···Cl, 6%, and O–Se···N, 2%).[19] The compounds comprised in this category mainly belong to the class of selenite and selenate anions, selenoxides, and derivatives thereof.

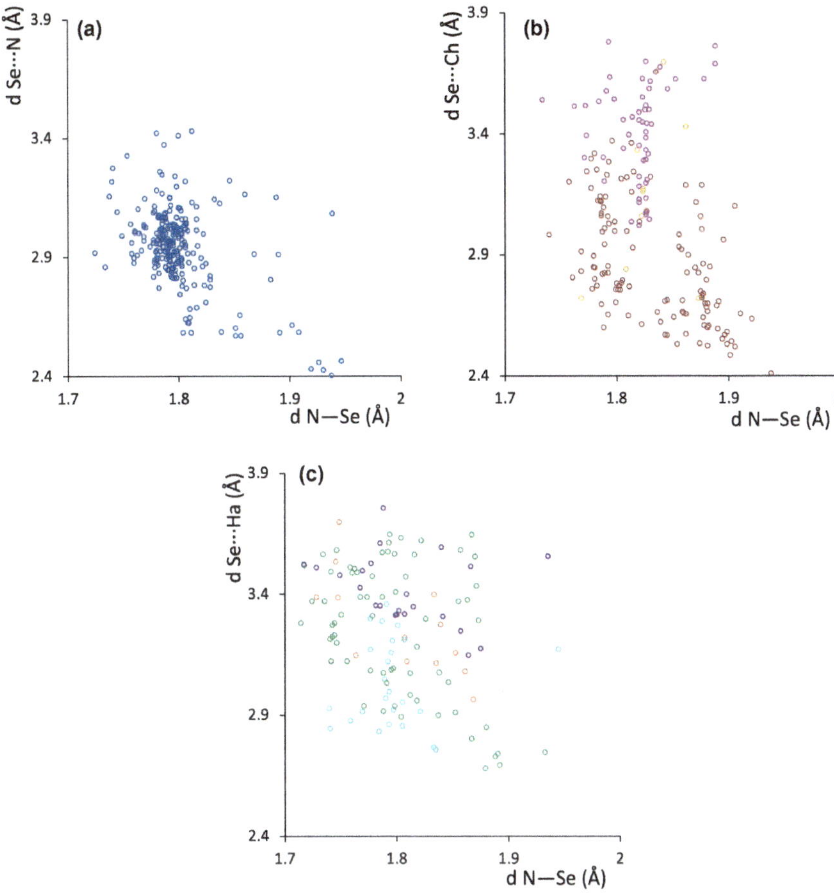

Figure 16.14 Scatterplots of $d_{\mathrm{Se}\cdots\mathrm{Nu}}$ *vs.* $d_{\mathrm{N-Se}}$ distances in the N–Se···Nu fragments for Nu = N (a); Ch (b); Ha (c). Nu colour codes: N = blue in (a); O = red, S = yellow, Se = magenta in (b); F = cyan, Cl = green, Br = orange, I = violet in (c).

Table 16.6 Occurrence, mean distances d [Å] (standard deviation, s), and δ values for N–Se···Nu fragments.

Nu	Occurrence	$d_{\mathrm{N-Se}}$ (s)	$d_{\mathrm{Se}\cdots\mathrm{Nu}}$ (s)	$\delta_{\mathrm{N-Se}}$ (%)	$\delta_{\mathrm{Se}\cdots\mathrm{Nu}}$ (%)
O	118	1.832 (0.05)	2.857 (0.23)	2.75	− 16.47
S	11	1.827 (0.03)	3.174 (0.34)	6.25	− 14.22
Se	57	1.816 (0.03)	3.412 (0.20)	7.71	− 10.22
N	253	1.800 (0.04)	2.945 (0.17)	5.89	− 14.65
Br	13	1.803 (0.05)	3.270 (0.20)	4.14	− 12.81
Cl	70	1.800 (0.06)	3.254 (0.27)	4.31	− 10.85
F	28	1.785 (0.05)	3.003 (0.19)	4.91	− 10.88
I	23	1.807 (0.05)	3.418 (0.02)	5.35	− 11.90

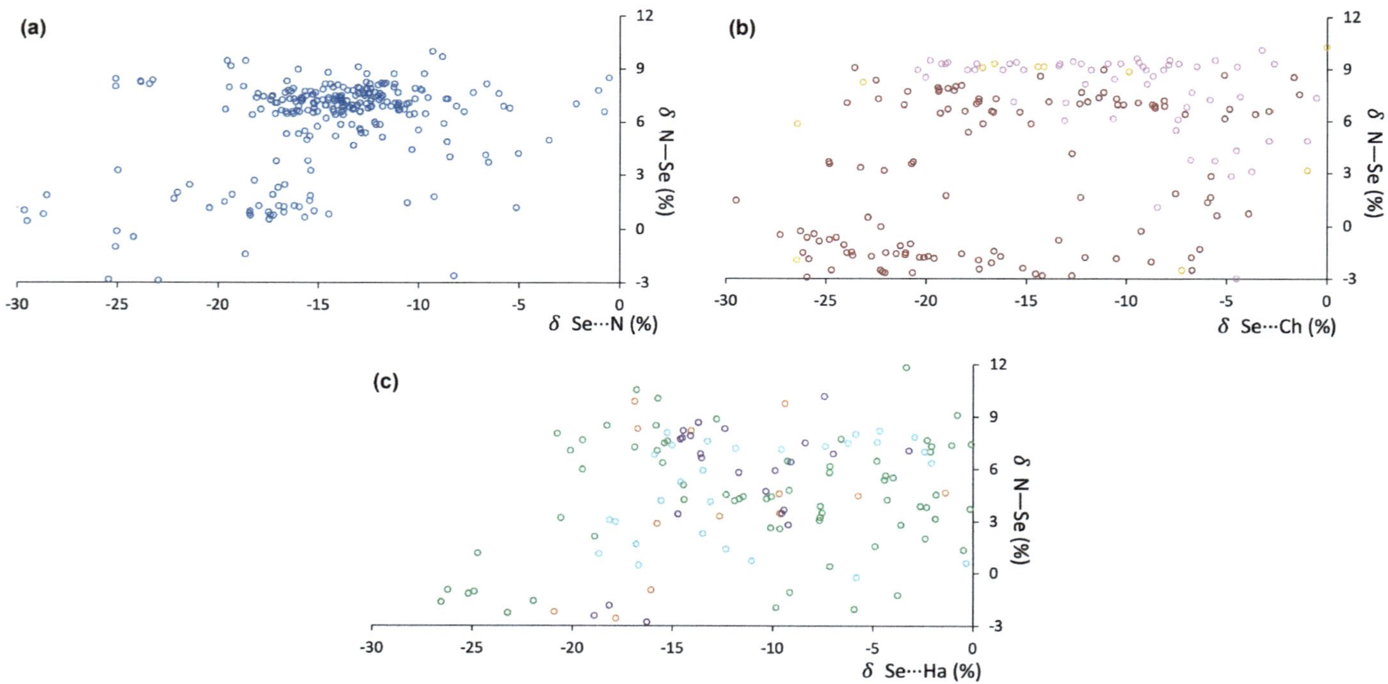

Figure 16.15 Scatterplots of $\delta_{N\text{–}Se}$ *vs.* $\delta_{Se\cdots Nu}$ calculated for the N–Se···Nu fragments for Nu = N (a); Ch (b); Ha (c). Nu colour codes: N = blue in (a); O = red, S = yellow, Se = magenta in (b); F = cyan, Cl = green, Br = orange, I = violet in (c).

Table 16.7 Occurrence, mean distances d [Å] (standard deviation, s), and δ values for O–Se···Nu fragments.

Nu	Occurrence	$d_{O\text{–}Se}$ (s)	$d_{Se\cdots Nu}$ (s)	$\delta_{O\text{–}Se}$ (%)	$\delta_{Se\cdots Nu}$ (%)
O	189	1.734 (0.07)	3.028 (0.24)	2.00	− 11.45
Se	13	1.740 (0.07)	3.621 (0.10)	2.22	− 4.71
N	4	1.721 (0.16)	3.252 (0.19)	2.00	− 5.74
Cl	13	1.697 (0.09)	3.378 (0.14)	4.74	− 7.45

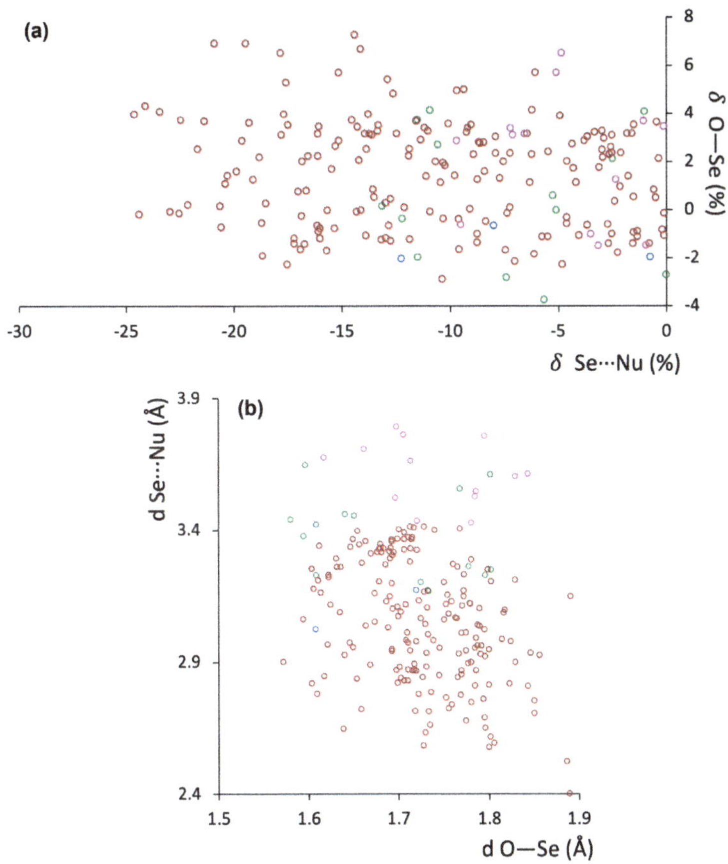

Figure 16.16 Scatterplots of $\delta_{O\text{–}Se}$ *vs.* $\delta_{Se\cdots Nu}$ (a) and $d_{Se\cdots Nu}$ *vs.* $d_{O\text{–}Se}$ (b) for the O–Se···Nu fragments. Nu colour codes: N = blue, O = red, Se = magenta, Cl = green.

The mean values calculated for the O–Se and Se···Nu distances in the fragments O–Se···Nu are summarized in Table 16.7, along with the mean of $\delta_{O\text{–}Se}$ and $\delta_{Se\cdots Nu}$ values. Figure 16.16 shows the scatterplots of $\delta_{O\text{–}Se}$ *vs.* $\delta_{Se\cdots Nu}$ values (a), and Se···Nu *vs.* O–Se distances (b). The fragments typically show short covalent O–Se and non-covalent Se···Nu distances mainly falling in the range 1.6–1.8 and 2.6–3.6 Å, respectively, following a tendency similar to that evidenced for the O–S···Nu fragments. The strength of the involved

ChBs is increased with respect to the lighter S-isologues with the stronger interactions involving O and Cl, as the ChB acceptors.

16.2.2.4 S–Se···Nu

The number of linear S–Se···Nu fragments amounts to 97, with Nu = N (29%), Ch (42%), and Ha (29%).[19] The compounds comprised in this group belong to the same classes examined in Section 16.2.1.4, with a selenium atom replacing a sulfur one, such as thiaselenazoles, selanyl-sulfanyl, thiaselane, and selenol derivatives.

The mean distances calculated within the fragments S–Se···Nu does not significantly depend on the nature of Nu, and therefore present very close values for the three Nu types [$d_{S\text{–}Se} = 2.234$ (0.05), 2.222 (0.03), and 2.219 (0.06) Å; $d_{Se\cdots Nu} = 2.84$ (0.25), 3.36 (0.26), and 3.16 (0.17) Å for Nu = N, Ch, and Ha, respectively]. As shown in Figure 16.17, where the scatterplots of

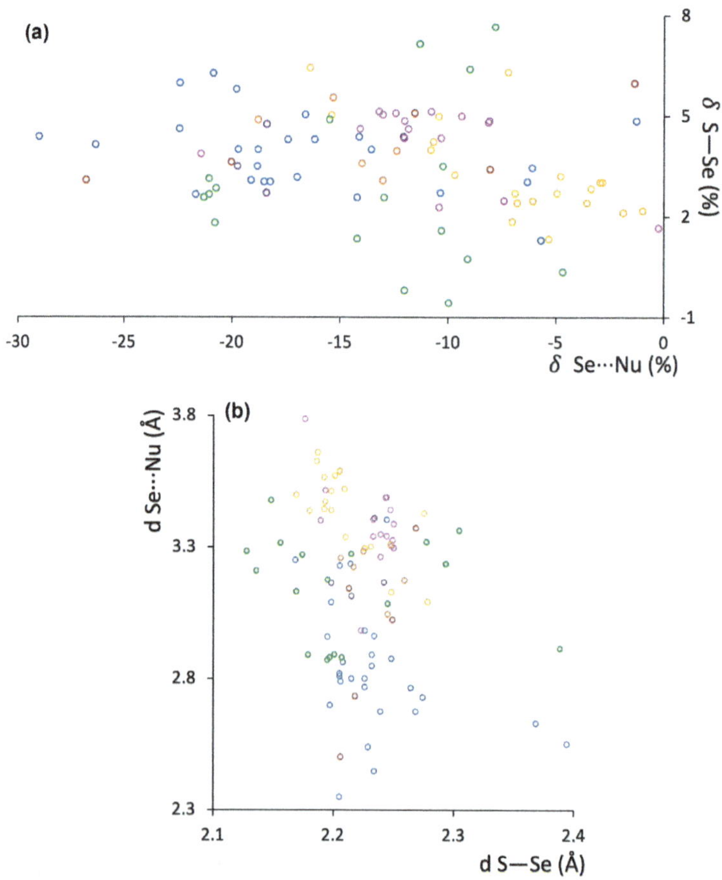

Figure 16.17 Scatterplots of $\delta_{S\text{–}Se}$ *vs.* $\delta_{Se\cdots Nu}$ (a) and $d_{Se\cdots Nu}$ *vs.* $d_{S\text{–}Se}$ (b) for the S–Se···Nu fragments. Nu colour codes: N = blue, O = red, S = yellow, Se = magenta, Cl = green, Br = orange, I = violet.

Table 16.8 Occurrence, mean distances d [Å] (standard deviations s), and δ values for S–Se···Nu fragments.

Nu	Occurrence	$d_{S\text{–}Se}$ (s)	$d_{Se\cdots Nu}$ (s)	$\delta_{S\text{–}Se}$ (%)	$\delta_{Se\cdots Nu}$ (%)
O	5	2.231 (0.03)	2.956 (0.34)	4.24	− 13.56
S	20	2.212 (0.03)	3.444 (0.16)	3.34	− 6.91
Se	16	2.231 (0.02)	3.385 (0.16)	4.26	− 10.93
N	28	2.234 (0.05)	2.839 (0.24)	4.40	− 17.71
Br	5	2.230 (0.02)	3.199 (0.09)	4.22	− 14.69
Cl	19	2.216 (0.07)	3.144 (0.20)	3.54	− 13.87
I	4	2.222 (0.02)	3.215 (0.13)	3.84	− 17.15

$\delta_{S\text{–}Se}$ *vs.* $\delta_{Se\cdots Nu}$ values (a) and Se···Nu *vs.* S–Se distances (b) are reported, the covalent S–Se distances fall in a narrow range, and the Se···Nu contacts show values comparable or even shorter that those found for the S–S···Nu isologues, despite the larger vdW radius of the Se atom compared to the S one.

The mean values calculated for the S–Se and Se···Nu distances in the fragments S–Se···Nu are summarized in Table 16.8, along with the mean of $\delta_{S\text{–}Se}$ and $\delta_{Se\cdots Nu}$ values calculated as described above (eqn (16.1) and (16.2)).

The data reported in Table 16.8 and Figure 16.17a confirm the small range covered by the S–Se elongations, with $\delta_{S\text{–}Se}$ values mainly falling between 0% and 5%. As expected, the mean $\delta_{Se\cdots Nu}$ values, ranging between −18% and −7%, reveal an increase in the strength of the ChB involved, as compared to the lighter sulfur isologues.

16.2.2.5 Se–Se···Nu

The number of linear Se–Se···Nu fragments involving a ChB amounts to 504.[19] Among them, moieties with Nu = N (10.7%), Ch (55.5%), and Ha (33.3%) are found, with only a single Se–Se···P and a single Se–Se···As fragment (0.5%). Belonging to this category are compounds with a Se–Se bond such as diselenides and polyselenides, and diselena-cycles and aza-cycles derivatives.

The mean distances calculated within the fragments Se–Se···Nu in the three Nu sub-categories present very similar values for the covalent Se–Se distances [$d_{Se\text{–}Se}$ = 2.336 (0.04), 2.343 (0.06), and 2.344 (0.05) Å, for Nu = N, Ch, and Ha, respectively]. The interaction Se···Nu shows mean values as large as 3.09 (0.16), 3.41 (0.27), and 3.24 (0.25) Å for Nu = N, Ch, and Ha, respectively, in line with the other R–Se···Nu systems. Figure 16.18 reports the scatterplots of Se···Nu *vs.* Se–Se distances for Nu = N (a), Ch (b), and Ha (c) showing spread regions of existence of fragments involving different acceptors without any loss of continuity.

The mean values calculated for the Se–Se and Se···Nu distances in the fragments Se–Se···Nu along with the mean values of the pertinent $\delta_{Se\text{–}Se}$ and $\delta_{Se\cdots Nu}$ are summarized in Table 16.9. Figure 16.19 shows the scatterplots of $\delta_{Se\text{–}Se}$ *vs.* $\delta_{Se\cdots Nu}$.

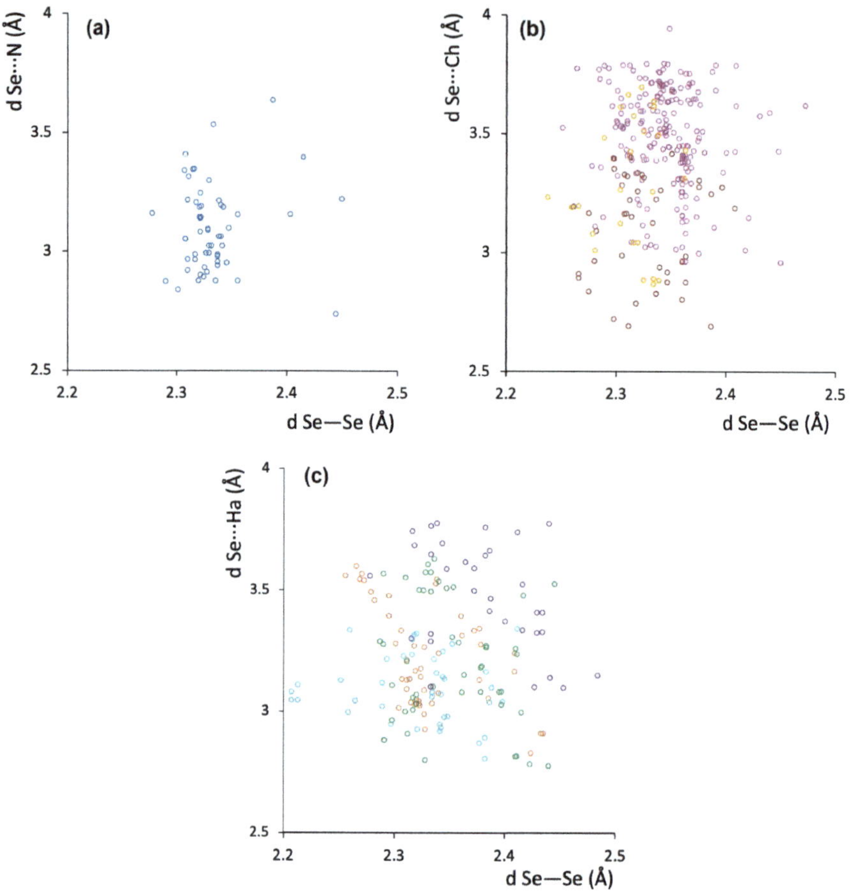

Figure 16.18 Scatterplots of $d_{\text{Se}\cdots\text{Nu}}$ *vs.* $d_{\text{Se–Se}}$ distances in the Se–Se···Nu fragments for Nu = N (a); Ch (b); Ha (c). Nu colour codes: N = blue in (a); O = red, S = yellow, Se = magenta in (b); F = cyan, Cl = green, Br = orange, I = violet in (c).

Table 16.9 Occurrence, mean distances *d* [Å] (standard deviations s in parentheses), and δ values for Se–Se···Nu fragments.

Nu	Occurrence	$d_{\text{Se–Se}}$ (s)	$d_{\text{Se}\cdots\text{Nu}}$ (s)	$\delta_{\text{Se–Se}}$ (%)	$\delta_{\text{Se}\cdots\text{Nu}}$ (%)
O	44	2.325 (0.04)	3.110 (0.22)	2.41	−9.05
S	26	2.313 (0.03)	3.285 (0.27)	1.91	−11.21
Se	209	2.350 (0.07)	3.487 (0.18)	3.53	−8.21
N	54	2.336 (0.04)	3.086 (0.16)	2.93	−10.56
Br	48	2.331 (0.04)	3.221 (0.21)	2.68	−14.12
Cl	46	2.352 (0.045)	3.211 (0.26)	3.65	−11.84
F	41	2.322 (0.05)	3.100 (0.14)	2.24	−8.13
I	33	2.381 (0.05)	3.480 (0.22)	4.88	−10.31

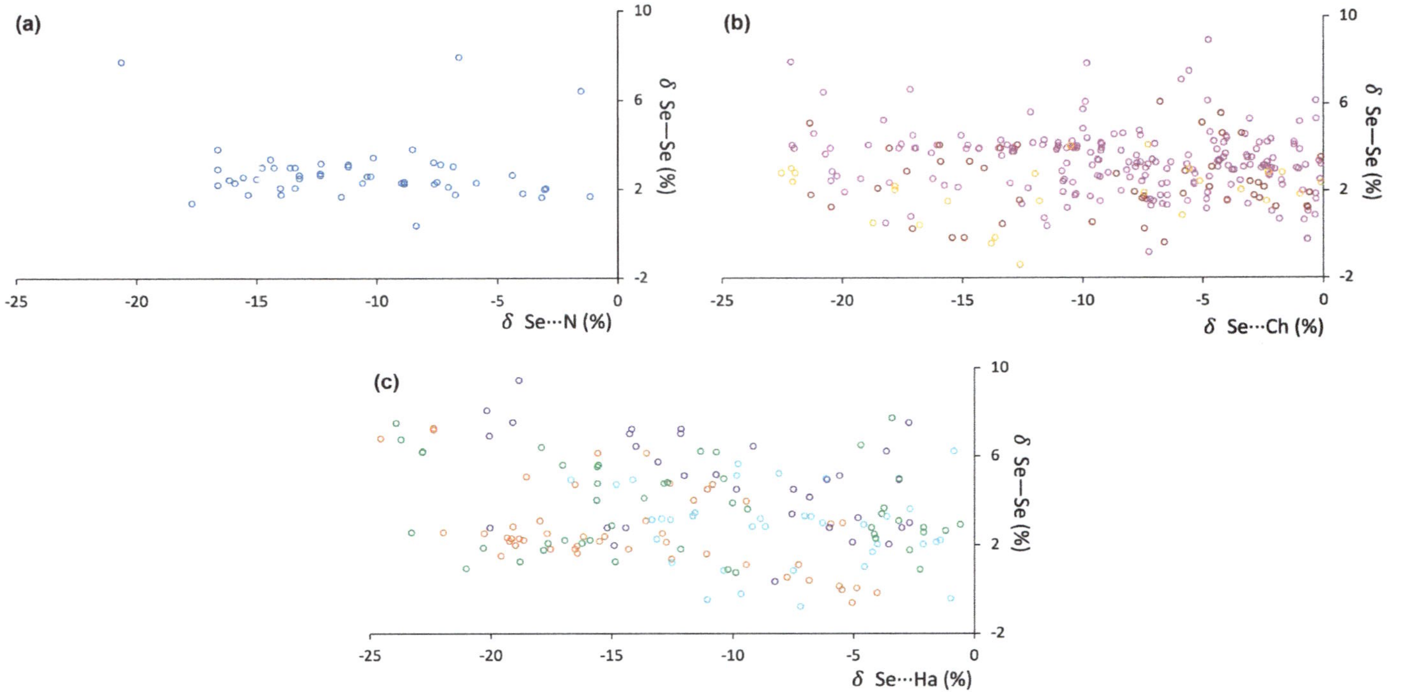

Figure 16.19 Scatterplots of $\delta_{\text{Se–Se}}$ *vs.* $\delta_{\text{Se}\cdots\text{Nu}}$ calculated for the Se–Se···Nu fragments for Nu = N (a); Ch (b); Ha (c). Nu colour codes: N = blue in (a); O = red, S = yellow, Se = magenta in (b); F = cyan, Cl = green, Br = orange, I = violet in (c).

Interestingly, the values reported in Table 16.9 and Figure 16.19 indicate that the ChB interacting in Se–Se···Nu fragments are intermediate in strengths between those found in S–S···Nu and S–Se···Nu ones.

16.2.3 R–Te···Nu Fragments

The CSD lists 2318 linear fragments of the type R–Te···Nu (R = B, C, N, P, As, O, S, Se, Te and Nu = N, P, As, Sb, O, S, Se, Te, F, Cl, Br, I).[19] Following the choice adopted in the previous sections, the survey here reported is limited to categories of fragments with R = C, N, O, S, with the addition of Te to include compounds with Te–Te bonds, in analogy with S–S and Se–Se isologues. The histogram plot in Figure 16.20 shows the occurrence of fragments R–Te···Nu sorted by R and Nu, with an evident prevalence of C–Se···Nu systems (1769, 76.3%) followed by Te–Te···Nu (172, 7.4%), N–Te···Nu (120, 5.2%), O–Te···Nu (98, 4.2%), S–Te···Nu (68, 2.9%), and Ha–Te···Nu (34, 1.5%) systems.

16.2.3.1 C–Te···Nu

The number of C–Te···Nu linear fragments amounts to 1769 with a prevalence for Nu = Ha and Ch (54.2% and 37.0%, respectively), the remaining 8.8% of occurrences being constituted by C–Te···N ones.[19] The compounds encompassed in this survey belong to various categories such as tellurophenes, tellurazoles, tellurafulvalente, telluroanhydrides, tellurocyanates, tellurides, and tellane derivatives.

The mean covalent and non-covalent bond distances calculated within the fragments C–Te···Nu show similar values [$d_{\mathrm{C-Te}}$ = 2.109 (0.03), 2.118 (0.03), and 2.127 (0.03) Å; $d_{\mathrm{Te\cdots Nu}}$ = 3.16 (0.22), 3.43 (0.39), and 3.57 (0.22) Å, for

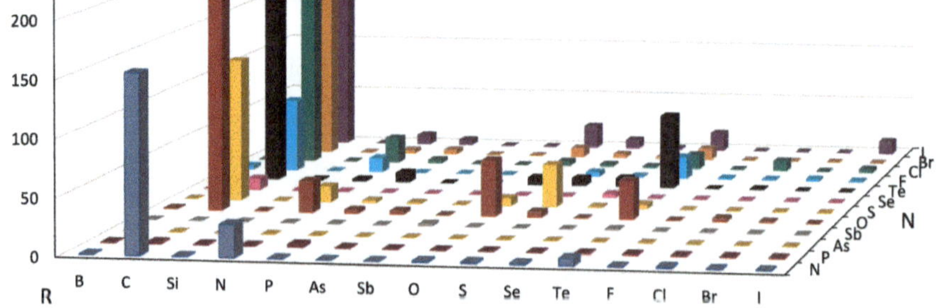

Figure 16.20 Linear R–Te···Nu fragment occurrence (R = B, C, Si, N, P, As, Sb, O, S, Se, Te, F, Cl, Br, I; Nu = N, P, As, Sb, O, S, Se, Te, F, Cl, Br, I) deposited at the CSD.

Nu = N, Ch, and Ha, respectively]. Despite the different vdW radii of the chalcogen atoms involved, non-bonding distances Se· · ·Nu and Te· · ·Nu fall in the same range and show very similar mean values. The scatterplots of Te· · ·Nu *vs.* C–Te distances for Nu = N (a), Ch (b), and Ha (c) are reported in Figure 16.21 and display wide distributions of the values, particularly for the C–Se· · ·Ha systems, as previously noticed for the selenium isologues (Figures 16.12c and 16.21c). Such wide distributions are in good agreement with the continuum from weak to strong Ch· · ·Nu interactions noted for C–Te· · ·I and C–Se· · ·Br systems.[23]

The mean values calculated for the C–Te and Te· · ·Nu distances in the fragments C–Te· · ·Nu are summarized in Table 16.10, along with the mean of $\delta_{\text{C–Te}}$ and $\delta_{\text{Te}\cdots\text{Nu}}$ values calculated, as described above (eqn (16.1) and (16.2)). Figure 16.22 shows the scatterplots of $\delta_{\text{C–Te}}$ *vs.* $\delta_{\text{Te}\cdots\text{Nu}}$.

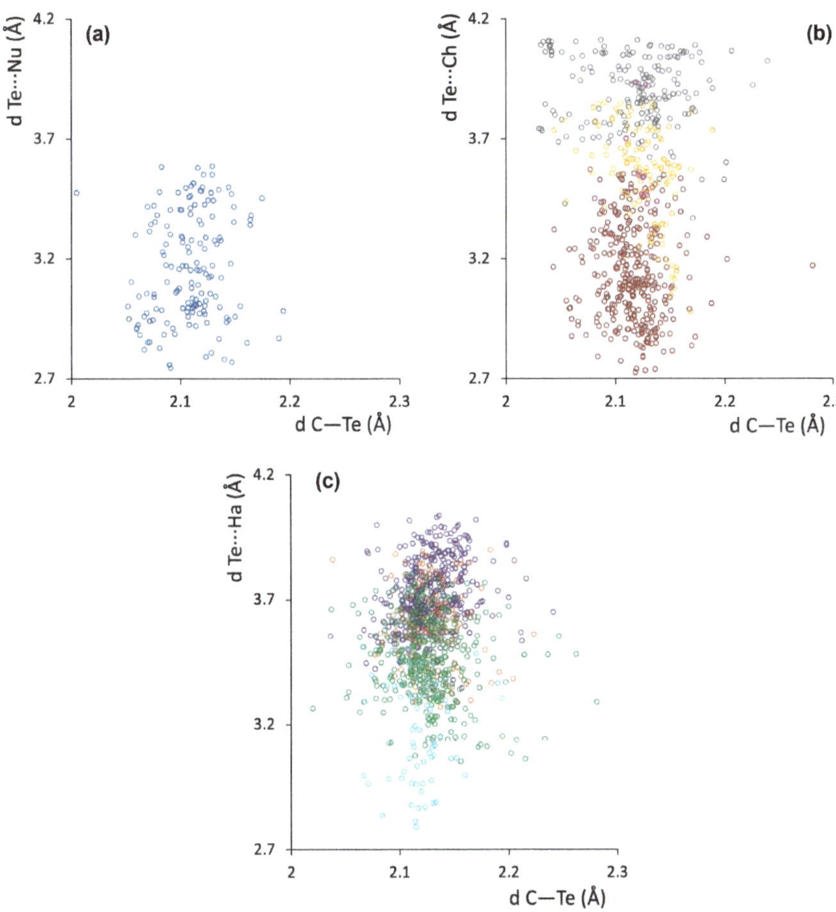

Figure 16.21 Scatterplots of $d_{\text{Te}\cdots\text{Nu}}$ *vs.* $d_{\text{C–Te}}$ distances in the C–Te· · ·Nu fragments for Nu = N (a); Ch (b); Ha (c). Nu colour codes: N = blue in (a); O = red, S = yellow, Se = magenta, Te = black in (b); F = cyan, Cl = green, Br = orange, I = violet in (c).

Table 16.10 Occurrence, mean distances d [Å] (standard deviation, s), and δ values for C–Te···Nu fragments.

Nu	Frequency	$d_{C\text{–}Te}$ (s)	$d_{Te\cdots Nu}$ (s)	$\delta_{C\text{–}Te}$ (%)	$\delta_{Te\cdots Nu}$ (%)
O	332	2.116 (0.03)	3.120 (0.17)	2.69	− 12.96
S	127	2.120 (0.04)	3.562 (0.20)	3.05	− 7.73
Se	10	2.125 (0.007)	3.602 (0.19)	3.15	− 9.03
Te	185	2.119 (0.04)	3.894 (0.15)	2.93	− 5.46
N	155	2.109 (0.03)	3.165 (0.22)	2.37	− 12.46
Br	165	2.129 (0.02)	3.588 (0.15)	3.33	− 8.25
Cl	352	2.126 (0.03)	3.460 (0.16)	3.09	− 9.39
F	66	2.119 (0.02)	3.153 (0.22)	2.78	− 10.89
I	376	2.128 (0.03)	3.729 (0.13)	3.30	− 7.69

The values of $\delta_{C\text{–}Te}$ vary in a range between 0% and 6% for almost all the fragments, without the loss of continuity found in the case of the isologues C–S···Nu (Figure 16.4b) and C–Se···Nu (Figure 16.13b). The $\delta_{Se\cdots Nu}$ values present values up to −13% indicating an increasing in strength of the ChBs involved on passing from C–Se···Nu to C–Te···Nu fragments.

16.2.3.2 N–Te···Nu

The number of linear fragments bearing a N–Te···Nu ChB amounts to 120 hits. These involve compounds usually corresponding to S- and Se-analogous, *i.e.*, tellurimides, amido-tellurium derivatives, and Te-containing heterocycles such as tellurazoles and fused polycyclic derivative thereof. Among the linear N–Te···Nu fragments, the Nu atom is a nitrogen in 28 hits (23.3%), a chalcogen in 44 (36.7%), and a halogen species in 48 (40.0%). The mean distances N–Te and Te···Nu calculated for the three sub-categories display very close values [$d_{N\text{–}Te}$ = 2.052 (0.05), 2.028 (0.07), and 2.009 (0.05) Å; $d_{Te\cdots Nu}$ = 2.932 (0.11), 3.19 (0.46), and 3.27 (0.28) Å for Nu = N, Ch, and Ha, respectively]. As noticed for the sulfur and selenium isologues, N–Te···Nu fragments show covalent and non-covalent distances notably shorter than those found in the analogous C–Te···Nu fragments, confirming the strengthening of the ChB donor ability on passing from carbon-bonded to nitrogen-bonded Te-donors. The involved distances fall in a range common to all fragments, as well as the relative $\delta_{Te\cdots Nu}$ and $\delta_{N\text{–}Te}$ values (Figure 16.23).

The mean N–Te and Te···Nu distances in fragments N–Te···Nu (Table 16.11) confirm the strengthening of the ChB on passing from selenium to tellurium with Te···Nu mean distances shorter than those found in the N–Se···Nu isologues, notwithstanding the larger Te vdW radium, and increased shortening $\delta_{Te\cdots Nu}$ reaching mean values of about −19% for Nu = O and N (Table 16.11).

16.2.3.3 O–Te···Nu

Linear O–Te···Nu fragments involving a ChB amount to 98 hits with a net prevalence of systems featuring a chalcogen as the ChB acceptor (O–Te···O, 51%;

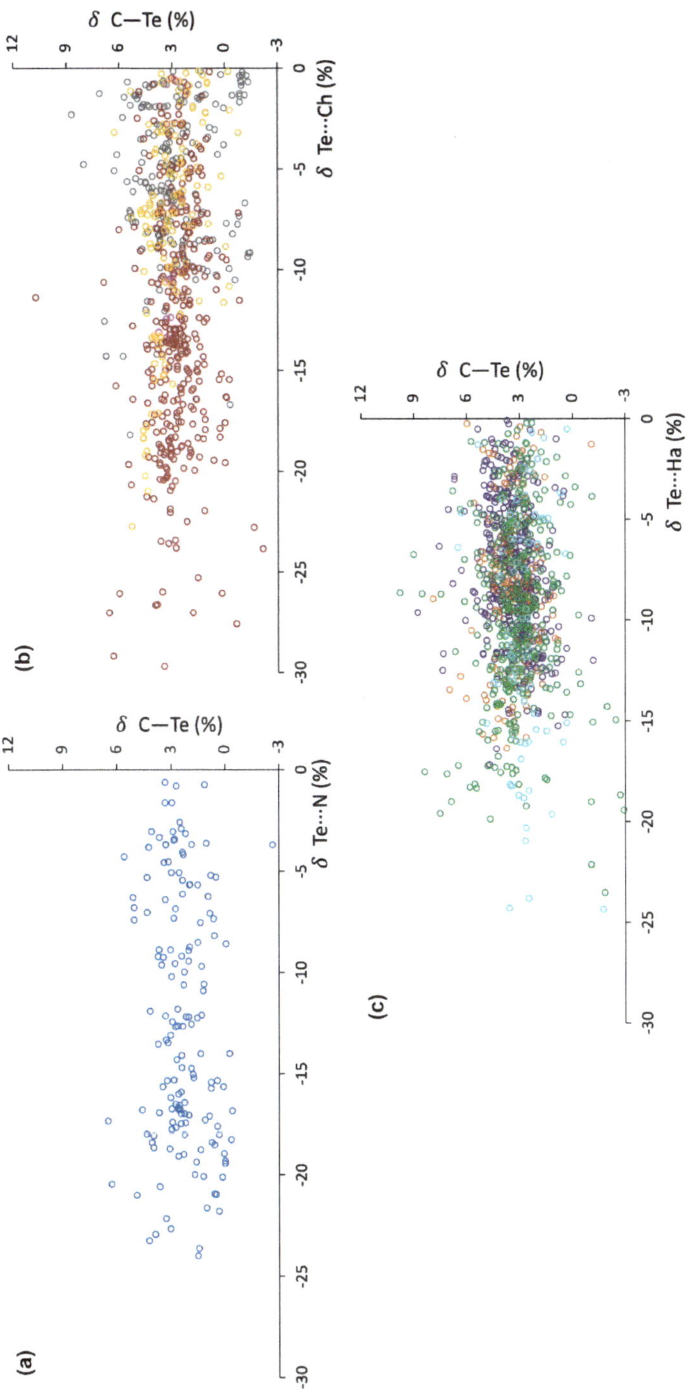

Figure 16.22 Scatterplots of $\delta_{C\text{–}Te}$ *vs.* $\delta_{Te\cdots Nu}$ calculated for the C–Te···Nu fragments for Nu = N (a); Ch (b); Ha (c). Nu colour codes: N = blue in (a); O = red, S = yellow, Se = magenta, Te = black in (b); F = cyan, Cl = green, Br = orange, I = violet in (c).

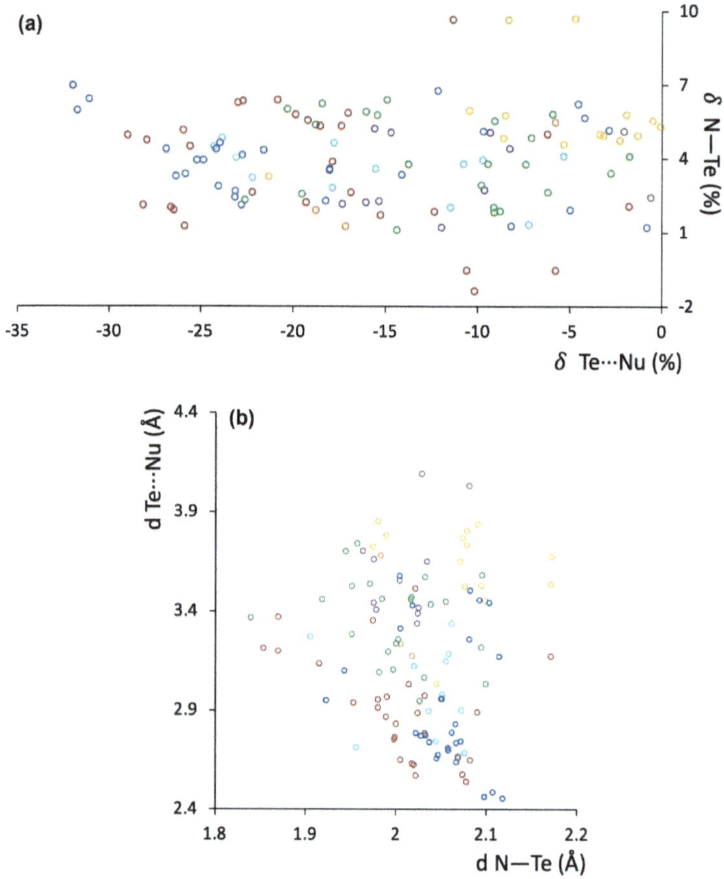

Figure 16.23 Scatterplots of $\delta_{N\text{–}Te}$ *vs.* $\delta_{Te\cdots Nu}$ (a) and $d_{Te\cdots Nu}$ *vs.* $d_{N\text{–}Te}$ (b) for the N–Te···Nu fragments. Nu colour codes: N = blue, O = red, S = sulfur, Te = black, F = cyan, Cl = green, Br = orange, I = violet.

Table 16.11 Occurrence, mean distances *d* [Å] (standard deviation, s), and *δ* values for N–Te···Nu fragments.

Nu	Occurrence	$d_{N\text{–}Te}$ (s)	$d_{Te\cdots Nu}$ (s)	$\delta_{N\text{–}Te}$ (%)	$\delta_{Te\cdots Nu}$ (%)
O	28	2.004 (0.07)	2.910 (0.26)	3.67	− 18.71
S	14	2.071 (0.06)	3.640 (0.21)	5.72	− 5.71
Te	2	2.054 (0.04)	4.065 (0.04)	3.76	− 1.33
N	28	2.052 (0.05)	2.932 (0.34)	4.00	− 18.77
Br	3	2.002 (0.02)	3.366 (0.28)	2.89	− 13.91
Cl	23	2.099 (0.06)	3.359 (0.22)	3.73	− 11.84
F	13	2.033 (0.05)	2.985 (0.23)	3.45	− 15.44
I	9	2.000 (0.03)	3.509 (0.14)	0.0337	− 13.13

O–Te···S, 7%, O–Te···Se, 7%, O–Te···I, 22%, O–Te···Cl, 8; O–Te···Br, 4%; O–Te···N, 1%).[19] The compounds comprised in this category mainly belong to the class of tellurium(IV) oxyanions, often showing complex polymeric structures.

Table 16.12 Occurrence, mean distances d [Å] (standard deviation, s), and δ values for O–Te···Nu fragments.

Nu	Occurrence	$d_{O\text{–}Te}$ (s)	$d_{Te\cdots Nu}$ (s)	$\delta_{O\text{–}Te}$ (%)	$\delta_{Te\cdots Nu}$ (%)
O	50	1.924 (0.07)	2.840 (0.33)	− 1.14	− 20.67
S	7	1.940 (0.04)	3.756 (0.03)	− 0.015	− 2.69
Te	7	2.090 (0.08)	3.835 (0.25)	2.46	− 6.91
Br	8	1.917 (0.004)	3.141 (0.03)	− 1.19	− 19.67
Cl	4	1.945 (0.04)	3.291 (0.25)	2.71	− 13.61
I	21	1.907 (0.017)	3.403 (0.13)	− 1.68	− 15.76

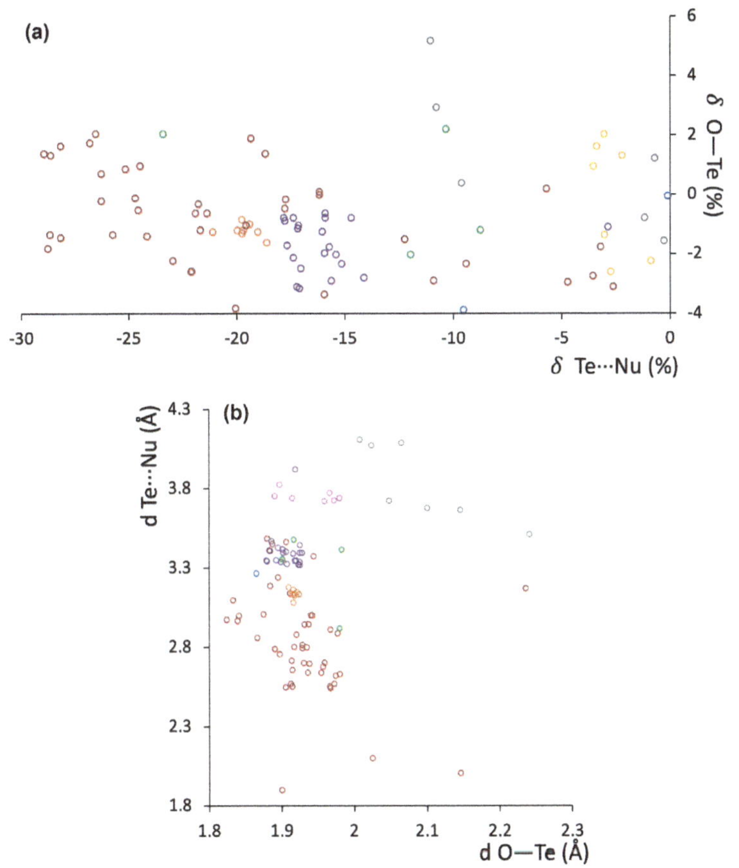

Figure 16.24 Scatterplots of $\delta_{O\text{–}Te}$ *vs.* $\delta_{Te\cdots Nu}$ (a) and $d_{Te\cdots Nu}$ *vs.* $d_{O\text{–}Te}$ (b) for the O–Te···Nu fragments. Nu colour codes: N = blue, O = red, S = sulfur, Te = black, Cl = green, Br = orange, I = violet.

The mean values calculated for the O–Te and Te···Nu distances in the fragments O–Te···Nu are summarized in Table 16.12, along with the delta $\delta_{O\text{–}Te}$ and $\delta_{Te\cdots Nu}$ mean values. Figure 16.24 shows the scatterplots of $\delta_{O\text{–}Te}$ *vs.* $\delta_{Te\cdots Nu}$ values (a) and Te···Nu *vs.* O–Te distances (b). The fragments display short covalent O–Te distances mainly falling in the range 1.8–2.0 Å with only a few cases, mainly O–Te···Te ones, featuring distances longer than 2.0 Å. The strength of the involved ChBs is increased with respect to the lighter sulfur and selenium isologues and to the analogous N–Te···Nu fragments, with the greater interactions regarding fragments involving O, Br, and Cl as the ChB acceptors.

16.2.3.4 S–Te···Nu

The 68 linear S–Te···Nu fragments found in the CSD are detailed in Table 16.13.[19] In these compounds Tellurium generally shows a metal character, being coordinated by the sulfur of thio- and dithiolates, thio- and dithiocarbamates, dithiophosphates and dithiophosphites, sulfonate, sulfides, and so on.

The S–Te···Nu fragments feature S–Te and Te···Nu values in the ranges 2.3–2.6, and 2.8–4.0 Å, respectively, and $\delta_{Te\cdots Nu}$ shortenings comprised between −10% and −25%, with mean values slightly shorter than those recorded for the S–Se···Nu isologues (Table 16.13 and Figure 16.25).

16.2.3.5 Te–Te···Nu

A total number of 172 linear Te–Te···Nu fragments involving a ChB can be found in the CSD. The compounds in this category are mainly di- and polytellurides derivatives.[19] The fragments mainly involve a Te–Te···Te interaction (67 items, 39%). The occurrences of Te–Te···Nu fragments along with the mean Te–Te and Te···Nu distances and $\delta_{Te–Te}$ and $\delta_{Te\cdots Nu}$ values are listed in Table 16.14.

The values reported in Table 16.14, as well as the data reported in the Te–Te and Te···Nu distances, and $\delta_{Te–Te}$ and $\delta_{Te\cdots Nu}$ scatterplots (Figure 16.26) show values well-suited with the presence of ChBs similar in strength to those found in the Se–Se···Nu fragments.

16.3 Conclusions

Notwithstanding the large number of scientific contributions on ChB interactions, a systematic survey of structural data was missing in the literature, being generally limited to selected groups of strictly related

Table 16.13 Occurrence, mean distances *d* [Å] (standard deviation, s), and δ values for S–Te···Nu fragments.

Nu	Occurrence	$d_{S–Te}$ (s)	$d_{Te\cdots Nu}$ (s)	$\delta_{S–Te}$ (%)	$\delta_{Te\cdots Nu}$ (%)
O	5	2.367 (0.05)	2.983 (0.08)	− 1.36	− 16.66
S	38	2.411 (0.04)	3.389 (0.15)	0.475	− 12.20
Te	7	2.419 (0.03)	3.697 (0.28)	7.80	− 10.27
Br	3	2.442 (0.005)	3.196 (0.13)	1.76	− 18.26
Cl	2	2.407 (0.03)	3.549 (0.07)	0.312	− 6.86
F	4	2.312 (0.02)	3.059 (0.27)	− 36.87	− 13.34
I	8	2.509 (0.04)	3.247 (0.08)	4.526	− 19.64

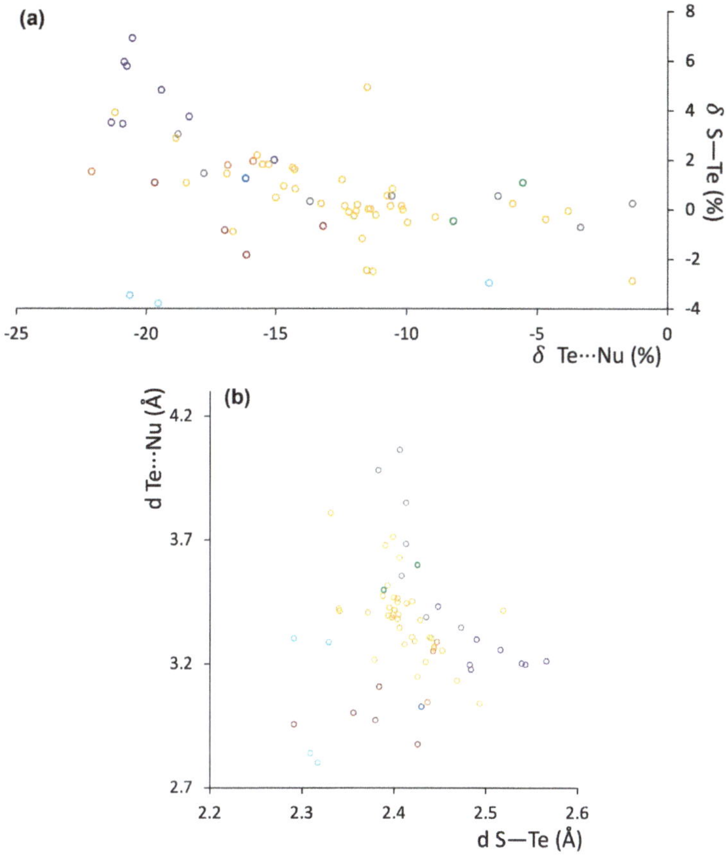

Figure 16.25 Scatterplots of $\delta_{S\text{–}Te}$ *vs.* $\delta_{Te\cdots Nu}$ (a) and $d_{Te\cdots Nu}$ *vs.* $d_{S\text{–}Te}$ (b) for the S–Te···Nu fragments. Nu colour codes: N = blue, O = red, S = sulfur, Te = black, F = cyan, Cl = green, Br = orange, I violet.

Table 16.14 Occurrence, mean distances d [Å] (standard deviation, s), and δ values for Te–Te···Nu fragments.

Nu	Occurrence	$d_{Te\text{–}Te}$ (s)	$d_{Te\cdots Nu}$ (s)	$\delta_{Te\text{–}Te}$ (%)	$\delta_{Te\cdots Nu}$ (%)
O	34	2.727 (0.06)	3.093 (0.21)	− 1.21	− 13.61
S	4	2.751 (0.04)	3.631 (0.03)	− 0.317	− 5.93
Te	67	2.774 (0.10)	3.753 (0.06)	0.510	− 8.92
N	6	2.697 (0.05)	3.300 (0.19)	− 2.29	− 8.58
Br	10	2.716 (0.06)	3.343 (0.23)	− 1.58	− 14.51
Cl	13	2.734 (0.06)	3.399 (0.25)	− 0.956	− 10.78
F	20	2.787 (0.16)	3.122 (0.23)	0.960	− 11.55
I	17	2.730 (0.06)	3.563 (0.22)	− 1.10	− 11.81

compounds. An analysis of all the structural data deposited at the CSD allows us to validate the assumptions generally accepted for ChB based on experimental or theoretical limited groups of compounds. The reported

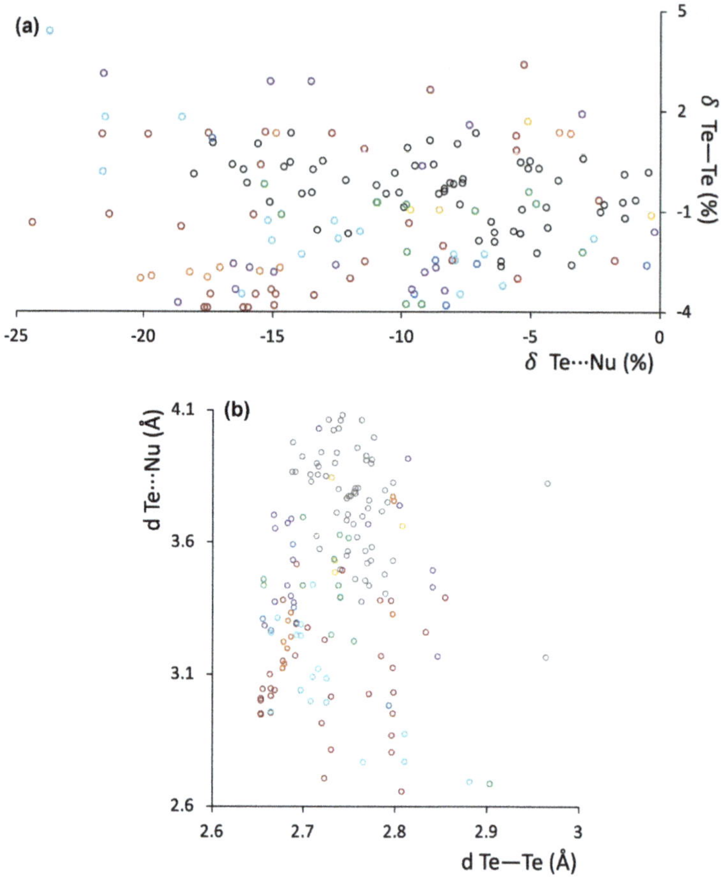

Figure 16.26 Scatterplots of $\delta_{Te\text{–}Te}$ *vs.* $\delta_{Te\cdots Nu}$ (a) and $d_{Te\cdots Nu}$ *vs.* $d_{Te\text{–}Te}$ (b) for the T–Te···Nu fragments. Nu colour codes: N = blue, O = red, S = sulfur, Te = black, F = cyan, Cl = green, Br = orange, I = violet.

structural data for the linear R–Ch···Nu fragments involving a ChB confirm a strengthening in the Ch···Nu interactions with the increasing of polarizability of the chalcogen atom (Te > Se > S). In fact, notwithstanding the increase of the vdW radius of the relevant Ch atoms, the Ch···Nu non-bonding distances fall in a similar range 2.4–3.6, 2.4–3.8, and 2.4–4.1 Å for Ch = S, Se, and Te, respectively (Figure 16.27).

On the contrary, the region of existence of the fragments featuring a different R atom covalently bonded to the chalcogen atom can be clearly identified for the different R–Ch···Nu systems by the different values displayed by the R–Ch distances. The increase in strength of the involved ChBs is better evidenced by the function $\delta_{Ch\cdots Nu}$ (eqn (16.2)) that normalizes the difference between the Ch···Nu distances and the sum of the relevant vdW radii. Figure 16.28 reports the $\delta_{Ch\cdots Nu}$ *vs.* $\delta_{R\text{–}Ch}$ values calculated for the R–Ch···Nu fragments where Ch = S (a), Se (b), and Te (c).

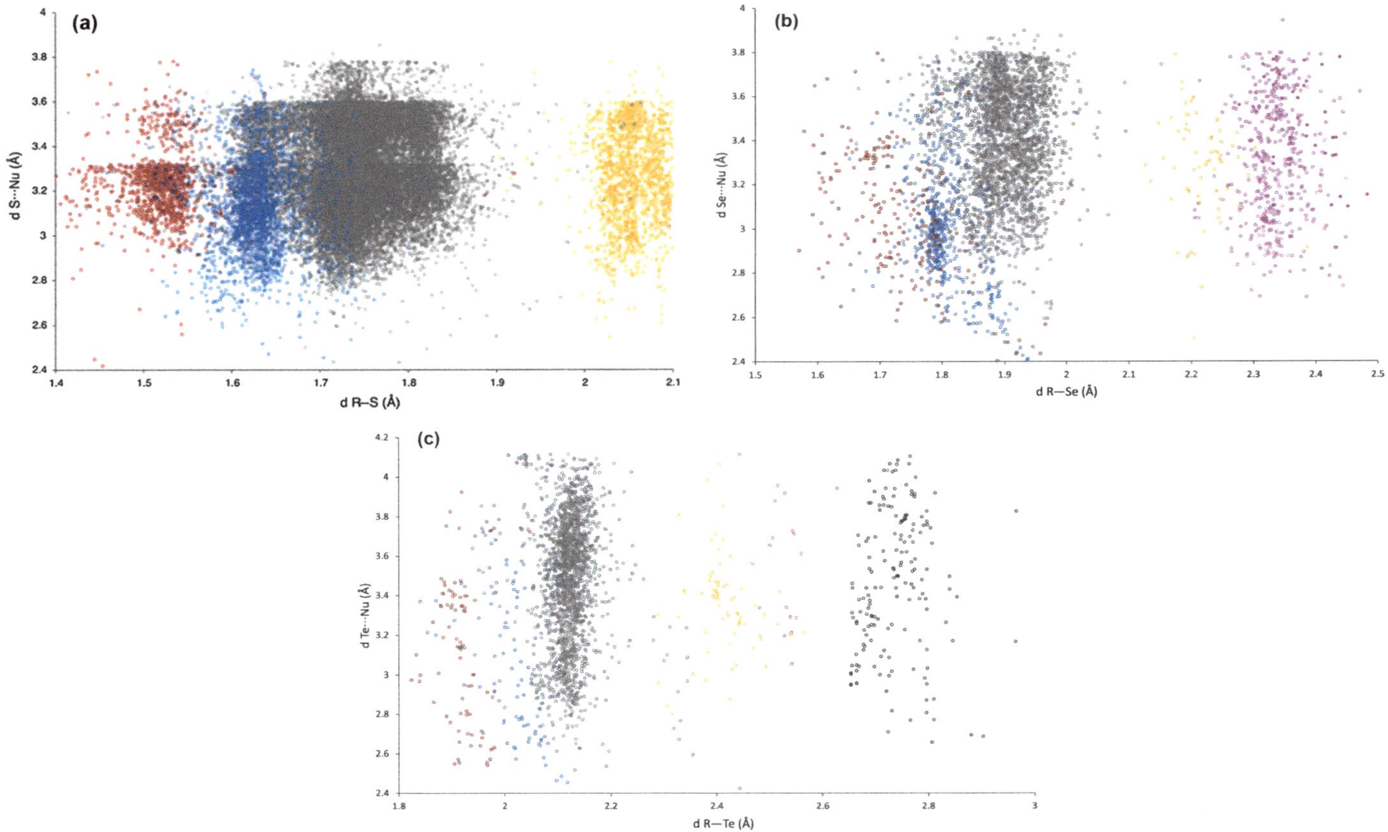

Figure 16.27 Scatterplots of $d_{\text{S}\cdots\text{Nu}}$ *vs.* $d_{\text{R–S}}$ (a); $d_{\text{Se}\cdots\text{Nu}}$ *vs.* $d_{\text{R–Se}}$ (b); $d_{\text{Te}\cdots\text{Nu}}$ *vs.* $d_{\text{R–Te}}$ (c) for the relevant R–Ch···Nu fragments. Nu colour codes: N = blue, O = red, S = sulfur, Se = magenta, Te = black, F = cyan, Cl = green, Br = orange, I = violet.

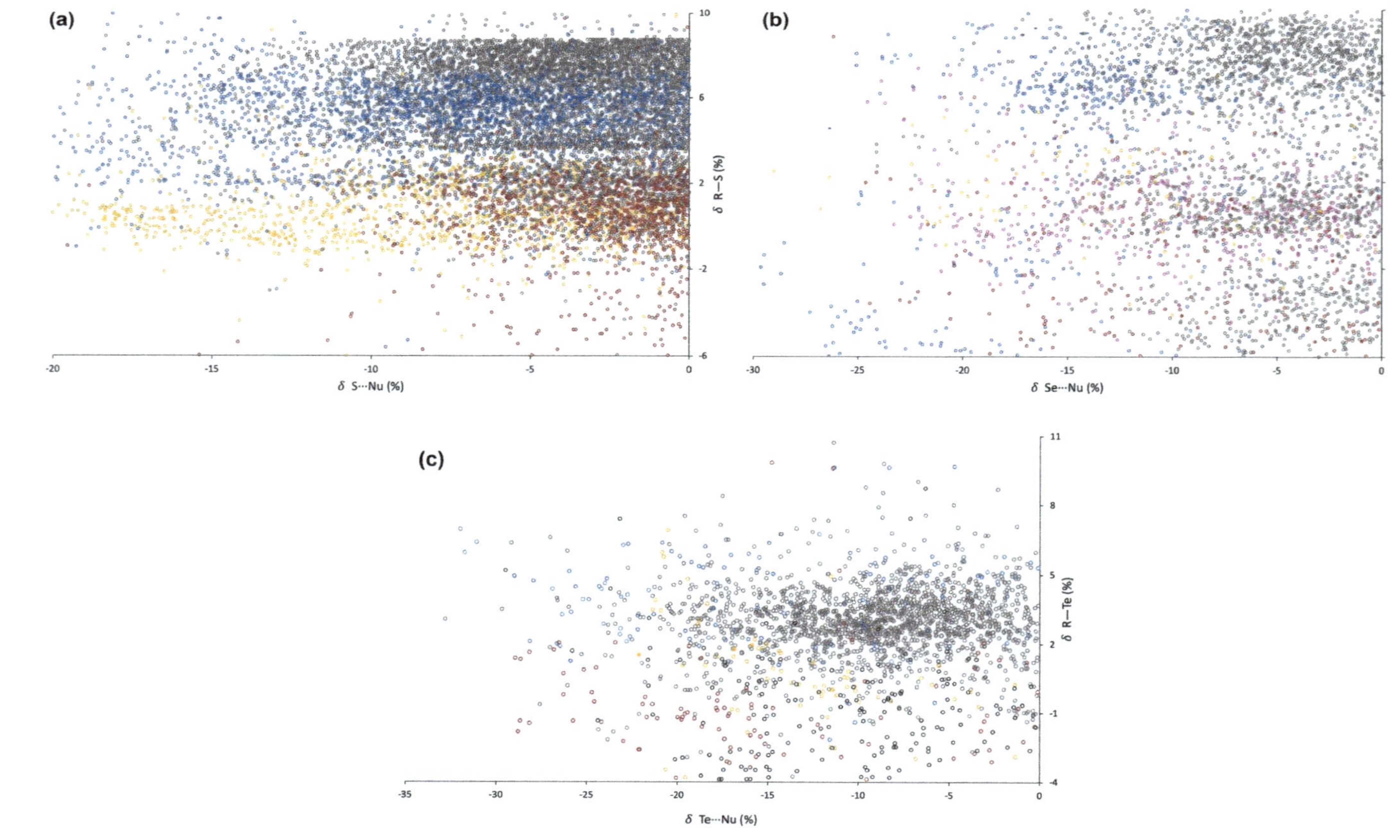

Figure 16.28 Scatterplots of the $\delta_{R\text{–}Ch}$ *vs.* $\delta_{Ch\cdots Nu}$ values calculated for the R–Ch···Nu fragments for Ch = S (a); Se (b); Te (c). Nu colour codes: N = blue, O = red, S = sulfur, Se = magenta, Te = black, F = cyan, Cl = green, Br = orange, I = violet.

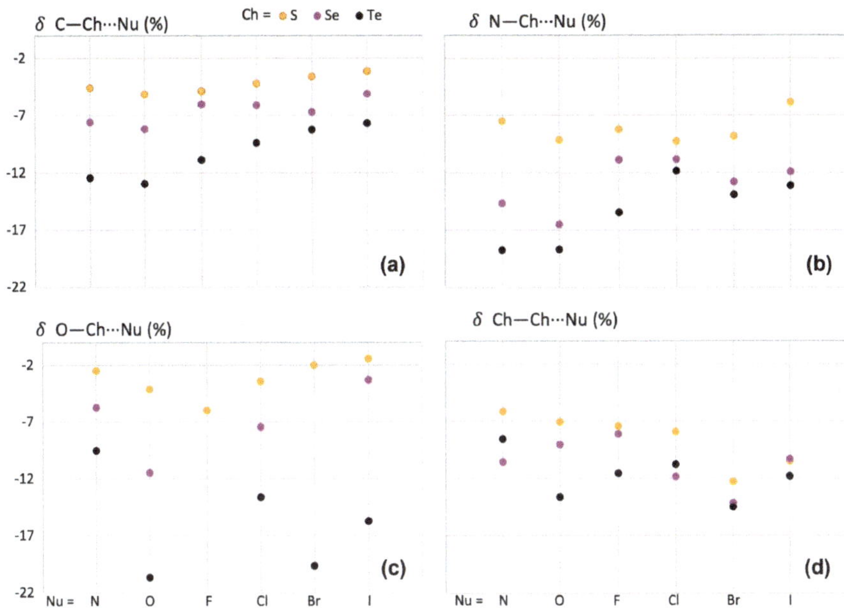

Figure 16.29 Mean $\delta_{\text{Ch}\cdots\text{Nu}}$ values calculated for the R–Ch···Nu fragments [R = C (a); N (b); O (c); Ch (S–S, Se–Se, and Te–Te) (d)]. Nu = N, O, F, Cl, Br, I. Ch colour codes: S = yellow, Se = magenta, Te = black.

The scatterplots of the $\delta_{\text{R-Ch}}$ *vs.* $\delta_{\text{Ch}\cdots\text{Nu}}$ values for the fragments R–Ch···Nu show similar variation ranges in the covalent distances R–Ch and an increasing in difference with respect to the sum of the involved vdW radii in the order Te > Se > S confirming a corresponding increasing in strength of the pertinent ChB interactions. The increase in the ChB strength on passing from S to Se and Te is well represented also by the mean values calculated for the function $\delta_{\text{Ch}\cdots\text{Nu}}$ for the different fragments with Nu = N, O, Ch (*i.e.*, S for S, Se for Se, and Te for Te), and Ha, reported in detail in Tables 16.1–16.14, and showed in Figure 16.29.

Less regular is the effect recorded for the fragments R–Ch···Nu on varying the nature of the R atom directly bonded to the Ch ones (Figure 16.30). The strength of the ChB increases when enhancing the electron-withdrawing character of the substituents, and a clear trend is observed on passing from C–Ch···Nu to N–Ch···Nu fragments for the all the examined systems and for all the involved Ch atoms (see Tables 16.1–16.14 and Figure 16.30). Fragments S–Ch···Nu also involve ChBs stronger than those found in C–Ch···Nu systems. On the contrary, a comparison between S–Ch···Nu and N–Ch···Nu fragments shows that the strength of the relative ChB interactions, usually higher when the chalcogen is covalently bonded to a nitrogen, can show increased values for the S–Ch···Ha systems, where the ChB acceptor Nu is a halogen species. With few exceptions, O–Ch···Nu fragments show the weakest ChB interactions, probably due to the nature of the involved compounds that are mainly chalcogen-oxides. The ChB formed by Se–Se···Nu

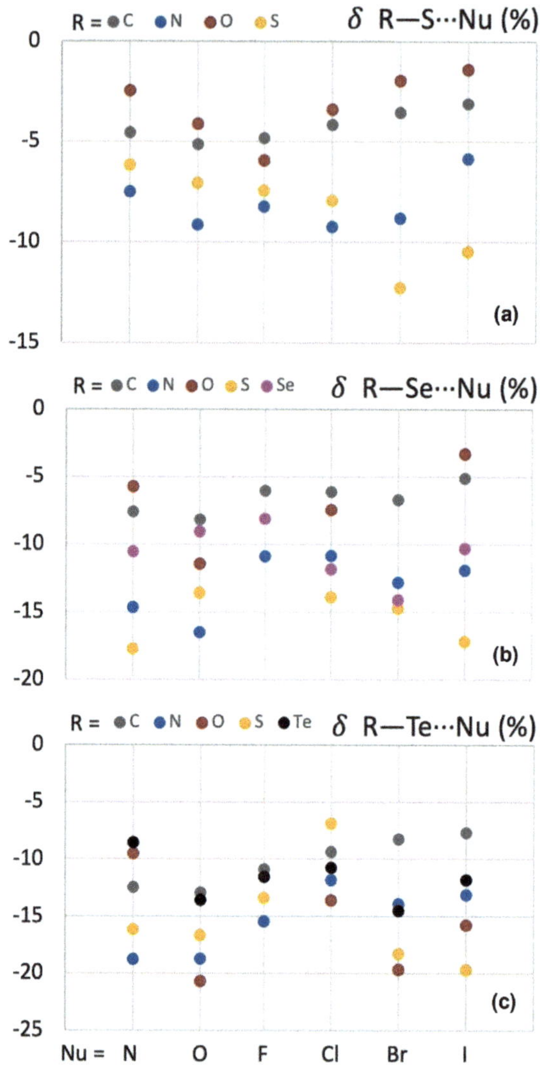

Figure 16.30 Mean $\delta_{\text{Ch-Nu}}$ values calculated for the R–Ch···Nu fragments [Ch = S (a); Se (b); Te (c)]. Nu = N, O, F, Cl, Br, I. R colour codes: C = grey, N = blue, O = red, S = yellow, Se = magenta, Te = black.

and Te–Te···Nu fragments display a strength very similar to those found in analogous C–Se···Nu and C–Te···Nu systems.

References

1. P. Scilabra, G. Terraneo and G. Resnati, *Acc. Chem. Res.*, 2019, **52**, 1313.
2. R. E. Rosenfield, R. Parthasarathy and J. D. Dunitz, *J. Am. Chem. Soc.*, 1977, **99**, 4860.

3. J. S. Murray, P. Lane, T. Clark and P. Politzer, *J. Mol. Model.*, 2007, **13**, 1033.
4. C. B. Aakeroy, D. L. Bryce, G. R. Desiraju, A. Frontera, A. C. Legon, F. Nicotra, K. Rissanen, S. Scheiner, G. Terraneo, P. Metrangolo and G. Resnati, *Pure Appl. Chem.*, 2019, **91**(11), 1889.
5. (a) The covalent radii utilized in the present work were taken from the work of B. Cordero, V. Gomez, A. E. Platero-Prats, M. Revés, J. Echeverrìa, E. Cremades, F. Barragàn and S. Alvarez, *Dalton Trans.*, 2008, 2832; (b) for atoms involved in single bonds, and from the work of P. Pyykkç and M. Atsumi, *Chem. - Eur. J.*, 2009, **15**, 12770, for atoms involved in multiple bonds.
6. K. T. Mahmudov, M. N. Kopylovich, M. F. C. Guedes da Silva and A. J. L. Pombeiro, *Dalton Trans.*, 2017, **46**, 10121.
7. L. Vogel, P. Wonner and S. M. Huber, *Angew. Chem., Int. Ed.*, 2019, **58**, 1880.
8. N. Biot and D. Bonifazi, *Coord. Chem. Rev.*, 2020, **413**, 213243.
9. M. Fourmigué and A. Dhaka, *Coord. Chem. Rev.*, 2020, **403**, 213084.
10. P. Politzer, J. S. Murray and T. Clark, *Phys. Chem. Chem. Phys.*, 2010, **12**, 7748.
11. V. Oliveira, D. Cremer and E. Kraka, *J. Phys. Chem. A*, 2017, **121**, 6845.
12. J. S. Murray, P. Lane, T. Clark and P. Politzer, *J. Mol. Model.*, 2007, **13**(10), 1033.
13. R. Bader, *Atoms in Molecules: A Quantum Theory*, Oxford University Press, USA, 1994, ISBN 978-0-19-855865-1.
14. M. C. Aragoni, M. Arca, F. A. Devillanova, F. Isaia and V. Lippolis, *Cryst. Growth Des.*, 2012, **12**, 2769.
15. N. Tarannam, R. Shukla and S. Kozuch, *Phys. Chem. Chem. Phys.*, 2021, **23**, 19948.
16. J. Zhang, W. Li, J. Cheng, Z. Liua and Q. Li, *RSC Adv.*, 2018, **8**, 26580.
17. D. Esrafili and R. Nurazar, *Mol. Phys.*, 2016, **114**, 276.
18. L. M. Azofra, I. Alkorta and S. Scheiner, *J. Phys. Chem. A*, 2015, **119**, 535.
19. CDS searches were made by ConQuest Version 2021.3.0 on Q_A–Ch···Q_B fragments by imposing a Q_A–Ch···Q_B angle between 160 and 180°. Ch···Q_B distances were imposed to be shorter than the sum of relevant van der Waals radii. Q_A = B, C, N, P, As, Sb, O, S, Se, Te, F, Cl, Br, I; Q_B = N, P, As, Sb, O, S, Se, Te, F, Cl, Br, I; Ch = S, Se, Te.
20. A. S. Lundemba, D. D. Bibelayi, P. A. Wood, J. Pradon and Z. G. Yava, *Acta Crystallogr.*, 2020, **B76**, 707.
21. A. J. Peloquin, C. D. McMillen, S. T. Iacono and W. T. Pennington, *ChemPlusChem*, 2021, **86**, 549.
22. P. Sanz Camacho, K. S. Athukorala Arachchige, A. M. Z. Slawin, T. F. G. Green, J. R. Yates, D. M. Dawson, J. D. Woollins and S. E. Ashbrook, *J. Am. Chem. Soc.*, 2015, **137**, 6172.
23. D. K. Miller, I. Y. Chernyshov, Y. Torubaev and S. Rosokha, *Phys. Chem. Chem. Phys.*, 2022, **24**, 8251–8259.

CHAPTER 17

Computational Methods to Study Chalcogen Bonds

MASSIMILIANO ARCA,*[a] GIANLUCA CIANCALEONI*[b] AND ANNA PINTUS*[a]

[a] Dipartimento di Scienze Chimiche e Geologiche, Università degli Studi di Cagliari, S.S. 554 bivio per Sestu, 09042 Monserrato, CA, Italy;
[b] Dipartimento di Chimica e Chimica Industriale (DCCI), Università di Pisa, Via Giuseppe Moruzzi, 13 – 56124 Pisa, Italy
*Emails: marca@unica.it; gianluca.ciancaleoni@unipi.it; apintus@unica.it

17.1 Introduction

According to the IUPAC definition,[1] a chalcogen bond (ChB) is a net attractive interaction R–Ch···A occurring between an electron-rich donor site A and an electron-poor chalcogen atom (Ch) included in a R–Ch molecule or ion (R = organic framework). Although chalcogen atoms can also behave as electron-rich species, the term chalcogen bond is reserved to the former case. The definition does not imply any model for the ChB, and the nature of the interaction is a current matter of debate.

In a molecular orbital (MO) view, the interaction between a ChB donor R–Ch and a ChB acceptor A to give a RCh···A assembly can be interpreted as a charge transfer from a LP of the Lewis base A to a virtual σ*-MO localized on a R–Ch bond. In the case of double-bonded sp^2-hybridized and hypercoordinated chalcogen species, a slightly different CT scheme occurs. In such cases, charge is transferred from the LP on A to a π*-MO, as reflected in the formation of remarkably bent ChB adducts.[2]

Chalcogen Chemistry: Fundamentals and Applications
Edited by Vito Lippolis, Claudio Santi, Eder J. Lenardão and Antonio L. Braga

Published by the Royal Society of Chemistry, www.rsc.org

In association with the diffusion of the model in the analysis of ChBs,[3] the anisotropy of the charge distribution at the chalcogen atoms was investigated in detail, so that the so-called σ-hole concept became common in publications on chalcogen-bonded systems.

In fact, nowadays ChB is considered a non-covalent yet directional interaction, related to other such sister interactions such as hydrogen bond (HB), pnictogen bond (PnB), and halogen bond (XB). This approach derives from the acknowledgment that even negatively-charged atoms feature one or more regions (σ-holes) of positive molecular electrostatic potential (MEP) on their surfaces that can direct the interaction with negatively-charged bond acceptors.[4] σ-Holes are typically oriented along the extensions of covalent bonds of the bond donors. In the case of ChB involving neutral interacting species, one or two σ-holes can be present, depending on the oxidation state and charge of the Ch atom (Figure 17.1). In the case of saturated chalcogen species, the largest charge depletion region is localized opposite the most electronegative substituent R.

In general, a CT MO-approach or a purely electrostatic interpretation are not singularly sufficient to describe appropriately σ-hole interactions. Accordingly, it is largely acknowledged that ChB derives from a combination of orbital, electrostatic, and dispersion contributions, whose relative weight depends on the nature of the interacting tectons and their substituents. A unifying view has been proposed by Kraka and co-workers, who classified ChB adducts as weakly (bond order $n < 0.1$), normally ($0.1 < n < 0.2$), or strongly ($n > 0.2$) bonded. The degree of the CT term would increase with n, so that strong complexes show a remarkable 3c-4e nature in the R–Ch···A three-body system.[2]

In this context, a key role in the interpretation of the bond is represented by theoretical quantum-mechanical (QM) calculations, which over the past two decades have been exploited systematically as a computational tool capable of interpreting and predicting the nature of ChB. A variety of theoretical approaches have been adopted, ranging from Molecular Mechanics (MM) *ab initio* Hartree–Fock (HF) and post-HF methods (most commonly according to the second order Møller–Plesset perturbative approach), to calculations based on the Density Functional Theory (DFT) and more computationally demanding methods such as couple cluster single-double (CCSD) and complete active space self-consistent field (CASSCF) techniques.

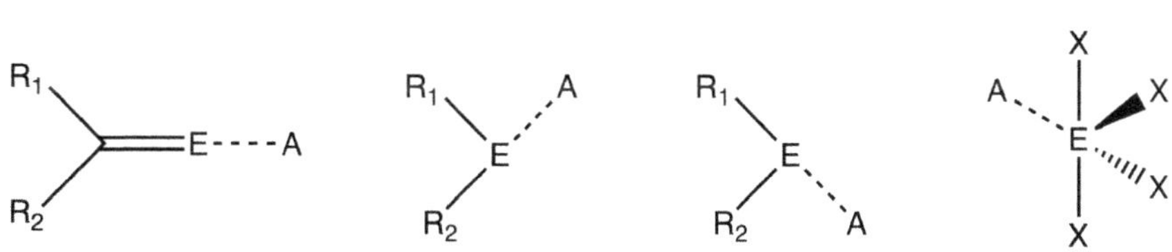

Figure 17.1 Common ChB motifs (Ch = chalcogen; A = ChB acceptor; X = halogen; R_1, R_2 = substituent).

In this chapter, the most commonly adopted methods will be reviewed, followed by a systematic classification of the model compounds described in the literature. In the final paragraph, the cooperativity and/or competition of ChB with other σ-hole interactions will be briefly discussed.

17.2 Computational Tools

The variety of computational tools available to interpret ChB interactions is well-known to theoretical chemists: different levels of approximation exist (HF methods or DFT, for example), and within each method, a large number of choices are available, such as the basis set and the functional. Only within the DFT method, a whole "Jacob's ladder" of different approximations is available, ranging from local-density approximation to fully non-local methods. Along with this, the choice of the computational details and the size of the system are also dictated by the computational resources currently available. Historically, a clear trend arises from an analysis of the literature: in the period 2002–2015, most of the results were based on MP2 methods, except when the size of the system made such a choice unpractical. This was due to the fact that, traditionally, DFT methods are not the best choice to analyze non-covalent interactions. Indeed, most of them do not contain specific dispersion physics, leading to the under- or over-estimation of the bond in noble gas dimers,[5] used as a test case. Anyway, between 2013 and 2015 some benchmark studies discovered the superior results of some DFT methods, in particular M06-2X, on the characterization of "σ-hole" interactions.[6,7] Indeed, Kozuch and Martin[6] demonstrated that the geometry of a family of halogen-bonded adducts optimized with the M06-2X functional showed a mean signed error (MSE) of −0.009 Å and a root mean square deviation (RMSD) of 0.028 Å, taking heavy correlated methods (as CCSD(T)) as a reference. Similarly, the dissociation energies showed a MSE of 0.04 kJ mol^{-1} and a RMSD of 1.80 kJ mol^{-1}. Bauzà and others[7] estimated that the average absolute error (AAE) for a family of chalcogen-bonded adducts optimized with the M06-2X functional was 0.046 Å with a mean absolute percentage error (MAPE) of 1.7%, also in this case taking the CCSD(T) results as a reference. As for the energy, the AAE was 3.2 kJ mol^{-1} and MAPE 8.3%. Better than M06-2X was only the MP2 method using resolution of the identity approximation (RI-MP2), with slightly lower values of AAE and MAPE. Even better are the double-hybrid functionals (such as SOS0-PBE0-2-D3(BJ) and revDSD-PBEP86-D3(BJ)),[8] but they are computationally demanding, and their use is recommended only for small systems or if large computational resources are available. Among the hybrid functionals, ωB97M-V and PW6B96-D3(BJ) reportedly work quite well.[8]

As for the basis set, it goes without saying that it has to be large enough, and generally a triple-zeta basis set gives enough confidence and reliability to make the basis set superimposition error (BSSE) not significant.[7] This has been verified also for other theoretical methods by Hamlin, Bickelhaupt, and co-workers.[9] Importantly, according to various results, the widely used

Grimme's correction for the dispersion energy (D3) tends to worsen the results and is not recommended. This happens for many reasons: first some DFT functionals already overestimate the binding energy, and the addition of another stabilizing contribution makes the overestimation even worse. More in detail, when a halide or a chalcogen are involved, the main cause of error does not lie in the functional, but in the self-consistent density.[10] When this happens, the errors can amount to up to 20 kJ mol^{-1}. They can be partially fixed by evaluating the functional on the HF density, decreasing the error to 4–8 kJ mol^{-1}, which is chemically acceptable. But geometry and dissociation energy are not the only useful parameters for ChB adducts. As mentioned in the Introduction, the nature of the ChB has been thoroughly explored in recent years. This debate is common to all σ-hole interactions, and in the case of the most explored one, XB, many studies tried to support the only electrostatic/polarization model.[11–14] In most cases, the main proof is a linear correlation between the electrostatic properties of the system (such as the value of the MEP in correspondence of the σ-hole) and the formation energy of the adduct. Nevertheless, most of the latest studies tend to acknowledge the importance of charge transfer.

Among the computational tools, the most popular are certainly the various decomposition schemes, which start from the interaction energy between the two fragments of the adduct (arbitrarily chosen) to isolate the different components of the bond (electrostatic interaction, orbital mixing, exchange/repulsion, and others). The most used ones are the canonical Energy Decomposition Analysis (EDA) and the Symmetry-Adapted Perturbation Theory (SAPT), which generally estimate comparable interaction energies and provide similar insights into the nature of the bond.[15] According to these analyses,[16,17] it is quite clear that when the interaction is very weak the most important contributions are electrostatics and dispersion, whereas for normal and strong ChBs the relative weight of induction and orbital mixing is larger.[18–20] The latter arises from the mixing of the np-type LPs of the Lewis base with the R–Ch σ* antibonding molecular orbital of the ChB donor, as confirmed by the Natural Bond Orbitals (NBO) results.[21] Another useful tool is the Quantum Theory of Atoms in Molecules (QTAIM), which is a topological study of the electron density developed by Bader.[22] This analysis is based on the partitioning of the molecular electron density based on zero-flux surfaces. As an output, a "bond path" is obtained, which is "a single line of maximum electron density linking the nuclei of two chemically-bonded atoms".[23] The "bond path" concept does not represent the chemical bond, as bond paths can appear also when bonding can be ruled out.[24] The maximum value along the bond path is called the bond critical point (BCP) and each BCP is characterized by some parameters, which are helpful in the characterization of the interaction. In particular, the local kinetic energy $G(\mathbf{r})$ and the local potential energy density $V(\mathbf{r})$ are generally examined, the sum of which is the energy density $H(\mathbf{r})$. From this analysis, the ChB emerges as a relatively strong interaction with some covalent character.[16] Interestingly, a correlation exists between the interaction energy of a ChB adduct and the values of $V(\mathbf{r})$ and $G(\mathbf{r})$ at the BCP.[12,25–29] When applied to adducts held together by multiple interactions, QTAIM also allows for the evaluation of the different contributions.

The ChB can also influence regions of the adduct far from the chalcogen atom, and these effects can be evaluated with the density deformation maps, which subtract the electron density of the non-interacting units from that of the adduct. These maps describe all the electronic effects due to the formation of the ChB, both on the atoms involved in the interaction (the chalcogen and the electron-rich atom) and on the rest of the molecule. For a quantitative result, the density deformation functions can be integrated along a chosen axis to derive the amount of charge that passes through the chosen point upon the formation of the ChB (Charge Displacement, CD, analysis).[30] The CD results can be used to predict the experimental Gibbs formation energy of ChB adducts[31] or to rationalize the activation of a Lewis acid by a Lewis base.[32] As for the latter, the method has been applied to phenylchalcogeniranium–olefin complexes ($PhSe^{+}\cdots$olefin) where the selenium atom can be involved in a ChB with the Lewis base. The integration of the CD method with the Natural Orbitals for Chemical Valence (NOCV) analysis allowed to isolate and quantify the Se ← olefin σ donation, the Se → olefin π back-donation, the propensity of the olefin to undergo a nucleophilic attack, and the effect of ChB on all these factors.[32]

17.3 Chalcogen Bond Models

The variety of molecular systems featuring ChB interactions is indeed very vast. In this paragraph, a brief examination of the interacting model compounds reported so far is discussed. These generally consist of two, or more rarely three, simple tectons featuring ChB interactions. Tectons are generally closed-shell neutral species, although few examples of interactions between neutral ChB donors and anions or ChB donor cations with neutral acceptors have been reported. In all cases, the ChB donor features an electrophile nature while the ChB acceptor must be an electron-rich species acting as a nucleophile. This implies that the donor species shows one, two, or three σ-hole(s) (Figure 17.2), depending on its oxidation state and chemical surrounding, orienting the interaction of the ChB acceptor.

In order to provide a systematic classification of the adducts, in the following different systems are categorized based on the interacting groups. For the sake of conciseness, chalcogen···chalcogen, chalcogen···pnictogen, and chalcogen···halogen interactions involving only two tectons will be discussed here.

17.3.1 Chalcogen···Chalcogen Interactions

Several papers have dealt with ChB interactions involving two chalcogen atoms (Ch···Ch′; Ch, Ch′ = O, S, Se, Te), the most electrophilic of which behaves as the ChB donor.[2] Most typically intermolecular interactions are

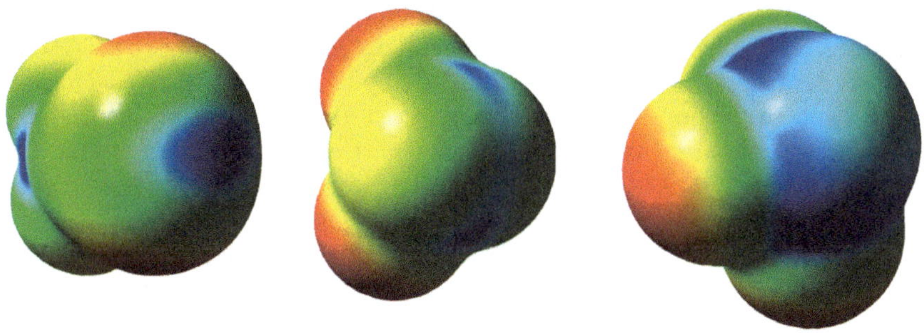

Figure 17.2 Molecular electrostatic potential (MEP) maps calculated for F_2CTe, F_2Te, and F_3Te^+ at the MP2/aug-cc-pVDZ-PP level. σ-holes on the chalcogen atom are mapped in a blue color.

involved in such studies, although intramolecular interactions have been investigated as well.[33,34] A first question arises on whether oxygen can be included among chalcogen species capable of forming ChB interactions. In fact, several papers reportedly demonstrated that ChBs would originate only from σ-holes of S, Se, or Te, and that oxygen would not behave as a chalcogen bond donor.[35–41] Varadwaj and co-workers explored theoretically at the DFT M06-2X and *ab initio* MP2 level the capability of the oxygen atom in the OF_2 molecule to behave as a ChB donor site in the interaction with a variety of species (CH_3F, CH_3Cl, CH_3Br, H_2CO, HFCO, HF, SO, CH_3CN, PN, HSCN, and HCN), concluding that "oxygen is indeed capable of forming a chalcogen bond".[36] Notably the authors also underlined the inadequacy of the ESP model in describing ChB interactions in the analyzed complexes.[36] The ability of oxygen to behave not only as an ChB acceptor but also as a ChB donor was further confirmed by subsequent papers, also supported by structural data.[42–44]

Looking at the most commonly encountered systems featuring ChB donor atoms heavier than oxygen, the nature of the interaction depends on the bond environment at the chalcogen donor. In a previously mentioned paper,[2] a set of 100 chalcogen bonded homodimeric adducts and complexes, formed starting from variously substituted neutral and cationic sulfur, selenium, and tellurium donors with N-acceptors and Cl^-, was described and interpreted at the DFT level.

When the chalcogen atom shows sp^2 hybridization, it features a single σ-hole oriented along the double bond axis (Figure 17.2).[2,45,46] A popular model of ChB donor of this type is represented by Ch=C=Ch′ molecules (Ch, Ch′ = O, S, Se, Te).[45–51] The interaction of Ch=C=Se (Ch = O, S, Se) with a variety of acceptors, including H_2S and H_2O, was investigated.[2,47,51,52] Due to the topology of the σ-hole, Ch=C=Se···Ch″H_2 adducts (Ch″ = O, S) displaying roughly linear C=Se···Ch″ moieties are formed. This notwithstanding, the molecular geometries, optimized at the MP2/aug-cc-pVTZ level, depend on the nature of Ch″ (C_{2v} and C_s point groups for Ch″ = O and S, respectively) and the BSSE-corrected energy of the interaction systematically decreases on passing from Ch″ = O to Ch″ = S, and from Ch = O to Ch = Se, so that the O=C=Se···OH_2 adduct is calculated to be the most stable and Se=C=Se···SH_2 the least stable among the considered assemblies (8.01 and 4.93 kJ mol^{-1}, respectively).[47] Related is a study on the interaction of CS_2 with formic aldehyde H_2C=O and its mono- and di-methyl and chloro derivatives, which eventually emphasizes the electrostatic contribution to the relatively weak interaction (<9.6 kJ mol^{-1}) by means of a SAPT decomposition of the interaction energy.[48,53] Ch=C=Ch′ compounds are also capable of forming bifurcated ChB when interacting with molecules displaying two adjacent ChB acceptor sites, such as $Ph_2N_2O_2$,[46] 1,2-dihydroxybenzene, or 1,2-dimethoxybenzene.[45] M06-2X/def2-TZVPPD AIM calculations showed that bifurcated interactions are closed-shell in nature, and that their strength increases with the atomic number of the interacting ChB atom Ch′ (*i.e.*, along the series S < Se < Te),[45] accompanied by an increasing CT, demonstrated at the NBO level, along the series.[45] The nature of the non-interacting atom Ch also affects the interaction, which is always larger for Ch = O. Notably, in the case of Ch=C=Ch′·$Ph_2N_2O_2$, the Ziegler–Rauk EDA[54] showed that the largest contribution is electrostatic in nature (43–55%).[46] The introduction of strongly electron-withdrawing groups X to a C=Ch moiety in a X_2C=Se ChB donor is expected to increase the electrophilic nature of the chalcogen atom. A different example of S···O interaction is represented by the adducts between SO_2, whose central S atom behaves as the ChB donor, and a variety of carbonyl donors, including $HCOCH_3$, $CH_3COOCOCH_3$, and $CH_3CONHCOCH_3$.[27] The presence of electron-withdrawing substituents at the CO Lewis donor reduces, as expected, the binding energy, while in amides the interaction is strengthened. The ChB interaction is attributed by the authors mainly to a CT orbital interaction involving the LP on the O atoms of the carbonyl donors and the antibonding π*-MOs on SO_2 (π-ChB) with binding energies as large as 36.0 kJ mol^{-1}.

ChB donors featuring a sp^3-chalcogen atom display two σ-holes (Figure 17.2),[55] whose entity is again dependent on the substitution, but also on the total charge. Several model compounds of this class have been reported in the literature.[55–59]

In a pair of papers, an extended set of ChB donors and acceptors were investigated, defining up to 32 different 'dimeric' model systems composed of a dimethylchalcogenoether $(CH_3)_2$Ch interacting with a CH_3Ch′Z acceptor (Ch, Ch′ = O, S, Se, and Te; Z = Me, CN).[18] Following different approaches, the authors concluded that Te-containing species are the most strongly bound (interaction energies in the range 12–25 kJ mol^{-1}), while O and S species show lower ChB energies (8–12 kJ mol^{-1}), with a trend of increasing strength along the Ch′ series Te > Se > S > O, largely attributed to dispersion contributions, while electrostatic attractive or repulsive interactions dominate only if at least one of the interacting atoms is oxygen or sulfur.[18] The dative contribution $p\rightarrow\sigma^*$ was found not to describe satisfactorily trends and details of the examined structures.[18,60]

CH_3SH and CH_3SF have been adopted by Scheiner for modeling the interactions with *N*-methylacetamide, which can be of ChB, HB, and tetrel bond type.[56] The introduction of the electron-withdrawing F-species increases the ChB interaction (O···S–F: 27 kJ mol^{-1}). A series of adducts between differently substituted XHSe acceptors and water were also studied, showing also in this case a strong dependance on the electron-withdrawing and donating capabilities of the X substituent.[61]

When charged sulfured compounds $CH_3SH_2^+$ and CH_3SHF^+ were considered, the interactions were dramatically enhanced, the counterpoise-corrected binding energies being as high as 163 kJ mol^{-1} for the O···S–F interaction in the CH_3SHF^+·*N*-methylacetamide adduct.[56] In SX_2 compounds (X = F, Cl, Br), MP2/aug-cc-pVTZ-PP QTAIM calculations showed that the S···O interaction with formaldehyde is mainly driven by electrostatic contributions, decreasing on passing from X = F to X = Br.[55] Notably, the terminal X species are themselves potentially involved in XB interactions (S–X···O), that were evaluated to be less favored as compared to ChB.[55] Related are HSX donors (X = H, F, Cl, Br, CH_3, CCH, CN, NC), whose interaction with HOCl was investigated (MP2/aug-cc-pVDZ, aug-cc-pVTZ).[57] The role played by electron-withdrawing X groups in activating ChB X–S···O interactions was confirmed, and attributed mainly to a CT from the LP on the O atom to a σ^*(S–X) MO.

When the chalcogen atom features a positive charge in trivalent sulfonium, selenonium, and telluronium cations of the type $(CH_3)_3Ch^+$,[2] three σ-holes are present on the Ch atom (Figure 17.2), located at the extension of the C–Ch bonds, and capable of orienting as many directional charge-assisted interactions (CAChB) with ChB acceptor species.[62] A CSD search showed that such interactions occur frequently.[62] In this context, 30 different complexes were theoretically investigated (PBE0-D3/Def2TZVP), among which 18 feature Ch···O interactions in the adducts formed between $(CH_3)_3Ch^+/F_3Ch^+$ (Ch = S, Se, Te) and three tripodal ChB acceptors containing terminal –CHO, –COOH, or $-N(O)H_2$ groups.[62]

The adducts formed by water with a series of different sulfur- and selenium-containing heterocycles were modeled at the B3PW91/6-31G(d,p) level of theory, showing that the σ-holes on the chalcogen atoms and the ensuing interactions can be selectively tuned by modifying the molecular frameworks and appropriately choosing and positioning substituents.[63]

Interestingly, the capability of chalcogen-rich species to form homodimers or oligomers based on ChB interactions has been occasionally investigated. Homodimers of chalcogenocyanic acid $(HChCN)_2$ (Ch = S, Se, Te) proved to be stable in a variety of relative orientations optimized at the MP2/aug-cc-pVTZ level, depending on the type of interactions established.[59] QTAIM analysis showed that the head-to-tail parallel orientations are supported by ChB Ch···Ch interactions, the electron density ρ at the bond critical points

(BCPs) increasing with the size of the chalcogen Ch, as reflected by the NBO analysis, displaying a regular decrease of the orbital interaction along the series Te > Se > S.[59] The corresponding binding energies, evaluated at the CCSD(T)/CBS level, fall between −20 and −44 kJ mol^{-1}, depending on the nature of the chalcogen atom Ch and the optimized dimer conformation.[59]

17.3.2 Chalcogen···Pnictogen Interactions

The computational study of ChB interactions involving pnictogen-containing electron donors started in the early 2010s, with a series of papers by Scheiner and co-workers.[64–69] The authors first studied the ChB of H_2S with ammonia,[64] finding that this interaction arises equally from induction and electrostatic components, with a minor contribution from dispersion energy.[66] The same authors also studied the influence of substitution at the H_2S molecule on the same type of interaction (namely YHS···NH_3; Y = F, Cl, Br, CH_3, NH_2, CF_3, OH, NO_2), noting a rough dependence of the interaction energy on the electron-withdrawing properties of the substituent Y.[65,66] Studies on the tunability of the S···N interaction depending on the nature of the Lewis base were performed on the adducts between $F_{3-n}H_nS^+$ cations (X = H, F; n = 1–2) and different sp and sp^3 bases,[70] revealing that while for the sp-hybridized ones (NCH, NCF, NCCl, NCCN, $NCCH_3$, NCOH, and N_2) the energy of the interaction decreases in the order $FH_2S^+ > F_2HS^+ > F_3S^+$, no single pattern was found for the variation of the interaction energy in complexes with sp^3-hybridized donor atoms (NH_3, NH_2F, NH_2Cl, NH_2CN, NH_2CH_3, and NH_2OH). Similar studies were performed on O=C=S in combination with different bases,[71] showing that complexes with the sp-hybridized nitrogen have linear chalcogen bonds whose interaction energies depend on the electron-donating or -withdrawing ability of the substituents, while in the case of sp^3-hybridized nitrogen bases the ChBs deviate slightly from linearity. A similar trend was observed in the case of SOXY (X, Y = F, Cl).[72]

The formation of chalcogen bonds through a π-hole rather than a σ-hole in the case of SO_3 was also explored by the same authors.[73] An important structural feature of these complexes is the distortion from planarity of the SO_3 molecule, so that in the corresponding adducts the deformation energy represents an important contribution to the interaction energy. The interaction energies depend in this case on the electron-donating ability of the P- and N-donor Lewis bases investigated, which in turn is improved by the introduction of electron-donating substituents and is larger for sp^3-hybridized than for the sp-hybridized bases.

Scheiner and co-workers[74] studied cationic systems at the MP2/aug-cc-pVDZ level, investigating the effect of charge, along with that of substitution, on S···N chalcogen bonds in XHS···NH_3 and X(Me)S···NH_3 systems (X = H, Me, NH_2, CF_3, OH, Cl, F, NO_2) and the corresponding cation dimers

$XH_2S^+\cdots NH_3$ and $X(Me)_2S^+\cdots NH_3$. They observed that the ionic complexes are several-fold more strongly bound as compared to the neutral ones.

17.3.3 Chalcogen···Halogen Interactions

Interactions of chalcogen-rich species with halogen-containing ChB acceptors are commonly encountered, and several papers investigated such interactions on both model compounds[26,36,57,75–80] and case studies.[41,81,82] The conceptually simplest cases involve the interactions of ChB donors with halides, a field of flourishing interest for anion recognition[80,83] and catalysis.[79] One of the few reports exploring theoretically ChB interactions between sp^2 ChB donors >C=Ch and the chloride anion dates back to 2009.[75] In particular, the interactions $X_2C{=}S\cdots Cl^-$ (X = H, F) and Ch=C=S $\cdots Cl^-$ (Ch = O, S), resulting in a contraction of the Ch=C bond distance for the latter compounds, were investigated in detail at the MP2/aug-cc-pVTZ level by means of a variety of approaches, including AIM theory and NBO analysis, that resulted in the acknowledgment of ChB as a sister non-covalent interaction to XB.[75] A larger number of donors featuring double-bonded Ch (Ch = S, Se, Te) donor atoms ($F_2C{=}Ch$, O=C=Ch, F–N=Ch) interacting with a chloride anion was examined later at the DFT level.[2]

Much larger is the number of papers involving the interaction between sp^3 ChB donors and halides or halogen-containing acceptors. Chalcogenides X_2Ch featuring electron-withdrawing substituents X are potentially ideal ChB donors for this purpose. In this context, the interaction of $(CN)_2Ch$ and X–Ch–Y (X, Y = Cl, F; Ch = S, Se) with chloride and fluoride anions was investigated, and MP2 calculations were capable of correctly predicting the arrangement of sub-units in the resulting complexes, particularly when X≠Y.[26] Notably, the charge-assisted ChB is classified as a strong interaction (with BSSE-corrected binding energies as large as 230 kJ mol^{-1} for $(CN)_2Se{\cdot}F^-$), with negative values of total electron density at BCPs, testifying for a partly covalent nature of the interaction.[2,26] Also $(C_6F_5)_2Ch$, $(C_6F_5)PhCh$, and Ph_2Ch diarylchalcogenides (Ch = S, Se, Te) were proposed as model ChB donors capable of forming stable interactions with the chloride anion.[79] The authors remarked that, consistently with experimental findings, DFT AIM/EDA investigations (M06-2X/aug-cc-pVDZ, with ECPs for Te) suggest perfluorophenyl substituents increase the strength of the interaction.[79] Moreover, in contrast to what it was previously reported,[2] on intensifying the interaction strength in these systems, the electrostatic contribution increases while the orbital term decreases.[79] As expected, the introduction of a positive charge on the ChB donor increases the electrostatic contribution of the interaction with halides.[2,77] This is the case of variously substituted cyclic thio- and selenopyrylium cations, whose interaction with the chloride anion was investigated at the DFT (BP86, B97), MP2, and CCSD(T)/CBS levels.[77]

Only a few papers explored the interactions of neutral ChB donors with neutral ChB acceptors featuring halogen atoms, focusing in particular on the interaction with the chlorine atom of hypohalous acids. $F_2C{=}Se$ forms adducts with HXO species (X = F, Cl, Br, I) held together by a Se···X ChB (with an almost linear optimized C=Se···X moiety), a Se···O ChB (See Section 17.3.1 above), or HBs.[76] While HBs do prevail in stabilizing the adducts, ChB involving Se···X interactions were calculated to range between −4.2 and 7.6 kJ mol^{-1} at the CCSD(T) level.[76] MP2/aug-cc-pVTZ calculations showed that the largest contribution to the ChB interactions arises from the electrostatic (from 36.3% to 45.4% on passing from X = F to X = I) and dispersion (from 51.4% to 41.9% on passing from X = F to X = I) components.[76]

Further examples of ChB in HChX derivatives have been reported.[78,84] The interaction of HChF model ChB donors (Ch = S, Se, Te) toward *N*-methylacetamide, adopted as a model protein environment, was investigated at the MP2/aug-cc-pVDZ level,[84] while in the ternary cyclic adduct XBr·HSX·PH_2X (X = F, Cl, Br, CN, NC, OH, NH_2, OCH_3), the Br···S interaction was interpreted as an XB, and the S···P interaction as a ChB.[78] Finally, a very specific case regards the interactions between F_2O and CH_3X compounds (X = F, Cl, Br), identified as authentic chalcogen bonds at least for X = F and Cl based on AIM analysis.[36]

17.3.4 Cooperativity and Competitivity of ChB in Comparison with Different σ-hole Interactions

As mentioned, ChB is a member of the growing "σ-hole" family, the progenitor of which is the halogen bond. Other members are the pnictogen and tetrel bonds. In all these cases, the MEP surface analysis reveals that the electrostatic potential is positive on the elongation of the X–Y bond, where X is an electronegative moiety and Y is the atom subject to the σ-hole. Generally, HB is stronger than other σ-hole interactions,[4] with some exceptions (see below). Regarding their nature, EDA schemes did not evidence a substantial difference among them, with the exchange term being the largest among the stabilizing contributions (40–50% of the attracting energy).[57] Anyway, there are also substantial differences, the most important one being the steric factor. In the case of XB, the approaching Lewis base lies on the prolongation of the only R–X bond, minimizing the steric hindrance between the LB and R. On the other hand, in R_1–Ch–R_2 molecules there will be some steric hindrance between the LB and the σ-hole originated by R_1 and R_2. Obviously, chalcogenone species $R_2C{=}Ch$ do not suffer this inconvenience, as in this case there is only one σ-hole (see above), if R is electron-withdrawing enough.

The competition between ChB, XB, and HB in differently bound E=C=E···OHX adducts (Ch = S, Se; X = Cl, Br) was studied, suggesting that while in these systems HBs and XBs are similar in strength, both are stronger than ChBs.[53] In $F_2C{=}Se$·HOX (X = halogen) systems, Se···O interactions compete with Se···H HB and Se···X XB, and C···O tetrel bonds.[76] Also in this case, ChB interactions were found to be the least stable ones. Moreover, the electrostatic energy is dominant in the H-bonded (Se···H–O interaction) and X-bonded (Se···X–O; see Section 17.3.3 above) complexes, although the polarization and dispersion contributions are also important, while the dispersion energy is comparable with the electrostatic contribution in tetrel-bonded (featuring C···O short contacts) and ChB (C=Se···O) complexes.[76]

An experimental and theoretical investigation was carried out by Caballero, Frontera, and co-workers on the chloride/bromide sensing abilty of chalcogenophene derivatives (C_3H_4Ch; Ch = S, Se, Te), such as C_6H_5–C_3H_4Ch, C_6F_5–C_3H_4Ch, and $C_6F_4(C_3H_4Ch)$,[80] potentially capable of ChB, HB, and π-interactions. A MEP surface analysis (carried out at PBE0-D3/def2-TZVPP optimized geometries) showed that the anions have a preference for ChB/HB rather than ChB/π interactions. The σ-hole interaction increases its contribution on passing from Ch = S to Ch = Te, so that in tellurophene derivatives ChB becomes more relevant than HB in stabilizing the halide adducts.[80]

Esrafili and co-workers compared at the MP2 level ChB and PnB interactions in $XHS\cdots PH_2Y$ and $H_2YP\cdots SHY$ adducts (X, Y = F, Cl, H, OH, OCH_3, CH_3, C_2H_5, NH_2, CCH, COH).[49,85] These studies report that the preferential formation of ChBs or PnBs depends on the different electron-donating or withdrawing nature of the X and Y substituents at the two monomers, as also confirmed by related studies.[86] ChBs were also compared to HBs by investigating the different isomers of radical cationic dimers formed by PH_2X and SHX (X = H, F), showing that in this case ChBs are stronger than HBs.[87] Related is a study at the DFT and CCSD(T) level on the S···N, C···H and S···π bonded complexes of CS_2 with different diazines, also showing that the chalcogen bonded and stacking complexes are more stable than the hydrogen bonded ones, and while in the S···N bonded complexes the principal attractive force is the electrostatic energy, in the remaining compounds the dispersion term is the dominant one.[88] A comparison between chalcogen and tetrel bonds was performed, again at the MP2 level,[89] on the dimers formed by F_2CSe and F_2CTe with HCN and NH_3, showing that the tetrel bond is stronger than the chalcogen bond for the F_2CSe dyads, while an opposite result was found for F_2CTe. Comparative studies between tetrel, pnictogen, chalcogen, and halogens bonds were also performed at the MP2 level by Scheiner and co-workers,[90] by investigating the interaction of different H_nY, MeH_nY, MeH_nF, and FH_nY Lewis acids ($n = 0$–3; Y = Br, Se, As, Ge) with ammonia, showing that different substitutions can reverse the strength order of the different interactions.

The interactions of 1,4-benzoquinone and its perfluoro-derivative toward HSF and HSCl compounds were analyzed theoretically (MP2/6-311++G**) with and without the presence of a halide X^- interacting perpendicular to the π-electron system of the quinone derivative.[91] The anion···π interaction was found to strengthen the ChB O···S interaction: a combined CT and QTAIM analysis indicated that the calculated CT values both in the anion···π and ChB interactions are slightly larger than the relevant ones in the isolated complexes, thus testifying for a cooperativity between the two types of interactions.[91]

A series of papers by Esrafili and co-workers analyzed these aspects at the MP2 level of theory, by studying the mutual effects between chalcogen bonds and halogen,[92–95] hydrogen,[25,92,95] and even aerogen[96] and lithium bonds[25] in linear trimeric systems, where the central molecule can play a dual role of Lewis acid and Lewis base simultaneously. A positive cooperative effect can be observed in all cases, arising from the mutual polarization of the interacting moieties, and the formation of the additional interaction results in a loss of electron density on the chalcogen atom and a simultaneous electron density accumulation on the pnictogen atom. The positive cooperative effect also shows a dependence on the nature of the substituent at the chalcogen-containing Lewis acid, the more electron-withdrawing groups generally enhancing the cooperation. The mutual positive cooperativity of S···N chalcogen bonds was also confirmed on XHS···NCHS···4-Z–Py trimers (X = F, Cl; Z = H, F, OH, CH_3, NH_2, NO_2, and CN; Py = pyridine).[97]

Other studies about ChB cooperativity rely on the investigation of potentially infinite linear adducts, inspired by solid-state structures,[98,99] or using *ad hoc* designed molecules.[45,100] For example, Dronskowski and co-workers published a paper about linear adducts $(XCN)_n$, where X corresponds to a halogen, a $Ch(CN)_2$ or a $Pn(CN)_3$ moiety, with the aim of comparing XB, ChB, and PnB interactions.[99] In all cases, the monomers form stable linear head-to-tail aggregates through X···N σ-hole interactions, and the stability of the chain increases as the central atom becomes heavier, coherently with the σ-hole framework. The adduct also becomes more stable as the number of interacting monomers increases, showing a cooperative effect. In the same papers, systems showing infinite chains held together by XB have been analyzed, showing that the degree of cooperativity is system-dependent and even anti-cooperativity effects can be found (for example 1,4-dioxane plus a halogen molecule). The reason for this cooperativity can be found in another paper by Esrafili and others, in which a slightly different chain is analyzed: $(OCCh)_{2-8}$, with Ch = S and Se.[100] For these systems, the ChB occurs between the oxygen atom and sulfur or selenium, again with a head-to-tail arrangement. First, the intermolecular Ch···O distances shortens as the length of the supramolecular chain increases, from an average value of 3.225 to 3.205 Å when Ch = S and from 3.244 to 3.224 Å when Ch = Se. The EDA reveals that the electrostatic contribution is additive and it has a negligible effect on the cooperativity, its increase depending exclusively on the shortening of the bonds. Conversely, the polarization contribution strongly affects the "non-additive energy" (E_{NA}, a measure of the cooperativity degree), which lies in the range 4–9% of the total E_{int}. Similar results are obtained with different supramolecular chains, and even mixing different interactions (HB or XB).[45,97,101,102]

Further cooperative effects between ChB and other interactions have been recently explored by Beer and co-workers.[44] In their paper, they synthesized a

3,5-bis-triazole pyridine structure covalently linked to benzo-15-crown-5 ether motifs. The remaining position on the triazoles was substituted with hydrogens, iodine, or tellurium, in order to measure the anion binding activity by HB, XB, or ChB. As a result, in all cases the binding was low (anion association constants $K_a < 50\ M^{-1}$), and even null in the case of tellurium. But upon adding $NaPF_6$ in solution, K_a exceeded 350, 5000, and 1000 M^{-1}, for R = H, I, and Te, respectively. The binding between the sodium cation and the crown ether makes the interaction stronger just because of electrostatics: the anion is interacting with a doubly charged species. Beyond this, DFT calculations demonstrated that also the MEP in correspondence with the σ-hole becomes much more positive upon complexation. The latter enhances the polarization of iodine and chalcogen atoms, strengthening the ChB.

17.4 Conclusions

In this chapter, the wide adoption of theoretical calculations as an invaluable tool to investigate, interpret, and predict ChB interactions has been explored. While a general agreement can be found in including ChB among non-covalent σ-hole interactions, it is undoubtable that the subtle balance of different contributions, such as charge-transfer, electrostatic, and dispersion factors, renders sophisticated calculations necessary to correctly model these aspects.

Over the past two decades, more than one hundred papers have been published on the theoretical analysis of ChB interactions. Among them, a large number deal with the modeling of two, three, or even more extended simple interacting model molecules or ions featuring one or more ChB interactions involving a chalcogen atom Ch on the ChB-donor and a nucleophile ChB-acceptor featuring a halogen (Ch···X), pnictogen (Ch···Pn), or even a different chalcogen atom (Ch···Ch′). This analysis of model systems, briefly summarized in this chapter, allowed us to shed a light on the most promising computational methods to approach theoretically the title systems. Significant efforts were also made to study the competitive and/or cooperative effects between ChB and other noncovalent interactions.

While initially the theoretical exploration of ChB made use almost only of post-HF perturbative methods, such as MP2, or multiconfigurational (CCSD) calculations, during the past decade an increasing role of DFT has emerged, with a remarkable prevalence of the functionals based on the meta-GGA approximation belonging to the Minnesota family, such as M06-2X, although DFT calculations employing the ever-green B3LYP hybrid functional, albeit not necessary providing the best accuracy, were often reported.

Whatever the theoretical level, computational tools based, just to cite a few, on the quantum theory of AIM, NBO, and energy decomposition techniques have been used jointly in the interpretation of ChB interactions.

While only simple models were presented here, QM calculations were also exploited to explore a larger number of specific and more complex compounds, with applications in catalysis, crystallography, spectroscopy, or biochemistry.

The framework of the research testifies for the growing importance of theoretical calculations in interpreting the nature of ChB and sister interactions and highlights the continuing effort of the computational chemists' community to model non-covalent interactions, thus shifting their focus from single molecules to increasingly more complex supramolecular aggregates.

References

1. C. B. Aakeroy, D. L. Bryce, G. R. Desiraju, A. Frontera, A. C. Legon, F. Nicotra, K. Rissanen, S. Scheiner, G. Terraneo, P. Metrangolo and G. Resnati, *Pure Appl. Chem.*, 2019, **91**, 1889.
2. V. Oliveira, D. Cremer and E. Kraka, *J. Phys. Chem. A*, 2017, **121**, 6845.
3. T. Clark, M. Hennemann, J. S. Murray and P. Politzer, *J. Mol. Model.*, 2007, **13**, 291.
4. P. Politzer, J. S. Murray and P. Lane, *Int. J. Quantum Chem.*, 2007, **107**, 3046.
5. G. A. Dilabio and A. Otero-de-la-Roza, *Rev. Comput. Chem.*, 2016, **29**, 1.
6. S. Kozuch and J. M. L. Martin, *J. Chem. Theory Comput.*, 2013, **9**, 1918.
7. A. Bauzá, I. Alkorta, A. Frontera and J. Elguero, *J. Chem. Theory Comput.*, 2013, **9**, 5201.
8. N. Mehta, T. Fellowes, J. M. White and L. Goerigk, *J. Chem. Theory Comput.*, 2021, **17**, 2783.
9. L. de Azevedo Santos, T. C. Ramalho, T. A. Hamlin and F. M. Bickelhaupt, *J. Comput. Chem.*, 2021, **42**, 688.
10. Y. Kim, S. Song, E. Sim and K. Burke, *J. Phys. Chem. Lett.*, 2019, **10**, 295.
11. M. H. Kolář and P. Hobza, *Chem. Rev.*, 2016, **116**, 5155.
12. P. V. Bijina and C. H. Suresh, *J. Chem. Sci.*, 2016, **128**, 1677.
13. T. Brinck and A. N. Borrfors, *J. Mol. Model.*, 2019, **25**, 125.
14. P. Politzer, J. S. Murray and T. Clark, *Phys. Chem. Chem. Phys.*, 2010, **12**, 7748.
15. O. A. Stasyuk, R. Sedlak, C. F. Guerra and P. Hobza, *J. Chem. Theory Comput.*, 2018, **14**, 3440.
16. F. De Vleeschouwer, M. Denayer, B. Pinter, P. Geerlings and F. De Proft, *J. Comput. Chem.*, 2018, **39**, 557.
17. L. de Azevedo Santos, S. C. C. van der Lubbe, T. A. Hamlin, T. C. Ramalho and F. Matthias Bickelhaupt, *ChemistryOpen*, 2021, **10**, 391.
18. C. Bleiholder, R. Gleiter, D. B. Werz and H. Köppel, *Inorg. Chem.*, 2007, **46**, 2249.
19. W. Zierkiewicz, J. Fanfrlík, P. Hobza, D. Michalska and T. Zeegers-Huyskens, *Theor. Chem. Acc.*, 2016, **135**, 217.
20. J. Fanfrlík, A. Přáda, Z. Padělková, A. Pecina, J. Macháček, M. Lepšík, J. Holub, A. Růžička, D. Hnyk and P. Hobza, *Angew. Chem., Int. Ed.*, 2014, **53**, 10139.

21. D. J. Pascoe, K. B. Ling and S. L. Cockroft, *J. Am. Chem. Soc.*, 2017, **139**, 15160.
22. R. F. W. Bader, *Acc. Chem. Res.*, 1985, **18**, 9.
23. C. F. Matta and R. J. Boyd, *An Introduction to the Quantum Theory of Atoms in Molecules*, Wiley-VCH Verlag GmbH & Co. KGaA, Weinheim, 2007.
24. L. E. Bickerton, A. Docker, A. J. Sterling, H. Kuhn, F. Duarte, P. D. Beer and M. J. Langton, *Chem. – Eur. J.*, 2021, **27**, 11738.
25. M. D. Esrafili, F. Mohammadian-Sabet and M. Solimannejad, *Struct. Chem.*, 2014, **25**, 1197.
26. E. Alikhani, F. Fuster, B. Madebene and S. J. Grabowski, *Phys. Chem. Chem. Phys.*, 2014, **16**, 2430.
27. L. M. Azofra and S. Scheiner, *J. Phys. Chem. A*, 2014, **118**, 3835.
28. M. D. Esrafili, S. Asadollahi and Y. Dadban Shahamat, *Struct. Chem.*, 2016, **27**, 1439.
29. A. Bauzá and A. Frontera, *ChemPhysChem*, 2020, **21**, 26.
30. G. Ciancaleoni, F. Nunzi and L. Belpassi, *Molecules*, 2020, 25.
31. G. Ciancaleoni, C. Santi, M. Ragni and A. L. Braga, *Dalton Trans.*, 2015, **44**, 20168.
32. G. Ciancaleoni, *ACS Omega*, 2018, **3**, 16292.
33. G. Sánchez-Sanz, I. Alkorta and J. Elguero, *Molecules*, 2017, **22**, 227.
34. V. D. P. N. Nziko and S. Scheiner, *J. Org. Chem.*, 2015, **80**, 2356.
35. S. Benz, J. López-Andarias, J. Mareda, N. Sakai and S. Matile, *Angew. Chem., Int. Ed.*, 2017, **56**, 812.
36. P. R. Varadwaj, A. Varadwaj, H. M. Marques and P. J. MacDougall, *Phys. Chem. Chem. Phys.*, 2019, **21**, 19969.
37. B. R. Beno, K. S. Yeung, M. D. Bartberger, L. D. Pennington and N. A. Meanwell, *J. Med. Chem.*, 2015, **58**, 4383.
38. A. Bauzá, T. J. Mooibroek and A. Frontera, *ChemPhysChem*, 2015, **16**, 2496.
39. A. J. Mukherjee, S. S. Zade, H. B. Singh and R. B. Sunoj, *Chem. Rev.*, 2010, **110**, 4357.
40. H. Huang, L. Yang, A. Facchetti and T. J. Marks, *Chem. Rev.*, 2017, **117**, 10291.
41. S. Benz, J. Mareda, C. Besnard, N. Sakai and S. Matile, *Chem. Sci.*, 2017, **8**, 8164.
42. P. R. Varadwaj, *Molecules*, 2019, **24**, 3166.
43. M. A. A. Ibrahim and E. M. Z. Telb, *ACS Omega*, 2020, **5**, 21631.
44. T. Bunchuay, A. Docker, U. Eiamprasert, P. Surawatanawong, A. Brown and P. D. Beer, *Angew. Chem., Int. Ed.*, 2020, **59**, 12007.
45. M. D. Esrafili and F. Mohammadian-Sabet, *Chem. Phys. Lett.*, 2015, **634**, 210.
46. S. Massahi, M. Ghobadi and M. Nikoorazm, *Theor. Chem. Acc.*, 2020, **139**, 162.
47. P. Ramasami and T. A. Ford, *J. Mol. Model.*, 2015, **21**, 35.
48. M. D. Esrafili and F. Mohammadian-Sabet, *Chem. Phys. Lett.*, 2015, **628**, 71.

49. M. D. Esrafili and H. Akhgarpour, *Mol. Phys.*, 2016, **114**, 1847.
50. J. E. Del Bene, I. Alkorta and J. Elguero, *Chem. Phys. Lett.*, 2019, **730**, 466.
51. P. Ramasami and T. A. Ford, *J. Mol. Struct.*, 2014, **1072**, 28.
52. F. T. Burling and B. M. Goldstein, *J. Am. Chem. Soc.*, 1992, **114**, 2313.
53. Q. Z. Li, R. Li, P. Guo, H. Li, W. Z. Li and J. B. Cheng, *Comput. Theor. Chem.*, 2012, **980**, 56.
54. T. Ziegler and A. Rauk, *Inorg. Chem.*, 1979, **18**, 1558.
55. L. Mo, Y. Zeng, X. Li, X. Zhang and L. Meng, *J. Mol. Model.*, 2016, **22**, 167.
56. S. Scheiner, *J. Phys. Chem. A*, 2015, **119**, 9189.
57. Y. Fang, A. Y. Li and F. Y. Ma, *J. Mol. Model.*, 2015, **21**, 61.
58. Y. Geboes, F. De Vleeschouwer, F. De Proft and W. A. Herrebout, *Chem. – Eur. J.*, 2017, **23**, 17384.
59. V. Previtali, G. Sánchez-Sanz and C. Trujillo, *ChemPhysChem*, 2019, **20**, 3186.
60. C. Bleiholder, D. B. Werz, H. Köppel and R. Gleiter, *J. Am. Chem. Soc.*, 2006, **128**, 2666.
61. R. Shukla and D. Chopra, *J. Phys. Chem. B*, 2015, **119**, 14857.
62. B. Galmés, A. Juan-Bals, A. Frontera and G. Resnati, *Chem. – Eur. J.*, 2020, **26**, 4599.
63. J. S. Murray, P. Lane and P. Politzer, *Int. J. Quantum Chem.*, 2008, **108**, 2770.
64. S. Scheiner, *J. Chem. Phys.*, 2011, **134**, 164313.
65. U. Adhikari and S. Scheiner, *Chem. Phys. Lett.*, 2011, **514**, 36.
66. U. Adhikari and S. Scheiner, *J. Phys. Chem. A*, 2012, **116**, 3487.
67. U. Adhikari and S. Scheiner, *Chem. Phys. Lett.*, 2012, **532**, 31.
68. S. Scheiner, *CrystEngComm*, 2013, **15**, 3119.
69. S. Scheiner, *Int. J. Quantum Chem.*, 2013, **113**, 1609.
70. M. D. Esrafili and F. Mohammadian-Sabet, *Chem. Phys. Lett.*, 2016, **645**, 32.
71. I. Alkorta, J. Elguero and J. E. Del Bene, *ChemPhysChem*, 2018, **19**, 1886.
72. L. M. Azofra, I. Alkorta and S. Scheiner, *J. Phys. Chem. A*, 2015, **119**, 535.
73. M. D. Esrafili and R. Nurazar, *Mol. Phys.*, 2016, **114**, 276.
74. U. Adhikari and S. Scheiner, *J. Phys. Chem. A*, 2014, **118**, 3183.
75. W. Wang, B. Ji and Y. Zhang, *J. Phys. Chem. A*, 2009, **113**, 8132.
76. X. Guo, Q. Li, B. Xiao, X. Yang, W. Li and J. Cheng, *RSC Adv.*, 2015, **5**, 52667.
77. D. Quiñonero, *Molecules*, 2015, **20**, 11632.
78. M. D. Esrafili, F. Mohammadian-Sabet and M. M. Baneshi, *Int. J. Quantum Chem.*, 2015, **115**, 1580.
79. L. Lu, Y. Lu, Z. Zhu and H. Liu, *J. Mol. Model.*, 2020, **26**, 16.
80. E. Navarro-García, B. Galmés, M. D. Velasco, A. Frontera and A. Caballero, *Chem. – Eur. J.*, 2020, **26**, 4706.
81. P. Wonner, L. Vogel, F. Kniep and S. M. Huber, *Chem. – Eur. J.*, 2017, **23**, 16972.
82. N. Tarannam, M. H. H. Voelkel, S. M. Huber and S. Kozuch, *J. Org. Chem.*, 2022, **87**, 1661.

83. A. Casula, P. Begines, A. Bettoschi, J. G. Fernandez-Bolaños, F. Isaia, V. Lippolis, Ó. López, G. Picci, M. Andrea Scorciapino and C. Caltagirone, *Chem. Commun.*, 2017, **53**, 11869.
84. M. Michalczyk, W. Zierkiewicz, R. Wysokiński and S. Scheiner, *Molecules*, 2019, **24**, 3329.
85. M. D. Esrafili and N. Mohammadirad, *J. Mol. Model.*, 2015, **21**, 176.
86. R. Shukla and D. Chopra, *Phys. Chem. Chem. Phys.*, 2016, **18**, 13820.
87. X. Li and A. Y. Li, *Chem. Phys. Lett.*, 2016, **659**, 126.
88. W. Zierkiewicz, J. Fanfrlík, M. Michalczyk, D. Michalska and P. Hobza, *Chem. Phys.*, 2018, **500**, 37.
89. X. Guo, Y. W. Liu, Q. Z. Li, W. Z. Li and J. B. Cheng, *Chem. Phys. Lett.*, 2015, **620**, 7.
90. W. Dong, Q. Li and S. Scheiner, *Molecules*, 2018, **23**, 1.
91. H. A. Samimi, M. D. Esrafili, F. Mohammadian-Sabet and H. Haddadi, *Mol. Phys.*, 2015, **113**, 1442.
92. M. D. Esrafili and S. Asadollahi, *J. Sulfur Chem.*, 2017, **38**, 83.
93. M. D. Esrafili and M. Vakili, *Mol. Phys.*, 2014, **112**, 2746.
94. Q. Zhao, *J. Mol. Model.*, 2014, **20**, 2458.
95. M. D. Esrafili, M. Vakili and M. Solimannejad, *Mol. Phys.*, 2014, **112**, 2078.
96. M. D. Esrafili, P. Mousavian and F. Mohammadian-Sabet, *Mol. Phys.*, 2019, **117**, 58.
97. M. D. Esrafili and N. Mohammadirad, *Mol. Phys.*, 2015, **113**, 3282.
98. A. Bauzá, D. Quiñonero, P. M. Deyà and A. Frontera, *CrystEngComm*, 2013, **15**, 3137.
99. J. George, V. L. Deringer and R. Dronskowski, *J. Phys. Chem. A*, 2014, **118**, 3193.
100. M. D. Esrafili and F. Mohammadian-Sabet, *Struct. Chem.*, 2015, **26**, 199.
101. M. D. Esrafili, S. Asadollahi and Y. D. Shahamat, *Can. J. Chem.*, 2016, **94**, 567.
102. M. D. Esrafili and A. Sadr-Mousavi, *Mol. Phys.*, 2017, **115**, 1713.

CHAPTER 18

Supramolecular Chemistry via *Chalcogen Bonding Interactions*

DEBORAH ROMITO,[a] PETER C. HO,[b] IGNACIO VARGAS-BACA*[b] AND DAVIDE BONIFAZI*[a]

[a] Institute of Organic Chemistry, Vienna University, Währinger Str. 38, 1090 Wien, Austria; [b] Department of Chemistry and Chemical Biology, McMaster University, 1280 Main St. W., Hamilton, Ontario L8S 4M1, Canada

*Emails: davide.bonifazi@univie.ac.at; vargasi@mcmaster.ca

18.1 Introduction

In recent years, chalcogen atoms have been increasingly employed in organic molecules for their ability to develop an "atypical" type of non-covalent attractive interaction with nucleophiles, which are now termed chalcogen bonds (ChBs).[1] Despite their electron lone pairs, chalcogens also embody a Lewis acidic role, analogous to the one observed for main group elements from groups 14–15 and 17–18 in the periodic table. More generally, these interactions are specific cases of secondary bonding, that have a distance longer than "primary" covalent bonds, but shorter than the sum of the van der Waals radii ($\sum r_{vdw}$) of the atoms involved.[2] The origin of this interaction resides in the anisotropic distribution of the electron density around covalently bonded atoms, which generates a region of positive electrostatic potential collinear but opposite to a σ-bond, and thus called a σ-hole.[3] This depletion of electrons leads to connections with areas with a higher electron

Chalcogen Chemistry: Fundamentals and Applications
Edited by Vito Lippolis, Claudio Santi, Eder J. Lenardão and Antonio L. Braga

Published by the Royal Society of Chemistry, www.rsc.org

density, such as lone-pair possessing atoms or anions. Since chalcogen atoms typically form two covalent bonds R–Ch, two σ-holes can be located opposite to these bonds (Figure 18.1a). Their magnitude is enhanced by increasing the electronegativity of the R substituents attached to Ch, either by increasing the polarizability of Ch itself.[4,5] However, several studies have proven that van der Waals dispersion forces and orbital mixing need to be considered[6,7] to fully rationalize the roots of ChBs. Specifically, a $n^2(A_2) \rightarrow \sigma^*(R_1\text{–}Ch)$ interaction takes place when R–Ch···A chalcogen bonds occur, where non-bonding electrons of electron-donating atom A (ChB acceptor) interact with the empty antibonding $\sigma^*_{(R_1-Ch)}$ molecular orbital located on the chalcogen atom Ch (ChB donor, Figure 18.1b).

This aspect supports the experimental evidence which states a strengthening of ChBs when the R–Ch···A angle becomes closer to 180°, thus maximizing the orbital overlapping. The combined effect of the three contributions allows us to anticipate that stronger ChBs will take place when heavier chalcogen atoms are employed in the interaction (Te > Se > S > O). By increasing the polarizability of the Ch atom, a synergic enhancement of the anisotropic charge distribution, and lowering of energy of the σ* molecular orbital is achieved, as well as a more favourable contribution from the dispersion forces.[8] The "exotic" nature of this non-covalent interaction led to a progressively significant investigation of ChBs as complementary interactions to the ubiquitous hydrogen bonding (HB) in several research fields.[9–13] Among the advantages in choosing chalcogen bonding to control the self-assembly, one can easily point out the bivalence of λ^2 and λ^4 chalcogens, whereas HBs are limited to a single interaction per atom, thus notably affecting the stability of the resulting self-assembled structure. Moreover, the strong orbital contribution to the bonding nature makes these interactions less sensitive to solvent medium, and tunable to bind to softer nucleophiles, while hydrogen bonding strictly relies on the electronegativity of the HB acceptor to be effective. Since its discovery, modest attention has been paid to the investigation of ChBs,[14,15] but at the turn of the millennium a large and growing volume of literature has focused on the potential of this interaction for controlling the structure and properties of matter.[16–21]

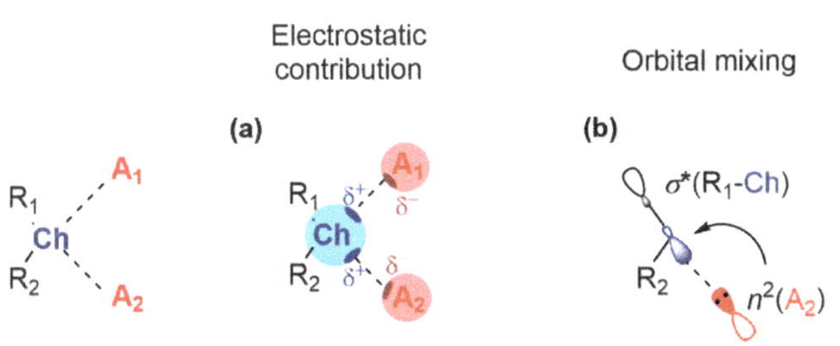

Figure 18.1 Schematic representation of chalcogen bonds and description of their nature by (a) electrostatic contribution[3] and (b) orbital mixing theory.[6] Ch = S, Se, Te; A is a chalcogen-bond acceptor.

Starting from a crystallographic survey in the 70s of intermolecular distances involving chalcogens, recent years have witnessed almost exclusive applications in the solid state, by using crystal engineering as a practical rational tool to build functional materials through stable multidimensional networks.[22,23] However, in the last few years incredible efforts have been carried out to explore the potentialities of ChBs in solution-based molecular recognition,[24–27] with remarkable results achieved in catalysis and sensing applications.[28] This chapter is, therefore, organized in two sections dealing with the most relevant studies which exploit ChBs for effective application in the solid state and in solution.

18.2 Chalcogen Bonding in the Solid State

ChBs have been chiefly employed to control the supramolecular arrangement in the solid state, aiming at the preparation of materials with diverse functions. This is possible by carefully choosing the molecular geometry and thus the resulting intermolecular recognition motif, which will guide the formation of precise supramolecular assemblies that have specific properties.[29] By looking into the Cambridge Structural Database (CSD), the molecular scaffold most largely investigated as a milestone for solid-state application is based on 1,2,5-chalcogenadiazole rings. Assembling through antiparallel Ch···N intermolecular contacts, four-membered $[Ch\cdots N]_2$ squares are created, which constitute a small but effective example of supramolecular synthons.[30,31] The lower polarizability of lighter S- and Se-containing materials prevent their association into infinite ribbon chains, which instead occurs with the unsubstituted Te-congener **10_1**.[32] The absence of steric constraints determines a persistent association through multiple strong Te···N bonds ($d_{Te\cdots N} \sim 2.7$ Å), being significantly shorter than the $\sum r_{vdw}$ of the atoms involved (3.61 Å). Adding small substituents to the molecular backbone does not dramatically influence the solid-state arrangement (as observed for the crystal structure of **10_2**),[33] but it determines a lengthening of the ChBs. On the contrary, the presence of sterically-hindered substituents closer to the $[Ch\cdots N]_2$ synthon (as for **10_3**) disrupts the supramolecular polymeric architecture, in favour of discrete dimeric species, still marked by short Te···N intermolecular bonds ($d_{Te\cdots N} = 2.696$ Å, in blue in Figure 18.2c). The molecular packing extends through Ch–halogen bonds ($d_{Te\cdots Br} = 3.683$ Å, in ochre in Figure 18.2c), involving the second σ-hole of Te atoms and one of Br atoms located on neighbouring molecules, as ChB acceptors.

Notably, Vargas-Baca took advantage of the synergistic robustness of multiple ChBs, along with the distorting steric effects arising by the ketone in **10_2**, to grow non-centrosymmetric crystals of **10_2** having second-order non-linear optical (NLO) properties.[33] The endurance of the $[Ch\cdots N]_2$ supramolecular synthon also allowed us to accomplish the first permanently porous chalcogen-bonded organic framework through the self-assembly of a triptycene tris-substituted with benzoselenadiazoles **10_4**.[34] Functionalizing triptycene scaffolds with chalcogen-containing rings previously proved to be

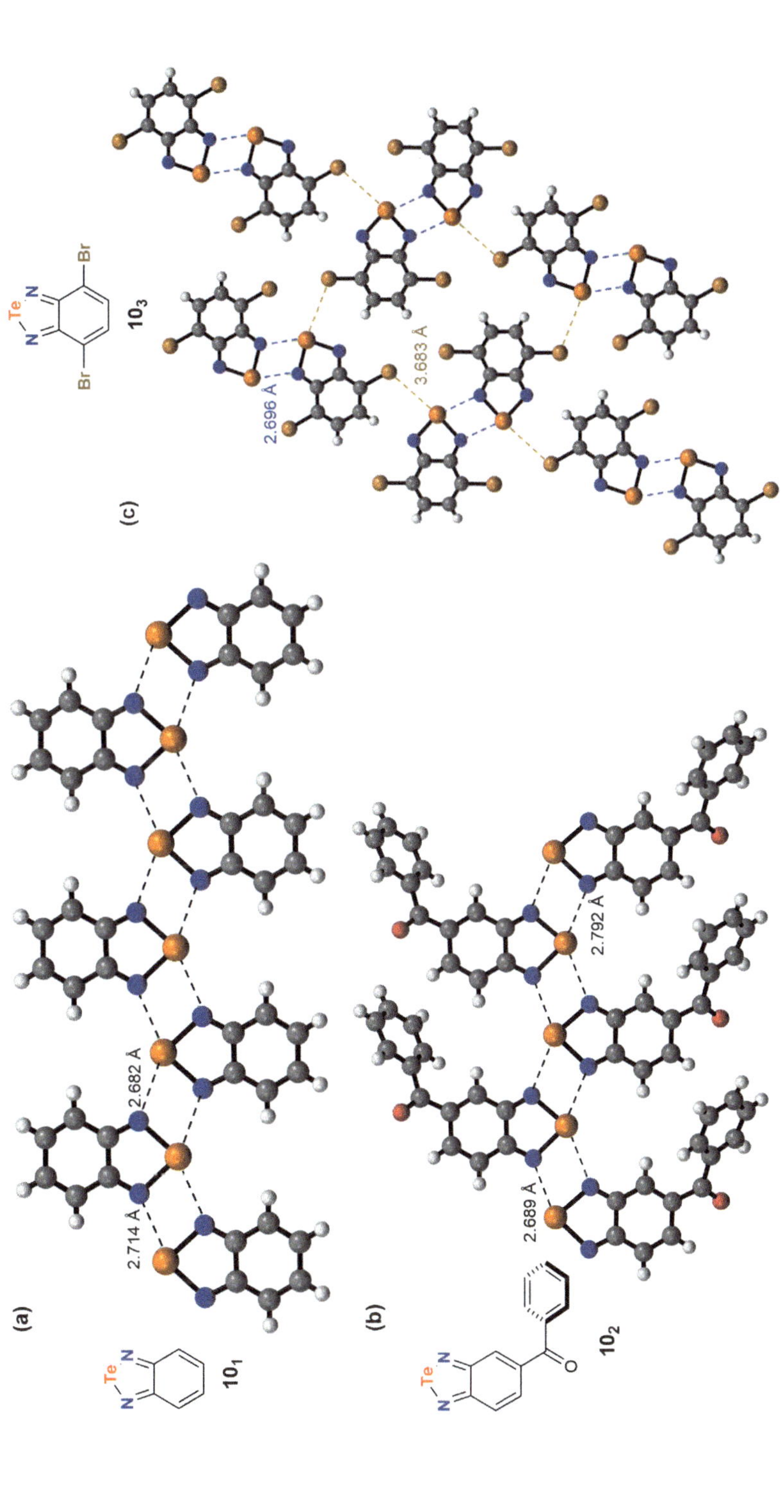

Figure 18.2 Infinite ribbon chains achieved with unsubstituted (a) $\mathbf{10_1}$ or bearing small substituent (b) $\mathbf{10_2}$; on the right: dimeric association of (c) $\mathbf{10_3}$ and solid-state arrangement through additional ChBs. Adapted from ref. 32 and 33 with permission from American Chemical Society, Copyright 2006 and 2010.

effective in the growth of highly-ordered crystal structures, with molecules of fullerene C_{60} fully incapsulated in supramolecular cages, held together by S···N ChBs.[35] The very recent study of Wang and co-workers described the formation of three-dimensional (3D) porous organic cages by means of combined Se···N ChBs and Se···π contacts. The non-solvated crystal structure is marked by one benzoselenadiazole ring seeing both Se σ-holes interacting with two neighbouring molecules through a symmetrical ChB ($d_{Se_2\cdots N_1} = 3.328$ Å; N_3–$Se_2\cdots N_1 = 157°$). The other two benzoselenadiazole units participate in the self-assembly as ChB acceptors, thus employing N atoms (Figure 18.3a). The molecular packing further extends in a 3D 4-connected $\{4^2{\cdot}8^4\}$ *lvt* topological framework, with each molecule acting in a 4-connected mode (outlined in red in Figure 18.3b). A second porous cage was also achieved in the solvated crystal structure incorporating THF molecules **10_4·THF**. Similarly to the case previously described, one benzoselenadiazole unit is simultaneously engaged as the ChB donor by means of both Se σ-holes ($d_{Se_1\cdots N_7} = 2.934$ Å, $d_{Se_1\cdots N_{10}} = 3.237$ Å), as well as ChB acceptor involving one N atom ($d_{N_2\cdots Se_4} = 3.269$ Å). Instead, another Se-containing moiety participates in the crystal structure by connecting to

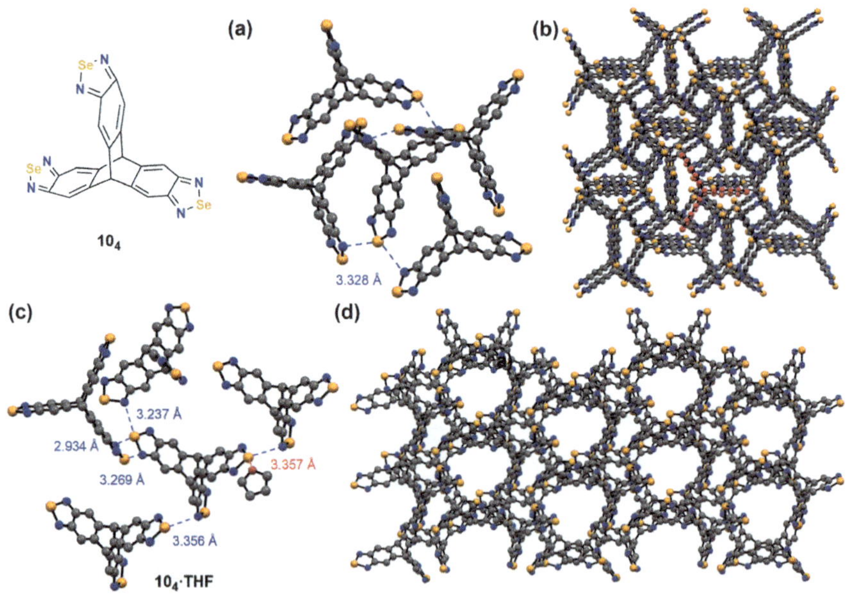

Figure 18.3 Crystal structure of tris-substituted triptycene **10_4** depicting (a) the intermolecular ChBs developed by each single molecule and (b) the 3D supramolecular architecture seen along the *c* axis, with one molecule highlighted in red as the 4-connected node; (c) intermolecular ChBs in the solvated structure of **10_4·THF**, with Se···N bonds indicated in blue and Se···O contacts in red; (d) resulting chalcogen-bonded porous organic cage, with THF molecules omitted to better highlight the porosity. Distances are expressed in Å, H atoms are omitted for clarity. Adapted from ref. 34 with permission from American Chemical Society, Copyright 2021.

two different ChB acceptors, one being a N atom of a neighbouring molecule ($d_{Se_2\cdots N_6}$ = 3.356 Å), the other being the O atom of THF ($d_{Se_1\cdots O_2}$ = 3.357 Å, Figure 18.3c). The combination of these interactions led to the formation of a supramolecular 3D porous tubular architecture, filled by THF molecules (omitted in Figure 18.3d to better show the porosity of the material). A very recent work further investigated the stability of this chalcogen-bonded porous framework, submitting it to solvent-exchange conditions. The persistent $[Se\cdots N]_2$ Ch-bonded motif allows the achievement of permanent porosity, thus promoting ChBs as effective supramolecular tools for building diverse porous framework materials.[36]

The occurrence of the persistent $[Ch\cdots N]_2$ supramolecular synthon can also be observed in salts of benzotelluradiazoles, prepared by employing KC_8 as the reducing agent in the presence of 18-crown-6 ether.[37] Differently from analogous S and Se congeners, whose anions were previously isolated as monomeric salts of $[K(THF)]^+$ and $[K(18\text{-crown-6})]^+$,[38,39] the crystal structure described in Zibarev's work displays the trimeric anion $[(\mathbf{10_1})_3]^{2-}$. This exotic species is constituted of a neutral **10₁** molecule, which is chalcogen-bonded to two benzotelluradiazalidyl radical anions through two pairs of Te···N interactions. Notably, anion $[(\mathbf{10_1})_3]^{2-}$ is diamagnetic and its covalent bonds are significantly longer than those of neutral $\mathbf{10_1}$ in the solid state ($d_{N\text{–}Te}$ = 2.019–2.096 Å *vs.* 1.987–2.014 Å). Three of the non-covalent Te···N contacts are noteworthy for being among the shortest ChBs reported in the literature so far ($d_{Te\cdots N}$ = 2.250–2.385 Å ≪ $\sum r_{vdw}$ = 3.61 Å, highlighted in blue in Figure 18.4). However, the fourth Te···N bond is remarkably longer ($d_{Te_2\cdots N_{12}}$ = 2.904 Å, N_{22}–$Te_2\cdots N_{12}$ = 141°), and the consequent asymmetry of strength among the ChBs determines a substantial deviation of $[(\mathbf{10_1})_3]^{2-}$ from planarity. Notably, asymmetry between chalcogen-bonded intermolecular contacts is hardly observed when $[Te\cdots N]_2$ recognition occurs, but it was already reported for salts of benzoselenadiazoles co-crystallized with a *N*-methyl cationic analogue.[40] Given the persistence of Te···N interactions, the 18-crown-6 ether stays out of the plane formed by the dianion species.

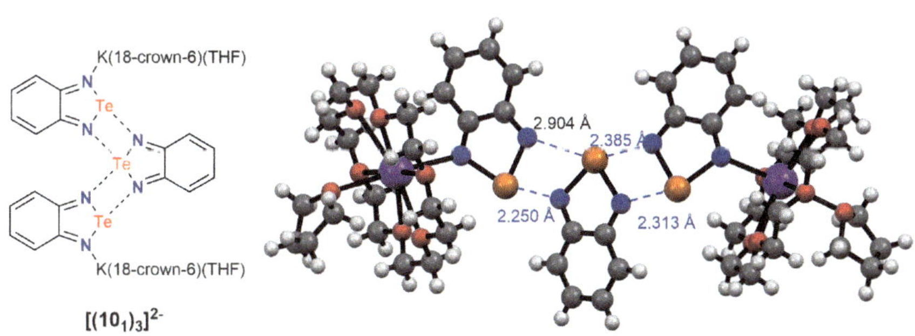

Figure 18.4 Schematic representation of dianion $[(\mathbf{10_1})_3]^{2-}$ and its Te···N ChBs, with the shortest ones highlighted in blue. Distances are expressed in Å. Adapted from ref. 37 with permission from the Royal Society of Chemistry.

In sharp contrast, a recent work of the same author used cyclic polyethers to govern the solid-state arrangement of 1,2,5-chalcogenadiazoles.[41] By analyzing the supramolecular structures of the co-crystals there described, the occurrence of several chalcogen bonds is depicted, as estimated by density functional theory (DFT) calculations and natural bond orbital (NBO) analysis. The theoretical results suggested the ambivalent Lewis nature of these organochalcogenides. Aside from a charge transfer (CT) from the lone pair located on an O atom into the $\sigma^*_{(Ch-N)}$ molecular orbital, the presence of a back CT was calculated, further proving the ability of 1,2,5-chalcogenadiazoles towards simultaneous σ-donor and σ-acceptor interactions.[42] Weaker contacts occur in the crystal structure of S-containing congener **10_5**, which displayed a S···O distance almost equal to the $\sum r_{vdw}$ of the atoms involved ($d_{S_1 \cdots O_2} = 3.309$ Å *vs.* $\sum r_{vdw} = 3.45$ Å, Figure 18.5a). As expected, the strongest Ch···O chalcogen bonds are observed when a more polarizable Te atom is present, **10_6**. Remarkably, the two resulting Te···O intermolecular distances are ~0.6 Å shorter than $\sum r_{vdw}$ of Te and O (Figure 18.5b).

Co-crystallization has proved to be an effective strategy to prepare diverse materials and thoroughly study the nature of chalcogen bonds, given the wide range of information which can be collected in relation to their directionality and strength with several species. One of the most significant contributions to the field was provided by Fourmigué and co-workers, firstly investigating intermolecular ChBs developed by organic selenocyanates.[43,44] With an amphoteric nature as ChBs donors (through Se atom) and acceptors (by means of N lone pair), these molecules exhibit a recurrent supramolecular motif, based on the regular repetition of highly directional Se···N≡C interactions in the solid state, leading to the formation of one-dimensional (1D) structures. *o*-Substituted bis(selenocyanomethyl) **10_7** and **10_8** were further employed by the same author as chelating agents towards halide anions in solution, as well as in the solid state.[44] It is noteworthy mentioning that this example constitutes the first published Se···X$^-$ non-covalent contact

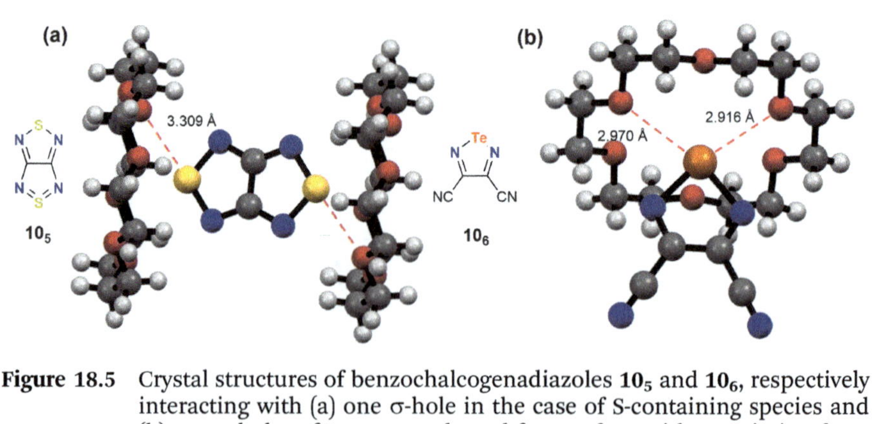

Figure 18.5 Crystal structures of benzochalcogenadiazoles **10_5** and **10_6**, respectively interacting with (a) one σ-hole in the case of S-containing species and (b) two σ-holes of Te atom. Adapted from ref. 41 with permission from American Chemical Society, Copyright 2020.

that employs a Se atom in its neutral state. Co-crystalline salts grown with Ph_4PCl, Ph_4PBr, and Bu_4NCl reveal both the presence of μ_2- and μ_4-halide structures, formed through multiple highly-directional ChBs. Specifically, Cl^- anions are chelated by a single molecule of **10_7** by utilizing all the SeCN moieties, which dispose following an *anti*-orientation among themselves, to maximize the degree of directionality with the ChB acceptor ($d_{Se\cdots Cl^-}=3.12–3.21$ Å $\ll \sum r_{vdw}=3.65$ Å; C–Se$\cdots Cl^- \sim 174°$, Figure 18.6a). In contrast, when Br^- anions are engaged, a double chelation is still observed with **10_7**, though involving two Se atoms belonging to two different molecules. Likely, the double chelation recognition governs the growth of a non-covalent polymeric architecture in the solid state ($d_{Se\cdots Br^-}=3.23–3.28$ Å $\ll \sum r_{vdw}=3.75$ Å; C–Se$\cdots Br^- \sim 176°$, Figure 18.6b). When using molecule **10_8** as the chelating agent, the co-crystals grow with two Se-containing molecules interacting with the same anion, which displays a μ_4-halide coordination. A tetrahedral conformation in observed with Cl^-, having Se$\cdots Cl^-$ ChBs slightly longer than the previous case described ($d_{Se\cdots Cl^-}=3.17–3.47$ Å; C–Se$\cdots Cl^-=168°$ for the longest bond, ~173° for the others, Figure 18.6c).

Instead, chelation of Br^- occurs in a so-called seesaw geometry, which is marked by a lengthening of the intermolecular Se$\cdots Br^-$ contacts ($d_{Se\cdots Br^-}=3.23–3.55$ Å $\ll \sum r_{vdw}=3.75$ Å; C–Se$\cdots Br^-$ among 173 and 178°, Figure 18.6d). To further evaluate how the Se$\cdots$N≡C could be tuned, the co-crystallization of a novel selenocyanate derivative with several ditopic Lewis bases was performed.[45] Mixing two equivalents of 1,3,5-tris-(selenocyanatomethyl)-2,4,6-trimethylbenzene **10_9** with one equivalent of tetramethylpyrazine (**TMP**), benzoquinone (**BQ**) or *p*-dinitrobenzene (**DNB**), a novel recurrent oligomeric system unites each co-crystal solid-state organization.

Notably, two out of three SeCN groups are involved in a Se$\cdots$N≡C bond with neighbouring molecules ($d_{Se\cdots N}=2.99–3.32$ Å $<\sum r_{vdw}=3.45$ Å), whereas the third one is engaged with one of the Lewis-base centres of ***p*DNB, TMP,** or **BQ**. As a result, every Lewis base acts as a bridging ChB acceptor among two Se-containing molecules (Figure 18.7), with the shortest intermolecular ChBs observed when **BQ** is used ($d_{Se\cdots O}=2.966$ Å $<\sum r_{vdw}=3.55$ Å, Figure 18.7c). Building on these results, the authors reported a very recent study focusing on the chelating role of chiral bis(selenocyanates) with tetraphenylphosphonium halides.[46] All crystal structures of ***anti*-10_{10}** as the ChB donor display a 1:1 stoichiometry with the halide, being bischelated by each Se atom through intermolecular Se$\cdots X^-$ distances significantly larger than those of the achiral species previously mentioned. This can be justified by the disfavouring role played by the methyl substituents on the benzylic positions, playing an effect on anti-isomers both as electron-donating and sterically hindering moiety ($d_{Se\cdots Cl^-}=3.26–3.30$ Å <3.65 Å; $d_{Se\cdots Br^-}=3.38–3.44$ Å <3.75 Å, Figure 18.8a and b). The chelating effect is even lower with I^-, because it is too big to satisfy the geometrical constraints of the *anti*-conformation ($d_{Se\cdots I^-}=3.60–3.67$ Å $<\sum r_{vdw}=3.88$ Å), thus determining a substantial deviation from 180° for one of the Se$\cdots$I angle bonds ($C_{12}–Se_2\cdots I_1=168°$, Figure 18.8c). Instead, a different outcome

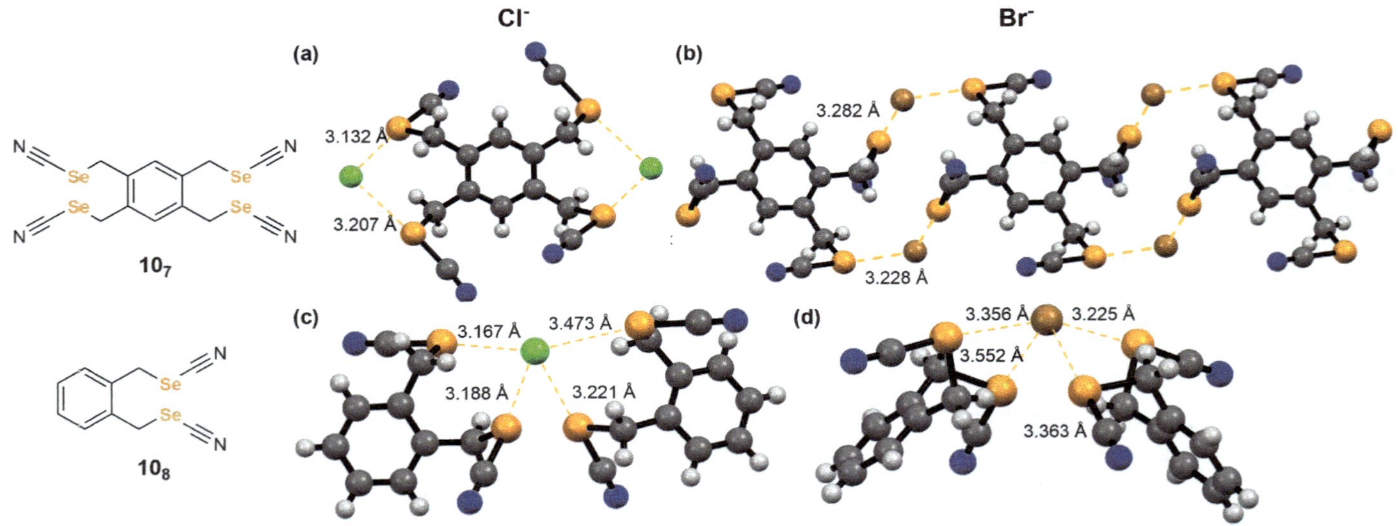

Figure 18.6 Co-crystals of μ_2-chelated (a) $\mathbf{10_7}$:Cl^- and (b) $\mathbf{10_7}$:Br^-, as well as μ_4-chelated (c) $\mathbf{10_8}$:Cl^- and (d) $\mathbf{10_8}$:Br^-. All distances are expressed in Å. All the respective cations are omitted for clarity. Adapted from ref. 44 with permission from American Chemical Society, Copyright 2019.

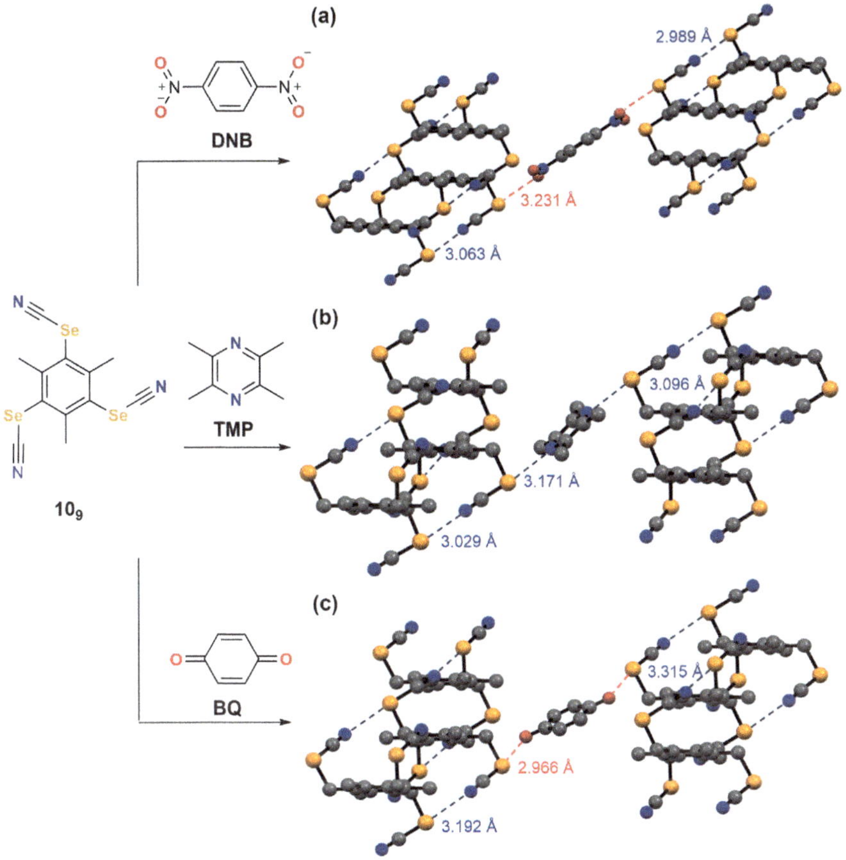

Figure 18.7 Schematic representation of ChB donor $\mathbf{10_9}$ and the crystal structures achieved when co-crystallized with ditopic Lewis bases (a) **DNB**, (b) **TMP**, and (c) **BQ**. Se···N ChBs are indicated in blue, Se···O in red, all distances are in Å. Adapted from ref. 45 with permission from The International Union of Crystallography, Copyright 2019.

is obtained when ***syn*-$\mathbf{10_{10}}$** is used as a chelating agent, starting with a 2 : 1 stoichiometry which determines a μ_4 square-planar environment for Cl^-. Moreover, the Se···Cl^- bonds are impressively short, even shorter than those reported for its achiral analogue ($d_{Se\cdots Cl^-} = 3.11–3.16$ Å $\ll 3.65$ Å; C–Se···Cl^- ~ 176°, Figure 18.8d).

Another study supporting this concept comes from a very recent publication, dealing with icosahedral carborane derivatives which are functionalized with –SeMe, –TeMe, and –I moieties to act as ChB or XB donors in the interaction with halide anions.[47] Given its 3D aromaticity, the substituents located on the more electronegative carbon vertices of carboranes undergo enhanced electron-withdrawing effects. This characteristic would strengthen the electron-positive character of heavy chalcogen and halogen atoms. As a result, the co-crystals of Se and Te congeners with Bu_4NX salts

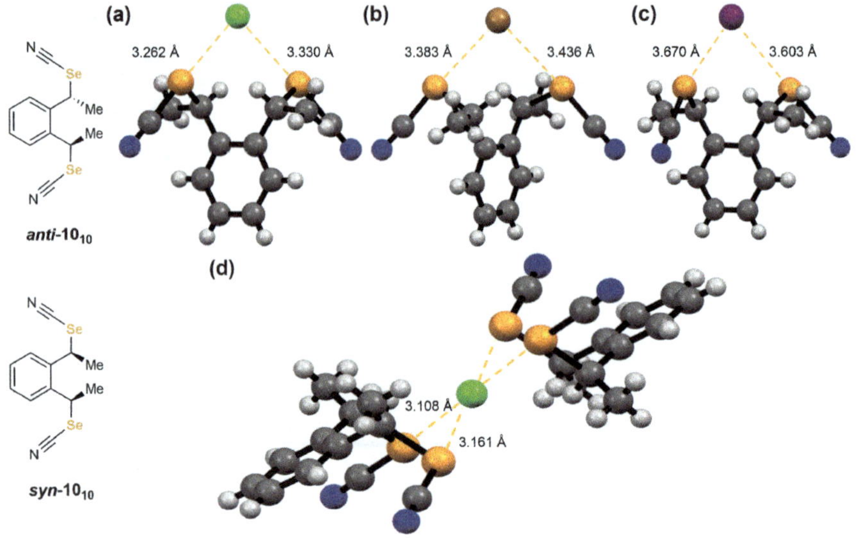

Figure 18.8 Crystal structures of the halide adducts of ***anti*-10$_{10}$** with (a) Cl^-, (b) Br^-, and (c) I^-, as well as (d) anionic moiety in ***syn*-10$_{10}$** with Cl^-. Distances are expressed in Å, cationic counterions are omitted for clarity. Adapted from ref. 46 with permission from the Royal Society of Chemistry.

display highly directional and short ChBs. Specifically, the salt **10$_{11_Se}$**:Br^- shows one of two Se atoms involved as a ChB donor with Br^-, by means of a non-covalent ChB of 3.681 Å ($<\sum r_{vdw} = 3.86$ Å), with a C–Se···Br^- angle of 173° (Figure 18.9a). Moving to heavier Te-containing co-crystals of **10$_{11_Te}$**:Cl^-, both chalcogen-containing moieties take part in the solid-state arrangement by bridging chloride anions, forming zigged-zagged 1D supramolecular chains (Figure 18.9b). Looking at the ChB distances in the molecular packing, the shortest Te···Cl bond is 3.120 Å long with an angle of C_{10}–Te_6···$Cl^- = 178°$, making this one of the strongest Te-based chalcogen-bonds reported so far.

Supporting the idea that the molecular scaffold embodies a crucial role in the activation of the σ-holes in chalcogen atoms, it is worth mentioning the example of alkyltelluroalkynyl derivatives published by Gleiter, which self-assemble in supramolecular nanotubes held by means of four Te···Te ChBs.[48,49] More recent examples describe the use of molecular synthons bearing electron-withdrawing groups, to activate the σ-holes along the C≡C–Ch bond.[50] Se-containing monomer **10$_{12}$** self-assembles into tetrahedral motifs through multiple Se···Se contacts, thus simultaneously employing each Se atom as a ChB donor and acceptor ($d_{Se\cdots Se} = 3.65$–3.73 Å $< \sum r_{vdw} = 3.8$ Å; C–Se···Se ~ 166°, Figure 18.10a). Notably, a tubular square-lattice network is formed in the solid state, in which six helical chains along the *c* axis of the crystal structure are present (represented in green in Figure 18.10b and c). At the centre of each nanotube, one can notice the presence of a second channel formed by two molecules arranged into a

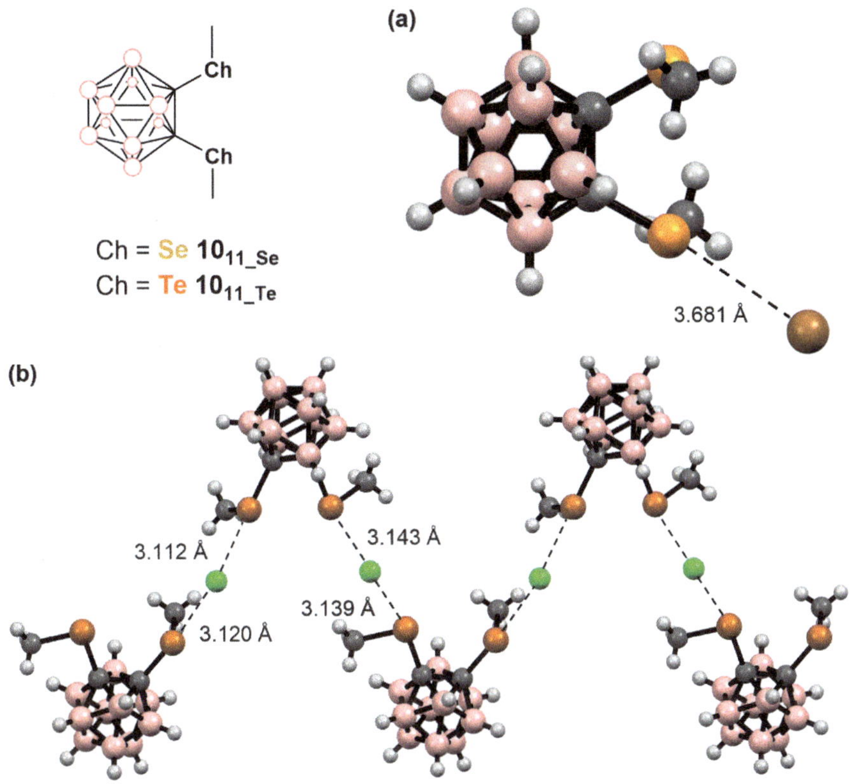

Figure 18.9 Schematic representation of icosahedral carborane derivative (pink spheres stand for B atoms) and solid-state arrangement of salts of Se-containing (a) $\mathbf{10_{11\_Se}}$ and Te-congener (b) $\mathbf{10_{11\_Te}}$, respectively. Bu_4N^+ cations are omitted for clarity.[47]

double helix through Se···Se bonds ($d_{Se\cdots Se} = 3.504$ Å for helix depicted in red, 3.471 Å for the helix highlighted in blue in Figure 18.10c).

Co-crystallizing similar Se- and Te-congeners with linear ditopic Lewis bases as 4,4′-bipyridine or 1,4-di(4-pyridyl)piperazine **bpi-pip**, afforded supramolecular polymers in the solid state, marked by short and directional ChB contacts. Instead, using the latter Lewis base with Te-congener $\mathbf{10_{13}}$ provided a significant change in the resulting molecular packing, as proved in the very recent work of Fourmiguè and co-workers.[51] The chosen ChB-donor monomer was opportunely designed to form a discrete and heteromolecular supramolecular architecture in the solid state. Specifically, two Te-containing sites were positioned facing the same side of the molecular backbone to prevent the development of the supramolecular organization into infinite chains. Instead, the building of supramolecular discrete "rectangles" with a series of ditopic Lewis bases was favoured (Figure 18.11). Three out of the four Lewis bases chosen for this study successfully achieved the predicted rectangular motif, driven by four intermolecular Te···N ChBs, being all substantially shorter than

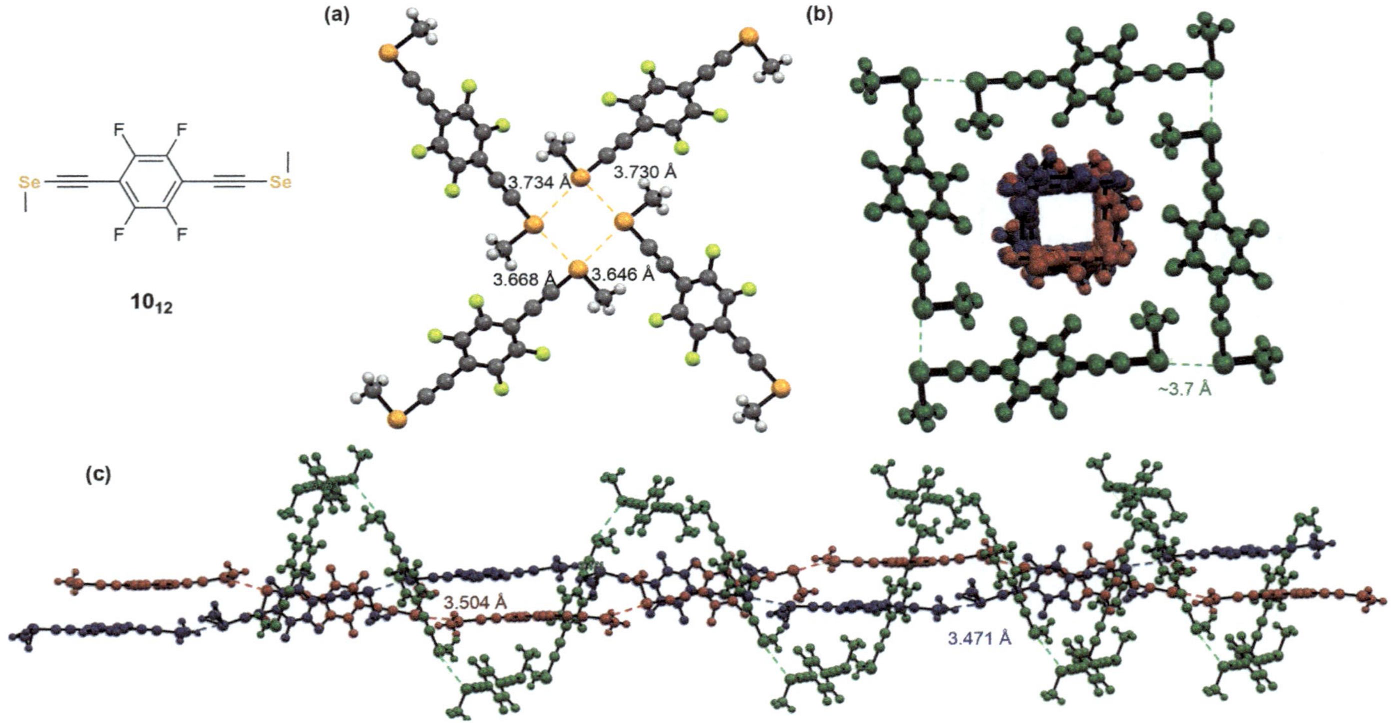

Figure 18.10 Supramolecular Se-containing nanotubes developed by $\mathbf{10_{12}}$ through (a) non-covalent tetrahedrals, generating (b) tubular square-lattice network, containing (c) an infinite double helix, respectively represented in blue and red. Distances are expressed in Å.[50]

$\sum r_{vdw}$ (3.61 Å) and highly directional (with an average value of C≡C–Te···N angle among 172 and 175°). It is noteworthy mentioning that two supramolecular rectangles are formed when **bpi-pip** is employed, since it crystallizes in two different conformations in the solid state (Figure 18.11c).

Similar to what is observed with these derivatives, even a slight but precise modification in the molecular backbone can significantly affect the resulting supramolecular organization. This concept was effectively proved by Bonifazi and co-workers, who first discussed the solid-state organization of benzochalcogenazoles in wiring-like supramolecular polymers.[52–54] By replacing the C atom proximal to the Ch-containing ring with a N atom led to the chalcogenazolo[5,4-β]pyridines (CGPs), with ChB donor and acceptor units located as second nearest neighbouring (SNN) atoms between themselves.[55] Designing the molecular scaffold in this way determines their supramolecular self-assembly into dimers through two frontal Ch···N interactions. To evaluate the persistence of the association, several substituents were introduced in the 2-position, either depicting electron-withdrawing or steric effects, and both Se- and Te-congeners were prepared (Figure 18.12). The latter ones persistently self-assemble into dimeric species, independently of the nature of the given functionalization, displaying a shortened Ch···N distance in relation to a highly electronegative substituent in 2-position ($d_{Te\cdots N}$ = 3.013 Å for **10_{15}-Pyr** and ~2.92 Å for **10_{15}-CF$_3$**). A completely different outcome is instead observed when the Se atom is employed as a ChB donor. Given its lower polarizability, the choice of the 2-substitution dramatically influences the solid-state association of Se-congeners. If the molecular scaffold does not bear a highly electronegative functionalization, dimerization would occur only through double H···N HBs, as displayed by 4-pyridyl substituted **10_{14}-Pyr** (Figure 18.12a). Increasing the electronegativity of the 2-substituent, a weak Te···N ChB involving the *N*-azole ring can be achieved ($d_{Te\cdots N}$ = 3.318 Å < 3.61 Å), supported by a H···N HB (Figure 18.12b). Only the use of the CF_3 moiety allowed the targeted dimerization through double Se···N contacts ($d_{Se\cdots N}$ = 2.963 Å < $\sum r_{vdw}$ = 3.45 Å, Figure 18.12c).

The careful choice of the substituent in the 2-position, respectively incorporating halogen bonding (XB) donor or acceptor units, led to the growth of supramolecular heteromolecular polymers in the solid state by co-crystallization with suitable complementary molecular modules.[56] Further studies from the same research group proved that the addition of a second functional group in the 5-position of the CGP scaffold, and tangling it with that borne in 2-position, affords supramolecular homomolecular ribbons through multiple combined ChBs and HBs.[57]

18.3 Chalcogen Bonding in Solution

Early investigations of intramolecular Se···N ChBs in solution were firstly reported by Tomoda in the 90s. The study examined 2-selenobenzylamines (**10_{16_X}**, Scheme 18.1).[58,59] Variable-temperature (VT) ^{1}H NMR spectra in CD_2Cl_2 and $CDCl_3$ showed that the strongest ChBs are formed by **10_{16_Cl}**, **10_{16_Br}** and

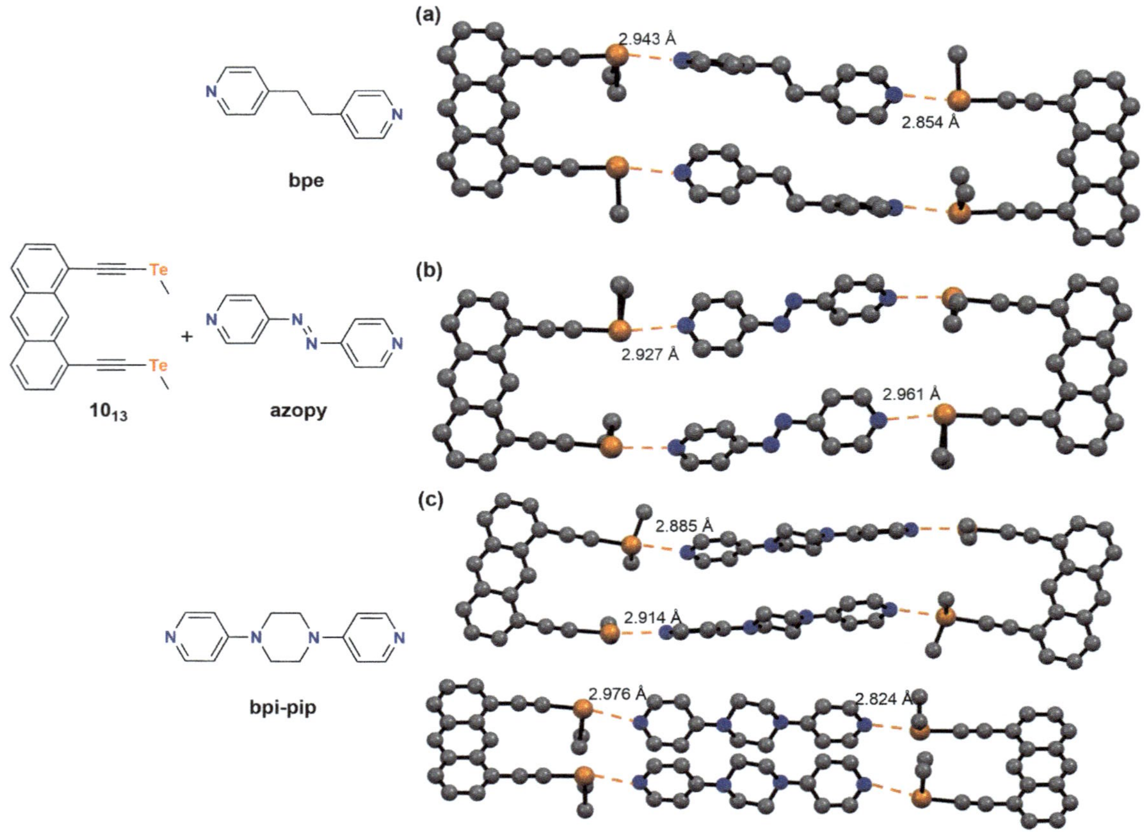

Figure 18.11 Supramolecular rectangles achieved by co-crystallization of $\mathbf{10_{13}}$ with (a) **bpe**, (d) **azopy** and (c) **bpi-pip**. Distances are expressed in Å, H atoms are omitted for clarity. Adapted from ref. 51 with permission from the Royal Society of Chemistry.

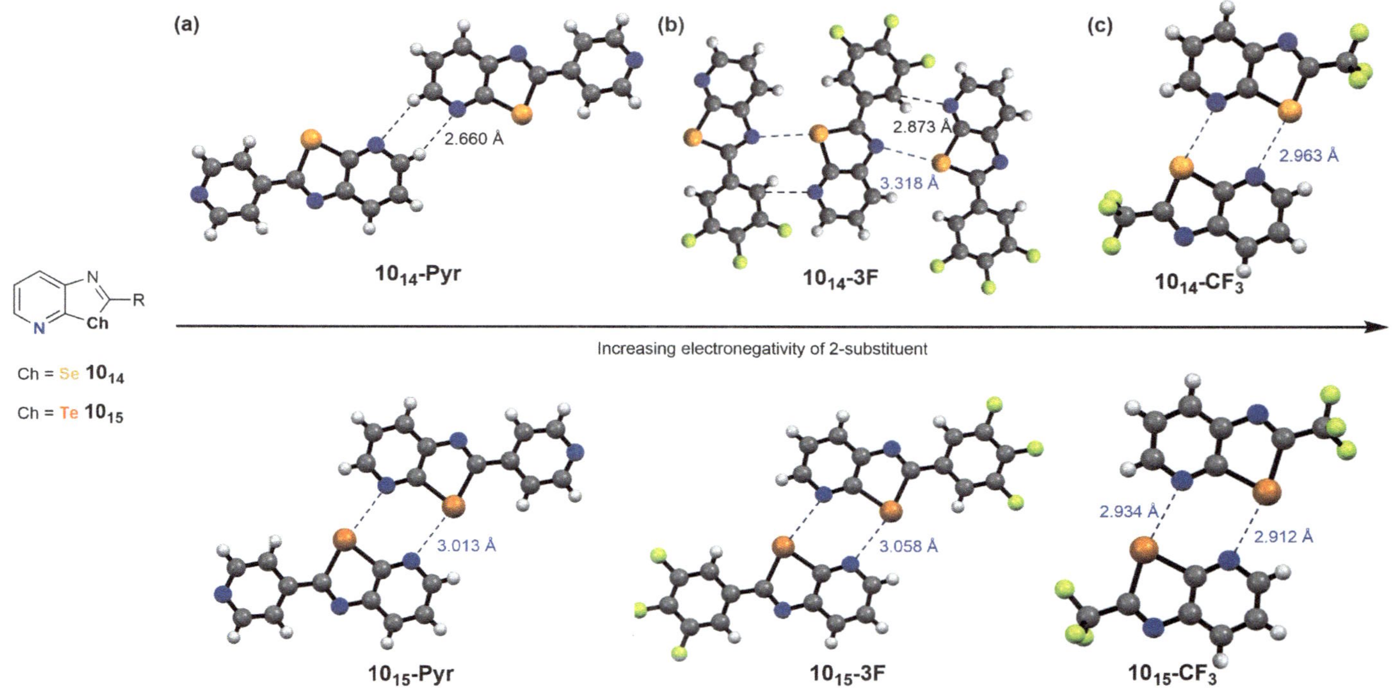

Figure 18.12 Crystal structures of CGP-derivatives (a) 4-pyridyl substituted **10_{14}-Pyr** and **10_{15}-Pyr**, (b) trifluorophenyl functionalized **10_{14}-3F** and **10_{15}-3F**, and (c) CF_3-containing **10_{14}-CF_3** and **10_{15}-CF_3**. Distances are expressed in Å. Adapted from ref. 55 with permission from John Wiley & Sons, Copyright © 2018 Wiley-VCH Verlag GmbH & Co. KGaA, Weinheim.

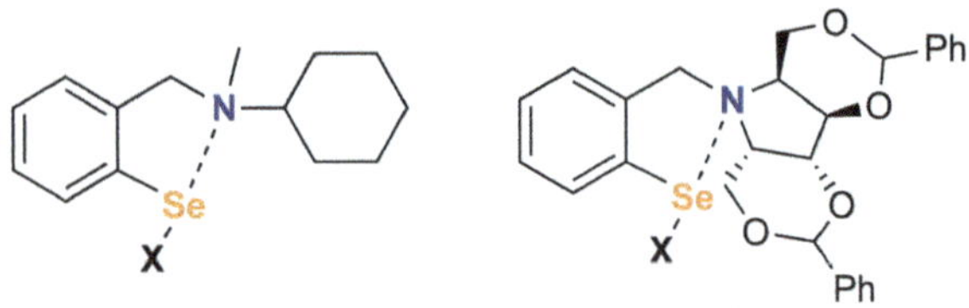

10$_{16}$ (X = Cl, Br, OAc, CN, SPh, Me) **10**$_{16\_Br^*}$

Scheme 18.1 Early examples of intramolecular Se···N ChBs studied in solution by Tomoda.[58–60]

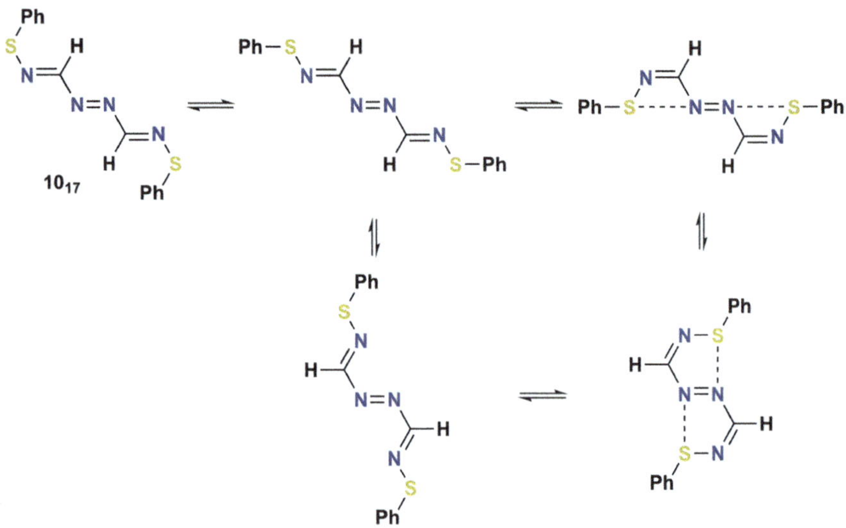

Scheme 18.2 Different chalcogen-bonded stabilized isomers observed in solution for **10**$_{17}$. Adapted from ref. 61 with permission from American Chemical Society, Copyright 1996.

10$_{16\_OAc}$ (>18.8 kcal mol^{-1}). These are followed by the ChBs in **10**$_{16\_CN}$ (12.4 ± 1.6 kcal mol^{-1}) and **10**$_{16\_SPh}$ (10.8 ± 1.1 kcal mol^{-1}), whereas the corresponding diselenide and methyl-substituted selenide (**10**$_{16\_Me}$) have the weakest interactions (<7.7 kcal mol^{-1}). The $J_{^{77}Se\cdots{}^{15}N}$ coupling was observed as Se···N ChB increased in strength. Taking advantage of the strength of the Se···N ChBs, Tomoda applied **10**$_{16\_Br^*}$ to the enantioselective synthesis of olefins.[60]

Also in the 90s, Chivers found intramolecular ChB between nitrogen and selenium or sulfur atoms in the structures of 1,8-dichalcogena-2,4,5,7-tetraza-octatrienes in solution. VT ^{1}H NMR spectroscopy of **10**$_{17}$ demonstrated the co-existence and interconversion of geometrical isomers (Scheme 18.2). Four independent resonances are observed at room temperature for the protons on the backbone, but these coalesce into a singlet at high temperatures in toluene.[61]

In 2016, Vargas-Baca reported evidence of auto-association of organoditellurides, R–Te–Te–R, in solution by ^{125}Te NMR measurements

(Scheme 18.3).[62] Ditellurides $\mathbf{10_{18a}}$ and $\mathbf{10_{18b}}$ undergo a low-field shift as the sample concentration increases in various solvents (hexane, toluene, CH_2Cl_2). The positive concentration coefficients for these two molecules are a result of supramolecular tellurium-centred supramolecular interactions. In contrast, sterically-hindered ditelluride $\mathbf{10_{18c}}$ and telluric acid experienced an opposite effect. In the telluric acid, the octahedral arrangement of oxygen atoms in the molecule prevents the chalcogen from interacting with other atoms. However, the influence of autoassociation is much smaller than the effects of solvent polarity and the conformation of the ditelluride bond.

In 2016, Bonifazi reported the UV-vis spectra in CH_2Cl_2 of pyridyl-substituted benzo-1,3-chalcogenazoles ($\mathbf{10_{19Ch}}$) to be greatly redshifted in comparison with the phenyl analogues.[53] The presence of the intramolecular Te···N interaction forces the heterocyclic moieties to adopt a planar conformation also in solution (Figure 18.13). This effect increases the

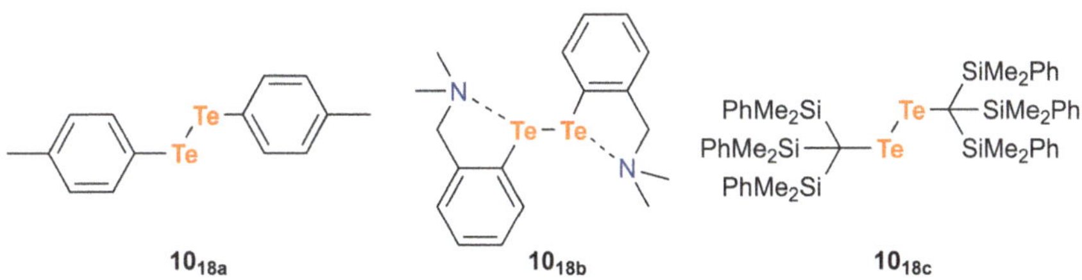

Scheme 18.3 Organo-ditellurides $\mathbf{10_{18a}}$ and $\mathbf{10_{18b}}$, and sterically hindered $\mathbf{10_{18c}}$.[62]

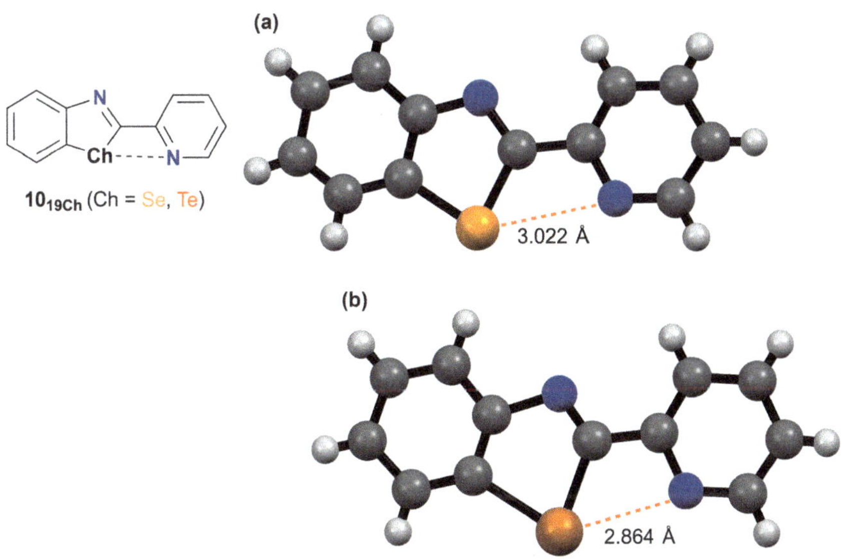

Figure 18.13 Intramolecular Ch···N displayed by $\mathbf{10_{19Ch}}$ in the solid state, which produces bathochromic shifts in solution. Adapted from ref. 53 with permission from John Wiley & Sons, Copyright © 2016 Wiley-VCH Verlag GmbH & Co. KGaA, Weinheim.

π-conjugation across the molecule resulting in a bathochromic shift of the electronic transitions.

Aiming at quantifying the strengths of a ChB interaction in solution, in 2017 Cockroft investigated ChB interactions in formamidochalcogenophenes using molecular balances by taking advantage of the dependency of the conformational equilibrium on the intramolecular ChB and solvent effects (Scheme 18.4).[63] The rotational conformational equilibrium constant around formamidothiophene molecules was probed by ^{19}F NMR at room temperature. The formamidothiophene $\mathbf{10_{20a}}$ and formamidoselenophene $\mathbf{10_{20b}}$ showed preference towards the closed conformers with short ChB (< -7.4 to -1 kJ mol^{-1}) even without an electron-withdrawing group, the Se···O ChB in formamidoselenophene $\mathbf{10_{20b}}$ is more favourable than the S···O ChB in $\mathbf{10_{20a}}$. For those molecules that could potentially have S···S ChB in thioformamidothiophene $\mathbf{10_{21}}$ were about 1.5 kJ mol^{-1} less favourable than those of S···O in formamidothiophene $\mathbf{10_{20a}}$. The conformational free energies are nearly solvent-independent suggesting minimal solvophobic, electrostatic, and dipolar effects on the ChBs.

In a very recent work, Kukushkin described ChB interactions between $(4\text{-}NC_5F_4)_2Ch$ $\mathbf{10_{22Ch}}$ (Ch = Se, Te) and platinum(II) complexes. The d_{z^2} electron density allows the cation to act as a ChB acceptor, consistent with the crystal structure displayed in Figure 18.14, with calculated bonding energies ranging from -7 to -12 kcal mol^{-1}.[64] The association of $\mathbf{10_{22Te}}$ in $CDCl_3$ solutions was studied by using NMR spectroscopy and NMR diffusion methods. Upon mixing the components, a downfield shift of the ^{195}Pt resonance and upfield shift of the ^{125}Te resonance were observed. Titration experiments in $CDCl_3$ yielded 1 : 1 binding constants of 56 and 79 M^{-1} and ΔG of -2.4 to -2.6 kcal mol^{-1}.

Compared to neutral Lewis bases, anions are better electron donors and form stronger ChBs in solution.[65] This is particularly useful as it can be applied to anion recognition.[4,19,21,66] As the field has been the subject of recent reviews,[19,21,24,25,65,66] we only highlight key contributions. The application of ChB to organocatalysis and organic synthesis is covered in Chapter 11.

Zibarev reported binding of thiophenolate anion to the selenium atom of 1,2,5-selenadiazole ($\mathbf{10_{23\_Se}}$).[67,68] The ^{77}Se NMR resonance of the adduct is shifted downfield by *ca.* 45 ppm compared to the free selenadiazole. UV-vis spectroscopy revealed partial dissociation in solution with high binding constants ($(7.4 \pm 1.0) \times 10^4$ and $(3.0 \pm 0.6) \times 10^3$) in THF and CH_3CN). Even

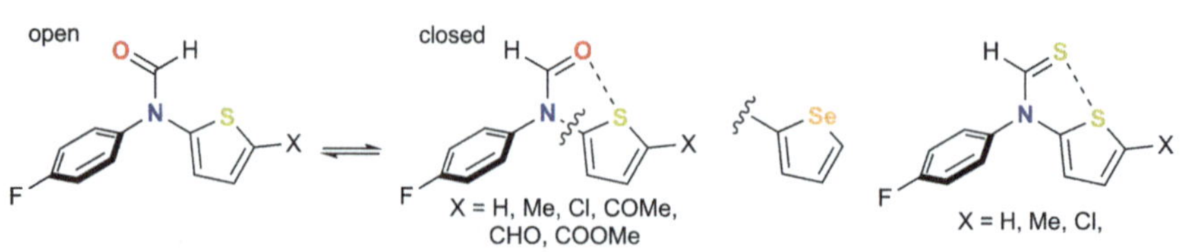

$\mathbf{10_{20a}}$ $\mathbf{10_{20b}}$ $\mathbf{10_{21}}$

Scheme 18.4 Examples of molecular balances studied by Cockroft.[63]

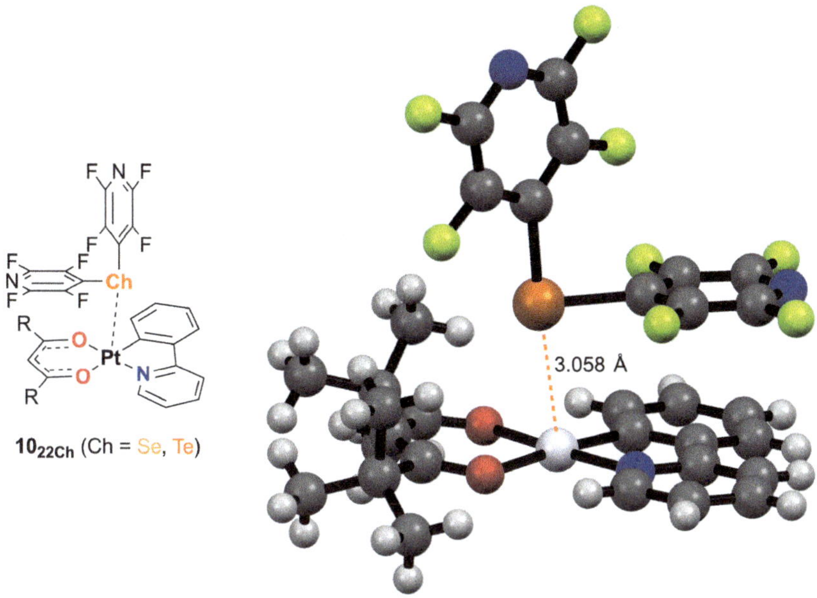

Figure 18.14 Molecular structure of $\mathbf{10_{22}}$, displaying a Te···Pt ChB in the solid state, as well as in solution. Adapted from ref. 64 with permission from American Chemical Society, Copyright 2021.

larger binding constants were obtained for the adducts of I^- ($(6.8 \pm 1.4) \times 10^5$), SPh^- and F^- (both $> 10^8$ with SPh^-) with 3,4-dicyano-1,2,5-telluradiazole $\mathbf{10_{23\_Te}}$ (Figure 18.15).[69,70] Further investigations have included binding of pseudohalides CN^- and $ChCN^-$ (Ch = O, S, Se, Te) to 1,2,5-chalcogenadiazoles (E = Te, Se, S).[71] The adducts have been crystallized with $[K(18\text{-crown-}6)]^+$ as a counterion. The $\mathbf{10_{23\_Te}}$ binding constants for $SeCN^-$ (600 ± 60 to 700 ± 40 M^{-1}) are larger than for SCN^- (503 ± 16 M^{-1}). These adducts are not formed by selenadiazole ($\mathbf{10_{23\_Se}}$) or thiadiazole ($\mathbf{10_{23\_S}}$), in agreement with thermodynamic calculations. The ^{77}Se NMR spectrum of $\mathbf{10_{23\_Te}}$-$SeCN^-$ displayed a positive chemical shift ($\delta = -321$ to -147 ppm) compared to $\mathbf{10_{23\_Te}}$. The $SeCN^-$ ion in $\mathbf{10_{23\_Te}}$-$SeCN^-$ can be displaced by F^- as suggested by ^{77}Se NMR.

In 2015 Taylor and Seferos reported a comprehensive investigation of ChB binding of benzotelluradiazoles to anions and quinuclidine in organic medium.[5] In the strongest case, *i.e.*, with the tetra-fluorinated benzotelluradiazole derivative, the magnitude of ChB-anion association constants follows the trend of Cl^- (130 000 M^{-1}) > Br^- (9800 M^{-1}) > NO_3^- (190 M^{-1}) > quinuclidine (96 M^{-1}). The interaction with I^- was too weak for an accurate measurement. The more polar solvents give smaller ChB association constants. Later in 2016, the same group reported anion binding by a bidentate chalcogen bond donor $\mathbf{10_{24}}$, consisting of two tellurophenes rings bridged by an ethynylene group (Scheme 18.5).[72] The Cl^- binding constant (2290 M^{-1}) is about one order of magnitude higher than that measured with the monodentate 1,5-bis(perfluoroaryl)-tellurophene ($\mathbf{10_{25}}$) (310 M^{-1}).

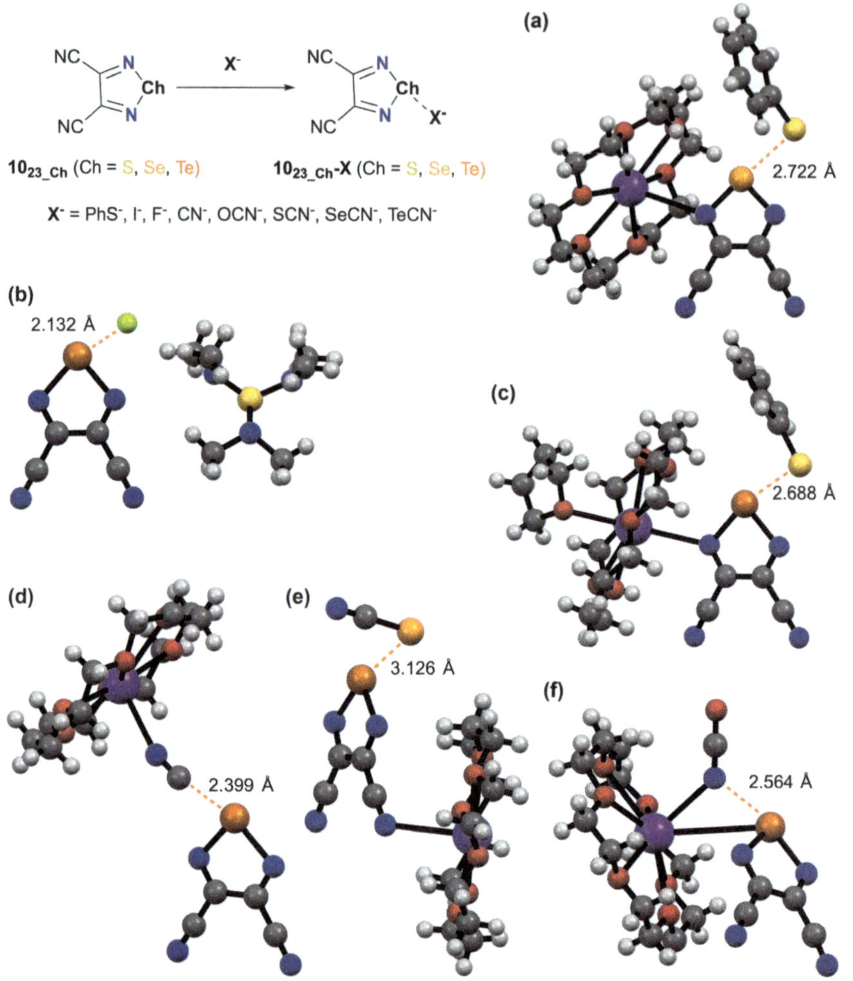

Figure 18.15 Schematic representation and structures of the adducts of 1,2,5-chalcogenadiazole $\mathbf{10_{23\_Ch}}$ (E = Se, Te) with anions (a) $\mathbf{10_{23\_Se}}$:$PhS^{-}[K(C_{12}H_{24}O_6)]^{+}$, (b) $\mathbf{10_{23\_Te}}$:$Cl^{-}[S(N(CH_3)_2)_3]^{+}$, (c) $\mathbf{10_{23\_Te}}$:$PhS^{-}[K(C_{12}H_{24}O_6)(C_4H_8O)]^{+}$, (d) $\mathbf{10_{23\_Te}}$:$CN^{-}[K(C_{12}H_{24}O_6)]^{+}$, (e) $\mathbf{10_{23\_Te}}$:$SCN^{-}[K(C_{12}H_{24}O_6)]^{+}$, and (f) $\mathbf{10_{23\_Te}}$:$OCN^{-}[K(C_{12}H_{24}O_6)$.[69–71]

In 2017, Beer inserted a 5-(methylchalcogeno)-1,2,3-triazole (Ch = Se, Te) motif in a rotaxane architecture as a ChB donor for anion binding.[73] The Te-containing compound ($\mathbf{10_{26TeMe}}$-rotaxane) displayed a greater propensity for binding anions of higher charge density (SO_4^{2-} (1130 M^{-1}) > F^{-} (360 M^{-1}) > Cl^{-} (127 M^{-1}) > Br^{-} (110 M^{-1}) > AcO^{-} (98 M^{-1}) > I^{-} (34 M^{-1})). In contrast, the selenium analogue ($\mathbf{10_{26SeMe}}$) did not bind anions to any significant extent; however, its ChB acceptor ability was enhanced by methylation of the triazoles (Scheme 18.6).

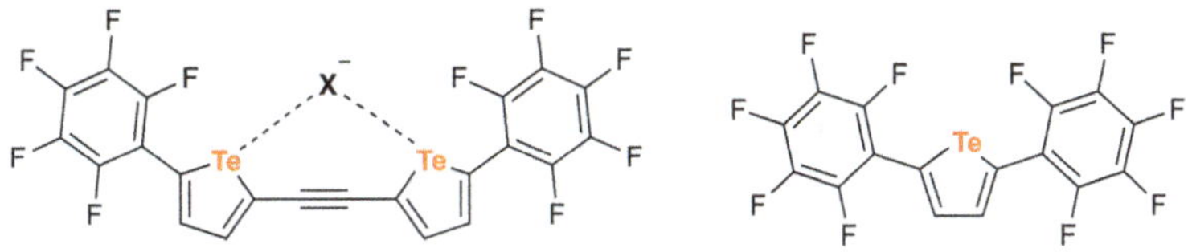

$\mathbf{10_{24}}$ X^- = Cl^-, Br^-, I^-, NO_3^-, TsO^-, BzO^- $\mathbf{10_{25}}$

Scheme 18.5 Anion-binding tellurophenes $\mathbf{10_{24}}$ and $\mathbf{10_{25}}$.[72]

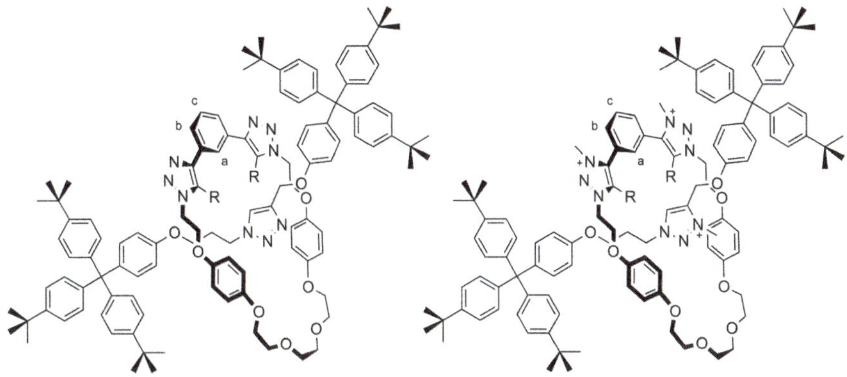

$\mathbf{10_{26R}}$-rotaxane (R = H, SeMe, TeMe) [$\mathbf{10_{26SeMe}}$-rotaxane]$^{3+}$

Scheme 18.6 Examples of chalcogen-bonded rotaxanes studied by Beer.[73]

The halide binding constants of [$\mathbf{10_{26SeMe}}$-rotaxane]$^{3+}$ were even higher (SO_4^{2-} (3531 M^{-1})>I^- (2084 M^{-1})>Br^- (983 M^{-1})>Cl^- (696 M^{-1})) than those measured with the neutral $\mathbf{10_{26TeMe}}$-rotaxane. In the heteronuclear ^{125}Te and ^{77}Se NMR study, upfield shifts of the signals were observed when binding to halides. In a follow up report, Beer reported in 2018 the anion binding behaviour of three bidentate chalcogen bond donors ($\mathbf{10_{27}}^+$ to $\mathbf{10_{29}}^+$) probed by ^{1}H NMR titration experiments in CD_3CN (Scheme 18.7).[74] The addition of tetrabutylammonium (TBA) chloride to a solution containing Te-based derivative $\mathbf{10_{27}}^+$ resulted in downfield shifts only in resonances of H_a and $TeCH_3$ protons suggesting the binding vicinity between the Te and ChB donor groups. Measurement of the association constants of $\mathbf{10_{27}}^+$ showed decreased anion affinities in the order of charge density ($H_2PO_4^-$ (411 to 1530 M^{-1})>Cl^- (652 M^{-1})>Br^- (503 M^{-1})>N_3^- (330 M^{-1})>I^- (305 M^{-1})>CH_3COO^- (269 M^{-1})), which revealed to be in full agreement with Taylor and Seferos' work.[5,72] Notably, the anion association constants decreased as the polarity of the solvent increases. The ^{1}H NMR spectrum of the Se-analogue ($\mathbf{10_{28}}^+$), displayed H_b down- and H_a up-field shifts upon addition of TBA chloride. Considering that the ^{77}Se resonance showed nearly no perturbations, these observations suggest that the anion is attached in the vicinity of the cationic pyridinium nitrogen atom. There was no evidence of Cl^- binding to the macrocyclic selenium ChB donor ($\mathbf{10_{29}}^+$).

Thermodynamic parameters for halide binding with the tellurium analogue were determined from variable temperature NMR in CD_3CN, which displays a strong enthalpic contribution that gradually increases for large halogen anions. On the other hand, a favourable entropic contribution is only observed with Cl^-, as it is negligible for Br^- and unfavourable for I^-.

In a parallel avenue, the group of Beer reported fluorescent dicarboxylate anion sensing using chiral halogen and chalcogen bonding receptors (**10_{30_I}**$^{2+}$ and **10_{30_SeMe}**$^{2+}$, Scheme 18.8), which exhibit different selectivity for stereo- and geometric dicarboxylate isomers compared to the hydrogen bonding analogue **10_{30_H}**$^{2+}$, shown to be non-selective.[75] The dicarboxylate anion binding properties for **10_{30_I}**$^{2+}$ and **10_{30_SeMe}**$^{2+}$ were investigated by ^1H NMR titration experiments in d_6-acetone/D_2O 85 : 15 v/v. The ChB donor showed enhanced fumarate/maleate discrimination over the HB and XB donors. The stability of ChBs to highly basic anions makes ChB hosts valuable alternatives to XB for anion recognition and sensing. On the other hand, **10_{30_I}**$^{2+}$ showed better enantioselectivity with tartrate than *N*Boc-glutamate.

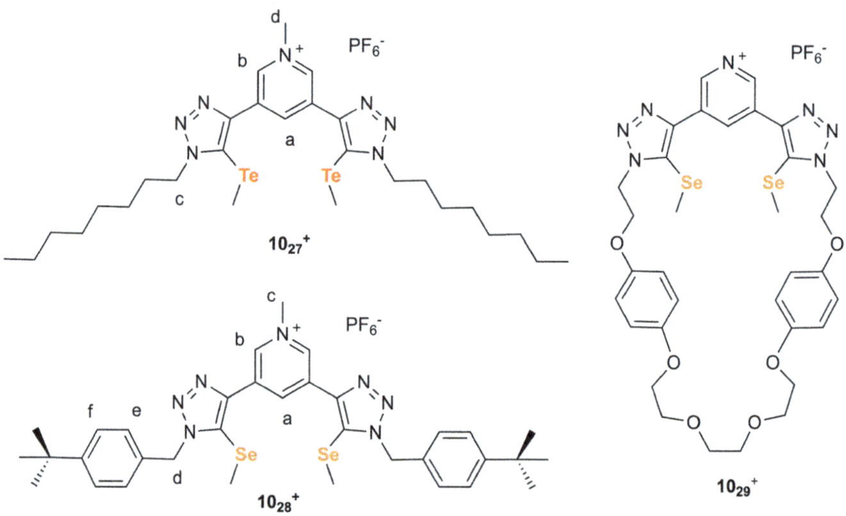

Scheme 18.7 Bidentate ChB donors investigated by Beer for anion recognition.[74]

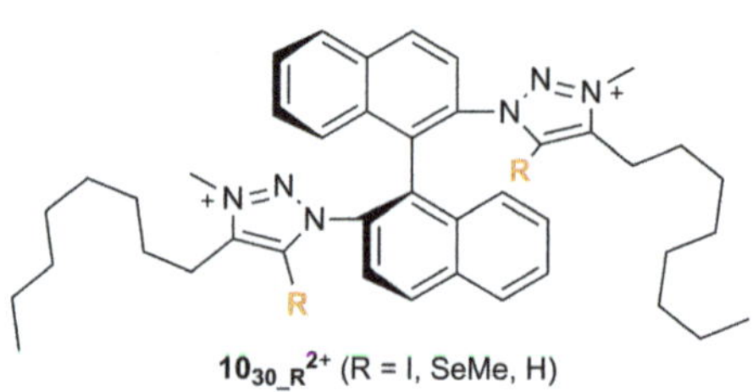

Scheme 18.8 Molecular structure of the chiral receptors **10_{30_R}**.[75]

In 2019, Beer increased the binding sites to four hydrophobic halogen and chalcogen bond donors into a foldamer containing hydrophilic functionalities (**10_{31_X}**).[76] In organic solvents, **10_{31_TeMe}** forms 1:1 receptor-anion adducts with SCN^- (K_a 492 M^{-1}), I^- (K_a 367 M^{-1}), ReO_4 (K_a 184 M^{-1}), and Br^- (K_a 169 M^{-1}). The anion binding was mainly enthalpy driven due to the ability of the σ-hole donors to form strong ChB interactions within the largely dehydrated core of the host complex. The ChB coordination was confirmed by the large downfield changes in ^{125}Te chemical shifts upon titration with TBAI. In water, the hydrophobic collapse drives the formation of 2:1 bound states, prevailing over the weak solvation of anion (I^- (K_a 730 M^{-1}) > SCN^- (K_a 116 M^{-1}) > ReO_4 (K_a 86.7 M^{-1}) > Br^- (K_a 0.0694 M^{-1}) 226 Cl^-; with the ClO_4^- being too weakly bound for K_a determination. In 2021, Duarte, Beer, and Langton reported a bidentate ChB donor chloride transporter based on an electron deficient telluromethyl-triazole motif (**10_{32_TeMe}**).[77] The transmembrane anion transport activity, with a low effective concentration to reach 50% of maximal activity in the HPTS assay ($EC_{50} = 0.1$ mol% or 0.032 ± 0.003 μM), in lipid bilayer vesicles, showed high selectivity for chloride, with high association constant ($K_a = 18\,543$ M^{-1}) and very negative free energy (−9.2 kcal mol^{-1}), over proton/hydroxide. It is worth noting that the iodo analogue (**10_{32_I}**) showed a record anion transporter activity ($EC_{50} = 0.01$ mol% or 0.003 ± 0.001 μM) reported to date with a high Cl^- association constant ($K_a > 10^5$ M^{-1}).[78] In a parallel study, Beer studied a series of acyclic anion receptors containing ChB donors integrated into a neutral 3,5-bis-triazole pyridine scaffold **10_{33_TeMe}**, as well as comparing with the XB analogue **10_{33_I}** (Scheme 18.9).[79] Donors **10_{33_X}** bearing EW perfluorophenyl (PFP) groups showed an enhanced association constant

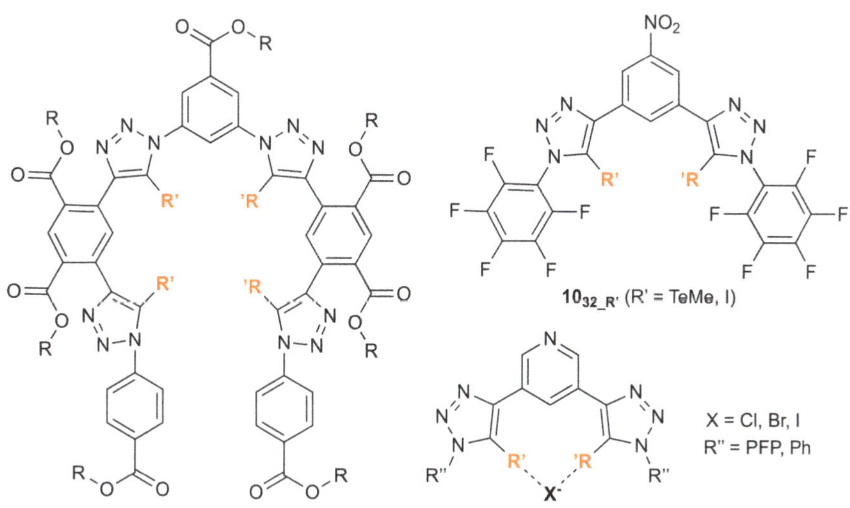

$10_{31\_R'}$ (R' = TeMe, I)

$10_{32\_R'}$ (R' = TeMe, I)

$10_{33\_R'}$ (R' = H, TeMe, Te-EWG, I)

Scheme 18.9 Ditopic and tetratopic anion recognition systems studied by Beer.[76–79]

in acetone for Cl^- (18 500 M^{-1} X = TeMe, 2920 M^{-1} X = I), Br^- (4620 M^{-1} X = TeMe, 4600 M^{-1} X = I), and I^- (1230 M^{-1} X = TeMe, 5230 M^{-1} X = I) relative to the donors substituted with phenyl groups (Cl^-: 590 M^{-1} X = TePh, 2920 M^{-1} X = I) (Br^-: 304 M^{-1} X = TePh, 1060 M^{-1} X = I) (I^-: 234 M^{-1} X = TePh, 1450 M^{-1} X = I). It was demonstrated that the ChB donor is highly sensitive to the local electronic environments, as $\mathbf{10_{33\_TeEWG}}$ shows an enhancement of selectivity for the lighter halides as the electron-withdrawing ability of the Te-bonded group increases.

In a seminal work, Matile reported in 2016 the electron deficient dithieno[3,2-*b*;2′,3′-*d*]thiophenes (DTTs, $\mathbf{10_{34}}$) for anion binding in solution and anion transport across lipid bilayers (Scheme 18.10).[80] An UV-vis study showed that Cl^- bindings caused hypo- and batho-chromism. The Job plot supported the formation of a 1 : 1 complex and suggested that $\mathbf{10_{34}}$ transported anions as monomers ($K_D = 1.13 \pm 0.03$ mM^{-1}; $K_a = 885$ M^{-1}). When embedded into large unilamellar vesicles, it transported Cl^- anions across the membrane with a transport activity (EC_{50}, effective concentration needed to reach 50% of maximal activity in the HPTS assay; from the Hill analysis of dose response curves) to be 1.9 ± 0.2 μM.

In 2020, Frontera and Caballero reported thiophene, selenophene, and tellurophene-based neutral receptors ($\mathbf{10_{35\_Ch}}$) for molecular recognition of anions (Scheme 18.11).[81] These receptors had no interaction with I^-, NO_3^-, AcO^-, $H_2PO_4^-$, and PF_6^-, as confirmed by ^1H NMR spectra. However, the resonance of the acidic proton next to the chalcogen of $\mathbf{10_{35\_S}}$ and $\mathbf{10_{35\_Se}}$ shifted downfield upon the addition of Cl^- and Br^-. As no perturbation was detected in the ^{77}Se NMR of $\mathbf{10_{35\_Se}}$, it was suggested that HBs are preferred with these anions. For Te-derivative $\mathbf{10_{35\_Te}}$, ChB with the anions (X = Cl^-, $K_a = 51.1$ M^{-1}; X = Br^-, $K_a = 12.7$ to 14.0 M^{-1}) in THF-d_8 was in competition

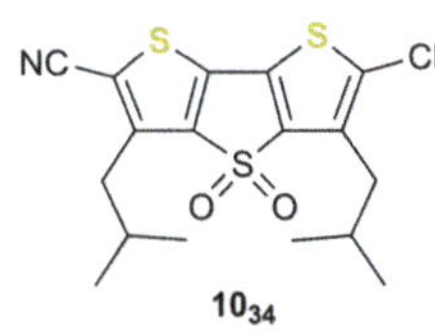

$\mathbf{10_{34}}$

Scheme 18.10 Molecular structure of the dithienothiophene $\mathbf{10_{34}}$, employed for anion transport across lipid bilayers.[80]

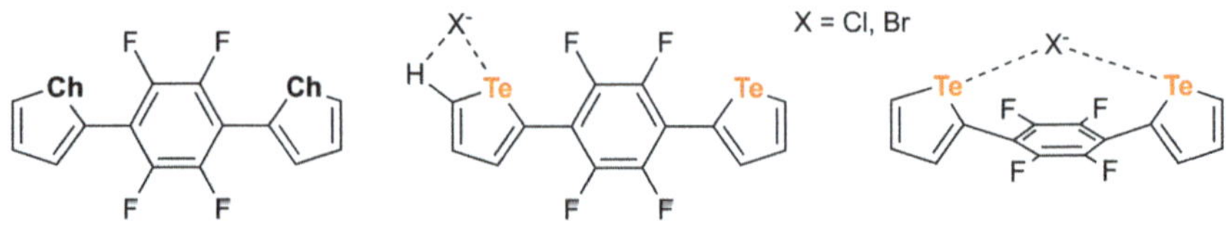

$\mathbf{10_{35\_Ch}}$ (Ch = S, Se, Te)

Scheme 18.11 Chalcogenophene-based neutral receptors for anion recognition.[81]

with hydrogen bonding interactions. The calculated binding energies are more negative for Cl^- (-3.56 to -1.5 kcal mol^{-1}) than Br^- (-1.71 to 0.3 kcal mol^{-1}) for both mono and bidentate recognition motifs.

In 2020, Gabbaï reported $[Mes(C_6F_5)TeMe]^+$ (**10$_{36}$**) and $[(C_6F_5)_2TeMe]^+$ (**10$_{37}$**) telluronium cations for Cl^- recognition and transport across phospholipid bilayers.[82] These telluronium cations showed enhanced Lewis acidity towards Cl^- in CH_3CN ($K_1 = (1.2 \pm 0.6) \times 10^4$ M^{-1} for **10$_{36}$** and $\geq 10^5$ for **10$_{37}$**, $K_2 = 260 \pm 80$ M^{-1} for **10$_{37}$**), whereas the neutral compounds are only weakly to non-interactive, and have better membrane transport ability (0.75% s^{-1} for **10$_{36}$**, 1.68% s^{-1} for **10$_{37}$**) compared to neutral $(C_6F_5)_2Te$ ($K_1 = 36$ M^{-1}, 0.40% s^{-1}). The V_{max} increased from 103.1 to 138.3 kcal mol^{-1} from the neutral compounds (9.2 to 39.2 kcal mol^{-1}). In a very recent report, the same group expanded the work to bis-telluronium dication **10$_{38}$** that functions as a bidentate Lewis acid towards Cl^- ions (Scheme 18.12).[83] Such dications **10$_{38}$** abstract chloride from Ph_3CCl with an equilibrium constant ($K_{rel} = 14.5$ M^{-1}), determined by UV-vis absorption spectroscopy, almost four orders of magnitude higher than its monofunctional analogue (**10$_{39}$** $K_{rel} = 1.58 \times 10^{-3}$ M^{-1}). ^{31}P NMR measurements upon addition of Ph_3PO show a greater deshielding of the phosphorus resonance for the dication ($\Delta\delta = 7.9$ ppm) than the monocation ($\Delta\delta = 6.1$ ppm) host. The dications are also found to activate the metal–chloride bonds in Ph_3PAuCl and [*cis*-$(Ph_3P)_2PtCl_2$] in transition metal catalysis.

The use of ChBs in the assembly of large structures in solution is by far less studied than the application of this interaction to anion binding and crystal engineering. In 2010, Vargas-Baca reported the discovery of macrocyclic tetramers of an iso-tellurazole *N*-oxide (**10$_{40_tBu}$**, Scheme 18.13).[84] In 2016 it was demonstrated that these are reversibly assembled by Te···O ChBs. In addition to the tetramer, the R = Ph analogue crystallizes as a cyclic hexamer and a helical polymer.[85] Variable concentration and continuous variations NMR (^1H and ^{125}Te) investigations of mixtures of **10$_{40_tBu}$** and **10$_{40_Ph}$** showed that the hexamers and tetramers coexist in solution with an equilibrium constant of 0.28 ± 0.01 dm$^{1.5}$ mol$^{-0.5}$, and $\Delta H = -16 \pm 1$ kJ mol^{-1} and $\Delta S = -88 \pm 6$ J mol^{-1} K^{-1}. The supramolecular tetramers formed adducts with C_{60} and κ^4Te coordination complexes with transition metal ions such as Cu(I), Pd(II), Pt(II), and Rh(III), while Au(I) forms a κ^1Te binuclear complex.[85–87]

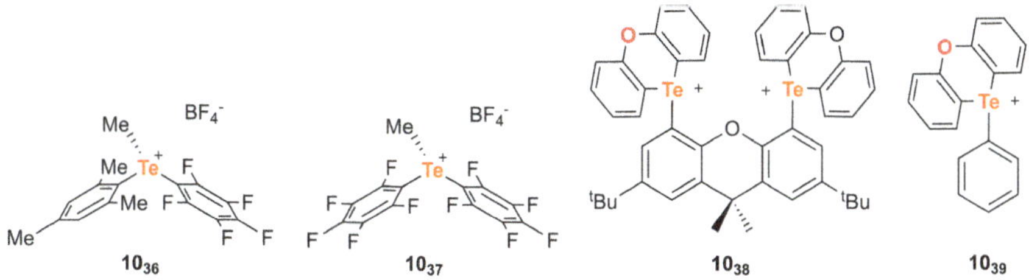

Scheme 18.12 Series of telluronium cationic species developed by Gabbaï for anion transport.[82,83]

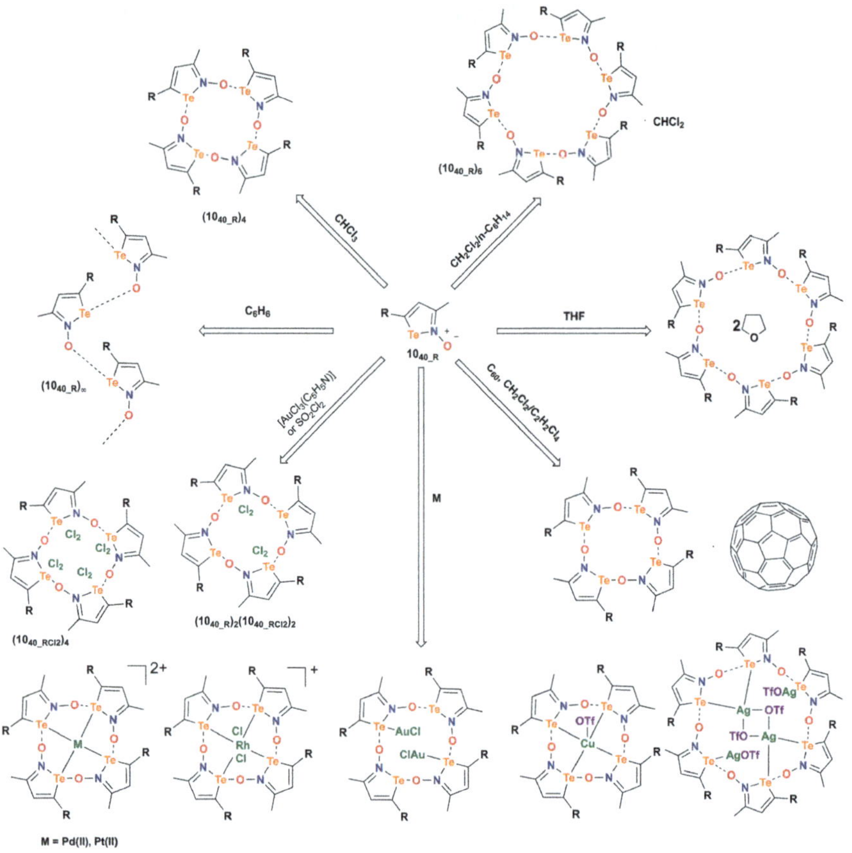

Scheme 18.13 Schematic representation of the several reversible supramolecular macrocycles achieved by $\mathbf{10_{40\_R}}$.[84–87]

The hexamers form Ag(I) complexes with two endocyclic and two exocyclic Ag(I) ions.[87] In the Pt(II) and Rh(III) complexes, the ^{125}Te–^{195}Pt and ^{125}Te–^{103}Rh ^1J coupling constant proved that these complexes are stable in solution.

Chlorination of $\mathbf{10_{40\_Ph}}$ yielded the λ^4Te dichloro derivative $\mathbf{10_{40\_PhCl2}}$, which retains the ability to aggregate by ChB forming macrocyclic tetramers. DFT calculations indicate that halogenation strengthens the ChB donor but also weakens the ChB acceptor abilities of the individual molecule. Scrambling of $(\mathbf{10_{40\_Ph}})_4$ with the chlorinated homotetramer $(\mathbf{10_{40\_PhCl2}})_4$ selectively yielded the 2 : 2 heterotetramer $(\mathbf{10_{40\_Ph}})_2(\mathbf{10_{40\_PhCl2}})_2$ which is the thermodynamically preferred array according to DFT calculations. In 2021, Vargas-Baca reported another supramolecular building block ($\mathbf{10_{41}}$) that features formal substitution of the oxygen atom of $\mathbf{10_{40}}$ with deprotonated (ortho, meta, and para)-hydroxyphenyl groups (Scheme 18.14).[88]

These molecules form strong ChBs in solution as shown by VT ^{1}H NMR, and supported by DFT calculations. X-ray diffraction revealed polymeric

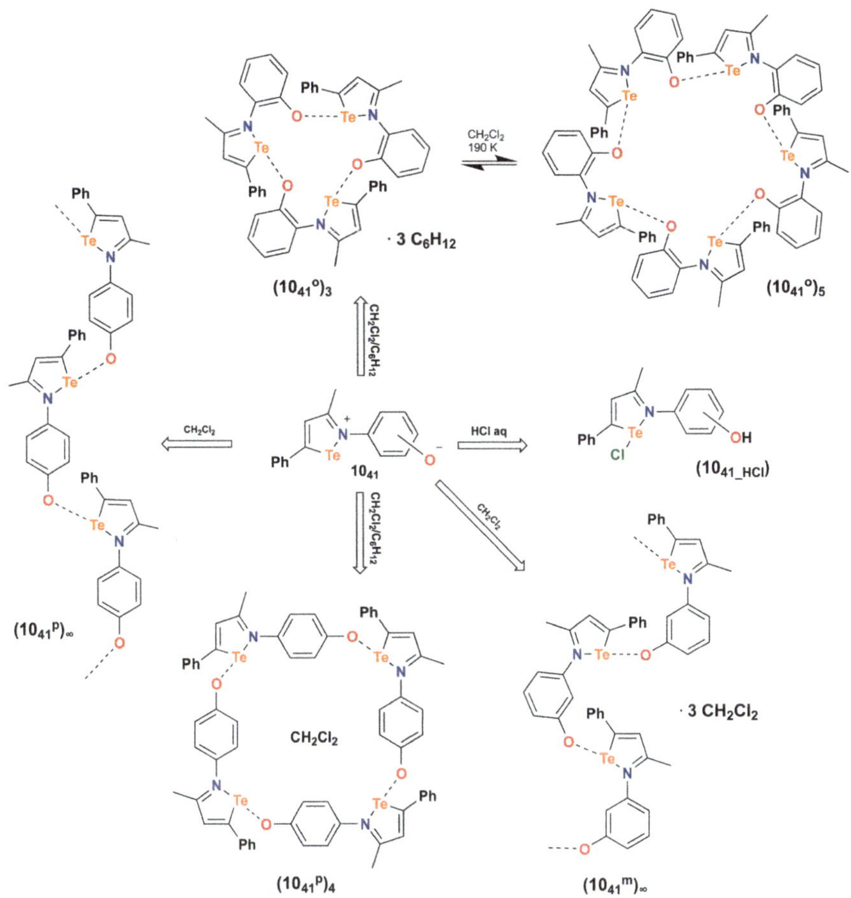

Scheme 18.14 Novel supramolecular assemblies achieved in the recent work of Vargas-Baca by $\mathbf{10_{41}}$.[88]

$(\mathbf{10_{41}})_{\infty}$ as well as trimeric $(\mathbf{10_{41}}^{\mathbf{o}})_3$ and tetrameric $(\mathbf{10_{41}}^{\mathbf{p}})_4$ aggregates with calixarene-like trimer and square structures, respectively. The VT ^{1}H NMR of the trimer further suggests the formation of the pentamer $(\mathbf{10_{41}}^{\mathbf{o}})_5$ at 190 K in solution in equilibrium ($\ln K = 6.1 \pm 0.4$) with thermodynamic parameters consistent with autoassociation ($\Delta H = -43.3 \pm 0.5$ kJ mol^{-1}, $\Delta S = -58 \pm 3$ J mol^{-1} K^{-1}).

In an earlier work, Laitinen and Oakley reported in 2008 the self-association of *N*-methyl benzotellurodiazolylium cations in solution.[89] The crystal structure of the cations consists of centrosymmetric dimers linked by the $[\text{Ch}\cdots\text{N}]_2$ supramolecular synthon with very short ChB distances (2.417(3) Å). Cyclic voltammetric studies in CH_3CN showed an oxidative event centred at $E_{1/2} = +0.168$ V, which anodically shifted when compared to that obtained for the sulfur (−0.351 V) and selenium (−0.308 V) analogues, known to be monomers in their crystals.

An outstanding example of the application of the supramolecular $[Ch\cdots N]_2$ recognition motif is the contribution of the group of Diederich, who prepared supramolecular dimeric capsules by functionalization of resorcin[4]arene cavitands with benzochalcogenadiazoles.[90] Mainly, the strength of their interaction was controlled depending on the chalcogen atom employed, with Te-based $\mathbf{10}_{\mathbf{42\_Te}}$ forming dimers independently on the solvent media. Interestingly, the S-congener $\mathbf{10}_{\mathbf{42\_S}}$ showed a different association behaviour, resulting in being solvent-dependent. The tremendous strength of multiple $[Ch\cdots N]_2$ squares was demonstrated by the 16 intermolecular Te···N bonds occurring in the solid state, all significantly shorter (2.6–2.9 Å) than the $\sum r_{vdw}$ (3.61 Å), regardless of the solvation effects (Figure 18.16a). The effectiveness of the polydentate ChB interaction was also proven by an ESI-MS titration study, which provided a dimerization constant of $\mathbf{10}_{\mathbf{42\_Te}}$ in the order of 10^7 M^{-1}. On the other hand, the lower tendency of S atoms to establish strong ChBs was confirmed by the observation of different supramolecular arrangements in the solid state by $\mathbf{10}_{\mathbf{42\_S}}$ when diverse solvents were used. Choosing aromatic solvents such as benzene or toluene afforded the formation of capsules, partially bonded through 12 S···N long contacts ($d_{S\cdots N}=3.0$–3.3 Å$<\sum r_{vdw}=3.35$ Å, Figure 18.16b). Instead, dichloromethane completely disrupted the capsular assembly, favouring the development of supramolecular polymers. Further studies of the capsule-association by ^{1}H NMR analysis allowed us to determine a dimerization constant that was three orders of magnitude smaller than that obtained with $\mathbf{10}_{\mathbf{42\_Te}}$. This work further inspired the use of chalcogen bonds to form leakproof, water-stable supramolecular capsules with benzoselenadiazole-containing resorcinarenes, cemented together through multiple Se···N ChBs.[91,92]

18.4 Conclusion

Our modern understanding of the nature of chalcogen bonding has allowed taking advantage of this interaction in multiple aspects of supramolecular chemistry. While all three stabilizing energetic contributions (electrostatic, polarization, and dispersion) favour tellurium centres to be the strongest ChB donors, the interaction strength can be modulated by inserting a large variety of ancillary groups featuring electron withdrawing abilities. Systematic investigations have identified building blocks with a reliable ability to form ChBs, prominent in this category are molecules that form antiparallel ChB pairs. ChBs bear similarities to hydrogen and chalcogen bonds, but offer clear advantages in some respects, namely their compatibility with nonpolar media and the absence of reactive hydrogen atoms. The characterization of chalcogen-bonded structures in the solid state will likely remain a major trust, but focus is shifting towards enabling and manipulating the properties of functional materials, *e.g.*, NLO crystals, porous frameworks, and semiconductors. In the last seven years, an unprecedented number of studies have not just shown ChBs in the liquid phase, but have demonstrated their application in anion recognition with binding constants as high

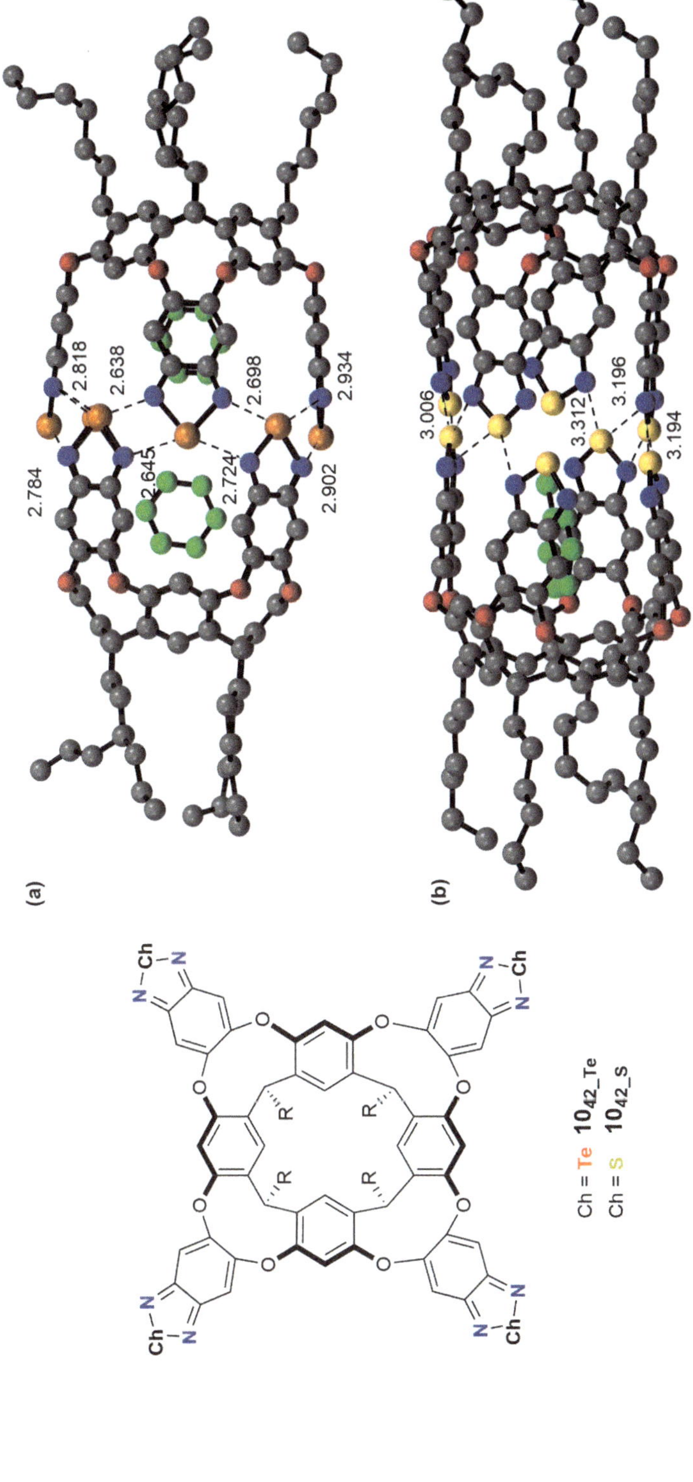

Figure 18.16 Schematic representation of resorcinarene-based scaffold and the resulting supramolecular capsules of (a) **10$_{42_Te}$** and (b) **10$_{42_S}$**. Solvent molecules of benzene and toluene are shown in green. Distances are expressed in Å, H atoms are omitted for clarity.[90]

as 10^6 M^{-1}, transport across membranes, and the construction of supramolecular capsules and macrocycles. Research in chalcogen bonding is beginning to bear fruit and is likely to yield major advances in the future especially in the development of new recognition motifs and applications in materials science.

Acknowledgements

The continuing support of the Natural Sciences and Engineering Research Council (NSERC) of Canada is gratefully acknowledged by PCH (PGSD) and IVB (DG RGPIN-2016-06452). DB gratefully acknowledges the EU through the funding scheme projects MSCA-RISE INFUSION (No. 734834), H2020-NMBP-2017 DECOCHROM (No. 760973), and MSCA-ITN-ETN STiBNite (No. 956923) and the University of Vienna for the financial support.

References

1. C. B. Aakeroy, D. L. Bryce, G. R. Desiraju, A. Frontera, A. C. Legon, F. Nicotra, K. Rissanen, S. Scheiner, G. Terraneo and P. Metrangolo, *Pure Appl. Chem.*, 2019, **91**, 1889.
2. N. W. Alcock, *Advances in Inorganic Chemistry and Radiochemistry*, Elsevier, 1972, vol. 15, pp. 1–58.
3. P. Politzer, J. S. Murray and T. Clark, *Phys. Chem. Chem. Phys.*, 2013, **15**, 11178.
4. J. Y. Lim and P. D. Beer, *Chem*, 2018, **4**, 731.
5. G. E. Garrett, G. L. Gibson, R. N. Straus, D. S. Seferos and M. S. Taylor, *J. Am. Chem. Soc.*, 2015, **137**, 4126.
6. D. J. Pascoe, K. B. Ling and S. L. Cockroft, *J. Am. Chem. Soc.*, 2017, **139**, 15160.
7. C. Bleiholder, D. B. Werz, H. Köppel and R. Gleiter, *J. Am. Chem. Soc.*, 2006, **128**, 2666.
8. T. Chivers and R. S. Laitinen, *Chem. Soc. Rev.*, 2015, **44**, 1725.
9. S. Sarkhel and G. R. Desiraju, *Proteins: Struct., Funct., Bioinf.*, 2004, **54**, 247.
10. A. J. Wilson, *Soft Matter*, 2007, **3**, 409.
11. N. V. Medhekar, A. Ramasubramaniam, R. S. Ruoff and V. B. Shenoy, *ACS Nano*, 2010, **4**, 2300.
12. S. Herbst, B. Soberats, P. Leowanawat, M. Lehmann and F. Würthner, *Angew. Chem.*, 2017, **129**, 2194–2197; S. Herbst, B. Soberats, P. Leowanawat, M. Lehmann and F. Würthner, *Angew. Chem., Int. Ed.*, 2017, **56**, 2162.
13. X. Du, X. Lu, J. Zhao, Y. Zhang, X. Li, H. Lin, C. Zheng and S. Tao, *Adv. Funct. Mater.*, 2019, **29**, 1902078.
14. T. Chivers, I. Krouse, M. Parvez, I. Vargas-Baca, T. Ziegler and P. Zoricak, *Inorg. Chem.*, 1996, **35**, 5836.

15. A. F. Cozzolino, P. J. W. Elder and I. Vargas-Baca, *Coord. Chem. Rev.*, 2011, **255**, 1426.
16. J. Fanfrlík, A. Přáda, Z. Padělková, A. Pecina, J. Macháček, M. Lepšík, J. Holub, A. Růžička, D. Hnyk and P. Hobza, *Angew. Chem.*, 2014, **126**, 10303.
17. K. T. Mahmudov, M. N. Kopylovich, M. F. C. G. da Silva and A. J. Pombeiro, *Dalton Trans.*, 2017, **46**, 10121.
18. R. Gleiter, G. Haberhauer, D. B. Werz, F. Rominger and C. Bleiholder, *Chem. Rev.*, 2018, **118**, 2010.
19. N. Biot and D. Bonifazi, *Coord. Chem. Rev.*, 2020, **413**, 213243.
20. S. Kolb, G. A. Oliver and D. B. Werz, *Angew. Chem.*, 2020, **132**, 22490–22495; S. Kolb, G. A. Oliver and D. B. Werz, *Angew. Chem., Int. Ed.*, 2020, **59**, 22306.
21. P. C. Ho, J. Z. Wang, F. Meloni and I. Vargas-Baca, *Coord. Chem. Rev.*, 2020, **422**, 213464.
22. P. Scilabra, G. Terraneo and G. Resnati, *Acc. Chem. Res.*, 2019, **52**, 1313.
23. M. Fourmigué and A. Dhaka, *Coord. Chem. Rev.*, 2020, **403**, 213084.
24. L. Vogel, P. Wonner and S. M. Huber, *Angew. Chem.*, 2019, **131**, 1896–1907; L. Vogel, P. Wonner and S. M. Huber, *Angew. Chem., Int. Ed.*, 2019, **58**, 1880.
25. K. Strakova, L. Assies, A. Goujon, F. Piazzolla, H. V. Humeniuk and S. Matile, *Chem. Rev.*, 2019, **119**, 10977.
26. C. M. Young, A. Elmi, D. J. Pascoe, R. K. Morris, C. McLaughlin, A. M. Woods, A. B. Frost, A. De La Houpliere, K. B. Ling and T. K. Smith, *Angew. Chem.*, 2020, **132**, 3734–3739; C. M. Young, A. Elmi, D. J. Pascoe, R. K. Morris, C. McLaughlin, A. M. Woods, A. B. Frost, A. De La Houpliere, K. B. Ling and T. K. Smith, *Angew. Chem., Int. Ed.*, 2020, **59**, 3705.
27. W. Wang, H. Zhu, L. Feng, Q. Yu, J. Hao, R. Zhu and Y. Wang, *J. Am. Chem. Soc.*, 2020, **142**, 3117.
28. A. Docker, C. H. Guthrie, H. Kuhn and P. D. Beer, *Angew. Chem.*, 2021, **133**, 22144–22149; A. Docker, C. H. Guthrie, H. Kuhn and P. D. Beer, *Angew. Chem., Int. Ed.*, 2021, **60**, 21973.
29. A. F. Cozzolino and I. Vargas-Baca, *Cryst. Growth Des.*, 2011, **11**, 668.
30. A. F. Cozzolino, I. Vargas-Baca, S. Mansour and A. H. Mahmoudkhani, *J. Am. Chem. Soc.*, 2005, **127**, 3184.
31. A. F. Cozzolino and I. Vargas-Baca, *J. Organomet. Chem.*, 2007, **692**, 2654.
32. A. F. Cozzolino, J. F. Britten and I. Vargas-Baca, *Cryst. Growth Des.*, 2006, **6**, 181.
33. A. F. Cozzolino, Q. Yang and I. Vargas-Baca, *Cryst. Growth Des.*, 2010, **10**, 4959.
34. W. Yang, R. Jiang, C. Liu, B. Yu, X. Cai and H. Wang, *Cryst. Growth Des.*, 2021, **21**, 6497.
35. S. Langis-Barsetti, T. Maris and J. D. Wuest, *J. Org. Chem.*, 2017, **82**, 5034.
36. B. J. Eckstein, L. C. Brown, B. C. Noll, M. P. Moghadasnia, G. J. Balaich and C. M. McGuirk, *J. Am. Chem. Soc.*, 2021, **143**, 20207.

37. N. A. Puskarevsky, A. I. Smolentsev, A. A. Dmitriev, I. Vargas-Baca, N. P. Gritsan, J. Beckmann and A. V. Zibarev, *Chem. Commun.*, 2020, **56**, 1113.
38. S. N. Konchenko, N. P. Gritsan, A. V. Lonchakov, U. Radiusd and A. V. Zibarev, *Mendeleev Commun.*, 2009, **19**, 7.
39. N. A. Pushkarevsky, E. A. Chulanova, L. A. Shundrin, A. I. Smolentsev, G. E. Salnikov, E. A. Pritchina, A. M. Genaev, I. G. Irtegova, I. Y. Bagryanskaya and S. N. Konchenko, *Chem. – Eur. J.*, 2019, **25**, 806.
40. G. Berionni, B. Pégot, J. Marrot and R. Goumont, *CrystEngComm*, 2009, **11**, 986.
41. E. A. Chulanova, E. A. Radiush, I. K. Shundrina, I. Y. Bagryanskaya, N. A. Semenov, J. Beckmann, N. P. Gritsan and A. V. Zibarev, *Cryst. Growth Des.*, 2020, **20**, 5868.
42. Y. Ishigaki, K. Asai, H.-P. J. de Rouville, T. Shimajiri, V. Heitz, H. Fujii-Shinomiya and T. Suzuki, *Eur. J. Org. Chem.*, 2021, 990.
43. H.-T. Huynh, O. Jeannin and M. Fourmigué, *Chem. Commun.*, 2017, **53**, 8467.
44. A. M. S. Riel, H.-T. Huynh, O. Jeannin, O. Berryman and M. Fourmigué, *Cryst. Growth Des.*, 2019, **19**, 1418.
45. A. M. S. Riel, O. Jeannin, O. B. Berryman and M. Fourmigue, *Acta Crystallogr., Sect. B: Struct. Sci., Cryst. Eng. Mater.*, 2019, **75**, 34.
46. H.-T. Huynh, O. Jeannin, E. Aubert, E. Espinosa and M. Fourmigue, *New J. Chem.*, 2021, **45**, 76.
47. M. Beau, S. Lee, S. Kim, W. S. Han, O. Jeannin, M. Fourmigué, E. Aubert, E. Espinosa and I. R. Jeon, *Angew. Chem.*, 2021, **133**, 370–374; M. Beau, S. Lee, S. Kim, W. S. Han, O. Jeannin, M. Fourmigué, E. Aubert, E. Espinosa and I. R. Jeon, *Angew. Chem., Int. Ed.*, 2021, **60**, 366.
48. D. B. Werz, R. Gleiter and F. Rominger, *J. Am. Chem. Soc.*, 2002, **124**, 10638.
49. R. Gleiter, D. B. Werz and B. J. Rausch, *Chem. – Eur. J.*, 2003, **9**, 2676.
50. (a) A. Dhaka, O. Jeannin, I. R. Jeon, E. Aubert, E. Espinosa and M. Fourmigué, *Angew. Chem.*, 2020, **132**, 23789; (b) A. Dhaka, O. Jeannin, I. R. Jeon, E. Aubert, E. Espinosa and M. Fourmigué, *Angew. Chem, Int. Ed.*, 2020, **59**, 23583.
51. A. Dhaka, O. Jeannin, E. Aubert, E. Espinosa and M. Fourmigué, *Chem. Commun.*, 2021, **57**, 4560.
52. A. Kremer, C. Aurisicchio, F. De Leo, B. Ventura, J. Wouters, N. Armaroli, A. Barbieri and D. Bonifazi, *Chem. – Eur. J.*, 2015, **21**, 15377.
53. A. Kremer, A. Fermi, N. Biot, J. Wouters and D. Bonifazi, *Chem. – Eur. J.*, 2016, **22**, 5665.
54. D. Romito, N. Biot, F. Babudri and D. Bonifazi, *New J. Chem.*, 2020, **44**, 6732.
55. N. Biot and D. Bonifazi, *Chem. – Eur. J.*, 2018, **24**, 5439.
56. N. Biot and D. Bonifazi, *Chem. – Eur. J.*, 2020, **26**, 2904.
57. N. Biot, D. Romito and D. Bonifazi, *Cryst. Growth Des.*, 2021, **21**, 536.
58. M. Iwaoka and S. Tomoda, *J. Am. Chem. Soc.*, 1996, **118**, 8807.

59. K. Fujita, M. Iwaoka and S. Tomoda, *Chem. Lett.*, 1994, **23**, 923.
60. K. Fujita, K. Murata, M. Iwaoka and S. Tomoda, *Tetrahedron*, 1997, **53**, 2029.
61. T. Chivers, I. Krouse, M. Parvez, I. Vargas-Baca, T. Ziegler and P. Zoricak, *Inorg. Chem.*, 1996, **35**, 5836.
62. P. J. W. Elder and I. Vargas-Baca, *Phys. Chem. Chem. Phys.*, 2016, **18**, 30740.
63. D. J. Pascoe, K. B. Ling and S. L. Cockroft, *J. Am. Chem. Soc.*, 2017, **139**, 15160.
64. A. V. Rozhkov, E. A. Katlenok, M. V. Zhmykhova, A. Y. Ivanov, M. L. Kuznetsov, N. A. Bokach and V. Y. Kukushkin, *J. Am. Chem. Soc.*, 2021, **143**, 15701.
65. M. S. Taylor, *Coord. Chem. Rev.*, 2020, **413**, 213270.
66. K. T. Mahmudov, V. A. Aliyeva, M. F. C. G. da Silva and A. J. L. Pombeiro, *Halogen Bonding in Solution*, Wiley-VCH GbmH, 1st edn, 2021, pp. 363–382.
67. E. A. Suturina, N. A. Semenov, A. V. Lonchakov, I. Y. Bagryanskaya, Y. V. Gatilov, I. G. Irtegova, N. V. Vasilieva, E. Lork, R. Mews, N. P. Gritsan and A. V. Zibarev, *J. Phys. Chem. A*, 2011, **115**, 4851.
68. A. Y. Makarov, I. G. Irtegova, N. V. Vasilieva, I. Y. Bagryanskaya, T. Borrmann, Y. V. Gatilov, E. Lork, R. Mews, W. D. Stohrer and A. V. Zibarev, *Inorg. Chem.*, 2005, **44**, 7194.
69. N. A. Semenov, A. V. Lonchakov, N. A. Pushkarevsky, E. A. Suturina, V. V. Korolev, E. Lork, V. G. Vasiliev, S. N. Konchenko, J. Beckmann, N. P. Gritsan and A. V. Zibarev, *Organometallics*, 2014, **33**, 4302.
70. N. A. Semenov, A. V. Lonchakov, N. P. Gritsan and A. V. Zibarev, *Russ. Chem. Bull.*, 2015, **64**, 499.
71. N. A. Semenov, D. E. Gorbunov, M. V. Shakhova, G. E. Salnikov, I. Y. Bagryanskaya, V. V. Korolev, J. Beckmann, N. P. Gritsan and A. V. Zibarev, *Chem. – Eur. J.*, 2018, **24**, 12983.
72. G. E. Garrett, E. I. Carrera, D. S. Seferos and M. S. Taylor, *Chem. Commun.*, 2016, **52**, 9881.
73. J. Y. C. Lim, I. Marques, A. L. Thompson, K. E. Christensen, V. Félix and P. D. Beer, *J. Am. Chem. Soc.*, 2017, **139**, 3122.
74. J. Y. C. Lim, J. Y. Liew and P. D. Beer, *Chem. – Eur. J.*, 2018, **24**, 14560.
75. J. Y. C. Lim, I. Marques, V. Félix and P. D. Beer, *Chem. Commun.*, 2018, **54**, 10851.
76. A. Borissov, I. Marques, J. Y. C. Lim, V. Félix, M. D. Smith and P. D. Beer, *J. Am. Chem. Soc.*, 2019, **141**, 4119.
77. L. E. Bickerton, A. Docker, A. J. Sterling, H. Kuhn, F. Duarte, P. D. Beer and M. J. Langton, *Chem. – Eur. J.*, 2021, **27**, 11738.
78. L. E. Bickerton, A. J. Sterling, P. D. Beer, F. Duarte and M. J. Langton, *Chem. Sci.*, 2020, **11**, 4722.
79. (a) A. Docker, C. H. Guthrie, H. Kuhn and P. D. Beer, *Angew. Chem.*, 2021, **133**, 22144; (b) A. Docker, C. H. Guthrie, H. Kuhn and P. D. Beer, *Angew. Chem., Int. Ed.*, 2021, **60**, 21973.

80. S. Benz, M. Macchione, Q. Verolet, J. Mareda, N. Sakai and S. Matile, *J. Am. Chem. Soc.*, 2016, **138**, 9093.
81. E. Navarro-García, B. Galmés, M. D. Velasco, A. Frontera and A. Caballero, *Chem. – Eur. J.*, 2020, **26**, 4706.
82. B. Zhou and F. P. Gabbaï, *Chem. Sci.*, 2020, **11**, 7495.
83. B. Zhou and F. P. Gabbaï, *J. Am. Chem. Soc.*, 2021, **143**, 8625.
84. J. Kübel, P. J. W. Elder, H. A. Jenkins and I. Vargas-Baca, *Dalton Trans.*, 2010, **39**, 11126.
85. P. C. Ho, P. Szydlowski, J. Sinclair, P. J. W. Elder, J. Kübel, C. Gendy, L. M. Lee, H. Jenkins, J. F. Britten, D. R. Morin and I. Vargas-Baca, *Nat. Commun.*, 2016, **7**, 11299.
86. P. C. Ho, R. Bui, A. Cevallos, S. Sequeira, J. F. Britten and I. Vargas-Baca, *Dalton Trans.*, 2019, **48**, 4879.
87. J. Wang, P. C. Ho, J. F. Britten, V. Tomassetti and I. Vargas-Baca, *New J. Chem.*, 2019, **43**, 12601.
88. P. C. Ho, V. Tomassetti, J. F. Britten and I. Vargas-Baca, *Inorg. Chem.*, 2021, **60**, 16726.
89. M. Risto, R. W. Reed, C. M. Robertson, R. Oilunkaniemi, R. L. Laitinen and R. T. Oakley, *Chem. Commun.*, 2008, 3278.
90. L. J. Riwar, N. Trapp, K. Root, R. Zenobi and F. Diederich, *Angew. Chem.*, 2018, **130**, 17506–17512; L. J. Riwar, N. Trapp, K. Root, R. Zenobi and F. Diederich, *Angew. Chem., Int. Ed.*, 2018, **57**, 17259.
91. F.-U. Rahman, D. Tzeli, I. D. Petsalakis, G. Theodorakopoulos, P. Ballester, J. Rebek Jr and Y. Yu, *J. Am. Chem. Soc.*, 2020, **142**, 5876.
92. Y.-J. Zhu, Y. Gao, M.-M. Tang, J. Rebek and Y. Yu, *Chem. Commun.*, 2021, **57**, 1543.

CHAPTER 19

Chalcogen Bond Donors in Organic Synthesis

TIM STEINKE AND STEFAN M. HUBER*

Fakultät für Chemie und Biochemie, Ruhr-Universität Bochum, Universitätsstraße 150, 44801 Bochum, Germany
*Email: stefan.m.huber@rub.de

19.1 Introduction

The noncovalent interaction between Lewis bases and Lewis acidic chalcogen atoms in organic or inorganic compounds is defined as chalcogen bonding (ChB).[1–7]

This interaction is closely related to hydrogen bonding (HB)[8] and halogen bonding (XB).[9,10] Hydrogen bonding in particular has been strongly investigated in recent decades,[11–13] followed by an increased research interest in halogen bonding.[10,14,15] However, chalcogen bonding has been mostly neglected although it has been known from nature for decades.[16–18]

One illustrative example shows the ability of chalcogens to act as Lewis bases (as already commonly known), and as Lewis acids: the complex of iodine with diphenyl diselenide.[19] This complex with a square geometry demonstrates the interrelated interactions of halogen bonding alongside chalcogen bonding (Figure 19.1). First of all, the complex contains two strong halogen bonds with intermolecular distances of 2.99 Å (3.88 Å is the sum of van der Waals radii[20]), in which the chalcogens act as Lewis bases. But also the second face of chalcogens is observed, in which the selenium centers act as Lewis acids and form noncovalent bonds with the Lewis basic region of iodine, with a distance of 3.59 Å (which is again shorter than the

Chalcogen Chemistry: Fundamentals and Applications
Edited by Vito Lippolis, Claudio Santi, Eder J. Lenardão and Antonio L. Braga

Published by the Royal Society of Chemistry, www.rsc.org

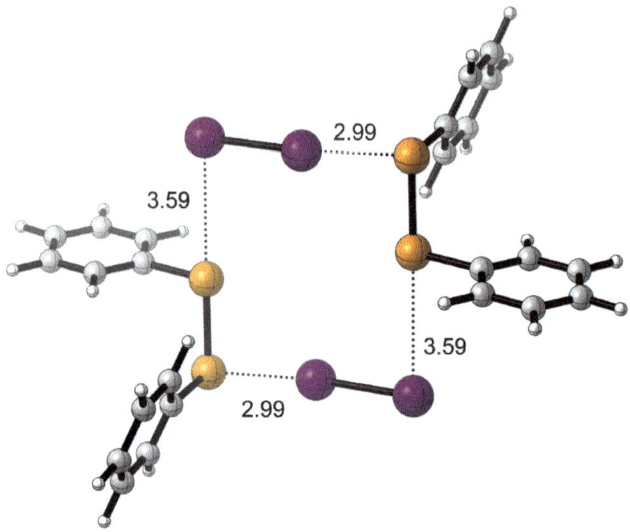

Figure 19.1 Crystal structure of the complex between diphenyl diselenide and iodine.[19,21]

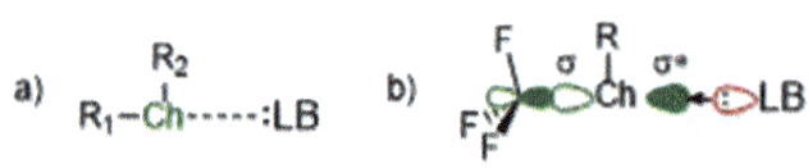

Scheme 19.1 (a) Chalcogen bond. (b) Orbital interaction between a Lewis base and a chalcogen bond donor (Ch = chalcogen).

sum of the van der Waals radii). Therefore, the complex nicely showcases the amphiphilic behavior of both halogens and chalcogens.

Chalcogen bonding can be described by three attractive contributions: (a) charge transfer,[16] (b) electrostatics,[22–25] and (c) dispersion.[26]

The first contribution refers to charge transfer[27] from a lone pair n of the Lewis base to the σ*-orbital of the R_1–Ch and/or R_2–Ch bond (Scheme 19.1b).[16] It is expected that this is the major contribution for strong chalcogen bonding. The electrostatic contribution is explained by the anisotropic electron distribution of the chalcogen due to electron-withdrawing substituents, including an electronegative region and, important for chalcogen bonding, two electron-deficient regions on the elongation of the R_1–Ch and R_2–Ch bonds.[22–25] This electron-deficient region is called "σ-hole"[2,28] and it can undergo interactions with Lewis bases. The last contribution, dispersion, is expected to be important for weak chalcogen bonding.[26]

For the strength of chalcogen bonding, several features of the chalcogen bond donor and the interaction partner have to be considered. The chalcogen bonding strength, first and foremost, depends on the nature of the used chalcogen. Easier polarization and less electronegativity of the chalcogen results in a more anisotropic electron distribution and also a lower lying R–Ch σ* orbital. Therefore, the chalcogen bonding strength increases with the atomic number (O < S < Se < Te). Other than that, the backbone attached to the chalcogen plays a central role as well, as electron-withdrawing groups, such as cationic or polyfluorinated backbones, typically provide stronger Lewis acids.[29,30] The interaction angle needs to be close to 180° to enable a strong interaction.[16,17,31,32] Finally, not only the donating ability of the chalcogen bonding donor is important, but also the basicity of the Lewis base. In contrast, in interactions in which the Lewis basicity of the chalcogens is exploited, the interacting Lewis acids approach perpendicular to the R_1–Ch–R_2 plane. All in all, these electronic origins and especially the characteristics are closely related to halogen bonding, although there is the crucial difference of a second substituent attached to the chalcogen, which on the one hand generates a second electrophilic region, and on the other hand is important for further steric properties.

Although chalcogen bonding is related to hydrogen bonding, there are some crucial differences between those two interactions. First of all, hydrogen bonding is obviously limited to one element. The interaction site for chalcogen bonding can be varied between the elements in their main group. This leads to a superior tunability of chalcogen bonding compared to hydrogen bonding.[29,33] Considering the HSAB principle,[34] hydrogen bonding is supposed to be based on hard Lewis acids. On the other hand, chalcogens are considered soft Lewis acids and should therefore show preferential binding properties to soft Lewis bases.

The interaction angle of chalcogen bonding is somewhat limited to approx. 160–180° whereas hydrogen bonding shows a bigger range of angles. The latter is explained by two facts: (a) the absence of filled p orbitals on hydrogen, and (b) the R–H σ^* orbital being comprised of the 1s orbital on the side of the hydrogen.

Investigations of chalcogen bonding have been relatively rare, especially compared to hydrogen and halogen bonding in recent decades.[7] However, a few applications of this noncovalent interaction have been carried out in the solid state, particularly in the field of crystal engineering.[1,35–40] Also, the use of intramolecular chalcogen bonding in synthesis was investigated long before applications of chalcogen bonding in solution (*e.g.*, for anion recognition and transport) were gaining attention.

In this chapter, we aim to present an overview of chalcogen bonding in organic synthesis. Therefore, we will provide examples for the use of intramolecular chalcogen bonding as well as for intermolecular applications.

19.2 Intramolecular Chalcogen Bonding

19.2.1 Electrophilic Catalysis Assisted with Chalcogen Bonding

First, we want to discuss applications of intramolecular chalcogen bonding in organic synthesis, as these works were also the pioneer applications for intermolecular Lewis acids.

After investigations of Tomoda[41,42] and Singh[43] on such intramolecular chalcogen bonding in organoselenium compounds, further applications emerged, as the potential of enantioselective methoxyselenenylations was discovered.[44] In these matters, diphenyl diselenide compounds with chiral pyrrolidine units in the ortho position to the selenium were synthesized and upon oxidation of the diselenides the corresponding selenenyl bromide species (**1**, Scheme 19.2) was obtained.[45] These compounds were used for the asymmetric methoxyselenenylation of *trans*-β-methylstyrene (**2**) and *trans*-anethole. The asymmetric induction is enabled by the rigidification of the reagent, achieved *via* an intramolecular chalcogen bond of the selenenyl to the chiral pyrrolidine unit. Medium to good yields were obtained of products **3** and **4**, however, up to 53% diastereomeric excess was obtained at −78 °C. Furthermore, the selenenyl bromide species (**4**) was investigated in ^{77}Se NMR experiments, and a downfield shift of the selenium was observed, indicating the intramolecular chalcogen bond.[45]

Whereas Tomoda *et al.* used the interaction of an amine nitrogen to selenium, Wirth and coworkers have chosen a different approach as they extended the concept to selenium oxygen interactions. In this matter, chiral secondary alcohols in the benzyl position were used with cationic selenium moieties in the ortho position.[46,47] The chirality induction by these compounds is once again due to the fixation of the conformation by the intramolecular chalcogen bond between the selenium and the oxygen of the alcohol, leading to a five-membered pseudo-heterocycle (**5**, Scheme 19.3). In the methoxyselenenylation of styrene (**7**, Scheme 19.3) at −78 °C a yield of 70% with a diastereomeric excess of 77% for **8** was obtained using reagent **5**. Lowering the temperature to −100 °C resulted in an increase of the diastereomeric excess to 83% with only a slight drop of the yield to 67%.

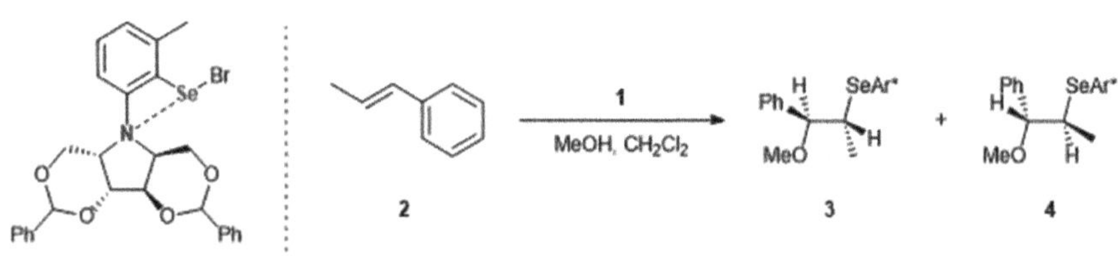

1

Scheme 19.2 Selenenyl bromide **1** employed in the enantioselective methoxyselenenylation.[45]

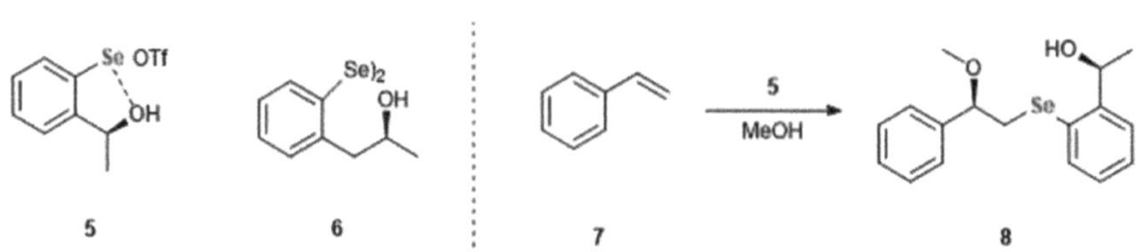

Scheme 19.3 Intramolecular chalcogen bonding induced enantioselective methoxyselenenylation of styrene (**7**).[46] OTf = trifluoromethanesulfate.

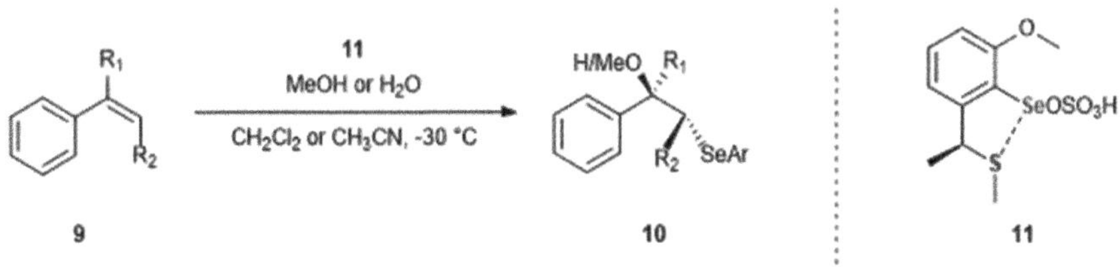

Scheme 19.4 Asymmetric selenomethoxylation and selenohydroxylation in the presence of **11**.[49]

With compound **6**, which forms a six-membered pseudo-heterocycle upon formation of the intramolecular chalcogen bond, the diastereomeric excess dropped to 60% whereas the yield dropped only slightly to 63%. The authors explain this decrease in the diastereomeric excess by the higher flexibility of the conformation of the formed six-membered ring.[46] Following this report, another publication of the group confirmed the chirality transfer based on the fixation of the conformation by intramolecular chalcogen bonding.[47] It should also be mentioned that these selenide compounds were attached to polymers which allowed the recyclability of the catalyst. Again, only modest yields were obtained of 44–72% but up to 80% enantiomeric excess was achieved.[48]

Another advancement on the use of intramolecular chalcogen bonding in organic synthesis was presented by Tiecco in 2002.[49] In this report, the interaction partners were again modified and sulfur from a thioether group was implemented as a Lewis base. The Lewis acidic selenium in these compounds is a selenenyl sulfate, which is obtained from the respective diselenide. Yields from 60 to 79% of **10** (Scheme 19.4) were obtained with diastereomeric ratios up to 98 : 2 in the selenomethoxylation and selenohydroxylation of alkenes (*e.g.*, **9**) at temperatures of only −30 °C with compound **22**. Previously, temperatures of −78 °C and −100 °C were necessary in order to obtain suitable diastereomeric ratios.

In summary, chiral diselenides have found valuable applications in asymmetric methoxy- and hydroxy-selenenylation of alkenes. The stereoinduction in the examples was based on intramolecular chalcogen bonds of selenium to different intramolecular chiral Lewis bases (*i.e.* Lewis bases attached to a chiral moiety), which leads to fixation of the conformation and therefore to the stereoinduction.

19.2.2 Nucleophilic Catalysis Assisted with Chalcogen Bonding

A different approach to use Lewis acidic chalcogens is to implement them into heterocycles, as these compounds can also form intramolecular chalcogen bonds to Lewis bases.[50–56] One of the first applications of this concept was presented by Romo in 2010, using the chiral isothiourea derivative (*S*)-HBTM (**15**) in a nucleophile-catalyzed aldol lactonization (NCAL)

(Scheme 19.5).[57] This catalyst was previously developed and applied by Birman *et al.* in kinetic resolutions of aryl-cycloalkanols[58] as well as of α-aryloxy- and α-arylthioalkanoic acids.[59,60] Bicyclic β-lactones (such as **13**) were obtained by Romo in yields up to 93% with an enantiomeric excess as high as 90%.[57] Similar results were obtained for tricyclic β-lactones with a slightly reduced yield of 85%, but 98% enantiomeric excess. The authors explained the high enantiomeric excess with an intramolecular $n_O \rightarrow \sigma^*_{C-S}$ interaction formed by the sulfur of (*S*)-HBTM and the remaining oxygen of the former carboxylic acid, fixing the conformation in the transition state (**14**).

An intramolecular chalcogen bond was also postulated in the intramolecular Michael-addition lactonization of enone carboxylic acid (**16**, Scheme 19.6) with tetramisole hydrochloride (**18**), reported by Smith

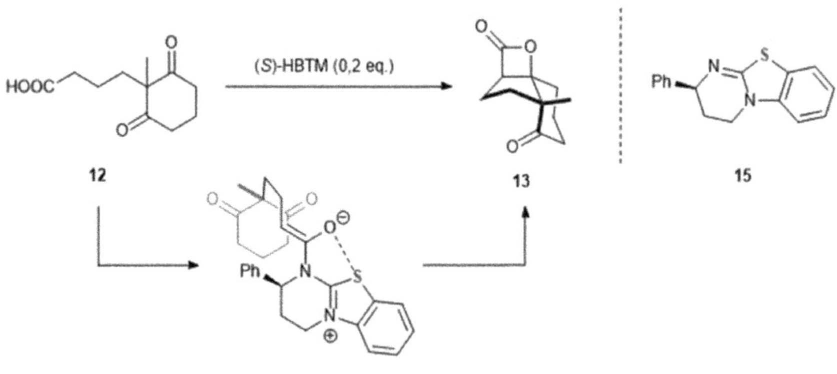

Scheme 19.5 Nucleophile-catalyzed aldol lactonization (NCAL) of **12** with catalyst **15**. The intramolecular chalcogen bond rigidifies the intermediate (**14**) and ultimately induces chirality.[57]

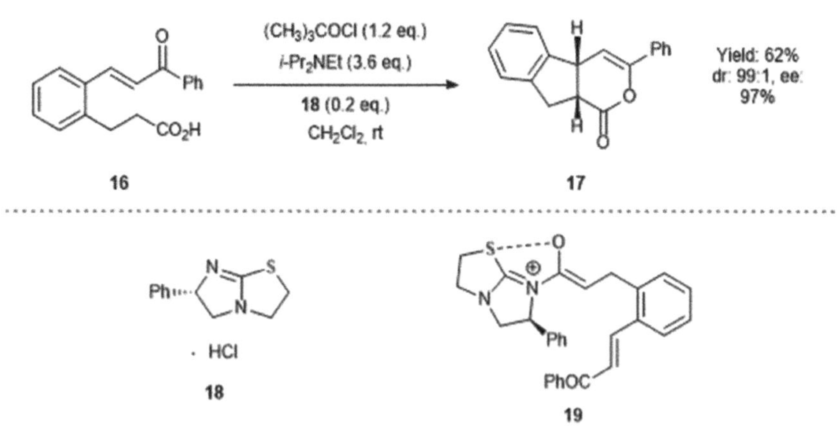

Scheme 19.6 Enantioselective intramolecular Michael-addition lactonization of enone carboxylic acid **16** by intramolecular chalcogen bonding in the presence of **18**.[61]

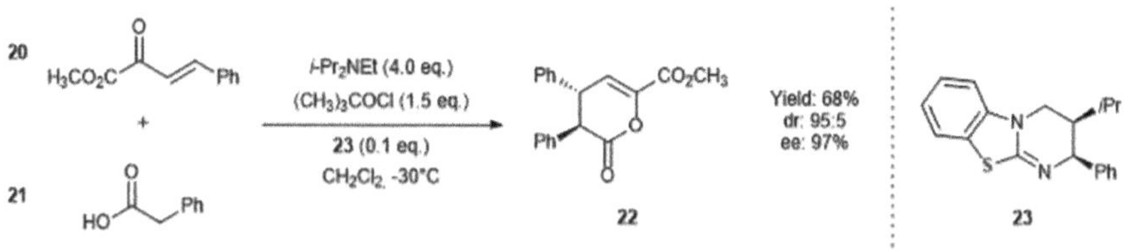

Scheme 19.7 Intermolecular Michael-addition lactonization by intramolecular chalcogen bonding applying catalyst **23.** Intramolecular chalcogen bond of the enolate and the sulfur of **18** which rigidifies the transition state **19** and induces chirality.[61]

et al. in 2011.[61] Employing 0.2 mol% of the latter compound in catalytic amounts along with the activating agent pivaloyl chloride, compound **17** was obtained in a reasonable yield of 62%. However, an impressive diastereomeric ratio of 99 : 1 was achieved, as well as 97% enantiomeric excess at room temperature. Appling isothiourea **23** (Scheme 19.7), on the other hand, increased the yield to 72% while the enantiocontrol decreased to a diastereomeric ratio of 80 : 20 and 70% ee. Several different derivatives with substituents attached to the benzene moiety of the enone and dihydrobenzofurane derivatives were then tested (employing again **18**) and yielded comparable diastereomeric ratios and enantiomeric excess.

Interestingly, in the intermolecular variant of the reaction using phenylacetic acid and a chalcone derivative as the starting material, catalyst **18** showed decent yield along with diastereomeric control. However, catalyst **23**, which was previously less suitable in the intramolecular Michael addition, is ahead of **18** in terms of yield, diastereomeric ratio, and enantiomeric excess (Scheme 19.7).

In both cases, the enantioselective induction was postulated to be based on the stabilizing intramolecular $n_O \rightarrow \sigma^*_{C-S}$ chalcogen bond with similar pathways. This pathway includes the *N*-acylation of the isothiourea by the formed anhydride of the carboxylic acid with the activating agent. After the deprotonation of the α-proton of the acyl ammonium transition state, the corresponding ammonium enolate is formed and the oxygen of the enolate, which is *syn* to the sulfur, features the intramolecular chalcogen bond (**19**, Scheme 19.6). Later in 2013, Smith and co-workers extended the use of isothioureas to the anellation of α,β-unsaturated acyl ammonium compounds with 1,3-diketones, β-ketoester, and azaarylketones using a modified HBTM derivative (**23**).[62] In the following years, several more applications of related nucleophilic catalysis, assisted by intramolecular chalcogen bonding in asymmetric organic synthesis, were reported such as an asymmetric [2 + 2] cycloaddition of arylacetic acids and arylacetic anhydrides with *N*-sulfonylaldimines,[63] the [3 + 2] cycloaddition to form 1,5-benzothiazepines[64] and the addition of 4-nitrophenyl esters to iminium ions.[65] However, all of these examples have one thing in common: sulfur is the Lewis acidic chalcogen that is used to support the nucleophilic catalysis with chalcogen bonding. The report of Smith and co-workers in 2020 was, to the best of our

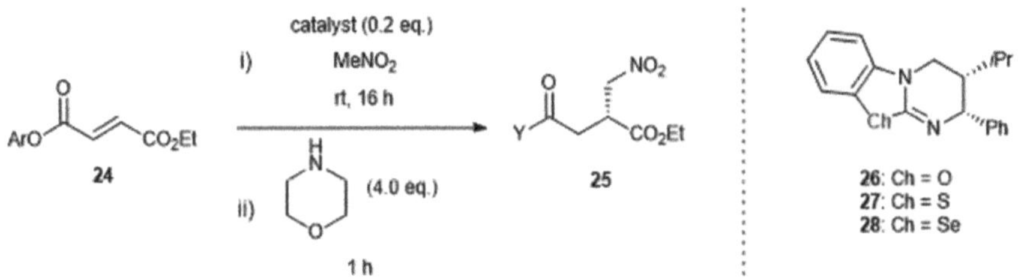

Scheme 19.8 HBTM derivatives with different chalcogens as a Lewis acidic center in a Michael addition of **24** and nitromethane to give **25**.[66]

knowledge, the first application of selenium as a Lewis acidic center in this regard.[66] In this report, the corresponding oxygen and selenium analogs (**26** and **28**, Scheme 19.8) of HBTM derivative **27** were synthesized and investigated *via* X-ray crystallography. In these experiments, the oxygen derivative of HBTM **26** does not feature an O···O contact and the compound shows an *anti*-coplanar conformation. The sulfur (**27**) and selenium (**28**) containing compound, on the other hand, feature a coplanar *syn* conformation due to the formed intramolecular S/Se···O chalcogen bond, which fixes the conformation of the molecule. The expected red-shifts are indeed observed in IR spectroscopic experiments along with downfield shifts of the carbonyl carbon in ^{13}C NMR for the sulfur- and selenium-containing compounds, which were not observed for the oxygen species. The three compounds were used for the kinetic resolution of secondary and tertiary alcohols. Due to the missing stabilization of the *N*-acyl intermediate, no acylation was observed for **26**, whereas **27** and **28** were suitable for the kinetic resolution and showed comparable results in terms of yield and selectivity. Finally, the three catalysts were applied in the Michael addition of nitromethane to α,β-unsaturated esters to investigate the catalytic activity and the enantiocontrol of the catalysts. As expected, **26** did not show any activity. The corresponding sulfur and selenium compounds were similar in terms of activity (both 55%) and enantiomeric ratio (95 : 5 for **27**, 96 : 4 for **28**). Kinetic observations *via* ^{19}F NMR experiments revealed a stronger acceleration of the reaction for the more Lewis acidic selenium-based compounds **28**.

All in all, intramolecular chalcogen bonding was used to rigidify intermediates in nucleophilic catalysis, in order to induce chirality. This interaction turned out to be crucial, as equivalent molecules that could not form the respective interaction also could not induce chirality. The Lewis acidic active site in this field was mostly limited to sulfur, however lately the first use of selenium centers was presented.

19.3 Intermolecular Chalcogen Bonding

After discussing several applications of intramolecular chalcogen bonding, we now aim to present the still very recent developments of intermolecular chalcogen bonding in organic synthesis. Intermolecular chalcogen bonding was successfully applied in solid state systems,[35–38,67] anion recognition,[68–71] and transport.[72,73] Especially since the very familiar interaction, halogen bonding, was successfully applied in organocatalysis,[14] the interest in applying chalcogen bonding in organocatalysis was obviously increasing. This subchapter will be divided into two parts, referring to the activated substrates. Whereas the first application of chalcogen bonding was achieved by Matile and co-workers in 2016 with the activation of neutral compounds,[74] we will start with the development in the activation of carbon–halide bonds. Subsequently, we will discuss the more recent applications of chalcogen bonding in further neutral compound activation.

19.3.1 Carbon–Halide Bond Activation by Chalcogen Bonding

The first activation of a carbon–halide bond was reported referring to the carbon–bromide bond in benzhydryl bromide (**29**), using cationic selenium-based chalcogen bond donors (**31**/**32**, Scheme 19.9).[75] The catalysts were based on a 1,3-bis(benzimidazolium)benzene scaffold with a trifluoromethane unit in the 2 position of the core, which freezes the C–N bond rotation between the central benzene and the benzimidazolium unit, resulting in two atropisomers of the catalyst, a preorganized *syn* atropisomer (**31**) and the corresponding *anti* atropisomer (**32**). With one equivalent **31**, a yield of 64% was observed within 140 h reaction time. The corresponding *anti* atropisomer **32** produced the expected smaller yield of 45%. Interestingly, the halogen bond donor analog (using bromine as halogen) was less active with 35% product formation after 140 h, indicating a superior activity of chalcogen bond donors compared to halogen bond donors (featuring elements of the same row). Compared to Matile and co-workers, who applied their chalcogen bond donors in catalytic amounts,[74] the Lewis acids in this report were used as an activator in stoichiometric amounts.[75] The use of several reference compounds excluded any other activation of the reaction by other interactions and traced the activation back to chalcogen bonding.

The same reaction was later used by Matile and co-workers as well, as they introduced cationic species of their previously used selenium-based catalysts (see below).[76]

After catalyst **31** was successfully applied in stoichiometric amounts, Huber *et al.* presented the catalytic application of the structurally strongly related catalyst **36** (Scheme 19.10) in the carbon–chloride bond activation of

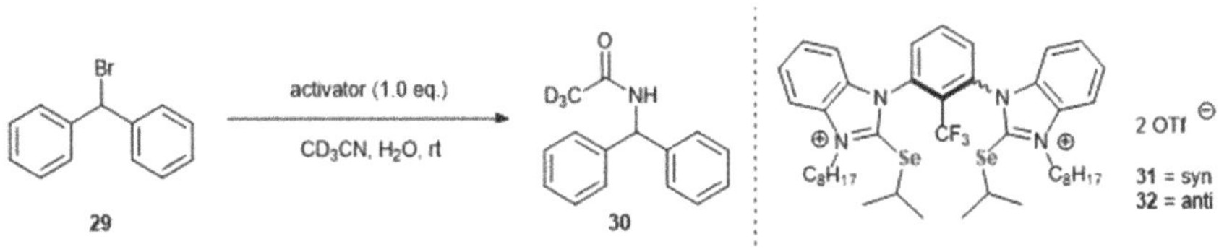

Scheme 19.9 Chalcogen bonding activated bromide abstraction of benzhydryl bromide (**29**).[75]

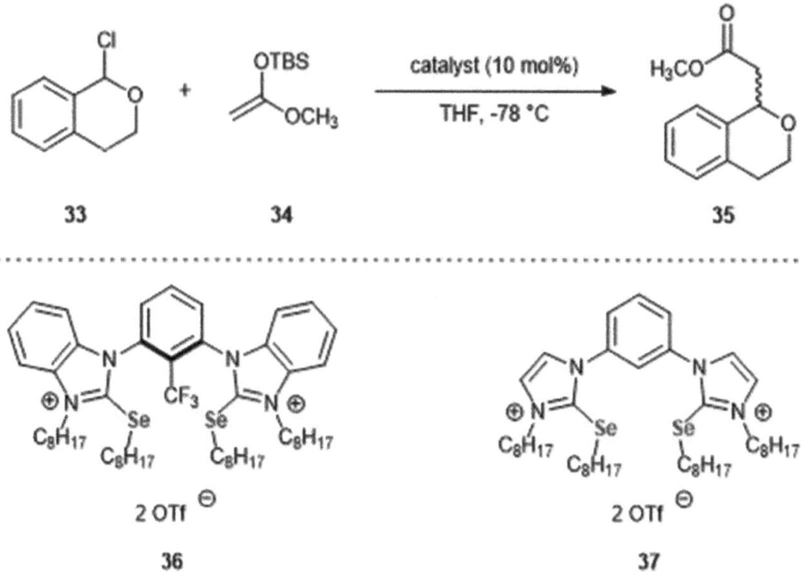

Scheme 19.10 Chloride abstraction of 1-chloroisochroman (**33**) catalyzed by selenium based chalcogen bond donors (**36** and **37**).[77]

1-chloroisochroman (**33**).[77] After 118 h, a yield of 92% was obtained with this catalyst. The corresponding *anti* atropisomer and the brominated halogen bonding analog were again less active with only 24% and 40% of **35** being obtained. These results indicate again the power of preorganization and the possible superior activity of chalcogen bonding over halogen bonding. Furthermore, the related 1,3-bis(imidazolium)benzene-based chalcogen bond donor (**37**) was introduced and applied in this reaction, however no catalytic activity of this compound was observed.

In 2018, Matile and co-workers used the same chloride abstraction of 1-chloroisochroman (**33**) for comparison of halogen, chalcogen, and pnictogen bonding catalysis.[33] In this publication, monodentate bis(pentafluorophenyl)tellurium and -selenium were used. After 55 h reaction time, a 48% yield of the product was determined for the tellurium-based compound, whereas only 6% yield was observed for the selenium-based analog. Since neutral chalcogen bond donors were used, which are expected to be less active than cationic species, but the reaction time was also halved, the great potential of applying tellurium as the Lewis acidic center is shown.

In 2020, the Huber group published another example that emphasizes the powerful properties of tellurium-based chalcogen bond donors.[78] Again, the chloride abstraction of 1-chloroisochroman (**33**) was used as a simple benchmark reaction. In this report, 1,3-bis(triazolium)benzene-based chalcogen bond donors (**38–40**, Figure 19.2), which were previously introduced,[79,80] were applied and a yield of 83% for product **35** was obtained after 12 h reaction time using catalyst **38**. This is a reduction of the reaction time by a factor of ten compared to the first published example of this reaction using chalcogen bonding.[77] The corresponding selenium-based catalyst **39** was less active, as only 15% yield was determined. Also, the use of weakly coordinating anions was promoted, as catalyst **38,** which contains the weak coordinating tetrakis(3,5-bis(trifluoromethyl)phenyl)borate ($BArF^4$) anion, showing faster acceleration in kinetic experiments compared to **40** (which uses the stronger coordinating triflate anion).

19.3.2 Neutral Compound Activation by Chalcogen Bonding

The first successful application of chalcogen bond donors in organic synthesis was presented by Matile *et al.* in 2016.[74] Quinoline derivatives were catalytically activated for their reduction by sulfur-based chalcogen-bond donors in the presence of a Hantzsch ester (Scheme 19.11). The neutral

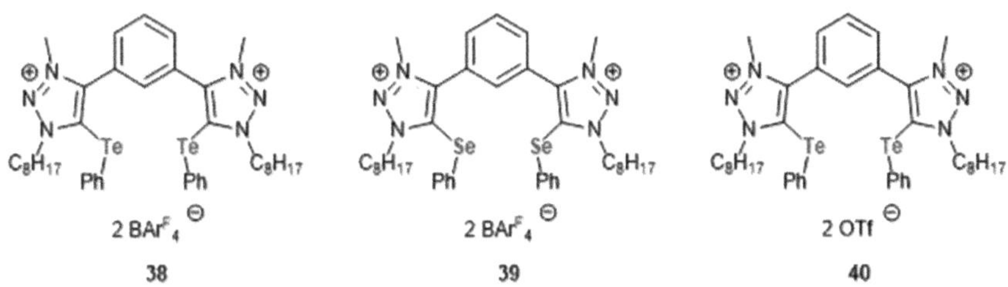

Figure 19.2 1,3-Bistriazoliumbenzene-based chalogen bond donors.[78] $BArF_4$ = tetrakis(3,5-bis(trifluoromethyl)phenyl)borate.

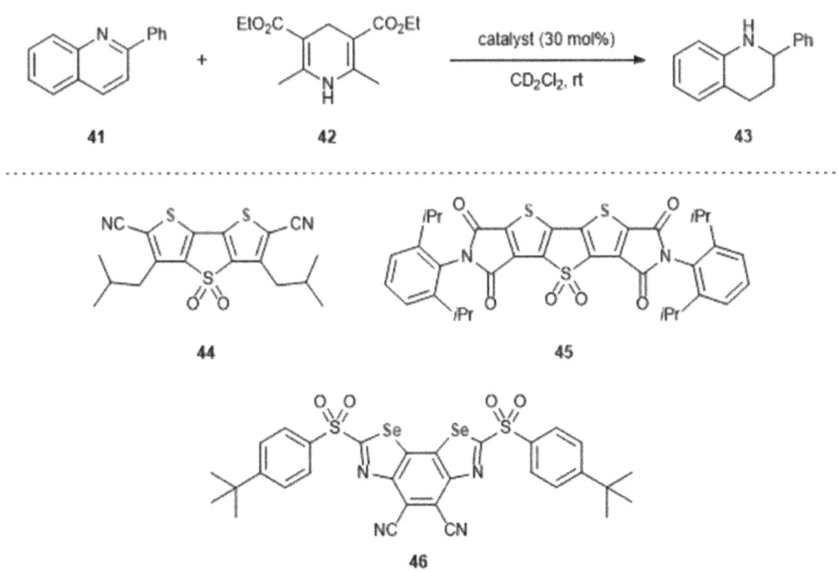

Scheme 19.11 Chalcogen bonding catalyzed reduction of 2-phenylquinoline (**41**) by Matile and co-workers.[74,81]

catalysts were based on the dithienothiophene (DTT) motif (**44**) that was already known from anion recognition and anion transport by the same group. Also, a modified bisimide motive of DTT (**45**, DDI) was applied.

With 30 mol% of the DTT-based catalyst **44**, 96% yield was obtained after 12 days. However, computational studies and fluorescence measurements with chloride as the Lewis base revealed the higher Lewis acidity of the DDI-based catalysts **45**. Those studies were confirmed by experimental data on catalysis, as an increased acceleration of the reaction was obtained with **45** and thus a similar yield was obtained after only 7 days with reduced amounts of 10 mol% of the respective catalyst.

After this first successful application of sulfur-based chalcogen bond donors in organocatalysis, the herein used catalysts were modified and the Lewis acidic center was changed to selenium.[81] With this modification, catalysts **46** was obtained, and the acceleration for the transfer hydrogenation of quinoline derivatives was improved from a rate enhancement of 1290[74] with **45** to 1.5×10^5 with **46**.[81]

In 2019, this reaction was used by the Huber group as a benchmark.[82] Cationic sulfur- and selenium-based ChB donors with different catalyst scaffolds were investigated, as again the preorganized 1,3-bis(benzimidazolium)-2-trifluoromethylbenzene scaffold (**47**, Figure 19.3) was applied along with the flexible 1,3-bis(benzimidazolium)benzene scaffold (**48**) and the corresponding 1,3-bis(imidazolium)benzene derivative (**49**). In the end, the 1,3-bis(imidazolium)benzene-based catalysts showed the lowest activity with 21% and 15% of **43** (Ch = S, Se) after 12 h at room temperature. The flexible catalyst $\mathbf{48^{Se}}$ and the preorganized catalyst $\mathbf{47^{Se}}$ produced comparable yields of 95% (for $\mathbf{48^{Se}}$) and 91% ($\mathbf{47^{Se}}$). Interestingly the *anti*-atropisomers also showed similar activities as their *syn*-atropisomers and moreover, for each scaffold, the sulfur-based catalysts were as active as their selenium-based analogs.

As the scope of substrates that are activated by chalcogen bonding grew, Wang presented the first activation of a carbonyl group in 2019.[83] Different β-ketoaldehydes (such as **51**, Scheme 19.12) were converted with indoles into seven-membered N-heterocycles in good yields. Several mono- and bidentate catalysts were used, however the best yield was obtained with catalyst **53**.

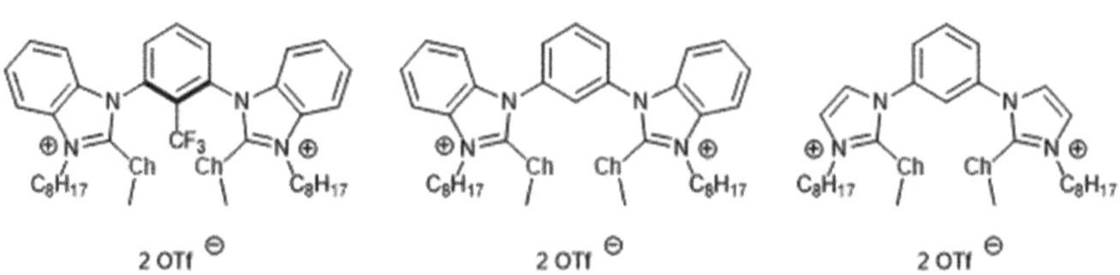

Figure 19.3 Chalcogen bonding catalysts used in the reduction of 2-phenylquinoline by Huber and co-workers.[82]

Adding the weakly coordinating anions tetrafluoroborate or tetrakis(3,5-bis(trifluoromethyl)phenyl)borate (both as their sodium salts) along with the drying agent sodium sulfate accelerated the reaction further, as the catalysts with weaker coordinating anions were formed *in situ*. The scope of substrates was investigated, and several electronically different indoles and also heterocyclic substituents were feasible for this reaction. Several reference experiments indicated an activation based on chalcogen bonding and confirmed the stability of the catalyst. Furthermore, certain additives which inhibit the catalysts were included to further stress the activation by chalcogen bonding.

A further report was presented by Wang and co-workers in 2020, in which similar catalysts were applied in a Rauhut–Currier type reaction, introducing a "dual chalcogen–chalcogen bonding catalysis approach" (Scheme 19.13).[84] First, the interaction of enone **54** and methanol was observed with catalyst **53**. This was further investigated using ^{13}C, ^{31}P, and ^{77}Se NMR spectroscopy, confirming the interaction between the chalcogen bonding donor and the Lewis bases. The goal of this study was the simultaneous activation of enone **54** and methanol by chalcogen bond donor **53** to push the reaction of the two substrates to a somewhat more intramolecular fashion. Therefore, the ratio between enone and methanol and the addition procedure proved to be crucial, as the absence or the excess of reaction partners, as well as alternating mixing orders led to inhibitions. However, using 15 mol% of catalyst **53** along with 0.6 equivalents of $NaBF_4$ and 5 equivalents of methanol in the presence of the drying agent Na_2SO_4 in dichloroethane, a conversion of 74% was determined.

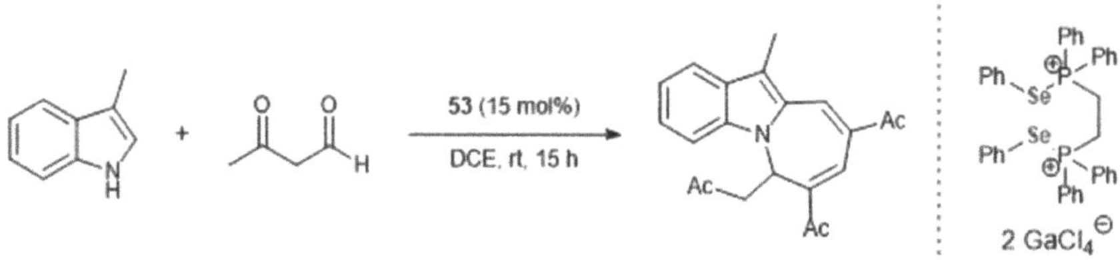

50 51 52 53

Scheme 19.12 Formation of seven-membered heterocycle **52** catalyzed by chalcogen bond donor **53**.[83]

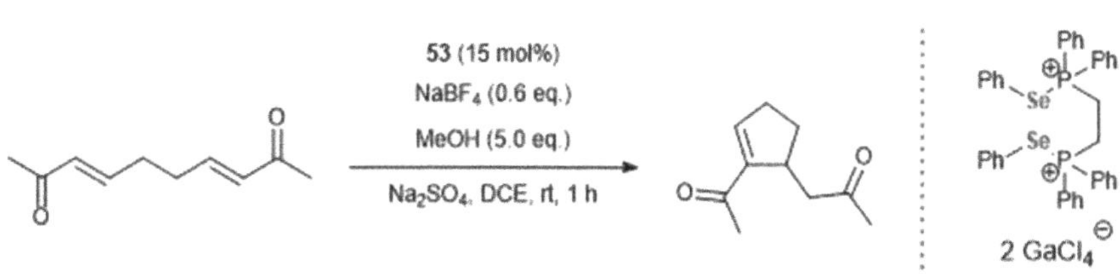

54 55 53

Scheme 19.13 Dual chalcogen bonding catalysis.[84]

Meanwhile, Huber and co-workers reported the activation of crotonophenone (**57**) for a Michael addition, applying a 1,3-bis(triazolium)benzene-based catalyst scaffold using tellurium, selenium, and sulfur as Lewis acidic centers (Scheme 19.14).[80] In this report, the distinct catalytic potential of the three chalcogens is illustrated. Tellurium-based catalyst **38** showed the strongest activation of the reaction, as over 95% of **58** was observed after 4 hours reaction time. With the selenium-based analog (**39**) 16% yield was obtained, whereas no activation of the reaction was possible for the sulfur-based catalyst **62**. Thus, the expectation of tellurium being the most active chalcogen center due to its polarizability is confirmed, as well as selenium being less active than tellurium but more active than the sulfur-containing compound. Furthermore, a counter anion effect was once again visible for the tellurium-based catalysts. After 4 h reaction time, the triflate compound **40** provided 57% yield, whereas the BF_4 containing analog **59** gave 46% yield. This is in line with the expected activity, as both catalysts are supposed to be weaker than the $BArF_4$ containing one. However, the catalyst loading could be reduced to 7.5 mol% for catalyst **38** without a big drop in activity (82% yield after 4 h). Interestingly, the corresponding halogen bond donor and hydrogen bond donor were not active in this reaction. Also other reference compounds were demonstrated to be inactive, showcasing the superior properties of the 1,3-bis(triazolium)benzene-based chalcogen bond donor. Finally, the substrate scope for the indoles was investigated using electron-rich and electron-poor indoles, as well as a sterically hindered one, and yields between 64–99% were observed.

Wang and co-workers then reported a chalcogen bonding catalyzed cyanosilylation of several cyclic, acyclic, α,β-unsaturated, and sterically hindered ketones with ppm-level catalyst loading (Scheme 19.15).[85] Whereas no

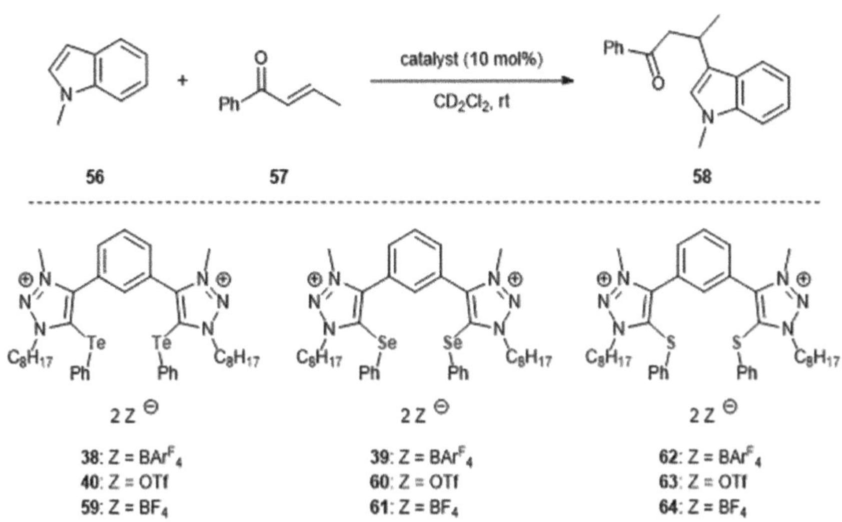

Scheme 19.14 Michael reaction of 1-methylindole (**56**) and *trans*-crotonophenone (**57**) catalyzed by 10 mol% chalcogen bonding donor.[80]

reaction took place in the absence of any catalyst, 94% yield was obtained after only 8 minutes at room temperature in dichloromethane by employing 15 ppm of catalyst **67** to the reaction (Scheme 19.15). As Lewis basic solvents may interfere with the catalyst, the reaction was also tested under neat conditions without any solvent. In this case, 94% yield of **66** was obtained after only 2 minutes with catalyst **67**. Lowering the catalyst load to 10 and 5 ppm was therefore possible. An analysis of the mixtures of catalyst **67** with either ketone **65** or trimethylsilylformonitrile indicated an interaction between the catalyst and the ketone as well as an interaction with the cyanide. The authors conclude that catalyst **67** binds both the ketone and the cyanide to form a very reactive species which undergoes cyanosilylation very fast.

Furthermore, the same group observed an interaction between Lewis acidic sulfur and selenium centers with a π-system in single-crystal structures of monodentate catalysts, which might allow the activation of π-systems. Indeed, an activation of vinyl-substituted indoles for a Diels–Alder [4 + 2] cycloaddition reaction was later realized.[86]

Besides these numerous examples for carbonyl activation, other functional groups were targeted by chalcogen bonding as well. One such report is the activation of *trans*-β-nitrostyrene (**69**, Scheme 19.16) in a Michael

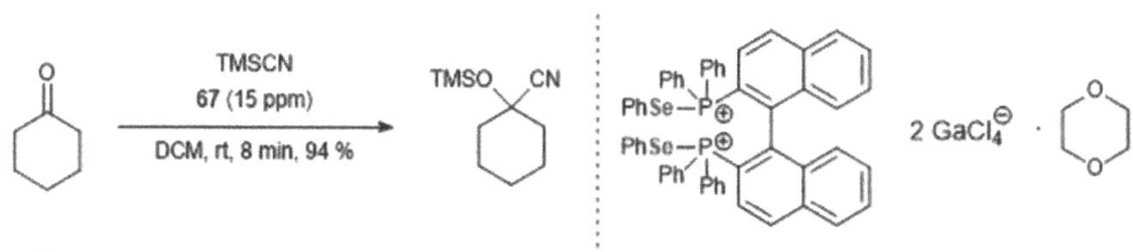

Scheme 19.15 Cyanosilylation of cyclohexanone (**65**) with a catalyst loading of 15 ppm of **67**.[85]

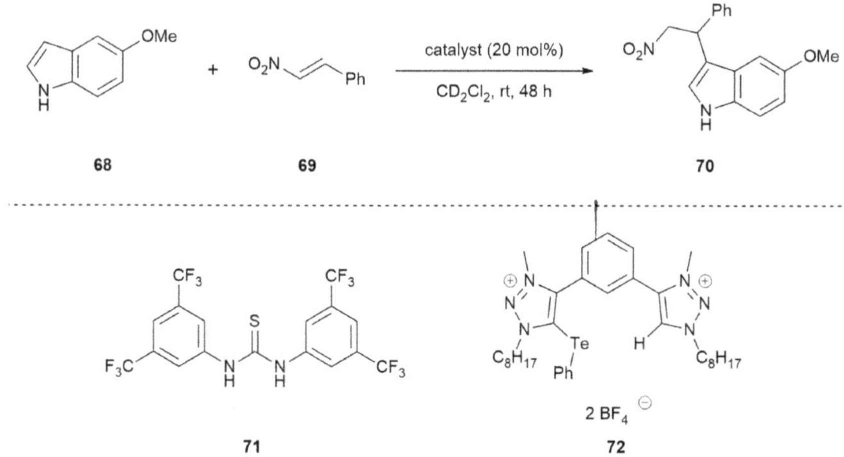

Scheme 19.16 Catalyzed nitro-Michael reaction by chalcogen bonding.[79]

addition by the Huber group in 2019.[79] This reaction between 5-methoxyindole (**68**) and *trans*-β-nitrostyrene (**69**), which was previously not catalyzed by halogen bonding, does not take place in the absence of any activating agent. Furthermore, widely used hydrogen bonding donor **71** was also inactive, along with further reference compounds. With chalcogen bonding catalyst **59** (Scheme 19.14), 78% of **70** was obtained after a reaction time of 48 h. The corresponding selenium- (**61**) and sulfur- (**64**) containing catalysts were also not active, stressing the superior Lewis acidic properties of tellurium compared to selenium and sulfur. Employing the weakly co-ordinating anion $BArF_4$ led to a stronger acceleration of the reaction. Complementary DFT calculations were performed, which indicated a two-point coordination of the chalcogen bonding catalyst to the oxygens of the nitro group. Employing the monodentate catalyst **72** in the reaction further confirmed this result, as a lower yield is obtained compared to **59**.

All these examples for the use of intermolecular chalcogen bonding in organic synthesis are based on divalent chalcogen species with tellurium(II) and selenium(II) centers. Only the report of Lenardão mentioned selenonium-based co-catalysts as ionic liquids in a Baylis–Hillman reaction.[87] However, recently the use of trivalent telluronium species was published independently by the group of Gabbai[88] and Mamane.[89] The former used mono- and bidentate telluronium salts (**73** and **74**) in the chloride abstraction of **metal**-chloride species, enabling different cyclizations.[88] A cyclization reaction (Scheme 19.17), previously studied in terms of halogen bonding,[90] was successfully activated by both **73** and **74**.[88] The chalcogen bonding donors abstract chloride from the transition metal, forming the active species of the gold(I) complex. Without the chalcogen

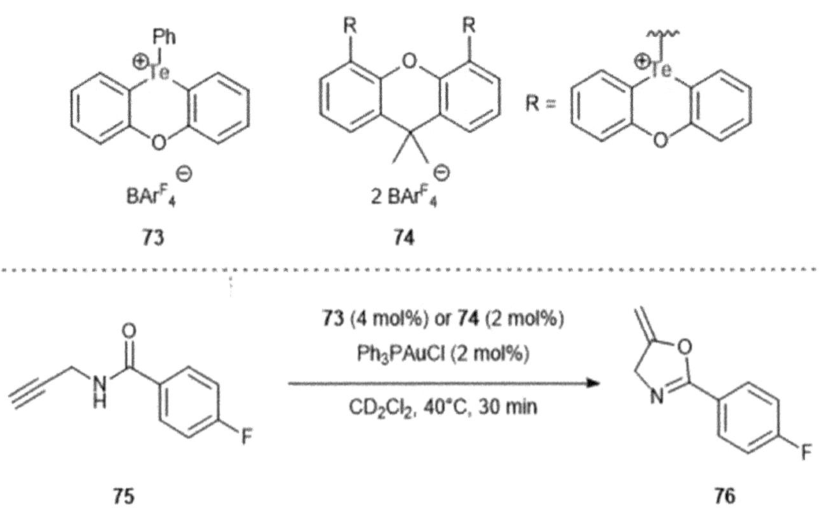

Scheme 19.17 Chalcogen bonding-based chloride abstraction yielding the formation of the active gold species for the cyclization of **75** to **76**.[88]

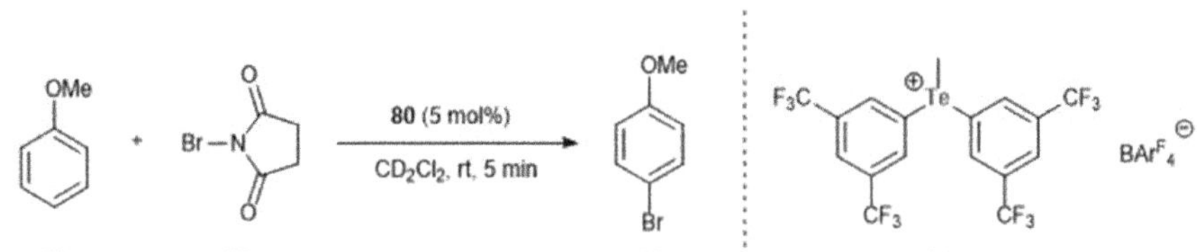

77 78 79 80

Scheme 19.18 Chalcogen bonding catalyzed bromination of anisol (**77**) by monodentate telluronium cation **80**.[89]

bonding donor, no reaction takes place. The bidentate activator **74** showed a superior activation of the reaction, as 95% conversion of **76** was detected after 30 minutes reaction time at 40 °C, with 2 mol% of **74** and the gold catalyst. For **73**, 4 mol% were used, but a lower conversion of only 36% was observed, stressing a positive cooperative effect of activator **74**.

Mamane and co-workers used the monodentate telluronium cation **80** (Scheme 19.18) for the bromination of anisol with *N*-bromosuccinimide (NBS).[89] This reaction also does not take place without the presence of the chalcogen bonding donor. Therefore, the authors postulate that the σ* orbital of the N–Br bond is lowered by the interaction of **80** with the carbonyl group of NBS. In this manner, a yield of 96% was observed using **80**. Several more telluronium species were tested, which vary in the counter anion and the substituents located at the phenyl rings. However, the depicted telluronium species is the most active in this report. Besides this reaction, a bromolactonization of different ω-unsaturated carboxylic acids and the condensation of Danishefsky's diene with imines were successfully catalyzed by chalcogen bonding using **80**. Furthermore, the same reaction was previously also catalyzed by the corresponding selenonium cations.[91]

19.4 Summary

In this chapter, we have presented several fascinating applications of chalcogen bonding in organic synthesis. Whereas the similar interactions of hydrogen and halogen bonding have been studied in more detail, chalcogen bonding,[16–18] known at least since the 1970s, has been less extensively investigated. After initial applications in solid-state chemistry,[1,35–40] the focus shifted more toward applications in solution.

In the 1990s, different diselenides and their related cationic species were used to rigidify reaction intermediates or reagents, in order to induce chirality. With this approach, excellent results were obtained in methoxyselenenylation reactions. Furthermore, intramolecular chalcogen bonding was used to induce chirality in nucleophilic catalysis in aldol lactonization and intramolecular Michael addition reactions. In these reports, sulfur was commonly used as a Lewis acidic interaction site, however, the first example with selenium as the Lewis acidic center was also provided lately.

Within recent years, several applications of intermolecular chalcogen bonding interactions in catalysis were presented, introducing different cationic and neutral, as well as mono- and bidentate chalcogen bond donors. As Lewis acidic sites, sulfur, selenium, and tellurium were employed. To date, several different functional groups were successfully activated for different types of reactions. Even though many challenges remain, these proof-of-principle type studies have clearly illustrated the potential of this quite unexplored interaction in organic synthesis, and it is foreseeable that its utilization will only pick up speed in the following years.

References

1. D. B. Werz, R. Gleiter and F. Rominger, *J. Am. Chem. Soc.*, 2002, **124**, 10638.
2. J. S. Murray, P. Lane, T. Clark and P. Politzer, *J. Mol. Model.*, 2007, **13**, 1033.
3. K. T. Mahmudov, M. N. Kopylovich, M. F. C. Guedes da Silva and A. J. L. Pombeiro, *Dalton Trans.*, 2017, **46**, 10121.
4. J. Y. C. Lim and P. D. Beer, *Chem*, 2018, **4**, 731.
5. S. Kolb, G. A. Oliver and D. B. Werz, *Angew. Chem., Int. Ed.*, 2020, **59**, 22306.
6. J. Bamberger, F. Ostler and O. G. Mancheño, *ChemCatChem*, 2019, **11**, 5198.
7. L. Vogel, P. Wonner and S. M. Huber, *Angew. Chem., Int. Ed.*, 2019, **58**, 1880.
8. E. Arunan, G. R. Desiraju, R. A. Klein, J. Sadlej, S. Scheiner, I. Alkorta, D. C. Clary, R. H. Crabtree, J. J. Dannenberg, P. Hobza, H. G. Kjaergaard, A. C. Legon, B. Mennucci and D. J. Nesbitt, *Pure Appl. Chem.*, 2011, **83**, 1619.
9. G. R. Desiraju, P. S. Ho, L. Kloo, A. C. Legon, R. Marquardt, P. Metrangolo, P. Politzer, G. Resnati and K. Rissanen, *Pure Appl. Chem.*, 2013, **85**, 1711.
10. G. Cavallo, P. Metrangolo, R. Milani, T. Pilati, A. Priimagi, G. Resnati and G. Terraneo, *Chem. Rev.*, 2016, **116**, 2478.
11. S. J. Connon, *Chem. – Eur. J.*, 2006, **12**, 5418.
12. A. G. Doyle and E. N. Jacobsen, *Chem. Rev.*, 2007, **107**, 5713.
13. M. S. Taylor and E. N. Jacobsen, *Angew. Chem., Int. Ed.*, 2006, **45**, 1520.
14. D. Bulfield and S. M. Huber, *Chem. – Eur. J.*, 2016, **22**, 14434.
15. L. C. Gilday, S. W. Robinson, T. A. Barendt, M. J. Langton, B. R. Mullaney and P. D. Beer, *Chem. Rev.*, 2015, **115**, 7118.
16. R. E. Rosenfield, R. Parthasarathy and J. D. Dunitz, *J. Am. Chem. Soc.*, 1977, **99**, 4860.
17. R. Weiss, C. Schlierf and K. Schloter, *J. Am. Chem. Soc.*, 1976, **98**, 4668.
18. R. Laitinen, R. Steudel and R. Weiss, *J. Chem. Soc., Dalton Trans.*, 1986, 1095.
19. S. Kubiniok, W. W. Du Mont, S. Pohl and W. Saak, *Angew. Chem., Int. Ed.*, 1988, **27**, 431.

20. A. Bondi, *J. Phys. Chem.*, 1964, **68**, 441.
21. C. Legault, C. Y., 1.0b, Université de Sherbrooke, 2009, http://www.cylview.org
22. J. G. Angyan, R. A. Poirier, A. Kucsman and I. G. Csizmadia, *J. Am. Chem. Soc.*, 1987, **109**, 2237.
23. N. W. Alcock, *Adv. Inorg. Chem.*, 1972, **15**, 1.
24. J. S. Murray, P. Lane and P. Politzer, *J. Mol. Model.*, 2009, **15**, 723.
25. H. B. Buergi and J. D. Dunitz, *J. Am. Chem. Soc.*, 1987, **109**, 2924.
26. C. Bleiholder, D. B. Werz, H. Köppel and R. Gleiter, *J. Am. Chem. Soc.*, 2006, **128**, 2666.
27. R. S. Mulliken, *J. Am. Chem. Soc.*, 1952, **74**, 811.
28. T. Clark, M. Hennemann, J. S. Murray and P. Politzer, *J. Mol. Model.*, 2007, **13**, 291.
29. M. Iwaoka and S. Tomoda, *J. Am. Chem. Soc.*, 1996, **118**, 8077.
30. M. Iwaoka and N. Isozumi, *Molecules*, 2012, **17**, 7266.
31. P. Politzer, J. S. Murray, T. Clark and G. Resnati, *Phys. Chem. Chem. Phys.*, 2017, **19**, 32166.
32. T. N. G. Row and R. Parthasarathy, *J. Am. Chem. Soc.*, 1981, **103**, 477.
33. S. Benz, A. I. Poblador-Bahamonde, N. Low-Ders and S. Matile, *Angew. Chem., Int. Ed.*, 2018, **57**, 5408.
34. R. G. Pearson, *J. Am. Chem. Soc.*, 1963, **85**, 3533.
35. D. B. Werz, T. H. Staeb, C. Benisch, B. J. Rausch, F. Rominger and R. Gleiter, *Org. Lett.*, 2002, **4**, 339.
36. J. H. Schulte, D. B. Werz, F. Rominger and R. Gleiter, *Org. Biomol. Chem.*, 2003, **1**, 2788.
37. P. C. Ho, P. Szydlowski, J. Sinclair, P. J. W. Elder, J. Kübel, C. Gendy, L. M. Lee, H. Jenkins, J. F. Britten, D. R. Morim and I. Vargas-Baca, *Nat. Commun.*, 2016, **7**, 11299 EP.
38. P. Scilabra, G. Terraneo and G. Resnati, *Acc. Chem. Res.*, 2019, **52**, 1313.
39. M. Gsänger, E. Kirchner, M. Stolte, C. Burschka, V. Stepanenko, J. Pflaum and F. Würthner, *J. Am. Chem. Soc.*, 2014, **136**, 2351.
40. M. Gsänger, D. Bialas, L. Huang, M. Stolte and F. Würthner, *Adv. Mater.*, 2016, **28**, 3615.
41. M. Iwaoka and S. Tomoda, *Phosphorus, Sulfur Silicon Relat. Elem.*, 1992, **67**, 125.
42. M. Iwaoka and S. Tomoda, *J. Chem. Soc., Chem. Commun.*, 1992, 1165.
43. K. Srivastava, T. Chakraborty, H. B. Singh and R. J. Butcher, *Dalton Trans.*, 2011, **40**, 4489.
44. S. Tomoda, K.-I. Fujita and M. Iwaoka, *J. Chem. Soc., Chem. Commun.*, 1990, 129.
45. K.-I. Fujita, M. Iwaoka and S. Tomoda, *Chem. Lett.*, 1994, **23**, 923.
46. T. Wirth, *Angew. Chem., Int. Ed.*, 1995, **34**, 1726.
47. T. Wirth, G. Fragale and M. Spichty, *J. Am. Chem. Soc.*, 1998, **120**, 3376.
48. L. Uehlin and T. Wirth, *Org. Lett.*, 2001, **3**, 2931.
49. M. Tiecco, L. Testaferri, C. Santi, C. Tomassini, F. Marini, L. Bagnoli and A. Temperini, *Chem. – Eur. J.*, 2002, **8**, 1118.

50. F. T. Burling and B. M. Goldstein, *J. Am. Chem. Soc.*, 1992, **114**, 2313.
51. P. Franchetti, L. Cappellacci, M. Grifantini, A. Barzi, G. Nocentini, H. Yang, A. O'Connor, H. N. Jayaram, C. Carrell and B. M. Goldstein, *J. Med. Chem.*, 1995, **38**, 3829.
52. P. Franchetti, G. Cristalli, M. Grifantini, L. Cappellacci, S. Vittori and G. Nocentini, *J. Med. Chem.*, 1990, **33**, 2849.
53. P. Franchetti, L. Messini, L. Cappellacci, M. Grifantini, P. Guarracino, M. E. Marongiu, G. Piras and P. L. Colla, *Nucleosides Nucleotides*, 1993, **12**, 359.
54. H. Li, W. A. Hallows, J. S. Punzi, V. E. Marquez, H. L. Carrell, K. W. Pankiewicz, K. A. Watanabe and B. M. Goldstein, *Biochemistry*, 1994, **33**, 23.
55. B. M. Goldstein, H. Li, W. H. Hallows, D. A. Langs, P. Franchetti, L. Cappellacci and M. Grifantini, *J. Med. Chem.*, 1994, **37**, 1684.
56. Y. Nagao, T. Hirata, S. Goto, S. Sano, A. Kakehi, K. Iizuka and M. Shiro, *J. Am. Chem. Soc.*, 1998, **120**, 3104.
57. C. A. Leverett, V. C. Purohit and D. Romo, *Angew. Chem., Int. Ed.*, 2010, **49**, 9479.
58. V. B. Birman and X. Li, *Org. Lett.*, 2008, **10**, 1115.
59. X. Yang and V. B. Birman, *Adv. Synth. Catal.*, 2009, **351**, 2301.
60. X. Yang and V. B. Birman, *Chem. – Eur. J.*, 2011, **17**, 11296.
61. D. Belmessieri, L. C. Morrill, C. Simal, A. M. Z. Slawin and A. D. Smith, *J. Am. Chem. Soc.*, 2011, **133**, 2714.
62. E. R. T. Robinson, C. Fallan, C. Simal, A. M. Z. Slawin and A. D. Smith, *Chem. Sci.*, 2013, **4**, 2193.
63. S. R. Smith, J. Douglas, H. Prevet, P. Shapland, A. M. Z. Slawin and A. D. Smith, *J. Org. Chem.*, 2014, **79**, 1626.
64. Y. Fukata, K. Asano and S. Matsubara, *J. Am. Chem. Soc.*, 2015, **137**, 5320.
65. J. N. Arokianathar, A. B. Frost, A. M. Z. Slawin, D. Stead and A. D. Smith, *ACS Catal.*, 2018, **8**, 1153.
66. C. M. Young, A. Elmi, D. J. Pascoe, R. K. Morris, C. McLaughlin, A. M. Woods, A. B. Frost, A. de la Houpliere, K. B. Ling, T. K. Smith, A. M. Z. Slawin, P. H. Willoughby, S. L. Cockroft and A. D. Smith, *Angew. Chem., Int. Ed.*, 2020, **59**, 3705.
67. P. C. Ho, J. Rafique, J. Lee, L. M. Lee, H. A. Jenkins, J. F. Britten, A. L. Braga and I. Vargas-Baca, *Dalton Trans.*, 2017, **46**, 6570.
68. G. E. Garrett, G. L. Gibson, R. N. Straus, D. S. Seferos and M. S. Taylor, *J. Am. Chem. Soc.*, 2015, **137**, 4126.
69. G. E. Garrett, E. I. Carrera, D. S. Seferos and M. S. Taylor, *Chem. Commun.*, 2016, **52**, 9881.
70. J. Y. C. Lim, I. Marques, A. L. Thompson, K. E. Christensen, V. Félix and P. D. Beer, *J. Am. Chem. Soc.*, 2017, **139**, 3122.
71. A. Docker, C. H. Guthrie, H. Kuhn and P. D. Beer, *Angew. Chem., Int. Ed.*, 2021, **60**, 21973.
72. S. Benz, M. Macchione, Q. Verolet, J. Mareda, N. Sakai and S. Matile, *J. Am. Chem. Soc.*, 2016, **138**, 9093.

73. L. M. Lee, M. Tsemperouli, A. I. Poblador-Bahamonde, S. Benz, N. Sakai, K. Sugihara and S. Matile, *J. Am. Chem. Soc.*, 2019, **141**, 810.
74. S. Benz, J. López-Andarias, J. Mareda, N. Sakai and S. Matile, *Angew. Chem., Int. Ed.*, 2017, **56**, 812.
75. P. Wonner, L. Vogel, M. Düser, L. Gomes, F. Kniep, B. Mallick, D. B. Werz and S. M. Huber, *Angew. Chem., Int. Ed.*, 2017, **56**, 12009.
76. S. Benz, C. Besnard and S. Matile, *Helv. Chim. Acta*, 2018, **101**, e1800075.
77. P. Wonner, L. Vogel, F. Kniep and S. M. Huber, *Chem. – Eur. J.*, 2017, **23**, 16972.
78. T. Steinke, P. Wonner, E. Engelage and S. M. Huber, *Synthesis*, 2021, **53**, 2043.
79. P. Wonner, A. Dreger, L. Vogel, E. Engelage and S. M. Huber, *Angew. Chem., Int. Ed.*, 2019, **58**, 16923.
80. P. Wonner, T. Steinke, L. Vogel and S. M. Huber, *Chem. – Eur. J.*, 2020, **26**, 1258.
81. S. Benz, J. Mareda, C. Besnard, N. Sakai and S. Matile, *Chem. Sci.*, 2017, **8**, 8164.
82. P. Wonner, T. Steinke and S. M. Huber, *Synlett*, 2019, **30**, 1673.
83. W. Wang, H. Zhu, S. Liu, Z. Zhao, L. Zhang, J. Hao and Y. Wang, *J. Am. Chem. Soc.*, 2019, **141**, 9175.
84. W. Wang, H. Zhu, L. Feng, Q. Yu, J. Hao, R. Zhu and Y. Wang, *J. Am. Chem. Soc.*, 2020, **142**, 3117.
85. L. Bao, X. Kong and Y. Wang, *Asian J. Org. Chem.*, 2020, **9**, 757.
86. X. Kong, P.-P. Zhou and Y. Wang, *Angew. Chem., Int. Ed.*, 2021, **60**, 9395.
87. E. J. Lenardão, J. D. O. Feijó, S. Thurow, G. Perin, R. G. Jacob and C. C. Silveira, *Tetrahedron Lett.*, 2009, **50**, 5215.
88. B. Zhou and F. P. Gabbaï, *J. Am. Chem. Soc.*, 2021, **143**, 8625.
89. R. Weiss, E. Aubert, P. Pale and V. Mamane, *Angew. Chem., Int. Ed.*, 2021, **60**, 19281.
90. J. Wolf, F. Huber, N. Erochok, F. Heinen, V. Guérin, C. Y. Legault, S. F. Kirsch and S. M. Huber, *Angew. Chem., Int. Ed.*, 2020, **59**, 16496.
91. X. He, X. Wang, Y.-L. Tse, Z. Ke and Y.-Y. Yeung, *Angew. Chem., Int. Ed.*, 2018, **57**, 12869.

CHAPTER 20

Structure, Bonding, and Ligand Chemistry of Macrocyclic Seleno- and Telluroethers

RISTO S. LAITINEN,*[a] RAIJA OILUNKANIEMI[a] AND WOLFGANG WEIGAND[b]

[a] Laboratory of Inorganic Chemistry, Environmental and Chemical Engineering, University of Oulu, P.O. Box 3000, FI-90014 Oulu, Finland; [b] Institut für Anorganische und Analytische Chemie, Friedrich-Schiller-Universität Jena, Humboldt Str. 8, 07743 Jena, Germany
*Email: risto.laitinen@oulu.fi

20.1 Introduction

During recent decades, the versatile chemistry of crown-ethers and their complexes has generated interest in related macrocycles of heavier chalcogen elements.[1,2] Several reviews have been published in recent years since 2000 describing the progress in synthetic methods for the preparation, structural characterization, and coordination chemistry of heterocyclic chalcogenoethers, which is largely due to contributions by Levason and Reid, by Pinto, and by Gleiter, and their co-workers.[3–16] Research literature describing progress in heterocyclic chalcogenoethers is very extensive concentrating mainly on thioethers, and a complete review is beyond the scope of this chapter. This discussion is restricted to macrocyclic seleno- and telluroethers of ring size above ten and containing unbranched aliphatic spacers. The study of macrocyclic chalcogen heterocycles helps to gain insight on the chalcogen bonding and its application in supramolecular

Chalcogen Chemistry: Fundamentals and Applications
Edited by Vito Lippolis, Claudio Santi, Eder J. Lenardão and Antonio L. Braga

Published by the Royal Society of Chemistry, www.rsc.org

chemistry and crystal engineering.[17–22] This chapter is therefore concerned not only with syntheses and molecular structures, but also with bonding, and solid-state intermolecular interactions of saturated selenoether and telluroether macrocycles.

20.2 Selenoethers

20.2.1 Preparation

The earliest reports on the preparation of $Se(CH_2)_4$ and $Se(CH_2)_5$ date back to the work of Morgan and Burstall[23,24] in 1929. This synthesis required the use of hydrogen selenide and hydrogen gas, and metallic sodium. The modified preparation involved the reduction of elemental selenium with sodium hydroxymethylsulfate(IV) and the treatment of the selenide thus formed with α,ω-$Br(CH_2)_mBr$ ($m = 4, 5$).[25] The same methodology was also applied in the preparation of 1,4-$SeO(CH_2)_4$ and 1,4-$SeS(CH_2)_4$. The related reaction involving lithium selenide and 1,2-dichloroethane afforded 1,4-$Se_2(CH_2)_4$.[26] Other reducing agents have also been utilized for the reduction of elemental chalcogen, such as sodium tetrahydridoborate,[27] lithium triethylhydridoborate,[28] lithium tetrahydridoaluminate,[29] and hydrazine hydrate-potassium hydroxide.[30] The number m of the ethylene groups in α,ω-$X(CH_2)_mX$ can be used to control the ring size. The known macrocyclic selenoethers have been listed in Scheme 20.1.

The irradiation of thiohydroxamic esters of n-(benzylseleno)pentanoic, -hexanoic, and -heptanoic acid ($n = 5–7$, as appropriate) at the NMR scale have afforded $Se(CH_2)_4$, $Se(CH_2)_5$, and $Se(CH_2)_6$ in good yields.[31]

Recently, a macrocyclic 1,15-$Se_2(CH_2)_{28}$ was prepared by the decomposition of $[PtCl_2\{[Se(CH_2)_6\}_2(CH{=}CH)\}_2]$, which had been synthesized from $PtCl_2$ and $Se\{(CH_2)_6CH{=}CH_2\}_2$.[32] The reaction of hydrogen selenide and formaldehyde afforded 1,3,5-$Se_3(CH_2)_3$.[33] This compound has also been prepared from Na_2Se and CH_2Cl_2.[34] In addition to 1,3,5-$Se_3(CH_2)_3$, the reaction afforded small

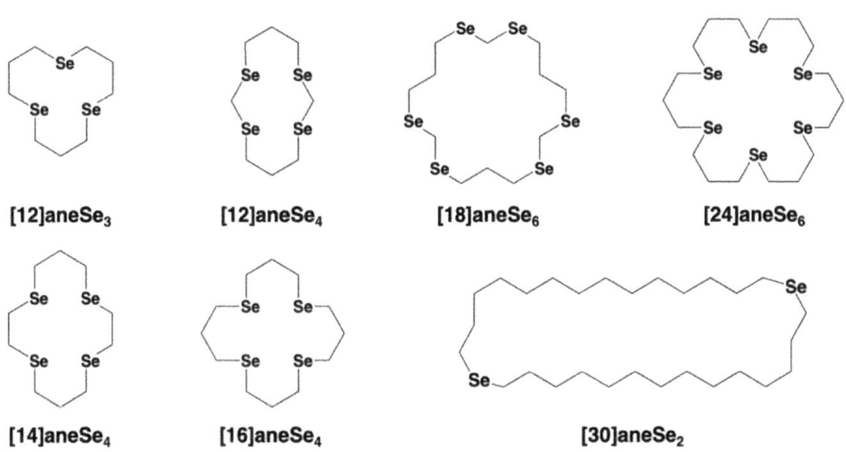

Scheme 20.1 Saturated, well-characterized selenoether macrocycles.

amounts of 1,3,5,7-$Se_4(CH_2)_4$. Macrocyclic poly-selenoethers 1,5,9-$Se_3(CH_2)_9$ and 1,5,9,13,17,21-$Se_6(CH_2)_{18}$ have been prepared, as shown in Scheme 20.2.[35]

The larger selenoether macrocycles have been prepared by reducing $NCSe(CH_2)_nSeCN$ ($n=2$, 3) by metallic sodium in liquid ammonia. The diselenolates, which are formed, are treated with α,ω-$Br(CH_2)_mBr$ ($m=1$–3; see Scheme 20.3).

The reduction of 1,2-$Se_2(CH_2)_4$ with $NaBH_4$ followed by the treatment with $Br(CH_3)_3Br$ afforded 1,6-$Se_2(CH_2)_8$.[37,38] The treatment of $[Pt_2(\mu\text{-}Se)_2(PPh_3)_4]$

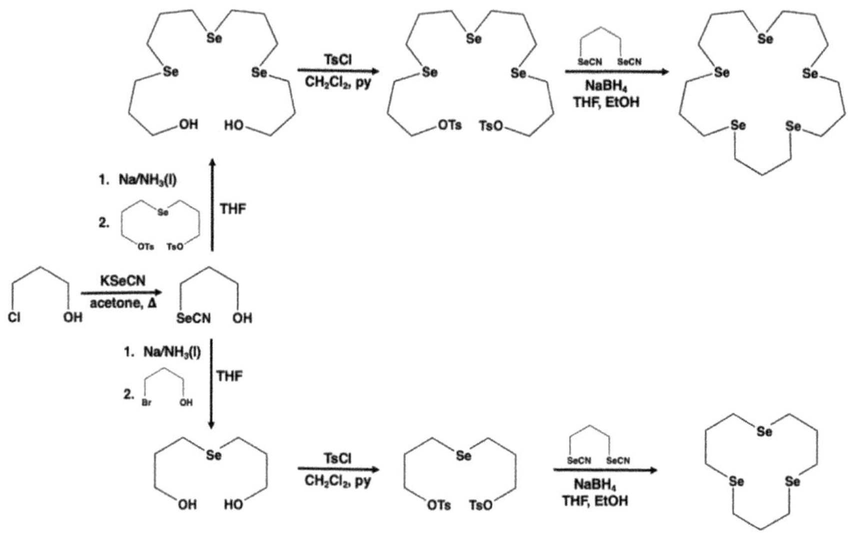

Scheme 20.2 Preparation of 1,5,9-$Se_3(CH_2)_9$ (**[12]aneSe$_3$**) and 1,5,9,13,17-$Se_5(CH_2)_{15}$ (**[20]aneSe$_5$**).[35]

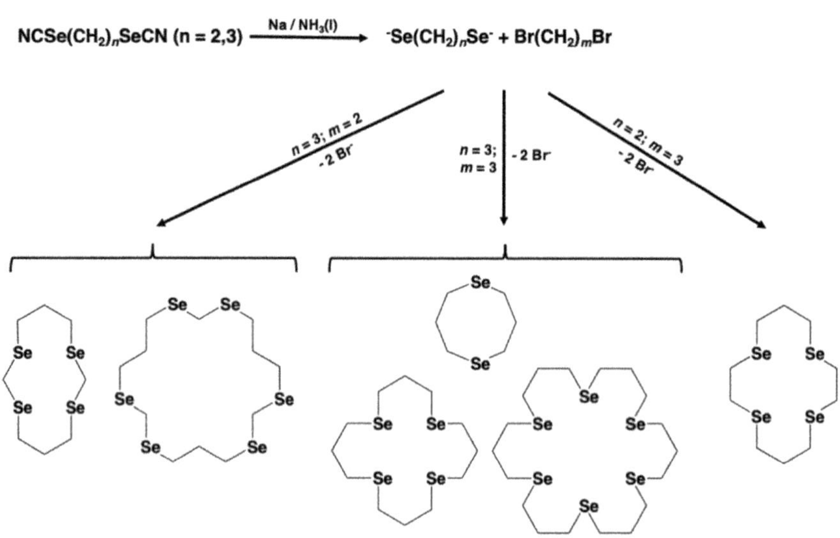

Scheme 20.3 Preparation of selenoether macrocycles.[36]

with the large excess of $Br(CH_2)_4Br$ afforded 1,6-$Se_2(CH_2)_8$ with an improved yield.[39] The compound was identified by positive-ion ESMS in MeCN/H_2O (50 : 50 v/v) solution.

20.2.2 Crystal Structures

The selenoether macrocycles shown in Scheme 20.1 are crystalline solids with well-established crystal structures, which have been determined by single-crystal X-ray diffraction. Their bond parameters are consistent with the values expected for single bonds [$r_{SeC} = 1.907(6)–1.98(2)$ Å, $r_{CC} = 1.471(14)–1.536(12)$ Å, $\alpha_{Se} = 91.5(4)–100.6(3)^\circ$, $\alpha_C = 109.9(4)–118.5(3)^\circ$].[32,36,40] The τ_{SeC} torsional angles are concentrated near to the values of 60°, 90°, and 180°, and the τ_{CC} torsional angles show mostly values near to 180°.

Most macrocyclic chalcogenoethers show columnar packing in the solid lattices due to Se···Se secondary bonding interactions of 3.457(14)–4.402(9) Å, as depicted in Figure 20.1. **[16]aneSe$_4$** is the most notable exception.[36]

20.2.3 Complexes

Small selenoether heterocycles $Se(CH_2)_4$ (**[5]aneSe**), 1,4-$Se_2(CH_2)_4$ (**[6]aneSe$_2$**), and 1,3,5-$Se_3(CH_2)_3$ (**[6]aneSe$_3$**) have long been known to react with HgX_2 (X = Cl, Br, I) to form complexes, the crystal structures of which have been determined.[42–44] $Se(CH_2)_4$ also forms a coordination polymer $[Au\{$**[5]aneSe**$\}_2{}^+]_n[AuI_2{}^-]_n$ upon the slow reaction with elemental gold and iodine.[45] $[Ag\{$**[6]aneSe$_3$**$\}_2][AsF_6]\cdot SO_2$ can be prepared quantitatively from 1,3,5-$Se_3(CH_2)_3$ and $AgAsF_6$ at −196 °C in liquid SO_2.[46]

The 1,5-$Se_2(CH_2)_6$ rings also react with a series of carbonyl complexes of transition metal carbonyl complexes forming $[MX(CO)_3\{$**[6]aneSe$_2$**$\}]$ (M = Mn, Re; X = Cl, Br, I) and $[M(CO)_4\{$**[6]aneSe$_2$**$\}]$ (M = Cr, Mo, W).[47] This selenoether also coordinates to the palladium[48] and silver[49] center.

The selenoether macrocycle **[16]aneSe$_4$** reacts with MCl_3 (M = Ga, In) in dichloromethane to afford $[MCl_2($**[16]aneSe$_4$**$)][MCl_4]$.[50] A small amount of acetonitrile was added to improve the solubility of the starting halides. The 1,5,9,13-$Se_4(CH_2)_{12}$ ring also reacts with a number of transition metal halogenides, such as $[Ru(DMF)_6Cl_3]$,[51] CoX_2 (X = Cl, Br) and MX_3 (M = Rh, X = Cl; M = Ir; X = Cl, Br),[52] MCl_2 (M = Pd, Pt),[53–55] and $Cu(OTf)_n$ (n = 1, 2; OTf = $CF_3SO_3^-$, triflate),[56,57] and $Hg(CN)_2$.[57] Thc bond parameters of the coordinated 1,5,9,13-$Se_4(CH_2)_{12}$ ring are close to those of the free molecule. The structures of some of the resulting complexes are exemplified in Figure 20.2.

The treatment of 1,5,9,13-$Se_4(CH_2)_{12}$ with $AsCl_3$ or MBr_3 (M = As, Sb, Bi) affords a 1 : 1 adduct.[58–60] The reaction of $GeCl_2$(dioxane), $GeBr_2$, and GeI_2 in dichloromethane with 1,5,9,13-$Se_4(CH_2)_{12}$ leads to the formation of $[(GeX_2)_2\{$**[16]aneSe$_4$**$\}]$ (X = Cl, Br, I).[61]

The direct reaction of 1,5,9,13,17,21-$Se_6(CH_2)_{18}$ (**[24]aneSe$_6$**) with $AsCl_3$ in dichloromethane generate $[(AsCl_3)_4\{$**[24]aneSe$_6$**$\}]$,[62] and that of $PdCl_2$ in

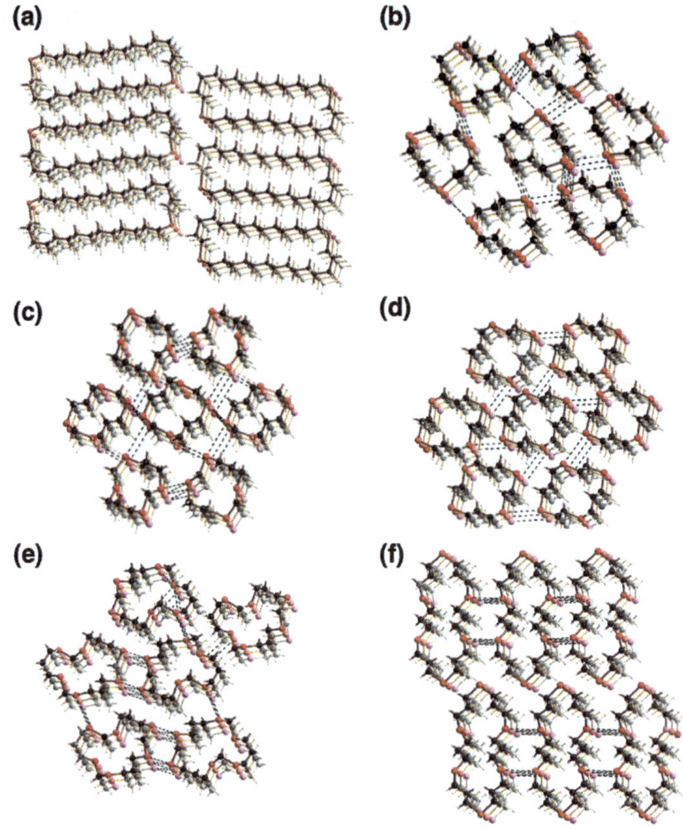

Figure 20.1 Columnar stacks in the solid lattices of (a) **[30]aneSe$_2$**;[32] (b) **[12]aneSe$_3$**;[40] (c) **[12]aneSe$_4$**;[41] (d) **[14]aneSe$_4$**;[36] (e) **[18]aneSe$_6$**;[36] and (f) **[24]aneSe$_6$**.[36]

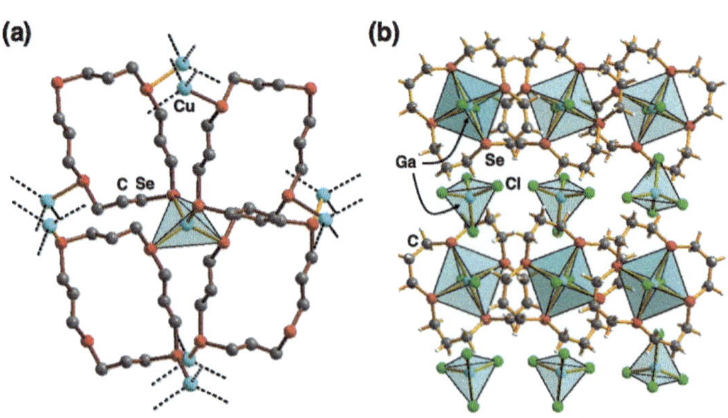

Figure 20.2 Examples of the molecular structures of 1,5,9,13-$Se_4(CH_2)_{12}$ **[16]aneSe$_4$** coordinated to the metal centers: (a) [Cu{μ_4-(**[16]aneSe$_4$**)}](OTf)[57] and (b) [$GaBr_2${**[16]aneSe$_4$**}][$GaBr_4$].[50]

acetonitrile yield $[Pd_2Cl_2\{$**[24]aneSe$_6$**$\}][BF_4]_2$ after the anion change with $NaBF_4$.[53] The crystal structures of both species are shown in Figure 20.3.

The selenium crown-ether analog 1,4,10,13-tetraoxa-7,16-diselenaoctadecane (**[18]aneO$_4$Se$_2$**)[63] provides rare ligation through soft selenium atoms to electropositive alkali metals,[64] some alkaline earth metals,[65,66] and rare-earth elements.[67,68] In addition to these complexes, the complexation of divalent selenium to a very electropositive metal is rare. The crystal structures of the three lanthanoid complexes of **[18]aneO$_4$Se$_2$** are shown in Figure 20.4.

The lanthanoid-selenium distances are 3.0971(7)–3.0884(8),[68] 3.1035(7),[67] and 2.8960(18)–2.9033(17) Å,[68] for lanthanum, ytterbium, and lutetium complexes, respectively. The relative bond lengths are consistent with the oxidation states and the lanthanum contraction. The Ce–Se bond distance of 3.0733(3) Å is also in agreement with these values.[69]

It is more to be expected that the soft selenium donor in **[18]aneSe$_2$** forms complexes with soft centers, as exemplified by the formation of the platinum complex, which was made by treating the ligand with $[PtX_2(NCMe)_2]$ (X = Cl, Br).[63] Platinum expectedly coordinates *via* selenium atoms forming a transannular Se–Pt–Se fragment (see Figure 20.5a). The antimony salt $[SbCl_2\{$**[18]aneO$_4$Se$_2$**$\}]_2[Sb_3Cl_{11}]\cdot MeCN$ was prepared by stirring the ligand with $SbCl_3$ in acetonitrile (Figure 20.5b).[70] The cation shows the oxygen–selenium macrocycle coordinated to antimony with the Sb–Se bond length spanning the range of 2.8657(14)–2.9187(15) Å. In addition, antimony in $SbCl_2$ interacts with the four oxygen atoms of the ring with the close Sb···O contacts of 2.763(9)–2.866(6) Å. Three lead salts $[Pb\{$**18aneO$_4$Se$_2$**$\}][X]_2$ ($X = BF_4^-$, PF_6^-, or NO_3^-) were prepared by the reaction of PbX_2 with **[18]aneO$_4$Se$_2$**.[71] The Pb–O distances in the three complexes span a range of 2.584(5)–2.717(4), 2.543(4)–2.664(3), and 2.777(8)–2.929(8) Å, respectively and the respective Pb–Se distances 3.1820(14), 3.173(1), and 3.174(2) Å.

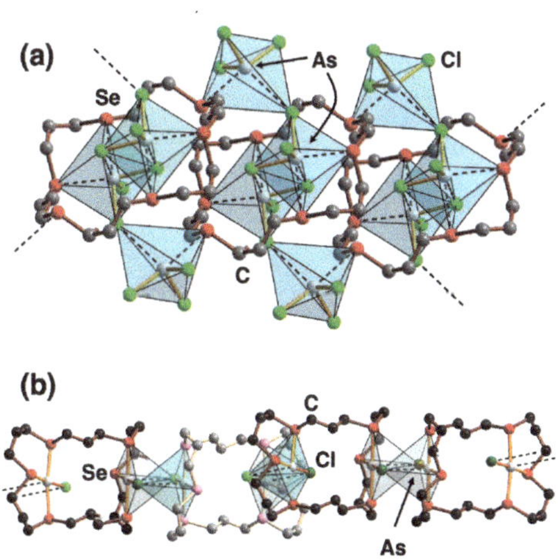

Figure 20.3 Crystal structures of (a) $[(AsCl_3)_4\{$**[24]aneSe$_6$**$\}]$[62] and (b) the $[Pd_2Cl_2\{$**[24]aneSe$_6$**$\}]^{2+}$ cation.[53]

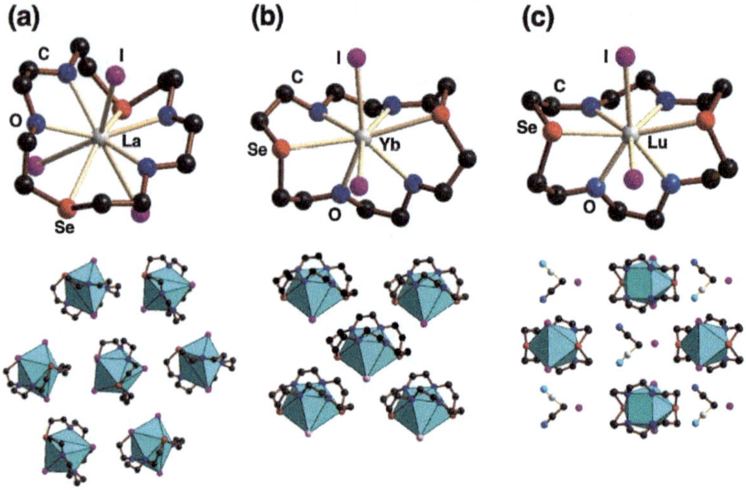

Figure 20.4 Molecular structures and the packing of the complexes of (a) [La{**[18]aneO$_4$Se$_2$**}(I)$_3$],[67] (b) [Yb{**[18]aneO$_4$Se$_2$**}(I)$_2$],[68] and (c) [Lu{**[18]aneO$_4$Se$_2$**}(I)$_2$](I)·2MeCN.[68]

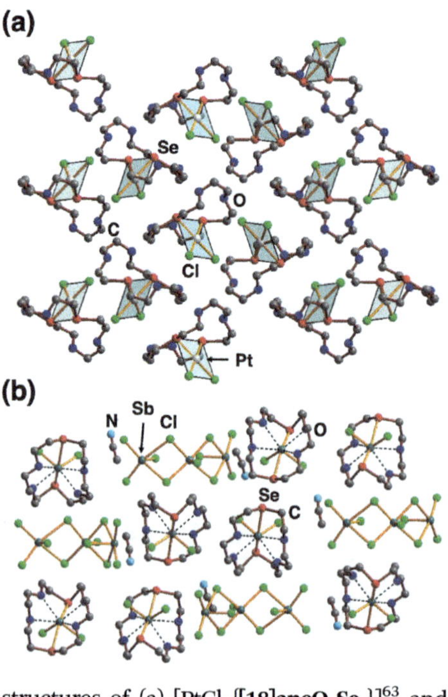

Figure 20.5 Crystal structures of (a) [PtCl$_2${**[18]aneO$_4$Se$_2$**}][63] and (b) [SbCl$_2${**[18]aneO$_4$Se$_2$**}]$_2$[Sb$_3$Cl$_{11}$]·MeCN.[70]

20.3 Telluroethers

20.3.1 Preparation

The known chemistry of saturated telluroethers $Te_n(CH_2)_m$ is much simpler than that of the related selenoethers, and these species have a significantly shorter history with the exception of $Te(CH_2)_4$[72] and $Te(CH_2)_5$,[73] the preparation of which was reported almost a century ago. The synthesis of 1,5,9-$Te_3(CH_2)_9$ has briefly been mentioned in connection with the preparation and structural characterization of the halogen derivative 1,1,5,5,9,9-hexchloro-1,5,9-tritelluracyclododecane, but no structural information was provided for this tritelluroether.[74] A systematic preparation of a number of macrocyclic telluroethers together with their structural and NMR spectroscopic characterization has only recently been reported.[75] The known telluroethers $Te_n(CH_2)_m$ have been listed in Scheme 20.4.

A systematic method to prepare telluroethers involves the reduction of elemental tellurium with $NaBH_4$ followed by the treatment with α,ω-$Br_2(CH_2)_m$ (see Scheme 20.5)[75] (for reviews of earlier literature, see ref. 3, 5, and 7–9). These reactions generally afford mixtures of different telluroethers which, however, can be separated by flash chromatography. The yield of the

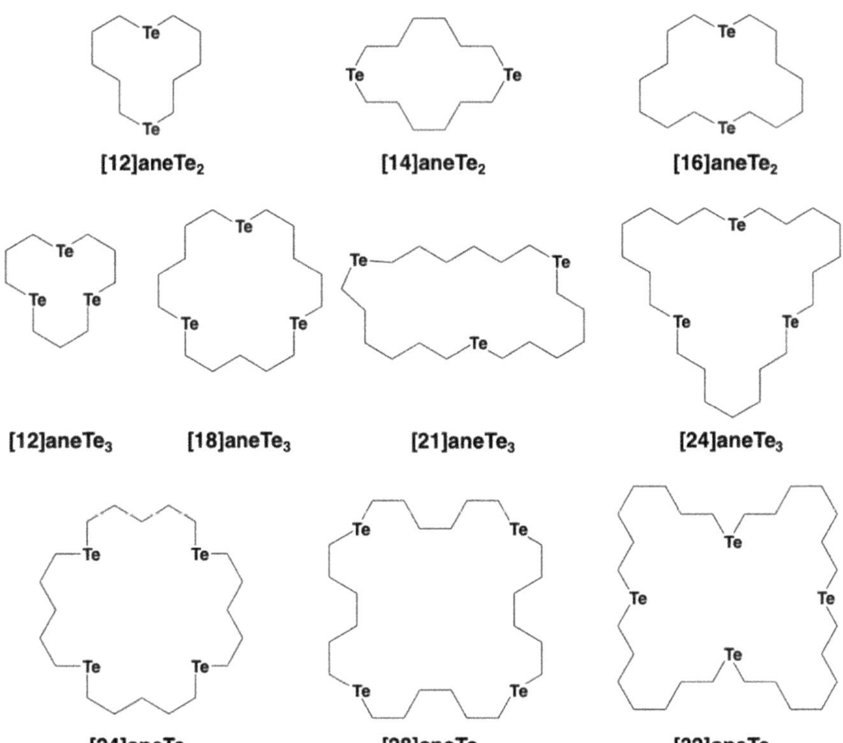

Scheme 20.4 Saturated macrocyclic telluroethers.

monotelluroether is the highest in each series and gets rapidly smaller as the number of tellurium atoms in the ring increases.[75] The PBE0-D3/def2-TZVPP calculations have shown that the Gibbs energy of formation of a given homologous series of heterocyclic rings from Na_2Te(c) and a given α,ω-dibromoalkane spacer is fairly constant, as exemplified by $[Te(CH_2)_5]_n$ ($n = 1$–4; see eqn (20.1)–(20.4)). Each reaction is thermodynamically equally probable and the relative yields depend on the rates of the individual reactions. The ring closure is kinetically favored in the case of unstrained rings and it is therefore understandable that the main product in most of these reactions is the monotelluroether $Te(CH_2)_m$ ($m = 3$–7). The stoichiometry of the different reactions exemplified in eqn (20.1)–(20.4) also favors the formation of telluroethers with fewer tellurium atoms.[75]

$$Na_2Te(c) + Br(CH_2)_5Br(solv) \rightarrow Te(CH_2)_5(solv) + 2\,NaBr(c)\ \Delta G_F = -241.6\ \text{kJ} \tag{20.1}$$

$$Na_2Te(c) + Br(CH_2)_5Br(solv) \rightarrow \tfrac{1}{2}\ 1,7\text{-}Te_2(CH_2)_{10}(solv) + 2\,NaBr(c)\ \Delta G_F = -217.6\ \text{kJ} \tag{20.2}$$

$$Na_2Te(c) + Br(CH_2)_5Br(solv) \rightarrow \tfrac{1}{3}\ 1,7,13\text{-}Te_3(CH_2)_{15}(solv) + 2\,NaBr(c)\ \Delta G_F = -214.3\ \text{kJ} \tag{20.3}$$

$$Na_2Te(c) + Br(CH_2)_5Br(solv) \rightarrow \tfrac{1}{4}\ 1,7,13,19\text{-}Te_4(CH_2)_{20}(solv) + 2\,NaBr(c)\quad \Delta G_F = -212.5\ \text{kJ} \tag{20.4}$$

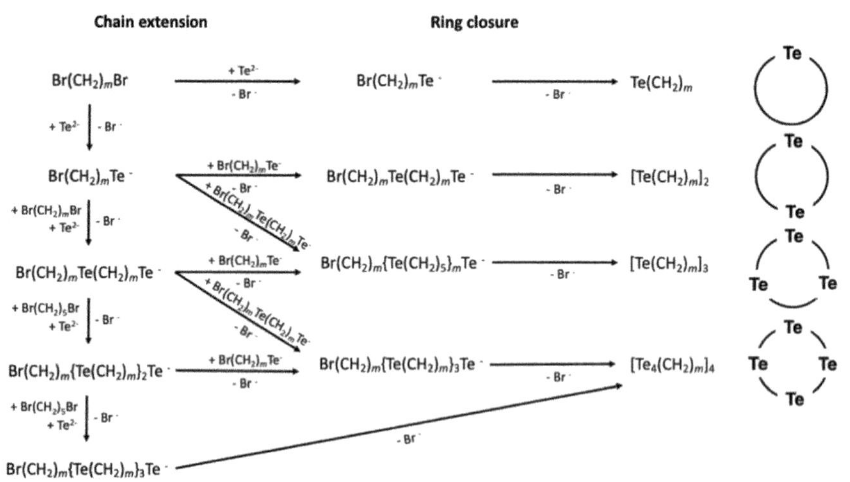

Scheme 20.5 Formation of cyclic telluroethers $[Te(CH_2)_m]_n$ ($n = 1$–4; $m = 3$–7).[75] Reproduced from ref. 75, https://doi.org/10.1002/chem.202002510, under the terms of the CC BY 4.0 license https://creativecommons.org/licenses/by/4.0/.

Hybrid macrocyclic thiotelluroethers **[11]aneS_2Te**, **[12]aneS_2Te**, and **[14]aneS_3Te** have been prepared in an analogous manner by the reaction of sodium telluride and an appropriate $Cl(CH_2)_nS(CH_2)_n[S(CH_2)_n]_mCl$ (n, $m=2$, 3) in THF/NH_3(liq) solution.[76,77] The products are light yellow and slightly soluble in organic solvents. The tellurium-containing crown-ether analog **[18]aneO_4Te_2**, which is similar to **[18]aneO_4Se_2**, was obtained in a good yield by the related reaction involving $Cl(CH_2)_2O(CH_2)_2O(CH_2)_2Cl$.[63]

20.3.2 Crystal Structures

Di-, tri-, and tetratelluroethers (see Scheme 20.4) are crystalline solids and their structures have been determined by X-ray diffraction methods.[75] 1,7-$Te_2(CH_2)_{10}$ (**[12]aneTe_2**) has been determined to be disordered with the ring molecules randomly assuming two different orientations. The intermolecular orientations between the disordered molecules have been analyzed with the solid-state PBE0-D3/pob-TZVP computations utilizing periodic boundary conditions (see Figure 20.6).[75]

There are four different alternatives for the relative orientations of neighboring molecules and the Te···Te secondary bonding interactions link the molecules in infinite chains. HOMO–LUMO overlap and a resulting charge transfer explains that result. These Te···Te secondary bonding interactions can be described best by σ-hole interaction/mutual polarization model.[78] In a similar way, the molecular structure of 1,8-$Te_2(CH_2)_{12}$ (**[14]aneTe_2**) also shows Te···Te interactions leading to infinite chains with a Te···Te distance of 4.0390(3) Å. As already observed for homologous thio- and selenoethers the Te containing molecules also form infinite tubular columns in the solid state (Figure 20.7a).[18–21,79–81]

Fascinating solid-state structures are formed when the number of Te atoms in the macrocyclic molecules increases. An important finding is that the complexity of the resulting supramolecular assembly grows as the number of Te···Te contacts increases as shown in Figure 20.7b and c. Moreover, it is well known that the chalcogen bonds become remarkably stronger for the heavier homolog tellurium.[18–21,79–83] This fact has a direct influence on the formation of even more complex structures when considering macrocycles containing four tellurium atoms. The prominent structural characteristics of compounds **[24]aneTe_4**, **[28]aneTe_4**, and **[32]aneTe_4** formed in the crystals are the extension of the coordination numbers of the tellurium atoms from two to six by developing four chalcogen bonds leading to a quasi-octahedral coordination sphere (Figure 20.8).

The closest Te···Te contacts in these three compounds have been determined in a range of 4.047(2)–4.066(2) Å. The intermolecular interactions between the macrocycles have been analyzed with the solid-state PBE0-D3/pob-TZVP computations utilizing periodic boundary conditions resulting in bond orders in a range of 0.11–0.17. The columnar packing of

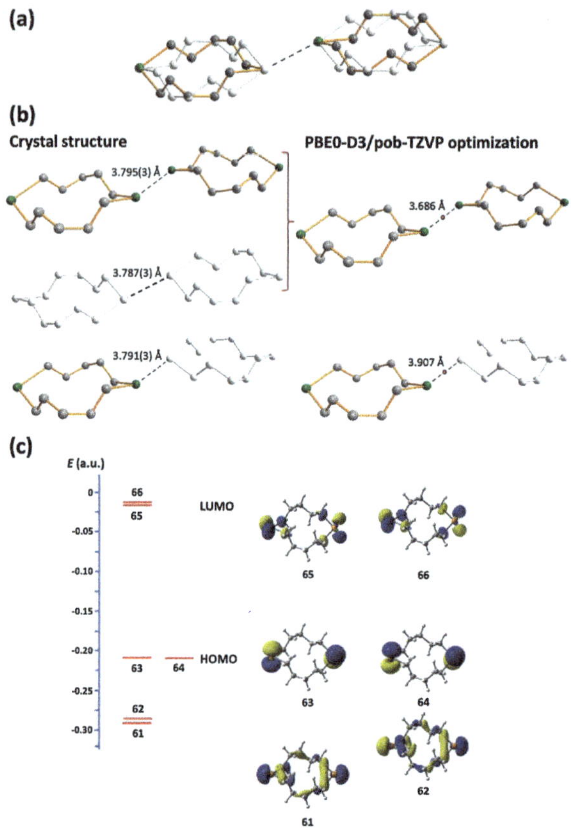

Figure 20.6 (a) Disordered crystal structure of 1,7-$Te_2(CH_2)_{10}$ (**[12]aneTe_2**); (b) comparison of the experimental intermolecular contacts between alternative orientations with those yielded by the solid-state PBE0-D3/pob-TZVP calculations involving the periodic boundary conditions (the red dots indicate the locations of the bond critical points (BCPs)); (c) the frontier orbitals of 1,7-$Te_2(CH_2)_{10}$ (**[12]aneTe_2**).[75] Reproduced from ref. 75, https://doi.org/10.1002/chem.202002510, under the terms of the CC BY 4.0 license https://creativecommons.org/licenses/by/4.0/.

compounds **[24]aneTe_4**, **[28]aneTe_4**, and **[32]aneTe_4** results in the formation of large voids, which are partially filled with solvent molecules. These observations showing disordered solvent molecules were similar to those reported by Gleiter and co-workers for the structurally related macrocyclic chalcogenethers.[21,84] In the case of **[24]aneTe_4**, the BET experiments using nitrogen as adsorptive gas confirm a surface area of 952 $m^2 g^{-1}$. The pore openings in **[24]aneTe_4**, **[28]aneTe_4**, and **[32]aneTe_4** are estimated approximately as 6.3, 8.0, and 8.7 Å, respectively (see Figure 20.8).[75]

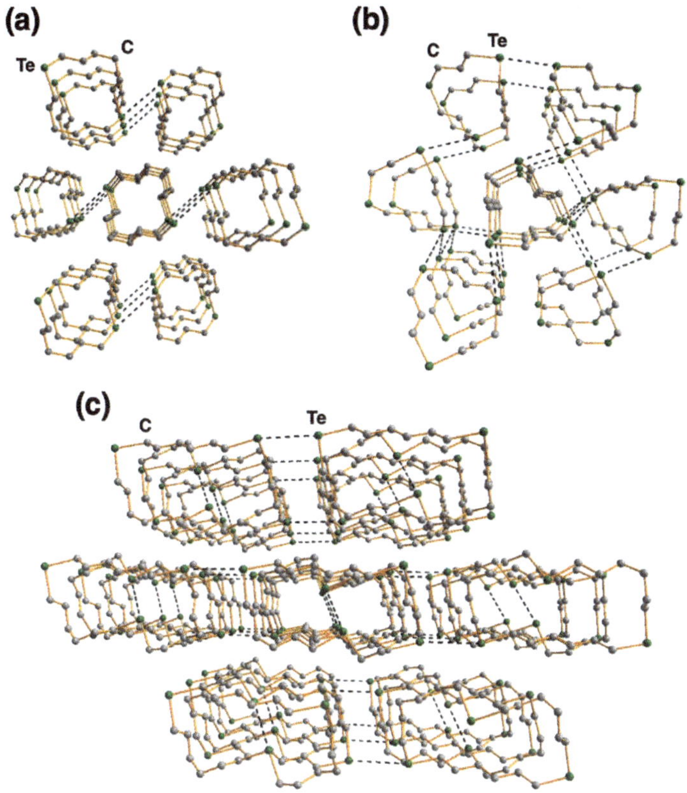

Figure 20.7 Columnar packing of (a) the 1,8-$Te_2(CH_2)_{12}$ (**[14]aneTe$_2$**) molecules; (b) the 1,5,9-$Te_3(CH_2)_9$ (**[12]aneTe$_3$**) molecules; and (c) the 1,8,15-$Te_3(CH_2)_{18}$ (**[21]aneTe$_3$**) molecules in the crystalline lattice.[75] Reproduced from ref. 75, https://doi.org/10.1002/chem.202002510, under the terms of the CC BY 4.0 license https://creativecommons.org/licenses/by/4.0/.

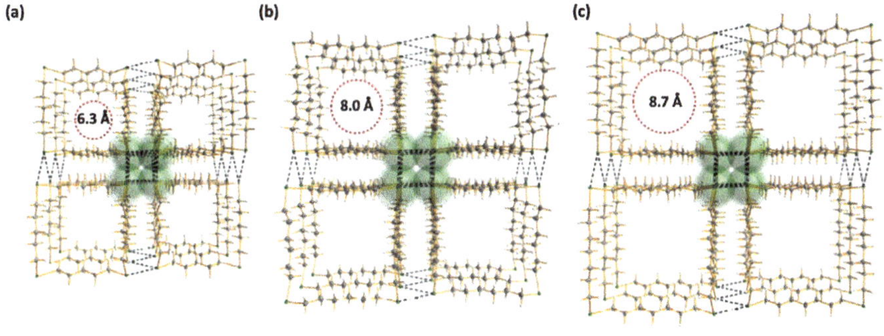

Figure 20.8 Molecular packing in the crystal and the pore openings in (a) **[24]aneTe$_4$**, (b) **[28]aneTe$_4$**, and (c) **[32]aneTe$_4$**.[75] Reproduced from ref. 75, https://doi.org/10.1002/chem.202002510, under the terms of the CC BY 4.0 license https://creativecommons.org/licenses/by/4.0/.

20.3.3 Complexes

There are very few reports on the formation and structures of complexes containing macromolecular telluroether ligands. The only information of telluroethers based on X-ray diffraction involves $Te(CH_2)_4$[85] and $Te(CH_2)_6$,[86] which have been determined as ligands in palladium and platinum complexes, respectively (see Figure 20.9). Since cyclic monotelluroethers $Te(CH_2)_m$ ($m = 4–7$) are liquids at room temperature like $Se(CH_2)_m$ ($m = 4–6$), the structures of the complexes provide the only way to establish the bond

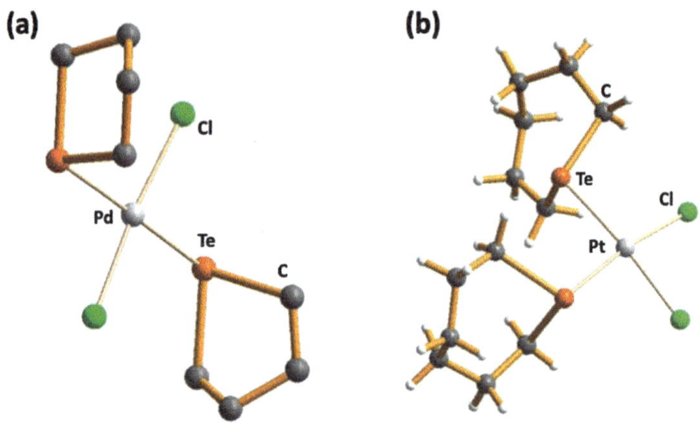

Figure 20.9 Molecular structures of (a) $[PdCl_2\{Te(CH_2)_4\}]$[85] and (b) $[PtCl_2\{Te(CH_2)_6\}]$.[86]

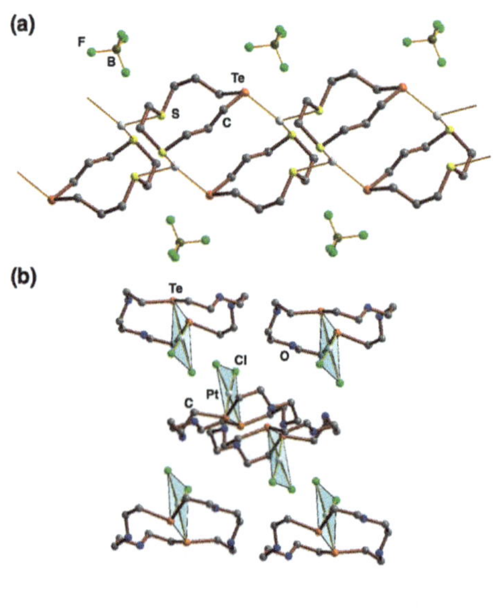

Figure 20.10 Crystal structures of (a) $[Ag\{$**[11]aneS₂Te**$\}]_n(BF_4)_n$[76] and (b) $[PtCl_2\{$**[18]aneO₄Te₂**$\}]$.[63]

parameters and conformations of these small heterocyclic telluroethers by X-ray crystallography.

The hybrid **[11]aneS$_2$Te** reacts with Ag(OTf) to afford a one-dimensional polymeric cation (see Figure 20.10a).[76] Refluxing the crown-ether analog **[18]aneO$_4$Te$_2$** with [MX$_2$(NCMe)$_2$] (M = Pd, Pt; X = Cl, Br) yields [MX$_2${**[18]aneO$_4$Te$_2$**}] complexes (see Figure 20.10b), which are mutually isomorphic but the packing is different from that of the related [MCl$_2${**[18]aneO$_4$Te$_2$**}].[63]

20.4 Summary and Outlook

The preparation, structural aspects, and solid-state intermolecular interactions of macrocyclic saturated selenoethers and telluroethers have been reviewed. One synthetic approach involves reduction of elemental selenium and tellurium or a suitable organic derivative and the treatment of the resulting anion with X(CH$_2$)$_n$X (X = Cl, Br).

Currently seven selenoether macrocycles, **[12]aneSe$_3$**, **[12]aneSe$_4$**, **[14]aneSe$_4$**, **[16]aneSe$_4$**, **[18]aneSe$_6$**, **[24]aneSe$_6$**, and **[30]aneSe$_4$** have been prepared and structurally characterized. In addition, a mixed crown-ether analog **[18]aneO$_4$Se$_2$** is known. **[16]aneSe$_4$** and **[24]aneSe$_6$** have formed a number of complexes with p-block halides and soft transition metal centers. **[18]aneO$_4$Se$_2$** behaves in a similar manner to crown-ether ligands and have ligated to alkali metals, alkaline earth metals, and rare earth elements. All complexes show a rare example of soft selenium coordinating to the hard Lewis acid.

By contrast to selenoethers, the chemistry of related telluroethers is much less developed. While ten macrocyclic telluroethers have been spectroscopically identified, X-ray structures are known only for **[12]aneTe$_2$**, **[14]aneTe$_2$**, **[12]aneTe$_3$**, **[21]aneTe$_3$**, **[24]aneTe$_4$**, **[28]aneTe$_4$**, and **[32]aneTe$_2$**. There is no published information about the macrocyclic telluroethers as a ligand in complexes, but there are a few structural reports of complexes containing hybrid thiotelluroether or oxotelluroether ligands. The development of coordination chemistry of telluroethers is a very worthwhile task for the future. Another promising goal might be to intercalate small gas molecules in the channels and cavities of the supramolecular constructs of **[24]aneTe$_4$**, **[28]aneTe$_4$**, and **[32]aneTe$_4$** in the solid state.

References

1. C. J. Pedersen, *J. Am. Chem. Soc.*, 1967, **89**, 7017.
2. G. W. Gokel, S. Negin and R. Catwell, in *Crown Ethers, Comprehensive Supramolecular Chemistry II*, ed. J. L. Atwood, G. Gokel and L. J. Barbour, Elsevier, Amsterdam, 2017, vol. 3, pp. 3–48.
3. W. Levason, S. D. Orchard and G. Reid, *Coord. Chem. Rev.*, 2002, **225**, 159.
4. M. Segi, *Sci. Synth.*, 2007, **39**, 1069.
5. A. J. Barton, A. R. J. Genge, N. J. Hill, W. Levason, S. D. Orchard, B. Patel, G. Reid and A. J. Ward, *Heteroatom. Chem.*, 2002, **13**, 550.

6. A. Panda, *Coord. Chem. Rev.*, 2009, **253**, 1056.
7. W. Levason and G. Reid, in *Handbook of Chalcogen Chemistry*, ed. F. A. Devillanova, RSC Publishing, Cambridge, 2007, pp. 81–106.
8. W. Levason and G. Reid, in *Comprehensive Coordination Chemistry II*, ed. J. A. McCleverty and T. J. Meyer, Elsevier, Amsterdam, 2004, pp. 399–410.
9. G. L. Sommen, in *Comprehensive Heterocyclic Chemistry III*, ed. A. R. Katritzky, C. A. Ramsden, E. F. V. Scriven and R. J. K. Taylor, Elsevier, Amsterdam, 2008, pp. 863–900.
10. W. Levason, G. Reid and W.-J. Zhang, *Dalton Trans.*, 2011, **40**, 8491.
11. C. N. Moorefield and G. R. Newcombe, *Prog. Heterocycl. Chem.*, 2020, **31**, 649.
12. A. K. Singh and S. Sharma, *Coord. Chem. Rev.*, 2000, **209**, 49.
13. W. Levason and G. Reid, *J. Chem. Soc., Dalton Trans.*, 2001, 2953.
14. W. Levason and G. Reid, *J. Chem. Res., Synop.*, 2002, 467.
15. V. K. Jain and R. S. Chauhan, *Coord. Chem. Rev.*, 2016, **306**, 270.
16. L. R. Gahan, *Coord. Chem. Rev.*, 2016, **311**, 168.
17. P. D. Boyle and S. M. Godfrey, *Coord. Chem. Rev.*, 2001, **223**, 265.
18. A. F. Cozzolino, P. J. W. Elder and I. Vargas-Baca, *Coord. Chem. Rev.*, 2011, **255**, 1426.
19. W.-W. du Mont and C. G. Hrib, in *Handbook of Chalcogen Chemistry: New Perspectives in Sulfur, Selenium and Tellurium*, ed. F. A. Devillanova and W.-W. du Month, RSC Publishing, Cambridge, 2nd edn, 2013, vol. 2, pp. 273–316.
20. T. Chivers and R. S. Laitinen, *Chem. Soc. Rev.*, 2015, **44**, 1725.
21. R. Gleiter, G. Haberhauer, D. B. Werz, F. Rominger and C. Leiholder, *Chem. Rev.*, 2018, **118**, 2010.
22. S. Kolb, G. A. Oliver and D. B. Werz, *Angew. Chem., Int. Ed.*, 2020, **59**, 22306.
23. G. T. Morgan and F. H. Burstall, *J. Chem. Soc.*, 1929, 1096.
24. G. T. Morgan and F. H. Burstall, *J. Chem. Soc.*, 1929, 2197.
25. J. D. McCullough and A. Lefohn, *Inorg. Chem.*, 1966, **5**, 150.
26. E. S. Gould and J. D. McCullough, *J. Am. Chem. Soc.*, 1951, **73**, 1105.
27. D. I. Klayman and T. S. Griffin, *J. Am. Chem. Soc.*, 1973, **95**, 197.
28. J. A. Gladysz, J. L. Hornby and J. E. Garbe, *J. Org. Chem.*, 1978, **43**, 1204.
29. H. Ishihara, M. Koketsu, Y. Fukuta and F. Nada, *J. Am. Chem. Soc.*, 2001, **123**, 8408.
30. E. P. Levanova, V. A. Grabel'nykh, A. V. Elaev, N. V. Russavskaya, L. V. Klyba, A. I. Albanov and N. A. Korchevin, *Chem. Heterocycl. Compd.*, 2012, **47**, 1345.
31. C. H. Schiesser and K. Sutej, *J. Chem. Soc., Chem. Commun.*, 1992, 57.
32. H. Joshi, S. Kharel, N. Bhuvanesh and J. A. Gladysz, *J. Organomet. Chem.*, 2018, **875**, 80.
33. L. Mortillaro, L. Credali, M. Mammi and G. Valle, *J. Chem. Soc.*, 1965, 807.
34. M. Russo, L. Mortillaro, L. Credali and C. De Checchi, *J. Polym. Sci.*, 1966, **4**, 248.

35. I. Cordova-Reyes, E. VandenHoven, A. Moihammed and B. M. Pinto, *Can. J. Chem.*, 1995, **73**, 113.
36. (a) R. J. Batchelor, F. W. B. Einstein, I. D. Gay, J.-H. Gu, B. D. Johnston and B. M. Pinto, *J. Am. Chem. Soc.*, 1989, **111**, 6582; (b) R. J. Batchelor, F. W. B. Einstein, I. D. Gay, J.-H. Gu, B. D. Johnston and B. M. Pinto, *J. Am. Chem. Soc.*, 1990, **112**, 1297.
37. H. Fujihara, R. Akaishi, T. Erata and N. Furukawa, *J. Chem. Soc., Chem. Commun.*, 1989, 1789.
38. H. Fujihara, R. Akaishi and N. Furukawa, *Tetrahedron*, 1993, **49**, 1605.
39. J. S. L. Yeo, J. J. Vittal and T. S. A. Hor, *Eur. J. Inorg. Chem.*, 2003, 277.
40. R. J. Batchelor, F. W. B. Einstein, I. D. Gay, J.-H. Gu, S. Mehta, B. M. Pinto and X.-M. Zhou, *Inorg. Chem.*, 2000, **39**, 2558.
41. G. Valle, G. Zanotti and M. Mammi, *Acta Crystallogr., Sect. B: Struct. Crystallogr. Cryst. Chem.*, 1978, **B34**, 2634.
42. C. Stålhandske and F. Zintl, *Acta Crystallogr., Sect. C: Cryst. Struct. Commun.*, 1988, **C44**, 253.
43. C. Stålhandske and F. Zintl, *Acta Crystallogr., Sect. C: Cryst. Struct. Commun.*, 1986, **C42**, 1449.
44. H. Hope and J. D. McCullough, *Acta Crystallogr.*, 1964, **17**, 712.
45. S. Ahrland, K. Dreisch, B. Norén and Å. Oskarsson, *Mater. Chem. Phys.*, 1993, **35**, 281.
46. H. Hofmann, P. G. Jones, M. Noltemeyer, E. Peymann, W. Pinkert, H. W. Roesky and G. M. Sheldrick, *J. Organomet. Chem.*, 1983, **249**, 97.
47. M. K. Davies, M. C. Durrant, W. Levason, G. Reid and R. L. Richards, *J. Chem. Soc., Dalton Trans.*, 1999, 1077.
48. N. R. Champness, W. Levason, J. J. Quirk and G. Reid, *Polyhedron*, 1995, **14**, 2753.
49. D. G. Booth, W. Levason, J. J. Quirk and G. Reid, *J. Chem. Soc., Dalton Trans.*, 1997, 3493.
50. K. George, M. Jura, W. Levason, M. E. Light, L. P. Ollivere and G. Reid, *Inorg. Chem.*, 2012, **51**, 2231.
51. W. Levason, J. J. Quirk, G. Reid and S. M. Smith, *J. Chem. Soc., Dalton Trans.*, 1997, 3719.
52. W. Levason, J. J. Quirk and G. Reid, *J. Chem. Soc., Dalton Trans.*, 1996, 3713.
53. R. J. Batchelor, F. W. B. Einstein, I. D. Gay, J. Gu, B. M. Pinto and X. Zhou, *Inorg. Chem.*, 1996, **35**, 3667.
54. N. R. Champness, P. F. Kelly, W. Levason, G. Rcid, A. M. Z. Slawin and D. J. Williams, *Inorg. Chem.*, 1995, **34**, 651.
55. W. Levason, J. J. Quirk and G. Reid, *Inorg. Chem.*, 1994, **33**, 6120.
56. R. J. Batchelor, F. W. B. Einstein, I. D. Gay, J.-H. Gu, B. M. Pinto and X.-M. Zhou, *J. Am. Chem. Soc.*, 1990, **112**, 3706.
57. R. J. Batchelor, F. W. B. Einstein, I. D. Gay, J.-H. Gu and B. M. Pinto, *J. Organomet. Chem.*, 1991, **411**, 147.
58. N. J. Hill, W. Levason and G. Reid, *Inorg. Chem.*, 2002, **41**, 2070.
59. A. J. Barton, N. J. Hill, W. Levason, B. Patel and G. Reid, *Chem. Commun.*, 2001, 95.

60. A. J. Barton, A. R. J. Genge, W. Levason and G. Reid, *J. Chem. Soc., Dalton Trans.*, 2000, 2163.
61. A. L. Hector, W. Levason, G. Reid, M. Webster and W. Zhang, *Dalton Trans.*, 2011, **40**, 694.
62. A. J. Barton, N. J. Hill and W. Levason, *J. Am. Chem. Soc.*, 2001, **123**, 11801.
63. M. J. Hesford, W. Levason, M. L. Matthews and G. Reid, *Dalton Trans.*, 2003, 2852.
64. M. J. D. Champion, W. Levason, D. Pugh and G. Reid, *Dalton Trans.*, 2015, **44**, 18748.
65. P. Farina, W. Levason and G. Reid, *Dalton Trans.*, 2013, **42**, 89.
66. W. Levason, D. Pugh, J. M. Purkis and G. Reid, *Dalton Trans.*, 2016, **45**, 7900.
67. M. J. D. Champion, P. Farina, W. Levason and G. Reid, *Dalton Trans.*, 2013, **42**, 13179.
68. P. N. Bartlett, J. D. Champion, M. E. Light, W. Levason, G. Reid and P. W. Richardson, *Dalton Trans.*, 2015, **44**, 2953.
69. A. C. Behrle, J. R. Levin, J. E. Kim, J. M. Drewett, C. L. Barnes, E. J. Schelter and J. R. Walensky, *Dalton Trans.*, 2015, **44**, 2693.
70. P. Farina, W. Levason and G. Reid, *Polyhedron*, 2013, **55**, 102.
71. P. Farina, T. Latter, W. Levason and G. Reid, *Dalton Trans.*, 2013, **42**, 4714.
72. G. T. Morgan and F. H. Burstall, *J. Chem. Soc.*, 1931, 180.
73. G. T. Morgan and H. Burgess, *J. Chem. Soc.*, 1928, 321.
74. Y. Takaguchi, E. Horn and N. Furukawa, *Organometallics*, 1996, **15**, 5112.
75. M. Rodewald, J. M. Rautiainen, T. Nicksch, H. Görls, R. Oilunkaniemi, W. Weigand and R. S. Laitinen, *Chem. – Eur. J.*, 2020, **26**, 13806.
76. W. Levason, S. D. Orchard and G. Reid, *Chem. Commun.*, 2001, 427.
77. M. J. Hesford, W. Levason, M. L. Matthews, S. D. Orchard and G. Reid, *Dalton Trans.*, 2003, 2434.
78. P. Politzer, J. S. Murray, T. Clark and G. Resnati, *Phys. Chem. Chem. Phys.*, 2017, **19**, 32166.
79. C. Bleiholder, R. Gleiter, D. B. Werz and H. Köppel, *Inorg. Chem.*, 2007, **46**, 2249.
80. C. Bleiholder, D. B. Werz, H. Köppel and R. Gleiter, *J. Am. Chem. Soc.*, 2006, **128**, 2666.
81. R. Gleiter, D. B. Werz and B. J. Rausch, *Chem. – Eur. J.*, 2003, **9**, 2676.
82. D. B. Werz, B. J. Rausch and R. Gleiter, *Tetrahedron Lett.*, 2002, **43**, 5767.
83. M. M. Karjalainen, C. Sanchez-Perez, J. M. Rautiainen, R. Oilunkaniemi and R. S. Laitinen, *CrystEngComm*, 2016, **18**, 4538.
84. D. B. Werz, R. Gleiter and F. Rominger, *J. Am. Chem. Soc.*, 2002, **124**, 10638.
85. T. Kemmitt, W. Levason, R. D. Olroyd and M. Webster, *Polyhedron*, 1992, **11**, 2165.
86. M. Rodewald, H. Görls, R. Oilunkaniemi, W. Weigand and R. S. Laitinen, *103rd Canadian Chemistry Conference and Exhibition*, Winnipeg 24–28/5/2020, 2020.

CHAPTER 21

The Long Story of Ebselen: From about One Century of its Synthesis to Clinical Trials

PABLO A. NOGARA,[a,b] MEIRE E. PEREIRA,[c]
CLAUDIA S. OLIVEIRA,[c] LAURA ORIAN[d] AND
JOÃO B. T. DA ROCHA*[a]

[a] Universidade Federal de Santa Maria, Departamento de Bioquímica e Biologia Molecular, Av. Roraima 1000, 97105-900, Santa Maria, RS, Brazil; [b] Instituto Federal de Educação, Ciência e Tecnologia Farroupilha (IFFAR), Campus Santo Augusto, 98590-000, Santo Augusto – RS, Brazil; [c] Instituto de Pesquisa Pelé Pequeno Príncipe, Av. Silva Jardim 1632, 80250-060, Curitiba, PR, Brazil; [d] Dipartimento di Scienze Chimiche, Università degli Studi di Padova, Via Marzolo 1, 35131 Padova, Italy
*Email: jbtrocha@yahoo.com.br

21.1 Introduction: A Brief History of Ebselen

Ebselen (2-phenyl-1,2-benzisoselenazol-3(2*H*)-one, also called PZ-51) is by far the most popular organoselenium compound and since the 1990s it has entered into several clinical trials.[1–9] Of particular therapeutic significance, Ebselen has been associated with some borderline beneficial effects in studies related to brain ischemia, stroke, and noise-induced deafness,[1,3,5] but so far it has not been approved to be used as a pharmaceutical agent to treat a specific disease or pathology. Despite this, the low toxicity of Ebselen to humans and its marginal beneficial effects in some clinical trials has given new momentum to the synthesis of organoselenium compounds. Here, we will present and

Chalcogen Chemistry: Fundamentals and Applications
Edited by Vito Lippolis, Claudio Santi, Eder J. Lenardão and Antonio L. Braga

Published by the Royal Society of Chemistry, www.rsc.org

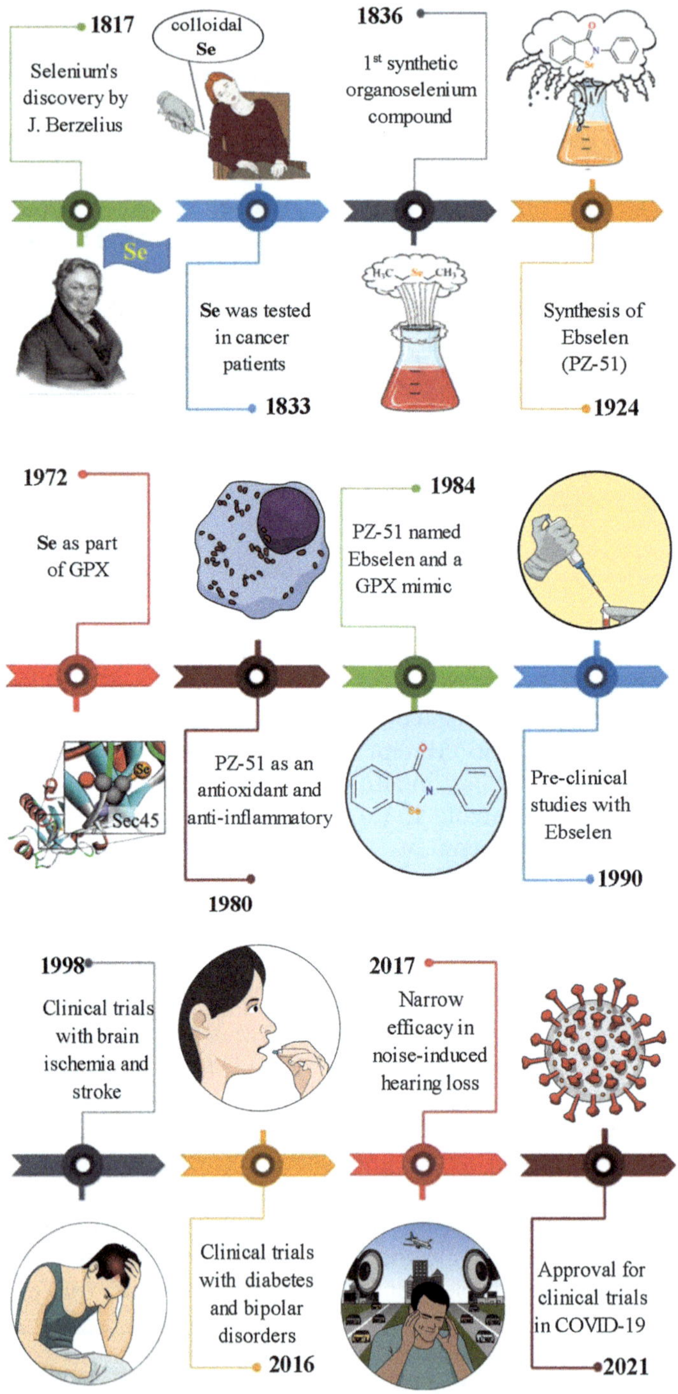

Figure 21.1 Historical overview of selenium, Ebselen, and its clinical trials.

discuss the published clinical trials and we will also illustrate the approved or ongoing clinical trials with Ebselen. In each case, we will briefly discuss the chemistry and biochemistry supporting the "authors' rationale" to test Ebselen in different pathologies. The authors considered the weak glutathione peroxidase-like properties of Ebselen to be the main molecular mechanism mediating Ebselen pharmacology. However, as will become evident, the biochemical behavior of Ebselen is much more complex than a simple low-molecular-mass compound capable of decomposing peroxides after interacting with endogenous thiol-groups in the body.

The history of Ebselen started in 1924, when its synthesis was first reported by Lesser and Weiss (Figure 21.1).[10] Since then, Ebselen has been little studied. Indeed, in the 1930s, Franke and colleagues published a series of studies demonstrating that selenium (Se) was the causative agent found in toxic grains.[11–14] They also demonstrated that the majority of the Se was in organic form and bound to the protein fraction of the toxic foodstuff.[12,15] In addition, the removal of Se from toxic protein hydrolysates by precipitation indicated a behavior similar to those of methionine and cystine.[12] These classical studies by Franke, Painter, and colleagues were the first indication that organic selenium (possibly the analogs of methionine and cysteine) could be found at toxic concentrations in plant proteins.

The demonstration that inorganic and organic Se forms could be highly toxic toward chicken and mammals created a fear of Se that was posteriorly named as "selenophobia" by Frost and Olson in 1972.[13,14,16] Indeed, in the 1930s, Franke and co-workers categorized as monstrosities the anatomical malformations caused by selenite (Se(VI) or SeO_3^{2-}) and selenate (Se(VI) or SeO_4^{2-}) in developing chickens.[14] The change in perspective regarding Se from a highly toxic to a beneficial element began after the classical study by Schwarz and Foltz in 1957.[17] In this short communication to the editor of the Journal of the American Chemical Society, Se either as part of a "factor" (Factor 3, an unspecified organic form of selenium, later demonstrated to be mainly selenomethionine[18]) or as selenite protected rats from a necrotic and lethal liver pathology caused by a diet deficient in vitamin E.[17,19] Then, the biochemical and nutritional interest in different organic forms of Se increased considerably. For instance, Schwarz and colleagues extensively investigated the potential protective properties of quite distinct organic forms of Se against the hepatotoxicity induced by a Se and vitamin E deficient diet.[20–25] Of particular biochemical and nutritional importance, they demonstrated that a variety of organo-selenides and -diselenides could furnish metabolically-protective selenium against the liver necrosis induced by vitamin E and Se deficiency.

However, the exact biochemical role of Se in mammals is unknown. The situation started to change in 1972–1973, when Dr Rotruck *et al.*[26,27] and prof. Leopold Flohé and colleagues[28] described that Se is an integral component of mammalian glutathione peroxidase (GPx). The subsequent clarification of the exact chemical nature of Se in bacteria and rat selenoenzymes as a selenocysteinyl residue[29,30] boosted the chemical, biochemical, nutritional, and pharmaceutical interest in the element.

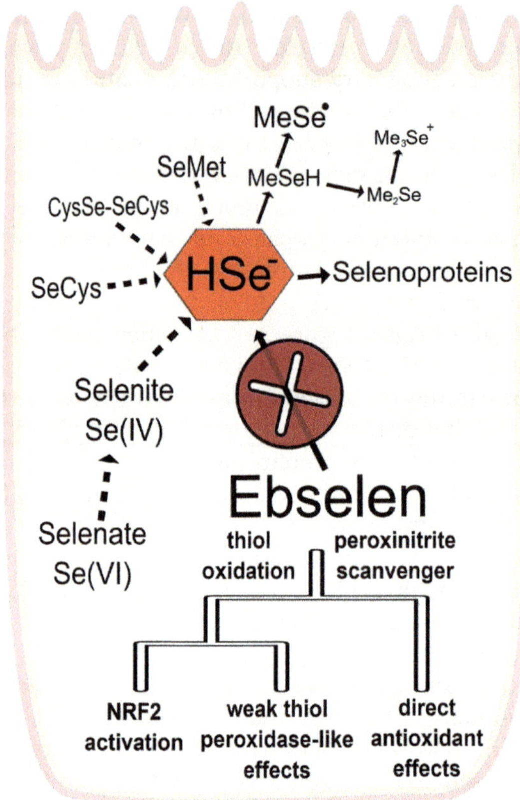

Figure 21.2 General metabolism of selenium in cells. Inorganic and organic selenium is converted to selenide (HSe^-) which is incorporated into selenoproteins. On the other hand, the Se atom of Ebselen is not available for selenoprotein synthesis (as indicated by the white and red X, Ebselen does not contribute to the inorganic selenide pool (indicated by HSe^- in the figure)). The Ebselen's plausive mechanism of action is due to its thiol-modifier effect and Nrf2 activation, modulating the transcription of antioxidant genes. Selenomethionine (SeMet) can be metabolized to methylselenol (MeSeH) which is further methylated to dimethylselenide (Me_2Se) and trimethylselenonium (Me_2Se^+). The conversion of MeSeH to its respective radical (MeSe•) is associated with Se toxicity.

It was under this context that pioneering researchers started to test the potential exploitation of Ebselen as a nutritional source of Se to "feed" the synthesis of the GPx enzyme.[31] In contrast to the researches expectation, the ^{75}Se-labeled Ebselen did not release the Se atom from the organic moiety; consequently, Ebselen could not be a source of Se for selenoprotein synthesis (Figure 21.2).[31–34] But, as cited by Dr Parham and Sies,[31] serendipitously they discovered some interesting antioxidant properties of Ebselen. The most famous of them was the capacity of Ebselen to weakly imitate the GPx activity[31,35] and still it has generally been assumed that the antioxidant properties of various organoselenium compounds are mediated by their capacity to form selenol (–SeH)-containing intermediate that could decompose hydrogen peroxide and organic peroxides.[36,37] However, in view of the instability of low molecular weight –SeH molecules (LMW-SeH) in physiological media, the GPx-like activity cannot be the main route involved in Ebselen's and other organoselenium compounds' beneficial pharmacological effects. In fact, the thiol modifier properties of Ebselen and other selenides is probably the main mechanism of their pharmacological action in the human body (Figure 21.3).

21.2 Ebselen Synthesis

The synthesis of Ebselen was first reported by Lesser and Weiss in 1924,[10] when the authors described the synthesis of benzothiaselenolone, benzodiselenolone, and many benzisoselenazolones (Figure 21.4). In this process, the conversion of 2,2′-diselanediyldibenzoic acid (DSBA) to a dichloride intermediate (**2**), using thionyl chloride, is a key step. DSBA was obtained by the selenenylation of the diazonium salt of the anthranilic acid, according to the method of Schoeller (1919).[40]

The Se-chloro-selenophenol-*o*-carboxylic acid chloride (**2**) is a very reactive compound that is able to react with ammonia, primary and aromatic amines (aniline) to form benzo[*d*][1,2]selenazol-3(2*H*)-one (**3**), 2-methylbenzo[*d*][1,2]-selenazol-3(2*H*)-one (**7**), and 2-phenylbenzo[*d*][1,2]selenazol-3(2*H*)-one (Ebselen), among other molecules. As reported by the authors, the compound **3** forms alkali salts (**5**) easily and can be used to obtain **7** and 2,2′-methylene-bis(benzo[*d*][1,2]selenazol-3(2*H*)-one) (**6**) by the addition of methyl iodide and methylene iodide, respectively. In addition, acetyl-derivatives (2-acetyl-benzo[*d*][1,2]selenazol-3(2*H*)-one (**4**)) is formed from **3** using acetic anhydride. Furthermore, the dichloride **2** reacts with sodium sulfide and selenide to give the corresponding 3*H*-benzo[*c*][1,2]thiaselenol-3-one (**8**) and 3*H*-benzo[*c*][1,2]-diselenol-3-one (**9**) derivatives.

As demonstrated by Lesser and Weiss, this synthetic route is a convenient method to prepare organoselenium compounds, especially benzisoselenazolones, including Ebselen (at that time it was not this). This same methodology was applied by Cantineau *et al.*[33] for the synthesis of the ^{75}Se–Ebselen labeled compound, which was used in the study that demonstrated that the ^{75}Se from ^{75}Se–Ebselen was not incorporated in selenoproteins in mouse liver,[42] *i.e.*, in Ebselen the ^{75}Se is not bioavailable for selenoproteins synthesis. Nowadays, many other approaches to obtain Ebselen and its derivatives have been reported in the literature (for a comprehensive review see Santi *et al.*[41]). For instance, the method described by Engman in 1989,[43] starting from *N*-phenylbenzamide and using *n*BuLi and elemental selenium (Se^0), and the route by Kumar in 2010[44] using *ortho*-iodo-benzamide, Se^0 and copper as a catalyst. Sancineto *et al.* in 2015,[45] used the DSBA converting it into diselenobisbenzamide, and cyclizing it with I_2, to obtain Ebselen. Here, we have focused only in the classical procedure of Lesser and Weiss to prepare Ebselen, but more information can be found in the recent literature.[41,46]

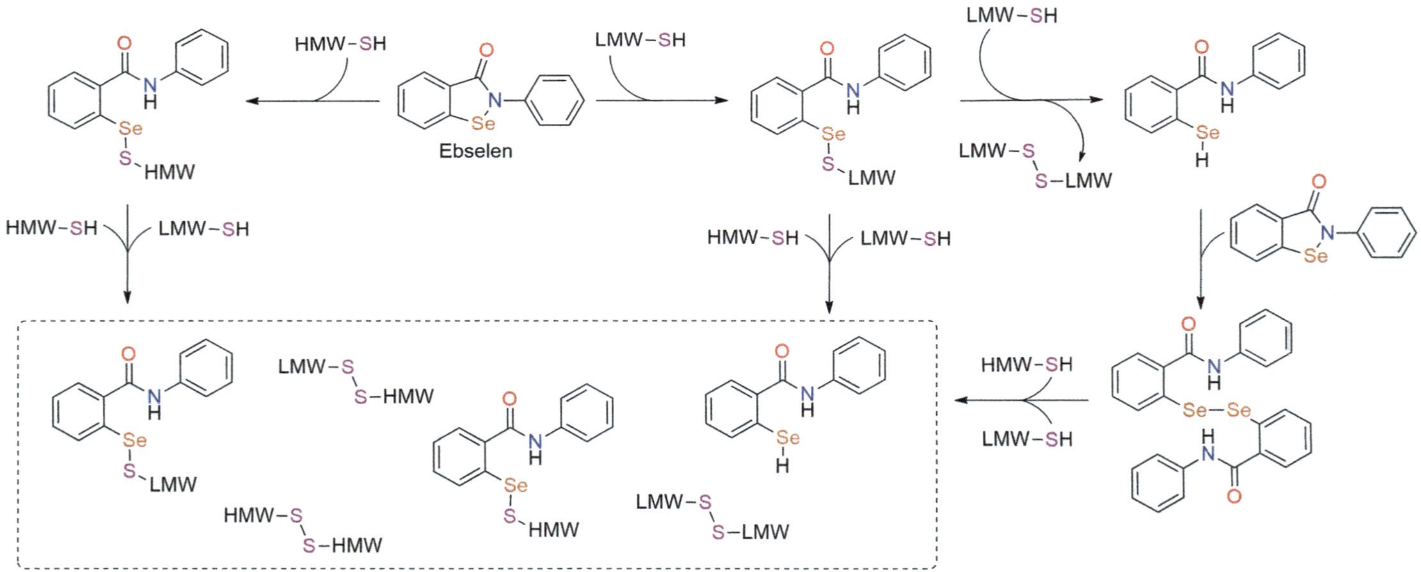

Figure 21.3 Possible Ebselen thiol-exchange reactions in the cell. Low- and high-molecular-weight thiol molecules (LMW-SH and HMW-SH, respectively) can be Cys, GSH, or thiol-containing proteins, such as albumin. Ebselen can also do exchange reactions with selenol-containing proteins (selenoproteins). For instance, thioredoxin reductase can interact with Ebselen and diselenide of Ebselen forming Ebselen–enzyme adducts (*via* Se–Se bond formation), which can sequentially release the selenol intermediate of Ebselen and the oxidized enzyme, for details see Zhao and Holmgren (2002)[38] and Zhao *et al.*, (2002).[39]

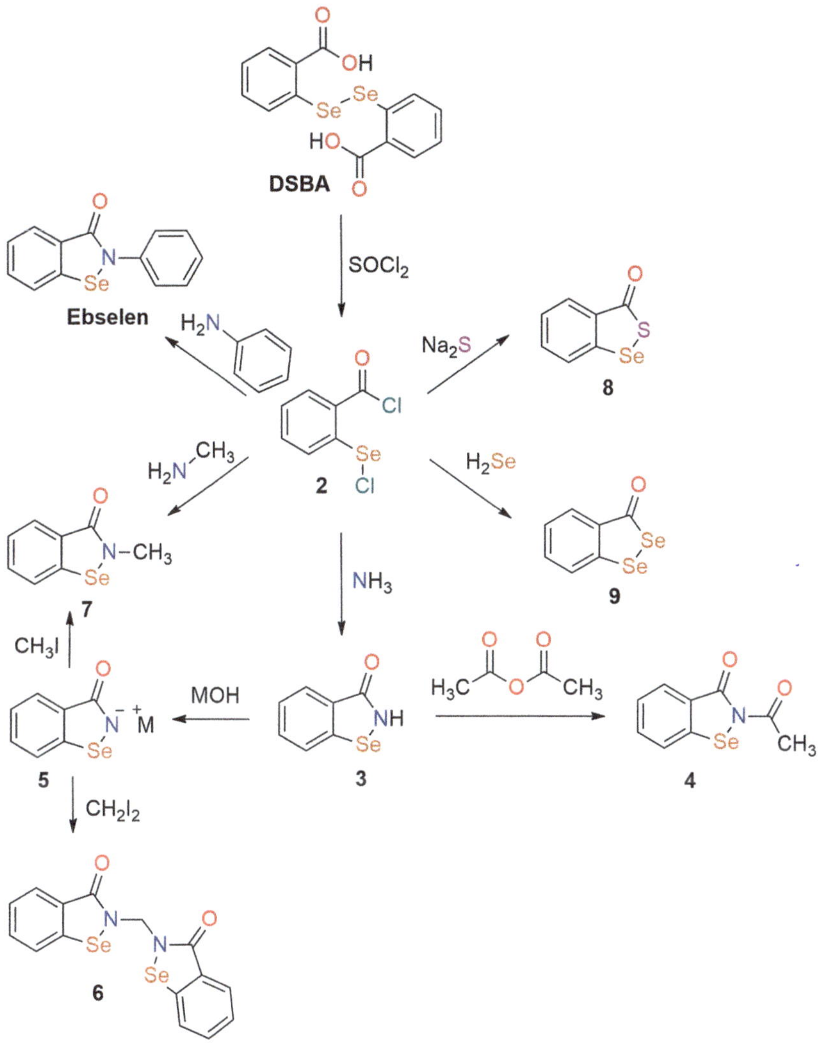

Figure 21.4 The classical synthesis of Ebselen by Lesser and Weiss in 1924. For an elegant and comprehensive review about the routes of Ebselen synthesis see Santi *et al.*, (2021).[41]

21.3 Clinical Uses of Ebselen

As commented in the Introduction, Ebselen was re-discovered as an antioxidant agent in the early 1980s (Figure 21.1).[31,32] Since then, the compound was tested in countless pre-clinical studies based on the rationale of Ebselen as a mimetic of GPx. Of note for the therapeutic application of Ebselen, the pre-clinical trials as a protective agent against brain and other organs ischemia have clearly indicated the beneficial effects of Ebselen in situations normally associated with redox unbalance and low levels of O_2 dissolved in blood fluids.[47–54] Those reported positive outcomes of Ebselen in mammals have motivated the indication of Ebselen as a potential therapeutic agent against brain pathologies associated with oxygen supply deficiency (see Section 21.3.1) by a Japanese pharmaceutical company in the 1990s (Figure 21.1 and Table 21.1). However, despite the borderline efficacy of Ebselen against brain ischemia, it was not approved by the Japanese governmental agencies for therapeutic use in brain pathologies associated with oxygen supply deficiency.

In the same vein, the beneficial effects of Ebselen against mammalian models of diabetes-induced pathologies[55–59] have stimulated the clinical investigation of Ebselen in diabetic patients (see Section 21.3.2; Figure 21.1 and Table 21.1). In the case of vascular reactivity in diabetes patients, Ebselen was not effective in a small clinical trial.[4] In contrast to the diabetes trial with Ebselen, pre-clinical studies and clinical studies have indicated the protective effects of Ebselen against noise-induced hearing pathologies,[60–67] but the clinical efficacy of Ebselen in volunteers was very narrow (see Section 21.3.3; Figure 21.1 and Table 21.1).[5,60] Pre-clinical studies have also indicated the potential antidepressant-like effects of Ebselen in rodents[68] and some "lithium mimetic activities" *in vitro* and *in vivo* animal models.[8,69,70] However, ongoing clinical trials have given only modest (if any) therapeutic effects of Ebselen as a substitute for lithium in bipolar disorders (see Section 21.3.4; Figure 21.1 and Table 21.1).[71] In short, although Ebselen is safe for human intake, the drug has not yet been approved for the treatment of specific diseases. In the following section, we will present details about the clinical trials already performed with Ebselen (Table 21.1).

21.3.1 Human Neuroprotective Effects of Ebselen: The Japanese Clinical Trials

As commented above, in the early 1980s studies demonstrated *in vitro* the antioxidant activity of Ebselen.[31,32,72] Later, in the early 1990s, the neuroprotective effects of Ebselen, explained by its antioxidant and anti-inflammatory activities, were demonstrated in pre-clinical studies.[50,54,73] Based on the results observed *in vitro* and *in vivo*, the neuroprotective effects of Ebselen were evaluated in three Japanese clinical trials. Several Japanese neurological and neurosurgical units participated in the clinical trials, which enrolled patients between June 1994 and December 1996. The three clinical trials were placebo-controlled and double-blind and the Ebselen treated patients received 150 mg of Ebselen, orally, twice a day for 14 days (Table 21.1).

In the first clinical trial "Ebselen and acute ischemic stroke", the authors had a window of 48 hours after stroke onset to begin the Ebselen or placebo treatment. The results demonstrated that Ebselen had small but significant neuroprotective effects 1 month after the stroke onset, but no difference was observed 3 months after the onset. Interestingly, when the patients were separated by the time of Ebselen treatment beginning, the improvement was significant in patients who started Ebselen within 24 hours of stroke onset but not in those who started treatment after 24 hours.[1] In the second

Table 21.1 Concluded Ebselen clinical trials.

Disease	Treatment	Type of study	Number of participants	Main findings	Country
Ischemic stroke[1]	Ebselen granules suspended in water (150 mg twice a day) or placebo was administered in a window of 48 h after onset and continued for 14 days.	Multicenter, double-blind, placebo-controlled clinical trial.	Ebselen: 151 Placebo: 149	The improvement was significant in patients who started Ebselen within 24 hours of stroke onset but not in those who started treatment after 24 hours.	Japan
Aneurysmal subarachnoid Hemorrhage[2]	Ebselen granules suspended in water (150 mg twice a day) or placebo was administered in a window of 48 h after onset and continued for 14 days.	Multicenter, double-blind, placebo-controlled clinical trial.	Ebselen: 145 Placebo: 141	Ebselen treatment reduced brain damage in patients with delayed ischemic neurological deficits after subarachnoid hemorrhage.	Japan
Acute middle cerebral artery occlusion[3]	Ebselen granules suspended in water (150 mg twice a day) or placebo was administered in a window of 12 h after onset and continued for 14 days.	Multicenter, double-blind, placebo-controlled clinical trial.	Ebselen: 43 Placebo: 56	Ebselen treatment reduced the volume of cerebral infarct areas only in patients that started the treatment within 6 h after the onset.	Japan
Type 1 and type 2 diabetes mellitus[4]	Ebselen (150 mg twice a day) or placebo, for 4 weeks.	Single-site, randomized, double-masked, placebo-controlled, crossover clinical trial.	Type 1 diabetes mellitus: 16 Type 2 diabetes mellitus: 10	Ebselen did not improve the oxidative stress profile, nor does it affect endothelium-dependent vasodilation in patients.	USA
Healthy participants with the intention to explore the results in bipolar disorders[6]	*Sleep and inositol study* Ebselen (600 mg twice a day) or placebo for two days *Emotional processing study*	*Sleep and inositol study* Single-site, randomized, double-blind, placebo-controlled, crossover clinical trial.	*Sleep and inositol study* *N*: 16 *Emotional processing study* Ebselen: 20	Ebselen decreased myo-inositol in brain regions associated with emotional processing, decreased slow wave	UK

Table 21.1 (*Continued*)

Disease	Treatment	Type of study	Number of participants	Main findings	Country
	Ebselen (600 mg) or placebo 3 times (every ~11 h)	*Emotional processing study* Single-site, randomized, double-blind, placebo-controlled, parallel clinical trial.	Placebo: 20	sleep, and changed performance in several emotional processing tasks (*e.g.*, increase the recognition of emotions).	
Bipolar disorder[7,8]	Ebselen (400 mg) or placebo in three doses given over 2 days.	Single-site, randomized, double-blind, crossover clinical trial.	*N*: 20	Ebselen treatment increased the recognition of facial expressions and decreased impulsivity, inhibited human brain IMPAse.	UK
Noise-induced hearing loss[5]	Ebselen (200, 400, or 600 mg) or placebo twice daily, orally, for 4 days.	Single-site, randomized, double-blind, parallel-group, placebo-controlled, phase 2a clinical trial.	Ebselen 200 mg: 22 Ebselen 400 mg: 20 Ebselen 600 mg: 21 Placebo: 20	Ebselen was effective at a dose of 400 mg in preventing a noise induced hearing loss.	USA
Mania or hypomania[9]	Ebselen (200 mg capsules) or placebo taken orally twice a day for 3 weeks.	Single-site, randomized, double-blind, parallel-group, placebo-controlled, phase 2a clinical trial.	Ebselen: 33 Placebo: 35	Ebselen did not improve the primary outcomes in the mania rating scales applied to patients. But it was effective in secondary outcomes.	UK

Japanese clinical trial "Ebselen and aneurysmal subarachnoid hemorrhage" the window to begin the treatment was 96 h after the onset. When the patients were stratified by the incidence or not of delayed ischemic neurological deficits (DINDs) caused by vasospasm, Ebselen treatment increased the percentage of patients classified as "good outcomes" two weeks and three months after the onset only in patients that had DINDs.[2] The result suggests that the neuroprotective effects of Ebselen occurred after ischemic insults, confirming the borderline beneficial effects of Ebselen previously demonstrated in stroke ischemic patients by Yamaguchi *et al.* (1998).[1] The third Japanese clinical trial "Ebselen and middle cerebral artery occlusion" had a window of 12 hours after onset to start the treatment. The patients treated with Ebselen within 6 hours after the onset presented a significant decrease in the volume of cerebral infarcted areas as determined by following up the patients by using computed tomography (CT) with the National Institutes of Health (NIH) Image program.[3]

The three Japanese clinical trials indicated that the effects of Ebselen seem to occur during the ischemic period. As elegantly reviewed by Rodrigo *et al.* (2013),[74] oxidative stress is implicated in the cellular brain injury observed after an ischemia and reperfusion process. Since the effects of Ebselen were better when the treatment started a few hours after the ischemic period, Ebselen might be carrying out its protective effects by scavenging the reactive species generated either directly (*via* its thiol-peroxidase-like activity, reacting with nitrogen reactive intermediates, *e.g.*, peroxynitrite)[75–78] or indirectly by activating antioxidant transcription factors.[79–82] For instance, Ebselen can increase the expression of antioxidant genes and can increase the translocation of the Nrf2 to the nucleus. The activation of the Nrf2 pathway stimulates the synthesis of important antioxidant elements.[36] As commented in previous sections, the reactivity of the selenol intermediate of Ebselen is too high under physiological conditions and possibly, the GPx-like effects of Ebselen are not the most important pathway for its neuroprotective effects. In fact, the thiol-modifier effects and the activation of electrophilic sensitive transcription factors are possibly the main routes of Ebselen neuroprotective effects.

21.3.2 Inefficacy of Ebselen Against Diabetes Mellitus' Oxidative Stress

Diabetes mellitus (DM) has an environment permissive for the development of endothelial dysfunction and atherosclerosis.[83] One mechanism by which these alterations occur is mediated by an exaggerated inflammatory response, which is associated with the generation of oxidative stress. The general picture of type 1 and 2 DM is marked by the chronicity and progressive deleterious alterations in various physiological functions (renal, vascular, cardiac, *etc.*).[84] Patients with DM can have a low concentration of endogenous antioxidants,[85] and supplementation with antioxidants (for instance, lipoic acid) has occasionally been associated with amelioration of some pathological consequences of DM.[86] It is observed that in rodents, the absence of GPx is associated with an increase in oxidative stress, reduced endothelium-dependent vasodilation, and uncoupling of endothelial nitric oxide synthase (eNOS). Accordingly, GPx1 knockout has been reported to participate in the endothelial dysfunction in experimental diabetes and Ebselen protected from GPx1 deficiency.[80]

Based on the above cited pre-clinical studies, a small double-masked, placebo-controlled, crossover clinical trial (NCT00762671) carried out in the Brigham and Women's Hospital, Massachusetts, USA evaluated the effects of Ebselen, as a GPx-mimetic in parameters of oxidative stress in type 1 and 2 DM patients.[4] However, the Ebselen treatment (oral intake, 150 mg twice a day for 4 weeks) did not improve the oxidative stress parameters or vascular function in DM patients. Possibly, the effectiveness of redox modulation by Ebselen is more pronounced at the beginning of the inflammatory process (which is followed by reactive species unbalance and further exacerbation of the inflammatory response), but not when the pathological consequences of inflammation and overproduction of pro-oxidant species have already settled. More specifically, the clinical trial with DM patients was carried out after a non-returning point of damage caused by inflammatory and oxidative damage. To some extent, this is in accordance with the findings of the Japanese clinical trials, where a better improvement was observed in patients who received Ebselen a few hours after the ischemic onset.

21.3.3 Ebselen as a Treatment for Hearing Loss

Animal studies indicated that the antioxidant system is pivotal to hearing manutention, as well as highlighting that oxidative stress is involved in cochlear pathologies, such as noise-inducing hearing loss.[87,88] Pre-clinical studies observed that rodents exposed to high noise presented a decrease in GPx activity, outer hair cell loss, stria swelling, and decrease in the auditory brainstem threshold; and the treatment with Ebselen reduced these alterations.[61,64] Thus, in the same line of the clinical trials mentioned in the previous sections, the GPx-like effect of Ebselen was evaluated in a noise-inducing hearing loss clinical trial (NCT01444846) (Table 21.1).[5]

This was a single-center, randomized, double-blind, placebo-controlled phase 2 clinical trial carried out at Florida University, USA. The healthy volunteers were enrolled between January 2013 and March 2014; and randomly assigned to receive placebo or Ebselen treatment (200, 400, or 600 mg), orally, twice a day. The volunteers treated with 400 mg of Ebselen had a significant improvement in the temporary auditory threshold shift (~68%) when compared with the placebo group. But no significant improvement was observed in volunteers treated with 200 and 600 mg of Ebselen,[5] which raised questions about the Ebselen therapeutic window that seems to be very narrow.

A recently published review study summarizes three clinical trials (not yet published) were the Ebselen has been tested in the Menieres' disease (inner ear disorder) and in the ototoxicity caused by tobramycin in cystic fibrosis patients.[60] Although the authors mentioned only previous results, the treatment with 400 mg Ebselen demonstrates borderline significative improvements in the patient's auditory threshold.[60]

21.3.4 Ebselen as a Lithium Mimetic: Clinical Trials for Bipolar Disorders

In 2013, a group from the Department of Pharmacology from Oxford University published a study describing the lithium-like activity of Ebselen *in vitro* and *in vivo*.[69] The motivation of the authors was based on the necessity of searching for new low-molecular-weight compounds that could mimic the molecular effects of Li^+ without the typical lithium side effects. The authors considered specifically the inositol monophosphatase (IMPase) as the main target of Li^+. For finding potential lithium mimetics, the authors screened the NIH Clinical Collection provided by the National Institutes of Health Molecular Libraries Roadmap Initiative (www.nihclinicalcollection.com). The molecules in the above libraries have already been tested in human clinical trials and could be safely repurposed in new pathologies. After an *in vitro* screening with the purified human IMPase, Ebselen was found to be an effective inhibitor of the enzyme with IC_{50} of 1.5 μM,[69] which is a therapeutically relevant concentration of Ebselen. The authors also demonstrated the effectiveness of Ebselen *in vivo*, using mice in behavioral and neurochemical models.[69]

Subsequently, the clinical trials with healthy volunteers indicated some behavioral (*e.g.*, decreased impulsivity in a standardized neuropsychological test and emotional processing) and neurochemical (a reduction in brain myo-inositol concentration) changes after the oral intake of large doses of Ebselen (3 or 6 capsules of 200 mg of Ebselen was given in three doses over 2 days, total doses 1.8 or 3.6 g of Ebselen) (Table 21.1).[6–8] Recently, the same group published the results of the phase 2a, single-center, 3 week, double-blind, and randomized clinical trial with 60 patients with bipolar disorder.[9] In this study, the authors used the formulation SPI-1005 (three capsules of 200 mg of SPI-1005, two times a day, total daily dose of Ebselen equal to 1.2 g), which has Ebselen as the active ingredient, and the equivalent placebo. The study accessed the Young Mania Rating Scale (YMRS) as the primary outcome; and the Altman Self-Rating Mania (ASRM) Scale and Clinical Global Impression-Severity Scale (CGI-S) as the secondary outcomes. The authors reported a borderline ($p = 0.054$) improvement in the YMRS scores and a significant ($p = 0.04$) improvement in the Clinical Global Impression Severity (CGI-S) scale in the Ebselen group after 3 weeks of treatment. The authors concluded that it would be important to test Ebselen in clinical trials as a monotherapy in acutely manic patients.[9]

21.4 Ebselen as a Promising Agent against Microorganisms: Data from Pre-clinical Models

Experimentally, the potential repurposing of Ebselen either alone or in combination with standard antimicrobial agents has been extensively investigated (for a review see Wang *et al.*[89] and Nogara *et al.*[90]). In fact, Ebselen has been reported as an anti-microbial agent against different infectious micro-organisms,[91–93] including *Trypanosoma cruzi*,[94] *Bacillus anthracis*,[95] *Aspergillus fumigatus*,[96,97] and the brain-eating ameba *Naegleria fowleri*.[98]

The interactions between Ebselen and Cys residues from critical proteins of the pathogens could be associated with the microbicidal effect. In fact, Ebselen presents anti-HIV-1 activity (EC_{50} of 4.3 μM) binding covalently to capsid proteins and forming a selenylsulfide bond with Cys residues, perturbing protein–protein interactions, virion assembly, and consequently, the HIV-1 maturation.[99] Recently, Ebselen demonstrated antiviral action against the severe acute respiratory syndrome coronavirus 2 (SARS-CoV-2), which will be discussed below.

21.5 Ebselen and Derivatives as Antivirals against SARS-CoV-2

Coronavirus disease (COVID-19) appeared at the end of 2019 in China and spread very quickly around the world. At the beginning of March 2020, the World Health Organization (WHO) officially confirmed the COVID-19 pandemic[100] and, approximately one month later, Jin and colleagues[101] reported that Ebselen is a potent inhibitor of the main protease (M^{pro}) from SARS-CoV-2, suggesting it as a promising antiviral drug. The study of Jin *et al.* (2020), which was followed by several other studies on the effects of Ebselen and its derivatives against SARS-CoV-2 *in vitro*,[102–104] motivated the same Company that tested clinically the effects of Ebselen against hearing loss to submit and approve a clinical trial of Ebselen against moderate and severe COVID-19 in patients. In the next two subsections, we will present some details about the approved clinical trials and the potential mechanisms of Ebselen effectiveness as an inhibitor of thiol-containing proteases of the virus (the main protease or M^{pro} and the papain-like protease or PL^{pro}).

21.5.1 Approved Clinical Trials Against Moderate and Severe COVID-19 Cases

Ebselen was approved by the United States Food and Drug Administration (FDA) to be tested in a phase 2, randomized, double-blind, placebo-controlled, dose escalation study for the evaluation of the safety and efficacy in moderate and severe COVID-19 patients (Table 21.2). However, the clinical trials are still in the recruiting phase, and we do not know its

Table 21.2 Ebselen (SPI-1005) clinical trials (phase 2) against COVID-19.[a]

Title	Status	Treatment	NCT no.
SPI-1005 Treatment in Moderate COVID-19 Patients	Enrolling by invitation	400 or 800 mg BID orally for 7 days	NCT04484025
SPI-1005 Treatment in Severe COVID-19 Patients	Enrolling by invitation	400 or 800 mg BID orally for 14 days	NCT04483973

[a]BID: *bis in die* (twice a day). Last access in 01 Feb 2022.

potential therapeutic effect yet. The dosage of Ebselen used is in the same range as in previous clinical interventions (Table 21.1). In the two approved, recruiting-phase clinical trials, Ebselen will be used in the SPI-1005 formulation, which is a classified formulation from the company Sounds Pharmaceutical. The SPI-1005 has already been reported to have very narrow effectiveness against loss hearing.[5] But in the company page (https://soundpharma.com/technology/) it is possible to read a very optimistic indication of SPI-1005 as an effective otoprotective therapeutic agent.

21.5.2 Potential Mechanisms of Ebselen Against SARS-CoV-2 *In Vitro*

Jin *et al.*[101] first demonstrated the *in vitro* inhibitory effects of Ebselen against SARS-CoV-2, which, to some extent, should be expected in view of the knowledge that the two proteases (M^{pro} and PL^{pro}) of closed related coronaviruses SARS-CoV (the etiologic agent of Severe Acute Respiratory Syndrome and Middle East Respiratory Syndrome, respectively) were sensitive to thiol-blocking or oxidizing agents.[105] In relation to Ebselen, Jin *et al.*[101] indicated an individual half-maximal effective concentration (EC_{50}) of 4.67 μM against the virus replication in Vero E6 cells and a half-maximal inhibitory concentration (IC_{50}) value of 0.67 μM against the purified M^{pro} enzyme *in vitro*.

The papain-like protease (PL^{pro}) from SARS-CoV-2 has also been reported to be inhibited by Ebselen *in vitro* (Table 21.3). In this sense, the inhibition of the two viral proteases by Ebselen is of paramount importance. In fact, the proteolytic activities of M^{pro} and PL^{pro} are essential for the completion of the virus cycle. The two proteases are involved in the maturation of the poliprotein chain into functional non-structural proteins (nsp).[106–108] Furthermore, the PL^{pro} can catalyze the deubiquitination of host proteins (*i.e.*, PL^{pro} has also a deubiquitinatinase activity) and disrupt the innate immune response against SARS-CoV-2.[109] Experimental and theoretical studies have demonstrated that M^{pro} and PL^{pro} inhibition by Ebselen can involve the formation of a covalent adduct of the type selenyl–sulfide bond (Eb–Se–S–Enz) between the Se atom of Ebselen and the thiol group from the catalytic cysteine (Cys) residue from the active site of the proteases (Figure 21.5).[101,110–112] However, as reported by Ma *et al.*,[113] it is highly expected from the chemical behavior of Ebselen in the biological media (for a critical review see ref. 114), the interaction of Ebselen with free thiol-groups is non-specific, occurring with the M^{pro} thiol-groups and other cysteine proteases. In fact, Chen *et al.*[115] using a biotin derivative of Ebselen (biotinylated Ebselen probe), identified that the compound was able to bind in 462 proteins from in HeLa cells. In this sense, this lack of specificity could lead to non-specific effects of Ebselen (see for an elegant critical review, Orian and Toppo[114]).

In the same study of Ma *et al.*,[113] it was also reported that under reducing conditions, such as in the presence of an excess of 4 mM 1,4-dithiothreitol (DTT) or 1 mM GSH, the inhibitory potency of Ebselen against M^{pro} decreased considerably (Figure 21.5). Here, the thiol molecules, like GSH, can

Table 21.3 IC_{50} (μM) of Ebselen and derivatives against the SARS-CoV-2 proteases.[a]

Entry	Molecule	M^{pro}	PL^{pro}	Ref.
1	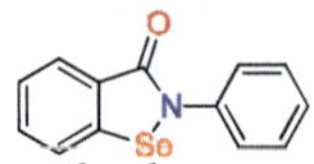 Ebselen	0.03–3.7	1.1–10.3	101–103 and 110
2	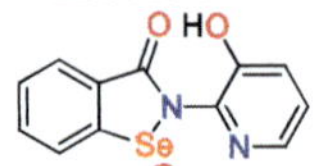	0.05	0.6	103
3	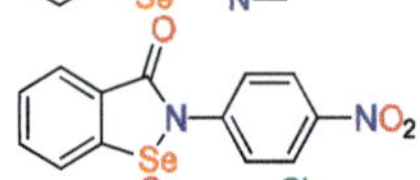	0.03	1.9	103
4		0.02	2.1	103
5		0.074	nd	121
6	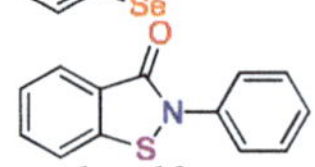 Ebsulfur	0.13–0.49	nd	102 and 121
7	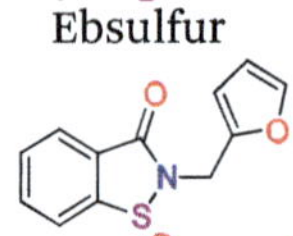	0.11	nd	121
8	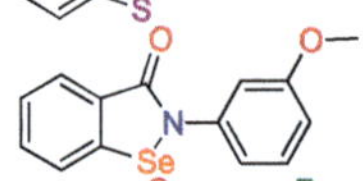	0.13	nd	122
9	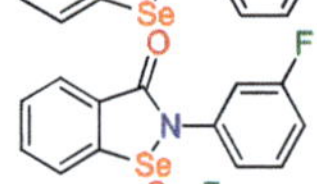	0.11	nd	122
10	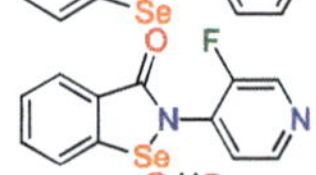	0.07	nd	122
11	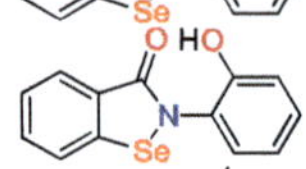	nd	0.2	104
12	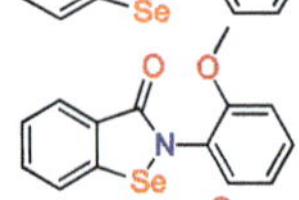	nd	0.2	104
13	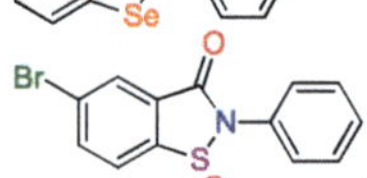	0.03	nd	102
14	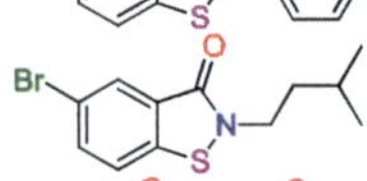	0.21	nd	102
15	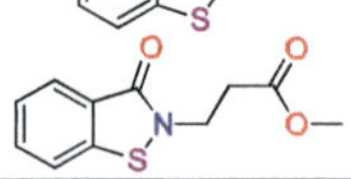	0.25	nd	102

[a]Not determined (nd).

displace Ebselen from the active site of M^{pro} by an exchange Rabenstein-like reaction,[116] forming the Ebselen–GSH adduct and releasing the free and re-activated enzyme.

Alternatively, the excess of GSH or DTT could decrease the interaction of Ebselen with M^{pro} by the law of mass action principle. In fact, the interaction of Ebselen-low-molecular weight thiol (sulfide) molecules with SARS-CoV-2 proteases are critical to predict the potential effectiveness of Ebselen as a protease inhibitor. In summary, the theoretical and *in vitro* studies of the Ebselen–Se–S–R (complex or adduct of Ebselen with low-molecular weight thiol-containing molecules) with M^{pro} and PL^{pro} are highly needed. Particularly, because Ebselen could induce a conformational change in M^{pro} after the formation of the adduct, leading to a permanent denatured state (a sort of aged enzyme). Hypothetically, the conformational change can impair the abstraction of Ebselen by GSH and the reactivation of the enzyme to the native state. As briefly commented in the previous paragraph, the thiol-containing molecules (free Cys, GSH, albumin, hemoglobin) can react with Ebselen before the interaction with the virus enzyme, forming the selenyl–sulfide adduct. These selenyl–sulfide intermediates (putative metabolites) can have lower inhibitory activity compared with free Ebselen (Figure 21.5). In fact, the energetics computed for the Ebselen metabolite (Se–S adduct) suggests it is a weak inhibitor compared with EbSe, in agreement with experimental data. In this case, the structural and electronic nature of the leaving thiol, during the exchange reaction, might affect the whole inhibition reaction.[111]

In short, the detailed study of this type of Rabenstein-like reaction using different low-molecular-weight and high-molecular-weight (proteins) thiol-containing species will be crucial to better understand the Ebselen absorption, distribution, and interaction with target and non-target thiol-containing proteins either in the human body or in target microorganisms. This interaction could considerably affect the stability of the Se–S complex with proteins, enhancing the enzymatic inhibition and reducing the rate of the detoxification process.[117]

Besides the M^{pro} and PL^{pro} inhibition, Ebselen has been demonstrated to target the replication and transcription complex (RTC) of SARS-CoV-2. The RTC is critical for the viral replication and transcripts new viral RNA molecules.[118] The Ebselen's inhibition of nsp13 ATPase and nsp14 exoribonuclease activities (IC_{50} of 292 nM and 3.18 μM, respectively) occurred simultaneous with the ejection of Zn(II) ions from conserved essential Zn(II) sites of nsp13 and nsp14 proteins.[119] Similar data were observed in the nsp10 and PL^{pro} proteins,[110] where Ebselen can bind to Cys residues coordinated to Zn(II), and, as a consequence, ejects the Zn(II) from the zinc-binding site, causing the protein denaturation.

The interaction between Ebselen and critical Cys residues in target proteins can be mediated by the Se···S interaction, followed by the covalent bond formation. The nucleophilic attack from thiol/thiolate of Cys to the electrophilic Se atom of Ebselen can lead to the cleavage of the N–Se bond and the concomitant isoselanzolone ring opening and formation of the selenyl–sulfide adduct.[111,112,120] In addition, in the case of the zinc-finger sites, which is composed by four Cys residues, the Se–S adduct can lead to

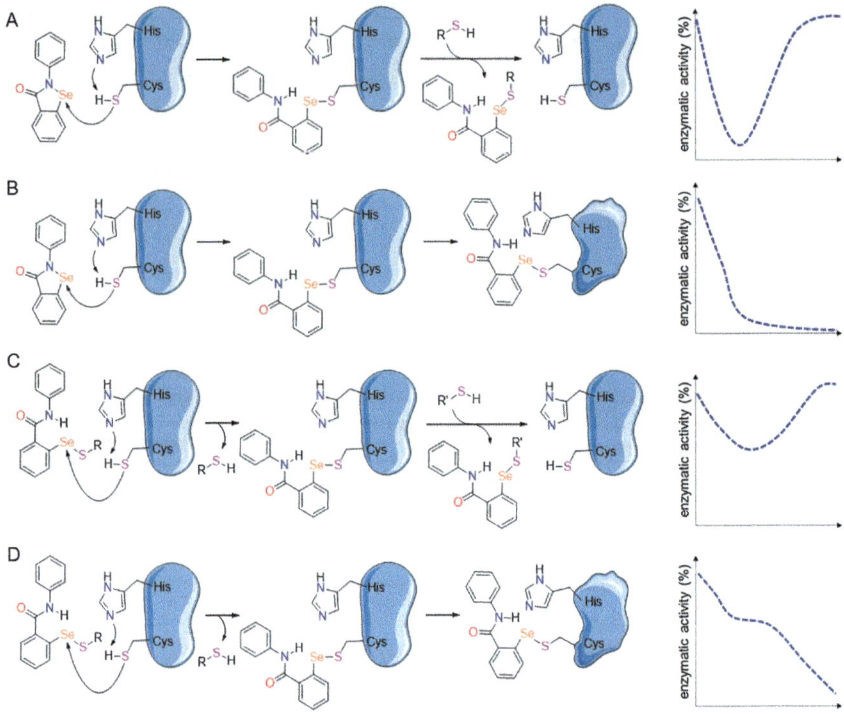

Figure 21.5 Potential mechanisms of SARS-CoV-2 proteases inhibition by Ebselen. (A) Ebselen binds covalently to the active site, however, the inhibition is reversible in the presence of a thiol-reducing agent. (B) Ebselen binds covalently to the active site, leading to a protein conformational change (denaturation or aging) and the inhibition is not reversible. The enzyme is "aged", *i.e.*, the conversion of the inhibited enzyme into an irreparable thermodynamic state. (C) Ebselen–selenylsulfide adduct can bind covalently to the active site *via* an exchange reaction, however, the inhibitory potency is possibly lower than that of free Ebselen and it is reversible in the presence of a thiol-reducing agent. (D) Ebselen–selenylsulfide adduct exchanges and binds covalently to the active site, leading to protein denaturation (aged enzyme). R and R′ can be any thiol-reducing agent or low-molecular-weight thiol molecules (LMW-SH), such as Cys, GSH, or DTT.

the Cys oxidation (S–S), by the nucleophilic attack of a vicinal Cys (Figure 21.6).

Taken all together, the isoselanazolone ring appears to be a good warhead for Cys residues inactivation, since several Ebselen analogs are M^{pro} and PL^{pro} inhibitors (Table 21.3). Many of the structural modifications occurs at the *N*-position, introducing different substituents (F, Cl, MeO, NO_2, OH) and rings (pyridyl and furyl). For instance, Sun *et al.*,[121] demonstrated that Ebsulfur is a good M^{pro} inhibitor and Ebselen analogous molecules containing furan substituent presented high M^{pro} inhibition (Table 21.3, entries 5–7). In

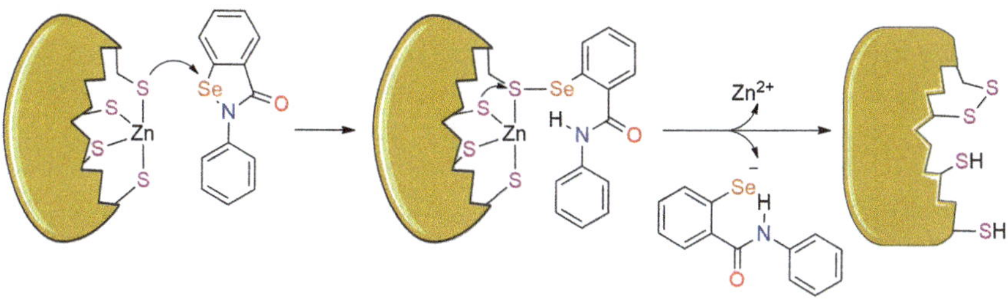

Figure 21.6 Possible mechanism of SARS-CoV-2 zinc-finger proteins inhibition by Ebselen. Although the formation of adducts between Ebselen and thiol groups from the protein is possible, here we represent only the so-called "Zn-ejection" effect of Ebselen.

addition, in the study of Qiao *et al.*,[122] three new Ebselen-like molecules (Table 21.3, entries 8–10) were found to be potent M^{pro} inhibitors because they inhibited the viral replication in infected HPAepiC cells (EC_{50} = 4.07, 19.91, and 8.23 μM, respectively), better than Ebselen itself (21.60 μM). In this sense, the design of Ebselen-like molecules with high selectivity toward viral proteins and with low binding affinity to endogenous thiol-containing molecules is of paramount importance.

Furthermore, besides the potential anti-SARS-CoV-2 effect of Ebselen, it could modulate negatively the exaggerated inflammatory response of the host, decreasing the formation of pro-oxidants and the release of inflammatory cytokines.[108,123] In short, in addition to impairing the virus replication during the first phase of COVID-19, Ebselen could potentially also modulate the exaggerated pro-inflammatory phase of the disease. Thus, although the non-specific interaction of Ebselen with different thiol molecules can have some undesirable toxic effects, the multi-targeting ability of Ebselen is possibly influenced by the capacity of interacting with different thiol-containing molecules and can have beneficial pharmacological implications. In fact, the multi-targeting ability of Ebselen could have some therapeutic effects since the SARS-CoV-2 has organotropism, infecting the respiratory tract and other organs, such as kidneys, liver, heart, and brain.[124]

21.6 Conclusions

Almost one hundred years have passed since the synthesis of Ebselen was reported for the first time and, of particular importance to the field of organoselenium compounds, this molecule is safe for human use. However, its therapeutic efficacy in human pathologies (noise-induced hearing loss and bipolar disorder)[5,9] resulted only in borderline beneficial effects. In addition, in the hearing loss study, the therapeutic effective dose (400 mg, twice a day) was close to the ineffective lowest (200 mg, twice a day) and highest (600 mg, twice a day) doses, indicating a very narrow therapeutic window. In the bipolar disorder study, the effectiveness of Ebselen in the primary outcome scale was marginal, indicating that large clinical trials are needed to confirm or not the effectiveness of Ebselen as a potential substitute to lithium. Despite the clinical failures of Ebselen, its thiol-modifier properties of SARS-CoV-2 proteases and its *in vitro* inhibitory properties against the virus replication have allowed its rapid approval for use in clinical trials with moderate and severe COVID-19 patients. Hopefully, these clinical trials will demonstrate the effectiveness of Ebselen as a therapeutic agent to treat coronaviruses.

Acknowledgements

The authors would like to thank the financial support by Coordination for Improvement of Higher Education Personnel CAPES/PROEX (no. 23038.004173/2019-93; no. 0493/2019). P. A. N. and J. B. T. R. were funded by CAPES (Edital 09-88887.505377/2020-00, 88887.511828/2020-00). C. S. O. and M. E. P. were supported by Pequeno Príncipe Complex. L. O. was funded by Università degli Studi di Padova, thanks to the P-DiSC (BIRD2018-UNIPD) project MAD^{3}S (Modeling Antioxidant Drugs: Design and Development of computer-aided molecular Systems).

References

1. T. Yamaguchi, K. Sano, K. Takakura, I. Saito, Y. Shinohara, T. Asano and H. Yasuhara, *Stroke*, 1998, **29**, 12.
2. I. Saito, T. Asano, K. Sano, K. Takakura, H. Abe, T. Yoshimoto, H. Kikuchi, T. Ohta and S. Ishibashi, *Neurosurgery*, 1998, **42**, 269.
3. A. Ogawa, T. Yoshimoto, H. Kikuchi, K. Sano, I. Saito, T. Yamaguchi and H. Yasuhara, *Cerebrovasc. Dis.*, 1999, **9**, 112.
4. J. A. Beckman, A. B. Goldfine, J. A. Leopold and M. A. Creager, *Am. J. Physiol. Circ. Physiol.*, 2016, **311**, H1431.
5. J. Kil, E. Lobarinas, C. Spankovich, S. K. Griffiths, P. J. Antonelli, E. D. Lynch and C. G. Le Prell, *Lancet*, 2017, **390**, 969.
6. N. Singh, A. L. Sharpley, U. E. Emir, C. Masaki, M. M. Herzallah, M. A. Gluck, T. Sharp, C. J. Harmer, S. R. Vasudevan, P. J. Cowen and G. C. Churchill, *Neuropsychopharmacology*, 2016, **41**, 1768.
7. C. Masaki, A. L. Sharpley, B. R. Godlewska, A. Berrington, T. Hashimoto, N. Singh, S. R. Vasudevan, U. E. Emir, G. C. Churchill and P. J. Cowen, *Psychopharmacology*, 2016, **233**, 1097.
8. C. Masaki, A. L. Sharpley, C. M. Cooper, B. R. Godlewska, N. Singh, S. R. Vasudevan, C. J. Harmer, G. C. Churchill, T. Sharp, R. D. Rogers and P. J. Cowen, *Psychopharmacology*, 2016, **233**, 2655.
9. A. L. Sharpley, C. Williams, A. A. Holder, B. R. Godlewska, N. Singh, M. Shanyinde, O. MacDonald and P. J. Cowen, *Psychopharmacology*, 2020, **237**, 3773.
10. R. Lesser and R. Weiß, *Ber. Der Dtsch. Chem. Ges.*, 1924, **57**, 1077.
11. K. W. Franke and V. R. Potter, *J. Nutr.*, 1935, **10**, 213.
12. E. P. Painter and K. W. Franke, *J. Biol. Chem.*, 1935, **111**, 643.

13. W. C. Tully and K. W. Franke, *Poult. Sci.*, 1934, **13**, 343.
14. K. W. Franke, A. L. Moxon, W. E. Poley and W. C. Tully, *Anat. Rec.*, 1936, **65**, 15.
15. K. W. Franke and E. P. Painter, *J. Nutr.*, 1935, **10**, 599.
16. D. V. Frost and O. E. Olson, *Crit. Rev. Toxicol.*, 1972, **1**, 467.
17. K. Schwarz and C. M. Foltz, *J. Am. Chem. Soc.*, 1957, **79**, 3292.
18. G. N. Schrauzer, *Pure Appl. Chem.*, 2006, **78**, 105–109.
19. K. Schwarz and C. M. Foltz, *J. Biol. Chem.*, 1958, **233**, 245.
20. K. Schwarz and A. Fredga, *J. Biol. Chem.*, 1969, **244**, 2103.
21. K. Schwarz and A. Fredga, *Bioinorg. Chem.*, 1974, **3**, 153.
22. K. Schwarz and A. Fredga, *Bioinorg. Chem.*, 1975, **4**, 235.
23. K. Schwarz, L. A. Porter and A. Fredga, *Bioinorg. Chem.*, 1974, **3**, 145.
24. A. Fredga, *Bioinorg. Chem.*, 1972, **2**, 47.
25. K. Schwarz and A. Fredga, *Bioinorg. Chem.*, 1972, **2**, 171.
26. J. T. Rotruck, A. L. Pope, H. E. Ganther, A. B. Swanson, D. G. Hafeman and W. G. Hoekstra, *Science*, 1973, **179**, 588.
27. J. T. Rotruck, W. G. Hoekstra, A. L. Pope, H. Ganther, A. Swanson and D. Hafeman, *Fed. Proc.*, 1972, **31**, 691.
28. L. Flohé, W. A. Günzler and H. H. Schock, *FEBS Lett.*, 1973, **32**, 132.
29. J. E. Cone, R. M. Del Rio, J. N. Davis and T. C. Stadtman, *Proc. Natl. Acad. Sci.*, 1976, **73**, 2659.
30. J. W. Forstrom, J. J. Zakowski and A. L. Tappel, *Biochemistry*, 1978, **17**, 2639.
31. M. J. Parnham and H. Sies, *Biochem. Pharmacol.*, 2013, **86**, 1248.
32. N. Noguchi, *Arch. Biochem. Biophys.*, 2016, **595**, 109.
33. R. Cantineau, G. Tihange, A. Plenevaux, L. Christiaens, M. Guillaume, A. Welter and N. Dereu, *J. Label. Compd. Radiopharm.*, 1986, **23**, 59.
34. A. Müller, H. Gabriel, H. Sies, R. Terlinden, H. Fischer and A. Römer, *Biochem. Pharmacol.*, 1988, **37**, 1103.
35. S. R. Wilson, P. A. Zucker, R. R. C. Huang and A. Spector, *J. Am. Chem. Soc.*, 1989, **111**, 5936.
36. N. V. Barbosa, C. W. Nogueira, P. A. Nogara, A. F. De Bem, M. Aschner and J. B. T. Rocha, *Metallomics*, 2017, **9**, 1703.
37. T. Wirth, *Angew. Chem., Int. Ed.*, 2015, 54, 10074.
38. R. Zhao and A. Holmgren, *J. Biol. Chem.*, 2002, **277**, 39456.
39. R. Zhao, H. Masayasu and A. Holmgren, *Proc. Natl. Acad. Sci. U. S. A.*, 2002, **99**, 8579.
40. A. Schoeller, *Chem. Ber.*, 1919, **52**, 1517.
41. C. Santi, C. Scimmi and L. Sancineto, *Molecules*, 2021, **26**, 1.
42. A. Wendel, M. Fausel, H. Safayhi, G. Tiegs and R. Otter, *Biochem. Pharmacol.*, 1984, **33**, 3241.
43. L. Engman and A. Hallberg, *J. Org. Chem.*, 1989, **54**, 2964.
44. S. J. Balkrishna, B. S. Bhakuni, D. Chopra and S. Kumar, *Acc. Chem. Res.*, 2010, **12**, 5394.
45. L. Sancineto, A. Mariotti, L. Bagnoli, F. Marini, J. Desantis, N. Iraci, C. Santi, C. Pannecouque and O. Tabarrini, *J. Med. Chem.*, 2015, **58**, 9601.

46. P. A. Nogara, C. S. Oliveira and J. B. T. Rocha, in *Organoselenium Chemistry*, ed. B. C. Ranu and B. Banerjee, De Gruyter, Berlin, 2020, pp. 305–346.
47. S. Ueda, T. Yoshikawa, S. Takahashi, Y. Naito, H. Oyamada, T. Takemura, Y. Morita, T. Tanigawa, S. Sugino and M. Kondo, *Adv. Exp. Med. Biol.*, 1990, **164**, 187.
48. J. F. Wang, P. Komarov, H. Sies and H. Groot, *Hepatology*, 1992, **15**, 1112.
49. T. Ishikawa, F. Yamamoto, T. Ohashi, Y. Shimada, K. Kagisaki, Y. Kumada, Y. Kito and Y. Kawashima, *J. Jap. Assoc. Thorac. Surg.*, 1995, **43**, 458.
50. D. A. Dawson, H. Masayasu, D. I. Graham and I. M. Macrae, *Neurosci. Lett.*, 1995, **185**, 65.
51. S. Knollema, J. W. Elting, R. M. Dijkhuizen, K. Nicolay, J. Korf and G. J. Ter Horst, *Neurosci. Res. Commun.*, 1996, **19**, 47.
52. M. Ozaki, M. Nakamura, S. Teraoka and K. Ota, *Transpl. Int.*, 1997, **10**, 96.
53. Y. X. Tan, W. H. Li, X. B. Tao, Y. Y. Jiang, W. P. Chen and B. Zhou, *Acta Pharmacol. Sin.*, 1997, **10**, 201.
54. T. Takasago, E. E. Peters, D. I. Graham, H. Masayasu and I. M. Macrae, *Br. J. Pharmacol.*, 1997, **122**, 1251.
55. S. V. Brodsky, O. Gealekman, J. Chen, F. Zhang, N. Togashi, M. Crabtree, S. S. Gross, A. Nasjletti and M. S. Goligorsky, *Circ. Res.*, 2004, **94**, 377.
56. O. Gealekman, S. V. Brodsky, F. Zhang, P. N. Chander, C. Friedli, A. Nasjletti and M. S. Goligorsky, *Kidney Int.*, 2004, **66**, 2337.
57. P. N. Chander, O. Gealekman, S. V. Brodsky, S. Elitok, A. Tojo, M. Crabtree, S. S. Gross and M. S. Goligorsky, *J. Am. Soc. Nephrol.*, 2004, **15**, 2391.
58. A. H. Bubolz, Q. Wu, B. T. Larsen, D. D. Gutterman and Y. Liu, *Am. J. Physiol. - Hear. Circ. Physiol.*, 2007, **283**, 2231.
59. J. Chen, H.-C. Park, S. Patschan, S. V. Brodsky, O. Gealikman, M.-C. Kuo, H. Li, F. Addabbo, F. Zhang, A. Nasjletti, S. S. Gross and M. S. Goligorsky, *Drug Discov. Today Ther. Strateg.*, 2007, **4**, 93.
60. J. Kil, E. E. Harruff and R. J. Longenecker, *Hear. Res.*, 2022, **413**, 108209.
61. E. D. Lynch, R. Gu, C. Pierce and J. Kil, *Laryngoscope*, 2004, **114**, 333.
62. A. Pourbakht and T. Yamasoba, *Hear. Res.*, 2003, **181**, 100.
63. T. Yamasoba, A. Pourbakht, T. Sakamoto and M. Suzuki, *Neurosci. Lett.*, 2005, **380**, 234.
64. J. Kil, C. Pierce, H. Tran, R. Gu and E. D. Lynch, *Hear. Res.*, 2007, **226**, 44.
65. E. D. Lynch, R. Gu, C. Pierce and J. Kil, *Hear. Res.*, 2005, **201**, 81.
66. E. Lynch and J. Kil, *Semin. Hear.*, 2009, **30**, 47.
67. S.-J. Kim, C. Park, A. L. Han, M.-J. Youn, J.-H. Lee, Y. Kim, E.-S. Kim, H.-J. Kim, J.-K. Kim, H.-K. Lee, S.-Y. Chung, H. So and R. Park, *Hear. Res.*, 2009, **251**, 70.
68. T. Posser, M. P. Kaster, S. C. Baraúna, J. B. T. Rocha, A. L. S. Rodrigues and R. B. Leal, *Eur. J. Pharmacol.*, 2009, **602**, 85.
69. N. Singh, A. C. Halliday, J. M. Thomas, O. Kuznetsova, R. Baldwin, E. C. Y. Woon, P. K. Aley, I. Antoniadou, T. Sharp, S. R. Vasudevan and G. C. Churchill, *Nat. Commun.*, 2013, **4**, 1332.

70. C. Barkus, J.-M. N. Ferland, W. K. Adams, G. C. Churchill, P. J. Cowen, D. M. Bannerman, R. D. Rogers, C. A. Winstanley and T. Sharp, *J. Psychopharmacol.*, 2018, **32**, 1018.
71. I. Antoniadou, M. Kouskou, T. Arsiwala, N. Singh, S. R. Vasudevan, T. Fowler, E. Cadirci, G. C. Churchill and T. Sharp, *Br. J. Pharmacol.*, 2018, **175**, 2599.
72. A. Müller, E. Cadenas, P. Graf and H. Sies, *Biochem. Pharmacol.*, 1984, **33**, 3235.
73. H. Johshita, T. Sasaki, T. Matsui, T. Hanamura, H. Masayasu, T. Asano and K. Takakura, *Brain Edema VIII*, 1990, vol. 51, p. 239.
74. R. Rodrigo, R. Fernandez-Gajardo, R. Gutierrez, J. Matamala, R. Carrasco, A. Miranda-Merchak and W. Feuerhake, *CNS Neurol. Disord. – Drug Targets*, 2013, **12**, 698–714.
75. H. Masumoto and H. Sies, *Chem. Res. Toxicol.*, 1996, **9**, 262.
76. H. Masumoto, R. Kissner, W. H. Koppenol and H. Sies, *FEBS Lett.*, 1996, **398**, 179.
77. H. Sies and H. Masumoto, *Adv. Pharmacol.*, 1996, **38**, 229.
78. A. Daiber, M. H. Zou, M. Bachschmid and V. Ullrich, *Biochem. Pharmacol.*, 2000, **59**, 153.
79. C. A. Wood-Allum, S. C. Barber, J. Kirby, P. Heath, H. Holden, R. Mead, A. Higginbottom, S. Allen, T. Beaujeux, S. E. Alexson, P. G. Ince and P. J. Shaw, *Brain*, 2006, **129**, 1693.
80. P. Chew, D. Y. C. Yuen, N. Stefanovic, J. Pete, M. T. Coughlan, K. A. Jandeleit-Dahm, M. C. Thomas, F. Rosenfeldt, M. E. Cooper and J. B. De Haan, *Diabetes*, 2010, **59**, 3198.
81. P. Chew, D. Y. C. Yuen, P. Koh, N. Stefanovic, M. A. Febbraio, I. Kola, M. E. Cooper and J. B. De Haan, *Arterioscler., Thromb., Vasc. Biol.*, 2009, **29**, 823.
82. A. F. de Bem, B. Fiuza, P. Calcerrada, P. M. Brito, G. Peluffo, T. C. P. Dinis, M. Trujillo, J. B. T. Rocha, R. Radi and L. M. Almeida, *Nitric oxide*, 2013, **31**, 20.
83. H. A. Hadi and J. Al Suwaidi, *Vasc. Health Risk Manag.*, 2007, **3**, 853.
84. O. Oguntibeju, *Int. J. Physiol. Pathophysiol. Pharmacol.*, 2019, **11**, 45.
85. F. A. Matough, S. B. Budin, Z. A. Hamid, N. Alwahaibi and J. Mohamed, *Sultan Qaboos Univ. Med. J.*, 2012, **12**, 5.
86. S. Ghibu, C. Richard, C. Vergely, M. Zeller, Y. Cottin and L. Rochette, *J. Cardiovasc. Pharmacol.*, 2009, **54**, 391.
87. J. Lautermann, S. A. Crann, J. McLaren and J. Schacht, *Hear. Res.*, 1997, **114**, 75.
88. D. Coling, S. Chen, L.-H. Chi, S. Jamesdaniel and D. Henderson, *Neurosci. Lett.*, 2009, **464**, 22.
89. J. Wang, P. Wang, C. Dong, Y. Zhao, J. Zhou, C. Yuan and L. Zou, *Future Med. Chem.*, 2020, **12**, 2141.
90. P. A. Nogara, C. S. Oliveira, M. E. Pereira, M. Bortoli, L. Orian, M. Aschner and J. B. T. Rocha, in *Redox Chemistry and Biology of Thiols*, ed. B. Alvarez, M. Comini, G. Salinas and M. Trujillo, Elsevier, 2022, pp. 1–20.

91. T. F. Kubiça, L. B. Denardi, É. S. Loreto, G. Zeni, C. Weiblen, V. Oliveira, J. M. Santurio and S. H. Alves, *Mycoses*, 2019, **62**, 428.
92. R. Vartak, S. Menon, M. Patki, B. Billack and K. Patel, *Eur. J. Pharm. Sci.*, 2020, **148**, 105323.
93. J. Padiadpu, P. Baloni, K. Anand, M. Munshi, C. Thakur, A. Mohan, A. Singh and N. Chandra, *ACS Infect. Dis.*, 2016, **2**, 592.
94. F. Irigoín, L. Cibils, M. A. Comini, S. R. Wilkinson, L. Flohé and R. Radi, *Free Radic. Biol. Med.*, 2008, **45**, 733.
95. T. N. Gustafsson, H. Osman, J. Werngren, S. Hoffner, L. Engman and A. Holmgren, *Biochim. Biophys. Acta – Gen. Subj*, 2016, **1860**, 1265.
96. K. M. Sakita, I. R. G. Capoci, P. C. V. Conrado, F. A. V. Rodrigues-Vendramini, D. R. Faria, G. S. Arita, T. C. A. Becker, P. de, S. Bonfim-Mendonça, T. I. E. Svidzinski and E. S. Kioshima, *Front. Cell. Infect. Microbiol.*, 2021, **11**, 684525.
97. A. C. Marshall, S. E. Kidd, S. J. Lamont-Friedrich, G. Arentz, P. Hoffmann, B. R. Coad and J. B. Bruning, *Antimicrob. Agents Chemother.*, 2019, **63**, e02281-18.
98. A. Debnath, A. T. Nelson, A. Silva-Olivares, M. Shibayama, D. Siegel and J. H. McKerrow, *Front. Microbiol.*, 2018, **9**, 414.
99. W. M. Mcfadden, A. A. Snyder, K. A. Kirby, P. R. Tedbury, M. Raj, Z. Wang and S. G. Sarafianos, *Retrovirology*, 2021, **18**, 1.
100. N. Chow, K. Fleming-Dutra, R. Gierke, A. Hall, M. Hughes, T. Pilishvili, M. Ritchey, K. Roguski, T. Skoff and E. Ussery, *MMWR. Morb. Mortal. Wkly. Rep.*, 2020, **69**, 382.
101. Z. Jin, X. Du, Y. Xu, Y. Deng, M. Liu, Y. Zhao, B. Zhang, X. Li, L. Zhang, C. Peng, Y. Duan, J. Yu, L. Wang, K. Yang, F. Liu, R. Jiang, X. Yang, T. You, X. Liu, X. Yang, F. Bai, H. Liu, X. Liu, L. W. Guddat, W. Xu, G. Xiao, C. Qin, Z. Shi, H. Jiang, Z. Rao and H. Yang, *Nature*, 2020, **582**, 289.
102. S. T. Thun-Hohenstein, T. F. Suits, T. R. Malla, A. Tumber, L. Brewitz, H. Choudhry, E. Salah and C. Schofield, *ChemMedChem*, 2021, **1**, e202100582.
103. M. Zmudzinski, W. Rut, K. Olech, J. Granda, M. Giurg, M. Burda-Grabowska, L. Zhang, X. Sun, Z. Lv, D. Nayak, M. Kesik-Brodacka, S. K. Olsen, R. Hilgenfeld and M. Drag, *bioRxiv*, DOI: 10.1101/2020.08.30.273979.
104. E. W. Tomczak, J. M. Tomczak and M. Talma, *Sci. Rep.*, 2021, **11**, 3640.
105. M. H. Lin, D. C. Moses, C. H. Hsieh, S. C. Cheng, Y. H. Chen, C. Y. Sun and C. Y. Chou, *Antiviral Res.*, 2018, **150**, 155.
106. W. Rut, K. Groborz, L. Zhang, X. Sun, M. Zmudzinski, B. Pawlik, X. Wang, D. Jochmans, J. Neyts, W. Młynarski, R. Hilgenfeld and M. Drag, *Nat. Chem. Biol.*, 2021, **17**, 222.
107. S. A. Amin, S. Banerjee, K. Ghosh, S. Gayen and T. Jha, *Bioorg. Med. Chem.*, 2021, **29**, 115860.
108. H. Sies and M. J. Parnham, *Free Radic. Biol. Med.*, 2020, **156**, 107.
109. S. Yan and G. Wu, *FASEB J.*, 2021, **35**, e21197.

110. K. Sargsyan, C. C. Lin, T. Chen, C. Grauffel, Y. P. Chen, W. Z. Yang, H. S. Yuan and C. Lim, *Chem. Sci.*, 2020, **11**, 9904.
111. A. Madabeni, P. A. Nogara, F. B. Omage, J. B. T. Rocha and L. Orian, *Appl. Sci.*, 2021, **11**, 6291.
112. P. A. Nogara, F. B. Omage, G. R. Bolzan, C. P. Delgado, M. Aschner, L. Orian and J. B. T. Rocha, *Mol. Inform.*, 2021, **40**, 2100028.
113. C. Ma, Y. Hu, J. A. Townsend, P. I. Lagarias, M. T. Marty, A. Kolocouris and J. Wang, *ACS Pharmacol. Transl. Sci.*, 2020, **3**, 1265.
114. L. Orian and S. Toppo, *Free Radic. Biol. Med.*, 2014, **66**, 65.
115. Z. Chen, Z. Jiang, N. Chen, Q. Shi, L. Tong, F. Kong, X. Cheng, H. Chen, C. Wang and B. Tang, *Chem. Commun.*, 2018, **54**, 9506.
116. P. A. Nogara, C. S. Oliveira, G. L. Schmitz, P. C. Piquini, M. Farina, M. Aschner and J. B. T. Rocha, *Biochim. Biophys. Acta - Gen. Subj.*, 2019, **1863**, 129284.
117. C. Santi, L. Sancineto, F. Mangiavacchi, C. Scimmi and S. Misra, 1st International Conference of Chemo and Bioinformatics, October 26–27, 2021. Kragujevac, Serbia, DOI: 10.46793/ICCBI21.020S.
118. P. V'kovski, A. Kratzel, S. Steiner, H. Stalder and V. Thiel, *Nat. Rev. Microbiol.*, 2021, **19**, 155.
119. T. Chen, C. Y. Fei, Y. P. Chen, K. Sargsyan, C. P. Chang, H. S. Yuan and C. Lim, *ACS Pharmacol. Transl. Sci.*, 2021, **4**, 898.
120. A. Parise, I. Romeo, N. Russo and T. Marino, *Int. J. Mol. Sci.*, 2021, **22**, 9792.
121. L.-Y. Sun, C. Chen, J. Su, J.-Q. Li, Z. Jiang, H. Gao, J.-Z. Chigan, H.-H. Ding, L. Zhai and K.-W. Yang, *Bioorg. Chem.*, 2021, **112**, 104889.
122. Z. Qiao, N. Wei, L. Jin, H. Zhang, J. Luo, Y. Zhang and K. Wang, *Bioorg. Chem.*, 2021, **117**, 105455.
123. C. V. Haritha, K. Sharun and B. Jose, *Int. J. Surg.*, 2020, **84**, 53.
124. V. G. Puelles, M. Lütgehetmann, M. T. Lindenmeyer, J. P. Sperhake, M. N. Wong, L. Allweiss, S. Chilla, A. Heinemann, N. Wanner, S. Liu, F. Braun, S. Lu, S. Pfefferle, A. S. Schröder, C. Edler, O. Gross, M. Glatzel, D. Wichmann, T. Wiech, S. Kluge, K. Pueschel, M. Aepfelbacher and T. B. Huber, *N. Engl. J. Med.*, 2020, **383**, 590.

CHAPTER 22

Chalcogen-containing Therapeutic Molecules

LUCA SANCINETO*[a] AND VANESSA NASCIMENTO*[b]

[a] Group of Catalysis and Green Organic Chemistry, Department of Pharmaceutical Sciences, University of Perugia, Via del Liceo 1, 06100, Perugia, Italy; [b] Departamento de Química Orgânica, Instituto de Química, Universidade Federal Fluminense, Campus do Valonguinho, CEP 24020-150, Niterói-RJ, Brazil
*Emails: luca.sancineto@unipg.it; nascimentovanessa@id.uff.br

22.1 Antiviral Activity

The identification of ebselen (compound **1**, Figure 22.1) as a plausible anti SARS-CoV-main protease inhibitor, within the article that can be considered the first drug repurposing approach ever reported in the context of COVID19,[1] lead to renewed interest around the use of selenorganic compounds as antivirals,[2,3] among other therapeutic applications.[4–8] From a historical perspective the first selenium-containing antiviral is selenazofurin (compound **2**), developed in 1983 by Kirsi, able to inhibit the replication of a plethora of DNA and RNA viruses with a potency which was far higher than that displayed by its sulfur analog, thiazafurin (not shown), and of the marketed, broad spectrum drug ribavirin.[9] Then, ebselen itself was found to be active as an anti-HIV compound due to its ability to block tat-mediated transcription[10] and covalently bind the capsid protein.[11] The group of Mlochowski and Giurg contributed a lot by preparing several diselenides and benzisoselenazolones endowed with antiviral activities.[12–14] Some of us, in 2015, reported the identification of DiSelenoBisbenzAmides (DiSeBAs, compounds **3**), diphenyldiselenides

Chalcogen Chemistry: Fundamentals and Applications
Edited by Vito Lippolis, Claudio Santi, Eder J. Lenardão and Antonio L. Braga

Published by the Royal Society of Chemistry, www.rsc.org

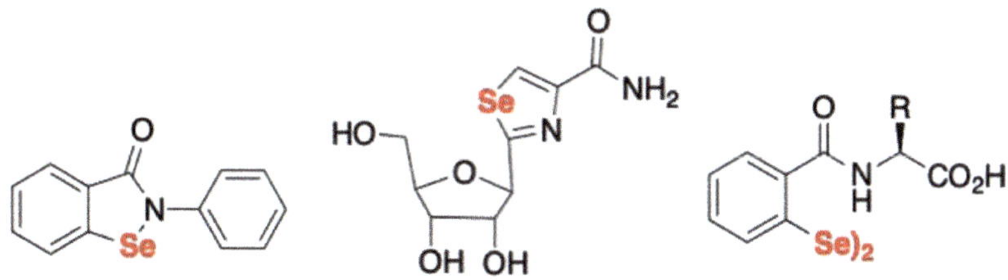

1 2 3

Figure 22.1 Ebselen (**1**), selenazofurin (**2**), and DiSeBAs (**3**).

substituted in the *ortho* position with an amide functionality, capable of exerting a broad and selective antiviral activity.[15] Their ability to inhibit viral replication is due to the inhibition of the key viral component NCp7, a zinc finger protein which is essential for the HIV to successfully duplicate.[16,17] Not only HIV-1 but also HIV-2 wild type and clinical isolates resistant to the marketed drugs are inhibited by DiSeBAs.[15] Several reviews,[3] books,[18] and book chapters[7,19] have been devoted to this topic, and as a result the authors decided to limit the discussion to the most salient and timely examples of selenium-containing compounds endowed with antiviral activities detailing all the examples of compounds endowed with anti-SARS-CoV-2 properties.

22.1.1 Ebselen and Derivatives

With their article, Jiang, Rao, and Yang solved the crystal structure of the SARS-CoV-2 core protein, the so-called Main protease (Mpro) illustrating the first drug repurposing approach ever reported in the context of COVID19. Mpro is an ideal drug target since it has no homolog in the human proteome, and it is well conserved among SARS strains. With a high-throughput screening approach, they identified several plausible candidates and among these, ebselen was reported to be the most potent, with a half maximal inhibitory concentration (IC_{50}) of 670 nM. The mechanism by which the protease is inhibited is irreversible, ebselen is indeed able to covalently bind Cysteine 145 blocking the active site. The anti-Mpro activity translates into an antiviral activity measured in a cellular context with a potency in the low micromolar range ($EC_{50} = 4.67$ μM).[1] The dynamic of Mpro inhibition was thoroughly studied through molccular dynamic simulations that highlighted two probable interaction sites, one is of course the catalytic pocket, the other, which recognizes the compound with higher affinity, is located in the region between the II and III domains, essential for Mpro dimerization.[20] It was also found that the inhibition can be fully reverted by the addition of reducing agents such as dithiothreitol and GSH.[21]

Later on, the computational analysis of the ebselen/Mpro complex formation was extended to all the plausible metabolites of ebselen, including oxidation products and adducts with cellular thiols.[22] Most probably, after the reaction with Mpro, the selenorganic framework of ebselen undergoes a

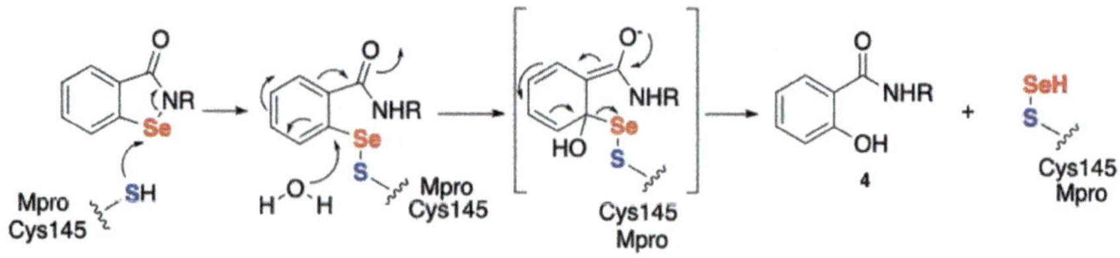

Scheme 22.1 Selenylation of Cys145.

nucleophilic aromatic substitution that leads to the selenenylated Cys145 and 2-hydroxy-*N*-phenylbenzamide (**4**) (Scheme 22.1).[23]

Not only Mpro, but also Zn-containing non-structural proteins (Nsp) 13 and 14 are inhibited with an IC_{50} of 292 nM and 3.18 μM, respectively. Interestingly the compound displays synergism with the newly approved anti-SARS-CoV-2 drug Remdesivir.[24]

Previously reported studies clearly suggest that ebselen is ideally suited to treat respiratory infections[25] and COVID19 in particular,[26,27] since it is able to reach the lungs in high concentrations and to reduce inflammation *in vivo*.[28–30] In addition, the hyperinflammation state coupled with thrombotic events that result in the death of SARS-CoV-2 infected patients[31] seems to be reverted by ebselen in studies carried out decades before COVID19.[32] Liver injury is observed in the more severe cases of COVID19 patients and ebselen may here again be of therapeutic benefit due to its capabilities of inhibiting liver injury induced by several stimuli.[33] All these features make ebselen a useful therapeutic option for the treatment of SARS-CoV-2 infection so much so that it is currently under clinical evaluation for moderate COVID19 patients (see clinicaltrials.gov). In addition, its scaffold, the benzisoselenazolone core, was used as a template for the preparation of anti-SARS-CoV-2 compounds.

In 2021, Weglarz-Tomczak and Brul *et al.* tested their collection of bensisolenenazolones for the ability to inhibit the papain-like proteases (PLpro) of both SARS and SARS-CoV-2. Compounds **5** and **6** (Figure 22.2) were identified as the most potent, but unfortunately, no anti-viral activity was ever reported for these interesting compounds.[34] In the same year Yang *et al.* reported the biological evaluation of a series of ebselen and ebsulfur (not shown) analogues bearing aromatic, heteroaromatic, and aliphatic substituents at the nitrogen atom. This classical medicinal chemistry approach led to the identification of compounds endowed with a better Mpro activity when compared to native lead (compounds **7–9**). It is worth noting that ebsulfur and its analogs are able to inhibit Mpro, covalently, although with a lower potency. This demonstrates the unique feature of the electrophilic selenium and the reactivity of the benzisoselenazolone core.[35]

Very recently, Rana *et al.* designed and synthesized a small library of benzisoselenazolones then tested them against recombinant SARS-CoV-2 Mpro through FRET-based screening identifying six compounds that inhibit proteolysis with nanomolar IC_{50} values (compounds **10–15**). Investigation of

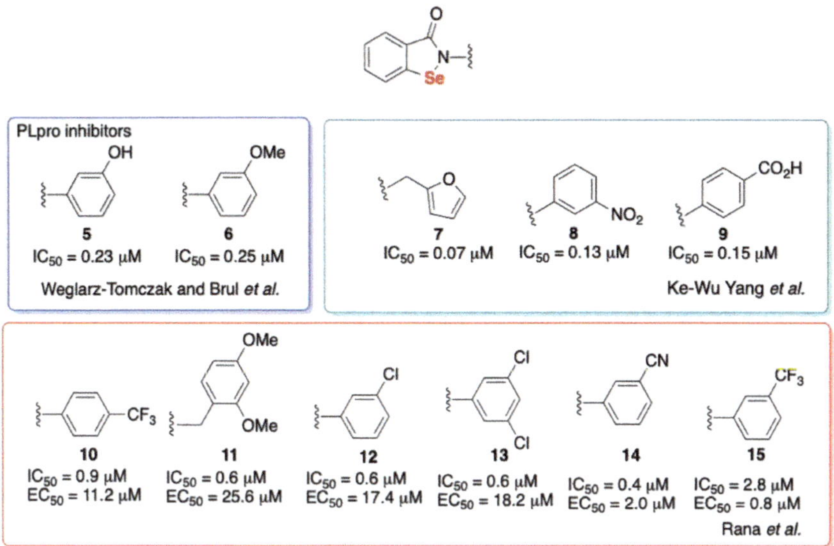

Figure 22.2 Ebselen-like structures tested in the context of COVID19.

the inhibition mechanism clearly revealed that these compounds can inhibit the protein either covalently or noncovalently. As an example, derivative **10** inhibits Mpro competitively. Compound **15** displayed an antiviral activity measured in Vero E6 cells with a nanomolar EC_{50} value (844 nM), which is far lower than the anti-Mpro activity probably indicating that the inhibition of the protease activity is only one of the steps inhibited by this really interesting compound. Interestingly, the antiviral activity was further confirmed in human lung epithelial cells and human-induced pluripotent stem cell-derived 3D lung organoids, cell constructs closer to the tissue actually infected by the virus.[36]

22.1.2 Diselenides and Selenides

In the frame of their study meant to identify anti-COVID19 agents, Weglarz-Tomczak and Brul *et al.* found not only benzisoselenazolones but also diselenides endowed with anti-PLpro activity. Compounds **16** and **17** (Figure 22.3) displayed an inhibitory potency comparable to that shown by ebselen-like compounds. Again, no antiviral activity measured in a cell context was ever reported for such compounds.[34]

Some of us, in the framework of ongoing efforts to develop pharmaceutically active compounds,[37–42] reported the introduction of a selenium framework into quercetin, already known to inhibit Mpro.[43] Compound **18** is indeed able to inhibit the proteolytic activity with a potency in the low micromolar range and such activity couples with a mild antiviral activity measured in Vero E6 cells ($EC_{50} = 192$ μM). The precise introduction of a *para*-methylphenylselenyl functionality in the C8 position of the quercetin

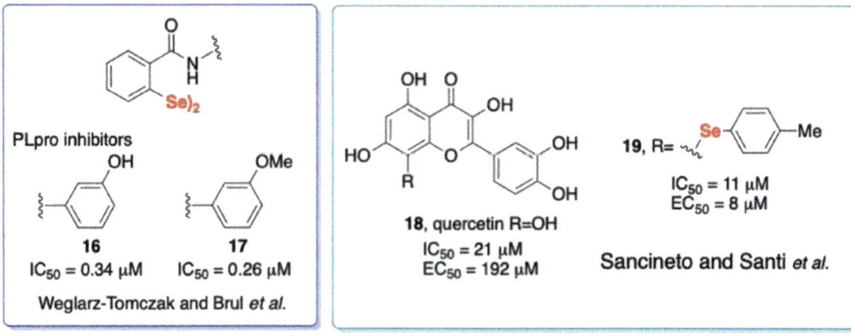

Figure 22.3 Diselenides and selenides as anti-SARS-CoV-2 agents.

scaffold led to compound **18** that has a higher anti-Mpro potency (11 μM) and pronounced antiviral activity (8 μM) without major cytotoxicity. Computational efforts clarified the binding mode of **18** into the Mpro active site where the selenium is engaged in hydrogen bond interaction with Gln189.[44]

22.2 Neurodegenerative Diseases

Neurodegenerative disease is a term used to designate a series of diseases that affect the nervous system, causing the death of neurons. Although there are some treatments that can minimize a few symptoms associated with these illnesses, unfortunately to date there is no treatment that will lead to a cure or even delay the progress of these diseases.[45] In this context, selenium organic compounds have been studied and show great potential in combating these disorders.[7,46] Several research groups have been dedicated to these studies for a long time. However, for this chapter the authors decided to focus on the most recent and representative examples in the area.

22.2.1 Ebselen and Derivatives

There are many articles in the literature that report the ebselen neuroprotective effects in a great number of *in vitro* and *in vivo* models of brain toxicity.[8,25,47–49]

Recently, Sucheck and co-workers published the synthesis of ebselen derivatives and their neuroprotective activity tests. Compounds 7 (Figure 22.2) and **20–23** (Figure 22.4) demonstrated a significant improvement in cell viability, at a 10 μM concentration, through an oxygen–glucose deprivation/reperfusion model of ischemia. Compounds **7** and **21** showed to be promising neuroprotective compounds also able to protect human neuroblastoma SH-SY5Y cells exposed to microglia-activated media. The 21 molecules prepared in the frame of this study (not shown) were tested for glutathione peroxidase (GPx)-like activity however the authors did not find any obvious correlation between this assay and neuroprotectivity.[50]

20, R_6=H **21**, R_6=H **22**, R_6=H **23**, R_6=F

Figure 22.4 Ebselen derivatives with neuroprotective activity.

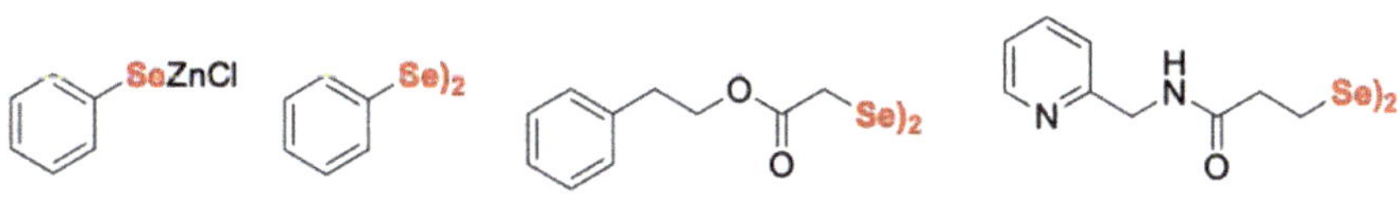

24 **25** **26** **27**

Figure 22.5 Structures of diselenides and PhSeZnCl.

A further approach for ebselen **1** and some of its derivatives was described by Hasnain and co-authors. In this study the potential of these compounds for the treatment of amyotrophic lateral sclerosis disease (ALS), a progressive neurological disorder, was demonstrated by a superoxide dismutase-1 (SOD-1) stabilization effect. Ebselen-based compounds showed better *in vitro* neuroprotection than edaravone (approved drug to combat this disease with limited therapeutic benefits) and with acceptable safety characteristics. *In vivo*, **1** delayed the onset of the disease in a transgenic ALS mice model.[51]

More recently, Wang *et al.* reported the potential of novel organoselenium molecules, based on ebselen, against Alzheimer's disease (AD), *via* multi-target action. In this work compound **23** exhibited good activity in inhibiting β-secretase 1 (BACE1) as well as a mimetic of the GPx enzyme. Also, it showed the best result in PC12 cell protection against insults caused by H_2O_2 or 6-hydroxydopamine, among other related activities. Through all the results, including mechanism studies, compound **23** can be considered a promising multi-target directed ligand candidate for the treatment of AD.[52]

22.2.2 Diselenides and Selenides

Due to high oxygen consumption and low levels of antioxidant enzymes, neuronal cells are quite susceptible to oxidative damage. This, in turn, is one of the main factors associated with neurodegenerative diseases such as Alzheimer's and Parkinson's. In this context, de Bem *et al.* investigated and compared the protective effects of three different diselenides and PhSeZnCl (compound **24**, Figure 22.5) against oxidative challenge in cultured neuronal HT22 cells. All compounds showed lower thiol peroxidase activity than bovine erythrocytes GPx enzyme. They protected the tested cells against glutamate-induced cytotoxicity – pre- or co-treatment. Compounds **24** and **25** compared to **26** and **27** showed higher thiol peroxidase and lower thiol oxidase activities and **25** and **26** demonstrated good activity to prevent peroxinitrite-induced dihydrorhodamine oxidation through pre-treatment. Besides, diphenyl diselenide, compound **25**, increasing the expression of antioxidant and pro-survival proteins was able to act as a redox-modulating molecule. According to the authors, these results are encouraging for conducting *in vivo* tests and, furthermore, they demonstrated that the activity of these compounds is not directly connected with their GPx and thiol oxidase activities, but rather may be related to the regulation of antioxidant defences through their thiol-modifying effects.[53]

In 2020, Soares and co-workers reported a study focused on the use of diphenyl diselenide **25** for the treatment of Huntington's disease (HD). This heredodegenerative disease affects the central nervous system and is characterized by the abnormal deposition and aggregation of proteins in the cell. Through *Caenorhabditis elegans* as an animal model for HD, the authors have demonstrated that the organoselenium compound is able to protect against polyglutamine (poly-Q) aggregation in muscle, reduce neuronal cells death, and maintain the neuronal function. Moreover, **25** showed effectiveness in combating the reactive oxygen species (ROS) and extended the lifespan and health span of wild type and PolyQ mutant worms. According to the authors, these protective results might be attributed to insulin pathway modulation through activation of DAF-16, heat-shock protein-16.2 (HSP-16.2), and superoxide dismutase-3 (SOD-3) in whole body tissues.[54]

Lin *et al.* in 2021, presented an extensive and promising study of the neuroprotective activity of 2-substituted 1,3-selenazole amide derivative **28** (Figure 22.6) on Aβ1–42 transgenic *Caenorhabditis elegans* CL4176 as a model of Alzheimer's disease. The results achieved showed that the worms fed with

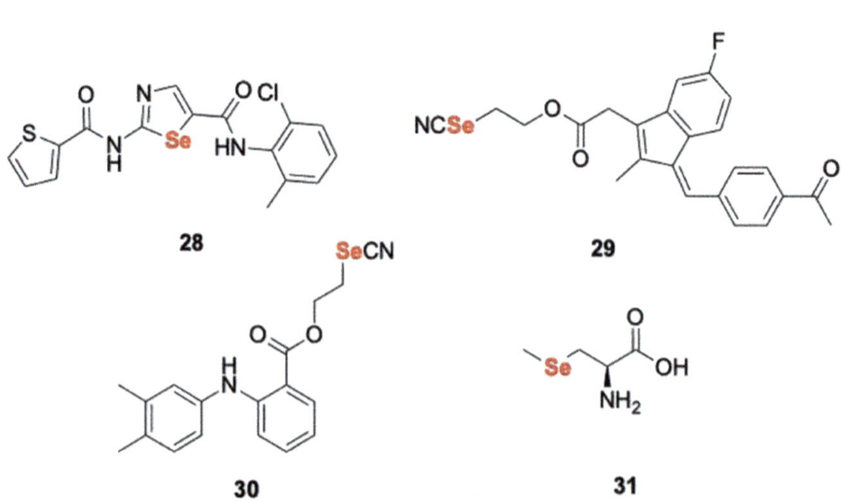

Figure 22.6 Structures of 2-substituted 1,3-selenazole amide and selenocyanate derivatives.

28 had a significant reduction in paralysis, decreased levels of toxic Aβ oligomers and Aβ plaque deposition as well as lower levels in the production of ROS when compared to worms that were not fed the compound. The results also suggested that the reduction in toxic protein deposition, observed in the selenazole administration, may be regulated at the post-transcriptional level, not transcription level. The protective effects of **28** may be involved in the HSF-1 signaling pathway in CL4176.[55]

Later, Bi's group described the synthesis of new organoselenium compounds based on the hybridization of nonsteroidal anti-inflammatory drugs (NSAIDs) skeleton and organoselenium moiety and their evaluation as anti-neuroinflammatory potential using lipopolysaccharide (LPS)-induced BV2 mouse microglia. Importantly, neuroinflammation is an immune response of the central nervous system to pathogens and is correlated with a series of neurodegenerative diseases. Among the synthesized compounds, derivatives **29** and **30** showed the best results in protecting cells against oxidative stress and inflammation. The first activity can be attributed to down-regulating NADPH oxidase 2 (NOX2) and its downstream targets and the second by the fact of the molecules are able to suppress the secretion of IL-18 and IL-1β by cells.[56]

Concerning naturally-occurring selenide-containing amino acids, selenomethionine **31** is the most studied in models of AD. In 2021, Shen and coworkers related the successful application of this seleno-amino acid in the amelioration of neuropathology and cognitive deficits in a triple-transgenic mice model of Alzheimer's disease (3×Tg-AD mice). After treatment with a selenium compound, through a proteomic study, it was possible to identify that 138 differentially expressed proteins (DEPs) were reversed in the diseased mice group. Also, the supplementation with **31** reduced the decline of cortical adenosine triphosphate (ATP) levels and synaptic structure damage in AD mice. According to the authors the main pathogenic phenomena of AD may be involved in energy metabolism disorders, abnormal amino acid metabolism, synaptic dysfunction, and oxidative stress. Selenomethionine was shown to be effective in these underlying mechanisms, ameliorating the pathological changes and symptoms of AD mice.[57]

In the same year, Song *et al.* proposed that selenomethionine **31** improved synaptic plasticity and cognitive function, through modulating Ca^{2+} influx mediated by *N*-methyl-D-aspartate acid receptors (NMDARs), in triple transgenic AD (3×Tg-AD) mice. The authors described in this study the potential of a seleno-compound in regulating selenoprotein K (SELENOK) levels and the equilibrium between synaptic and extrasynaptic NMDAR expression in AD mice. This was an important finding as it furnishes insight into mechanisms by which Se-compounds act in synaptic deficits in AD.[58]

Acknowledgements

The authors thank FAPERJ (E-26/202.911/2019) for financial support. The international network SeSRedCat is acknowledged.

References

1. Z. Jin, X. Du, Y. Xu, Y. Deng, M. Liu, Y. Zhao, B. Zhang, X. Li, L. Zhang, C. Peng, Y. Duan, J. Yu, L. Wang, K. Yang, F. Liu, R. Jiang, X. Yang, T. You, X. Liu, X. Yang, F. Bai, H. Liu, X. Liu, L. W. Guddat, W. Xu, G. Xiao, C. Qin, Z. Shi, H. Jiang, Z. Rao and H. Yang, *Nature*, 2020, **582**, 289.
2. E. J. Lenardão, C. Santi and L. Sancineto, *New Frontiers in Organoselenium Compounds*, Springer International Publishing, Cham, 2018.
3. I. Di Leo, F. Messina, V. Nascimento, F. G. Nacca, D. Pietrella, E. J. Lenardão, G. Perin and L. Sancineto, *Mini-Rev. Org. Chem.*, 2019, **16**, 589.
4. D. Radomska, R. Czarnomysy, D. Radomski and K. Bielawski, *Int. J. Mol. Sci.*, 2021, **22**, 1009.
5. N. V. Barbosa, C. W. Nogueira, P. A. Nogara, A. F. de Bem, M. Aschner and J. B. T. Rocha, *Metallomics*, 2017, **9**, 1703.
6. C. W. Nogueira and J. B. T. Rocha, *Arch. Toxicol.*, 2011, **85**, 1313.
7. P. A. Nogara, C. S. Oliveira and J. B. T. Rocha, *Organoselenium Chemistry*, De Gruyter, 2020, pp. 305–346.
8. H. Chuai, S.-Q. Zhang, H. Bai, J. Li, Y. Wang, J. Sun, E. Wen, J. Zhang and M. Xin, *Eur. J. Med. Chem.*, 2021, **223**, 113621.
9. J. J. Kirsi, J. A. North, P. A. McKernan, B. K. Murray, P. G. Canonico, J. W. Huggins, P. C. Srivastava and R. K. Robins, *Antimicrob. Agents Chemother.*, 1983, **24**, 353.
10. M. Baba, *Antiviral Res.*, 1997, **33**, 141.
11. S. Thenin-Houssier, I. M. S. de Vera, L. Pedro-Rosa, A. Brady, A. Richard, B. Konnick, S. Opp, C. Buffone, J. Fuhrmann, S. Kota, B. Billack, M. Pietka-Ottlik, T. Tellinghuisen, H. Choe, T. Spicer, L. Scampavia, F. Diaz-Griffero, D. J. Kojetin and S. T. Valente, *Antimicrob. Agents Chemother.*, 2016, **60**, 2195.
12. M. Pietka-Ottlik, P. Potaczek, E. Piasecki and J. Mlochowski, *Molecules*, 2010, **15**, 8214.
13. M. Pietka-Ottlik, M. Burda-Grabowska, M. Woźna, J. Waleńska, R. Kaleta, E. Zaczyńska, E. Piasecki and M. Giurg, *Arkivoc*, 2017, **2017**, 546.
14. M. Giurg, A. Gołąb, J. Suchodolski, R. Kaleta, A. Krasowska, E. Piasecki and M. Piętka-Ottlik, *Molecules*, 2017, **22**, 974.
15. L. Sancineto, A. Mariotti, L. Bagnoli, F. Marini, J. Desantis, N. Iraci, C. Santi, C. Pannecouque and O. Tabarrini, *J. Med. Chem.*, 2015, **58**, 9601.
16. N. Iraci, O. Tabarrini, C. Santi and L. Sancineto, *Drug Discovery Today*, 2018, **23**, 687.
17. L. Sancineto, N. Iraci, O. Tabarrini and C. Santi, *Drug Discovery Today*, 2018, **23**, 260.
18. E. J. Lenardão, C. Santi and L. Sancineto, *New Frontiers in Organoselenium Compounds*, Springer International Publishing, Cham, 2018, pp. 99–143.

19. C. Santi, F. Marini and E. J. Lenardão, *Organoselenium Compounds in Biology and Medicine*, Royal Society of Chemistry, Cambridge, 2018, pp. 35–76.
20. C. A. Menéndez, F. Byléhn, G. R. Perez-Lemus, W. Alvarado and J. J. de Pablo, *Sci. Adv.*, 2020, **6**, eabd3045.
21. C. Ma, Y. Hu, J. A. Townsend, P. I. Lagarias, M. T. Marty, A. Kolocouris and J. Wang, *ACS Pharmacol. Transl. Sci.*, 2020, **3**, 1265.
22. P. A. Nogara, F. B. Omage, G. R. Bolzan, C. P. Delgado, M. Aschner, L. Orian and J. B. Teixeira Rocha, *Mol. Inf.*, 2021, **40**, 2100028.
23. K. Amporndanai, X. Meng, W. Shang, Z. Jin, M. Rogers, Y. Zhao, Z. Rao, Z.-J. Liu, H. Yang, L. Zhang, P. M. O'Neill and S. Samar Hasnain, *Nat. Commun.*, 2021, **12**, 3061.
24. T. Chen, C.-Y. Fei, Y.-P. Chen, K. Sargsyan, C.-P. Chang, H. S. Yuan and C. Lim, *ACS Pharmacol. Transl. Sci.*, 2021, **4**, 898.
25. C. Santi, C. Scimmi and L. Sancineto, *Molecules*, 2021, **26**, 4230.
26. H. Sies and M. J. Parnham, *Free Radicals Biol. Med.*, 2020, **156**, 107.
27. C. V. Haritha, K. Sharun and B. Jose, *Int. J. Surg. Pathol.*, 2020, **84**, 53.
28. C. Duong, H. J. Seow, S. Bozinovski, P. J. Crack, G. P. Anderson and R. Vlahos, *Am. J. Physiol. Cell. Mol. Physiol.*, 2010, **299**, L425.
29. F. Petronilho, D. Florentino, F. Silvestre, L. G. Danielski, D. Z. Nascimento, A. Vieira, L. A. Kanis, J. J. Fortunato, M. Badawy, T. Barichello and J. Quevedo, *Inflammation*, 2015, **38**, 1394.
30. S. Yatmaz, H. J. Seow, R. C. Gualano, Z. X. Wong, J. Stambas, S. Selemidis, P. J. Crack, S. Bozinovski, G. P. Anderson and R. Vlahos, *Am. J. Respir. Cell Mol. Biol.*, 2013, **48**, 17.
31. P. Mehta, D. F. McAuley, M. Brown, E. Sanchez, R. S. Tattersall and J. J. Manson, *Lancet*, 2020, **395**, 1033.
32. S. Gladilin, H.-J. Bidmon, A. Divanach, G. E. Arteel, O. W. Witte, K. Zilles and H. Sies, *Arch. Biochem. Biophys.*, 2000, **380**, 237.
33. T. Koyanagi, M. Nakamuta, M. Enjoji, H. Iwamoto, K. Motomura, H. Sakai and H. Nawata, *Int. J. Mol. Med.*, 2001, **7**, 321.
34. E. Weglarz-Tomczak, J. M. Tomczak, M. Talma, M. Burda-Grabowska, M. Giurg and S. Brul, *Sci. Rep.*, 2021, **11**, 3640.
35. L.-Y. Sun, C. Chen, J. Su, J.-Q. Li, Z. Jiang, H. Gao, J.-Z. Chigan, H.-H. Ding, L. Zhai and K.-W. Yang, *Bioorg. Chem.*, 2021, **112**, 104889.
36. S. Huff, I. R. Kummetha, S. K. Tiwari, M. B. Huante, A. E. Clark, S. Wang, W. Bray, D. Smith, A. F. Carlin, M. Endsley and T. M. Rana, *J. Med. Chem.*, 2022, **65**, 2866.
37. V. Nascimento, N. L. Ferreira, R. F. S. Canto, K. L. Schott, E. P. Waczuk, L. Sancineto, C. Santi, J. B. T. Rocha and A. L. Braga, *Eur. J. Med. Chem.*, 2014, **87**, 131.
38. L. Sancineto, N. Iraci, S. Massari, V. Attanasio, G. Corazza, M. L. Barreca, S. Sabatini, G. Manfroni, N. R. Avanzi, V. Cecchetti, C. Pannecouque, A. Marcello and O. Tabarrini, *ChemMedChem*, 2013, **8**, 1941.
39. L. Sancineto, N. Iraci, M. L. Barreca, S. Massari, G. Manfroni, G. Corazza, V. Cecchetti, A. Marcello, D. Daelemans, C. Pannecouque and O. Tabarrini, *Bioorg. Med. Chem.*, 2014, **22**, 4658.

40. D. Krasowska, N. Iraci, C. Santi, J. Drabowicz, M. Cieslak, J. Kaźmierczak-Barańska, M. Palomba, K. Królewska-Golińska, J. Magiera and L. Sancineto, *Molecules*, 2019, **24**, 2914.
41. L. Sancineto, M. Piccioni, S. De Marco, R. Pagiotti, V. Nascimento, A. L. Braga, C. Santi and D. Pietrella, *BMC Microbiol.*, 2016, **16**, 220.
42. L. S. Galant, M. M. Braga, D. de Souza, A. F. de Bem, L. Sancineto, C. Santi and J. B. T. da Rocha, *Free Radical Res.*, 2017, **51**, 657.
43. O. Abian, D. Ortega-Alarcon, A. Jimenez-Alesanco, L. Ceballos-Laita, S. Vega, H. T. Reyburn, B. Rizzuti and A. Velazquez-Campoy, *Int. J. Biol. Macromol.*, 2020, **164**, 1693.
44. F. Mangiavacchi, P. Botwina, E. Menichetti, L. Bagnoli, O. Rosati, F. Marini, S. F. Fonseca, L. Abenante, D. Alves, A. Dabrowska, A. Kula-Pacurar, D. Ortega-Alarcon, A. Jimenez-Alesanco, L. Ceballos-Laita, S. Vega, B. Rizzuti, O. Abian, E. J. Lenardão, A. Velazquez-Campoy, K. Pyrc, L. Sancineto and C. Santi, *Int. J. Mol. Sci.*, 2021, **22**, 7048.
45. B. N. Dugger and D. W. Dickson, *Cold Spring Harbor Perspect. Biol.*, 2017, **9**, a028035.
46. C. W. Nogueira, N. V. Barbosa and J. B. T. Rocha, *Arch. Toxicol.*, 2021, **95**, 1179.
47. A. C. Ruberte, C. Sanmartin, C. Aydillo, A. K. Sharma and D. Plano, *J. Med. Chem.*, 2020, **63**, 1473.
48. F. Martini, S. G. Rosa, I. P. Klann, B. C. W. Fulco, F. B. Carvalho, F. L. Rahmeier, M. C. Fernandes and C. W. Nogueira, *J. Psychiatr. Res.*, 2019, **109**, 107.
49. A. Kumawat, S. Raheem, F. Ali, T. A. Dar, S. Chakrabarty and M. A. Rizvi, *J. Phys. Chem. B*, 2021, **125**, 1531.
50. A. D. Landgraf, A. S. Alsegiani, S. Alaqel, S. Thanna, Z. A. Shah and S. J. Sucheck, *ACS Chem. Neurosci.*, 2020, **11**, 3008.
51. K. Amporndanai, M. Rogers, S. Watanabe, K. Yamanaka, P. M. O'Neill and S. S. Hasnain, *EBioMedicine*, 2020, **59**, 102980.
52. L. Qu, L. Ji, C. Wang, H. Luo, S. Li, W. Peng, F. Yin, D. Lu, X. Liu, L. Kong and X. Wang, *Eur. J. Med. Chem.*, 2021, **219**, 113441.
53. L. S. Galant, J. Rafique, A. L. Braga, F. C. Braga, S. Saba, R. Radi, J. B. T. da Rocha, C. Santi, M. Monsalve, M. Farina and A. F. de Bem, *Neurochem. Res.*, 2021, **46**, 120.
54. F. Bicca Obetine Baptista, L. P. Arantes, M. L. Machado, A. F. da Silva, L. Marafiga Cordeiro, T. L. da Silveira and F. A. A. Soares, *Metallomics*, 2020, **12**, 1142.
55. H. Wang, Y. Yue, H. Zhao, H. Wu, K. Jiang, S. Li, M. Zhao and F. Lin, *Neurotoxic. Res.*, 2021, **39**, 841.
56. Y. Han, Y. Lu, X. Li, X. Niu, A. K. Chang, Z. Yang, X. Li, X. He and X. Bi, *Int. Immunopharmacol.*, 2021, **101**, 108377.
57. X. Du, Q. Shi, Y. Zhao, Y. Xie, X. Li, Q. Liu, J. Iqbal, H. Zhang, X. Liu and L. Shen, *ACS Chem. Neurosci.*, 2021, **12**, 1112.
58. Z.-H. Zhang, C. Chen, S.-Z. Jia, X.-C. Cao, M. Liu, J. Tian, P. R. Hoffmann, H.-X. Xu, J.-Z. Ni and G.-L. Song, *Antioxid. Redox Signaling*, 2021, **35**, 863.

CHAPTER 23

The Discovery of Glutathione Peroxidases: Milestones in Understanding the Biological Role of Selenium und Sulfur

LEOPOLD FLOHÉ[a,b]

[a] Department of Molecular Medicine, University of Padova, Viale G. Colombo 3, I-35121, Padova, Italy; [b] Universidad de la República, Facultad de Medicina, Departamento de Bioquímica, Avda, General Flores 2125, 11800, Montevideo, Uruguay
Email: l.flohe@t-online.de

23.1 The Glutathione Scene Before Glutathione Peroxidase

The co-substrate of glutathione peroxidases (GPx), glutathione (GSH), was discovered in the 19th century by de Rey-Pailhade and named "philothion".[1] A partial elucidation of its composition was achieved in the early 1920s by Hopkins.[2] The final characterization as the atypical tripeptide γ-glutamyl-cysteinyl-glycine required the evidence of the chemical synthesis by Harrington and Mead[3] and du Vigneaud and Miller[4] in the 1930s. The biosynthesis of GSH was then elucidated in the 1940s by Waelsch and Rittenberg[5] and Bloch and Anker[6] (for a recent review see Forman *et al.*).[7] If one searches the internet for functions of glutathione, one can learn that it is a "key detoxifying molecule" or "the mother of all antioxidants", both terms were primed early in the history of GSH and are not entirely adequate.

Chalcogen Chemistry: Fundamentals and Applications
Edited by Vito Lippolis, Claudio Santi, Eder J. Lenardão and Antonio L. Braga

Published by the Royal Society of Chemistry, www.rsc.org

Instead, I would recommend reading the greatly written review on the early GSH history by the late Alton Meister.[8] Although it was published in 1988, it is still a most valuable source to find almost-forgotten knowledge and explanations of cloudy characteristics.

The image of GSH as a detoxifying agent can be traced back to the 19th century. In 1879 Baumann and Preusse,[9] and in the same year Jaffe,[10] reported the formation of mercapturic acids in dogs poisoned with brombenzene and chlorbenzene. Of course, the involvement of GSH in the formation of mercapturic acids, which are S-conjugates of these poisons with *N*-acetyl-cysteine, was not yet known, since GSH had not yet been discovered. The first hints of the involvement of GSH in a similar detoxification reaction were reported by Nakashima in 1934 in his studies on the mechanism of cataract formation upon naphthalene exposure: the content of GSH dramatically declined upon naphthalene administration in the liver, lens, and muscle.[11] The final elucidation of the mercapturic acid pathway was not achieved before the late 1950s[12,13] and the early 1960s,[14] when Boyland and co-workers demonstrated the presence of a naphthalene epoxide in rat liver slices that first became conjugated with glutathione and subsequently processed to cysteinyl-naphthalene and finally to *N*-acetyl-cysteinyl-naphthalene, the mercapturic acid. In the meantime, we have learned that the first step of mercapturic acid formation is generally catalyzed by a realm of glutathione *S*-transferases that catalyze the glutathionylation of a huge variety of compounds comprising epoxides, halogenated alkanes, aromatics, and olefines. Gamma-glutamyl-transpeptidase removes the γ-glutamyl residue, a dipeptidase splits the cysteinyl–glycine bond, and *N*-acetylation of the cysteinyl residue then yields the final mercapturic acid. The end product is usually, but not always, less toxic and more readily excreted *via* the bile than the mother compound. This pathway is, however, not restricted to exogenous poisons. It equally applies to endogenous toxins such as hydroxynonenal or orthoquinones. Moreover, the enzymes catalyzing the initial step, the glutathione *S*-transferases, are essential for biosynthesis of steroid hormones, peptido-leukotrienes and other eicosanoids, and play an important role in *cis–trans* isomerization during amino acid catabolism,[15] contributing to redox regulation[16] and can even mimic glutathione peroxidase[15] or transfer glutathione peroxidase activity to a peroxiredoxin.[17] In short, some of the glutathione *S*-transferases may have evolved to cope with frequent and unfavorable xenobiotic exposure, the majority, however, serve completely different purposes, and, accordingly, GSH as a substrate of these enzymes is not just a detoxifying agent.

The antioxidant function of GSH is closely linked to the development of antimalarial drugs. In 1926 acute hemolysis was reported as a side effect of primaquine treatment.[18] Related research became a hot topic during the Korean war (1950–1953). Seemingly healthy American soldiers, in particular those of African origin, responded to antimalarial drugs such as pamaquine or other aminoquinolines with an acute hemolytic crisis.[19] The crisis was usually accompanied by formation of "Heinz bodies", *i.e.*, membrane-attached particles primarily containing denatured hemoglobin. These phenomena were attributed to an enzymatic defect in red blood cells,[20] which could be identified as a deficiency in glucose-6-phosphate dehydrogenase (G6PDH).[21] A hallmark of G6PDH deficiency proved to be an instability of the GSH levels in the red blood cells, since the pentose phosphate shunt with its key enzyme G6PDH provides the NADPH that is required for regeneration of GSH from its oxidized form GSSG. Accordingly, prognostic tests for GSH stability under different challenges were widely performed to screen for aminoquinoline sensitivity.[22]

In parallel, research was going on in the Mediterranean countries on a disease evidently known for centuries: favism. In 1905 about 0.5% of the Sardinian population had experienced a hemolytic crisis after ingestion of fava beans (*Vicia faba*).[23] In an excellent review, Arese and de Flora described that favism is also caused by glucose-6-phosphate dehydrogenase deficiency.[24] It occurs in different ethnic groups in the Mediterranean area, particularly in Sardinia and in the near east, but also in India, China, and Africa. The disease is due to different mutations of the G6PDH gene, but their clinical manifestations are very similar. The worldwide incidence of genetically impaired G6PDH is estimated to range around 400 000, which, however, only results in clinical favism if challenged with oxidant drugs or fava beans.[25] In the case of favism, the culprits are the autoxidizable compounds divicine and isouramil contained in the fava beans as glycosides, and the incidence of severe clinical outcomes largely depends on eating and cooking habits and the intestinal flora. Yet individuals susceptible to fava bean ingestion are equally sensitive to aminoquinoline-based antimalarials, sulfonamides, and the like.[25] Also the risk of favism may be monitored by GSH stability tests.[26] In most cases of genetic G6PDH deficiency, the mutations affect the stability of the enzyme, which explains the restriction of the symptoms to the red blood cell. Some aspects of the disease, in particular the self-limitation of the hemolytic crises and the refractory period thereafter, still remain unsolved.[24]

Favism-like conditions have meanwhile been observed in other situations that are associated with impaired NADPH supply or utilization. The deficiency in NADPH-dependent GSSG reduction by glutathione reductase is mostly due to limited alimentary FAD supply and appears seldom caused by a genetically-modified enzyme.[27] Genetic deficiencies of the enzymes catalyzing GSH biosynthesis are equally rare and possibly associated with additional pathological phenomena distinct from hemolysis.[28] Severe cases of genetic alteration of the GSH biosynthesis may also affect tissues with active protein synthesis.[28]

If one reads related publications of the 1950s or earlier, one may suspect it is GSH itself, which protects the cell against intoxication and oxidant challenge. The sheer abundance of GSH in cells (2–10 mM) and the low redox potential of the GSH/GSSG couple ($E_{m7} \sim -240$ mV) made people believe that keeping GSH in the reduced state guaranteed the wellbeing of the cell. This is not the case.[29] As outlined above, the detoxification depends on *S*-transferases, and also the antioxidant function of GSH hardly works without enzymatic support. Radiation chemists discussed the GSSG formation to result from a combination of two glutathionyl radicals.[30] Of course, this is theoretically possible. However, the low levels of thiyl radicals and their promiscuity in reactions with a multitude of cellular compounds make this kind of GSSG formation a highly unlikely event. Similarly, the formation of mixed disulfides between GSH and proteins is not just the consequence of an altered redox potential of the GSH system.[31] Typically, it is a two-step process initiated by thiol oxidation to a sulfenic acid derivative of GSH or a protein SH and followed by the reaction with a thiol. The reversal of protein disulfide formation is the domain of redoxins (proteins characterized by a CysxxCys motif). In the case of *S*-glutathionylated proteins the catalytic reductant is usually a glutaredoxin.[32]

The reactions of GSH typically depend on the dissociation of its –SH group. It is the thiolate form that reacts. However, the precise pK_a of the GSH thiol is still debated because of the presence of functional groups with similar dissociation behavior. The very first interpretation of the titration curve of GSH yielded a pK_a value for its –SH group of 9.62 at room temperature.[33] A later ^{13}C-NMR-based re-investigation of the problem relying on the pH-dependent chemical shift of the carbon atom nearest to the dissociating group yielded a value of 9.2.[34] Whatever the precise value is, the pK_a of the thiol function of GSH is appreciably higher than that of any naturally occurring low molecular weight thiol. Accordingly, the reactivity of GSH with oxygen or hydroperoxides is comparatively low. GSH appears to be designed by nature not to react spontaneously at any relevant speed with any physiological component. If it reacts at all, it will scavenge the most abundant radical, which is molecular dioxygen, and ultimately produce H_2O_2, likely mediated by interim superoxide formation.[35] Accordingly, GSH supported alloxan-induced hemolysis,[36] and it induced hemolysis when added to washed erythrocytes,[37] and it also induced loss of volume control and contractibility of isolated mitochondria,[38] an effect later shown to be caused by oxidative membrane damage.[39] In conclusion, when deprived of its enzymatic environment, GSH is not a good antioxidant.

In short, both the elimination of xenobiotics and the balance of oxidative challenge are not achieved by GSH alone, but depend on enzymatic catalysis. However, the enzyme family most important for GSH-dependent hydroperoxide detoxification and redox regulation still had to be discovered (see Section 23.2).

23.2 From the Discovery of Glutathione Peroxidase up to its Identification as the First Selenoprotein in Mammals

In 1957 Gordon C. Mills described a peroxidase that protected hemoglobin from oxidative destruction.[40] The enzyme was claimed to be highly specific for GSH and, unlike catalase or any other known peroxidase, proved not to be inhibited by cyanide or azide.

This glutathione peroxidase, now known as GPx1, was initially reported to catalyze the reaction

$$2\ GSH + H_2O_2 \rightarrow GSSG + 2\ H_2O \tag{23.1}$$

and, thus, had been able to immediately solve the enigma of GSH being a pro-oxidant in model systems, where it is deprived of its enzymatic environment, and the clearly antioxidant role it has in the intact cell. However, the Mills discovery first had to overcome a biochemical dogma. Almost by definition, peroxidases had to be iron porphyrin proteins, and, in consequence, GPx was not at all taken seriously; it was even publically declared not to exist at all (Gerald Cohen 1930 – 2001, personal communication).[41] Supporting information in the 1960s was very scarce. Cohen and Hochstein demonstrated that the hemoglobin in H_2O_2-exposed erythrocytes did not suffer until the last trace of GSH was consumed and therefore concluded that indeed GPx is the most important enzyme of H_2O_2 metabolism in the red blood cell.[42] In the late 1960s and early 1970s the substrate spectrum of GPx was enlarged to hydroperoxides of lipids[43–45] and DNA[46] according to eqn (23.2).

$$2GSH + ROOH \rightarrow GSSG + H_2O + ROH \tag{23.2}$$

R may be methyl, ethyl *t*-butyl, a more or less complex lipid residue, a thymine residue in DNA or NO. It should, however, be mentioned that the early experiments were performed with liver extracts, which may contain more than one isozyme of the GPx family. Later studies, however, confirmed that GPx1 also reacts with a variety of organic hydroperoxides.[47] In the early 1970s, the first large-scale preparation of GPx1 from bovine blood[48] confirmed that Mills was right in his conclusion that GPx is not a heme protein, nor did it contain any other known prosthetic group.[49] Moreover, it could be demonstrated that the oxidation of GPx1 by H_2O_2 is faster than that of catalase, until then considered the fastest enzyme for the catalysis of bimolecular reactions.[47]

The poor appreciation of GPx did not change very much, until a link to selenium biochemistry was established. In the year of the first description of GPx, Klaus Schwarz discovered that selenium was an essential trace element for mammals.[50] However, it was 15 years before the two lines of research started to fertilize each other. In 1972, Hoekstra's group in Wisconsin reported that erythrocytes from selenium-deficient rats tend to hemolyzc upon exposure to H_2O_2. However, unlike in favism, the H_2O_2 sensitivity of these erythrocytes was not ameliorated by glucose and their GSH level did not decline.[51] The authors correctly concluded that not the provision of reduced GSH, but its utilization, was impaired and the activity of GPx was found to be dramatically reduced. Yet even their full paper, *inter alia* describing the properties of GPx1 isolated from sheep blood, left some questions unanswered.[52] GPx1 was known to be a homotetramer.[48] The selenium content of the sheep enzyme, however, was reported to be about 2 gram-atom per mol of tetrameric GPx.[52] Either GPx was a real selenoprotein and the preparation was not pure, or the determination of selenium was too unprecise, or it was a protein with a dissociable selenium-containing cofactor, or it was not a homotetramer. The minimum value of 4 gram-atom of selenium per tetrameric GPx expected for a real homotetrameric selenoprotein was first reported in March 1973 by the group in Tübingen[53] and published two months later.[54] In 1978, a selenocysteine residue firmly integrated into the peptide chain was identified as the catalytic moiety[55,56] and later confirmed by X-ray analysis[57] and sequencing.[58]

Detailed kinetic analysis[47] revealed that the catalytic mechanism consists of a sequence of independent bimolecular reactions. First, the selenocysteine residue is oxidized by ROOH to a selenenic acid derivative according to eqn (23.3):

$$\text{E–Se}^- + \text{ROOH} + \text{H}^+ \rightarrow \text{E} - \text{SeOH} + \text{ROH} \tag{23.3}$$

Whether this step of the catalysis involves a typical enzyme/substrate complex remains a semantic problem. Certainly, the hydroperoxide has to meet the selenium in an adequate orientation, and the wide variety of organic hydroperoxides which are accepted by the enzyme argues against a specific enzyme/substrate complex. The selenenic acid nature of the oxidized enzyme could so far not be verified experimentally, since the selenic acid reacts with a backbone nitrogen in the absence of a reducing substrate.[59] However, the selenenic acid form is the only theoretical possibility. Definitely, the selenocysteine residue is oxidized, as its replacement by cysteine or serine lowers or abolishes the activity, respectively. The formation of an intramolecular Se – S bond, as originally been discussed,[53] can be excluded since all cysteine residues in the three-dimensional structure are too far away from the selenium.[60] Finally, the homologous sulfenic acid in GPx mutants with the selenocysteine residue replaced by cysteine was verified by mass spectrometry.[59]

In the second step, the selenenic acid reacts with GSH to form a Se–glutathionylated intermediate (eqn (23.4)). In view of the pronounced specificity of GPx1 for GSH, we here have to postulate the formation of an enzyme/substrate complex. Unlike in Michaelis–Menten-kinetics, the complex formation must be slower than the formation of the intermediate.

$$\text{E} - \text{SeOH} + \text{GSH} \leftrightarrow [\text{E} - \text{SeOH} \cdots \text{GSH}] \leftrightarrow \text{E} - \text{Se} - \text{SG} + \text{H}_2\text{O} \tag{23.4}$$

In the final step (eqn (23.5)), the ground-state enzyme is regenerated, again by a specific reaction of the intermediate with GSH:

$$\text{E} - \text{Se} - \text{SG} + \text{GSH} \leftrightarrow [\text{E} - \text{Se} - \text{SG} \cdots \text{GSH}] \leftrightarrow \text{E} - \text{Se}^- + \text{H}^+ + \text{GSSG} \tag{23.5}$$

In short, the glutathione peroxidase reaction is a typical case of "Zwischenstoff-Katalyse" (catalysis by the formation of intermediates),[61] which is not uncommon in enzymology and evidently the rule in the case of peroxidases and superoxide dismutases.

23.3 Toward Solving the Mystery of the Glutathione Peroxidase Efficiency: The Orian Proposal

The kinetic analysis of GPx1 has not only disclosed the basic reaction scheme, but also surprising rate constants. The rate constant k'_{+1} for the reaction of the reduced enzyme with H_2O_2 (eqn (23.3)) was determined to be $6\times10^7\ M^{-1}\,s^{-1}$, if calculated per subunit, and the rate constants for the analogous reaction with organic hydroperoxides were only a bit smaller.[47] After the detection of selenium in GPx1, these extremely high rate constants were tentatively attributed to the magic catalytic power of the trace element. However, this assumption was soon falsified. When the selenocysteine in GPx1[62] or GPx4[63] (see Section 23.4) was replaced by cysteine, the remaining activity was still appreciably larger than the non-catalyzed reaction of H_2O_2 with any low molecular mass fully dissociated thiol ($<50\ M^{-1}\,s^{-1}$).[64] (Experimental data for the oxidation of free selenocysteine by H_2O_2 do not exist; an estimation based on density functional theory (DFT) yielded a value of only $35.4\ M^{-1}\,s^{-1}$.)[65] A naturally-occurring GPx of *Drosophila* even reduced phosphatidylcholine hydroperoxide with the impressive rate constant of $\sim1\times10^6\ M^{-1}\,s^{-1}$.[66] Also another family of thiol-dependent peroxidases, the peroxiredoxins, reach rate constants for thiol oxidation by hydroperoxides up to $10^7\ M^{-1}\,s^{-1}$.[67] These observations clearly prove that it is not the selenium that facilitates the fast catalytic hydroperoxide reduction. The rates with selenium catalysis may be a bit higher, but also the thiols or selenols in glutathione peroxidases (see Section 23.4 for GPx isozymes) and peroxiredoxins react up to five orders of magnitude faster than those of low molecular thiol compounds, in which the thiol function is freely accessible. Speculations about the accessibility and pK_a values of the enzymes' thiols or selenols cannot possibly explain their catalytic efficiency. At best, they may modulate the catalytic efficiency of individual enzymes. The reason for the unusual chalcogen reactivity has to be sought in the architecture of the active site.

Already the first X-ray analysis of GPx1 had seen the selenium in the close vicinity of a tryptophan and a glutamine residue.[57] In these residues, which together with the selenocysteine form a catalytic triad, the selenium was believed to be activated by hydrogen bridges. The functional relevance of the triad components could indeed be verified by site-directed mutagenesis.[63] The triad was later enlarged to a tetrad by including an essential asparagine residue.[68] However, the conclusions therefrom remained ambiguous. When Toppo *et al.*[69] reinvestigated the structure of different GPxs, a dense network of hydrogen bridges was confirmed between the active site residues, but none arising from the selenol or the thiol, which is in line with the weak tendency of these elements to sustain stable hydrogen bridges. A plausible interpretation of the unusually high reaction rates was finally achieved by an *ab initio* calculation based on an active site model and the density functional theory.[59]

The active site model that Laura Orian calculated consisted of 7 amino acid residues. It comprised the strictly conserved Trp/Asn/Phe motif (residues 136 – 138, residue numbering of rat cytosolic GPx4), the highly conserved Gln81, Sec46, which in some instances is replaced by Cys, the mostly conserved Gly47, and a non-conserved Gly just to eliminate a negative charge close to the reaction center. Whenever feasible, the suggestions of the DFT calculations were verified by mass spectroscopy. Since the active site cysteine is often replaced by cysteine, the calculation was performed for both possibilities. Irrespective of the level of theory applied and the chalcogen involved, the residues spontaneously adopted positions that were practically identical to those known from X-ray analysis. (It should be mentioned here that for X-ray analysis the active site Sec was usually replaced by a non-reactive residue. The only case of X-ray analysis of a real seleno-GPx found the selenium over-oxidized to a seleninic acid.)[57] The model contains all tetrad residues previously shown to be essential for full GPx activity by site-directed mutagenesis[63,68] and, thus, may be considered representative of most members of the glutathione peroxidase family. It further revealed that neither the selenium nor the sulfur in a homologous position is involved in hydrogen bridges with the tetrad residues. This model, which represents the ground state, *i.e.*, the reduced, enzyme, was then used to calculate the energetic profile of the entire catalytic cycle (eqn (23.3)–(23.5)). H_2O_2 was chosen as the oxidizing substrate, while, for technical reasons, the reductant was changed to a simpler molecule than GSH (ethanthiol). Accordingly, quantitative data on the reductive part of the catalytic cycle are less reliable. The first outcome of these calculations was frustrating. The selenocysteine always remained not dissociated. This clearly conflicted with p*K* calculations and the observation that reduced glutathione peroxidases were readily inactivated by alkylating agents,[53,63] which requires a dissociation of the selenol function. Yet Laura Orian had a simple, but bright idea: She inserted a water molecule into the reaction center and started the calculations again. Now, the selenocysteine (or cysteine) became dissociated (Figure 23.1). However, the proton of the selenol group did not leave the reaction center but migrated *via* water to the ring nitrogen of the Trp in the tetrad. This position of the dislocated proton is favored, because the original proton at the ring nitrogen is engaged in a hydrogen bridge with the carboxamide function of Gln81.

The importance of the charge separation became obvious in the next step, which is the oxidation of the chalcogen (eqn (23.3) and Figure 23.2). When the substrate H_2O_2 was introduced at a position previously demonstrated to allow a reaction with the selenium by molecular dynamics simulation,[70] the adduct of the charge-separated ground state enzyme and H_2O_2 decayed instantly without any measurable activation energy. The negatively-charged selenium (or sulfur) attacks one oxygen of the peroxide bond, while the dislocated proton shuttling back *via* water reacts with the other oxygen to form water as an ideal leaving group. By these reactions, which happen concertedly, the peroxide bond is cleaved, and an oxidized intermediate containing a selenenic or sulfenic group, respectively, is formed.[59]

In a more recent report, the model was further simplified by replacing the Trp by an indol ring and the Gln and Asn residues by formamide.

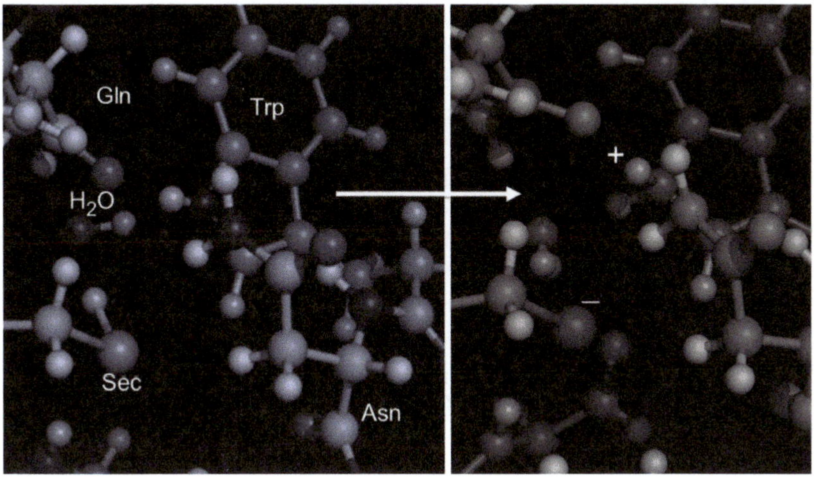

Figure 23.1 Snapshots of an original density functional theory movie of a GPx active site model. The left panel shows the starting situation (associated selenol), with the tetrad residues, and the required water molecule; the right panel demonstrates the final result, where the proton of the selenol shuttled to the ring nitrogen of the active site Trp. For details see the text (see Section 23.3). Adapted from ref. 59 with permission from Elsevier, Copyright 2015.

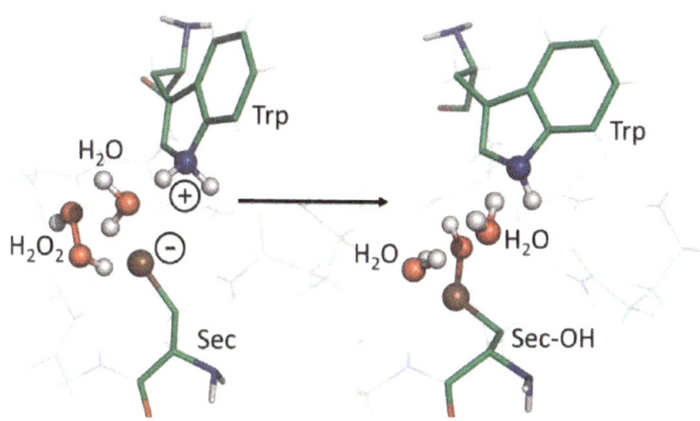

Figure 23.2 Barrier-less decay of the zwitterionic H_2O_2–GPx complex according to ref. 59. The delocalized proton shuttles back via water and hits one oxygen of the peroxide bond, while the selenolate attacks the other oxygen. In this concerted reaction, the peroxide bond is cleaved and a selenenic acid derivative and an additional water molecule is formed. For sake of clarity only the atoms directly involved in the peroxidatic process are shown as balls. Adapted from ref. 59 with permission from Elsevier, Copyright 2015.

The outcome of the calculations was similar. Also in this case, a water molecule was required to make the zwitterionic form. The activation energies for the decay of the H_2O_2 complex, however, were higher than in the

7-amino-acid model, adopting values that were clearly different from zero and higher for sulfur than for selenium.[71] These findings reveal that the protein moiety surrounding the reaction center significantly contributes to lowering the activation energy. Such contributions of more remote residues should be particularly relevant for the reductive part of the catalytic cycle, in which larger substrates such as GSH or thioredoxin (Trx) have to interact with the reaction center.[72]

The mechanistic proposal of Orian and her co-workers is the first one that complies with all experimental findings determined with the native enzymes. It explains their strange kinetics, which in the case of the selenium-containing enzymes is always a ping-pong mechanism with infinite maximum velocities and infinite Michaelis constants. This kinetic pattern is the opposite of the more common Michelis–Menten-type kinetics, which assumes fast complexation of the enzyme by its substrate(s) and a rate-limiting decay of the complex. In the case of the GPxs, we may assume that the small substrate H_2O_2 readily finds its ideal position, likely with the experimentally accessible rate constant k_1', which comes close to that of the net forward association of the enzyme with H_2O_2, *i.e.*, k_{+1}. Yet the barrierless decay of the enzyme/H_2O_2 complex is evidently even faster, which implies that this complex would never accumulate and, accordingly, Michaelis–Menten-type saturation kinetics are not observed. Of course, one might assume that a GPx becomes saturated by GSH in the reductive steps (eqn (23.4) and (23.5)). However, it appears feasible that it takes more time for the flexible substrate GSH to establish multiple hydrogen and salt bridges, which are required to adopt its ideal position,[72] than to react with an Se – OH or Se – S within the complexes. In line with this interpretation, GPxs, which are reduced by the more rigid substrate thioredoxin (see Section 23.4), tend to show saturation kinetics.[73]

After having seen that the first step of the GPx catalysis is qualitatively identical for selenocysteine- and cysteine-containing GPxs, we reasoned that the proton-shuttling mechanism outlined above could also be responsible for an unusually high SH oxidation by H_2O_2 in other protein families. Orian therefore subjected models of various unrelated proteins to the same DFT calculations. Dalla Tiezza *et al.*[74] could indeed verify that proton dislocation from the thiol group with generation of a zwitterionic structure is possible in the glycolytic enzyme glyceraldehyde phosphate dehydrogenase (GAPDH), in the bacterial hydroperoxide sensor OxyR, and in peroxiredoxins. The proton shuttles to different acceptors, to Thr or His in GAPDH, to a histidine in OxyR and to the oxygen of a Thr (or Ser) in the peroxiredoxins. Water molecules may be involved or not. The principle, however, remains the same: The nucleophilicity of the selenol or thiol is greatly enhanced due to dissociation and, thus, facilitates a nucleophilic attack on one oxygen of the peroxide bond. The dislocated proton, when shuttling back, exerts an electrophilic attack on the other oxygen, whereby the O – O bond is split. Evidently, a dual attack on the peroxide bond is always required to cleave the peroxide bond faster than observed in non-enzymatic H_2O_2 reduction.

23.4 The Diversification Within the Glutathione Peroxidase Family

Not all members of the glutathione peroxidase family are selenoproteins, nor are they all glutathione peroxidases. Moreover, glutathione peroxidase activity has been found associated with unrelated enzyme families such as glutathione-*S*-transferases and peroxiredoxins (see Section 23.1). Typically, hydroperoxide metabolism is catalyzed by more than a dozen enzymes, which co-exist in one species. This multiplicity points to distinct roles of hydroperoxides and the metabolizing enzymes, as will be exemplified below. This paragraph, however, will only focus on those proteins that, according to sequence, belong to the glutathione peroxide family. In humans, 8 members of the family have been identified. GPx1–GPx4 and GPx6 are selenoproteins (SecGPxs), and GPx5, GPx7 and GPx8 have the active site selenocysteine replaced by cysteine (CysGPxs). In rodents, GPx6 is a CysGPx. SecGPxs are common in vertebrates, but have also been detected in non-vertebrate animals, viruses, and protozoa.[75] SecGPx are absent in terrestrial plants and yeasts, since they lack the machinery to synthetize selenoproteins. GPxs differ in substrate specificity, tissue distribution, regulation of their synthesis, and physiological functions. The field has been amply reviewed.[75–81] I shall here only briefly compile the role of the individual GPxs in mammals and the basic diversifications in other species.

23.4.1 Mammalian SecGPxs

23.4.1.1 GPx1

GPx1, the enzyme discovered by Mills,[40] is the most widely distributed enzyme of the family. Its role has been seen in the defense against hydroperoxide challenge. In line with this view, its gene could be knocked out without any serious phenotype. The GPx1$^{-/-}$ mice, however, were highly susceptible to diverse oxidative challenges such as exposure to bacterial lipopolysaccharides, macrophage activation, and viral infection. Its biosynthesis is of low priority, which means that it declines quickly in selenium deficiency. Deficiency of GPx1 is discussed as the major contributor to the manifestation of Keshan disease. More recently, GPx1 has also been known to contribute to regulatory processes. Overexpression of GPx1 in rodents caused insulin resistance and obesity.[78,82] It also dampens TNFα-induced NF-κB activation[83] and is believed to generally influence signaling cascades that are controlled by redox-sensitive protein phosphatases.[76]

23.4.1.2 GPx2

GPx2 was discovered as a gastro-intestinal isozyme by Fong Fong Chu and colleagues[84] but also occurs in other tissues, often in malignant ones. In the colon, it declines from the crypt grounds toward the lumen and is believed to regulate the delicate balance of proliferation at the crypt ground and apoptotic cell shedding at the lumen.[79] GPx2 inhibits inflammation and early stages of carcinogenesis but promotes tumor growth in a mouse xenograft model.[85,86] Among the GPxs, GPx2 ranks highest in the hierarchy of selenoproteins. GPx2$^{-/-}$ mice are vital but susceptible to inflammatory stimuli.[87]

23.4.1.3 GPx3

GPx3 was first described as an immunologically distinct enzyme by Takahashi and Cohen in 1986.[88] It is primarily synthetized in the kidney and secreted into the circulation. It is often used as an indicator of kidney disease or to monitor the selenium status, since it responds to selenium restriction as fast as GPx1. Its physiological role is unclear. It is often claimed to protect the extracellular space against oxidative challenge. This assumption, however, conflicts with the extremely low extracellular concentration of GSH or the alternate substrate thioredoxin (Trx) and the absence of any suitable substrate regeneration system in this compartment. Knock-out mice are largely asymptomatic. More recently, however, an increased platelet aggregation in GPx3-deficient mice has been reported.[89] This finding could be explained by the oxidative activation of cyclooxygenases due to an increase in peroxide tone.[76]

23.4.1.4 GPx4

Gpx4 was first isolated as a protein that inhibits lipid peroxidation and called PIP (for peroxidation inhibiting protein), then renamed phospholipid hydroperoxide glutathione peroxidase (PHGPx) and finally GPx4 (for a historical review see).[80] The isozyme is unique in several respects: (1) Its gene is expressed in 3 different forms. (2) It is apparently specialized to reduce complex lipids, even when integrated into biomembranes. (3) It still prefers GSH as a reducing substrate, but can also interact with protein thiols. (4) It is the only one of the family that so far proved to be of vital importance in mice.

Most recently, also a clinical correlate of mutations in the Gpx4 gene have been reported. It was proved to be the Sedaghatian-type spondylometaphyseal dysplasia,[90] which was first described in 1980[91] as an extremely rare neonatally lethal form of spondylometaphyseal dysplasia characterized by severe metaphyseal chondrodysplasia with mild limb shortening, platyspondyly, cardiac conduction defects, and central nervous system abnormalities. From the location on the Gpx4 gene in the 21 cases compiled in,[92] it can be concluded that all Gpx4 expression forms (see below) are affected. As far as investigated, the mutations have to be homozygous to cause the clinical symptoms. Residual enzyme activities could usually not be assessed. It can, however, be inferred that the activities of the mutated enzymes are significantly impaired, although the mutations do not change the catalytic tetrad.[92] The clinical symptoms associated with the Gpx4 mutations *inter alia* point to a role of GPx4 in the development of bones and brain, which are not easily explained by our present knowledge of GPx4 functions.

23.4.1.4.1 The Cytosolic Expression Form of GPx4. Cytosolic Gpx4 is the one that was proved to be essential. It reduces the hydroperoxy groups of complex lipids, including those in cholesterol, and inhibits all kind of lipoxygenases. This appears to be its essential function. It has recently attracted considerable interest as the key regulator of ferroptosis, a novel form of regulated cell death that is characterized by iron-catalyzed lipid peroxidation.[80,93,94] When overexpressed, Gpx4 blocks interleukin-1-dependent NF-κB activation more efficiently than GPx1, which strongly suggests the involvement of a lipid hydroperoxide in this signaling cascade.[95] Its pronounced affinity to biomembranes evidently resides in positively-charged residues, which surround the reaction center in a way that the negatively-charged heads of the phospholipids direct their hydroperoxy groups toward the active site selenium.[96]

23.4.1.4.2 The Nuclear Expression Form of GPx4. The use of an ATG located in the first intron of the GPx4 gene yields the nuclear expression form. In consequence, it differs from the other two expression forms in the N-terminal amino acid sequence, which contains the nuclear import signal.[97] Gene disruption specifically blocking the nuclear expression yielded vital offspring. Only the protamine compaction in epididymal spermatozoa appeared impaired, which however did not affect fertility.[98] As previously suggested by Godeas *et al.*[99] the nuclear GPx4 here oxidizes SH groups of protamine. This is not an antioxidant function, but rather the H_2O_2 is used to stabilize the chromatin *via* thiol oxidation.

23.4.1.4.3 The Mitochondrial Expression Form of GPx4. The mitochondrial expression form results from the use of an upstream TGA as the start codon. It thus has an N-terminal extension, which contains the mitochondrial import signal. After mitochondrial import, which is associated with the cleavage of the extension, the protein is identical with the cytosolic form.[100] Mitochondrial GPx4 is preferentially expressed in the testis and here mostly in round spermatids.[101] In mature spermatozoa, it has lost its enzymatic activity and forms the sheath that surrounds the mitochondrial helix.[102] Mechanistically, this metamorphosis of the mitochondrial GPx4 results from a loss of GSH in late spermiogenesis. Under the conditions of GSH shortage, GPx4 reacts with its selenenic acid form with a surface-exposed SH of GPx4 itself[103] and other cysteine-rich proteins[104] as alternate substrates. These processes are clearly the opposite of an antioxidant function; hydroperoxides are used to generate a structural protein network that is indispensable for normal sperm function. Accordingly, specific knockout of the mitochondrial GPx4 yielded vital offspring, which only displayed male infertility. This surprising function of mitochondrial GPx4 explained infertility in life stock grazing on selenium-deficient soils, in laboratory animals fed a selenium-deficient diet, and possibly in humans.[105]

23.4.1.5 GPx6

The expression of GPx6, which is a selenoprotein in men, is restricted to the olfactory system. It is only known by sequence. Its biological role has not yet been identified.

23.4.2 Mammalian CysGPxs

23.4.2.1 GPx5

GPx5 is an extracellular protein that appears to be restricted to the male reproductive tract. Therefore, a role in sperm function has been suggested. Inverse genetics, however, yielded vital and fully fertile offspring. Its biological function, thus, remains to be established.

23.4.2.2 GPx7

GPx7 was discovered in 2004 by Utomo *et al.* as a non-selenium glutathione peroxidase.[106] It is interesting, since it shows a change in donor substrate specificity toward a redoxin-type protein (proteins characterized by a CysxxCys motif), in this case protein disulfide isomerase (PDI). The reduction by GSH occurs with a rate constant (12.6 $M^{-1} s^{-1}$) that roughly complies with a non-catalyzed reaction. In contrast, the oxidized enzyme accepts PDI with a rate constant near $10^4 M^{-1} s^{-1}$.[107] The enzyme is located at the luminal site of the endoplasmic reticulum and believed to reduce superfluous H_2O_2 produced by Ero 1α (ER oxidoreductin 1α) or to be directly involved on oxidative protein refolding.[107,108] Gpx7 also oxidizes the "glucose-regulated protein 78" (GPR78, also called mortalin) *via* heterodimer formation and thereby enhances its chaperone activity. Knockout of Gpx7 results in animals with multiple organ dysfunctions, increased risk of carcinogenesis and reduced life expectancy.[109]

23.4.2.3 GPx8

Like GPx7, GPx8 is also associated with the endoplasmic reticulum, however as a transmembranal protein. It also is a weak GSH peroxidase, but reacts with PDI. The endoplasmic peroxidases are believed to complement each other in the complex process of oxidative protein folding.[106] Dampening of fibroblast growth factor and insulin signaling in Hela cells by GPx8 has also been reported.[110] GPx8 is one of the rare examples of a GPx, in which the typical tetrad is modified; the tetrad's Gln is mutated to Ser.

23.4.3 The Non-mammalian 2-Cys-gutathione Peroxidases

Members of the glutathione peroxide family had been detected in plants and protozoa, but were considered to be poor GPxs, before it had been noticed that they prefer substrates distinct from GSH. The first example of a thioredoxin-specific "glutathione" peroxidase was an enzyme from *Plasmodium falciparum*, which practically was devoid of any GPx activity, but efficiently reduced hydroperoxides with reduced thioredoxin ($Trx(SH)_2$).[111] In the meantime, it has become clear that this type of enzyme prevails in protozoa, plants, bacteria, yeasts, and insects. A systematic comparison of related sequences revealed that these GPx-type proteins, which in functional terms are thioredoxin peroxidases, disclosed some special characteristics: This subfamily contains a flexible loop that harbors a second conserved cysteine residue[66] (2-Cys-GPx). This cysteine residue forms an intramolecular disulfide bridge with the selenenic acid form, which is generated as in other GPxs. The disulfide formation, however, affords a huge motion of the domain that contains the second cysteine. In this respect, this type of mechanism mimics that of atypical 2-Cys-peroxiredoxins[73] (eqn (23.6) and (23.7)).

$$\text{2-Cys-GPx(SH)}_2 + \text{ROOH} \rightarrow \rightarrow \text{2-Cys-GPxS}_2 + \text{ROH} + \text{H}_2\text{O} \quad (23.6)$$

$$\text{2-Cys-GPxS}_2 + \text{Trx(SH)}_2 \rightarrow \rightarrow \rightarrow \text{2-Cys-GPx(SH)}_2 + \text{TrxS}_2 \quad (23.7)$$

The connection with thioredoxin with its pleiotropic roles in cellular regulation and metabolism shifts the interest in these enzymes far beyond the protection against oxidative challenges. Evidently, this subfamily of the GPxs may affect and regulate a wide range of thioredoxin-dependent events. A special case of H_2O_2-sensing was reported by the group of Michel Toledano[112] (Figure 23.3): a GPx-type protein of yeast (yeast GPx3 or Orp1), which proved to be a typical thioredoxin peroxidase, efficiently senses H_2O_2 and, as oxidized enzyme, can form a heterodimer with the transcription factor Yap1. Finally, *via* disulfide reshuffling, the transcription factor is oxidized and thereby activated. The system is regulated by thioredoxin in a dual way: (1) the reaction of the oxidized Orp1 with Yap1 will likely only happen when the regular thioredoxin peroxidase cycle is limited by thioredoxin shortage. (2) The reduced transcription factor is regenerated by thioredoxin. The teleological meaning of this reaction scheme is seen in rendering specificity to H_2O_2, which by itself would be too unspecific to act as useful signaling molecule, whereas the oxidized GPx can interact with the transcription factor *via* specific protein/protein interaction. So far this kind of transcriptional regulation has been observed only once with GPx-type proteins. It appears to be more common with peroxiredoxins.[113]

23.5 Conclusions

The first discovery of a glutathione peroxidase (GPx1) clarified that glutathione itself does not act as an efficient antioxidant *in vivo*, but as a substrate of an antioxidant enzymatic system. In erythrocytes, this system links

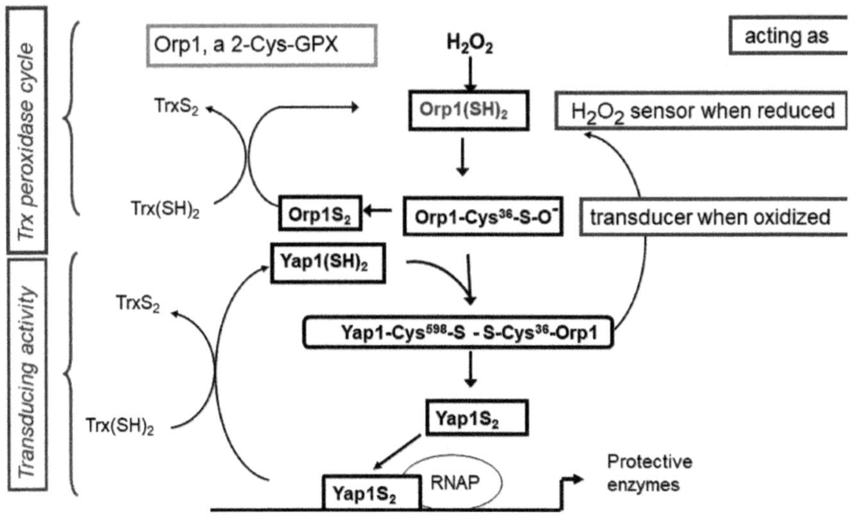

Figure 23.3 A glutathione peroxidase as H_2O_2 sensor. Orp1, which is a GPx-type thioredoxin peroxidase of yeast (yeast GPx3), can react with the transcription factor Yap1. In shortage of reduced thioredoxin ($Trx(SH)_2$), it forms a heterodimer with the transcription factor. After disulfide reshuffling, the latter is oxidized and activated for biosynthesis of protective proteins. The process is modulated and reversed by reduced thioredoxin. The scheme was prepared by the author based on data reported in ref. 109.

NADPH generation by the pentose phosphate shunt *via* glutathione reductase to H_2O_2 detoxification by GPx1. The insight enabled the understanding of various hematological disorders such as favism and drug-induced hemolysis.

Over the decades, new forms of GPx were discovered. They differ in sequence, regulation of biosynthesis, tissue distribution, subcellular compartmentation, substrate specificity, and biological function. In vertebrates and some other animals, the family of glutathione peroxidases comprises enzymes that contain a selenocysteine residue, which together with Trp, Asn, and Gln form the catalytically active tetrad. With very few exceptions, the tetrad consisting of Sec (or Cys), Trp, Asn, and Gln is conserved in the entire family. In 4 of the 8 mammalian GPxs the active site selenocysteine may be replaced by cysteine. This substitution is also observed in other species and regularly in plants.

The basic enzymatic function of GPxs is the reduction of hydroperoxides by thiols. In a first step, the (seleno)cysteine selenol (or thiol) reacts with H_2O_2 or another hydroperoxide to yield a seleninic or sulfenic acid derivative, respectively. The efficient cleavage of the peroxide bond requires a dual attack, which is achieved by a water-mediated proton shuttling mechanism. The selenol (thiol) dissociates, but the released proton remains bound to the active site in an unstable position, preferentially at the ring nitrogen of the

Trp in the tetrad. The selenolate (thiolate), due to its increased nucleophilicity, attacks one oxygen of the peroxide bond, while the delocalized proton, when shuttling back, exerts an electrophilic attack on the other one. Thereby, rate constants for chalcogen oxidation by hydroperoxides of >10 000 000 $M^{-1} s^{-1}$ can be achieved. This first step of the catalytic cycle appears to be common to all GPxs. The difference between SecGPxs and CysGPxs is only a quantitative one. Variations in hydroperoxide specificity are determined by accessibility and residues surrounding the reaction center.

The mechanistic studies on the enzymatic chalcogen oxidation in GPxs and other protein families have greatly improved the general understanding how a cysteine or selenocysteine becomes a "peroxidatic" one. It is the dual attack on the peroxide bond, which is only possible in the peculiar architecture of specialized enzymes.

The reductive part of the catalytic cycle of GPxs is more variable. GPx1 is highly specific for GSH. Mammalian GPx3 reportedly also accepts thioredoxin and glutaredoxin. GPx4 can react with surface-exposed thiols of proteins. The endoplasmic isozymes GPx7 and GPx8 show high affinities and rate constants for protein disulfide isomerase. A large subfamily of GPxs, the 2-Cys-GPxs, which prevail in protozoa, insects, bacteria, yeast, and plants, use redoxins (proteins with a Cys-xx-Cys motif) as reductants. Their reaction scheme, thus, resembles that of atypical 2-Cys-peroxiredoxins.

With the awareness that hydroperoxides are not only toxic, but physiological intermediates or signaling molecules, a fundamental enlargement of the biological roles of GPxs became obvious. They are now widely re-defined as regulatory proteins that remove hydroperoxides, which modulate signaling cascades. The most common interference with signaling processes is the prevention of oxidative inactivation of redox-sensitive protein phosphatases. Other regulatory interferences are the prevention of oxidative activation of lipoxygenases. They may also sense, *i.e.*, react with, H_2O_2 and transfer the oxidation equivalents directly to a transcription factor or modify the activity of a target enzyme. Which of the multiple possibilities is exploited, depends on the biological context, *inter alia* on its location. Mammalian GPx4 may be quoted as a convincing example: its cytosolic expression prevents lipid peroxidation and, thus, ferroptosis; in the nucleus it supports chromatin compaction; the mitochondrial enzyme undergoes a metamorphosis. It becomes part of the mitochondrial sheath in the mid piece of spermatozoa and, thus, is essential for male fertility.

References

1. J. de Rey Pailhade, *Compt. Rend, Acad. Sci.*, 1888, **106**, 1683.
2. F. G. Hopkins, *Biochem. J*, 1921, **15**, 286.
3. C. R. Harrington and T. H. Mead, *Biochem. J*, 1935, **29**, 1602.
4. V. du Vigneaud and G. L. Miller, *J. Biol. Chem.*, 1936, **116**, 469.
5. H. Waelsch and D. Rittenberg, *J. Biol. Chem.*, 1941, **139**, 761.

6. K. Bloch and H. S. Anker, *J. Biol. Chem.*, 1947, **169**, 765.
7. H. J. Forman, H. Zhang and T. J. Kavanagh, in *Glutatione*, ed. L. Flohé, CRC Press, Boca Raton, London, New York, 2019, p. 3.
8. A. Meister, in *Glutathione*, ed. D. Dolphin, R. Poulson and O. Avramovic, Wiley, New York, 1988, vol. 3, Part A, p. 1.
9. E. Baumann and C. Preusse, *Ber. Dtsch. Chem. Ges.*, 1879, **12**, 806.
10. M. Jaffe, *Ber. Dtsch. Chem. Ges.*, 1879, **12**, 1092.
11. T. Nakashima, *J. Biochem.*, 1934, **19**, 281.
12. M. M. Barnes, S. P. James and P. B. Wood, *Biochem. J*, 1959, **71**, 680.
13. H. G. Bray, T. J. Franklin and S. P. James, *Biochem. J*, 1959, **71**, 690.
14. E. Boyland, G. S. Ramsay and P. Sims, *Biochem. J*, 1961, **78**, 376.
15. B. Mannervik and B. Sjödin, in *Glutathione*, ed. L. Flohé, CRC Press, Boca Raton, FL, 2019, p. 175.
16. K. D. Tew, in *Glutathione*, ed. L. Flohé, CRC Press, Boca Raton, FL, 2019, p. 201.
17. Y. Manevich, in *Glutathione*, ed. L. Flohé, CRC Press, Boca Raton, FL, 2019, p. 143.
18. W. Cordes, Annual Report Unite Fruit Co. (Med. Dept.), 1926, vol. 15, p. 66.
19. R. S. Hockwald, J. Arnold, C. B. Clayman and A. S. Alving, *J. Am. Med. Assoc.*, 1952, **149**, 1568.
20. A. S. Alving, P. E. Carson, C. L. Flanagan and C. E. Ickes, *Science*, 1956, **124**, 484.
21. P. E. Carson, *Fed. Proc.*, 1960, **19**, 995.
22. E. Beutler, *Blood*, 1959, **14**, 103.
23. C. Fermi and P. Martinetti, *Ann. Ig. Sperim.*, 1905, **15**, 75.
24. P. Arese and A. De Flora, *Semin. Hematol.*, 1990, **27**, 1.
25. K. D. Belfield and E. M. Tichy, *Am. J. Health-Syst. Pharm.*, 2018, **75**, 97.
26. A. Szeinberg, C. Sheba, N. Hirshorn and E. Bodonyi, *Blood*, 1957, **19**, 603.
27. E. Beutler, in *Glutathione*, ed. L. Flohé, H. C. Benöhr, H. Sies, H. D.Waller and A. Wendel, Georg Thieme Publishers, Stuttgart, 1974, p. 109.
28. P. Boivin, C. Galand and J. F. Bernard, in *Glutathione*, ed. L. Flohé, H. C. Benöhr, H. Sies, H. D. Waller and A. Wendel, Georg Thieme Publishers, Stuttgart, 1974, p. 146.
29. C. Berndt, C. H. Lillig and L. Flohé, *Front. Pharmacol.*, 2014, **5**, 168.
30. R. Rink, in *Glutathione*, ed. L. Flohé, H. C. Benöhr, H. Sies, H. D. Waller and A. Wendel, Georg Thieme Publishers, Stuttgart, 1974, p. 206.
31. L. Flohé, *Biochim. Biophys. Acta*, 2013, **1830**, 3139.
32. C. H. Lillig and C. Berndt, *Antioxid. Redox Signaling*, 2013, **18**, 1654.
33. W. N. Pirie and K. G. Pinhey, *J. Biol. Chem.*, 1929, **84**, 321.
34. G. Jung, E. Breitmeier, W. A. Günzler, M. Ottnad, W. Voelter and L. Flohé, in Glutathione, ed. L. Flohé, H. C. Benöhr, H. Sies, H. D. Waller and A. Wendel, Georg Thieme Publishers, Stuttgart, 1974, p. 1.

35. H. P. Misra, *J. Biol. Chem.*, 1974, **249**, 2151.
36. C. S. Rose and P. György, *Blood*, 1950, **5**, 1062.
37. N. R. Brownlee, J. J. Huttner, R. V. Panganamala and D. G. Cornwell, *J. Lipid Res.*, 1977, **18**, 635.
38. D. Neubert and A. L. Lehninger, *J. Biol. Chem.*, 1962, **237**, 952.
39. L. Flohé and R. Zimmermann, in *Glutatione*, ed. L. Flohé, H. C. Benöhr, H. Sies, H. D. Waller and A. Wendel, Georg Thieme Publishers, Stuttgart, 1974, p. 245.
40. G. C. Mills, *J. Biol. Chem.*, 1957, **229**, 189.
41. L. Flohé, *Antioxidants*, 2020, **9**, 1274.
42. G. Cohen and P. Hochstein, *Biochemistry*, 1963, **2**, 1420.
43. C. Little and P. J. O'Brien, *Biochem. Biophys. Res. Commun.*, 1968, **31**, 145.
44. B. O. Christophersen, *Biochim. Biophys. Acta*, 1969, **176**, 463.
45. C. Little, *Biochim. Biophys. Acta*, 1972, **284**, 375.
46. B. O. Christophersen, *Biochim. Biophys. Acta*, 1969, **186**, 387.
47. W. A. Günzler, H. Vergin, I. Müller and L. Flohé, *Hoppe-Seyler's Z. physiol. Chem.*, 1972, **353**, 1001.
48. L. Flohé, B. Eisele and A. Wendel, *Hoppe-Seyler's Z. physiol. Chem.*, 1971, **352**, 151.
49. L. Flohé, E. Schaich, W. Voelter and A. Wendel, *Hoppe-Seyler's Z. physiol. Chem.*, 1971, **352**, 170.
50. K. Schwarz and C. M. Foltz, *J. Am. Chem. Soc.*, 1957, **79**, 3292.
51. J. T. Rotruck, W. G. Hoekstra, A. L. Pope, H. E. Ganther, A. Swanson and D. Hafemann, *Fed. Proc.*, 1972, **31**, 691.
52. J. T. Rotruck, A. L. Pope, H. E. Ganther, A. B. Swanson, D. G. Hafeman and W. G. Hoekstra, *Science*, 1973, **179**, 588.
53. L. Flohé and W. A. Günzler, in *Glutathione*, ed. L. Flohé, Georg Thieme Publishers, Stuttgart, 1974, p. 132.
54. L. Flohé, W. A. Günzler and H. H. Schock, *FEBS lett.*, 1973, **32**, 132.
55. A. Wendel, B. Kerner and K. Graupe, in *Functions of Glutathione in Liver and Kidney*, ed. H. Sies and A. Wendel, Springer, Heidelberg, 1978, p. 107.
56. J. W. Forstrom, J. J. Zakowski and A. L. Tappel, *Biochemistry*, 1978, **17**, 2639.
57. R. Ladenstein and O. Epp, in *Oxidative Damage and Related Enzymes – EMBO Workshop 1983*, ed. G. Rotilio and J. V. Bannister, Harwood Academic Publishers, Chur, London, Paris, New York, 1984, p. 363.
58. W. A. Günzler, G. J. Steffens, A. Grossmann, S. M. Kim, F. Ötting, A. Wendel and L. Flohé, *Hoppe-Seyler's Z. physiol. Chem.*, 1984, **365**, 195.
59. L. Orian, P. Mauri, A. Roveri, S. Toppo, L. Benazzi, V. Bosello-Travain, A. De Palma, M. Maiorino, G. Miotto, M. Zaccarin, A. Polimeno, L. Flohé and F. Ursini, *Free Radicals Biol. Med.*, 2015, **87**, 1.
60. R. Ladenstein and O. Epp, *Life Chem. Rep.*, 1984, **2**, 363.
61. W. Ostwald, *The Nobel Lecture in Chemistry*, NobelMedia Ag, Stockholm, 1909.

62. C. Rocher, J. L. Lalanne and J. Chaudiere, *Eur. J. Biochem.*, 1992, **205**, 955.
63. M. Maiorino, K. D. Aumann, R. Brigelius-Flohé, D. Doria, J. van den Heuvel, J. McCarthy, A. Roveri, F. Ursini and L. Flohé, *Biol. Chem. Hoppe-Seyler*, 1995, **376**, 651.
64. C. C. Winterbourn and D. Metodiewa, *Free Radicals Biol. Med.*, 1999, **27**, 322.
65. B. Cardey and M. Enescu, *Chemphyschem*, 2005, **6**, 1175.
66. M. Maiorino, F. Ursini, V. Bosello, S. Toppo, S. C. Tosatto, P. Mauri, K. Becker, A. Roveri, C. Bulato, L. Benazzi, A. De Palma and L. Flohé, *J. Mol. Biol.*, 2007, **365**, 1033.
67. M. Trujillo, G. Ferrer-Sueta, L. Thomson, L. Flohé and R. Radi, *Subcell. Biochem.*, 2007, **44**, 83.
68. S. C. Tosatto, V. Bosello, F. Fogolari, P. Mauri, A. Roveri, S. Toppo, L. Flohe, F. Ursini and M. Maiorino, *Antioxid. Redox Signaling*, 2008, **10**, 1515.
69. S. Toppo, L. Flohé, F. Ursini, S. Vanin and M. Maiorino, *Biochim. Biophys. Acta*, 2009, **1790**, 1486–1500.
70. F. Fogolari, A. Corazza, S. Toppo, S. C. Tosatto, P. Viglino, F. Ursini and G. Esposito, *J. Biomed. Biotechnol.*, 2012, **2012**, 303.
71. M. Dalla Tiezza, F. M. Bickelhaupt, L. Flohé and L. Orian, *ChemPlusChem*, 2021, **86**, 525.
72. F. Ursini, M. Maiorino, R. Brigelius-Flohé, K. D. Aumann, A. Roveri, D. Schomburg and L. Flohé, *Methods Enzymol.*, 1995, **252**, 38.
73. L. Flohé, S. Toppo, G. Cozza and F. Ursini, *Antioxid. Redox Signaling*, 2011, **15**, 763.
74. M. Dalla Tiezza, F. M. Bickelhaupt, L. Flohé, M. Maiorino, F. Ursini and L. Orian, *Redox Biol.*, 2020, **34**, 101540.
75. L. Flohé and R. Brigelius-Flohé, in *Selenium. Its Molecular Biology and Role in Human Health*, ed. D. L. Hatfield, U. Schweizer, P. A. Tsuji and V. N. Gladyshev, Springer, New York, 4th edn, p. 211.
76. R. Brigelius-Flohé and L. Flohé, *Antioxid. Redox Signaling*, 2020, **33**, 498.
77. D. E. Handy and J. Loscalzo, in *Glutathione*, ed. L. Flohé, CRC Press, Boca Raton, FL, 2019, p. 67.
78. H. Steinbrenner and L.-O. Klotz, in *Glutathione*, ed. L. Flohé, CRC Press, Boca Raton, FL, 1919, p. 81.
79. A. P. Kipp, in *Glutathione*, ed. L. Flohé, CRC Press, Boca Raton, FL, 2019, ch. 6, pp. 95–109.
80. M. Maiorino, A. Roveri and F. Ursini, in *Glutathione*, ed. L. Flohé, CRC Press, Boca Raton, FL, 2019, p. 111.
81. R. Brigelius-Flohé and M. Maiorino, *Biochim. Biophys. Acta*, 2013, **1830**, 3289.
82. J. P. McClung, C. A. Roneker, W. Mu, D. J. Lisk, P. Langlais, F. Liu and X. G. Lei, *Proc. Natl. Acad. Sci. U. S. A.*, 2004, **101**, 8852.
83. C. Kretz-Remy, P. Mehlen, M. E. Mirault and A. P. Arrigo, *J. Cell Biol.*, 1996, **133**, 1083.

84. F. F. Chu, J. H. Doroshow and R. S. Esworthy, *J. Biol. Chem.*, 1993, **268**, 2571.
85. R. Brigelius-Flohé and A. Kipp, *Biochim. Biophys. Acta*, 2009, **1790**, 1555.
86. A. Banning, A. Kipp, S. Schmitmeier, M. Lowinger, S. Florian, S. Krehl, S. Thalmann, R. Thierbach, P. Steinberg and R. Brigelius-Flohé, *Cancer Res.*, 2008, **68**, 9746.
87. S. Krehl, M. Loewinger, S. Florian, A. P. Kipp, A. Banning, L. A. Wessjohann, M. N. Brauer, R. Iori, R. S. Esworthy, F. F. Chu and R. Brigelius-Flohé, *Carcinogenesis*, 2012, **33**, 620.
88. K. Takahashi and H. J. Cohen, *Blood*, 1986, **68**, 640.
89. R. C. Jin, C. E. Mahoney, L. (Coleman) Anderson, F. Ottaviano, K. Croce, J. A. Leopold, Y. Y. Zhang, S. S. Tang, D. E. Handy and J. Loscalzo, *Circulation*, 2011, **123**, 1963.
90. A. C. Smith, A. J. Mears, R. Bunker, A. Ahmed, M. McKenzie, J. A. Schwarzentruber, C. L. Beaulieu, E. Ferretti, FORGE Canada Consortium, J. Majewski, D. E. Bulman, F. C. Celik, K. M. Boycott and G. E. Graham, *J. Med. Genet.*, 2014, **51**, 470.
91. M. R. Sedaghatian, *Am. J. Med. Genet.*, 1980, **6**, 269.
92. D. M. Cheff, A. R. Muotri, B. R. Stockwell, E. E. Schmidt, Q. Ran, R. V. Kartha, S. C. Johnson, P. Mittal, E. S. J. Arnér, K. M. Wigby, M. D. Hall and, S. K. Ramesh, www.preprint.org, 2021, DOI: 10.20944/preprints202104.0105.v1.
93. J. P. Friedmann Angeli, V. E. Kagan and M. Conrad, in *Glutathione*, ed. L. Flohé, CRC Press, Boca Raton, FL, 2019, ch. 8, p. 129.
94. I. Ingold, C. Berndt, S. Schmitt, S. Doll, G. Poschmann, K. Buday, A. Roveri, X. Peng, F. Porto Freitas, T. Seibt, L. Mehr, M. Aichler, A. Walch, D. Lamp, M. Jastroch, S. Miyamoto, W. Wurst, F. Ursini, E. S. J. Arnér, N. Fradejas-Villar, U. Schweizer, H. Zischka, J. P. Friedmann Angeli and M. Conrad, *Cell*, 2018, **172**, 409.
95. R. Brigelius-Flohé, B. Friedrichs, S. Maurer, M. Schultz and R. Streicher, *Biochem. J.*, 1997, **328**, 199.
96. G. Cozza, M. Rossetto, V. Bosello-Travain, M. Maiorino, A. Roveri, S. Toppo, M. Zaccarin, L. Zennaro and F. Ursini, *Free Radicals Biol. Med.*, 2017, **112**, 1.
97. H. Pfeifer, M. Conrad, D. Roethlein, A. Kyriakopoulos, M. Brielmeier, G. W. Bornkamm and D. Behne, *FASEB J.*, 2001, **15**, 1236.
98. M. Conrad, S. G. Moreno, F. Sinowatz, F. Ursini, S. Kolle, A. Roveri, M. Brielmeier, W. Wurst, M. Maiorino and G. W. Bornkamm, *Mol. Cell. Biol.*, 2005, **25**, 7637.
99. C. Godeas, F. Tramer, F. Micali, A. Roveri, M. Maiorino, C. Nisii, G. Sandri and E. Panfili, *Biochem. Mol. Med.*, 1996, **59**, 118.
100. M. Arai, H. Imai, D. Sumi, T. Imanaka, T. Takano, N. Chiba and Y. Nakagawa, *Biochem. Biophys. Res. Commun.*, 1996, **227**, 433.
101. M. Maiorino, J. B. Wissing, R. Brigelius-Flohé, F. Calabrese, A. Roveri, P. Steinert, F. Ursini and L. Flohé, *FASEB J.*, 1998, **12**, 1359.

102. F. Ursini, S. Heim, M. Kiess, M. Maiorino, A. Roveri, J. Wissing and L. Flohé, *Science*, 1999, **285**, 1393.
103. P. Mauri, L. Benazzi, L. Flohé, M. Maiorino, P. G. Pietta, S. Pilawa, A. Roveri and F. Ursini, *Biol. Chem.*, 2003, **384**, 575.
104. M. Maiorino, A. Roveri, L. Benazzi, V. Bosello, P. Mauri, S. Toppo, S. C. Tosatto and F. Ursini, *J. Biol. Chem.*, 2005, **280**, 38395.
105. L. Flohé, *Biol. Chem*, 2007, **388**, 987.
106. A. Utomo, X. Jiang, S. Furuta, J. Yun, D. S. Levin, Y. C. Wang, K. V. Desai, J. E. Green, P. L. Chen and W. H. Lee, *J. Biol. Chem.*, 2004, **279**, 43522.
107. M. Maiorino, V. Bosello-Travain, G. Cozza, G. Miotto, A. Roveri, S. Toppo, M. Zaccarin and F. Ursini, *Free Radicals Biol. Med.*, 2015, **83**, 352.
108. V. D. Nguyen, M. J. Saaranen, A. R. Karala, A. K. Lappi, L. Wang, I. B. Raykhel, H. I. Alanen, K. E. Salo, C. C. Wang and L. W. Ruddock, *J. Mol. Biol.*, 2011, **406**, 503.
109. T. Ramming and C. Appenzeller-Herzog, *Intern. J. Cell Biol.*, 2013, **2013**, 180906.
110. V. Bosello-Travain, H. J. Forman, A. Roveri, S. Toppo, F. Ursini, R. Venerando, C. Warnecke, M. Zaccarin and M. Maiorino, *Free Radicals Biol. Med.*, 2015, **81**, 58.
111. H. Sztajer, B. Gamain, K. D. Aumann, C. Slomianny, K. Becker, R. Brigelius-Flohé and L. Flohé, *J. Biol. Chem.*, 2001, **276**, 7397.
112. A. Delaunay, D. Pflieger, M. B. Barrault, J. Vinh and M. B. Toledano, *Cell*, 2002, **111**, 471.
113. L. Flohé, *Free Radical Res.*, 2016, **50**, 126.

CHAPTER 24

Chalcogen-containing Protein and Nucleic Acid Derivatives – Synthesis and Applications

MICHIO IWAOKA

Department of Chemistry, School of Science, Tokai University, 4-1-1 Kitakaname, Hiratsuka, Kanagawa 259-1292, Japan
Email: miwaoka@tokai.ac.jp

24.1 Introduction

When sulfur (S)-, selenium (Se)-, or tellurium (Te)-containing chalcogen compounds are administrated to living organisms, they are *Absorbed*, *Distributed* to the whole body, *Metabolized* in cells, and then *Excreted* to the environment. Through this ADME cycle, the chalcogen compounds are assimilated into various functional biomolecules and dissimilated to small organic or inorganic chalcogen compounds.[1–3] The production of various chalcogen-containing biomolecules *in vivo* can be either an active process, which is strictly controlled by genomic information, or a passive random process without genetic control. In any case, these chalcogen-containing biomolecules, in particular Se- or Te-containing ones, attract much attention of researchers in the fields of chemical biology, metallomics, and medicinal science as they exhibit interesting biological activities, such as antioxidant activity, anticancer activity, cell toxicity, *etc.*[4,5]

In the meantime, new chemical as well as molecular biology techniques have been developed in the last few decades, which made it possible to introduce S, Se, or Te atoms artificially at any specific positions of

Chalcogen Chemistry: Fundamentals and Applications
Edited by Vito Lippolis, Claudio Santi, Eder J. Lenardão and Antonio L. Braga

Published by the Royal Society of Chemistry, www.rsc.org

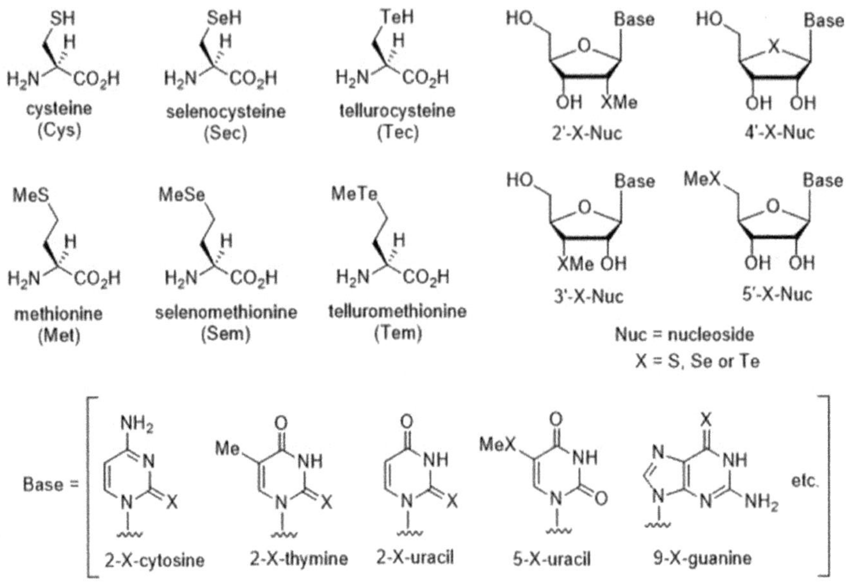

Figure 24.1 Representative S-, Se-, or Te-containing molecular fragments, which can be found in proteins and nucleic acids.

natural biomolecules. The chalcogen analogues thus obtained showed unique biological activities that are different from those exhibited by the original biomolecules.[4,6,7] In addition, various small organic molecules modelling chalcogen-containing enzymes and nucleic acids have also been designed and synthesized. These organochalcogen compounds are now utilized as enzyme mimics to probe the reaction mechanisms. Their possible applications to medicinal science are also elaborately being examined.[8,9]

Molecular structures of representative chalcogen-containing amino acids and nucleoside units, which can be found in proteins and nucleic acids, are collected in Figure 24.1.

This chapter focuses on recent advances in the chemical synthesis and applications of chalcogen analogues of proteins and nucleic acids. In Section 24.2, naturally-occurring chalcogen-containing proteins and nucleic acids are overviewed. In Sections 24.3 and 24.4, the synthesis and applications of chalcogen analogues of natural proteins and nucleic acids, respectively, are summarized. Finally, perspectives of the relevant research areas are presented in Section 24.5.

24.2 Natural Products Containing Chalcogen Atoms

24.2.1 Proteins

Chalcogen-containing proteins are abundant in nature. Most of them are sulfur-containing proteins but selenium-containing proteins can also be found. In these proteins, the S and Se atoms are involved as cysteine (Cys), methionine (Met), selenocysteine (Sec), or selenomethionine (Sem) (see Figure 24.1). Among these, Cys, Met, and Sec are proteinogenic amino acids, which are incorporated into proteins at the fixed positions through translation of the corresponding genetic codons, and play significantly important roles in many biological functions of proteins as their positions along the amino acid sequences are strongly conserved during the evolution. On the other hand, selenomethionine (Sem) is taken into proteins, instead of Met, through a passive process when inorganic selenium nutrients, such as selenite (SeO_3^{2-}) and selenate (SeO_4^{2-}), are ingested into microbes[10] and plants.[11] Thus, this peculiar amino acid is randomly incorporated in proteins.

Tec- and Tem-containing proteins are not found in natural proteins usually, probably due to the strong cytotoxicity of tellurium. Under particular conditions however, Te would be incorporated in a protein as Tem.[12,13]

24.2.1.1 *Sulfur-containing Proteins*

Since the S atom of Cys can change its oxidation state easily, the Cys residues in proteins participate in various redox biological processes through the transition among the variable oxidation states. Usually, Cys exists as either a reduced thiol (CysSH) or an oxidized disulfide (CysSSCys) state. CysSH can be reversibly modified to CysSNO, CysSSG, CysSSH, *etc.* in cells by reacting with various reactive species, such as nitric oxide (NO) and glutathione (GSH) (see Figure 24.2).[14] In the catalytic cycles of cysteine protease enzymes, CysSH can also be oxidized with reactive oxygen species (ROS) to a highly reactive sulfenic acid (CysSOH) reversibly. However, its overoxidized products, *i.e.*, sulfinic acid ($CysSO_2H$) and sulfonic acid ($CysSO_3H$), are so stable

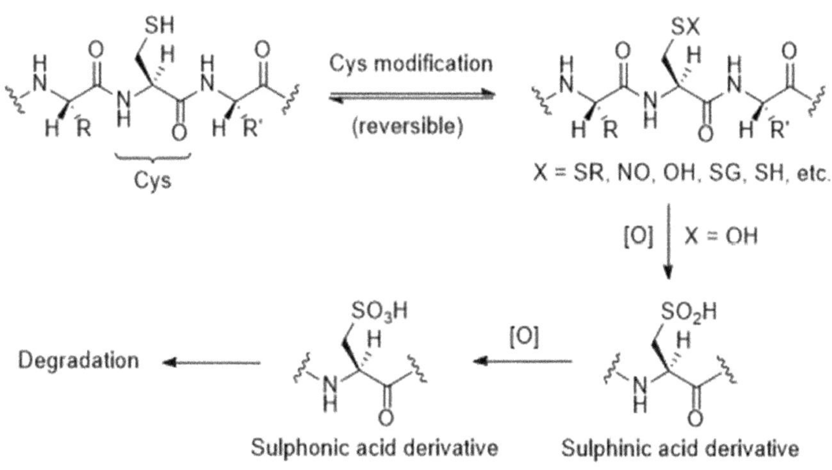

Figure 24.2 Reversible post-translational modifications (PTMs) of Cys in cells and the degradation pathways.

that they cannot be transformed back to CysSH with thiols, except for the enzymatic reduction of $CysSO_2H$ with sulfiredoxin.[15] $CysSO_2H$ and $CysSO_3H$ would be generated during degradation processes of proteins *in vivo*.

Met possesses a sulfanyl (or methyl sulfide) S atom, which can be oxidized to sulfoxide with ROS. The generated methionine sulfoxide (Met(O)), which is harmful to cells, can be reduced back to Met enzymatically with methionine sulfoxide reductases (MsrA, MsrB, *etc.*).[16] Interestingly, among these enzymes, MsrB1, which catalyzes the selective reduction of (*R*)-Met(O) to Met, is one of Se-containing enzymes (see Table 24.1).[17]

Cys residues in proteins are known to play several roles, including stabilization of protein structures by disulfide (SS)-bond formation, stabilization of metal (Fe, Zn, Cu, *etc.*) complexes as a ligand and active centres of particular enzymes, such as cysteine proteases and peroxiredoxins. In addition to these conventional roles, novel aspects of the biological functions of Cys are recently being explored.

24.2.1.1.1 Multiple SS-Bonded States. Multiple SS-bonded states of SS-rich proteins have been focused on regarding protein polymerization (*e.g.*, blood clotting).[18] Fibrinogen and α2-macroglobulin are secreted from cells to the blood stream and play a role in blood-clotting. These proteins have more than 10 SS-bonds but about 10–70% of them are reduced on average during blood circulation. Thus, the number of possible SS-bonded states should be very large.[19] It was recently suggested that this complex multiple SS-bonded state of the proteins can dynamically respond against stimuli from outside.[18]

24.2.1.1.2 Proteome Analysis of Thiol Redox States. There are a huge number of Cys-containing proteins coexisting in a cell or in the organelles. The thiol groups of these proteins receive a variety of post-translational modifications (PTMs) in response to external stimuli as well as environmental changes (see Figure 24.2).[14,20] Recently, the methods to quantify the ratios of these thiol modifications by mass spectrometry have been developed and utilized for the proteome analysis of thiol redox states in a whole cell.[21,22] Most of the thiol modifications are reversible, except for the overoxidation to sulfinic and sulfonic acids. These PTMs play roles in the regulation of the catalytic activity of S-containing proteins, controlling the protein–protein interaction, regulation of protein conformation, allosteric regulation, or switches of protein functions, *etc.*, in cells.

24.2.1.1.3 Chalcogen Bonds. The formation of S···X (X = O, N, S) chalcogen bonds has also recently been focused on as a new function of sulfur in proteins (see Figure 24.3).[23–25] S···X chalcogen bonds are weak nonbonding interactions characterized as a weak coordination of the lone pair of the X atom to the σ hole of the S atom.[26,27] It has been reported that these weak interactions would influence structures and biological functions of proteins.[28] Relating to this weak nonbonding interaction, Se···X

Table 24.1 Representative Se-containing proteins in humans.

Representative selenoenzymes	Enzymatic reactions	Catalytic cycles	Ref.
Glutathione peroxidase (GPx)	$ROOH + 2GSH \rightarrow ROH + H_2O + GSSG$ (R = H or Lipid)	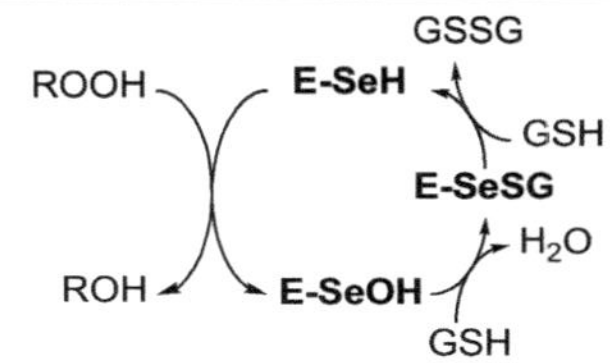	35
Thioredoxin reductase (TrxR)	$Trx(S{-}S) + NADPH + H^+ \rightarrow Trx(SH)_2 + NADP^+$	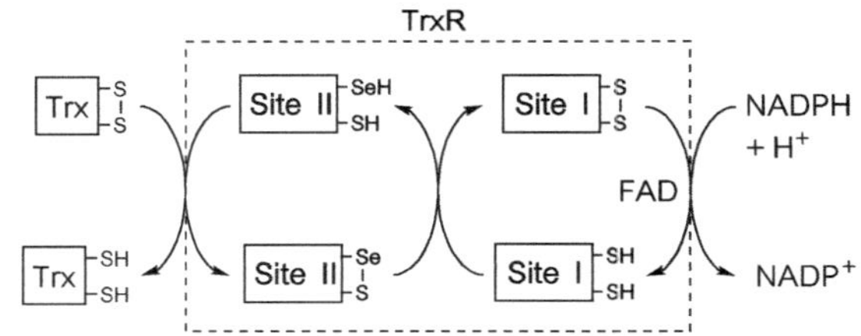	36
Methionine-sulfoxide reductase B1 (MsrB1)	(R)-$Met(O) + Trx(SH)_2 \rightarrow Met + Trx(S{-}S) + H_2O$	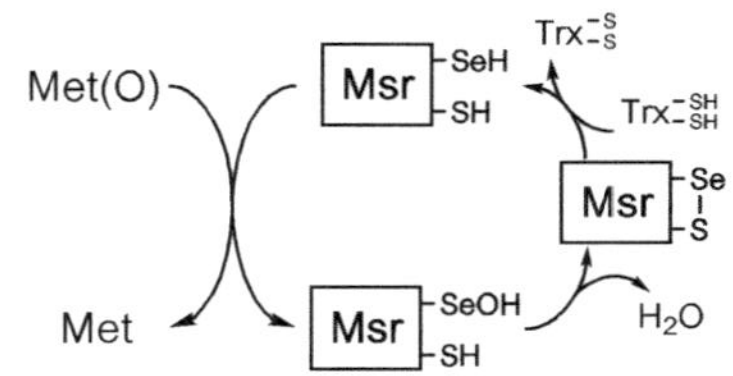	17
Type 1 iodothyronine deiodinase (DIO1)	Thyroxine (T_4) + (2RSH) → T_3 + (HI + RSSR) (The cofactor is not identified. Here, RSH is tentatively assigned.)	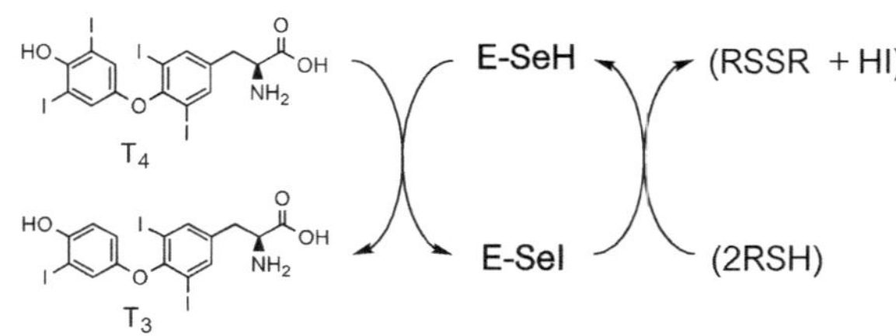	37

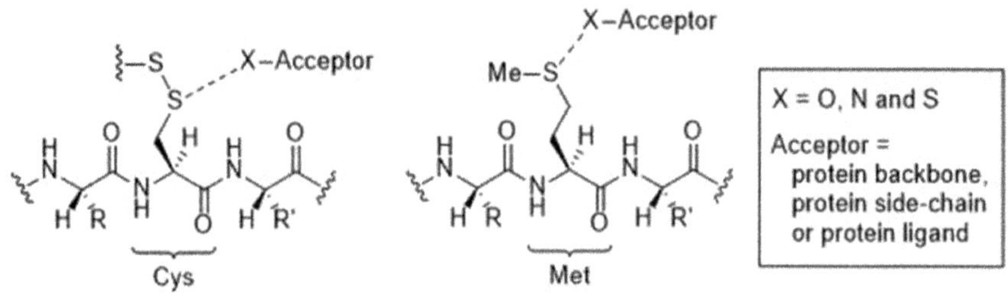

Figure 24.3 S···X chalcogen bonds in proteins.

chalcogen bonds[29] and Ch···π (Ch = S, Se, Te) interactions[30] are also elaborately studied recently.

24.2.1.2 Selenium-containing Proteins

In contrast to the fact that sulfur is involved in most proteins, selenium is found in a limited number of proteins. For example, only 25 selenoproteins have been identified in the human genome.[31] Representative selenoproteins are listed in Table 24.1 along with their enzymatic functions. Although the number of selenoproteins is small, they play biologically important roles and selenium deficiency causes fatal conditions for people. To introduce Sec to proteins *in vivo*, a series of specific enzymes participate, such as specialized elongation factor (SelB), SECIS (selenocysteine insertion sequence)-binding protein 2 (SBP-2), 43 kDa RNA-binding protein (Sep43), *O*-phosphoserine-tRNA:selenocysteine-tRNA synthase (SepSecS), *etc.*[32] Similar Sec-expression systems are commonly deployed for a wide range of living beings from fungi to human, but this kind of system does not generally exist in plants.[33] The introduction of Sec to proteins is a laborious job for life. Despite this, life has adopted Sec, not Cys, at the active centre of several oxidoreductase enzymes, which are indeed indispensable for maintaining life activities. This would be due to the characteristic redox properties of selenium.[34]

At the active state of selenoenzymes, Sec exists as a reduced selenol (SecSeH), which should be deprotonated to a selenolate ($SecSe^-$) under physiological conditions because of a low pK_a value of the selenol (5.2).[38] Since $SecSe^-$ is a highly reactive species, it would be stabilized at the active sites of selenoenzymes. For example, the $SecSe^-$ of glutathione peroxidase (GPx), a representative antioxidant selenoenzyme, is suggested to be thermodynamically stabilized by two NH···Se hydrogen bonds with nearby glutamine (Gln) and tryptophan (Trp) residues.[35] For thioredoxin reductase (TrxR), another representative selenoenzyme, the $SecSe^-$ in the active site II forms a metastable selenosulfide bond (Se–S) with the adjacent Cys residue. The resulting unreactive state can be activated to $SecSe^-$ with a dithiol unit in the active site I.[36] Selenium can change its oxidation state during the catalytic cycles, like the case of S, but the transition takes place much easier. Thus, selenoenzymes maintain the redox homoeostasis in cells against the oxidative stress by removing ROS directly and indirectly.

24.2.2 Nucleosides and Nucleic Acids

Like proteins, nucleosides and nucleic acids frequently contain S and Se atoms. These chalcogen atoms usually replace the oxygen atoms at the 5′-position of the ribose or at the 2-position of uracil, thymine, and cytosine nucleobases (see Figure 24.1). Representative examples are shown in Figure 24.4. The biological functions of these chalcogen analogues are already implicated in various biological reactions, such as an increase in the nuclease resistance, a decrease in the toxicity, and an enhancement of the translational efficiency.[5] In contrast, Te-containing nucleosides have not been observed *in vivo* probably due to the high reactivities and toxicities.[39,40] In the following subsections, naturally-occurring thio- and seleno-nucleosides are briefly overviewed.

24.2.2.1 Sulfur-containing Nucleosides and Nucleic Acids

5′-Deoxy-5′-methylthioadenosine (MTA, **1**) is the first thionucleoside, which was isolated from yeast and other organisms.[41] MTA (**1**) is derived from *S*-adenosylmethionine (SAM) in cells and has versatile biological functions. 2-Thiouridine (S^2U, **2**), 4-thiouridine (S^4U), 2-thiocytidine (S^2C), and 2-thiothymidine (S^2T) are also found in a wide range of living organisms.[42] Among these, S^2U (**2**) is characteristic as it is present in the wobble position of tRNA and plays crucial roles in the translational functions.[43,44] Thiouridine **2** can be modified *in vivo* at the 5-position of uracil to produce 5-methylaminomethyl-2-thiouridine (mnm^5S^2U, **3**), *etc.*

24.2.2.2 Selenium-containing Nucleosides and Nucleic Acids

Several selenonucleosides, such as 2-selenouridine (Se^2U, **4**) and 5-methylaminomethyl-2-selenouridine (mnm^5Se^2U, **5**), have been found in certain bacterial tRNAs.[45] They are formed *in vivo* from the corresponding thiouridines (**3**, *etc.*) using monoselenophosphate as a selenium donor. The process is catalyzed by a specific enzyme, tRNA 2-selenouridine synthase.[46] The selenouridines exist at the wobble position of the anticodon loop and would augment translational efficiency.[47]

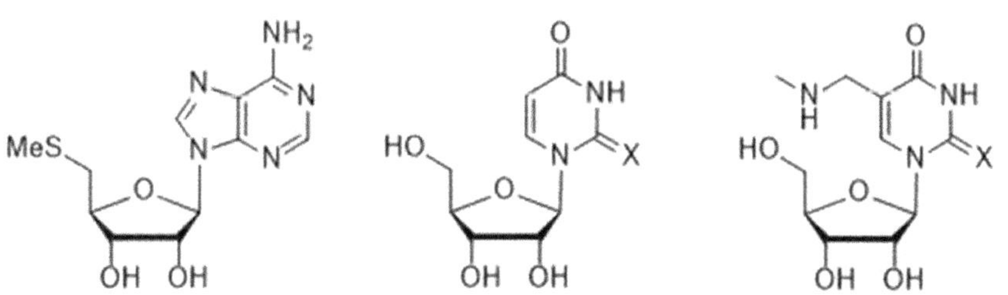

MTA (**1**)

s^2U (**2**) (X = S)
Se^2U (**4**) (X = Se)

mnm^5S^2U (**3**) (X = S)
mnm^5Se^2U (**5**) (X = Se)

Figure 24.4 Examples of naturally occurring S- or Se-containing nucleosides.

24.2.3 Other Natural Products

S, Se, and Te chalcogen atoms are present *in vivo* in various forms. For example, a variety of S-containing natural products are known, such as thiamine, biotin, lipoic acid, taurine, *etc.* The biological functions of these organosulfur compounds are well known. On the other hand, only few Se- and Te-containing natural products have been characterized to date. Suzuki first found 1β-methylseleno-*N*-acetyl-D-galactosamine as a major urinary selenium metabolite.[48,49] The other Se-containing mono- and di-saccharides were also characterized in staple crops, such as wheat, rice, and maize.[50] Selenoneine, a selenium analogue of ergothioneine, was discovered by Yamashita as a strong antioxidant in tuna blood.[51] Se and Te are also present as rather simple inorganic or organic compounds, such as $SeMe_2$, $TeMe_2$, $SeMe_3^+$, and $TeMe_3^+$. Interestingly, *E. coli* and some bacteria are known to harvest energy from inorganic tellurite (TeO_3^{2-}) and tellurate (TeO_4^{2-}) *via* the reduction into non-toxic Te nanoparticles.[52–55]

24.3 Chalcogen Analogues of Proteins

24.3.1 Synthesis of Selenium-containing Protein Analogues

The methodologies to introduce a Se atom(s) at a specific position(s) in a protein (or an enzyme) have been explored since around 1990. As a result, chemical transformation of Ser to Sec at the active site of a serine protease using a selenating reagent,[56,57] biological substitution of Met with Sem by expression of a Met-containing protein in culture medium containing Sem, instead of Met, using a methionine auxotrophic strain of *E. coli*[58] and molecular biological expression of Sec-containing proteins utilizing recombinant genes including SECIS and all specific enzymes required for Sec expression[59] have been developed. The Met-to-Sem substituted proteins thus obtained are widely utilized to solve the phase problem in protein structure determination by single crystal X-ray analysis.[60] Using the genetic methodologies, proteins labelled with ^{77}Se isotopes were also obtained, and their structures and reactions were investigated by ^{77}Se NMR.[61] However, it is difficult to find a supply of the ^{77}Se-enriched materials, which are required for the synthesis.

In the meantime, entirely chemical synthesis of a Sec-containing protein, which has a Sec residue(s) at any desired position(s), has become possible since around 2000 thanks to the development of the native chemical ligation method using Sec at the ligation point (Sec-NCL).[62–64] The Sec-containing peptide, which is utilized for Sec-NCL, is available from the solid-phase peptide synthesis (SPPS) using Sec derivatives, such as Boc–Sec(Bzl)–OH, Fmoc–Sec(PMB or MPM)–OH, *etc.*[65] By applying this chemical synthesis method, various proteins containing Sec, instead of Cys, were synthesized (see Table 24.2).[6,66,67] These selenium analogues disclosed interesting features in their oxidative folding pathways as well as the biological functions. Furthermore, since the Sec residue introduced by Sec-NCL can be transformed to alanine (Ala) or other proteinogenic amino

Table 24.2 Examples of Sec-substituted analogues of natural proteins.

Natural proteins	Number of amino acid residues	Positions of SS bonds or Cys residues	Sec-analogues	Ref.
Endothelin-1 (human)	21	Cys1–Cys15, Cys3–Cys11	Cys1–Cys15, **Sec3–Sec11**	72
α-Conotoxin (marine snail)	12	Cys2–Cys8, Cys3–Cys12	**Sec2–Sec8**, Cys3–Cys12	73
			Cys2–Cys8, **Sec3–Sec12**	
			Sec2–Sec8, **Sec3–Sec12**	
Ribonuclease A (bovine pancreas)	124	Cys26–Cys84, Cys40–Cys95, Cys58–Cys110, Cys65–Cys72	Cys26–Cys84, Cys40–Cys95, Cys58–**Sec110**, Cys65–Cys72	63
Glutaredoxin 3 (engineered)	82	Cys11–Cys14 (active site)	**Sec11**–Cys14	74
			Cys11–**Sec14**	
			Sec11–Sec14	
Glutaredoxin 1 (*E. coli*)	85	Cys11–Cys14 (active site)	**Sec11**, Ser14	75
Trypsin inhibitor or aprotinin (bovine pancreas)	58	Cys5–Cys55, Cys14–Cys38, Cys30–Cys51	Cys5–Cys55, Cys14–**Sec38**, Cys30–Cys51	64
			Sec5–Cys55, **Sec14**–Cys38, Cys30–Cys51	76, 77
			Sec5–Cys55, Cys14–Cys38, Cys30–Cys51	77
			Cys5–Cys55, **Sec14–Sec38**, Cys30–Cys51	
Insulin (bovine)	21 (A-chain) 30 (B-chain)	Cys6A–Cys11A, Cys7A–Cys7B, Cys20A–Cys19B	Cys6A–Cys11A, **Sec7A–Sec7B**, Cys20A–Cys19B	78, 79
Insulin (human)	21 (A-chain) 30 (B-chain)	Cys6A–Cys11A, Cys7A–Cys7B, Cys20A–Cys19B	**Sec6A–Sec11A**, Cys7A–Cys7B, Cys20A–Cys19B	80
Ferredoxin-1 (*T. elongatus*)	97	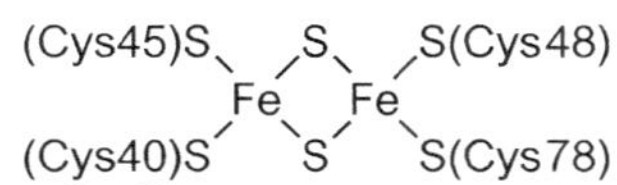 Cys86	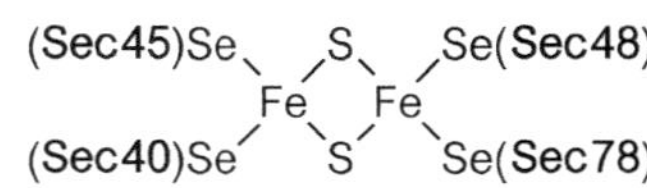 Cys86	81
Hirudin (leech)	65	Cys6–Cys14, Cys16–Cys28, Cys22–Cys39	**Sec6–Sec14**, Cys16–Cys28, Cys22–Cys39	82
			Cys6–Cys14, **Sec16–Sec28**, Cys22–Cys39	
			Cys6–Cys14, Cys16–Cys28, **Sec22–Sec39**	
			Sec6–Cys14, **Sec16**–Cys28, Cys22–Cys39	

acids by deselenization,[68,69] the selenium analogues can be utilized as synthetic intermediates for large natural proteins.[70,71]

Herein, from recent chemical syntheses of Sec-substituted proteins, diselenide (SeSe)-substituted insulin (selenoinsulin), Sec-substituted ferredoxin (selenoferredoxin), and SeSe-substituted hirudin (selenohirudin) are exemplified below.

24.3.1.1 Selenoinsulin

Insulin is a small protein (MW = 5800) consisting of two peptide chains (A- and B-chains), which are crosslinked by two SS bonds. The folded structure of insulin is stabilized by the interchain SS bonds and one SS bond residing in the A-chain. Recently, selenium analogous of insulin, which have a SeSe bond in place of one of the three SS bonds, were synthesized, and their unique features have been revealed. The first selenoinsulin was synthesized by Arai *et al.* by replacing a SeSe bond for the SS bond between Cys7A and Cys7B (see Figure 24.5A).[78] They prepared the Sec-containing A- and B-chains of bovine pancreatic insulin *via* common SPPS with an Fmoc strategy. The C7U mutants of A- and B-chains thus obtained were mixed in a buffer solution under optimal conditions to afford [C7U^A,C7U^B] bovine insulin in a yield up to 27%. This simple method of two-chain folding, called "native chain assembly", was later applied to the chemical synthesis of wild-type insulin too.[83] The selenoinsulin obtained was almost identical to the wild-type[84] in the molecular structure (see Figure 24.5C) and the cellular activity to induce phosphorylation of Akt and GSK2β through an insulin receptor on the cell membrane. However, a significant enhancement of the resistance against insulin degrading enzyme (IDE) (*ca.* 8 times in comparison with the wild-type) was found, suggesting the long-lasting nature of the selenoinsulin in a blood stream.[78] It was inferred that slight structural perturbations (or the changes of hydrogen-bonding networks) caused by replacing a SeSe bond for the SS bond near the molecular surface made the affinity change to IDE. Later, Weiss[79] synthesized the same selenoinsulin *via* a different synthetic strategy and investigated its functions. However, the IDE resistance was not so high as Arai reported. The controversy is open for future assessments.[85]

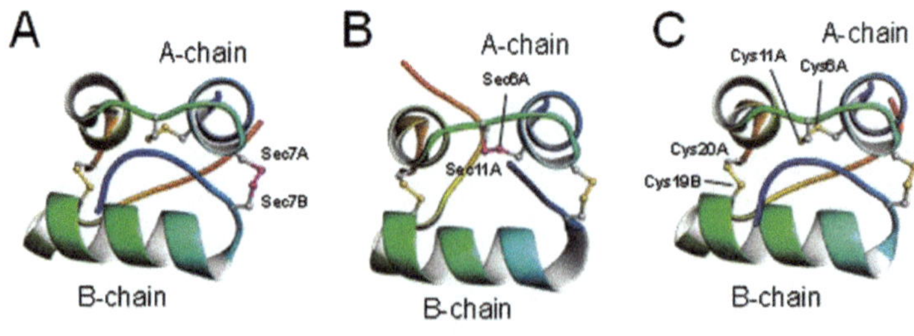

Figure 24.5 Molecular structures of selenoinsulins. A: [C7U^A,C7U^B] bovine insulin (PDB code: 5azz).[78] B: [C6U^A,C11U^A] human insulin (PDB code: 6h3m).[80] C: Wild-type bovine insulin (PDB code: 2bn3).[84]

Another example of selenoinsulin was reported by Metanis and Weiss,[80] who synthesized a selenium analogue of human insulin with a SeSe bond in place of Cys6A–Cys11A SS bond, which is located in the hydrophobic core of the native fold (see Figure 24.5B). The synthesis was achieved by reacting the native B-chain disulfonated at the Cys7B and Cys19B sidechains with the [C6U,C11U] mutant of A-chain, which was obtained from the Fmoc-SPPS method. The obtained [C6U^{A},C11U^{A}] insulin exhibited enhanced stabilities against guanidine denaturation and protease digestion due to the presence of the SeSe bond in the hydrophobic core.

24.3.1.2 Selenoferredoxin

Ferredoxins are electron-transfer proteins containing an iron–sulfur cluster, such as [2Fe–2S], [3Fe–4S], and [4Fe–4S], which is coordinated to thiolates of the cysteine residues. Takei *et al.* recently succeeded the total synthesis of ferredoxin-1, in which all four Cys residues engaged in the coordination are replaced by Sec.[81] The obtained selenoferredoxin (see Table 24.2) possessed a similar folded structure to the native protein, but the selenium analogue exhibited slightly decreased electron transfer activity.

24.3.1.3 Selenohirudin

Hirudin is an inhibitor of thrombin. Metanis recently reported the synthesis of selenohirudins, in which one of the three native SS bonds is replaced with a SeSe bond, and examined the effects of the SeSe substitution on the folded structure as well as the biological function.[82] The three selenium analogues (see Table 24.2) possessed a similar structure and the activity to the wild-type. Interestingly, the substitution accelerated the folding rate significantly. They also found that the selenium analogue of hirudin with the substitution of two Cys with Sec at the non-native position also possessed similar properties to the wild-type, suggesting that the Cys to Sec substitution approach is a useful strategy to investigate their effect on the folding, structure, and activity.

24.3.2 Applications of Selenium-containing Protein Analogues

Selenocysteine-containing peptides, so-called selenopeptides, have been utilized to model the active-site structures of selenoenzymes.[86–89] These small analogues of selenoproteins were applied as enzyme mimics as well as the mechanistic probes for selenoenzyme's functions. Two interesting points have been investigated recently using such selenopeptides.

24.3.2.1 Prevention of E–SeOH Degradation

In the normal GPx cycle (see Table 24.1), E–SeOH rapidly reacts with glutathione (GSH) to produce the metastable selenenyl sulfide intermediate (E–SeSG). If this reaction is disrupted by any reasons, E–SeOH would degrade and lose the enzymatic activity. It has long been believed that the over-oxidation of E–SeOH to the stable seleninic acid (E–SeO_2H) is an important haven to avoid the degradation of E–SeOH because E–SeO_2H can be reversibly converted back to E–SeOH by reacting with GSH (see Figure 24.6, path **a**).[90] This is in sharp contrast to the corresponding cysteine sulfinic acid (Cys–SO_2H) that cannot be reduced back to the sulfenic acid (Cys–SOH) ordinarily (see Figure 24.2). This difference in the reactivity can be an advantage of Sec being applied at the active centre of redox enzymes instead of Cys.

Orian *et al.* recently proposed a novel path for escaping the degradation of E–SeOH (see Figure 24.6, path **b**).[91] In this path, E–SeOH is dehydrated to a semi-stable selenenyl amide intermediate (E–SeN<) with a five-membered ring, which can be sent back to the GPx main cycle by reacting with GSH. The possible formation of a larger E–SeN< intermediate with an eight-membered ring was also proposed. The presence of this second escaping path has been endorsed by Goto,[89] who devised a cradle-type bulky protecting group and utilized it to reproduce the molecular environment around the active site of GPx. At the centre of the cradle, the formation of Sec–SeOH was characterized at −20 °C by NMR spectroscopy. With warming the solution up to 0 °C they observed the formation of a selenenyl amide with a five-membered ring in addition to the degradation product. The selenenyl amide could be converted to the selenenyl sulfide by the reaction with cysteine thiol. Interestingly, the Sec–SeOH could not be oxidized to Sec–SeO_2H by excess H_2O_2 in the cradle, suggesting that path **b** would be more effective than path **a** for E–SeOH to escape the degradation. A similar cyclic selenenyl amide was previously proposed as a possible precatalyst of GPx mimics.[92] On the other hand, an analogous cyclic sulfenyl amide intermediate (E–SN<) is known to be involved in the catalytic cycle of protein tyrosine phosphatase 1B (PTP1B).[93,94]

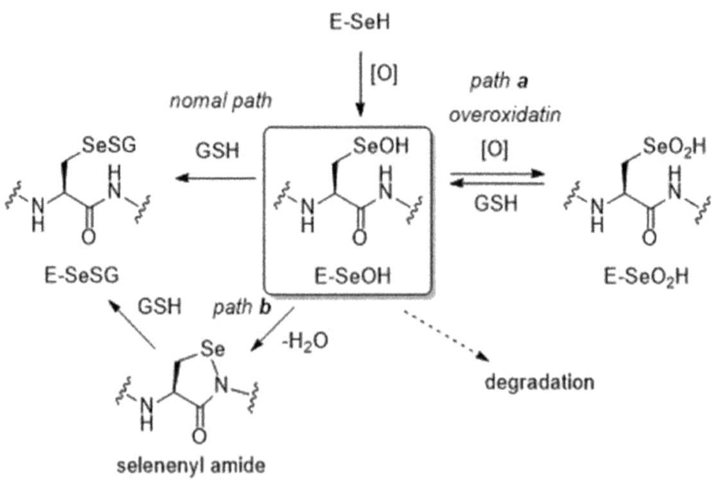

Figure 24.6 The paths for the highly reactive selenenic acid intermediate (E–SeOH) of GPx to escape the degradation.

Thus, the formation of a cyclic chalcogeneyl amide with a main-chain amide group would be a common process of a highly reactive E–ChOH (Ch = S or Se) intermediate to escape the degradation.

24.3.2.2 Activation of the Se–S Bonded State

Another interesting proposal was recently given for the activation process of the selenenyl sulfide (Se–S) state of thioredoxin reductase (TrxR(Se–S)).[95,96] This Se–S bonded state is frequently involved as a metastable intermediate in the catalytic cycles of various selenoenzymes (see Table 24.1). To enhance the total catalytic activities of those selenoenzymes, it is necessary to break the Se–S bond reductively to produce an activated selenolate intermediate ($E–Se^-$). However, this process does not proceed efficiently because the electrophilicity of Se is generally greater than that of S. Thus, a thiol nucleophile preferentially attacks the Se, not the S atom, of the Se–S bond. This results in a non-productive thiol exchange reaction of E–SeSR (attack **a**) (see Figure 24.7A). At the active site of selenoenzymes, however, a thiol nucleophile smoothly attacks the S atom (attack **b**) selectively due to the site-specific binding or the

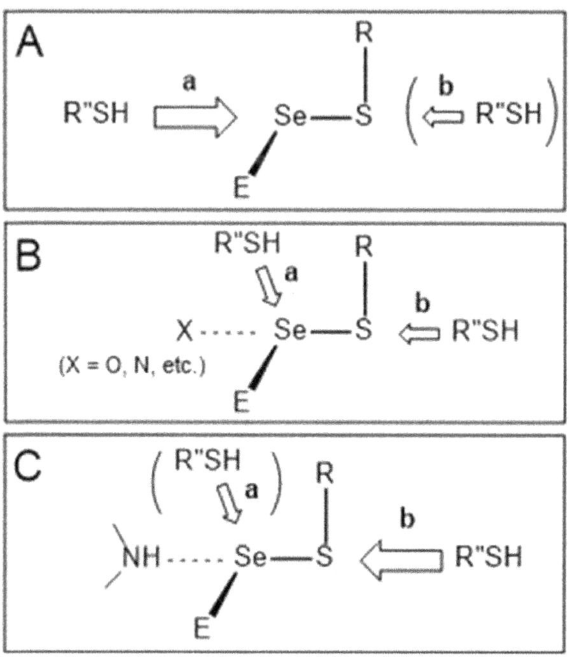

Figure 24.7 The electronic and steric control of the reactivity of a Se–S bond. (A) Inherent reactivity of a Se–S bond. Attack **a** is more favourable than attack **b**. (B) Effects of Se···X chalcogen bond. Attack **a** is more favourable electronically, while attack **b** is more favourable sterically. (C) Effects of NH···Se hydrogen bond. Attack **b** is both electronically and sterically more favourable than attack **a**.

suitable positioning of the thiol substrate. From a viewpoint of molecular design of efficient selenoenzyme mimics, it has long been discussed how to control the direction of the nucleophilic attack of a thiol not to the Se but to the S atom of the Se–S bond.[97]

Tomoda and Iwaoka previously proposed that a nonbonded Se···N interaction (a so-called Se···N chalcogen bond) sterically hinders a thiol attack at the Se atom, allowing the preferential attack at the S atom (see Figure 24.7B, attack **b**).[98] On the other hand, Mugesh[97] pointed out that the formation of such a chalcogen bond electronically increases the electrophilicity of the Se atom, leading a thiol to the undesired direction (attack **a**). In a new model, it is proposed that the formation an NH···Se hydrogen bond plays an important role to control the reactivity of the Se–S bond (see Figure 24.7C).[95,96]

At the active centre of TrxR, a histidine residue (His) is present in proximity to the active centre. The roles of this His residue were previously considered to be a base to deprotonate a thiol substrate to enhance the nucleophilicity or an acid to protonate the thiolate anion that is eliminated from the Se–S bond through a nucleophilic attack of the thiol.[36] In addition to these functions of His, the new model proposed that His forms an NH···Se hydrogen bond in the mixed Se–S intermediate, which is transiently formed after a nucleophilic attack of the first thiol to the Se–S bond, to invert the reactivity of the Se–S bond so that the second thiol preferentially attacks the intrinsically less electrophilic S atom.[96] A similar umpolung effect of the NH···Se hydrogen bond may be possible at the active site of the other selenoenzymes.

Although a species with a Se–S bond is thermodynamically stable, it can be converted to a reactive thiolate or selenolate rather easily by the reaction with a thiol or other reductants. Thus, selenenyl sulfides are interesting targets to study.[99] One recent successful application is the chemical synthesis of proteins *via* native chemical ligation (NCL) utilizing *N*-selenoethyl cystine (SetCys, **6**) at the ligation point as shown in Figure 24.8.[100]

24.3.3 Synthesis and Applications of Tellurium-containing Protein Analogues

Although no telluropeptides and telluroproteins have been found in nature, tellurium-containing protein analogues have been synthesized artificially.

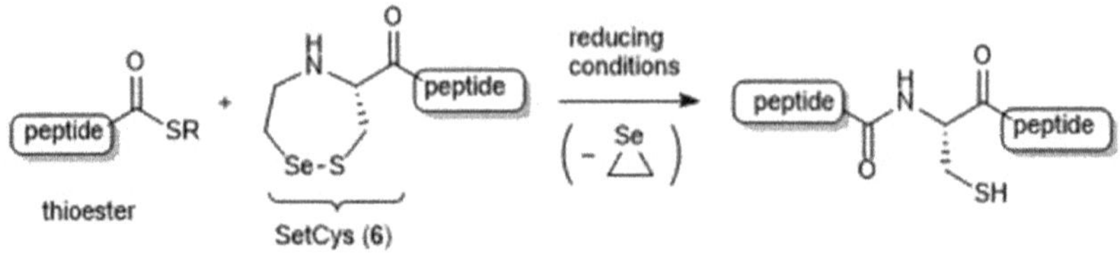

Figure 24.8 Application of the Se–S chemistry to the chemical synthesis of proteins.

These tellurium analogues exhibited interesting chemical and biological properties. However, the reports on the synthesis of telluropeptides and telluroproteins are still limited.

Liu and co-workers succeeded the synthesis of a Tec-containing protein for the first time by chemically-transforming the active-site Ser residue of subtilisin with a tellurating reagent (NaHTe).[101] The obtained tellurosubtilisin showed a very high GPx-like activity to catalyze the reaction between H_2O_2 and 3-carboxy-4-nitrobenzenethiol. The same group also reported the incorporation of Tec into glutathione transferase (GST) by applying a cysteine-auxotrophic system for the recombinant GST that has Cys instead of Ser at the active site.[102] The telluro-GST possessed a high GPx-like activity.

Singh recently investigated the reactivities of Tec under a variety of redox conditions and demonstrated that simple Tec derivatives can exhibit a significant GPx-like activity.[103]

24.4 Chalcogen Analogues of Nucleic Acids

Various types of thio-, seleno-, and telluro-nucleosides modified at the 2′-, 3′-, 4′-, and 5′-positions, the nucleobase or the phosphate bond have been synthesized to date for the purposes of the structural biology study and the pharmaceutical applications. These chalcogen analogues have already been reviewed in several notable articles.[5,104–107] In this section, recent examples are summarized focusing on their biological applications.

24.4.1 Synthesis and Applications of Sulfur-containing Nucleic Acid Analogues

The sulfur atom incorporated in nucleosides and oligonucleotides exhibits interesting effects on the structure as well as the resistance against nuclease enzymes. Thus, these sulfur analogues are potential targets of drug discovery, mechanistic probes, photosensors, *etc.* Representative man-made thionucleoside and thionucleotide fragments are shown in Figure 24.9.

4′-Thionucleic acids have long been targets of organic synthesis.[108] Oligonucleotides containing 4′-thioribonucleosides (**7**) or 2′-deoxy-4′-thionucleoside (**8**) were prepared by Bellon[109] and Hancox,[110] respectively. Since these S analogues showed a significant stability in biological tissues and made

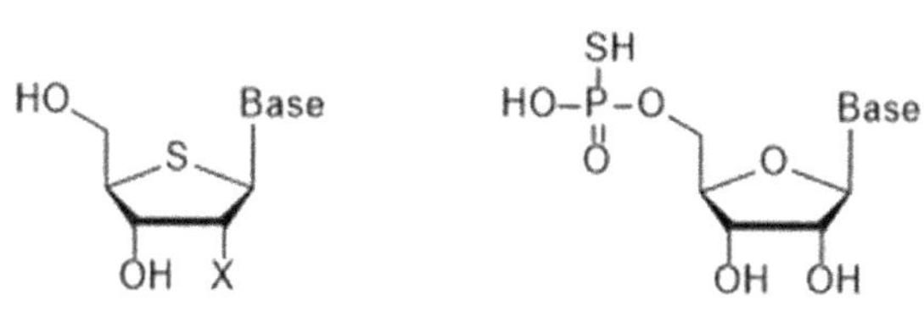

7 X = H
8 X = OH

9

Figure 24.9 Representative synthetic thionucleoside and thionucleotide fragments.

thermally stable duplexes, their applications to RNA aptamers and RNA interference were also attempted.[111] On the other hand, Minakawa succeeded with the stereoselective introduction of a nucleobase at the anomeric position of these skeletons.[112,113]

Sulfur modifications in the phosphate ester bond of RNA are applicable as nucleic acid medicines. Several oligonucleotides, which consist of ribonucleotides with a phosphorothioester linkage (**9**), have already been approved as antisense drugs, such as inotersen[114] and volanesorsen,[115] or small interfering RNA (siRNA), such as givosiran.[116] Thus, thionucleosides and thiosucleotides are promising targets of pharmaceutical applications.

24.4.2 Synthesis and Applications of Selenium-containing Nucleic Acid Analogues

Various selenosugars have been developed for applications as potential glycosidase inhibitors and antioxidants.[106,107] In the meantime, substantial efforts to synthesize selenonucleic acids have also been exerted elaborately to date. Figure 24.10 collects representative selenonucleoside and selenonucleotide skeletons, which were recently developed.

4′-Selenonucleosides (**10**) were synthesized by several research groups by applying seleno-Pummerer-type reactions.[117–120] However, the yields were moderate, and the stereo control of the α/β selectivity in the nucleobase introduction was generally difficult compared to the oxygen and sulfur analogues. Further modifications have been applied to the 4′-selenonucleoside skeletons.

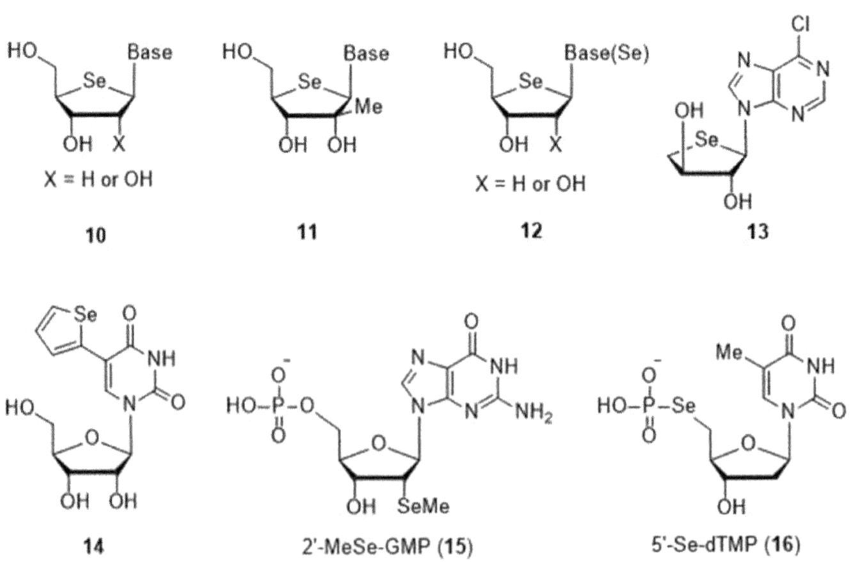

Figure 24.10 Representative synthetic selenonucleoside and selenonucleotide fragments.

Thus, 2′-methyl-4′-selenonucleosides (**11**) as anti-hepatitis C virus agents,[121] various nucleosides (**12**) with a Se-modified nucleobase[122] and 4′-selenothreonucleoside (**13**) with a novel selenothreose skeleton[123] were recently prepared.

As for the applications of selenonucleic acids, Nuthanakanti synthesized a 5-selenophene-modified uridine derivative (**14**) and incorporated it into oligo-RNA as a conformation-sensitive fluorescent probe and an anomalous X-ray diffraction label.[124] Moroder synthesized 2′-methylselenoguanosine (2′-MeSe-GMP, **15**)-containing RNAs and investigated the oxidation behaviours.[125] Conlon synthesized phsophoroselenolate-modified DNAs, which contain a 5′-selenonucleotide fragment, *i.e.*, 5′-Se-dTMP (**16**).[126] Incorporation of the 5′-selenonucleotide unit facilitated the X-ray crystallography without having an impact on the molecular conformation.

24.4.3 Synthesis and Applications of Tellurium-containing Nucleic Acid Analogues

Telluronucleosides (see Figure 24.11) were also synthesized and have already been applied to the preparation of oligonucleotides. These tellurium analogues were used for X-ray crystallography and electron-microscope imaging of nucleic acids.[5]

Braga synthesized 5′-tellurouridine derivatives (**17**) and investigated the GPx-like antioxidant activities.[127] Huang first succeeded in the synthesis of oligonucleotides containing telluronucleoside units by solid-phase synthesis.[128] In this study, 2′-tellurouridine derivatives (**18**) were utilized. Similarly, Huang also employed 2′-deoxy-5-tellurothymidine (**19**) to synthesize oligonucleotides and analyzed the structures of the obtained telluro-DNA by X-ray crystallography and scanning tunnelling microscope (STM) imaging.[129] Munchen synthesized a series of 3′-triazolyl-5′-thio-, seleno-, and telluro-thymidines and investigated the antioxidant and antiproliferative bladder carcinoma activities.[130] Among these chalcogeno-azidothymidines, the tellurium variant (**20**) exhibited significant antitumor effects towards human breast cancer cell lines.[131]

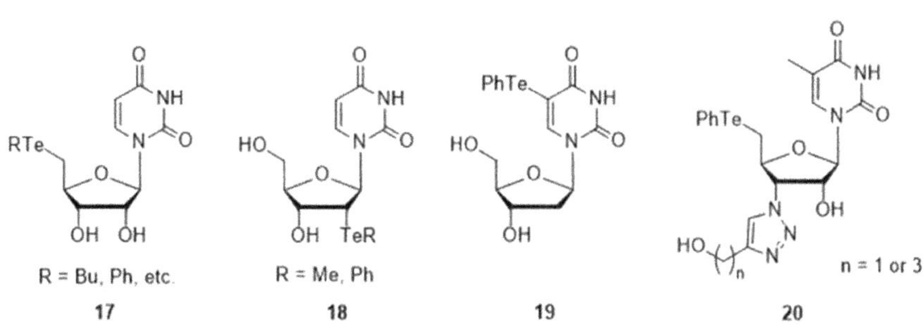

Figure 24.11 Representative synthetic telluronucleoside fragments.

24.5 Perspectives

Chalcogen atoms, particularly S and Se, are ubiquitously found in living organisms in the forms of amino acids, peptides, proteins, carbohydrates, nucleic acids, and other organic and inorganic compounds. Since these chalcogen-containing biomolecules have specific biological functions and some of them are indispensable for life, various chalcogen analogues of biomolecules have been synthesized and investigated for biological applications to date. The introduction of heavier chalcogen atoms, such as Se and Te, in place of O and S would change the structures, electronic states, folding reactions, interactions to the substrates, and biological functions from a little to a significant extent. Thus, these chalcogen analogues can be applied for imaging of the static and dynamic conformations, probing the reactions in the cellular environment, and developing new drugs. Some of the recent examples were summarized in this chapter. In the relevant research areas, alteration of biological functions of natural proteins and nucleic acids by replacing chalcogen atoms, creation of new chalcogen-modified biomolecules and development of chalcogen-containing therapeutic agents, especially Se and Te drugs, are the current research objectives to be investigated.

References

1. R. J. Huxtable and W. M. Lafranconi, *Biochemistry of Sulfur*, Plenum Press, New York, 1986.
2. K. T. Suzuki, *J. Health Sci.*, 2005, **51**, 107.
3. Y. Ogra, *Anal. Sci.*, 2009, **25**, 1189.
4. L. Moroder and H.-J. Musiol, *J. Pept. Sci.*, 2020, **26**, e3232.
5. M. Kaur, A. Rob, J. Caton-Williams and Z. Huang, in *ACS Symposium Series*, American Chemical Society, 2013, vol. 1152, p. 89.
6. N. Metanis and D. Hilvert, *Curr. Opin. Chem. Biol.*, 2014, **22**, 27.
7. N. Metanis, J. Beld and D. Hilvert, in *PATAI'S Chemistry of Functional Groups*, John Wiley & Sons, Ltd, 2011.
8. B. Guan, R. Yan, R. Li and X. Zhang, *Int. J. Nanomed.*, 2018, **13**, 7473.
9. B. Zare, M. Nami and A. R. Shahverdi, *Biol. Trace Elem. Res.*, 2017, **180**, 171.
10. W. Jin, C. Yoon, T. V. Johnston, S. Ku and G. E. Ji, *Molecules*, 2018, **23**, 2860.
11. M. Gupta and S. Gupta, *Front. Plant Sci.*, 2017, **7**, 2074.
12. N. Budisa, B. Steipe, P. Demange, C. Eckerskorn, J. Kellermann and R. Huber, *Eur. J. Biochem.*, 1995, **230**, 788.
13. Y. Ogra, T. Kitaguchi, N. Suzuki and K. T. Suzuki, *Anal. Bioanal. Chem.*, 2007, **390**, 45.
14. N. J. Day, M. J. Gaffrey and W.-J. Qian, *Antioxidants*, 2021, **10**, 499.
15. S. G. Rhee, W. Jeong, T. S. Chang and H. A. Woo, *Kidney Int.*, 2007, **72**, S3–S8.

16. A. Gennaris, B. Ezraty, C. Henry, R. Agrebi, A. Vergnes, E. Oheix, J. Bos, P. Leverrier, L. Espinosa, J. Szewczyk, D. Vertommen, O. Iranzo, J.-F. Collet and F. Barras, *Nature*, 2015, **528**, 409.
17. H.-Y. Kim and V. N. Gladyshev, *PLoS Biol.*, 2005, **3**, e375.
18. D. Butera and P. J. Hogg, *Nat. Commun.*, 2020, **11**, 5468.
19. P. J. Hogg, *Molecules*, 2020, **25**, 5729.
20. J. Duan, M. J. Gaffrey and W.-J. Qian, *Mol. BioSyst.*, 2017, **13**, 816.
21. M. Y. Lim, J. O'Brien, J. A. Paulo and S. P. Gygi, *J. Proteome Res.*, 2017, **16**, 4217.
22. J. Duan, T. Zhang, M. J. Gaffrey, K. K. Weitz, R. J. Moore, X. Li, M. Xian, B. D. Thrall and W. J. Qian, *Redox Biol.*, 2020, **36**, 101649.
23. M. Iwaoka and N. Isozumi, *Molecules*, 2012, **17**, 7266.
24. M. Iwaoka, in *Noncovalent Forces*, Springer International Publishing, Cham, 2015, pp. 265–289.
25. K. Kříž, J. Fanfrlík and M. Lepšík, *Chem. Phys. Chem.*, 2018, **19**, 2540.
26. M. Iwaoka, S. Takemoto, M. Okada and S. Tomoda, *Chem. Lett.*, 2002, 132.
27. M. Iwaoka, S. Takemoto and S. Tomoda, *J. Am. Chem. Soc.*, 2002, **124**, 10613.
28. M. Iwaoka and N. Babe, *Phosphorus, Sulfur Silicon Relat. Elem.*, 2015, **190**, 1257.
29. O. Carugo, G. Resnati and P. Metrangolo, *ACS Chem. Biol.*, 2021, **16**, 1622.
30. M. Bortoli, S. M. Ahmad, T. A. Hamlin, F. M. Bickelhaupt and L. Orian, *Phys. Chem. Chem. Phys.*, 2018, **20**, 27592.
31. J. C. Avery and P. R. Hoffmann, *Nutrients*, 2018, **10**, 1203.
32. B. Anne-Laure and C. Laurent, *Antioxid. Redox Signal.*, 2015, **23**, 775.
33. S. V. Novoselov, M. Rao, N. V. Onoshko, H. Zhi, G. V. Kryukov, Y. Xiang, D. P. Weeks, D. L. Hatfield and V. N. Gladyshev, *EMBO J.*, 2002, **21**, 3681.
34. H. J. Reich and R. J. Hondal, *ACS Chem. Biol.*, 2016, **11**, 821.
35. O. Epp, R. Ladenstein and A. Wendel, *Eur. J. Biochem.*, 1983, **133**, 51.
36. Q. Cheng, T. Sandalova, Y. Lindqvist and E. S. J. Arnér, *J. Biol. Chem.*, 2009, **284**, 3998.
37. S. Mondal, D. Manna, K. Raja and G. Mugesh, *ChemBioChem*, 2020, **21**, 911.
38. M. J. Axley, A. Böck and T. C. Stadtman, *Proc. Natl. Acad. Sci.*, 1991, **88**, 8450.
39. Y. Ogra, R. Kobayashi, K. Ishiwata and K. T. Suzuki, *J. Anal. At. Spectrom.*, 2007, **22**, 153.
40. A. Kobayashi and Y. Ogra, *J. Toxicol. Sci.*, 2009, **34**, 295.
41. H. G. Williams-Ashman, J. Seidenfeld and P. Galletti, *Biochem. Pharmacol.*, 1982, **31**, 277.
42. E. G. Mueller, *Nat. Chem. Biol.*, 2006, **2**(4), 185.
43. P. Ajitkumart and J. D. Cherayil, *Microbiol. Rev.*, 1988, **52**, 103.
44. R. Kambampati and C. T. Lauhon, *Biochemistry*, 2003, **42**, 1109.
45. G. Leszczynska, M. Cypryk, B. Gostynski, K. Sadowska, P. Herman, G. Bujacz, E. Lodyga-Chruscinska, E. Sochacka and B. Nawrot, *Int. J. Mol. Sci.*, 2020, **21**, 2882.

46. M. Sierant, G. Leszczynska, K. Sadowska, P. Komar, E. Radzikowska-Cieciura, E. Sochacka and B. Nawrot, *FEBS Lett.*, 2018, **592**, 2248.
47. T. C. Stadtman, *FASEB J.*, 1987, **1**, 375.
48. Y. Ogra, K. Ishiwata, H. Takayama, N. Aimi and K. T. Suzuki, *J. Chromatogr. B: Anal. Technol. Biomed. Life Sci.*, 2002, **767**, 301.
49. Y. Kobayashi, Y. Ogra, K. Ishiwata, H. Takayama, N. Aimi and K. T. Suzuki, *Proc. Natl. Acad. Sci. U. S. A.*, 2002, **99**, 15932.
50. F. Aureli, L. Ouerdane, K. Bierla, J. Szpunar, N. T. Prakash and F. Cubadda, *Metallomics*, 2012, **4**, 968.
51. Y. Yamashita, T. Yabu and M. Yamashita, *World J. Biol. Chem.*, 2010, **1**, 144.
52. J. Theisen, G. J. Zylstra and N. Yee, *Appl. Environ. Microbiol.*, 2013, **79**, 3171.
53. S. M. Baesman, T. D. Bullen, J. Dewald, D. Zhang, S. Curran, F. S. Islam, T. J. Beveridge and R. S. Oremland, *Appl. Environ. Microbiol.*, 2007, **73**, 2135.
54. D. H. Kim, M. G. Kim, S. Jiang, J. H. Lee and H. G. Hur, *Environ. Sci. Technol.*, 2013, **47**, 8709.
55. T. Horiike, O. Otsuka, Y. Tanaka, T. Terahara, C. Imada and M. Yamashita, *J. Gen. Appl. Microbiol.*, 2019, **65**, 246.
56. Z. P. Wu and D. Hilvert, *J. Am. Chem. Soc.*, 2002, **112**, 5647.
57. X. Ren, P. Jemth, P. G. Board, G. Luo, B. Mannervik, J. Liu, K. Zhang and J. Shen, *Chem. Biol.*, 2002, **9**, 789.
58. H. Walden, *Acta Crystallogr.*, 2010, **D66**, 352.
59. C. Z. Chung, C. Miller, D. Söll and N. Krahn, *Curr. Protoc.*, 2021, **1**, e54.
60. W. A. Hendrickson, *Q. Rev. Biophys.*, 2014, **47**, 49.
61. S. Rozovsky, in *ACS Symposium Series*, American Chemical Society, 2013, vol. 1152, pp. 127–142.
62. M. D. Gieselman, L. Xie and W. A. van Der Donk, *Org. Lett.*, 2001, **3**, 1331.
63. R. J. Hondal, B. L. Nilsson and R. T. Raines, *J. Am. Chem. Soc.*, 2001, **123**, 5140.
64. R. Quaderer, A. Sewing and D. Hilvert, *Helv. Chim. Acta*, 2001, **84**, 1197.
65. S. Flemer, *J. Pept. Sci.*, 2015, **21**, 53.
66. R. Mousa, R. Notis Dardashti and N. Metanis, *Angew. Chem., Int. Ed.*, 2017, **56**, 15818.
67. K. Arai and M. Iwaoka, *Molecules*, 2021, **26**, 195.
68. N. J. Mitchell, L. R. Malins, X. Liu, R. E. Thompson, B. Chan, L. Radom and R. J. Payne, *J. Am. Chem. Soc.*, 2015, **137**, 14011.
69. X. Wang, L. Corcilius, B. Premdjee and R. J. Payne, *J. Org. Chem.*, 2020, **85**, 1567.
70. J. Sayers, P. M. T. Karpati, N. J. Mitchell, A. M. Goldys, S. M. Kwong, N. Firth, B. Chan and R. J. Payne, *J. Am. Chem. Soc.*, 2018, **140**, 13327.
71. J. Sayers, R. J. Payne and N. Winssinger, *Chem. Sci.*, 2018, **9**, 896.
72. S. Pegoraro, S. Fiori, S. Rudolph-Böhner, T. X. Watanabe and L. Moroder, *J. Mol. Biol.*, 1998, **284**, 779.

73. C. J. Armishaw, N. L. Daly, S. T. Nevin, D. J. Adams, D. J. Craik and P. F. Alewood, *J. Biol. Chem.*, 2006, **281**, 14136.
74. N. Metanis, E. Keinan and P. E. Dawson, *J. Am. Chem. Soc.*, 2006, **128**, 16684.
75. G. Casi, G. Roelfes and D. Hilvert, *ChemBioChem*, 2008, **9**, 1623.
76. N. Metanis and D. Hilvert, *Angew. Chem., Int. Ed.*, 2012, **51**, 5585.
77. N. Metanis and D. Hilvert, *Chem. Sci.*, 2014, **6**, 322.
78. K. Arai, T. Takei, M. Okumura, S. Watanabe, Y. Amagai, Y. Asahina, L. Moroder, H. Hojo, K. Inaba and M. Iwaoka, *Angew. Chem., Int. Ed.*, 2017, **56**, 5522.
79. B. Dhayalan, Y.-S. Chen, N. B. Phillips, M. Wain, N. K. Rege, A. Mirsalehi, M. Jarosinski, F. Ismail-Beigi, N. Etanis, M. A. Weiss, D. Y.-S Chen, D. Arosinski, M. A. Weiss, N. B. Phillips, M. Swain, D. KRege, A. Mirsalehi and F. Ismail-Beigi, *Chem. - Eur. J.*, 2020, **26**, 4695.
80. O. Weil-Ktorza, N. Rege, S. Lansky, D. E. Shalev, G. Shoham, M. A. Weiss and N. Metanis, *Chem. - Eur. J.*, 2019, **25**, 8513.
81. T. Takei, T. Ando, T. Takao, Y. Ohnishi, G. Kurisu, M. Iwaoka and H. Hojo, *Chem. Commun.*, 2020, **56**, 14239.
82. R. Mousa, T. Hidmi, S. Pomyalov, S. Lansky, L. Khouri, D. E. Shalev, G. Shoham and N. Metanis, *Commun. Chem.*, 2021, **4**, 30.
83. K. Arai, T. Takei, R. Shinozaki, M. Noguchi, S. Fujisawa, H. Katayama, L. Moroder, S. Ando, M. Okumura, K. Inaba, H. Hojo and M. Iwaoka, *Commun. Chem.*, 2018, **1**, 26.
84. M. H. Nanao, G. M. Sheldrick and R. B. G. Ravelli, *Acta Crystallogr.*, 2005, **D61**, 1227.
85. J. A. Karas, J. D. Wade and M. Akhter Hossain, *Chem. Rev.*, 2021, **121**, 4531.
86. A. P. Lothrop, G. W. Snider, E. L. Ruggles, A. S. Patel, W. J. Lees and R. J. Hondal, *Biochemistry*, 2014, **53**, 654.
87. T. Takei, Y. Urabe, Y. Asahina, H. Hojo, T. Nomura, K. Dedachi, K. Arai and M. Iwaoka, *J. Phys. Chem. B*, 2014, **118**, 492.
88. S. Shimodaira, T. Takei, H. Hojo and M. Iwaoka, *Chem. Commun.*, 2018, **54**, 11737.
89. R. Masuda, R. Kimura, T. Karasaki, S. Sase and K. Goto, *J. Am. Chem. Soc.*, 2021, **143**, 6345.
90. H. Ungati, V. Govindaraj, M. Narayanan and G. Mugesh, *Angew. Chem., Int. Ed.*, 2019, **58**, 8156.
91. L. Orian, P. Mauri, A. Roveri, S. Toppo, L. Benazzi, V. Bosello-Travain, A. De Palma, M. Maiorino, G. Miotto, M. Zaccarin, A. Polimeno, L. Flohé and F. Ursini, *Free Radical Biol. Med.*, 2015, **87**, 1.
92. M. Iwoaka, in *ACS Symposium Series*, American Chemical Society, 2013, vol. 1152, pp. 163–177.
93. A. Salmeen, J. N. Andersen, M. P. Myers, T. C. Meng, J. A. Hinks, N. K. Tonks and D. Barford, *Nature*, 2003, **423**, 769.
94. R. L. M. Van Montfort, M. Congreve, D. Tisi, R. Carr and H. Jhoti, *Nature*, 2003, **423**, 773.

95. S. Shimodaira and M. Iwaoka, *Phosphorus, Sulfur Silicon Relat. Elem.*, 2019, **194**, 750.
96. K. Arai, T. Matsunaga, H. Ueno, N. Akahoshi, Y. Sato, G. Chakrabarty, G. Mugesh and M. Iwaoka, *Chem. – Eur. J.*, 2019, **25**, 12751.
97. B. Kanta Sarma and G. Mugesh, *J. Am. Chem. Soc.*, 2005, **127**, 11477.
98. M. Iwaoka and S. Tomoda, *J. Am. Chem. Soc.*, 1994, **116**, 2557.
99. A. Hamsath and M. Xian, *Antioxid. Redox Signaling*, 2020, **33**, 1143.
100. V. Diemer, N. Ollivier, B. Leclercq, H. Drobecq, J. Vicogne, V. Agouridas and O. Melnyk, *Nat. Commun.*, 2020, **11**, 2558.
101. S. Mao, Z. Dong, J. Liu, X. Li, X. Liu, G. Luo and J. Shen, *J. Am. Chem. Soc.*, 2005, **127**, 11588.
102. X. Liu, L. A. Silks, C. Liu, M. Ollivault-Shiflett, X. Huang, J. Li, G. Luo, Y.-M. Hou, J. Liu and J. Shen, *Angew. Chem., Int. Ed.*, 2009, **48**, 2020.
103. K. Satheeshkumar, S. Raju, H. B. Singh and R. J. Butcher, *Chem. – Eur. J.*, 2018, **24**, 17513.
104. P. Gunaga, H. Moon, W. Choi, D. Shin, J. Park and L. Jeong, *Curr. Med. Chem.*, 2004, **11**, 2585.
105. P. K. Sahu, S. D. Naik, J. Yu and L. S. Jeong, *Eur. J. Org. Chem.*, 2015, **2015**, 6115.
106. C. Kona, J. Shimabukuro, M. Kiso and H. Ando, in *Organoselenium Compounds in Biology and Medicine*, Royal Society of Chemistry, 2018, p. 223.
107. F. Mangiavacchi, I. F. C. Dias, I. Di Lorenzo, P. Grzes, M. Palomba, O. Rosati, L. Bagnoli, F. Marini, C. Santi, E. J. Lenardao and L. Sancineto, *Pharmaceuticals*, 2020, **13**, 1.
108. N. Minakawa, S. Hoshika, N. Inoue, Y. Kato and A. Matsuda, *Front. Org. Chem.*, 2005, **1**, 79.
109. L. Bellon, F. Morvan, J. L. Barascut and J. L. Imbach, *Biochem. Biophys. Res. Commun.*, 1992, **184**, 797.
110. E. L. Hancox, B. A. Connolly and R. T. Walker, *Nucleic Acids Res.*, 1993, **21**, 3485.
111. S. Hoshika, N. Minakawa and A. Matsuda, *Nucleic Acids Res.*, 2004, **32**, 3815.
112. T. Naka, N. Minakawa, H. Abe, D. Kaga and A. Matsuda, *J. Am. Chem. Soc.*, 2000, **122**, 7233.
113. N. Minakawa, Y. Kato, K. Uetake, D. Kaga and A. Matsuda, *Tetrahedron*, 2003, **59**, 1699.
114. M. D. Benson, M. Waddington-Cruz, J. L. Berk, M. Polydefkis, P. J. Dyck, A. K. Wang, V. Planté-Bordeneuve, F. A. Barroso, G. Merlini, L. Obici, M. Scheinberg, T. H. Brannagan, W. J. Litchy, C. Whelan, B. M. Drachman, D. Adams, S. B. Heitner, I. Conceição, H. H. Schmidt, G. Vita, J. M. Campistol, J. Gamez, P. D. Gorevic, E. Gane, A. M. Shah, S. D. Solomon, B. P. Monia, S. G. Hughes, T. J. Kwoh, B. W. McEvoy, S. W. Jung, B. F. Baker, E. J. Ackermann, M. A. Gertz and T. Coelho, *N. Engl. J. Med.*, 2018, **379**, 22.

115. J. L. Witztum, D. Gaudet, S. D. Freedman, V. J. Alexander, A. Digenio, K. R. Williams, Q. Yang, S. G. Hughes, R. S. Geary, M. Arca, E. S. G. Stroes, J. Bergeron, H. Soran, F. Civeira, L. Hemphill, S. Tsimikas, D. J. Blom, L. O'Dea and E. Bruckert, *N. Engl. J. Med.*, 2019, **381**, 531.
116. Y. Y. Syed, *Drugs*, 2021, **81**, 841.
117. Y. Inagaki, N. Minakawa and A. Matsuda, *Nucleic Acids Symp. Ser.*, 2007, **51**, 139.
118. J. K. Watts, B. D. Johnston, K. Jayakanthan, A. S. Wahba, B. M. Pinto and M. J. Damha, *J. Am. Chem. Soc.*, 2008, **130**, 8578.
119. L. S. Jeong, D. K. Tosh, O. K. Hea, T. Wang, X. Hou, S. Y. Ho, Y. Kwon, K. L. Sang, J. Choi and X. Z. Long, *Org. Lett.*, 2008, **10**, 209.
120. J. Yu, J.-H. Kim, H. W. Lee, V. Alexander, H.-C. Ahn, W. J. Choi, J. Choi and L. S. Jeong, *Chem. – Eur. J.*, 2013, **19**, 5528.
121. H. Lee, D. B. Jarhad, J. Yu, C. Lee and L. S. Jeong, *J. Org. Chem.*, 2019, **84**, 14414.
122. X. Zhou, W. Yan, C. Zhang, Z. Yang, P. Neubauer, I. A. Mikhailopulo and Z. Huang, *Catal. Commun.*, 2019, **121**, 32.
123. M. Iwaoka, Y. Hiyoshi, S. Arai and T. Ito, *ACS Omega*, 2021, **6**, 17621.
124. A. Nuthanakanti, M. A. Boerneke, T. Hermann and S. G. Srivatsan, *Angew. Chem., Int. Ed.*, 2017, **56**, 2640.
125. H. Moroder, C. Kreutz, K. Lang, A. Serganov and R. Micura, *J. Am. Chem. Soc.*, 2006, **128**, 9909.
126. P. F. Conlon, O. Eguaogie, J. J. Wilson, J. S. T. Sweet, J. Steinhoegl, K. Englert, O. G. A. Hancox, C. J. Law, S. A. Allman, J. H. R. Tucker, J. P. Hall and J. S. Vyle, *Chem. Sci.*, 2019, **10**, 10948.
127. A. L. Braga, W. A. Severo Filho, R. S. Schwab, O. E. D. Rodrigues, L. Dornelles, H. C. Braga and D. S. Lüdtke, *Tetrahedron Lett.*, 2009, **50**, 3005.
128. J. Sheng, A. E. A. Hassan and Z. Huang, *Chem. – Eur. J.*, 2009, **15**, 10210.
129. J. Sheng, A. E. A. Hassan, W. Zhang, J. Zhou, B. Xu, A. S. Soares and Z. Huang, *Nucleic Acids Res.*, 2011, **39**, 3962.
130. T. S. Munchen, M. S. Sonego, D. de Souza, L. Dornelles, F. K. Seixas, T. Collares, B. C. Piccoli, F. D. Á. da Silva, J. B. T. da Rocha, N. Quoos and O. E. D. Rodrigues, *ChemistrySelect*, 2018, **3**, 3479.
131. A. M. O. Rocha, F. S. S. Sousa, V. M. Borba, T. S. Munchen, J. G. Leal, O. E. D. Rodrigues, M. G. Fronza, L. Savegnago, T. Collares and F. K. Seixas, *Bioorg. Med. Chem. Lett.*, 2020, **30**, 127365.

CHAPTER 25

Small Organoselenium Catalysts as a Potential Manipulator for Redox Homeostasis and Proteostasis

KENTA ARAI

Department of Chemistry, School of Science, Tokai University, Kitakaname, Hiratsuka-shi, Kanagawa 259-1292, Japan
Email: k-arai4470@tokai-u.jp

25.1 Introduction

Proteins are biopolymers consisting of 20 amino acids. The amino acids are sequentially elongated in the ribosome *via* translation and transcription from genetic information to synthesize polypeptide chains. In many cases, the synthesized polypeptides undergo oxidative folding in the endoplasmic reticulum (ER), during which cross-linking of the disulfide (SS) linkage(s) between cysteine (Cys; C) residues in the peptides is coupled with the formation of a unique three-dimensional structure (*i.e.*, conformational folding) to exert biological functions[1] (path a in Figure 25.1). It is well known that polypeptides can fold spontaneously to some extent under oxidative conditions *in vitro* according to the primary sequence information.[2] *In vivo*, various enzymes cooperatively promote oxidative folding and regulate unfolding and subsequent degradation to effectively maintain proteostasis.[3] The most representative ER-resident enzymes are protein disulfide isomerase (PDI) and more than 20 types of family proteins (PDIs).[4] Although the detailed

Chalcogen Chemistry: Fundamentals and Applications
Edited by Vito Lippolis, Claudio Santi, Eder J. Lenardão and Antonio L. Braga

Published by the Royal Society of Chemistry, www.rsc.org

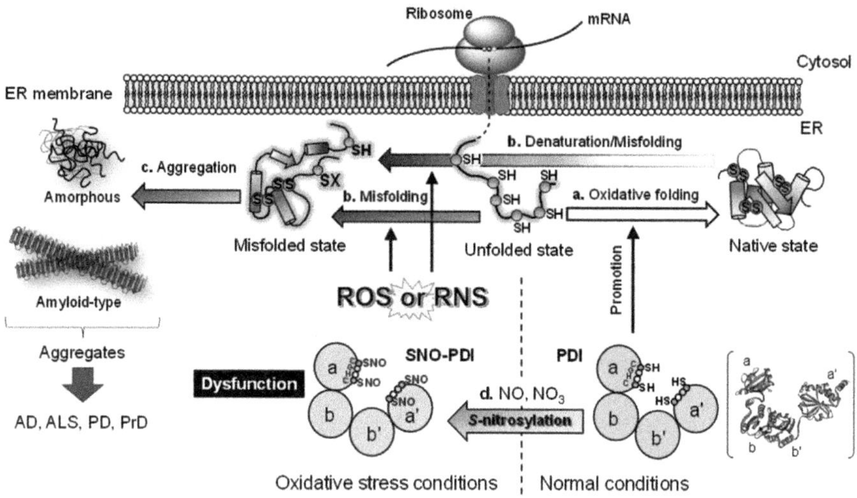

Figure 25.1 Protein misfolding induced by redox imbalance in the ER.

functions and molecular mechanisms of many of them are still unknown, they all consist of thioredoxin (Trx)-like domains with a redox quartet sequence called the CXXC motif as the active center, which regulates folding and unfolding while promoting SS-related reactions (*i.e.*, SS-formation, SS-reduction, SS-isomerization, see Figure 25.10 in Section 25.4.1) in the substrate proteins.

Under oxidative stress conditions, reactive oxygen species (ROS) and reactive nitrogen species (RNS) oxidize proteins in the ER (path b in Figure 25.1), inducing misfolding, oligomerization, and aggregation of proteins (path c in Figure 25.1). This eventually leads to undesired cell death and fatal diseases such as neurodegenerative diseases, diabetes, and atherothrombotic brain infarction. Furthermore, it has been reported that lipid peroxidation (LPO) of the plasma membrane under oxidative stress conditions is closely related to the accumulation of misfolded proteins in the ER and the induction of ER stress, and that lipid peroxidation products and protein aggregates are increased in neuronal cells in neurodegenerative diseases.[5]

Additionally, it has been reported that cysteinyl thiol (SH) groups in two CGHC motifs of PDI, which are the redox active centers, are readily modified to nitrosothiol (SNO) under a nitrosative stress (path d in Figure 25.1), resulting in a significant decrease in the ability of PDI as a catalyst for SS-related reactions.[6] Moreover, PDI has a U-shaped structure composed of four tandem Trx-like domains (a–b–b′–a′), and its inner surface is a hydrophobic environment.[7] Therefore, PDI can capture structurally immature proteins, which expose hydrophobic regions, into the hydrophobic cavity and thus exhibit chaperone-like activity by inhibiting protein–protein intermolecular associations that lead to aggregation. Because the opening and closing of the U-shaped structure is coupled with the redox of the CGHC, *S*-nitrosylation in CGHC results in a decrease in not only the catalytic activity for oxidative folding but also the capturing capability toward substrate proteins.[6,8] Consequently, the ER is equipped with various antioxidative and antinitrosative factors to remove toxic ROS and RNS generated during the metabolic process of molecular oxygen (O_2).[9]

Glutathione (GSH), a monothiol compound, is one of the most ubiquitous and abundant small molecule antioxidants in cells, which can reduce ROS and function as an effective reducing agent for converting *S*-nitrosylated cysteine (SNO-Cys) in PDIs and other ER-resident proteins to the original Cys.[10] Additionally, hydrogen peroxide (H_2O_2), which is a representative ROS produced as a metabolite of O_2 during protein folding in the ER, is reduced to water by the functions of antioxidative enzymes, such as peroxiredoxin (Prx) 4 and glutathione peroxidase (GPx) 7/8 with Cys residues in the active center.[9] The dysfunctions of these biomolecules are accompanied by a redox imbalance in the ER, leading to excessive oxidative and nitrosative stress in the ER and various diseases due to protein misfolding and aggregation. In this context, the following chemical strategies may be applied to maintain proteostasis in the ER.

(a) Maintenance of normal redox balance in the ER by alleviating oxidative stress.
(b) Conversion of SNO-Cys residues into Cys to avoid undesired protein misfolding.
(c) Direct control of correct protein folding using an alternative molecule to PDI.

In this chapter, we review compounds that were recently reported as antioxidant catalysts and protein folding enhancers, focusing on small molecules with molecular weights below 1 kDa that have Se atom(s) as a redox center, and discuss their potential pharmacological applications in proteostasis.

25.2 Aspects as Peroxidase Mimics

25.2.1 Seleno-antioxidative Enzymes

Selenium, an essential micronutrient for mammals, is incorporated as selenocysteine (Sec; U), which is a selenium analog of Cys and known as the 21st amino acid; it functions as a redox center in several important oxidoreductases. The selenol (SeH) group mainly exists as a deprotonated selenolate (Se^-) species in biological environments owing to its higher acidity than that of the SH group. Thus, the high nucleophilicity of Se^- enhances the catalytic activity of selenoenzymes, enabling the effective reduction of substrates. Among the glutathione peroxidase (GPx) family, GPx1 and GPx4 are representative cytosolic seleno-antioxidative enzymes, and

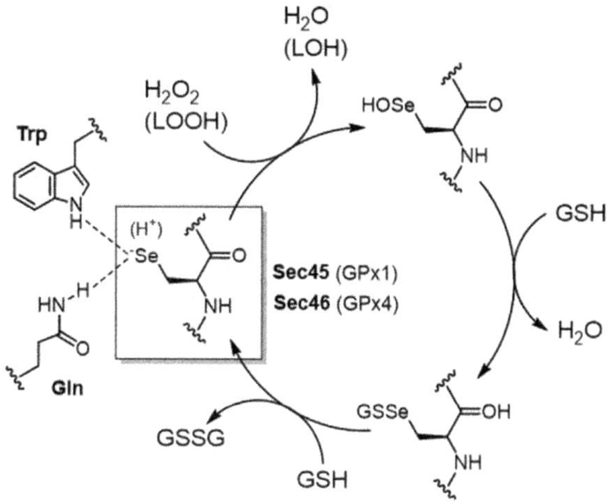

Figure 25.2 Catalytic mechanisms of GPx1 and GPx4 for the reduction of hydroperoxides (ROOH). It should be noted that this cycle represents a typical major one and other complecting bypaths are presented.

their functions and molecular mechanisms have been the focus of the development of artificial antioxidative catalysts.[11] In the case of GPx1, Sec45 as the active center reduces H_2O_2 to H_2O, whereas the SeH group in Sec is oxidized to selenenic acid (SeOH) (Figure 25.2). The SeOH then reacts with the coexisting GSH and is converted to a mixed selenenyl sulfide (Se–S) intermediate, which further reacts with GSH to produce oxidized GSH (GSSG) and regenerate the reactive SeH group. GPx4, a phospholipid peroxidase, catalyzes the reduction of lipid peroxides (LOOH) as a substrate to the corresponding alcohols (LOH) *via* a similar catalytic cycle. In both GPx1 and GPx4, the presence of a catalytic triad structure consisting of Sec, glutamine (Gln), and tryptophan (Trp) has been reported. These three amino acid residues interact with each other in the intermediates produced during the catalytic reaction to exert a remarkable catalytic activity.[12]

Although these selenoenzymes are not found in the ER, they have similar peroxide-reducing activities to those of ER-resident peroxidases, such as GPx7/8 and Prx4, which are not selenoenzymes. Therefore, selenium-containing compounds, which can imitate selenoenzyme functions, may be promising drug candidates for modulating the redox balance in the ER.

25.2.2 Ebselen, Diphenyl Diselenide, and their Derivatives as a GPx1 Model

In the development of GPx-like catalysts, based on the knowledge that the excellent antioxidant activity of GPx1 is generated by the high reactivity of the SeH group and weak interaction network in the triad structure, small molecules with highly polar functional groups spatially arranged near the Se atom(s) have been designed and synthesized.[12]

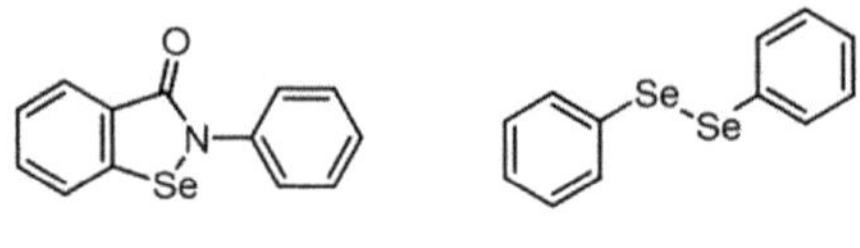

Ebselen (**1**) Diphenyl diselenide (**2**)

Figure 25.3 Ebselen and diphenyl diselenide as typical aromatic antioxidants.

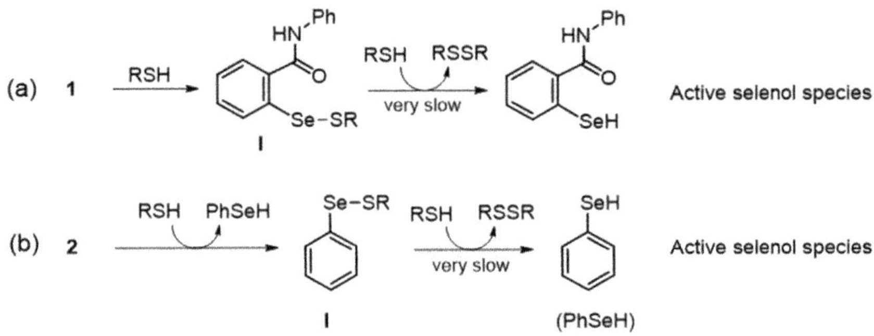

Figure 25.4 Generation of active selenol species against H_2O_2 *via* an intermolecular bond exchange reaction with thiol co-substrate. (a) Activation of ebselen *via* Se–N cleavage. (b) Activation of diphenyl diselenide *via* Se–Se cleavage.

Ebselen (**1**, Figure 25.3), with a five-membered isoselenazoline ring fused to a benzene skeleton facilitated this trend in the research field. The catalytic cycle of ebselen for the reduction of peroxides has been found to consist of several different pathways; detailed descriptions can be found in recent reviews[13,14] and in another chapter of this book. Depending on the reaction conditions, the catalytic cycle initiated by the cleavage of the Se–N bonds is one of the major pathways. Ebselen can be converted into an Se–S intermediate (**I**) with an *N*-benzamide moiety by reaction with the coexistent thiol (RSH), and the generated **I** subsequently reacts with another RSH to cleave the Se–S bond and produce the disulfide (RSSR) and active SeH species (Figure 25.4a). The resulting SeH species reduces H_2O_2 to H_2O with its own oxidation to the SeOH form, which can undergo intramolecular dehydration condensation to regenerate **1**. Similarly, diphenyl diselenide (**2**, Figure 25.3) and its derivatives, which have been reported as a second-generation aromatic GPx mimic, can also catalyze the reduction of H_2O_2 to H_2O through a catalytic cycle similar to that of **1**. The formation of an active SeH form *via* the Se–S intermediate is one of the key reaction steps (Figure 25.4b), although a series of **2** also has multiple bypasses in the GPx-like catalytic cycle.

To liberate the active selenol from the Se–S intermediate, RSH must preferentially attack the S atom of the Se–S bond (Figure 25.4). However, the liberation of the SH group by the nucleophilic attack of RSH on the Se atom is thermodynamically and kinetically more favored.[15] Consequently, the molecular design, which allows the preferential attack of RSH on the S atom in the Se–S intermediate, has been one of the main topics in the development of aromatic GPx mimics. It has been reported that the derivatives of **1** and **2**, which have substituents such as methyl, hydroxy, and amino groups, peptides, and halogens at the *ortho*- or *para*-position(s) with respect to the Se atom, efficiently generate SeH groups in the catalytic cycle and exhibit remarkable GPx-like activity *in vitro* because of their steric and/or electronic substituent effects or bonding/non-bonding interactions with the Se atom.[12,14]

Simultaneously, biologists have examined the antioxidant effects of organoselenium compounds in cells and animals. Unfortunately, there are presently no concrete prospects for the clinical application of antioxidant therapy.[16] Nonetheless, with experimental evidence, it has been reported that ebselen (**1**) exerts a protective effect on neuronal cells by exerting direct antioxidant activity and/or promoting the expression of antioxidant enzymes.[17,18] Furthermore, it has recently been reported that **2** has similar effects.[19,20] Additionally, ebselen has been reported to suppress the production of amyloid β and hyperphosphorylation of the tau protein in brain cells, thereby improving or preserving cognitive function in Alzheimer's disease.[21,22] Recently, we found that ebselen intrinsically possesses a catalytic function that promotes oxidative protein folding (see Section 25.4). Further evaluation of aromatic GPx mimics, in which PDI-like activity is also included, will be an important future task in the field of drug discovery against misfolding diseases.

25.2.3 Aliphatic Diselenide-based GPx Models

Although not as abundant as aromatic selenium compounds, aliphatic selenium compounds have also been reported as potent antioxidants (Figure 25.5). The aliphatic diselenides, such as 3,3′-diselenodipropionic acid (**3**),[23] selenocystamine (**4**),[24] and selenocystine (**5**),[25] exhibit good GPx-like activity *via* a similar catalytic cycle to that of aromatic diselenides, which involves the Se–Se bond cleavage as a rate-limiting step. Particularly, compound **3**, which has low toxicity toward cells, has been found to possess not

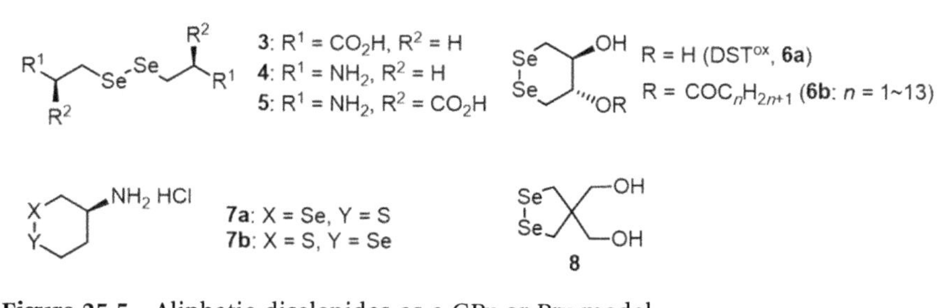

Figure 25.5 Aliphatic diselenides as a GPx or Prx model.

only GPx-like functions but also radical scavenger and radio protection effects, based on both *in vivo* and *in vitro* studies.[26]

Recently, this author's group also reported aliphatic cyclic diselenides (series **6**) as a new class of peroxidase mimic.[27] The Se–Se bond is reductively activated by the coexisting dithiol co-substrate to the active diselenol ([SeH, SeH]) form, which is subsequently oxidized back to the original [Se–Se] form by reducing H_2O_2 to H_2O (Cycle A in Figure 25.6). On the other hand, during LPO induced by 2,2′-azobis(2-amidinopropane)dihydrochloride (AAPH) using lecithin/cholesterol liposomes as a model cell membrane, compound **6** can directly reduce LOOH to LOH and inhibit the lipid radical chain reaction (Cycle B in Figure 25.6). Thus, **6** can exhibit GPx1- and GPx4-like activities through switchable catalytic cycles depending on the substrate. As a general trend, highly hydrophilic compounds such as **6a** and monoester **6b** with short alkyl chains ($n \leq 2$) exhibit high GPx1-like activity *via* Cycle A in an aqueous medium. In contrast, hydrophobic compound **6b** with long alkyl chains ($n \geq 5$) showed high GPx4-like activity *via* Cycle B, possibly due to their high membrane affinity. Remarkably, the 50% inhibitory activity ($IC_{50} = 10.7 \pm 2.9$) of **6b** with lauroyl monoester ($n = 11$) against LPO was better than that ($IC_{50} = 29.3 \pm 5.9$) of edaravone, which has been clinically applied as an anti-LPO agent. Furthermore, a series of monoesters with long alkyl chains ($n = 7$, 9, 11, and 13), which are less toxic to cultured cells, effectively inhibited AAPH-induced LPO. Recently, we also found that water-soluble cyclic selenenyl sulfides (series 7), which were prepared as a model compound of the active center of thioredoxin reductase, a cytosolic selenoenzyme, can exert a good hydroperoxidase-like function *via* similar catalytic mechanisms to cycle A for series **6** both in test tubes and cells.[28] Recently, Mugesh *et al.* reported that a five-membered cyclic diselenide **8**

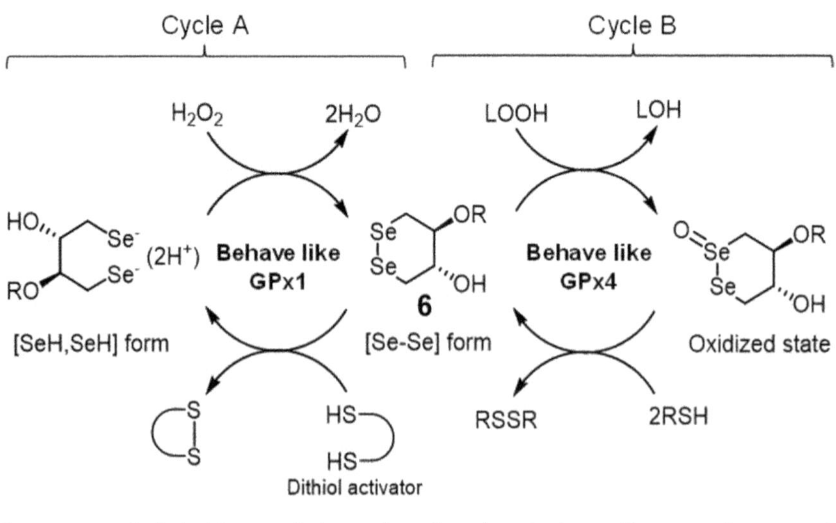

Figure 25.6 Switchable catalytic cycle of series **6** depending on the peroxide substrates.

also exhibited hydroperoxidase-like activity in a similar manner to **6**. Interestingly, when Prx2 and catalase, which are major hydroperoxidases in erythrocyte cells, were chemically inhibited, the compound administered in the cells was found to function as an alternative catalyst for those enzymes and to effectively suppress oxidative stress-induced erythrocytosis.[29]

25.2.4 Selenide-based GPx Models

Several aromatic and aliphatic selenides (R^1SeR^2) have frequently been reported as GPx mimics (Figure 25.7). The catalytic cycle of selenides for the reduction of peroxides is quite different from that of diselenide compounds. A selenide can directly reduce the substrate peroxides (H_2O_2 or LOOH), and the generated selenoxide ($R^1Se(=O)R^2$) is subsequently reduced back to the original selenide with the oxidation of RSH as a co-substrate to RSSR (Cycle A in Figure 25.8).

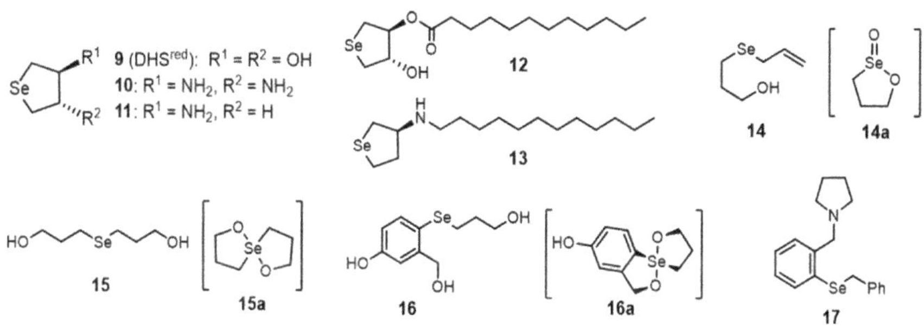

Figure 25.7 Selenide compounds as a GPx model.

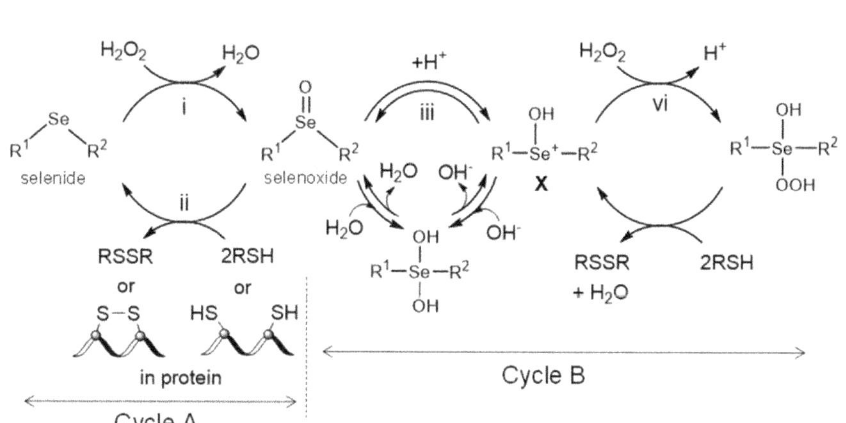

Figure 25.8 GPx-like catalytic cycles of selenides. Cycle A: Simple redox cycle involved a selenide and a selenoxide mainly observed in aqueous medium. Cycle B: Another catalytic cycle involved over oxidized intermediate, hydroxy perhydroxy selenane, as a reactive species toward thiols.

Iwaoka *et al.* previously reported that five-membered cyclic selenides (**9–11**) with various highly polar functional groups reasonably exhibited GPx-like catalytic activity in an aqueous medium *via* Cycle A, as shown in Figure 25.8.[30,31] Because the reduction process of peroxides by selenide (path i) is the rate-limiting step in Cycle A, the magnitude of the GPx-like catalytic activity of the selenides in water strongly correlates with the reducing capability of the selenide, that is, the energy level of the HOMO surrounding the Se atom. Additionally, dihydroxy selenolane (DHS^{red}: **9**) has been reported to have a radical-scavenging ability to protect cells against radiation-induced oxidative stress.[32] Amphiphilic derivatives of **9** such as monoester **12**, as well as diselenide **6b** (see previous section), can infiltrate liposomes and effectively reduce LOOH to LOH *via* path i in Cycle A, inhibiting radical-induced LPO.[33] Furthermore, amphiphilic selenides **12** and **13** could also modulate the expression of antioxidant selenoenzymes, such as the GPx family in cultured cells to protect against oxidative stress induced by free radicals or radiation.[34,35] Notably, the oxidized (selenoxide) form of **13** can preferentially oxidize SH groups in proteins to SS bonds *via* path ii in Cycle A without the oxidation of coexisting small thiols such as GSH or DL-dithiothreitol (DTT^{red}), owing to the interaction of the long-chain alkyl groups with solvent-exposed hydrophobic regions of proteins.[36] Thus, selenide **13** can recognize proteins as substrates and behave like a hybrid mimic of both the PDI and GPx family, possibly being a promising drug candidate for both the regulation of the redox balance and protein folding in the ER.

Back *et al.* reported several selenides such as hydroxyalkyl allyl selenide (**14**),[37] di(3-hydroxypropyl)selenide (**15**),[38] and phenolic derivative of **15** (**16**)[39] as GPx mimics. Although these selenides can also be oxidized to the corresponding selenoxide form by peroxides, the selenoxide forms are subsequently converted into cyclic seleninate ester (**14a**) or spirodioxyselenuranes (**15a** and **16a**) and exhibit strong GPx-like catalytic activities *via* different catalytic cycles from Cycle A in Figure 25.8. On the other hand, Braga *et al.* reported that sub-cycle (Cycle B) accompanies the general catalytic cycle involving selenide and selenoxide (Cycle A).[40] In methanol, aromatic selenoxide derived from **17** can be further oxidized to the corresponding hydroxy perhydroxy selenane (paths iii and vi in Figure 25.8), which rapidly oxidizes RSH to RSSR with its own reduction back to the selenoxide form *via* cation X. Cyclic selenides (**10** and **11**) were also found to exhibit high antioxidant activity in less polar solvents than water, such as methanol, through Cycle B as the major catalytic cycle.[41]

25.3 Aspects as S-denitrosylase Mimics

Nitric oxide radical (NO) and dinitrogen trioxide (N_2O_3) modify cysteinyl SH groups in the PDI active center and in its substrate proteins, causing enzyme dysfunction, protein denaturation, and protein misfolding, which in turn lead to fatal oxidative stress, neurodegeneration, *etc.*[42] Therefore, the development of artificial catalysts that promote denitrosylation of the cysteinyl SNO group in proteins is an important task for biological chemists.

GPx1 promotes *in vitro* S-denitrosylation of *S*-nitrosothiol (RSNO), such as nitrosoglutathione (GSNO [Figure 25.9]) (*i.e.*, reductive regeneration of SH), as well as reduction of H_2O_2.[43] Therefore, the compounds developed as models for GPx1 should intrinsically possess S-denitrosylase-like activity. Indeed, diselenide compounds, such as **4** and **5**, and their polymer- and metal-supported materials, are useful as NO-release catalyst from RSNO and are promising for application in antithrombotic therapy utilizing antiplatelet adhesion properties of NO.[44] However, the inhibitory activity of diselenide compounds against protein misfolding induced by S-nitrosylation has not been evaluated even *in vitro*. Recently, we developed compound **19**, which is a Pro–His dipeptide-conjugate of 1,2-diselenane-4-amine (**18**) as a parent compound and catalyzes the S-denitrosylation of GSNO and *S*-nitrosoproteins (SNO-proteins).[45] Similar to the GPx1-like catalytic cycle (Cycle A in Figure 25.6) of compound **6**, compounds **18** and **19** are activated by the coexisting thiol (RSH) to the [SeH,SeH] form, which subsequently reduces SNO-Cys residues in GSNO and SNO-protein to original Cys residues with the simultaneous generation of nitrosyl (HNO) (Figure 25.9). Compound **19** undergoes a conformational change in the peptide backbone upon reductive ring-opening in the Se–Se bond, forming a γ-turn, which seldom exists in natural short peptides, and presumably an NH···Se hydrogen bond between the imidazole side chain of His and the Se atom, as shown in Figure 25.9. Therefore, the [SeH,SeH] form is thermodynamically stabilized and its productivity in the reaction solution is increased, resulting in **19** exhibiting significantly better S-denitrosylase-like activity than **18**. In addition, compound **19** can inhibit the protein misfolding caused by S-nitrosylation and promoted correct oxidative folding. These results may provide a useful strategy for the development of artificial molecules to prevent the dysregulation of the redox balance and proteostasis in the ER under nitrosative stress conditions.

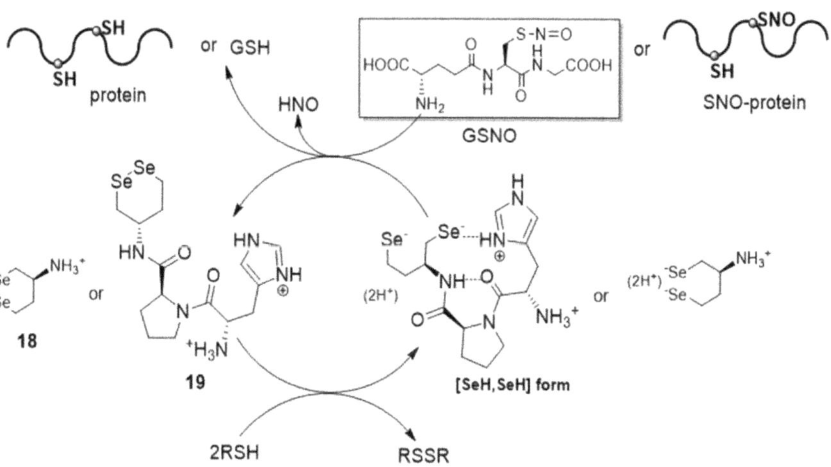

Figure 25.9 Catalytic cycle of a cyclic diselenide with Pro–His dipeptide unit as a S-denitrosylase mimic.

25.4 Aspects as Foldase Mimics

25.4.1 Protein Disulfide Isomerase

PDI, which catalyzes oxidative folding in the ER, contains two Trx-like domains with a CGHC motif. These active centers exist in equilibrium between the [S–S] (PDI^{ox}) and [SH,SH] (PDI^{red}) forms in the ER (Figure 25.10). PDI^{ox} randomly introduces SS bonds into the substrate proteins *via* an intermolecular SS-exchange reaction (path i→ii in Figure 25.10). In contrast, PDI^{red} transiently cleaves the non-native (mispaired) SS bonds and rearranges the SS patterns (path iii→iv→v in Figure 25.10) to gain SS linkages found in the native state. The CGHC motifs in PDI possess crucial physicochemical properties that effectively catalyze these S–S-related reactions in substrate proteins. The pK_a (*ca.* 6.7) of the active SHs in PDI^{red} is lower than that of normal cysteinyl SHs (*ca.* 8.3) owing to the general base effect of the adjacent His side chain.[46] This means that the SH groups predominantly exist as nucleophilic thiolate (S^-) anions in the ER (pH = *ca.* 7.2), which is advantageous for catalyzing SS-isomerization initiated by the nucleophilic attack on the S atom in SS bonds. Additionally, the SS-reduction potential ($E^{\circ\prime}_{SS} = -180\,mV$) of the CGHC motifs is higher than those of the SS bonds formed between ordinary Cys residues in proteins.[47] This means that PDI^{ox} functions as a strong SS-forming reagent and that PDI^{red} can favorably catalyze SS-isomerization rather than SS-reduction (path iii→vi in Figure 25.10) because of the thermodynamic property of the PDI^{red} form, which is less likely to cyclize to PDI^{ox}. In this context, to develop a prominent PDI mimic, it is essential to design a redox molecule that can imitate the physicochemical properties of the PDI active center.

25.4.2 Selenium-based Protein Disulfide Isomerase Mimics

In addition to commercially available thiol compounds such as GSH, cystamine, and DTT^{red}, numerous SH-based reagents, which imitate the physicochemical properties of the PDI active center, have been developed and applied as oxidative folding promoters (please refer to a previous review[48] for the representative reagents). Although these reagents positively promote oxidative folding effectively, the addition of the mM scale, which is approximately 10–100 fold with respect to the substrate protein, is required to achieve reasonable folding yields and rates. In contrast, PDI exerts sufficient catalytic activity for *in vitro* oxidative folding even when approximately 1/800–1/7 molar equivalents with respect to the substrate protein is applied.[49]

As described in Section 25.2.2, organodiselenide compounds can be converted into the corresponding SeH forms *via* an intermolecular bond exchange reaction with coexisting thiols, as shown in Figure 25.4. This reductive cleavage of the Se–Se bond is reversible, whereas the forward reaction (*i.e.*, generation of the SeH form) is thermodynamically unfavorable. Conversely, the Se–Se bond has an extremely higher electrophilicity than the

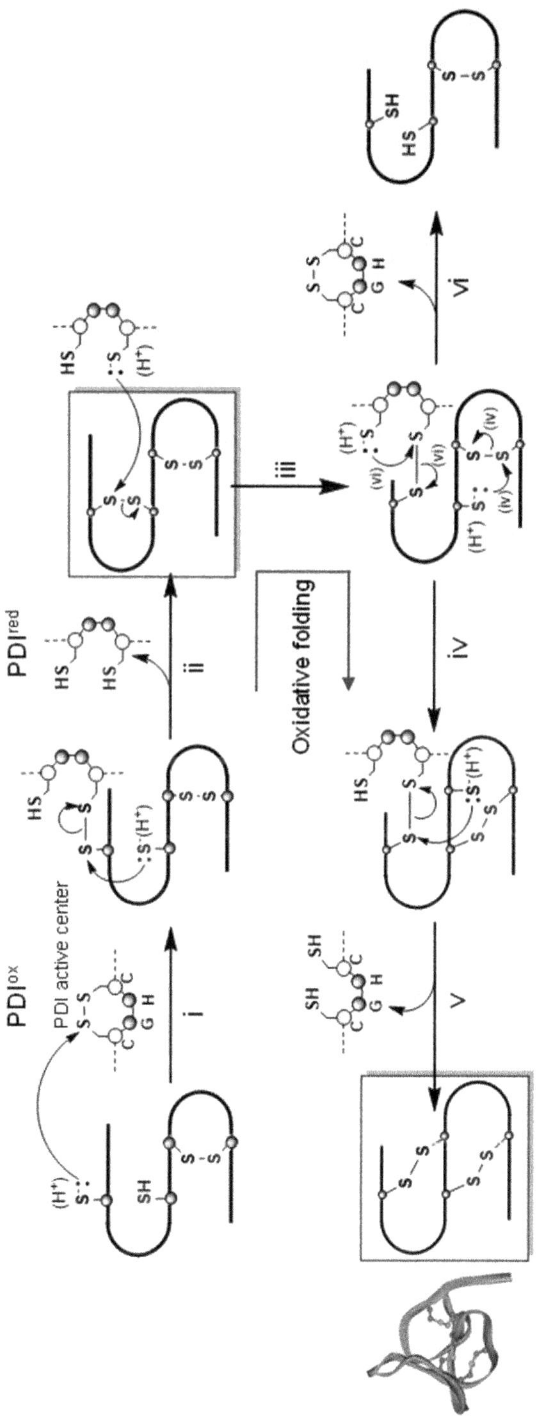

Figure 25.10 Catalytic mechanisms of PDI during the promotion of oxidative folding.

S–S bond, and hence the generation of the SeH group by the initial attack of SH on a Se atom in the Se–Se bond is effective from the viewpoint of the kinetics.[50] More notably, the pK_a of SeH is much lower than that of SH, and SeH almost exists as the Se^- species, which is highly nucleophilic toward the S–S bond in substrate proteins, under physiological pH conditions, indicating that SeH compounds have more ideal physicochemical properties to mimic the active center of PDI than SH compounds.

To improve the efficiency of oxidative folding by artificial reagents, Hilvert and co-workers reported selenoglutathione (GSeSeG; **20**), a selenium analog of GSSG, as a first diselenide-based folding reagent, into which the concept of Se–Se/SeH chemistry is incorporated[51] (Figure 25.11). They reported that, in the oxidative folding of both reduced ribonuclease A (RNase A) and bovine pancreatic trypsin inhibitor, GSeSeG added as an oxidant effectively promoted the reaction more than GSSG, which has been widely used as a typical SS-forming reagent (oxidant) in oxidative folding, in terms of both the reaction rate and folding yield under aerobic conditions. This result may be due not only to the potential advantage of the Se–Se compound as an SS-forming reagent, but also because GSeSeG consumed in SS-formation can be regenerated by O_2 dissolved in the reaction solution (*i.e.*, $4GSeH + O_2 \rightarrow 2GSeSeG + 2H_2O$) to promote SS-formation in a catalytic manner. Furthermore, GSeH ($GSe^- + H^+$) can efficiently promote SS-isomerization because of its superior nucleophilicity compared to GSH. In addition to thiols, NADPH can reduce GSeSeG to reactive GSeH by the function of glutathione reductase, and the resulting GSeH can repair a misfolded protein with non-native S–S bonds.[52] GSeSeG as a folding reagent (or catalyst) has also been successfully applied to other diverse proteins.[53] Additionally, Hilvert *et al.* reported that selenocystamine (**4**) also accelerates oxidative folding, whereas its catalytic ability is slightly lower than that of GSeSeG.[54] Furthermore, it has also been reported that when **4** is administered at low concentrations (1–10 μM) to *E. coli* lacking DsbA, which is a typical folding enzyme containing a CXXC motif in the periplasm of gram-negative bacteria, **4** can function as an oxidative folding catalyst instead of DsbA.[55] Later, Metanis *et al.* reported low-molecular diselenide compounds (**21–23**) as an oxidative folding catalyst, which can readily be synthesized from inexpensive starting materials, and showed that their capability as a folding promoter is similar to or slightly higher than that of GSeSeG.[56] Additionally, our group recently reported that ebselen, which is the most representative GPx1 mimic and convertible to the reactive SeH form in the presence of GSH (Figure 25.4a), also reasonably accelerates the oxidative folding of a reduced protein.[45]

Furthermore, we previously demonstrated that a catalytic amount (1/4 equivalent to total protein thiols) of the five-membered ring diselenide (**24**) significantly accelerated the oxidative folding of reduced RNase A in a buffer solution containing GSH and GSSG as redox reagents at pH 7.5.[57] The Se–Se reduction potential $\left(E'^{\circ}_{SeSe}\right)$ of compound **24** (−265 mV) is significantly higher than that of the six-membered ring analog (**18:** −365 mV), owing to the high

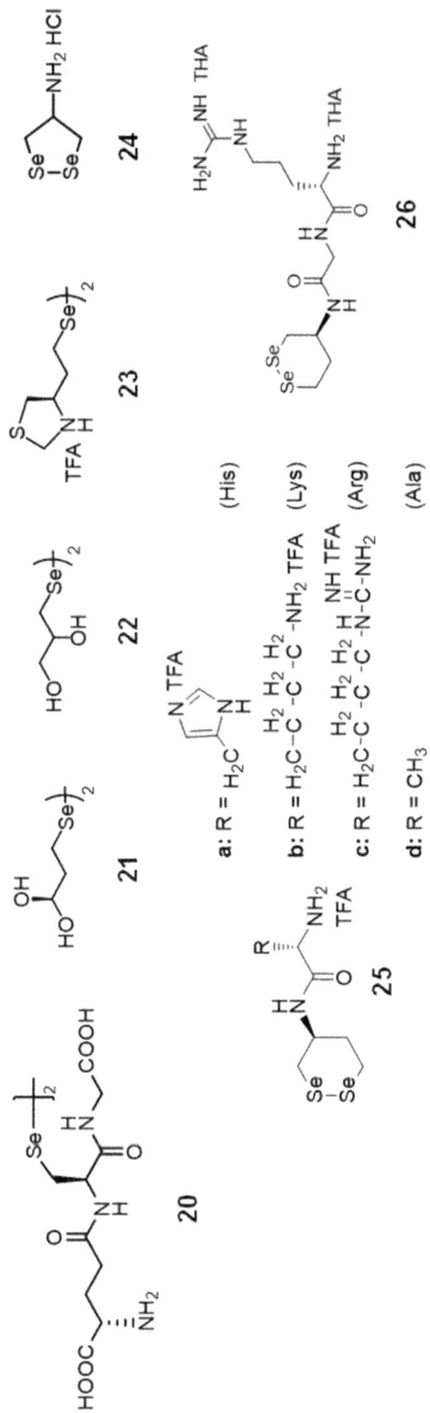

Figure 25.11 Selenium-based folding reagents and catalysts.

ring strain of the five-membered ring structure; thus, **24** functions as a better SS-forming and SS-isomerizing catalyst than **18** during oxidative folding. Furthermore, we recently reported that the His conjugate (**25a**) with compound **18** has a significantly higher $E'^{\circ}_{\mathrm{SeSe}}$ value than the parent compound (**18**), owing to the thermodynamic stabilization of the [SeH,SeH] through the intramolecular NH···Se hydrogen bond between the imidazole side chain and the Se atom.[58] Consequently, it was found from the evaluation using hen egg white lysozyme (HEL) as a model substrate that compound **25a** has a higher catalytic activity for both oxidative folding and repairing of misfolded form with non-native S–S bonds than not only other basic amino acids, but also non-basic amino acid conjugates (**25b–d**). Similarly, compound **19**[45] (Figure 25.9) and arginine-conjugated compound with a Gly spacer (**26**),[59] which have a high $E'^{\circ}_{\mathrm{SeSe}}$ value, also showed an oxidative folding catalytic activity that was comparable to or slightly higher than that of **25a**.

25.4.3 Redox Catalysts Including Chaperon-like Activity

PDI not only catalyzes SS-related reactions in oxidative folding, but also functions as a molecular chaperone that inhibits the association between immature proteins. The length of the open side in the hydrophobic cavity of PDI is 15–30 Å, which can accommodate substrate proteins or their parts.[60] Currently, there are no reports on the development of middle- and high-molecular redox compounds, which can incorporate substrate proteins in a supramolecular manner similar to that of PDI. Nevertheless, it has been reported that several SH-based compounds with surfactant moieties,[61] aromatic substituents,[62] or urea-type functional groups,[63,64] which can associate with proteins *via* hydrophobic interactions and salt bridges, have an inhibitory effect against protein aggregation in addition to their SS-redox and isomerization activities. However, to achieve sufficient enhancement of the oxidative folding rate and inhibition of aggregation, high concentrations (1–10 mM) must be added.

Although selenium-based oxidative folding catalysts with an aggregation-inhibiting capability have not been reported, this author's group recently reported that compound **25a** efficiently inhibited the aggregation of HEL induced by thermal denaturation despite adding low concentrations (0.3–0.5 mM). Additionally, redox buffer solution containing compound **25a** (0.2 mM) and GSH (1.0 mM) not only accelerated the oxidative folding of a high concentration of HEL (0.7 $\mathrm{mg\,mL^{-1}}$), but also suppressed the competing aggregation and oligomerization of HEL. Consequently, the final folding yield was 55%, compared to 25% when GSSG 0.2 mM was used instead of **25a**. However, although PDI exhibits excellent chaperone-like activity at concentrations equal to or less than the substrate protein concentrations, the artificial compound should be added at least two-fold more than the concentration of the substrate protein. The development of organic redox molecules that more effectively capture substrate proteins and promote accurate folding is an important issue in the quest for artificial proteostasis control in cells.

25.5 Summary and Future Challenges

Organoselenium compounds have been suggested to possess various pharmacological effects, such as antioxidant, anticancer, anti-fibrolytic, and antiparasitic activities due to the unique chemical properties of the selenium center and have attracted considerable attention in medicinal chemistry. In this chapter, we focused on ROS, RNS, and protein misfolding in the ER, and discussed the potential of selenium compounds as a regulating factor for proteostasis. Numerous studies have suggested that organic selenium molecules as antioxidants may be sufficiently effective in inhibiting the structural transformation of normal proteins into pathogenic proteins caused by the disruption of the redox balance in the ER. Furthermore, some organoselenium compounds with antioxidant activity, such as ebselen, can also catalyze the oxidative folding of proteins and repair of the misfolded form. Additionally, it has recently been shown that some of these organoselenium compounds can also inhibit protein–protein interactions, inhibit protein aggregation, and contribute to the correct structural maturation of proteins. However, further biological evaluations using living cells and animals are necessary to assess whether these compounds have effective activity against misfolding disease. Additionally, there are countless redox factors in cells. To avoid undesired side reactions in organelles other than the ER, a molecular design that enables the selective activation of selenium compounds and expression of appropriate antioxidants and PDI-like activities in the ER will be the key to future drug development for ER-related disorders.

References

1. M. J. Feige, *Oxidative Folding of Proteins: Basic Principles, Cellular Regulation and Engineering*, RSC Publishing, London, 2018.
2. C. B. Anfinsen, *Science*, 1973, **181**, 223.
3. M. Matsusaki, S. Kanemura, M. Kinoshita, Y. H. Lee, K. Inaba and M. Okumura, *Biochim. Biophys. Acta, Gen. Subj.*, 2020, **1864**, 129338.
4. S. Kanemura, M. Matsusaki, K. Inaba and M. Okumura, *Int. J. Mol. Sci.*, 2020, **21**, 9351.
5. K. Iuchi, T. Takai and H. Hisatomi, *Biology*, 2021, **10**, 399.
6. T. Uehara, T. Nakamura, D. Yao, Z. Q. Shi, Z. Gu, Y. Ma, E. Masliah, Y. Nomura and S. A. Lipton, *Nature*, 2006, **441**, 513.
7. C. Wang, W. Li, J. Ren, J. Fang, H. Ke, W. Gong, W. Feng and C. C. Wang, *Antioxid. Redox Signaling*, 2013, **19**, 36.
8. C. Wang, J. Yu, L. Huo, L. Wang, W. Feng and C. C. Wang, *J. Biol. Chem.*, 2012, **287**, 1139.
9. T. Konno, E. P. Melo, J. E. Chambers and E. Avezov, *Cells*, 2021, **10**, 233.
10. M. Deponte, *Antioxid. Redox Signaling*, 2017, **27**, 1130.
11. R. Brigelius-Flohé and M. Maiorino, *Biochim. Biophys. Acta, Gen. Subj.*, 2013, **1830**, 3289.
12. M. Iwaoka and K. Arai, in *Organoselenium Compounds in Biology and Medicine*, RSC Publishing, London, 2017, pp. 197–222.

13. M. Govindasamy, *Curr. Chem. Biol.*, 2013, **7**, 47.
14. D. Bhowmick and G. Mugesh, *Org. Biomol. Chem.*, 2015, **13**, 10262.
15. B. K. Sarma and G. Mugesh, *J. Am. Chem. Soc.*, 2005, **127**, 11477.
16. C. Santi, C. Scimmi and L. Sancineto, *Molecules*, 2021, **26**, 4230.
17. M. Aras, M. Altaş, S. Meydan, E. Nacar, M. Karcıoğlu, K. T. Ulutaş and Y. Serarslan, *Int. J. Neurosci.*, 2014, **124**, 771.
18. C. Barkus, J.-M. N. Ferland, W. K. Adams, G. C. Churchill, P. J. Cowen, D. M. Bannerman, R. D. Rogers, C. A. Winstanley and T. Sharp, *J. Psychopharmacol.*, 2018, **32**, 1018.
19. R. L. Quispe, M. L. Jaramillo, L. S. Galant, D. Engel, A. L. Dafre, J. B. Teixeira da Rocha, R. Radi, M. Farina and A. F. de Bem, *Redox Biol.*, 2019, **20**, 118.
20. L. S. Galant, J. Rafique, A. L. Braga, F. C. Braga, S. Saba, R. Radi, J. B. T. da Rocha, C. Santi, M. Monsalve, M. Farina and A. F. de Bem, *Neurochem. Res.*, 2021, **46**, 120.
21. G. L. Godoi, L. de Oliveira Porciúncula, J. F. Schulz, F. N. Kaufmann, J. B. da Rocha, D. O. G. de Souza, G. Ghisleni and H. L. de Almeida, *Neurochem. Res.*, 2013, **38**, 2359.
22. Y. Xie, Y. Tan, Y. Zheng, X. Du and Q. Liu, *J. Biol. Inorg. Chem.*, 2017, **22**, 851.
23. A. Kunwar, B. Mishra, A. Barik, L. B. Kumbhare, R. Pandey, V. K. Jain and K. I. Priyadarsini, *Chem. Res. Toxicol.*, 2007, **20**, 1482.
24. B. Mishra, A. Barik, A. Kunwar, L. B. Kumbhare, K. I. Priyadarsini and V. K. Jain, *Phosphorus, Sulfur Silicon Relat. Elem.*, 2008, **183**, 1018.
25. B. G. Singh, P. P. Bag, F. Kumakura, M. Iwaoka and K. I. Priyadarsini, *Bull. Chem. Soc. Jpn.*, 2010, **83**, 703.
26. A. Kunwar, P. Bansal, S. J. Kumar, P. P. Bag, P. Paul, N. D. Reddy, L. B. Kumbhare, V. K. Jain, R. C. Chaubey, M. K. Unnikrishnan and K. I. Priyadarsini, *Free Radicals Biol. Med.*, 2010, **48**, 399.
27. K. Arai, Y. Sato, I. Nakajima, M. Saito, M. Sasaki, A. Kanamori and M. Iwaoka, *Bioorg. Med. Chem.*, 2021, **29**, 115866.
28. K. Arai, T. Matsunaga, H. Ueno, N. Akahoshi, Y. Sato, G. Chakrabarty, G. Mugesh and M. Iwaoka, *Chem. – Eur. J.*, 2019, **25**, 12751.
29. G. Chakrabarty, S. K. NaveenKumar, S. Kumar and G. Mugesh, *ACS Chem. Biol.*, 2020, **15**, 2673.
30. F. Kumakura, B. Mishra, K. I. Priyadarsini and M. Iwaoka, *Eur. J. Org. Chem.*, 2010, 440.
31. K. Arai, F. Kumakura, M. Takahira, N. Sekiyama, N. Kuroda, T. Suzuki and M. Iwaoka, *J. Org. Chem.*, 2015, **80**, 5633.
32. A. Kunwar, P. Verma, H. N. Bhilwade, M. Iwaoka and K. I. Priyadarsini, *Mutat. Res., Genet. Toxicol. Environ. Mutagen.*, 2016, **807**, 33.
33. M. Iwaoka, N. Sano, Y. Y. Lin, A. Katakura, M. Noguchi, K. Takahashi, F. Kumakura, K. Arai, B. G. Singh, A. Kunwar and K. I. Priyadarsini, *ChemBioChem*, 2015, **16**, 1226.
34. P. Verma, A. Kunwar, K. Arai, M. Iwaoka and K. I. Priyadarsini, *Toxicol. Res.*, 2016, **5**, 434.

35. P. Verma, A. Kunwar, K. Arai, M. Iwaoka and K. I. Priyadarsini, *Biochimie*, 2018, **144**, 122.
36. K. Arai, K. Moriai, A. Ogawa and M. Iwaoka, *Chem. – Asian J.*, 2014, **9**, 3464.
37. T. G. Back and Z. Moussa, *J. Am. Chem. Soc.*, 2003, **125**, 13455.
38. T. G. Back, Z. Moussa and M. Parvez, *Angew. Chem., Int. Ed.*, 2004, **43**, 1268.
39. N. M. R. McNeil, D. J. Press, D. M. Mayder, P. Garnica, L. M. Doyle and T. G. Back, *J. Org. Chem.*, 2016, **81**, 7884.
40. V. Nascimento, E. E. Alberto, D. W. Tondo, D. Dambrowski, M. R. Detty, F. Nome and A. L. Braga, *J. Am. Chem. Soc.*, 2012, **134**, 138.
41. K. Arai, A. Tashiro, Y. Osaka and M. Iwaoka, *Molecules*, 2017, **22**, 354.
42. I. Gusarov and E. Nudler, *Mol. Cell*, 2018, **69**, 351.
43. Y. Hou, Z. Guo, J. Li and P. G. Wang, *Biochem. Biophys. Res. Commun.*, 1996, **228**, 88.
44. T. Yang, A. N. Zelikin and R. Chandrawati, *Small*, 2020, **16**, 1907635.
45. R. Mikami, S. Tsukagoshi, Y. Oda and K. Arai, *ChemBioChem*, 2022, **23**, e202100394.
46. H. C. Hawkins and R. B. Freedman, *Biochem. J.*, 1991, **275**, 335.
47. J. Lundstroem and A. Holmgren, *Biochemistry*, 1993, **32**, 6649.
48. W. J. Lees, *Curr. Opin. Chem. Biol.*, 2008, **12**, 740.
49. J. S. Weissman and P. S. Kimt, *Nature*, 1993, **365**, 185.
50. N. Metanis, E. Keinan and P. E. Dawson, *J. Am. Chem. Soc.*, 2006, **128**, 16684.
51. J. Beld, K. J. Woycechowsky and D. Hilvert, *Biochemistry*, 2007, **46**, 5382.
52. S. Shimodaira, Y. Asano, K. Arai and M. Iwaoka, *Biochemistry*, 2017, **56**, 5644.
53. J. Beld, K. J. Woycechowsky and D. Hilvert, *J. Biotechnol.*, 2010, **150**, 481.
54. J. Beld, K. J. Woycechowsky and D. Hilvert, *Biochemistry*, 2008, **47**, 6985.
55. J. Beld, K. J. Woycechowsky and D. Hilvert, *ACS Chem. Biol.*, 2010, **5**, 177.
56. P. S. Reddy and N. Metanis, *Chem. Commun.*, 2016, **52**, 3336.
57. K. Arai, H. Ueno, Y. Asano, G. Chakrabarty, S. Shimodaira, G. Mugesh and M. Iwaoka, *ChemBioChem*, 2018, **19**, 207.
58. S. Tsukagoshi, R. Mikami and K. Arai, *Chem. – Asian J.*, 2020, **15**, 2646.
59. R. Mikami, S. Tsukagoshi and K. Arai, *Biology*, 2021, **10**, 1090.
60. L. Wang, X. Wang and C. Wang, *Free Radicals Biol. Med.*, 2015, **83**, 305.
61. M. Potempa, M. Hafner and C. Frech, *Protein J.*, 2010, **29**, 457.
62. D. J. Madar, A. S. Patel and W. J. Lees, *J. Biotechnol.*, 2009, **142**, 214.
63. S. Okada, M. Matsusaki, K. Arai, Y. Hidaka, K. Inaba, M. Okumura and T. Muraoka, *Chem. Commun.*, 2019, **55**, 759–762.
64. S. Okada, M. Matsusaki, M. Okumura and T. Muraoka, *Molecules*, 2021, **26**, 879.

CHAPTER 26

Speciation of Organic and Inorganic Selenium in Food and the Environment

PABLO HUGO PACHECO

Instituto de Química San Luis (INQUISAL-CONICET), Chacabuco y Pedernera, 5700, San Luis, Argentina
Email: ppacheco@unsl.edu.ar

26.1 Introduction

The toxicity, bioavailability, and metabolic pathways of elements are tightly connected to the chemical forms in which they occur. Thus, particular species of elements can have a different impact on living organisms.[1] The International Union of Pure and Applied Chemists have defined a chemical species as the "Specific form of an element defined as to its isotopic composition or oxidation state, and/or complex or molecular structure." The speciation of an element is the "distribution of an element amongst defined chemical species in a system" while speciation analysis is the "analytical chemistry: analytical activities of identifying and/or measuring the quantities of one or more individual chemical species in a sample".[2]

Selenium may act as a quasi-essential micronutrient for living organisms. Selenium is involved in several physiological and metabolic processes. Se intake in humans is often low and very seldom excessive, and its bioavailability depends also on its chemical form, with organic Se as the most available after ingestion. In this chapter the analytical methods developed for the speciation of this chalcogen element in food and the environment

Chalcogen Chemistry: Fundamentals and Applications
Edited by Vito Lippolis, Claudio Santi, Eder J. Lenardão and Antonio L. Braga

Published by the Royal Society of Chemistry, www.rsc.org

will be comparatively discussed. The bioavailability of a given element to an organism depends on the species in which it enters the body. "Bioavailable fraction is the fraction of an element that is absorbed and that reaches the systemic circulation in order to be distributed to organs and tissues, where it can eventually become bioactive".[3] Effective supplementation requires the use of bioavailable forms of the elements. Selenium is an essential element, recently attracting more attention due to its antioxidant and anticancer properties. Nutritionists recommend increasing Se intake, especially in regions where environmental selenium levels are low.[4] Therefore, selenium supplementation should be carried out very carefully, considering appropriate doses not only for total element concentration, but also for its particular chemical forms. Selenium may be present in foods as inorganic compounds: selenate (Se(VI)) and selenite (Se(IV)) or as organic compounds: selenomethionine (SeMet), selenocysteine (SeCys), selenocystine ($(SeCys)_2$), and their methylated derivatives. Literature reports indicate that SeMet is in the long term more effective in preventing selenium deficiency than other seleno-organic compounds. On the other hand, inorganic selenium salts may more rapidly compensate for selenium deficiency, though they have a higher chance of becoming toxic in the long term.

Selenium is introduced into the environment by both natural processes and by human activity. Selenium can bioaccumulate by means of the food chain, and selenium can be at significant levels for higher animals.[5] Selenium distribution and speciation depends on a variety of physical, chemical, and biological processes, such as volcanic activity, combustion of fossil fuels, weathering of rocks, soil leaching, groundwater transport, plant and animal uptake, and release, adsorption, and desorption, chemical or biological reduction, and oxidation as well as mineral formation.[6] Bioavailability of selenium is governed by the selenium species present in soil and therefore factors that affect the selenium dynamics (*i.e.* mobility, solubilization, and transformations) in the soil should be considered.[7]

Speciation of selenocompounds requires (i) separation with an appropriate technique, (ii) destruction of the organocompound, and (iii) selenium quantification.[8] The principal analytical approach to selenium speciation has been based on the fractionation of biological extracts by chromatography (or electrophoresis) while specifically monitoring selenium by inductively coupled plasma mass spectrometry (ICP MS).[5] The latter allows one to isolate the Se-specific signal, to decrease considerably the detection limits, to calculate the recovery from the chromatographic columns, and to implement reasonable measures of quality control. The increase in the chromatographic resolution, the introduction of multidimensional separation approaches and the growing sensitivity of ICP MS detection owing to collision cell and triple quadrupole mass spectrometers have resulted in the increasing number of unidentified peaks in HPLC and ICP MS chromatograms. The absence of retention time standards has rendered electrospray mass spectrometry indispensable when identifying new selenium species.[9]

26.2 Speciation Analysis

26.2.1 Sample Treatment

The successful extraction of Se compounds will depend on whether they are water soluble, protein bound, or polysaccharide bound. Liquid extraction is the most conventional approach to extract Se species and a variety of different solutions have been used to extract species from selected matrices.[10] The identification and quantitative determination of selenium species in foods or biota presents several challenges. The extraction and the separation steps must be carefully considered in order to maintain the integrity of seleno-species. Extraction conditions must be chemically mild but sufficiently efficient to liberate the species from the matrix. Sampling and storage affect the reliability of the results since volatilization, adsorption, inter-conversion of species, precipitation, or contamination may change the Se composition of the sample.

In sequential extraction, a series of reagents is applied to the same sample to sub-divide the total metal content. The "vigor" of the treatment generally increases through the steps of the procedure, from initially mild conditions (*e.g.*, shaking with water, a salt solution, or dilute acetic acid) to the use of much harsher reagents (*e.g.*, hot mineral acid).[11] Sequential treatment with enzymes combined with conventional techniques may also improve the extraction efficiency. This strategy was applied to Se-enriched yeast material[12] using Tris–HCl buffer for water-soluble fractions, followed by Tris–HCl buffer with SDS for solubilization of the protein fraction and, finally, concentrated HNO_3 for the dissolution of the resulting solid residue. After the three steps, the total Se recovery reached 91.5%.

Ultrasonic-Assisted Solvent Extraction (UAE) has gained popularity in the last few years as an attractive alternative to Microwave Assisted Extraction (MAE) techniques as it does not require high temperatures, pressures, or harmful chemicals.[13] A solid sample is placed into a solution and is exposed to ultrasonic energy, which results in the formation of imploding bubbles, leading to high localized temperatures and pressures. In this way compounds or elements can be extracted from the sample matrix. Ultrasonic methods also increase the rate of reaction in sequential extractions and assist with the homogenous suspension and distribution of slurries.

Selenium in animal tissues is not present as selenite or selenate but bound within proteins mainly as SeCys and/or SeMet and normally requires enzymatic hydrolysis with protease to release these species for analysis. The low recoveries of Se from tissues after the extraction and separation can be due to the loss of labile SeCys and the presence of other Se species at low concentrations.[14] SeMet is relatively stable, easy to extract with typical recoveries of 70–90%. SeCys is not stable during hydrolysis and artifacts will be formed, often elemental selenium. Thus, SeCys containing proteins must be protected by derivitization before hydrolysis.[2] Traditional Se enzymatic digestion for speciation is a time-consuming approach, far from being quantitative and with a risk of selenium inter-conversion. Those procedures with enzymes have been performed with incubation in a bath at 37 °C or with bath sonicaton, the latter being less time-consuming than the former.[8]

In vitro gastrointestinal digestion was used to assess the bio-accessibility of Se compounds from fish samples (swordfish, sardine, and tuna).[15] The digestion method was studied in gastric juice at 37 °C on a water bath, with shaking for a few hours and then adding intestinal juice for a few more hours under the same conditions.

Attention is focused on the application of liquid–liquid micro-extraction (LLME) methods and the use of environmentally friendly extraction solvents such as those formed by the combination of quaternary ammonium salts with hydrogen bond donors (carboxylic acids, polyhydric alcohols).[16]

Solid-phase extraction (SPE) has been used as a separation and/or pre-concentration technique prior to the determination of selenium species due to the need for accurate measurements for Se species at extremely low levels.[17] This widespread use of SPE is due to its well-known advantages, high enrichment factor, good recovery, fast phase separation, low consumption of organic solvents, time saving with the possibility of automation, sampling in the field, suitability for combination on-line or off-line with different detection techniques and low cost. Appropriate enhancements in sensitivity and adequate elimination of interferences and sample matrix have been achieved with this procedure. In addition, SPE requires minimal sample manipulation and it preserves chemical species and avoids modifications in their distribution. A wide variety of distinct SPE-sorption materials have been used for this task, including different kinds of polymers and silica gel-based sorbents, metallicoxides and hydroxides, powdered activated carbon, nanomaterials, and biological substrates such as several types of bacteria and amino acids. Solid-phase microextraction is also considered, especially those procedures involving the application of nanoparticles as the solid phase.[18] Nanomaterials were widely applied to enhance analytical procedures in trace element speciation due to their strong adsorption capacity, high catalytic ability and excellent optical performance. Nanomaterials can significantly improve the existent sampling, pretreatment, and pre-concentration techniques.[19] Ideally, nanomaterials possess a large specific surface area, high adsorption capacity, and fast sorption kinetics compared with conventional sorbents. Chen *et al.*[20] reported the use of sulfonated polystyrene-coated magnetic nanoparticles (Fe_3O_4@PSS MNPs) for selenium speciation analysis in selenium-enriched yeast cells. Wang *et al.*[21] developed a UV-TiO_2 photocatalysis reduction device (UV-TiO_2-PCRD) to achieve the online reduction of Se(VI) under UV irradiation in a flow system. Pre-concentration procedures generally involve coprecipitation with different hydroxides and metal chelates, dispersive liquid–liquid microextraction (DLLME) and hollow fiber liquid phase microextraction (HF-LPME), cloud point extraction (CPE), isotachophoresis, and anion exchange chromatography.[17]

26.2.2 Separation of Seleno-species

There are three distinct steps in undertaking speciation analysis, (1) species extraction and concentration, (2) separation by liquid chromatography (LC) or gas chromatography (GC), and (3) ICP MS measurement. LC is the most commonly used technique for trace selenium speciation analysis. LC has the advantage of performing separations of nonvolatile selenium species; thus, it has generally better versatility than GC.

As far as separation is concerned, Se speciation has been carried out with different chromatographic approaches. Figure 26.1 shows different strategies for selenium analysis according to the target seleno-specie.

Ion-exchange chromatography (IEC) uses the mechanism of exchange equilibra between a stationary phase containing surface ions and oppositely-charged ions in the mobile phase. IEC may be used in either of two separation modes: cation or anion-exchange, both of which have been employed in selenium speciation analysis.[5] Anionic chromatographic columns have

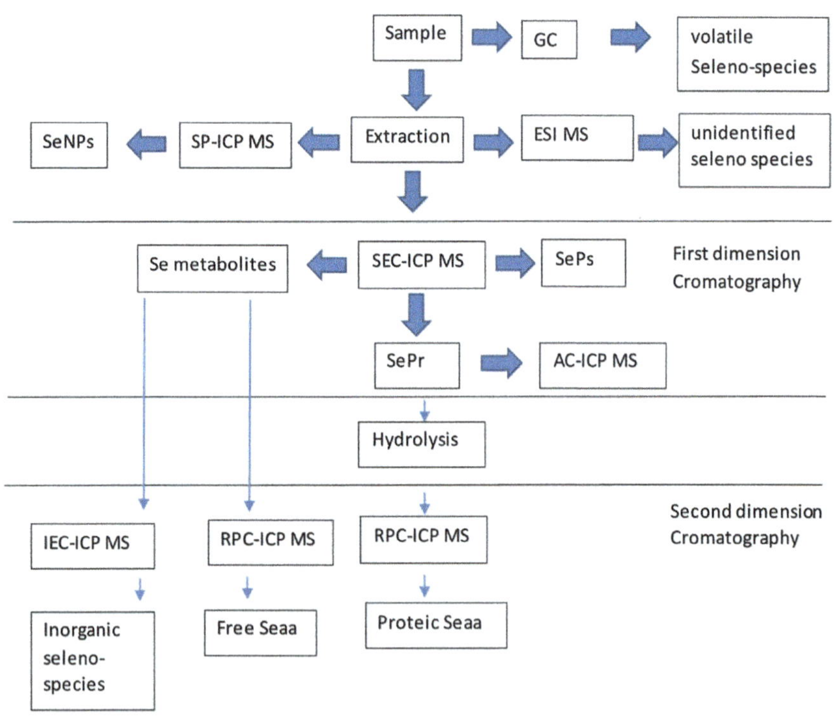

Figure 26.1 Schematic diagram of analysis strategies for selenium species determination. GC, gas chromatography; ESI MS, electrospray ionization mass spectrometry; SP, single particle; ICP MS, inductively coupled plasma mass spectrometry; SeNPs, selenium nanoparticles; SEC, size exclusion chromatography; SePs, selenized polysaccharides; SePr, seleno proteins; AC, affinity chromatography; IEC, ion exchange chromatography; RPC, reverse phase chromatography.

been used to separate Se(IV) and Se(VI).[22] Organic cationic forms of Se, such as tri-methyl-selenonium (TMSe), dimethylselenocysteineselenonium, methylselenomethionineselenonium, Se-methylselenocysteine (Se-Me-SeCys), and SeMet have been separated with cationic columns.[23]

Separation of seleno-amino acids has been achieved by reverse phase columns (Figure 26.1). Separation by reverse phase chromatography (RPC) is based on the different interactions between neutral non-polar analytes and aliphatic carbonyl chains of 18 or 8 carbons. Since seleno-amino acids are charged molecules they need to be neutralized by an ion-pair reagent. Seleno-amino acids were determined in extra virgin olive oils employing a C8 chromatographic column and trifluoroacetic acid (TFA) as the ion pair reagent.[24]

Size-exclusion chromatography (SEC) is often used in selenium speciation to characterize the various selenoproteins encountered from different sample matrices. SEC is a separation mode in which the separation is according to the hydrodynamic volume of the analytes, and the separation takes place in a porous non-absorbing stationary phase with pores of approximately the same size as the analyte's effective solution dimensions. Unlike many of the other modes of HPLC, interaction between the stationary phase and the analyte is minimized for effective size separation. SEC does not generally provide the resolution required for the correct identification of the various selenium species when using the ICP MS detector. SEC has been employed successfully as a screening separation methodology for use with other modes of LC as shown in Figure 26.1.[5]

Single chromatographic steps and/or multidimensional strategies can be applied for purification of protein extracts in order to characterize a given metal–protein complex. The simplest strategy as a first screening test of the presence of a metal in a molecular mass range is the combination of size exclusion (SEC) on-line to ICP MS (Figure 26.1). This strategy has been conducted to monitor the elemental distribution patterns of several elements in whole milk and mussel tissues.[25] Seleno-amino acids can then be separated from extraneous material by SEC and identified and quantified using cation and C18 reverse phase chromatography. This approach has been used to identify selenium species in a range of selenium enriched food stuffs. Selenium in grain plants such as wheat and barley may be present in a variety of complex species. Some are readily extracted into water (*e.g.*, Se-methylselenocysteine, γ-glutamyl-Semethylselenocysteine) while other species also require the use of enzymatic extraction with protease or acid hydrolysis to release selenium species.[2] The fractionation of Se-containing proteins in the rat serum was achieved by affinity chromatography by using the HEP and BLUE columns and two mobile phases composed of 0.05 and 1.5 mol L^{-1} NH_4CH_3COOH, respectively.[18]

Capillary electrophoresis (CE) is discussed as a complementary technique to GC and LC, and it is a powerful tool in element speciation with a high separation capability and environmentally friendly nature due to the use of aqueous buffer solutions with moderate pH and its extremely low reagent and sample consumption. Separation by CE is usually faster than that by LC, and therefore is potentially a rapid and highly efficient separation technique. However, some problems are presented when CE is coupled to ICP MS due to liquid flow incompatibility, *i.e.*, the liquid flow is one order of magnitude lower in CE than in ICP sample introduction system.[26] Because the packed columns will lead to poor reproducibility, thus capillary or multi-capillary GC is usually used to provide superior separation power and better detection limits. Volatile organic Se compounds of dimethylselenide (DMSe) and dimethyldiselenide (DMDSe) in environmental and biological samples were determined by Ghasemi *et al.*[27] using a headspace hollow fiber protected LPME combined with capillary GC–MS. It is noted that although GC attracts particular attention due to its high efficiency and simplicity of coupling, however, in contrast to GC, HPLC has the ability of dealing with non-volatile compounds without any derivatization in some cases, and thus extending the range of application.[26]

26.2.3 Selenium Determination

Elemental speciation usually requires coupling of two techniques: a technique to separate the element's chemical forms of interest and a sensitive detection method to provide detection of the analyte at low levels. ICP MS detection offers high sensitivity and relative ease of interfacing with common chromatographic and separation techniques. ICP MS has several advantages for general speciation analysis, as well as selenium speciation analysis, over more traditional detectors. These advantages include multi-element and multi-isotope detection and high sensitivity with a wide linear dynamic range. The majority of the chemical elements have a dynamic linear detection range of 4–11 orders of magnitude using the ICP MS. In ICP MS, the high efficiency of atomization and ion formation of ICP is coupled with the specific and sensitive detection capability offered by mass spectrometry. ICP is a high-energy ionization source; the plasma typically operates at temperatures of 5000–10 000 K at atmospheric pressure. The argon plasma is generated in a quartz torch under the conditions of a powerful (600–1800 watts) radio frequency electromagnetic field (27–40 MHz). Plasma is formed when a spark is applied to the argon stream in the quartz torch. Free electrons within the plasma are accelerated by the induced magnetic field and bombard other gas atoms, which causes further ionization and sustains the plasma. Sample introduction is by means of a carrier argon gas flow through the central tube in the quartz torch. For aqueous samples, the liquid is first nebulized into an aerosol, a finely dispersed liquid mist, before being carried into the plasma torch. An aerosol facilitates a uniform sample introduction into the plasma and vaporization, atomization, and ionization of the analyte occur nearly simultaneously. ICP is a "hard" ionization source; thus, ions of elemental atoms generated from the plasma are sampled by the mass spectrometer.[5]

A basic understanding of the nebulizer function and the types of nebulizers is necessary to successfully interface LC, or for that matter, CE, to the

ICP MS. Nebulization, as described previously, is the process to form an aerosol, that is, to suspend a liquid sample into a cloud of droplets less than 100 μm in size. There are numerous nebulizers available commercially for use with ICP MS systems. Pneumatic designs, both concentric and cross flow, are the most popular for LC interfaces. Interfacing can be readily accomplished with LC–ICP MS; mobile phase flow rated used for analytical and small-bore column dimensions are in the range of 0.2–1.5 mL min^{-1} which is compatible with most conventional nebulizer systems employing the concentric or cross-flow nebulizer and a standard spray chamber. To minimize possible peak broadening from excessive post-column volume, the transfer line from the LC column outlet to the nebulizer is kept to a minimum length. The standard concentric nebulizer delivers only 2% of the sample entering the nebulizer to the plasma. The microconcentric nebulizer combined with a cyclonic spray chamber was found to be the optimal sample introduction system taking chromatographic peak shape, sensitivity, and LODs into account.[28]

Many HPLC techniques make use of organic modifiers in the mobile phase; therefore, the organic solvent content introduced into the ICP must also be a consideration. Large volumes of organic solvent introduced into the plasma can result in its instability. This problem can be minimized by desolvating the sample aerosol before it reaches the plasma. A cooled spray chamber is generally used to condense the solvent vapor, although membrane dryers and Peltier condensers have occasionally been used for this purpose. Also, acetonitrile is relatively incompatible with ICP MS, so methanol is the organic modifier of choice for plasma/signal stability. Reduction of the liquid volume introduced into the plasma can be accomplished by flow splitting after the analytical column or with the use of small-bore or microbore HPLC columns.

The potential of ICP MS methods and strategies for screening multiple heteroatoms (*e.g.*, S, P, Se, metals) in proteins and their mixtures and extraordinary capabilities to tackle the problem of absolute protein quantifications, *via* heteroatom determinations, selenium and iodine are also naturally present in some proteins making them easily detectable by ICP-MS. ICP-MS is an extraordinary tool to screen metal-containing fractions in biological samples.[25,29] The advent of ICP MS has made it relatively easy to measure metalloid species. Increasingly health and environment regulations have been written around knowledge of metalloid species rather than total concentrations. In addition, speciation measurements are being used to gain an understanding of the effectiveness of food or food supplements in supplying a particular element in a nutritional or active form. The active ingredient in selenium vitamin tablets, often either SeMet or selenate (SeO_4^{2-}).[2]

The major drawback of using ICP MS is that the co-elution of chemical species with the same mass as the species of interest can essentially mask the species of interest thereby leading to an overestimation of the species concentration. When measuring Se concentrations by ICP MS, carbon enhancement in the ICP plasma and isobaric inferences from chloride-, sulfur-, bromine, and oxygen-containing matrices need to be considered. The use of dynamic reaction cell (DRC) or collision cell minimizes or eliminates these interferences.[14] The most abundant Se isotope is ^{80}Se. At mass 80, however, the determination of Se is not possible with single quadrupole ICP-MS because of the strong interference from the argon dimer (Ar_2) formed in the argon plasma. Methane was applied as the reaction gas in the DRC to eliminate the Ar-based spectral interferences while monitoring the ion signals of the ^{78}Se, ^{80}Se, and ^{82}Se isotopes on-mass.[28]

Selenium nanoparticles (SeNPs) synthesized by microorganisms, such as bacteria, fungi, or yeast were demonstrated to possess antibacterial, antiviral, and antioxidant properties. The process of the Se(0) nanoparticle formation is based on the reduction of a toxic selenite or selenate to the less toxic (for the host organism) elemental selenium through the intra- or extracellular formation of SeNPs with a typical spherical shape and a diameter of 50–400 nm. Single Particle (SP)-ICP MS is able to discriminate (detect and quantify) dissolved *versus* particulate forms of the element in a sample, and to provide information about the element mass content per particle. Moreover, if additional information about their composition, shape, and density is available, the size of the particles can be obtained, as well as their number and mass concentration.[30]

Quality control is critical in speciation analyses, which highlights the use of certified reference materials (CRMs), whenever possible, to improve the reliability of the analytical results. In the absence of a suitable CRM, physically different methods should be employed for comparison.[13] Numerous certified reference materials exist for the measurement of trace elements in biological materials, but few materials are certified for metalloid species analysis. Although ESI-MS is the best approach for identification of selenium compounds, ICP MS offers the highest sensitivity and offers the best quantitative methodology when reference compounds are available.[5]

The hyphenation of chromatography to atomic absorption or atomic fluorescence spectrometry (AAS or AFS) is still a reliable and low-cost alternative for routine laboratories.[8] Concerning total determination with hydride generation from all the selenium species, only Se(IV) forms the volatile selenium hydride, H_2Se. For this reason transformation of the different Se compounds into Se(IV) prior to hydride formation is mandatory. Concerning destruction/quantification of the Se organocompounds, this can be done by on-line procedures with the aid of microwave or ultraviolet heating (LC-UV(MO)-HG-AAS/AFS). Selenoproteins were also separated by gel electrophoresis in bands that later were excised for selenium quantification by ETAAS.[31]

Hydride generation (HG) coupled with atomic spectrometric techniques is a sensitive analytical tool for the determination of trace levels of Se. The popularity of HG, nearly 30 years after its introduction as an analytical technique, is still increasing because of its ability to differentiate different oxidation states of Se through simple procedures, without the use of chromatographic separation. By applying suitable separation techniques, elemental species which form the corresponding hydrides can be quantitated. The separation techniques include selective hydride formation based on pH adjustment, buffer, reaction media, and matrix modification.[32] The speciation analysis of selenium (Se(IV) and Se(VI)) could also be achieved without chromatographic separation, owing to the relative inertness of Se(VI) without photocatalysts in photo-cold vapor generation process. Under UV irradiation, only selenite could be transformed to its volatile species at low temperature and determined subsequently, and the total inorganic selenium (selenite + selenate) could be detected by using a boiling water bath together with TiO_2, thus the concentration of selenate could be calculated from the difference between selenite and total selenium.

26.2.4 Identification of Novel Seleno-species

Hyphenated techniques which use ICP MS do not provide structural information about species thus the identification of unknown species is not possible. Mass spectrometers with soft ionization techniques such as electrospray used in conjunction with hyphenated ICP MS techniques are required to unequivocally identify the structures of unknown species (Figure 26.1). Mass spectrometry discriminates molecules according to their mass–charge ratio (m/q) by applying different magnetic fields varying the voltage of a quadruple. Different mass spectrometer designs or tandem configurations reach different selectivity and sensibility. Ordinary, complimentary techniques, such as LC–MS^n should be used to identify unknown species. Identified species can then be isolated in sufficient quantities for use as standards in subsequent analyses. Quadrupole, time of flight (TOF), or sector field (SF) mass spectrometers have all been employed with the ICP and general speciation analysis.

The increase in the chromatographic resolution, the introduction of multidimensional separation approaches and the growing sensitivity of ICP MS detection owing to collision cell and triple quadrupole (QQQ) mass spectrometers have been resulting in the increasing number of unidentified peaks in HPLC and ICP MS chromatograms.[9] The sensitivity of ESI MS has long been regarded as inferior to ICP MS. The comparison is actually difficult because, in contrast to ICP MS, the sensitivity of ESI MS is not only compound-dependent but also is dependent on thc compound purity at the moment of its arrival at the detector whereas that of ICP MS is largely not.

At the attainable chromatographic resolution, there is a severe problem of co-elution of selenium species which can be solved only by high resolution mass spectrometry capable to describe the peak composition in terms of the selenocompounds present. Studies open with a discussion on the growing role of high-resolution ESI-MS techniques in selenometabolomic and selenoproteomic studies and how these techniques have started surpassing ICP MS for speciation studies. For instance, the use of LC–TOF-MS and LC–Orbitrap-MS instruments has allowed the acquisition of metabolomic fingerprints containing 200 Se-containing compounds in selenized yeast.[33] The authors predict a secondary role of ICP MS in Se speciation as a complementary technique for determining extraction and column recoveries. Elucidation of Se species and metabolites in Se supplements is still a topic of much research. In terms of selenium-protein determination, a proteomics approach based on 2D gel electrophoresis followed by LC–ESI-Orbitrap MS/MS enabled the identification of 19 Se-containing proteins in yeast to investigate the degree of the Se/S replacement.

ESI-MS, which has been considered for a long time as a technique complementary to ICP MS and used to confirm or enable the species identification, is becoming a standalone tool for speciation analysis. It can be used in direct (infusion) mode or coupled to HPLC sample introduction and is able to produce data on tens or hundreds of element species in a single mass spectrum. The use of ESI-MS as a standalone technique for speciation analysis imposes conditions on its performance, especially in terms of resolution and mass accuracy. Not only should the target species ion be separated from the ions of other species of the same element, but it should also be separated from any other ion produced by the ionization of other concomitant molecules.[33] Figure 26.1 shows different strategies for selenium analysis according to the target seleno-species.

26.3 Seleno-species in Food

Food chemistry is involved in elucidating both the composition of the raw materials and final products, as well as the changes which occur in food during production, storage, and processing. The analysis of food products may be directed to the assessment of food quality and authenticity, the control of a technological process, the determination of nutritional values, and the detection of molecules with a possible beneficial or a toxic effect on human health.[34]

Selenium yeasts are effective and safe sources of selenium and represent a better absorbed form of this element.[35] Animals unlike plants and microorganisms cannot synthesize SeMet themselves. However, they are capable of synthesizing SeCys from both inorganic and organic selenium compounds. The latter transformation occurs as a result of the transfer of selenium atom from SeMet to serine, similarly as in the case of sulfur amino acids, methionine, and cysteine.[9] To assess selenium yeast quality, a LC–ICP MS method was applied to determine seleno-amino acids and related oxidation products, selenate, selenite, and other species relatable to the quality and/or accuracy of the labeled selenium ingredients[36]

In addition to traditional forms such as Se salts, amino acids, or selenium-enriched yeast supplements, next-generation selenium supplements, with a lower risk for excess supplementation, are emerging. These are based on selenium forms with lower toxicity, higher bioavailability, and controlled release, such as zerovalent selenium nanoparticles (SeNPs) and selenized polysaccharides (SPs).[37] The widespread selenium yeast supplement was demonstrated to contain zerovalent selenium nanoparticles.[30] The use of elemental Se in the form of NPs for different biological applications has gained a lot of attention lately, because of several proposed advantages compared with other forms of Se, including higher bioavailability and lower toxicity. Another next-generation group of selenium supplements was proposed under the form of SPs, which are extra- and intra-cellular polysaccharides (EPS and IPS, respectively) modified with selenium.[37] They were mainly investigated for their strong antioxidant properties. In the case of SPs, the molecular weight can be determined with SEC,[38] and the presence and quantification of Se can be confirmed by ICP MS.

In plants, selenium is metabolized by mechanisms similar to those of sulfur metabolism. The uptake and translocation of selenate in plants is significantly faster than that of selenite. Selenate is taken up by the plants *via* high affinity sulfate transporters in the roots and assimilated into SeCys and SeMet.[7] Se-enriched Allium group vegetables such as garlic, onion, and ramps have been mainly the subject of several studies. Apart from the total Se uptake, enrichment treatments normally undergo certain metabolic changes that determine the final product as well as its translocation and accumulation in different plant tissues.[39] A member of the Brassicaceae family, Indian mustard (*Brassica juncea*) is one of the most studied plants in terms of Se accumulation because of its fast-growing cycle and high biomass. The resistance to excessive Se has been related to the formation of organo-selenium compounds in plants that cannot be incorporated into proteins, like Se–Me–SeCys.[24] Determination of this compound can be performed by LC–ICP MS. The ability of these plants to convert selenium into volatile species can be assessed by GC–ICP MS.[40]

In meat, fish, as well as offal and cereals, selenium often occurs in combination with proteins, thus these products with high protein content are typically characterized by a higher selenium content.[35] SeMet, SeCys, and selenite in chicken eggs were determined by a sample treatment protocol including defatting, protein denaturation, and carbamidomethylation to achieve complete protein digestion and to avoid SeCys losses. Quantification was carried out by LC–ICP MS after quantitative isolation of the selenium-containing fractions by SEC.[41]

26.4 Seleno-species in the Environment

Selenium is introduced into the environment by both natural processes and by human activity. Weathering of cretaceous marine shale rock has been known to produce high levels of selenium in the soils of many areas in the western region of the United States. Irrigation of seleniferous soils for crop production in arid and semi-arid regions is a major source of selenium introduction into the environment. These agricultural activities in some areas of the western United States have caused high levels of selenium to be released into local aquatic systems. Another major source of selenium from human activity is by the acquisition, processing and combustion of fossil fuels. Selenium is a trace element present in coal, coal conversion products, oil, oil shale, and their waste by-products. Selenium leachate or run-off from these disposal areas often contaminate local aquatic systems. Selenium concentrations can rapidly increase in fish, water fowl, and other aquatic organisms. Selenium can bioaccumulate by means of the food chain, and selenium can be at significant levels for higher animals. Biomagnification of selenium, which is the accumulation of progressively greater concentrations by successive trophic levels of the food chain, can range from 2 to 6 times between the primary producer (the plants and algae) and the lower consumer (the invertebrates and fish).[5] The assimilation of inorganic Se by living organisms, such as bacteria, yeast, or plants, results in the natural synthesis of organoselenium compounds which, depending on the nature of the molecules and the biological species, can be incorporated into proteins, stored, or eliminated by volatilization.[9] Figure 26.2 shows the selenium cycle in the environment, and the location of selenium species in soils, water, air, living organisms, plants, and food.

26.4.1 Seleno-species in Water

Dissolved inorganic selenium can be found in natural waters as selenide(II), as colloidal elemental Se(0), as selenite ion Se(IV), and as selenate anion Se(VI). Organic forms of Se that may be found in the aqueous environment are volatile (methyl selenides) or non-volatile (trimethyl selenonium ion, seleno-amino acids, and their derivatives). Seleno-species can be determined in water by non-chromatographic methods involving HG and atomic spectrometries. As only Se(IV) is reduced to SeH_2 with $NaBH_4$, the other species should be converted into Se(IV) in order to generate SeH_2, by performing an analysis using three separate aliquots, *i.e.* (a) with no further treatment: direct determination of Se(IV), (b) after reduction of Se(VI) to Se(IV) with HCl or HBr/KBr–Br_2: the sum of Se(IV) and Se(VI), Se(VI) thus being determined by difference, and (c) mineralization of organic matrix by UV-irradiation or wet digestion followed by reduction to Se(IV)-determination of total Se. The difference between the total Se and the sum of Se(IV) and Se(VI) is attributed to organic selenium compounds. However, it has been reported recently that the presence of dimethylselenium (DMeSe) and dimethyldiselenium (DMeDSe) compounds also contributes to Se signal in HG methods.[32]

Inorganic species (selenite and selenate) can be easily separated by AXC, which could not be easily accomplished with RPC. For example, the inorganic selenium (Se) species, selenite, selenate, and selenocyanate, in sea and rain water were separated by AXC using a gradient of 0.1 M NaOH as the mobile phase and determined by (HG ICP-DRC-MS) with instrumental detection limits of 0.15, 0.27, and 0.19 ng Se per L, respectively.[10]

Hydrogen selenide by itself is not expected to exist in the aquatic environment since the Se(0)/H_2Se couple falls even below the H^+/H_2 couple. Aqueous solutions of H_2Se are actually unstable in air due to its decomposition into

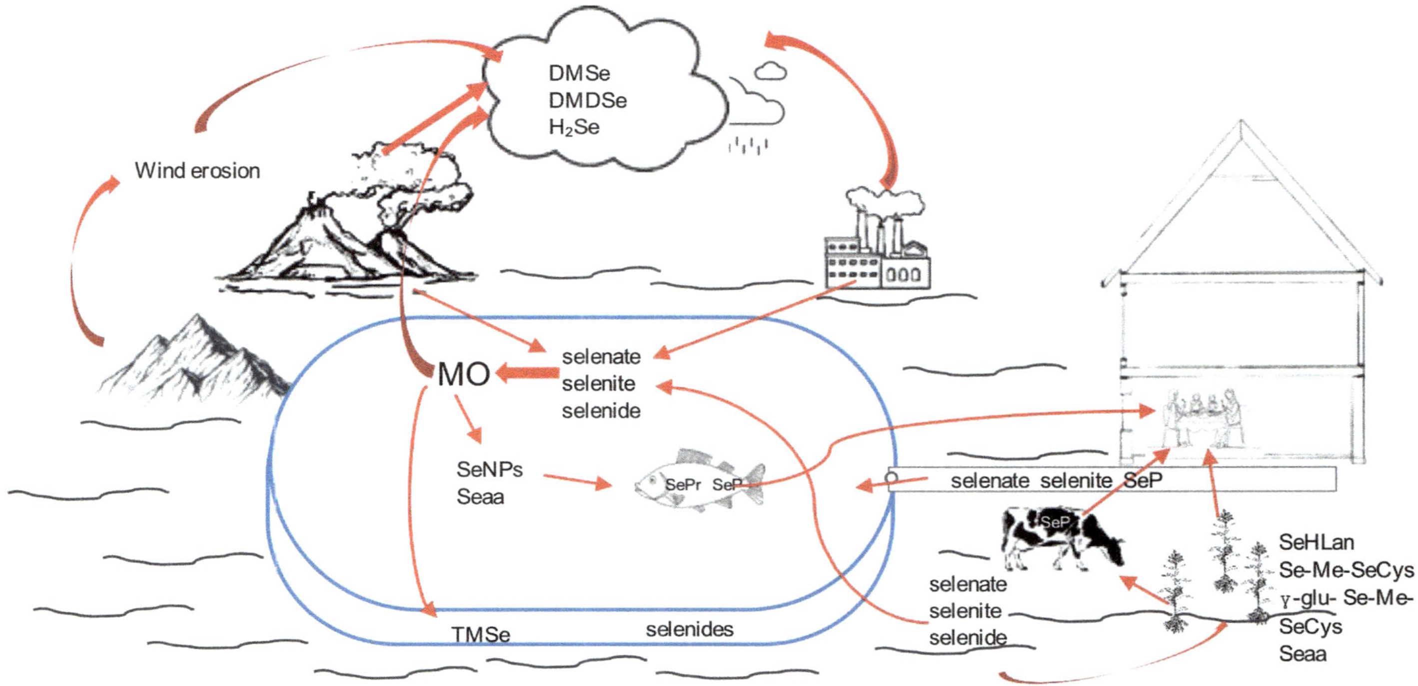

Figure 26.2 Selenium cycle in the environment. DMSe, dimethylselenide; DMDSe, dimethyldiselenide; SeNPs, selenium nanoparticles; SePr, seleno-proteins; SeP, selenized polysaccharides; Seaa, seleno-amino acids; Se–Me–SeCys, seleno-methyl-selenocysteine; γ-glu–Se–Me–SeCys, γ-glutamyl-semethylselenocysteine; SeHLan, selenohomolanthionine.

elemental selenium and water.[42] The precipitation of selenium as heavy metal selenides can be an important factor affecting the cycling of the element in soils and natural waters. The evaluation of the role of elemental selenium in the speciation of this element is a critical point, due to the difficulty in the detection of Se(0) at environmental concentrations.

Several inorganic species of Se (−2, 0, +4, and +6) and organic species (monomethylated and dimethylated) have been reported in aquatic systems. Both $HSeO_3^-$ and SeO_3^{2-} would be present in natural waters. Under mild oxidizing conditions, $HSeO^{3-}$and SeO_3^{2-} are the major species, while HSe^- would be the dominant species at pH higher than 4 and strong reducing conditions.[43]

Selenosugars have been confirmed to be important urinary selenium metabolites, while selenohomolanthionine (SeHLan) is mainly detected in some Se-accumulating plants and yeast.[26]

26.4.2 Seleno-species in Soil and Sediments

Selenium concentrations in soils vary widely throughout the world.[7] Although the normal range of selenium in soil ranges between 0.1 and 2 μg Se per g, the selenium content in soil varies from 0 to 100 μg Se per g throughout the world and can reach up to 1200 μg Se per g in selenium rich soils. Soils containing less than 0.05 μg Se per g are considered selenium-deficient, whereas those containing more than 5 μg Se per g soil are termed seleniferous soils. In soils and sediments, selenium can be present in the soluble, exchangeable, organic matter-bound, sulfide, carbonate, and oxide fractions of sequential extraction procedures. Selenate, a stable form of selenium in the soil and water environment,[44] is water soluble and thus readily absorbed by plant roots.

Selenium in marine sediments is usually associated with effluent discharges from mining industries or coal fired power plants. Overflow from ash dams of the coal-fired power station into the lake leads to gradual methylation of the selenium present in the sediments to volatile dimethyl selenide (DMSe).

26.4.3 Seleno-species in the Atmosphere

The atmosphere is an important transient reservoir of selenium (Se). According to recent evaluations of the global Se budget, approximately 13 000–19 000 tons of Se is cycled through the troposphere annually.[45] Natural sources of atmospheric selenium include crustal erosion (windblown) resulting in suspended soil and dust, biological production of volatile organoselenium compounds, volcanoes, and aerosol particles associated with sea spray,[43] oxidation from the surface of seas, oceans, and soils. Metal refining and coal combustion have been proposed as the major sources of anthropogenic atmospheric emissions. Three major species of Se are generally emitted, which include particulate Se, volatile inorganic Se (Se(0),

SeO_2, and H_2Se), and organic Se methylated Se, *e.g.*, dimethyl selenide, DMSe ($(CH_3)_2Se$), and dimethyl diselenide, DMDSe ($(CH_3)_2Se_2$).

Recent studies have shown that bacteria, soil microbes, microbes in sewage sludge, and plants are also capable of producing DMSe and DMDSe and most of the Se gas reaching the atmosphere will be DMSe. This might be attributed to the rapid sorption and instability of DMDSe relative to DMSe. Finally, like most metals, Se is returned to the Earth's surface through wet and dry depositional processes.

References

1. M. Marcinkowska and D. Barałkiewicz, *Talanta*, 2016, **161**, 177.
2. W. Maher, F. Krikowa, M. Ellwood, S. Foster, R. Jagtap and G. Raber, *Microchem. J.*, 2012, **105**, 15.
3. C. Thiry, A. Ruttens, L. De Temmerman, Y. J. Schneider and L. Pussemier, *Food Chem.*, 2012, **130**, 767.
4. H. M. Crews, *Spectrochim. Acta, Part B*, 1998, **53**, 213.
5. C. B'Hymer and J. A. Caruso, *J. Chromatogr. A*, 2006, **1114**, 1.
6. L. Wu, *Ecotoxicol. Environ. Saf.*, 2004, **57**, 257.
7. S. L. Wadgaonkar, Y. V. Nancharaiah, G. Esposito and P. N. L. Lens, *Crit. Rev. Biotechnol.*, 2018, **38**, 941.
8. J. L. Capelo, C. Fernandez, B. Pedras, P. Santos, P. Gonzalez and C. Vaz, *Talanta*, 2006, **68**, 1442.
9. K. Bierla, S. Godin, R. Lobinski and J. Szpunar, *TrAC, Trends Anal. Chem.*, 2018, **104**, 87.
10. W. Wang, Z. Chen, D. E. Davey and R. Naidu, *J. Liq. Chromatogr. Relat. Technol.*, 2010, **33**, 1151.
11. J. R. Bacon and C. M. Davidson, *Analyst*, 2008, **133**, 25.
12. H. Chassaigne, C. C. Chéry, G. Bordin and A. R. Rodriguez, *J. Chromatogr. A*, 2002, **976**, 409.
13. E. M. Kroukamp, T. Wondimu and P. B. C. Forbes, *TrAC, Trends Anal. Chem.*, 2016, **77**, 87.
14. R. Jagtap and W. Maher, *Microchem. J.*, 2016, **124**, 422.
15. A. I. Cabañero, Y. Madrid and C. Cámara, *Anal. Chim. Acta*, 2004, **526**, 51.
16. K. L. LeBlanc, P. Kumkrong, P. H. J. Mercier and Z. Mester, *Sci. Total Environ.*, 2018, **640–641**, 1635.
17. C. Herrero Latorre, J. Barciela García, S. García Martín and R. M. Peña Crecente, *Anal. Chim. Acta*, 2013, **804**, 37.
18. R. Clough, C. F. Harrington, S. J. Hill, Y. Madrid and J. F. Tyson, *J. Anal. At. Spectrom.*, 2019, **34**, 1306.
19. F. Xu, J. Hu, J. Zhang, X. Hou and X. Jiang, *Appl. Spectrosc. Rev.*, 2018, **53**, 333.
20. B. Chen, B. Hu, M. He, Q. Huang, Y. Zhang and X. Zhang, *J. Anal. At. Spectrom.*, 2013, **28**, 334.
21. Q. Wang, J. Liang, J. Qiu and B. Huang, *J. Anal. At. Spectrom.*, 2004, **19**, 715.

22. H. C. Mehra and W. T. Frankenberger Jr, *Chromatographia*, 1988, **25**, 585.
23. R. J. Kraus, S. J. Foster and H. E. Ganther, *Anal. Biochem.*, 1985, **147**, 432.
24. S. Torres, R. Gil, M. F. Silva and P. Pacheco, *Food Chem.*, 2016, **197**, 400.
25. J. Bettmer, M. Montes Bayón, J. Ruiz Encinar, M. L. Fernández Sánchez, M. D. R. Fernández de la Campa and A. Sanz Medel, *J. Proteomics*, 2009, **72**, 989.
26. X. Yu, C. Liu, Y. Guo and T. Deng, *Molecules*, 2019, **24**, 926.
27. E. Ghasemi, M. Sillanpää and N. M. Najafi, *J. Chromatogr. A*, 2011, **1218**, 380.
28. B. Klencsár, S. Li, L. Balcaen and F. Vanhaecke, *TrAC, Trends Anal. Chem.*, 2018, **104**, 118.
29. Y. He, Y. Zhang, C. Wei, C. Li, Y. Gao and R. Liu, *Appl. Spectrosc. Rev.*, 2014, **49**, 492.
30. J. Jiménez-Lamana, I. Abad-Álvaro, K. Bierla, F. Laborda, J. Szpunar and R. Lobinski, *J. Anal. At. Spectrom.*, 2018, **33**, 452.
31. U. Sidenius and B. Gammelgaard, *Fresenius' J. Anal. Chem.*, 2000, **367**, 96.
32. A. R. Kumar and P. Riyazuddin, *Int. J. Environ. Anal. Chem.*, 2007, **87**, 469.
33. K. Bierla, G. Chiappetta, J. Vinh, R. Lobinski and J. Szpunar, *Front. Chem.*, 2020, **8**, 612387.
34. P. Dugo, T. Kumm, F. Cacciola, G. Dugo and L. Mondello, *J. Liq. Chromatogr. Relat. Technol.*, 2008, **31**, 1758.
35. M. Kieliszek and S. Błazejak, *Molecules*, 2016, **21**, 609.
36. K. M. Kubachka, T. Hanley, M. Mantha, R. A. Wilson, T. M. Falconer, Z. Kassa, A. Oliveira, J. Landero and J. Caruso, *Food Chem.*, 2017, **218**, 313.
37. D. Constantinescu-Aruxandei, R. M. Frîncu, L. Capră and F. Oancea, *Nutrients*, 2018, **10**, 1466.
38. J. Wang, Q. Li, A. Bao, X. Liu, J. Zeng, X. Yang, J. Yao, J. Zhang and Z. Lei, *Carbohydr. Polym.*, 2016, **152**, 70.
39. K. Pyrzynska, *Food Chem.*, 2009, **114**, 1183.
40. J. Meija, M. Montes-Bayón, D. L. Le Duc, N. Terry and J. A. Caruso, *Anal. Chem.*, 2002, **74**, 5837.
41. E. Lipiec, G. Siara, K. Bierla, L. Ouerdane and J. Szpunar, *Anal. Bioanal. Chem.*, 2010, **397**, 731.
42. M. Pettine, T. J. McDonald, M. Sohn, G. A. K. Anquandah, R. Zboril and V. K. Sharma, *Trends Environ. Anal. Chem.*, 2015, **5**, 1.
43. V. K. Sharma, T. J. McDonald, M. Sohn, G. A. K. Anquandah, M. Pettine and R. Zboril, *Environ. Chem. Lett.*, 2015, **13**, 49.
44. M. Yasin, A. F. El Mehdawi, C. E. Jahn, A. Anwar, M. F. S. Turner, M. Faisal and E. A. H. Pilon-Smits, *Plant Soil*, 2014, **386**, 385.
45. H. Wen and J. Carignan, *Atmos. Environ.*, 2007, **41**, 7151.

CHAPTER 27

Sulfur and Selenium in Nutrition Biology

MAURIZIO TOMMASINI

Italian National Order of Biologist, Via Icilio 7, 00153, Rome, Italy
Email: info@mauriziotommasini.it

27.1 Sulfur

Sulfur is the seventh most abundant element in the human body, it has been estimated that the total amount of sulfur in the body of a reference male of 70 kg is about 140 g.[1] Sulfur is present in the composition of two amino acids (SAA, sulfur amino acids), methionine (Met) and cysteine (Cys). Humans are unable to produce Met and Cys *de novo* and require the intake of both from dietary sources. Met is one of the nine essential amino acids in humans while Cys is considered conditionally essential since it is usually produced in the necessary amounts *via* the metabolism of Met. The sulfur amino acids are the primary sources to incorporate sulfur in a vast array of molecules, many of which are key metabolic intermediates.

27.1.1 Sulfur in the Human Organism

Met and Cys are raw materials for protein synthesis. Met is the initiating amino acid in the synthesis of human proteins,[2] and Cys has a critical role in the protein structure since Cys side-chain thiol groups allow the formation of inter- and intra-chain disulfide bonds, found predominantly in extracellular proteins where they contribute to the stability of the folded state in the ever-changing extracellular environment.[3]

Chalcogen Chemistry: Fundamentals and Applications
Edited by Vito Lippolis, Claudio Santi, Eder J. Lenardão and Antonio L. Braga

Published by the Royal Society of Chemistry, www.rsc.org

Sulfonation is the transfer of a sulfonate group ($-SO_3^-$) from the donor 3′-phoasphoadenosine-5′-phosphosulfate (PAPS), synthesized from sulfate in the cytosol, to a hydroxyl or amino group.[4] Many endogenous and xenobiotic molecules may undergo sulfonation which is an important step in the activation or deactivation of xenobiotics,[5] modulation of the function of hormones and catecholamines and structure and function of many macromolecules[6] such as glycosaminoglycans, proteoglycans, and galactoglycerolipids.[7]

Met is converted by L-methionine-*S*-adenosyltransferase (MAT) to *S*-adenosyl-methionine (SAM), a molecule that acts as a universal methyl group donor to a wide variety of acceptors. When the methyl group of SAM is transferred to the target molecule, in a reaction catalyzed by SAM-dependent methyltransferase, *S*-adenosylhomocysteine (SAH) is produced which is then hydrolyzed to homocysteine (Hcy) and adenosine by SAH hydrolase. Homocysteine can be remethylated to Met by the folate-dependent methionine synthase (MTR), a process that links sulfur to one carbon metabolism,[8] or by betaine-homocysteine methyltransferase (BHMT), closing the *methionine cycle* which occurs in most cell types. Met is catabolized through *trans*-sulfuration which converts homocysteine to cysteine by the sequential actions of cystathionine β-synthase (CBS) and cystathionine γ-lyase (CSE), an irreversible process restricted to the liver, kidney, intestine, and pancreas.[9] When Met intake is high there is a rapid interconversion to SAM which inhibits folic acid synthesis and activates CBS thus reducing remethylation and increasing *trans*-sulfuration and the production of Cys and GSH. Conversely, a low dietary intake of Met activates folic acid synthesis and inhibits CBS activity which in turn increases the remethylation of homocysteine to compensate for the dietary deficiency.[10] Vitamin B_{12}, folic acid, riboflavin, pyridoxine, niacin, and panthotenic acid are co-factors in various stages of the methionine cycle, a notable example of the role of vitamins in cell metabolism.[11]

Taurine is a sulfonic acid, an amino acid not used in protein synthesis, abundant in almost every tissue in the body, particularly the brain, muscle, liver, and kidneys. Taurine is mainly synthesized from Met and Cys *via* cysteine sulfinic acid decarboxylase (CSD).[12] Taurine has been proposed to act as an antioxidant, intracellular osmolyte, membrane stabilizer, neurotransmitter, and there is an increasing body of evidence that taurine depletion may lead to a wide range of pathological conditions.[13]

Glutathione (GSH, γ-glutamylcysteinylglycine) is the most abundant nonprotein thiols in the cell, a majority – up to 80% in the cytosol, the remaining mainly in the mitochondria. GSH is produced in two steps: in the first step, which requires the enzyme glutamate–cysteine ligase (GCL), Cys and glutamine combines to form γ-glutamylcysteine; in the second step glycine is added to the C-terminal of γ-glutamylcysteine, a condensation catalyzed by glutathione synthetase.[14] Both steps require ATP, the first reaction is the rate limiting step in the synthesis of GSH with the availability of Cys and a critical factor.[15] The thiol group on the Cys residue has a redox potential of −240 mV which makes it able to transfer electrons to a variety of molecules, a fact that allows GSH to perform a crucial role in neutralizing free radicals whose unchecked action could lead to severe damage to key cellular components and eventually to cell death.[16] GSH is essential in scavenging free radicals, in the maintenance of the overall redox state of the cell, in thiol protection, in nucleic acid and protein synthesis, in signal transduction, cytokine production, and in mitochondrial performance.[17]

Coenzyme A (CoA) is a molecule constructed from adenosine triphosphate, pantothenic acid, and Cys.[18] CoA is a ubiquitous essential cofactor utilized by a wide range of enzymes, a key molecule in major metabolic pathways, a carrier of activated acetyl groups essential for acetylation, and one of the central substrates that controls the balance between anabolic and catabolic reactions in the cell.[19] Lipoic acid,[20] thiamin[21] (vitamin B_1), and biotin[22] (vitamin H or factor H of the B_2 complex) are other important sulfur containing molecules acting as coenzymes in carbohydrate and amino acid metabolism, fatty acid biosynthesis, and gluconeogenesis.

Iron–sulfur proteins contain iron–sulfur clusters that consist of iron atoms, two or more, bridged by sulfur ligands provided by the enzymatic cleavage of Cys.[23,24] These proteins, which are present in the mitochondria, cytosol, endoplasmic reticulum, and nucleus, serve mainly as electron carriers and are of vital importance in biochemical tasks related to respiration, iron homeostasis, and gene expression.[25] Metallothionein (MTs) are a family of low molecular weight proteins, rich in Cys residues that are not engaged in disulfide bonds.[26,27] MTs have potent metal-binding and redox capabilities and are involved in Zn homeostasis and in the immune response, in metabolic regulation, and in protection against oxidative stress and heavy metals such as cadmium, copper, and mercury:[28–30] constant low-level exposure to heavy metals is known to increase the production of certain MTs, especially in the kidney, making them valid biomarkers in the medical and environmental fields.[31,32] There is evidence that MTs play a role in tumor formation, progression and drug resistance and may be utilized as biomarkers for cancer diagnosis.[33]

Hydrogen sulfide (H_2S) is produced from Cys through two distinct desulforation paths, one requiring the mediation of CBS and CSE and the other under the control of cysteine aminotransferase (CAT) and 3-mercaptopyruvate sulfurtransferase (MST). Hydrogen sulfide is the intermediate or end product of metabolic pathways in gut microbiota whose presence is associated with higher H_2S concentration in the intestinal tract and in plasma.[34] Hydrogen sulfide is a highly diffusible gas, a reactive species that requires tight regulation of synthesis and metabolism, whose signaling role through two major mechanisms, interaction with protein metal centers[35] and protein persulfidation,[36] is involved in the regulation of several cellular processes.

27.1.1.1 Sulfur Deficiency

A low intake of Met may impair protein synthesis, may affect folic acid synthesis, may decrease the organism capability to handle reactive and toxic chemical substances and reduce the availability of SAM in several methylation processes with alteration in patterns of DNA methylation.[37] Met deficiency may cause DNA hypomethylation, a condition associated with extensive damage to DNA which can possibly result in mutations and carcinogenesis.[38] A decrease in the dietary intake or in the endogenous production of Cys may affect GSH metabolism,[39] change the cellular redox state, and alter vital redox-sensitive signaling pathways.[40]

Hyperhomocysteinemia (HHCy) is a metabolic anomaly that can result from defective synthesis of folic acid, Hcy remethylation, or *trans*-sulfuration.[41] Amongst the many causes for this condition, the most common are an inadequate activity of methylene tetrahydrofolate reductase (MTHFR)[42] – an enzyme necessary for the synthesis of 5-methyltetrahydrofolate, a co-substrate for homocysteine remethylation to methionine – or any causes of vitamin B_{12}, B_6, or folate deficiency.[43] HHCy and accumulation of HCy in organ and tissues is associated with an increased risk of cardiovascular disease (CVD) and stroke, hypertension, kidney disfunction, Alzheimer's disease, and inflammatory bowel disease,[44] probably through the production of reactive oxygen species (ROS) and reactive nitrogen species (RNS) that may cause endothelial dysfunction that promotes blood vessel injury and atherosclerosis.[45] Acute and chronic stressful conditions are also associated with HHCy; in the latter case the overproduction of HCys, C-reactive protein, and inflammatory cytokine, can act with insulin resistance in elevating CVD risk.[46] HHCys is likely linked to reduced Met intake and bioavailability through modulation of the enzymes participating to remethylation and *trans*-sulfuration pathways, suggesting that maintenance of Met homeostasis is a critical factory in the regulation of protein balance in body tissues[47] as evidenced by the negative correlations between lean body mass and HHCys observed in kwashiorkor,[48] anorexia nervosa,[49] and bariatric surgery even after vitamin supplementation.[50]

27.1.1.2 Sulfur Excess

Usually, an excessive dietary intake of sulfur amino acid is stored as GSH[51] or excreted in the urine as sulfate.[52] High intake levels of Met in patients with impaired liver functions results in hyperammonemia and neurological dysfunctions.[53] In normal subjects, a high intake of Met – up to 5 times the normal – may contribute to the development of HHCys an effect that is mitigated by an adequate supply of vitamins B_6, B_{12}, C, and folic acid.[54] In animals, excess dietary levels of SAA may cause growth depression and show toxicity at levels higher than 5 times the normal.[55]

27.1.2 Sulfur Dietary Recommendations

There isn't a clearly defined and documented requirement of elemental sulfur, but a value of roughly 250 mg per day has been proposed,[56] an amount usually exceeded in a normal diet, with any excess readily excreted in the urine. Met and Cys are by and large the primary source of sulfur in the diet. Met is an essential amino acid in humans while the physiological requirement for Cys can be met by dietary Cys or by an excess of dietary Met. An adequate level of Cys intake may reduce the requirements of dietary Met by sparing its utilization in essential metabolic processes. The RDA for an adult of the combined sulfur amino acids has been estimated to be around 13 mg per kg of body weight: for a reference male adult of 70 kg this amounts to 1.1 g of Met/Cys per day, a requirement usually met by an average Western diet that may include up to 3.6 g of Met/Cys per day.[57] Higher values are suggested for infants and in early childhood due to a deficiency of the cystathionaze enzyme required for the production of both Cys and Tau from Met *via* the transmethylation pathway.[58]

Studies on older individuals, evaluating the SAA necessary to remain in positive nitrogen balance, point to higher requirements, from 2.1 to 3.0 kg per day,[59] but these findings were not confirmed in studies using amino acid oxidation.[60] There are no studies that take into account the role of SAA in providing S for sulfonation and for the synthesis of essential metabolic intermediates, but it has been established that certain drugs (Acetaminophen, Salicylamide, *etc.*) require sulfates for their excretion and can significantly reduce the availability of Met/Cys.[61] In older individuals the use of these drugs, a low-level inflammation, with the accompanying alteration of the redox state, and a low protein intake may lead to an adequate SAA intake that may be 2 to 3 times higher than those suggested by the WHO.[62]

Table 27.1 reports the sulfur amino acid requirements according to the literature.[63–67]

27.1.3 Sources of Sulfur in the Diet

The intake of inorganic sulfur in the standard Western diet is estimated to range from 0.2 to 1.5 kg per day, with minute quantities coming from sulfur-containing preservatives and additives.[68] The SAA in animal and plant proteins is the main source of sulfur in the diet. Animal tissues usually have a mean protein concentration twice as high as that of plants with a larger proportion of SAA amino acids (4.48% per gram of proteins) and a higher ratio of Met to Cys (approximately 3 : 1): egg white is one of the richest sources of S in the diet. Colostrum milk and maternal milk are rich in SAA

Table 27.1 Daily sulfur amino acid requirements (methionine + cysteine) (mg kg^{-1}) according to different estimations.

	Infants	Preschool children	Adults
Millward[63]	29	19	16
Di Buono[64]			13
Dewey[65]	27		
Young[66]			13
FAO/WHO/UNU[67]	58	27	13

and taurine which allow for an adequate daily intake of these nutrients despite a deficit of the CSE enzyme in the neonatal period.[69]

Data on the SAA content of foods suggest that Met and Cys content are limiting essential amino acids in many plant foods therefore individuals following strictly vegan diets may incur the risk of S deficiency:[70] soybeans have a content of SAA amino acid that is half that of egg white but a high protein content (33%) that may provide an adequate intake of Met and Cys.[71]

GSH is particularly abundant in fruits and vegetables and contribute up to 50% of the daily intake, meat contributes less than 25%, while dairy, cereals, and roots are devoid of this nutrient. Taurine is abundant in intracellular muscular tissue and is mainly supplied by the intake of meat and fish: subjects on strictly vegan diets run the risk of developing taurine deficiency, due to the total absence of taurine in plant foods.[72]

Plant belonging to the *Brassicaceae*, *Euphorbiaceae*, and *Liliaceae* families – cabbage, cauliflower, cassava, garlic, and onion – contain several organosulfur compounds such as allylsulfur molecules and glucosinolate-derived isothiocyanates that *in vivo* have a protective action against pathogens and pests.[73] The contribution of this compound to the daily intake of sulfur is negligible but a habitual consumption may have health benefits due to their therapeutic properties as antithrombotic and anticancer agents.[74]

27.1.4 Sulfur Supplementation

Plants and animals may benefit from sulfur supplementation and poultry diets are routinely enriched with Met/Cys to enhance growth. It is probable that a balanced omnivorous diet may guarantee an adequate daily intake of sulfur, sufficient to meet the needs of an average individual, while certain populations may require targeted supplementation.[75]

Vegan individuals may benefit from vitamin B_{12} supplementation;[76] Met deficiency can be caused by a chronic dietary folate insufficiency and may be alleviated by folate supplementation;[77] acute stress patients such as burn patients[78] may have an increased demand for Met; individuals on severe weight-losing diets, septic patients,[79] and cancer and HIV positive patients[80] may benefit from supplementation with *N*-acetylCysteine; older individuals, who often decrease their food intake at the expense of protein, and individuals in therapy with drugs that reduce the availability of Met/Cys, may benefit or need supplementation with SAA in amounts exceeding 2–3 times the minimal recommended daily intake. Taurine supplementation may be necessary in pre-term infants and in individuals with conditions that cause intestinal malabsorption.[81]

27.1.5 Methionine, Cancer, and Aging

Met-poor diets have been suggested to play an important role in the control of cancer growth because of the high requirements for essential amino acid shown by cancer cells: *in vitro* and *in vivo* studies[82] on Met-deficient culture media or with synthetic Met-free diets have shown selective killing of Met-dependent cancer cells, limited tumor growth, and a reduction in tumor volume.[83,84]

Met restriction, without malnutrition, is associated to metabolic health[85] and lifespan extension in several model organisms, from yeast to rats, and can mimic the effects of dietary and protein restriction on longevity, regardless of calorie intake.[86] A Met-restricted diet may increase longevity through signaling pathways involving mTOR,[87] insulin, and GH/IGF, through an increase in the production of hydrogen sulfide and polyamines and a reduction in the production of SAM, with a decrease in oxidative stress due to inhibition of the production of reactive oxygen species in mitochondria and an increase of autophagy.[88]

27.2 Selenium

Selenium (Se) was discovered in 1817 by Swedish chemist Jacob Berzelius while investigating the cause of illness among workers at a sulfuric acid manufacturing plant. Accurate sample analysis identified in some sediments from the plant a new element, with properties similar to sulfur and tellurium, which was called selenium, after the Greek moon goddess Selene.[89]

For a long time, selenium was regarded as a toxic element, especially after it was found to cause poisoning in livestock in areas with a high content in the soil,[90] an occurrence that had already been described by Marco Polo, in China in the 13th century, reporting a necrotic hoof disease in horse and camel feed containing *Astragalus danicus*, a plant known to accumulate large quantities of selenium.

The role of selenium as an essential trace element was established by Klaus Schwartz and Calvin Foltz,[91] who investigated the effects of a selenium deficient diet in rats that led to necrotic liver formation; other studies showed that in calves and lambs selenium deficiency was associated with a form of myopathy known as "white muscle disease".[92] These classic studies led to a widespread use of selenium supplementation in livestock with positive economic impacts in the economy of several countries.

The focus of selenium studies shifted to the field of molecular biology in 1973, when selenium was found to be present in the active center of the enzyme glutathione peroxidase,[93] as a constituent of the amino acid selenocysteine (Sec). This discovery sparked a flurry of studies that have led to the identification of several other enzymes that include the element in their active centers, the so-called selenoproteins, which play many important biological roles, chief amongst them the control of redox balance in cells, with effects on immunity, fertility, hormone balance, and the aging process.

27.2.1 Selenium in the Environment

Total soil selenium concentrations range from 0.01 to 2 mg kg^{-1} with a mean of 0.4 mg kg^{-1}. Higher concentrations, up to 1200 mg kg^{-1} are found in soil derived from seleniferous parent materials originating from sedimentary rocks, widespread in part of the United States, Canada, South America, China, and Russia.[94] Lower concentrations of Se are associated with soils derived from igneous rocks, in area with elevated erosion rates and limited atmospheric deposition. The total selenium content depends on a host of factors, soil type and texture, origin and geological history, organic matter content, weathering degree and rainfalls, natural Se emission and anthropogenic emission, from fossil fuel combustion, non-ferrous metal production and manufacturing, and agricultural activities such as irrigation and long-term use of organic and chemical fertilizers.[95]

Selenium in soil is found in inorganic form with different oxidation states [selenite (SeO_3^{2-}) and selenate (SeO_4^{2-})] with solubility and mobility raising with increasing oxidizing conditions and poor correlation between total soil Se and available soil Se. Organic forms of selenium are usually found in biological matters, and include methylated and unmethylated Se-amino acids, volatile forms of dimethyl selenide and dimethyl diselenide and other forms yet to be indagated.[96]

Plant takes up both inorganic and organic Se species and accumulate the element in their organs with large variation between genera and species; hyperaccumulator species thrive on Se-rich soils and tolerate high concentrations of selenium, up to 1000 μg Se per g d.wt., which can be toxic to grazers eventually. Hyperaccumulators store selenium in the form of methylselenocysteine (MeSec) and selenocystathionine, while non-accumulators store selenium mainly as selenomethionine (SeMet).[97]

27.2.2 Selenium Absorption, Metabolism, and Excretion

The selenium in food, either of inorganic or organic origin, is absorbed in the duodenum, cecum, and colon: Sec and SeMet are absorbed by active transport system, while inorganic form, mainly selenate, are absorbed by passive transport (anion changers of the SLC26 gene family).[98] Most organic and inorganic form are well absorbed – absorption range from 70 to 90% – except for selenite, which usually has an absorption lower than 60%.[99] In the enterocytes all the absorbed selenium is converted to hydrogen selenide (H_2Se) through a different cascade of reactions and transported to the liver linked to LDL and VLDL fractions, albumin, and alpha-globulin. In the liver the hydrogen selenide is converted to selenophosphate (SePhp) which reacts with tRNA-bound serinil residues to give Sec-bound tRNA from which Sec[100] can be co-translationally-incorporated in selenoproteins, at loci encoded by specific UGA codons.[101]

Selenium is transported in the plasma, in the form of selenoprotein P (SePP),[102] a heparin-binding glycoprotein that delivers selenium to organs where apolipoprotein E receptor 2 or megalin are expressed;[103] glutathione peroxidase 3 (Gpx 3) is another selenoprotein abundant in plasma but does not appear to transport selenium for specific uptake by cells. When selenium intake is extremely high a small molecule form of selenium[104] can transfer the element *via* a nonspecific pathway.

Selenium is excreted in urine, in feces, by exhalation in breath, and through loss of hair and skin cells. Selenium excretion in urine in increased with a high selenium intake: the major metabolite in urine is a selenosugar, β-methylseleno-*N*-acetyl-D-galactosamine, synthesized in the liver.[105] The volatile compound dimethyl selenide is excreted through breath only at high selenium intakes, and in the case of selenium poisoning is responsible of the unpleasant breath and body odor that are typical symptoms of the condition. The non-absorbed selenium from food is incorporated into the bile and the pancreatic and intestinal secretions and is eliminated in the feces.[106]

27.2.3 Selenium in Human Organisms

Selenium is an essential micronutrient, incorporated as Sec in the active site of several proteins where it usually performs catalytic redox reactions. To date, the human selenoproteome is encoded by 25 genes:[107] glutathione peroxidases (GPx), thioredoxin reductases (TXNRD), iodothyronine deiodinases (DIO), and selenoprotein P (SePP) have been extensively indagated and their role in human physiology has been described in detail.[108] Selenoproteins serve quite diverse functions: redox signaling, control of the reduced state of Cys residues in cytosolic and mitochondrial proteins, removal of hydrogen peroxide, repair of oxidatively-damaged proteins, control of cytoskeleton assembly, formation and quality control of structural disulfide bonds in proteins, thyroid hormone activation and inactivation, selenoprotein synthesis, Se transport, protein folding, and ER-associated degradation.[109]

A vast body of studies has clearly shown that selenium has a double-edged behavior, with problems associated with both deficiency and excess.

27.2.3.1 Selenium Deficiency

Severe selenium deficiency may occur in areas with selenium-poor soil. Three endemic diseases associated with selenium deficiency are:

1. *Keshan disease* is a severe form of juvenile cardiomyopathy with a high mortality rate, named after Keshan County, in Northeast China, where symptoms were first described. The incidence of the disease fluctuates with the seasons, a fact that suggests that viral infection by a particular strain of coxsackie virus is an important cofactor. Supplementation with sodium selenite can prevent the development of the disease even in a subject infected with the virus.[110]
2. *Kashin-Beck disease* (KBD) is a type of osteochondropaty mainly found in a diagonal belt from northeast to southwest China, and also in Mongolia, Siberia, and North Korea. The cause of KBD is multifactorial, with selenium deficiency predisposing the target cell to oxidative damage. Other cofactors contributing to the development of the disease are iodine deficiency, the prolonged consumption of mycotoxins from fungi on storage grains and the presence of organic matter – humic and fulvic

acids – in drinking water.[111] Selenium supplementation has been reported to prevent the disease in some studies but has had no significant effect in others, showing that KBD requires other factors, linked to specific regions and to families: a polymorphism in the GPx 1 gene has been indicated as a potential genetic risk factor.[112]

3. *Myxedematous endemic cretinism* is prevalent in iodine and selenium poor areas of central Africa rich in thiocyanate. The disease is induced by thyroid atrophy resulting in mental retardation, short stature, and goiter.[113]

Low levels of selenium in the body is associated with infertility in men. GPx 4 has vital function for spermatogenesis: deletion of the mitochondrial isoform of the protein is associated with impaired sperm quality and severe structural abnormalities in the midpiece of the spermatozoa.[114] Se deficiency in pregnancy negatively affects the development of the embryo[115] and is associated with an increased risk of pre-eclampsia[116] while an adequate Se intake is associated with increased gestational length and a decreased risk of preterm delivery.[117]

The thyroid gland is the organ with the highest concentrations of selenium in the body and expresses several selenoproteins that are involved in the protection of the gland from oxidative stress and in the fine regulation of the metabolism of thyroid hormones (TH).[118] On a selenium deficient diet the thyroid gland is preferentially supplied and retains selenoprotein activity[119] but it is unclear the role of Se in the modulation of peripheral TH metabolism. In several epidemiological studies a low selenium status appears to be associated with an increased risk of thyroid disease[120] and dietary intake of Se is associated with sub-clinical hypothyroidism,[121] but other studies found no significant correlations.[122]

Low serum Se is associated with low levels of natural killer cells and Se deficiency is associated with increased susceptibility to infections and to inflammation-associated cancers. Research conducted in cell cultures and animal models indicates that Se plays an important role in the optimal function of immune cells, in innate immunity, and in B-cell dependent antibody production.[123] Deficiencies in Se and other micronutrients are common in individuals infected by HIV.[124]

The brain is another organ in which selenium supply is prioritized for normal development and functions. Se deficiency can nonetheless adversely affect the functioning of the nervous system[125] and is frequently observed in individual suffering of depression, anxiety, and Alzheimer's disease, and increases the risk of seizure in epilepsy.[126]

Many studies have shown that low Se status, resulting from low intake, is associated to a higher risk of developing breast, prostate, lung, and colorectal cancer.[127] Several mechanisms of the effect of Se on cancer have been proposed, including the antioxidant effects through the action of selenoproteins, the regulation of cell cycle and apoptosis, the modulation of angiogenesis and extracellular matrix, histone deacetylase inhibition, and alteration of DNA damage and repair mechanism.[128]

27.2.3.2 Selenium Excess

An excess of Se can be toxic to an organism in both organic and inorganic form. The toxicity of Se is related with competitive inhibition between selenium and sulfur: selenium may substitute sulfur amino acids while inorganic compounds displace sulfur during the synthesis of mercapturic acid and the reaction of selenites with thiol groups, leading to the synthesis of distorted, dysfunctional enzyme and protein molecules which cause anomalies in the biochemical functioning of the cell. Symptoms of acute Se poisoning include hypotension, tachycardia, neurological disorders, fever, and pulmonary edema, while chronic Se toxicity (known as selenosis) is characterized by nail fragility, hair loss, skin lesions, joint pain, and garlic odor in breath due to excretion of excess Se as dimethyl selenide.

Several studies have shown an association between selenium supplementation and an increased risk of type 2 diabetes in populations that have a sufficient or good Se status.[129] Results from several animal studies indicated that selenium supplementation and selenoproteins may interfere with insulin action and glucose homeostasis. The proposed mechanism is that daily doses near the tolerable upper intake level may cause selenium to accumulate in pancreatic tissue, exerting oxidative stress that over time damages beta-cells.[130] In an intervention study Se supplementation is associated with an increased risk of squamous cell carcinoma,[131] while supplementation of Se and vitamin E in Se repleted men (SELECT study)[132] is associated with an increased risk of type 2 diabetes and prostate cancer.[133]

27.2.4 Selenium Dietary Recommendations

There is a U-shaped relationship between Se intake/status and its effects on human health, with adverse outcomes related both to deficiency and excess.

Evaluation of Se intake is a difficult task due to relevant variation of the Se content of foodstuff and partial or incomplete Se concentration data in the food composition table. Individual dietary intake ranges from 40–50 μg per day in many European countries to well over 300 μg per day in seleniferous area of South America and China.[134]

The recommended dietary allowance (RDA) is based on the amount of selenium necessary to maximize the activity of GPx 3 in serum. The RDA for adults, with no difference between males and females, is 55 μg per day. The RDAs in pregnancy and breastfeeding are 60 μg per day and 70 μg per day, respectively (Table 27.2). An intake of 55 μg per day is considered sufficient to support an adequate synthesis of the 25 selenoproteins but there may be differences between individuals, gender, and races. EFSA suggests a slightly higher value with the adequate intake (AI) set at 70 μg per day for adults and 85 μg per day for breastfeeding women.[135] A dosage of 200–300 μg per day is considered therapeutic and is widely used in intervention studies.[136]

The Food and Nutrition Board (FNB) of the US Institute of Medicine set the tolerable upper intake level (UL) for selenium at 400 μg per day in adults,

Table 27.2 Recommended Dietary Allowance (RDA) of selenium for individuals in different age groups and genders.[136]

Age group	Male (μg per day)	Female (μg per day)	Pregnancy (μg per day)	Lactation (μg per day)
Birth to 6 months	15[a]	15[a]		
7 to 12 months	20[a]	20[a]		
9 to 13 years	40	40		
14 to 18 years	55	55	60	70
19 to 30 years	55	55	60	70
31 to 50 years	55	55	60	70
50 to 71 years	55	55		
Above 71 years	55	55		

[a]Adequate Intake (AI), a recommended intake value based on observed or experimentally determined approximations or estimates of nutrients used when an RDA cannot be determined.

based on the prevention of early signs of chronic selenium toxicity such as hair and nail brittleness and loss. Populations that live in areas with selenium-rich soil can tolerate a daily intake of 800 μg per day, a value considered the NOAEL (no observed adverse effect level) for Se, while the LOAEL (lowest observed adverse effect level) is set at 900 μg per day.[137]

The ability of some people and/or populations to tolerate low or high Se intake may be related to the presence of polymorphism in selenoproteins genes, genes involved in Se metabolism and Se excretion and detoxification. High levels of Se intake may have beneficial health effects in populations exposed to toxic elements such as mercury, arsenic, and cadmium.[138]

The effect of Se intake may depend by the speciation of Se in foodstuff:[139] it has been suggested that the organic form of Se, SeMet, may be more toxic than the inorganic form as they can be incorporated in tissue proteins rather than be excreted rapidly but several studies with supplementation of Se-yest, rich in SeMet, up to 800 μg per day for long periods have been carried out without indication of the toxic effects. Data from animal studies indicate that the inorganic forms of Se, selenite and selenate, are more acutely toxic than the organic form while elemental Se is quite nontoxic with an oral LD50 of 6700 mg kg^{-1}.[140]

27.2.5 Selenium Biomarkers

The assessment of Se nutritional status is a challenge since it depends on several factors, such as Se content of the soil, Se incorporation in plant and foodstuff, bioavailability from food sources, dietary intake, and health state of the individual.

Se status is a function of dietary intake, retention, metabolism, and excretion. Predicting Se status from food consumption is difficult due to the lack of precision of food composition tables and the huge geographical variation of Se content in soil. Plasma Se levels can be used to evaluate Se intake when SeMet is the predominant form in the food consumed.[141]

Plasma Se concentrations are affected by several factors, such as geographical location, gender, age, race, BMI, smoking status, and alcohol consumption, nonetheless a systematic review has recommended the use of plasma Se concentrations as a reliable biomarker in supplementation studies with adults. The optimal value of Se in plasma has been estimated at 90–120 $\mu g L^{-1}$, a level sufficient to saturate selenoproteins in this fraction: deficiency sets in at levels below 85 $\mu g L^{-1}$.[142]

The concentration of Se in toenails provides integrated measurements of long-term exposure to Se and has a direct relationship with SePP and SeCys and an inverse relationship with the inorganic form. Toenails are used in large epidemiological studies for their slow growth, ease of collection and low susceptibility to external contamination.[143]

The measurement of selenoproteins directly reflects Se function. GPx 1 is a good biomarker in the case of low Se status and is effective in evaluating Se supplementation due to its rapid response, GPx 3 can identify the risk of nutritional deficiency, while SePP is a useful biomarker of Se status in populations with relatively low Se intake, but not in populations with high intake. A recent study has observed a discrepancy between serum Se level and GPx activity suggesting that other factors may impact the activity of this enzyme, such as genetic polymorphism.[144]

27.2.6 Sources of Selenium in the Diet

Selenium is accumulated in the human organism mainly through ingestion. The amount of selenium in the diet depends on several factors: the soil on which crops are grown, the fodder fed to animals, the food consumed, and the way in which food is processed.[145]

The content of Se in the soil is a major factor driving the amount consumed and is related to deficiency and toxicity in some regions. The selenium content of the soil is usually in the range of 1–1.5 μg Se per g, with values as high as 5.0 μg Se per g in seleniferous soils. The bioavailability of Se is another major factor with organic forms (SeMet, SeC, and methylated derivates) that appear to be more readily available than the inorganic forms selenite and selenate. The bioavailability of Se is reduced in the presence of heavy metals and sulfur, is increased in the presence of vitamin A, C, and E. Cooking and thermal processing can lead to significant loss of selenium while the matrix of the food does not appear to have an important role in Se bioavailability.[146,147]

Bread and cereals are major sources of selenium in the diet, with a content that ranges widely, from 0.01 to 30 $mg kg^{-1}$, with predominance of SeMet (55–85%). Meat is another major source of Se with high levels in the offals, 0.93 $mg kg^{-1}$ in the liver, and lower levels in the muscle, usually 0.2 $mg kg^{-1}$, susceptible to an increase up to 0.6 $mg kg^{-1}$ when cattle are fed with Se-enriched yeast; the predominant form in meat is SeMet (50–60%) and SeCys (20–30%). The selenium content of fish ranges from 0.1 to 5.0 $mg kg^{-1}$, is higher in some marine fish such as cod and tuna, with SeMet (30–70%) and selenite and selenate (12–45%) as the predominant form. The selenium content of eggs, milk, and dairy products varies widely and is increased by supplementation of the animal's diet. Fruit and vegetables usually contain small amounts of selenium but certain species such as onion, garlic, broccoli, and other *Brassica* species are good accumulators and when grown in soil rich in Se can reach a content up to 150–300 mg kg^{-1}; the predominant species are Se-methylselenocysteine and γ-glutamyl-Se-methylselenocysteine.[148] Brazil nuts, the fruit of *Bertholletia excelsa*, a plant from the Amazon, has the highest content of Se in non-enriched food and can reach levels up to 600 mg kg^{-1}, mainly in the form of SeMet; a single nut grown in an area with Se rich soil may provide ∼100 μg of the mineral, more than the RDA for adults, but a nut grown in soil poor in Se usually provides 10 times less, a reminder of the inherent difficulties in evaluating Se intake from food.[149]

Bread and cereal products cover ∼50% of the daily intake of selenium, meat and fish account for ∼35%, fruit and vegetables provide ∼10% each, and water and drinks contribute 5–25% of selenium in the diet.

27.2.7 Selenium Supplementation

Se deficiency has increased over recent years, with a negative impact on public health. Recent estimates suggest that nearly 1 billion people may have a limited Se intake[150] and this number is going to grow in the future because of predicted climate changes that will probably reduce the Se content of the soil in agricultural areas.[151]

The development of crops with a high content of Se using biofortification intervention has been secured in several countries (Finland,[152] Malawi,[153] Tibet, and China[154]) where Se deficiency has been endemic. Factors that must be considered in the planning strategies of biofortification are the Se species to be used, the mode of Se fertilization, and crop species.[155] Se has no essential metabolic role in higher plants, is incorporated in SeMet and SeCys *via* the metabolic pathway of sulfur, and can be toxic in excess, causing the production of malformed proteins.[156] Lower concentrations of Se have shown beneficial effects, particularly in hyper-accumulator plants such as several species of the *Brassica* and *Allium* genus.[157]

Selenium supplements are available in both inorganic and organic form. Sodium selenate has the best absorption but is rapidly excreted in urine, before being incorporated into proteins; sodium selenite has a lower absorption, about 50%, but better retention; SeMet is nearly completely absorbed, nearly 90%, but only a third is converted to free SeMet.[158] Different Se species contribute in a different manner to the maintenance of an adequate Se status: in intervention studies, the organic form increased blood Se concentrations more effectively than the inorganic form, but selenate and selenite appear to increase plasma GPx activity more effectively than the organic form.[159,160]

Observational studies have shown that low Se status is correlated with a higher risk of developing cancer,[161] metabolic disease,[162] cardiovascular disease (CVD),[163] viral infections,[164] anomalies in thyroid-hormone metabolism,[165] and low fertility.[166] Intervention trials with selenium supplementation have given mixed results, with the outcome of the research dependent on the Se species used, the dose taken, the bioavailability of the element, baseline Se status and, in studies about cancer, the type of cancer and stage of cancer lesions.[167] Recent reviews concluded that there is no convincing evidence suggesting that Se supplements may prevent cancer in humans,[168] no convincing evidence the Se supplements may be used in the primary prevention of CVD[169] or in the treatment of Hashimoto's thyroiditis[170] and no clear evidence to recommend Se supplementation of critically ill patients.[171]

27.3 Sulfur, Selenium, and the Microbiota

The human digestive tract is inhabited by a host of microorganisms including bacteria, virus, fungi, and protozoa, collectively called microbiota. The composition of the microbiota varies along the enteric tract with a central core of bacterial phylum – *Bacteroidetes* and *Firmicutes* – always present and with individual variation dependent on age, diet, health status, geographic location, and other influencing factors.[172]

An elevated production of hydrogen sulfide by sulfur-reducing bacteria in the colon[173] may be linked to the development of ulcerative colitis, Crohn's disease, and irritable bowel syndrome, with lesion to colonocytes and to the intestinal mucosa even though sulfur-reducing bacteria are found in individuals that do not develop such conditions, an indication that other environmental and genetic factors are at play.[174,175]

Consumption of food rich in selenium – brazil nuts – may regulate gut microbiota and increase the production of selenoproteins.[176] Studies on animals have shown the Se supplementation can modulate the composition of gut microbiota for protection against intestinal dysfunction,[177] while Se-deficient diets may adversely affect the gut barrier function and alter the intestinal and immune response.[178,179] These observations indicate the need for randomized clinical trials necessary to elucidate the connection between the S and Se content of the diet, the composition of the intestinal microbiota, and its effects on human health.

References

1. G. B. Forbes, Body composition, in *Present Knowledge in Nutrition*, ed. E. E. Ziegler and L. J. Filer, ILSI Press, Washington, DC, 7th edn, 1996.
2. J. T. Brosnan and M. E. Brosnan, *J. Nutr.*, 2006, **136**(6), 1636S–1640S.
3. S. F. Betz, *Protein Sci.*, 1993, **2**(10), 1551.
4. C. D. Klaassen and J. W. Boles, *FASEB J.*, 1997, **11**(6), 404.
5. F. C. Kauffman, *Drug Metab. Rev.*, 2004, **36**(3–4), 8233.
6. A. W. Leung, I. Backstrom and M. B. Bally, *Oncotarget*, 2016, **7**(34), 55811.
7. Y. S. Yang, C. C. Wang, B. H. Chen, Y. J. Hou, K. S. Hung and Y. C Mao, *Molecules*, 2015, **20**(2), 2138.
8. G. S. Ducker and J. D. Rabinowitz, *Cell Metab.*, 2017, **25**(1), 27.

9. M. V. Martinov, V. M. Vitvitsky, R. Banerjee and F. I. Ataullakhanov, *Biochim. Biophys. Acta*, 2010, **1804**(1), 89.
10. M. Di Buono, L. J. Wyke, D. E. Cole, R. O. Ball and P. B. Pencharz, *J. Nutr.*, 2003, **133**(3), 733.
11. D. S. Froese, B. Fowler and M. R. Baumgartner, *J. Inherited Metab. Dis.*, 2019, **42**(4), 673.
12. H. Ripps and W. Shen, *Mol. Vision*, 2012, **18**, 2673.
13. S. Schaffer and H. W. Kim, *Biomol. Ther.*, 2018, **26**(3), 225.
14. H. Zhang and H. J. Forman, *Semin. Cell Dev. Biol.*, 2012, **23**(7), 722.
15. S. C. Lu, *Mol. Aspects Med.*, 2009, **30**(1–2), 42.
16. H. J. Forman, H. Zhang and A. Rinna, *Mol. Aspects Med.*, 2009, **30**(1–2), 1.
17. C. Gaucher, A. Boudier, J. Bonetti, I. Clarot, P. Leroy and M. Parent, *Antioxidants*, 2018, **7**(5), 62.
18. R. Leonardi, Y. M. Zhang, C. O. Rock and S. Jackowski, *Prog. Lipid Res.*, 2005, **44**(2–3), 125.
19. F. Pietrocola, L. Galluzzi, J. M. Bravo-San Pedro, F. Madeo and G. Kroemer, *Cell Metab.*, 2015, **21**(6), 805.
20. A. Gorąca, H. Huk-Kolega, A. Piechota, P. Kleniewska, E. Ciejka and B. Skibska, *Pharmacol. Rep.*, 2011, **63**(4), 849.
21. S. Manzetti, J. Zhang and D. van der Spoel, *Biochemistry*, 2014, **53**(5), 821.
22. R. J. McMahon, *Annu. Rev. Nutr.*, 2002, **22**, 221.
23. R. Lill and S. A. Freibert, *Annu. Rev. Biochem.*, 2020, **89**, 471.
24. J. J. Braymer, S. A. Freibert, M. Rakwalska-Bange and R. Lill, *Biochim. Biophys. Acta, Mol. Cell Res.*, 2021, **1868**(1), 118863.
25. A. Sheftel, O. Stehling and R. Lill, *Trends Endocrinol. Metab.*, 2010, **21**(5), 302.
26. P. Coyle, J. C. Philcox, L. C. Carey and A. M. Rofe, *Cell. Mol. Life Sci.*, 2002, **59**(4), 627.
27. E. Carpenè, G. Andreani and G. Isani, *J. Trace Elem. Med. Biol.*, 2007, **21**(1), 35.
28. M. Vasák, *J. Trace Elem. Med. Biol.*, 2005, **19**(1), 13.
29. P. Dziegiel, B. Pula, C. Kobierzycki, M. Stasiolek and M. Podhorska-Okolow, *Adv. Anat., Embryol. Cell Biol.*, 2016, **218**, 1.
30. N. Thirumoorthy, K. T. Manisenthil Kumar, A. Shyam Sundar, L. Panayappan and M. Chatterjee, *World J. Gastroenterol.*, 2007, **13**(7), 993.
31. L. Chen, L. Ma, Q. Bai, X. Zhu, J. Zhang, Q. Wei, D. Li, C. Gao, J. Li, Z. Zhang, C. Liu, Z. He, X. Zeng, A. Zhang, W. Qu, Z. Zhuang, W. Chen and Y. Xiao, *J. Biol. Chem.*, 2014, **289**(32), 22413.
32. S. Ganguly, E. Taioli, B. Baranski, B. Cohen, P. Toniolo and S. J Garte, *Cancer Epidemiol., Biomarkers Prev.*, 1996, **5**(4), 297.
33. M. Si and J. Lang, *J. Hematol. Oncol.*, 2018, **11**(1), 107.
34. A. Giuffrè and J. B. Vicente, *Oxid. Med. Cell. Longevity*, 2018, **2018**, 6290931.
35. R. Pietri, E. Román-Morales and J. López-Garriga, *Antioxid. Redox Signaling*, 2011, **15**(2), 393.

36. M. R. Filipovic, J. Zivanovic, B. Alvarez and R. Banerjee, *Chem. Rev.*, 2018, **118**(3), 1253.
37. D. M. Townsend, K. D. Tew and H. Tapiero, *Biomed. Pharmacother.*, 2004, **58**(1), 47.
38. M. J. Hoffmann and W. A. Schulz, *Biochem. Cell Biol.*, 2005, **83**(3), 296.
39. F. Jahoor, *Nutr. Rev.*, 2012, **70**(3), 176.
40. W. Dröge, *Philos. Trans. R. Soc., B*, 2005, **360**(1464), 2355.
41. K. S. McCully, *Am. J. Pathol.*, 1969, **56**(1), 111.
42. D. S. Froese, M. Huemer, T. Suormala, P. Burda, D. Coelho, J. L. Guéant, M. A. Landolt, V. Kožich, B. Fowler and M. R. Baumgartner, *Hum. Mutat.*, 2016, **37**(5), 427.
43. J. Kim, H. Kim, H. Roh and Y. Kwon, *Arch. Pharmacal Res.*, 2018, **41**(4), 372.
44. C. Tinelli, A. Di Pino, E. Ficulle, S. Marcelli and M. Feligioni, *Front. Nutr.*, 2019, **6**, 9.
45. C. van Guldener and C. D. Stehouwer, *Semin. Thromb. Hemostasis*, 2000, **26**(3), 281.
46. D. M. Cummings, D. E. King, A. G. Mainous and M. E. Geesey, *Eur. J. Cardiovasc. Prev. Rehabil.*, 2006, **13**(2), 180.
47. Y. Ingenbleek, *Nutrients*, 2017, **9**(9), 1035.
48. A. G. M. Abdel Gader, S. H. Abdullah and A. Y. Kordofani, *East. Mediterr. Health J.*, 2009, **15**(6), 1432.
49. J. Wilhelm, E. Müller, M. de Zwaan, J. Fischer, T. Hillemacher, J. Kornhuber, S. Bleich and H. Frieling, *J. Neural Transm.*, 2010, **117**(4), 521.
50. N. Komorniak, M. Szczuko, B. Kowalewski and E. Stachowska, *Obes. Surg.*, 2019, **29**(11), 3735.
51. R. L. Goul and R. Pazdro, *Nutrients*, 2019, **11**(5), 1056.
52. D. E. Cole and J. Evrovski, *Crit. Rev. Clin. Lab. Sci.*, 2000, **37**(4), 299.
53. E. A. Phear, B. Ruebner, S. Sherlock and W. H. Summerskill, *Clin. Sci.*, 1956, **15**, 93.
54. P. J. Garlick, *J. Nutr.*, 2006, **136**(6), 1722S.
55. D. H. Baker, *J. Nutr.*, 2006, **136**(6), 1670S.
56. Y. Ingenbleek and H. Kimura, *Nutr. Rev.*, 2013, **71**(7), 413.
57. Y. Ingenbleek, *J. Nutr.*, 2006, **136**(6), 1641S.
58. J. Viña, M. Vento, F. García-Sala, I. R. Puertes, E. Gascó, J. Sastre, M. Asensi and F. V. Pallardó, *Am. J. Clin. Nutr.*, 1995, **61**(5), 1067.
59. S. G. Tuttle, S. H. Bassett, W. H. Griffith, D. B. Mulcare and M. E. Swendseid, *Am. J. Clin. Nutr.*, 1965, **16**, 229.
60. N. K. Fukagawa, Y. M. Yu and V. R. Young, *Am. J. Clin. Nutr.*, 1998, **68**(2), 380.
61. J. H. Lin and G. Levy, *J. Pharm. Sci.*, 1983, **72**(3), 213.
62. M. E. Nimni, B. Han and F. Cordoba, *Nutr. Metab.*, 2007, **4**, 24.
63. D. J. Millward, A. Fereday, N. Gibson and P. J. Pacy, *Am. J. Clin. Nutr.*, 1997, **66**(4), 774.
64. M. Di Buono, L. J. Wykes, R. O. Ball and P. B. Pencharz, *Am. J. Clin. Nutr.*, 2001, **74**(6), 756.

65. K. G. Dewey, G. Beaton, C. Fjeld, B. Lönnerdal and P. Reeds, *Eur. J. Clin. Nutr.*, 1996, **50**(1), S119.
66. V. R. Young and S. Borgonha, *J. Nutr.*, 2000, **130**(7), 1841S.
67. FAO/WHO/UNU, Report of a joint FAO/WHO/UNU expert consultation, World Health Organization, Geneva, 1985.
68. T. H. J. Florin, G. Neale, S. Goretski and J. H. Cummings, *J. Food Compos. Anal.*, 1993, **6**(2), 140.
69. M. E. McNally, S. A. Atkinson and D. E. Cole, *J. Nutr.*, 1991, **121**(8), 1250.
70. Y. Ingenbleek and K. S. McCully, *Nutrition*, 2012, **28**(2), 148.
71. M. Friedman and D. L. Brandon, *J. Agric. Food Chem.*, 2001, **49**(3), 1069.
72. M. F. McCarty, *Med. Hypotheses*, 2004, **63**(3), 426.
73. I. D. Nwachukwu, A. J. Slusarenko and M. C. Gruhlke, *Nat. Prod. Commun.*, 2012, **7**(3), 395.
74. N. Miękus, K. Marszałek, M. Podlacha, A. Iqbal, C. Puchalski and A. H. Świergiel, *Molecules*, 2020, **25**(17), 3804.
75. M. C. van de Poll, C. H. Dejong and P. B. Soeters, *J. Nutr.*, 2006, **136**(6), 1694S.
76. D. R. Bakaloudi, A. Halloran, H. L. Rippin, A. C. Oikonomidou, T. I. Dardavesis, J. Williams, K. Wickramasinghe, J. Breda and M. Chourdakis, *Clin. Nutr.*, 2021, **40**(5), 3503.
77. C. H. Halsted, *Clin. Chem. Lab. Med.*, 2013, **51**(3), 457.
78. Y. M. Yu, J. F. Burke and V. R. Young, *J. Trauma*, 1993, **35**(1), 1.
79. W. Druml, G. Heinzel and G. Kleinberger, *Am. J. Clin. Nutr.*, 2001, **73**(5), 908.
80. R. Breitzkreuz, S. Holm, N. Pittack, M. Beichert, A. Babylon, J. Yodoi and W. Dröge, *Res. Hum. Retroviruses*, 2000, **16**, 203.
81. H. S. Geggel, M. E. Ament, J. R. Heckenlively, D. A. Martin and J. D. Kopple, *N. Engl. J. Med.*, 1985, **312**, 142.
82. D. Wanders, K. Hobson and X. Ji, *Nutrients*, 2020, **12**(3), 684.
83. R. M. Hoffman, Y. Hoshiya and W. Guo, *Methods Mol. Biol.*, 2019, **1866**, 7.
84. X. Durando, E. Thivat, P. Gimbergues, E. Cellarier, C. Abrial, M. Dib, O. Tacca and P. Chollet, *Bull. Cancer*, 2008, **95**(1), 69.
85. E. K. Lees, E. Król, E. L. Grant, L. K. Shearer, C. Wyse, E. Moncur, A. S. Bykowska, N. Mody, T. W. Gettys and M. Delibegovic, *Aging Cell*, 2014, **13**(5), 817.
86. V. L. Malloy, R. A. Krajcik, S. J. Bailey, G. Hristopoulos, J. D. Plummer and N. Orentreich, *Aging Cell*, 2006, **5**(4), 305.
87. B. B. Lee, A. Kaya and V. N. Gladyshev, *Ann. N. Y. Acad. Sci.*, 2016, **1363**, 116.
88. M. Kitada, Y. Ogura, I. Monno, J. Xu and D. Koya, *Biomedicines*, 2021, **9**(2), 130.
89. J. Trofast, *Chem. Int.*, 2011, **33**(5), 16.
90. K. W. Franke, *J. Nutr.*, 1934, **8**, 597.
91. K. Schwarz and C. M. Foltz, *J. Am. Chem. Soc.*, 1957, **79**, 3292.

92. O. H. Muth, J. E. Oldfield, L. F. Remmert and J. R. Schubert, *Science*, 1958, **128**, 1090.
93. J. T. Rotruck, A. L. Pope, H. E. Ganther, A. B. Swanson, D. G. Hafeman and W. G. Hoekstra, *Science*, 1973, **179**, 588.
94. F. M. Fordyce, *Essentials of Medical Geology*, Springer, Dordrecht, 2013, 375.
95. H. Hartikainen, *J. Trace Elem. Med. Biol.*, 2005, **18**, 309.
96. L. H. Winkel, B. Vriens, G. D. Jones, L. S. Schneider, E. Pilon-Smits and G. S. Bañuelos, *Nutrients*, 2015, 7(6), 4199.
97. M. Schiavon and E. A. Pilon-Smits, *New Phytol.*, 2017, **213**(4), 1582.
98. L. A. Daniels, *Biol. Trace Elem. Res.*, 1996, **54**, 185.
99. M. Roman, P. Jitaru and C. Barbante, *Metallomics*, 2014, **6**(1), 25.
100. X. Xu, B. A. Carlson, H. Mix, Y. Zhang, K. Saira, R. S. Glass, M. J. Berry, V. N. Gladyshev and D. L. Hatfield, *PLoS Biol.*, 2007, 5(1), e4.
101. J. Donovan and P. R. Copeland, *J. Mol. Biol.*, 2010, **400**, 659.
102. K. E. Hill, R. S. Lloyd and R. F. Burk, *Proc. Natl. Acad. Sci. U. S. A.*, 1993, **90**, 537.
103. R. F. Burk and K. E. Hill, *Biochim. Biophys. Acta*, 2009, **1790**(11), 1441.
104. K. E. Hill, S. Wu, A. K. Motley, T. D. Stevenson, V. P. Winfrey, M. R. Capecchi, J. F. Atkins and R. F. Burk, *J. Biol. Chem.*, 2012, **287**, 40414.
105. T. Jäger, H. Drexler and T. Göen, *Arch. Toxicol.*, 2016 May, **90**(5), 1069.
106. S. Bügel, E. H. Larsen, J. J. Sloth, K. Flytlie, K. Overvad, L. C. Steenberg and S. Moesgaard, *Food Nutr. Res.*, 2008, **52**, 10.
107. G. V. Kryukov, S. Castellano, S. V. Novoselov, A. V Lobanov, O. Zehtab, R. Guigó and V. N. Gladyshev, *Science*, 2003, **300**(5624), 1439.
108. V. M. Labunskyy, D. L. Hatfield and V. N. Gladyshev, *Physiol. Rev.*, 2014, **94**(3), 739.
109. S. Gromer, J. K. Eubel, B. L. Lee and J. Jacob, *Cell. Mol. Life Sci.*, 2005, **62**(21), 2414.
110. Y. Shi, W. Yang, X. Tang, Q. Yan, X. Cai and F. Wu, *Front. Pediatr.*, 2021, **9**, 576916.
111. K. Wang, J. Yu, H. Liu, Y. Liu, N. Liu, Y. Cao, X. Zhang and D. Sun, *Semin. Arthritis Rheum.*, 2020, **50**(2), 366.
112. Y. M. Xiong, X. Y. Mo, X. Z. Zou, R. X. Song, W. Y. Sun, W. Lu, Q. Chen, Y. Z. Yu and W. J. Zang, *Osteoarthritis Cartilage*, 2010, **18**(6), 817.
113. J. B. Vanderpas, B. Contempré, N. L. Duale, W. Goossens, N. Bebe, R. Thorpe, K. Ntambue, J. Dumont, C. H. Thilly and A. T. Diplock, *Am. J. Clin. Nutr.*, 1990, **52**(6), 1087.
114. M. Schneider, H. Förster, A. Boersma, A. Seiler, H. Wehnes, F. Sinowatz, C. Neumüller, M. J. Deutsch, A. Walch, M. Hrabé de Angelis, W. Wurst, F. Ursini, A. Roveri, M. Maleszewski, M. Maiorino and M. Conrad, *FASEB J.*, 2009, **23**(9), 3233.
115. J. Pieczyńska and H. Grajeta, *J. Trace Elem. Med. Biol.*, 2015, **29**, 31.
116. M. Xu, D. Guo, H. Gu, L. Zhang and S. Lv, *Biol. Trace Elem. Res.*, 2016, **171**(2), 283.

117. M. Barman, A. L. Brantsæter, S. Nilsson, M. Haugen, T. Lundh, G. F. Combs, G. Zhang, L. J. Muglia, H. M. Meltzer, B. Jacobsson and V. Sengpiel, *Br. J. Nutr.*, 2020, **123**(2), 209.
118. J. Köhrle, F. Jakob, B. Contempré and J. E. Dumont, *Endocr. Rev.*, 2005, **26**(7), 944.
119. L. Schomburg, C. Riese, M. Michaelis, E. Griebert, M. O. Klein, R. Sapin, U. Schweizer and J. Köhrle, *Endocrinology*, 2006, **147**(3), 1306.
120. Q. Wu, M. P. Rayman, H. Lv, L. Schomburg, B. Cui, C. Gao, P. Chen, G. Zhuang, Z. Zhang, X. Peng, H. Li, Y. Zhao, X. He, G. Zeng, F. Qin, P. Hou and B. Shi, *J. Clin. Endocrinol. Metab.*, 2015, **100**(11), 4037.
121. G. R. G. Andrade, B. Gorgulho, P. A. Lotufo, I. M. Bensenor and D. M. Marchioni, *Nutrients*, 2018, **10**(6), 693.
122. C. D. Thomson, S. K. McLachlan, A. M. Grant, E. Paterson and A. J. Lillico, *Br. J. Nutr.*, 2005, **94**(6), 962.
123. J. C. Avery and P. R. Hoffmann, *Nutrients*, 2018, **10**(9), 1203.
124. P. K. Drain, R. Kupka, F. Mugusi and W. W. Fawzi, *Am. J. Clin. Nutr.*, 2007, **85**(2), 333.
125. H. Steinbrenner and H. Sies, *Arch. Biochem. Biophys.*, 2013, **536**(2), 152.
126. R. Pillai, J. H. Uyehara-Lock and F. P. Bellinger, *IUBMB Life*, 2014, **66**(4), 229.
127. M. Vinceti, C. M. Crespi, C. Malagoli, C. Del Giovane and V. Krogh, *J. Environ. Sci. Health, Part C: Environ. Carcinog. Ecotoxicol. Rev.*, 2013, **31**(4), 305.
128. D. Radomska, R. Czarnomysy, D. Radomski, A. Bielawska and K. Bielawski, *Nutrients*, 2021, **13**(5), 1649.
129. M. Laclaustra, A. Navas-Acien, S. Stranges, J. M. Ordovas and E. Guallar, *Environ. Health Perspect.*, 2009, **117**(9), 1409.
130. L. E. Fridlyand and L. H. Philipson, *Ann. N. Y. Acad. Sci.*, 2005, **1066**, 136.
131. A. J. Duffield-Lillico, E. H. Slate, M. E. Reid, B. W. Turnbull, P. A. Wilkins, G. F. Combs Jr, H. K. Park, E. G. Gross, G. F. Graham, M. S. Stratton, J. R. Marshall and L. C. Clark, *J. Natl. Cancer Inst.*, 2003, **95**(19), 1477.
132. E. A. Klein, I. M. Thompson Jr, C. M. Tangen, J. J. Crowley, M. S. Lucia, P. J. Goodman, L. M. Minasian, L. G. Ford, H. L. Parnes, J. M. Gaziano, D. D. Karp, M. M. Lieber, P. J. Walther, L. Klotz, J. K. Parsons, J. L. Chin, A. K. Darke, S. M. Lippman, G. E. Goodman, F. L. Meyskens Jr and L. H. Baker, *JAMA*, 2011, **306**(14), 1549.
133. D. L. Hatfield and V. N. Gladyshev, *Mol. Interventions*, 2009, **9**(1), 18.
134. World Health Organization, Vitamin and mineral requirements in human nutrition, in *Report of a Joint FAO = WHO Expert Consultation*, Bangkok, Thailand, World Health Organization, Switzerland, 2nd edn, 2004.
135. EFSA Panel on Dietetic Products, Nutrition and Allergies (NDA), *EFSA J.*, 2014, **12**(10), 3846.
136. Institute of Medicine, *Dietary Reference Intakes for Vitamin C, Vitamin E, Selenium, and Carotenoids*, The National Academies Press, Washington, DC, 2000.

137. J. Donadio, G. Duarte, P. Borel, S. Cozzolino and M. M. Rogero, *Nutr. Rev.*, 2021, **79**(11), 1259.
138. M. P. Rayman, *Hormones*, 2020, **19**(1), 9.
139. D. G. Barceloux, *J. Toxicol., Clin. Toxicol.*, 1999, **37**(2), 1452.
140. National Research Council (US) Subcommittee on Selenium, *Selenium in Nutrition: Revised Edition 7, Effects of Excess Selenium*, National Academies Press (US), Washington (DC), 1983.
141. G. F. Combs Jr, *Nutrients*, 2015, **7**(4), 2209.
142. A. P. Kipp, D. Strohm, R. Brigelius-Flohé, L. Schomburg, A. Bechthold, E. Leschik-Bonnet and H. Heseker, *J. Trace Elem. Med. Biol.*, 2015, **2**, 195.
143. J. A. Satia, I. B. King, J. S. Morris, K. Stratton and E. White, *Ann. Epidemiol.*, 2006, **16**(1), 53.
144. S. M. Müller, C. Dawczynski, J. Wiest, S. Lorkowski, A. P. Kipp and T. Schwerdtle, *Nutrients*, 2020, **12**(3), 676.
145. A. Schubert, J. M. Holden and W. R. Wolf, *J. Am. Diet. Assoc.*, 1987, **87**(3), 285.
146. J. W. Finley, *Nutr. Rev.*, 2006, **64**(3), 146.
147. C. Thiry, A. Ruttens, L. D. Temmerman, Y. J. Schneider and L. Pussemier, *Food Chem.*, 2012, **130**(4), 767.
148. M. Rayman, *Br. J. Nutr.*, 2008, **100**(2), 254.
149. B. R. Cardoso, G. B. S. Duarte, R. Reis and S. M. F. Cozzolino, *Food Res. Int.*, 2017, **100**(Pt 2), 9.
150. G. F. Combs Jr, *Br. J. Nutr.*, 2001, **85**(5), 517.
151. G. D. Jones, B. Droz, P. Greve, P. Gottschalk, D. Poffet, S. P. McGrath, S. I. Seneviratne, P. Smith and L. H. Winkel, *Proc. Natl. Acad. Sci. U. S. A.*, 2017, **114**(11), 2848.
152. G. Alfthan, M. Eurola, P. Ekholm, E. Venäläinen, T. Root, K. Korkalainen, H. Hartikainen, P. Salminen, V. Hietaniemi, P. Aspila and A. Aro, *J. Trace Elem. Med. Biol.*, 2015, **31**, 142.
153. E. J. M Joy, A. A. Kalimbira, D. Gashu, E. L. Ferguson, J. Sturgess, A. D. Dangour, L. Banda, G. Chiutsi-Phiri, E. H. Bailey, S. C. Langley-Evans, R. Murray Lark, K. Millar, S. D. Young, L. Matandika, J. Mfutso-Bengo, J. C. Phuka, F. P. Phiri, J. Gondwe, E. L. Ander, N. M. Lowe, P. C. Nalivata, M. R. Broadley and E. Allen, *Trials*, 2019, **20**(1), 795.
154. Z. Wu, G. S. Bañuelos, Z. Q. Lin, Y. Liu, L. Yuan, X. Yin and M. Li, *Front. Plant Sci.*, 2015, **6**, 136.
155. R. D'Amato, L. Regni, B. Falcinelli, S. Mattioli, P. Benincasa, A. Dal Bosco, P. Pacheco, P. Poietti, E. Troni, C. Santi and D. Businelli, *J. Agric. Food Chem.*, 2020, **68**(14), 4075.
156. R. Chauhan, S. Awasthi, S. Srivastava, S. Dwived, E. A. H. Pilon-Smits, O. P. Dhankher and D. Rudra, *Crit. Rev. Environ. Sci. Technol.*, 2019, **49**, 1937.
157. M. Schiavon, S. Nardi, F. dalla Vecchia and A. Ertani, *Plant Soil*, 2020, **16**, 1–26.
158. M. P. Rayman, H. G. Infante and M. Sargent, *Br. J. Nutr.*, 2008, **100**(2), 238.

159. E. N. Bermingham, J. E. Hesketh, B. R. Sinclair, J. P. Koolaard and N. C. Roy, *Nutrients*, 2014, **6**(10), 4002.
160. A. Giacosa, M. A. Faliva, S. Perna, C. Minoia, A. Ronchi and M. Rondanelli, *Nutrients*, 2014, **6**(3), 1251.
161. W. C. Willett, B. F. Polk, J. S. Morris, M. J. Stampfer, S. Pressel, B. Rosner, J. O. Taylor, K. Schneider and C. G. Hames, *Lancet*, 1983, **2**(8342), 130.
162. N. Wang, H. Y. Tan, S. Li, Y. Xu, W. Guo and Y. Feng, *Oxid. Med. Cell. Longevity*, 2017, **2017**, 7478523.
163. X. Zhang, C. Liu, J. Guo and Y. Song, *Eur. J. Clin. Nutr.*, 2016, **70**(2), 162.
164. M. A. Beck, O. A. Levander and J. Handy, *J. Nutr.*, 2003, **133**(5 Suppl 1), 1463S.
165. L. Schomburg and J. Köhrle, *Mol. Nutr. Food Res.*, 2008, **52**(11), 1235.
166. I. H. Qazi, C. Angel, H. Yang, E. Zoidis, B. Pan, Z. Wu, Z. Ming, C. J. Zeng, Q. Meng, H. Han and G. Zhou, *Antioxidants*, 2019, **8**(8), 268.
167. X. Cai, C. Wang, W. Yu, W. Fan, S. Wang, N. Shen, P. Wu, X. Li and F. Wang, *Sci. Rep.*, 2016, **6**, 19213.
168. M. Vinceti, T. Filippini, C. Del Giovane, G. Dennert, M. Zwahlen, M. Brinkman, M. Pa Zeegers, M. Horneber, R. D'Amico and C. M. Crespi, *Cochrane Database Syst. Rev.*, 2018, **1**(1), CD005195.
169. K. Rees, L. Hartley, D. Day, N. Flowers, A. Clarke and S. Stranges, *Cochrane Database Syst. Rev.*, 2013, **2013**(1), CD009671.
170. E. J. van Zuuren, A. Y. Albusta, Z. Fedorowicz, B. Carter and H. Pijl, *Cochrane Database Syst. Rev.*, 2013, **6**, CD010223.
171. M. Allingstrup and A. Afshari, *Cochrane Database Syst. Rev.*, 2015, **2015**(7), CD003703.
172. L. K. Ursell, J. L. Metcalf, L. W. Parfrey and R. Knight, *Nutr. Rev.*, 2012, **70**(Suppl 1), S38.
173. S. Lewis and S. Cochrane, *Am. J. Gastroenterol.*, 2007, **102**(3), 624.
174. D. Dordević, S. Jančíková, M. Vítězová and I. Kushkevych, *J. Adv. Res.*, 2020, **27**, 55.
175. S. C. Mitchell, *Adv. Food Nutr. Res.*, 2021, **96**, 123.
176. A. C. Creedon, E. S. Hung, S. E. Berry and K. Whelan, *Nutrients*, 2020, **12**(8), 2347.
177. Q. Zhai, S. Cen, P. Li, F. Tian, Ji. Zhao, H. Zhang and W. Chen, *Environ. Sci. Technol. Lett.*, 2018, **5**, 724.
178. Q. Zhai, Y. Xiao, P. Li, F. Tian, J. Zhao, H. Zhang and W. Chen, *Food Funct.*, 2019, **10**(9), 5398.
179. E. A. Franzosa, A. Sirota-Madi, J. Avila-Pacheco, N. Fornelos, H. J. Haise, S. Reinker, T. Vatanen, A. Brantley Hall, H. Mallick, L. J. McIver, J. S. Sauk, R. G. Wilson, B. W. Stevens, J. M. Scott, K. Pierce, A. A. Deik, K. Bullock, F. Imhann, J. A. Porter, A. Zhernakova, J. Fu, R. K. Weersma, C. Wijmenga, C. B. Clish, H. Vlamakis, C. Huttenhower and R. J. Xavier, *Nat. Microbiol.*, 2019, **4**(2), 293.

Subject Index